The Story of Corn

The Story of Corn

Betty Fussell

*I believe in the forest, and in the meadow, and
in the night in which the corn grows.*

—HENRY THOREAU, "Walking" (1862)

NORTH POINT PRESS

FARRAR, STRAUS AND GIROUX

NEW YORK

North Point Press
A division of Farrar, Straus and Giroux
19 Union Square West, New York 10003

Distributed in Canada by Douglas & McIntyre Ltd.
Printed in the United States of America
Designed by Mia Vander Els
First published in 1992 by Alfred A. Knopf
First North Point paperback edition, 1999

Library of Congress Cataloging-in-Publication Data
Fussell, Betty Harper.
The story of corn / Betty Fussell. — 1st North Point pbk. ed.
p. cm.
Originally published: New York : Knopf, 1992.
Includes bibliographical references and index.
ISBN 0-86547-545-8 (alk. paper)
1. Corn. 2. Corn—America—History. 3. Corn—America—Folklore.
4. Indians—Folklore. 5. Folklore—America. I. Title.
[SB191.M2F87 1999]
633.1′5′0973—dc21   *98-42378*

Frontispiece: Maya mural painting of a corn stalk with human heads in the form of the Tonsured Maize God, from east Temple Rojo, c. A.D. 700–900, at Cacaxla in the province of Tlaxcala, Mexico. The elongated heads show grains of yellow corn topped by tassels of corn silk. In J. B. Carlson's Center for Archaeoastronomy Technical Publication No. 7 (1991). Photo copyright © 1992 by Bob Sacha

Part title collages created by Arlene Lee

FOR MY GRANDMOTHERS

Ellen Josephine Culver Kennedy
1857–1946

Carrie Hadassah Erskine Harper
1866–1955

ACKNOWLEDGMENTS

Many thanks to many people, not only to those named in the text, but to legions of others, among them: Eloise Alton, Robert Bird, Ruth Bronz, Judith Brown, Millie Byrne, Jack Carter, Keith Crotz, Phyllis Drozd, Mary Mills Dunea, Meryle Evans, Charlotte Green, Katherine Hamilton-Smith, Tiff Harris, Stan Herd, Carrie Hollister, Josefina Howard, Marcia Keegan, Barbara and Justin Kerr, Eli Klein, Kornelia Kurbjahn, Jan Longone, Katie MacNeil, Deborah Madison, Clark Mangelsdorf, John McElhaney, Steve Miller, Nicholas Millhouse, Edith Monroe, Alicia Rios, Alice Ross, Trevor Rowe, Cynthia Rubin, Drew Schwartz, Eugene Sekaquaptewa, Sarah Morgan Snead, Duff Stoltz, Terry Wilde and to archivists of collections, libraries, museums across the land.

Muchas gracias to friends in Peru: Marco Bustamante, Maria Elena Gonzalez Buttgenbach, Federico Perez Eguren, Luis Antonio Guerra, Betty Leyva, Hilda Linares, Ricardo Sevilla Panizo.

Thanks for untold mercies to George Andreou, Josephine Caruso, Don and Eleanor Crosier, Matthew Culligan, Roger Feinstein, Harry Ford, Sam Fussell, Tucky Fussell, Barbara Hatcher, Leah Holzel, Arlene Lee, Gloria Loomis, Patricia Rucidlo, Elisabeth Sifton, Fiona Stevenson, Kendra Taylor, Mia Vander Els.

Thanks to Gregory Thorp, who transmutes corn madness into photographic art.

And thanks, in memoriam, to Lloyd Wescott, who planted the seeds of this book in his stand of New Jersey corn.

Contents

ONE *A Babel of Corn*

CORN MAD   3
THE POPCORN CONNECTION   9
WHAT IS THE CORN?   15
CORN, U.S.A.   21

TWO *Seeds of Life*

The Language of Myth   29

THE SEED OF BLOOD   29
CORN CANNIBALS   38
THE CIRCLES OF CHACO CANYON   47

The Language of Science   59

SOLVING THE CORN MYSTERY   59
IN THE NAME OF DIVINE PROGRESS   67
THE CORN WAR   76
THE CORN RACE   86

THREE *The Daily Round*

The Language of the Seasons   99

THE SACRED CORN OF INCA AND AZTEC   99
THE DESERT FARMERS OF THE SOUTHWEST   113
BUFFALO BIRD WOMAN'S GARDEN   121

The Language of the Machine   133

THE OCEAN-DESERT OF NEBRASKA   133
THE MARCH OF MECHANICAL TIME   143
"CORN SICK" LAND   154

FOUR *Flesh and Blood*

The Language of Food   167
FROM PIKI TO CORNFLAKES   167
GREEN-CORN COOKING   176
ASH COOKING   195
CORNMEAL COOKING   209

The Language of Drink   249
CHICHA IN PERU, MOONSHINE IN ARKANSAS   249

The Language of Commodities   265
SQUEEZING THE MOLECULE   265

FIVE *The Sacred Round*

The Language of Ceremony   281
DANCING UP THE CORN   281
GATHERING IN THE CORN   286
TEARING OUT THE HEART   291
DRINKING DOWN THE SUN   296

The Language of Carnivals and Kings   301
HOOPESTON'S SWEET-CORN MAGIC   301
COURTLAND'S CORNHUSKERS   304
THE WORLD'S ONLY CORN PALACE   311

SIX *Closing the Circle*

THE NIGHT OF SHALAKO   323
THE BLESSINGWAY   326

Selected Bibliography   335
Index   345

A
Babel
of
Corn

(overleaf) In a Classic Maya image of fertility, corn sprouts from the belly of a human sacrifice. Drawn from a carving on Stela 11 at Piedras Negras, Guatemala, A.D. *731.* In S. G. Morley's *The Ancient Maya* (1946)
In a contemporary American advertising image, a corn ear becomes a pencil to record bankbook savings for farmers who buy hybrid seed from Lynks Seeds in Marshalltown, Iowa. Courtesy Lynks Seeds

Corn Mad

"At the end of the third quarter," the cabin steward announces over the loudspeaker, "the Cornhuskers are leading twenty-one to nineteen." The passengers burst into cheers. This United Airlines steward knows how to milk his audience. He got laughs at the beginning of the flight from O'Hare Airport in Chicago to Lincoln, Nebraska, when he asked us to don our Mae Wests "in the unlikely event that we should land in water." Now he congratulates our pilot on back-to-back perfect landings: "Let's show him our appreciation, folks." We burst into applause. We have just landed in the city named for America's favorite president, in the state that is the exact center of the United States.

My cousin Eleanor and her husband, Don, meet me at the airport dressed in Cornhusker red because on football Saturdays *tout* Lincoln paints the town red. "The State Capitol is one of the top ten Wonders of the World," says my cousin, "that's what they say." I've seen most of the classical seven but never the Capitol nor Nebraska nor my cousin, although I'm here for a reunion of cousins, the grandchildren of Parks Ira and Ellen Culver Kennedy, whose eight children were born in Nebraska. Most of the grandchildren are now grandparents themselves, most live in the Midwest, and all but one are strangers whom I've never met. I've been to Halicarnassus and the country of Peru, but never to the center of America, which boasts four Perus, one of them a small town in Nebraska where my mother taught country school. As a migrant bi-coastal American, old enough to be a grandmother, I have come at last to seek and to find, as my Grandmother Kennedy would have said, the center of my family, my country, my culture—founded and sustained on the cultivation of corn.

When I traveled around the United States in the early 1980s looking for the genus and the genius of American food, I discovered that American cooking was rooted in corn. But because I was looking then for the diversity of our hybrid cuisine, I focused on ports of entry around the perimeter—Boston, New Orleans, Santa Fe, Seattle. I spent less time in the center because it was the last land settled and the most Anglo-Saxon in its settling. Nebraska belonged entirely to buffalo and the hunters who pursued them until the last century, when pioneering folk like Grandfather Kennedy pushed west from Ohio and Iowa to turn the sod and dig wells and plant corn.

Nebraska wasn't even a state until 1867, when my grandfather was eleven years old. By the time he married another Scotch-Irish Presbyterian in 1880 and exchanged his "soddie" for the first of many frame houses he would build himself, the farming revolution had already begun to turn his small acreage in Wilcox into a notch in the world's Corn Belt. He would recognize the Wilcox I saw a century later, because it is still no more than a crossroads with a water-tower and a silo and a cluster of white frame houses, but the scale of the plane geometry of green that I saw from the air flying into Lincoln would have discombobulated him considerable.

My Grandfather Kennedy, who was born in 1856 in Benton County, Iowa, and died in Los Angeles, California, in 1947, experienced in his ninety-one years a revolution on land no less extraordinary than that in the air. "A lot of North Americans forget," Joel Garreau has reminded us in *The Nine Nations of America* (1981), "that the Breadbasket was the *last* frontier," and that the transformation of the Great American Desert into the World's Breadbasket took place only within the last hundred years.

A land and its people founded on corn.
Photo by Gregory Thorp

By the time my grandfather pushed west from Iowa into Nebraska, he had John Deere's "singing" Moline plow of wrought iron and steel to turn the sod and a single-row horse-drawn planter to deposit seed and dried manure in one drop. By the 1930s, when he left farming and well digging—he had dug the town well in Wilcox—for carpentry and retirement in California, he wasn't needed as a farmer anymore. His hands and his horses had been replaced by tractors and combines. A bushel that had taken him two to three hours to produce now took three minutes. The very seed he saved from open pollinated fields to plant again each year had undergone revolution. By the year he died, virtually all Midwestern corn grew not from farmer's seed saved year to year but from "factory-produced" hybrid seed. That seed nearly pentupled the amount of corn my grandfather could have harvested; he averaged perhaps 25 bushels an acre, but today he would average 120. Within the span of a single life, the pioneer farmer and his seed went under as quickly as the prairie sod, the plow and the horse.

"Sing a song of popcorn
When the snowstorms rage
Fifty little round men
Put into a cage.
Shake them till they laugh and leap
Crowding to the top;
Watch them burst their little coats
Pop!! Pop!! Pop!!"

—NANCY BYRD TURNER,
"A Popcorn Song" (1988)

While nearly all the Kennedy grandchildren were born on a farm or ranch (I on a California orange ranch), none of them farms today. Most have moved into cities, like the rest of America's farm population, which in the 1980s dropped another 11 percent. Today 97 percent of us live on what 3 percent of us farm. Small wonder that, in the age of service industries, corn is not always a major topic of conversation, even in Nebraska. When Cousin Eleanor introduced me to a friend as an Eastern writer, the friend asked with excitement, "What subject?" I told her, "Corn," and she groaned. "Why don't you write about something glamorous, something sexy?"

My days of innocence—before I understood the true sexiness of corn, before I learned the power of hidden corn—were nostalgic days, when Eleanor would tell me how her mother had canned hominy and her father had saved seed corn. He would take one kernel each from selected ears, wrap the kernels in a wet sheet to germinate and then throw out the ears of any kernels that had failed to sprout. They were days when I ate large helpings of caramel corn, scalloped corn and cornhusker bread, which Eleanor made from a can of cream-style corn, a can of whole corn kernels, two boxes of cornbread mix and simple nostalgic things like six eggs, a pint of cream and a cup of butter.

They were the days before I went corn mad. Even then, however, I could see that no matter what I touched, it turned to corn. Corn was everywhere.

Family snapshots in Nebraska showed cousins in their youth standing in cornfields higher than their heads. When I left Lincoln to drive through canyons of corn in Kansas, Iowa, Ohio, Illinois, I stumbled on corn folk I never knew existed. There was the cornhusking champion in Kewanee, Illinois—home of the Cornhuskers' Museum; the champion Corn Collector displaying his wares in Courtland, Kansas; the Omaha Indian and his wife at Macy, Nebraska, who hickory-smoked corn; the retired corn canner in Hoopeston, Illinois—home of the National Sweetcorn Festival; the pioneer hybrid-seed producer, Henry Wallace's brother James, in Des Moines, Iowa. Corn made the whole world kin.

Neither my Grandfather Kennedy nor my Grandfather Harper, a Kansas farm boy, ever went east of the Alleghenies or west of California. Their fathers on both sides were born and buried in the Midwest and fought Indians in between. Both generations shared territorial imperatives that ended with the oceans on each side, and they shared the politics of America First. My grandfathers would have been astonished to learn that today a farmer's income in the Corn Belt is tied, directly or indirectly, as Garreau has pointed out, "to the weather in Siberia." Today poke your finger anywhere in corn country and you strike corn one hundred, one thousand, ten thousand miles away.

Bantus in 1954 in East Cape Province, South Africa, flail cobs with poles to remove kernels, which they will grind into "mealies." Photo by A. M. Duggan-Cronin

The living plants at the Corn Belt center have roots that reach deep into the past and extend wide across the present. Following roots means exploring our earliest corn cultures in what is left of Indian territory in scattered patches in North Dakota, Oklahoma, New Mexico, Arizona. On the Second Mesa, I found Hopi grandmothers who grew blue corn and made piki bread as their ancestors had done centuries before a handful of Italians and Spanish beached their ships on a tropical isle, bit into the little yellow cakes offered them by the natives and wondered what this peculiar substance was. To find out what it was and where it came from, I had to explore the steamy rainforests of Mexico and the icy highlands of Peru. I had to talk in broken Spanish to Zinacotecans in Chiapas and Quechuans in Cuzco, places where the same breeds of corn had grown for two or even seven thousand years, and where corn bred and sustained sacred rites that still bind people to one another and to the land that nourishes spirits as well as bodies.

The corn culture of the Western world circled the globe a good four centuries ago, wherever the Spanish, Italians and Portuguese were foolhardy enough to go next. Today one can see—at a Polenta Festival in Gubbio, Italy, or in villages by the Great Wall of China outside Beijing—on both sides of the Eastern Hemisphere, rooftops, houses and walls that are compact with ears of drying corn. So quickly did the American corn revolution of this century spread that not only the Great American Desert was transformed, but also barren lands in Egypt and South Africa, India and China; today China, after the United States, is the world's largest producer of corn.

Geographic spread was only the beginning. Every item in Cousin Eleanor's cornhusker bread came from corn—not just the corn kernels but the eggs, cream and butter; even the baking powder, sugar and salt were touched indirectly by corn. My very flesh was compacted of corn and, at Eleanor's table, was beginning its own geographic spread. Everything that went into my mouth—including the fresh fruit and vegetables if they had been sprayed with insecticide—had been touched in one way or another by corn. To enter a supermarket was to step into cornland—all the items but the fresh fish having made contact with some product or by-product of processed corn, be it syrup, oil or starch in one of its multifoliate, mostly inedible forms.

Edible corn for humans, whether fresh, canned, frozen or in the form of cornmeal, makes up less than 1 percent of the American corn market, a tiny amount that nonetheless adds up to about three pounds of corn per person per day, a pound per meal in foodstuffs derived from converted corn. We are but one product of the Great Corn Conversion Chain that converts sun and water into animal, vegetable, mineral and synthetic. From the more than 200 million metric *tons* of corn that the United States produces each year, 85 percent is converted into cows, hogs and chickens in the proportion of 60 million cows, 100 million hogs and 4 billion chickens. As an index of corn's super conversion powers (double those of wheat), one bushel of corn in a mixed feed bag translates into 15 pounds of retail beef, 26 pounds of pork and 37 pounds of poultry. And this is still corn in a form we can recognize: fodder, silage, shelled grain or fibrous by-products.

Far more pervasive is the world of hidden corn, in which corn's living flesh is converted alchemically into industrial gold. Before my Grandfather Kennedy was born, the local gristmills of the East were already being replaced by Colgate's roller mills in Jersey City, which by 1844 extracted starch from corn in the process now called wet-milling. Twenty years later, starch was converted into syrup and sugar and, soon after that, into dextrose and the roasted starch called dextrin that turns corn into glue. After a century and a half of playing with the chemical constituents of corn, America has created an invisible and awesome corn network.

We expect to find corn oil in mayonnaise and even soap, but aren't apt to think of it in paint and insecticides. While corn syrup or sugar is not unthinkable in a jar of peanut butter, it is downright worrisome in baby foods, chewing gums, soft drinks, canned and frozen vegetables, beer and wine, crackers and breads, frozen fish and processed meats like hot dogs and corned beef hash. It is mind-boggling in cough drops, toothpaste, lipstick, shaving cream, shoe polish, detergents, tobacco, rayon, tanned leather, rubber tires, urethane foam, explosives and embalming fluid.

When we get to invisible cornstarch, it's cause for national alarm. Just how much so, Howard Walden told us in *Native Inheritance* (1966), when he pointed to corn's indispensability in World War II or indeed in any war. Starch or dextrin touches all dehydrated foods, powdered and granulated foods, aspirin tablets, antiseptics, the surgeon's gloves and sponges, shotgun shells, dynamite, cloth fabrics, adhesives, cigarettes, books and magazines, every paper product but newsprint, penicillin and sulfa, molded plastics, oil-well drills, battery dry cells, aluminum in every form from wire to airplanes, and metal molds of every type from gun carriages to tanks.

The list goes on. Think of the distilling industries, from whiskey to ethanol, and you must think corn. Think of all the metal and oil industries, and you must think corn. Think not only of what's in the supermarket but of the materials of which the store is built and the machinery that moves both you and the food in and out of the supermarket, and you must think corn. Think of the greenhouse effect at the end of the green revolution, and you must think how green *was* the corn. For the revolution in chemical farming that turned green plants into industrial plants hinged on corn.

One dip into Nebraska and my green corn days were over. If corn was everywhere and everything was corn, how could I find the perimeter or center of the maze? Where to begin? As the subject possessed me, my apartment dispossessed me by the mounting boxes of books, photographs, paintings, jewelry, rugs, recipes, ephemera and kitsch, bags of meal and grits, stacks of blue and white tortillas and dried ears of purple, black and green kernels. I found that corn, like Captain Ahab's monstrous whale, possesses its pursuers until it drives them mad. But I wasn't alone. America is full of corn madness.

The townsfolk of Mitchell, South Dakota, have built the world's only corn palace of kernel mosaics and minarets, like a prairie Xanadu. A field artist in Kansas has planted corn, sunflowers and sorghum to replicate, from the air, Van Gogh's *Sunflowers*. A noted geneticist in Massachusetts pats the leaves of his baby corn plants and asks them to talk to him nicely. A photographer in Vermont, when I telephone to ask about corn pictures, immediately puts down the phone and lets out a long "Whooooopeeeee!" A Nobel Prize winner in Long Island has said that the best thing about studying corn is that you become part of the subject. Corn breeds its own poets, lunatics and lovers. After five years of corn madness, I can't think of anything sexier than corn. Or more dangerous.

THE POPCORN CONNECTION

T KEEP a grain of sanity, I turned to food. What did corn mean to me? Popcorn. When I think of my Grandfather Harper, I think of popcorn. Or rather, I smell popcorn, with rivulets of real butter trickling down through the red-striped bag of hot roasted kernels, fresh from the electric popper that was a prominent feature of the lobby of the movie theater in California we attended every Saturday afternoon in the 1930s. When I think of my son in the 1960s, sprawled on the living-room sofa of our house in suburban New Jersey watching Saturday-afternoon football, I smell popcorn, heavily buttered and lightly salted in a large wooden bowl that I bought on my honeymoon in Cape Cod.

In the 1980s, with my son long gone and the honeymoon over, I still pop corn the old-fashioned way, in a cast-iron skillet, holding the lid slightly ajar while I shake the skillet back and forth over a high flame until the pop-pop-popping quiets and only an occasional late-bloomer explodes against the lid. I upend the skillet over the wooden bowl, dark now after four decades of oiled salads and buttered popcorn that gave luster to its intended function as a quahog chopping bowl. How curious that my grandfather in the West who died long before my son was born in the East should be linked to him, a continent and decades apart, by popcorn.

Popcorn is a truly indigenous fast finger-food that links all ages, places, races, classes and kinds in the continuing circus of American life. Popcorn is the great equalizer, which turns itself inside out to attest to our faith that color is only skin deep, and class superfluous. Popcorn connects toddlers and grannies, hard-hats and connoisseurs, moviegoers and sports fans, beer drinkers and tee-totalers, sharecroppers and industrialists, cops and robbers, even cowboys and Indians. Popcorn bags the past with the present and evokes even in old age the memories of childhood. We eat it not because it's good for us, which it is, but because it's joke food, and Americans love joke food. We eat popcorn for fun.

Each year every man, woman and child of us eats an average of fifty-six quarts, popped, from a total of 900 million pounds of kernels, unpopped. We

"The first thing I saw when I got into his house was a console TV entirely covered with popcorn. Above it, suspended from the ceiling, was a popcorn bra and a popcorn football."
—SPALDING GRAY, "Travels Through New England" (1986)

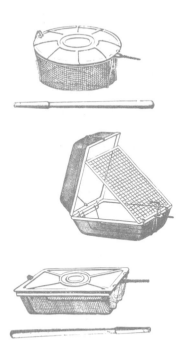

Corn poppers advertised in trade catalogue of Fletcher, Jenks & Company, 1889. From the Collections of Henry Ford Museum and Greenfield Village

The decorative possibilities of popcorn, exploited by the Aztec, are evident in this "tasselseed" mutant, grown and popped by Paul Mangelsdorf and R. G. Reeves in their 1939 researches into the origin of corn. Today, this popcorn "bouquet" decorates the Botanical Museum of Harvard University. Botanical Museum, Harvard University

love to pop corn because it explodes with the noise and violence of our language, our streets, our practical jokes. The late lamented puppet Ollie, despite his single dragon tooth, was addicted to it. Texas Junior Leaguers can't roast turkey without it. "You fill the cavity with kernels," Ruth Bronz explains, "seal it and bake it until the turkey blows its ass off." My cousin Don of Nebraska couldn't date college girls without it, in the days when they bought grocery sacks of popcorn and gallon jugs of root beer for a blanket party in the park. "Popcorn and root beer and it was impolite to belch," he remembers, "but you had to or it would have blown your head off."

Nebraska, I learned, is our number one popcorn state, although it feuds with Indiana for that title. Together they grow 65 percent of all processed kernels. Typically America has turned its favorite joke food into a $2-billion-a-year industry. That gives popcorn status as a cultural institution. Chicago is the home of the Popcorn Institute; Marion, Ohio, of the Popcorn Museum; Columbus, Ohio, features major Cracker Jackiana in its Center for Science and Industry; and men across the nation, vying for the title of Cracker Jack King, acquire "rare" Cracker Jack prizes with the same avidity Japanese show in acquiring Van Goghs.

In the last few years, popcorn sales have exploded, skyrocketing 25 percent in a single year. At a recent national foods show, I found new "fun flavors" like yogurt and jalapeño; innovative coatings like Jell-O and chocolate; fashion colors like black and blue (although the kernels always pop white); and "fun packaging" like disposable poppers, hot-air poppers, popcorn-on-the-cob. Merchandising is the name of the popcorn game and that, too, happened within the lifetime of my grandfathers.

If corn is the prime model of America's industrialization of crops at the production end, popcorn is the model at the distribution end, where packaging is all. By nature, popcorn has an edge, for each dull kernel comes equipped with its own snowy-white display. Popcorn's eye appeal is what Spanish soldiers and priests in the wake of Cortez noted when they saw Aztec virgins place "over their heads, like orange blossoms, garlands of parched maize, which they called *mumuchitl*." This was "a kind of corn which bursts when parched and discloses its contents and makes itself look like a very white flower," wrote Father Bernardino de Sahagún in the 1560s, relating how Aztecs honored the god who protected fishermen and other water workers by scattering these white blossoms as if they were "hailstones given to the god of water." There was precedent for Thoreau's calling popped corn "a perfect winter flower hinting of anemones."

"Parched corn" is what Governor John Winthrop called it when he observed the natives of Connecticut stirring it in hot embers until it "turned almost the inside outward, which will be almost white and flowery." Parched corn is what Quadequina, Chief Massasoit's brother, is said to have brought in a deerskin bag to the first Thanksgiving feast. Parched corn is what Benjamin Franklin described a century later, when he approved the method of throwing grains into sand heated in an iron pot until "each Grain burst and threw out a white substance of twice its bigness." Four centuries later, an Englishman encountering parched or popped corn for the first time exclaimed, "How extraordinary—fluffed maize."

While all kinds of corn pop to some degree, not all will turn inside out to become light, white and fluffy. The two major types of popcorn grown commercially—"rice," with sharply pointed kernels, and "pearl," with smooth rounded crowns—contain such a high proportion of hard starch that their prime feature is popping expansion or, as the admen say, "poppability." As scientists say, expansion depends upon moisture within the grain that turns into steam when the kernel is heated, but which is held within the starch-protein matrix until it suddenly explodes. In popcorn types where the starch-to-protein ratio is somewhat low, steam leaks out without exploding. Poppability is best when the kernel's moisture is between 13.5 and 15.5 percent and when the kernel is heated in such a way that the water vapor can escape. For this reason, when you pop corn in a skillet, you should cover the skillet only partially with a lid or the kernels will reabsorb the moisture and become tough. Texture is the essence of popcorn flavor. "Ideally, it should be crisp but not tough," stated *Consumer Reports* (June 1989). "It should shear off cleanly and compress easily as you chew, but not pack the teeth too much. Once chewed, it should be easy to swallow, leaving few crumbled bits in the mouth."

Today popcorn is so easy to swallow that *Consumer Reports* has rated fifty-one brand-name varieties and has observed that sales have increased more than 50 percent in the last decade through the magic combination of a microwave in every kitchen and a TV in every bedroom. Americans eat 70 percent of their popcorn at home, and they microwave 60 percent of that for couch potatoes. Today's surge of popularity

Moche pottery corn poppers (A.D. 100 to 800) from the northern coast of Peru, where the Moche had developed extensive irrigation canals for farming. Field Museum of Natural History, Neg. #A96665, USA.

In his No. 1 Wagon of 1893, Charles Cretors put his steam-powered peanut roaster and corn popper machine on wheels and into the hands of street vendors, who found a ready audience for the theatrics of automated popping. C. Cretors & Co., Chicago

began, however, in World War II, when home-fronters sacrificed candy for our boys over there and chewed popcorn instead. In movie-theater lobbies in the 1940s we purchased corn popped in streamlined machines named "Majestic" and "Hollywood." Today we run to our microwaves during commercials.

If the earliest method of popping was simply to toss kernels into the embers of a fire, the method of popping in a sand-filled pot is as ancient as prehistoric Peru. A three-legged container with an open mouth on one side, an *olla canchera,* kept the heat in and let the steam out. While colonists substituted iron for clay, natives continued to pop corn in clay as late as this century. Buffalo Bird Woman of the Hidatsa tribe in North Dakota described parching hard yellow corn in the hot sand of "a clay pot of our own make," until "all the kernels cracked open with a sharp crackling noise; they burst open much as you say white man's popcorn does."

New England colonists improvised a popper by punching holes in a sheet-iron container shaped like a warming pan, a rare version of which was a pierced cylinder that revolved on an axle, like a squirrel cage, in front of the fire. In 1847, in his *Memoir on Maize, or Indian Corn,* D. J. Browne listed two "Modern Modes of Popping Corn," one of them by heating the kernels in a buttered or larded frying pan and the other by "a very ingenious contrivance" of recent invention:

> It consists of a box made of wire gauze, with the apertures not exceeding one twentieth of an inch square, and is so constructed that the corn can be put within it, without being burnt, and can be held over a hot fire made either of wood or coal. The carburetted hydrogen gas, produced within the box by the decomposition of the oil in the corn, is prevented from explosion in a similar manner as *fire-damp,* in mines, is prevented from explosion by the safety-lamp.

While his chemical analysis and analogy may give us pause, Browne's is one of the first descriptions of the wire-basket corn popper that became a staple of American hearths until popcorn machines put popcorn vendors on every street corner.

Charles C. Cretors of Chicago was the first to develop a steam-driven machine in 1885 to pop corn in such volume that it could be sold from street wagons propelled first by hand, then by horse and finally by gasoline motor. Within the decade, Frederick William Rueckheim, also of Chicago, "improved" the popcorn he sold from his corner stand by sweetening it with molasses and calling it—from the same slang pool that gave us "swell" and "topnotch"—Cracker Jack. There was nothing new about sweetening parched corn with a native sweetener; Indians had been doing it for millennia. Nor was there anything new about using molasses candy to stick together toasted almonds or other kinds of nuts. With molasses candy, Rueckheim and his brother Louis stuck popcorn together with marshmallows and other sweets before they settled on popcorn and peanuts. Their inventiveness was not in cooking but in marketing, once they packaged their candy after its success in the 1893 World's Columbian Exposition in Chicago.

At first the brothers sold their product in tins and barrels, until Henry Eckstein joined them in 1902 and developed a moisture-proof sealed box that contained the first "green stamp," a printed coupon redeemable for adult prizes like clothing and sporting goods. In 1912, the Rueckheims replaced coupons with children's prizes: miniature books, magnifying glasses, tiny pitchers, beads, metal trains—"a prize-in-every-package." In 1916 they capitalized on the fame of that year's naval Battle of Jutland by putting F.W.'s grandson Robert on the box in his sailor suit (with his pet dog Bingo) and calling him Jack the Sailor. In the 1930s they added presidential medals, movie-star cards, Mystery Club prizes. During World War II, Cracker Jack went to war, and the company was commended for "high achievement in the production of materials needed by our armed forces." In the 1960s the company was sold to Borden and moved into the modern age with the foil bag, the big party-pack tub, safety-tested toys, and automated plant machinery that pops eighteen to twenty tons of corn every day.

In the 1930s just plain popcorn had become popular enough to encourage a rival to the Cretors company in the person of William Hoover Brown of Marion, Ohio, known then as "Shovel City" from the success of the Marion Steam Shovel Company. Brown founded his Wyandot Popcorn company after he married Ava King, daughter of the shovel company head, whose family mansion was frequented by President Warren Harding, born and buried in Marion. At Harding's burial in 1923, the King mansion lodged for the occasion Henry Ford, Harvey Firestone and Thomas Edison. The conjunction was fortuitous in the history of the popcorn industry, for Brown adapted Ford's assembly-line method to mass-produce popcorn, which he then distributed by trucks, equipped with Firestone tires, to chain theaters, which showed the movies that Edison had helped to invent.

Sailor Jack and Bingo in 1925 salute their prize package. Courtesy Borden, Inc.

A former county agriculture agent in Indiana, Orville Redenbacher scrutinizes ears of the hybrid Snowflake variety he developed into his trademarked Gourmet Popping Corn. Orville Redenbacher and the Popcorn Institute, Chicago

In the 1960s the kernel itself was streamlined by a county extension agent in Indiana named Orville Redenbacher, who developed a hybrid popcorn seed he called "Gourmet Popping Corn." "It's a Snowflake variety," says Orville, with "higher popping volume" than other kinds—one ounce pops to a quart—and higher reliability—fewer of those unpoppable kernels the industry calls "old maids." Redenbacher put his smiling bumpkin face on a label and marketed his gourmet kernels from the back of a car until Hunt-Wesson bought up the company in 1974. Six-million-dollar television campaigns, plus clever microwave marketing, have made Orville's face and name ubiquitous.

The super-marketing of a Redenbacher leaves little room for popcorn farmers like Michael Della Rocco in Melrose, New York. A traditional crop farmer who had run into hard times, Della Rocco began experimenting in 1985 with popcorn types that might grow in New York's short season. "I wanted something I could grow and market so that I could take control of my own destiny," he says. "I wanted to have my name on it." The name was Dellwood Farm, but it's tough work going up against nationally known brands. "Look, I'm convinced my product is good, but you're looking at a farmer, and farmers aren't known for their marketing abilities."

Who knows anymore what a farmer is? It's as hard to imagine TV Time Gourmet Microwave Pop actually growing in dirt fields as it is to imagine popcorn, our clown of corns, at the center of the story of corn in America. Popcorn, I knew, linked my grandfather to my son, but I never dreamed that popcorn was the connection between Pop-a-Cobs for Couch Kernels and the cobs, no bigger than half a thumbnail, that were uncovered exactly one year before the summer on Cape Cod when I married a Harvard graduate student of English and bought a quahog chopping bowl to put our popcorn in.

It seems that on a hot August day in 1948, another pair of Harvard graduate students, one in anthropology and the other in botany, were digging in an abandoned rock shelter 165 feet above the San Augustin plain in Catron County, New Mexico. They were exploring Bat Cave, a site known to have been occupied by cave dwellers practicing primitive agriculture on the rim of an ancient lake three thousand years ago. In order to join his anthropologist friend Herbert Dick, the botanist Earle Smith had asked his professor in Cambridge for enough cash to pay for a round-trip bus ticket and a sleeping bag—$150. His professor was Paul Mangelsdorf, then director of Harvard's Botanical Museum, who later wrote, "Seldom in Harvard's history has so small an investment paid so large a return."

What they discovered in the Bat Cave's layers of trash, garbage and excrement, which had accumulated over two thousand years, were "766 specimens of shelled cobs, 125 loose kernels, 8 pieces of husks, 10 of leaf sheath, and 5 of tassels and tassel fragments." The deeper they dug, the smaller and more primitive the cobs, until they reached bottom and found tiny cobs of popcorn in which each kernel was enclosed in its own husk, the "pod popcorn" that Mangelsdorf identified as the genetic ancestor of a modern Mexican brown-skinned popcorn called Chapalote. Among these prehistoric kernels they found six that were partly or completely popped. Mangelsdorf and his colleague Walton Galinat took a few unpopped kernels and dropped them into a little hot oil to prove that two or three millennia later they could *still* pop. The Bat Cave find was, as Mangelsdorf said, "a landmark discovery." Not only was this the oldest known corn in America but, as later discoveries proved, popcorn was the oldest known corn in the world.

What Is the Corn?

For thousands of years before corn was popcorn, corn was grass. A century ago, Whitman asked, "What is the grass?" Fifty years ago, looking at the green heart of America where leaves of corn had replaced grass, Paul Mangelsdorf asked, "What is the corn?" The answer depends on who wants to know—in what mother tongue and in which mother culture. "*Mamaliga,*" says the Romanian. "Stir-about," says the Irishman. "*Makkal,*" says the East Indian. "*Welschkorn,*" says the Amish. "*Masa, our daily bread,*" says the Mexican. "*Tâ'-a,* the Seed of Seeds," says the Zuni.

What kind of corn? my botanist father would have asked the questioner. To a botanist corn is maize or, better yet, *Zea mays,* one of three related grasses in the tribe Maydeae, a member of the family Gramineae. Since maize hybridizes so readily, it has more varieties than any other crop species. There are thousands of varieties of corn, so many that taxonomists, engaged in what is today called "systematics," group these varieties loosely into three hundred races for the Western Hemisphere alone. Earlier, when corn seemed simpler than it does today, textbooks divided it into six kinds: dent, flint, flour, sweet, pop and waxy. But today we talk about "racial complexes." Northern Flints, Great Plains Flints and Flours, Pima-Papago, Southwestern Semidents, Southwestern 12-row, Southern Dents, Derived Southern Dents, Southeastern Flints, Corn Belt Dents are the nine major racial complexes of corn in the United States, excluding popcorns and sweet corns.

Corn in what form? the Quechuan in Peru would ask. If you mean parched kernels, the answer would be *kancha hara*. If boiled kernels, *mote*. If young ears for roasting, *choklo*. If cut kernels cooked in corn husks, *shatu*. If corn soaked in lye to remove their hulls, *mote lyushki*. If ears boiled and dried, *chochoka*. If fermented corn, *hara togosh*. If corn fermented and dried for corn beer, *shura*.

Corn at what time of year? the Iroquois would ask. He would use a different Onondaga word for "corn" in each of its stages or aspects: She is dropping or planting the corn. She is doing first-corn-hoeing. She is doing second-corn-hoeing. The corn pollen is being shed. The corn is in the milk. She is plucking the green corn. She is plucking the ripe corn. She is husking corn. He is making a string of corn. The corn is hung over the pole.

Corn for what purpose? the Iowan would ask. If the whole plant is to be chopped up and used for animal fodder, the answer would be silage. If hybridized for planting, seed corn. If fermented for distilling, corn mash. If used for feed and other manufacturers, cash crop.

Corn in what culture? the mythographer would ask. Corn to the ancient Maya was embodied in One Hunahpu, one of the twin heroes who defeated the Lords of Death. Corn to the Inca was embodied in Manco-Paca, son of the Sun and founder of the dynasty of the Royal Lords of Cuzco. Corn to the Totonac of Central America was Tzinteotl, wife of the sun. Corn to the Aztec was the goddess Xilonen and the god Quetzalcoatl. Corn to the Chippewa was Mondawmin. Corn to the Pawnee was the Evening Star, the mother of all things, who gave corn to the people from her garden in the sky.

From the beginning of corn in Mesoamerica, its languages have been as varied as the races and languages of the peoples who grew it. The Taino that Columbus heard the Arawaks of the West Indies speak was but one of two thousand languages that had developed among the diverse peoples of the Western Hemisphere. In the centuries after Columbus, the Old World struggled to incorporate into languages rooted in Greco-Roman or Anglo-Saxon the flora and fauna of peoples who spoke in unknown tongues. The first act of the explorers was an act of linguistic imperialism—the naming of plants. If, as some historians have said, the New World was an invention of the Renaissance, shaped by the European imagination, so too the Old World's perception of the

New was shaped by European crops. Explorers and colonists alike spoke the language of wheat, oats, barley, rye and millet. Looking for similar grains here, they found none. They found instead a monster for which they had neither language nor understanding.

Characteristically, the first time the plant appeared in written form in the Old World, it appeared in imperial Latin: "*maizium, id frumenti genus appellant.*" So wrote Peter Martyr d'Anghiera of the grain we now call "maize." Peter was a cleric at the court of Barcelona in May 1493, when Columbus returned from his first voyage. Peter wrote excitedly to his patron, Cardinal Sforza, in Rome to tell him everything Columbus said of the strange islands of the Arawaks. "There were dogs that never barked," Columbus reported of the fauna, and of the flora, "All the trees were as different from ours as day from night, and so the fruits, the herbage, the rocks, and all things."

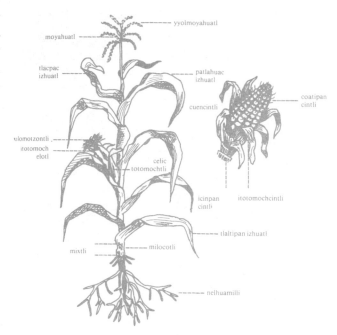

The naming of corn parts in the dialect of the Indians of Acatlán in Chilapan de Álvarez, Guerrero, one of dozens of corn dialects current in Mexico today. Museo Nacional de Culturas Populares, Mexico City

When the Arawaks had pointed to heaped ears of grain and the cakes they made from its meal, Columbus heard the word *mahiz*. Columbus amazed the court with details of its tapered ear longer than the span of a human hand and thick as an arm, its grains "affixed by nature in a wondrous manner and in form and size like garden peas, white when young." He delighted them with tales of black seeds which, when split open, revealed interiors whiter than snow. "*Fracta candore nivem exuperant,*" Peter wrote Cardinal Sforza about popcorn.

Peter's notes weren't published until 1511, in *De orbe novo decades*, and by that time Europeans were rendering the word *mahiz* in every way, in every language. While Columbus's journal of the first voyage circulated among many during this time, the original manuscript was eventually lost. Today, we must rely on abridgments by later chroniclers to learn that on November 6, 1492, Columbus sent two of his men, Rodrigo de Jerez and Luiz Torres, with a pair of native guides, into the interior of the island to search for the large Asian city and its emperor that Columbus expected to find. While they found no emperor, his men did find a city of a thousand people, who showered them with gifts, including "a sort of grain like millet, which they called *maize* [or *mahiz*], which it very well tasted when boiled, roasted, or made into porridge." Another version speaks of the grain "which the Indians called *maiz* but which

In a symbolic conjunction of New and Old Worlds in Florida in 1564, Chief Athore of the Timucua Indians displays to René de Laudonnière the fruits of field and forest before a column erected by Jean Ribault at his Huguenot colony. The scene was rendered by Jacques Le Moyne de Morgues in gouache on vellum. Collection of the late James Hazen Hyde, in the Print Collection, the New York Library

the Spanish called *panizo."* *Panizo* was the term Spaniards applied generically to the grains they knew, such as wheat, millet, barley, oats, sorghum.

In the next two decades, Spaniards and Portuguese spread variants of the name "maize" as rapidly as they spread variants of its seed through Europe, Africa, the Middle East and Far East. In 1516, when Portuguese sailors finally reached the Asian coast Columbus had sought, they deposited maize seeds in China. So quickly did these seeds take root wherever they were dropped, even in the most extreme climates, that patriots of Africa, Turkey, India and China alike claim corn as native. "So completely has India now appropriated the Makkal [maize]," wrote a nineteenth-century botanist, "that few of the village fathers would be found willing to admit that it had not always been with them as it is now, a staple article of diet." I know many Chinese today who insist, with some heat, that China was the onlie begetter of corn.

There are good reasons for confusion about the place of corn's origin. In the sixteenth and seventeenth centuries, as herbalists began to incorporate the world of newly discovered plants into their *Theatrum Botanicum,* as John Parkinson called his monumental book in 1640, plants were what you called them. An early Parisian botanist, Jean Ruel (or Ruellius), in his *De Natura Stirpium* (1536), called maize *Turcicum frumentum.* For centuries thereafter, Europe called maize Turkish wheat. Ruel may have thought the grain originated in Asia Minor, then under Turkish rule, or he may have wanted to distinguish "savage" maize from civilized wheat.

To this day, the French sometimes call maize *blé de Turquie,* despite the admonitions of one of their own botanists, Alphonse de Candolle, who as long ago as 1890 threw up his hands at the persistence of this misnomer. To call America's native grain *blé de Turquie,* he argued, was as silly as to call America's native bird *coq d'Inde* or "turkey." The English herbalist John Gerarde in his 1597 *Herball* was similarly exasperated, but he cited the authority of the ancients to show off his Latin and Greek, unaware that no ancient had ever heard of the plant:

Turkey wheat is called of some *Frumentum Turcicum*, and *Milium Indicum. Strabo, Eratostenes, Onesicritum, Plinie* and others, haue contended about the name heereof, which I minde not to rehearse, considering how vaine and friuolous it is: but leauing it vntill such time as some one *Oedipus* or other shall bewraie any other name therof that hath been described, or known of the old writers. . . . In English it is called Turky corne and Turky wheate: the inhabitants of America . . . do call it *Maizium* and *Maizum* and *Mais*.

But the English inhabitants of America never took to the word "maize." Rather, they adopted the term "Indian corn" to distinguish it from British corn, meaning all the grains known in Europe before Indians came into their ken. "Their Corn is of two sorts," John Harris reported in his *Voyages & Travels* (1748), "*English* Wheat . . . and Maize, or *Indian* Corn, which . . . grows in a great Ear or Head as big as the Handle of a large Horsewhip."

"Corn" in Old English meant a small worn-down particle, as, say, a grain of sand or salt; by extension, "corn" meant cereal grain in general and wheat in particular. When Biblical Ruth "stood in tears amid the alien corn," she was standing in a field of Mesopotamian wheat. Germans compounded the confusion by calling maize *Welschkorn,* after a botanist labeled it thus in 1539, saying it was "without doubt first brought to us by merchants from warm lands of fat soils." To the Turks during the period of the Ottoman Empire, on the other hand, maize, which they called *kukuruz,* signified barbarism. The Ottoman sultans erected a grain wall between their own people and those they ruled: the Turks ate wheat, while their Balkan and Hungarian subjects ate maize, which did nothing to elevate the status of "Turkey corn" in the rest of Europe. Portuguese muddled the names further by confusing maize with millet or sorghum. To this day the Portuguese term for corn is *milho,* a close derivative of *milhete,* "millet," and the South African name for maize, "mealies," comes from *milho.*

Contemporary anthropologists have made us aware that different societies live in different worlds, "not merely the same world with different labels attached." The Old World attached wheat labels to colonize an alien corn that was far more alien than it knew. Thirty years before the first English colonists landed in Plymouth, Gerarde's *Herball* set the tone of Europe's attitude toward New World corn for years to come:

Turky wheate doth nourish far lesse than either Wheate, Rie, Barly or Otes. . . . The barbarous Indians which know no better, are constrained to make a vertue of necessitie, and think it a good food; whereas we may easily iudge that it nourisheth but little, and is of hard and euill digestion, a more conuenient foode for swine than for man.

From John Gerarde's Herball, or Generall Historie of Plantes *(1597).* Rare Books and Manuscripts Division, The New York Public Library, Astor, Lenox and Tilden Foundations

One hundred and forty years after the English landed, Josiah Atkins of Connecticut, following General Washington through Virginia in 1781, lamented that their want of provisions reduced them to eating coarse Indian meal. "This is what people live on chiefly in these parts & what they call Hoe cakes. However, we not being used to such bread nor such country . . . all these together make my trials unsupportable." So unsupportable, it seems, that Atkins died in Virginia that same year, at age thirty-two. The hoe-cake natives, on the other hand, thought British wheat vastly inferior to their own grain and were horrified when these aliens fed sacred Indian corn—stalk, leaf, husk and cob—to their cows, chickens and pigs.

From a wheat culturist's point of view, corn was peculiar indeed. First was the plant's ungainly height, which rivaled a man's and sometimes exceeded it, reaching twenty feet. Next was the clumsy arrangement of its seeds, clamped together into one or more thick clubs halfway down the stalk, instead of the numberless graceful tassels waving atop stalks of "orient and immortal wheat." Third were the bastard colors and shapes produced by a profligate copulation. Fourth was the cloistering leaf sheath that kept seeds chaste as nuns, *and* as infertile, until man violated the cob. Finally, corn was common, cheap and therefore vulgar.

From a corn culturist's point of view, these very peculiarities made corn a godsend and, to the Indian, literally divine. It grew tall with outstretched arms like its brother, man. For its brother's use, it sacrificed every part of its body. Its abundant seeds, massed and well anchored in a central stem, could be eaten as fruit, vegetable or cereal, and at every stage of their growing, green or mature, raw or roasted, whole or ground. It stored its own grain in a waterproof wrapper on the stalk and did not have to be harvested when ripe. In addition, it grew with great speed and abundance and with half the labor of the foreign grains.

From the point of view of nature, corn was unnatural. Botanists today call it a "hopeless monster," for it cannot reproduce itself without man's help. Not only does the husk wrap the seeds so tightly that they cannot disperse, but the seeds are so tightly massed that, if a shucked ear happens to be buried in earth, the young shoots die from overcrowding. On the other hand, since corn adapts more readily than other grains to extreme climates and varied soils, it grows in more places than any other of the world's grains. As an energy converter, it surpasses all—in yield, speed and edibility of the total plant.

This drawing made from a pottery vessel from the Chicama Valley, an early agricultural site (2500 to 1800 B.C.) on the north coast of Peru, renders a detailed botanical understanding of the corn plant: the internodes of the stalks, corn ears ending in silks, tassels with a central spike and branches. From Lehman and Doering, *The Art of Old Peru,* courtesy Paul Mangelsdorf

From the point of view of culture, corn is unique among the world's staples because the earliest corn we have found was man-made, a cultural artifact. Since it cannot reproduce unaided, its relation to man is without parallel. To its native cultivators, corn coexisted with man on equal terms, as if corn and men grew up together as children of earth and sky. "When you're working with a wheat plant, who cares?" asks a contemporary botanist, Garrison Wilkes. "But when you're dealing with a corn plant, it's different. It's of human height, and you can look it in the eye. It's one on one."

To English yeomen who settled Plymouth in hopes of bettering themselves, wheat stood for upward mobility from barley-and-oats peasantry. Wheat was a matter of class, and corn was beyond the pale. When Edward Johnson wrote home in 1654 to praise the *Wonder Working Providence of Sions Savior in New England,* he focused on the new egalitarianism of wheat: "Now good white and wheaten bread is no dainty, but even ordinary man hath his choice." When Preacher Edward Taylor looked for a metaphor to encompass the ineffable qualities of his Lord, he didn't speak of barley cakes or corn pone. God's Bread of Life was ground from "the Purest Wheate in Heaven, his deare-dear Son" to make "Heavens Sugar Cake." And when Mistress Anne Bradstreet poeticized "corn" in her elegy on the death of a grandchild, you can be sure she didn't mean the Indian corn outside her door but the grain of British poetry:

> *And Corn and grass are in their season mown,*
> *And time brings down what is both strong and tall.*

Corn, U.S.A.

How much the naming of corn mattered I discovered at the second reunion of the Kennedy clan, this time in Oklahoma City. I found that Oklahoma was the only place in the country where you could go to the National Cowboy Hall of Fame and the National Hall of Fame for Famous American Indians on the same day. Of course this is the only place in the country that even has a Cowboy or Indian Hall of Fame. It is also the only place in the country that has a town named Corn.

"The only way you git to Corn is if that's where you're going," Flo says. "Corn's not a place you just drive through." Flo Ratzluff Richardson is Corn born and bred, as were her parents, retired now from their farm in the "country," as she calls it, meaning the land around Corn. The town of Corn, including a Main Street two blocks long, is what I would call country. A few white bungalows line up like opposing baseball teams on either side of Main Street behind opposing churches, Mennonite Brethren on one side and Calvary Baptist on the other, flanked by the Corn Historical Society Museum and Post Office, the Dutch Touch Cafe, Albert Odefehr American Legion, Veal's Grocery, Corn Town Hall and a tall white grain elevator labeled "CORN" at the end of town. Beyond that is nothing but red earth and flat green fields. Red earth gave the state its name, from the Choctaw *humma* for red and *okla* for people. Green fields gave the town its name of Corn, but not for maize.

Corn is a Mennonite community of Dutch-German wheat farmers who spelled Corn with a "K" when they first settled here in 1893, because *Korn* is German for "grain" and the grain they raised was wheat. Earlier they had raised wheat in the Ukraine, where Catherine the Great had given their ancestors both land and religious freedom in the eighteenth century and where a century later Alexander II had taken both away. Migrating through Canada to Kansas and finally to Oklahoma, after the Indian Appropriation Act of 1889, they built a church of sod, then frame, then brick. Today their church of poured concrete is the center for Mennonite missionaries from around the world.

I got to Corn—population 483—by driving due west from Oklahoma City on I-40 for a couple of hours, then south on Route 54 until a watertower spelled "CORN" in green letters above a faded sign, "CORN BIBLE ACADEMY, A Christian High School." Mennonite Dirk Bouma, owner of the Dutch Touch Cafe, got to Corn from Holland. His wife, Shabnum, got here from India. The entire population of Corn drops in to Dutch Touch at least three or four times a day to eat *verenikas* and *schnetkas* with their coffee and to keep in touch with the day's events. No jail, no police, no saloon—everyone keeps an eye out for, and on, his neighbor. "You sacrifice privacy," says James Resneder, at thirty-four sole owner and operator of the *Washita County Enterprise*, where Flo works part-time. "But I like living in a small town," he says. "I like being able to pick up the phone, get a wrong number and talk anyway because you know who's on the line."

Flo likes the family gatherings, listening to stories of how folks used to come to town on Saturdays and park in the middle of the street to sell their eggs and get their haircuts. "I've always had a real soft spot for the old days when the whole family did everything together," she says. "It was a closeness."

Four miles south on Route 54 there's an even closer community, about half as big as Corn, named Colony. "We're two hundred and one, countin' the dogs," says the town mayor, John Kauger, whose German parents also came here from Russia but as Lutherans, not Mennonites. To the left of John's house are horses, to the right are cows, and dead ahead are three churches and a small adobe Indian museum founded by his daughter Yvonne. Colony took its name from John Seger's Indian Colony, built in 1872, when this was Cheyenne-Arapaho land in the midst of Indan Territory.

Inside John introduces me to his wife, Alice, from Mississippi, and to a slender dark-haired girl who is dusting the chairs and the Indian paintings on the walls. She is Mary Haury, whose great-grandfather was an Arapaho chief at the time John Seger was a farmer working with the Bureau of Indian Affairs and trying to persuade Arapaho hunters to stay put and learn to farm. After the Dawes Act of 1887 reduced tribal homelands to allotments of 160 acres, and after the Land Run of 1889 brought in land-hungry ranchers, Seger built with the help of the Arapahos and in defiance of the Bureau a schoolhouse encircled by cottages—Seger's Indian Colony. The schoolhouse is a two-story, red-brick colonial affair of a certain grandeur, which it retains even now when its windows are boarded and its walls crumbling in the dusty oak grove where Cheyenne-Arapahos gather for their annual powwow in the fall. Even then, Seger's reign was brief, for he was caught in the crossfire between white Bureau chiefs and Cheyenne chiefs, and in 1905 was fired from the school. When he died in 1928 at the age of eighty-two, his friend Joe Creeping Bear delivered the eulogy:

> *Dear Mr. Seger is gone, his darling form no more on earth is seen.*
> *He is gone to live just o'er the sea on a shore that is ever green.*

Corn and Colony, the one named for wheat and the other for Indians, named the contradictions as America's midsection got settled and unsettled by a Babel of opposing tribes. "There's some animosity between Corn and Colony," says James Resneder of the *Washita County Enterprise*. "The land down there was poorer, so the farmers were poorer, until they found there was an underground river and they could irrigate." Colony is on the edge of a seven-mile-wide river that runs underground from Canada through central Oklahoma. Now while Corn grows wheat, Colony grows peanuts and prospers, but the Arapahos have gone. "Only two families left," says Mary Haury, "and so few gathered

"I took up a conversation with a gorgeous country girl wearing a low-cut cotton blouse that displayed the beautiful sun-tan on her breast tops. She was dull. She spoke of evenings in the country making popcorn on the porch. . . . And what else do you do for fun? I tried to bring up boyfriends and sex. . . . What do you do on Sunday afternoons? I asked. She sat on her porch. The boys went by on bicycles and stopped to chat. She read the funny papers, she reclined on the hammock. What do you do on a warm summer's night? She sat on the porch, she watched the cars in the road. She and her mother made popcorn."

—JACK KEROUAC, *On the Road* (1957)

for the powwow last year that they didn't even set up the tepee they used to call Big Church."

"When I was growing up, there were about thirty Indians to one white," says John, who graduated in 1932 from the first class of the high school that replaced Seger's schoolhouse. One of his classmates, Paul Bitchknee, served with the Army's 45th Division in World War II and when he landed in Sicily, John recalls, he raised his hand and said, "Columbus, I have come to return your visit."

Oklahoma is full of anomalies like the conjunction of Corn and Colony, and so is its history, like so much of history in this part of the world. When the fathers of my grandfathers crossed the Mississippi with a blueprint of the Promised Land in their heads, killing Indians when they had to in the name of the Peaceable Kingdom, they were driving roughshod over the graves of a civilization that was cradled eight thousand years before Moses was cradled in the bulrushes of the Nile. A hundred years before Christ was cradled in Bethlehem, the Hopewell Indians of Ohio had begun to build the giant circular earthworks of their ceremonial centers and to sculpt finely wrought effigies in copper and mica to bury in their domed mounds. A thousand years before the Kennedy clan reunited in Oklahoma, the Moundbuilders of the Mississippi had built from Maya blueprints the four-square tabernacle they named Cahokia, the largest temple complex in the world, larger by three acres than the Great Pyramid of Giza. When my grandfathers' fathers set out across the wilderness it was less virgin, as one historian says, than widowed.

From my cousins I learned that my Grandmother Kennedy was an eighth-generation Culver, descended from Edward and Ann, married in Dedham, Massachusetts, in 1638, to produce progeny named Zebulun, Ebenezer, Hezekiah, names sacred to the Biblical cultivators of wheat. Culvers may have been among the batch of English who landed on Cape Cod in 1620 and who fell on the stored grain they found there as if it were manna from Heaven, rather than the hard-won labor of civilized hands. Six years earlier, Captain Thomas Hunt had plundered the Nausets on Cape Cod not only of corn but of seven of their men, whom he sold as slaves in Spain. As heathen, these were assuredly agents of Satan and therefore fair game.

"The whole earth is the Lord's garden," Governor John Winthrop would tell Edward and Ann Culver at Massachusetts Bay, "and He hath given it to the sons of Adam to be tilled and improved by them." One of Satan's slaves who had escaped from Spain to England had learned enough English to help the colonists improve the Lord's garden in Plymouth, after he was sent back to Massachusetts, but Squanto too was an anomaly. The trick of fertilizing fields with alewives or other fish he had learned in England, where land was scarce. The native blueprint for corn was different.

"For Americans," Daniel Boorstin wrote, "discovery and growth were synonymous from the very beginning." From the very beginning, these words meant radically different things to the people of wheat and to the people of corn. From the very beginning, the wheat people's discovery of this new land was synonymous with the growth of a new language that fostered territorial exploration, discovery and change—the language of science. But what did "from the very beginning" mean? To the sons of Father Adam it meant one thing, to the sons of Mother Corn another. Those who had inhabited this land for at least twenty thousand years, who spoke the language of myth and metaphor and enacted it in ceremony and art, meant different things by "discovery," "growth" and "beginning."

From the very beginning of my search for the center of my own people, in Nebraska, in Oklahoma, I fell into the hotchpotch confusions registered in Arapaho Colony and Mennonite Corn. Not just conflicting worlds but self-contradictory ones, fractured and displaced. Improving the Lord's garden took its toll and I had ghosts to appease, including the ghost of a dead mother who did not find paradise in an orange grove in California and went corn mad for real. The Way West for some was the Trail of Tears for others, for settlers as well as natives, and I saw no way to reconcile the two. Still, there was my own discovery of the land's center. Still, there was the growth of corn. A beginning.

A mile west of the wheatfields of Corn I found a stand of Indian corn grown by a man who looked like my Grandfather Kennedy, farmer A. C. Hinz, whose gnarled hands and eyes blue as his overalls testified to eighty-three years of Oklahoma winds and weathers. Born in a dugout on this same land, Hinz grows at the far end of his wheatfields a little corn just for roastin' ears, he says. "If you want a good-tastin' roastin' ear, you use big yellow dent, the old-fashioned corn." Farmer Hinz has kept his dent seed pure for fifty years. "I don't detassel," he explains, "but corn, it's funny how it interbreeds so quickly, just that *poll*-en [poll as in pole] fallin' on the silks."

On the way to his cornpatch he points to a dam that beavers have built from good-sized boulders that fell in the creek, beavers weighing two hundred pounds, he says, big enough to move rock. He points to a cottonwood tree split in two by lightning. "That's a historical tree," he says. "Used to be Indian camps all along here and the cowboys and Indians were feudin', stealin' a few of each other's cattle, so the Indians tied a few boys to the tree, scalped 'em and burnt 'em, so the government sent in the Fort Wayne cavalry and killed them some Indians." That was in his father's day. The radio in his pickup truck is playing

The coupling of oil derricks and corn stalks, in a bas-relief on a public building in Chickasha, Oklahoma, is only one of the incongruities in the settling of Oklahoma. Photo by Gregory Thorp

a country song that warns, "Don't call him cowboy, Until you make him ride." That was in my day.

Not until I step into the red earth of his greening field do I begin to put the two together, his past and my present, my past and the country's present. The smell—dank and musty, yet fresh like the smell of corn silks when you first rip off the husks—is overpowering. "It's shooting for tassel," he says. And I see that this squadron of green ten feet tall, armed with outstretched blades, is helmeted by plumes of pale lavender, some malformed with bulbous flowers, others with hermaphroditic ears and tassels entangled in the same husks. Where a leaf joins a stalk I see a secondary shoot of leaves and, within, a delicate ear the length of my finger, lined with rows of tiny pointed kernels white as baby's teeth, each shooting a fine silken hair from its center, hairs that seem to grow as I watch, that seem to stretch for release from their tight sheath to burst like pale sea-green anemones into a cloud of pollen. Shooting for tassel.

Stripped of its green cloak, the silk lies as glossy and abundant as pubic hair. The sexiness of corn. Old World paintings of Danaë and Jove, coming in a shower of gold. New World photos of Culvers and Kennedys, coming to turn corn into cash. In farmer Hinz's cornfield in Corn, U.S.A, the corn silks smell of wet, fecund earth the color of dried blood. And I know that the growth of corn is not only up but down, down into the muck where roots stretch toward the darkness, toward the germinating seeds of life which are sown, as both Christians and Indians knew but knew differently, in blood.

TWO

Seeds
of
Life

(overleaf) A Mixtec drawing depicts a cornfield made fertile by ritual offerings of blood from above and a corpse below, curled like a seed in the earth. The germinating Goddess of Earth and of Maize is personified as "Maize Stalk Drinking Blood," in the Codex Vindobonensis (Codex Vienna), late fifteenth century. British Museum *Olaf Krans portrays* Women Planting Corn *in a settlement of Swedish religious dissenters at Bishop Hill, Illinois, c. 1875–1895. To achieve symmetrical rows, a pair of men move a string across the field while the women poke holes at evenly spaced intervals. Shown is Stanley Mazur's 1939 rendering of Krans's original painting.* Courtesy The National Gallery of Art, Washington, D.C.

The Language of Myth

THE SEED OF BLOOD

BLOOD IS THE COLOR of the feast of martyred San Sebastian in the valley of Zinacantán near San Cristobal, in Chiapas, on January 21, "the very day of very day," as the Zinacantecs say. I stood on a hill by a shaman's cross festooned with pine and looked into a lake of red. On top of hair coiled at each ear like the hair of Gauguin's Tahitian beauties, women wore folded shawls in colors that burned the eyes—aqua, deep pink, red. Men wore flat wide-brimmed skimmers that burst into streamers of multicolored ribbons over pinstriped tunics of white and red. The tassels of their headcloths were exclamation points of fuchsia. The long black ponchos of the elders were banded with red at neck and shoulders and their turbans were tasseled with red. Two dozen elders, each with a stave of new wood, sat in a row at one side of the square, facing the church front covered in flowers—cloudbursts of white calla lilies, flames of gladioluses and chrysanthemums.

A mariachi band followed a procession of men and boys, carrying daisies and tapers, into the church. Inside, red and green banners spanned the nave beneath a ceiling painted blue, above a floor green with pine needles. Kneeling amid trays of white tapers, clustered families chanted before the tiny figure of San Sebastian on the altar, half swallowed by a wall of flowers and half suffocated by the scent of burning copal. Here and there a body sprawled face down, overcome by poche, the drink of fermented sugar cane ritually dispensed by shot glass from plastic demijohns. Outside, people flocked to stands of tamales and popcorn, Fanta and Pepsi, hovered around great round black fry pans, bubbling with strips of meat and ballooning frybread. A large bull, four legs tied together, waited with open eyes for the knife that would signal the beginning of the feast.

". . . all the historian can do is to reconstruct a myth based on his own selection of fact."
—LÉVI-STRAUSS

Suddenly, from a group of wooden crosses, a band of forest creatures, dangling scarlet penises, sprang into the center, dancing to the cries of beasts and birds. There was Jaguar in a spotted baggy jumpsuit next to the black-masked Spooks. With hand puppets of monkeys and squirrels, they taunted the crowds with obscene gestures. There were Spanish lords and ladies, men in drag, comically elegant in lace and velvet. There was Raven flapping his wings of painted boards and pointing his conical beak. His beak held an ear of corn, for Raven, or *K'uk'ulchan* in the language of Tzotzil, is none other than Quetzalcoatl, the Plumed Serpent of the Toltec, the god of wind and rain who first brought corn to man.

The dancers enacted their version of the story of San Sebastian, left for dead in the woods but saved by wild animals in order that he might captain the Spaniards and reign as patron saint of the Zinacantecs. Here Sebastian was Maya, and in his victory over death the Zinacantecs celebrated their triumph over the conquistadors, reaffirming in present ritual those collective memories that the Maya call dreams and Europeans call history and myth.

While the dancers danced, a group of men mounted horses and raced one by one through a leaf-covered chute. Like jousting knights, they charged with staves, which they aimed at the heart of the saint painted on a stick hanging between poles. It is said that only a pure Zinacantec can pierce his heart, for the Zinacantecs have absorbed the saint into their own pantheon of martyrs, holy martyrs like Quetzalcoatl, who absorbed earlier gods of corn.

Maize gods native to Central and South America were far more ancient than Christian saints or the crucified God whose image the Spaniards planted in maize fields red with the blood of conquest. For these Maya descendants, the association of maize with blood is as old as the oldest Maya memory, as old as the first planted seed. As their culture evolved, ancient Maya fertilized seeds of corn with the sacrificed blood of their enemies and the blood of their own kings. For the Maya a single kernel of corn is symbolic of what Christians symbolize by the holy cross—the tragic and monstrous truth that the seed of life is death.

Today, in the Maya ruins of Palenque in the Yucatán jungle, the Temple of the Foliated Cross reveals in its carvings what Christians call the Tree of Life. For the Maya, it is the World Tree in the shape of a cross, where the crosspiece or branches are formed by leaves and silk-topped ears of corn, each ear a human head. The corn sprouts from a trunk of blood rooted in the head of the Water-Lily Monster that floats on the primal waters of the Underworld. Here out of the monster's mouth a god is born—God K, the Young Lord, the Maize God.

The Foliated Cross at Palenque depicts the cosmos as a leafy corn plant, through which the life-blood of the universe flows, uniting the World Tree at the center with the kings on either side, the Underworld with the heavens, the living with the dead. Neg. No. 320486. Courtesy Department of Library Services, American Museum of Natural History

Today, Maya descendants in Chiapas, Yucatán and Guatemala live double lives between Christian and Maya crosses, Christian Axle-trees and Maya World Trees, Christian saints and maize gods, Catholic priests and shamans— lives colored by blood. "People, when they are dying, save their corn, which has beautiful grains," a modern Chorti told an interviewer in 1972. "They look for those with beautiful white grains, with black corn, with red corn. Because they say that that is the blood of Jesus Christ." In a kernel of red corn, a contemporary Maya sees not only the cosmic globe but a drop of blood that condenses all of human history into a single germ of life.

Above a tributary of Rio Motagua in western Honduras stands Copán, which Michael Coe calls "one of the loveliest of all Classic Maya ruins" and which the nineteenth-century archaeologist John Lloyd Stephens once bought for $50. The price included the perfectly preserved Ball Court, the Temple of the Hieroglyphic Stairway and four figures carved in green volcanic tuff that are among the finest pieces of Maya art—the busts of the Young Maize God.

One of these, carved around A.D. 775, in the Late Classic period, is a delicate boy whose upraised and downturned palms curiously suggest the hand gestures of Buddha Sàkyamuni. This Young Lord once adorned the temple-palace of Yax-Pac, last of the Maya kings. When the king entered the

Here the Young Corn God (shaped as a mold whistle in the Late Classic period) embodies the rhythm of regeneration as he dances between earth and sky, with the Witz or Mountain Monster beneath him and signs of the Moon Goddess in the leaf curls above his head. Private collection, photo © Justin Kerr

innermost sanctum of his acropolis, he walked through the mouth of a gigantic serpent, over its stone fangs and through its gullet, to reach the sacrificial altar within. Here from his penis and earlobes the king drew his own blood to fertilize the Young Maize Gods "growing" from clefts in the stone heads of the Cauac-Monsters that ornamented the cornices of the temple's exterior. A Maya would have seen in the face of the Young Lord an ear of corn: in his hair, corn silk; in his gesturing hands, waving leaves of corn; in his closed eyes, the silent teeming life within the plant. Even today this carved stone evokes the living plant. When the palace fell into ruin, wrote the authors of *The Blood of Kings* (1986), "the Maize Gods fell across the East Court like a ripened crop that was not harvested."

So subtle and complex is the ancient Maya language of corn, carved in stone, painted on walls and pottery and screen-folds made of beaten bark, that only in recent years have its mysteries begun to be decoded. We now see that the Maya Maize God, like the medieval Christian God, stands at the center of a cluster of images and symbols that evolved slowly but took primary shape in the third to ninth centuries after Christ, a period rich in Christian saints and Maya maize gods. Rich also in Maya script which recorded the history and destiny of a people as expressively as Christian Scripture. Maya hieroglyphs, once we can read them, may help us learn what "discovery," "growth" and "beginning" meant to a civilization built on the symbolic as well as the physical potency of maize.

Until 1952, epigraphers tended to misread Maya hieroglyphs by oversimplifying them. Ever since Bishop Diego de Landa in the mid-sixteenth century sketched an "alphabet" of twenty-nine signs, scholars had quarreled over their meanings by construing them as either exclusively a phonetic or exclusively an ideographic system. The breakthrough came when they were read as both—as phonetic letters or syllables *and* as ideograms or rebuses. Anagogic symbols were as typical of the New World as Aristotelian logic was of the Old, a logic compelled to divide what the mythic mind unites.

Divide and destroy, some might say, in view of Bishop Landa's torching in 1562 of the thousands of heathen texts—red and black drawings on bark paper bound in covers of jaguar skin—that recorded the Maya world. Only four such texts escaped the flames: three codices that we name for the European cities which house them—Madrid, Paris, Dresden—a fourth, the Grolier, which is the only one to remain in Mexico. Until recently, what we knew about the ancient Maya we knew principally from the encyclopedic work Landa proceeded to write about the life of the natives in post-Conquest Yucatán, *Relación de las cosas de Yucatán* (1566). Ironically, modern editions of Landa's work are illustrated with the blasphemous images of the Maya books that escaped his burnings.

Not only the books but the monuments of Classic sites are now read in a new way, as we see that Maya glyphs record history as well as myth, the history of dynasties contemporary with the early Christian martyrs in Rome and the close of the Han dynasty in China. When Linda Schele and Mary Ellen Miller explored the dynasties and rituals of Maya art in their *Blood of Kings,* they found that the Maize God depicted in the great carved Temple of the Foliated Cross was but a mask of the historical king who built the major temples of the Sun, Cross and Foliated Cross at Palenque in the seventh century.

Lord Chan-Bahlum, or Snake-Jaguar, who stands like a giant on the left of the Foliated Cross, became king in A.D. 684, after the death of his father, Lord Pacal, who stands on the right. Chan-Bahlum was believed to incarnate in mortal form a higher Lord, the Young Lord of Maize. At the same time, he stood as an embodiment of the Cross itself, the World Tree as a living corn plant, bearing human heads, rooted in the Kan-cross Water-Lily Monster, symbolic of the raised-field method of agriculture, with its ecology of canals, water lilies, birds, fish and corn plants, all flourishing together, though only with the aid of man. The Tree is crowned with Celestial Bird, or Vision Serpent Bird, symbolic of the vision quest by which the king communicated, on behalf of his people, with the Otherworld of gods and ancestors. Pacal offers to his son the Personified Bloodletter, by which Chan-Bahlum will draw blood painfully from his penis in order to evoke the beneficence of the gods and sustain the social order of man and the agricultural order of nature.

The Foliated Maize God, as he is called, is another form of the Tonsured Maize God, discerned by Karl Taube in the canoe images carved on bones found in the temple mausoleum of Tikal in Guatemala. The forehead of the god who sits in the center of the canoe has been shaved and the head flattened and elongated. The tonsuring separates a short brow fringe from his tufted top, as if his head were a husked corn ear and his hair, corn silk. On his head, he wears the Jester God, a personification of royal and sacred power that Maya kings wore as a crown of jade, shaped in three points like a jester's cap and signifying the Hero Twins of the *Popol Vuh*. His wrist at his forehead suggests

SACRAMENTS:
Guatemala City, 1775
"Nor do the Indians come to Mass. They do not respond to announcements of the bell. They have to be sought out on horseback in villages and fields and dragged in by force. Absence is punished with eight lashes, but the Mass offends the Mayan gods and that has more power than fear of the thong. Fifty times a year, the Mass interrupts work in the fields, the daily ceremony of communion with the earth. For the Indians, accompanying step by step the corn's cycle of death and resurrection is a way of praying; and the earth, that immense temple, is their day-to-day testimony to the miracle of life being reborn. For them all earth is a church, all woods a sanctuary.
—EDUARDO GALEANO, *Memory of Fire II: Masks & Faces* (1987)

Stingray
Paddler   iguana   spider   dead   parrot   Kankin   Jaguar
                     monkey   king   dog   Paddler

The Tonsured Maize God in
the middle of the canoe on his
journey to the Underworld.
From Linda Schele and Mary
Ellen Miller, *Blood of Kings*
(1986)

death, imaged by the sinking of his canoe into the waters of the Underworld. He is both Maize God and king, for the bone carvings celebrate the life of the historical Ruler A, buried with a royal treasury of pottery and jade in the mausoleum built after his death in A.D. 735.

Everywhere the phallus-shaped head of the Tonsured Lord is associated with fertility, corn and blood. Glyphs of the god show corn kernels, or "corn curls," as Maya iconographers say, affixed to his head or hands. Often the curls sprout corn leaves and cobs, like those of the Foliated God. We see both forms often in the imagery of Late Classic plates and vessels which depict the severed head of the Maize God floating in a pool of blood. The curving tassel at the back of his head suggests corn tassels, as his beaded necklace suggests drops of blood. On the rim of such plates are Kan-crosses, which bear both the phonetic value of the Kan sign and its ideographic value as the sign for Day, suggesting the color yellow, ripeness, maize.

The life cycle of maize was the great metaphor of Maya life, the root of its language, its rituals and its calendar. We now see that the many configurations of the Maize God evolved from the seed of life embodied in the Kan sign. Kan is only one of the twenty named days of the Maya calendar, but wherever the Kan sign appears in conjunction with a god, it refers to crops and the powers for good and evil that affect them. Kan is also the syllable *wah,* which denotes bread, tortilla, tamale. Bowls holding Kan signs may represent offerings of maize, and therefore blood offerings and other precious things—jade beads, drops of blood. The beaded streams that fall from the hands of the Tonsured Lord and king in Stela 1 at Yaxchilan suggest drops of blood falling as dynastic seed.

Sometimes from the Kan or kernel there sprouts a serpent's head with a single eye at the top. The Madrid Codex, which deals largely with farm ceremonies, includes a cartoon sequence of corn plants growing like serpents from godheads, from seeds in the earth, from Kan signs and from Ik, the sign of Breath or Life. As the serpentine leaves and stalks of the plant represent the fertility associated everywhere with the snake, so the kernel represents the originating breath we name the soul. In the Maya language, the word for "serpent" is a homonym of the word for "sky," and both are pronounced "Kan."

The earliest Maize Gods of the Olmec, named gods I and II by the iconographers, prefigure by a thousand years the later Maize Gods of the Maya, which culminate in the many guises of the Plumed Serpent. Like a many-leaved plant of corn, they spring from the same "Seed-corn Dot." The Olmec saw corn as a three-part plant, its vegetation sometimes sprouting from the round Dot, sometimes incorporated into the Bracket, a sign abstracted from the paired fangs and vegetation that bracket the mouth of God I. From the cleft head of God II maize sprouts from the cleft itself, from the forehead, from the eyes or mouth—sometimes as a phallic cone (Banded Maize), sometimes as a dot with three feathers (Feathered Seed), sometimes as a fleur-de-lis (Three-pronged Vegetation). Like a plant the symbols of the three-part seedling develop into the three-pointed headdress of the Maya Jester God, uniting the primal cleft gods of the Olmec at La Venta with the Foliated Maize Lords of the Maya at Copán.

Through such complex figurations of vegetation and violence, blood and maize, the Maya expressed the symbiotic exchange between corn and man, each imaging the other. The image of the tonsured and foliated head of the Maize God severed from its body is at the same time a phallic ear of corn stripped from its stalk. The ritual perforation of the penis with one of the sacred bloodletting instruments—the obsidian blade, stingray spine and flint knife—at the same time enacts the husking of the cob with sharpened wood or bone. At Palenque, the historical Lord Chan-Bahlum saw in the severed head of the Maize God a mirror image of his own tonsured head, in the pool of blood his own life stream, rising as sap through the World Tree that connects the Cauac Monster of the Underworld to the Celestial Bird of the Heavens.

What the Maya read as sacred, we read as art. Today, a fine version of the Beheaded God, incised in the lid of a cache vessel used to catch the blood in these perforating rituals, rests peaceably despite its horrific message in the Art Museum of Princeton University. Three bloodletting lancets sharpened from stone and bone, together with the signs of Kan-cross and bone-beads, mark the undulating wave patterns in the background as blood, on which the severed head appears to float.

Although I spent most of my adult life in the town of Princeton, I never saw this vessel and if I had, its figurations would have signified nothing to me. They would have meant little more, or so I imagine, to my Uncle Roy, my father's younger brother, when he attended the Presbyterian Theological Sem-

maize sprouting from seed-corn dot

maize sprouting from bracket

vegetation sprouting from the corners of the mouth

maize sprouting from head cleft

banded maize

feathered seed corn

three-part seedling as three-pointed headdress

Redrawn from Peter David Joralemon, "A Study of Olmec Iconography" (1971).

A drawing of the cache vessel lid (Early Classic Maya, A.D. 350 to 500) shows the decapitated head of the Maize God in profile beneath three bloodletting instruments—obsidian blade, stingray spine and flint knife— on top of a symbolic bloodletting bowl, marked with the sign for the sun. The Cleveland Museum of Art; lent anonymously; drawing by Lin Deletaille

inary at Princeton before setting off as a missionary for the jungles of Brazil. Had he seen the vessel, he might have thought of John the Baptist, but it's doubtful that he would have thought also of the corn he had husked as a boy in Kansas. And when he sang hymns for the jungle heathen, accompanied by his wife on her portable organ, it's doubtful that he would have seen any connection between his sacrificed God and theirs. He had brought them the Word of God by way of Calvin, a Word that severed the head of the Creator God from His creation and that damned at once the World, the Flesh and the Devil. He brought words and worlds unintelligible to savages who believed that it was now as it was in The Beginning, that men and gods were created one flesh, one blood, from the same vital substance of corn.

Maya Corn Language, from Tamales to Corn Gods

| *"curled" balls* | *"notched" balls* | *wrapper and ball* | *ball and wrapper* | *ball + wrapper = KAN sign = bread* | *KAN sign + frame + tripod = the DAY sign of the calendar* |

Above left, drawings represent actual tamales, or balls of corn dough with fillings marked as "curled" or "notched." The dough and the wrapper of the tamale metamorphose into the Kan sign, which means bread or sacred offerings, and into the Kan glyph for Day. Below, corn kernels, ears and foliage are integrated in different ways with human heads and beaded streams to signify sacred corn gods and blood sacrifice.

     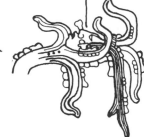

| *corn ears as human profile* | *corn curls and foliage* | *corn curl and foliage sprout from head* | *corn ear extends head* | *beaded streams of water* | *beaded streams of blood* | *corn god as blood sacrifice* |

At far left, a large ball of corn dough is unwrapped on a bed of banana leaves for a contemporary Yucatec communion feast. At near left, God K (from a West Court panel in Copán) holds a bowl of corn dough as an offering and sits on a bed of corn leaves, as if he were himself the communion tamale. Based on Karl Taube, "The Maize Tamale in Classic Maya Diet, Epigraphy, and Art" (1989)

CORN CANNIBALS

WHEN MY UNCLE ROY landed in São Paulo, the Tupinambas who had once occupied a two thousand-mile strip of Brazilian coast had given up the cannibal ways reported with a certain relish by Amerigo Vespucci in his *Mundus Novus* (1504–5), when he first encountered the savages of Brazil. "I saw salted human flesh suspended from beams between the houses, just as with us it is the custom to hang bacon and pork," he wrote, and added that the cannibals, for their part, wondered "why we do not eat our enemies and do not use as food their flesh which they say is most savory."

For two centuries, accounts of the cannibalism of the New World evoked the zeal of priestly converters even as they justified the greed of conquistadors. Even as late as this century, when the works of Father Bernardino de Sahagún were first published in English, a new wave of missionaries, Protestant and Catholic, was fired by his account of the cannibal stews of the ancient Aztec. After they had slain a captive, Sahagún reported, they cut him in pieces and offered the best piece of dark meat (a thigh) to Moctezuma. Then, at the house of the captor, "They made each one an offering of a bowl of stew of dried maize, called *tlacatlaolli*. . . . On each went a piece of the flesh of the captive."

How do Americans today respond when they learn that cannibalism was practiced by many of the tribes of our own continent? As late as 1756 there were reports of the Iroquois around Ontario eating captives taken in war, which jibed with Jesuit accounts from the sixteenth century. Jesuits who had witnessed the torture and ritual deaths practiced by the Iroquois and Huron on each other noted that the torturers seemed to act not in rage but in gentleness when they first burned different parts of the live victim, thrusting flaming brands down his throat and red-hot embers into his eyes, then cutting off several of his members, a foot here, a hand there, and finally the head "to carry it to the Captain Ondessone, for whom it had been reserved, in order to make a feast therewith." Commoners were content to make a feast of the trunk and whatever part they could carry home. When the feast was over, the Jesuits observed, "we encountered a Savage who was carrying upon a skewer one of [the victim's] half-roasted hands."

This drawing from the Florentine Codex underlines Sahagún's account of cannibal stews, eaten ceremonially with and without dried maize. Neg. No. 292757, Department of Library Services, American Museum of Natural History

None of the priests, adventurers, missionaries and criminals who busied themselves over four centuries with saving, civilizing and exterminating the brutes could have been expected to connect the cannibalism of "perfidious Savages," as Captain John Smith called them, with the cannibalism planted in the godhead of Christ. Whenever Christians celebrated the Eucharist, they consumed their Lord cannibalistically, but they had so long substituted wheat and grapes, in the form of bread and wine, for the primal connectedness of flesh and blood that the vegetative origins of cannibalism were as easily obscured as vows of chastity and poverty. How could Christians be other than horrified by savages who put a vegetable on equal footing with man and who believed that the primal being whose body had become and was now the world, flesh of their flesh, was incarnate equally in an ear of roasted corn and a half-roasted human hand?

"Then they surrounded those who danced, whereupon they went among the drums. Then they struck the arms of one who beat the drums; they severed both his hands, and afterwards struck his neck, so that his head flew off, falling far away. Then they pierced them all with iron lances, and they struck each with iron swords. Of some they slashed open the back, and then the entrails gushed out. Of some they split the head; they hacked their heads to pieces; their heads were completely cut up...."

—From Bernardino de Sahagún, in the Florentine Codex, quoted by Nigel Davies, *The Aztecs* (1973)

The cannibal ways of the New World sprang from the strength and persistence of an ancient planting culture that had kept its roots intact. High civilizations, as we define them today, began when man with great daring risked burying wild seeds in the earth at one time of year in the hope that green plants would appear in the same place at a later time. By that one act of imagination, Paleolithic man radically altered his relation to earth and to time. In his dawning awareness that the fertility of the earth was related to the alternating light and darkness of the skies, he began to tame by his own selection the wildness of plants he found edible, just as he began to tame by his own selection the wildness of animals he found useful.

That, at any rate, is the scenario of modern science that divorces history from myth to call the one "fact" and the other "fiction." According to the time projections of history, for nearly two million years men hunted, fished and gathered before they planted seed. Then twenty thousand years ago or more, peoples hunting mammoths and giant bison crossed the narrow land-bridge of the Bering Straits and wandered south between two massive sheets of ice that once covered the waterways of Chicago. Archaeologists who have recently uncovered rock paintings in the caves of Pedra Furada in Brazil, and who believe that the charcoal found there came from cooking fires, claim a much earlier date. Indisputably Paleo-Indians were in the Western Hemisphere twelve thousand years ago, from Alaska to Patagonia, roasting over campfires

*"Corn is the only veg-
etable we eat that is
made entirely of seeds,
like a pomegranate. To
eat corn on the cob is to
eat life, like fish roe or
caviar, in which we
cannibalize the future in
the instant."*
—SAMUEL WILSON,
*The Inquiring
Gastronome* (1927)

the game they killed with the flaked stone points that have been found at Onion Portage in Alaska, Clovis in New Mexico, Ayacucho in Peru.

In the context of two million years of non-planting, the three agricultural matrices of the world may be said to have begun at roughly the same time, in 9000 to 5000 B.C. In each matrix, man's crucial act was to tame a wild grass into a domestic grain that could feed an increasingly settled and expanding community. The cultivation of wheat is thought to have begun in Mesopotamia in 9000 to 7000 B.C., the cultivation of rice in southwest Asia in 6000 to 4500 B.C. Three major discoveries by archaeo-botanists in the past forty years give evidence of the cultivation of corn in Mexico, between the highlands of Tehuacán and the tropics of Yucatán, in 6000 to 5000 B.C.

In a remarkable find in the 1960s, Richard MacNeish, encouraged by Paul Mangelsdorf, explored the caves of Tehuacán in the arid highlands of south-central Mexico, near Oaxaca and the ruins of Mitla, and uncovered 24,186 specimens of maize in a total of five caves. More than half of these were whole cobs, so that for the first time an evolutionary growth of seven thousand years could be arranged in sequence. The smallest of the soft cobs were even tinier than those found in New Mexico's Bat Cave and could be dated to 5000 B.C. At first, a jubilant Mangelsdorf announced that this was the wild corn he had known they would find somewhere, but his "discovery" was not unlike Columbus's "discovery" of India. Today, most archaeo-botanists believe that these prehistoric cobs are the earliest known relics of *cultivated* corn and that sometime around 2300 B.C. corn underwent the major evolutionary changes that made it the primary staple of Mesoamerica a thousand years later.

A recent discovery of fossil corn pollen in Oaxaca thought to be nine thousand years old (but pollen dating is extremely speculative) confirms Mexico as the seedbed of corn. So too does an earlier discovery of pollen samples taken by drill cores from depths of six to sixty-nine feet beneath the Palacio de Bellas Artes in the heart of Mexico City. The earliest remains of corn thus far found in South America are in Colombia and suggest a date around 3000 B.C., a thousand years earlier than the arrival of corn in the highlands of Peru. These "discoveries" of modern science confirm what native Americans have always known, that the New World was new only in the sense of Prospero's reply to Miranda, " 'Tis new to thee."

From their first seedbeds in Tehuacán and Yucatán, primitive forms of corn spread north, east and south: north to the Woodland native tribes of the Mississippi and Ohio valleys, east to the Arawaks and Caribs of the Antilles, south to the rainforest tribes of the Amazon and the highland tribes of the Andes. Wherever corn went, the local ecology—rainfall, soil fertility, weed growth—

determined how the crop was grown and what it was grown with. Slash-and-burn, the shifting-field farming called *milpa,* was the earliest and is still the chief method of corn cultivation in the tropics of Mexico and Central and South America. It was a method ideally suited to tropical lands because slashed and burned jungle growth could quickly regenerate itself and renew earth exhausted after two or three years of corn plantings. In recent decades, however, archaeo-botanists have discovered more and more evidence of intensive raised-field farming in ancient Mexico and South America. They have discovered how vast and intricate were the canals dug to create and to irrigate raised fields, which were fertilized organically by water plants and organisms, by water animals and water birds.

As agriculture developed, Mesoamerican Indians grew corn in partnership with plants domesticated earlier—squash, pumpkin, and chili—in a "seed agriculture." Peruvian Indians, on the other hand, grew corn in conjunction with earlier tubers like potatoes and oca in a "vegetable agriculture." The connecting link between the two agricultures, north and south, was corn.

Wherever corn went, civilization followed. When the Olmec erected two thousand years ago the earliest known ceremonial center in this hemisphere at La Venta in northern Tabasco, they connected the action of planting seeds in the earth with the movement of planets and stars in the sky. As corn grew, so did narratives and ceremonies that expressed man's new relation to the organic world when he entrusted his life to buried seeds as well as to migratory animals. Men felt a kinship to animals because they moved as men did across the earth's surface. The myths of the earliest hunting cultures dramatize the struggle between antagonists who are equal, hunter and hunted, alike independent and mobile creatures who rely on skill and speed to outmaneuver each other. Plants, in contrast, stay put. The myths of planting cultures reveal a new terror, an abandonment of human consciousness and immersion in the primeval ooze, the formlessness and flux beneath the earth's crust. Here all creatures breed, grow and decay in the same generative mud. Here man experiences more primally "the voracity of life, which feeds on life," as Joseph Campbell put it, and "the sublime frenzy of this life which is rooted (if one is to see and speak truth) in a cannibal nightmare."

Ritual cannibalism is as central to early planting cultures, Campbell found, as the image of a primal being whose body is the universe. Because for subsistence planters, survival depends directly upon the facts of organic life, their myths express and their ceremonies ritualize the dependency of that life upon death: in the plant world it is vegetable death that generates and sustains new life; in the world of man it is chosen death, self-sacrifice, that generates and sus-

Life from death imaged as the Tree of the Middle Place growing from the belly of a human sacrifice, representing the primal earth goddess. The quetzal bird on top devours the serpent, symbolic of transformation and rebirth. From the Dresden Codex, Maya Postclassic Period, c. A.D. 1300, redrawn by S. G. Morley, *The Ancient Maya* (1946)

tains social order. The tragedy for the conscious animal called man is that pain and death are not only inevitable but necessary. In the Western World, Nietzsche's birth of tragedy begins with the birth of corn.

In the Western World, a plant rather than an animal was the supreme symbol of the life-death connection. Because the New World domesticated few animals in comparison to the Old, its high civilizations retained their planting-culture origins, their primal sense of the mutual dependence of humankind and plant-kind, the interchange of blood with sap and flesh with grain. For the same reason, these civilizations adhered to cannibalistic rituals; animal flesh could not substitute for human flesh because animals did not cultivate plants, only people did.

At the same time, wild animals survived in the New World in unprecedented abundance across vast stretches of tundra, plain, prairie and desert, and their flesh sustained migratory human tribes. One of the oddities of the New World in contrast to the Old is that the ancient conflict between nomad and planter persists to this day. Joseph Campbell spoke of "the culture tides" that for centuries flowed in opposite directions after a few nomadic groups first turned to planting. Settlers widening their village centers to tame more land were in constant conflict with nomads invading those centers to take what others had so conveniently gathered. This cultural dichotomy was shaped and maintained by the topographical extremes of the American landscape. "The great problem of the cultural geography of America," Leo Frobenius said, "lies in the question of how, between the two prodigious negative fields of the northern hunting nomads and the southern forest and water nomads, the American high civilizations could have originated and developed at all."

Develop they did, as those of the Old World did, from the beginning of the Word. "This is the beginning of the Ancient Word" is the way the Sacred Book of the Quiché Maya begins, the Council Book, the *Popul Vuh*. This book, the codices and the post-Conquest *Book of Chilam Balam* are all collections of texts like the Bible or the Book of the Dead. They are both myth and history for a civilization that did not divorce the two but told the history of the gods and the mythos of mankind in a single narrative, as if the Old Testament began not with the genesis of man but with the genesis of the gods and "the emergence of all the sky-earth":

> *the fourfold siding, fourfold cornering,*
> *measuring, fourfold staking*
> *halving the cord, stretching the cord*

in the sky, on the earth,
the four sides, the four corners,

as if the gods were measuring out the entire universe as a cornfield.

The narrative begins with gods and ends with humans, in a collection of stories about the gods who prepared sky-earth before planting within its four sides and four corners the men and women of corn. The literatures of Mesoamerica attempt to chart time as well as place and so recount a series of creations, a fourfold staking of "world ages" from which man emerges at the end. In the *Popul Vuh,* we find that the gods were lonely and wanted someone with enough understanding to talk to but not enough to rival them. At the end the situation is reversed, and we find man worrying about how to keep in touch with the gods. Male and female alike share in the creation of humanity, which is at once a sowing of seeds and a dawning of light, connecting sky-earth by corn.

Corn also connects creation to sacrifice. The twins One Hunahpu and Seven Hunahpu are playing on the ball court when they are summoned by the Lords of Death who have already triumphed over their fathers. The boys are sent to the House of Gloom, given lighted pine sticks and rolled tobacco and told that they must bring them back whole at dawn. But alas, they burn the sticks for light and smoke the tobacco, and so at dawn they are killed. The Lords throw the head of One Hunahpu into a barren tree and instantly the tree is covered with yellow gourds as large as a man's head, the fruit of the calabash tree. A girl called Blood Woman, daughter of the Lord of the Underworld (named Blood Gatherer), sees the fruit and wonders if she should pick it. Then a voice in the branches tells her that the fruit is mere bone and asks her if that is what she wants. I want it, she says. As she reaches up, the bone spits in her palm and tells her that through his spit she will put flesh on the bone of his son. When her father sees that she is pregnant, he calls her a whore and commands the owl Keepers of the Mat to kill her and to bring back a bowl that contains her heart. The girl persuades the owls to fill the bowl with red sap from the croton tree, which looks just like blood, and she escapes to the Upperworld.

Here the girl, her stomach swollen, comes to the

"And so then they put into words the creation,
  The shaping
Of our first mother
  And father.
Only yellow corn
  And white corn were their bodies.
Only food were the legs
  And arms of man.
Those who were our first fathers
  Were the original men.
Only food at the outset
  Were their bodies."
    —*Popul Vuh,* Monro Edmundson
    translation (1971 edition)

The Young Maize God of Copán (Maya Late Classic, c. A.D.775), who is also One Hunahpu of the Popul Vuh. Reproduced by courtesy of the Trustees of the British Museum

house where the mother of the twins still lives. She calls the woman Mother and tells her that the sons of One Hunahpu are inside her. The woman, Xmucame, is outraged by this claim because she believes that her sons are dead. To test the girl, she tells her to fill a big net with corn from the field. In the field, the girl sees but a single corn plant and cries aloud, "Where will I get a netful of corn?" She begs for help from the guardians of the field—Generous Woman, Harvest Woman, Cacao Woman, Cornmeal Woman. She then stands before the tall green plant, takes the red silk of the ear in her hand and pulls it out, leaving the single ear untouched. When she lays out the silk in the net as if the strands were ears of corn, the net is filled. The animals of the field come running, take the net from the girl and parade down the path to the grandmother.

When she sees the full net, the old woman is astonished. "Where is this from?" she asks and storms down the path to the field, but when she gets there she finds only the single stock of corn as before and the place where the net was on the ground. "The sign is there," she says. "I see that you really are my daughter-in-law and that my grandsons will be soothsayers." And so Blood Woman gives birth to twins, whose fathers could not conquer the Lords of Death by strength but did so by guile. The spittle of One Hunahpu was at once corn seed and the flesh of his progeny, as his severed head was the head of the Young Maize God whose death brings the return of the Sun.

Now it becomes time for the grandsons, Hunahpu and Xbalanque, to play ball with the Lords of the Underworld, but they leave with their grandmother "a sign of their word," a sign that they will return: in the center of her house, each boy plants an ear of corn on the earthen floor; when the corn dries, they tell her, it is a sign of their death and when the corn sprouts, it is a sign of their life. In the Underworld, the boys outwit the Lords of Death and by guile recover the head of One Hunahpu and use it as a ball in the ball court. But since they have outwitted both the gods and monster animals sent to defeat them, the boys predict that the Lords of Death will build a great stone oven and burn them there. This death they choose and so they leap into the oven. They have told the seers Xulu and Pacam to grind their bones on a stone the way corn is ground and to sprinkle the bone meal in the river. When the seers do so, the ground bone sinks and after the fifth day the boys reappear—first as catfish, then as vagabonds who dance in various masks the dance of the heart sacrifice. With each sacrifice they bring themselves back to life, and the Lords of Death are so ravished by the dance that they too beg to be sacrificed. So the boys do as they ask, but do not bring them back to life.

When the boys burn in the oven, their grandmother cries out and burns copal in their memory, for the corn plants they have planted in her house have dried up. When the corn plants sprout again, the grandmother rejoices and gives them names: Middle of the House, Middle of the Harvest, Living Corn, Earthen Floor, in order that they will never be forgotten. So today the Quiché Maya burn copal in the pair of shrines (the *uinel*) they place by their cornfields, and after praying that the seeds will sprout again, they pass the ears through the smoke and place them in the center of the house in harvest time. These ears are to be neither eaten nor used as seed corn, Dennis Tedlock has reported, for they are "the heart of the corn" that does not die but remains alive throughout the year, "between the drying out of the plants at harvest time and the sprouting of new ones after planting."

Today, the Quiché use the *Popul Vuh* as a divining book, like the Chinese *I Ching,* because it provides a way of seeing daily events within the unchanging cycles of sky-earth as they were measured at "The Dawn of Life" in "Our Place in the Shadows," when the dawning of seeds and the sowing of stars began. For the people of corn, dawning and sowing are metaphors of each other, for their stories also tell of the "one dawn for all tribes" in the birth of the gods who become sun and moon, Venus and Mars, charted in their cyclic course by the 20-day, 260-day and 52-year cycles of the Maya and Toltec calendars.

One of the twenty named days of the Maya calendar, like the day-sign Kan, is Net. When the old grandmother sees the imprint of the girl's net in the earth by the corn plant, she takes it as a sign that the evening star that rose on the day named Death will reappear as the morning star on the day named Net. The twin maize gods are also aspects of the evening star, the star-mask of Quetzalcoatl, whose heart at his fiery death was transformed into the bright planet that descends into the Underworld in cycles that parallel the cyclic life of corn—and of man. In the *Popul Vuh,* man who sprouts like a seed in the womb sees his birth as the dawn of life; man whose corpse decays in the earth sees his death as the dawn of spirit. Like the twin gods, like the Plumed Serpent, he is both seed and star.

The authors of the *Popul Vuh*—who called themselves "daykeepers" or "mother-fathers of the Word," in imitation of the first "mother-father of life"—recounted also the pilgrimage of the Quiché lords to the East—to the kingdom of the Plumed Serpent. They returned with the Council Book they call "The Light That Came from Across the Sea," which enabled them to found the pyramids and palaces of Rotten Cane. This was the historical citadel of Utatlán, near Chichicastenango on the Pacific side of Guatemala, which the Quiché-speaking Maya founded around A.D. 900, after the abandonment of Palenque and Copán and after the invasions of Nahuatl-speaking Toltec from

Tabasco spread the name and power of the Plumed Serpent. In the *Popul Vuh,* the first men of corn were created in the darkness in the East—at Tula Zuyua, Seven Caves, Seven Canyons—evoking both the historical capital of the Toltec in Tula and the far older capital of Teotihuacán in central Mexico. Here beneath the Pyramid of the Sun are seven underground chambers and here, the *Popul Vuh* says, began the split between the Nahuatl languages of Mexico and the many languages of the Maya, a Pyramid of Babel. In the *Popul Vuh,* as in the Book of Exodus, the story of man is one of exile and wandering, after the golden age at Tula, at the dawn of the many tribes of Mesoamerican civilization in their rising and falling.

Utatlán, which rose as other Maya citadels fell, flourished until 1524, when the Spanish soldiers of Pedro de Alvarado slaughtered most of the inhabitants. Thirty years later, "amid the preaching of God, in Christendom now," a few survivors translated Maya hieroglyphs into the Roman alphabet in order to preserve the *mythistory* of their lost kingdom. The Quiché manuscript survived until the beginning of the eighteenth century, when a Franciscan friar at Chichicastenango, Francisco Ximenez, copied down the text and translated it into Spanish. While the Quiché original was lost, Ximenez's manuscript was published in the mid-nineteenth century, first in Vienna and then in Paris, before it worked its way back across the Atlantic to the Newberry Library in Chicago (a city named, by the way, for the wild onions that grew there after the ice retreated and left fields for the planting of corn).

The daykeepers of the *Popul Vuh* adopted the foreigners' alphabet as a mask,

In this Mexican Huichol yarn painting of The Five Sacred Colors of Maize, *by Guadeloupe and Ramón Medina Silva, each corn stalk represents a sacred corn maiden watched over by the dove, Our Mother Kukuruku.* The Fine Arts Museums of San Francisco, Gift of Peter F. Young

like the Zinacantec mask of San Sebastian, for the invaders' Word did not displace their own, any more than my Uncle Roy's Blood of the Lamb displaced the natives' Seed of Blood. As symbolic lambs absorbed actual human sacrifice in the Old World, so in the New World symbolic corn absorbed cannibal rites, but without a corresponding loss of vision. Corn remained both sentient and sacred. Even today a Maya farmer in Chichicastenango will warn that white and yellow corn left alone on the porch will copulate. If a farmer feels the earth quake as if a giant had turned over, he will comfort the corn in the field and tell it not to be afraid. To the north and south of the seedbed at Teotihuacán, the Word dawned wherever the seed of corn was sown.

THE CIRCLES OF CHACO CANYON

FAR TO THE NORTH and west of ancient Teotihuacán, the present ruins of Chaco Canyon in New Mexico are best seen from above, for only then can you grasp the scale of Pueblo Bonito. The weirdness you get right away, by driving north from Gallup toward Navaho country through miles of flat scrub desert before turning east on a straight and narrow track to nowhere. Chaco Canyon is mercilessly dry, hot, barren and seemingly empty even of ruins, for at a distance the layered stones are the color of sand and cliff, and not until you trace on foot the contours of ruined walls do you begin to sense the achievement of the Anasazi civilization that flowered here in A.D. 1100, just as the Normans "discovered" England, and that withered around A.D. 1300, not long after Marco Polo "discovered" China.

I "discovered" Chaco Canyon on my way back from the sixty-second Inter-Tribal Ceremonial at Red Rock State Park, outside Gallup, in August 1983. Tribes from all over the country were camped in permanent white concrete tepees during four days of parades, rodeos, frybread and dances. Every night the dancers paraded in a giant circle beneath a cliff of red rock silken as skin. When the circle was complete, they sat in the shadows cast by three pine fires burning in the center where the dancing began. There were a couple of thousand spectators, and those of us who had failed to adjust to ceremonial time were numbed by events like the Navaho Corn Dance, a boringly gentle pantomime of women grinding corn on their metates and brushing and braiding each other's hair. The final event, however, featuring "The Aztec Flyers" from Mexico City, made up for any languors. First a lone dancer climbed to the top of a dangerously tall pole and, while he played a flute, his feet beat out a rhythm on a platform small as a drum. Soon he was enclosed at the top by four feathered dancers, who swayed like a squash blossom until the blossom burst and each plumed body dove head first into space, spinning with ropes tied to their ankles in an ever-widening circle until they touched the ground.

I saw this circle carved in earth when I looked at Pueblo Bonito, the largest of the eight "Great Houses" of Chaco Canyon, where the Anasazi built a five-story apartment building of eight hundred rooms to house more than a thousand people. In the center of the crescent-shaped enclosure are nothing but circles, thirty-two of them, outlining the excavations of large and small kivas. The kivas began as pit houses dug into the earth for shelter, storage and burial, a practical cradle-to-grave womb, but their circular shape became potent as they evolved into ceremonial centers. For every kiva was and is today among descendant Pueblo Indians the vital link between the not yet born, whom we call the dead, and the undead, whom we call the living. The word "kiva" means "world below," and the ladder that men climb to reach the opening in the roof is also the ladder by which man emerged, with other forms of life, from the Earth Mother below—not a sudden fall from a heavenly garden but a slow ascent from a watery kingdom to one dry as sand.

Directly below the roof opening of the kiva is a sunken fire pit, for in the First of the Four Worlds, in the cosmogony of Mesoamerica and the Southwest, life begins with fire. Next to the pit is a small hole called by the Hopi *sipápuni,* "the path from the navel." This is the umbilical cord that attaches man in life and death to the Place of Beginning in the Underworld, in the earth's womb, even as he climbs toward the sky. For the Pueblo people, the center of the kiva is the navel of the universe, "the center of centers, the navel of navels," the sacred place from which the people of corn emerge and return.

What is now the American Southwest was once the northern edge of the Olmec corn circle as it spread slowly from Chihuaha into New Mexico through the Mogollon culture that emerged around 100 B.C., followed by the Hohokam and finally the Anasazi culture. By A.D. 400 the Anasazi were planting corn in the clefts of impossible canyons and dwelling in the overhang of improbable cliffs in the place where Colorado, Utah, Arizona and New Mexico now join to make "Four Corners." Here they developed irrigation systems based on seasonal rains and floods, covering an area as large as Ireland, to make the desert bloom at Mesa Verde, Aztec, Canyon de Chelly and Chaco Canyon. Builders as well as planters, they constructed massive buildings of fitted stone and dug in an ever-widening circle the vast underground chamber of the Great Kiva. At the end of the thirteenth century, suddenly, like the Maya at Palenque, they abandoned their settlements for reasons unknown and migrated to the high plateaus of Zuni and Hopi and to the valley of the Rio Grande.

Until a couple of decades ago, archaeologists believed that the climate at the time of the Anasazi must have been different for planters even to attempt to develop a canyon now as barren and waterless as Chaco. But recent studies have shown that the climate was the same; it's the farming that was different. The Anasazi, unlike the Hohokam of southern Arizona, had no perennial streams to tap for irrigation. But they turned disaster to advantage by a complex system of desert farming, in which they channeled flash floods as they poured over the cliff faces into a network of dams and diversion walls, canals and dikes, trenches and gates, and thereby distributed the water gradually into fields laid out in waffle patterns. The system we now call "flood farming" exhibits such sophisticated engineering that we are only now beginning to credit its intention.

At the time the Anasazi laid out their waffle fields, invading Athapaskan hunters from the Canadian Northwest swept down across the Plains and called these farmers *anasazi,* or "ancient enemy." The Zuni descendants of the Anasazi called the Athapaskan hunters *apache,* or "enemy." After a group of the nomadic hunters learned from the planters in the Pajarita Plateau in northern New Mexico how to plant corn, the Pueblo people began to call them the *apaches* of *navahu,* or "enemies of planted fields." While both Pueblos and Navahos plant cornfields today, the cultural tides are primordial and the enmity remains.

The myths of the Pueblo planters are divided between tales told as fiction and stories of The Beginning told as historical truth. "In The Beginning there were only two: Tawa, the Sun God, and Spider Woman, the Earth Goddess" is the way one Hopi creation story begins. From earth and spit, Spider Woman created twin boys whose job was to keep order in the world; next she created all plants, birds and animals; finally, from earth-balls mixed with the four colors—yellow, red, white and black—she created the First People of the First World. Here as in Maya creation, the First People emerge gradually, but now they emerge specifically from a circular center, from "Earth mother earth navel middle place." Their path is upward through the four womb worlds of Mother Earth, through the Second World, which froze into ice, through the Third, which was flooded with water, into the Fourth, which is the Middle World, the here and now of the four language groups of Tewa, Keresan, Hopi and Zuni, united as children of corn.

In his translation of Zuni creation stories, Frank Cushing began with the five elemental facts of Pueblo life—air, earth, water, fire and corn.

I once heard a Zuni priest say: "Five things alone are necessary to the sustenance and comfort of the dark ones among the children of earth.

> *"The sun, who is the Father of all.*
> *"The earth, who is the Mother of men.*
> *"The water, who is the Grandfather.*
> *"The fire, who is the Grandmother.*
> *"Our brothers and sisters the Corn, and seeds of growing things."*

So Cushing related what the Zuni priest told him in 1884 of The Beginning.

In The Beginning was watery darkness from which the Sun Father rose to

"Grass thus became as milk to the creatures of the animal kingdom, and corn became the milk for mankind."
—FRANK WATERS,
Book of the Hopi
(1963)

give light and life. From his skin he rolled balls of flesh that became Earth Mother and Sky Father. When Earth and Sky embraced each other, Earth Mother conceived in her four cavernous wombs the first men and other creatures. When Sky Father spread out his hand, in the lines and wrinkles of his palm were yellow grains to guide their children. From a bowl of foam, Earth Mother created the boundaries of the Earth, but the parents were afraid to let their children go and so kept them in Earth Mother's womb. From foam Sun Father then created twin boys, warriors, whose bows were rainbows and whose arrows, thunderbolts. With their bolts they split the cave wombs of Earth Mother and gave the children of darkness a tall ladder, which they put against the cavern roof so that they might climb into the cave of twilight and finally into the light of the Sun. So bright was the light that "they fell grasping their eye-balls and moaning."

The priest-chiefs then taught them the sacred words and rituals, divided them into clans, and gave each clan a magic medicine. They were told they must wander for many generations until they reached the "Middle of the World." In their wanderings they came on "The Place of Misty Waters," and here their own seed clan found a rival group called Seed People, who challenged the clan to show its powers. The clan planted prayer plumes, and for eight days it rained and new earth washed into the valleys. Then for eight days the Seed People danced and sang and at the end the prayer plumes had turned into seven corn plants.

"These," said the Seed People, "are the severed flesh of seven maidens, our own sisters and children." They introduced their eldest sister, who was yellow corn, and so on down the line through blue, red, white, speckled and black, until the youngest, who was sweet corn. Each corn color came from a place, yellow from the wintry North, blue from the watery West, red from the summery South, white from the light-giving East, speckled from the cloudy heavens, black from the womb caves. And so the clans joined to form the one Corn-clan and their maidens danced the dance of the Beautiful Corn Wands.

As the Corn-clan continued to wander, they heard flute music from the Cave of the Rainbow and sent two chief warriors to investigate. In the cave, they found the God of Dew and Great Father of the medicine priests, *Pai'-a-tu-ma,* who showed them new dances to take back to their people. At night the God of Dew led his flute players to watch the corn maidens dance and, as they watched, the flute players lusted after the maidens of corn, "more beautiful than all human maidens," until the corn maidens passed their hands over their own bodies and vanished, "leaving only their flesh behind."

The people mourned and sent the eagle to look for them, then the sparrow hawk, then the crow, who had a sharp eye for the maidens' flesh, but even he could find no trace. The God of Dew found them at last in the Land of Everlasting Summer and brought them back, but they could stay only part of the year. "As a mother of her own blood and being gives life to her offspring," said the God of Dew, "so have these given of their own flesh to you." From the beginning of the new Sun each year, they should treasure their flesh; and when winter winds and water bring new soil, they should bury their flesh. Then when the corn maidens blew rain clouds from their homes in Summer-land, green corn plants would spring from the heart of the Earth Mother, the Land of the Zuni. Thus, said the priest, was born *Tâ'-a,* the Seed of Seeds.

Rain is the salvation and drought the enemy of these desert peoples, placed in the middle between the Spider Woman below and the Plumed Serpent above. Hopi creation myth and purification rites come together in stories that tell of famine, where a young hero must descend into the Underworld to find Spider Woman and by his bravery bring back the seeds of life. In one story, after the boy is aided by Spider Woman, he is tested by the God of All Life Germs, who whips him with yucca and willow, then gives him a prayer plume and a bundle of seeds. Finally, through a cleft in the deepest rock of the Underworld, the god reveals his splendor, haloed by his wives.

Upon their heads were terraced rain clouds; many kinds of flowers and jewels adorned them, besides every seed of corn such as they have sent us for food. There was white, red, yellow, blue and black; another kind was speckled red and yellow. Such colors! They were like precious stones.

The god commands the boy and others to serve as priests. They are to dip their hands in clay and make a print on the rock as a sign they will show no fear when flogged. After being whipped, each youth holds in his hand five different-colored grains of corn. "Plant one for the hot wind, one for the field rat, one for the kachina, and two for yourselves," the god tells them, as he departs, leaving his mask and robes. One day while they chant his prayers, a giant snake appears, crowned with feathers—Palulkon, the Great Plumed Serpent—who tells them that the bravest must return the mask and robes to the god. But the Death God has deceived them and stolen the mask. From now on, the Plumed Serpent commands, the bravest each year must don the god's mask at the feast of Powamu, the time of purification, when Hopi children even today are initiated by ceremonial whipping into the rites of the kachina, when men impersonate the Hopi gods.

For Anasazi descendants, Earth Mother and Corn Mother are exchangeable masks, both represented by perfect ears of corn whose tips end in four kernels—"Corn Mothers." At the birth of a Hopi child, his Corn Mother is cradled beside him for twenty days while the house is kept dark. Before dawn on the last day, the child's aunts arrive, each carrying a Corn Mother in her right hand, to bless the child and present it to its Sun Father as he rises in the East. The child has dawned in the earth, the child is sown in the sky, for earth-sky are the parents of this child in spirit as well as in flesh. For the Zuni, the masked impersonators of the Corn Mothers appear at the winter solstice six days before the Fasting and six days after the night of Shalako, when the gods each year return. Even in their seasonal absence, the Corn Mothers are always present, either as sacred objects at the hearth or as masked dancers in the kiva.

Far to the north of the Olmec circle of corn, however, where the Iroquois descendants of Woodland tribes once stretched from New York to Ohio and from the Great Lakes to Canada's Georgian Bay, star gods vied with corn gods wherever hunting vied with planting. In a Senecan creation story (recorded in 1883 by Jeremiah Curtin), the First People fell to earth from the sky.

Senecan artist Ernest Smith's oil painting in 1936 of Sky Woman, *based on the Woodland creation story.* Rochester Museum & Science Center, Rochester, N.Y.

Originally, people lived above, in the center of the Blue, and in the middle of their village grew a tree with blossoms giving light. A woman dreamed of a man who told her that that tree should be uprooted. "A circle must be dug around it," he said; "then a better light will come." So the people cut around their tree and, sinking, it disappeared. Darkness fell, and the chief, becoming furious, ordered the woman pushed into the hole. Down, down she fell. Below there was only water, with waterfowl and aquatic animals at play. Hell-diver, however, brought up mud, and Loon then sent all the members of that tribe down for more. "Put the mud on Turtle's back," he said. Beaver flattened it with his tail. Then Fishhawk brought the woman down and, work continuing, the earth increased; bushes presently appeared, and soon the woman gave birth to a girl.

Very quickly the child grew. And when a

young woman, she was one day strolling, enjoying the animals and birds, when she met a nice young man. With their union, day and night came. At daybreak, she would go to meet him; at twilight, return home. One evening, she looked back and saw a big turtle walking where the man had just been. She thought: "A turtle has deceived me." At home, she told her mother and said, "I am going to die. You must bury me and cover me well. From my breasts will grow two stalks, on each of which an ear will appear. When ripe, give one to each of the boys I am to bear." She gave birth to twins, died, and was buried. And that was the origin of maize.

A better-known story of its origin, told by an Ojibway shaman of the Great Lakes to Henry Rowe Schoolcraft in the 1820s (and incorporated by Longfellow into *The Song of Hiawatha*), eliminated Earth Mother altogether in favor of a male named Mondawmin, who descends from the sky and wrestles with a young hunter named Wunzh, who is undergoing the ceremonial fast that will show him his spirit-guide. After the third day, the sky-youth Mondawmin surrenders, asking that his garments be stripped away so that he can be buried naked in the earth. He tells Wunzh to keep the spot clear of weeds and lay on fresh earth once a month. In secret the boy tends the grave and watches as green plumes come through the ground. When at last he brings his father to the spot, they see a tall, graceful plant with bright silken hair and golden arms. "It is my friend, Mondawmin," he says. "We need no longer rely on hunting alone. For this I fasted, and the Great Spirit heard."

"All alone stood Hiawatha,
Panting with his wild exertion,
Palpitating with the struggle;
And before him, breathless, lifeless,
Lay the youth, with hair dishevelled,
Plumage torn, and garments tattered,
Dead he lay there in the sunset.
And victorious Hiawatha
Made the grave as he commanded,
Stripped the garments from Mondamin,
Stripped his tattered plumage from him,
Laid him in the earth and made it
Soft and loose and light above him. . . ."
—HENRY WADSWORTH LONGFELLOW,
Book V: "Hiawatha Fasting,"
The Song of Hiawatha (1855)

Although the Ojibway became planters, they retained their hunting mythos. In this story of exclusively male struggle, triumph and expertise, a Father-Spirit, wrestler-athlete and farmer-chief displace Earth Mothers, pregnant women and corn maidens and make them irrelevant. In the Senecan story, by contrast, the dynamic of complementary creative powers, male and female, resembles the planting mythos of the Pueblo peoples. When Earthwoman mates with Turtle-man, their union generates day and night, birth and death, and sustenance. The corn that sprouts from the dying woman's breasts will furnish milk for her sons, for if wild grass is milk for animals, cultivated grass is milk for man. But still, the first woman falls out of the Blue into darkness, is pushed down a hole and deceived by a turtle. Not a great role for an Earth Mother.

Longfellow's Hiawatha, on the other hand, so preempts all creative roles for himself that one suspects the Ojibway story has been polluted by patriarchal Victorian germs. Earth Mother Nokomis is reduced to a nanny, while Hiawatha and Mondawmin are transfigured into the sort of soft-focus image captioned, in my grandmother's Bible, as "The Light of the World." So Christ-like is Longfellow's Indian that my Uncle Roy could have presented him without qualm to the Brazilian heathen as a model of muscular Christianity that would put to scorn idolators of both the Corn Mother and the Virgin Mary.

Nonetheless, the boy who planted corn by the shores of Gitche Gumee is descended from the Young Maize Lords in the tropics of Copán, Palenque and Tula, just as the earth navel of the Pueblo peoples is connected to those earlier centers where civilization emerged. The Hopi tell of a legendary Red City of the South, a ceremonial center said to have been built by the original kachina people before a serpent destroyed the city and the hero-twins fled northward. Some Hopi who know Maya signs say that the Red City was Palenque and that here the clans were named.

At Chaco Canyon all the migrating clans of the ancestral Anasazi left their stories incised in stone, as well as in legends. On a cliff wall of pictographs you can see the clan signatures of Snake, Sun, Sand, Bear, Coyote, Lizard, Eagle, Water, Parrot, Spider and Bow. You can see the red handprints of the brave and the footprints of the wanderers. You can see the Guardian Snakes of the six directions and you can see too the Hump-backed Flute Player, who has left his mark from the bottom tip of South America to the top of Canada, bearing in the hump on his back the life-giving seeds of corn and bean and squash. It was the Flute Player who led the migrating bands to the ends of the earth in each of the four directions before they found their home in "Earth mother earth navel middle place," the sacred center of each kiva, of each village and of the universe. It was the Flute Player who led the feathered dancers of the Aztec Flyers at the Inter-tribal Dances at Red Rock. If red was the color of the seed of blood, flute music was the sound that set the feathered seed in motion.

Of all the marks at Chaco Canyon the most awesome figure of motion is a spiral carved into the cliff behind a pair of rock slabs weighing two tons each, positioned on top of a 480-foot butte that commands the full horizon. This is the spiral that a young artist, Anna Sofaer, began to puzzle over in 1977. It took her nineteen trips and an unprecedented melding of photography, archaeology, astronomy and physics to persuade specialists in each of these fields that she had discovered a unique sun-and-moon calendar for planting and ceremonies.

During the summer solstice, June 21, at high noon, the pair of behemoth rocks so control the sun's shadow that, as it moves, it bisects the spiral vertically, moving downward like a dagger from top to bottom. At the winter solstice, December 21, the arrangement of the stones divides the sun's shadow into two shadows that bracket the spiral, leaving its center empty of light. Sofaer was able to prove that the site of the Sun Dagger records not only the alternating positions of the solstices of the sun, but also the analogous positions of the moon. Her discovery brought to light the earliest known calendar in the New World that tracks both the twelve-month cycle of the sun and the nineteen-year cycle of the moon.

The site of the Sun Dagger would have been a spot as sacred as the kiva to the Anasazi because the spiral is a form of the circular maze that expresses in spatial terms the temporal myth of Emergence, compressing into one symbol both the birth from the earth-wombs and the migratory Path of Life. The Hopi call this circular maze *Tapu'at,* or Mother and Child, for the concentric spirals enfold the child just as the straight line at the entrance of the maze leads the child out through an umbilical cord. At the same time, the cross formed by the intersection of straight lines symbolizes the Sun and centers the four cardinal

The three rock slabs of the Anasazi calendar at Chaco Canyon (A.D. 950 to 1150) are arranged to channel a dagger of light at summer solstice through the center of a spiral petroglyph, carved on the rock face behind the stones. Photos by Karl Kernberger, the Solstice Project

The circular form of seasons and calendars shaped also kivas, cities and fortresses, linking the Americas north to south. Seen from above when it was excavated in 1917, the circle of the Great Kiva at Aztec, New Mexico (right), is echoed by the circular walls of the fortress of Sacsahuaman in Cuzco (opposite page, left), built 300 years later. Both echo the circularity of an ear of corn (opposite page, right), shown here in cross sections of two varieties: the 12-rowed is from the north, the 8-rowed is from Peru and named for Cuzco. (The Great Kiva) Neg. No. 119748, photo E. H. Morris; (Sacsahuaman) Neg. No. 288012, photo Charles H. Coles; Courtesy Department Library Services, American Museum of Natural History. (Corn) Courtesy Paul Mangelsdorf

points that will guide the unborn on his Path. At the site of the Sun Dagger in Chaco Canyon, the sun's shadow literally crosses the spiral, marking the place of Emergence from the literal earth. Circles within circles—calendar, kiva, corn. Maize within maze.

"For a people so intensely agrarian for so many centuries of their existence, all life does result from happenings within the earth, from the union of earth, water, and sun," explains Alfonso Ortiz, the noted Tewa anthropologist born at San Juan Pueblo, where his father taught him to throw a pinch of white cornmeal to the sun every morning upon awakening. A Tewa is at home in the universe because all its elements are members of his family and he of theirs. Family is the root metaphor that relates Pueblo man intimately both to the outer galactic fires of Father Sun and to the inner volcanic fires of Mother Earth. Wherever he is, he is at home. Earth navels are like airports, a Tewa once told Ortiz. No matter where airplanes fly off to, they always return. Emergence and return. "The word of mythology," as Levi-Strauss said, "is round."

The word of my own migrating clan was neither round nor square because my clan had long ago lost its sense of intimacy with the elements and with the music of the spheres that set them in motion. Earth roots we had left behind in our temporary encampments on the prairies, and "star light, star bright" was

not a guide but a child's wish-rhyme. The heavenly Father was not the sun that rose and set daily, but a stern avenger who damned without mercy and forever. My mother took it to heart. Uprooted and displaced, she lost her way in a maze unlit by the sun's path, pushed into darkness with no way out and no power on earth, or under the earth, to comfort her. Hers was a maze like the ones my father kept in his classroom laboratory for mice and rats.

On the darkest day of the year, by chance I visited another Anasazi site. Between blizzards, I had made my way with difficulty over ice-covered mesas to reach a small point on the map called Aztec, on the border of New Mexico and Colorado. I wanted to see the largest of all kiva navels, the Great Kiva, sunk in the center of the Anasazi ruins which, like those of Chaco Canyon, were once high-rise apartments. Excavated and now restored, the main altar room of the kiva is forty-eight feet in diameter and the ninety-ton roof is held by four elephantine pillars of masonry resting on four huge disks of stone. A raised stone altar in the center is flanked on either side by a stone-lined pit. These were used for the magic fire ceremony Hopis call the Ya Ya Ceremony. *"Yah-hi-hi! Yah-hi-hi!"* cried the chief of the Fog Clan when the god of the animals, Somaikoli, appeared, for Ya Ya gained its power from the animal world. The power was so great that men might jump naked into the fire-pits and come out unscathed. But some began to use the powers for evil, and now Ya Ya is practiced only by sorcerers and witches who kill others to prolong their own lives.

The Great Kiva was eerily dark and deserted, but it had been so fully restored that it seemed phony, and I was quite put out by the canned flute music and the amplified drums. I climbed back outside for a breath of reality and was hit by a bright sun that shot through a rent in the clouds. By its light a pair of archaeo-astronomers, armed with surveying instruments, were attempting to track the winter-solstice shadows across some vertical stones. That too disappointed. No spirals, no sun daggers. Noon passed and nothing happened. When I went back down into the kiva, however, for a final look, something had. The kiva was as empty as before, but this time the flute and drum got to me. There on the center of the altar was a freshly killed bird, its feathers still warm.

The Language of Science

SOLVING THE CORN MYSTERY

WHEN PAUL MANGELSDORF ASKED, "What is the corn?" he answered, "Corn is a mystery." For the scientist, mystery is not a condition but a prelude to action. "And mysteries are there to be solved," Mangelsdorf asserted, "as surely as mountains are there to be scaled." Corn's "mystery" has meant one thing to those who speak the discursive language of science and quite another to those who speak the symbolic language of myth.

My botanist father solved the conflict between science and myth by filing them in separate compartments. The Bible had solved one mystery of origin, and Darwin the other. My father was not a speculative man, and he took Darwin's *Origin of Species* with the same good faith with which he took God's Book of Genesis. God created and Darwin evolved. It was the scientist's job to describe and improve the plants and animals of God's garden, while measuring the mountain of knowledge that was there to be scaled. Although he was a Bible Belt fundamentalist, my father believed it was knowledge, not faith, that moved mountains, for in his world there was no mystery that a little midnight oil and mental elbow grease couldn't solve.

The first botanists to examine the mysterious plants of the New World, however, undertook their task with a fuller measure of wonder. "This Corne is a marvelous strange plant, nothing resembling any other kind of grayne," wrote Henry Lyte in *A New Herbal* of 1619, "for it bringeth forth his seede cleane contrarie from the place whereas the Floures grow, which is against the nature and kinds of all other plants." Corn was marvelous strange, that is, in its sex life, for the seeds of the plant grew at one place and the flowers at another. As we would now say, the ear was distant from the tassel. But to understand the implications of this sexual division, the first botanists had first to "discover" that plants in

The corn plant as imagined in the mid-16th century by an artist who has never seen one, for he improvises tassels, blossoms, symmetrical leaves and kerneled ears without husks.
Courtesy Paul Weatherwax

general had a sex life. Early in the seventeenth century, Camerarius in Germany, experimenting, as it happened, with the corn plant, determined that plants have sexual parts that correspond to the human sperm and egg.

Europe's discovery of the New World coincided with a new way of seeing the natural world, in which the medieval way of reading the world symbolically as God's poem was replaced in the Renaissance by a textual analysis of nature's prose. The new scientists began to apply human analogues to plants, as Indian myth makers had done, but for different ends. Where Indians had given corn gender and genetic relationship to make all God's creatures one, Europeans now used human analogies to separate the plant world from the human one and to reconstitute plants as an autonomous family. It was less the invention of new tools for observation, like the microscope, that led plant scientists to believe they were discovering "the thing in itself" than it was the growth of a new language in which to systematize their "discoveries" of new mountains to scale. To the question "What is the Corn?" the Zuni who even today answers, "Our brothers and sisters," is worlds apart from the contemporary morphologist who answers, "A polystichous diploid with unisex inflorescences and rigid rachilla."

It was Swedish botanist Linnaeus who, in the eighteenth century, determined to marshal plants into families labeled with the logic of Latin and Greek, but his names were as arbitrary as anthropologists' names for Maya gods. Linnaeus christened corn first with a Greco-Latin generic name, *Zea,* which means a wheatlike grain, and then with a Latinized specific name, *mays,* from its Taino original *mahiz,* which means "life-giver." For modern taxonomists, corn is a member of the grass family Poaceae, which divides into the tribe Andropogoneae (corn, sugar cane, sorghum and teosinte), then into the subtribe Maydeae (corn, teosinte and tripsacum) and finally into the clan *Zea,* which includes the species *Z. mays,* of which domesticated maize is a subspecies.

What was and is jabberwocky to me was a constant delight to my label-loving father, who was never happier than when he had a blackboard and chalk at his command. He would point out to his botany students that the seed of this grain called *mays* should more properly be called a fruit, in which the germ is embedded in nutrient flesh. Those nutrients he would itemize in the language of chemistry: for an average dried kernel of corn, 72 percent starch, 10 percent protein, 4.8 percent oil, 3 percent sugar, 8.5 percent fiber, 1.7 percent ash. Unlike most fruits, which convert starch to sugar, he explained, corn reverses the process and converts sugar to starch. The oil is in the germ and the rest of the nutrients are in the surrounding endosperm, composed of both hard (horny) starch and soft starch. Soft starch occupies the crown and central part, hard starch the sides and back. The ratio of hard to soft starch varies with each kind of corn, and we label them accordingly. Different pigments, such as blue, red, black, white, are literally only skin-deep, the pigmentation being limited to a thin layer (aleurone) covering the endosperm just within the thinner outer layer of the hull (pericarp). Yellow pigment is the only one that resides in the endosperm.

After thus anatomizing the kernel, my father would set up a laboratory experiment to show how corn grows. He would put a few kernels in a saucer of water in a warm room and in five or six days they would sprout. He would point to the shield, the scutellum, on the rim of the germ, which absorbs water into the germ's tip and channels it to the tiny shoot growing upward and the tiny shoot growing downward. Through a microscope he would point out the minuscule hairs covering the root, designed to adhere to soil and absorb water. Once you plant the kernel, he would explain, it will develop almost immediately a set of temporary (seminal) roots that branch sideways before the permanent roots develop underground to form a tight circle a foot or two in diameter. As the stalk grows, brace roots form another circle above ground, a kind of ballet dancer's tutu—a metaphor my father would not have used.

My father would quote straightforward texts like *Corn and Corn Growing* (1949), by Henry A. Wallace and Earl N. Bressman, who stated that corn grows primarily on air and water. While 97 percent of the elements in the kernel comes from air and only 3 percent from soil, corn demands a quantity of water because it sweats, or "transpires," as they said, heavily. They noted that an acre of corn plants on a hot July day in an Iowa cornfield might transpire up to 18 tons of water each day, or 720 tons of water per acre during the growing season, the equivalent of seven inches of rainfall. Wallace and Bressman used the metaphor of an unfolding telescope to explain how the stalk grows in nodes, about eight to twenty nodes per stalk. The section between nodes (the internode) is where growth takes place, one leaf for each node on alternate sides of the stalk. When the internode slides out of its leaf sheath, it makes a noise loud enough for farmers to swear that they can "hear the corn grow."

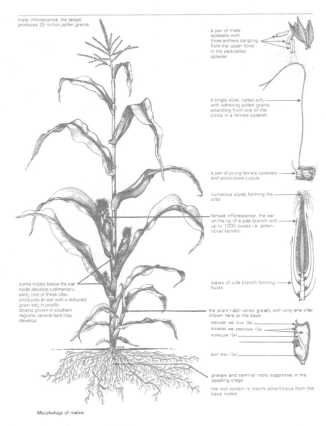

An anatomical drawing of a corn plant by contemporary geneticist Walton C. Galinat, detailing male and female parts. Courtesy Walton C. Galinat

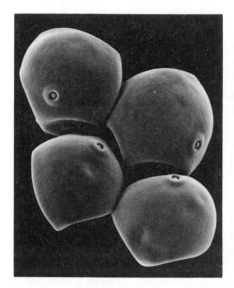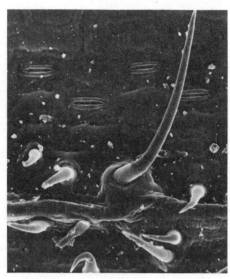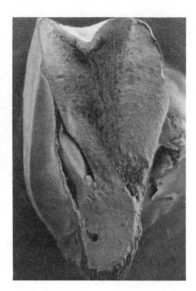

The electron micrograph reveals the sexuality of corn parts in a new perspective. Pollen grains (magnified x 1210), resemble mammaries. On a leaf's surface (magnified x 770), stomates look like mouths and spines like penile horns. The kernel, in cross section (magnified x 25) is a womb that cradles and feeds the embryonic germ. Pollen and leaf photos by Dennis Kunkel; kernel photo from the USDA Agricultural Research Service

During his own student days, my father would have read earlier textbooks like Frederick Sargent's *Corn Plants* (1899), in which Sargent charted the corn from infancy to old age. "For the baby leaves," he said, "a tube forms the snuggest sort of cradle." The upper part of the leaf unrolls as it grows, enclosing the rolled blade of a younger leaf, and because the edges of the blade grow more than the center, the edges get a wavy look. Through their green coloring matter (chlorophyll), the leaves suck up sunlight and carbon dioxide from the air and mix it with water to form sugar and starch.

When he took up the sex life of the plant, in a chapter titled with Victorian propriety "Provision for Offspring," Sargent explained that the flowers of the corn plant, divided into male and female, grow in spikelets on different parts of the plant, an unusual arrangement peculiar to the subtribe Maydeae. The male flowers grow in spikelets in a tassel (inflorescence) at the top of the stalk, where the spikelets are really miniature husks enclosing "the tender parts." These are a pair of flowers in each spikelet and in each flower a trio of anthers to hold and release the pollen, formerly called *farina fecundas*. Each tassel produces about two thousand grains per anther, or 14 to 18 million grains per plant. Like sperm, pollen production is supernumerary rather than cost-efficient. Since each kernel needs but one pollen grain to fertilize it, the plant overrun is about twenty thousand pollen grains per kernel. As Sargent said, "a very generous margin for mishaps has been allowed."

The female flowers, or "mother cells," he explained, are located in pairs (although only one will develop) along the tiny pubescent cobs that will turn into ears of corn. In Corn Belt corn, there is usually one ear per plant. Each cob grows within a leaf sheath at a node about halfway down the stalk. The flowers grow into kernels, four hundred to eight hundred or more per ear: one part of each flower encloses the ovary (pistil) and another part sends out a single strand of silk (style). In this remarkable arrangement, each single kernel must be fertilized separately by means of its single silk. Timing is everything, for the kernels at the base must send out silks from the tip of their leaf cocoon at the very moment the pollen is ready to drop. It usually takes the silks two to four growing days to emerge from the tip in a feathery green spray. The surface of each silk is hairy and sticky in order to catch that single but crucial dot of pollen dust. Once caught, each nuclear dot of sperm divides itself and becomes twins, as if following the plot of Maya myth. One twin forms a tube within the silk so that its brother twin can slide down the six- or eight-inch length of the silk to reach the virgin embryo sac at the base. While one twin fuses with the egg to create the embryo, which then becomes the kernel's germ, the other twin creates the endosperm. Corn mating takes place within twenty-four hours, and within a couple of days a farmer will know that conception has occurred, because the silks at the tip will then change color from green to reddish brown.

"By forming a judicious mixture with the gourdseed and the flinty corn, a variety may be introduced, yielding at least one third more per acre, on equal soil, than any of the solid [flint] corns are capable of producing, and equally usuable and saleable for export."

—JOHN LORAIN, in a letter to the Philadelphia Agricultural Society, dated July 21, 1812

Corn Belt corn takes about fifty days to mature after fertilization, Sargent said, ripening in five stages: milk, when the starch is still in fluid form; soft dough, when the starch is "soft and cheesy"; hard dough, when the starch is firm; glazed, when the hull is complete; and ripe, when the kernels are fully matured. Because kernels are attached to the cob in paired rows, corn rows on the ear are usually (there are a few variant types) even in number. A farmer running his hand over the outside of a green husk can tell whether an ear is ripe merely by the feel. If he feels separate rows, it's not fully ripe; if he feels separate kernels, it's over the hill; if he feels a smooth "heft," it's just right.

My father felt at home with the teleology of the scientific language of our founding fathers because it was based on utility. "What is not useful," said Cotton Mather in the late seventeenth century, "is vicious." Cotton Mather is important to the history of corn because he was one of the first of the preacher-scientists in this country to experiment with corn in the field. In 1716 he wrote from Boston to a friend to describe the effects of cross-pollination:

My friend planted a row of Indian corn that was colored red and blue; the rest of the field being planted with corn of the yellow which is the most usual color. To the windward side this red and blue row so infected three or four whole rows as to communicate the same color unto them; and part of ye fifth and some of ye sixth. But to the leeward side, no less than seven or eight rows had ye same color communicated unto them; and some small impressions were made on those that were yet further off.

Soon after, a personal enemy of Mather, Paul Dudley, made a similar observation, noting that a high fence could prevent the admixture of color resulting from this "wonderful copulation" of corn.

To a utilitarian like Benjamin Franklin, who conducted his own experiments in corn copulation, corn's sex life was precisely what made corn a most useful medium for studying and improving plants. Franklin was the man who produced the first commercial crop of broom corn (a distant relative of maize) and who founded the American Philosophical Society. One outgrowth of this Society is today's vast network of American agricultural societies that have, in effect, reinvented corn in our time.

Many pioneers of the new science in the colonies experimented with corn breeding, applying both Latin and colonial English to Indian corn in order to analyze principles of corn reproduction. Such was James Logan, secretary to William Penn, who wrote up his experiments in Latin in 1739 to validate the notion that male tassels were essential to female silks. Such was John Lorain, also of Pennsylvania, who saw the benefits of mixing different strains of corn from different parts of the country to improve yield. "They do not mix minutely like wine and water," he wrote in *Nature and Reason Harmonized in the Practice of Husbandry* (1825). "On the contrary like the mixed breeds of animals, a large portion of the valuable properties of any one of them . . . may be communicated to one plant, while the inferior of one or the whole may be nearly grown out."

From the beginning, what interested the newcomers to corn was mixing the breeds to further their utilitarian aims of providing the greatest number of ker-

Corn copulation sometimes produces deformities, like these multibranched cobs. Courtesy Paul Weatherwax

nels for the largest number of people. Instructive is Joseph Cooper's account in 1808, in the initial volume published by the Philadelphia Agricultural Society:

> In or about the year 1772 a friend sent me a few grains of a small kind of Indian corn, the grains of which were not larger than goose shot.... These grains I planted and found the production to answer the description, but the ears were small and few of them ripened before frost. I saved some of the largest and earliest and planted them between rows of the larger and earlier kinds of corn, which produced a mixture to advantage; then I saved seed from stalks that produced the greatest number of largest ears, and first ripe, which I planted the ensuing season, and was not a little gratified to find its production preferable, both in quantity and quality to that of any corn I had ever planted.

From the beginning, the language of the newcomers reflected different agricultural values from those of the natives, who were concerned less with productivity than with purity. Indian tribes had carefully selected seed to preserve purity of color and type, because each color had sacred meaning as well as food meaning. From each kind and from each crop, Indians selected perfect ears and saved these for seed, as their ancestors had done for thousands of years. A Hidatsa woman named Buffalo Bird Woman, born along the Knife River in North Dakota in 1839, described in detail the selecting methods of her tribe:

> When I selected seed corn, I chose only good, full, plump ears; and I looked carefully to see if the kernels on any of the ears had black hearts. When that part of a kernel of corn which joins the cob is black or dark colored, we say it has a black heart. This imperfection is caused by plucking the ear when too green. A kernel with a black heart will not grow.... When I came to plant corn, I used only the kernels in the center of the cob for seed, rejecting both the small and the large grains of the two ends.

She also described how they planted each variety separately, in fields away from each other. "We Indians understood perfectly the need of keeping the strains pure," she said. "We Indians knew that corn can travel." They knew that adjoining rows of yellow and white corn would produce mixed ears. "We Indians did not know what power it was that causes this," she said. "We only knew that it was so."

White men wanted to know what power it was because they wanted to control it. The white man's approach was founded on the principle that crossbreeding is better than inbreeding. Instead of a conservator of blooded lineage, the white man became a broker of mixed marriages, exercising control over every stage of the breeding process. As a result, he founded a new corn dynasty that was clean contrary to the native one. Over two centuries of experimentation, he developed a highly specialized method of crossing two sets of inbred strains, not once but twice, so that each hybrid seed today comes equipped with a pedigree of four named grandparents and two parents, in a listing that constitutes a kind of Debrett's Peerage of Hybrid Corn.

"Hybrid," however, is another term that breeds confusion. When my grandfathers began to farm as boys, they controlled pollination much as the Indians did, isolating different strains of corn in separate plots to prevent random crossings and to create inbred strains. Since the winds might blow from each and every corn tassel 14 million grains of pollen, some bastardy was inevitable and occasionally beneficial in open pollination. By the time my grandfathers were grown men, however, they had learned to cross the inbred strains of different kinds of corn deliberately, in order to get larger and hardier ears. These artificial hybrids they would keep pure by selecting and saving the best seed at each harvest. What they gained in genetic uniformity, however, they lost in reproductive vigor. To solve this problem, the modern corn breeder must return, at each planting, to seed produced from the first crossing of inbreds. Modern hybrid corn is generated not by the farmer but by the hybrid seed industry, a child of the industrial revolution of corn.

A remarkable figure who tried to mediate between the Indian's and white man's conflicting corn languages was a seedsman in the early decades of the century named George F. Will, who with his colleague George E. Hyde wrote *Corn Among the Indians of the Upper Missouri* (1917). Born in Bismarck, North Dakota, in 1884, George Will knew the Mandan, Arikara and Hidatsa at first hand, since they traded at his father's store and greenhouse. When his father put out a seed catalogue in 1887, the first seed corn he listed was a squaw flint named "Ree," short for Arikara. Father and son perceived the value of the early-maturing varieties that had been developed by the tribes of the upper Missouri, and the Wills experimented with them to provide good seed for their neighbor farmers in the northwest.

Trained at Harvard in archaeology, ethnology and botany, Will was uniquely equipped to recount Indian Corn Ceremonies and the Sacred Character of Corn at the same time he analyzed Indian agricultural methods and compiled a staggering list (with photographs) of the hundreds of corn varieties still grown in 1917 by tribes from the Northwestern Mandans to the Southwestern Pueblos. As one contemporary admirer put it, "He used what he

"The great scientific weakness of America today is that she tends to emphasize quantity at the expense of quality—statistics instead of genuine insight—immediate utilitarian application instead of genuine thought about fundamentals."

—HENRY A. WALLACE AND WILLIAM L. BROWN, *Corn and Its Early Fathers* (1956)

learned from the American Indian for the benefit of every Northern Plains citizen." Not every modern seedsman or scientist was as wise in seeing that the plant he took such pride in improving was one that was already highly bred. In the words of geneticist Walton Galinat, "The American Indians were not simply the first corn breeders. They created corn in the first place."

In the Name of Divine Progress

"THE DESTINY of the nation is in the hands of the farmers," Edward Enfield wrote in 1866 in *Indian Corn: Its Value, Culture, and Uses.* This was at a time when farmers like my grandfathers governed the destiny and the imagery of the United States, not only by numbers—they outnumbered industrial workers by more than two to one and tradesmen by more than four to one—but by their sense of divine mission. When Enfield said farmers, he meant corn farmers, since the corn crop was five times greater than the crop of wheat and other cereals, and all other major vegetable crops, put together.

During the crisis of the Civil War, corn for the first time became big business, with a total of nearly 840 million bushels a year. At the end of the war, Enfield exhorted farmers to increase production so that the national *average* would be an unheard-of fifty bushels an acre instead of the usual twenty or thirty. Neither he nor my grandfathers could have imagined that exactly a century later, in 1966, the year Henry A. Wallace died, the national average per acre was more than 73 bushels, producing a total of 5 billion bushels a year valued at over $6 billion. My own family abandoned farming at the very moment Wallace staged the farming revolution that brought this about. In less than thirty years, from the 1920s to the 1950s, Wallace fulfilled the manifest destiny of American farmers by "industrializing" corn breeding and thus laying the foundations of modern American agribusiness. "No plant has changed so fast in so short a time as has corn," Wallace wrote, "in the hands of the white man."

Change, in the lexicon of the white man, was synonymous with improvement, and corn's natural variability made it a prime subject for "improvement" by human hands. It was corn, not wheat or potatoes, that became as crucial as oil to a nation hellbent on converting natural elements to mechanical ends. "Corn," boasted a contributor to the American Society of Agronomy's *Corn and Corn Improvement* (1979), "has achieved a higher level of industrial utilization than any other cereal grain."

More than a century ago, Enfield envisaged a New Jerusalem where agriculture and industry would march hand in hand to the sweet music of machinery. Of this new-age farmer he wrote,

> The discordant clatter of machinery that shocks the ears of other men is to him the sweetest of music; for it starts the long dormant corn from the crib, gives new activity and interest to butter and beef, and infallibly prognosticates a new top to the Sunday carriage, a silk gown for the wife, a suit of clothes for the little boy, and a new dress for the baby.

Enfield hailed the farmer as a closet manufacturer: "It may indeed be said that the farmer, in a broad and important sense, is himself a manufacturer, for, like the latter, he is essentially a creator of values."

The organic corn plant shared in this ennoblement by becoming an industrial plant. So Frederick Sargent in his *Corn Plants* (1901) praised corn: "The plant is like a factory, with the discouraging sign 'no admittance.' " From the way in which roots, stems and leaves are organized into "a singularly perfect system," he went on, "we may see that our self-building food-factory is governed by advanced business methods." Remember that this was a corn plant Sargent was describing:

> From the start, the policy pursued is to devote at once as much as possible of the product of manufacture to building additions to the establishment and to insuring its future safety. It is as if there were a wise and enterprising manager in charge of its affairs. This same spirit of enterprise which leads these plants to take fullest advantage of their opportunities appears also in the establishment of what we may call "branch factories" [or tillers].

Sargent's personification was as mythic as the *Popul Vuh,* but instead of a pair of enterprising twins in the Underworld we now had a factory foreman inside each cob.

The twin heroes of the new industrial myths were Energy and Power, generated in new form when agriculture, science, business and industry ganged up on Mother Corn. "Decade after decade, beginning in 1780," Wallace wrote in *Corn and Its Early Fathers* (1956), "the progress of American civilization was measured by the western expansion of the corn acreage." Progress and Civilization, another set of twins, were measured not just by plant expansion, that is, corn and factory acreage, but by increased efficiency of production. Since the industrial farmer produced his merchandise by employing the minimum amount of human labor to convert the maximum amount of solar energy, the rate of conversion was crucial. Where the Indian required twenty hours of hand labor for each bushel of corn, the Corn Belt farmer in 1956 required only six minutes. The story of the conversion of a native Indian corn into the world's most efficient industrial crop was, for Wallace, "one of the great and vital romances of all time."

This romance took its tone from evangelists in the latter half of the nineteenth century who preached Divine Progress through the medium of corn. One of the earliest preachers was Wallace's grandfather, "Uncle Henry" Wallace, a United Presbyterian minister in Iowa in 1862, about the time my own Great-Grandfather Harper, a minister of the same cloth, was preaching God and raising corn in the same neck of the woods. The spirit of these preacher-farmers still invests the corporate offices of Pioneer Hi-Bred International, in Des Moines, where I found three generations of hybrid seed growers. Pioneer is the company Henry Wallace founded in 1926 by selling forty-nine shares for $100 a share. Today, Pioneer sells 645 million pounds of seed corn a year, worth $500 million, from a complex of offices, laboratories, greenhouses, fields, processing plants and research stations operating in ninety countries around the world.

In 1914, tall corn was an index of progress and national destiny, accomplished here at the hands of Andrew Engstrom in Junction City, Kansas. Joseph J. Pennell Collection, Kansas Collection, University of Kansas Libraries

So recent is the hybrid-corn revolution that some of the founding members of Wallace's company are still there, still fired by the idea of doing the world some good. "The Idea," Wallace wrote in a 1932 booklet describing his original company, was to "improve corn by controlling its pollination." The possibilities seemed infinite: "The best hybrids of the future will be so much better than the best hybrids of today that there will be no comparison." When I spoke to James Wallace, Henry's brother, a modest, dignified man in his eighties, his belief in the Idea was undiminished. "I don't know how much more we can improve it," James said in his Iowan drawl, "but we spend several million dollars a year working at it and that ought to produce something." One thing it produces is seven to ten new hybrids each year, totaling more than five hundred now, which helps the company retain 35 percent of the U.S. market. But James is concerned with more than business. He worries that the Idea has not solved the problem of how to feed the world. Such worries, however, are mere static in the current of optimism that lights up the signs on his office wall: "The Three-Year Plan of Pioneer Overseas Corporation is not letting negativism creep into any of the operating units. WE CAN DO IT is the buzz word."

"We Can Do It" was the unofficial motto of *Wallace's Farmer,* the paper founded by James's father in 1895 under the rubric "Good Farming—Clear Thinking—Right Living." Not even the precepts of positive thinking, however, could blot out the boom-or-bust crisis after World War I, when in 1920 corn prices collapsed from $1.70 a bushel to 67 cents. The Depression that hit Wall Street later hit the farmers first. The early 1920s were such bad farm years, James recalled, that they may have hastened his father's death, in 1922, while he was serving as Secretary of Agriculture under Warren Harding. Certainly the farm depression hastened the departure of my family from their farms. When young Henry Wallace became Franklin D. Roosevelt's Secretary of Agriculture in the depths of the Depression, Kansas Republicans like my folks dubbed Wallace a Judas who had betrayed the farmer and sold his soul to a Democrat. And when, in the forties, he formed the Progressive Party, they knew he was the only thing worse than a Democrat—a Red.

Iconographic portrait of Henry A. Wallace with symbols of farming progress and prosperity: a twin-siloed barn loaded with corn silage over the winter and a jeep loaded with corn-fed chickens. Courtesy Pioneer Hi-Bred International, Inc.

From the beginning, Wallace's Idea was as radical as it was pragmatic. By applying the principles of mass production and marketing to the plant world, Wallace turned agriculture into business. George Mills, a reporter for *The Des Moines Register*, summed it up: "He looked like a sort of mystic, which he was, but he was a mystic who made an awful lot of money." That was part of the romance. As brother James explained him, "Henry was the first one to make a business out of the hybridization of corn."

Henry Wallace's curiosity about corn breeding had begun when, as a boy, he met George Washington Carver, the man who developed a more perfect peanut. From Carver, who believed that "God was in every plant and rock and tree and in every human being," Wallace learned to respect the way American Indians treated the plant world, even as he was altering that world forever.

In 1900, the revolution in plant breeding that was to make a businessman of Wallace and a botanist of my father was exploding. A trio of European botanists rediscovered Gregor Mendel's "laws" of inheritance, which he had drawn from genetic experiments in 1866. Mendel's "laws" were reinforced by Darwin's experiments with corn that led to his theory of "hybrid vigor," published in 1876 in *The Effects of Cross and Self Fertilization in the Vegetable Kingdom*. A year later, the United States produced its own revolutionist in William James Beal, whose systematic experiments with corn crossings caused one senator to accuse Beal of "pimping for the tassels."

After studying with Louis Agassiz at Harvard, Beal went on to Michigan Agricultural College and there produced the first controlled crosses of maize (scientists are wont to call corn "maize") by a technique of "castrating" the plant, or detasseling male spikelets, in order to "fix" the fathers. His first step was to create inbred strains by the process of "selfing," which was as cumbersome as safe sex and almost as simple. The breeder guaranteed corn chastity until the proper moment for copulation by slipping one paper bag over the tassel and another over the ear. At the right moment, he shook the pollen collected on top onto the ear below. His second step was to systematize the crossing of inbred strains by putting the pollen of one strain of corn on the silks of another strain. This was easy to control by making one row of corn female and the other male: that is, the breeder first detasseled a row of corn and then harvested the ears of that same row. His third step was to cross-breed the progeny yet again, for he discovered that doubling the crossings disproportionately increased yield, turning eight-rowed Indian corn into twenty-four-rowed hybrid corn. Beal's triumph was memorialized in a plaque erected by his successor at Michigan, P. G. Holden. "Near this spot in 1877, Beal became the first to cross-fertilize corn for the purpose of increasing yields through hybrid vigor," the plaque reads. "From his original experiment has come the Twentieth Century miracle—hybrid corn."

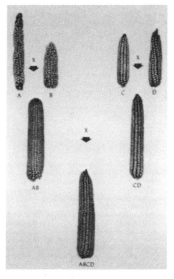

Double-cross hybrid corn is the offspring of four grandparents and two parents. From Wallace and Brown, *Corn and Its Early Fathers* (1956)

"Miracle" was the word an English botanist used in 1956 when he thanked American corn for saving Europe from starvation during and after the Second World War. From Beal's time throughout this century, the miracle workers were principally men like the Iowa teacher P. G. Holden, pragmatists who crossed theoretical science with applied agronomy in America's land-grant colleges.

Holden was a corn evangelist in the Wallace vein, and his mission was to educate Iowa farmers. First he motivated them to want to improve their corn crops and then he supplied them with the means to do it. To preach the gospel of longer, larger, smoother, more perfect ears, he sent out Seed Corn Gospel Trains, which would make seventeen stops a day. To spark the farmers' competitive spirit, he promoted the idea of the Corn Show, which was a county-fair beauty contest for corn. In 1893 at the World's Columbian Exposition in Chicago, he crowned the grand-prize winner "the world's most beautiful corn." This was the corn that changed the face of the American continent—Reid's Yellow Dent.

The Reid hybrid had begun accidentally in 1847, when an Ohio farmer named Robert Reid moved to Illinois and brought with him a reddish strain of corn which had been grown for generations in Virginia by the family of Gordon Hopkins. But when Reid planted this corn in Illinois, it did so poorly that he replanted some of his corn hills with a yellow flint from the Northeast, grown for centuries by the Indians. Over the years, Robert and his son James continually selected the best ears of this crossing of Gordon Hopkins Gourdseed corn and Little Yellow Flint to get the best qualities of each. They ended with their prize-winning beauty corn.

The romance of hybrid corn sometimes overlooks the evolution of the purebred strains that preceded the white man's "miracle." "One of the best things that happened for North America, and the world," wrote David Christensen in the 1989 *Seed Savers Exchange* was "the appearance in northwest Mexico about 700 AD of a soft flour starch which came in an eight-row strain." Because this strain proved to be highly productive, adaptable to cooler climates and easy to grind, it moved north through Anasazi territory and up the Missouri to

Reid's crossing of eight-rowed hard-starch northern flint (left) with many-rowed soft-starch southern gourdseed (right) produced Corn Belt dent, father of commercial corn around the world. From Wallace and Brown, *Corn and Its Early Fathers*, Michigan State University Press

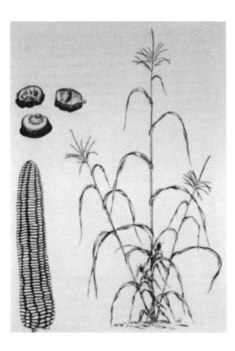

southern Canada. This Pueblo corn became the prototype of the many flint and flour corns of the upper Missouri tribes and, eastward, of New England Woodland tribes. Since the soft-flour starch component did not germinate well in cold climes, it took a few centuries of natural adaptation and selection for the strain to become the hard-starched Northern Flint that is "the more important of the two racial ancestors of today's Corn Belt Dents."

Ironically, one reason for the momentous success of Reid's Yellow Dent was that it came along soon after white settlers had taken over the major corn-growing territory of Indian tribes in the Northeast, at the end of the Black Hawk War in 1832. Further, the opening of the Erie Canal in 1825 had opened new lands for settlement and developed new markets for corn-on-the-hoof. In Cincinnati hogs were processed for oil in what became known as the "land whale" era, and then in Chicago, when railroads superseded canals, hogs were processed for every possible edible part. Increased corn production through Reid's Dent helped save the Union during the Civil War, and after the war a more perfect union of the industrial North and agricultural South was contained in the union of Northern flint and Southern gourdseed.

Saltzer Seed Catalogue of 1901. Courtesy Library of University of California at Davis

Reid's Yellow Dent was the medium through which Holden was determined to "carry the message of better farming, better living and better corn not merely to one hundred, one thousand, ten thousand, or even one hundred thousand farmers," as Wallace wrote, "but to all the farmers in the central Corn Belt." But Wallace, both a corn-show veteran and an ag-school graduate, wasn't buying it. He began his own backyard experiments in 1913 after reading one of the papers of George Harrison Shull, a farm boy from Indiana who was double-crossing inbred corn at the Carnegie Institution Station for Experimental Evolution at Cold Spring Harbor in Long Island (the place where Nobel Prize winner Barbara McClintock continues her work on corn today). Unbeknownst to Shull, another geneticist was working nearby, at the Connecticut Experiment Station, near New Haven, on the estate of Eli Whitney. In one of the buildings where Whitney had devised a method for the mass production of guns, as important to the North as his cotton gin to the South, Edward Murray East devised the first hybrid corn to retain full propagative vigor, known now as the Burr-Leaming double-cross. Shull and East did not meet until 1908 and then none too happily, but they are the twin heroes of modern factory hybrids.

In remembering Henry's unending experiments, James Wallace recalled watching him "in the early morning standing in his bare feet out in the small corn in his backyard reading his morning paper, just looking at the corn in between times and reading the paper in between times and observing how it was doing." Henry was convinced that beauty in corn was only skin-deep. "Looks mean nothing to a hog," said Henry Wallace, who initiated the Iowa Corn Yield Contest and in 1924 won it with a misshapen red-kerneled hybrid of his own named Copper Cross. From then on, Wallace's mission was to sell Corn Belt farmers on the Idea that more is best and then sell them the seed. With James and a young corn breeder named Raymond "Bake" Baker, Henry founded a company to sell seed corn. From then on, corn farmers depended upon the specialized knowledge and equipment that characterize today's industrial farms.

Over lunch in Pioneer's corporate dining rooms, "Bake" Baker recalls how hard it was in the early days to persuade farmers whose fathers and grandfathers had grown open-pollinated corn to switch to hybrid seed. With hybrids each farmer had to detassel alternate rows of corn, which might involve five thousand to eight thousand tassels per acre. By the 1930s corn detasseling for hybrid-seed companies had become as much a Corn Belt ritual as the husking bee had been earlier. During the 1940s, women and girls took over the detasseling brigades as farm boys went to war; and after the war detasseling was as common a form of coed student employment as waiting tables elsewhere.

A hybrid grown in Metropolis, Illinois, in 1917 to honor the Red Cross of World War I. The grower was Dr. J. T. Cummins, dentist. Lake County (Ill.) Museum, Curt Teich Postcard Archives

But if hybrid seed made planting more work, it made harvesting easier. For the first time corn ears could be harvested mechanically because breeding produced ears that stood up straight at a uniform height on each stalk, supplanting ears that bent with the weight of maturity. "The farmer was bent over all day picking up corn," Baker remembers. "He wasn't so sure about yield but he sure knew about bending over." Baker now concludes, "After fifty years there's no optimum ear, but we keep working to put more and more good characteristics together." His checklist includes Moisture, Stay Green, Test Weight, Seedling Vigor, Grain Quality, Ear Height, Plant Height, Early Stand Count, Number Stalk, Dropped Ears, Wind Resistance—to name a few.

Hybrids have changed the nature of farming, says Donald Duvick, head of the Plant Breeding Division, who has been with Pioneer for a mere thirty years. During this time the genetic capacity for grain yield has increased at the rate of one bushel per acre per year. Besides yield, plants have improved in root strength, resistance to stalk rot and stalk lodging, resistance to premature death, to barrenness, to the attack of the second-brood corn borer—Duvick runs out of breath. As a result, the factory farm has replaced the family farm, with every year fewer and fewer farmers working larger and larger farms.

Such displacement concerns William Brown, geneticist and coauthor with Wallace of *Corn and Its Early Fathers,* but after forty-five years as a Pioneer executive Brown still brims with Pioneer spirit. "This is the most exciting time in the history of botany or biology," he says. "The reason is molecular genetics." Through the techniques of genetic engineering that were first developed in the 1940s, he explains, we can now move genes from one organism to another without carrying any extraneous materials. In other words, you don't have to back-cross corn to get rid of side effects you don't want. You can move genes at will, make up a high-lysine strain, or one resistant to drought or to saline soils or to pests like corn-ear worms. It's open-ended, he says, but we won't see major results until around the year 2000, a short time to wait in the life of "the amazing grass called corn."

Brown sustains Pioneer's zeal for improving corn breeds, but he too admits disappointment in improving men's lives: "Where the need is greatest, as in Africa, we've had least success." Elsewhere, they've had almost too much success. Supplied with American hybrid seed, China and other countries are now becoming competitive with the U.S. Corn Belt. In 1986, China produced 65,560,000 metric tons of corn, second only to America's 209,632,000. "We must improve productivity by reducing yield along with production costs," Brown says, "to get a greater net return." The Idea of ever-increasing yield has boomeranged, for we're now depleting our soils to grow ever greater surpluses that the government has to pay for. "We need programs to get out of production," Brown believes, "since the major income of farmers today is not what they sell but what they get from the goverment, and when the public learns that the big farms get close to a million a year for growing surplus, it will raise serious questions."

After the snazzy corporate headquarters of Pioneer, I visited the other end of the seed-corn scale in El Paso, Illinois, where the major buildings are a silo, the Corn Belt Motel, and the house and barns of Pfister Hybrid Corn. Here Lester Pfister, born in 1897 to tenant farmers, became the Horatio Alger of hybrid corn in the 1930s, in a story worthy of *Reader's Digest,* where it duly appeared. Young Lester had to quit school when his father died, but as a hired hand he began to test different strains of corn and record the results. He took samples to the agent at the County Poor Farm, where P. G. Holden had set up a system for testing and comparing the yields of local farmers. The agent was impressed with Pfister's records and asked him to do a three-year testing of Woodford County corn. In 1922 Wallace had awarded a prize to a high-yield Woodford strain brought in by an illiterate dirt farmer named George Krug. Krug had crossed a Nebraska strain of Reid corn with Iowa Gold Mine to make, as Wallace said, "the highest-yielding strain of old-fashioned nonhybrid yellow corn ever found in the Corn Belt." For the next five years Pfister worked to inbreed Krug corn, despite the ridicule of neighbors who laughed at

"For ten years, up until 1935, Lester Pfister's neighbors in El Paso, Illinois, were convinced that he wasn't quite right in the head. They couldn't understand why any sane individual should spend hours in a field under the boiling sun taping paper bags on corn tassels."

—GEORGE KENT, "A Farmer Bags a Million Dollars," *The Reader's Digest,* September 1938

Today, paper-bagged fields of corn to produce hybrid seed are commonplace. Courtesy Pioneer Hi-Bred International, Inc.

his fields of paper bags, covering row after row of corn tassels, and despite the painstaking selection that reduced 388 ears to 4. With these four he experimented for the next five years with crossings, going deeply into debt, feeding his family on cornmeal mush and fueling his stove with dried cobs.

Threatened with foreclosure in 1933, Pfister at last hit upon a double-crossed hybrid that outyielded the original Krug so substantially that word, and corn, began to spread. In another five years he was grossing $1 million a year, had invented a detasseling machine and was named "the outstanding corn breeder of the World" by *Reader's Digest* and *Life*. Selling most of the business in 1941, he kept the parent company and family farm, now run by his son Dan. "We're fifty years old," Dan says, as we talk in his backyard next to the shelling and drying shed, "but we've proved our quality." When I visit in September, they are working around the clock. "You've got to pick seed corn when the sun is shining and moisture is in the high thirties," Dan explains. He points to his new seed-corn harvester, an eight-row beauty with a special trailer that cost $128,000. "Look at that sucker go," he says, watching as it eats up the rows.

Dan's excitement comes from machinery, as he guides me through the shed where elevators take two hundred bushel loads of ears up to shellers (six hundred bushels an hour) and onto giant dryers with automatic temperature controls. I find the heavy, sweet smell of drying corn overpowering, but Dan is in full flush talking about the latest engineered product—Pfister's Kernoil. "Kernoil could revolutionize the feed industry," he says, since "feed efficiency" rises 4 or 5 percent with livestock fed on corn that contains 60 percent more oil in the germ. "My mind can't get away from the possibility of starting something like Dad did," Dan says. "You have a breakthrough, you're the only one doing it and pretty soon you establish a megatrend—you reinvent the world."

THE CORN WAR

To REINVENT THE WORLD is the dream of technologic man, but the question is—in which language? In the course of corn's reinvention in this century, war broke out among the second generation of plant breeders in whose hands the destiny of farmers lay, a war that accelerated as descriptive botanists like my father were displaced by geneticists who took giant leaps from taxonomy to morphology to cytology. The Corn War, as it has been called, is as much a product of post–World War II America as the Cold War, and springs in part from isolationist America's sudden discovery of the rest of the world. If in the first half of the century botanists quested for the Perfect Ear, in the second half they quested for the Primal Ear. The war over the origin of corn was part of the growing pains of the revolution in biology that gave us a new common language in DNA and of the revolution in outlook that gave Kansas a common corn language with Siberia and Peru. But not before shots were fired and opposing corn worlds reinvented.

After the Second World War, the two opposing generals in the burgeoning Corn War left their farms in Kansas and Nebraska to become world contenders. One became an internationally known geneticist at Harvard and the other won a Nobel Prize. The Nobel Prize went to George Wells Beadle, born in 1903 in Wahoo, Nebraska, and educated at the University of Nebraska's College of Agriculture and then Cornell. The Harvard prize went to Paul Christoph Mangelsdorf, born in 1899 in Atchison, Kansas, home of the famed corn carnivals at the century's turn, where buildings, streets and people were draped in husks, kernels and cobs.

Beadle was a new-wave geneticist concerned with how genes and chromosomes work within the cell. In researching the process of cell division (meiosis) in maize, he discovered that inherited defects in maize pollen were related to the behavior of chromosomes during cell division. Later he studied gene change in materials like the fruit fly and bread mold, becoming a co-winner of the 1958 Nobel Prize in Physiology or Medicine for his discovery that "genes act by regulating definite chemical events." Later still, he became president of the University of Chicago, but when he retired he returned to the corn trenches and remained there until his death in June 1989, just two months before Mangelsdorf's.

When Beadle began research in the 1930s, botanists agreed that corn was both the most productive of the grasses and the one least able to reproduce itself. No one agreed on how it had got that way. Beadle revived nineteenth-century speculation that corn might have evolved from a kindred grass, teosinte, when he began to experiment in his Stanford lab on teosinte seeds collected from Mexico and Central America, where annual teosinte was still found in the wild. When he heated the seeds, he found that their kernels "exploded out of their fruit cases indistinguishable from popped corn." The story of modern corn, he proposed in 1939, began with the *accidental popping* of

Paul Christoph Mangelsdorf in 1979, still plumbing the mysteries of corn in his garden at Carol Woods near Chapel Hill, North Carolina. Courtesy Mary Eubanks and Clark Mangelsdorf

such wild seeds. The problem with teosinte as a hypothetical ancestor of maize, however, was that the plant was so different from the earliest known maize that breeders couldn't reproduce its evolutionary steps.

In the same year, 1939, a fellow Midwestern farm boy, trained in taxonomy, fired the first shot in the Corn War from his post at the Agricultural Experiment Station of Texas A&M. Paul Christoph Mangelsdorf was destined for the ministry, his mother believed, and she joked that his initials stood for "Preach Christ Mangelsdorf." When I visited Mangelsdorf in his retirement home in Chapel Hill, North Carolina, in his eighty-fifth year of corn preachments, he joked that his initials stood for "Pod Corn Maize." Pod corn, in which each kernel on the ear is enclosed in its own tiny husk, is the earliest form of corn discovered in archaeological remains. And a wild pod corn, Mangelsdorf had declared in 1939 and continued to declare until his death in 1989, is the origin of cultivated corn.

Where Beadle had worked primarily in the laboratory, Mangelsdorf worked primarily in the field. After studies at Kansas State Agricultural College, the Connecticut Agricultural Experiment Station in New Haven, and Harvard, Mangelsdorf had continued the labors of earlier corn breeders like Edward East. At Texas A&M, Mangelsdorf gained fame for breeding a popular sweet corn named Honey June and for discovering a sterile strain of corn (named "T" for Texas) that seemed to be the answer to a hybrid-corn farmer's prayer because it ended the need for detasseling. Unfortunately, the strain was not resistant to the corn-leaf blight that in 1970 swept the Corn Belt and finally alerted farmers and breeders to the danger of reducing the wide world of corn to a handful of commercial types. Mangelsdorf did his most important field work, however, in the wilds of Mexico, Guatemala and Peru, collecting and organizing thousands of corn varieties into "races of corn," while he searched for the Primal Ear.

When I met him, Mangelsdorf was still a handsome man with a mane of white hair and an unabated mania for corn. Leaning on a walker, he still bred corn plants in his backyard and collected artifacts to add to the collection of almost five hundred corn objects he and his wife had donated to North Carolina State University at Raleigh. A glass case in the botany building there displayed the most valuable—a seventeenth-century Chinese corn ear carved from ivory, a Japanese corn netsuke, a Meissen pudding mold with indented ears, a Zapotec funerary urn with corn headdress. Stored unseen were box after box of majolica, 1930s pottery, carnival glass, British porcelain, tin molds, copper molds, ironstone and plastic—corn shapes cloned in the least likely materials. This was a form of corn madness I recognized.

Mangelsdorf's form of corn language, however, I did not. As a corn breeder, he had experimented with a third kindred grass, *Tripsacum*, and with his Texas colleague Robert Reeves had in 1939 formulated a "tripartite theory" to prove that "the ancestor of cultivated corn was corn." He speculated that there must once have been a wild corn, now extinct, and that a hydrid of this wild pod popcorn mated with *Tripsacum* to become the parents of teosinte. He concluded that a gene mix of these three related grasses evolved into our modern races of corn. Mangelsdorf preached his trinitarian gospel with such passion that after his comprehensive *Corn: Its Origin, Evolution, and Improvement* (1974), he became to the general public a Corn Father on the order of Henry Wallace.

Mangelsdorf's theory polarized the tight world of corn geneticists, seducing some and outraging others. Even today a Mangelsdorf loyalist such as Umesh Banerjee of North Carolina Central University will speak of conspiratorial enemies who know but refuse to acknowledge the *truth* for reasons of politics and revenge. From a Beadle loyalist, on the other hand, the mere mention of Mangelsdorf provokes gunfire. "It's all bombast, this crazy theory," shouts Hugh Iltis of the University of Wisconsin at Madison. "Mangelsdorf dominated the scene with it for forty years and almost all of it is *wrong!*"

What was right about Mangelsdorf and what gave juice to his theory was his pursuit after the Second World War of the connections between botany, archaeology and anthropology that a few adventurers—like Junius Bird in Peru—had initiated in the decade before the war. The work of Edgar Anderson and of Paul Weatherwax, a botanist at Indiana University, explained to the layman what such crossovers could do. Studying Pre-Columbian ceramics in Central and South America, Weatherwax found that actual corn ears had often been used for the molds and thus early forms of corn were recorded in clay. Mangelsdorf too enlarged his experiments in the lab by encouraging archaeologists in the field, beginning with the pair of graduate students who discovered those prehistoric cobs of pod popcorn in Bat Cave in 1948.

Mangelsdorf was certain that he had won the Corn

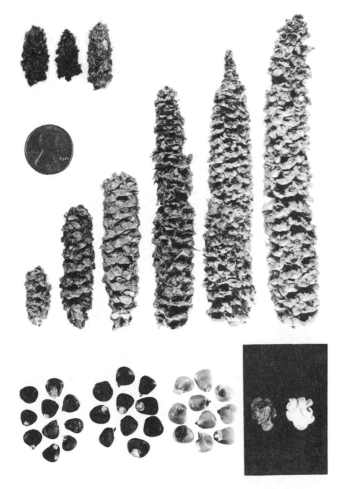

Cobs found in three levels at Bat Cave are shown next to a penny to indicate actual size. At bottom right, a prehistoric popped kernel (dark) is paired with a kernel from the same level that Mangelsdorf popped in 1948. Courtesy Paul Mangelsdorf

War when the first radio-carbon dating at Bat Cave suggested that the earliest layer of corn was seven to eight thousand years old. Although the date was revised, a discovery in 1949 supported botanists who thought the solution to corn's origins lay in archaeology. In that year, Richard MacNeish was digging for potsherds in the La Perra Cave in northeastern Mexico, near Tamaulipas, where he hoped to uncover some connection between the Moundbuilders of the lower Mississippi Valley and more ancient Mexican tribes. At La Perra he found eighty-seven cobs of cultivated prehistoric corn in a layer that he dated roughly between 250 B.C. and A.D. 150. The majority of cobs were primitive forms of a living Mexican race of popcorn called Nal-Tel, and some of the cobs had been charred, suggesting that the kernels had been popped on the ear. Some cobs showed evidence of battering and since stone mortars were also found in the cave, it seemed reasonable that its inhabitants ground meal from popped corn. Although this corn was not as old as the Bat Cave cobs, the find suggested that Mexico was the place to begin digging in earnest.

Mangelsdorf helped MacNeish get funds for further digs in the same area, particularly after he examined some grains of fossil pollen shown him in 1950 by a Yale botanist, Paul B. Sears. These came from two hundred feet below the present level of Mexico City, which had been built on the site of an ancient lake. (Sears had learned that engineers preparing for the city's first skyscraper had taken core samples in a couple of locations, and he got the cores for study.) When Mangelsdorf scrutinized nineteen "large" grains of pollen estimated to be sixty thousand years old, "as old as the early stages of the Iowan advance of the Wisconsin Ice Sheet" in the late Pleistocene era, he believed that he had found indisputable proof of the existence of wild corn. He was wrong. Nothing in the Corn War was beyond dispute.

The climactic discovery for MacNeish came in a different area of Mexico, in caves of the Tehuacán Valley north of Oaxaca, in a region dry as a desert but with seasonal rains and underground springs. In 1960, after digging through the debris in thirty-nine caves in the area, MacNeish came upon a big one, covered with a layer of goat dung. It took his workers four days, digging straight down, before they uncovered a single corncob—the size of a cigarette butt. When they uncovered more, MacNeish wired Mangelsdorf, "We've hit corn."

With new funding, Mangelsdorf and his wife joined the dig, which covered five caves and an unprecedented find of 24,186 specimens, dating from seven thousand to five hundred years ago. These remain the earliest cobs yet discovered, and Mangelsdorf was convinced that some of the cobs were wild. "Here was corn telling its own history," wrote Mangelsdorf, as he fell into a reverie that suggested Cortez discovering the Pacific:

As I stood with MacNeish in the San Marcos Cave looking down on the alluvial terraces I could see in my mind's eye wild corn plants growing here and there, never in thick stands but in favored spots. I could visualize the prehistoric cave-dweller, primarily a hunter of small game and a food-gatherer, bringing back small ears of wild corn to his shelter, picking off the small, flinty kernels, much too hard for him to chew even with his excellent teeth, and exposing them one by one to the glowing coals of his fire. After absorbing heat for several minutes, the kernels would have exploded, transforming the stony little grains into tender, tasty morsels. . . .

Like earlier corn evangelists, Mangelsdorf took to lecturing on "A Botanist's Dream Come True." The ears of wild maize, as he conjured them, were about an inch long and as thin as a pencil. The top half was a tassel and the bottom half a cob, enclosed by the leaves we now call husks, with eight rows of brown or orange kernels, seven kernels to a row, each kernel enclosed entirely by chaff. Here were all the distinctive characteristics, he said, of modern corn.

In 1978, however, an extraordinary discovery in the enemy camp forced Mangelsdorf to capitulate on the tripartite theory. Only two years after Mangelsdorf published *Corn*, a young Mexican botany student at the University of Guadalajara, Rafael Guzman, spent his Christmas vacation looking for a form of wild teosinte thought to be extinct. He did so because of a Christmas card sent by Hugh Iltis of Wisconsin to young Guzman's botany professor in Guadalajara. Convinced that teosinte was the true ancestor of corn and hearing that a perennial teosinte had once been collected near Ciudad Guzman in 1910 but was now thought to be extinct, Iltis fantasized a *Zea perennis* on his greeting card and labeled it "extinct in the wild." Guzman's Guadalajara professor challenged her students to find the original.

In the heart of the mountains near Jalisco, not far from the earlier collection site, Guzman came on a large stand of grass and sent seeds to Iltis. Planting the seeds in his herbarium, Iltis found that the grown plants had the same number of chromosomes as maize, and that the student had discovered a new species, which Iltis labeled *diploperennis,* or perennial

Rafael Guzman discovers a probable ancestor of corn, perennial teosinte, growing wild at La Ventana in the Sierra de Manantlan of southwestern Mexico. Courtesy Hugh H. Iltis

teosinte. Like annual teosinte, this species could easily crossbreed with corn and produce fertile progeny. That made wild perennial teosinte, with its capacity to reseed itself, the primary candidate for the ancestor of corn. Later, the teosinte subspecies now labeled *parviglumis* was found to have the same specific proteins as maize. Iltis, not unnaturally, hailed perennial teosinte as "the botanical breakthrough of the twentieth century."

Mangelsdorf, not unnaturally, began to experiment with the seed as soon as he could get his hands on it, backcrossing it with primitive Mexican popcorns native to the same area. Earlier studies of pollen and chromosome structures had already ruled out *Tripsacum* as a progenitor, so the question now was how teosinte was related to corn. In the August 1986 issue of *Scientific American,* Mangelsdorf surrendered—but only halfway: "As I now see it both modern corn and annual teosinte are descended from the hybridization of perennial teosinte with a primitive pod-popcorn." From perennial teosinte, he concluded, modern corn got its good root system, strong stalks and disease resistance, while from primitive cultivated corn came modern corn's distinctive ear, tenacious cob and multiple paired rows of kernels. Because the chromosomes of the two parents were not perfectly aligned, new genes were exchanged through hundreds of thousands of crossings. "The ultimate result," Mangelsdorf wrote, "was a gene pool so extensive and so rich in variation that almost any kind of corn could evolve from it through natural and artificial selection."

Still, Mangelsdorf would not give up his faith that a hypothetical "wild corn" was the remote ancestor of all the Maydeae subtribe and their crossings, until wild corn was "eventually swamped by cultivated corn and became extinct." "The mystery of corn," he repeated, "has essentially been solved." "The domestication of corn is no mystery, and needs no mystique," countered Iltis. "The origin of corn is still a mess," sighed Richard Schultes of Harvard's Botanical Museum.

Actually the mystery of origin and domestication simply shifted ground. "Everybody agrees that domestic corn began about ten thousand years ago and that the place was Mexico and that present-day corn came, one way or another, from the interbreeding of teosinte and maize," says Garrison Wilkes at the Boston campus of the University of Massachusetts. The question is which way, rather than which one first—a question as problematic as the question of source; for as Paul Weatherwax determined thirty years ago, domestication is the real mystery of corn and whatever wild plant was ancestral, "domestication produced a sudden and profound change."

The nature of that change has drawn new battle lines, and while some allies have changed ground, others have changed sides. Walton Galinat, once a post-doctoral student and associate of Mangelsdorf, now embraces Beadle's theory but remains loyal to Mangelsdorf the man. "He got mad on Mangelsdorf," Banerjee mutters, "and joined the Beadles in revenge." When I visited Galinat at the Waltham Cornfield Station attached to the University of Massachusetts, I found anything but a vengeful man. In his late sixties, Galinat has the shy, sly smile of Walter Mitty. He wears funny hats in his cornpatch, drives a rattle-trap car with the license plate "CORN," and has gone corn-mad in his own fashion. "Corn is my religion," he says, "and this laboratory is my church."

His laboratory is also a cluttered fieldhouse stuffed with packets of seeds meticulously labeled to keep track of the two thousand pedigreed varieties planted in the adjoining two-acre plot. In the middle of the field he has painted a pair of giant eyes beneath Japanese balloons called Terror-Eyes, designed to scare off birds immune to the alarms that sound every three minutes. Near a greenhouse, an old meat locker serves as a germ plasm bank for some four thousand varieties, including the world's oldest viable corn seed. Upstairs in the "headhouse," his office door sports a brass cob-shaped knocker, his curtains are printed with gene gender signs, his desk bears a lusterware corn pitcher holding a bouquet of tiny real cobs and his walls are a gallery of corn productions, reinvented by his illustrative and breeding arts. He's bred corn two feet long, the world's largest. He's bred square ears of corn—as airplane food, he jokes, because they won't roll off the plate. He's bred red-white-and-blue "Old Glory" corn for events like the Bicentennial, with blue-dotted kernels made to look like stars.

Walton Galinat's monster eyes are designed to terrorize corn-eating birds.

At home in nearby Waban, his wife, Betty, a weaver, covers her spindle with a corn-shaped cloth, knits him sweaters with corn motifs and dusts the clutter of corn artifacts in the china cabinet they call the Corn Morgue—which takes up where Mangelsdorf left off. Galinat shows me pictures of the Corn Tassel Lady, Miss Margo Cairns, who launched a campaign in 1955 to make the tassel our national floral emblem, but failed because it wasn't what most people call a flower. He shows me scrapbooks of corn postage stamps, corn wrapping paper, corn wallpaper. He points to bookshelves of works like Carl Sandburg's *How to Tell Corn Fairies When You See 'Em.* "I think Galinat would almost *be* corn," says a Harvard compeer, "if he wasn't chained to the animal kingdom." Galinat's wife says, "Sometimes I tell him I'm going out, and I put on his hat and coat just to see if he notices." It's tough playing second fiddle to corn.

When I ask him why corn scientists battle so fiercely about their theories, Galinat replies, "The corn plant and ear are so beautiful morphologically and so important economically that they arouse high feeling, not to mention ego systems. People are more indifferent to poison ivy or skunk cabbage." Galinat relates to corn like a citified Hopi. "It shows you everything," he says. "It's like watching TV." As he walks among his nurslings, he asks, "See the baby corn just coming up, saying hello to the world? They're having a chance to express their genes. We put questions to them and then they talk to us here in the field." He fingers a green shoot. "I like to touch a plant when I see it. It's like shaking hands with a person." He admits that his vegetable love has provoked some antagonists to accuse him of the "unfettered imagination" of poetry—or worse. "Galinat is completely stupid," Hugh Iltis explodes, "and Mangelsdorf is crazy like a fox."

"Iltis," warns another veteran of the Corn War, "is the most volatile man I've ever met." Hugh H. Iltis brought his volatility with him from Moravia, Czechoslovakia, when he escaped in 1939 with his botanist father, one step ahead of Hitler. Iltis is the Sir Richard Burton of the plant world, scaling the Andes

Enlarged closeup of a cob of sweet corn with kernels removed shows how rows of kernels are paired. To Iltis, a radical sex change transformed and condensed the paired male spikelets of a teosinte tassel into the paired female kernels of a corncob. Courtesy Hugh H. Iltis

in search of a wild potato and discovering a wild unknown tomato. As head of the University of Wisconsin's Herbarium, he is passionate about conserving what nature has put on the earth, especially the tropical rainforest of the Sierra de Manantlan, in southwestern Mexico, where perennial teosinte was discovered wild.

To explain how teosinte could have evolved by natural selection into corn, Iltis devised a theory as flamboyant as his person, the Catastrophic Sexual Transmutation Theory, better known as CSTT. To bridge the wide gap between plants as different as teosinte and corn, Iltis argued for an "evolutionary jump," one giant step or sudden transformation instead of gradual evolution. Perennial teosinte, he concluded in 1983, underwent a major sex change in the process of becoming corn. Essentially he argued that the male tassel spike at the end of a lateral branch of teosinte was transmuted into the female ear of corn attached to the central stem. A radically telescoped branch became the shank at the base of the corn ear, and the teosinte leaves became the husks on the ear. By radical contraction and condensation, the double rows of kernels on a teosinte spike were transformed into the several paired rows of kernels on a corncob.

Galinat took issue with Iltis on the abruptness and the means of change, arguing for a gradual transformation over one or two centuries in a three-phase evolution controlled by man. In Galinat's view the hard spike of teosinte was transmuted without sex change into the soft cob of corn by condensation, which "liberated" kernels from the separate pods of

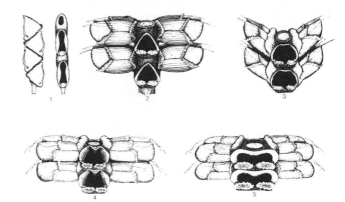

To Galinat, transformation was gradual as the process of condensation "liberated" paired kernels by forcing them outward from their coverings in the hard teosinte spike (figure 1 on the left). His drawing shows cross sections of different paired kernel rows. Courtesy of Walton C. Galinat

the teosinte spike. This change was a simple inheritable gene trait that was then "fixed" by human selection, so that what seemed to be a natural monstrosity was actually the deliberate creation of man.

To Galinat, the argument stems from the opposing languages of taxonomy and plant breeding. The taxonomist sees plant structures as discrete objects, created through natural selection and ordered in vertical lineages, with a place for every variant and every variant in its place. The plant breeder, on the other hand, looks for relatedness in "horizontal" clusters, in a continual process shaped by "the mind, hand and eye of man." What the taxonomist sees as a honeycomb of niches the plant breeder sees as a meandering stream with tributaries. Since a breeder can produce by whim or caprice objects that are nonsense to the taxonomist, he calls them catastrophic. "Iltis," Galinat says with a smile, "he's the catastrophe."

The catastrophe to Major Goodman, a botanist at North Carolina State University, is that both theories conflict with current archaeological and biosystematic evidence, which suggests that "maize and teosinte were as different seven thousand years ago" as they are today. While there is much evidence for the early use of maize, there is none for teosinte. How corn came to be remains a mystery, despite Edgar Anderson's fiat fifty years ago, "The history of corn is now an exact science."

To Garrison Wilkes, the catastrophe is the loss of common goals in research and the commodification of plant genes. "We've lost the leadership in plant breeding," he laments; "we've lost the edge." He lists the old who are winding down and laments the young who have forsaken research for technologic development in molecular biology. That was the field that gave Beadle a bright vision of hope when he accepted his Nobel Prize. "Our rapidly growing knowledge of the architecture of proteins and nucleic acids is making it possible, for the first time in the history of science," he said, "for geneticists, biochemists, and biophysicists to discuss basic problems of biology in the common language of molecular structure." But in that common language new wars are stirring between those who would reinvent a brave new world of corn and those who would preserve what Henry Wallace called "the world of forgotten corns."

THE CORN RACE

THE WORLD of genetic manipulation, which followed mechanical manipulation, moved too fast for my father. Although he lived to be ninety-one, not dying until 1983, he never set foot in an airplane. He preferred to move more slowly. When he moved from Kansas to California, he had already lived through two rapid farm revolutions—one the change from horse to engine, the other the change from open-pollinated to hybrid seed—and that was quite enough. Subsequent revolutions in chemical and genetic engineering moved with such speed that, like the airplane, they were beyond his ken. He applauded Wallace's statement that the hybrid revolution was "as dramatic and important as the history of the automobile," but the genetic revolution of gene pools, germ banks, biotech protoplasts and opaque-2 mutants was as unthinkable as a trip to the moon. In only sixty years, the hybrid revolution that began with Reid's Yellow Dent in his childhood had spread so rapidly that by his death Reid's corn was not merely atavistic, but nearly extinct.

In that short span, corn became the substance and symbol of a new conflict between the prophets of divine progress and those of demonic decline. Science was discovering the diversity of corn's many races at the very moment it was eliminating most of them, so that the very notion of corn races suggested also a race against time. Even forty years ago the speed with which corn had moved from a plant of infinite variety to a "monoculture" hybrid alarmed the very men who had sped the change. Henry Wallace saw that in the first half of this century, man's exploitation of corn had altered the plant irrevocably and, in evolutionary lingo, more "catastrophically" than in all the millennia over which it had evolved. "When Reid's Yellow Dent swept the Corn Belt from 1890 to 1920, it destroyed thousands of [forgotten corns]," he

Problems of classification arise from the wide diversity of corn, revealed here in front and back views (and some cross sections) of corn kernels taken from a few major varieties of corn available in 1902. Yearbook U.S. Department of Agriculture, 1902

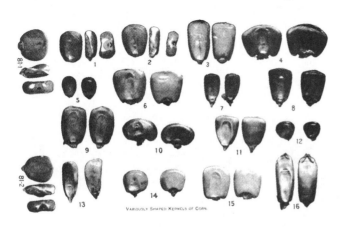

VARIOUSLY SHAPED KERNELS OF CORN.

wrote in 1956. "When hybrid corn swept the Corn Belt from 1930 to 1950, it destroyed most of what remained."

The Second World War, which sped genetic change through its need for high grain production, also alerted people, in the midst of waste and extinction, to the frailty of what already existed. In 1943, Wallace and a group of scientists including Paul Mangelsdorf met with representatives of the Mexican government and the Rockefeller Foundation to start a research program in corn and wheat that would help to solve Mexico's chronic food shortages. Scientists at the same time embarked on a vast inventory of existing corn varieties, first in Mexico and Central America and finally in the Western Hemisphere, the first time such a project had been undertaken anyplace in the world.

The ancient Peruvian and Mexican device of molding actual corn ears in clay has helped modern scientists to classify corn races. This Peruvian ceramic doubled ear (shown front and back), originally mistaken for a fossilized specimen, was classified by Mangelsdorf as a type of prehistoric pod corn. Courtesy Paul Mangelsdorf

The first problem was to create a new system of corn classification. The geneticists Edgar Anderson and Hugh Cutler had already started to revise Sturtevant's classification of corn by endosperm texture, the basis of the Big Six (pop, dent, flour, flint, sweet and waxy). They had decided to group varieties of corn by related genetic characteristics such as kernel shape, ear shape, row number. These groups they called "races." "Classifying the races of maize is comparable to classifying the races of man," Mangelsdorf wrote, "and the task is no less formidable and no less fraught with hazards."

For thirty years North and South American botanists took to the field to collect samples from every farm in every region under study. They grouped the samples into sets, planted together members of each set in order to eliminate strays, then saved the best ears with their seeds for future breeding. The list filled twelve encyclopedic volumes, beginning with *The Races of Maize in Mexico* in 1951–52. They arranged the thousands of varieties that they had found into roughly two hundred and eighty races, two hundred and ten of those unique to South America (most of them in Peru), forty to all of Europe, thirty to Mexico and twenty to the United States and Canada.

There were large problems of duplication, not to mention national pride. Mangelsdorf recalls that he was nearly lynched in Mexico City when he speculated aloud that maize might have originated in Peru, and was nearly lynched in Cuzco when he speculated that Peru's Gigante Race might have originated in Mexico. Since lineage is what interested the classifiers, they organized races into family trees. Peru and the upper Andes were especially valuable areas

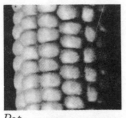

Pop

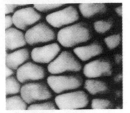

Flint

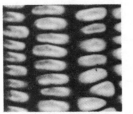

Dent

Flour

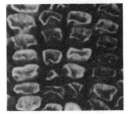

Sweet

Classification of corn by the type of starch in the kernels of each. From *Maize*, CIBA-GEIGY Agrochemicals Monograph (1979)

because corn strains there had not been hybridized. The result was "a virtually complete inventory of the corn of this hemisphere."

Since any classification is arbitrary, some geneticists now argue that corn races were caused as much by geographic isolation as by natural hybridization and that they should be called racial complexes. Revisionists like Bruce Benz have rearranged the earliest races (Ancient Indigenous, Precolumbian Exotics, Prehistoric Mestizos, Modern Incipient and Poorly Defined) into complexes divided by locales: for example, Mexican Narrow Ear Complex, which divides into Balsas-West Mexico Alliance, Isthmian Alliance, Mixe Alliance and Unaffiliated.

Whatever the linguistic manipulations whereby Poorly Defined becomes Unaffiliated, the achievement of the original fieldwork was unparalleled. The classification of races helped to explain the ancestral roots and routes of corn in the Americas and put them in the context of a world map. That map is based on the Big Seven: Northern Flints (found in the northern United States, southern Canada, central Europe), Corn Belt Dents (covering most of the temperate regions), Cateto Flints (southern South America, southern Europe), Mexican Dents, Cuban Flints, Caribbean Flints and Tusons (all found in tropical and semitropical regions). In Mangelsdorf's scheme, these races evolved from six major ancestral groups: Palomero Toluqueño and Chapalote-Nal-Tel (Mexico); Pira Naranja (Colombia); and Confite Morocho, Chullpi and Kculli (Peru).

The earliest corn to reach what is now the United States, according to Galinat, was a pre-Chapalote type. This was a twelve- to fourteen-rowed ear with brown kernels, each kernel enclosed in its own "weak pod" or husk. Because of longer days and shorter growing seasons north of Mexico, the spread of corn northward was delayed a few millennia until a combination of selection and irrigation helped to foster "day-neutral" varieties.

If we try to trace the ancestry of our own Corn Belt Dents, we find that they are composed of 25 percent Northern Flints (traced to a hybrid of Chapalote) and 75 percent Southern Dents (traced to Palomero Toluqueño). The key family name for Northern Flints is Maiz de Ocho, the soft-flour eight-rowed Mexican hybrid that appeared in the American Southwest around A.D. 700. The earliest Northern Flints, found in Ohio and upstate New York, date to A.D. 1040. By natural selection, their tight husks, which in desert climes protected the ears from ear worm and borer pests, evolved into looser husks to prevent kernels from molding in the rainy weather of the north. Southern Dents arrived later, since the earliest remains date only to A.D. 1500, and they probably arrived first in Florida, Virginia and Louisiana by sea routes from Mexico or the Caribbean. Two types of Southern Dents, Gourdseed and Shoepeg, are close kin to Pepitilla, all three derived from Palomero Toluqueño, of southern and coastal Mexico.

Sweet Corns

THE ANCESTRY and evolution of sweet corn tell a more complicated story. In ancient times, native Americans thought of two forms of corn as sweet. The most common was any type of corn that was young, or "green," and therefore sweet because the immature kernels had not yet converted their sugar to starch. The other type was ripe corn that was born sweet by genotype. This is the type that Galinat traces to the ancestral race from Peru called Chullpi. A descendant of this type, Confite Morocho, still grows in the high Andes, relatively unchanged from its appearance in ceramic replicas made one thousand years ago. The ear has a white cob with eighteen to thirty rows of deep-yellow kernels, which are extremely shriveled in their upper half.

The Andeans used this corn as a source of sugar before the post-Columbian introduction of sugarcane and honeybees. Peruvians today snack on parched and dried sweet-corn kernels (*kcancha*), just as we snack on the kernels of Cuzco flour corn we eat today as "corn nuts." Peruvians use descendant strains of Chullpi for their national drink, fermented and otherwise, called *chicha*. A Chullpi descendant arrived in Mexico several centuries before Columbus, and this Maiz Dulce made its way into the American Southwest by A.D. 1200 or 1300, the date of the earliest sugary kernels found in New Mexican caves.

Another type of sweet corn, called Papoon, reached New England from Mexico by way of the upper Missouri River. This one had a red cob (from the P1 gene named for "purple plant color"), which Indians used for dye. A letter written in 1779 in Plymouth, Massachusetts, by one "Plymoutheus" notes that "the core was a bright crimson, and after being boiled, and the corn taken off, if the core was laid in contact with any linen (the table cloth or napkin), it communicated an indelible stain." Papoon was the forerunner of hundreds of later varieties that crossed with types of Northern Flint, and one of the progeny was Darling's Early, born in Connecticut in 1844 and remembered now only as a parent of Golden Bantam.

The taste for sweet corn accelerated in the mid-nineteenth century "for culi-

"Corned Indian" and a "Corned Maid" from a catalogue for Parker and Wood Seeds and Tools. From the collection of Cynthia Elyce Rubin, photograph by Carleton Palmer

nary purposes," in the words of the *The Prairie Farmer* of May 22, 1856, at a time when that organ was still dividing all corn into two major types, yellow and white. A variety called Early Sweet Corn, *The Prairie Farmer* explained, had recently been introduced into Massachusetts from Virginia by a Captain Richard Bagnall and, when boiled green, this type could be preserved in hermetically sealed tin cans. "Preserved in this manner, you have apparently a fresh dish of corn at any season in the year." Another type which crossed northern Papoon with Southern Dent was called Old Colony, and this became a favorite variety for canning for the next fifty years. Ironically, the push to create new sweet corns, which today are synonymous with fresh corn-on-the-cob, began with the creation of the can.

Stowell's Evergreen, introduced in 1853 in New Jersey, was named for its ability to keep moist for nearly a year without a can—if the husks were tied at the ends with a piece of string and laid on shelves in a cool place. The desire to taste "green" corn year-round by developing and preserving sweet-corn types is revealed in the records of the U.S. Patent Office, which lists only six varieties of sweet corn in 1858, twelve in 1866, thirty-three in 1884 and sixty-three in 1899. The turn of the century ushered in the modern craze for sweet corn, although its cultivation was limited largely to the Northeast. Country Gentleman, a white sweet corn, became immensely popular after its introduction in 1890, but it was soon overtaken by a yellow sweet corn introduced commercially by the Burpee Company of Philadelphia in 1902 as Golden Bantam. By 1911 *Burpee's Annual* ("The Plain Truth about the Best Seeds that Grow") listed twenty-one "Select Strains of Sugar Corn," which included names like Black Mexican and Howling Mob. By 1953 the offering of sweet corns was so extensive it required subheads. But most of these would be irrelevant within the decade because of the takeover by the supersweets that began in the 1950s when J. R. Laughman, a geneticist at the University of Illinois, developed a mutant from a three-way cross and called it by the unlovely name of "shrunken 1." To botanists sweet corn is "a mutant defective" because its sweetness derives from a recessive gene, *su* (short for "sugary"). This gene prevents the full conversion of sugar into starch that occurs in other types as the endosperm develops. Because the endosperm of sweet corn contains many more sugars than starch grains, as the kernels dry they become wrinkled and somewhat translucent and appear shrunken or shriveled. In his *Garden Book* of 1810, Thomas Jefferson recorded entries at Monticello for "shriveled corn," which we would call sweet corn. Since their sugary condition reduces the total amount of food they can store, the kernels of sweet corn are more susceptible to mold than those of other kinds, and are therefore frailer and less productive. For that very reason, perhaps, their sweetness was all the more precious to native eaters and growers of corn.

In explaining the way corn genes work in relation to sweetness, Robert L. Johnston of *Johnny's Selected Seeds* in Albion, Maine (a good source of heirloom seeds), uses the metaphor of a "recipe." The nucleus of each corn cell, he explains, contains a recipe for all elements of its growth. The vocabulary consists of "words" or "sentences" that we call chromosomes, which are formed in turn of individual "letters," or genes. The chromosomes are made of chemical strands (DNA) that in corn form different combinations of four nitrogenous bases, plus a phosphate and a sugar. These are the basic ingredients of the recipe, and the genes spell out how much to add when, as in "add a pinch of salt," or in this case "add a teaspoon of sugar." The degree of sweetness depends upon how many teaspoons of sugar are added to each type.

Today's sweet corn genotypes are so sweet that seed catalogues classify them by sugar percentages: "Sweet" has 5 to 10 percent sugar, "Sugar-Enhanced" has 15 to 18 percent and "Supersweet" has 25 to 30 percent. Sweet (Su) types are bred for flavor and crispness, while Sugar-Enhanced (SE) types balance sweetness with juiciness, or good "poppy," as they say in the trade. So rapidly have the types multiplied that the trade has been left speechless: after Kandy Korn, Peaches and Cream, Sugar Buns and Snow Queen, they are reduced to names like Phenomenal and Incredible. Among supersweet (Sh) types, the mutant Sh2 is less vigorous than regular sweet corn, but may be five times as sweet and stay that way for two weeks longer because the sugar-to-starch conversion has been that long delayed.

Seed Movers and Savers

LIFE FOR GENETIC REVOLUTIONISTS is sweetened by any kind of corn. As a raw material, corn has always been at the forefront of genetic research because, like the fruit fly, its variability makes it a favorite medium for experimentation. As genetics became the message of modern botany and biology, corn became the medium.

What made corn so variable was its ability to take unpredictable leaps, as was shown in the molecular experiments of Barbara McClintock, once described as "the Greta Garbo of genetics" because she preferred to live and work alone. As a graduate student of botany at Cornell in the early 1920s, at the same time as George Beadle, McClintock devised a method for studying the morphology of maize chromosomes. For forty years she was a woman alone in a field of men, many of whom ridiculed her experiments at the Cold Spring Harbor Laboratory when she concluded that some genetic systems, such as those that control multiple colors in an ear of Indian corn, behave unpredictably rather than by Mendelian rules. She found that these genes were transposable and mobile—now called "jumping." But her theory so violated genetic dogma that she had to wait until 1983, when she was eighty-one, before the Nobel Prize validated her "jumping genes" and put them into the current mainstream of "evolutionary jumps" and genetic mobility.

Corn varieties delineated by Matthieu Bonafous in his History Naturelle, Agricole et Economique de Maïs *(1836).* Courtesy Library of the University of California at Davis

Research, too is jumping. "Things we thought would take years to develop are taking months," said Virginia Walbot, a researcher at Stanford University, in 1986. "Everything is moving very quickly." She referred to a new high-speed method of altering corn cells by electric pulses. Her team had applied a technology of electricity (electroporation) to classic breeding techniques that was "100 times more efficient . . . for altering the genetic characteristics of plants" and that would soon "generate whole plants from single corn cells." A new breakthrough, wrote *The New York Times,* in custom-designed crops.

More than a century ago, the French botanist Matthieu Bonafous had threnodized corn's native ability to adapt to hostile climes: "One sees maize growing in the sandy plains of Nouvelle-Jersey; in the land of Carthagenes in Colombia, too humid for the *froment* and *orge* to succeed; upon the icy *ingrates* situated between Trevise and Bassan; in the schist terrain at the foot of the Apennines." Today the icy steppes of Russia grow an early-maturing, cold-resistant corn; the dry earth of Africa grows high-lysine, opaque-2 and QPM (quality protein maize); the humid jungles of Thailand grow downy-mildew–resistant corn; the insect-plagued fields of France grow corn-borer–resistant corn. Every month brings news of the newest biotech innovation: a gene-altered corn plant that grows its own pesticide to combat predators or one that grows its own antibiotic to combat disease.

Yet Henry Wallace in 1956 had warned corn breeders and researchers in their rush for the new to remember the Hopi: "For thousands of years, the Hopi or their ancestors have lived with corn with a depth of feeling which no white man can fully understand." That feeling is reflected in the language of mind and heart, he said, as well as in the ceremonies of daily life. The Hopi focus not on change but on "preparing," repeating what was in order to make ready for what will be. "The white man, on the contrary," Wallace observed, "rushes with maximum speed to change his machinery, his surroundings, his way of life and his attitudes, thereby creating obsolescence, waste and frustration, as well as rich, new values."

A catastrophe in 1970 lent credence to Wallace's view. Before that date, scientists had applauded the low annual disease rate in corn, ranging from 2 to 7 percent, and the localized nature of corn diseases. In 1970—the same year that brought a Nobel Peace Prize to Norman Borlaug for his development of a high-yield semidwarf high-lysine wheat that greened the world from Mexico to India—brought an epidemic to American corn that wiped out 13 to 25 percent of the crop. The genetic uniformity of commercial hybrid corn had made it susceptible to a new mutant strain of southern corn-leaf blight, *Helminthosporium maydis*. The ravages of the epidemic dramatized the dangers of sacrificing diversity to yield. Taxonomists and breeders alike now qualify "genetic improvement" with words like "genetic erosion" and "genetic vulnerability." More and more geneticists agree that the extinction of local corn varieties and primitive races in an effort to "improve" them is analogous, as Garrison Wilkes says, "to taking stones from the foundation to repair the roof."

In 1966, the same foundation that inventoried corn races organized an International Maize and Wheat Improvement Center. On a site just north of Mexico City, where Texcoco Indians had once created the garden capital of the Aztecs, the Center (CIMMYT) built a storage vault in which to preserve its seed collections and use them as a germplasm bank for both a global wheat and a global maize "improvement system." Currently the bank holds about 12,500 entries for maize, "more thoroughly collected than that of any other major food crop," grouped into thirty-seven gene pools, based on similar characteristics and environmental adaptations.

Originally the main idea was to help disseminate Latin American germplasm to Third World countries in order to end world hunger. More corn for more mouths. By 1971 the program was so large that a research consortium (CGIAR) was funded by the United Nations and the World Bank. Five centuries after the beginning of the Columbian Exchange, Iowa's massive Corn Belt has been cloned in southeastern Brazil, northeastern Argentina, the Danube Basin from Germany to the Black Sea, the Po Valley of northern Italy, eastern South Africa and the plains of Manchurian China. At the same time, improvement was redefined: conservation is now seen to be as important as change.

Within the United States, we now have a national network for distributing and preserving seed in the form of a twin storage program, both a checking and a savings bank. In 1954 the National Academy of Sciences and the National Research Council set up the Committee on the Preservation of Indigenous Strains of Maize to continue the collecting work begun in Mexico. This committee evolved into the National Seed Storage Laboratory in Fort Collins, Col-

"One of the major paradoxes of modern agriculture and particularly of corn production is that the short-term goal of high productivity runs directly counter to the long-term genetical health of the crop."
—"Irony and Paradox in the Corn Field," *Missouri Botanical Garden Bulletin* (July/August 1982)

orado. Ironically, it took over the function of what was previously called the Bureau of Plant Industry, organized in 1898 to regulate and encourage the introduction of new plants. By 1970 the Bureau's inventory listed three hundred and fifty thousand "introductions," but since there was no parallel system of preservation, 98 percent of those introductions were lost.

At Fort Collins, the National Plant Germplasm System houses, in addition to corn, other basic crops such as cotton, soybean, sorghum and tobacco. Today the bank holds about two hundred and fifty thousand genetically different samples from two thousand crop species, each sample containing around five thousand seeds, to total one and a quarter-billion seeds from every major world crop. The lab's director, Steve Eberhart, explains that seeds as they are received are tested for viability, then dried, hermetically sealed and stored in refrigerated vaults at zero degrees (Fahrenheit), where they retain viability for twenty to one hundred years. The next step is cryogenic storage, with temperatures at −196 degrees (Centigrade). Like animal breeds, seeds arrive with papers documenting their parentage, and these are stored in a data base with the acronym GRIN (Germplasm Resources Information Network), which keeps track of seed in all the U.S. gene banks.

Working seed collections, as opposed to storage collections, are located in four Regional Plant Introduction Stations, each specializing in a major crop. The place to go for corn is Ames, Iowa, at the North Central Plant Introduction Station sponsored jointly by the Department of Agriculture and Iowa State University. The station's job is to provide seeds for research and to keep track of the movement of germplasm in other countries that might affect what's going on in the States. Since the stations are geared to large-scale operations, domestic gardeners and others who want to grow a particular strain are encouraged to use networks like the Seed Savers Exchange.

The Seed Savers Exchange (SSE) at Heritage Farm in Decorah, Iowa, began in 1975 when Kent Whealy and his wife, Diane, heeded warnings by Garrison Wilkes and other breeders about genetic erosion. They began to look for gardeners who had continued to grow heirloom vegetables and fruits in their own gardens and then organized them into an annual seed exchange. What started as "a six-page newsletter to twenty-nine people copied on an unguarded Xerox machine at Boeing Aircraft in Wichita, Kansas," Kent explains, has become an annual *Winter Yearbook,* listing nearly eight hundred members and more than eleven thousand varieties of seeds. The organization is nonprofit and members must grow seeds to get seeds, since the notion is exchange: "We are making our seeds available to any gardener who will help us maintain them."

In the pages of *Seed Savers Exchange,* members exchange information on organic gardening, on hand-pollinating corn the old-fashioned way, on maintaining selected lines of corn, even on composing corn pictures of ducks and owls with kernels colored purple, pink and brown. In 1985 *SSE* compiled and published the first comprehensive inventory of the U.S. and Canadian garden seed industry since 1903. Overall, only 3 percent of vegetable varieties available then survive today in the Seed Storage Laboratory in Fort Collins. While in 1987 the number of newly recovered varieties increased, still there were fifty-four fewer seed companies than two years earlier. "Think what it would cost, in terms of time and energy and money, to develop this many outstanding varieties," Whealy has written. "But they already exist. All we have to do is save them."

Another seed savers' group evolved from a gardening and nutrition project labeled Meals for Millions, on the Tohono O'odham Indian Reservation southwest of Tucson, Arizona. Here Gary Nabhan, an ethnobotanist, and Mahina Drees, an extension agent, were working with tribal members to revitalize the land. In 1982 Nabhan and Drees, with their spouses, founded Native Seeds/SEARCH in order to collect, spread and save native seeds throughout the Southwest and northern Mexico. Although corn makes up much of their inventory, they also concentrate on corn's sister crops of beans, squashes, chilies, herbs and related wild plants that grow in these low desert and high semi-desert regions.

In their quarterly *Seedhead News,* you can learn of the Salt River Indian Community's restoration of native crops to fields long abandoned, of endangered monuments such as the eighteen prehistoric irrigation canals of the Hohokam that run through the Park of the Four Waters near Phoenix, and of current planting celebrations of the Pimas and O'odhams, described in the voices of celebrants. "When a man goes into a cornfield he feels he is in a holy place, that he is walking among Holy People, White Corn Boy, Yellow Corn Girl, Pollen Boy, Corn Bug Girl, Blue Corn Boy, and Variegated Corn Girl," a Navaho named Slim Curley said in 1938. "If your fields are in good shape you feel that the Holy People are with you, and you feel buoyed up in spirit when you get back home. If your field is dried up you are down-hearted because the Holy People are not helping you."

In the 1950s Carl and Karen Barnes of Turpin, Oklahoma, were ordinary farmers who saw the need to establish a living seed bank of open-pollinated varieties before they disappeared. Now the Barneses have developed a network of some three thousand corn growers in fifty states who work to maintain breed purity as Indian cultures have always done. In the 1990s the Garst & Thomas Hybrid Corn Company in Coon Rapids, Iowa, a town that boasts the

"We're like a society that's eating its seed corn instead of planting it."
—WALTER FRIEDMAN, Department of Economics, Harvard University (January 1989)

world's largest seed-corn processing plant, has started the Kernel Club to fund a Seed Corn Museum at the base of a five-story sculpture that will be the world's largest ear of corn. In 1927 Roswell "Bob" Garst had bought a bushel of hybrid seed corn from Wallace's new Hi-Bred Corn Company and parlayed it into today's multimillion-dollar seed-corn industry. Within sixty years hybrid corn has turned from monster prodigy to museum piece. "We have found that most people outside the seed-corn industry," a Kernel Club member explained, "have no idea what seed corn is, or why it is needed."

The corporate offices of Pioneer Hi-Bred International still display in a glass case an ear of Hopi Indian corn that is labeled "Grown from seed by Henry Wallace, gift of Hopi Indians, planted very dry." Henry Wallace's plea to save forgotten corns and to listen to forgotten voices has not gone unheeded. High-powered geneticists now speak of "plant sensitivity" and of the intimate "cultivator-to-plant relationship" among native growers of corn. Good thoughts are good for the plant, the Hopi believe, and bad thoughts are bad. That's a Hopi definition of maize improvement. "It is not scientifically feasible to demonstrate that *thought,* that *loving* a corn plant, will improve it or its progeny," Wallace wrote, but "all sincere plantbreeders use this very approach in some modified way." For the Hopi, time is not an antagonist, but rather a friend to the races of corn and men, who wax and wane together in the recurrent rhythm of the seasons.

THREE

The Daily Round

*(overleaf) Cornfields are the
organic and festal source of life
in the village of Secota, depicted
by John White and engraved by
Theodor de Bry in 1590, for
Thomas Harriot's* Briefe and
true report of the new found
land of Virginia.
From the collections of the
Library of Congress
*A rotating silage cutter on a
modern corn combine cuts up
ears and stalks for fodder.* Photo
by Gregory Thorp (1988)

The Language of the Seasons

THE SACRED CORN OF INCA AND AZTEC

TIME IS OUT OF JOINT when you stand in the nave of the Church of Santo Domingo in Cuzco, high in the Andes of Peru, and imagine these massive slabs of Inca stone covered with sheets of gold. Where the cross of the Christian Son now stands you imagine the high altar surmounted with "the image of the Sun on a gold plate twice the thickness [of the other gold sheets]," as Garcilaso de la Vega described it, "with a round face and beams and flames of fire all in one piece . . . so large that it stretched over the whole of that side of the temple from wall to wall." In Santo Domingo, the crucified Son is shadowed by the blaze of the Sun that was.

The Temple of the Sun that dazzled the eyes and hardened the steel of Francisco Pizarro's swordsmen when they entered Cuzco in 1533 adjoined a garden of herbs and flowers, lizards and butterflies and birds, a small patch of quinoa and a great field of corn, in which every single natural thing was hammered from silver and gold, including the human figures of the gardeners and their spades and hoes. The corn was copied with exact realism, down to the "leaves, cob, stalk, roots, and flowers," Garcilaso has told us, as he described how "the beard of the maize husk was done in gold and the rest in silver, the two being soldered together."

The Inca called this entire temple and garden complex *Coricancha,* "the golden quarter." *Cancha,* meaning "enclosure," is also the Quechua name for toasted kernels of corn, each kernel a miniature garden of gold. The association of corn kernels with the globe of the sun and gardens of gold was natural to a people who called gold "the sun's sweat." And yet, while they honored gold, they were unable to understand the Spaniard's visceral hunger for it. An

This Inca figure of cast silver, holding five ears of corn and wearing a vegetative costume, may represent a harvest-festival dancer. Neg. No. 328548, photo Boltin, courtesy Department of Library Services, American Museum of Natural History

PRIMER DEGENERACIOWS
VARIVIRACOCH

TRAVAXO
3ARATARPVMITAN

TRAVAXA
CHACRAMĀTAPISCO

TRAVAXA
3ARACARPAÑACOMVC

The cycle of planting, harvesting and storing, from Poma de Ayala's almanac.

old drawing shows an Inca handing a bowl of gold nuggets to a Spaniard and asking, "Do you eat this gold?" The Spaniard answers, "We eat this gold."

The misunderstanding was in their guts but deeper than hunger, for the Inca and Spaniard ordered their lives by different visions of the sacred, and the clash of Holy Sun with Son was as inevitable as it was bloody. In Peru as in Mexico, with the Inca as with their contemporaries, the Aztec, it was not gold but corn that was sacred as an emblem of the life-giving sun. The movement of the sun dictated the rhythms of daily life and organized the outposts of empire for the Inca, whose very name meant "Lords of the Sun." Although they had domesticated a few mountain animals, theirs was a planting culture and corn was its lifeblood.

Like the Aztec, the Inca inherited the agricultural wisdom of the many ancient planting cultures that had preceded them. Fifteen hundred years before the Inca empire commanded the Andes from Ecuador to Chile, corn was being cultivated in the highlands and on the coast. Potatoes were cultivated in the highlands at least eight thousand years ago, but the underground tuber lacked the cachet of sun-gilt corn. If gold was the sun's sweat, corn was the sun's gold, but exactly when Peru first domesticated corn is even more problematic than corn dates in Mexico, because archaeology in Peru is relatively young.

Darwin in 1859 became the first scientist to describe archaeologic corn there, when he found "heads of indian corn" mingled with shells on San Lorenzo Island near the port of Callao. In the 1860s, George Squier, then United States Commissioner to Peru, found corn in the contents of a tomb he uncovered in the ruins of the oracular shrine at Pachacamac, built in the first millennium after Christ on a headland five hundred feet above the sea. Buried with six mummies, including a fisherman with his lines and sinkers, a woman with spindle and thread, a girl with a makeup kit of pigments and cotton dabs and a baby with a seashell rattle of pebbles, were six earthenware pots of foodstuffs like peanuts and maize. Not until 1899 did Max Uhle uncover the ceramic corn pots at Huaca del Sol, buried between A.D. 300 and 600 in the Moche Valley south of Trujillo. Not until 1911 did an enterprising Yalie, Hiram Bingham, uncover the corn terraces of Machu Picchu. The story of corn in Peru has only begun to be told.

In 1989 a group of American archaeologists unearthed in the Casma Valley on the northern coast, at Pampa de las Llamas-Moxeke, a complex of storage chambers, stepped pyramids and temples ten feet high, built thirty-five hundred to thirty-eight hundred years ago, two thousand years before the great Mayan temples of Palenque, three thousand before the pyramids of the Aztec.

Pampa was as old as the great pyramids of Egypt. While corn was not found at Pampa, there was evidence of other cultivated crops such as peanuts, beans, sweet potatoes and manioc. So far, the earliest evidence of corn in Peru has been found in the highlands around Ayacucho, where Richard MacNeish discovered in the cave of Pikimachay (or Drunken Flea) corn remains that are three to four thousand years old and animal remains that may indicate aborigines hunted there sixteen thousand years ago. On the coast, where dry sands preserve well, the earliest evidence of corn has been the prehistoric cobs discovered at Huaca Prieta, a late pre-ceramic site of the Chavin culture at the mouth of the Chicama Valley, which dates from around 1050 B.C.

In the 1920s, Julio C. Tello, Peru's first native archaeologist, claimed that corn originated in Peru separately from Mexico, a claim Mangelsdorf welcomed to support his theory of wild corn. Evidence was sought in what appeared to be a petrified ear of corn found in Cuzco that botanists labeled *Zea antigua* until they discovered, on breaking it open, that it was made of clay and contained three small pellets—it was a child's rattle. Today the consensus of most archaeo-botanists is that corn migrated to South America from Mesoamerica along sea and land routes, but in Peru the extremes of the landscape, from coastal deserts to high sierras to Amazonian jungles, created a unique culture of corn.

Corn founded the culture that for three millennia dug irrigation canals and terraced the steep cliffs of the Andes, shaped and painted clay, wove fine cotton and alpaca, hammered gold and molded feathers, carved desert rock to chart the stars and morticed stone to "harness" the sun. During this time, not only did Peru evolve more diverse races of corn than any other country (Mangelsdorf distinguishes forty-eight), but its planters also evolved methods and rituals of agriculture in these formidable topographies that are still in effect today. While the Inca were but the last of Peru's rich civilizations, it was their awesome if fatal genius for organization that made permanent impress on the land, incorporating disparate ecologies into the "Land of the Four Quarters" and binding disparate peoples by chains of corn.

Francisco Pizarro's conquistadors reported corn plants "tall as soldiers' pikes" when they landed in 1528 at the Gulf of Guayaquil on the northern coast, but detailed knowledge of Peru's corn culture comes not only from Spanish chroniclers like Father Bernabé Cobo, but from two mestizos who knew Cuzco from the inside. One was Guaman Poma de Ayala, whose *El primer nueva corónica y buen gobierno,* written between 1583 and 1613 in a jumble of Indian dialects and queer Spanish, provides us with an Inca Farmer's Almanac. While his drawings melded mythic and historic time, juxtaposing Adam and

Photo Logan, courtesy Department of Library Services, American Museum of Natural History

A Moche stirrup-spout water jar in the shape of the fanged creator-god, Viracocha, whose fertility is figured in corn children miniatures of himself. The corn was molded from actual ears related to the highland race called Cuzco. Neg. No. 45663, photo J. D. Wheelock, courtesy Department of Library Services, American Museum of Natural History

Eve in the Garden with contemporary battles and earthquakes in Peru, his calendar showed clearly how the Inca organized his daily life around the seasons of the sun. Like the Christian calendar, the Inca calendar was based on a year of twelve months, and the word *quilla,* for month, suggests that the primary counting language of the Inca, the system of knotted cords called *quipu,* originated in an agricultural ordering of time, numbered by the recurrent cycles of planting, irrigating, weeding and harvesting corn.

The other mestizo was Garcilaso de la Vega, "El Inca," as he called himself proudly, son of a Spanish nobleman and an Inca princess. Born in Cuzco in 1539 and educated royally, Garcilaso left for Spain when he was twenty-one to claim his share of his father's estate. Denied it by reason of his bastardy, Garcilaso turned soldier and then scholar when he retired to Andalusia to finish his *Comentarios reales de los Incas,* four years before his death in 1616. In that same year Shakespeare died, and wrote no more royal commentaries on England and Rome. In that same year my ancestral Kennedys had already turned Presbyters in Scotland and my ancestral Culvers were plotting to set forth from England as the Children of Israel had done, in Exodus to the Wilderness. Even had Garcilaso's account been translated and made available to them, they would have spurned it. They had the new King James Bible for travel guide. Only in these latter, more secular days do we value the uniqueness of Garcilaso's early account of the New World, written in a European language, but by a native American.

When Garcilaso describes the farming of the Inca, he reveals in every sentence the Inca need for *system.* The Inca divided their empire into four quarters, each quarter into provinces, each province into communities, or *ayllus.* Their first act in conquering new territories was to make the land arable for corn. Their engineers dug irrigation channels and leveled fields to distribute the waters. They terraced impossibly steep mountains by a stair-step system of masonry walls filled with earth, "the platforms being flattened out like stairs in a staircase," diminishing in depth as they ascended. Once the land was arable, they divided it into thirds, one part for the Sun (or the Church, we would say), one for the king and one for the commoners. They planted corn in the fields each year without letting them lie fallow because they knew how to water and fertilize them. Along with corn, instead of planting beans they planted quinoa, an indigenous grain which Garcilaso called "a seed rather like rice," and which we know to be rich in protein.

The Inca monarchy was as absolute as the setting and rising of the sun, and the Inca believed that their emperors were direct descendants of the Sun and

His many brides—Mama Quilla (Mother Moon), Pacha Mama (Mother Earth), Mama Sara (Mother Corn). Commanded by the Sun, as they said, the first Inca left their island home in Lake Titicaca to travel through the high valleys of the altiplano until Manco Capa's golden staff should sink into the ground at the first blow. On the hill of Huanacauri above the Valley of Cuzco, the king's staff sank, and there they built the Temple of the Sun and planted fields of corn. "The keynote to all this activity was the search for good farmland," Paul Weatherwax commented, "and, at the dawn of modern history, the Valley of Cuzco was the center of the greatest diversity of maize types and the most highly specialized agriculture to be found anywhere in the New World."

It was also the most highly controlled. The Inca plowed in August, planted in September and harvested in May. After the proper planting ceremonies, sprinkling cornmeal and corn beer, the emperor himself would dig the first hole with a plow of gold, not unlike a city mayor in the States wielding his ceremonial shovel on Arbor Day. Garcilaso described the week of first planting as a time of great festivity, when the nobles strapped on their gold and silver breastplates, donned feather headdresses and sang songs of praise to the Sun and his Lords. Between the city of Cuzco and the hilltop fortress of Sacsahuaman was a terrace dedicated to the Sun and reverenced as the body of Mama Huaco (mother-sister-wife of Cuzco's founding Inca), reserved for exclusive tillage by those of royal blood. These lords set the rhythm of their plowing to songs that played with military and sexual puns on the word *hailli,* which was their victory cry when they crushed the enemy, furrowed the earth, ravished a woman. At harvest time they turned the maturing of corn into a puberty rite for their young knights, dressing them in parrot feathers and ear plugs, kilts and breechclouts, and allowing them alone to harvest this field of sacred corn and bring it to the temple, crying their version of hallelujah, *"Hailli, hailli!"*

Since this was a hieratic military society, the hierarchies were strict. At the bottom of the social order were the plowmen, *yanca ayllu,* or "worthless ones." These were the ones who worked the fields in common, but always in strict order: first the plots belonging to the Sun; then plots belonging to the king and state, which included land cultivated for the poor—widows, orphans, the disabled; finally their own communal plots. Their plow was a stick, the *taclla,* about two feet long and four fingers thick, with a flat front and a rounded back. Two sticks were lashed to the main shaft above the pointed end to serve as a footrest, and the other end was fashioned into a handle. A band of seven or eight men worked in a group, planting their plows in unison to the beat of the songs. A band of women followed them with clod-breakers or hoes; the *lampa* was a short-handled stick lashed to a doughnut-shaped stone or, later, to a broad blade of bronze.

The rhythm of the plowing was dictated by the rhythm of the songs. Garcilaso recalls that a choirmaster in the cathedral in Cuzco once composed a part-song for the feast of Corpus Christi in which eight of Garcilaso's mestizo schoolmates, plows in hand, mimed the action of plowing while singing a *hailli* in praise of God, the Spanish God that the converted Inca called Pachacamac, "He who gives life to the universe."

Each plowman was given a measure of land called a *tupu* for growing corn. If he had children, each boy was given a *tupu* and each girl half a *tupu*. Noblemen were given land in proportion to their retinue—wives, concubines, children, servants. In Cuzco and the highlands, they fertilized the soil with human manure, taking care first to dry and pulverize it. On the southern coast the manure used was "the dung of seabirds"; the offshore islands where birds flocked in great numbers seemed like "the snowy crests of a range of mountains," so thick was the guano. Each guano island was divided into village allotments, and trespassers were rigorously punished. In other parts of the coast they manured the land with the heads of sardines, which they would burn in small pits made by stakes and thereafter plant two or three grains of corn in each pit.

The system of storage was part of the system of tribute. Mixing clay with straw, they constructed *pirua,* or bins of different size, which they arranged in rows within the granary. They left alleys between the rows so that the bins could be emptied and filled as needed. By making a small hole in the front of each bin they could see at a glance how much grain remained. The harvests belonging to the Sun and to the king were stored in the same granaries but kept separate. The seed corn, *marca moho,* which they reserved for next year's crop, had its own special place in the granary and its own symbolic place in the heavens, in the constellation of the Pleiades, which they called "The Star of the Overflowing Grain Bins." These seeds were given out each year by the landowners to plowmen for their communal allotments.

In his *Historia del Nuevo Mundo,* the good Catholic Father Bernabé Cobo admires the totality of control the Inca emperor exerted over his subjects. Cobo's history, begun during his sojourn in Cuzco in 1609–13, was not finished until 1653 and not published until 1890, by which time other totalitarian empires had fallen and new ones were soon to be born. The acts of sowing, tilling and harvesting, Father Cobo assures us, were for the Inca acts of worship in which the entire community joined. The king and his nobles joyfully worked their plows of gold and only then did they sit down to the banquets and fiestas prepared for the celebration of the Sun on Field Planting Day. The commoners, of course, continued the job until the harvest was in, dividing their communal plots into sections where each laborer worked with his wives, children and hired hands if he could afford them. A poor man with no helpers simply took longer to work his allotment.

"When a field is to be broken up, they have a very loving, sociable, speedy way to dispatch it; all the neighbors, men and women, forty, fifty or a hundred, do joyne and come in to help freely; with friendly joyning they break up their fields and build their forts."

—ROGER WILLIAMS,
A Key into the Language of America (1643)

Cobo was struck by the communality of the farming. No man was granted any land except at the king's favor, and even that was divided equally among his kinsmen for sowing and harvesting in common. "When it was time to sow or cultivate the fields, all other tasks stopped," Cobo writes, and if there was a war on or some other emergency, "the other Indians of the community themselves worked the fields of the absent men without requesting or receiving any compensation beyond their food."

Marketing was also a communal exchange, since corn was the major unit of currency in the Inca barter system. In the central markets of village or town, Cobo explains, Indian women would exchange raw or cooked foods for corn-bread or for kernels of corn, which served as money:

The Indian women put all their goods, or part of them if they are fruits or things of this nature, in small piles arranged in a row; each little pile is worth about one-half to one real; if it is meat, it is cut up in pieces of equal value, and likewise with the rest of their goods. The Indian woman who comes to buy with her maize instead of money sits very slowly next to the one selling and makes a small pile of maize with which she plans to pay for what she is buying. This is done without exchanging a single word. The one who is selling looks at the maize, and if she thinks it is too little, she does not say a word or make any sign. She continues looking at it, and while she continues in this pose, it is understood that she is not satisfied with the price. The one buying has her eyes on the seller, and during the time she sees her uncommitted, she adds a few more grains to her pile of maize, but not many. If the seller remains inflexible, she adds again and again to the pile, but always in very small amounts, until the one selling is happy with the price and declares her approval—not with words, which from the beginning to the end are never spoken, even if the transaction lasts half an hour, but rather by deeds. The seller reaches with her hand

and brings the maize toward her. They never stop to think about whether
or not these exchanges are in keeping with the principle of just and equi-
table commerce.

The tradition of assiduous bargaining did not end with the Inca, as any traveler to Peru today knows well. Nor did the ancient methods of farming. Anyone traveling along the Urubamba River through the Vilcanota Valley, once
thick with runners and boatmen who carried jungle fruits from the Amazon
to the banquet tables of the Cuzco lords, knows why it is called still the Sacred
Valley of the Incas. To travel there in June, as I did, is to travel through a time
warp lit by a harvest landscape golden with corn and wild Scotch broom.
Warmed by the sun, protected from the snow-covered peaks that glitter against
an ice-blue sky, the valley is fragrant with orange trees and lemon trees, red
tomatoes and green peppers, geraniums and Pisonay trees that flame their
color against hills of gold and ochre squares.

These are *tendales,* where rows of husked ears of corn have been laid flat on
the fields in precise geometrical design to dry in the sun, squares of white alternating with red, deep yellow or startling purple. Drying ears poke from niches
and corners of the squat adobe houses, some strung from a rafter, others heaped
in baskets in a courtyard. Ahead of me by the side of the road three men run
barefoot uphill, leaning forward against straps that bind their brows to the
mountainous loads of cornstalks on their backs. On the hill behind them I
glimpse a woman's red petticoated skirt and her tall felt hat, a boy tending his
sheep, a pair of donkeys circling and circling a threshing floor, and in the distance a solitary man and woman bent unmoving in the field as if frozen in time.

Time is permanently out of whack in this rude land. Today's villagers still
cultivate the steep terraces that mount the ruins of ancient Pisac, a place famed
today as it was in the time of the Incas for the quality of its large-kerneled corn.
And on the cliffside adjoining the sinister ruins of Pisac's fortress, you can see
still a multi-leveled Inca granary covered so smoothly with yellow plaster that
it looks like a modern urban high-rise. At Ollantaytambo, where the gorge
narrows and the rapids begin, the village still conforms to the grid that Incas
laid around a central square, each cobbled street cut down the center to channel water. At nearby Yucay, stone troughs still bring water down zigzag terraces from the glaciers of the Urubamba Cordillero far above to irrigate the
valley below. High in the Cordillero, the terraces that crown ghostly Machu
Picchu seem still to breathe with life from the smoke of little piles of burning
grass and brush set afire by plowmen who now tend the ruins of this Sun Tem-

ple for tourists instead of for the Chosen Virgins of the Inca. Four centuries disappear in an eye-blink.

Armed with Garcilaso, on my first trip to Peru I came deliberately for the festivities at winter solstice, but I didn't expect ghosts to come to life at Machu Picchu the way they did on June 21, the day of the solstice itself. Having managed to climb, despite myself, the footholds carved by Inca in the vertiginous cliff of Huayna Picchu that thrusts above the site like a fang of the god Viracocha, I lay flat on a rock to peer over the edge. Across the central greensward of Machu Picchu far below, I saw a procession of white-robed figures, streaming banners of red and gold. They formed a circle, some four hundred of them, then moved slowly toward the great carved stone Intihuatana, the "Stone to which the Sun was Tied" on this day only, as the sun's rays crossed it at dawn. Later I learned from one of the white-robed sun worshippers that they come from around the world to meet here every twelve years at the solstice. "Thees ees magnetic center of earth where all forces converge," the man said in his distinctly East Indian accent. "Soon will end Age of Aquarius to begin Golden Age, when all will live in piss and hominy."

I knew what he meant, but I couldn't help thinking of why the emperor's virgins were chosen in the first place. They were commoners separated from their families as small girls, sequestered and trained in the domestic arts until

*"the Toltecs did not in
fact lack anything
no one was poor or had a
shabby house
and the smaller maize
ears they used as fuel
to heat their steam baths
with"*
—From the Florentine
Codex, Book 3,
Chapter 3

puberty, then presented to the emperor at the great solstice festival in Cuzco. He chose the beauties as servants or concubines for himself or as gifts to friends and reserved the rest for sacrifices to the Sun. Machu Picchu and Cuzco evoke such ghosts at every stone corner, but I preferred a quieter ghost, of a woman I had seen in the necropolis museum at Paracas on the southern coast, by the guano-coated islands of Ballestis, an ancient fertilizer factory. She was a mummy of leathery skin preserved for two thousand years, sitting cross-legged in a large ceramic bowl with the loom and spindles of her weaving. Next to her a tiny boy, who had, according to the sign, died of tuberculosis, lay curled around his little pipe, his pet cuy (guinea pig), four llama hooves, a bit of yucca and a few kernels of corn. I found considerable comfort in that corn.

LIKE THE INCA, the Aztec were empire builders, erecting in the swampy lagoons of the Valley of Mexico in 1345 the white city of Tenochtitlán, founded like Cuzco on sacred corn. Also like the Inca, the Aztec were cultural arrivistes, a tribe of barbarian Mexica (pronounced Me-*sheek*-a) from Aztlan, to the west, who wandered into the valley some thousand and a half years after the Olmec erected the metropolis of Teotihuacán in its cornfields and a thousand years after the Toltec raised in the same valley the powerful city of Tula with the help of the Plumed Serpent, Quetzalcoatl, who discovered the twin arts of agriculture and writing.

When I visited the National Museum of History in the castle of Chapultepec and looked east across the smog-filled bowl of Mexico City, built on the ruins of Tenochtitlán and Tlatelolco, toward Texcoco on the eastern perimeter where now lie hidden the world's accumulated seeds of corn, I found I was standing on the place where the last Toltec ruler hanged himself, where Aztec emperors carved their portraits in the rock and where only a century ago Maximilian set up house with Carlotta. Such are the palimpsests of history in Mexico, where corn was born and farming began.

The Aztec inherited an intensive method of agriculture, necessary to the Mesoamerican peoples who did not use domestic animals for planting and therefore used neither plow nor wheel for cultivating the soil extensively. While the hot wet jungles of the Yucatán favored the simple slash-and-burn method, the Valley of Mexico required different treatment because it was a swamp of fresh- and salt-water lakes. Planters there around 300 B.C. began to build artificial islands they called *chinampas* and we call "floating gardens." You can see their contemporary forms in the gardens of Xochimilco and their ghostly outlines in aerial photographs. The Franciscan Juan de Torquemada saw them in the sixteenth century and described how the Indians in their fresh-water lagoons "without much trouble plant and harvest their maize and greens, for all over are ridges called chinampas; these are strips built above water and surrounded by ditches, which obviates watering."

The Aztec like the Inca were masters of organization, and they elaborated the ridges into a system of dams, aqueducts and dikes to control the waters during their annual flooding as well as to provide for irrigation, cultivation and transport all at once. They divided the city into four quarters and each quarter into subdistricts based on clans, or *calpulli*, who tilled their land communally. Through the marshes, they dug a network of canals and built between them mounds of water plants covered with thick layers of lake-bottom mud. The mounds they held in place by hurdles of wickerwork, often three hundred feet long by fifteen or thirty feet wide, large enough to support reed houses and willow trees. They could grow crops year-round without the land lying fallow because they fertilized it with human dung, sold to farmers by city sanitation engineers, and renewed the land as needed with fresh mud. The mounds provided movable nurseries for seedlings of corn, which they transplanted as needed, and the canals provided transport for farmers and merchants alike.

The network was a marvel of city planning that at its peak in the fifteenth century supported an urban population of more than a million and a half people, three-fourths of whom were farmers. The first major public work of the Aztec was a broad raised-stone dike, which was both a highway and a dam to create reservoirs of fresh water for ever-increasing areas of cultivated land. How dependent the city of Tenochtitlán was on its gardens of corn is evident from records of successive years of frost and drought beginning in 1450, when famine was so severe that families sold their children for a basket of corn: four hundred cobs for a girl, five hundred for a boy.

The majority of farmers were hereditary clan members, with rights and duties to the state, but beneath them were *mayeques* who belonged to no clan and consequently became sharecroppers working for others. Each farm family paid tribute to church and state with goods and services, and when a neighborhood was required to supply firewood, say, to a god's temple, each family would share the cost by contributing "one large mantle, four small ones, a basket of shelled maize, and 100 cobs." So states a pre-Conquest tax list ordered by Moctezuma and recorded in pictures on thick sheets of maguey (since annotated in both Nahuatl and Spanish), whereby we have a fair notion of the annual tribute of corn paid to the imperial granaries. In this Tribute Roll, you can see two storage bins, or *troxes,* of wide planks or wicker plastered with mor-

Long before the Aztec, the Zapotec founded their civilization upon corn at Monte Albán, A.D. 300 to 900, near Oaxaca. Here a Zapotec funeral urn depicts the corn god, Pitao Cozobi, with stylized ears on his headdress. A cob with rows of kernels topped by a spike was typical of the prehistoric corn ears found in the Tehuacán Valley near Oaxaca. Courtesy Departamento de Arqueología, Museo Nacional de Antropología, Mexico City

From the Codex Mendoza

tar, each topped with a bean and a kernel of corn. Each bin held eight thousand to ten thousand bushels, and the annual tribute due the state was twenty-eight *troxes* of maize, twenty-one of beans, the like number of a seed called *chian,* and eighteen of a mixture of greens and grain amaranth that they called *huauhtli.* Additional tribute was levied for the support of local temples, schools, garrisons and indigents.

Corn planting began in March and continued through May, depending on the breed, in time for the heavy rains of summer. Seed corn from the previous season, after its blessing by the goddess of maize, Chicome Couatl, was soaked in water while the farmer prepared the ground with his digging stick, which resembled the *taclla* of Peru. He hilled the earth in small hummocks in rows about a yard apart, made a small hole in the top of each and dropped in a few kernels from the cloth bag over his shoulder. He hoed the ground but two or three times during the growing season. Mid-July, when each stalk bore two or three young ears, he removed all ears but one that the green corn might be made into special cakes for the festival of Xilonen, the goddess of young corn. By mid-August he bent each stalk over just below the maturing ear and left it to dry. In September, after the harvest sacrifice to the god of ripe corn, Cinteotl (*cintli* means "maize" in Nahuatl; *teotl* means "god"), the farmer tied the ears in bundles, shelled some for his house and took the rest to the communal storage bins.

The Spanish calculated that the average yield per acre was sixteen bushels. As William Bray speculated in *Everyday Life of the Aztecs* (1968), an Aztec family of five would have needed to till three acres of land to provide enough food for themselves, apart from their tribute fields. The combination of a temperate climate and sophisticated farming methods ensured that a family could produce what they needed by working only half a year at farming. As a result, the Aztec had plenty of leisure for the arts, for building, for the festivals and ceremonies that punctuated their daily round, and for the rituals of war. "The material foundations for Aztec imperialism," one archaeologist has said, "were established by the farmers who had conquered the swamps."

We know details of Aztec farming, and other aspects of daily life, from the writings and drawings of Father Bernardino de Sahagún, a Franciscan who arrived in Mexico from Spain in 1529, learned the Nahuatl language and spent the rest of his life chronicling what he heard and saw in *La Historia general de las cosas de Nueva España,* not published until 1829. What he heard he transcribed with great care in both Nahuatl and Spanish—a lexicographer's method that created by mere accumulation a rhythm as incantatory as a planting song. Here is his account of the farmer in Book 10 of the Florentine Codex:

The farmer [is] strong, hardy, energetic, wiry, powerful. The good farmer, the [good] field worker [is] active, agile, diligent, industrious: a man careful of things, dedicated—dedicated to separate things; vigilant, penitent, contrite. [He goes] without his sleep, without his food; he keeps vigil at night; his heart breaks. He is bound to the soil; he works—works the soil, stirs the soil anew, prepares the soil; he weeds, breaks up the clods, hoes, levels the soil, makes furrows, makes separate furrows, breaks up the soil. He sets the landmarks, the separate landmarks; he sets the boundaries, the separate boundaries; he stirs the soil anew during the summer; he works [the soil] during the summer; he takes up the stones; he digs furrows; he makes holes; he plants, hills, waters, sprinkles; he broadcasts seed; he sows beans, provides holes for them—punches holes for them, fills in the holes; he hills [the maize plants], removes the undeveloped maize ears, discards the withered ears, breaks off the green maize stalks, thins out the green maize stalks, breaks off the undeveloped maize ears, breaks off the nubbins, harvests the maize stalks, gathers the stubble; he removes the tassels, gathers the green maize ears, breaks off the ripened ears, gathers the maize, shucks the ears, removes the leaves, binds the maize ears, binds the ears [by their shucks], forms clusters of maize ears, makes necklaces of maize ears. He hauls away [the maize ears]; he fills the maize bins; he scatters [the maize ears]; he spreads them, he places them where they can be reached. He cuts them, he dismembers them. He shells them, treads on them, cleans them, winnows them, throws them against the wind.

The bad farmer is a shirker, a lukewarm worker, a careless worker; one who drops his work, lazy, negligent, mad, wicked, noisy, coarse, decrepit, unfit—a field hand, nothing but a field hand; a glutton—one who gorges himself; nothing but a beggar—stingy, avaricious, greedy, niggardly, selfish. . . . He is lazy, negligent, listless; he works unwillingly; he does things listlessly.

Sahagún's doubled litanies of Nahuatl and Spanish are rendered by a voice and ear so tuned to the

This earliest recorded drawing of raised-field agriculture, from the Drake Manuscript, *or* Histoire Naturelle des Indes, *ca. 1586, shows a West Indian farmer using a pointed stick to plant seeds of different plants together. On the right, three corn stalks abut two bean stalks. The translated inscription reads: "The Indian making his garden sows several kinds of grain for his food to make it appear that he is working hard and also to please his fiancée and sufficient to feed his wife and children, the soil being so fertile as to bear fruit all the time."* Courtesy Pierpont Morgan Library

This sequence from the Florentine Codex shows farmers planting, cultivating, harvesting and storing corn. Negative Nos. 1738–40, 1743, courtesy Department of Library Services, American Museum of Natural History

high liturgical language of his own church that he makes even the most hum-drum activities of growing and eating corn sound like corn crucifixion and res-urrection: "For it was said that we brought much torment to it—that we ate it, we put chili on it, we mixed salt with it, we mixed saltpeter with it; it was mixed with lime. As we troubled our food to death, thus we revived it. Thus, it is said, the maize was given new youth when this was done."

Describing the many different kinds of corn that the Aztec farmer planted, Sahagún translates each as a particularized jewel. At the apex is white maize, a small ear "hard, like a copper bell—hard, like fruit pits," clear like a seashell, like a crystal, "an ear of metal, a green stone, a bracelet—precious, our flesh, our bones." This was the maize whose perfect seeds the farmer selected for planting, whose growth Sahagún turns into a planting primer and an ode:

> Then it is gathering moisture; then it swells; then the grain of maize bursts; then it takes root. Then it sprouts; then it pushes up; then it reaches the surface; then it gathers moisture; it really flies. Then it forks, it lies dividing; it spreads out, it is spreading out. And they say it is pleasing. At this time it is hilled, the hollow is filled in, the crown is covered, the earth is well heaped up.
>
> Also at this time beans are sown or cast. They say that at this time this [maize] once again begins to grow, also begins to branch out. Then it reaches outward; then it spreads out; then it becomes succulent. Once again at this time it is hilled. Then the corn silk develops; then the corn tas-sels form. At this time, once again it is hilled; it is, they say, the hatching of the green maize ear. Then an embryonic ear forms. Then the green maize ear begins to form; the green maize ear shines, glistens, spreads glistening.
>
> [The kernels] spread forming little droplets; it is said that they spread taking form. Then [the kernels] form milk; they spread forming milk. Then the surface [of the kernels] become evened. Then it becomes the *nix-*

tamal flower; now it is called *chichipelotl*. Then [the milk] thickens, at which time it is called *elotl*. Then, at this time, it begins to harden; it turns yellow, whereupon it is called *cintli*.

Sahagún's minute attention to the naming of each moment in the birth, ripening and death of each corn plant conveys as powerfully as any transcription can the personal identification of ancient corn growers with their corn. His litanies are like prayers to Xilonen and Xochipilli, the green corn gods, to Cinteotl and Chicome Couatl, the ripe corn gods. Sahagún evokes as few others can the symbolic power of Chicome Couatl, "the Demeter of old Mexico, the Mother Goddess," as Erich Neumann has called her, whose son was the Sun. In the Aztec calendar which governed the daily round of Aztec life, chaos comes again every fifty-two years, when the Great Mother, Earth Mother, Terrible Goddess, Snake Woman, reasserts her power, swallowing all in darkness in order that her son, god of ripe corn, god of the heavens, may be born again. In the doubled columns of Sahagún's transcriptions we can hear the doubled rhythms of apocalypse and resurrection that underlie the metamorphic changes of both Sun and Son.

THE DESERT FARMERS OF THE SOUTHWEST

FARMING IN the Southwest desert of what is now the United States was far more difficult than in the Sacred Valley of the Inca or in the marshes of Tenochtitlán. Yet for three centuries before Christ was born in a desert across the seas, the Hohokam were making their stretch of the Sonora desert bloom with squashes, lima and tepary beans, tobacco and, at the time of Christ's birth, corn. Although earlier desert cultures like the Mogollon had cultivated corn in the mountainous borderlands of southern Arizona and New Mexico, corn as a civilizer of the Southwest desert began with the new group called by their Pima descendants the *Huhugam O'odham,* or the Hohokam, meaning "the vanished ones," who developed in the first millennium after Christ the most diverse plantings of any of the tribes and one of the largest canal systems of North America.

The Hohokam were farmers, despite the fact that the desert climate was as hostile then as it is now. Besides the usual hazards of sandstorms and flash floods, temperatures ranged, then as now, from 7 degrees to 119 degrees Fahrenheit and rainfall averaged less than ten inches a year. Fortunately, there was the floodplain of the Gila River, formed by the Salt and Verde rivers flowing toward the Colorado. With stone hoes these farmers began to dig pit houses and wide shallow canals. If you visit Arizona's Park of the Four Waters today, you can see the remains of eighteen prehistoric irrigation canals, which run parallel to the modern canal that now waters downtown Phoenix. Phoenix itself is built over the ruins of 1,750 miles of Hohokam canals. Along the Gila south of Phoenix, you can see the extensive ruins of Snaketown, where the Hohokam flourished until A.D. 1450, covering 240 acres with their dwellings and irrigated fields.

As Mesoamerican cultures bred new strains of corn resistant to drought and found new ways of irrigating it, a Reventador popcorn, a Chapalote flint and the eight-rowed sixty-day flour corn we call Maiz de Ocho reached Mexico's northern frontier. Along with adaptive corn breeds came Mesoamerica's planting sticks and, no less importantly, their planting songs. In their many different languages, all the Southwestern peoples knew the power of corn song. "It is late July, the moon of rain," Ruth Underhill wrote in *Singing for Power: The Song Magic of the Papago Indians of Southern Arizona* (1938). "Now planting can begin." Every man, she explained, placed his field at the mouth of a wash where, after a torrent, the earth would be soft enough to puncture with his digging stick. In each hole he dropped four kernels and, kneeling, spoke to the seed so that there would be no misunderstanding: "Now I place you in the ground. You will grow tall. Then they shall eat, my children and my friends who come from afar." While the corn grew, night after night the farmer walked around his field, "singing up the corn." "There is a song for corn as high as his knee," Underhill explained, "for corn waist high, and for corn with the tassel forming."

> *The corn comes up;*
> *It comes up green;*
> *Here upon our fields*
> *White tassels unfold.*
>
> *The corn comes up;*
> *It comes up green;*
> *Here upon our fields*
> *Green leaves blow in the breeze.*

There was a song for the blue evening when the corn tassels trembled, for the wind when the corn leaves shook, for the fear the corn felt when the striped woodpecker struck at its heart, for the green time when the tassels waved for joy and for the harvest time when the corn was embraced lovingly by the har-

"During the long Indian tenure the land remained undefiled save for scars no deeper than the scratches of cornfield clearings or the farming canals of the Hohokams on the Arizona desert."
—STEWART UDALL,
The Quiet Crisis (1963)

vester's arms. "Sometimes," Underhill said, "all the men of a village meet together and sing all night, not only for the corn but also for the beans, the squash, and the wild things." Songs and prayers were an essential part of cultivating wild things, and while no one has proved that loving a corn plant, as Henry Wallace said, will improve it or its progeny, the sense that corn was sacred did it no harm.

It was not lack of desert song but a disastrous series of droughts and floods that "stressed the Hohokam system," as Gary Nabhan puts it in *Enduring Seeds,* and depopulated their valleys a century before the first Europeans set foot in the Southwest. The first to come were three Spanish slave-catchers and a Moorish slave named Estevan de Dorantes, who had been shipwrecked on the Gulf coast of Texas in 1535. Before they worked their way back to Mexico, the Spaniards were told by the Indians they encountered of a kingdom of rich cities to the north. A later foray headed by Estevan and a Franciscan priest, Fray Marcos de Niza, got the slave killed for his trouble, but the priest's tales of imagined gold were enough for Francisco Vásquez de Coronado and his troops to saddle their horses and set out for the "Seven Cities of Cíbola." In 1540 they found, instead, the cornfields of the Zuni.

Zuni Pueblo is on the New Mexico side of the Arizona border, due south of Gallup and north of Bat Cave. Farther north the Navaho reservation straddles the two states, adjoining the Hopi to the west and the Pueblo group to the east. To reach Zuni Pueblo you must climb through rocky canyons and forests of piñon and juniper before hitting a barren plateau. Few Zuni grow corn here today, but theirs is the only one of the eighty pueblos conquered by Coronado that has retained its ancestral land, land they call the Middle Ant Hill of the World. When they do grow corn, largely for ceremonial purposes, they grow it as their ancestors did.

The corn of their ancestors impressed Pedro de Castañeda, who wrote about his trip with Coronado in *Relación de la jornada de Cíbola* (a copy of his original manuscript, made in 1596, rests in the New York Public Library). "The Indians plant in holes, and the corn does not grow tall, but each stalk bears three and four large and heavy ears with 800 hundred grains each," Castañeda wrote, "a thing never seen in these regions," by which he meant the regions around Mexico City. "In one year they harvest enough for seven years."

Even a century ago the methods of Zuni corn farming impressed another recorder, Frank Hamilton Cushing, who first led a Smithsonian expedition to the Pueblo in 1879 and discovered there a handful of Presbyterian missionaries working hard to cultivate a tribe of seventeen hundred unsaved souls. As an ethnographer and translator of the distinctive Zuni language, Cushing was as

This Mimbres pottery food bowl from New Mexico (c. A.D. 1000–1150) depicts six men with planting sticks, and a feline, in a cornfield. When a man died, a food bowl was inverted over his face like the domed sky of the Underworld, and a "kill hole" was punched in the center to allow his spirit to emerge, just as his ancestors had once emerged from the Underworld. Photo by John Bigelow Taylor, *Within the Underworld Sky* (1981)

Frank Hamilton Cushing in Zuni regalia around 1882. National Anthropological Archives, Smithsonian Institution

remarkable as Sahagún had been among the Aztec. His Smithsonian bureau chiefs, however, found him remarkable in ways they did not approve. They thought he'd gone native, and he had. The Easterner who had come but to observe aboriginal ways remained four and a half years to become a Priest of the Bow and First War Chief, before his assimilation was cut short at the age of forty-three when he choked on a fishbone.

From Cushing we have learned in precise detail how Zuni and other desert farmers made the sand sprout corn. When a young Zuni wished to mark land as his own, Cushing told us, he looked for the mouth of an arroyo and carefully "lifted the sand" with his hoe—the nature of Southwest farming is condensed in the phrase. After lifting little mounds at intervals around his proposed field, he built them into embankments called "sand strings." At each corner he placed a rock to establish his claim to this land for his lifetime and, after his death, his clan's. Significantly, the first month of spring, the month of planting, was called the month of "Lesser Sand Storms."

Before he could plant, the farmer had to extend his flood control into an irrigation system. Upstream of his chosen arroyo, he drove forked cedar branches in a line across the dry streambed and built a dam with branches, rocks and earth. Downstream, he built more barriers on either side. Finally he looked for a ball of clay, which he buried at the side of the bed where he wanted rain freshets to flow, then packed earth over the clay to form a long embankment that angled into the arroyo. The buried ball of clay was important symbolically as well as practically, for the ball represented a wooden cylinder used in a Zuni game called *Ti'-kwa-we,* or Race of the Kicked Stick. Here two opposing teams, running at full speed, competed over a twenty-five-mile course by kicking ahead of them two small cylinders of wood. The clay ball provided a like cylinder for the water gods, encouraging them to race with the Kicked Stick and to push their waters ahead of them with like force and speed.

By this method of earth banking, the farmer built a network of smaller barriers across his field and created an irrigation system, directing any freshets that came and catching any rain that fell. To encourage the rain, he sought the

Corn-Priest of his clan to prepare a plumed prayer stick and a cane of wild tobacco. The priest knelt in the new field facing the east and implored the god-priests of earth, sky and cavern "not to withhold their mist-laden breaths, but to canopy the earth with cloud banners, and let fly their shafts little and mighty of rain, to send forth the fiery spirits of lightning, lift up the voice of thunder whose echoes shall step from mountain to mountain bidding the *mesas* shake down streamlets." That the streamlets might become torrents and feed the earth seeds, the priest "this day plants, standing in the trail of the waters, the smoke-cane and prayer-plume." Not quite the incantatory repetition of Sahagún, but Cushing's prose shares its Latinate root. Through the overlay of his romance-derived tongue, we glimpse the stone-hard reality of Zuni land.

The Zuni farmer had to allow a full year for the rains to deposit loam. Only then did he plant rows of sagebrush along the western boundaries of his field to catch the fine dust and sand blown during the month of the Crescent of the Greater Sand Storms. Usually there was but a single rainstorm in the spring and its waters, channeled by the embankments, helped redistribute and tamp down the soil until it was at last ready for the ceremonies of planting. In winter the seed had been blessed by the Corn-Matron, who had sprinkled a tray of selected kernels with the black powder of corn-soot (corn smut, we call it) and a mixture of yellow paint, yellow flowers and corn pollen, so that the grains became bright yellow in token of their strength. She then stored them in a pouch made from the whole skin of a fawn.

A cornfield laid out in waffle patterns and protected by corn-stalk hedges with the Zuni pueblo beyond, seen from the southwest. From Cushing's *Zuni Breadstuff,* reprinted 1974

A Zuni cornfield sports a network of traps to scare crows. In the foreground, a man-sized "watcher of corn sprouts" lolls his tongue at a caught crow. Neg. No. 273529, Courtesy Department of Library Services, American Museum of Natural History

On May 1, when the farmer heard the Sun Priest call from the housetops, he sharpened his planting stick of juniper. The base of his stick was forked with a branch stump that he could use as a brace for his foot. He brought with him to the fields a plumed prayer stick and, from a pouch, six different-colored kernels of corn, which a corn-matron sprinkled with water to bless with "rain." When the farmer arrived at the field, he dug four holes equidistant from the center, each of them a cardinal point, and in each planted a corn kernel of the right color—red for the South, white for the East, yellow for the North, blue for the West. He dug two more holes and planted a white kernel for the Sky regions and a black kernel for the world Below. In lines extending from each of the four directions, he planted rows of corn until all the kernels in his pouch were gone. After he returned home, he fasted and prayed for four days.

Now the farmer could plant his corn. Taking with him a lunch of piki bread and a seed bag, the planter dug holes four to seven inches deep, alongside last year's rows. A boy followed him and dropped in each hole twelve to twenty kernels, covering them with sand. Where the broken stalks of last year's rows were thin, the planter reinforced them with twigs of greasewood or sagebrush to catch drifting soil blown by the wind. The country was so dry, Cushing explained, that seeds had to be planted deep to gain protection from the underlying loam.

The planters made a network of crow traps. They erected a number of cedar poles, topped them with prickly leaves and strung between them ropes made from split yucca leaves. On these they hung rags, strips of hide, moss, old bones, anything that would sway in the wind. They also made small hair nooses in which a hair thread was baited at each end with a corn kernel, contrived to choke two crows at once. Boys were set to fashion human scarecrows with faces of painted rawhide, eyes of cornhusk-balls, teeth of cornstalk strips, hair of black horsetails and lolling tongues of red leather. These were the "watchers of corn sprouts," and soon they were joined by crows that had been caught and hung beak downward to warn their fellows off the corn.

When the kernels had sprouted, it was "leaf-lifting" time. The farmers pulled up all but four or five of the best shoots and killed the white grubs near the root. Then it was "hoeing time" or "staving time," when they dug out weeds with their ancient hard wood scythes or with the white man's hoe of hand-wrought iron. The work was hard and when the men went home at sunset, the women had prepared a feast with bowls of smoking stew red with chili and with baskets of waferbreads colored red, yellow, white and blue. After the second or third hoeing of the summer, the men hilled the corn with a pickaxe made from an elk's scapula or a broad stone and left the corn until harvest.

In autumn, they picked all the corn that was still too green to show signs of ripening and carried it to a hill. Here they dug a deep funnel-shaped pit, several feet in diameter at the base, and on the windward side a hole about two feet in diameter opening into the interior of the pit. Through the hole in the top, they threw in dried grass, leaves and wood until the pit was full, then fired the pit so that it would burn all night. When the coals were glowing, they threw in green cornstalks and all the unhusked ears of corn, plugging the top and the draft hole with green stalks, then mounding earth over the whole. After a day and night, the mound and the stalk plug were removed and steam shot hundreds of feet in the air like a geyser. By afternoon the mass had cooled and they could shovel out the corn through the draft hole and carry it into the village, where they husked the golden-brown ears and braided them, to hang from the rafters of their houses.

While the remaining corn in the field ripened, they built little huts in which children and old men would keep watch against coyotes and burros as well as crows. Any burro so foolhardy as to venture into a cornfield after its first beating regretted it. Cushing recalled a burro nicknamed "Short-horn," which had had its ears shaved, its tail and tongue cut short, some teeth pulled and its left eye put out. "The Zunis, and probably most other Indians," Cushing explained, "are touchy on the subject of their breadstuff."

When frost turned stalks to gold and shucks to feathers, the corn was picked and carried into town for husking. Women formed the husking bees, shucking the cobs in great numbers, selecting ears to string on threads of yucca fiber and carrying baskets of cobs to the roofs to dry next to heaps of chili peppers. Dried corn was stored in the corn room or granary of each house, in the center of which four sacred objects were kept: a perfect ear of yellow corn, a bifurcated ear of white corn, a bunch of black corn-soot and an ear blessed by a Seed-Priest in the sacred Salt Lake, "Las Salinas." The salt ear and the soot ear formed the couch or "resting place" of the "Father and Mother of corn-crops," the yellow ear for the Father and the white for the Mother. Each year the Corn-Matron presented the first new corn to its corn parents in a ceremony called "Meeting of the Children," reenacting the return of the lost corn maidens and their welcome by the Seed-Priests of the Zuni of former times. Each year repeated the cycle of the seasons and retold the narratives that gave them voice.

At Acoma Pueblo (visible on the clifftop in the distance), a farmer in 1945 harvests his corn with horse and wagon, but the corn stalks are planted as of old. Photo by Ferenz Fedor, courtesy Museum of New Mexico, Neg. No. 100471

Today Zuni and other Pueblos still farm by a version of the ancient system now called waffle gardens, in which they build mud walls four inches high to enclose small beds four feet wide by twelve to twenty feet long, dividing them into two-by-two-foot squares to hold rainwater and water they pour in by hand. Today as of old they plant in May, "under a waxing moon, so as to grow with the moon," as ethnobotanists reported in 1916, for "under a waning moon the seeds cease growing." And still they harvest in late September and early October, "after the watermelons have been taken." The day the Summer chief takes office, Alfonso Ortiz reports of contemporary Tewa, he begins the cycle anew by "bringing the buds to life," and then it is time for the medicine men to go to the sacred center of the village in dead of night to "reseed mother earth navel," to reach deep into the ground and waken her to life for the new year.

Still today the Zuni keep their varied strains of colored corn pure by cultivating them in fields isolated from each other, and still they keep back some seed year after year, to maintain the bloodlines of their own pueblo. "We could buy other seed, and perhaps better, from white people; or we could get seed from other pueblos; but the old men do not want that," a native of San Ildefonso told the 1916 ethnobotanists. "They want to keep the very corn of the pueblo, because the corn is the same as the people." Still they plant their corn with planting sticks and tend it by singing. "A planting stick is special," George Blue-eyes told some Navaho students in 1979 at Point Community School in

Chinle, Arizona. "You must finish your planting and put it away before the last quarter of the moon. . . . You should sing after planting in the first four holes."

There are more planting songs than cornfields now, for three centuries after Spanish adventurers brought horses and swords to the Southwest, white American engineers brought dams and similar improvements. The Salt River Project effectively ended Pima farming in 1920 when it diverted water from Pima lands. The pattern is familiar. At that date there were still more than forty-eight thousand native American farmers in the United States, half of whom owned the land they farmed. In 1982 the number had dropped to no more than seven thousand. Each year there are fewer to say with George Blue-eyes that "planting the old way is still best."

"Corn planted by tractor may take two weeks to come up," he explained. "Mine might be up in four days. Wind can blow tractor-planted corn right out the ground. Mine is strong. The plow rolls the ground over on itself across the whole field. Too much soil dries out. The stick digs just enough for each seed. The ground underneath stays wet." He might have added that the roar of a modern combine is nothing like a song, so how can it keep the corn from misunderstanding, how can it keep the tassels from trembling, keep the wind from its tender leaves, the woodpecker from its heart, until it is gobbled up by the ravening harvester?

BUFFALO BIRD WOMAN'S GARDEN

"OFTEN IN SUMMER I rise at daybreak and steal out to the cornfields; and as I hoe the corn I sing to it, as we did when I was young." So said Buffalo Bird Woman in *Waheenee: An Indian Girl's Story* (1921), after a long life devoted to the growing of corn. She'd been born in 1839 (just seventeen years before my grandfather Parks Ira Kennedy was born in Benton County, Iowa) to her father, Small Ankle, and her mother, Want-to-be-a-Woman, in one of the villages at Knife River, North Dakota. The year 1839 was one of grief for the Hidatsa, who had lost half their tribe two years earlier in the "smallpox year," so that they numbered but eight hundred. The Knife River region of the upper Missouri River had been the homeland of these Plains Villagers for more than a thousand years, beginning with the Mandan and later the Hidatsa and Arikara (known as the Three Affiliated Tribes), but their Northern Plains ancestors had cut a wide swath from Montana to Wisconsin for more than nine thousand years. Here around A.D. 900, about the time Monte Albán was being abandoned in western Mexico and the great city of Tula was rising in the east, the Plains Villagers of the barbaric north were beginning to plant seeds as well as to hunt wild game.

According to Sioux legend, it was Buffalo Cow Woman who first brought the Sacred Pipe to her people and the four different-colored grains of maize. These sprang from the milk that dropped from her udder when she kicked up her hind legs and departed, "so that maize and the buffalo were given together to be the food of all the red tribes." Cheyenne legend created a pair of culture heroes to express this dual culture. Corn was represented by Sweet Root Standing, whose other names were Rustling Leaf, Rustling Corn Leaf, Sweet Medicine. His twin was Standing on Ground, named also Erect Horns and Straight Horns.

But these legends are relatively recent, for corn rites among the tribes of the upper Missouri were older than buffalo rites. Long before the white man brought horses and guns to the Plains, the red tribes were far more dependent upon corn than buffalo meat. The Hopewells at the confluence of the Ohio, Mississippi and Illinois rivers had already come and gone, leaving behind them a continent-wide network of trade, fine ornaments of copper and mica, engravings of bone, magnificent pottery and the great earth circles of their ceremonial centers and burial mounds, all founded on a knowledge of agriculture. The Moundbuilders had planted corn up and down the Mississippi and Missouri rivers, and the myths of origin of the later Plains people reflected the ancientness of corn.

Here men were hunters and women were gardeners, and the Plains Villagers prospered as traders of both fur and corn, building prototypes of the sod houses of my grandfathers in their circular earth lodges, and farming the rich

In Florida as in Virginia, men used wooden "mattocks" to cultivate the ground while women used a "pecker" to make holes for planting. Jacques Le Moyne's version, as engraved by De Bry, gives the women Botticelli hair and the men Michelangelo muscles. From the Collections of the Library of Congress

river bottoms that snaked through the plains. To the east, the Woodland ancestors of Sioux, Algonquian and Iroquois from Minnesota to Maine were also planting the hardy flint corns as they adapted to snow and ice. To the southeast, corn supported the massive temple mounds of the Mississippians, the ritual fires and Great Sun king-priest of the Natchez. Wherever corn grew, north and south, it was planted with sticks and songs.

One of the best early descriptions of Indian corn planting was provided by Thomas Hariot in 1588, when he reported on the coastal Indians of the new-found land of Virginia, where land was easier to dig than in the frozen plains of the North or the desert of the Southwest. As a wheat grower, he puzzled over the lack of native plows:

> The ground they never fatten with muck, dung, or anything, neither plow or dig it as we in England but only prepare it in a sort as followeth: A few days before they sow or set the men with wooden instruments made almost in the form of mattocks or hoes with long handles, the women with short peckers or parers, because they use them sitting, of a foot long and five inches in breadth, do only break the upper part of the ground, to raise up the weeds grass and old stubs of cornstalks with their roots. The which after a day or two days drying in the sun, being scraped up into many small heaps, to save them the labor of carrying them away, they burn to ashes. . . . Then their setting or sowing is after this manner. First, for their corn, beginning in one corner of the plot with a pecker they make a hole wherein they put out four grains, with care that they touch not one another (about an inch asunder), and cover them with the mould again.

A few years later, Samuel de Champlain, reporting on the Iroquois along the St. Lawrence River, was also struck by the oddity of planting with sticks. "In place of ploughs, they use an instrument of hard wood, shaped like a spade," he wrote:

> Planting three or four kernels in one place they then heap up about it a quantity of earth with shells of the signoc. . . . Then three feet distant they plant as much more, and this in succession. With this corn they put in each hill three or four Brazilian beans which are of different colours. When they grow up they interlace with the corn which reaches to the height of from five to six feet; and they keep the ground very free from weeds.

Another early French traveler, Gabriel Sagard, noted in 1632 how Huron men began to prepare the ground a year ahead by girdling trees in the spring so that they would die over the winter, after which they burned the underbrush to make "oak openings" for planting. The women would then clean the ground and with a stick dig round holes at intervals, in each of which they sowed nine or ten grains of corn that had been soaked in water.

If in the Northeast clearing the fields was man's work, planting them was woman's work, unlike in the Southwest, where farming was largely a matter of hydrodynamics. Across the North, the survival of the tribe through corn was entrusted to the women, and the status of planting was equal to hunting.

"The quantity of corn destroyed, at a moderate computation, must amount to 160,000 bushels, with a vast quantity of vegetables of every kind. . . . I flatter myself that the orders with which I was entrusted are fully executed, as we have not left a single settlement or a field of corn in the country of the Five Nations."
—MAJOR GENERAL JOHN SULLIVAN, *Journals of the Military Expedition of Major General John Sullivan against the Six Nations* (1779, published 1887)

Among the Iroquois of New Hampshire the women of each settlement elected annually one of their number to direct their work. Her name meant literally "Corn Plant, Its Field Female Chief." "In the summer season we planted, tended and harvested our corn, and generally had our children with us," Mary Jemison, who had been taken captive by the Iroquois, later told James Seaver, who recounted it in *A narrative of the life of Mrs. Mary Jemison* (1862). The labor was not severe, she said, and they had "no masters to oversee or drive us, so that we could work as leisurely as we pleased."

The yeoman masters of wheatfields were unaccustomed to this division of labor and misunderstood the methods of planting along with their social meaning. "The women as is the custom with the Indians do all the drudgery," John Bradbury wrote of the Mandan in 1811, noting that the squaws were excellent cultivators despite the fact that they were so destitute of implements that they hoed their corn "with the blade bone of buffalo, ingeniously fixed to a stick for that purpose." Alexander Henry, traveling through the Mandan and Hidatsa settlements in the early nineteenth century, better understood the cultural differences when he wrote in his *Journal,* "Let it not be thought that this work was mere drudgery. Every woman had the company of some of the young girls, and the gardens were close enough together to permit of friendly intercourse. The women usually sang as they worked, and there are preserved great numbers of field-songs which were sung only in the gardens." Besides which, young men were always hanging about on the pretext of protecting the gardens from prowling war parties and that seemed "to detract in no way from the pleasure of the girls."

We know much about these gardens because of an unusual and invaluable collaboration between the Hidatsan Buffalo Bird Woman and the Presbyterian Gilbert L. Wilson, a minister who turned anthropologist after visiting Fort Berthold in 1906, when Buffalo Bird Woman was sixty-seven. Fifty years earlier, Buffalo Bird Woman had watched the construction of that fort next to her cornfields in a bend of the upper Missouri called Like-a-Fishhook. Here the small band of survivors of both smallpox and roving Dakota warriors had migrated, only to be rounded up in 1874 into the Fort Berthold Indian Reservation, where they were compelled to abandon the communal lands they held sacred in deference to the white man's consecration of private property. This was part of the United States government's reformist policy instituted by the Dawes Act of 1887, and formulated by the founder of the Carlisle School for Indians as: "Kill the Indian and save the man."

Buffalo Bird Woman refused to be saved. While her brother Wolf Chief learned English and became a storekeeper, Buffalo Bird Woman kept the

"The Mandans and Manitaries cultivate very fine maize without ever manuring the ground, but their fields are on the low banks of the river . . . where the soil is particularly fruitful. . . . They have extremely fine maize of different species."

—ALEXANDER
PHILIP
MAXIMILIAN,
Maximilian's Travels
(1833)

Buffalo Bird Woman appears on the right, with her son, Edward Goodbird, in the middle and her husband, Son of a Star, on the left. Photo by Gilbert Wilson in 1906. Minnesota Historical Society

Indian tongue and the gardening traditions practiced for centuries by Hidatsa women. Fortunately, Gilbert Wilson was as unusual in his way as Buffalo Bird Woman was in hers, for he wanted to learn about Indian life from the Indian point of view. Over the course of a decade and more, with Buffalo Bird Woman's son Edward Goodbird acting as interpreter, Wilson received from her a comprehensively detailed account of her life at Knife River Village beginning in 1846, about the time my grandfather's father, Roswell Kennedy, was applying the plow to the prairie sod of Indian territories south of the Dakotas and west of the Missouri. In 1917 Wilson published as a doctoral thesis the work now titled *Buffalo Bird Woman's Garden*—one of the most complete and articulate records we have of native agricultural life in the mid-1800s.

Buffalo Bird Woman, whose grandmother was Soft-white Corn, explained to Wilson how they had always scorned the hard and dry prairies for the bottomlands, which they worked with hoes instead of plows. "I think our old way of raising corn is better than the new way taught us by white men," she said. Just the year before, her corn had taken first prize at an agricultural county fair: "I raised it on new ground; the ground had been plowed, but aside from that, I cultivated the corn exactly as in old times, with a hoe."

Because her memory was sharp, Buffalo Bird Woman's descriptions of the old times were detailed and precise. When her grandmothers cleared a new

The seasonal round expressed in the corn language of the Senecans:

onä′o‘	corn
waeeyŭnt′to’	she plants
ohwĕⁿo‘dadyiĕ’	it is just forming sprouts
oga″hwäodaⁿ	it has sprouted
otgaä′häät	the blade begins to appear
otga′äähät	the blade has appeared
deyuähä′o	the blade is already out
ogwäⁿ′dääodyiĕ’	the stalk begins to appear
ogwäⁿ′dää′e’	the stalk is fully out
oge″odadyie’	it is beginning to silk
owäⁿ′dǎ’	the ears are out
o’geot	it has silked out
ogwäⁿdŭ′äe’, ogwäⁿ′däⁿe‘	the tassles are fully out
ono″gwaat	it is in the milk
dĕju′göⁿsäät	it is no longer in the milk
oweäⁿdäädyĕ’, owĕⁿdādyĕ’	the ears are beginning to set
onĕ′oda′dyiĕ’	the kernels are setting on the cob
hadi′nonyoⁿcos	they are husking
yestä′änyoⁿnyano’	she is braiding
dŭstaⁿ′shoni	it is braided
gasdäⁿt′shudoho’	it is hung over a pole
ganoⁿ′gadi‘	it is strung along a pole

field, they dug corn hills with digging sticks of ash and worked between the hills with hoes of bone lashed to wooden handles. At the time of Buffalo Bird Woman's girlhood, they worked with iron axes and hoes they bought from traders, although they still used rakes which they had fashioned from bent wood or from the antlers of black-tailed deer. Once the fields were cleared, they burned the dry grass, felled willows and brush and distributed the ash. This was the old method of slash-and-burn.

The old women knew it was time to begin planting when they saw wild gooseberry bushes leafing in the woods in May. First they loosened the soil of the old hills with their hoes, then planted six to eight seeds half an inch deep in a circular pattern in each hill. They planted the hills in rows about four feet apart and the same distance between hills, for if the hills were so close together that the growing plants touched, "we called them 'smell-each-other,'" Buffalo Bird Woman explained, "and we knew that the ears they bore would not be plump nor large." In her family's garden, they planted nine rows of twelve-hilled corn, alternating with beans and surrounded by squash, in an area 180 yards long and 90 yards wide. This garden was in the keeping of her grandmothers Turtle and Otter, and of their daughters Corn Sucker, Red Blossom, Strikes-Many-Women and Want-to-be-a-Woman, all of whom were wives of Small Ankle.

When the corn was about three inches high, they called it "Young-bird's feather-tail corn," from the shape of its sprouting leaves. They weeded the hills carefully with their hands and hoed between the hills. When the plants began to blossom, they hilled the earth further by building it up around the roots to protect them. They made stick-figure scarecrows, covered in buffalo robes to ward off crows, and built a watching platform to protect the young corn from thieves and to encourage it with "watch-garden songs." "We cared for our corn in those days as we would care for a child," Buffalo Bird Woman said, "for we Indian people loved our gardens, just as a mother loves her children; and we thought that our growing corn liked to hear us sing, just as children like to hear their mother sing to them."

Most of the watch-garden songs were love songs, she said, sung by the young girls who sat on the platforms, doing their "needlework" with porcupine quills and teasing passing boys with "love-boy" songs. When the girls were young, around twelve, they sang,

> *You bad boys, you are all alike!*
> *Your bow is like a bent basket hoop;*
> *You poor boys, you have to run on the prairie barefoot;*
> *Your arrows are fit for nothing but to shoot up into the sky!*

Older girls elaborated on the theme for older boys:

> *You young man of the Dog society, you said to me,*
> *"When I go to the east on a war party, you will hear*
>   *news of me how brave I am!"*
> *I have heard news of you;*
> *When the fight was on, you ran and hid!*
> *And you think you are a brave young man!*
> *Behold you have joined the Dog society;*
> *Therefore, I call you just plain dog!*

Before their numbers were reduced, even settled agricultural tribes like the Hidatsa were accustomed to leave their villages and crops after the first hoeings for the summer buffalo hunt. According to George Will, in *Corn Among the Indians of the Upper Missouri* (1917), the hunt lasted through July for the Mandan, Hidatsa and Arikara, who returned to their villages in time for the green-corn harvest of August. This was a season of great feasting, said Buffalo Bird Woman. If they were working in the fields they ate whenever they got hungry, in little booths made of willow. They broiled buffalo meat on the coals and boiled green corn just shelled from the cob with freshly shelled beans and ate them with spoons made from the stems of squash leaves.

"The first corn was ready to be eaten green early in the harvest moon, when the blossoms of the prairie golden rod are all in full, bright yellow; or about the end of the first week in August," she said. Some of the corn they roasted and some they boiled in a pot "as white people do." She could tell by looking at the corn on the stalk if it was ripe enough to boil. "The blossoms on the top of the stalk were turned brown, the silk on the end of the ear was dry, and the husks on the ear were of a dark green color." Today, she lamented, young folks are not very good gardeners, for they have to open up the ears to look at the kernels. "In old times, when we went out to gather green ears, we did not have to open their faces to see if the grain was ripe enough to be plucked!"

Green-corn season lasted about ten days, she said, and every family dried and stored green corn for winter. When the later corn was fully ripened in September, the men went hunting to get buffalo meat for the husking feasts of the ripe-corn harvest. Now the whole community went to work, each family plucking the ears and piling them in a huge heap in the middle of their field. Next day the field's owner would notify the crier of some society like the Dog or Fox, and he would cry from the roof of the lodge, "All you of the Fox society come hither; they want you to husk." "As we approached the fields we began to sing, that the girls might hear us," Water Chief said, joining in Buffalo Bird Woman's recollections. "We knew that our sweethearts would take notice of our singing." The youths and their sweethearts would be dressed to the nines for the husking feast of a side of fresh buffalo roasted over the fire next to a steaming pot of meat and corn. The pile of unshucked ears might be four feet high and twenty feet long, but the young men worked fast, particularly near the pile of a pretty girl.

As they husked the smaller ears, they saved the best big ones to braid into strings of fifty to sixty ears. "A braid," Edward Goodbird remembered, "was long enough to reach from the thigh around under the foot and up again to the other side of the thigh." A husker tested a new braid with his foot by holding the ends in his hand. "Unless this was done a weak place in the string might escape notice and the braid break, and all the others would then laugh." There was a lot of laughter at husking season because it was courting season. "Boys and young men went to the husking bees because of the fun to be had; they wanted to see the girls!"

Ponies carried the braided corn and bags of ears to the family's drying stage, where the braided strings were hung "like dead snakes" over the railings. Loose corn was spread evenly over the stage, and after eleven days it was usually dry enough for threshing. Next spring, they selected their seed corn from the best ears of the many braided strings. Usually they took out five strings of soft white, half a string of soft yellow and ten ears of the type of sweet corn they called *ma'idadicake,* or "gummy" corn, for making cornballs. These were favorites among the Mandan and Hidatsa, George Will said, "made of pounded sugar corn mixed with grease" and tasting something like peanut butter.

Buffalo Bird Woman shelled her seed corn carefully, she said, using only the kernels in the center of the cob. Since the seed corn remained fertile for at least two years, in seasons when the crop was good her family would put up seed to last a second year and built up a good business of selling seed corn to less provident families. "The price was one tanned buffalo skin for one string of braided seed corn." When Buffalo Bird Woman recalled these events for Wilson in 1912, she still sold seed corn and seed beans, but the units of exchange had altered. "A handful of beans, enough for one planting, I sell for one calico," she said, "enough calico, that is, to make an Indian woman a dress, or about ten yards."

For threshing, they built an enclosed booth of buffalo-cow hides under the drying stage, removing some planks so that they could push the corn through. Three women, each with a flail of ash or cottonwood, would sit facing the pile of corn now in the booth and beat the dry cobs so that the kernels flew into the air but were caught by the walls of the booth. They spread a tent cover just outside the booth for the threshed cobs, and at the end of the day they shelled off every clinging grain, "for we wasted nothing." At sunset when the air was still, they carried the cobs outside the village, piled them up and fired them. Buffalo Bird Woman's job was to stay and watch the fire to prevent mischievous boys

"In my tribe in old times, some men helped their wives in their gardens. Others did not. Those who did not help their wives talked against those who did, saying, 'That man's wife makes him her servant!'

"And the others retorted, 'Look, that man puts all the hard work on his wife!'

"Men were not all alike; some did not like to work in the garden at all, and cared for nothing but to go around visiting or to be off on a hunt.

"My father, Small Ankle, liked to garden and often helped his wives. . . . My father said that that man lived best and had plenty to eat who helped his wife. One who did not help his wife was likely to have scanty stores of food."

—Wolf Chief, as told in 1910 to Gilbert L. Wilson, *Buffalo Bird Woman's Garden* (1917)

from playing mudballs. This was a game in which boys put a ball of wet mud on a green stick and stuck it into the fire so that some burning coals would stick to the mud. Then they would snap the fireballs through the air at each other, like snowballs. Next day Buffalo Bird Woman would squeeze pieces of the crusted corncob ash into little balls and carry them home, wrap them in dried buffalo-heart skins and store them in baskets to use as seasoning during the winter.

Each morning, they would winnow the grain from the previous day's threshing by holding their baskets aloft and pouring the grain into the wind. Then they would store the grain temporarily in round containers called "bull boats" within the lodge. In a good year the corn would last until the next harvest, and in a bad year there might be some left from the previous harvest. In any case "there were elk and buffalo and antelope to be had for the hunting."

Digging the cache pits for winter storage was also woman's work. The pit was shaped like a jug, wide at the bottom and narrow at the mouth. Big ones were deep enough to require a ladder. Once dug, the pit was lined with a bluish water grass that would not mold. Then they laid a floor of dry willow sticks on the bottom of the pit and covered it with a thickness of dried grass and then a circular skin. They next secured a thick layer of grass on the inside of the walls with willow sticks, against which they laid rows of braided ears with the tips of the ears pointed inward, the husk ends outward to protect against moisture.

When the braids were four rows deep, they pushed in enough shelled corn to make an even layer and then coiled a string of dried squash in the middle of it, covering it with corn to protect the squash from moisture. They continued layering until the pit was filled, storing at least four strings of squash and thirty or more strings of braided corn. On top they fitted a thick buffalo hide which had been cut to measure, and covered it with grass and puncheons split from small logs and laid flat side down in a trench so that they wouldn't roll. Over the puncheons, more grass, another hide and then earth topped with ashes and refuse to conceal it "from any enemy that might come prowling around." Small cache pits they dug just outside the lodge for family storage, one pit for yellow corn, another for white, another for dried squash and other vegetables, a fourth inside for dried berries and valuables.

Drawing by Edward Goodbird, in Buffalo Bird Woman's Garden *(1917)*

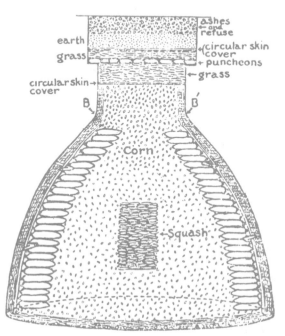

The first two crops planted on new ground were the best, Buffalo Bird Woman said, and when the yields grew less, they let the land lie fallow for two years, planting their crops elsewhere, since they never used all their fields at one time. Once the corn was husked, they let their horses eat the standing fodder and any husks left in the field, so that few vestiges of corn lingered until spring. They removed any dung the horses had dropped, for they found it attracted worms, insects and foreign weeds like sharp thistle and mustard with black seeds. "Now that the white men have come and put manure on their fields," she remarked, "these strange weeds have become common."

When Wilson asked if young men ever worked the fields, Buffalo Bird Woman laughed long and hard. "Certainly not! The young men should be off hunting, or on a war party," she said. "Also they spent a great deal of time dressing up to be seen of the village maidens." Buffalo Bird Woman had a strong sense of what was fitting and not fitting. It was not fitting for young men to help women in the fields. It was fitting only for old men, too old to hunt or go to war. While it was fitting for her son to learn new ways in the white man's school and read books and raise cattle, it was not fitting for her. "But for me, I cannot forget our old ways," she told Wilson.

Today you can visit the Knife River Indian Villages, north of Bismarck, North Dakota, as part of the National Park System, and catch glimpses of the old ways. You can examine reconstructed earth lodges and watch a historical pageant showing the arrival of Lewis and Clark in 1804, where they picked out Toussaint Charbonneau and the maiden Sacagawea to guide them farther west. You can learn from Gerard Baker, a National Park Ranger who is a Hidatsa, that the corn he plants in his garden grows from seeds planted for three hundred years and that every spring he asks a tribeswoman to come over from Fort Berthold to bless them. "That's the way it was done in the old culture," Baker says.

Today, there are white men too who are working to save some part of the old culture of the Mandan and Hidatsa. "One of the rarest of all corn," says seed-saver Charles Hanson, "is the 'Mandan little corn,' or midget blue flour corn." This corn, which produced the earliest roasting ears in green-corn season, had disappeared forever, he was told by an aged Indian near Fort Yates. Hanson persevered in his search and a decade later heard through an Indian trader of an old Oglala woman who had kept the gardening traditions of her mother. Hanson found the garden, the woman and one perfect little ear of midget blue. "It was enough," he reported in the *Seed Savers Exchange*; "today there is plenty of seed to keep the variety for posterity."

When the seed-saver Gary Nabhan visited the tribal headquarters of the

Mandan at New Town a few years ago and looked at the waters behind Garrison Dam which had swallowed the old tribal fields, he thought of John Bradbury's wonderment in 1811 at the elegance of Mandan cornfields: "I have not seen, even in the United States, any crop of Indian corn in finer order or better managed than the corn about the three villages." The old ways were extinct, Nabhan feared, until he saw a group of children eating cornballs and tracked them to a contemporary Mandan. "For cornballs, you need soft corn, flour corn, to make it," Vera Bracklin explained. "Mine is corn from my mother-in-law. I can't raise it here, but there's a white lady who gives me room to plant out at her place." Buffalo Bird Woman had said that in her time cornballs were a good present for a woman to give to her daughter to take to her husband and his parents and sisters. Today Vera Bracklin sells cornballs door-to-door to help support her family.

The old ways take new turns as the seasons come and go, but some ways do not come round again. "No one cares for our corn songs now," Buffalo Bird Woman lamented seventy years ago.

Sometimes at evening I sit, looking out on the big Missouri. The sun sets, and dusk steals over the water. In the shadow I seem again to see our Indian village, with smoke curling upward from the earth lodges; and in the river's roar I hear the yells of the warriors, the laughter of little children as of old. It is but an old woman's dream. Again I see but shadows and hear only the roar of the river; and tears come into my eyes. Our Indian life, I know, is gone forever.

The Language of the Machine

The Ocean-Desert of Nebraska

THE SONGS of the Oglala are gone forever from Beaver Valley near St. Edward in Boone County, Nebraska, on the eastern edge of the Sandhills, but the beavers are not. Even in the fall when the pin oaks are turning red, the milo a deep rust and the cornfields ash-blond, the hills roll like green waves above water hidden underground. Paradoxically, the prairies of Nebraska in what was once called the Great American Desert are honeycombed with water in the Oglala aquifer. This geologic core at the center of America was the alluvial tip of the Ice Age glacier that covered the top half of the globe and, when it retreated, left Nebraska with the largest aquifer in the United States—good for beavers, good for corn.

"Nebraska has no sublime scenery. No lake, cliff, cave or cataract," James Butler wrote in 1873 when he addressed his pamphlet *Nebraska: Its Characteristics and Prospects* to "progressive, *calculating* men, who love nothing so well as the logic of facts and figures, to prove how they can better their condition." As if in compensation, unsublime Nebraska named its towns Harvard, Oxford, Hebron, Cairo and Peru, in defiance of a landscape that suggested none of these places. The rolling waves of corn that replaced a sea of grass in the prairie center of America have a prospect and a rhythm of their own that defy more civilized sites, as they defy the logic of facts and figures.

Prairie rhythms are better caught by America's poets than by its men of calculation:

> *America was beavers,*
> *Buffalo in seas,*

Cornsilk and the johnnycake,
Song of scythes and bees.

So sang Robert Tristram Coffin in *Primer for America* (1943), evoking a pastoral like Buffalo Bird Woman's. Their corn songs, as it happened, bracketed the decade that wiped out the prairie center and turned green pastures and beaver valleys into the valley of the shadow of death. "As far as the eye could see in the Dust Bowl there was not a tree, or a blade of grass, or a field; not a flower or a stalk of corn, or a dog or a cow, or human being—nothing at all but gray raw earth and a few far houses and barns, sticking up like white cattle skeletons on the desert," Ernie Pyle wrote in *Home Country* (1947). "It was death, if ever I have seen death."

The rhythm of the prairies alternated between fecundity and drought, between heat blasts and blizzards, unmitigated by the communality of the Indians to whom this harsh and unforgiving land was native. The white man was there to go it alone, to challenge the wild and, if need be, the God who created it:

You see them heat waves out there on the prairie? Them's the fires of hell, licking round your feet, burning your feet, burning your faces red as raw meat, drying up your crops, drawing the water out of your wells! You see them thunderheads, shining like mansions in the sky but sprouting fire and shaking the ground under your feet? God is mad, mad as hell!

Nebraska might lack sublime scenery, but ordinary farmers like this one quoted in the WPA volume on *Nebraska* in 1939 acknowledged the sublimity of its weathers, which tested faith and toughened hands and hearts.

In 1873 James Butler, as a man of his time, saw only heroic challenge in the white man's duty to transform the "vast ocean-desert" of Nebraska, as he called it, into "Nebraska the garden of the west." While the romantic ethos remains a century later, the economics have changed, and I had to work hard to locate a native Nebraskan who still wants to prove himself by running a family farm. Although every one of my Corn Belt cousins had been born on a farm, not a one of them was now a farmer. "I went to college so I wouldn't *have* to marry a farmer," laughs my cousin Eleanor, whose farm-born husband moved quickly into the petroleum business. Today the path to betterment leads straight off the farm because Nebraska's garden of the west, like other Corn Belt gardens, suffers from an excess of success.

The economic paradox was made clear when Duane Choat, one of St.

Edward's progressive native-son farmers, pointed to his corncribs stuffed to the bursting point with bumper harvests, three years in a row. Duane is both a progressive and a *calculating* man, and when he threw his arms wide as if to embrace with pride the three circular cribs by his barn, he groaned aloud. The logic of facts and figures, once the pride of the Nebraska farmer, is now his despair, for his farm like many another suffers from a mortal disease called "chronic excess capacity."

Duane is a heavyset farmer with a slow, deadpan sense of humor that takes some getting used to. He married his high-school sweetheart, Addis, a girl born on the far side of town on the farm of her parents and grandparents. It's not far to the far side of town, which consists of a barbershop, a beauty parlor, a gas station, a bank, a lumber company, a church, the City Cafe, the Bottle Barn Liquor Store and the St. Edward Beavers—a club. Addis worked in a factory for three years inspecting hypodermic needles, but now she raises Alida, Ann Marie and Wayne. Wayne at fifteen was raising prize-winning sheep and helping his father run the farm. When I spoke to Duane three years later, Wayne had won a regents' scholarship to the University of Nebraska to study agricultural engineering. "He's interested in farming," Duane said, "but he's going to want an income. He doesn't particularly want to work for nothing."

The rhythm of the prairies in 1987 created by the logic of physics and math.
Courtesy Pioneer Hi-Bred International, Inc.

Duane and his three brothers farm the land they inherited from their father. "Where did your father come from?" I ask, looking for origins. "Four miles up the valley," Duane replies. Duane's portion was 560 acres, but to make an income he's pooled his land with his brothers' to total 3,000 acres. Duane's facts and figures are a litany of costs. First the pests—corn borers, red spiders, corn-root worm—"gotta be watchin' all the time." Then the chemicals. Duane and his brothers have their pesticide license but they must still hire a specialist to find out what chemicals they need before they hire an "aerial applicator" (an airplane, to the rest of us) to spray them at a cost of $8 to $10 an acre. Then the machines. The many machines they need to fertilize, cultivate and harvest their crop are costly to buy and costly to run. Duane's combine got a little rattle under the seat and repairs came to $1,369.12. He shows me the bill with the same mixture of pride and despair evoked by his bursting corncribs.

A mechanical leviathan swallowing up the rows, six at a clip. Photo by David Plowden

His two combines condense a century of machine lingo and costs in their multiple parts: low-profile dividers, sloped gather sheets, individually mounted row units with long gathering chain lugs, center-pivot system, transversely mounted threshing cylinder, two-stage high-velocity air cleaner, cage sweep, distribution augers, trailing auger, center-mounted bin, swivel unloading tube. With the lingo I'm all at sea, but the sight and sound of one of these leviathans swallowing the rows in its steel-tusked jaws is, if not sublime, as awesome as an imagined herd of mastodon thundering across the glaciers.

Duane's planting year moves not to the songs of Buffalo Bird Woman but to the rhythmic clank of his machines. In November and December he fertilizes the ground with a "chisel" applicator mounted on a tractor, which injects anhydrous ammonia directly into the earth, three to fifteen rows at a time. He uses about two hundred pounds of fertilizer per acre, which costs 10 to 16 cents a pound and occupies five days on the tractor. He can't fertilize until the ground temperature has dropped to 50 degrees or below to prevent leaching into the groundwater. But he must fertilize in the fall rather than the spring because he doesn't want the tractor to compact the earth at planting time. After the "chisel" has done its work, a "shredder" pulled by a tractor on a PTO (power take-off) drive cuts up the stalks left from the harvest. Then a disk attachment mixes the top layer of soil with the stalk residue to make a compost four to five inches deep. "What we're doing," Duane explains, "is we're taking care of a big garden—if you helped your mother in her garden you'll know how to take care of your cornfield."

Last year he took care of 343 acres of cornfield because, under the government farm program, he's taken out between 10 and 50 percent of his crop base for the last ten years. He begins planting about April 25, but first he must turn the ground and weeds over with the disk, which is a cultivator "with a lot of little shovels on it." His planter does four rows at a time, but some planters can do up to sixteen. By a series of sprockets, the drive wheel controls how fast the seed is released into the furrows dug by the planter as it goes. A packer wheel then packs the ground over the seed. Duane's fields should be done within a week and a half, because if the corn isn't in by May 5, he says, "research shows that we'll lose a bushel of corn per day per acre."

At planting time he does his first spraying of herbicides and insecticides. The herbicide, a lasso-attrex mix, is sprayed by pressure through a nozzle attached to "saddle tanks" on the tractor. The insecticide—he uses about seven pounds per acre of a brand named Counter—goes on dry. He's used to defending his chemicals: "We're as much concerned as anyone about the chemicals, but if you have head lice you can get a chemical at the drugstore that is worse than anything we put on our fields." Like many traditional farmers, he believes that the big chemical companies are working with the EPA for the general good. "They're going hand in hand in the same direction," he says, and meantime he needs insecticides to fight the first broods of the root worm and root borer that attack root and stalk in May and June, then the second broods in August when the root worm becomes a beetle and when the borer attacks the shank of the ear so that it falls plumb off.

During the growing season, he continues to use the cultivator to throw dirt up around the plants for root protection and to make furrows for irrigation. He

"December 25: A lovely Christmas morning. The ground is frozen hard, yet the sun is shining nice. They are picking corn. I will read awhile in the Testament."
—From the diary of H. Greene, an Ohio farm woman in 1887 (a manuscript in the Schlesinger Library, Cambridge, Mass.)

irrigates by a gravity system that pumps well water at a cost of $10 to $30 an acre, depending on the rain. There's no outlay on scarecrow devices because there aren't enough birds to attack the corn, but there are plenty of other "maize thieves" to deplete the crop. The woods shelter marauding raccoons, squirrels and deer, and beavers will come up from the river to make a haul.

Duane's combines begin harvesting around September 20 when corn moisture is down to 25 percent. His pair of combines are twenty years old now, but a new one would cost him $100,000 or $150,000 for such added refinements as air-conditioned cabins and monitors that run like digital computers. When his corn yield is 150 to 200 bushels per acre, he can harvest ten thousand bushels, or about fifty acres a day. The combines run on diesel fuel, consuming four to eight gallons an hour, with fuel costs running from 60 cents to more than $1 a gallon. Annual upkeep on all his major machines—four tractors, two or three pickup trucks, two combines—begins at $4,000. "If an engine goes, just double the costs." Finally, Bahlen wagons and trucks haul the harvested corn into his series of grain bins, small ones that hold three thousand bushels and big ones that hold ten thousand.

The logic of facts and figures demonstrates that for today's farmer storing corn is as crucial as growing it. "High yield for high net income is the aim," Duane says. "Highest yield with the driest moisture gives you the net." Fourteen percent moisture is the key for his large reserves of corn, some three or

Ever-rising yield demands high-rise grain elevators that dwarf the small prairie towns that build them. Photo by Gregory Thorp

four years old, which he keeps in the glass-lined steel storage tanks that make his old-fashioned corncribs look like farm toys. These tanks are automatically monitored to balance temperature with humidity and keep it constant. New corn goes into a holding tank for wet grain, which starts at 35 percent moisture and is then air-dried for permanent storage. "Keep it dry to keep the insects out and clean to keep the rodents out and I suppose it would keep for ten years," Duane says, "but I wouldn't want to try it."

In his basement office a big computer taps into the head office of Pioneer Hi-Bred International in Ioway. In the Corn Belt, Iowa is always "Ioway." Pioneer devised its computer program principally for big-scale land managers who run farms for absentee landlords, Duane explains, but since he sows annually about nineteen kinds of seed corn, so that some

October 13: *corn. Dig well.*
October 29: *I finished corn crib, sawed wood, moved pigs . . .*
November 2: *Went farm husked 17 shocks.*
November 3: *Clark—Adams and I husked corn. prayer meet.*
November 5: *cold. fix corn crib.*
[Undated]: *. . . Downs [?] and I husked 42 bu. corn.*
November 10: *144 bu. upstairs 172 bu. downstairs.*
November 18: *Warm and pleasant . . . wheat not thrashed corn not husked.*

> —From Joseph Burt-Holt's diary when he homesteaded in Minnesota in 1859, part of the Holt-Messner Family file at Schlesinger Library, Cambridge, Mass.

will mature early and some late, he needs a computer to feed him the latest data on high-yield improvements. At the same time, organizations like the National Corn Growers Association encourage him with annual prizes to increase his yield to the maximum. He won a first in 1984 for 204 bushels, his best yield ever. Yield, however, is both the chief symptom and cause of the disease called "chronic excess capacity."

Duane is caught between loyalty to the old ethos and the illogic of the new economics. Chronic excess had not been envisaged by Nebraska's progressive pioneers, who were once so desperate simply to survive that they burned their crops for fuel. Today the government doesn't burn crops to support prices, but it puts them away. Duane supports the "grain reserve program," in which the government pays the farmer to store grain in his own bins. "The government does a good job," he says. "Our margins are so close we need a system to help control production, we need an organization and somebody has to pay the bill. We *all* do. We have tough times, but we're doing okay." Okay, however, may not be enough to keep farmers down on the farm. "As long as we have the land, we'll be growing corn—there's not much else you can do with it," he laughs, "but we'll have to grow our farmers too. With the present economy the young ones are in trouble and if they can't survive by farming, they'll leave for other jobs."

Around midnight, late for a farmer, he plays me a tape made by the American Agricultural Movement, a grass-roots group that protests corporate

Winslow Homer's The Last Days of the Corn Harvest *in* Harper's Weekly, *December 6, 1873. Courtesy New-York Historical Society, New York City, Neg. No. 64980*

takeover of the land. Only now, and in whispers, does Duane confess to frustration at what motivated the farmers to organize in the first place—the desire to control their own destinies. "We don't join the co-ops because we want to know who's controlling the co-ops," he says quietly. "We want to be self-sufficient, but you have to get cash to pay taxes on the land. You can't just go your own way and grow your own food the way we used to."

Those who took destiny in their hands in the 1930s and, like my cousins, left farming for good are not nostalgic about it. Having experienced the hardships, they welcomed the change. Change is still the Corn Belt credo as long as it can be made to suggest progress, prosperity and betterment in the spirit of the Optimist's Creed that hangs on Cousin Eleanor's bathroom wall: "Promise Yourself to talk health, happiness and prosperity to every person you meet, to wear a cheerful countenance at all times and give every living creature you meet a smile."

Even without her family farm in Hildreth, Cousin Eleanor kept the faith of our grandfathers and grandmothers who had invaded the land in order to change it. When Eleanor recalled her own farm days, she remembered the climactic moments of change, like the arrival of electricity in 1934. For the first time they had a water pump and a semi-attached bathroom and running water in the kitchen. "People who talk about the good old times forget the mother who died in childbirth, the babies' graves in the cemeteries, the stress and loneliness of pioneer women who came out here and might not see their husbands for weeks at a time," Eleanor says. She remembers how long it took the pair of horses to grind corn into meal between turning stones, how many jars of corn exploded because they hadn't been processed right, how often the cream wouldn't churn to butter because it wasn't cold enough. Today, her mother's butter churn sits in a corner of the parlor in Eleanor's neatly decorated house in Lincoln, holding a large-leafed houseplant.

But a new generation is questioning whether change is always an unmitigated good. Milan Moore, a young man in his thirties, uses words like "generational change" to describe what is happening today to Nebraska farmers. Milan, with his wide-set eyes and sincere voice, sitting with his quiet wife in their bungalow in Sutherland, himself exemplifies generational change. He is a farmer who doesn't farm at all. After graduating in animal science at the University in Lincoln, he found the demand by farmers for information on new chemicals and new equipment so large that he started his own consulting business.

"People come to me for information on federal regulations, environmental protection, chemical use, that sort of thing," Milan explains. "There are so many regulations—you have to be trained and licensed as a private applicator or commercial applicator to put any chemicals on the land. We have a chemical so toxic that one-sixteenth of an ounce per acre is enough to kill the weeds of the entire area, so it's critical that you make no mistake in how many drops to put in the tank. Farmers don't want to become chemical experts or business managers or lots of things they're required to be today. So they hire an animal nutritionist to formulate rations, a crop consultant to determine fertilizers, a chemical consultant to determine crop rotations, because all they want to do is *till the soil.*"

The big change for Nebraska farmers, he believes, began in the 1950s with the introduction of "center pivot irrigation." For the first time the aquifer of the Sandhills could be efficiently tapped by a sprinkling system in which a long mobile pipe, attached to a well in the center of a field, rotates in a half-mile-wide circle uphill and down to cover about 130 acres. With pivot irrigation, farmers could cultivate not only more land, but rolling land. Between the mid-1970s and 1980s, Nebraska increased its cropland by one million acres, to make corn by far the country's largest irrigated crop. Corn, Milan says, now occupies 19 percent of all irrigated acres in contrast to 4 percent for wheat and 4 percent for soy.

Banks were eager to lend money for land expansion, Milan continues, and the younger generation of farmers, unlike their fathers, were eager to borrow. Rapid expansion drove up land prices crazily in the 1970s and when they dropped by half in the 1980s, borrowers were in trouble. Farmers who'd borrowed heavily to buy more land had to buy more machines to produce more and more crops to pay the interest, but the more they produced the more they lowered their crop prices. Milan believes a major change must come in the economics of agriculture if any small farmers are to survive. It's the Catch-22 of the farmer, who needs, as Joel Garreau has said in his *Nine Nations of America*, "more and more land to pay for his machinery, the machinery that keeps on coming bigger and more expensive in order that one man can till more and more land." This is the rhythm Carl Sandburg caught in 1928 in "Good Morning, America":

"There are three great crops raised in Nebraska. One is a crop of corn, one a crop of freight rates, and one a crop of interest. One is produced by farmers who by sweat and toil farm the land. The other two are produced by men who sit in their offices and behind their bank counters and farm the farmers. The corn is less than half a crop. The freight rates will produce a full average crop. The interest crop, however, is the one that fully illustrates the boundless resources and prosperity of Nebraska."
—Farmer's Alliance Bulletin (ca. 1890)

We raise more corn
to feed more hogs
to buy more land
to raise more corn
to feed more hogs
to . . .

I heard an acceptance of change, without either the white man's optimism or his despair, in the talk of Frank and Alice Saunsoci on the Omaha reservation at Macy, up by Sioux City northeast of Beaver Valley. The Omaha were an agricultural tribe led by Big Elk and Iron Eye when they lost their territory in the Kansas-Nebraska Act of 1854 and the Homestead Act of 1862. "We used to have our big harvest festival, our big powwow, after the Buffalo Hunt," Alice says, "a hundred years ago." They still have an annual powwow in August, but there are only two hundred or so Omaha on the reservation to do the Gourd Dance and the War Dance and to protest the squatters' takeover of Black Bird Bend—a hundred years ago.

While looking for Alice, I decided to grab a bite in the corner grocery opposite the Umo Ha Ta'Paska school. I ate a pizza burrito heated in the Deli-Express microwave and then found Alice's husband, Frank, in front of the pinball machines watching some friends play Pac-Man. Frank took me to their house to see their cornfield because the Saunsocis are known hereabouts for their smoked corn, which they make every year from the deep-purple corn in their back lot. "It has no name," says Frank. "It's just the corn we grow, from seed handed down for generations." Some corn they parch, some they cook for hominy and some they smoke, to use in soups and stews for special occasions like the powwow. They build a fire with red elm—slow-burning, good for flavor—place a window screen over the coals and spread a couple of gallons of boiled sweet-corn kernels over the screen to smoke. Their four sons would like to keep on with the cornfields, Frank says, "to see every year the corn grow," and perhaps to acknowledge like their ancestors the blessings of Mother Corn. But change is afoot. Last year a French boy spent a month with them on a student exchange program, after their eldest son had spent a scholarship month in Nice. "Already the old things are being lost," the Pawnees' Eagle Chief had said almost a century ago. "It is well that they should be put down, so that our children, when they are like white people, can know what was their fathers' ways."

"Improvement in Illinois was snail-paced at first, for it came in on ox-teams," James Butler wrote in detailing Nebraska's "Progress of Settlement" after the Homestead Act. "It entered Iowa on steamboats, and was therefore long confined to the banks of navigable rivers. Its advent into Nebraska was on locomotives, which, plying on iron rivers, that render all prairies navigable, leave no corner of them untouched." Butler was so inspired by the muse of Nebraskan transport that only epic simile could express, in whatever fractured rhythms, his song of progressive betterment. "As much as steam is swifter than a steer, as much as railroads are more pervasive than rivers, so much more rapid and ubiquitous development than that of Illinois and Iowa may we expect to behold in Nebraska."

THE MARCH OF MECHANICAL TIME

IN THE 1990s we can behold in Nebraska the results of the radical changes that took place more than a century ago in the 1860s, at the close of the Civil War. The decade in which Nebraska achieved statehood was the decade that inaugurated America's industrial revolution of the farm. The prairies incited revolution because radical change was required to turn sod into corn, but revolution cut far deeper than sod in furrowing men's minds. Planting, harvesting and preserving—the organic cycle of agriculture plotted by seasonal change—were transformed by machinery that commandeered time and geared it to an ever swifter man-made clock. As homesteaders displaced Indians and penned them in reservations, the worship of technology displaced the worship of Father Sun and Mother Corn and organic myths and metaphors gave way before the newly deified Machine.

In that one decade, the Homestead Act and the Land Grant Act of 1862 both delivered land free to farmers, and the new agricultural colleges told them what to do with it. In that one decade the Indian Territory that was Nebraska was reorganized as a state (in 1867) and linked east to west by the completion of the transcontinental railroad (in 1869). In that one decade, a number of scientific theories proposed earlier in Europe confirmed America's faith that the laws of nature were ours to control and improve. Justus von Liebig's *Organic Chemistry in its Application to Agriculture and Physiology* (1840) became known to American farmers through journals such as the *Albany Cultivator,* and the farm chemical industry began. Darwin's *Origin of Species* (1859), together with Louis Pasteur's discovery of the germ (1861), brought the visible and invisible world of organisms to the attention of farm breeders, and the genetic industry began. By the 1860s German agricultural experiment stations had become the

"Attacking your opponent for being well educated is an old rube campaign tactic aimed at fetching the know-nothing vote, but to work successfully it has to come from a candidate who majored in corn at the state university. Coming from a Yale man, it is absurd."

—RUSSELL BAKER, on presidential candidate George Bush, in "The Ivy Hayseed," *The New York Times,* June 1988

model for newly founded land-grant colleges, and the union of agriculture, science and industry was cemented at the source.

It's small wonder that within that one decade the number of Nebraska farms grew from 2,789 to 12,301. But these were not subsistence farms, for from the beginning they were geared to industrial products. In 1844 Colgate & Company of New Jersey established a plant to extract starch from corn by the new method of steel-toothed rollers that Orlando Jones had introduced into England for wheat. The wet-milling industry began. After Isaac Winslow sold in 1843 a dozen tin cans of Maine corn to Samuel S. Pierce of Boston, he was finally issued Patent No. 35,346 in 1862 for an "Improved Process of Preserving Green Corn." The canning industry began.

And so began the industry of farm machinery. The *U.S. Patent Office Report of 1860* issued new patents for corn shellers, cornhuskers, corn cultivators, corn-shock binders, cornstalk shocking machines, cornstalk cutters, corn cleaners, corn and cob crushers, seed drills, corn harvesters, rotary harrows, corn and cob mills, smut machines and hundreds of corn planters. Nebraska farmers, on the cutting edge of the last frontier, turned sod into farmland in the heady atmosphere of industrialism expressed in the 1860 *Report*: "Agriculture is experiencing the truth taught in the history of all other manufactures—that machinery is, in the long run, the best friend of the laborer." The ruling metaphor for farmers for the next century and a half was manufacturing and the machine, and the Corn Belt farmer was unique in that he was the first to labor under its dominion.

In Nebraska as elsewhere in the Corn Belt, half of those laborers were German. In the 1850s the prairies had been swept by a second wave of immigrants from northern Europe. "To a bedrock of Anglo-Saxonism," as Garreau put it, "was added Germans, Swedes, Germans, Norwegians, Germans, Finns, Germans, Ukrainians, Germans, Poles, and Germans." With their ready cash and strong work ethic, Germans added scientific thoroughness to Yankee ingenuity. They brought new vigor to farmer-soldiers exhausted by civil war to help turn the prairies themselves into "vast armies . . . of gleaming corn." The military metaphor was current. "That so soon after turning their backs upon the field of battle, they should exhibit to the world a countless array of harvest fields stretching over a thousand hills and valleys, and covering a land redeemed by their valor and now embellished by their toil," wrote Edward Enfield of the citizen-soldiers of America in 1866 in *Indian Corn,* "—this indeed is a moral spectacle instructive to the world, and more to be prized than all the material prosperity and affluence which it indicates."

Material prosperity, however, was the trumpet and drum to which these

The Randall and Jones Double Hand Planter, popular in the 1850s. Smithsonian Institution, Photo No. 42113

Christian Soldiers marched, firm in their belief that, like religion, "science is an unfailing source of good," as James Dana said, in which "every new development is destined to bestow some universal blessing on mankind." In the last half of the century, four new languages of science fused—the languages of genetics, chemistry, mechanics and economics—to explode in a cannonade of technology that propelled pioneer farming into commercial farming and rocketed the family farm into corporate industry. The end of the century heralded the birth of modern American agribusiness.

That birth had to occur in the lowly mangers of the prairies because the prairies required the first major revolution of the plow since the time of the Pharaohs. On prairie sod iron plows were nearly as useless as wooden ones. "The prairie breaker," which had been developed in the 1820s from Jethro Wood's 1814 iron plow, weighed 125 pounds with the moldboard alone, and made slow, heavy and hard work for the three or four yoke of oxen required to pull it. In addition, its moldboard clogged in the dense loam and the farmer had to scrape it constantly with a paddle. In 1833 a blacksmith in Lockport, Illinois, named John Lane realized that highly polished iron or steel was needed to scour the furrows and prevent clogging, but he was unable to market his moldboard. Four years later, a more entrepreneurial blacksmith, John Deere of Grand Detour, Illinois, mounted a steel share from a broken band saw onto a wrought-iron moldboard and scoured the loam so cleanly it "sang." By the end of the 1840s, Deere had built a plant at Moline, Illinois, and a decade later was mass-producing ten thousand "singing plows" a year. By the 1860s, after a foundry in Wyandotte, Michigan, introduced the Bessemer prócess of converting pig iron to steel, Deere's improved steel plows had taken over the prairies, replacing popular cast-iron models like the Eagle and the Michigan Double-Plow. The same decade brought in specialty plows like the "rod breaker," the stirring, ditching or paring plows, rigs like the sulky or riding plows, and soon the American Rotary, or disk plow.

Harrows followed the same swift evolution. At first, the pioneer farmer used a hoe or a simple brush harrow, dragged by an animal across the fields. In the 1840s he replaced these with the giant rake called the Geddes, a triangular hinged harrow made of wood with iron teeth, and in the 1860s with one made of iron and steel. Soon the Geddes was overtaken by the Nishwitz rotary disk harrow, which first pulverized the soil and then with wooden rollers or clod-crushers packed it down.

The same period brought new developments in

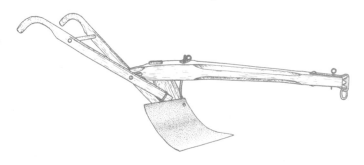

John Deere's "singing plow" in 1837, with its steel share to scour the prairie loam. Courtesy Smithsonian Institution Photo No. 42647

Nancy Hendrickson planting corn with a "stabber" on her farm near Mandan, North Dakota. Photo courtesy State Historical Society of North Dakota

corn planters. Until 1850, wrote R. Douglas Hurt in *American Farm Tools* (1982), American farmers planted corn as the Indians had first taught them, with a hoe or pointed "dibble stick." From the vantage point of New England, Thoreau could write as late as the 1840s, "This generation is very sure to plant corn and beans each year precisely as the Indians did centuries ago and taught the settlers to do, as if there were a fate in it." But Thoreau didn't reckon with Nebraska, where a pioneer required an axe to dig a hole in the sod. The first improvement on the Indian stick was a hand planter called the "stabber" or "jobber," which dropped seeds from a canister attached to a pair of boards hinged at the bottom like jaws and mounted on a pair of iron blades that were able to pierce the ground. The farmer now had only to jab the planter into the earth and open its jaws to deposit a few kernels of corn.

Improved tools for wheat farmers in America had developed half a century earlier, in conjunction with the wheat cultures of Britain and Europe. But the mechanical seed drill for planting wheat that was patented by Eliakim Spooner in 1799, for example, was of little use for corn. Corn could not be sown broadcast but had to be planted kernel by kernel. The first successful mechanical corn planter was not developed until the 1850s by George W. Brown at Galesburg, Illinois. His was a two-row horse-drawn vehicle which dropped seed from a mechanism attached to the wheels, but the mechanism was hand-operated. By the 1860s he had improved the rig by adding a pair of furrow openers or "shoes" in front and, at the rear, a "dropper seat" so that a boy could sit in front of the driver for more accurate dropping. As the farmer drove across the field at right angles to parallel lines previously marked, the boy worked the handle of the seed box to drop seed over each intersection.

Earlier pioneers had adapted the Indian method of planting their corn hills in a checkerboard pattern, or "check-rows," by dragging a four-runner sled over a field in first one direction and then another, so that they could cultivate between the rows. Thomas Jefferson described the method to his friend Charles Willson Peale when he wrote from Monticello on March 21, 1815, "We have had a method of planting corn suggested by a mr. Hall which dispenses with the plough entirely." Jefferson explained how the ground was marked off in squares and a grain of corn planted and manured at each intersection. Since Mr. Hall had taken a patent on the process, Jefferson was glad that he "has given me a right to use it, for I certainly should not have thought the right worth 50.D. the price of a licence."

Later farmers paid the price. In 1857 Martin Robbins of Cincinnati patented the first corn planter to drop seed automatically at the intersections of check-rows by the device of a chain that was fitted with iron buttons at regular intervals and was staked at each side of the field. In 1864 John Thompson and John Ramsay of Illinois substituted a knotted wire for the chain and thus created the standard check-row planter for the next forty years and beyond. Where the Western prairie bordered the Great Plains and was even harder to dig than in the Midwest, farmers developed the "lister planter," which dug deeper by means of a double moldboard plow to which a seed canister was attached.

By the 1880s the vision of material prosperity had altered the tone of America's moral spectacle, and lawsuits proliferated as fast as machines. George Brown sued Charles Deere and his partner Alvah Mansur for their version of the automated planter called the "Deere Rotary Drop." As Brown's threats flew, so did Deere's flyers:

> We shall continue to supply the trade with rotary drop planters, greatly improved, differing wholly from and therefore not infringing this old, abandoned and worthless device contained in Mr. Brown's reissued patent No. 6384, and continue as heretofore, to fully guarantee all parties who have bought or may buy, not only corn planters but any other implement from us, against all loss or damage, by reason of any claims or threats of Mr. Brown or any other man.

By 1885 Deere & Mansur had at least outblustered Mr. Brown sufficiently to add a fertilizer attachment, a stalk cutter and a sulky hay rake to their planter. A decade later they had developed the "edge drop," which regulated the number of kernels dropped in each hill by measuring a kernel's thickness rather than its length or breadth. Interestingly, the basic wire check-row system remained the chief method of planting corn until the advent of the corn combine in the 1930s revolutionized both planting and harvesting.

Still, planting large acreages had to await the evolution of cultivators, from a man wielding a hoe to one riding a sulky cultivator behind two or three horses that could till two rows at once. Usually, corn farmers hoed their crop four times during the growing season and, hoeing by hand, a farmer might have to spend six days per acre chopping out weeds. A one-shovel plow pulled by a horse was in common use by the 1840s, followed in 1856 by a "walking straddle-row cultivator," designed by George Esterly of Heart Prairie, Wisconsin, with movable shovels to cultivate two rows at a time. In the 1860s the sulky cultivator put the farmer on a seat behind a pair of horses, and he could now weed and ride with speed. Using three or four horses, he could weed fifteen acres a day with "as much ease and comfort as a day's journey in a buggy." With refinements—and in 1869 there were 1,900 patents issued for them—this cul-

"For reasons I never knew, perhaps it was nothing more complicated than pride of workmanship, farmers always associated crooked rows with sorry people. So much of farming was beyond a man's control, but at least he could have whatever nature allowed to grow laid off in straight rows. And the feeling was that a man who didn't care enough to keep his rows from being crooked couldn't be much of a man."

—HARRY CREWS, *A Childhood* (1978)

The industrial evolution of corn harvesters, from the first one, labeled "primitive," to the latest progressive combine of 1892, published by the United States Patent Office. Courtesy Alfred A. Knopf, Inc.

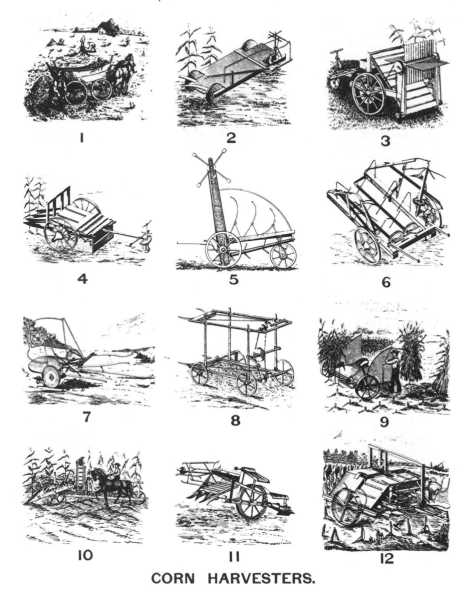

CORN HARVESTERS.

tivator remained standard until the International Harvester Company in 1924 changed plowing, planting and cultivating at once with a gasoline-powered tractor.

Although machinery for harvesting wheat appeared as early as 1831 with McCormick's Reaper, harvesting corn was another kettle of porridge. First the gasoline tractor had to be born and then corn had to be bred to produce only two ears per stalk and at a standardized height. Until then, corn continued to involve more hand labor than any other grain. Although the time required to produce one bushel of corn dropped from four and a half hours in 1855 to forty minutes in 1895, the only time-savers were in plowing and planting. Ears and stalks still had to be cut by hand. Where Indians had used cutters of stone and bone, colonists had at first used swords, sickles and knives. They then improvised a "corn knife" from a scythe blade (using an eighteen-inch blade and a six-inch handle), and eventually a knife manufacturer attached a knife-like blade to a boot so that the farmer could kick his cornstalk rather than bend to cut it. It was still back-breaking work.

One of the earliest methods of pioneer harvesting was the "corn sled," a wooden triangular platform with runners and blades extending on both sides to cut the stalks. Gleaners might follow to pick ears from the "down rows," or a couple of men might ride the platform to gather the stalks as they fell and bind them into shocks. With improvements like two-row cutters and chains to guide the stalks, two men could cut four and a half acres a day instead of one and a half with a corn knife.

While colonists learned from the Indians how to cut and bind stalks together to let the ears dry in the field, colonists had to devise new methods of harvesting to use green stalks and leaves for animal fodder. One of their earliest methods involved "fodder pulling" and "topping," in which a farmer cut off all the leaves just below the maturing ears and bundled them into fodder stacks. Next he topped the stalks just above the ears and added those to the stacks. He could then leave the ears to dry and harvest them when he was ready, but leaf removal tended to impede the full ripening of the ears and the method was abandoned early in the nineteenth century.

Corn "shocking" replaced "pulling" as the most efficient way to harvest, store and dry the ears, leaves and stalks in a single operation. As described by Nicholas Hardeman in *Shucks, Shocks and Hominy Blocks* (1981), a farmer would mark out a square of stalks for each shock, then twist the stalks from each of the four corners into a double-arched brace called a "horse," "gallows" or "gallus." With his heavy corn knife, he then cut and stacked corn in the openings around the brace, slanting the stalks from bottom to top to make a tepee. When the shock was five or six feet in diameter, he tied the "waist" with a band of twisted cornstalks or twine. Sometimes he used a wooden pole with cross-arms or a three-legged brace, some kind of "wooden horse," to help build his shocks.

The first successful mechanical binder didn't appear until 1892, when A. S. Peck of Geneva, Illinois, devised a horse-powered machine that passed down each side of a corn row and fed stalks into a pair of cutters. A chain carried the

stalks back to a binder that bundled them with twine and dropped them in the fields for stacking. Increasingly, farmers used the binder to cut green corn for fodder, which they would run through a shredder or chopper and blow through an elevator into a silo. Since the binder could harvest seven to nine acres per day, two or three shockers were needed to keep up with it, which incited inventors to combine cutting, binding and shocking mechanically. In 1888 A. N. Hadley devised a mechanical "shocker" in the form of a rotating table behind the cutter that gathered the stalks vertically and enabled a binder to tie them on the platform.

The corn shock was a storage method designed to shed rain reasonably well and thus prevent spoilage among the ears of drying corn. When a farmer was ready to pick and shuck ears already dried on the stalks, his family would work as a team, each person removing a bundle of stalks. If, however, a farmer preferred to pick the ears before cutting and shocking the stalks, he would snap off the ears while walking behind a wagon equipped with a "bangboard." As a picker tossed his ears against the board, they would fall into the wagon. Inventors came up with various mechanical pickers—cutters, rollers, prongs—but nothing proved superior to the farmer's hands until the increased power of the gasoline tractor altered the entire sequence.

Until this century, farmers had improved on the husking pegs of the Indians only by substituting iron for bone. With metal pegs, a skilled husker could husk about an acre of shocked corn per day, or one and a half acres of unshocked corn rows. Still, husking was hard on the hands and those who didn't wear heavy husking gloves poured tallow on their cracked hands at night, or a patented cornhusker's lotion that is still sold today. Huskers were hopeful when Jonathan Cutler in Putney, Vermont, patented the first mechanical husker in 1837, which worked with a pair of rollers attached to a wagon, but his invention didn't take off. Fifty years later an inventor came up with a husker-shredder combination that stripped the ears and fed the husks, along with the stalks, through rotating shredding knives that pulverized them. The shredded fodder was then channeled through a funnel and onto a stack while the ears were carried by an elevator to a bin or wagon. This device could eat up to a thousand bushels a day, but it was too costly for the ordinary farmer.

Shelling by hand was as time-consuming as husking, and mechanical corn shellers, large and small, became standard household equipment during the nineteenth century. Colonists had at first simply imitated Indians in scraping kernels from their cobs with seashells or bones. But by the 1820s hand-turned shellers like the "Little Giant" had become popular. By turning a spiked wheel, a farmer could strip off the kernels and drop them into a basket or tray while the cobs fell through a separate opening. In the 1840s the cast-iron Burrall sheller became standard: this required one person to crank the wheel and another to feed ears into the spout. Soon the horse-powered "Cannon" could shell a hundred bushels a day and, when steam replaced horsepower, the Sandwich sheller could shell two hundred to three hundred bushels an hour.

As tools "improved," so did storage. From the beginning, colonists had favored "corn houses," "barnes," "cratches" or "cribbes" over the underground pits favored by many of the Indians, because the grains that Europeans knew were more subject to mold than corn was and were best preserved in dry lofts, sheds and barns. They noted, however, that tribes from Canada to Florida also built corn houses on poles above the ground, sometimes square and sometimes round, with thatched or bark roofs for protection. In the humid Southeast, Indians made corncribs of cane or saplings, which one of De Soto's companions in 1542 described by the Indian word *barbacoa:* "a house with wooden sides, like a room, raised aloft on four posts, and has a floor of cane." The corn within was thus "barbecued" by smoke from fires lit beneath the platform, which helped to dry the ears more quickly and to ward off insects.

Colonists likewise constructed V-shaped corn houses of rough slatted boards, spaced so that plenty of air could circulate between them for drying the ears stored within. These were the corncribs still seen in backwoods farms in rural America, cradling the corn like babies in a manger. While the origin of

Round wood-slat cribs, heaped with dried corn and backed by a pair of elevators, in Casselton, North Dakota, ca. 1940. Courtesy State Historical Society of North Dakota

the word is obscure, the first written reference to "corncrib," Keith Roe has told us in *Corncribs: In History, Folklife, and Architecture* (1988), comes from the town records of Brookhaven, Long Island, in June 6, 1681, and suggests that corncrib theft was as serious as cradle robbing: "I, Hannah Huls, through inadvertance and passion, defamed Nathanell Norten, of this towne, by saying he had stollen Indian corn out of my fatther daiton's [Samuel Dayton's] his corn cribb."

Even the round silos that have come to symbolize the Midwestern farm were adapted from Indian granaries. In the early seventeenth century the English explorer Henry Hudson reported seeing a great quantity of maize stored from the year before in a corn house "well constructed of oak bark and circular in shape with the appearance of being built with an arched roof." A century later, Father Joseph Lafitau found Senecan "granaries of bark in the form of towers, on high ground, and they pierce the bark on all sides, to allow the air to penetrate and prevent the grain from moulding."

The significant change in silos did not come until 1875, when America imported a French version for the large-scale use of green fodder, or ensilage. These silos, built of wood or brick, were made airtight, since their function was to ferment the green leaves and stalks, like pickles in a crock, in order to make the fodder easier for animals to eat and digest. The first silo in Nebraska was built as an experiment at the Agricultural College farm of the University in 1882. Ensilage on this scale demanded more efficient cutters than the old-fashioned "fodder sled" or the hand-operated stalk cutter that used a hinged knife like a paper cutter. But power cutters that could fill the silos had to wait for the advent of the gasoline engine.

The wait was short, for the turn of the century brought the internal combustion engine and a new revolution of power. Charles Hart and Charles Parr of Iowa built the first successful gasoline tractor in 1901, but it was Henry Ford's first tractor, the Fordson in 1917, that made the combustion engine available to the American farmer at a price most could afford—$397. International Harvester followed with the Farmall tractor in 1924, the first to be fitted with removable attachments to make an all-purpose machine that would outdo the steam-driven wheat combines for reaping and threshing in use since the 1880s. Once again, the more difficult process of harvesting and shelling corn delayed the advent of a practicable corn combine until the 1930s. In 1936 Allis Chalmers took the lead by adding a corn-head attachment to its Gleaner and calling it the ALL-CROP Harvester, but the onset of World War II delayed production until post-war peace.

After the war, corn "picker-shellers"—one homemade version in Ohio was called the "cornfield battleship"—came into vogue because they were far cheaper than combines with corn-head attachments, though combines were more versatile and efficient. Once artificial dryers and steel bins replaced the old-fashioned cribs, however, they demanded the mass harvesting of shelled corn that only combines could do. With the development of airtight Harvestore silos, those great blue towers that dot the land from Wisconsin to Mississippi, corn did not even require drying before storing because the silos prevented mold in high-moisture corn by replacing oxygen with nitrogen or carbon dioxide.

As silos multiplied, so did the need for corn to fill them, and the corn was now supplied by a combine that could devour four, six, eight, even twelve rows at a clip and thousands of bushels in a day. Like some fire-breathing dragon, the combine demanded an ever-greater supply of fodder, contributing mightily to the malaise of chronic excess supply. As the diesel power of the combine put an end to horsepower and steam power, the revolution of the Machine, like other revolutions, threatened to consume its perpetrators, displacing the family farmer as the pioneer had displaced the Indian.

In 1986, when I talked to farmers like Everett Johnson at the North Central Corn Husking Contest in Courtland, Kansas, where old-time huskers gathered to compete for sport, I found that the march of time had brought a change of tune. When Everett spoke of his machines, it sounded less like reveille than like taps:

A combine costs maybe a hundred and twenty thousand dollars to buy and has a life of probably six to eight years at the most. Then there's the cost to run it—burns up ten gallons an hour—plus labor, plus repairs. Now, thirty years ago, you could tear down and build up your own tractor, but today there's so much electronic stuff you don't dare touch it. They trade 'em in in a year to take the depreciation. The machines are rusting, the houses are gone. Pretty soon all we got to do is import some Indians.

"Corn Sick" Land

WHEN COLONISTS, rather than Indians, were the imports, Captain John Smith at Jamestown in the spring of 1609—"this starving time," as he called it—wrote that "for one basket of corn they would have sold their souls." And so, without knowing it, they did. For two centuries they rationalized their territorial imperatives as God's will. "What is the right of a huntsman to the forest of a thousand miles over which he has accidentally ranged in quest of prey?" John Quincy Adams asked rhetorically in 1802. "Shall the fields and vallies, which a beneficent God has formed to teem with the life of innumerable multitudes, be condemned to everlasting barrenness?"

Toward the close of the third century after Jamestown, with the huntsman-planter now stripped of both forest and corn and enduring his own "starving time," economic imperatives were claiming dominion not just over fields and valleys but over minds and hearts as well. Addressing a conference of Friends of the Indians in 1896, the president of Amherst College, Merrill E. Gates, proclaimed:

> We have, to begin with, the absolute need of awakening in the savage Indian broader desires and ampler wants. To bring him out of savagery into citizenship we must make the Indian more intelligently selfish before we can make him unselfishly intelligent. We need to *awaken in him wants.* In his dull savagery he must be touched by the wings of the divine angel of discontent. . . . Discontent with the teepee and the starving rations of the Indian camp in winter is needed to get the Indian out of the blanket and into trousers,—and trousers with a pocket in them, and with a *pocket that aches to be filled with dollars!*

The nature of the white man's mission may have changed, but the moral zealotry of his language did not. The civilization of the intelligently selfish must convert the savage to the divinity of discontent. Corn, as always, was the catalyst in this divine alchemy, and the question, as always, was how many baskets of corn would it take to fill a pocket with the dollars it was aching for?

Despite a century tuned to ever-ampler wants by ever-louder choirs of discontent, there are small towns in America's corn country that do not ache for dollars. Some of the very small towns of Nebraska have slept so soundly that they seem to have awakened no wants at all. Wilcox is that kind of town—a Main Street, a grain elevator, a watertower and Dale's Cafe. My mother, Hazel Kennedy, was born in Wilcox on Christmas Eve, 1887, and I'm told that her father dug the town's first well. Today, Sunday dinner at Dale's Cafe is still served at noon and at century-old prices. You get a plate of crisp-fried chicken with mashed potatoes, boiled corn, white bread, iceberg lettuce and iced tea for $3. When I ask for the rest room, I am sent through the kitchen into a back room with a 1930s refrigerator and battered sofa, where an elderly man sits reading the funnies next to a hairy clump of dog asleep on the rug. The man points me to a clothes closet, and behind a door at the back of the clothes are a toilet and sink, flushed still, no doubt, by the water my grandfather dug. I would not have been surprised to see an Indian in his blanket and teepee out back.

Wilcox is one hundred miles west of Lincoln and next door to the equally small and unchanged crossroads of Hildreth, where Cousin Eleanor was born. Not far south is Red Cloud, where Willa Cather's upstairs bedroom is still papered in red and brown roses in the neat white frame house the Cathers lived in. After a century of land-shattering change elsewhere, Wilcox, Hildreth and Red Cloud stand like prehistoric prairie dolmens. To find a chunk of native prairie grass you have to go to a museum park like the Homestead National Monument near Beatrice, where little patches of big blue-stem, buffalo grass, blue gamma, Indian grass, switch grass and sand lovegrass have been "restored" for people who have never seen the like.

Not far north, at Grand Island on the Platte River, the changes of the land have been compacted in the Stuhr Museum of the Prairie Pioneer. Here an authentic steam engine and coach of the Nebraska Midland Railroad will take you by a real buffalo herd, a Pawnee Indian earth lodge, the log cabins of a pioneer settlement and the toy windmills and upright houses of Railroad Town, including the transported Grand Island cottage where Henry Fonda was born in 1905. Time has moved so fast on the prairie that one generation's wild is the next generation's Disney World and the ache of discontented pockets is at war with the ache of nostalgic hearts. In 1970 American nostalgia had produced so

"How beautiful to think of tough lean Yankee settlers, tough as gutta-percha, with a most unsubduable fire in their belly, steering over the Western Mountains, to annihilate the jungle, and to bring bacon and corn out of it for the posterity of Adam."
—THOMAS CARLYLE

many local farm museums and "living history" farms that a national association was formed, ALHFAM (Association of Living Historical Farms and
Agricultural Museums), headquartered in the National Museum of American
History in Washington, D.C. Even twenty years ago the family farm was
ancient history.

"When the rich, black, prairie corn lands of the Central West were first broken up it was believed that these were naturally inexhaustible lands and would
never wear out," wrote C. B. Smith, from the Office of Farm Management,
Bureau of Plant Industry, in the *Yearbook of the U.S. Department of Agriculture*
for 1911. That belief was ill-founded, he said even then, for farmers had
planted crop after crop of corn on the same land year after year, and as a result
had decreased yield, increased insects and exhausted the land. "The land," he
said, "was 'corn sick.'" The remedy, he said, was to rotate corn with restorative
crops like clover or alfalfa and to fertilize it with barnyard manure.

"How much corn must be grown per acre to make it profitable?" he asked,
at a time when the average cost per acre was $14.63 and the price per bushel
42.4 cents. With a cost-and-profit chart he easily demonstrated that yield was
crucial to profit, since a yield of thirty-five bushels an acre would return a profit
of but 21 cents, where a yield of forty bushels would bring a profit of $2.33.
When the twentieth century began, government officials had to persuade
farmers that what mattered most was yield, rather than tall corn with beautiful ears. At the close of this century, government officials still urge greater and
greater yield, despite the farmers' consensus that overproduction is the enemy.
In the last half of the 1980s America's farmers produced the largest grain surplus in the world's history, and yet every one of the corn farmers I talked to
agreed that in one way or another the simplistic equation of yield with profit
has made the land corn sick.

*A bleak scene in the "corn sick"
land of Sherman County,
Kansas, around the turn of the
century.* Courtesy Kansas State
Historical Society, Topeka

"We got a billion bushels piled up in Ioway alone," George Mills tells me, as we look at a green wall of corn behind his house on the south fringe of Des Moines. "One hundred thirty-five bushels to the acre, finest corn we've ever had." I've driven from Nebraska to Iowa, reversing the western trail of my grandparents, because Iowa is still the Tall Corn State. Mills can recall a time when Iowa Shriners convening in 1912 in Los Angeles brandished cornstalks and sang, "We're from Iowa-y, Iowa-y, That's where the tall corn grows." Mills was not a farmer, but a political writer with *The Des Moines Register* for thirty years, before he retired to pay more attention to his backyard corn and to write a book on the decline of the Bible Belt from the 1930s to the 1980s.

"Iowa is in the middle of the biggest plain this side of Jupiter. Climb on to a roof-top almost anywhere in the state and you are confronted with a featureless sweep of corn as far as the eye can see. It is 1,000 miles from the sea in any direction, 600 miles from the nearest mountain, 400 miles from skyscrapers and muggers and things of interest, 300 miles from people who do not habitually stick a finger in their ear and swivel it around as a preliminary to answering any question addressed to them by a stranger."

—BILL BRYSON, "Fat Girls in Des Moines" (1988)

Mills's reporting career coincided, as he sees it, with the rise and fall of the Bible Belt farm. "The old definition of a farmer was a man with one hundred and sixty acres who kept some hogs and some cows and whose wife kept chickens and sold eggs and took their milk to the milk station," he says. "Now the average farmer in Warren County tills eight hundred acres because you can barely make a living on three hundred. The land hasn't gone away but the farmer has." 1988 research figures show that although our national farm acreage has remained constant at 340 million acres, our average farm size has tripled while the number of farms has been halved, from 5.9 million in 1945 to 2.2 million in 1985. More significantly, farming today occupies only 2 percent of America's entire labor force.

Even at the dawn of the century, when enthusiasm for equating the farm home with a place of business was undimmed, there were warnings that profitability depended on size. "One of the first and most important factors having to do with profitable farming, as in all other lines of business, is the size of the enterprise," J. S. Cates, of the Office of Farm Management, wrote in the 1915 *Yearbook of the U.S. Department of Agriculture*. "Despite the much-talked of idea of 'a little farm well tilled,' actual records from thousands of farms covering pretty well the whole United States go to show that little farms do not often make big profits, and that as a rule the profits from farming vary directly with the size of the business."

Fifteen years later, at the onset of the Depression, little farms were going under wholesale, and the man who had done most to turn corn farming into a business, Henry Wallace, tried to save the small farmer by limiting production through the Agricultural Adjustment Act of 1933. After the act was declared unconstitutional, Wallace came up with the "Ever-Normal Granary Plan," which became the basis for government crop subsidies euphemistically called "government price-support loans." The government guaranteed the farmer a loan price and if the market price of corn fell below it, the farmer turned his corn over to the government for storage and the government paid him the difference.

Through this device, the government by controlling acreage could still encourage higher yields. The most the farmer produced, the larger his collateral. But at best it was a stopgap measure, conceived under the duress of the Depression and enlarged by the crisis of World War II. After the war, the benefits of Wallace's strategy were undone both by the unprecedented yields of his own hybrid corn and by the unprecedented use of chemical fertilizers. The herbicide 2,4-D, developed during the war as a chemical weapon for potential use against enemy food crops, was introduced in 1947 as a weapon against enemy weeds. Since that time, America's average yield per farm acre has increased annually by 2 percent.

Wallace's device, of course, was also a political subterfuge by which capitalism's nominally free-market economy could be manipulated behind the scenes. Talking to Mills, I begin to understand why my farming family excoriated Roosevelt and Wallace as twin Antichrists. To my kinfolk, self-sufficiency was divine and anyone who interfered with that God-given right was doing the devil's work. George Mills's view, on the other hand, is more historical, and he sees Wallace as caught between wars. "You call upon your farmer to vastly increase production as part of national defense, but when the war's over it's hell to scale down." In 1920 after the First World War, Iowa land sold for $254 an acre and corn for $1.70 a bushel. Six months later, corn sold for 67 cents—a disastrous deflation that wasn't "corrected" until World War II. "Same damn thing happened after the Vietnam War in the 1970s—land prices tumbled and so did corn," Mills says. "When politicians get mixed up in farm economics, hold on to your hat."

But in Ioway the politics and economics of corn are one and the same, says Lauren Soth, another silver-haired *Register* man whose editorials on corn politics over the past forty years won him a Pulitzer Prize. He recounted how overproduction after World War I had been increased by mechanization. With the advent of tractors, the need for oats to feed ten million horses and mules disappeared; so as more land was tilled, it was planted with corn. Soth had seen how overproduction after World War II was increased by commercial fertilizers, when specialization removed livestock and manure from the farms and

put cows and hogs into commercial feeding lots. By the 1960s he had seen how farmers had to turn their farms into cash crops in order to pay for the petroleum that ran their tractors and the anhydrous ammonia that increased the yields of a continuous corn crop. Today corn farming, he says, consumes 44 percent of all chemical fertilizers used in the States and half of all farm pesticides and herbicides. Now 95 percent of corn and soy acreage uses the synthetic chemicals that are demanded by intensive continuous cropping.

Smith in the 1911 USDA *Yearbook* warned, "Continuous corn culture has no place in progressive farming." Soth in a 1986 *Register* editorial uttered the same warning for opposite reasons: "The hard fact that the USDA top command and the Fertilizer Institute cannot face is that agriculture in this country is substantially overexpanded. So is the agribusiness superstructure that serves it." As one who remembers the Dust Bowl well, Soth wants the government and the farmer to retire eroded land and to restore it with grass and trees, to reduce the use of chemicals severely and to practice crop rotation, as Smith advised eighty years ago.

But Americans don't believe in what they can't see, and the superstructure of American agribusiness that controls the production of corn is as invisible and pervasive as the industrial products of corn. Measured by output per labor unit, the business of producing grain on farms is larger than any non-farm business in this country, and has been since the crop explosion at the end of the 1970s put America for the next five years in control of 70 percent of the world's trade in grain. Between the 1960s and 1980s, while inputs on American farms increased

A boy with his shovel contemplates the fruits of his labors in western Kansas, in a post–World War I economy where ever more corn brought ever less cash. Courtesy Kansas State Historical Society, Topeka

from $50 billion to $80 billion, gross outputs increased from $235 billion to $450 billion. But 80 percent of that output is produced by only 15 percent of all United States farms. In the heart of the Corn Belt itself, the average farm size is but 294 acres, so that three-quarters of its farm households must generate "outside" income off the farms in order to keep them.

The problem, Soth believes, lies in men's minds. The industrial model simply will not work for corn. "Growing corn is not like manufacturing automobiles," he says. "The small-enterprise farm is an industry, but it's also a living place, and they get in the way of each other." The free-enterprise market of supply and demand doesn't work for corn, because you can't control the supply side the way you can control it with cars or shoes—nature interferes. So, he continues, we have to rely on some kind of government program, which may suit the big corporations that are in it only for money, but not small farmers for whom farming is a way of life.

Today's government program for corn, already limited by the facts that it is voluntary and applies only to acreage, still operates on the basis of the antiquated loan system set up by Wallace in 1938, a half-century ago. Even the language is antiquated and obfuscatory. A farmer who goes into the program volunteers not to plant a certain percentage of his corn acres, which is called his "corn base." He gets a "loan," which is a government-guaranteed "target" price, for what he produces on the planted acres. "Congress picks the target price out of the air," says Soth. "That price is Congress's idea of what they think the farmer this year ought to get—it's a wholly political decision." The difference between the target price and the market price is the "deficiency payment" that the government pays the farmer that year. In 1986, for example, the target price was $3.03 while the market price was only $1.25, so that government subsidies for the year totaled $30 million. "It's ironic that Reagan, that luminous champion of the free-enterprise market and frugal federal budget," Soth says, "not only paid this sum but bragged about it."

Soth maintains, as my own family had vehemently confirmed, that small farmers don't like taking money from the government instead of from the market. Nor do they like government control. If the government would exert mandatory controls over both crop acreage and production, they could create a marketing quota. But more government control doesn't sit well with farming families who remember what brought their ancestors to the prairies in the first place. "I loved seeing things grow, being my own boss and working together as a family," says an Iowa farmer who recently lost his farm and is now a door-to-door salesman. "Farming is a calling," says another, "not a living."

Many ex-farmers like my cousin Don still believe that the farmers who are losing out today are those incapable of making the change to total mechanization and computerized farming. "We don't need farmers, we need businessmen," Don says with the fervor of his youth. Milan Moore, on the other hand, speaks for a younger generation when he says that the small farmer has simply been squeezed out by corporate business. Big corporations mean absentee landlords, managerial companies and ever-more-specialized factory functions that manipulate and displace organic ones, like industrial feedlots for cattle and hogs. Size is the determinant. "I think there'll have to be a major change in the economics of agriculture before it's all controlled by large companies who have to get a large return on their money or they won't get involved," says Milan. Farmers with small acreage would be satisfied with smaller profit, he believes, in order to be more self-sufficient. The remedy as he sees it lies in price control, where the government would initiate a reverse graduated price to decrease production: "Say four dollars a bushel for the first ten thousand bushels, then three for the next ten thousand, and so on down the line."

In the 1980s other voices began to question the rigid logic of facts and figures outlined a century ago. In 1989 specialists gathered by Wes Jackson of the Land Institute met in Salinas, Kansas, for a symposium on "The Marriage of Ecology and Agriculture." They assessed the costs of industrial farming in a larger context than crop economics, including the costs of land erosion and groundwater loss and of the pollution of both elements. "The pioneers who ripped away the virgin sod of the prairie to plant wheat and corn committed sacrileges for which man has yet to atone," Jackson said, "and the penalties have been erosion, droughts and dust bowls."

In November 1989, Richard Rhodes, in a *New York Times* editorial, voiced the concerns of all those who have adopted the term "sustainable agriculture" to replace the "chemical agriculture" of agribusiness. "We ought to place the blame for chemically intensive agriculture where it belongs: in Washington, with the politics of farm subsidies," Rhodes wrote. Collectively, agribusiness, the largest industry in America, has redefined a farmer as "someone who launders money for a chemical company." He cited as typical a farming family near Kansas City, Missouri, who grossed $152,090 in 1986. After the chief expense—feed for their cattle and hogs—they paid $22,345 for fertilizers and pesticides and $17,910 for machines and fuel. With the government subsidy of $11,000 to idle 10 percent of their land, they cleared $19,000. In short, more than $41,000 of their gross went to agribusiness. Rhodes's plea was for a federal farm program that would limit production, encourage crop rotation and biological controls, and remove dependence on chemicals.

The Summer-Autumn 1989 issue of *The Journal of Gastronomy* contained a number of articles by different writers addressing the costs to food and consumers of current methods of farming. "Farming isn't manufacturing," said the noted restaurateur Alice Waters of Chez Panisse. "If our food has lacked flavor . . . that may be because it was treated as dead even while it was being grown." We are a nation of industrial eaters, warned the noted essayist Wendell Berry. "The industrial eater is, in fact, one who does not know that eating is an agricultural act, who no longer knows or imagines the connections between eating and the land, and who is therefore necessarily passive and uncritical—in short, a victim." Most tellingly, Frances Moore Lappé was concerned that we have no democratic language of agriculture, but rather only the "free-market" economic language of industry, which assumes that crops are commodities, based on private property and the accumulation of resources, on a division between labor and ownership and on control over processing and distribution. "Cheap food has been made possible at the farmer's expense, not at the expense of the food processor or distributor, who receives 75 percent of every dollar spent on food," she wrote. "The economic miracle of today's cheap American food has involved a colossal transfer of income and capital from producers to middlemen—to the agricultural equivalents of Wall Street arbitrageurs and bond sellers."

In 1989 the Board on Agriculture of the National Research Council for the first time in its history put out a volume called *Alternative Agriculture,* which applied the logic of facts and figures to the costs of our shortsighted farm policies since the 1930s. "In effect, a high target price subsidizes the inefficient, potentially damaging use of inputs [fertilizers, pesticides, and irrigation]," the Council concluded. "It also encourages surplus production of the same crops that the commodity programs are in part designed to control, thus increasing government expenditures."

At the same time, the National Research Council was aware that any significant decrease in American corn production will have not just a national but a global effect. In 1988 the United States, with 4.1 billion bushels of corn in storage, held 73 percent of the world's supply. Meanwhile the world's demand for corn has been rising by about 200 million bushels a year. Oddly enough, China has become the world's second largest supplier of that demand and in China today one can see more visibly than elsewhere the political effect of corn. In China one can see how closely the alarming increase in the world's population is tied to the increased production of corn.

When I went to China in October 1987, the moment I shook the city dust of Beijing from my heels, I was in corn country. Corn was everywhere. The bike paths, where kernels were laid to dry, were calligraphic strokes of yellow.

Tile rooftops, eaves, ledges were looped with corn. Golden-red ears were stacked like logs against walls, hung in bunches from trees, strung from poles like laundry, heaped in mountains walled by drying stalks. I might have been in Mexico or Peru. Not a tractor in sight, just men, women, burros and donkeys, toiling under their golden burden as they have done for four centuries, ever since corn found its way to China along the trade routes, sometime between the 1520s and 1560s.

The introduction of corn revolutionized the agricultural life of China and helped produce a population explosion that began in the seventeenth century, quadrupled between the eighteenth and nineteenth centuries and is now out of sight. To the Chinese, corn was and remains poor man's food because it can be grown where their favored crops of rice and wheat cannot. "We think of corn as an auxiliary staple," said a young country-faced woman named Ren Rui, a crop researcher with the Chinese Academy of Agricultural Sciences in Beijing. Ren Rui, whose husband is a physicist, studied for a year at Ames, Iowa, and knows well the wide world of corn. Corn now forms 22 percent of China's staple crops, she tells me, yielding 14.3 kilograms per *mou* (a *mou* is one-sixth of an acre). China's commercial crop, which began in the 1950s with seed imported from the United States, is grown in Heilongjiang province in the northeast, in Manchuria, which with its Corn Belt climate is now the second home of Corn Belt Dent.

Bouquets of corn carpet a village square in China in the 1970s. From *Scenes of China,* Hong Kong Beauty View Publications

Corn has been important historically to China; it helped to extend and sustain the power of the Manchu invaders who overthrew the last of the Ming emperors in the mid-seventeenth century. Corn enabled the overcrowded Yangtze delta farmers to cultivate inland hills where other crops would not grow, as it enabled laborers in the mountains of Yunnan and Szechwan to become full-scale farmers for the first time. By the end of the eighteenth century, corn was the primary food crop in southwest China and today it still provides a goodly portion of the food energy for north-central China, although those who can afford rice or wheat eat corn only when desperate. "We once *had* to eat corn," Ren Rui says, "so we don't want to eat it anymore." Today, from a

"Let there be more corn and more meat and let there be no hydrogen bombs at all."
—PREMIER NIKITA KHRUSHCHEV, on his first visit to Roswell Garst's farm in Coon Rapids, Iowa, in September 1959. After Khrushchev was deposed, he devoted himself to writing his memoirs, puttering in his garden, and growing corn.

crop yield of 65,560,000 metric tons (in 1986), they can afford to feed 60 percent of their corn to their animals and export most of the rest to Russia and Japan.

As American farmers have learned that corn in Kansas is tied to the weather in Siberia, so they know that their corn dollars are tied to the value of the yuan in Manchuria. The language of agriculture, like the language of economics and ecology, is worldwide no matter what the national politics of its constituencies. While Chinese farmers have been sowing more land with corn, American farmers have been sowing more cornfields with soybeans. In 1987 the dollar value of American exports of soy topped both corn and wheat by $1 billion—a new variation on the post-Columbian Exchange. After a century and a half of exporting the Industrial Ideal along with our industrialized crops, the United States is beginning to look not only outward but inward, seeing perhaps for the first time that the cure for a corn-sick land begins at home.

At long last we are listening to the native huntsman-planter, who has never lost his sense that man belongs to the land, not the land to man, and that a nation's health as well as a planet's requires that earth, sky, plant, animal and man be treated as one. Such are the complexities that, near the dawn of a new century, are altering our corn politics and our corn-pone opinions, as in Mark Twain's pronouncement, "You tell me whar a man gits his corn pone, en I'll tell you what his 'pinions is." I doubt that Twain knew the Chinese had been eating corn pone for as long as our colonists had, but he knew a thing or two about cultural arrogance and humility, as he knew the savagery of those who would mistake their corn-pone opinions for God's will and their greed, in this or any other starving time, for divine discontent.

One of fifty houses of sun-dried corn in a village in the Erlang Mountains in Sichuan Province in 1985. Photo by Harald Sund

FOUR

Flesh
and
Blood

(overleaf) Pueblo women at San Ildefonso grind corn from coarse meal to fine on a triple metate, to drumbeat and song, in this painting by Gilbert Atencio in 1962. From the Dale Bullock Collection, New Mexico State Records Center

A corn commodities train speeds through a rural landscape of villages, silos and windmills, embodied in products using corn sugars. Corn Refiners Association, Inc. booklet, 1980s

The Language of Food

FROM PIKI TO CORNFLAKES

"CORN IS LIFE," said Don Talayesva, Sun Chief of the Hopi in 1942, "and piki is the perfect food." Most of America has never heard of the fast finger-food called piki, but all the world has heard of cornflakes. "Piki are the original cornflakes," the Hopi matriarch Helen Sekaquaptewa told me, and indeed, the more I learned about North American native cooking, the more I saw that cornflakes are the white man's piki. They are also the white man's aboriginal fast food—"In two jiffies a flavory meal to satisfy the hungriest man," read a Kellogg's ad in the 1930s. "In a bowl, flakes race with milk to the finish, beyond which the food turns and collapses to paper, to the cardboard, soaked, from which it had been poured," wrote the food critic Jeff Weinstein in *The Village Voice*. "A bowl of cornflakes ignited by milk is a meal automatically clocked, constructed to speed you up and out." Cornflakes speed our racing engines, and therein lies the difference between the Hopi's view of corn as life and the white man's view of corn as high-octane fuel.

From the start, the food languages of native and colonial America were as disparate as the cultures from which they came. Any organic food culture is rooted in the place where it was born and bred, but white Americans are all transplants, migrating to this place, these weathers, this ecology, with a load of culinary baggage packed elsewhere. The subtlety and sophistication of native corn cooking was integral with corn, life and language on this continent. "The terms in Zuni for every phenomenon connected with corn and its growth are so numerous and technical," Frank Cushing wrote a century ago, "that it is as difficult to render them into English as it would be to translate into Zuni the terminology of an exact science." Because the primary language of corn, in its thousands of dialects, was more often spoken than written, was performed rather than abstracted, the language of wheat subdued the language of corn as quickly as swords subdued arrows. Many of corn's secrets were misunderstood and lost, literally and figuratively, in translation. Like the forgotten seeds from which it sprang, however, native corn cuisine did not die but evolved and hybridized. While much of it dwindled into a few standardized forms, much remained hidden, like the prehistoric cobs in Bat Cave, awaiting rediscovery.

For more than eleven centuries, piki sustained and still sustains life in Old Oraibi on Second Mesa, the oldest continuously inhabited village in the United States, as old as the earliest settlement of Aztec or Inca and the center of the ten thousand Hopi who occupy the small chunk of northeast Arizona called Hopiland—the land of "the people of peace." Old Oraibi, however, is barely visible, so fully does it blend into the pink wrinkled skin of the mesas, as the mesas themselves blend into the desert horizon of sand and sky.

In contrast, Battle Creek, named for the people of war, in central Michigan near the kingdom of Ford in Dearborn, shouts from the rooftops its name and cereal number. On the side of his four-story factory of 1906, William Keith Kellogg, the father of cornflakes, had "THIS IS BATTLE CREEK" painted in block letters to make "Kellogg's of Battle Creek" a chivalric vaunt like "William of Burgundy" or "Timon of Athens." If the original stone Sanitarium of his brother, Dr. John Harvey Kellogg, the father of health zealots, is now "just a heap of rubble—passed away, dead," as a local Battle Creekian put it, the grandiose "San" of 1904 (now the Federal Center) still proclaims its glory fifteen stories tall and twenty Ionic pillars wide.

The difference in language is not just a matter of style but of organic substance. Although the Hopi language was officially outlawed in the 1910s and the Hopi land officially shrunk to 501,501 acres in the 1940s, even today the Hopi are the American natives who continue to speak most fully the ancient language of corn, because their daily lives are still intimate with the growing, cooking and ceremonies of corn that gave meaning to the lives of their ancestors. Theirs is also the last place on the continent where the art of transforming blue corn into the tra-

Dr. John Harvey Kellogg, with his brother Will K., applied Barnum hoopla to "health" and sold the world on flaked corn. Courtesy Battle Creek Sanitarium, Battle Creek, Michigan

ditional blue wafer bread is still handed down by the grandmothers to the mothers and daughters of the clan.

Helen Sekaquaptewa, born in Old Oraibi ninety-three years ago, is one of the honored grandmothers and the reigning matriarch of the Eagle clan, whom everyone calls "Grandma." Her life, like the life of every Hopi from cradle to grave, has been shaped by corn. For nineteen days after childbirth, her mother fasted, while Helen's paternal grandmother placed beside the new baby two perfect corn ears, a long one to represent the mother and a smaller one for the baby. Each day for twenty days her mother marked each of the four walls of her room with white cornmeal in five horizontal lines to prepare for the naming ceremony. Then both mother and baby were washed with soapweed and the baby's face rubbed with white cornmeal. As the grandmother blessed the baby with the pair of corn ears, she put a pinch of cornmeal in the baby's mouth and said, "This is your food, which the earth mother will give you all your life."

When Grandma dies, her kinsmen will dust her face with cornmeal and cover it with a cotton mask in which holes have been cut for eyes and mouth. They will fold her body in a fetal curve, wrap it in a blanket, tie it with yucca strips and bury it deep in the earth. They will cover her grave with a slab of sandstone and rocks, in which a greasewood stick will provide a way out for the spirit to depart on the third day to the spirit world. At the far end of Second Mesa you can see the rocks and sticks of Hopi graves clustered at the foot of Corn Rock, which juts from the flat mesa like a thick and phallic double ear of corn, symbolic of eternal as well as daily life.

Grandma's earliest memory was of her mother taking her to the "training rocks" beyond the village, where she and her playmates were each given a few kernels of corn and a stone to grind them with in the small hollows of the rock. Now I could see the shallow depressions where the girls ground corn next to grooves where boys learned simultaneously to file the shafts of arrows and to flirt with girls. "Grinding is good for you," Grandma said, since grinding was for girls what running and hunting were for boys, a form of exercise as well as a way to get food. At puberty, girls had to grind corn ceremonially for four days, just as boys had to run all night to place cornmeal and prayer sticks on distant shrines and return before the rise of the morning star.

Even today when a Hopi girl is betrothed she takes, as Grandma did seventy-three years ago, a pan "full of fine white cornmeal" and a basket of blue piki bread to the house of her groom-to-be. From then until the day of her wedding, the women of the village grind corn for the wedding feast while the men weave the bride's wedding robes from white cotton, which formerly they both grew

"*The Navaho chose the long ear of yellow corn because it was soft and easy to shell and easy to grind, but the Hopi, he chose the shortest, hardest ear. It was tough. It would survive.*"

—George Nasoftie, in "The Hopi Migrations" by Tony Hillerman, in *Arizona Highways* (September 1980)

A Hopi woman lifts a sheet of piki from the hot stone. The bowl of batter sits next to the row of folded wafers on the mat. Courtesy Arizona State Museum, the University of Arizona, Helen Teiwes, photographer

and spun. The last four days before the wedding, the bride grinds corn in the house of the groom in a room set apart, shades drawn, talking to no one. Each day her husband's kinswomen bring her a quart of corn, white the first day, blue the second and third. "After the first grinding I handed the corn out and waited while it was roasted and passed back to me to be ground real fine," Grandma remembered. On the fourth day she ground unparched blue corn for the special wedding cake called *someviki*. After the wedding, the bride returned to her mother's home to grind more corn to "pay" for the weaving of her wedding robes, and only then would the groom move into her home.

Grandma's son Eugene took me to the village of Shipaulovi to watch a group of women in the house of Frieda Youhoena make piki in preparation for the "payback" feast of a recent bride. I had read many accounts of the sacred piki stone, of how the men would quarry a two-foot granite slab two to three inches thick, how their wives would rub and polish it with fine pebbles until it was smooth as glass, how they would season it with watermelon seeds, ground and roasted until black, which they would then rub over the surface of the stone until its blackness shone like a mirror. So I was startled to find a young woman in jeans and harlequin glasses, kneeling against a piece of foam rubber over a smoky fire in the back shed that served as Frieda's piki room.

"I'm just a novice," laughed the daughter of Rex Pooyouna, a famous moccasin maker, as she chatted with her husband. He sat on a low bench while she fed a small but intense wood fire beneath the stone placed on a foot-high hearth. She dipped her right hand into a large plastic bowl of blue batter, rubbed off excess batter along the rim, then smeared her hand quickly over the top of the stone, from the far edge to the near in successive arcs, dip and smear, dip and smear, until the stone was completely covered. On top she put a sheet of already cooked piki in order to soften and fold it. She folded both sides into the middle and then folded the wafer in thirds from the bottom up. She added the folded piki to a pile of wafers at her side before she stripped the new layer from the stone. Every now and then she cleaned the stone with a rag of soft leather and greased it from a bowl of fat rendered from sheep brains.

"Piki's so expensive now," she said. "I wanted to learn to make it because it costs so much and because I thought it was time. I have three girls and I want them to learn." At first, she burned her fingers constantly on the stone, which is heated to 400 degrees, but she had learned to use the heel of her hand and her fingertips had toughened as she had picked up speed. "I'm still so slow," she groaned. "It will take me two or three hours to finish this amount of batter." Although Frieda had already stored four boxes, she would need hundreds more piki wafers for the ceremony and the women in her kitchen were making tubfuls of silky blue batter.

Because of the way the corn has been prepared before the batter, piki will keep indefinitely if properly stored. First the corn is cracked, then ground coarsely, then roasted, then ground fine. It's the roasting that preserves, as Indians learned in ancient times when making their "traveling" bread or "journeying" cake, the little bag of parched meal they carried with them on their hunts and migrations. In Grandma's home they always kept plenty of piki on hand for guests, she told me, some white, some blue, some lavender—the color blue-corn meal turns when heated without any alkali. But the alkaline of wood ash not only keeps the color true blue but also lends flavor, leavening and nutritional value to the meal. Ash cooking, as we shall see later, begins the alphabet of America's corn cuisine.

The Hopi make many other blue-corn dishes—boiled dumplings called *mumuozpiki,* pancakes called *saquavikavike,* fried grits called *huzrusuki,* which any traveler can eat in the restaurant at the Hopi Center—but piki remains their quintessential food. "No man will marry a girl unless she can make piki," they said in Grandma's day, but that was then. Now Hopi girls buy hand mills and electric blenders to grind most of their corn, and many a fine finger will never have touched a piki stone. But if a Zuni or Pueblo girl wants to learn to make wafer bread, she must go to Hopiland and begin by grinding meal by hand as fine as the powdered corn Grandma's mother taught her to put on her face to make her skin as smooth and silky as a piki stone.

Such wafer breads were once a staple among the Pueblo people, called *mowa* by the Tewa, *buwa* by the Rio Grande Tewa and *he'-we* by the Zuni, for whom a *he'-we* stone was "the rock's flesh," to be anointed with hot piñon gum and cactus juice. In *Zuni Breadstuff* (1920), Frank Cushing devoted a chapter to the wafer's many colors and kinds: a salty one flavored with lime yeast; a rich one made with milk, tasting like "cream biscuits"; a red one of red-corn kernels, leaves and shoots, sweetened like "London sugar wafers"; a delicious fermented one sweetened by saliva, baked in layers between hot stones, called "buried-bread broad *he'-we.*" Stored wafers which became dry and flaky they might turn into a batter and cook again as "double-done *he'-we,*" like zwieback, or they might eat them as chips or flakes. Grandma said that the Hopi liked to crumble piki and eat it with milk and sugar as a breakfast cereal. Like cornflakes.

Cornflakes! The Kellogg brothers no more invented cornflakes in Battle Creek in 1898 than the Hopi invented corn in Oraibi in 898. The century that invented the adman has been conned by its own hokum. We believed our tribal chiefs who said, "Publicity is life and cornflakes the perfect food." In 1986 I participated in a cornflake ritual that condensed the personal breakfast lives of four generations, a ritual performed over the last eighty years by six and a half million people from around the world at the sacred fount of the Kellogg Company and which on April 11 ended forever. It was, as Jeff Weinstein put it, The Last Cornflake Show.

There had been thirty thousand of us that week waiting to tour the cornflake plant, waiting in the anteroom beneath a world map that flagged twenty-two Kellogg's plants in seventeen countries and fifteen languages, from Japanese to Finnish. We stood by the thirty-pound silver National Corn Trophy that bore an enamel portrait of the "Sweetheart of the Corn," the original cornflake girl of 1907, whose image evolved along with the art and science of advertising to put Kellogg's factory version of piki into the mouths of babes in Rangoon and Patagonia. We stood in our hairnets, whole families of us, including bearded men encircled in hairnets from chin to eyebrows, waiting by posters that recapitulated a century's myths of pep, health and digestion, waiting to be touched by the magic of machines that, among the 6 million cereal packages they produced each day, turned 48,700 bushels of corn grits into 70,000 cartons of cornflakes.

My interest was personal. The culinary rites of my family table sprang directly from Battle Creek at the turn of the century, which had enshrined Clean Colons, Purity and Postum, sanctified Horace Fletcher's "thirty-two chews per bite" and elevated colonic irrigation to lustration and atonement. My grandparents rejoiced in the Bible Belt union of health foods and high colonics that paved the way for the union of engineer, preacher and adman in the Kellogg boys. While Dr. John sold health cures at the "San," brother Will sold health food at the factory. They appealed to the puritan need to convert flesh, and the grain that fed it, into the transparent wafer of the Eucharist. Or into Elijah's Manna, as C. W. Post first christened his rival brand of cornflakes before he renamed them Post Toasties.

Fabricating their own myths of origin, the new breed of zealots believed they were inventing what they were simply industrializing. Processing corn grits by

"It is highly necessary that the bowels move freely. If they do not completely evacuate at least two or three times a day you are constipated. It is wise to clean them once or twice a week with an herb enema."

—JETHRO KLOSS,
Back to Eden (1939)

roasting, drying, tempering, flavoring, flaking and retoasting to make a form of transportable, long-keeping and instantly digestible finger-food, in both powdered and wafer form, was of course ancient cooking history. What the industrialists contributed, along with their machines, was not culinary methods but techniques of marketing—the Big Sell.

Since mid-century, health moralists had been pounding cereal grains into granules on their kitchen tables to make them digestible as water. Dr. James Jackson had cracked wheat into "Granula" (made from water mixed with toasted graham flour), which C. W. Post would promote as "Grape Nuts" and Dr. John Kellogg as "Granola." Dr. John had ground wheat into finer granules for his coffee substitute "Caramel Cereal," which Post labeled first "Monk's Brew" and then "Postum." Soon the Kelloggs began to crank a gummy mess of boiled wheat through a pair of rollers on their kitchen table in an attempt to rival Henry Perky's newly patented shredded wheat. In the lucky accident endemic to myths of origin, the Kelloggs "invented" flaking when they forgot a batch of boiled wheat for a couple of days and found that the "tempering," despite mold, resulted in a perfect flake.

Local myths ignored the facts that England had introduced the roller mill to America after the Civil War and that the brewing industry had already been rolling raw corn grits into flakes (Cerealine or Crystal Malt Flakes) to use for malt. When the Kelloggs turned to corn, their homemade flakes were at first thick, tasteless and rancid, until they decided to flake hulled corn with commercial rollers and to flavor it with salted and sugared brewer's malt. In 1898 their Sanitas Toasted Corn Flakes had the edge, but within a decade a cornflake boom had generated 108 rivals, who leapt to exploit the milling industry's "improved" rollers, which automatically degermed corn to remove the oil.

In 1906, after the first of many court battles with his brother, W. K. Kellogg set up his own company and began to exploit the new century's appetite for sugar and showbiz. W. K. was nothing if not a showman. He threw his signature cornflakes onto streetcar signs, shop windows, billboards. He fashioned nine-foot ears of corn and giant cornflake boxes into walking ads. He sent canvassers door-to-door with cases of free samples, snared shopping housewives with gimmicks like "Wink Day" when a wink brought a prize from their local grocer, dazzled the jaded in Broadway's Times Square with the world's largest electric sign, in which a smiling boy declared, "I want KELLOGG'S." In downtown Chicago, he improved the sign when a change of lights on the word "KELLOGG'S" changed a boy crying "I want" to a boy smiling "I got." With the same Barnum hoopla, the current Kellogg Company ended its family tours with brass bands, a cavorting Tony the Tiger and a costumed Rice Krispies trio

A typical Kellogg's ad in 1913. The box displays the Kellogg logo, "The sweetheart of the corn," and advises the consumer to heat before serving. Courtesy Historical Pictures/Stock Montage

handing out balloons in the guise of Snap! Crackle! and Pop! Despite the death of both brothers at ninety-one, still locked in litigation and brotherly hate, the Kellogg name, branded on $3 billion worth of products each year, lives on.

By harnessing mechanized power to old-fashioned flaked corn, the Kelloggs introduced cornflakes to a world that, five centuries after Columbus, still identifies corn with field corn, the food of pigs and peasants but not of civilized tables. "The hardest part of selling corn to Europeans is getting them to taste it," Peter Schubeline said recently. An Americanized Swiss, Schubeline retired from nuclear physics to found the Unicorn Sweetcorn Company in Strasbourg, France, in order "to colonize Europe with sweet corn." Europe is a few centuries late. The reports of early French explorers like Dumont de Montigny fell on deaf ears. Navigating the Mississippi in 1753, Montigny was impressed by the high degree of Indian culture evidenced in their corn cookery: "forty-two styles, each of which has its special name." A century later, however, genteel snobs like Harriet Martineau and Frances Trollope set the European tone of contempt. "Everything eats corn from slave to chick," Martineau complained of the American South. "They eat Indian corn in a great variety of forms," Trollope acknowledged of the American North, "but in my opinion, all bad."

Early colonists in North America struggled to translate corn into wheat because their lives depended on it; but while wheat was simple, corn was complex. Not only was corn both cereal and vegetable, but there were many different kinds of corn, each demanding different preparation, whether raw or cooked, fresh or dried, pureed or popped. A modern European encyclopedia of *The Food of the Modern World* still divides all corn into two parts, *indentata* and *saccarahata:* the first is "maize or Indian corn," used for a cereal, and the latter is sweet corn, used for a vegetable.

Today we know that different Indian tribes grew many different varieties of corn and varied their cooking methods according to type. The Chickasaw named corn types by their cooking uses; the Tewa in the Southwest organized their universe and their palates by corn color; Buffalo Bird Woman, like all Woodland peoples, distinguished corn types by both color and texture, discriminating hard and soft white, hard and soft yellow, and gummy. Roger Williams, searching for *A Key into the Language of America* in 1643, found it in the culinary language of the Narragansett, who used different nouns for dozens of differently prepared corns, not to mention verbs for grinding and parching. Trying to grind Zuni words into English, Frank Cushing noted different terms for corn on the ear, in the grain, cracked, skinned, rubbed, broken into coarse meal, reduced to fine meal, ground to exceedingly fine flour— naming hundreds of corn dishes.

While modern Americans divide corn roughly between sweet corn and cornmeal—between a fresh vegetable and a dried cereal—ancient Americans divided corn by season and ceremony. Their basic distinction was between early-ripe and late-ripe, between corn harvested at the beginning and at the end of the growing season, between the green-corn harvest of early summer and the ripe-corn harvest of the fall. Beyond the realm of season was their ash cooking, which might be called lye or lime cooking.

"I ought here to describe to my English readers what this same Indian Corn *is. . . . It is eaten by man and beast in all the various shapes of whole corn, meal, cracked, and every other way that can be imagined. It is tossed down to hogs, sheep, cattle, in the whole ear. The former two thresh for themselves, and the latter eat* cob *and all. It is eaten, and is a very delicious thing, in its half-ripe, or* milky *state."*

—WILLIAM COBBETT, *A Year's Residence in the United States of America* (1819)

There is virtually no corn preparation today that doesn't have some ancient analogue. Indians, after all, have a recipe backlog of a few thousand years. They roasted, steamed, parched, dried and pickled fresh corn on the cob. They cut kernels from the cob for succotash, soups and stews, just as they "milked" fresh cobs by scraping, grating and pureeing the kernels to make puddings, dumplings and breads. They dried fresh kernels to preserve them whole or parched them to eat as snacks or ground them to make what they called cold meal. Their subsistence crop, however, was ripe corn as a cereal grain, preserved for winter feeding until spring came around again. Their ancient methods of milling corn produced degrees of refinement from coarsely cracked grain to meal ground fine as powder; their methods of preserving produced diversity from roasting to toasting, smoking and parching.

Ash cookery, which was once the major building block of ripe-corn cookery, evolved differently in North America than in Mesoamerica. Just as the first corn plant required man's help in planting, so corn as a staple food required man's help in processing, because unprocessed corn lacks certain nutrients essential to man. Early planters instinctively made up for the protein deficiencies of both corn and beans by planting and cooking them together. Then they discovered "how to unlock the nutritional potential of maize," as Solomon Katz of the University of Pennsylvania describes it, by processing the grain with alkali. The great civilizations of Mesoamerica with their large populations, Katz asserts, could not have been built or sustained without the knowledge of alkali-processed corn.

In Mexico, divided between jungle and desert, the Indians processed corn chiefly with slaked lime. The processed kernels they called *nixtamal,* which they ground wet to make what the Spaniards called *masa* or dried to make *masa harina. Masa* was the dough they turned into their staple breadstuffs of tortillas and tamales. North of Mexico, however, where forests were deep, Indians processed corn chiefly with lye made from wood ashes, to make what we generally call whole-kernel hominy and what our Southwest called posole. Because the North American colonists tended to look on corn as deformed wheat, they misunderstood the nutritional function of ash cooking and saw it only as a form of milling, since the ash bath removed the hulls. In America hulled corn, or hominy, became a minor branch of corn cooking and today is nearly extinct. As soon as they could, our colonists imported their own wheat-milling devices to remove the hulls of corn and did not think to apply chemical processing until they wanted to turn corn into commercial starch.

Their misunderstanding was no different from the rest of the world's. After Columbus, in countries where corn was imported to become a major staple, men suffered the disease we now know as pellagra, because they did not import a knowledge of alkali processing along with the corn. Recently Professor Katz, a "biocultural evolutionist," studied fifty-one Indian societies in the Americas and concluded that all of them that were dependent upon corn had devised some form of alkali processing, to compensate for corn's lack of niacin. Certainly such processing was as ancient as the lime-soaking pots of 100 B.C. discovered at Teotihuacán, a discovery which has led many to believe that corn is the oldest chemically processed grain in the world.

Cooking is by definition an ephemeral language, but we have many witnesses in the past to the glories of the urban cuisines created by Aztec and Inca and to the country cuisines of those farther north. We also have witnesses in the present, women like Grandma Helen Sekaquaptewa and her friend Louise Udall (mother of Congressman Morris Udall) who have tried to bridge the gap between languages and cultures by mutual respect: "I am talking," Grandma said. "She is writing." Even if our texts are garbled, there is still a tie that binds Grandma's piki to Kellogg's cornflakes, Tex-Mex tortillas to New England johnnycakes, popcorn to Cracker Jack and metates to General Mills in a biocultural evolutionary succotash that could have happened nowhere else but here.

GREEN-CORN COOKING

"PEOPLE HAVE TRIED and they have tried," says our *Prairie Home Companion,* Garrison Keillor, "but sex is not better than sweet corn." For the Indians "sweet" meant the first tender cobs that were "green" and "in the milk," rather than a specific sugary breed, and the sexiness of sweet corn was sung and danced at the fertility feasts called Green Corn Busks. The types of corn that were planted to be harvested young, or "early-ripe," varied by tribe and ecology. In the climes of North Dakota, the Hidatsa planted two crops of corn to be harvested green, one that would be ready by the first week of August and another that would be ready in late September when the other corn was getting hard. In far warmer Virginia, the early-ripe corn, as described by Robert Beverley, was of two types: a smaller ear, about the size of "the Handle of a Case Knife," which was ready to eat by mid-May, and the larger ear, "thick as a Child's Leg," ready a fortnight later. Down in Florida among the Chickasaw and Choctaw, the earliest corn harvested was "little corn" or "six weeks' corn," which was a popcorn; then came the soft dent corns, harvested in time for the Green Corn Dance, and finally the flint or hominy corns for late harvesting. In general, each stage of ripeness of each type of corn—which controlled all those variables of size, texture, flavor, hardness, softness, sweetness and starchiness—dictated the way the particular corn was cooked. Buffalo Bird Woman found that hard yellow corn and soft yellow corn made the best "green roasting ears," but she was less choosy about green boiling ears. "For green corn, boiled and eaten fresh, we used all varieties except the gummy," she said, "for when green they tasted alike." (The gummy, ironically, was a type of sugar corn—or sweet corn—that the Hidatsa used exclusively for making "sweet cornballs.")

Benjamin Franklin in the spring of 1785 neatly summarized the differences between green- and ripe-corn cookery when he outlined for the French, suffering famine at the time, the virtues of importing and planting Indian corn. Franklin noted that when English farmers first reached America they were wont to "despise & neglect the Culture of Mayz" because they were Wheat farmers, but that they inevitably turned to Mayz when they observed "the Advantage it affords to their Neighbours, the older Inhabitants":

> 1. The Family can begin to make use of it before the time of full Harvest; for the tender green Ears stript of their Leaves and roasted by a quick Fire till the Grain is brown, and eaten with a little Salt or Butter, are a Delicacy.
> 2. When the Grain is riper and harder the Ears boil'd in their Leaves, and eaten with Butter are also good & agreable Food. The green tender Grains, dried [may] be kept all the Year; and mixed with green Haricots also dried, make at any time a pleasing Dish, being first soak'd some hours in Water, and then boil'd.

He was right thus to summarize the three basic ways green corn was prepared by Indians and settlers alike until our own century: roasting, boiling and

"Corn balls *were a favorite article of food among the Mandans and Hidatsas. . . . One was made of pounded sugar corn mixed with grease. . . . Another kind of corn ball was made of pounded corn, pounded sunflower seed, and boiled beans. It tasted like peanut butter and was called* opata *by the Mandans."*

—GEORGE F. WILL,
Corn Among the Indians of the Upper Missouri (1917)

drying. Although drying was eventually replaced by canning, it was once the chief means of preserving green corn. "Sweet corn is gathered before it is ripe, and dried in the sun," John Bradbury wrote in 1811 to explain the food ways of the Arikara to the English. "It is called by the Americans green corn or corn in the milk." Roasting and boiling, on the other hand, were the ways to enjoy fresh corn-on-the-cob, and even the odd wheat lover was occasionally seduced by its erotic mystique. "Roasting Ears," William Cobbett averred in *A Treatise on Indian Corn* in 1828, "are certainly the greatest delicacy that ever came in contact with the palate of man." If the older inhabitants had the edge on emollients—"When boiled green with rich buffalo marrow spread on it," an English traveler reported, "it is very sweet and truly delicious"—colonists decided that butter was no more a bad substitute for marrow than sweet corn for sex.

Corn-on-the-Cob

"FIRES ARE BLAZING in all directions around which gather merry groups to feast on boiled and roasted ears," Henry Boller wrote of North Dakota tribes feasting at Green Corn time in the 1860s. Whether the ears were steam-baked in an underground pit lined with green husks, or boiled in a bark pot or roasted in the embers, what Indians prized was sweetness. When they roasted green ears they usually husked them first so that the kernels acquired a caramel taste and color in the embers. Buffalo Bird Woman said they would lay a "fresh

Georgia O'Keeffe: A Portrait—Eating Corn, 1920/1922, by Alfred Stieglitz, 1864–1946. National Gallery of Art, Washington, D.C., Alfred Stieglitz Collection

ear on the coals with the husk removed" and roll the ear back and forth to roast it evenly. If any kernels popped with a loud noise, they would joke that it was corn stolen from a neighbor's garden. They would roast ripe corn as well as green until frost, when it lost its freshness, she said, but they had a trick to restore it: "We would take the green corn silk of the same plucked ear and rub the silk well into the kernels of the ear as they stood in the cob."

Settlers adapted the Indian method of roasting corn, husks off, in their hearth fires. Even as late as 1847, Mrs. T. J. Crowen in her *American Lady's Cookery Book* instructed the housewife: "Strip off all the husk from green corn, and roast it on a gridiron, over a bright fire of coals, turning it as one side is done. Or, if a wood fire is used, make a place clean in the front of the fire; lay the corn down, turn it when one side is done; serve with salt and butter." Long after gas and electric broilers had replaced wood hearths, Irma Rombauer followed the same ancient tradition in her 1943 *Joy of Cooking*. To roast corn, husk the ears, she said, spread them with melted butter and turn them under a heated broiler to brown.

Roasting green corn in the shuck, however, had its own traditions. The Iroquois roasted corn in quantity with the husk on, digging a trench for the fire and placing a stick lengthwise over the embers. "The ears of green corn were then leaned against the stick on both sides and turned from time to time until they were roasted," F. W. Waugh observed in 1916, adding that children were warned to eat green corn from the whole cob rather than break it in pieces, lest they be gobbled up by a woodland monster made entirely of legs.

Our modern practices of roasting with or without husks vary from one man's grill to another man's microwave and from one ethnic pocket to another. In

Settlers in Emmons County, North Dakota, in 1915 roast corn in hot coals "Indian style" by rolling the ears back and forth. Courtesy State Historical Society of North Dakota

Roast Corn Man, *Orchard and Hester Streets, May 3, 1938.* Berenice Abbott, photographer. Museum of the City of New York

Alabama, A. L. Tommie Bass, an Appalachian herbalist, prefers to roast his corn not only in the shuck but inside foil as well: "I generally cut off the part that fits to the stalk and one inch off the silk end and wrap it up in foil. You put it in the oven or in ashes. Of course, if you don't use foil, the ashes don't get through the shuck. Most folks boil the corn on the cob, but I can't eat it on account of having no teeth. I takes the corn off." I know a lot of city folk who soak their corn in the shuck for ten to twenty minutes in cold water before putting them on their grills to prevent the husks from burning and imparting a scorched taste. I know other folk like Juanita Tiger Kavena who, Hopi-style, deliberately scorch their husked ears over live coals or a gas-stove flame to make *twoitsie,* or Scorched Corn Stew.

Street vendors around the world husk corn to flavor it before grilling. Iranians soak their charcoal-roasted corn in heavy brine and call it *balal; balal* vendors, I'm told, are as popular in Iranian towns as pretzel vendors in New York or chestnut vendors in Paris. Southeast Asians brush their grilled corn with salted coconut milk, which they call *kao pot ping.* Chinese shake on soy sauce and Mexicans chili pepper. Best of all, Peruvians grill the big-kerneled corn they call *choclo* over roadside braziers, then wrap it in grilled cheese.

Nowadays all this happy diversity goes for naught when it comes to *boiling* fresh corn-on-the-cob. Here even cultural relativists become shrill with authority, myself included, in dictating the One True Way. Dictation begins, as in John Doerper's *Eating Well* (1984), with the demand for speed: "Make sure to have as little distance as possible between the corn patch and the kettle." Indians solved the timing problem by putting the kettle, or earth oven, *in* the cornpatch. Because we learned decades ago that corn sugar begins to convert to starch the moment the ear is plucked, and because the cornpatch for city folk may be a state or a continent away from the kettle, our primary sweet-corn ritual is a mad dash to the kitchen with whatever corn we've found.

Our ancestors were far more particular. A Wisconsin friend remembers that her farming grandmother first stripped back the husk at the top to test the kernels with her thumbnail. If milk oozed out, she threw that corn to the pigs because it was already too *old*. Today, even with supersweet corns, each breed and batch will differ in its ripeness and sweetness. It's wise to nibble a few kernels raw to determine precise cooking times. If there's no starchy taste at all, the less heat the better to retain sweetness and crispness. With the supersweets we've even learned to eat corn raw.

Still, the full erotic thrust of corn-on-the-cob comes only when the corn is at least as warm as human flesh, so a dip into a hot bath will do it no harm if the dip is quick. The formulaic three-minute bath, which has been standard in cookbooks for twenty years, is as antediluvian for modern sweet corn as the hand plow. Worse is the prescription for a three-minute boil *after* the kettle of water has returned to the boil. Worst of all is the advice, from no less an authority than James Beard, to start corn in a skillet of *cold* water and then bring it to the boil. Since heat speeds the conversion of corn sugar to starch, the longer the cooking time the tougher, starchier and soggier the corn. Jasper White, of Jasper's restaurant in Boston, who grew up with corn on a New Jersey farm, recommends no more than a second for the youngest, most sugary ear and no more than a minute for a slightly riper one.

The cooking times for our elders were much longer because their corn was very different. In nineteenth-century cookbooks, twenty-five or thirty-five minutes was standard for flint and dent corns that had a high degree of starch and tough hulls. But cooks of that time followed an Indian precedent for retaining sweetness which we have abandoned: boiling corn in the husk. As usual, Eliza Leslie's *Directions for Cookery in Its Various Branches* (1837) was exact in explaining how to boil Indian corn with the inner husks on:

> Corn for boiling should be full grown but young and tender. When the grains become yellow it is too old. Strip it of the outside leaves and the silk, but let the inner leaves remain, as they will keep in the sweetness. Put it into a large pot with plenty of water, and boil it rather fast for half an hour. When done, drain off the water, and remove the leaves.

Not until the turn of the century did our cookbooks advise housewives to remove all the leaves before boiling. In her 1878 *Dinner Year-Book,* Marion Harland suggested tying the inner leaves with thread around the tip of each ear to present a smooth appearance, but by 1903 she said simply in her *Complete Cook Book:* "Strip husk and silk from the ear and put over the fire in plenty of boiling water, slightly salted." *The White House Cook Book* (1902) explained why the new practice of denuding corn before boiling caught on: "The corn is

The Hot Corn Seller, *1840/1844, a watercolor drawing by Nicolino Calyo, 1799–1884.* Museum of the City of New York, Gift of Mrs. Francis P. Garvin in memory of Francis P. Garvin

much sweeter when cooked with the husks on, but requires a longer time to boil." Since corn is also sweeter when cooked without salt (salt toughens the kernels), many cookbooks advised housewives to add, instead of salt, a little sugar and milk to the water.

The English, for whom Indian green corn remained an exotic novelty, wavered between husking and not husking, but their concerns were often as extra-culinary as William Cobbett's in the voluntaries he performed as Peter Porcupine in *A Treatise on Indian Corn* (1828). To prepare green ears for cooking, he observed,

> there are none of those cullings and pickings and choosings and rejectings and washings and dabblings and old women putting on their spectacles to save the caterpillars from being boiled alive; there are none of those peelings and washings as in the case of potatoes and turnips, and digging into the sides with the knife for the eyes, the maggots and the worms, and flinging away about half the root, in order to secure the *worst* part of it, which is in the middle; none of those squeezings and mashings and choppings before the worthless mess can be got upon the table. . . . Nature has furnished this valuable production with so complete a covering, that washings from the purest water cannot add to its cleanness.

Nature, he assumed, had devised this covering for the sole purpose of sanitation, for he concluded: "The husk being stripped off, it is at once ready for the pot."

In her recipe for "Boiled Indian Wheat or Maize" in her *Book of Household Management* (1861), Mrs. Isabella Beeton called "the ears of young and green Indian wheat" "one of the most delicious dishes brought to table," but her concern was more for the growing than the cooking of it. She regretted that it was so rarely seen in Britain and wondered that it was not invariably cultivated in the gardens of the wealthy. She followed Eliza Leslie in directing that "the outside sheath" and "waving fibres" be removed, but since Beeton mentioned no inner leaves, it's hard to know whether she stripped or retained them. We do know that Eliza Acton stripped them all in her *Modern Cookery* of the same date, asserting without truth that Americans picked and ate their corn "at a less sweet and delicate stage" and therefore needed to retain the inner leaves.

Significantly, it was the French or French-trained who first halved the customary half-hour boil and who exploited the husk for both cooking and serving. "If it seems desirable to strip off the inner husk just before sending to the table," Oscar Tschirky of the Waldorf wrote in 1896, "this must be done very

Jules Starbuck Bourquin, eating corn in Horton, Kansas, 1910. Kansas Collection, University of Kansas Libraries

quickly and the corn covered with a clean napkin or cloth to prevent the escape of heat." Escoffier was more explicit in his 1921 *Guide culinaire:* "Take the corn when it is quite fresh and still milky, and cook it in either steam or salted water; taking care to leave the husks on. When cooked, the husks are drawn back so as to represent stems, and the ears are bared if served whole. This done, set the ears on a napkin, and send a hors-d'oeuvre dish of fresh butter to the table with them." By the time an American edition of Escoffier's *Guide* came out, however, in 1941, American practice had so changed that the editor noted, "In the United States, the silk and husks are removed, and corn-holders are sent to the table with the ears of corn."

The proper way to eat corn had been a subject of dispute from the mid-nineteenth century on. Sam Beeton had noted in his wife's original *Household Management* that he had been introduced to green Indian wheat when visiting America and "found it to combine the excellences of the young green pea and the finest asparagus." That may be why Mrs. Beeton directed that it be "sent to table with a piece of toast underneath, as one might serve asparagus, to absorb any drops of water." Sam, however, had further worries. He confessed that "he felt at first slightly awkward in holding the large ear with one hand, whilst the other had to be employed in cutting off with a knife the delicate green grains." To simply gnaw the cob with one's teeth was unthinkable, and Americans worried the subject according to their pretensions to refinement.

Corn holders for the cob were a late refinement that appeared after Frederick Stokes in 1890 had given a full account in *Good Form: Dinners Ceremonious and Unceremonious* of how and how not to eat corn-on-the-cob:

> Corn may be eaten from the cob. Etiquette permits this method, but does not allow one to butter the entire length of an ear of corn and then gnaw it from end to end. To hold an ear of corn, if it be a short one, by the end, with the right hand, and bite from the ear is good form. A little doily, or very small napkin, is sometimes served with corn to fold about the end of a cob that is to be grasped by the hand, but this arrangement is as inconvenient as it is unnecessary. Good form disallows it.

Whether good form would have disallowed the corn holders Sarah Tyson Rorer recommended in her *New Cook Book* (1902) we do not know, since her concern was less with manners than digestion: "To eat, score each row of grains right through the centre. Then spread the corn on the cob lightly with butter, dust it with salt, and with the teeth press out the centre of the grains, leaving the hulls on the cob. Fresh, carefully cooked corn, eaten in this way, will not

"Some people take the whole stem and gnaw [the kernels] out with their teeth . . . it really troubles me to see how their wide mouths . . . ravenously grind up the beautiful white, pearly maize ears."
—FREDERICK BREMER, visiting America in 1850

produce indigestion." As an afterthought, she mentioned corn holders in the form of small wooden or silver handles, which were pressed into the cob ends, and also wire frames for the same purpose, "rather clumsy, but convenient." Even at that date she warned cooks not to remove husks, for they "prevent the sweetness from being drawn out in the water."

Corn holders were not sufficient to satisfy Emily Post in her *Etiquette: The Blue Book of Social Usage* (1928). "Corn on the cob could be eliminated so far as ever having to eat it in formal company is concerned, since it is never served at a luncheon or a dinner," she warned, "but, if you insist on eating it at home or in a restaurant, to attack it with as little ferocity as possible is perhaps the only direction to be given, since at best it is an ungraceful performance and to eat it greedily or smearingly is a horrible sight."

The original drawing by John White (in 1585–87) for Thomas Hariot's Briefe and True Report of the New Found Land of Virginia *shows a native couple eating corn. "Their meate is Mayz sodden . . . of verye good taste, deers flesche, or of some other beaste, and fishe. They are verye sober in their eatinge, and trinkinge, and consequentlye verye longe liued because they doe not oppress nature." The only visible ingredients here are kernels of cut corn and perhaps beans.* Courtesy of the Collections of the Library of Congress

Milking the Cob

Raw kernels cut from the cob and mixed with a variety of ingredients was the basis of Indian succotash (from the Narragansett word *m'sick-quatash,* with variants *sukquttahash* and *msakwitash*). The word meant "fragments," or the jumble of ingredients that the Spanish call *olla podrida* and American-Chinese call chop suey. The Indian jumble always included corn, for Roger Williams tells us that the Narragansett word also meant "boild corne whole."

Green corn and green beans were natural partners for the pot because they were planted and plucked together, so that succotash suggested spring feasts. John Bartram described a great feast among council chiefs in which the climactic dish was "a great bowl, full of Indian dumplings, made of new soft corn, cut or scraped off the ear, then with the addition of some boiled beans, lapped well up in Indian corn leaves." Cushing called Zuni succotash "the delicacy of the year," made into a kind of thick soup-stew that mingled milky kernels with little round beans and bits of fresh meat, thickened with sunflower seeds or piñon-nut meal.

Since corn and beans were twin staples for reasons of nutrition as well as of taste, succotash became an important American one-dish meal, to which colonists added salt pork when venison or other game was unavailable. By the mid-nineteenth century, salt pork was the third staple ingredient, evidenced in "How to Make Succotash" from the *Western Farmer and Gardener,* a recipe which D. J. Browne included in his *Memoir on Maize or Indian Corn* (1846). Instead of sunflower-seed thickening, the Western Farmer boiled the cobs.

> To about half a pound of salt pork, add 3 quarts of cold water, and set it to boil. Now cut off 3 quarts of green corn from the cobs; set the corn aside, and put the *cobs* to boil with the pork, as they will add much to the richness of the mixture. When the pork has boiled, say half an hour, remove the cobs and put in 1 quart of freshly-gathered, green, shelled beans; boil again for fifteen minutes; then add 3 quarts of corn and let it boil another fifteen minutes.

By the turn of the century the dish had declined into a mixture of salt pork, dried beans and dried corn for winter, seasoned with sugar and thickened with flour. Charles Murphy in *American Indian Corn* (1917), commented without irony: "This old Indian dish was probably served originally without the seasoning or thickening that is used by modern cooks." Fortunately, some of our own modern cooks have resuscitated the dish by returning to fresh ingredients, seasoned with interesting new possibilities like caraway seeds and yogurt.

Wherever corn grows around the world, each locale has added its own particularized fragments to the pot, so that in West Africa we find an "okra-corn mix-up" seasoned with hot Guinea peppers, in East Africa a "coconut corn curry" flavored with poppyseeds and peanuts. A French succotash, sampled in Giverny, mixed sweet corn with grated onion and—hold on to your hats— caviar. A Peruvian one mixed corn with onions and red chilies, thickened with evaporated milk and garnished with hard-cooked eggs. As for sweet kernels raw, I have found them sprinkled over everything from broiled oysters to chicken breasts; I have found them folded into whipped cream as a garnish for fish, layered into salads from lentils to seafood, mixed into crunchy salsas of chilies and tomatillos or into creamy sauces of mustard and mayonnaise. Unexpectedly I even found them, on a hot summer's day in Umbria, mixed into angel-hair pasta in a cream sauce flavored with lemon rind.

The pride of Indian green-corn cooking, however, was neither whole sweet kernels nor whole cobs, but pounded kernels and milked cobs: Indian milk came not from cows or sheep, but from corn. Indians used mussel or oyster shells, or "the half of a deer's jaw with the articular portion or ramus broken off," as Waugh detailed it, to scrape the milky kernels from the cob before they

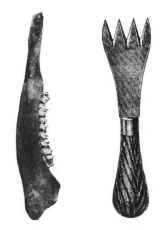

(left) A Seneca scraper for green corn fashioned from the jawbone of a deer. New York State Museum, Albany
(right) A nineteenth-century corn scraper fashioned from iron and wood. The Bettmann Archive

"I prod the nerveless novel succotash."
—JOHN BERRYMAN,
Homage to Mistress Bradstreet (1956)

pounded them into liquid. The Senecan name for a deer-jaw scraper of this sort was "green corn," and using it a Senecan housewife would say, "I am letting the deer chew the corn first for me."

"Ripening green corn, with the grain still soft, was shelled off the cob with the tip of the thumb or with the thumbnail, so as not to break open the kernels," Buffalo Bird Woman said. They then pounded the kernels in a mortar to make such delicacies as *naktsi'*, or "buried-in-the-ashes-and-baked corn bread." They would enclose the pulp an inch or two thick in green husks "overlapping like shingles," then bake the loaf in the ashes. For this bread, no fat or seasoning was added, she said, "the pounded green corn pulp was all that entered into it."

The Seneca called a similar bread "green corn leaf bread," Chief Gibson said, in which the corn might be pounded with beans, berries, apples, chestnuts or venison and enclosed in an "envelope" of corn leaves tied with basswood bark. The Cherokee called it "dogheads," or *di-ga-nu-li*. Two recipes for it appeared as late as 1933 in the brief *Indian Cook Book* put out by the Indian Women's Club of Tulsa, Oklahoma. Despite echoes of "hush puppies," the name "dogheads" seems to come from the squared shape of bread wrapped like tamales in green husks. The Zuni made a similar green-corn dish, mixing the pulp with pounded green squash and sunflower seeds, then boiling the mixture until "thick and gelatinous like curdled milk." However unappetizing Cushing's description, he declared that "no green-corn cookery of civilization can boast anything more delicious."

Such were the native origins of colonial creamed corn and green-corn pudding, which *The Genesee Farmer* of 1838 acknowledged were "derived from the tawny natives among whom our ancestors squatted down two hundred years ago." To enrich their green-corn "milk" from pounded corn, colonists added cow's milk, just as Indians had added "hickory cream" from pounded nuts. ("It is as sweet and rich as fresh cream," William Bartram had reported, "and is an ingredient in most of their cookery, especially homony and corn cakes.") For proper milking of the cob, instructions were exacting, as in the *The Buckeye Cookbook* (1883): "Cut with a sharp knife through the center of every row of grains, and cut off the outer edge; then with the back of the blade push out the yellow eye [the germ], with the rich, creamy center of the grain, leaving the hull on the cob." For "stewed corn" the *Buckeye* author added to a quart of this corn "milk" a cup of rich cow's milk, seasoned with salt, pepper and butter, and then steamed in a covered tin pail inside a kettle one-third full of boiling water.

Housewives soon found that grating corn was much quicker than slicing and squeezing, and corn graters became a standard kitchen item, for a grater could be made simply by pounding nails through a board in even rows. A friend from

Norfolk, Connecticut, John Barstow, tells me that his family has been grating corn this way for three generations, replacing graters as needed in high corn season. They found that plunging the cobs first into boiling water and then into cold water helps to set the milk and make the kernels easier to grate.

For settlers accustomed to the puddings, custards and thick sauces created by their wheat and dairy products, corn pulp had the advantage of being both thickener and liquid at once, with the added kicker that it was naturally sweet. Sophisticated urbanites like Eliza Leslie in 1837 turned the basic green-corn pudding into a rich baked custard, adding to twelve grated ears of corn a quart of rich milk, a quarter-pound fresh butter with sugar and eggs, and served with a sauce of whipped butter, sugar and nutmeg. As the century proceeded, the pudding was converted into "a veritable soufflé and incomparable," as Marion Harland said in 1903, but by the 1920s canned creamed corn had replaced freshly grated in Fannie Farmer's *Boston Cooking-School Cook Book,* and her soufflé was thickened with flour. Even Escoffier, in his 1934 *Ma Cuisine,* made corn soufflé with a can of creamed corn, noting, "If fresh corn is not on hand, excellent canned or frozen kinds are to be found on the market."

Southerners had their own ways with creamed corn. What they called fried corn was simply "young, tender green corn" that is "milked" from the cob, as *The Picayune's Creole Cook Book* (1901) said, and mixed Creole-style with browned minced onions or country-style with ham fat or butter and cooked in a skillet until thickened. Sarah Rutledge in *The Carolina Housewife* (1847) showed how readily colonists turned Indian green-corn "breads" into English "pies." One of her favorites was "green-corn pie with shrimp," in which she lined a baking dish with a batter of corn pureed with eggs and tomato pulp, then filled it with shrimp—or stewed chicken, veal, ham "or whatever you choose to make your pie of"—and topped it with another layer of batter.

Southerners were especially fond of turning green corn and eggs into a batter to make fritters, frequently called "corn oysters" because of their shape. "Grate the corn while green and tender," Sarah Rutledge advised, before adding egg yolks and whites beaten separately, thickened slightly with butter and flour, seasoned well and shaped into "oysters" with a spoon to fry in hot butter. New Englanders, on the other hand, chopped real oysters into their batter, mixing them with pureed kernels in "Pagataw pancakes." Texans didn't bother cutting off the kernels. They simply battered the cob whole and deep-fried it like chicken-fried steak.

There were dozens of variations on the fritter theme of battered corn, sometimes sweetened for dessert and sometimes peppered as a savory accompaniment to meat. Their continuing popularity inspired Irma Rombauer in her

"I remember my mother's fried corn. She cut it off the cob real thin, and she'd put her some butter in one of those old iron skillets and put the corn in that hot butter and just stir it till it thickened. Lots of times she'd put it in the oven and it would brown on the top and the bottom. There's nobody could cook it like she did."

—INEZ TAYLOR, quoted in *The Foxfire Book of Appalachian Cookery* (1984)

1943 *Joy of Cooking* to include an account written by a man whose father had told his children "of a fritter tree he was going to plant on the banks of a small lake filled with molasses, maple syrup or honey, to be located in our back yard. When one of us children felt the urge for the most delectable repast, all we had to do was to shake the tree, the fritters would drop into the lake and we could fish them out and eat fritters to our hearts' content." Today corn fritters fall from fritter trees in most countries of the world, flavored with cumin and coconut in India to make *makai nu shaak,* with minced leeks and carrots in Korea to make *kwang-ju,* and with chilies and ground shrimp in Indonesia to make *perkedel djagung.*

Corn soups and chowders also exploit the double blessing of green corn as both vegetable and thickener. The simplest green-corn soup was furnished by the Westminster Presbyterian Church ladies of Minneapolis in *Valuable Recipes* (1877): "Two quarts of milk, ten ears of corn scraped down, season highly." Today the same simple formula of liquid, scraped corn and seasoning appears in the sophisticated bowls of Alice Waters's Chez Panisse as "corn soup with garlic butter," or of Stephan Pyles's Routh Street Cafe as "grilled corn soup with ancho chile and cilantro creams." As for chowders, corn appears in an astonishing number of them, including seafood and shellfish chowders of New England and chile and tomato chowders of the Southwest.

The classic continuum in green-corn cookery was what was called interchangeably creamed corn and stewed corn, and here the white man reversed Indian tradition. Since Midwesterners canned whole-kernel and creamed corn as soon as they were farmed, cookbooks extolled the can. In recommending "stewed corn" for the month of October in her *Dinner Year-Book* (1878), Marion Harland advised, "Green corn, even in city markets is both indifferent and dear at this season. We do better, therefore, to fall back upon the invaluable canned vegetables that have made American housewives almost independent of changing seasons." Not independent, however, of rituals more appropriate to fine wines than canned corn: "Open a can of corn one hour before it is to be cooked," Harland said, even though she proceeded to cook the contents, in hot water to cover, for thirty minutes before adding milk, butter and flour and cooking fifteen minutes more.

Although French exponents like Ranhofer and Escoffier enriched with béchamel sauce their "corn stewed with cream," their *"crême de maïs, dit Washington,"* their "cream of green corn à la Hermann" (fancified with chicken forcemeat, egg yolk and butter) and their "cream of green corn à la Mendocino (with crawfish butter and shrimps), they did not eschew the can. Nor did Irma

Corn canners in New England scrape raw kernels from ears of green corn picked on the site. The Bettmann Archive

Rombauer, who used canned and fresh corn interchangeably in *Joy of Cooking,* giving preference to the No. 2 can. The craft of making a fresh green-corn pudding "a luscious dish" is not lost but made difficult, she said, "because corn differs with the season." Picked early, what the Indians called early-ripe, the corn may be so watery as to require flour; picked late, what the Indians called late-ripe, additional cream may be needed. When corn is just right it should look, after grating, "like thick curdled cream." Buffalo Bird Woman herself could not have been more precise.

Contemporary logo of John Cope, driers of sweet corn in Rheems, Pennsylvania.

Corn Dried, Pickled and Parched

CANNED CORN must have been a boon to prairie households longing for the sweetness of summer in mid-December. Indians had assuaged this longing by the art of drying, an art that lingered into the twentieth century but finally succumbed to the can. Today, sweet corn is dried in quantity only in the Southwest and in Pennsylvania Dutch country. In Rheems, Pennsylvania, four generations of the Cope family have been drying corn, first over wood-burning stoves and now in high-tech dehydrators, but John Cope's Food Products is the only supplier of any size left in the States. And that's a pity, for drying intensifies corn's sweetness and lends it a delicious caramel taste.

"Every Hidatsa family put up a store of dried green corn for winter," Buffalo Bird Woman said, and the first step was to pick the corn when the husks were a particular shade of dark green. "Sometimes I even broke open the husks to see if the ear was just right; but this was seldom, as I could tell very well by the color and other signs I have described. . . . Green corn for drying was always plucked in the evening, just before sunset; and the newly plucked ears were let lie in the pile all night, in the open air." In the morning she brought them to the lodge, where kettles of water boiled over the coals of the fire. After husking the ears—"a pretty sight; some of them were big, fine ones, and all had plump, shiny kernels"—she dropped them into the kettles and, when the corn was half cooked, lifted the ears onto the green husks "with a mountain sheep horn spoon" and then laid them in rows on the drying stage to dry overnight.

The second morning, she laid a skin tent cover on the floor of the lodge and began to shell the corn. "I had a small, pointed stick; and this I ran, point forward, down between two rows of kernels, thus loosening the grains." She shelled the rows one at a time with her thumb or loosened them with the stick to "shell them off with smart, quick strokes of a mussel shell held in my right hand." The kernels were spread on the drying stage outside, covered at night with old tent skins, and left about four days until they were fully dried. Finally she winnowed out the chaff and put the kernels in sacks for the winter.

The Hidatsa word for dried corn meant "treated-by-fire-but-not-cooked." The method of half cooking the corn by boiling or roasting was practiced everywhere; Arikara and Huron women also roasted their green ears before a fire, or in sand heated by fire, as Gabriel Sagard noted in *Le Grand Voyage du Pays des Hurons* (1632). They shelled and dried the kernels on bark sheets, then stored the corn with a quarter-portion of dried beans, ready to boil in the caldron with a little fresh or dried fish. As open fires evolved into kitchen stoves, tribes roasted and dried their corn indoors. Shawnee tradition held as late as 1932, when Roberta Lawson in *The Indian Cookbook* explained how she dried a pan of scraped kernels in a slow oven and baked them for an hour before crumbling the mass on a "drying board in the sun" to complete the drying.

The best corn for drying was late-ripe sweet corn, rather than early-ripe. Juanita Tiger Kavena, in *Hopi Cookery* (1980), specified that "the later crop" of sweet corn is the one to bake and dry on the cobs for winter use. Baking helps the corn retain its nitrogen, potassium and minerals, she claimed, while storing the whole cob helps protect it from insects. That settlers struggled to learn the right stages of early- and late-ripe is clear from Peter Kalm's journal of November 4, 1749, while he was traveling in New York state: "From Mrs. Vischer with whom I had lodgings, I learned one way of preparing corn.... They take the corn before it is ripe, boil it in a little water, allow it to dry in the sun and preserve it for future use.... The younger the corn is when picked, the better it is provided, however, that it is not too young."

In the Southwest, Spanish settlers called dried green corn *chicos*. Cleofas Jaramillo recalled how they made chicos at her family hacienda in Arroyo Hondo, New Mexico, almost a century ago: first they roasted white unhusked

Drying corn in Bismarck, Dakota Territories. Photo by D. F. Berry, State Historical Society, North Dakota

corn in the oven, then shelled and cracked the kernels on a metate and finally tossed them in a basket "to allow the air to blow away the skin." When ready to eat the corn, they would soften it and mix it with onions, garlic and red chilies fried in lard, or add it to stews with beef or venison jerky. In Peru, Spaniards developed one of the glories of criollo cooking from the dried purple corn that gives its name to the dish *mazamorra morada*. Boiled in water until the liquid is as royally purple as the corn, the corn is discarded and replaced with a variety of dried fruits, while the liquid is thickened with sweet-potato starch, sweetened with the dark sugar called *chancaca* and flavored with cinnamon, clove and lemon.

As late as 1908 drying corn was still so commonplace in American farm households that when Sidney Morse compiled his *Household Discoveries* he listed four different ways "To Dry Corn":

Cut the corn raw from the cob and dry it thoroughly in pans in an oven. This gives a finer flavor than when it is partly boiled.

Or dip green corn on the ear in boiling water, remove, and hang up the ears until dry in a room where there is a free circulation of air.

Or . . . place the ears in a colander over a kettle of steaming water, and steam a half hour or more. Split the kernels with a sharp knife, scrape out the pulp and dry it on clean tines or earthenware platters. . . .

Or shave off the kernels with a sharp knife, scrape the remaining pulp from the cobs, and lay on earthenware platters. Sprinkle ½ teacupful of sugar to each 3 quarts of corn, stir well and place in a medium hot oven for ten minutes, but do not scorch or brown it. . . .

It should be dried as quickly as possible as it deteriorates with exposure. Store in tight jars or boxes in a dry place. When required for use soak it in lukewarm water.

Morse gave equally detailed instructions for canning corn, but they are so tedious as to suggest why the commercial product was popular. Clearly both the new-fashioned canning methods and old-fashioned drying methods were practiced simultaneously for some time.

Methods of preserving other sorts of vegetables and fruit were also applied to corn. Charles Murphy in his *American Indian Corn* (1917) suggested smoking whole cobs in a tight wooden box with sulphur, as one smoked apricots to preserve their color, before immersing the cobs in a large crock of water and sealing. A more common practice was to pickle corn in layers of salt in a stone jar as one might put up cabbage to make sauerkraut. A nice recipe for pickled corn, collected in *Pioneer Cookery Around Oklahoma* (1978), affords a glimpse of the pioneer life of one Annie Henthorn, born in 1864, who advised her fellow corn picklers to keep "the crock jar" in a cool place like "the back bedroom," where in winter you can "push the ice off it" to dip out corn as needed.

A curious combination of pickling and canning appears in *The Picayune's Creole Cook Book* under the heading "canned corn," or *Du Maïs en Conserve*. First boil the corn on the cob "until no milk will exude if the grains are pricked with a needle." Then cut the kernels off and pack them in cans or jars in two-inch layers of corn alternating with one-inch layers of salt. Pour melted lard over each can and heat the cans to 80 to 100 degrees in a bain-marie, hermetically sealed.

Everywhere ladies pickled corn in the vinegar-sugar preservative characteristic of the sweet-sour relishes labeled "chowchow," in which corn kernels mingled with many other vegetables. Typical is a relish found in the Official Recipe Book of the *Patriotic Food Show* held in St. Louis, Missouri, in 1918, which called for twelve ears of corn, four cups cabbage, four green bell peppers, three cups chopped onions, one cup sugar, two cups white vinegar and one cup flour, seasoned with salt, turmeric, Tabasco, celery and mustard seed, the whole boiled together.

Parched dried corn was a branch of cooking unto itself and one of the most important for the Indian because parching cooked, flavored and preserved corn simultaneously. Most Americans today know parched corn only through whole-kernel snacks like "corn nuts" or popcorn, but Indians pounded the parched kernels into a fine powder to make their all-important "journeying" bread or "traveling food." The high culinary art by which they created "a highly concentrated and nutritious substance, capable alone of sustaining life throughout extended journeys," as Frank Cushing said, as easy to transport as it was quick and easy to prepare, might give pause to those who identify high cuisine or high culture exclusively with their own time and arts.

Both early- and late-ripe corn were used, but especially favored was "the little corn of the nature of popcorn, which was the first to mature." Benjamin Franklin in explaining parched corn in his "Mayz" pamphlet was clearly describing a popcorn of this type:

An Iron Pot is fill'd with Sand, and set on the Fire till the sand is very hot. Two or three Pounds of the Grain are then thrown in and well mix'd with the Sand by stirring. Each Grain bursts and throws out a white substance of twice its bigness. The Sand is separated by a Wire Sieve, and return'd into the Pot, to be again heated and repeat the Operation with fresh grain.

At the end of Prohibition, the parched dried corn now called "Corn Nuts" was originally sold as "Brown Jug," the proper snack to go with beer.

That which is parch'd is pounded to a Powder in Mortars. This being sifted will keep long fit for Use. An Indian will travel far, and subsist long on a small Bag of it, taking only 6 or 8 Ounces of it per day, mix'd with water."

This is the meal the Narragansetts called *nokehick,* of which Roger Williams had personal experience denied to the urban and urbane Franklin:

Nokehick: Parch'd meal, which is a readie very wholesome food, which they eate with a little water, hot or cold; I have travelled with neere 200 of them at once, neere 100 miles through the woods, every man carrying a *little Basket* of this at his *back,* and sometimes in a hollow *Leather Girdle* about his middle, sufficient for a man for three or four daies.

With this readie provision, and their *Bows* and *Arrowes,* are they ready for *War,* and *travell* at an *houres* warning. With a *spoonfull* of this *meale* and a *spoonfull* of water from the *Brooke,* have I made a good dinner and supper.

The Seneca called it "burnt corn" and the Natchez "scorched corn." Antoine Le Page Du Pratz in his 1758 *Histoire de la Louisiane* reported in detail how the grain was ground and then "scorched in a dish made expressly for the purpose," where it was mixed with ashes and shaken until it attained "the red color which is proper." Once the ashes were removed, it was pounded in a mortar with the ashes of dried beans until reduced to a coarse meal. The meal was then crushed finer and dried in the sun. To help preserve it, they exposed the meal to the sun from time to time, and when they ate it, they mixed it with twice the amount of water.

Parching corn in a vessel made for the purpose was the earliest form of pot cooking in the Americas. Pre-Inca Peruvians devised a pot with a small opening on one side and a handle on the other to shake over the fire so that the kernels would pop but not pop out. When iron replaced pottery, the Zuni used a black shallow roaster, according to Cushing, which they filled with dry sand and heated over the fire. They stirred corn kernels in with a bundle of hardwood sprigs until "a judicious shaking of the toasting vessel brought all the kernels to the top, whence they were easily separated from the sand." The hot sand

"Americans have never taken a proper pride in this great native cereal Indian corn. . . . I can personally speak of the sustaining power of Indian corn, for in the early days of our war of the rebellion I was a prisoner of war in Richmond, being captured at the battle of Bull Run. I was fortunate enough to be one of the first three officers who escaped from that prison. . . . We were eleven days reaching the Potomac River, and during all that long and dreary tramp through the woods and swamps of Virginia, subject to intense excitement, we had nothing to eat but raw corn, gathered at long intervals, which fortified us in our race for liberty."

—CHARLES MURPHY,
American Indian Corn (1917)

both cooked the kernels evenly and prevented them from popping out. (The use of such vessels was not limited to the Americas. A modern anthropologist in Upper Burma found Chin tribes there in the 1940s parching corn in a *pei-bung,* "the traditional grain-roasting pot, shaped rather like a beer pot but with a hole in one side." From this they made a type of "journeying corn" from grain that had been dried and soaked, then roasted and pounded with salt or honey to carry on hunting parties.)

While parching in itself imparted a deliciously sweet and nutty flavor, often other seasonings were added. Sometimes the Zuni parched their corn in heated salt rather than sand, Cushing said, and sometimes they flavored the ground meal with pounded and fermented licorice root to make a sweet gruel that was "ever the favorite at the Zuni evening feast." The Iroquois similarly flavored their parched meal with maple sugar or dried cherries, and we know that the Inca flavored their journeying corn with pounded dried vegetables and hot peppers. So tasty was journeying corn that J.G.E. Heckewelder, reporting in 1822 the *History, manners, and customs of the Indian nations which formerly inhabited Pennsylvania and the neighbouring states,* issued a warning: "Persons who are unacquainted with this diet ought to be careful not to take too much at a time, and not to suffer themselves to be tempted too far by its flavor; more than one or two spoonfuls at most at any one time or at one meal is dangerous; for it is apt to swell in the stomach or bowels, as when heated over a fire."

Akin to parched corn was the smoked corn Dumont de Montigny found among the Natchez of Louisiana in 1753. But the Natchez favorite was dried green corn that was parched and pounded to make *farine froide,* as Montigny called it, cold meal. The Indians called it *boota copassa* and its purpose was clear from the names they gave their lunar months: the third moon (May) was "Little Corn," the seventh moon (August) "Great Corn" and the eleventh moon (January) "Cold Meal."

Dehydrated powdered corn was for the Indians both food and drink. *"Sagamite"* was the word the Iroquois used, recorded by a French Jesuit father in 1638–39: "As for drinks, they do not know what these are—the sagamite serving as meat and drink." *"Pinole"* was the Aztec word as recorded by Bishop Diego de Landa in 1566: "They also toast the maize and then grind and mix it with water into a very refreshing drink, putting into it a little Indian pepper or cacao." Since they weren't accustomed to drinking water straight, he noted, they added water to maize as it was being ground, cooked and flavored in various ways during the day, and so had food and drink at once.

Such drinks are still common today throughout Central and South America and in our own Southwest, generally under the names *pinole* and *atole.* Cleofas

Jaramillo gave a typical recipe for *pinole,* saying it tasted something like malted milk: "Toast to golden brown a cup of *maiz concho,* dry sweet corn. Grind into fine powder. Dissolve in fresh milk, sweeten with maple sugar, and mix." Phyllis Hughes, in her *Pueblo Indian Cookbook* (1977), mixed toasted and ground blue or white corn with cinnamon and sugar in hot milk. Richard and Wendy Condon, in *!Ole Mole!* (1988), recorded an exotic variant made with toasted cocoa, dark-brown sugar and dried jasmine flowers, all beaten into cold water until foamy, the foam then scooped off and mixed with hot corn *atole.* In the Southwest, *atole* more often than not refers to toasted and finely ground blue-corn meal, whether it is dissolved in water to make a mush or a beverage. Because it is easily digestible, it has remained a favored nourishment for invalids, and was once prescribed as a "very light and nutritious" drink for the sick by a New Englander, Maria Parloa, who called it "corn tea."

ASH COOKING

Hominy

Hom-in-y man, hom-in-y man,
Com-in' around on time!
Hom-in-y man, hom-in-y man,
Com-in down the line!
Two quarts for fif-teen cents,
One quart for a dime!
Hominy! Beautiful hominy!

EARLY IN THE CENTURY, the hominy man sang his cantos of mutability throughout the land. From the days of his youth in Portland, Oregon, James Beard remembered how "the hominy man passed through our neighborhood in his little horse-drawn cart twice a week," selling horseradish as well as hominy. Near Portland, Maine, Arnold Weeks remembered the hominy man in "his horse-drawn wagon driving through Town selling his hulled corn to the housewives by the quart." In New Orleans, the *Picayune* remembered him as the "Grits man," who was "as common a figure in the streets of the old French quarter as the 'ring man,' the 'bottle man,' or the 'Cala woman.' " He had a great tin horn, nearly three feet long, the *Picayune* recounted, at which sound housekeepers rushed to their doors to bargain for their dainty breakfast cereal. But even in 1901 the Grits man, like the Cala woman, was "fast becoming a memory of other days."

Certainly few Americans today under the age of thirty have any memory of hominy, either in the form of Southern grits or Southwestern *posole* or in the canned form that was a staple of the Depression 1930s but is now a supermarket rarity. The kernels of the hominy we bought were always white, about the size of chickpeas but slightly softer, packed in water turned chalky from starch. We heated them in their liquid with butter and salt and served them instead of rice or potatoes or canned creamed corn. To a family of Fletcherizers they were a boon. They were firm but tender and a snap to chew and swallow since the kernels had no skins. In those days skins were thought to offend delicate constitutions.

When corn was the year-round staple for the Indians, no corn was referred to so often as hominy, nor so confusingly. Each Indian tribe called it by a different name, or by several names: what was *rockahomonie* and *sagamite* to the Algonquian was *atole* to the Mexica, *sofki* to the Creek, *tanlubo* and *tafala* to the Choctaw. English colonists muddled words and meanings when they lumped under the single word "hominy" any number of preparations made from the fully mature kernels of any late-ripe corns but principally of the thick-skinned flints that Northern tribes called "hommony corn." ("Iroquois Hominy Corn" was eight-rowed, fourteen inches long by 1½ inches round, and creamy white in color.) Perversely, French colonists preferred the word "sagamite," which Louisiana Creoles transmuted into *la saccamité*, so beloved in Creole households that grandmothers would say to the young and restless, "Tempi, pour toi! La Saccamité te raménera!"—"Never mind! Hominy will bring you back!"

A farm family and hired hands in Illinois fill up on a bowl of hominy or grits. Photo by Grant Heilman Photography, Inc.

Colonists of all languages struggled mightily to translate the Indian dishes made with this corn into the thick wheat gruels, pottages or puddings they called frumenty, loblolly and hasty pudding. What wheat-conditioned colonists failed to see was that how the corn was hulled was as important as how it was cooked. Indians distinguished between two kinds of hulled corn: one was made of kernels broken into coarse pieces in a mortar, then cleared of skins by winnowing; the other was cleared of skins by boiling the kernels in lye. Traveling in Virginia in 1849, William Strachey detailed the first process of winnowing and described the resulting pottage: "The growtes and broken pieces of the corne remayning, they likewise preserve, and by fannying away the branne or huskes in a platter or in the wynd, they lett boyle in an earthen pott three or four howres, and thereof make a straung thick pottage." The method was unchanged a half century later when the anthropologist Alanson Skinner took "Notes on the Florida Seminole" (1913):

> The meal [once taken from the mortar] is first sifted through an open-mesh basket and then winnowed by being tossed into the air, the breeze carrying away the chaff, while the heavier, edible portion of the corn falls back into the flat receiving basket. In this condition the meal is mixed with water and boiled to make sofki.

The second method was what settlers eventually called lye hominy, in which the hull was loosened and removed by the chemical action of wood ash or some other alkali. Anthropologists were no wiser than colonists in assuming that the ashes were there principally for flavoring or for softening the hulls to make them easier to remove. "The ripe corn is usually hulled for cookery purposes," Waugh wrote of Iroquois hominy in explaining the function of lye. "The first step in hulling is to add sifted hardwood ashes to a pot of water in the proportion of about one double handful to three quarts of water." Once the lye is dissolved by boiling and is strong enough "to bite the tongue when tasted," the corn is boiled until it swells and the skin can be slipped off when pressed between the fingers. The corn is then drained in a washing basket and washed in several tubs of water or under a running stream until the hulls are all removed, "a process which is assisted by friction against the twilled sides of the basket and by rubbing with the hands."

Eastern settlers hulled corn by both methods after cracking and pounding their corn in the hollowed log mortars and wooden pestles they called interchangeably "hominy blocks" and "samp mills." But throughout the nineteenth century, American cooks north and south labored valiantly, and hopelessly, to squeeze the rich nomenclature of native corn dishes into the narrow confines of hominy, samp and—worst of all—grits. Anglo-Saxon *grytt* for bran and *greot* for ground had melded in "grist," which colonists applied generically to dried, ground and hulled grain. The New Orleans *Picayune* only confused matters when it called hominy "the older sister of grits," since it was the Indians who taught Creoles to thresh the hulls from dried yellow corn until the

"Ginny cracked corn and I don't care
Ginny cracked corn and I don't care
Ginny cracked corn and I don't care
My master's gone away."
—American folk song from the South

"My squaws have fine mat, big wigwam, soft samp."

—JOHN GREENLEAF
WHITTIER,
Passaconaway (1866)

"Grits is grits. There's not a whole lot of range to them."

—ROY BLOUNT, JR.

grains were white. Grits might be yellow if the hull was left on, the *Picayune* specified, but "the daintier preparation" was white with the hull off. Plain hulled corn was "big hominy"; grits ground superfine were "small hominy," "without which no breakfast in Louisiana is considered complete." Lye hominy, on the other hand, was identical with samp as "an old-fashioned Creole way of preparing hulled corn," that is, with the lye from wood ashes. The virtue of lye hominy, the *Picayune* claimed, was that the alkali reduced the oil in the corn and therefore made samp in hot Louisiana summers "a splendid summer food."

In the North, samp (from the Narragansett *nasaump*, or unparched corn, beaten and boiled) came to be identified with coarsely ground corn however it was hulled. Thomas Hazard, already nostalgic in the 1870s for the hasty puddings and pottages of his Rhode Island childhood, wrote lovingly of "the samp—coarse hominy pounded in a mortar—and the great and little hominy, all Indian dishes fit to be set before princes and gods." And best of all, the hulled corn of old—"None of your modern tasteless western corn, hulled with potash, but the real, genuine ambrosia, hulled in the nice sweet lye made from fresh hard oak and maplewood ashes."

Grits for many reasons became strongly identified, as they are today, with the South. When Eliza Leslie of Philadelphia confronted corn terms in her *New Receipts for Cooking* (c. 1854), she defined samp as skinned corn pounded fine, hominy as "perfectly white" corn skinned with lye, and Carolina grits as "small-grained hominy," sometimes called "pearl hominy." This apparently was the white grits ground superfine that the *Picayune* called small hominy and that the rest of the country began to call hominy grits—everywhere but in Charleston, South Carolina, that is, where the phrase *hominy* grits can still provoke otherwise sweet-mouthed ladies to profanity. There grits are grits, or as a recent cookbook title has it, *Grits R Us.*

In the backwoods South, in the bayou Deltas, in the hills of Appalachia, in the garden patches of former slaves, the rural poor continued to grow and cook corn in Indian ways. "Transforming dried corn into hominy seemed a most miraculous process" to Edna Lewis, growing up in Freetown, Virginia, watching her Aunt Jennie Hailstalk make hominy in the big iron pot on the back of the cookstove. "In the old days two-thirds of the people were raised on hominy," Tommie Bass remembers of Appalachian Alabama. The best hominy was made with Hickory King corn and hickory ashes. "The hominy was actually healthy for you because lye," Tommie said, "would clean your liver off, and your stomach." You had to be careful, however, with your wood ash. "Mother wouldn't let Daddy around the fireplace when she was fixing to make lye, because she was afraid he would spit in the fire when he was smoking!" You had also to be careful with store-bought lye. "Folks began to use Red Devil lye instead of ash. Once the Red Devil people told you how to make hominy, but not now. The Red Devil lye is a poison. It will kill you right dead, if you take too much."

A favorite Southern dish was hog and hominy, a colonized version of the universal Indian dish described by William Biggs when he was captured by the Kickapoos in 1788: adopted by the tribe, Biggs was given an Indian bride who made a wedding dish of *"hominy,* beat in a mortar, as white as snow, and handsome as I ever saw, and very well cooked. She fried some dried meat, pounded very fine in a mortar, in oil, and sprinkled it with sugar." Gentrified, the dish became New Orleans' grillades and grits (pounded smothered steak, fried with onion and tomato and served with grits on the side) and Charleston's grits and liver pudding, a favorite Sunday-night supper of calf's liver with grits on the side. Liver pudding is a name hung over from the days when women made scrapple of cornmeal mush and "the head, liver and feet of a hog," as Lettie Gay said in *200 Years of Charleston Cooking* (1930). Gay put grits or hominy with everything, including a delectable recipe for butter-sautéed shrimp and hominy recommended by "a charming old gentleman seventy-eight years old [who] avers that as far back as he could remember he has eaten shrimp with hominy for breakfast every morning during the shrimp season and shrimp salad for supper every Sunday night."

Outside the South, the status of hominy and grits steadily declined, except for a brief resurgence in the 1890s. Recently hog and hominy has acquired new chic in the form of pozole (or posole), a Mexican version of Indian corn and game still so popular south of the border that every region has its own specialty. In Guerrero, from Acapulco through Taxco, Thursday is "green *pozole* day," Rick and Deann Bayless have told us in *Authentic Mexican* (1987). They define pozole as "half-cooked hominy" made from dried white field corn, boiled in slaked lime, washed and then "deflowered" by picking out the small, hard, pointed germ at the bottom of each kernel so that the kernel will splay out, or "flower," as it cooks. Green pozole is green from a sauce of pumpkin seeds, tomatillos, green chilies and herbs. Pozole Guadalajara-style is one familiar to our own Southwest, made with hominy, pork and chicken, flavored with garlic and ancho chilies and garnished with salsa and

Advertisement for State Fair.
Photo by Holmboe Studio, State Historical Society, Bismarck, North Dakota

cheese. While ideally this dish is as hearty as a French *pot-au-feu,* I have had many a pozole best described by Herb Walker's "Indian Rock Soup," which calls for hominy plus vegetables and three or four smooth clean rocks the size of potatoes to be boiled in a kettle, the rocks later washed and saved for future use.

One Indian hominy use that has almost completely disappeared is a baked hominy bread. Curiously, this is the one corn dish of which Mrs. Frances Trollope approved: she wrote that hominy flour (by which she meant white hulled corn finely ground) "mixed in the proportion of one-third, with fine wheat, makes by far the best bread I ever tasted." Southern recipes often mixed wheat and rice flours with boiled hominy and eggs to make a "custard-bread" they called "owendaw." In *The Carolina Housewife,* Sarah Rutledge mixed "two tea-cups of hot hommony" with butter or lard, eggs, milk and cornmeal; when cooked, she said, it looked like "a baked batter pudding" and tasted like a delicate custard. A century later the recipe resurfaced in the Junior League's *Charleston Receipts* (1950) as "Mrs. Ralph Izard's Awendaw," but during the interval many mixtures of hominy and grits with other corn and wheat meals appeared in such works as Célestine Eustis's *Fifty Valuable and Delicious Recipes Made with Corn Meal for 50 Cents* (1917)—"Pour les Pauvres." The ancient ancestry of Eustis's recipes might suggest that Indian palates were no less discriminating than the palates of the poor.

Nixtamal

HOWEVER COMPLEX the hominy cooking of the North, it was rough country cooking compared to the cuisine of Mexico that reached its apex in Moctezuma's court. The very name Moctezuma means "elite Aztec food," and the lords of Moctezuma, whatever else they did, ate royally. If sauce was the root of France's royal cuisine, bread was the root of the Aztec's, the twice-cooked bread of myriad shapes and kinds made from the highly processed corn called *nixtamal.* In Nahuatl, the language of the Aztec, *nextamalli* meant tamales of maize softened in wood ashes and *tenextamalli* meant tamales of maize softened in lime. Of both kinds there was a plenitude that put the northern world of hominy, samp and grits to shame.

Ordinary folk survived frugally on a diet of maize gruel, with the plainest of tortillas and tamales mixed with chili and beans, but the noble and rich feasted with an extravagance that the Sun King at Versailles might have envied. Every day, Bernardino de Sahagún wrote in his *History,* the majordomo set out for his king two thousand dishes, and when the king had eaten, the food was divided among the lords and then the various ranks of ambassadors, princes, high priests, warriors, courtiers, artisans and multiple hangers-on. That every meal was a feast was confirmed by Bernal Díaz del Castillo in *The True History of the Conquest of Mexico.* What with the "cooked fowls, turkeys, pheasant, native partridges, quail, tame and wild ducks, venison, wild boar, reed birds, pigeons, hares and rabbits, and many sorts of birds," Díaz wrote, what with the fruits of all kinds and the two thousand jugs of frothed-up cacao, and what with the numbers of servants and bread makers and cacao makers, Moctezuma's expenses, Díaz calculated, "must have been very great."

To illustrate the quality of the king's personal service, Díaz wrote,

while Moctezuma was at table . . . there were waiting on him two other graceful women to bring him tortillas, kneaded with eggs and other sustaining ingredients, and these tortillas were very white, and they were brought on plates covered with clean napkins, and they also brought him another kind of bread, like long balls kneaded with other kinds of sustaining food, and 'pan pachol' for so they call it in this country, which is a sort of wafer.

They also placed before him three gilded tubes of *liquidambar* mixed with the herbs they called *tabaco,* and after the music and dancing, he took a brief smoke and fell asleep.

In the midst of this dazzlement, Díaz mentioned rumors that Moctezuma was fond of eating "the flesh of young boys," but the chronicler assured us that once Cortez censured the practice, Moctezuma readily gave it up since he had so much else to eat. The Aztec were not hungry for flesh but for maize, for they regarded everything other than maize, as Sophie Coe explained in her article "Aztec Cuisine" (1985) as a sauce or an auxiliary to maize. The Aztec were able to construct an entire food chain on maize because they processed it with ashes or lime and thereby compensated for the lack of game or domesticated animals more abundant to the north and south. As Sophie Coe showed, the high art to which the Aztec developed ash or lye cookery, and its nutritional significance, should disarm any who claim that human sacrifice was the means by which "the protein-starved Aztec higher classes got their minimum daily requirement." The assumption that the lords of Moctezuma were protein-starved because they lived chiefly on corn restates the earliest colonial prejudices against Indian grain.

One need but listen to the music of Sahagún when he lists what the Tortilla Seller does with tamales in the marketplace:

[He sells] salted wide tamales, pointed tamales, white tamales, fast foods, roll-shaped tamales, tamales with beans forming a seashell on top, [with] grains of maize thrown in; crumbled, pounded tamales; spotted tamales, white fruit tamales, red fruit tamales, turkey egg tamales; turkey eggs with grains of maize; tamales of tender maize, tamales of green maize, adobe-shaped tamales, braised ones; unleavened tamales, honey tamales, beeswax tamales, tamales with grains of maize, gourd tamales, crumbled tamales, maize flower tamales.

Quite distinct from tamales, and even more varied, were the little cakes called *tlaxcalli*, or, as the Spaniards called them, tortillas.

All this baroque splendor went for naught to those convinced of the cultural barbarism of the Indian. That attitude prevented the true gift of the Aztec, the akaline processing of corn, from being understood. Only in the last three decades has modern science discovered its biochemical function, and only recently have we come to value a process described by Francisco Hernández as early as 1651, when he reported on the uses of plants found in Nueva España. *Atolli,* he said, was made by combining eight parts water with six parts maize, plus lime, cooked together until soft, the maize then ground and cooked until thick. In the language of biochemistry, the addition of slaked lime, or calcium hydroxide, altered radically the protein potential of corn. Nonetheless, the bad name corn had with the first colonists was extended in the eighteenth century in Europe and elsewhere when corn became increasingly the fodder of the poor and an apparent source of disease. Peasants who lived on corn throughout the winter came down with "corn sickness," as it was called, until an Italian in 1771 named it the disease of "rough skin," or pellagra.

By the end of the nineteenth century, Egypt and other African lands were suffering this sickness in epidemic proportions, but it was little reported in the United States until an epidemic broke out in 1906 in a black insane asylum in Alabama. "The characteristic symptoms are very raw rashes and skin sores, sore mouths, sore joints and muscles, accompanied by dizziness, nervousness, and not infrequently, insanity," C. C. and S. M. Furnas wrote in *Man, Bread and Destiny* (1937). In 1915 the U.S. National Institute of Health appointed Dr. Joseph Goldberger to investigate the sickness that had a name but no proven cause. For a century it had been medical opinion that corn contaminated with

An Aztec education manual lists the tasks a father should teach his son, and a mother her daughter. The age of the child and difficulty of the task are graded by the number of tortillas, from half a tortilla for a child of three to six years, to two tortillas for a child over ten (continued on opposite page).

ergotic rot or a similar plant disease was the probable cause of pellagra. Then once Louis Pasteur's discovery of the infectious nature of bacteria took hold, scientists looked for the cause of pellagra in contagion from animals such as sheep. Goldberger tested the theory of an infectious cause by scratching his skin with pus from pellagra sores. No results. Finally he experimented with the diet of eleven prisoners at the Rankin Prison Farm of the Mississippi State Penitentiary. He fed six on cornbread, grits and sow-back. To the diets of five, he added milk, butter, lean meat and eggs. None of the five got pellagra, but the other six did. Later he experimented with dogs and found that "dog pellagra" could be prevented with yeast extract and liver.

Perhaps because the discovery of vitamins was as recent as 1911, or because the belief that man could live on corn and salt pork alone was as old as the slave trade, Goldberger failed to convince fellow scientists that pellagra was caused by a deficient diet. Even long after Goldberger's death in 1929, champions like the Furnases of the new science of nutrition had to shout to be heard: "There is no doubt that an adequate diet is a sure-fire preventive. Good food—no pellagra; Goldberger proved it." Fortunately, Goldberger had proved it to the American Red Cross, who distributed yeast to Southern sharecroppers during the Depression and encouraged them to grow vegetables. By 1933 pellagra deaths had dropped by a third.

While early nutritionists were concerned with "digestibility" and "dyspepsia," nutritionists now began to speak of "biological value" and "protein balance." They were discovering that proteins in our bodies were rebuilt from proteins in our foods, but that plant proteins were incomplete. Ranking the biologic or protein value of cereals in descending order of merit, the Furnases put wheat first, then barley, rye, rice, corn and oats. Conclusion? "Corn-eating peoples are definitely more susceptible to [pellagra] than those who do not have access to this gift of the Aztecs." They still could not explain why the Aztecs themselves, and other corn peoples, did not suffer from pellagra.

The proteins needed in the human diet are made up of amino acids, and one of them, the acid called niacin, is "crucial to the metabolic processes of every cell in the body," as Harold McGee wrote in *On Food and Cooking* (1984), for without it "our tissues begin to degenerate." Corn lacks niacin, and while the human body can synthesize niacin from an amino acid named tryptophan, corn also lacks tryptophan; anyone living on corn needs compensatory amino acids from another source, such as beans, which happen to be rich in the three acids corn lacks—niacin, tryptophan and lysine. But Indians did not have to rely wholly on an external supplement for their corn, for the necessary amino

At bottom a girl grinding corn on a metate with a griddle (comal) in front of her is now old enough to make the two tortillas appropriate to her task. Codex Mendoza, Neg. Nos. 120209/10, 2 10321, Courtesy Department of Library Services, American Museum of Natural History

A nineteenth-century engraving of natives selling corn in the streets of Cape Coast Castle, Ghana, at a time when pellagra was common. North Wind Picture Archives

acids are not so much absent in corn as locked in and therefore unavailable as nutrients. The protein in corn is of four kinds—albumins, globulins, glutelin and zein. Adding an alkali changes the balance among these, decreasing the zein (which has the least lysine and tryptophan) and increasing the availability of these two acids in the glutelin. At the same time, alkali aids the conversion of tryptophan to niacin and thereby releases niacin for the body's use. In their study of fifty-one Indian societies in the Americas, Solomon Katz and his colleagues observed that no matter which alkali they used (lime yields calcium hydroxide, wood ash potassium hydroxide, and lye sodium hydroxide), there was nearly a "one-to-one relationship between those societies that both consume and cultivate large amounts of maize and those that use alkali treatment." The Indians who used lime in processing corn—all of them in Mesoamerica and the American Southwest—had the additional nutrient benefit of the calcium in lime.

In our Southwest the secrets of processing were inherited and transmitted by the blue-corn cuisine of the Anasazi and their descendants, the Tewa, Zuni and Hopi. Rather than first cooking corn in lime and then washing it off, the Tewa method was to mix wood ashes or finely ground lime directly into their ground meal. In this way they retained the sacred color of the blue corn. Because this corn is rich in anthocyanin dye, Katz explained, it is highly sensitive to color changes that occur with heat or with any changes in the acid content (measured by pH). The lower the pH of the corn, the pinker it turns; the higher, the bluer it remains. Anyone who has cooked with blue-corn meal knows that sometimes the meal will turn pink by the mere addition of water, or lavender, brown or gray in a hot pan, or greenish blue with the addition of baking soda. Baking soda, however, has neither the flavor nor the mineral content of the wood ash used by the Indians. "Culinary ashes," Juanita Tiger Kavena called them: Creek and Seminole favored hickory, the Navaho juniper and the Hopi the four-winged saltbush, or *chamisa,* especially when green. When burned, green chamisa has an unusually high mineral content of potassium, magnesium and sodium. The Hopi chop down chamisa bushes, which grow in clumps like sagebrush, then stack them in a pyre and fire them until they've burned to ash. Once the ashes are collected in tubs, sifted, and stored in airtight containers, they are ready to do three things at once: improve nutrition, heighten flavor and preserve the sacramental color blue.

Frank Cushing also observed that when the Zuni fermented this ash-meal, preferably by chewing the grain and then spitting it out, they produced a leavening that he called "lime-yeast": "The most prized leaven was chewed *sa'-ḳo-we* mixed with moderately fine meal and warm water and placed in little narrow-necked pots over or near the hearth until fermentation took place, when lime flour and a little salt were added. Thus a yeast, in nowise inferior to some of our own, was compounded." Although this yeast had the remarkable property of changing the color of the flour during cooking to a beautiful green or blue, color was secondary in Cushing's view to the yeast's remarkable power of leavening. To this yeast he attributes not only a "greater variety and nicety of cookery," but also a host of cooking appliances—a *batterie de cuisine* of yucca sieves, meal trays, bread plaques, enormous earthenware cooking pots, pygmy water boilers with round stone covers, polished baking stones, bread bowls, carved pudding sticks—and even a separate place in which to use them, "which we may without exaggeration term the kitchen."

In the Zuni kitchen, lime-yeast leavened cornbreads from the rudest to the finest. Shaped into thick cakes and set to rise, "fire loaves" (*mui-a-ti-we*) were baked over hot coals, and "ash-bread" (*lu-pan-mu'-lo-ko-na*) was buried under hot ashes. The Zuni's finest delicacy, "salty buried-bread" (*k'os-he-pa-lo-kia*), made of a sticky paste spread on husks between layers of stones, was buried in the hearth floor, sealed with mud and baked overnight. Often it was flavored with "dried flowers, licorice-root, wild honey, or more frequently than any of these, masticated and fermented meal." Made this way as "sweet buried-bread" (*tchik-k'we'-pa-lo-kia*), it was usually cooked in an earthenware mush-pot lined with husks and was "sweet like our own Indian pudding, which it exactly resembled in taste." Boiled very thick and placed between stones laid out in the cold, it resembled a much-prized if "exceedingly coarse ice-cream."

As we know from Hopi wedding breads, the Zuni were not alone in prizing saliva-fermented meal. Major Powell, exploring Hopi villages in 1875, found

(Below left): Women mixed corn with saliva to start fermentation for both beer and bread. Girolamo Benzoni, traveling in Central and South America in 1541–55, shows three steps in the making of corn liquor.
(Below right): Benzoni shows three stages of breadmaking: boiling corn with lime, grinding the husked corn to make fresh dough, patting the dough into small cakes to cook on the comal, or griddle. Courtesy Special Collections, New York Public Library

their daintiest dish to be "virgin hash." "This is made by chewing morsels of meat and bread, rolling them in the mouth into little lumps about the size of a horse-chestnut, and then tying them up in bits of corn husk" to be boiled like dumplings. He found it curious that "the tongue and palate kneading must be done by a virgin" and that anyone bent on settling an old score might do so by offering to his enemy a "virgin hash" that had been secretly "made by a lewd woman."

Many explorers had commented on the Indian practice of fermenting corn in different forms. Champlain had noted "corn rendered putrid in pools or puddles," which Waugh explained as "stinking corn," left to rot for a couple of months in stagnant water before important feasts. Sagard was notably appalled: "They also eat it roasted under the hot cinders, licking their fingers while handling these stinking ears, as though they were bits of sugar cane, notwithstanding that the taste and odor are vile, and more infectious than the filthiest gutters."

But the practice was as ancient as it was honorable. Garcilaso had observed it in Cuzco, as Sahagún had in Tenochtitlán, where the Aztec soured gruel as we sour milk for yogurt. Lime-water helped preserve *masa* and prevent it from souring, but sometimes the Indians deliberately fermented their dough until it developed "an agreeable sourness," wrote Hernández, and this they mixed with a double portion of fresh corn flour (as we would mix in a sourdough "starter") and cooked it with salt and chili to make a bread they called *xocoatolli*.

We now know that in addition to its agreeable taste, fermented dough or gruel provided the same nutritional advantages as alkali, by increasing lysine and niacin through the production of yeast cells. Masticating the grain first sped fermentation just as malting does in making beer; since an enzyme in saliva (ptyalin) will digest starch and turn it into malt sugar (maltose), adding chewed corn to meal was like adding mash to a brew. Chewed corn started the enzymatic action that would not only sweeten but leaven the meal with carbon dioxide as maltose fed the yeast. Cushing was accurate in his discovery of "lime-yeast."

Such were the sophistications that made the world of Aztec breads the wonder of the New World. Tamales were more notable for their fillings, while tortillas were praised for the variety and wit of their shapes. Spanish commentators were as awestruck by their configurations as Americans today by the con-

A contemporary tortilla maker in Mexico, using a trough metate and a comal on the open fire. From *El Maíz*, Museo Nacional de Culturas Populares (1987 ed.)

tents of a Paris boulangerie. Thus Francisco Hernández described tortillas:

> They make with the palms of their hands thin tortillas of medium circumference, which are immediately cooked in a *comal* placed over the coals. This is the most common and frequent way to prepare maize bread. There are those who make the tortillas three or four times larger and also thicker, they also make of the dough balls like melons, putting them to cook in a pot over the fire and at times mixing in beans; they eat this with pleasure, as they are very smooth, easily digestible and taste good. Some make these breads a palm long, and four fingers thick, mixed with beans and roast them on a *comal*. But for the important Indians they prepare tortillas of sifted maize, so thin and clean they are almost translucent and like paper, also little balls of sifted maize, which despite their thickness are quite transparent, but these things are only for the rich and for princes.

Not even the princes of imperial China could have been displeased with tortillas thin as paper or little balls translucent as dim sum. Such seventeenth-century descriptions balanced more moderate nineteenth-century ones, after the post-Columbian conquest had taken its toll. Here is John Stephens writing of a village in Guatemala near the Honduran border, in his *Incidents of Travel in Central America, Chiapas & Yucatán* (1841):

> The whole family was engaged in making tortillas. . . . At one end of the *cucinera* was an elevation, on which stood a comal or griddle, resting on three stones, with a fire blazing under it. The daughter-in-law had before her an earthen vessel containing Indian corn soaked in lime-water to remove the husk; and, placing a handful on an oblong stone curving inward, mashed it with a stone roller into a thick paste. The girls took it as it was mashed, and patting it with their hands into flat cakes, laid them on the griddle to bake. This was repeated for every meal, and a great part of the business of the women consists in making tortillas.

Even a century later, a great part of the business of the women of Mesoamerica was to make tortillas, but few do so today. The culture in which tortilla-making was held in high esteem as an art worthy of princes and potentates went the way of Louis XIV's *pièces montées*. In Mexico today the remotest village has its long line of women waiting with plastic bags to take home their daily supply from the local tortilla "factory-shop." Regional differences, however, still abound. Oaxaca, known for Mexico's finest and thinnest tortillas,

Corn kernels boiled with slaked lime, ready to have their skins removed (for nixtamal*) and then ground to make the dough (*masa*) for tortillas. This bucketful is enough to supply a Mexican family of six to eight with tortillas for a single day.* Courtesy Centro Internacional de Mejoramiento de Maíz y Trigo, Mexico City

boasts thirty different kinds. There are thick yellow ones in Guadalajara and platter-sized ones topped with garlic in Tabasco and blue and red ones in the central highlands.

To gain some insight into what the business of women must once have been in making tortillas, three times a day, every day, one can consult Diana Kennedy's fifteen-page instruction in *The Art of Mexican Cooking* (1989). Here a few paragraphs must serve: You must begin by slaking lime. In the market buy a lump of *cal* (calcium oxide), break off a piece the size of a golf ball, crush it, pour on enough cold water to make it sizzle like Alka Seltzer. When it stops, strain the liquid into your pot of dried corn and water. The water should taste just acrid enough to "grab your tongue." (You can buy slaked lime in powdered form in hardware or gardening stores, but only in formidable quantities. For two pounds of dough you'll need as little as a tablespoon of lime.)

Simmer your pot of corn about fifteen minutes and as the mixture heats, the kernel hulls will turn yellow. Let the kernels stand in the liquid overnight, then drain, rinse, and rub off the skins between your fingers so that the kernels are snowy white. Cook the kernels too long and your dough will be "tacky"; use too much lime and your tortillas will be bitter.

Now send the wet corn to your local mill to be ground, and if you don't have one, grind the corn yourself through a plate-style grain mill (*molino de maíz*), or dry it with paper towels and grind it in a food processor. Tortilla dough must be used as soon as made, so now work in water to get a very smooth dough, softer than Play-Doh but not sticky. Of course your tortilla press is ready and your griddle, a heavy black cast-iron *comal,* hot. Place one end of the griddle over a burner with medium high heat and the other over a burner with medium low heat (or on a single burner, know where the hot spot is in the center and the cooler spots around the edge).

Without proper temperatures, the tortilla will be tough and won't puff properly (the puff is from moisture evaporating). Put the pressed tortilla on the cooler part of the griddle and in fifteen or twenty seconds, after its edges have loosened, flip it over onto the hotter part of the griddle. After twenty to thirty seconds more, when the underside shows a few brown spots, flip the tortilla back and let the top puff and the bottom speckle for another fifteen to twenty seconds. Remove it and wrap it in a towel to keep it from drying out. You have now made a single tortilla.

Once you have made many tortillas, you can wrap them around almost anything—in Mexico, high-protein items that go back to pre-Hispanic days: fish eyes (*tortillas de ojos*); crisply fried maguey worms (*gusanos de maguey*); the dried and toasted egg skeins of water bugs (*moscos de pajaro*); toasted locusts doused in chili; salamanders, iguanas, rattlesnake meat; the lake algae or slime that the Aztec called *tecuitlatl* and the Spaniards "cheese of the earth." Or, instead of using tortillas as wrappers, you can layer them in casseroles and pies, or grind them up and reshape them into dumplings.

Fry them as chips and you have Fritos. Fritos are to tortillas what cornflakes are to piki, and the founding of the commercial Frito company is a comic echo of the Kelloggs' blare. In San Antonio in 1932, a man named Elmer Doolin bought a five-cent package of corn chips at a small cafe, liked what he ate and tracked down the Mexican who made them. For $100, Doolin bought the Mexican's "recipe" for *tortillos fritos* and his manufacturing equipment, an old potato ricer. Doolin handed over both to his mother, whereupon Mrs. Daisy Dean Doolin and son began to turn out ten pounds of fritos a day until they could afford to move to Dallas and expand. Expand they did through World War II, when they joined with H. W. Lay potato chips to fill America's eternal hunger for the salty, the crisp and the portable.

Should you wish to turn your *masa* into a tamale, more work is at hand. To get a spongier dough, you must boil the corn in a larger quantity of lime-water and grind the hulled kernels less fine. You may want to dry the hulled kernels and then grind them to fine grits for the *tamales cernidos* of central Mexico. Or you may want to whip in quantities of melted lard for the *tamales colados* of Yucatán. You may want to add beaten eggs and herbs, or chili-sprinkled frogs, wild cherries, peanuts, black olives, pork rind and hot or sweet peppers, sour orange juice and shredded chicken; or sweeten them with honey, cinnamon, raisins and *manjar blanco* candy, or dried prunes, chocolate and pumpkin seeds. "To sum up," Diana Kennedy wrote, "the cooking of corn in Mexico with all its elaborations and ramifications is, and always has been, within the realm of the highest culinary art, beyond that of any other country." I don't know any country cousin foolhardy enough to dispute that.

CORNMEAL COOKING

Murder at the Mill

THE CORNMEAL of my youth came from the supermarket in a round cardboard box with a smiling Quaker on the outside. It would keep forever, like baking soda, like Kool-Aid, like soap powder. My family of ex-farmers numbered the supermarket and the grains therein as evidence of God's grace, for which we thanked Him thrice daily in the blessings before each meal. Only much, much later did I realize I'd been accessory to a crime.

Even though my family had raised corn with their hands, like the rest of the nation they had wheat in their heads, and the American conflict between an invading wheat culture and a native corn culture was most clearly expressed in the process by which Americans turned grain into meal. The mother wheat culture had been emotionally combusted in 1851 by what Lewis Mumford called the machine age's "cock-crow of triumph," the Crystal Palace Exhibition of London, where Queen Victoria brooded like a mother hen over her new mechanical chicks. Among her brood were iron roller mills that were to revolutionize the milling of wheat in the motherland and therefore the milling of corn in the "colonies." Although England had lost her colonial land in America, she still ruled the minds of colonial descendants forever determined to transform the crude native stuff into British wheat. Now the alchemy was to be done by milling.

"The improvement in corn milling is by the adoption of the roller reduction process similar to that used in wheat milling, but requiring much greater power, and by this process a flour is produced quite as impalpable as the best grades of wheat flour," wrote the authors of *The Book of Corn* in 1904. They were largely professors of agriculture at land-grant colleges, and they rejoiced in the promise of the new when "corn will yet be the spinal column of the nation's agriculture." That corn had been the flesh and bone of native agriculture for several millennia went unnoticed. At the dawn of the new century, machinery would do the trick:

> Through the use of very ingenious machinery the chit, or germ, is mechanically removed from each grain of corn before it passes into the rolls, by this means removing all but a trace of oil from the meal or flour. The product by this process loses some of the distinctive corn flavor, but the loss is more than offset by the gain in keeping quality, and corn flour can now be used or shipped under the same conditions as wheat flour.

The milling history of wheat had been very different from that of corn because for two thousand years wheat had been ground by wheel. Rotary mills, which first appeared in the Middle East around 800 B.C., worked by crushing grains between a pair of thick round stones. Around the first century B.C., the Romans "improved" their mills by attaching a horizontal millstone with gears

Preparation for "a wedding" in the Drake Manuscript, or Histoire Naturelle des Indes (ca. 1586), shows a woman pounding or grinding corn in a basin metate or mortar, a hunter with a hare, and a basket of tortillas. The inscription reads, in part: "This woman beats the wheat kernel in a wooden mortar and produces a very white flour from which they make very good and very nourishing bread." Courtesy Pierpont Morgan Library

to a vertical wheel, so that the stones could be turned "mechanically" by the power of slave, animal, water or wind. From then on, there was no essential change in milling for almost two millennia.

Before the rotary mill, the Old World had crushed and pounded its grains with mortar and stone just as New World inhabitants had done. The stone saddle quern of ancient Egypt was the same saddle-shaped stone with a stone rolling pin used in ancient Mesoamerica, but the New World kept to the principle of the mortar and pestle and evolved its own "improvements" on handpowered mills. Around the time of Christ, the "basin" metate—a vessel shaped like a bowl in which corn was ground by rotating a simple round stone, or "one-handed mano"—was replaced by the "trough" metate, open at one or both ends so that the grinder could push back and forth the "two-handed mano." In the eleventh century, with the flowering of Toltec and Anasazi, the "trough" was replaced by "slab" metates, set at an angle within a "metate bin," a slab-lined box designed to catch the flour as it was ground. The bins were set three in a row in a grinding room, where the grinders had devised a cornmeal production line, the first woman crushing the meal coarsely and passing it on for successively finer grindings.

Such milling refinements worked more easily with the soft-flour varieties of corn grown in Mexico and the Southwest than with the hard flints of the North. These demanded pounding, rather than grinding, with wood mortars and pestles. "Each house had its permanent wood mortar, set firmly into the earth floor, with a heavy wooden pestle fitting into it," Will and Hyde reported

A trough metate, found in Argentina, is shaped like a fanged and double-headed beast. Catalogue No. 233636, National Anthropological Archives, Smithsonian Institution

of the Indians in the upper Missouri. Small stone mortars, they noted, were used only for seasonal migrations when mobile mortars were essential. "Hominy blocks" were still a household fixture when F. W. Waugh reported on the Iroquois in 1916. They made a block of black or red oak and placed it upside down just outside the door, next to a pestle sometimes six feet long of maple, ironwood, ash or hickory. To make a block, they felled a tree and, selecting a section of trunk with uniform diameter, cut it to a height convenient to the pounder. Next they hollowed out the top, about a foot deep, by burning the wood and then scraping out the ash. To make the pestle, they rounded it on the bottom end and cut it broader at the top to act as a weight and give it more force. One to four women might pound at once, bringing "the pestles down smartly one after the other," and they often made the task into a competitive game.

Whether the mortar was of stone or wood, it had the social and religious significance of corn itself. The mortar was an object of reverence as the act of grinding was an act of worship and of social bonding. The Aztec buried the umbilical cord of a newborn girl under the metate, as they buried a boy's cord under a shield and arrows. Even in 1909, when Frank Speck reported on the Yuchi Indians of the Southeast, he noted that the dooryard mortar seemed to be "an important domestic fetish," for "the navel string of a female child" was laid beneath it "in the belief that the presiding spirit will guide the growing girl in the path of domestic efficiency."

"Domestic efficiency," however, was not an Indian but a colonial term. Men geared to efficiency would have trouble understanding a modern Pueblo woman who can say that a woman never tires of grinding corn. "She can grind it for three or four hours straight, because while she grinds, the men come and sing," explains Agnes Dill of Isleta Pueblo. "And in the grinding songs they tell you almost what to do. And you have to grind to the beat, to the rhythm of the songs." An efficiency expert might mistake this for a method of improved milling by controlling motion and morale. Not so the Tewa. Corn grinding like corn dancing is a way of getting in tune with the primal energy of life. "The grinding song may tell you first of all that what you're handling is very sacred," says Paul Enciso of Taos Pueblo, "and that you've got to put yourself in tune with that spirit of what you're doing, so it doesn't become a chore to you, but it becomes part of you."

How much a part of you, and how much the sacred was part of the secular, are evident in Frank Cushing's account of the "crooning feast" of the Zuni, which was a kind of milling bee. A gaggle of young girls and boys met at a large house, Cushing said, to grind and sing and dance to "sounding drum, shrieking flute, clanging rattles, and the wailing, weird measures of the chant." About eight girls at a time would step to the milling trough and pass the toasted kernels from right to left, reducing the coarse meal to flour as fine as pollen: "Not only did they move their molinas up and down in exact time, but at certain periods in the song—shifting the stone from one hand to the other—passed the meal from trough to trough in perfect unison." The crooning party continued late into the night, Cushing added, "as it usually does in the well-to-do families of Zuni."

In 1902, a pair of Hopi women, Ruth and Mabel Honani, grind their corn on slab metates set within metate bins in a grinding room. The grass brush in the left bin is used to sweep the meal together. Photo by J. H. Bratley, National Anthropological Archives, Smithsonian Institution

The well-to-do families of colonial settlers, on the other hand, would have seen in such grindings cause not for singing but for groaning. Any descendant of those settlers who has tried to grind corn on a metate with a mano knows why Hopi grandmothers say that grinding makes women strong. She quickly learns that grinding is an exact and exacting skill; and knows with what relief her own ancestral grandmothers in the East watched the crew unloading from the hold of the *Fama* at Fort Christina in New Amsterdam in 1644, "3 large saws for Saw Mill, 8 grindstones, 1 pr. stones for Hand Mill, 1 pr. large stones for Grist Mill."

At first our colonial grandmothers made do with native mortars, but they soon "improved" the efficiency of the hominy block of the woodland Indians by attaching the pestle to a tree limb, or "sweep," to make a "sweep-and-mortar mill." By the eighteenth century, every cabin and clearing had a samp mill or two, said Nicholas Hardeman, and sailors traveling along Eastern shores in a heavy fog could locate land by the "thump, thump, thump" of the mills. Imported stones for gristmills were a luxury limited to settlements like New Amsterdam. At the gristmill, women could transfer their domestic grinding efficiency to a professional miller, who used water instead of woman power.

Toward the end of the eighteenth century, the women learned of a new "improvement": a Londoner in Blackfriars had substituted steam for river

Mrs. Little Crow pounds corn in a hominy block in North Dakota. Courtesy State Historical Society of North Dakota

power to drive his wheat mill. Still, the principle of crushing grain between rotary stone wheels held firm until the first iron rollers were devised in Switzerland in 1834 and until Henry Simon in England built the first complete roller-mill plant in 1878. This change was far more radical because, combined with more efficient bolting, it promoted in the name of industrial efficiency the genocide of the living germ. Modern industrial roller mills for corn are grooved in such a way that when the rollers mesh they first fragment and then process the parts: they shear open the kernel, scrape out the starch and crush the germ for oil. Thus developed the first "ingenious machinery" lauded by the 1904 *Book of Corn* for its ability to remove the germ and therefore "all but a trace of oil from the meal or flour."

While the progressivists of *The Book of Corn* admitted some loss of "distinctive corn flavor" in the process, the loss was minor in comparison to "the gain in keeping quality," which would make corn flour as mobile as wheat. The future was theirs: "The economic possibilities of the corn crop are only beginning to be understood." No sounding drum or shrieking flute for this milling bee, but certainly the sound of money. What was not understood by the wheat thinkers, however, was the far greater loss to corn than to wheat when the germ was removed. Today we know that a typical wheat berry is roughly 85 percent endosperm to 2 percent germ, whereas a typical dent corn kernel is about 82 percent endosperm to 11.5 percent germ. Of the total oil in the corn kernel, 83.5 percent is in the germ. The taste is in the oil: remove the germ and you remove the taste. Anyone who samples freshly ground whole-kernel cornmeal can tell first by smell and then by tongue the difference between dead and living corn.

In mainstream America, my grandparents' generation was the last to know, without conscious effort, the taste of meal ground from living corn, which they took for granted and never missed when it was gone. In their farming days they had hauled corn to a local mill by wagon, but they were as happy to trade their sacks of meal for Quaker boxes as they were to exchange Kansas mules for Ford V-8's. For a family of cornflake Fletcherizers, the sweetly nutty flavors and varied crunchy textures of freshly ground cornmeal were quite beside the point. The point for mainstream America was not corn, but wheat.

Even in the midst of our current gastronomic and health-food revolutions, American eaters remain faithful to wheat. In 1982 Americans consumed an average per capita of 114 pounds of wheat flour, plus 2.9 pounds of wheat cereal, in contrast to 11.1 pounds of corn flour and grits, and 2.3 pounds of corn cereal. While the number of people who search out local mills that sell freshly ground cornmeals has increased since the whole-food generation of the 1960s, such mills are even rarer than organic farms. Consumers can scarcely demand what they have never known, and the majority of Americans outside the South and Southwest are as innocent of newly ground cornmeal as of newly laid eggs.

Fortunately, however, Americans are quirky and there's usually a rebel around the bend. Today, just as there's a new crop of farmers rescuing endangered corn species, so there's a new crop of millers rescuing cornmeal flavor along with the moribund milling art. "We're not just selling flour and cornmeal," says Tim McTague at Gray's Gristmill in Adamsville, Rhode Island. "We're selling history." Young McTague learned his milling arts from John Hart, eighty-eight, who with his father kept Gray's Mill going for a century; it is now the oldest continually operating mill in the nation, with the original building dating back to 1675. The current 1870s building looks like a gray-shingled garage, but that's appropriate to the power source that replaced the old millrace of the Westport River with the drive shaft of a 1946 Dodge truck.

In the age of the motorcar, a man grinds corn between a pair of stones in a handmill.
Courtesy Utah State Historical Society

Gray's Gristmill is just down the road from Paul Pieri's farm, which grows the same type of white flint corn—called White Cap today—that was cultivated originally by the Narragansett. Though this type of corn was once common throughout Rhode Island, today in all of New England there are but thirty acres of White Cap, and twenty of these are Paul's. Some of the others are Robert Wakefield's at the University of Rhode Island, where Wakefield has long championed the purity and preservation of the breed. This is a matter of some moment to the Society for the Propagation of the Jonnycake Tradition in Rhode Island, loyal promoters of stone-ground Rhode Island White Cap

Flint corn and producers for twenty years of the *Jonnycake Journal* (spelled their way).

These minutemen have marshaled three local gristmills to carry on the Jonnycake Tradition, one of them at Uskepaugh in West Kingston, where Paul Drumm, father and son, are selling history at Kenyon's Grist Mill, in operation since 1886. Son Paul first alerted me to the importance of millstones made of Narragansett granite, so hard that it flakes the grain paper-thin instead of crushing it, and to the importance of the bible of Rhode Island millers and jonnycake makers, *The Jonny-Cake Papers of 'Shepherd Tom,'* written by Thomas Robinson Hazard for *The Providence Journal* in the 1870s.

Hazard (1797–1886), in chapters called First, Second and Third Bakings, was a practitioner of what I call the Cornmeal Lament, a subgenre of English pastoral. He was blowing his pipes at a time when imported French burrstones were in vogue: between 1877 and 1900, America imported three thousand a year. French burrstones, made of a porous quartz from La-Ferté-sous-Jouarre on the Marne River, were thought to be the finest stones for wheat because the cut faces of the stones kept their edge and seldom needed dressing. Some, but not Shepherd Tom, believed that burrstones were also best for corn because they did not expel the oil but left it in the meal. These were fighting words to Tom:

The 32-foot overshot water-wheel at Falls Mill, built in 1873 in Belvidere, Tennessee, to power a cotton and wool factory has been restored by John and Jane Lovett to grind corn and other grains. Courtesy John and Jane Lovett

The idea that a burr stone can grind meal even out of the best of Rhode Island white corn, that an old-fashioned Narragansett pig would not have turned up his nose at in disgust, is perfectly preposterous. Rushed through the stones in a stream from the hopper as big as your arm, and rolled over and over in its passage, the coarse, uneven, half-ground stuff falls into the meal box below, hot as ashes and as tasteless as sawdust.

Since jonnycake meal should feel soft and flat rather than harsh and round, Tom stated, the only millstones fit to grind it were the fine-grained stones of Narragansett granite found at Hammond's Mill, on the site of the elder Gilbert Stuart's snuff mill, above the head of Pettaquamscutt pond. Here the beloved miller tends his meal as some Raphael his palette or Canova his chisel:

See the white-coated old man now first rub the meal, as it falls, carefully and thoughtfully between his fingers and thumb, then graduate the feed and raise or lower the upper stone, with that nice sense of adjustment, observance, and discretion that a Raphael might be supposed to exercise in the mixing and grinding of his colors for a Madonna, or a Canova in putting the last touch of his chisel to the statue of a god, until, by repeated handling, he had found the ambrosia to have acquired exactly the desired coolness and flatness—the result of its being cut into fine slivers by the nicely balanced revolving stones—rather than rolled, re-rolled, tumbled, and mumbled over and over again, until all its life and sweetness had been vitiated or dispelled.

Milling is so much a lost art that a brief explanation of how a gristmill works may help: the miller feeds grain through a hopper (by means of a leather "shoe," agitated by a device called the "noisy damsel") into a hole in the center of a flat "runner" stone that rests on a lower "bed" stone. As the grain falls into the hole, only the upper stone turns. The grain is ground between the faces of both stones, which have been cut, or "dressed," in radiating grooves, or "furrows," that act like scissor blades when they come together. The cut grain, funneled through the furrows to the outer edges of the stones, is then channeled down a spout to a bin below. How closely the stones fit together and how fast they are turned determines how much friction is generated and how much heat. "If you get the rocks too close together, it'll burn. You can smell it," says George Oswald, who has restored a gristmill in Lexington County, South Carolina. "When the rocks are dull, or you grind too close, it heats up."

An interior view at Falls Mill of the wheels and belts that transmit power from the turning waterwheel to the runner and bed grinding stones. Courtesy John and Jane Lovett

The first enemy of finely milled corn is heat and the second is quantity milling. Shepherd Tom's was not the only Cornmeal Lament for the days of yore before the huge burr millstones, driven by steam, were geared to mass production. In his *American Indian Corn* Charles Murphy complained of how little the manner of grinding corn was understood and how much it mattered: "The steam-mill with its huge burr mill-stones, in the hands of a miller whose ambition is quantity, and who knows nothing of quality, cuts and burns all the life out of the meal, and leaves it a heavy, dead mass. The cook may use all the customary appliances for making light and palatable bread, but all in vain." The capacity of the old horse-driven millstones was about three bushels an hour, he explained, and in the hands of a miller whose ambition was quality, the "meal came cool and lively, with a 'grain' perceptible to the touch."

Millers and farmers knew that the oil in corn rendered it more liable than wheat to spoilage, so they milled corn in small quantities to avoid long storage. Indians had avoided spoilage by parching or toasting the meal they stored for winter, but they were also skilled in applying the right amount of heat to each kind and grind of corn. When commercial roller mills were first introduced in order to grind corn in previously unimagined quantities, millers began to dry corn in commercial kilns to speed the drying. That brought new Laments, among them Maria Parloa's in her *Kitchen Companion* (1887). She regretted that the sweet-flavored meal of old had passed away: ". . . the time for drying has been reduced more and more, until now the grains of the corn meal are as hard as the grains of hominy. . . . Then, too, the meal is ground much finer than formerly. All these changes in the meal have damaged it considerably, and it is almost impossible to get the moist, sweet corn-bread of years gone by." Since the finer grind alters proportions, Parloa suggested reducing the quantity of "modern" cornmeal by an eighth when following old recipes for cornbreads.

A favorite villain of current Laments is the heat presumed to be generated by steel roller mills as opposed to cool stones. Edward Behr, however, in a recent investigation of wheat mills, discovered that the temperatures of steel roller-milled flour were apt to be lower than those produced by electric-powered millstones. The major difference, Behr concluded, lay not in the mechanical process but in the miller—his knowledge, skill and care. The older miller was more apt to begin with a high-quality flavorful grain that had been locally grown and dried, and he was more apt to treat it with care in the milling because he knew the meal would be eaten fresh. "There is no substitute," Behr stated, "for freshly stone-ground corn."

Dedicated to that proposition is the Society for the Preservation of Old Mills, run by Fred Beals in Mishawaka, Indiana. In the hope of restoring and preserving "old mills and other Americana now passing from the present scene," the society's quarterly, *Old Mill News,* is a mine of practical and historical information on gristmills past and present in Canada and the United States. As "Keeper of the Computerized Mill List," Beals compiled a representative list of about 114 working mills that produce stone-ground cornmeal today, most of them in the South.

Old mills, like other anthropological relicts, condense social as well as milling history. Such is the mill at Philipsburg Manor, one of the Sleepy Hollow Restorations, at Upper Mills on the Hudson River above Tarrytown, New York. The mill was built in 1682 for the manor of Frederick Flypse, about forty years after the *Fama* had unloaded its grist stones in New Amsterdam and after Flypse struck it rich by shipping goods and marrying wealthy widows. Such is the War Eagle Mill on the banks of the War Eagle River southwest of Eureka Springs, Arkansas. Built in the 1830s, it was burned to prevent capture by the Union Army during the Civil War, rebuilt, burned again in the 1920s, then restored in 1973 by the Medlin family as the only working undershot waterwheel in the United States.

Thanks to a few history buffs and milling rebels, I have a better chance today of tasting the ambrosial meal of Shepherd Tom or the moist, sweet cornbread of Maria Parloa than I did in my youth. Even my supermarket offers an alternative to the smiling Quaker in the profile of a sober Indian. Since 1960 an Indian head has stamped the meal of Arrowhead Mills in Hereford, Texas, where Frank Ford set up a thirty-inch stone grinder in an abandoned railroad car in order to grind organically grown whole-grain wheat and corn. Although Ford now uses a "hammer mill" instead of stone, the function of the rotating hammers is to shatter the grains into flour without generating heat. "It has been an interesting struggle," Ford says, "against the flow of junk-food madness in this nation."

No matter how good the original grain nor how careful the milling, the problem of freshness remains. Whole-grain cornmeal must be kept in a cold place, refrigerated or frozen, to keep fresh. Costly stone-ground meals on the shelves of gourmet of health-food stores are no better than the storekeeper's awareness of the frailty of living grains. No matter how it's been ground, stale meal is stale. Not until we learn to treat milled corn as if it were as perishable as fresh sweet corn, or butter, or eggs, will we give corn the respect that is its due as "the spinal column of our nation's agriculture." If we can resurrect the dead art of milling, we may give new life to a murdered grain.

Chef Louis Romain, at the stove of the Anthony Wayne Hotel in Hamilton, Ohio, in 1927, shows Princess Dawn Mist of the Blackfoot tribe "how to create American hominy out of Indian cornmeal."
The Bettmann Archive

Sad Paste and Ash Powders

ONCE THE CORN was ground to meal, the question was what to do with it. For wheat eaters, corn was a punishment. Rebecca Burlend, who emigrated to Pike County, Illinois, in the frontier days of 1848, complained when she was reduced to making bread out of corn:

> As our money was growing scarce, [my husband] bought a bushel of ground Indian corn, which was only one-third the price of wheaten flour. . . . Its taste is not pleasant to persons unaccustomed to it; but as it is wholesome food, it is much used for making bread. We had now some meal, but no yeast, nor an oven; we were therefore obliged to make sad paste, and bake it in our frying pan on some hot ashes.

In frontier America, as in colonial America, any form of bread made with corn instead of wheat was the sad paste of despair. How sad is reflected in the lowliness of the names—pone, ashcakes, hoe-cakes, journey-cakes, johnny-cakes, slapjacks, spoonbreads, dodgers—all improvised in the scramble to translate one culture's tongue and palate into another's. Names got muddled by region and recipe as much as samp, hominy and grits and for the same reason: the desperate attempt of a wheat culture to order by its own canon the enormous variety of pastes, batters and doughs cooked by native grinders of corn.

From the start, colonists put interchangeable labels on the generic native ash-cake, baked in an open fire, which seemed to the people of iron griddles and pots an appalling reversion to Stonehenge. Words slithered in the mud of translation, Narragansett *nokehick* becoming "no-cake" and "hoe-cake"; "jour-neying cake," "Shawnee cake" or "every John's no-cake" becoming "jonny" or "johnnycake." At the same time, these Anglicizations of hoe-cake and johnny-cake took hold early and hung on late as symbols, like Yankee Doodle, of rebel identity. In 1766, just after the Stamp Act, Benjamin Franklin fought back when *The London Gazatteer* opined that "Americans, should they resolve to drink no more tea, can by no means keep that Resolution, their Indian corn not affording an agreeable or easy digestible breakfast":

> Pray, let me, an American, inform the gentleman, who seems ignorant of the matter, that Indian corn, take it for all in all, is one of the most agreeable and wholesome grains in the world; that its green leaves roasted are a delicacy beyond expression; that samp, hominy, succotash, and nokehock, made of it, are so many pleasing varieties; and that johny or hoecake, hot from the fire, is better than a Yorkshire muffin.

A century later, Joel Chandler Harris dubbed cornmeal the cause of America's political greatness: "Real democracy and real republicanism, and the aspirations to which they give rise, are among the most potent results of corn meal, whether in the shape of the brown pone or the more delicate ashcake, or the dripping and juicy dumplings."

Still, for those who actually cooked the stuff, cornmeal was hard going. Not only was corn obdurately hard to pound even to coarse meal, but the meal refused to respond to yeast. No matter how they cooked it, in iron or on bark or stone, corn paste lay flat as mud pies, which particularly saddened the Dutch. Accustomed to tile stoves, the Dutch were better bakers than the hearth-cooking British and had cause to be frustrated by the incapacity of Indian wheat to make the light cookies and yeasted doughnuts in which they took pride. In New Netherlands in the seventeenth century, refined wheat flour was often scarce and so valued that authorities prohibited Dutch bakers from bartering their breads and cookies for Indian pelts. Records show that a baker was fined a considerable sum after "a certain savage" was seen leaving his bakery "carrying an oblong sugar bun."

Feinschmeckers fond of sugar buns understandably had trouble with pone. Pone, like samp, was stamped Indian by its name (from Algonquin *oppone* and Narragansett *suppawn*) and ashcake origin. Isaack deRasieres, a Dutchman visiting Plymouth Colony, was more generous than most colonists in describing Indian bread as "good but heavy." Heaviness was a constant colonial complaint, which cooks sought to remedy by mixing cornmeal with the more finely ground flours of rye or wheat—when they could get them. In the first American cookbook of 1796, Amelia Simmons typically mixed wheat flour with Indian meal in her batters and cooked her Indian pudding in a British pudding bag. "Some think the nicest of all bread is one third Indian, one third rye, and one third flour," said Lydia Child in *The Frugal Housewife* (1829), of what was usually called "thirded bread." Typical again was her use of a British name, "bannock," for her flat cake of Indian corn.

Names were as hybrid as the pastes and doughs, as one can see just by skimming the dozens of names Eliza Leslie gave her recipes in *The Indian Meal Book* (1847). We can learn more about mid-nineteenth-century Eastern establishment cookery from Leslie than from anyone else because no one was more detailed or more discriminating than she, but we can also learn how impossible names are and how little they matter. Her "common hoe-cake" was baked on a griddle, her "plain and nice johnny cakes" on a board and her "Indian pone" in an oven. The hoe-cake was a stiff dough of coarse meal, patted into cakes the size of a saucer and laid on a hot griddle over the fire. "Where griddles are not," she hastened to explain, the cakes may be baked "on a board standing nearly upright before the fire, and supported by a smoothing-iron or a stone placed against the back." With a wood fire, she continued, the cakes could be wrapped in paper and laid on the hearth, covering the paper with red-hot ashes. This was what our early settlers did, and "the custom is still continued by those who cannot yet obtain the means of cooking them more conveniently." The dough for "plain johnny cake" she also smoothed onto a stout, flat board (such as a piece of the head of a flour barrel), placed "nearly (but not quite) upright" on a smoothing-iron before the fire, an improvement over the hoe that "in some parts of America it was customary to bake it on." "The cake must be very well baked," she warned, "taking care that the surface does not burn while the inside is soft and raw." To make a "plain" one into a "nice" one, she added a little molasses, butter and ginger to the batter and greased the board with fresh butter.

Terms, ingredients and methods remained in flux for two centuries and should make cautious any naïve epistemologists who seek the "definitive" and "authentic." *Mrs. Curtis's Cookbook* (1909) contained an encyclopedic number of johnnycakes and hoe-cakes, including Southern hoe cake No. 1 (which called for a greased hoe) and a "genuine johnny cake" distinct from a "Rhode Island johnny cake." Fighting words to the Society for the Propagation of the Jonnycake Tradition, for whom "genuine" can mean only Rhode Island. Fighting words to the residents of Newport County, Rhode Island, for whom "genuine" can mean only Newport County. Fighting words, stoves and fists to the Rhode Island Legislature in the 1890s, which fixed by law the spelling of "jonnycake" and would have legislated the recipe as well, had Newport County on the west of Narragansett Bay not deadlocked with South County on the east.

Thirty years later they were still at it. Representative Benjamin Boyd cudgeled his opponents with rhyme: "South County mush, stick the coffee in it and a pig would think it slush." Representative James Caswell of South County took it in snuff: "I did not come here to be insulted—especially by anyone who has been reared on Newport County hick feed." The issue of the cornmeal war was texture. Newport County added milk to meal to make a thin, crisp griddlecake about five inches in diameter; South County scalded the meal in boiling water to make a thick cake about three inches in diameter. "We don't scald all the pep out of it, and we never wash the griddle," said Boyd. Caswell refused to negotiate: "Over my way, we scald the meal and we wash the griddle." The fight itself is now part of the Jonnycake Tradition, celebrated twice annually in Rhode Island with May and October jonnycake cookoffs.

I had had my own battle with johnnycakes and lost handily. I followed Sarah Rutledge's recipe for "corn journey or johnny cake" in *The Carolina Housewife*: two tablespoons "cold hommony" mixed with one tablespoon butter, one egg, one cup milk and enough corn flour to make a batter "just so stiff as to be spread upon a board, about quarter of an inch thick." I had just such a board, twelve by fourteen inches, which I propped against an iron trivet in front of an open fire, as directed: "Put the board before the fire, brown the cake, then pass a coarse thread under it, and turn it upon another board, and brown the other side in the same way." Simple? No. The batter must be just so moist that it sticks to the board, and the board must be just so close that the batter will not burn or dry out and slide into the ashes. Forget the coarse thread; the problem was to keep the batter on the board. The first cake slid off whole into the ashes. The next cake stuck to the board in patches, some burned, some gummily raw. Choked with smoke, front red with heat and back blue with cold and stiff from bending, I had new respect for my ancestral grandmothers and sympathy for their corn complaints.

Like "johnnycake," the word "corn-dodger" was another compound neologism, as H. L. Mencken says, that colonists invented as a way of coping with "an entirely new mode of life." "Dodger" suggests that artful bakers took to substituting corn for honest wheat buns. Nineteenth-century cookbooks suggest that it was simply one more cornmeal dough made "just so stiff" that it could be shaped by hand and baked on tin sheets in a Dutch oven. Eventually dodgers took on a particularized shape, detailed by Harriett Ross Colquitt in *The Savannah Cook Book* (1933), in which she told the cook to form the dough "into long, round dodgers with the hand (about four or five inches long and one inch in diameter)." Corn sticks molded in cast-iron pans with corn-ear depressions replaced the dodger in common parlance, after Wagner Ware in the 1880s began to manufacture Tea-size, Junior-size and Senior-size "Krusty Korn Kobs Baking Molds."

"Come, all young girls, pay attention to my noise;
Don't fall in love with the Kansas boys,
For if you do, your portion it will be,
Johnnycake and antelope is all you'll see."
—From "Kansas Boys," in *A Treasury of Nebraska Pioneer Folklore* (1966)

The Southern hush puppy is simply a latter-day dodger, despite the regional mystique that puppies and dodgers are to ordinary corn bread what brandy is to wine. Because the puppy batter adds minced onions, French-style, some have attributed its origin to the Croquettes de Maïs cooked by the Ursuline nuns at the founding of their New Orleans convent in 1727, but there's no record of the word before 1918. And the thing itself is no more than a corn fritter fried with fish in the happy cafe style of Little Pigs in Asheville, North Carolina, whose customers every day consume a thousand puppies, shaped by an ice cream scoop.

These are the cornmeal civilities that Thoreau abjured when he baked unleavened corn loaves in the fire at Walden Pond. Publishing *Walden* in 1854, around the same time as Eliza Leslie's *Receipts*, he exulted in his return to culinary innocence, to the "primitive days and first invention of the unleavened kind"—the primal ashcake, the original staff of life. Even so, he encountered problems with pure corn:

> Bread I at first made of pure Indian meal and salt, genuine hoecakes, which I baked before my fire out of doors on a shingle or the end of a stick of timber sawed off in building my house; but it was wont to get smoked and to have a piny flavor. I tried flour also; but have at last found a mixture of rye and Indian meal most convenient and agreeable.

". . . so Jim he got out some corn-dodgers and buttermilk, and pork and cabbage and greens—there ain't nothing in the world so good when it's cooked right. . . . "

—MARK TWAIN,
The Adventures of Huckleberry Finn
(1885)

Thoreau's city knowledge was evident not only in his mixture of meals but in his preserving methods, when he acknowledged that he kept his ashcakes "as long as possible by wrapping them in cloths." Such urban adhesions in no way inhibited his pastoral Lament that "fresh and sweet meal is rarely sold in the shops, and hominy and corn in a still coarser form are hardly used by any." For primitive purity, he decided to eschew not only leaven but salt: "I do not learn that the Indians ever troubled themselves to go after it." Nor did they trouble themselves to go after rye. Thoreau was a native not of Walden but of Concord, and that made all the difference.

The distinctions between wheat and corn continued to make all the difference in the South, where it was wheat biscuits for the gentry and corn pone for the slaves. Wheat meant "baker's bread," as Huckleberry Finn said, "what the quality eat—none of your low-down corn-pone." A contemporary descendant of freed slaves recalls that on weekdays her Mammy made ashcakes in the fireplace "outen meal, water and a little pinch of lard," but on Sundays they were special, "on Sundays day wuz made outen flour, buttermilk an' lard." Sharecroppers black and white limited the distinction to breakfast, William Bradford Huie said in a collection of reminiscences, *Mud on the Stars* (1942), but it was absolute because biscuit for breakfast meant dignity. "A Garth Negro or white cropper would relish corn pone for dinner or supper, but to have had to eat it for breakfast would have broken his spirit," Huie said. "Corn pone for breakfast among croppers is like a patch on the seat of the britches for a man, or drawers made of flour sacks for a woman among the landowning whites."

Native corn eaters of the Southwest, whose caste status did not depend upon wheat, nonetheless incorporated wheat into their cornmeal pastes as they incorporated the Madonna into their Corn Mothers. A recipe for a contemporary Navaho cake, in *Traditional Navajo Foods and Cooking* (1983), is a true child of the hybrid cuisine engendered when wheat met corn. It is a ceremonial cake baked by the entire community of women at "the Fourth Dawn of the Kinaada," and they bake it as of old in an earth oven fired by piñon and juniper.

A nineteenth-century engraving after the painting by Alfred Kappes, Hoe-cake and Clabber *(thickened sour milk).* North Wind Picture Archives

They mix fifty pounds of yellow corn, roasted and ground to the consistency of "Cream of Wheat," with fifty pounds of stone-ground whole-wheat flour and twelve pounds of sprouted wheat, dried and ground. After sweetening the mixture with brown sugar and raisins, they pour it onto the bottom layer of cornhusks that line the pit, then draw a sacred pattern with the pollen of the four corn colors before covering the whole with a layer of husks. Next they shovel on a layer of sand and hot coals and let the cake bake overnight. When they uncover it, they cut it into eight giant pieces until only the center, the heart, remains. This they cut in four pieces, to represent the cardinal directions, and these they put back in the ground to feed Mother Earth. "Steamed and moist, slightly sweet, chewy, with the full flavor of the grains, nothing could taste as good on a cold, sleepy dawn," the community of Navaho women agree, four centuries after the community of women at Plymouth struggled to feed their men with a sad paste baked in the irons and ashes of hearths far distant from their native ground.

WHAT DISMAYED Plymouth women most about cornbread was that no matter how much you kneaded it, or how much yeast you added, the loaf would not rise to any baking occasion. Wheat, in contrast, was king of the mountain because, more than any other grain, wheat was made for yeast. A wheat dough kneaded with yeast would rise slowly like a balloon and stay there under fire. Because of the high gluten content of its proteins, wheat dough could be made so elastic that as the yeast slowly produced carbon dioxide bubbles, each bubble could be trapped within a gluten parachute to make a light and airy dough. Not so corn. The low gluten content of its proteins kept the dough from absorbing the bubbles of fermenting yeast, which simply dissipated into thin air. If corn dough were to be made lighter, it would have to work quickly to trap bubbles—like the air bubbles of whipped eggs or the gas bubbles of a carbonate that could be produced chemically by combining an acid with an alkali.

Wood ash, lime and lye, all the alkalis employed in Indian ash cookery, furnished just such a leavening, as Cushing saw, for their cakes and breads of corn. Colonists, improvising their cooking over campfire and hearth, learned to adapt and refine the ash powders of Indian meal for use with their own wheat and rye. When they leached wood ashes in a pot, the colonists appropriately called the liquid "pot-ash" (a potassium carbonate), which they mixed with fat to make soap, or with an acid like sour milk or molasses to make a quick leavening for dough. Potash they soon refined in both word and substance into "pearlash," and then into the more easily prepared aerated salt they named "saleratus," a sodium bicarbonate known today as baking soda.

The evolution was speedy, but because these were wheat bakers they first applied the quick source of bubbles to wheat. In 1796 Amelia Simmons used pearlash for her gingerbread cakes of wheat, leavened with "three small spoons of pearlash dissolved in cream." In 1805 J. B. Bordley, in his *Essays and Notes on Husbandry and Rural Affairs,* prescribed "a tea spoonful of salt of tartar heaped, or any other form of pot or pearl ash" for a "handy-cake," a bread made with soured milk and wheat. In 1825, by the time of her second edition of *The Virginia Housewife,* Virginia Randolph called the quick bread she made of soda and wheat flour "soda cakes."

A new world of quick wheat breads was opened by Indian ash powders meant for corn. We can see a strange crossbreeding of Old World yeast and New World ash powder in Eliza Leslie's "Indian Meal Preparations, &c." in *New Receipts* (1854), where mixed batters of wheat and corn got mixed leavenings of yeast and pearlash or some other powder. Leslie's batter for "Indian slap-jacks" of yellow meal mixed with flour she leavened with "three large table-spoonfuls of strong fresh yeast" and "a level teaspoonful of pearlash, soda, or sal-eratus, dissolved in warm water." After the dough had risen from the yeast, the cook was to "add the dissolved pearlash to puff it still more."

The refinement of ash powders led directly to a cosmos of cakes, the glory of the American kitchen in the second half of the nineteenth century. The "Madison cake" proposed by Eliza Leslie, however, was a bewildering hybrid of a traditional British spice cake, made of wheat flour enriched with the usual butter, eggs, raisins and brandy, *and* an American mush cake, made of yellow cornmeal stirred with "a hickory spaddle," which she explained is "like a short mush-stick, only broader at the flattened end." In addition to eggs, she leavened the cake batter with "a large salt-spoon of saleratus, or a small tea-spoonful of soda" combined with half a pint of sour milk, and she explained to the housewife how the new powder worked: "For this cake the milk must be sour, that the saleratus or soda may act more powerfully by coming in contact with an acid. The acidity will then be entirely removed by the effervescence, and the cake will be rendered very light, and perfectly sweet."

"Lightness" and "sweetness" were Cushing's words for Zuni cakes made with the addition of "lime-yeast" added to meal mixed with acid saliva. Sahagún applied the same words to Aztec cakes made with lime and fermented meal and other combinations of alkali and acid. Ironically, although acid-alkali compounds were as native to corn cookery as yeast fermentation to wheat, Britain rather than the colonies was the first to commercialize ash powders for both grains. Between 1835 and 1850, Britain began to manufacture baking-powder compounds on a large scale and they quickly caught on, not just for the baker's ease and speed but from a new general distrust of yeast. Louis Pasteur's discovery in 1857 that yeast was full of living organisms meant to the popular imagination that yeast was full of germs and therefore harmful. Scientists as respectable as Professor Eben Horsford of Harvard warned that yeast organisms were poisonous molds and suggested that a baker avoid these poisons and speed his work by premixing bread flour with sodium bicarbonate and "dry phosphate of lime" (extracted from bones) to make a "self-raising flour." At the same time, extremists in the American health movement that produced Sylvester Graham and the Kellogg brothers were fulminating, like Thoreau, against leavening of any kind as a corruption of the staff of life. "Rotted by fermentation or poisoned with acids and alkalis," the staff of life had well nigh become "the staff of death," gloomed the proponents of the Boston Water Cure in 1858. By 1872 Boston purists like S. D. Farrar had narrowed their focus to devil "saleratus, that abomination of the household, or cream-of-tartar, its companion in guilt."

Americans who wanted flat pones to be light as wheaten cakes had to do something and typically overdid it. If one leavening was good, three were better, as in the cornmeal cakes offered in *The Genesee Farmer* (1847): "Mix two quarts of corn meal, at night, with water, and a little yeast and salt, and make it just thin enough to stir easy. In the morning stir in three or four eggs, a little saleratus and a cup of sour milk . . . bake three-quarters of an hour, and you will have light, rich honeycomb cakes." Without wheat, the yeast would do nothing for the meal overnight, but American nineteenth-century cookbooks

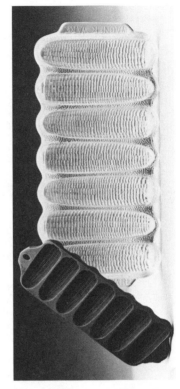

The ceramic corn molds used anciently in Mexico and Peru to cast corn ears in clay for sculpture are the ancestors of these modern pans used to shape corn batter into edible "corn sticks." The castiron pan is labeled Griswold Krispy Cornorwheat Stick Pan #262 from Erie, Pennsylvania. The glass one is Wagner Ware. From the collection of Meryle Evans, photo by David Arky

"A very common mode of aerating bread in America is by the effervescence of an acid and an alkali in the flour. The carbonic acid gas thus formed produces minute air-cells in the bread, or, as the cook says, makes it light. When this process is performed with exact attention to chemical laws, so that the acid and alkali completely neutralize each other, leaving no overplus of either, the result is often very palatable. The difficulty is, that this is a happy conjunction of circumstances which seldom occurs."

—CATHARINE BEECHER and
HARRIET BEECHER STOWE,
The American Woman's Home (1869)

are full of recipes that mash one culture's cookery with another's. Perhaps no bread tasted like bread to wheat eaters unless flavored by yeast.

By the turn of the century, an improved understanding of the chemistry of foods eliminated yeast for corn and demonstrated why cornmeal was better suited to quick-acting acid-soda mixtures and to small forms like biscuits and muffins. In *Fifty Recipes,* Célestine Eustis included a great variety of breads, cakes, biscuits and muffins leavened entirely with saleratus, soda, baking powders and eggs. Not a word about yeast.

Until baking powders were readily available and understood, housewives and especially Southern ones leavened their corn batters primarily with eggs. Just as they turned green-corn puddings into fritters and soufflés, so they turned puddings and breads into spoonbreads. Mrs. Bliss of Boston offered what we now call spoonbread in her *Practical Cook Book* (1850) under the name "Indian puffs." Her proportion of one quart of milk to eight tablespoons of meal and six eggs, "beaten as light as possible," suggests how much more easily Americans turned their corn in the direction of British puddings than British breads.

Americans were not alone in redeeming sad paste with mixed flours and leavenings. After the post-Columbian Exchange, Europe developed a wide range of baked corn goods that Britain knew nothing of. The Basques developed a huge cornmeal cuisine, with fine breads like *gâteau au maïs* (a sweetened bread with eggs and butter) and *méture au potiron basquais* (a pumpkin cornbread), cookies like *biscuits au maïs* (with wheat and corn flour mixed), and muffins like *taloa* (shaped flat and split open). In Agen they made deep-fried cornsticks to serve with oxtails and prunes in a dish called *queue de boeuf à l'agenaise et au milhas,* an echo from the time corn supplanted millet (*milhas*) by deed if not by word. Lombardy made a *pan Giallo* or *pan di Mais* of wheat and corn flavored with olive oil and shaped into wheels weighing two kilos. Ireland made a cornmeal soda bread called *paca, pike, yalla male* bread, said to be named for the ship *Alpaca* which brought cornmeal to Ireland during the potato famine of 1846. Romania made a cornbread with sour milk and dill called *alivence,* baked in cabbage leaves and served with sour cream.

England alone remained cut off from the possibilities of corn, for reasons that had more to do with its caste system than its agriculture. Indian paste was too base for the high monarchy of wheat. Franklin knew what he was about when he challenged Yorkshire muffins with Yankee hoe-cakes, and Joel Chandler Harris when he saw democracy rise from pone.

You Say Polenta, I Say Mush

WHEN I WAS A CHILD, I said *mush*. On a winter night when frost whitened the oranges outside our house and smudge blackened the curtains within, I said *mush*. Of all the belly-warming, heart-soothing, mind-comforting pottages, puddings, pablums and purees that I spooned down my childhood gullet, mush was the most. Mush was the heritage of our prairie past, the prerogative of a family palate honed to reduce all textures to mush. Mush, in our household, was the Platonic form to which all food aspired, and my mouth was ever open to it.

I little dreamed that halfway around the globe little Bantu children in Africa were comforted with *putu,* Venetian bambinas with *polenta,* petites filles in Gascony with *armottes,* junior Romanians with *mamaliga,* pocitas madrileñas with *milha* and tiny Transylvanians with *puliszka.* All these were names for cornmeal mush, and because mush was comfort food wherever eaten, each nation claimed it for its own.

Even now that I have put away childish things, I have never put away mush. I have simply incorporated a lifetime's travels within its wide and comforting embrace, adding Mexican jalapeños, Italian Parmesan, Danish butter, Ceylonese cinnamon, French crème fraîche. I don't mean at different times to different servings of mush, I mean all at once. I mean muddling peppery-hot, salty, sour, sweet and creamy all together in the same mush bowl to make one ambrosial pap.

The classic way to eat mush, of course, is with milk. Milk and mush was one of the earliest yokings in the hybrid cuisine of the colonials and remained one of the most cherished. A British army doctor during the War of 1812, tending captured soldiers in the prison hospital at Halifax, remembered how the prisoners cried incessantly for *"mush and milk,"* and since none was to be had, "their lamentations took the tone of despair."

A German artist, Rufus A. Grider, remembered in his 1888 memoirs of Dutch life in the Schoharie Valley the "Mush & Milk Dish" of his childhood which the Hollanders called "Mush Sapahn." Mush and Milk symbolized togetherness, not just the cold ingredients in the bowl, but the whole family eating communally from a large pewter dish:

A nineteenth-century engraving of a colonial gentleman, Eating Hasty Pudding.
The Bettmann Archive

> MUSH was prepared in the fall & winter of the year.—it was boiled in the afternoon & about one hour before meal time poured from the Iron Pot into the Pewter Dish & set in a cold place, cooling stiffens it. Near meal time the House Wife made as many excavations as there were guests—piling or heaping up the Centre, & filling the hollows with COLD MILK . . . as many PEWTER Table Spoons as milk Ponds were supplied. After Grace was said by the head of the family—Every one began to diminish the bank & increase the size of his White Lake by feeding on its banks and Centre—but there were limits beyond those no one could go—if for instance any one tapped his neighbors MILK POND it was ill manners—if children did so—the penalty was FINGER CLIPS.

Once the starving times of the seventeenth century were past, the manner and manners of eating mush and milk took on the significance of art and ritual. In New York, Grider transformed mush and milk into a Dutch genre scene, in his landscape of mush hills with milk ponds at the center, surrounded by squalling children in finger clips. In New England, Shepherd Tom in his *Jonny-Cake Papers* saw the mush-and-milk ritual of his childhood in the early nineteenth century as an index of class:

> Curious enough, I can remember when the eating of no-cake and milk was considered somewhat a test in Narragansett of good breeding. To be eaten gracefully, no-cake must be placed very carefully on the top of the milk, so as to float, and a novice, in taking a spoonful of it to his mouth, is very liable to draw his breath, when the semivolatile substance enters his throat in advance of the milk and causes violent strangling or sneezing.

For pioneers, tested by survival rather than breeding, the dish called "Dry Mush and Milk," in an 1861 issue of *The Nebraska Farmer,* was recommended to adults and children alike as a cure-all better than snake oil: the paper cited a student who, suffering from poor health, for a term of eleven weeks "ate *only* mush and milk for breakfast, dinner, and supper," and became first in his class.

Wherever mush was eaten, the mush mystique was sustained by associating the gruel with British pudding or Italian polenta. On March 26, 1722, the phrase "Indian Pudding" appeared in print for the first time in Boston's *New England Courant,* where Benjamin Franklin was working for the editor, his brother James, and noted with amusement "that at a certain House in Edgar Town, a Plain Indian Pudding, being put into the Pot and boiled the usual Time, it came out a blood-red Colour, to the great Surprise of the whole Family."

The name "Indian pudding" evoked the world of British pottages and gruels that had been upgraded in the seventeenth- and eighteenth-century homeland to puddings and pies. The staple cereal or milk pottage of medieval times, stirred with a stick in a pot over the fire, underwent radical change during the century when yeomen and bondsmen left for the colonies. For a "modern" British cook like Hannah Glasse in 1747, pottages were *hors de combat.* She preferred enlightened puddings, baked in a crust or boiled in a bag, at once avoiding the tedium of stirring and encouraging the pleasure of enrichments. For the colonies, suffering from the usual cultural lag, the term "hasty pudding" was a wheat-culture euphemism intended to redeem mush made of barbarous corn.

For Joel Barlow, hasty pudding was "A name, a sound to every Yankee dear," because it came to symbolize Yankee rebellion. A New Englander, Barlow was traveling through the Savoy in 1793 as Minister Plenipotentiary to France when he came on a steaming bowl of *polente au baton* in a small inn at Chambéry and was straightway seized by homesickness and the muse. Given to writing epics like *The Vision of Columbus,* he now tore off several hundred mock-epic lines on the subject of mush. Although Barlow was so avid a revolutionist that John Adams thought not even Tom Paine "a more worthless fellow," Barlow reveals his British heritage in a burlesque indebted to Alexander Pope's *Rape of the Lock.* "The soft nations round the warm Levant" call it *Polenta* Barlow said approvingly, "the French, of course, *Polente."* But his fellow Americans embarrassed him by the vulgarity of their tongue: "Ev'n in thy native regions, how I blush/ To hear the Pennsylvanians call thee *Mush!"* Its true name, Barlow explained, came from the haste with which the pudding was cooked, served and eaten, which proves he'd never actually stirred mush in a pot, because the one fact about mush is that you can't hurry it. The ladies of Deerfield, Massachusetts, aptly described "slow mush" in their recipe for "hasty pudding" in *The Pocumtuc Housewife* (1805): first you soften sifted meal in cold water, then add it to boiling water until thickened. "Then stand over the kettle and sprinkle in meal, handful after handful, stirring it thoroughly all the time. When it is so thick the pudding stick stands up in it, it is about right." Or, as *Ye Gentlewoman's Housewifery* put it as late as 1896, "beat like Mad with the Pudding Stick."

The stick was as essential as the iron pot to colonial and later kitchens. *Prac-*

"Father and I went up to camp,
Along with Captain Goodwin;
And there we saw the men and boys,
As thick as hasty pudding."
—A verse of "Yankee Doodle" used in 1767 in the ballad opera by Andrew Barton *The Disappointment; or, The Force of Credulity*

A wooden paddle (21 inches long) for stirring corn was carved by Iroquois in western New York in 1918. Milwaukee Public Museum

tical Housekeeping (1884) specified "a hard wood paddle, two feet long, with a blade two inches wide and seven inches long to stir with," and "stir-about" was the name the Irish gave to the pudding when England shoveled cornmeal into Ireland during the potato famine. Meanwhile American natives used two kinds of sticks for their mush. Among the Iroquois, one was a long narrow stick, often carved the length of the handle to exploit phallic suggestions. The other was a wide paddle, often carved with a heart-shaped hole in the center of the blade, to drain water from breads boiled in bark pots like dumplings. Even in this century, F. W. Waugh found that stirring paddles were in frequent use by Iroquois housewives for their "puddings" of beans, pumpkin, acorns, chestnuts and other native ingredients mixed with corn.

As soon as she could, the American housewife left behind the rustic stir-about pudding to make the batter pudding in vogue in the Age of Enlightenment. Comically, Indian pudding came to mean mush cooked *with* milk, either baked or boiled *English*-style in a pudding bag. While Amelia Simmons offered no hasty pudding recipe, she gave a trio of recipes for making "a nice Indian pudding." The simplest was softened with milk, sweetened with sugar and boiled in the strong cloth of a pudding bag for twelve hours. The nicer one was baked with eggs and butter, sugar or molasses, and spice. The nicest of all was baked like a custard, in which a small proportion of fine meal thickens the milk, along with a quantity of eggs, molasses, spice and a half pound of raisins.

Eliza Leslie went to town on Indian Puddings, from her first *Seventy-Five Receipts* (1828), when she elaborated the nicest Indian above with cinnamon, nutmeg, lemon and the finely chopped suet traditional in British mincemeat, to make a boiled pudding served with a drawn butter sauce of nutmeg and wine. In her *New Receipts* (1854) she offered a sumptuous "baked corn meal pudding," substituting butter for suet and emphasizing the rind "of a large fresh orange or lemon grated," plus a pound of Zante currants or sultana raisins, which suggests a British plum pudding. She patriotically balanced this one, however, with a "pumpkin Indian pudding," adding stewed pumpkin to much the same ingredients, all to be boiled in the bag.

In contrast to the urbane Leslie of Philadelphia, New Englanders remained apostles of austerity on the order of Lydia Child, who counseled that a hasty pudding made of rye grain and West India molasses would benefit any whose "system is in a restricted state" from dyspepsia. Both Catharine Beecher and Marion Harland provided recipes for "An Excellent Indian Pudding without Eggs (A Cheap Dish)," and Fannie Farmer's Indian pudding in 1896 was a strict survival mush of milk, meal and molasses, from which "ginger may be omitted." The same pudding without eggs and stiffened with rye flour became

Boston brown bread, steamed in a pot or mold (a one-pound baking-powder box or a five-pound lard pail "answers the purpose," said Farmer) instead of in a pudding bag.

Not that baking batters in an oven or steaming them in a pot was a British invention. Cushing described the Zuni version of a double-boiler used when a kind of steamed mush-bread was made from little balls of dough spread over a yucca sieve:

> A large pot half-filled with water was set over the fire, inside of which a smaller vessel, partially filled with water and weighted with pebbles to keep it steady, was placed. Upon this smaller part was laid the sieve or screen holding the balls of dough, the larger pot then being covered with a slab of stone and kept boiling until the dumplings were thoroughly cooked by steaming.

Fortunately today, many American chefs who are rediscovering native ingredients are also restoring sensuous pleasures to the minimalist version of Indian pudding purveyed by Fannie Farmer. One of my favorites is offered by a contemporary native Indian in a form that would have done no discredit to the luxurious palate of Eliza Leslie. The cornmeal dessert contributed by Ella Thomas Sekatau of the Narragansetts to Barrie Kavasch's *Native Harvests* (1979) puts cheap dishes to shame. She poured into her white cornmeal thick cream, maple syrup, berry juice and fresh berries, nutmeg and nut butter and enriched it with three lightly beaten eggs.

The other way colonists conferred dignity on mush was to call it polenta. Virginia Randolph's directions in her *Virginia Housewife* (1824) "To Make Polenta" were similar to the *Pocumtuc's* "hasty pudding," except that Randolph sliced the cooked mush when cold, layered it with butter and cheese in a deep dish and baked it in a quick oven—Italian-style. We know that she thought of it as Italian because she follows it with recipes for "macaroni" and "vermicelli." Fifty years later, Marion Harland included polenta in her *Dinner Year-Book*, turning it out on a plate to cool, then cutting it into squares and frying them yellow-brown. "This is a favorite dish with the Italian peasantry," she explained, "who generally eat it without frying." At the turn of the century, Célestine Eustis outlined four polenta recipes, including "alla Toscana," which initiated *les pauvres* in the polenta mystique taught today to rich Americans by expensive Italian cooking teachers. Here was the ritual pouring of the meal in a slow trickle into boiling water, the constant stirring, then the holding of the saucepan over the hottest part of the fire until the meal was detached from the

"I sing of food by British nurse designed
To make the stripling brave and maiden kind,
Let Pudding's dish most wholesome be thy theme,
And dip they swelling plumes in fragrant cream."
—BENJAMIN FRANKLIN KING, JR., *Hasty Pudding* (ca. 1875)

bottom so that it would turn out easily onto a board, then the cutting of it into slices with a wire or string. Eustis's *pauvres* may not have the cachet of Italian peasants in the American imagination, but the mush is the same.

So is the cornmeal, for there are even fewer stone-ground mills in Italy than in the United States. One of the last is the picturesque mill of the Scotti family outside Bergamo; the family grow heritage strains of corn for a mill that has been grinding since the early 1500s. As for continuous stirring in the traditional copper polenta pot, most Italian housewives I know buy blocks of packaged precooked polenta from their supermarkets. Busy American housewives, following British tradition, long ago displaced stir-about mushes with steamed ones, which were not only easier on the cook, but made a smoother, moister and more evenly cooked mush. Today the double-boiler (and microwave) has displaced both stirring sticks and pudding bags, so that the busiest housewife can make lump-free polenta, suppawn or mush: just moisten the meal first with cold water in the top of a double boiler, stir it until it comes to the boil over direct heat, add boiling water to the mush, place the top over the bottom of the boiler, clap on the lid and steam it for an hour.

A good reason for the polenta mystique, apart from the chic of copper pots and peasant cooking, is that Italians during four hundred years developed a fine corn cuisine of their own. When Columbus first brought corn to the continent, Italy was as suspicious as France or England of the grain they called *granoturco*. But since it was the cheapest of all grains to grow, they soon called corn by the ancient Roman name for pottage, *puls,* or *pulmentum,* and thus polenta. By the eighteenth century, corn had become a hedge against famine for the peasants of the northern provinces and, for the nobility, a version of

Nineteenth-century farmers in Middle Europe shelldried corn by "threshing" it as if it were wheat. Courtesy Library, University of California at Davis

pastoral. Among the Palladian villas a "polenta cult" developed, the subject of songs, poems and Manzoni's famous nineteenth-century romance *I Promessi Sposi.*

"Polenta is indeed a very simple food—cornmeal mush," Waverley Root wrote in *The Food of Italy* (1971), but like the simple wheat food called pasta, Italy subjected it to a multitude of regional spins. "Traviata," the wits called it, because of its ability to couple with any condiment. Since Italians knew nothing of Indian ash cooking, the condiments were important for nutrition and may be one reason that polenta was so often mixed with cheese, as in Piedmont's *polenta grassa,* in which polenta is layered with fontina, or Lombardy's *polenta cunscia,* in which it is mixed with Parmesan and garlic butter.

If Emilia-Romagna became the province of pasta, as Waverley Root said, Lombardy became the province of polenta. Here bowls or squares of polenta were sprinkled with truffles, flavored with lemon peel or orange water, accented with dried figs and walnuts, smoothed with roasted garlic and olive oil. In Trentino a black corn made "moorish polenta," enhanced with sausage, wild mushrooms, salt cod and cheese. It was the Veneto, however, which took polenta to new culinary heights in *polenta pastizzada,* layering the mush with minced veal, mushrooms and cockscombs in a wine-tomato-butter sauce seasoned with salt pork. Or in a sweetened *smejassa,* in which polenta was dried and crumbled with cookies, then baked like an Indian pudding with milk and molasses, raisins, pine nuts and candied citron.

The pride of the Veneto, however, was its famed *polenta con osei*—"fresh polenta, migratory birds, wine from the cellar and joyful folk," in the words of an ancient proverb. The birds were tiny grilled songbirds, wrapped in a baby blanket of crisp fat, each cradled in its juices on a square of polenta. Or sometimes the entire spit was brought to the table, Root said, "the chaplet of little birds skewered together on it, their beady eyes fixed reproachfully on the diner, a sight which has been known to indispose Anglo-Saxons." Venetians were more apt to be indisposed by *not* seeing beady eyes, and so gave a joke name to plain polenta, *polenta e osetiti scapai,* meaning "polenta and the little birds that got away." Cornmeal not only made mush, but also strengthened pastas, pizza crusts, soups, gnocchi, breads, cookies and cakes, among them a delicate confection flavored with maraschino liqueur and romantically titled *amor polenta.*

*While a woman in the foreground stirs her pot of mush (*mamaliga, puliszka, polenta*) with a mush paddle, a woman at the rear prepares to cut with a string the mush turned out on the table.* Courtesy Library, University of California at Davis

The French too developed a rustic cornmeal cuisine, but one far less familiar to Americans because each region gave its own vernacular twist to the mush theme. Perigord called it *las pous,* Paula Wolfert has said, because the porridge as it cooks sputters like a person softly sputtering in his sleep. The Savoy, where Barlow had his homesick meal, became the heart of French cornmeal cooking during the last years of the eighteenth century, when the French Revolution put peasant fare on the overturned tables of noblemen. Because Savoy mush (now called polenta) is stirred with a long stick for a long time, it is *polenta au baton.* The Savoyards serve it like the Italians with game and cheese, dress it in sauces, turn it into soups and pastas and into an unusual *polenta en pilaf.* This treats coarse grits like rice, adding the grits to sautéed onions, steaming them in chicken broth and garnishing them with prunes and bacon or snails and hazelnuts.

Romania was so taken by cornmeal that it included the comforting sound of "mama" in the mush it called *mamaliga.* This was the mush that made its way to delis like Ratner's on Manhattan's Lower East Side, where it had "somehow crossed an ocean and a continent to become a staple for generations of Romanian peasants and their Jewish neighbors," wrote a Romanian descendant in *The New Yorker* (January 15, 1990). "Mamaliga, which I had always thought of as quintessentially Rumanian, was, of course, sweet corn, maize—the New World's gift to the Old."

Cornmeal is still a country staple of Romania and Hungary, beloved especially by the gypsy tinsmiths who called themselves *kalderash,* the name given to little corn cakes flavored with cumin and coriander. Northern Hungary and Transylvania lived on the mush they call *puliszka,* whether made into simple mush-and-milk puddings or stuffed with ricotta or beat into sweetened cakes. "Poor people made it with water instead of milk," George Lang told us in *The Cuisine of Hungary* (1971), "and they drank the milk with it, making an entire meal of it."

Cornmeal mush became a staple too in Africa, wherever the sixteenth-century Portuguese landed their ships. Corn spread also from the north to the interior along Arab trade routes, but after 1517 it was grown principally around the slave-trade ports of the West in order to supply cheap fodder for slaves during transport. " 'Tis generally observ'd, that *Indian* corn rises from a crown to twenty shillings betwixt February and harvest, which I suppose is chiefly occasion'd by the great number of *European* slave ships yearly resorting to the coast," wrote John Barbot, traveling between Ghana and Dahomey in the late seventeenth century. In eastern Africa the maize meal called *posho* in

In the corn-slave conjunction, Thomas Branagan's plan for a slave ship in his 1807 The Penitential Tyrant *seems to disclose a giant ear of corn with paired rows of slaves for kernels.* Rare Book Room, Library of Congress

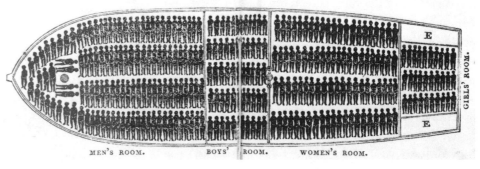

Plan of an African Ship's lower Deck, with Negroes in the proportion of not quite one to a Ton.

MEN'S ROOM.   BOYS' ROOM.   WOMEN'S ROOM.   GIRLS' ROOM.

Swahili was so common by the end of the nineteenth century that colonials used it as a verb: an accountant for the Imperial British East Africa Company in 1895 wrote that they had so many stores between Lake Victoria and the coast he was "in a position therefore to 'posho' caravans right through," meaning to supply them with meal. But not until this century did cornmeal become a staple of the eastern and central regions of Africa and so essential today throughout Africa, Laurens Van der Post has written, that "millions of Africans now eat it in the belief that it is something inherited from their own earliest beginnings."

Africans continued to grind corn, as Amerindians did, between flat stones and stone rollers or pounded it in wooden mortars. They too celebrated ripe-corn harvest festivals, such as Ghana's "Hooting at Hunger," celebrated today with the dish *kpekple,* a mush stew with fish and palm nuts. In South Africa mealie-meal has become the national porridge, despite the fact that when boiled simply with water, as Copeland Marks complained, "it comes out tasting like old inner tubes."

The cornmeal belt was a two-way stretch along the slave-trade routes that crossed the Atlantic: the Caribbean got its own back in an African form. Barbadian cornmeal mixed with okra in a dish called *coo coo* is African *frou frou.* West Indian *konkies* of coconut and sweet potatoes, a treat for Guy Fawkes Day, is African *kenkey.* West Indian *fungie* comes from Zaire, where they dump mush into a bowl, shape it into a smooth ball with the fingers, flip it against the side of the bowl to make it round, then tear off a small chunk and indent it with the thumb to use as a spoon for sauce. Where the cornmeal belt stretched north from the Caribbean into our Southern states, the African connection surfaces in the memories of slave descendants. One woman recalled the "meat tit" which they hung in their cabin to put meat skins in until they got enough for the mush pot, just as their African ancestors had done:

> I would put some of my meat skins in that pot and boil them and go out to that garden and crop some of them onion. I would make that pot of soup good and put some corn meal in. I had to feed all them children with that stuff. I'm like towards eleven or twelve years old. And I had to see that the children get something to eat or they would perish to death.

In America, a Southerner's meat tit was a Northerner's scrapple, and what bonded the childhood mush memories of one race to another, whether they had come as farmers from Saxony or slaves from Zaire, was the comfort of mush to empty guts. The Pennsylvania Dutch mixed it with scraps left over from

A Ghanaian woman uses a wooden paddle to mix her cornmeal dough of sofki. Courtesy Centro Internacional de Mejoramiento de Maíz y Trigo, Mexico City

butchering and called it "poor-do." In Germany and Holland they had mixed such scraps with buckwheat, and immigrants at first substituted cornmeal with reluctance until mush-and-hog scrapple became a dish as cherished as mush-and-milk. For the poor of the rural South, mush and hog offal had always been a special treat, like the "meat tit." A country woman in Appalachia today, a Mrs. Mann Norton, remembered for *Foxfire* how they cooked the scrapple of her youth:

> Take th' head, an' take th' eyeballs out, an' th' ears, an' cut down in there. Then y'got all th' hairs off of it. Y' put it in a big pot an' cooked it til th' meat just turned loose of th' main big bone.
>
> Y' lifted them bones out, an' laid your meat over in there an' felt of it with your hands t' see if they wasn't no bones in it. Then y' strain yer liquid through a strainer so th' little bones'd come out. Put'cher liquid back in a pot, and put that mashed meat back in that liquid. Put'cher sage an' pepper in there. Then y'stir it til it got t' boilin'. Then y' stick plain corn meal in there til its just plumb thick. Then y' pour it up in a mold, an' cut it off 'n fry it, an' brown it. Tastes just like fish.

And I remember the mush my Grandmother Harper made during the Depression for Sunday supper. She poured it hot into my Orphan Annie bowl and I made a well in the center with my Mickey Mouse spoon and filled it with a large pat of melting butter, then sprinkled the top with cinnamon and sugar and baptized it with cold milk. Hot mush, cold milk. The drama was there. I never asked for seconds because even before I'd finished my bowl I was dreaming of breakfast—thick slices of cold mush fried crisp on both sides in hot butter and slathered in syrup which, when my grandmother wasn't looking, I would lick from the plate with my tongue. Mush is still my favorite four-letter word.

One Man's Smut Is Another Man's Truffle

PERHAPS NO FOOD PLANT has ever been utilized more fully for edible, drinkable, inedible and unthinkable uses than corn, but all these uses depend upon language and the potency of names. What is "devil's corn" to a New Jersey farmer is *huitlacoche* (pronounced wheat-la-COH-chay) to a Mexican and *nanha* to a Hopi. These three terms name the common but visually monstrous smoky black-blue fungus, akin to a mushroom, that grows on corn and that American farmers call "corn smut." The Hopi thought it a delicacy and gathered it when young and tender for fried nanha; the fungus was parboiled for ten minutes, then sautéed in butter until crisp. Although older Hopi savored the fungus, few younger ones have tasted it, and today children use it to blacken faces in a game of tag.

Cushing said that for the Zuni corn-soot symbolized the "generation of life." In the Zuni granary, "a bunch of unbroken corn-soot" was laid each year next to a perfect ear of corn, dipped by a Seed-Priest in the sacred salt lake, Las Salinas. During the ceremony of the "Meeting of the Children," commemorating the return of the lost Corn Maidens, the Corn-Matron dusted the kernels of newly gathered and shelled corn with the old bunch of soot so that the seed might grow and reproduce. For the Zuni, Cushing added, such ceremonies were "testaments of faith, proving the infallibility of his 'medicine,' or fetishism, and of his practice of religion." By its very deformity, corn-soot was endowed with medicinal and miraculous powers.

For the typical American farmer, on the other hand, corn smut was and remains a disease to be eradicated, and it is seldom distinguished from the aflatoxin that attacks field corn in the form of a mold invisible to the eye and is wholly destructive to corn. Corn smut, on the contrary, is a highly visible spore, *Ustilago maydis,* that grows only on sweet corn and has been eaten for centuries, just as truffles have. But one man's medicine is another man's poison. Listen to the language of Edward Enfield in his *Indian Corn* (1866):

> The principal disease of this cereal appears in the form of a dark spongy growth, sometimes of a blue black or purple tinge, that occasionally shows itself on the stalk or leaves, but is more apt to take the place of the blighted ear. This substance increases gradually in size, sometimes reaching six or seven inches in diameter, and is generally regarded as a rank and luxuriant species of fungus [caused by bruises inflicted on the young plant by careless cultivation]. The bleeding that occurs from these wounds results in the formation of the dark morbid substance above described.

The modern scientific language of Paul Weatherwax doesn't carry the jolt of Enfield's bleeding wounds, but still presents a formidable obstacle to appetite: "The galls, which consist of hypertrophied host tissue mixed with fungal filaments, are at first greenish white in color throughout, but they darken from the inside outward and finally turn into sooty masses of black spores held together by remnants of the host tissue." The French called this *goitre du maïs,* Matthieu Bonafous reported in his *Histoire Naturelle, Agricole et Économique du Maïs* (1836), noting that most botanists considered it a form of mushroom. "I have tasted this substance several times without harm," he added. "I have seen cows eat a considerable quantity with impunity."

Soot, smut, goiter—none of them pretty names, least of all *huitlacoche,* which means "raven's shit." A story about Corn Soot Woman, translated by Ruth Benedict in her *Tales of the Cochiti Indians* (1931), deals with the name problem:

> One of the women [in the grinding house] heard somebody crying. She said, "Listen, somebody is crying." Just then the door opened and Corn Soot Woman came in crying. She said, "Nobody likes me to be with the corn they are to grind. I am fat but nobody has any use for me." The head woman of the [corn grinding] society said to Corn Soot Woman, "Why are you crying?" "I am crying because they don't ever put me among the good ears. I am not rotten." The head woman of the society said, "Don't ever separate her from the good corn. She is fat; that is why she is what she is. She is the mother of the corn soot and you must put her in with the good corn whenever you shell it, in order that they too may be fat, as she is." They gave her a new name, *Ioashknake* (shuck), and they gave the soot a ceremonial name, *Wesa.*

Bonafous's accurately baroque drawing of a corn "mushroom" in 1836 belies its less attractive names of corn soot, smut, goiter, raven's shit. Courtesy Library, University of California at Davis

If the Aztec gave her an irreverent name in *huitlacoche,* they relished the substance, as Mexicans do today in every kind of preparation from a hearty peasant soup to French-style crepes. When *huitlacoche* is renamed "Mexican truffle," even Americans become attentive, just as when fish eggs are called caviar. Since corn truffles grow only seasonally on unsprayed sweet corn, regrettably few Americans have had a chance to taste them outside Mexico. But Mexican farmers cultivate *huitlacoche* in special plots, so that it can be sold fresh in season and canned or frozen when not.

To give Americans a tasting and corn smut a new name, the James Beard House in New York City presented in 1989 an all-*huitlacoche* dinner of seven courses, conceived by Josefina Howard of Rosa

Mexicano restaurant. *Huitlacoche* appetizers were followed by *huitlacoche* soup, crepes, tortilla torte, salad and a startling but triumphant ice cream. Sautéed simply in butter, the corn truffle tastes delicately earthy and smoky like a mushroom, but because it is always attached to small kernels, it has also the sweetness and crunch of fresh corn. For mushroom and truffle lovers, corn truffles are a gastronomic joy. To American corn farmers, smut is smut.

Americans have, however, embraced without cavil other traditional uses of the corn plant that the ingenuity of native growers wrested from it. One was a major use of cornstalks, before the introduction of sugarcane, as corn sugar. Ben Franklin, pitching the sweetness of corn in his spiel to the French, focused on the juice in the stalk:

> The Stalks pressed like Sugar-Canes yield a sweet Juice, which being fermented and distill'd yields an excellent Spirit, boiled without Fermentation it affords a pleasant Syrop. In Mexico, Fields are sown with it thick, that multitudes of small Stalks may arise, which being cut from time to time like Asparagus are serv'd in Deserts, and their sweet Juice extracted in the Mouth, by chewing them.

Mexican bottled fruit drinks use corncobs for corks. Photograph © 1986 by Ignacio Urquiza, from *The Taste of Mexico* by Patricia Quintana; Stewart, Tabori & Chang, New York

Many early explorers mentioned cornstalk sugar, including Ralph Lane in Roanoke in 1585: "And within these few days we have found here maize, or Guinea wheat, whose ear yieldeth corn for bread, four hundred upon one ear; and the cane maketh very good and perfect sugar."

For both natives and invaders the uses of corn for food and drink overlapped with medicinal and household uses. The Iroquois, after extracting juice from the stalk, turned the juice into a medicinal lotion for wounds and the stalk into a medicine bottle to contain it, plugging the cut section at both ends. The pith they made into a sponge to absorb liquids, and from the dried stalks they cut floats for fishing lines, while boys used the stalks for playtime spears. Colonists were quick to absorb corn into their own traditions of herbal medicine, as John Josselyn did in 1672, citing the uses of "Indian wheat": *To ripen any Impostume or Swelling. For sore Mouths. The New Englands Standing Dish. . . .* It is hotter than our

Wheat and clammy, excellent in *Cataplasm* to ripen any Swelling or impostume," he explained. "The decoction of the blew Corn, is good to wash sore Mouths with." Roger Williams recommended it as a laxative: "If the use of it were knowne and received in England (it is the opinion of some skillfull in physic) it might save many thousand lives in England, occasioned by the binding nature of the English wheat, the Indian Corne keeping the body in a constant moderate loosenesse." English opinion, on the contrary, complained of the binding nature of Indian corn, along with its general indigestibility. This was a prejudice revived by lady nutritionists like Mrs. Mary Lincoln, who warned in 1887 that corn "is heating for persons with weak digestion, and should not be given to scrofulous children, or to invalids when there is any inflammatory condition of the system."

Corn silks brewed into tea, on the other hand, was a colonial medicine that remained in good standing. Herbalist Tommie Bass recommends it today to fellow Appalachians as a good kidney medicine. "If you just want to make a dose, you know, right quick, take about as many silks as on a good ear of corn and put it in a tea cup and run hot water over it and let it sit about ten or fifteen minutes and then drink it." Corn parched and ground as a substitute for coffee was once almost as common as corn tea, for both health and economy. "Dyspepsia Coffee," Libbie Thompson called it in the *Kansas Home Cook-Book* (1874); she diluted regular coffee with a mixture of cornmeal browned with molasses in the oven. During the Great War, corn coffee was a patriotic sacrifice of the home front, served up by Charles Murphy as "War Coffee":

> Mix together, thoroughly rubbing with the hands until the mixture resembles soft brown sugar, two quarts of bran, one quart of corn meal, and a large cup of molasses. Spread the mixture in a large dripping pan, brown it in a slow oven, stirring it often with a long-handled spoon until it is a rich seal brown. Do not let it scorch or burn. When done, put in jars like regular coffee.

Anthropologists were always keen to cite the medicinal uses of corn among their studied tribes and were often struck by the way in which every part of the corn plant was made a specific medicine against some ailment. Among the Tewa at Santa Clara, rubbing a warmed ear of corn with the foot was said to help swollen glands in the neck. At San Ildefonso Pueblo, corn pollen was recommended for palpitations of the heart and corn smut for women suffering from irregular menstruation. The Oklahoma Cherokee practiced preventive medicine with new corn silk in order to ward off green-corn fever: after wading into a stream at dawn, the medicine man filled a container of green ears and corn silk with water, stirred the contents counterclockwise while chanting four times, then presented the liquid to the patient for drinking or bathing.

Among Nebraska pioneers corn was used in ritual ways to prevent or cure warts, an alarming number of them summarized in *A Treasury of Nebraska Pioneer Folklore:*

> Steal a grain of corn and destroy it so that its owner will never have it again, and your warts will leave.
>
> Rub the warts with a kernel of corn and feed the corn to a rooster. Be sure the rooster gets it and your warts will disappear.
>
> Touch seventeen different kernels of corn to each wart, then feed the corn to the chickens. If they eat the corn, the warts will disappear.
>
> Bury a small bag of corn. When the corn decays, the warts will leave. Cut a corn cob crosswise and take out the white pulp center. Put this on the wart. Then fill the inside of the cob with soda and add a drop of vinegar to boil the soda away. Hold this over the warts. Repeat two or three times a day until cured.

The number of variants would suggest that warts were as common in Nebraska as corn, and as hard to get rid of.

For the Indian household, no part of the corn plant went to waste: leaves, stalks, husks and cobs were transformed into mats, trays, cushions, hammocks, scrub brushes, moccasins, feather holders, bottle stoppers, combs and back scratchers. Settlers followed aboriginal example in contriving their own uses, such as cornshuck mattresses. After the Civil War, these were a sign of hard times, wrote Elizabeth Woolsey Spruce (1873–1969) in a "corn memoir" telling how her grandmother prepared dowries for her daughters in south Texas:

> Since cotton was too expensive to use for mattresses, they used cornshucks instead. The hard end of the shuck was cut off. In one hand a thin layer of shuck was held and split all the way down with a table fork. . . . When they had sufficient for a bed it was placed in a bed tick without tacking. Of course it took days and days but what was time for, in those reconstruction days, except for hard work by every one? This mattress was always placed under the feather bed. There were no bedsprings.

The discomfort of cornhusk mattresses figured in a great deal of American humor, as in Josh Billings's description of a "modern slat bottom" bed with two

Cornhusk dolls made by Indians of the Six Nations. Milwaukee Public Museum, Neg. No. 425333

mattresses, "one cotton, and one husk, and both harder, and about az thick az a sea biskitt. Yu enter the bed sideways and kan feel every slat at once, az eazy az yu could the ribs ov a grid iron."

A more unusual use for cornhusks was that described by Mattie Huffman, one of the voices from the Kansas frontier collected by Joanna Stratton in *Pioneer Women* (1981). Mattie remembered how her aunt gathered cornshucks, cut them crosswise and wound her knitting yarn on them, twelve inches thick on each one, in order to dye the yarn: "When it was taken out, if the dye was red, for instance, next to the shuck would be the natural white color of the yarn, next to that a pink shade, and next or on the outside it would be red. This was the manner of making what they called clouded yarn." Clouded-yarn knitters in Kansas echoed weavers in Mexico, who primed the threads of their looms in corn gruel. Because of the corn, they kept their children away while weaving, explained a contemporary Maya woman in Chiapas: "If they put their heads in our loom, they will be big eaters and will want to drink a lot of corn gruel."

Indian craftsmen relied on corn products in even more unusual ways. In pre-Columbian Peru, south coast Indians ground corn kernels to make a glue for their feather masks: they first covered a human skull with finely woven cotton and then glued feathers to the cotton. In post-Columbian Mexico, artists ground up the inner core of cornstalks and mixed the stuff with a natural glue to make a material, like papier-mâché, for sculpting figures. After covering the sculpted paste with cloth, they painted it to make realistic saints and crucified Christs which, because they were light, were favored for carrying in religious processions.

For Indian and colonist alike, dried corncobs furnished a natural supply of fuel, but a new-fashioned twist is the invention of the Dovetec Corn Heater by Carroll Buckner of North Carolina, to replace wood. His corn stove holds seventy pounds of cobs, fed into the fire by an auger. "It takes years to grow a tree," Buckner says, "and only four months to grow a crop of corn." More than a century ago, *The Prairie Farmer* of March 20, 1869, advised its readers to burn corncobs to make "cob coal," not as a fuel but as a medicine for hog cholera:

> Collect your cobs in a pile and burn them until they are thoroughly charred, then wet them out. Sprinkle brine or salt on the coals and let your hogs eat all they want. . . . Last winter I neglected to give my hogs cob coal and the result was, in the spring, they took the cholera; ten of them had it bad. I raked up the cobs in the lot and charred them well, then salted the coal freely. They all ate of the coal but three that were too far gone to get up. They died; the rest got well.

A more usual use of cob coal was to "cure" hogs another way, smoking their slaughtered parts to flavor and preserve them for food. New Englander Amelia Simmons recommended corncobs for smoking bacon, and the tradition continues in the cob-smoked maple-cured hams from Enosberg Falls, Vermont, where "Puffer's Cob-Smoked Ham" has made Puffer Lumbra a household name.

Country folk everywhere remember one use of the corncob that was later supplanted by store-bought paper. "All the farmers, black and white, kept dried corncobs beside their double-seated thrones," Harry Crews wrote in *A Childhood* (1978), "and the cobs served the purpose for which they were put there with all possible efficiency and comfort."

A more public use of the American corncob was for pipes. The corncob-pipe center of the world is Washington, Missouri, where the Buescher Corn Cob Pipe Company turns out twelve thousand to eighteen thousand pipes a day to sell in seven thousand outlets in fifty-seven countries around the world. Today they cost $1.95; twenty years ago they cost 10 cents, and more than a century ago, when Henrick Tibbe of Holland started the company, they cost but a penny a pipe. Tibbe was an emigrant woodworker in 1869 when he decided to adapt his native Dutch pipe to corn country by shaping a cob on his lathe and applying a plaster-of-paris mixture to the outside. Today, his grandchildren

Interior of corncob ("meerschaum") pipe factory, Washington, Missouri, July 26, 1926. Photo Pierce W. Hangge, courtesy Missouri Historical Society

run the company and grow a special hard-cob hybrid for their pipes made according to family tradition: The cobs are aged for two years before they are sawed into "pipe bowl blanks"; then a "cob boring bit" drills a hole in the center of the blank and the blank is turned and shaped on a lathe; the indentations are filled with plaster of paris, which is finally sanded and shellacked "to produce a high luster while still preserving the texture and appearance of the corn cob."

If corncobs turned into pipes are as American as General MacArthur, striding through the waters of Manila Bay with a smoking cob clamped in his teeth, corncobs and cornshuck mattresses symbolize other unforgettable, if less heroic, moments in the American imagination. What reader of William Faulkner's *Sanctuary* can forget the way Temple lay in bed the night before her rape in the corncrib?

> There was no linen, no pillow, and when she touched the mattress it gave forth a faint dry whisper of shucks. . . . in the silence Tommy could hear a faint, steady chatter of the shucks inside the mattress where Temple lay, her hands crossed on her breast and her legs straight and close and decorous, like an effigy on an ancient tomb.

Faulkner cuts from the cornshuck innocence of Temple to the trial of Popeye, at which the district attorney offers as criminal evidence a corncob that appears to have been dipped in dark brownish paint. "You have just heard the testimony of the chemist and the gynecologist. . . . this is no longer a matter for the hangman, but for a bonfire of gasoline."

If a corncob in a psychopath's hand could be an instrument of dark defilement, so cornmeal, in a housewife's hand, could be an instrument of cleansing. Sidney Morse's *Household Discoveries* (1908) counted the ways: Rub cornmeal on calcimined walls to clean them, spread a cornmeal and oxalic-acid paste on straw goods to clean them, clean Panama hats with dampened cornmeal applied with a stiff nailbrush, clean white kid gloves by rubbing your gloved hands together in fine cornmeal, improve the luster of furs by browning cornmeal and while still hot rubbing it through the fur with a flannel cloth, clean wallpaper by making a dough of rye flour dipped in cornmeal to use as a rubbing cloth, clean blood or mud stains with cornstarch, make hand soap from cornmeal and powdered borax.

Just plain corn kernels were also used in extraordinary ways. If the French learned to force-feed geese with kernels poured through a funnel down the throat, to make their livers fat (or their *foies gras*), so the Spanish learned to force-feed heretics with dried raw kernels and water, and made them sit naked in the sun until their bellies burst. So Americans learned to tempt fish with kernels as bait. Louis de Gouy records an old Oregon game law that permitted fishermen the use of corn kernels but made it a misdemeanor to use canned corn. Corn fishing and hunting were part of the tradition of the West, as described by Richard Ford in a story named for Great Falls, Montana, in which the narrator tells how his father baited his lines with corn to catch fish in springtime: "He used yellow corn kernels stacked onto a #4 snelled hook, and he would rattle this rig-up along the bottom of a deep pool below a split-shot sinker, and catch fish." In the fall, they used corn as a "hunger-line" to catch ducks: "We would set out his decoys to the leeward side of our blind, and he would sprinkle corn on a hunger-line from the decoys to where we were. . . . And after a while, sometimes it would be an hour and full dark, the ducks would find the corn, and the whole raft of them—sixty sometimes—would swim in to us." Corn is used as bait even in Eastern cities, where poisoned corn kernels are strewn along city roofs to diminish the encroaching population of pigeons. Even in the animal kingdom, one bird's meat is another one's poison.

The totemic power of corn, for good or evil, has survived wherever people live close to the earth. A Southern black woman, who contributed an oral history to *When I Was Comin' Up* (1982), saw death in a cornshuck: "I seen a token, a great big, black bear, reachin' his mouth into a shuck of corn—and then my momma passed in three days. That was a token." In Haiti, corn retains demonic power, and the cornmeal that is ritual medicine is also ritual poison of a particularly potent kind. A recipe devised by the *voudon* La Bonte was outlined by Wade Davis in *Serpent and Rainbow* (1985): Take wood ground from a particular healing tree, leg-bone shavings and other decayed matter from a human cadaver, add white sugar, basil leaves, seven drops of rum, seven drops of claïrin and a small amount of cornmeal and mix well. Cornmeal and corn kernels figure not just in concoctions and incantations, as when the *mambo* traces a cabalistic design with meal on a coffin or gravel, but in rites of possession, as in Davis's powerful evocation of a woman possessed by the spirit of corn:

Then, on the steps of the church, the scene turned into an epiphany. A healthy peasant woman, dressed in the bright-blue-and-red solid block colors of Ogoun, the spirit of fire and war, swirled through the beggars possessed by her spirit. Over her shoulder was slung a brilliant red bag filled with dry kernels of golden corn. She twirled and pranced in divine grace, and with one arm stretching out like the neck of a swan she placed a small pile of corn into each of the begging bowls. When she was finished, her bag empty, she spun around to the delight of all and with a great cry flung herself from the steps of the church.

The power of the corn spirit that possessed Aztec priests as they once twirled and pranced on the steps of their temples resides still in the kernels of corn with which contemporary "daykeepers" of the Quiché Maya in Guatemala divine the future. The daykeeper passes his hands over the kernels and divides them into lots, then finds his auguries by counting the day numbers and names of the 260-day cycle of the ancient calendar. The Zuni too measured time, past and future, by kernels of corn. They measured lifetimes by "generations of corn," counting from the time a boy received from his father his first planting seed to the time he in turn gave seed to his son to plant. To number our seven ages of man, they counted seven ages of corn. In the numberless generations of corn possessed by the corn spirit, even the darkly monstrous corn-soot could take its place as a maker of miracles, when it was called "generator of life."

The Language of Drink

CHICHA IN PERU, MOONSHINE IN ARKANSAS

WHEN I STOOD in the great cathedral on the Plaza de Armas in Cuzco, my eye was seized not by the altar of beaten silver and the monstrance of solid gold, but by a large canvas of the Last Supper painted by an Indian converted in the seventeenth century to the white man's gods. Here the Inca Jesus and his twelve attendant lords sat before a table laden with the two chief elements of an Inca feast: not bread and wine, but *cuy* and *chicha,* or guinea pig and corn beer.

Corn was both bread and wine in the New World and particularly in Peru, for fermented corn did for the Inca what ash-processed corn did for the Hopi and lime-processed corn did for the Aztec. Not only did corn fermented from mash and malt make the drinkers of the brew happy in ways familiar to all drug-consuming cultures, but the enzymes of the ferment made the brew nutritious. In ancient Peru, the brewing of beer was held so sacred that it was entrusted to the hands of virgin priestesses, whose Christian equivalent would be the nuns of Mary. Even in modern Peru beer is a form of sacrament and no one drinks it before first pouring out a few drops on the ground to honor Mama Sara, the corn goddess, just as no one builds a house without putting a miniature bull and *chicha* jar on the roof, for luck and life.

My first taste of home-brewed *chicha de jora* sent me in search of a recipe for it. The earliest I found had been written by Girolamo Benzoni, in his *Historia del Mundo Nuevo* (1565), when he described how the Chosen Women of the Sun enhanced the potency of their brew by mixing the corn malt with saliva:

The women, taking a quantity of grain that seems to them sufficient . . . and having ground it, they put it into water in some large jars, and the women who are charged with this operation, taking a little of the grain, and having rendered it somewhat tender in a pot, hand it over to some other women, whose office it is to put it into their mouths and gradually chew it; then with an effort they almost cough it out upon a leaf or platter and throw it into the jar with the other mixture, for otherwise this wine would have no strength. It is then boiled for three or four hours, after which it is taken off the fire and left to cool, when it is poured through a cloth, and is esteemed good in proportion as it intoxicates.

Everywhere I went I came upon *chicha. Chicha* is the soda pop as well as the beer of Peru, and in one form or another everybody from nursing babes to toothless grannies drinks it. Children go for unfermented sweetened *chicha blanca,* which is made from white corn and sprinkled with cinnamon, or *chicha morada,* which is made from purple corn and looks and tastes like a strawberry milkshake. Adults move on to *chicha de jora,* made from sprouted and fermented corn, which produces a pale yellow-white brew with a two-inch head of foam and which tastes a bit like English barley water mixed with light pilsner, or like a shandygaff of ale mixed with cider or milk, once a country staple in both England and America. *Chicha picante* is the same brew zapped with lemon and the hot pepper sauce called *aji. Chicha frutillada* adds fresh fruit, such as strawberries, and *punchy* adds a shot of Pisco brandy to create a potent boilermaker.

A nineteenth-century engraving by A. de Neuville of chicha drinkers in Cuzco, Peru.

In Peru today, women are still the brewers and every woman knows how to make home brew, whether for a *chicheria,* or beer joint, or just for her family and neighborhood friends. Long ago Peruvian women learned to share the work, and every rural village will sport above some doorway a red or white flag, nowadays usually plastic, which means "Here's where you get fresh *chicha* today." You may be sharing an earth-floored room with a few chickens and pigs, but like a pub the mood is convivial in proportion to the size of the glass consumed, and a quart-sized *caporal grande* costs but a dime.

My recipe search in Cuzco took me one day to a padlocked gate of sheet iron, beyond which was a dirt yard cluttered with antique ceramic *chicha* jars and scraggly plants native to pre-Inca Peru. The yard was enclosed by adobe walls covered with frescoes of the sun and moon and scenes of Inca life in the style of a Peruvian Grandma Moses. This was the studio-classroom of Professor Chavez Bellon, an elderly man in a baggy suit with a big smile and a lifelong enthusiasm for all things Incan. He swept me and a handful of students in his wake to his local, a *chicheria* called Rosas Pata, on Ciro Alegria. He led us through a small adobe house into a back room, so black with smoke from the brazier in the corner that I could barely make out the pair of women stooped

"[Chicha] *precipitates them into incest, sodomy and homicide, and there is rarely a drinking bout that is not mixed with heathen rites, the Devil often being visibly present, disguised in the person of an Indian.*"

—FATHER GARCIA MARCOS, quoted by Father Antonio de la Calancha in *Coronoica Moralizada del Orden de San Augustin en el Peru* (1638)

over the boiling black-iron pots. Near them, a batch of corn mash fermented in a straw basket, awaiting the addition of cold water from a long wooden trough. Next to the trough stood a bucket of foaming *chicha,* ready to pour into the glasses of the men in the adjoining room, where the air was dense with smoke and the benches with beer drinkers.

Professor Chavez did his best to explain the old and the new way of making *chicha.* Old-style *chicha* was made simply by letting ground corn and water slowly ferment, but the new way is to speed the process by adding yeast and sugar, plus a little trigo (a type of wheat). Since then, every Peruvian I've met in Cuzco and elsewhere has had his own contradictory way with *chicha,* some adding quinoa and barley, others the heavy dark-brown sugar called *chancaca,* others insisting on straw baskets for filtering and buried earthen jars for fermenting. The closest I came to a "definitive" *chicha* for the would-be home brewer was from Felipe Rojas-Lombardo, the late Peruvian-born chef of Manhattan's Ballroom restaurant, known for its fine tapas:

Take dry corn, then make it wet to sprout it. Keep it moist with burlap (or some other loose covering) until the corn grows two- to three-inch sprouts. Then spread honey on the *maíz de jora* and put it in the sun for a day. Forget how it looks, it's a mess. Just scrape it up, put it in a pot with water and add some dry soaked corn that is not sprouted, along with a piece of *chancaca* or more honey. Bring to the boil slowly and boil for five hours, adding water as it evaporates. (My mother always added cinnamon and clove.) Take it off the fire and when it is fully cool, strain it through cheesecloth [a grass-lined basket was traditional; any metal other than stainless steel will ruin the fermenting process]. Store the liquid in a ceramic pot or jar in a dark place for a week or two. In the Andes, they bury the pot in the ground, leaving just the long neck of the pot above the earth, covered with a cloth, because the earth maintains an even cool temperature. The longer it stays, the stronger it gets.

I tried many times but the mess stayed a mess—souring, rotting, molding, yes, but fermenting? No. My advice is to leave it to the *chicheria* women. If not everybody makes it, everybody drinks it, said the Cuzco guide who pointed out the *chicha* grooves cut in the stones of the fortress of Sacsahuaman, and who now sat me with me in Cuzco's best-known *chicheria,* La Cholla on Calle Pumacurco, where a baby dunked her head in her mother's *caporal grande* to the sound of her father's homemade-fiddle music. But for the chatter of Spanish and Quechuan voices, I might have been listening to a hootenany in Arkansas, and a few months later I was doing just that, picking up the beat of Paralee and her Swinging Strings with other folks who appreciated home brew.

A colonial drinker taking his morning dram. North Wind Picture Archives

IT'S A LOT EASIER to find a *chicha* maker in Peru than an old-time moonshiner in Arkansas, but they share a brewer's knowledge of corn. If he is making good whiskey, an Arkansas moonshiner will begin with what he calls corn beer. In this country, whiskey originally was "pure corn," just like old-time *chicha*, without additional sugar or yeast. Mesoamerican Indians for centuries brewed beer from sprouted corn and cornstalks, as they distilled liquors from native plants and secretions like fermented honey to make a kind of mead, agave (or maguey) to make *pulque*, saguaro cactus to make a wine, peyote cactus to make a narcotic. Alcoholic beverages flourished wherever domesticated plants flourished, as dual signs of civilized life, but wheat-culture colonists brought their own ancient heritage of brewing techniques with them and had to learn to substitute corn malt for barley in the New World.

"Wee made of the same [mayze] in the countery some mault, wherof was brued as good ale as was to be desired," Thomas Hariot wrote from Virginia in 1588. By 1620 Captain George Thorpe, who had set up a distillery at Berkeley Plantation on the James River, could write to a friend in London that he now made "so good a drink of Indian corn as I protest I have divers times refused to drink good strong English beer and chosen to drink that." Captain Thorpe, however, was unusual in his open-mindedness and was repaid for his generosity by being scalped two years later by Indians who had drunk too much of his corn.

In England, of course, the brewhouse was as important an adjunct to the farm as the dairy, and small beer was the drink of choice for every member of the household just as soft drinks are with us. On the farm, brewing was woman's work. When Richard Bradley, a professor of botany at Cambridge

University, undertook in 1736 to instruct *The Country Housewife and Lady's Director* in the Management of a House, and the Delights and Profits of a Farm, his first concern was "Instructions for managing the Brew House, and Malt-Liquors in the Cellar; the making of Wines of all sorts." The housewife's traditional brewing tasks provoked a Virginia colonist a century earlier to scold lazy American housewives for dereliction of beer duty. The title John Hammond gave his work in 1656 suggests that women were much on his mind: *Leah and Rachel, or, the Two Fruitful Sisters, Virginia and Maryland: Their Present Condition, Impartially Stated and Related.*

Beare is indeed in some places constantly drunken, in other some nothing but Water or Milk, and Water or Beverige; and that is where the good-wives (if I may so call them) are negligent and idle; for it is not want of Corn to make Malt with, for the country affords enough, but because they are slothful and careless; and I hope this Item will shame them out of these humours; that they will be adjudged by their drinke, what kind of House-wives they are.

Spirits, or *aqua vitae,* were another matter. Distilling was man's work, and the early Scottish and Irish monastic tradition of distilling spirits from barley malt to make what the Celts called *usquebaugh,* or water of life, played an important role in the history of corn. It also, as I discovered with shock, signified in the history of my own American life. When I set out to find a moon-

A family in 1894 celebrates the Fourth of July with copious bottles of bourbon.
The Bettmann Archive

shiner who would give me a recipe To Make Corn Likker at Home, I had no notion that I would uncover the stills of my own teetotaling ancestors.

It wasn't easy to find Frank and Paralee Weddington in Butler Hollow near Beaver Town, total inhabitants fifty, in the wooded hills of northern Arkansas near Eureka Springs, but I got a little help from friends. The house we came to was a ramshackle weathered cabin, with rockers on the porch, barking dogs in the yard and a bluebottle tree out front. A bluebottle tree is one you make by putting empty blue glass bottles on the branches of a leafless tree to catch the sun and prettify the yard. Frank, bright-eyed and toothless, was tucked up in his overalls behind the big woodstove in the parlor. At eighty-five he doesn't hear or remember so good, but Paralee at a mere seventy-six, with short-cut hair and square-cut jaw, remembers for both of them. Since they've been married sixty years and have two daughters, six grandchildren and five great-grands to show for it, they have much to remember.

I asked Frank how he got started moonshining. "There wasn't work and I had to do something and I made a livin', " he said. "Back in the twenties, times was hard." Paralee took over: "Him and all his folk was always crazy about whiskey and I guess they didn't have no money to go buy it, so they went to makin' it. He growed up with boot jacks—a boot jack is a still," she explained. "We'd been married just nine months, Betty, the nineteenth of October and next July twenty-fourth he got caught and was in jail for eight months, so we spent our first anniversary with him in jail." Frank was caught just four days, I realized, before I was born.

It was a one-man operation, Frank told me. "I done everythin' meself." So when he gave me his whiskey recipe, I knew it'd been tested. Just as with *chicha,* however, there were an old and a new way of making corn whiskey. The old way was "pure corn" and slow fermentation. The new way was "sugar corn," with quantities of brown sugar substituted for grain because it was quicker and cheaper. For the purist, "sugar likker" was a corruption, but for the Prohibition moonshiner it was more like necessity. Frank's recipe was for "sugar corn":

You git your barrel, whatever you're goin' to put your mash in. If you're goin' to put up a whole lot, git you a sixty-gallon barrel, and put fifty pounds of chops [the chopped corn] in it and fifty pounds of sugar, and a package of yeast foam from the store. Stir that all up together and let it set three days till your still goes to pot and settles down. Put it on the far [fire] and go to burnin' it. Put somethin' under thar to catch your whiskey in. You start to boilin' your copper pot and the water come down to the boiler over into a barrel of cold water, so you got it coiling around and around and your whiskey comes out here. You got to watch it, when it gits down kinda weak, you're supposed to quit. Doublin' back, we call it, you catch a little and put it in your boiler for the next batch. Doublin' back, boy, it's pure dee alkeehol then.

Most whiskey is made from a combination of mash (cornmeal of ground unsprouted kernels mixed with water and then fermented) and malt (cornmeal of sprouted kernels, dried and ground) from white flint, not yellow hybrid, corn. A breed

A classical copper pot still of the kind used by distillers like George Washington at Mount Vernon, and by moonshiners everywhere, is now the property of the Bureau of Alcohol, Tobacco and Firearms, on permanent loan to the Oscar Getz Whiskey Museum in Bardstown, Kentucky. Courtesy Department of the Treasury, Bureau of Alcohol, Tobacco and Firearms

like Holcomb Prolific was favored by the Georgia moonshiner quoted by *Foxfire,* who explained how Georgians sprouted corn: In winter they soaked it in a barrel of warm water for a day, then drained it and put it in a tub by the stove; each day for four or five days they covered the corn with warm water for fifteen minutes, then drained and turned it. Soon they'd have a good malt with shoots two inches long. In summer, they put corn in tow sacks in the sun, sprinkling the sacks with warm water once a day and flipping them. They had to watch it, however, for if the corn got too hot, it would go "slick."

They would take the sprouted corn to be ground at the mill or they'd use a sausage grinder at home. They'd need about 1½ bushels of corn sprouted for malt to put with 9½ bushels unsprouted for mash. They'd grind the mash fine, then boil it in water in batches and put it in the mash barrels. "It's like a woman making biscuits," said one Tennessee moonshiner. "If she don't know how to mix that dough in the bread bowl over thar, when she puts 'em in the pan, they ain't no count. You don't make likker in a outfit. You make it over thar in the mash barrels."

In each of his mash barrels, the Georgia moonshiner first thinned the mash by sticking a mash stick upright and adding water until the stick fell to the side. Next he added a gallon of malt to each barrel and a double handful of raw rye over the top. After five days the mixture would develop a thick foamy cap about two inches deep, a suds-and-blubber foam called a "blossom cap." Once the alcohol had eaten up the cap, this "distiller's beer" was ready to run. You had to catch it at just the right moment, he warned, or it would turn to vinegar. Both distiller and revenuer could tell by the "smack," the feel of the beer when you dipped your fingers in it, whether the liquid had the right stickiness or tackiness to begin distillation.

All a still needs is a firebox attached to a cooker that is attached to a condenser. The simplest kind was a "ground hog" or "hog still," built directly on the ground with a furnace of mud and rocks built around it so that the flames could wrap around the pot and exit through a backside flue. Vapors would collect in a barrel on top that was connected to a condenser, so that fermenting and distilling could be done in one operation and the still could be easily built and concealed in a creek bank or hillside. The most common still was a pot still, in which a copper pot shaped like a turnip with a tall neck could be made airtight with a lid and fired by a furnace below. Like fermenting, firing had to be watched to keep the temperature steady at 173 degrees Fahrenheit, the point at which alcohol vaporizes. Higher heat meant more water and impurities in the condensed liquor.

For a "singling" run, as it was called, you could expect to get fifty gallons of weak alcohol from eight bushels of corn. "You run them singlin's 'til as long as they got any strength in 'em, then you put your hands under the worm [the distilling coil], rub 'em and inhale," explained a distiller from Blairville, Georgia. "When they smell right sour, the alkihol's out." That's when the whiskey, as they say, "breaks at the worm." Moonshiners pride themselves on sensing the moment of break. "Lots of times I could tell when the bead broke because it changed that twist," said another Georgian. "I'd say, 'It's broke, boys,' and go over and test it and it'd be dead." To bring the liquor up to strength, you'd have to run the singlings through again for a doubling, what Frank called "doublin' back," and thus "pure dee alkeehol." From fifty gallons of singlings, you'd get maybe 16 to 20 gallons of "doublin' likker," or about $2\frac{1}{2}$ gallons to a half bushel of cornmeal.

During Prohibition some hillfolk devised the thumper keg, which speeded up the doubling by distilling the vapors through a keg of beer inserted between the cooking pot and condenser. When the hot steam hit the beer, it began to bubble or thump. Beneath the condenser, the distiller put a funnel lined with a double layer of clean cloth and a double handful of clean hickory coals to act as a filter and remove the oil slick called "bardy grease." In the early days moonshiners were more likely to use a "bootleg bonnet," or felt hat, tacked across the top of the keg beneath the worm.

To test the strength of his doublin' likker, a moonshiner would proof it by shaking a sample in a glass vial and then looking at the bead. "If it's high

An 1830s temperance pamphlet from Salem, Massachusetts, depicts "The Demons in the Distillery," equipped with a classic copper pot still.
North Wind Picture Archives

proof—say 115 to 120—a big bead will jump up there on top when you shake it," explained one man. "If the proof is lower, the bead goes away faster and is smaller." When a doubling comes through the worm, the first shots are as high as 150 or 160 proof, but it gets weaker as it continues until it gets below 90 proof, when it breaks at the worm, and from then on is called "the backin's," which can be mixed with singlings for the next run. The ideal 100-proof pure corn makes a bead about the size of a BB shot. That confusing word "proof" was a British import, and it meant the proportion of alcohol it took to make a small heap of gunpowder burn with a steady blue flame. Whiskey that was about half alcohol and half water "proved," by means of the gunpowder test, to be the best proportion for human palates and bloodstreams; the British settled on 57.1 percent to mean 100 percent fit for drinking, whereas Americans rounded off the figure to 50 percent. In the colonies, 100 proof meant, as it still means, exactly half alcohol.

Two variations on pure corn liquor involved souring and aging. To get a sour mash whiskey, a distiller took the "slops" left over in the cooker after a run and added it to a new batch of mash to repeat the fermenting process. "Sloppin' back" was like adding a sourdough starter to flour and water when making bread to give a little more character to the dough. Unaged whiskey was called "white lightning," for the liquid was as clear and colorless as vodka when it came fresh from the still. The Scotch whiskey makers who invaded this country in the eighteenth century were accustomed to aging whiskey in wooden barrels or casks, since the wood in symbiosis with the liquor helped to filter out impurities as the liquid evaporated and to ripen its flavor through the tannin and other extractives of the wood. The liquid also took on color and tasted as smooth and sweet as the Scotch whisky they brewed back home from barley malt.

Legend differs on which Scotsman in which county of Kentucky first aged his corn whiskey in a newly charred white oak cask to create the distinctive taste of American bourbon, but the chief contenders are both Baptists. According to Gerald Carson in *The Social History of Bourbon* (1963), the honors are usually given to the Reverend Elijah Craig, a capitalist-entrepreneur/preacher, who in 1789 included in his watery activities distilling, as well as baptizing, convenient to his gristmill, papermill and clothmill business at the Big Spring Branch of North Elkhorn Creek in Georgetown of Scott County. But two years earlier, in Bourbon County, originally part of Virginia rather than Kentucky, the Reverend James Garrard, who combined preaching with tavern-keeping, was charged with selling locally distilled whiskey without a license. In neither case are there any records of charred casks.

"Here's to Old Corn Likker,
Whitens the teeth,
Perfumes the breath,
And makes childbirth a pleasure."
—North Carolina folk saying

A. W. Thompson's drawing for Harper's Weekly, *December 7, 1867, captioned "Illicit distillation of liquors—Southern mode of making whisky," exemplifies Moonshine Romance.* Picture Collection, The Branch Libraries, The New York Public Library

In the next two decades the bluegrass counties of northern Kentucky, blessed with good limestone water, produced excellent whiskey at such a rate that Fayette County alone numbered 139 distilleries and the state 2,000. Before this time and until the Revolutionary War, rum was the distilled drink of choice of all the colonies, and New England was the pot still that condensed molasses and slaves into money and rum. New England ship owners who traded rum for slaves on the Guinea Coast, then slaves for molasses in the West Indies, finally turned molasses into rum. By 1750, the sixty-three rum distilleries in Massachusetts consumed 1,500 hogsheads of molasses a year. After the war, the distilling center moved from New England to the South as whiskey replaced rum and as bourbon, with its echoes of pro-French and anti-English sentiment, became the whiskey of choice.

The name "bourbon," however, didn't take hold until the mid-nineteenth century (1846 saw the first printed use), when home distilling began to move off the farm and into the factory. "Old Bourbon" became synonymous with good aged whiskey, a class drink. Old Crow, for example, was the name given by W. A. Gaines & Co. in 1856 to the bourbon they produced by the refined methods for souring mash developed some decades earlier by James C. Crow, a Scottish-born physician. When a Kentucky paper-maker named Ebenezer Hiram Stedman recalled an annual fishing week at Elkhorn in the 1830s that attracted the area's richest and finest, he remembered that they had "alwais The Best of old Bourbon" and that the bank president "alwais Kept his Black Bottle in the Spring and the mint grew Rank and Completely Hid the Bottle." Stedman, no bourbon snob, proclaimed the virtues of even the most ordinary whiskey:

> Every Boddy took it. It ondly Cost Twenty Five Cents pr. Gallon. Evry Boddy was not Drunkhards. . . . A Man might Get Drunk on this Whiskey Evry Day in the year for a Life time & never have the Delerium Tremers nor Sick Stomach or nerverous Head Achake. . . . One Small drink would Stimulate the whole Sistom. . . . It Brot out kind feelings of the Heart, Made men sociable, And in them days Evry Boddy invited Evry Boddy That Come to their house to partake of this hosesome Beverage.

After the Civil War, the increasing popularity of this hosesome Beverage turned whiskey-making into a major Southern industry. While men like Colonel Edmund Hayes Taylor, Jr., a nephew of General Zachary Taylor of the Mexican War, retained home quality in manufacturing his Old Taylor, Hermitage, Carlisle and Old Sinner brands in Columbia, Kentucky, others were less scrupulous. Grocers would bottle bulk whiskey from the spigot and with impunity slap on fraudulent labels like "Old Crow Hand Made Sour Mash Bourbon Whiskey." For a long time, the government's sole interest lay not in supervising labels that claimed aged bourbon where there was none, but in collecting taxes on the basis of alcoholic content. Not until 1896 did congressional hearings reveal that, of the 105 million gallons sold annually as Old Bourbon, only 2 million was genuine—the rest being "blended" with everything from ethyl alcohol to prune juice—leading to enactment the following year of the Bottled-in-Bond Act, which guaranteed federal government supervision of bourbon bottling at the source, demanding 100-proof liquor aged at least four years and distilled at the same facility. And not until 1964 did the government legally define bourbon as a whiskey containing at least 51 percent corn and aged in new charred oak barrels. By this time, Bourbon County itself produced no bourbon at all and was, in fact, teetotaling dry.

Kentucky bourbon was one thing, Tennessee whiskey another. The oldest registered distillery in the country is in the Cumberland Mountains southeast of Nashville, in Lynchburg, Tennessee, home of the Jack Daniel Distillery. It began on the east side of Mulberry Creek in a log cabin built by Davy Crockett in 1811, by the graveyard of the Bethel Baptist Church. On the north side of Mulberry Creek was a deep gorge of limestone cut by a stream. This runs from Cave Spring through Daniel's Hollow, and this is where Jack Newton Daniel set up his distillery around 1866 at the ripe age of eighteen. His nephew Lemuel Motlow joined him a few years later, and Motlow's descendants carry on the company and the legend today.

Although Kentucky bourbon and Tennessee whiskey share water from the same clear limestone springs for brewing and the same new-charred white oak barrels for aging, the difference, say Jack Daniel's producers, is in the filtering. In Tennessee the distilled liquor is filtered through charcoal made from local hard-sugar maple wood. Once called "Old Lincoln County Process," the leaching process begins in the rickyard, where the charcoal is made from maple sticks, stacked in ricks six feet high and burned to the ground, after which the charcoal is ground fine and packed twelve feet deep in vats in the mellowing house. You can watch the entire process if you take a tour of the distillery today, at the end of which you'll be handed a glass of lemonade for your trouble. Like Kentucky's Bourbon County, Lynchburg, Tennessee, is bone-dry.

By the time I found the Weddingtons in Carroll County, Arkansas, America had had four centuries in which to develop a complete lexicon of comic whiskey talk, from the beginning a dialect of subversion against the Establishment—religious, political, social. For the language of moonshine, there's no better compendium than Joseph Earl Dabney's *Mountain Spirits: A Chronicle of Corn Whiskey from King James' Ulster Plantation to America's Appalachians and the Moonshine Life* (1974). A journalist from South Carolina, Dabney knew how to find and talk to moonshiners like the Weddingtons all over the South, and while they are now a dying generation their language lives on.

Whiskey slang from the beginning was a language of comic burlesque, ridiculing those who made and enforced the rules. "Moonshine" is eighteenth-century English slang for white brandy smuggled at nighttime from France and Holland onto the coasts of Kent and Sussex to avoid the liquor tax. "Bootlegger" is nineteenth-century American slang for the smuggler who concealed flasks of spirits in his boottops in order to trade with the Indians after the colonies expressly forbade it. "Bootlegger" was given new meaning in 1871 with the first federal excise tax on whiskey, when bootleggers would conceal revenue stamps in their boots to stick on a bottle when challenged.

That challenge, dear to America's rebel heart, created a literary genre I call Moonshine Romance, which pits a legendary Outsider against the Feds or—even better—a pair of Outsiders against each other, Hunter and Hunted.

Whether fact or legend, moonshine tales are compelling in the way that man-against-beast tales are, as in Faulkner's *The Bear* or Hemingway's *The Old Man and the Sea,* because the antagonists confront the same primal wild and share the mutual respect of those crafty enough to survive it. For Prohibition moonshiners the only disgrace was getting caught. When I told Paralee that I'd seen an old still at the Heritage Center in Berryville, labeled "Last Moonshine Still Captured in this Area," she let out a hoot. "Why that was Frank's old still the sheriff tore apart," but not at the time Frank got caught and put in jail. "They didn't ketch ya without somebody'd snitched on ya," Paralee explained:

I guess the revenoo men had heard about Frank because they came looking for him and soft-soaped this real old man, telling him they wanted five gallon a whiskey. Frank didn't want to trust 'em. He's mousey—or whaddaya call it?—foxey. He's hard to slip up on and hard to fool. But Frank got 'em the whiskey and they paid him off and then two or three weeks later they came back when we was out at a beer joint over by White River bridge and these fellers pulled their coats back and said they was Federal men. That's when he stayed in jail. After that he really had to know a person afore he'd trust 'em. But they didn't git the still that time, they just got him for sellin'. They found the still several times after that and busted it up but didn't ever ketch him again.

While we were talking, the Weddingtons' daughters dropped by with their instruments, Helen with a guitar, Laverne a mandolin and June a bass. Paralee handles the fiddle, electric organ and harmonica, the last two at the same time. The walls were full of photos of Paralee and her Swinging Strings, along with prizes for fiddlin', singin' and jiggin'. It was part of her heritage, Paralee explained: "Fiddlin' I learned from my daddy. He just loved music, just like me. I don't know how my momma put up with him. But I learned to jig on my own 'cause I like the

Josh Billings, in Everybody's Friend *(1874), spoofs both Darwin and Whiskey at once.* AMS Press

DARWIN THEORY AND WHISKEE THEORY.

Man waz kreated a little lower than the angells—

—and haz bin gitting a little lower ever sinse. (11)

music. None a my ancestors ever had a squeakin' joint but I got my knees mashed up against the dashboard in a car accident, so I got arthritis bad and a cruel old Arthur he is. Frank, once in awhile, he'll git up and cut off a limb or two jiggin'." And sure enough, as Paralee and the girls lit into their fiddlin' and singin', Frank hitched up his overalls and started jiggin' by the far.

Since I was double Scotch-Irish like Paralee Kirk, I felt at home among these descendants of the typical eighteenth-century American frontiersman, whom Dabney called "a dogmatic Presbyterian, a hard drinker and a contentious cuss who carried his long Pennsylvania rifle with him at all times." But not until I learned about the great migration of Ulster Presbyterians first to Pennsylvania and then south through the Shenandoah Valley to the Appalachians and Blue Ridge Mountains did I discover my home was here. I knew that a number of Harpers and Kennedys had come to America from County Tyrone in Northern Ireland, but I didn't know when or why until I hit the moonshine trail.

In 1608 when Ulster's two clan chieftains, the Earl of Tyrone and the Earl of Tyrconnel, joined their fellows in the "Flight of the Earls" to the Continent after the disastrous defeats at Kinsale, three million acres of land were left for the Crown's disposal. James I decided to plant Protestant Scotsmen there to tame Catholic Irishmen and to foster trade with England. But the Scots thrived so well in the woollen and whiskey trades, combining their native whiskey skills with those of Irish poteen makers, that England in 1640 imposed an "Act of Excyse" in the form of a punitive liquor tax. The Ulsters resisted and moonshining flourished, but as "rack-renting" Scottish landlords screwed up their rents, the Presbyters took off for America, where Pennsylvania offered both tolerance and land.

Between 1717 and 1776, some four hundred thousand Ulster Presbyterians worked their way south along the "western frontier" of Virginia, the Carolinas, Tennessee and Georgia. Fractious by temperament, they were as zealous fighting Indians as they'd been fighting British revenuers. Fired by their clansman Patrick Henry at the outbreak of the Revolution, they served General Washington at the rear while he commanded the van; their defeat of the Tory militia at the Battle of King's Mountain in South Carolina in 1780, when the Presbyters rallied to the Biblical cry of "The sword of the Lord and of Gideon," was a turning point in what George III called the "Presbyterian war." These were the men whom a minister of the Church of England traveling in South Carolina in 1766 called "a set of the most lowest vilest crew breathing—Scotch-Irish Presbyterians from the north of Ireland." And these were the men whom the Philadelphian Benjamin Rush compared to "the barbarous and indolent Indian"—a wild and disorderly people who grew nothing but corn.

"No nation is drunken where wine is cheap; and none sober, where the dearness of wine substitutes ardent spirits as the common beverage. It is, in truth, the only antidote to the bane of whisky."
—THOMAS JEFFERSON

These were my kind of people. A batch of Presbyterian ministers named Kennedy, the Reverends Gilbert, Robert, Thomas and John, all emigrated to America from Scotland or Ireland in the 1730s, some settling in Pennsylvania, some moving south to Virginia. My great-great-grandfather James Kennedy, born in Virginia in 1796, moved on to Kentucky and then Conneaut, Ohio. On the other side, a batch of Presbyterians named Harper—George, Andrew, William and James—all born in Ulster, emigrated in the 1780s to Pennsylvania, some moving into Virginia and some to Ohio. My great-great-grandfather James Harper settled at 20-Miles Stand, the first stagecoach station outside Cincinnati.

These were the settlers who as quick as they grew corn grew whiskey, as the best and cheapest way to transport their grain. Since they could distill eight pounds of corn into one gallon of whiskey, a mule could carry two eight-gallon kegs better than twenty-four bushels of corn. And besides, the price was right. During the Revolutionary War, good Monongahela whiskey sold for a dollar a gallon in Philadelphia and was a stabler currency than the coin of the realm. When Alexander Hamilton as Secretary of the Treasury decided to bail out the debts of the Revolutionary War by laying an excise tax on whiskey, he forgot that he was dealing with troops skilled in moonshining. Presbyterians were nothing if not Protestants and they protested vehemently the excise enacted in 1791 by the Congress of the new government they had fought for in order to end such taxes. It was West against East and Tom the Tinker's men against the Feds.

The Whiskey Rebellion came to a head in 1794 at Bower Hill near Pittsburgh, where rebel forces commanded by a former Revolutionary War officer, James McFarlane, stormed the mansion of General Neville, in charge of the revenue collectors. After McFarlane was killed while charging the hill, the rebels fired the mansion and, calling themselves "white Indians," marched on to Pittsburgh. They were joined by seventeen hundred insurgents while Washington hastily rounded up thirteen thousand army troops. But it was a wintry November and the whiskey army petered out before it reached the enemy. A handful of ragged prisoners were captured and driven on foot across the Alleghenies to the streets of Philadelphia to demonstrate the power of the new federal government. Jefferson repealed the unpopular tax eight years later, but the excise did much to spread the network and stiffen the resistance of illicit cottage whiskey farmers.

The whiskey interregnum lasted until 1862, when Congress needed to bail itself out from debts incurred in the Civil War. With the establishment of a Commissioner of Internal Revenue and a group of deputies to collect it, the

"A Recipe for Temperance Tipple:

Take one bushel corn-meal, 100 pounds of sugar, two boxes of lye, four plugs of tobacco, four pounds of poke root berries and two pounds of soda. Water to measure and distill."

—From the Asheville *Gazette News,* quoted by Joseph Earl Dabney, *Mountain Spirits* (1974)

Temperance crusaders in 1879 clean out the leading saloon of Fredericktown, Ohio. Culver Pictures, Inc.

excise tax rose in two years from twenty cents on a gallon of whiskey in 1863 to two dollars in 1865. Then there was the temperance movement, which made gains after the war with the founding of the Prohibition Party in 1869 and the Women's Christian Temperance Union in 1874. If taxes stimulated the sons of some illicit whiskey makers to new efforts, temperance turned the sons of others into staunch "tee (for emphasis) total" abstainers. The same language and culture that produced "teetotalers" also produced moonshiners and "pure dee alkeehol." There were purists on both sides of the barroom fence, and the quarrel between teetotalers and moonshiners was to some degree a family quarrel between Ulster folk who had kept to the old ways in the Appalachian hills and those who had moved on to the progressive frontiers of the prairies.

By the time Prohibition took off, after America's next big war—the World War now called First—my family was in thrall to a pair of household divinities, preacher Billy Sunday and axe-wielder Carry Nation. When Sister Carry swung her axe in the saloons of Medicine Lodge, Kansas, in 1899 she was pushing the heritage of her "white Indian" ancestors in a new direction. But there was no getting around it. Despite their Sunday preachments, my grandparents, praise the Lord, came from a long line of moonshiners. In the culture of corn, moonshiners and teetotalers were alike stubborn dissenters, brewed in the same mash, run through the same worm, but distilled into different bottles. Each had made his own accommodations to the ironies of heritage, and now it was my turn. After all, if an Inca painter in Cuzco three hundred years ago found it natural to compose a Last Supper of guinea pig and corn beer, surely I could come to grips with a United Presbyterian family in California who made a Last Supper of crackers and Welch's grape juice.

The Language of Commodities

SQUEEZING THE MOLECULE

INFUSED WITH THE SPIRIT of the brand-name age, my family shouted hallelujah in daily worship of the processed box, bottle and can. Presbyters who believed that wheat thins and grape juice could be transubstantiated into His Body and Blood could believe anything, anything but the equation of the "natural" with the "good." Fallen nature required redemption, and Christ was the soul's processor. We praised the manufacturers who, in imitation of Him, worked miracles of transformation. Instead of turning water into wine, they turned corn into starch, sugar and oil. Like the Indians, we too were the people of corn, but our corn was synthesized at the molecular level and extruded into the marketplace as the coin of the realm.

Today the corn synthesizers can transmute substances at will. "The primary message," the National Corn Growers Association declared in 1987, was that "anything made from a barrel of petroleum can be made from a bushel of corn." These are corn farmers, mind you, representing eighteen thousand growers in forty-six states, headquartered in St. Louis, Missouri. Many come from preacher stock, and today they are preaching the message of ethanol and biodegradables to tribes who have run out of gas in a wasteland of boxes, bottles and cans. Corn is a cash crop, say the Corn Growers, to be pumped aboveground as crude oil is beneath. Joining the chorus are the nation's nine giant wet-millers, the kings of corn processing known as the Corn Refiners Association, the archdeacons of pumping and squeezing, who proclaim, "Corn refiners constantly search for new technologies and more uses which will squeeze even more value out of every bushel of American corn."

Today, the way to squeeze money out of corn is to squeeze the molecule and manipulate its structure in the tee (for total) commodification of corn, in which

"The most promising new market for corn starches is as a raw material for the production of many industrial chemicals which are today made from petroleum feedstocks. As petroleum supplies dwindle or become less reliable, the importance of having an abundant source of basic industrial chemicals takes on new proportions, and corn industry scientists are at work on new systems for producing industrial necessities from the versatile corn plant."

—Corn Refiners Association pamphlet (1987)

farmers are replaced by marketers, millers by refiners and machinists by biochemists. From the perspective of industrial farmers, corn is of value only as it can be transformed chemically into whatever commodity is most in demand. The methodology of nineteenth-century chemistry underlies the refining of corn as it does the refining of oil: analyze the system, fragment the parts, quantify the parts and synthesize at will. In such a system organic plants are but raw material for the processor, who fabricates products for which the manufacturer creates demand. Equally, consumers are raw material for processing and consumption by the manufacturer. The consumer becomes the consumed.

In this interchange, chemists rule. "Chemists of the Kernel," the CRA trade bulletin *Corn* hails those who control the chemical milling of corn, known as wet-milling because the process begins with wetting and squeezing the corn. Corn is steeped in a solution of water and sulphur dioxide in order to separate each kernel into its components—hull, germ and starch. Indian ash processing, as we've seen, helped to remove the hull by chemical action but kept the rest of the kernel intact. The function of modern industrial processing is to break up the interior matrix in order to separate germ from starch and starch from gluten, so that the kernel chemists can go to work. "In general, corn refiners manufacture *components* of consumer products," *Corn* says, "rather than the products themselves."

In the abracadabra of corn chemistry, starch is the magic component of consumer products because starch, in turn, is a raw material for processing. In 1987, out of 971 million bushels of processed corn, 128 million became industrial starch, 290 million became ethyl alcohol, and 303 million became a sweetener for beverages. All these are transformations of starch, and one of the ironies of industrial history is that cornstarch was originally a substitute for wheat.

Starch for all kinds of purposes had been leached from wheat and rice by the simplest water solution from the time of ancient Egypt and China. In the sixteenth century, American colonists imported wheat starch in some quantity to powder their wigs and starch their collars, until they developed wheat- and later potato-starch factories of their own. Not until the nineteenth century was there any major change in starch extraction, when in 1840 the Englishman Orlando Jones patented a method which used an alkaline as catalyst. Within four years, Colgate & Company had applied the patent to corn in

its wheat-starch factory in Jersey City. An employee, Thomas Kingsford, was so struck by the results that he set up a cornstarch factory of his own in Oswego, New York.

This extracted cornstarch was first sold culinarily as a flour, superior to finely ground wheat flour for the baking of cakes. The fact that starch did not have the same properties as flour presented problems for the manufacturer in promoting his product. A recipe pamphlet of 1877 explained to the housewife why some of the recipes called for a little wheat flour to be added to the cornstarch: "This is done because the Oswego Corn Starch is so rich in all its parts, that it will not hold together in cakes, biscuits, etc., without the aid of flour. Still, if you wish to make them entirely of Corn Starch, do so, but use a little less butter and sugar than is mentioned in these recipes."

Wright Duryea, who had worked as a millwright, opened his own starch factory in 1854, coining the word Maizena for his starch. By 1891 he had the largest starch factory in the country, grinding up seven thousand bushels of corn a day. A Duryea plant engineer of the time, Frederick Jeffries, has given us a good working-man's account, in *Grinding Corn as I Have Seen It* (1942), of how starch was made in these early plants, with their wooden tubs and starch tables and profligate waste:

There was not a motor or electric light in the place. Corn was put in wooden tanks, covered with warm water, and nature was allowed to do its worst. Considerable of the corn was shovelled, as the tanks were flat bottomed. It was then ground. Then came the sieving and washing on silk shakers. The resulting liquor went to wooden tubs, where it was treated with caustic soda and allowed to settle. The water went to the sewer, carrying, aside from the solubles, quantities of insoluble starch and gluten.

Now to the starch tables—which were twelve feet wide and gave little trouble. When settled, the water, together with the gluten remaining in the starch, was decanted to the sewer. The starch, with fresh water added, again was put in solution with revolving sweeps, and the operation repeated. This settling or "topping" as it was called, was done three times. . . . All of the steepwater went to the sewer. No germs were separated, so there was neither oil nor cake. The fiber or slop as it left the end of the shakers was combined with the run from the ends of the tables, or the gluten. No settling or pressing was done. The material "as is" was run into a number of large bins [where] the water leached through the burlap to the ground. In a week or so the "starch feed," as it was called, would have leached to a point where it was solid enough to shovel. Then it was sold to local farmers who came for it in wagons.

How much of the corn was lost no one knew. In the starch-washing they made sure to get rid of the gluten. As to how much starch also went to the sewer, no one knew. When the feed could not be sold, it went to the sewer. No doubt, at times, one-half of the corn was lost.

Starch tables of the 1930s or earlier, described by an engineer of the Duryea plant. Courtesy Corn Refiners Association, Inc.

As day follows night, merchandising followed manufacturing and brought riches to new starch entrepreneurs like Gene Staley. Staley was a farm boy who had done a turn with the Royal Baking Powder Company of Chicago before 1897, when he decided to repackage bulk starch, sold at two cents a pound, into one-pound packages he could sell for seven cents a pound. When miffed starch manufacturers cut off his supply, he purchased the Cream Starch trademark and set up business in Baltimore. His Corn Products Company (which in 1906 he renamed Corn Products Refining Company) was the father of today's leviathan Corn Products Company International. By 1922, *The Staley Journal* reported the success of a company that understood the world value of a trademark: "Since 1912 we have ground 35 million bushels of corn. If all that corn were to be loaded into freight cars the train would be 400 miles long. Our products are shipped to every part of the globe. On the streets of Cairo, it is not uncommon to see a fallaheen wearing a Staley starch bag for apparel. In Constantinople, many a rain barrel displays the Staley brand."

STARCH IS STILL EVERYWHERE, especially in every variety of manufactured product using sugar. Historically, the first attempts to turn corn in any quantity into sugar began when corn syrup became a substitute for molasses after the Molasses Act of 1733. Legend, however, prefers to credit Napoleon's sweet tooth, aching during the Continental System blockade of 1806, which prevented West Indian cane from reaching French shores. Motivated by Napoleon's offer of one hundred thousand francs to anyone who could create sugar from a native plant, a Russian chemist, K. S. Kirchhof, discovered that he could obtain a clear, viscous syrup by adding sulfuric acid to heated potato starch. Cornstarch would do the same.

Contemporary language explains that the acid, or enzymes, from malt or mold alter the molecular structure of the starch, just as the enzymes in saliva do when they convert starch into sugar so that the body can digest it. The result is glucose, or as we now call it, dextrose, meaning "right-handed sugar" (right being "the direction in which the molecules rotate a beam of polarized light"). Fructose, which is chemically identical with dextrose, is molecularly different and should be called "left-handed sugar." It should also be called super-refined sugar, since it is no more "natural" than ethyl alcohol and derives from the same chemical conversions.

In America the industrial conversion of cornstarch to sugar began with the Union Sugar Company in New York in 1865 and accelerated with the American Glucose Company in Buffalo a decade later. Chemistry was again the key.

In the language of corn refining, once the starch matrix has been separated from its protein gluten, the starch is converted by chemical action (an acid or enzymes, or both, are added to starch suspended in water) into "simple" sugar, called a "low-dextrose solution." Sweetness and texture (crystal or syrup) are controlled at every point to produce different products, depending upon how much starch is digested by the acid or enzyme. "We now know that starch consists of long chains of glucose molecules," Harold McGee has explained in *On Food and Cooking* (1984), "and that both acid and certain plant, animal, and microbial enzymes will break this chain down into smaller pieces and eventually into individual glucose molecules." By the same initial process through which the Hopi made "virgin hash," our modern corn refiners make glucose, maltose, dextrose and fructose.

The larger the number of these long glucose chains in the molecule, the more viscous the syrup, a quality important to the baking and candy industries because it prevents graininess and crystallization. Without corn syrup, no easy-to-make chocolate fudge. The more complete the digestion of starch, the sweeter the syrup, because the rate of glucose and maltose is higher. Maltose is a "double-unit" sugar produced, as in brewing, by enzyme-manipulated starch. By manipulating the glucose units with an enzyme derived from— unlikely as it sounds—*Streptomyces* bacteria, the refiner can get a supersweet fructose called High Fructose Corn Syrup (HFCS). Today, this is where the king's share of cornstarch goes, because this syrup is the sweetener of choice (as long as it is cheaper than price-supported cane sugar) for the soft drink, ice cream and frozen dessert industries.

Although supersweet fructose tastes about twice as sweet as ordinary sugar, we do not as a result consume half as many soft drinks or ice cream cones. On the contrary, American sweetness consumption spirals ever upward, so that today, out of the 148 pounds of sweetness consumed per person per year ($1/3$ pound per day), 45 pounds come from corn syrup. Like new breeds of supersweet corn, supersweet syrup intensifies desire and invisibly and silently manipulates the molecules of the eater.

If we take a single corn kernel of No. 2 Yellow Dent through the refining process start to finish, we can see what the mind of modern industrial man hath wrought in squeezing every drop of juice from nature's molecules and in devising uses for each drop squeezed. A kernel arrives at the plant in a load of 51 metric tons, where it passes through mechanical cleaners on its way to the steep tanks. Here the kernel soaks for 36 to 48 hours in 120-degree (Fahrenheit) water with sulfur dioxide, which toughens the germ and softens the protein in its starch matrix. The steepwater, which is highly nourishing, is then concentrated by evaporation and sold as a component in antibiotics (steepwater provided the mold for penicillin in World War II), vitamins, amino acids, fermentation chemicals and—above all—animal feeds.

The softened kernel now goes to the degerminating mills, which tear it apart, loosening the germ and sloughing off the bran. This slurry of mixed parts is then sluiced into flotation tanks, where a centrifugal hydrocyclone separates the lighter germ from the heavier parts, which can then be filtered, ground and shaken. The germ is squeezed for oil (by expellers and solvent extractors), filtered and refined into edible oil and coarser "soap stock." The remainder, known as "corn oil cake," is ground into meal to be added, along with the ground hull, to corn gluten feed and gluten meal.

After a final shake, the lighter gluten is separated from the heavier starch in a high-speed centrifuge, and the gluten (70 percent protein) becomes the major component of feedstuff for dairy and beef cattle, poultry, hogs and sheep. (In poultry feed, the pigmentation of yellow corn, from a pro-vitamin called xanthophyll, is what turns chicken skin and egg yolks Perdue yellow.) The protein zein is, in turn, separated from the gluten, and the zein is then used by the pharmaceutical industry to make medicines in tablet form.

We are left with the industrial heart of the kernel—the starch. Some of it goes to the dryers to be marketed for home cooking, for paper and textile mills, metal casting and sand molds, laundry starches and—in the form of dextrins—

A Siamese-twin train of corn and Coke advertises the High Fructose bond for the Corn Products Refining Company. Courtesy Archer Daniels Midland

for adhesives. Dextrin is simply roasted starch, and its properties were discovered, according to legend, when a fire broke out in a potato-starch factory in Dublin in 1821, at the very moment the locals were celebrating a royal progress of King George IV with quantities of Irish whiskey. In attempting to man the fire pump, six of the men fell into a tank of starch-water, hot from the fire, and emerged glued together. The superglue properties of heated starch were thus tested and proved. Even unroasted starch is so hydrophobic that it has become an essential dusting agent for every variety of manufactured goods from plastic cups to fur coats.

The larger part of the starch is converted to syrup by acid, enzymes, heat or a combination of the three. In the first stage of conversion, 15 to 25 percent is converted to dextrose after the purified starch slurry has been thinned by means of an alpha-amylase enzyme or acid. In the second stage, the starch converts to 97 or 98.5 percent dextrose in reaction to a glucoamylase enzyme, over a period of seventy-two hours at 140 degrees Fahrenheit. After it is evaporated, crystallized, dried and remelted, the starch molecule may then be "cleaved," "modified" or "cross-linked" to produce different forms of corn sugars and syrups. The degree of conversion is measured by its "D.E.," or dextrose equivalent. More than half the dextrose output goes to manufacturers of breads, other baked goods and breakfast cereals. Because dextrose caramelizes easily, it is used as a coloring agent and sweetener in baked goods and drinks like cola, root beer and whiskey, not to mention dry mixes for everything from iced tea to salad dressing.

Refiners, in their own inimitable language, list the seventeen basic properties of corn syrup valued by "food industry scientists":

bodying agent (to improve "mouth feel" of chewing gum), browning reaction (color to bakery products), cohesiveness (for frostings and icings), fermentability (brewing beer), flavor enhancement (jams and jellies), flavor transfer medium (dried fruit drink powders), foam stabilizer (ice cream), lowers freezing point (frozen or dried eggs), stabilizes moisture (frozen fish), moisture absorption (marshmallows), sweetness and nutrition (almost all food products), osmotic pressure (candy), sugar crystal prevention (pie fillings), ice crystal prevention (frozen fruit), sheen (canned fruit), viscosity (pancake and waffle syrup).

The family of corn syrups includes hydrol, or corn sugar molasses, a dark, viscous syrup useful in animal feed and in drugs; lactic acid, a colorless syrup useful as a preservative and flavorer for everything from pickles to mayonnaise; and sorbitol (dextrose plus hydrogen), an emulsifier that shows up in toothpaste and detergents as well as processed edibles. But the premium syrup is high-fructose corn syrup, first commercialized in 1967 by the Clinton Corn Processing Co. of Clinton, Iowa, which patented *Isomerose* (named for the enzyme xylose isomerase, which converts glucose to fructose). By 1972, the company had increased the sweetness from 14 to 42 percent fructose, to make it equivalent to ordinary sugar. As sugar prices rose, food and beverage industrialists began to replace more and more sucrose with "Isosweet." Within four years, production of the supersweet syrup jumped from two hundred thousand

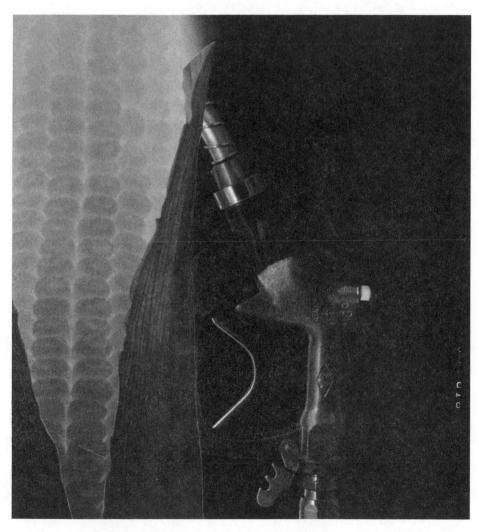

Corn fuels cars by way of ethanol. Archer Daniels Midland

to two and a half billion pounds a year, and within the decade it had become a major component of all major soft drinks. Today, HFCS can be made 25 percent sweeter than sugar (*Isomerose 600* contains 60 percent and *Isomerose 900* contains 90 percent fructose) and in crystalline form is an important rival to saccharin in the sugar-substitute industry.

This is the industrial alembic that makes corn and petroleum kin, for the same plants that produce high-fructose corn syrup also distill ethanol. In the 1970s, America's simultaneous oil crunch and pollutant anxiety revived the idea of fueling cars, as well as animals and people, with corn. Ethanol, or ethyl alcohol distilled from corn, began to be used both as a lead-free octane booster and as a replacement for regular leaded gasoline. Although the possibilities of fueling engines with ethyl alcohol had been contemplated from the beginning of car manufacturing, cheap oil made such experiments irrelevant until corn surpluses equaled oil shortages. Within six years, between 1978 and 1984, ethanol production jumped from 10 million to 430 million gallons, to make 4.3 billion gallons of the ethanol-blended gasoline called gasohol (10 percent ethanol to 90 percent gasoline). Those who laughed in the mid-1970s at the very word "gasohol" were sobered in the mid-1980s by the considerable tax benefits Congress granted in 1985 to gasohol producers.

Because it takes 56 bushels of corn to produce $2\frac{1}{2}$ gallons of ethanol, corn growers could also boast that ethanol had raised the value of corn by three to seven cents a bushel. In 1984 the future of ethanol looked so bright that the General Accounting Office projected a production of one billion gallons of ethanol in 1990 to generate a net income for farmers of $3.59 billion. In 1990, however, the costs of energy consumption and environmental pollutants in the simple growing of corn clouded the picture, and enthusiasm shifted to nurturing the environment with biodegradable plastics.

For the past decade, starch makers have been experimenting with ways to make plastics photodegradable: that is, composed of chemicals sensitive to ultraviolet rays, which help to break up the tightly bound "polymers," or chains of hydrocarbon molecules, that characterize plastic. In the spring of 1989 the St. Lawrence Starch Company of Canada introduced Ecolyte plastic, in which the addition of cornstarch and a vegetable oil for oxidation would allow plastic trash bags, for instance, to disintegrate along with the trash. While the starch granules are eaten by bacteria, the oxidizer reacts with salts in the soil to break up the polymers into little bits of litter. Environmentalists skeptical of the hard sell of corn refiners point out that these remnant shards do not turn into soil, as advertisers suggest, but into permanent plastic dust.

In Decatur, Illinois, Archer Daniels Midland—"the world's largest commercial practitioner of biotechnology"—has been a major producer and promoter of cornstarch degradable plastics through its subsidiary A. E. Staley Manufacturing. The story of wet-milled cornstarch comes full circle in the "Castle in the Corn Fields," as the Staley plant is called, once the dominion of the farm boy turned starch salesman and now the mother lode of high-fructose corn syrup and corn plastics. Significantly, another major producer of degradable plastic bags is Mobil Corporation, father of Hefty bags, which neatly completes the refiners' equation of a barrel of petroleum with a bushel of corn.

In a curious linguistic conversion, the word "feedstock," which once referred to the substance of field corn consumed by hogs, has now become a metaphor for industrial corn consumed by chemistry. "Corn is a high-tech industrial feedstock which can switch gears to become a biodegradable plastic or an extremely hard, lightweight building material," says a research scientist at Purdue University. "Corn is the only biomass feedstock worth considering in the chemical industry," says the president of Bio En-Gene-Er, in Wilmington, Delaware. In a decade when corn is but a subcategory of bio en-gene-ering, the watchwords "Onward and Upward" of the Iowa Corn Promotion Board seem as quaintly innocent as a Presbyterian hymn.

With Karo, Kingsford's and Mazola, of the Corn Products Refining Company, a whole new cuisine was born, based on the promotion of processed corn products that were embraced with rapture by my parents' generation, true believers as they were in the onward and upward of process. As Kellogg extended his reach from Battle Creek, Michigan, to fill my cornflake bowl, so Edward Thomas Bedford, a founder of New York Glucose and first president of CPRC, infiltrated all our meals with his thickeners, sweeteners and emollients in the name of "purity and refinement." Aseptic purity is the appeal on the title page of a 1920s *Corn Products Cook Book:* "You are cordially invited to inspect our Refineries where Karo, Kingsford's and Mazola are manufactured

Brand names for "the three great products of corn" became household words in the early 1900s through ads and free recipe booklets published by the Corn Products Refining Company.

under the most hygienic and sanitary conditions. From the time the yellow kernels of Corn are unloaded from the cars at our Refineries until the Pure Food Products made from them reach the Consumer in the form of Karo, Kingsford's and Mazola, they are not touched by human hands." CPRC knew how to exploit the popular confusion of cleanliness with godliness in all parts of the American home, but particularly in the kitchen. In the era of domestic engineers, when kitchens aspired to the conditions of chemical labs, Juliet Corson admonished her students in her New York Cooking School, "Remember that the best cook always has the cleanest kitchen." Fifty years later, in 1934, the same message was headed "Points to be considered in marketing" in the manual of my seventh-grade home economics class: "Insist upon a clean grocery store and meat market; Insist upon clean, courteous clerks."

To those who'd worked the soil with their hands, hand-free products signified progress and ease. Ready-mixed baking powders and baking mixes were a boon, not a corruption. Because cornstarch was an essential preserving element in commercial baking powder, to keep the powder dry, the new world of Bisquick breads, Aunt Jemima pancakes and Pillsbury ready-mix cakes was built on invisible starch. Cornstarch, which fine palates today view as an abomination, we viewed as a refinement over vulgar flour when we came to thicken our sauces, puddings, cakes and pies. Cornstarch, in fact, got its current bad name from its ubiquitous use in the 1930s Depression as a quick, cheap thickener of Anglo-American pudding and Chinese-American sauce. Anglos had earlier turned French blancmange, thickened classically with pounded almonds, into a common pudding thickened with flour. Americans translated English blancmange into cornstarch pudding, preferably flavored with chocolate, and so it appeared in my home economics manual, with 1/4 cup cornstarch thickening two cups milk. Americanized Chinese embraced cornstarch as a cheap substitute for water-chestnut powder or rice flour in sauces, reserving rice flour for puddings and cakes. A 1936 *Chinese Cook Book,* compiled by M. Sing Au of Philadelphia, gave three typical recipes for "Chinese gravy," each thickened by cornstarch.

Karo began as a cheap substitute for other syrups, but it soon created a dentist's fantasy of new candies and desserts. In our pantry there were always two bottles of Karo, one light and one dark. Crystal White Karo went into unthinkable tooth-dissolves like Karo Pie: "Line pie plate with rich crust, and fill two-thirds full with Karo (Crystal White). Stir in lightly without touching the paste two teaspoons Kingsford's Cornstarch." While this *Corn Products* recipe even in the reading may set teeth aquiver, note that pecan pie, still beloved today, didn't become part of the American repertoire until Karo (Red Label, dark like molasses) came on the market.

The great advantage of Karo of both colors was that it didn't crystallize, and so formed a stable and simple sugar base for candies like fondants that were otherwise tricky in the extreme. Fudge making, which became a freshman rite in Eastern women's colleges in the 1890s, was a Karo offshoot. So were subcategories like my childhood favorite, peanut butter fudge (a recipe for it well-thumbed in my home-ec manual), and my grandmother's favorite, divinity fudge (sometimes called sea foam), made with eggwhites, nuts, dates and of course Crystal White Karo. Karo also substituted for molasses in chewy caramels and pulled taffies. A recipe for saltwater taffy in a *Products Cook Book* managed to incorporate Karo and Kingsford's for the taffy making and Mazola for the taffy pulling.

"The half billion bushels of corn utilized in corn sweetener production represents over 4.5 million acres of planted land, which is more than the number of acres planted to corn in the entire state of Ohio in 1985.

More than one million corn growers produce raw material for the U.S. sweetener market. The equivalent of twenty-five acres from every Illinois and Iowa corn grower's crop will end up in corn sweeteners in 1985."

—National Corn Growers Association promotion materials, 1985

In 1910, when the current vegetable oils were olive, cottonseed, poppy, coconut and palm, a chemist at CPRC found a way to refine corn oil for cooking. The first customers were a group of sardine packers in Maine, who substituted the cheaper corn oil for olive in their packs. When the trademark "Mazola" moved into Boston grocery stores the following year, customers resisted buying until company demonstrators went into the stores and deep-fried french fries and onion rings to prove the superiority of corn to bacon grease and fish oil. Later, corn oil had to compete with peanut, soy and sunflower oil, but corn helped form and maintain the standard for a pure, colorless and minimally flavored vegetable oil for cooking and salads.

Mazola was born the same year as Hellmann's mayonnaise, made by a New York deli man from Germany, Richard Hellmann. He bottled his wife's mayonnaise in glass jars and put a "Blue Ribbon" label on them, as Frederick Pabst had done with beer. Like other food entrepreneurs of his time, Hellmann was a good salesman, able to persuade housewives to buy ready-made dressing the way Heinz persuaded them to buy ready-made pickles. Although Hellmann's today is made with soy oil, corn oil was the original base for its emulsion of oil, vinegar and eggs. In our house, Hellmann's mayonnaise, marketed west of the Mississippi as Best Foods, was the altarpiece of our dining table, removed only for ice-box protection between meals. With its proud blue-ribboned front, the

Richard Hellmann's original mayonnaise in 1912 wore a prominent blue ribbon to signify a winner. Courtesy CPC International Inc.

jar stood like a good angel to the right of my father's plate, since he could do without salt and he could do without sugar but he could never do without mayonnaise. At breakfast he spooned it onto his scrambled eggs; at lunch he slathered it top and bottom on the bread of his bologna sandwich. At dinner he scooped it onto his canned pineapple and Jell-O salad, onto his mashed potatoes and sometimes onto his Swiss steak.

He was, as I see it now, a mayo freak, gratefully processed by Hellmann salesmen like Paul Price, whose wife in 1937 made a chocolate mayonnaise cake, thickened with walnuts and dates, flavored with chocolate and made tender and moist by mayonnaise. For Hellmann's seventy-fifth birthday, a New York pastry chef updated the classic Hellmann version to make it suitable for microwaving. Behold—an American cuisine of molecular manipulation, start to finish. As a botanist of the old school and a cook of the old-fashioned range, my father knew little and cared less about molecules by the time he joined his ancestors. But I know how much he cared about mayonnaise, and I know he would have loved that cake.

The Sacred Round

The Language of Ceremony

DANCING UP THE CORN

CORN, BOTH NATURAL AND REDEEMED, invested with ceremony the tables of American settlers, but not their myths. These remained tied to the language of wheat, as formed by the English of the King James Bible. When my grandfather Doctor Charles Sumner Harper went west from Kansas, to leave farming for osteopathy, he landed in Greeley, Colorado, not far geographically from the mythical origins of the Tewa in the Lake of Emergence near Pagosa Springs, but light-years away from his own mythical origins in Adam's Eden. The geography of his tribe was entirely Old World, at the same time their migration across oceans, mountains and deserts rejected old worlds and old faiths for new. His journeying, like his language, was linear and progressive. Look onward and upward, never backward or downward. Upward was the direction of the pearly gates. Downward was the direction of the fallen earth, downward was the mouth of Hell.

My grandfather prayed in words, because dancing, in his language, was a sin. So was the body, so was all matter. Flesh and blood were of the earth, earthy, corruptible and corrupted, not God's but Devil's work, awaiting the lightning blasts of the Apocalypse and the trumpets of Armageddon. The split between body and soul was as complete as between man and God, Indian and white. In their westward migration, my tribal Presbyterians were aliens among the heathen and fundamentalists among the heretic. As self-willed exiles, they could not comprehend a people whose gods were rooted in this particular earth, both patch and globe, as strongly as stalks of corn. As self-appointed elect, they could not comprehend a people whose vision was egalitarian and inclusive, a vision which allowed Saints and Corn Mothers to live happily side by side.

My search for corn became a search to reconcile opposites that in my heritage were mutually exclusive. Even as my tribal ancestors mutated their "harvest homes" to cornhusking bees, their midsummer eves to sweet-corn feasts, their church steeples to corn palaces, still their division between corn savages and wheat saints seemed immutable. Even as I traveled from Nebraska to Cuzco, seeking connection, I was still my grandparents' child, preconditioned to square the circle of ceremony and myth by the Biblical language in which I was born. What good is an Indian peacepipe to a person taught not to smoke? Corn no more than tobacco figured in the ancient language of my heritage, but it was at the center of the land in which I lived. That challenge was one my ancestors had faced crossing the Atlantic. Now I would have to make my own crossings, searching in ritual and ceremony for the connections between opposites that elude reason and defy logic. I was a stranger in this land, no less than my ancestors, and my search for the hidden roots of corn was a search for a homeplace, a search ultimately for myself.

Hano Clowns—Koshare *of the Hopi, painted in watercolor by L. Kewanyouma.* Museum of Northern Arizona

Watching the midsummer corn dance at one or another pueblo in New Mexico, I felt the shades of my ancestors rising in wrath. Not only was this a pagan fertility rite but the feast day of an abominated Catholic saint. The descendants of the people of corn, however, had the logic of ceremony to connect opposites in a world altogether different, where time stands still and upward and downward are the same. In a desert as ancient as the sands of Samaria and Judah, this summer as every summer, the many tribes of the New Mexican Pueblos dance the rain down and dance the corn up. They dance while diesel trucks and cars speed past them on the superhighway between Albuquerque and Taos, while jets thunder overhead and space satellites circle far above, bouncing back signals to TV aerials that sprout like cornstalks on the roofs of their adobes.

At first, all I could take in at the pueblo of San Domingo, which holds the largest dance, were these weird conjunctions: the leafy shrine in the central plaza enclosing a life-sized image of Saint Dominic behind offerings of corn stew and Wonder Bread; the white leggings for winter dances crocheted by a devout Methodist lady in shrewd trade for turquoise and pots; the country-western blare from the carnival rides clashing with muffled drumbeats from the kiva; the gowned monk in a baseball cap laughing at the koshari clowns grabbing sno-cones and popcorn. And everywhere, as my grandfather had waited for the Second Coming, hundreds of spectators, natives and tourists, waiting hour after hour in folding chairs, under cowboy hats and sun umbrellas, for the coming of the dancers.

Restless and impatient for the dancing to begin, I remembered what Alfonso Ortiz had said about the Tewa traditions of his own pueblo of San Juan. "Westerners think of tradition as something static, frozen, opposed to progress," he told me. "For Indians tradition has always meant change, something living and organic—the Buffalo Dance now includes not only animals like buffalo and snakes but spacemen and rock stars." For the Tewa, who gather from around the country each summer to dance the sacred round of their ancestors, he said, the dance is a way of ordering their universe by a rhythmic beat, tuned to the energy of a universe that wires the galaxies of outer space to the germ within each single grain of corn.

When the Tewa dance, they pound the same "earth mother earth navel middle place" that their remote forefathers pounded in the courtyards of La Venta, Monte Alban and Copán. In the ancient rhythms of their dance, they invoke the presence not only of tribal ancestors but of Blue Corn Woman and White Corn Maiden, the summer and winter daughters of Mother Corn. They dance a universe of correspondences in a language so different from Latin and Greek that translation falters. Instead of the abstract signs of my root alphabet, the languages of the New World, though often carved, painted, sculpted and woven, were chiefly spoken, sung, chanted and danced, in the body language of ritual and ceremony.

It takes a poet to translate dance, a poet like D. H. Lawrence, who witnessed the corn dance at San Domingo in 1927. He described the koshari as "blackened ghosts of a dead corn-cob, tufted at the top," and the dancers of the sprouting corn as "some sort of wood tossing, a little forest of trees in motion, with gleaming black hair and gold-ruddy breasts . . . and the deep sound of men singing . . . like the deep soughing of the wind, in the depths of a wood." Others, like Captain John G. Bourke of the Third U.S. Cavalry in 1884, recorded the scene as amateur anthropologists. He noted the clowns painted black and white with bands of otter fur across their chests, cedar sprigs around

Young corn dancers at Cochiti Pueblo in 1935 are eloquent, even in a formalized camera pose, of the rhythmic language of corn. Photo by T. Harmon Parkhurst, courtesy Museum of New Mexico, Neg. No. 55189

their ankles and cornshucks woven into the whitewashed topknots of their hair. In the pine branch that each dancer waved, he found a symbol of green life, and in the gourd of dried corn kernels that each dancer shook a symbol of rattling rain.

A century later, I saw nothing but signs and symbols of corn. Painted corn-stalks embraced the portal of the white-plastered church, invigorated the legs and thighs of the koshari, claimed kinship with the sun on "the cane." The cane was the sun's emblem carried by the leader of the dancers—a pole which bore on top a kilt, hung with eagle feathers and fox tail, surmounted by a gourd of seeds and a bright crest of parrot and hummingbird feathers. This was the ancient *huitziton,* the crest of the Aztec war god Huitzilopochtli, and the pole was the ancient serpent-staff of the Feathered Serpent.

Every motif had meaning in a language that voiced the unity of corn, man, sun and rain, but I couldn't decipher it without instruction. To submit to the boredom of eternal waiting in heat and dust while nothing happened, to the boredom of repeated shufflings and chants, I had to unshuck my ancestors. I had to be told that all dances, even winter dances, speak to rain because rain

Men and women, dancing together at San Ildefonso in the mid-thirties, choreograph the energies of earth, sun, corn, and rain. Courtesy Museum of New Mexico, Neg. No. 3666

mediates between sun and earth, between the father and mother of corn. I had to learn that as a summer dance, the Corn Dance is both a harvest celebration and a prayer for the new year, a feast of thanksgiving for the crop of ripe corn just reaped and a prayer for rain for next year's crop.

What exhausted me, Ortiz said, energized the dancers. Since time and space no longer count, the dancer is more refreshed at the end of the day than when he started, for he receives power from the energy of earth and sun, power that makes the rain come, power that is choreographed in circle and line. The circle is the kernel, earth navel, womb, globe. The line is rainfall, lightning, irrigation. To a stranger like myself, the rhythms of the dance changed unaccountably and continually, like desert clouds, like a secret language. I began to see how varied and subtle was this unknown tongue that had to be felt in muscle and bone before it could be seen and heard.

A koshari makes zigzag motions with his hands to imitate lightning; then with sweeping motions, palms down, he moves like sheets of rain. Others shake their gourds from sky to earth to send the rain rattling down. Male dancers bare to the waist wear white cotton kilts, embroidered with cloud and rain symbols, tied with the long tasseled rain sash. Hanging behind each dancer is a fox skin, tied to his right knee is a rattle of turtle shell and deer hooves, across his chest a bandoleer of seashells, on his head a cluster of parrot feathers to bring rain from the south, over his moccasins a skunk fur to ward off evil— all reminders of his brothers the animals, whose lives he shares. Female dancers wear black *mantas* bare on one shoulder, tied with a red and green sash. Their feet are bare, their heads covered with *tablitas,* thin boards cut in the geometry of mesas and clouds and painted with signs of zigzagged lightning, sun and stars. Everything is coded, including the colors: yellow for sun, green and blue for crops, black for thunderclouds, red for fire, white for purity.

The dancers dance in two groups, alternately, divided like the moieties of the pueblo, into Summer People and Winter People, Turquoise and Squash, the male bodies painted with blue-gray clay for Turquoise and with yellow for Squash. The koshari of the Turquoise, naked but for loincloths, are striped horizontally, skull to toe, in black and white; the Squash are painted vertically, one half the body white with black spots and the other half ochre. Jangling their belts of bone or metal bells, the koshari are licensed clowns like the spirits of All Hallow's Eve who mediate between living and dead, gods and men, kivas and moieties, fear and laughter. In Old World language, they are Masters of the Revels, healers and tricksters, who at one moment rescue a fallen toddler and at the next tumble a girl for a kiss.

"He partakes in the springing of the corn, in the rising and budding and earing of the corn. And when he eats his bread, at last, he recovers all he once sent forth, and partakes again of the energies he called to the corn, from out of the wide universe."
—D.H. LAWRENCE, "Dance of the Sprouting Corn," *Mornings in Mexico* (1927)

Men and women dance face to face in opposite lines, then join as couples, the man ahead, the woman behind. All day they dance in alternation, Turquoise and Squash, under the dipping banner of the Sun. All day the band of singers, a massed choir of fifty or seventy-five men, calls the rain from the four quarters of the earth to thunder here. Houses are open all day to anyone hungry for ham and hominy, slices of watermelon, adobe bread, fruit-filled pies, soda pop and Kool-Aid. Toward the end of the day the two groups join at last to dance as one, five hundred dancers together, old men, young girls, fat men, athletic boys, matrons with babies, wrinkled crones, white hair and black hair shaking together, bare feet and moccasins stamping in unison as thunderheads mount in the east. In midsummer the clouds mount every afternoon over the Sangre de Cristo mountains, potent with the fire that spoke to Moses in the burning bush, but that here, in deserts half a world away, speaks in forked lightning to fertilize the seed deep in the womb of Mother Corn.

GATHERING IN THE CORN

To THOSE OF US whose ancestors crossed the Atlantic with their sacred texts packed among the apple seeds and iron pots in their sea chests, the summer solstice is an odd time to celebrate the new year. My Christian calendar puts the birth of the celestial sun, Holy Son and new year at the winter solstice. It's true that the Christian calendar bears vestiges of pagan planting calendars in both midwinter revels and spring resurrections. God's son is related, after all, to the wheat-and-barley son of the Great Mothers Demeter and Ishtar, whose province was spring. But after the Great Fathers dethroned them and in due time amalgamated the calendars of Roman emperor and pope, the new year and its symbolic babe were born in the dead of winter. In the New World, where Mother Corn continued to reign, summer is the beginning of the new year symbolized by corn and celebrated in the Green Corn Busk.

"The only way the Peaux-Rouges helped the Pelerins was when they taught them to grow corn (mais). The reason they did this was that they liked corn with their Pelerins. . . . In 1623, after another harsh year, the Pelerins' crops were so good that they decided to have a celebration and give thanks because more mais was raised by the Pelerins than Pelerins were killed by Peaux-Rouges."

—ART BUCHWALD,
    explaining le Jour de Merci Donnant
    (Thanksgiving) to the French

Midsummer for my Grandfather Harper, when he was still raising sugar beets and corn and peaches in Edgerton, Kansas, was a season of hard labor that was not to be celebrated until the end of November, when the corn was in

Theodore De Bry's engraving of Florida Indians shows women brewing and men expelling the purgative "black drink" during the purification ceremonies of the Green Corn Busk. Neg. No. 324297, courtesy Department of Library Services, American Museum of Natural History

the cribs. At that time, to be sure, the citizens of Edgerton sang hymns of thanksgiving in little white steepled churches and celebrated harvest abundance by eating fat turkeys. This ceremony, however, was an Old World "harvest home" they adapted to an alien corn. New World harvests were celebrated twice, first when green corn was gathered in midsummer and then when ripe corn was gathered in the fall. When Chief Massassoit and his men brought wild game and popcorn to the tables of the Pilgrims in Massachusetts, my ancestral fathers thanked God that the heathen were brought to His table for the salvation of their souls. That the Pilgrims themselves were guests, rather than hosts, at a seasonal corn feast as old and as sacred as Abraham and Isaac was literally unthinkable to those whose history and geography had been charted by Biblical maps.

Both my Grandfather and Grandmother Harper were born and raised in the fertile crescent of the Ohio and Mississippi rivers, which watered corn for the natives a couple of thousand years before they watered corn for the white man in parcels labeled Ohio, Illinois, Indiana, Iowa, Missouri. In Ohio, my grandmother was born not far from the world's largest serpent effigy mound, 5 feet high and 1,254 feet long, built by the Adenas in Adams County, Ohio, perhaps thirty centuries ago, at the time King Solomon built his temple in the kingdom of Israel. Ohio farmers tried to plow the mound flat, but the serpent with the egg in its mouth was obdurate. It was as native to Ohio as the serpent with an apple in its mouth to Eden.

Crossing the Mississippi River at St. Louis on their wedding trip from Ohio to Kansas, my grandparents could have seen the pyramidal earthworks called Monks Mound at Cahokia, covering sixteen square acres. But if they had, they would have seen a reminder of Moses' escape from the Pharaohs rather than vestiges of the high civilization of the Mississippians a millennium ago. And when they asked the Lord's blessing over the white starched cloth of their first Thanksgiving table, they would not have known that Woodland tribes from Florida to the Great Lakes had helped to shape that blessing of turkey and corn by the Green Corn Ceremony marking their new year.

Numbers of colonial travelers and settlers described that ceremony, each interpreting it by his own bent. Robert Beverley saw in the Creek ceremony in Virginia in 1705 a time of feasts, "War-Dances, and Heroick Songs." A later Creek chieftain, who learned to write English and adopted the unlikely pen name of Anthony Alexander M'Gillivray, concentrated on the fasting—which gave the ceremony the name "busk." While none were permitted to eat corn from April until July or August, "boskita" meant a three-day period of fasting accompanied by purging with the "black drink" of snakeroot, senneca or cassina, "which they use in such quantities as often to injure their health."

The explorer William Bartram was taken with the communality of the busk among the Creeks of Florida and Georgia in the late 1770s, and with their "proper and exemplary decorum."

It commences in August, when their new crops of corn are arrived to perfect maturity: and every town celebrates the busk separately, when their own harvest is ready. If they have any religious rite or ceremony, this festival is its most solemn celebration.

When a town celebrates the busk, having previously provided themselves with new clothes, new pots, pans, and other household utensils and furniture, they collect all their worn-out cloaths and other despicable things, sweep and cleanse their houses, squares, and the whole town, of their filth, which with all the remaining grain and other old provisions, they cast together into one common heap, and consume it with fire. After having taken medicine and fasted for three days, all the fire in the town is extinguished. During this fast they abstain from the gratification of every appetite and passion whatever. A general amnesty is proclaimed, all malefactors may return to their town, and they are absolved from their crimes, which are now forgotten, and they restored to favour.

On the fourth morning, the high priest, by rubbing wood together, produces new fire in the public square, from whence every habitation in the town is supplied with the new and pure flame.

Then the women go forth to the harvest field, and bring from thence new corn and fruits, which being prepared in the best manner, in various dishes, and drink withal, is brought with solemnity to the square, where the people are assembled, apparelled in their new cloaths and decorations. The men having regaled themselves, the remainder is carried off and distributed amongst the families of the town. The women and children solace themselves in their separate families, and in the evening repair to the public square, where they dance, sing and rejoice during the whole night, observing a proper and exemplary decorum: this continues three days, and the four following days they receive visits, and rejoice with their friends from neighbouring towns, who have purified and prepared themselves.

Where many white men found Christian echoes in this native ritual of purgation, purification, burial and resurrection, Thoreau found in Bartram's description a moral exemplum dear to his puritan heart. "Would it not be well if we were to celebrate such a 'busk,' or 'feast of first fruits,' as Bartram describes?" he asked. He wished to imitate not the feast but the fast, the stripping away of old clothes, old pots, old fires, old habits, old lives, a cleansing within and without to burn with a purer flame. He observed that "the Mexicans" practiced a similar purification every fifty-two years and concluded, "I have scarcely heard of a truer sacrament, that is, as the dictionary defines it, 'outward and visible sign of an inward and spiritual grace,' than this, and I have no doubt that they were originally inspired directly from Heaven to do thus, though they have no biblical record of the revelation."

No Biblical record certainly, but another record of revelation was at hand in the liturgical calendar the Aztec (Thoreau's "Mexicans") had inherited from preceding civilizations of Toltec and Maya, the shared calendar of the New World, which ordered ritual time according to the patterns of change they saw in the heavens and on earth. This was the great Calendar Round, based on fifty-two-year cycles, in which each new round was inaugurated by the ritual of fire-drilling. The same calendar timed the green-corn ceremonies in New England as in Tenochtitlán, reflecting how ancient and widespread were ceremonies of purgation, fire-drilling and gift-giving, how ancient and widespread the masking of a priest in the garments of a god.

Since the Woodlanders were hunters as well as planters, they needed shaman-priests to keep Sky-World and Earth-World in harmony. Their corn

"We give thanks to our sustainers,' says the Senecan male leader during their Green Corn Thanksgiving. In their ceremonial march the female leader holds an armful of corn and a cake of corn bread and leads her band in a march around a kettle of corn soup."
—From JOHN PINKERTON'S *Collection of Voyages and Travels* (1812)

Iroquois masks made of corn-husks and worn by the Husk Face Society during midwinter festivals, when their dances were prayers for the rebirth of green corn in mid-summer. Photo Ed Castle

ceremonies were different from those of the Pueblo planters: for the Pueblo peoples, corn was before time was; for the Woodland tribes, there was a time before corn. Hunter rituals are star treks, measuring the cycle of years, evoking the time when people fell to the earth like arrows from the sky. Planter rituals are earth-born and -bound, measuring the cycle of seasons, re-creating the emergence of people from the earth like seedlings.

Shamans were central to the green-corn dances of the Woodland Indians, which focused on fire rather than rain and connected earth fires to astral ones. The shaman's trance journey or vision quest lasted ritually eight days and nine nights, because that is the time it takes the planet Venus to make its west-to-east transit from morning to evening star. We can see from George Catlin's description in the 1830s of the green-corn dance of the Minataree, a small tribe on the Knife River in the upper Missouri, how the medicine men or priests orchestrated the event. Catlin called the dance "a ceremony of thanksgiving to the *Great Spirit,*" and the priests "doctors." First the doctors tested the corn's ripeness, then arranged for the corn to be boiled in a central kettle around which they danced with a larger circle of warriors. After the doctors sacrificed a few symbolic ears of corn on a pyre of sticks, they drilled new fire and incited the "unlimited licence" of surfeit and excess that characterize our own Thanksgiving feasts.

In this Woodland ritual, time and space are ordered in a linear narrative of four days, and meaning depends upon what happens in which order. The corn dance of the Pueblo peoples orders time differently, since everything happens within the round of a single day, from dawn to dusk, and in a single place. Its action is iconographic, like a Navaho sand painting, rather than like the pictographic narratives of Midewiwin scrolls or the recited narratives of Hiawatha.

Hiawatha is typical of the Woodland shaman-hero whose destiny unfolds in symbolic segments of time. For seven days he fasts to induce the vision quest; on four successive days he wrestles with Mondawmin until the god is overcome, dead and buried, and finally resurrected in the form of corn. The story tells of the sacrifice of an individual hero who concentrates in his own person the energy of the group and acts on their behalf. For the Pueblo tribes, on the other hand, the individual must submerge his unique identity and power with the inhuman energy of vegetative life. This is the force fused by the rhythms of the group, men and women, acting in concert—dancing rather than wrestling.

Among Woodland groups, the blood sacrifice common to earlier planting rituals continued to a surprisingly late date. George Will noted the ancient Pawnee midsummer tradition of sacrificing a human to the gods of the Sky-World: "At the summer solstice a human sacrifice was made to the Morning Star—a maiden captured from some hostile tribe usually being the victim." The Skidi Pawnee, he claimed, "kept up these human sacrifices until well on into the nineteenth century." Even where human sacrifices were not kept up, their meaning was part of the "sacrament" Thoreau attributed to divine revelation. Had he known that the Green Corn Busk was kin to the wholesale slaughter of the Mexica, he might have had second thoughts about the ceremony's exemplary decorum. Certainly the example of the shaman Quetzalcoatl was not the outward sign of an inward grace that my grandparents had in mind when they offered prayers of thanksgiving to the Lord God of Israel for turkey and corn.

After the girl's death, her skin was flayed and worn by a dancing priest who impersonated the flayed god, Xipe Totec, a god like Xilonen of spring and regenerating seed. Here the Florentine Codex depicts a priest being vested by his acolytes in a flayed skin and feathered ornaments. Neg. No. 292758. Photo Llane L. Bierwers, courtesy Department of Library Services, American Museum of Natural History

TEARING OUT THE HEART

THE CREEK BUSKS that Thoreau extolled for enlightenment an Aztec priest would have deplored for ignorance and barbarism. To end the long fast with no more than a feast of corn was to miss the whole point. Where was the god in the boiling pot? The only proper end for a corn festival celebrating great generating nature was the killing and eating of the god, or goddess. At the midsummer festival celebrating the goddess Xilonen (*xilotl* meant "green corn"), Aztec priests selected a young girl pretty enough to wear the robes and crown of the goddess. For a month, women danced in her honor, shaking their hair loose over their breasts like corn silk, to encourage the corn to multiply and be fruitful. At the end of the month, Torquemada reported, "The girl was placed on the sacrificial stone, and with her death ended the celebration, and with it the day."

So much for decorum. It's as if we were to end our celebration of Miss America with her dismemberment on the boardwalks of Atlantic City. But slaughtering, like dancing and feasting, has its own rules of propriety which vary with time and place; and while today Americans sanction wartime sacrifice on a global scale, individual dispatchings distress us. That our Woodland tribes did not always or even usually end their ceremonies in human sacrifice says less about their moral nicety than their geographic and cultural distance from the centers of urban civilization in Mesoamerica. To the citizens of Tenochtitlán, the Creek and Iroquois were as Goths and Gauls to the Romans; and as Romans inherited the glory that was Greek, so the Aztec inherited the grandeur that was Toltec. Tenochtitlán was the heart of civilized life in a vast continent, and the pious Mexica symbolized that heart in a real and palpitating organ torn from living flesh.

This horrific statue of Coatlicue, dug from the site of Tenochtitlán and now in the Museo Nacional de Anthropologia in Mexico City, embodies the cannibal relation of life and death in the central skull at her belly, the serpent skirt, the necklace of human hearts and hands and, surmounting all, the twinned serpent heads that form her face. The serpent heads are the streams of blood that gushed from her neck when she was beheaded by her sons. Neg. No. 333855, courtesy Department of Library Services, American Museum of Natural History

The greatest sacrament of the year for the Aztec, as Easter for the Christian, was the harvest celebration of the ripe-corn goddess and Earth Mother, Coatlicue, or Chicome Coatl. This was sacred theater, dramatizing primitive narratives embedded in the Green Corn Busk the way Christ's Passion dramatized earlier narratives of sacrifice told in Israel, Egypt and Babylon. When Father Sahagún described the ritual death of the maiden enacting the corn goddess at the September festival, Sahagún's imagination was informed by the ritual death of his own Lord. Sir James Frazer, translating Sahagún, spoke of the Aztec breaking their fast "as good Christians at Easter partake of meat and other carnal mercies after the long abstinence of Lent." Frazer's imagination was also informed by the fertility rituals celebrating one or another incarnation of the serpent goddesses of the Old World, and it is difficult not to read Coatlicue as a powerful manifestation of the Terrible Mother, in her skirt of serpents, necklace of skulls and shirt of flayed skin.

A moon goddess and mother of all the gods, Coatlicue had many guises: sometimes she appeared as the god of ripe corn, Cinteotl (*cintli* for "maize" and

and *teotl* for "god"); sometimes as her daughter Xilonen; sometimes as her son Xochipilli—"the Prince of Flowers," god of young love, spring and procreation, like Adonis or Osiris. Like other Great Mothers, Coatlicue had her son for lover, and spectators at the sacred ball game of the Aztec cried out to Xochipilli, "the great adulterer," when a player drove his ball successfully through the stationary stone ring at the side of the court. The mother's son who must be slain and reborn so that the green world may spring again to life is so familiar in Old World symbols and rites that it's easy to overlook the crucial feature in Mesoamerican rites—the visual shape of an ear of corn, at once phallus, face and knife.

Among Coatlicue's many sons, the Aztec chose for patron Huitzilopochtli, whom the Aztec associated with fire, sun and war. In his name the Aztec, with their genius for organization, institutionalized sacrifice on an epic scale. In his honor Moctezuma initiated the "flowery war," waged against six city-states to the east for the sole purpose of capturing suitable victims for sacrifice. Other peoples would not do because Huitzilopochtli could not abide the coarseness of their savage and barbarous flesh. "Their flesh is as yellowish, hard, tasteless tortillas in his mouth," an informant told Father Duran. Captives from the city-states, on the other hand, "come like warm tortillas, soft, tasty, straight from the griddle."

By the thousands, heads were chopped and hearts sundered on the round Sun Stone of Tenochtitlán, the very Calendar Stone that charted the course of the stars and the count of days, the flickering of fire, the breathing of winds, the flight of birds, the slithering of snakes, the undulation of waters and the ceaseless metamorphoses of matter and spirit. The rhythm of life engraved in the stone was also incarnate in the god whose mask the priest wore when he ripped from the victim's breast a heart pulsing to the same beat. At that moment the priest impersonated the greatest of the corn gods because he was the most mortal—the Feathered Serpent, Quetzalcoatl, twin to the god of war.

The Feathered Serpent was a descendant of the Maya god Kukulkan, legendary founder of the city of Chichen Itza, who sat at the center of the cross-shaped tree and embodied in his emblems of sprouting maize, fish, salamander and vulture the four elements of earth, water, fire and air. His day sign was Ik, the day of breath, and for the Maya he was

Carved probably in 1497, the great calendar stone of the Aztecs, 13 feet across, related human hearts directly to the sun at its center, "heaven's heart," on which the sacrificial victim was placed. Encircled by a pair of knife-tongued serpents, the stone dictated daily ritual in the present by the recorded history of the past. Neg. No. 318168, courtesy Department of Library Services, American Museum of Natural History

the god of resurrection, transmuted to the Creek and northern tribes as *the great master of breath,* the Great Spirit. The Toltec renamed him for the green-plumed quetzal bird of the rainforests of Guatemala and Chiapas and for the water serpent of the tropics. Sacred as jade, Quetzalcoatl became the priest-king who founded their city of Tula on the high plateaus, taught his people the arts, conceived their calendar and gave them maize. He was a mediator between gods and men, sky and earth, a mortal and immortal hero who sustained the sun with his fire-drill and the corn with his blood.

For more than a thousand years, from the rise of the empire of Teotihuacán in A.D. 250 to the fall of the Aztec empire in A.D. 1521, Quetzalcoatl ruled the cradle of civilization in the New World. He brought corn to man after he had transformed himself into a black ant to steal a grain from the red ants who lived in the heart of Food Mountain. To the later Aztec, his earlier reign in Tula was legendary and paradisal, a golden age of abundance in which pumpkins were so large "a man could hardly embrace one with his arms" and ears of corn so plentiful that "those that were not perfect were used to fuel the fires."

In the Codex Borbonicus, Quetzalcoatl, with a serpent in his hand, dances before his enemy Tezcatlipoca. Neg. No. 332117, photo Logan, courtesy Department of Library Services, American Museum of Natural History

In his time, it was said, every ear of corn was as strong and firm as a man.

Sahagún described his temple at Tenochtitlán as encrusted with precious metals and jewels: the room to the east was covered with gold, to the west with turquoise and emeralds, to the south with white seashells and to the north with silver. Another temple was lined entirely with four different colors of quetzal feathers, yellow, blue, white and red. The sacred jewels and colors reappear in the altars of the Hopi kivas, as Major John Wesley Powell noted in 1875: "In the niches was kept the collection of sacred jewels—little crystals of quartz, crystals of calcite, garnets, beautiful pieces of jasper, and other bright or fantastically shaped stones, which, it was claimed they had kept for many generations."

What set Quetzalcoatl apart from other creator sun-corn-rain gods in the New World was that he was also human; his flesh was mortal and he suffered as a man. To my ears his story was all too familiar. He prayed and fasted, did penance, was tempted and sinned. Thrice he was tempted and thrice fell. First vanity undid him. When his dread enemy Tezcatlipoca gave him a mirror, Quetzalcoatl was so appalled at his ugliness that he allowed Tezcatlipoca to dress him in a mask of turquoise and a mantle of feathers. Next gluttony undid him. He gorged on a meal of corn, beans, chili, tomatoes and wine mixed with honey. At first he refused the wine, but after tasting one drop he drank five gourds of it and called for his sister Quetzalpetlatl, who also drank. Then lust undid him, for they lay together in carnal sin.

When Quetzalcoatl woke to what must have been a stupendous hangover, he lay four days and nights in a stone casket, then threw himself into the flames of a funeral pyre and was consumed. His ashes rose as a flock of birds which bore his heart high into the heavens to become the morning and evening star, Venus. Now every night he journeys through the Underworld, repeating his first journey when he defeated the Dead Land Lord by piecing together the scattered bones of his father from which man was made. For the Aztec, man was made of bone ground into meal, stuck together with maize and moistened with penitential blood.

In the person of Quetzalcoatl, Aztec priests reen-acted in bloody sacrifice the mystery of metamor-

On a stepped pyramid altar, an Aztec priest holds the victim's husked heart aloft and offers it to his Royal Lord the sun, to sustain life for mankind at the source. From the Florentine Codex. Neg. No. 1731, courtesy Department of Library Services, American Museum of Natural History

phosis, "the fall" into matter and rebirth into spirit sung by the poets of many testaments in their psalms. The shape-changing of Quetzalcoatl sprang directly from the metaphoric relation of man to corn. The ritual act of tearing out the heart was figured in the husking of corn, which Erich Neumann has interpreted in Jungian terms: "The husking of the corn, the heart, is castration, mutilation, and sacrifice of the essential male part; but at the same time it is birth and a life-giving deed for the benefit of the world or mankind." The phallic shape of corn makes husking a logical, if startling, symbol of castration. But startling only to us, not to the Aztec, for whom the phrase "to bear a child" was interchangeable with three phrases of death and violence: "to die in childbed," "to take a prisoner," "to be sacrificed as a prisoner." The Aztec who glorified war and conquest did so in part to harvest captives as he harvested fields of corn, to sustain life at the source. The paradox of equating child-birth with prisoner-death, like the paradox of equating blood with corn, came easily to men whose world depended not only on the substance of corn but on its symbolic configurations, where rows of kernels appeared to be both protected and imprisoned by their husks, where husking appeared to be both castration and insemination, where corn-eating was both cannibalism and sacrament, a mystic communion with the sacrificed god.

Quetzalcoatl endured through centuries and civilizations because the metaphor worked as a figure of self-transformation: a carnal heart purified into the "precious eagle-cactus fruit" of the deified heart that blazes nightly in the transit of Venus. While Huitzilopochtli conquered the outer world of matter, Quetzalcoatl conquered the inner world of spirit, of penitence, of atonement for the guilt man was born with, or as my grandparents would have said, the guilt man was born for. Quetzalcoatl embodied what in other circumstances my grandparents would have called a change of heart.

Drinking Down the Sun

If I found connection in the rites of Quetzalcoatl, I lost it in Cuzco at the feast of Inti Raymi, celebrated at the summer solstice since the founding of the city and the Inca empire half a millennium ago. I found corn everywhere, but overwhelmingly in the form of drink. For a week, villagers come down from the mountains and up from the valleys to parade daily through the streets. For a week women in their Mad Hatter hats and Wizard of Oz skirts, behind tables improvised into bars with bottles of Pisco and rum, cry "Punchy, punchy," and "Chicha morada, chicha bianco, chicha picante, chicha chicha!" Cuzco's Plaza de Armas—enclosed by the spotlit fronts of the cathedral, the Church of La Compania and the Moorish balconies of the arcades—is a maelstrom of flutes, drums and harps, brass bands and conch shells, men in feathered capes and pointy caps, masked troops of children in top hats and wedding veils, youths in cloaks of Spanish moss waving yucca stalks, women in bowlers carrying plumed batons—everyone in costume, everyone dancing. To an American tourist like me, deafened by firecrackers and popped like a cork from the pushing, pressing, crushing, pickpocketing multitude, the festival is a combination Fourth of July, New Year's Eve and Mardi Gras, fused into one big drunken bash.

Inti Raymi so shocked the Spaniards that they forbade it, but to so little avail that after four centuries the Peruvian government officially restored it, in 1930, under the name "Indian Fiesta." In the Southern Hemisphere, the month of June ("Raymi") is the time of harvest, which belongs to the Sun ("Inti") and to corn. At the summer solstice, the reigning Lord of the Sun tallied the loads of grain, gold, feathers and coca leaves collected as tribute, arranged the marriages of young knights, distributed the year's chosen virgins, reviewed the year's oracles and punished the priests who had guessed wrong, numbered the different-colored flocks of llamas and ended the fast with a communion meal of *yahuar sancu,* little balls of raw corn mixed with the fat and blood of sacrificed llamas.

Sancu was a vestige of the cannibalism practiced in "ancient times," Garcilaso confessed, when tribes were so addicted to human flesh that they buried their dead in their stomachs: "As soon as the deceased had breathed his last, his relatives gathered round and ate him roasted or boiled, according to the amount of flesh he still had: if little, boiled, if much, roasted." Such practices, he explained, were more common to Indians of the jungle, where produce grew spontaneously, than to those of his native Cuzco, where men were required to plant maize and vegetables and so got themselves civilized. Where the civilizing power of the Inca did not reach, there were still barbaric tribes like the Anti, said to have come "from the Mexican area," who ate human flesh as a sacred thing—"without cooking it or roasting it thoroughly or even chewing it." Such tribes violated decorum. In his own enlightened times, Garcilaso continued, *sancu* was made only "by bleeding and not by killing the victims." First they half-baked the bread "in balls in dry pots, for they had no ovens," then added to the dough the blood of children between five and ten, "obtained from between the eyebrows, above the nostrils."

Only at the spring and fall harvest festivals, which resembled the fasting and feasting rites of the Aztec, did the Inca use corn for bread. Since they had other staples—quinoa, potato, cassava—they commonly ate corn in whole kernels rather than ground into meal. And because cornbread was sacred, "only pure young girls could grind the meal for it." These were the "wives of the Sun," who on the eve of both festivals worked all night grinding the corn flour to make bread in "little paste-balls the size of an apple." The main job of the pure young girls, however, was to brew *chicha* for massive consumption at the feast. When the Inca king broke his fast at dawn on Inti Raymi, he greeted the Sun, his father, with two golden vessels of the brew. Inviting the Sun to drink first, the king poured the sacred beer from the vessel in his right hand into a gold basin, from which it flowed through an underground channel into the fortress-temple of the Sun, so that "it disappeared as though the god-star had really drunk it." The Inca drank from the other vessel, then divided the rest among the gold goblets of his nobles, believing that the sanctified liquid "communicated its virtues to the prince and to all those of his blood."

"Each home kept a Corn Mother (sara-mama) of corncobs, wrapped in fine textiles, enclosed in a shrine. Since she represented the strength or weakness of the corn crop, if she was 'strong' during the year, she remained; if 'weak,' she was destroyed and replaced. After harvest, a new sara-mama was made from the most fruitful of the corn plants. The ritual is observed to this day."

—LYNN MEISCH,
*A Traveler's Guide
to El Dorado and
the Inca Empire*
(1977)

At the Sun temple, priests brought in lambs of the "Peruvian sheep" for sacrifice. The first sacrificed was always black, for this color was held most sacred because most pure and unmixed. (Even a white llama has a black snout.) After a priest slit the llama's belly to read its organs for auguries, large numbers of creatures were slaughtered and their flesh roasted on spits in the public squares for all to eat with the *sancu* bread: for "all Peru was one people during this feast." This custom, Garcilaso observed, led some Spanish commentators to assert that the Inca and their vassals took Holy Communion as Christians do. Garcilaso, however, was more cautious: "We have simply described what the Indians did and leave each reader to draw what parallel he wishes."

Christians who observed worshippers washed in the blood of the llama might well draw the parallel for themselves, particularly if they heard any hymns to the creator god Viracocha. One such hymn has come down to us through the knotted-rope language of the *quipu,* used not only to number flocks and crops, but to convey the formulaic phrases of Inca rituals:

> *peacefully, safely*
> *sun, shine on and illumine*
> *the Incas, the people, the servants*
> *whom you have shepherded*
> *guard them from sickness and suffering*
> *in peace, in safety.*

Pastoral metaphors of shepherding were possible to these highlanders because they had long ago domesticated llamas and found them proper for sacrifice, along with the usual pretty youths.

Blood sacrifice was as potent for Inca crops as for Aztec, but where the Aztec saw symbolic potential in "husking the corn," the Inca fixed on unhusked corn, young and green, with wrappers intact. Virgin purity was essential and, at earlier festival times, chosen youths of both sexes were kept in a pound, next to the llama pound, in the Field of Pure Gold before the Coricancha. While the llamas were slit open alive, the youths were first made drunk on *chicha* and then strangled before their throats were cut. Many pious parents, particularly in Cuzco, offered young daughters of their own free will, but more were purposefully "not very watchful with their daughters," Father Cobo wrote. "On the contrary, it is said that they were happy to see them seduced at a very early age."

Farmers used blood in practical as well as symbolic ways to fertilize their fields, just as they used guano. As late as the nineteenth century, a traveler in

Peru observed that on the feast day of San Antony, the natives of Acobamba, east of Lima, celebrated the day by staging a fierce fight in their plaza. The men divided themselves into two parties and battled until some fell wounded or dead. "Now the women rushed forth amongst the men, collecting the flowing blood and guarding it carefully," wrote the author of *Travels in Peru, 1838–42.* "The object of the barbarian fighting was to obtain human blood, which was afterwards interred in the fields with a view to securing an abundant crop."

The early Spanish Fathers were bothered less by the volume of bloodletting than by the volume of drinking. "Indians drank in the most incredible manner: indeed, drunkenness was certainly their commonest vice," Garcilaso commented. In his own day that vice had been mitigated by the shame Indians were made to feel from the good example set by the Spaniards. "If the same good example had been given them as regards all vices," Garcilaso the Inca added, "today all the Indians would be apostles and would be preaching the Gospel."

If the Aztec institutionalized wholesale sacrifice in his corn rites, the Inca institutionalized drunkenness. During Inti Raymi, feasting was followed by a drinking bout ritualized in a series of toasts. The invitation to drink was always extended with a pair of goblets so that host and recipient could drink equally: "The king has invited you and I have come in his name to drink with you." The bout lasted for nine days, the point being, as Father Cobo lamented, "to drink until they cannot stand up." To this end they preferred the strongest *chichas* and added other strong potions to "knock them out more rapidly."

If excess was the point, from what I saw at the festival of Inti Raymi in 1987, not much had changed. By the end of the week, back streets and hills were littered with bodies and here and there a bus that had toppled over the edge of a road. The body count was particularly heavy along the windy road leading to the fortress-temple of Sacsahuaman, which shares the view over Cuzco with a megalithic White Christ. Nowadays, the official climax of the week's festivities is a pageant performed inside the ruins of the Sun's House, built of cyclopean stones "so large that nobody who saw them would think they had been laid there by human hands."

The stones of Sacsahuaman become bleachers for spectators watching the procession of the Inca during the feast of Inti Raymi. Photo by Robert Frerck, Woodfin Camp & Associates

A native Cuzqueña celebrates Inti Raymi today as her ancestors did, by drinking chicha.
Photo by Mireille Vautier, Woodfin Camp & Associates

The pageant had been laid by too many human hands, at least for the hundreds of foreign tourists who dozed on bleachers, while local performers enacting the Inca and his voluminous court proclaimed in Quechua, hour after hour, the glories of empire. Interest picked up when a real llama was lugged up the altar steps and the Inca lifted his knife, but the victim disappeared through a trapdoor before a drop of blood was shed. All about us, however, on the Field of Spears, where blood aplenty had been shed during the Great Rebellion of 1536, the hilltop seethed with life. The field was a giant earth oven, a *huatia*, with a hundred smoking chimneys, where hundreds of families camping here for the week baked in the earth their potatoes and oca roots and corn.

The real harvest festival was here, in Chuqui Pampa, not in our segregated bleachers, where scruffy bands of boys, scampering out of range of the guards, begged and fought over our tourist lunchboxes of Jell-O and hard-boiled eggs. The festival was in Chuqui Pampa and back at the Plaza de Armas. There the dancing and drinking had never stopped. There thousands still snaked past the rows of alpaca-sweater sellers, the display of the vampire bat and Brazilian tarantula, the small boys peeing at the curbs. There I sought refuge with a glass of European beer inside Chez Victor. I liked the taste of *chicha,* but the decorum of excess defeated me. At the restaurant I was divided only by a pane of glass from one *chicha*-bombed celebrant, who stood for an hour with his Inca face pressed against the window. Looking in as I looked out, he looked through me as if I were glass, as if centuries were gone in smoke, and he would stand there always, fixed and motionless as stone.

The Language of Carnivals and Kings

HOOPESTON'S SWEET-CORN MAGIC

THE CORN RITES past and present of America's native civilizations kept me at a distance, but even those of the Midwest made me feel like a tourist. At 10:00 A.M. on Labor Day weekend, at the corner of Main and Sixth streets in Hoopeston, Illinois, I found men still unloading plastic folding chairs from the back of Ford pickup trucks for their womenfolk to sit on. In Hoopeston men wear their bib hats stiff and high; women wear their hair—colored blonde, white or blue—the same way. Men wear their jeans low and loose, cinched tightly below formidable bellies. Women wear their pastel polyester pants high, stretched tight across bellies worn proudly, even post-menopausally, as emblems of fertility. On Main Street, at the crossroads of what was once the north-south Chicago, Eastern & Illinois railroad line and the east-west Nickelplate line, Hoopeston's 7,092 inhabitants were gathering to celebrate their forty-third Annual National Sweetcorn Festival. Forty-three years is an eon out here, and the rites of this local festival are as decorously prescribed as those of the Pueblo, Creek, Aztec or Inca.

All over Illinois—not to mention Kansas, Nebraska, Ohio, Indiana, Missouri, Pennsylvania, Michigan—small towns, like an archipelago in a landlocked sea of corn, celebrate sweet-corn festivals. In Illinois I could have gone to Mendota for its thirty-ninth National Sweet Corn Festival, to Morris in Grundy County for its thirty-eighth, to Mount Vernon for its fourteenth Sweetcorn & Watermelon Festival, to Kewanee for its Annual Corn & Hog Days, to Bloomington, Urbana, Delavan, Warrenville, El Paso, Arcola. No town is too small to be without its Grand Parade, carnival rides and demolition derbies. I chose Hoopeston because it calls itself the Sweet Corn Capital of the World, not for corn in the field but corn in the can.

Hoopeston is built of cans. It was founded in 1871, in prairie between the Vermilion and Iroquois rivers, when a railroad was being laid along the north-south Hubbard Trail to Chicago. The Chicago, Danville & Vincennes line opened the following year, the same year as the founding of East Lynn next to Hoopeston and of the Sunbonnet Club of Young Ladies within Hoopeston. Shortly Hoopeston would have the first Illinois Canning Company, developed exclusively to can corn for the railroad to transport. Today Main Street divides the canneries of the north from those of the south, with names that have metamorphosed into Stokely-Van Camp, Green Giant and Pillsbury. To the north, the American Can Company supplies the cans and, to the south, the Food Machinery Corporation supplies the machinery for canning. In corn-pack season, when fifteen thousand acres of sweet corn in a thirty-mile radius around Hoopeston are ripe, Hoopeston cans around the clock seven days a week. As one festival booster, John McElhany, said, "It's a corn town."

John's dad, Clyde, is a dapper man of seventy-seven who's helped the Jaycees run the festival since 1938. I found Clyde joking with Pat Musk, a big woman with a Gertrude Stein haircut, who runs the National Sweetheart Pageant that began with a competition for Miss Sweetcorn. Now it's a "training pageant" for Miss America, said Pat, who is also town organist. Pat and Clyde go back a ways. "We used to smoke behind the barn together many years ago," Clyde said. "Yep, she played it on the organ out behind the barn." Pat heaved with laughter like a leaky bellows.

John was running the twenty-four-hour scavenger hunt this year—seven teams and 325 items—"some difficult, some embarrassing." He ran through the menu of events: the "I Like the Sound of America" pageant presented by the children's choir of the Church of the Nazarene, the Horseshoe Pitching Contest, the Horse and Pony Show, the "Festival of the Stars" Talent Show, Square Dancing with Merle Ponto and Mike Powell Callers, the Brownback Family Magic Act, and of course the parade. Baton twirling is big at the Hoopeston parade. The thin, the fat, the shapely, the squat, toddlers barely out of diapers—all strutted, swung, twiddled and dropped their batons in time to a thousand-and-one trombones. Local politicians like "Jerry Stout—Your Small Town Alternative" worked the crowd. The Shriners' Royal Order of Cooties invaded in motorized minicars like a plague of crickets. But the heart of the parade was cars, slick and penile, adorned with living dolls in organdy ruffles or glittering sequins, emerging like corn silks from the cob, waving smiles and bosoms for the title of National Sweetheart.

The entire festival celebrated engines and wheels: the Bigwheel/Tricycle Races, the Dual-Demolition Derby, the Sweetcorn Cruise, the Tractor Pull and the climactic King Krunch—a monster truck that stomped cars. Even the feast of "FREE HOT BUTTERED SWEET CORN" served to twenty-two thousand people a day, like a Midwestern potlatch, celebrated the machinery that can husk and steam some fifteen tons of corn for assembly lines of people, equipped with roasting pans, dish pans and plastic garbage bags to catch the Niagara fall of ears from the steamer. The Kids' Corn-Eating Contest was won by a freckled boy with no front teeth who turned his gums into a machine, "cleaning the cobs" round and round like a lawnmower. "Don't swallow," his brother yelled at him, "just chew." In the most casual talk cars figured: a man handed his busted shoe to the repair man with the words, "Blew out my tire."

The Kids' Corn-Eating Contest at Hoopeston's Annual Sweet-corn Festival. Photo by John McElhany

Machines are nature's way in Hoopeston; everyone works at one time or another for the canneries. "I used to work at Stokely cleaning smut," said the prim director of the Vermilion County Museum. "Awful stuff, you'd scrape it off and it'd get all over your glasses." "There are few hand jobs left now," said Bill Nichols at the present Stokely plant, and most of those are done by seasonal help, migrant Mexicans. But there is work running the machines that can Stokely's ten thousand acres of sweet corn, after they have been automatically husked and cut, cleaned for three hours, cooked for twelve minutes and cooled for five. Machines turn out 1.7 million cases of corn, one-third of it creamed.

Clyde Timmons, a well-kept man of sixty-six now retired, spent his life in the industry, as did his father before him. "I was pokin' cans at thirteen," he says, "for twenty-five cents an hour." He worked every "corn-pack" in high school and college until he came back from the army and went to work as a field man for Milford Canning, in charge of contracting and harvesting the sweet-corn crop. "Each company," he explained, "pays the farmer a contract price and furnishes the seed to control variety and timing, so that they can harvest everything during a four- to six-week period." When corn was all "hand snapped," they used to import Kentuckians and Missourians and mules to go with them to get the harvest in: "Each company had three hundred and fifty or so head of mules in their barns. At two-thirty every morning the workers would hitch a team of mules to a wagon that'd hold three ton a corn, they'd move out to the fields, jerk the corn to throw up into the wagon, then bring it back into the streets here with all the kids in town hollering, 'Git me a roastin' ear.' Each company had a muleskinner. He'd spend the whole year buying and selling the mules, too expensive to keep year-round, so they'd sell 'em after corn-pack and start again in the spring, buying 'em up for farming and jerking."

Mules were replaced by trucks in the late 1940s, hand snappers by mechanical harvesters in the mid-1950s. Hoopeston is the home of the first successful mechanical harvester, called the FMC Sergeant after the man who developed it in Iowa but didn't have "the finances" to continue. "We were among the first to put them in the field," Timmons said. In 1975 Alvin Slosser mounted a four-row harvester on a New Idea Uni-Harvester, which pulled a truck behind it, and that became the Byron Sweet Corn Harvester of today. When corn was shucked by hand, the shucker was given one brass coin per bucket, which he could exchange for money at the end of the day. But once William Sells developed a husker using iron rollers and then a feed machine called the "Peerless," the only major work left to automate was cutting kernels from the cob.

Originally whole ears of corn were canned, the first by Captain Isaac Winslow of Maine, who sold a dozen cans to S. S. Pierce of Boston in 1840. Three decades later, Barker's Semi-automatic Plunger Cutter paved the way for Welcome Sprague to create a cutter with rotating knives and an automatic silker. Sprague and Sells got together to form the company that became Food Machinery, with the aim of manufacturing all the machinery needed for canning corn. Soon a proper steaming kettle was invented by J. F. Rutter, who also devised a way to keep corn white in the can without bleach. In those days they boiled corn for three to four hours.

Today Hoopeston's canning companies may count on the fact that the taste for canned corn in the United States has not changed for the last twenty-five years: we still consume forty-two to forty-five million cases annually, of which Hoopeston produces eighty-one million cans. It is still a housewife's boon, said Timmons. "You just open it, heat it and dump it out." The voice of pragmatism in the rhythm of the machine and the ceremony of the can. I thought of the crowd's cheers when King Krunch stomped on cars in a monster sacrifice of metal. Where tin lizzies were enthroned with tin cans, King Krunch was Coatlicue. In the festival booklet labeled "Hoopeston's Sweet Corn Magic," an ad for one cannery boasted, "We built our plants right beside the cornfields in the heartland of America." Only to an outsider would that seem a contradiction.

COURTLAND'S CORNHUSKERS

THE LITURGIES OF INDUSTRIALISM boomed over the car radio on Sunday morning of that same weekend as I drove through minuscule Illinois towns, headed for Hog Days in Kewanee. A former Marine, now a born-again evangelist of the Church of God, pummeled me with questions: "Did you know that Jesus had short hair, was a taxpayer and dined with the wealthy as well as the poor?" I had been driving for miles through six-foot walls of corn, so mechanically uniform they looked like a Marine's dream of flat-topped Christian soldiers presenting arms to a short-haired Jesus. No wonder Khrushchev flipped when he saw the scale of these battalions, divisions, armies of corn, a rural proletariat regimented to military-industrial perfection. The paradox of the Midwest is that its farms and farmers are children of the industrial revolution, at once progressivist and conformist, like a capitalist Jesus—a paradox not lost on a Russian communist viewing for the first time America's machine-tooled factories of corn.

As an ex-Presbyterian, I saw the paradox of rural America celebrating Labor Day, a rite of the urban proletariat, with country corn festivals and churchgoing. For paradox, take Delavan, Illinois, where all but a handful of its two thousand inhabitants were attending the town's six churches before enjoying a communal barbecue in a cook-shed decorated with corn. First the eating, then the Frog Jumping Contest, Needle in the Straw, Tomahawk Throwing, Nail Driving, Corn Shelling, Lip Sync Contest and Cow Patty Throw—all played in the birthplace of factory corn. In Delavan in 1847, the corn that Robert Reid planted, and his son James crossbred, laid the foundations of America's corn industry—Reid Yellow Dent.

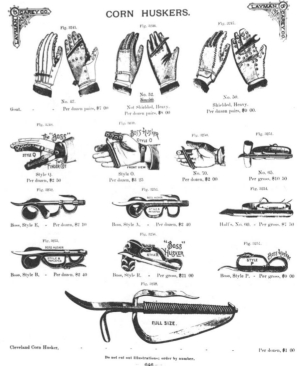

Cornhusking tools from the trade catalog of Laymen & Carey Co., ca. 1890. From the Collections of the Henry Ford Museum and Greenfield Village, Neg. No. 103449

Kewanee, while far from urban, was more proletarian than Delavan or Hoopeston. At Kewanee the bars outnumber the churches, and beer is not sequestered in a tent. The bars were jumping along the Main Street midway and so was the World's Largest Pork Chop Barbecue, surrounded by mountains of deep-fried onion rings, foot-long corn dogs, lemon shake-ups for the temperate, deep-fried elephant's ears with maple glaze and funnel cakes with apples and candied cherries for the fatties. There's nothing genteel about Kewanee, Hog Capital of the World. The Mud Volleyball Tournament was as satisfying as a hog wallow and the geriatric square dancers as invigorating as a troop of dervishes. I missed the Hog Jog Stampede, the Hog Calling Contest and the Kiddie Tractor Pull, but I found Kewanee's secret treasure, the National Corn Huskers' Hall of Fame, and an ex-cornhusking champ, Bill Rose.

Bill, a tall, barrel-chested man of seventy-nine

with hog-sized hands and thumbs like cigars, was selling raffle tickets for an afghan to benefit the Kewanee Historical Society, which devotes a corner of its building to the Hall of Fame. Bill pointed to a display of husking paraphernalia—the wrist hook, thumb hook, palm hook—and a wall of photos. "Here we're lined up waiting for the bomb to go off," he explained a *Life* photo of the 1935 national in Indiana. "Eighteen wagons, contestants from nine states, over one hundred thousand spectators mashed the corn down. There's Henry Wallace, he's the one started it in 'twenty-four in Des Moines, Ioway. I got fourth place that day in Indiana, got second in the state, won the county." Bill was off and running, recalling a world that ended as abruptly in 1941 as the pony express in 1861. For cornhusking, not only a necessity but a national sport in the 1930s, stopped when the war ended the contests and the combines ended husking by hand.

> Years ago we planted with a wire with buttons, about three foot apart, so the planter would click and plant a kernel. Then we'd husk two rows at a time and the ears'd hang down. Now they plant the corn so thick we pick only one row at a time and you grab the ears different and got to hold onto 'em. They're bred for combines 'cause that's what's doin' the pickin'. Then a man'd do good to pick a hundred bushels a day, or about two acres. Now a combine picks eight hundred bushels an hour.

Cornhusking champions trained in the field, where they learned to break, strip and throw an ear against the "bangboard" of the wagon in a single motion.
Photo by Grant Heilman
Photography, Inc.

In 1922 Henry Wallace challenged Midwestern farmers to improve the efficiency of their cornhusking (not just quantity but clean ears), Bill said, and offered a prize of fifty dollars for the best husking record for a day. The first contest took place in Des Moines and was won by Louis Curley, who picked 1,644 pounds. The following year Wallace doubled the prize and instituted rules: Wagons must be double-boxed, two boards high and twenty-six inches deep, with a forty-foot-square bangboard. Heats were to be eighty minutes, with a weight deduction for husks and a triple deduction for gleanings. County winners would go to the state contest, and state winners to the national. Farm magazines became sponsors, donating prize money and crowning husking queens. Churches got into the act, providing food and tents for husking banquets. Entertainment by bands and barbershop quartets, automobile and farm machinery exhibits, educational displays like Indian wickiups turned the contests into community fairs.

"In our opinion," Wallace said in 1923, after a horse named Zev had won the Kentucky Derby, "it should give the spectators as many thrills as the race between Zev and Papyrus or the Yale-Harvard football game." Here was a corn competition far more thrilling than the National Tall Corn Sweepstakes held annually at the Iowa State Fair, even though a world record was set in 1942 with a stalk of 26 feet, 10$\frac{1}{8}$ inches. Almost overnight champion cornhuskers became national sports heroes, sought by promoters to endorse everything from steak restaurants to insurance policies. About the time one husker was signed for a promised Hollywood movie about cornhusking, the farm papers required contestants to sign a pledge against paid promotions.

Wallace, himself no slouch at media promotion, used the pages of the *Farmer* to critique the husking techniques that were hotly debated by men and boys over suppers of creamed canned corn. Whetting their appetites for the match to come, Wallace compared the pinch-and-twist versus the free-swing methods of a pair of major contenders in the December 1923 issue of his *Farmer*:

> Both Rickelman and Paul use thumb hooks, one a Kees and the other a Clark. Their method of husking, however, is totally different. Rickelman uses almost invariably the pinch and twist husking method, grasping the ear at the butt with the left hand with the thumb up. The thumb hook on his right hand brushes the husks aside so that he can grasp the ear and give the twist which usually breaks the ear quite clean of husks. Occasionally, however, the Rickelman method results in a slip-shucked ear.... Paul uses the standard hook method, but he has a free and easy, rhythmical swing which makes his husking rather prettier to watch than the Rickelman husking.

By the 1930s the contests were covered by *Newsweek* and *Life,* by radio and newsreel, and President Franklin D. Roosevelt fired the opening shot from the White House. In 1940, the Nubbin Derby or Battle of the Bangboards, as NBC's *National Farm and Home Hour* called it, drew a record crowd of one hundred sixty thousand people, before the war ended contests already doomed by the advent of harvesting machines.

Machines may have supplanted the hands but not the hearts of veteran huskers. A few old-timers began to revive local contests in 1970, Bill told me, and revived the national in 1975 in Oakley, Kansas, purely for sport. Today

"At 12:45, when a five minute warning bomb exploded (upon signal from President Franklin D. Roosevelt in the White House) 160,000 people—more than have ever filled a football stadium—were throbbing with pre-kickoff excitement. The 18 contestants, each stationed in his own plot of land, waved aside photographers, radio announcers and reporters. Up floated an American flag, the starting bomb burst; and the 13th World Series of the Corn Belt started with a bang."

—From *Newsweek,* November 21, 1936

nine states—Kansas, Nebraska, Iowa, Indiana, Illinois, Ohio, South Dakota, Minnesota and Missouri—compete in the national, the host town and state varying from year to year. This year it was Oakley's turn again, Bill said, and I could watch the husking contenders warm up in the North Central Kansas Corn Husking Contest that takes place every September in Courtland, Kansas.

Courtland is just across the border from Hildreth, Nebraska, and is inhabited mostly by Johnsons and Thomsons, three generations of them, who run the show. Courtland is the kind of small town that wears its sporting heart on its shop windows: "Eat the Bluejays," says Gero's Cafe; "Bank the Panthers," says the Swedish-American Bank; "Flush the Bluejays, sink 'em," says Tebow Plumbing. I found the husking field on Highway 36, just three and a half miles east of Courtland. Horses and wagons were drawn up by privies labeled "HE" and "HER," with the footnote "Not Responsible for Accidents." A sign on a backboard said, "If you think nobody cares, just miss an interest payment on a couple of 8800 John Deere Combines." A sheet labeled "What We Named Our Horses" listed names like " 'Damn You Earl'—always called 'Damn You' by our children." When I pulled in, Esther Thomson was announcing the winners of the Patched Overalls and Jeans Contest—four divisions: largest, smallest, most unusual and most decorative.

Most of the folks were standing around waiting, waiting for the drivers to hitch up the wagons, waiting for the heats of each division to be announced, waiting for the starting gun to go off and then to go off again at the end of thirty minutes, waiting for the loads to be weighed and the results posted, waiting for the combine to finish off the picked rows, waiting in line to eat a porkburger and a lard-crust rhubarb pie and to drink a lot of lemonade to counter the hot dry heat. But no one's in a hurry in Courtland and waiting leaves plenty of time to talk. I talked to oldsters and youngsters and unhusked a whole new world of corn.

"This is my thirtieth contest," said Vernon Erickson, who is seventy-five and makes all the trophies for the contest by hand. His big competition is Gene Walker, a mere sixty-nine. "Gene won last year so Vernon's hot to win today," said Gene's wife. Gene said his best score was twenty-four pounds a minute, "but the corn is heavier this year and that slows you down." He showed me how a right-handed husker grabs the ear with his left hand, rips it open with the hook, shucks and tosses it in one motion with his right. It's the left hand that wears out quickest, and the men do exercises to strengthen it. As a boy he shucked from September to April, six days a week, with maybe two feet of snow on the cornfields. "You got ten cents a bushel and if you shucked a hundred bushels, ten bucks a day—ideal for a boy who wanted to get out and hustle."

As seniors, they don't compete with the age group designated Men's National, twenty-one to sixty-four. Of the seven divisions at each contest, this is the toughest, for the men have to pick for a full half-hour. Youths and seniors pick only for ten, women for twenty. "Ideal age is twenty to thirty but if you didn't grow up with it, it's kinda hard," said one husker, "gotta be able to do it blindfold." Waiting for all the heats of all divisions to end was like waiting for the Pueblo corn dances to begin. The contest drawled slow as speech through the hot afternoon, and I decided I could wait for the awards banquet that night to find out who'd won.

At seven o'clock sharp I joined the chowline with three hundred others in the high school gymnasium of a nearby hamlet, Scandia. Checked tablecloths stretched from one end of the basketball court to another, awaiting our paper plates of dry turkey, wet stuffing, mashed potatoes and corn. After the Doxology, the piano thumped out "Alexander's Ragtime Band." Everyone I met who was not a Johnson or Thomson was an Olson or Peterson. Across from me Orville Peterson, at eighty, had just competed for the fifty-eighth time. Next to him Roy Hubbard, at nine, had competed for his third and won first place in the sixteen-and-under division. Roy's father, Don, had won a prize for his horses—Most Popular Team—and Don began the evening's entertainment by reading Edgar Guest's "A Heap o' Livin'." The poetry reading was followed by a barbershop quartet, a clog dance and a skit featuring Dumb Bunny, Agra Bear and Farmerette. Vernon's trophies were almost anticlimactic, but I found out that Bob Ferguson, a "super husker" from New Sharon, Ioway, and a three-time national winner, had won today.

Afterward, as I drove to a distant motel through a landscape totally black, totally flat, the horizon exploded. "We have a little bit of everything out here— floods, droughts, tornadoes," some Thomson had told me. "A little twister the other day hit Harding and leveled the town, you wouldn't believe the debris." I believed it now, watching pitchforks of lightning stab ricks of clouds, booming like Armageddon, obliterating on my radio *Big Joe's Polka Show* and *Your Radio Bible Class.*

I caught up with Bob Ferguson again at the Iowa state contest, held at the Living History Farms in Des Moines. I also caught up with Herb Plambeck, who's been broadcasting farm news on the radio for fifty years, with time out for reporting World War II and Vietnam, and for serving as assistant to former Agriculture Secretary Earl Butz. Herb, an intense, worrying man of seventy-nine, is the guiding spirit behind the husking revival and founder of the National Corn Huskers Association and its newsletter, *Shucks*. In 1984, Herb started the first National Corn Throwing Contest and devised the rules: two throws each, one underhand and an optional overhand. That same year saw the first Champion of Champions Contest, limited to former national champs and won by John Jackson, husking 509½ pounds gross in twenty minutes. These are the kinds of statistics Herb keeps in his head, together with memories of former heroes like Carl Seiler of Oneida, Illinois, who hung up his hook after setting a world record in 1932, in deep mud in an ice storm. Ferguson's competition today, Herb told me, was another national champ, Tony Polich of Johnston, Iowa, age seventy-three to Bob's sixty-two. "It'll be that close today, a whisker's gonna tell the story," one husker predicted, but Bob won handily and set a new world record of 867½ pounds in thirty minutes.

The theme song of all this cornhusking revival is, There's been a change. "In the old days a couple would pick together and put the baby in the back of the wagon," said Herb. "The whole family picked or the harvest wouldn't get done. Tried to get it done by Thanksgiving, get the house banked up to insulate against the cold, get the produce stored in the root cellar, get the livestock penned. Then it was three hundred and sixty-five days you had work to do. Now it's corn, soybeans and Florida."

There's been a change, said Joe Anholt, a former national champ from Fort Dodge: "We'd stay out of school six to eight weeks every fall to get the crop done. They don't allow that anymore." There's been a change, said Ernest Heidecker of Lakota: "Nobody's got cattle anymore so nobody's using silage. Most people store corn shelled, 'cause last year we shelled one hundred and five thousand bushels from six hundred acres and if you did that with ear corn we'd a had to have cribs from here to Kansas and back."

Poster for the Missouri State and National Championship meets at Marshall, Missouri, in October 1987.

There's been a change, said Gary Kupferschmid, of Mediapolis, Iowa, who capitalized on change in 1981 by founding International Corn Hook Collectors. With more than a hundred members, they've expanded into all corn-related software and hardware, from corn sacks and shellers to Shawnee pottery, and now call themselves Corn Items Collectors. "About ten of us have got maybe three hundred to three hundred and fifty of them cornhooks, but I've probably got the biggest collection," Gary said, pointing to his display boards of antique pegs and

hooks, with rare beauties like the "twin-spurred palm hook" or the "iron-mesh thumb stall."

But some things don't change, like the "two-buckle thumb hook" made since 1929 by Raidt Manufacturing of Shenandoah, Iowa, and used today by Tony Polich. Or like the husking mitts and gloves that Raidt sold for twenty years after the bottom dropped out of his business, in 1948, when they were selling forty thousand dozen pairs of gloves and a hundred thousand palm hooks a year.

Some things don't change, like the voices of American men talking corn-husking as if it were baseball or football or golf. "Good huskers are born, not made," said Ray Oroke of Oskaloosa. "It's rhythm and stamina," said another. "It's breathing and concentration," said a third. "You just work hard as you can," said Bob Ferguson. "Hard work, clean living and a fierce determination to win," said Bill Rose. Winning takes the gumption of Vernon Erickson, the trophy maker of Scandia, who in 1978 lost the middle finger of his left hand in a sawing accident before the state meet and defended his title with a taped stub. Winning takes the passion of Herb Plambeck, still broadcasting in the pages of *Shucks* the glory of champions who contend for honor on the fields of corn.

THE WORLD'S ONLY CORN PALACE

GLORY IS WHAT TOOK ME to South Dakota in search of Mitchell, home of the World's Only Corn Palace, built to the greater glory of corn before the sport of husking champions was born. In late October, I picked up I-90, which runs straight across the state from Sioux Falls on the east to Rapid City on the west, where the four stony heads of Mount Rushmore look down on Crazy Horse Monument, Calamity Park and Custer State Park. If those faces could look to the southwest, beyond Buffalo Gap National Grassland and Badlands National Park, they would see Wounded Knee. If they could scrunch toward the east, nothing on these limitless barren plains, burnished to a shine by iced winds straight from the Pole, would interrupt their vision for 250 miles until they saw, in disbelief, the Kubla Khan domes and minarets of the Corn Palace.

Mitchell's Corn Palace of 1904 carried on the tradition begun by Sioux City in 1887.

After hours of bleached corn stubble and withered sunflowers, lit now and then by the startling blue eye of a pond, I was glad to reach Mitchell. The plains howl with the ghosts of travelers stranded after dark, and I was cheered to see, in the utterly deserted crossroads of North Main Street, the little Christmas-tree lights that outlined the only remaining corn palace from a century ago, when such palaces signaled the triumph of King Corn over the last stand of Mother Corn. By that time, Mother Corn had already lost the prairies and was about to lose the plains.

"We are strangers in the land where we were born."

—INSHTAMAGA OF THE OMAHA

It was hard to believe that the loss was so recent and so complete. The Indians of the United States had made their last stand in the rightly named Badlands but three generations ago. When South Dakota became a state in 1889, the Dakotas—last of the Louisiana Purchase lands to become a territory—were still half Indian. Just a year later came the massacre of the Sioux at Wounded Knee, the murder of Sitting Bull and the death of the Ghost Dance that promised the buffalo would come back and the invader depart forever. The year before, when the corn was in, my grandparents took a honeymoon trip upriver from their new home in Edgerton, Kansas, to Sioux City, Iowa, sixty miles south of Sioux Falls, to celebrate the Grand Harvest Jubilee Festival and take a gander at Sioux City's third Palace of Corn.

The new corn boom of the industrialites had already foretold the destiny of the continent and persuaded poet-prophets like Walt Whitman, traveling through prairies and plains a decade earlier, that one day they would feed the world. It was booster time, when new settlements vied to become towns, county seats and railroad centers, when booster talk turned hotels into People's Palaces and booster enterprise turned crops into Corn Palaces. It was the time when the nineteenth century changed gears, and Sioux City typified the speed of change. In less than a decade a farm town of seven thousand became a meat-packing metropolis of thirty thousand. The formula was corn plus railroads. What is striking in the sudden rise of these Midwestern cities is not the quick change from rural to urban but the fact that the urban preceded and shaped the rural before a field was planted or an ear of corn canned. In marveling at the way an entire continent packaged itself, Daniel Boorstin has written that the development of the American West is "one of the largest monuments to *a priorism* in all human history . . . a relic of the young nation's need to make a commodity of its land, and hastily to map and sell it, even before it was explored or surveyed." In the same booster spirit, the nation made a commodity of its crops; they were packaged and sold before planted. This despite the realities of a landscape biblically plagued by grasshoppers and locusts, droughts, tornadoes, prairie fires, white blizzards of snow and black blizzards of dust.

When the fathers of Sioux City got together in 1887 to give thanks for the rain that had spared them from the drought afflicting their neighbors, they didn't do a ceremonial dance as the Sioux would have done. These champions of liberty, who had fled Old World monarchies in the name of independence and democracy, built a palace, a palace that would sell their town as a center of commerce and culture inferior to none. "Palacing" their town was a way of packaging it. "St. Paul and Montreal can have their ice palaces, which melt at the first approach of spring," noted *The Sioux City Daily Journal* of August 21, 1887, "but Sioux City is going to build a palace of the product of the soil that is making it the great pork-packing center of the northwest."

The Sioux City Corn Palace of 1891, the fifth and last before Mitchell, South Dakota, took on the palace glory. Photo courtesy State Historical Society of Iowa—Special Collections

They would build a palace entirely of corn to celebrate a "Sioux City Thanksgiving and Harvest Festival and Corn Jubilee." The image of a corn palace inspired the citizens of Iowa when America was consciously searching for a symbol of national identity, and what better symbol for the union of science, commerce, agriculture and art than corn? Corn was no longer Indian wheat or Indian anything. Corn was 100 percent Made in the USA and therefore a prime symbol for a new nation, indivisible, that defined itself not by the generative life that bound its tribes but by the commodified life of its products. Since corn was the ground of the entire industrial empire that sprang from the prairies and plains, corn was appropriated symbolically to convey the white man's aspirations and achievements. Corn was the glue that bound one small community to another as tracks were laid across the plains.

These newest of New World settlers were too new to despise corn as inferior to wheat, for here the mere planting of corn was a symbol of white man's progress. The settlers were unaware that a thousand years before the white man set foot on the Plains, the Plains Indians were planters as well as hunters and that horses were not used in numbers until the middle of the eighteenth century. They knew nothing of the Woodland tribes, who had extended the Hopewell culture up the Missouri, of Sioux subcultures like the Mandan and Hidatsa who were tending their gardens of corn in North Dakota at the moment General Custer was hunting for Crazy Horse and gold. Did the white man conquer the Indian or did the Indian absorb the white man? Perhaps the battle of Wounded Knee did not so much mark the end of an empire as transform it into a new ceremonial mask in which Corn Mother became Corn King.

The corn palace as state booster, trumpeting Corn Belt science and art, was a mainstay at national expositions well into the 20th century. At San Francisco's Panama-Pacific Exposition of 1915, the North Dakota building enlightened the world with corn. Courtesy State Historical Society of North Dakota

According to *The Sioux City Journal,* 1887 was the year Sioux City went "corn crazy." The architect W. E. Loft was commissioned to turn the northwest corner of Fifth and Jackson streets into "the Eighth Wonder of the World." He designed a hundred-foot cupola, thatched with green stalks, from which flying buttresses sprang to the four corner towers, representing the four-square states of Iowa, Dakota, Nebraska and Minnesota. The Sioux City and Pacific Railroad brought in 15,000 bushels of yellow corn, 5,000 bushels of variegated corn, 500 pounds of carpet tacks and 3,360 pounds of nails to cover the frame. Iowans were energized by the discovery of a new art material, available to all: "Everyone was experimenting with grain as a medium of artistic expression." Every man could be a Michelangelo in corn.

The entire skin of the palace inside and out, every pinnacle, cupola and turret, was made of colored corn and other prairie grains, like sorghum and oats. The interior was a gallery of corn mosaics, executed by the Ladies' Decorative Association, creating allegorical scenes and panoramas not with gaud, tinsel or precious metals but with "an ear of corn, a handful of grasses, a bunch of weeds, a wisp of straw—the materials of nature's own painting." Rendered in corn, Hiawatha's Mondawmin joined hands with Demeter in a frieze around the dome. The Iowa State Seal was depicted in corn and cattails. Millet's *Angelus* and a map of the United States were alike figured in cereals. Even Ceres herself, the Great Mother of the ancient civilization of wheat, was robed in satiny husks on a stairway of yellow kernels and given a cornstalk scepter in homage to the King.

President and Mrs. Grover Cleveland, Cornelius Vanderbilt, Chauncey Depew and a party of "135 eastern capitalists from Boston and New York" led a hundred thousand visitors to celebrate the new empire. Through the corn-decorated arches of the streets marched a parade of industrial machines *and* a whooping procession of two hundred Omaha, Sioux and Winnebagoes in war paint, feathers and little else. The "fastidious are requested not to blush," warned the *Journal,* as the redmen performed their traditional corn dances in native undress. In the evening, fortified by "corn juice on tap even though Iowa is a prohibition state," the entire community dressed in corn costumes of husks, leaves and cobs to dance through city streets lit by seven thousand colored gas jets, past storefronts blazoned with harvest scenes, and into the armory for Iowa's first Corn Ball.

So successful was the first Jubilee Palace that Sioux City built four more in four years; the fifth, a cross between Moscow's St. Basil Cathedral and Washington's National Capitol, expired in a puff of glory, but floods and economic panic aborted a sixth. As an embodiment of progress, each successive palace had to top the last, with ever more voluptuous renderings of the royal treasury of corn: "Above the arch was a spacious balcony bounded at each end by stately turrets which were flanked by minarets overlaid with wild sage and white corn, giving the appearance of a chased silver column divided into diamond sections by bars of ivory." Visitors were tripled and quadrupled by publicity stunts like the Corn Palace Train, which toured the Eastern seaboard to receive the blessings of newly inaugurated President Benjamin Harrison and accolades from *The New York Times*—"Everything used in the decorations except the iron nails is the product of Iowa cornfields and the whole train is a marvel of beauty."

Corn palaces were the Midwest's answer to Eastern hauteur in the cultural line of literature, music and art. The Ladies of the Lilac History Club in Sioux City won first prize in the 1890 palace with a miniature corn library: "The walls of the library booth were adorned with [corn] pictures—a portrait of Dante, a winter scene, and a country maid with an apron of flowers. . . . Upon a table were quill pens of cane and oat straw, a corn lamp, a gourd inkwell, and several corn husk blotters." Nor were the dramatic arts neglected. The palace

The World's Only Corn Palace as it looks today, with its annual panels of corn and permanent domes of fiberglass.

of 1891 featured the balcony scene from *Romeo and Juliet* in white corn, the lovers in cornhusks and corn silk, shown at Romeo's moment of parting down a white corn ladder. In a medley of arts worthy of grand opera and lit by ten thousand of the city's first electric lights, King Corn was crowned in front of the palace, surrounded by Knights of Pythias mounted on steeds, while the Pullman Band of Chicago vied with the Ladies' Cornet Band of Clinton and a Mexican band played "Hail, Columbia." Proudly the city announced that the cost of decoration was $8,700 and of advertising $4,000. By happy coincidence the man crowned King Corn was James E. Booge, president of the Corn Palace Association.

In 1892 in South Dakota, Mitchell with upstart

cockiness took up where Sioux City left off. Mitchell had begun as a railroad town, founded by Alexander Mitchell, president of the Chicago, Milwaukee and St. Paul line. But after a decade it had no more than a few dirt streets and wooden sidewalks in a land so hostile that the U.S. government provoked gamblers to settle it: "The government bets you 160 acres of land against $18.00 [or three cents an acre] that you will starve to death before you live on it five years." The government's wager seemed a sure thing after January 12, 1888, when a gale of forty miles an hour blew in a blizzard that reached 45 below zero and stayed there, leaving cattle frozen in the fields and snow banks sixteen feet high, which people froze their ice cream with on the following Fourth of July.

Mitchell decided to erect a Corn Palace in this unpromising landscape because it wanted to outdo its rival, Pierre, to become the state capital. "Sioux City has abandoned her Corn Palace," said L. O. Gale, town druggist and jeweler. "Why not build one here?" Fifty-nine days later, after every man, woman and child pitched in with labor and cash ($3,700 total), the Palace opened at Fourth and Main on September 28, 1892, at 4:30 P.M. with a May Pole Dance, as announced. "Sixteen of Mitchell's loveliest ladies attired in bewitching costumes will take part, and in their graceful evolutions will circle about and surround the living personification of Ceres, goddess of grain." The grain of fabled Greece and Rome had turned to corn.

True to form, the following year's palace was far grander, corn "nubbings" and husks now imitating stained-glass windows, Grecian scrolls, a New England kitchen, a bridal chamber and the Black Hills Crystal Cave. The Phinney (Iowa) State Band and the Santee Indian Band warred for listeners already distracted by a flower dance, a German parade, a reunion of Civil War veterans and a large Indian wedding. The *Mitchell Gazette* touted the benefits of the Palace as (1) aesthetic—a happy escape from work, (2) pragmatic and educational—new ideas for farmers and (3) promotional—effective advertising to show corn farmers from other states that South Dakota was a good place to grow corn. They were wrong about number three. After 1893, a severe drought canceled both palaces and crops for six years.

Undaunted, Mitchell organized a Corn Palace Commission in 1902 and has held a festival and exposition the last week of September every year since, even after they lost the contest for state capital to Pierre. What they won was a more unique distinction, which attracted celebrities as grand as John Philip Sousa in 1904, hired at the cost of $7,000 a day for a six-day engagement. When Sousa saw the mud streets of the town, he refused to get off the train unless he was paid—in cash, in sacks brought to his carriage. "Whatever you gentlemen lack in judgment," he is supposed to have said, "you certainly make up in nerve."

The natural union of corn and railroads was celebrated by Atchison, Kansas, at their corn carnival of 1897 when they crowned corn king. Courtesy Kansas State Historical Society, Topeka

Mitchell nerve made for survival and Mitchell's palace survived where Sioux City's did not. In 1921 Mitchell made the building permanent and changed annually the theme of the exterior panels, which became a kind of corn mirror of the times. Egyptian motifs one year gave way to patriotic themes during the Great War, to scenes celebrating "Allied Victory" in World War II, to vignettes of "Relaxin' in South Dakota" in the 1950s and of "75 Years of Mitchell Progress" in the 1980s. For nearly a century, Mitchell's palace headliners mirrored the evolution of American entertainment. The minstrel shows and military bands of the first decade were replaced by revues and variety shows like Ernie Young's Golden Girls in the 1920s, by the big swing bands of Jimmy Dorsey and Paul Whiteman in the 1930s and by comedians like Red Skelton and Bob Hope in the 1970s.

From Jim Sellars, building manager, I learned that an arsonist had burned one of the palace's domes ten years ago, so its minarets are now fiberglass and its foundations brick and concrete. The outside panels, however, are renewed yearly to the tune of a thousand pounds of nails, a hundred thousand cobs of corn, two thousand bushels of sorghum, murdock, buffalo grass, broom grass, wild oats, milo, sudan grass, slough grass and "just plain weeds." Designs are projected onto tar paper, then nailed on wood panels numbered by color for the segments of corn. Mitchell farmers mix their palettes in the cornfield, breeding corn by color as the Indians did, to achieve pink, orange, calico, brown, green, smoky gray and black. Corn must be picked at precisely the right point of moisture and cobs must be cut to precisely the right thickness, so that the corn segments won't curl as they dry. "If they're cut right, the kernels won't blow out even in a high wind," Sellars explained, "and in Mitchell the wind blows quite hard." Outside, wind erosion is abetted by pigeons and squirrels; inside, erosion is abetted by kids who find ready ammunition in kernels during the heat of a basketball game.

Tossing kernels is also a tradition stemming from the days of corn carnivals at the turn of the century, as the empire of King Corn spread through Midwestern towns. Railroad towns that couldn't afford a palace created corn carnivals, such as the one Atchison, Kansas, staged in 1897 under the banner, "King Corn is supreme." Atchison's carnival was the brainchild of the editor of *The Atchison Globe,* and was heartily supported by the Burlington and later the Atchison, Topeka and Santa Fe railroad lines. "When one is separated by five hundred miles from a lake-shore, by as many from the mountains, and by a thousand miles from the ocean," wrote a *Topeka Journal* newsman, "it becomes a merit to conceive new forms of enjoyment dependent on none of these."

New forms were dependent, however, upon the wheels of spring-wagons,

The high point of the Atchison corn carnival of 1897 was the elephantine erection of the Burlington Route. Courtesy Kansas State Historical Society, Topeka

The rage for corn costumes spread to other occasions, such as this 1912 May Day revel on the greenswards of the University of Illinois at Urbana, where Corn Belt maidens of Xilonen danced around a May Pole "in high glee as in the old days of 'Merrie England.' Courtesy University Library, University of Illinois at Urbana-Champaign

buggies, surreys and excursion trains. The corn carnival moved the old agricultural county fair into town. City merchants promoted the new street fairs and conjured the mixed carnival bag that provided city pleasures by means of country products. One of the chief pleasures was masquerading as corn people in suits, dresses and hats of shredded husks and tassels, a garb less surprising to Indians than to Easterners. "The Corn Milliner of Kansas," Mrs. H. J. Cusack, gained notoriety when she sent a dainty corn bonnet to the White House for President McKinley's wife. The carnival combined the mimes and mummeries of New Orleans's Mardi Gras with the genteel floats that would soon grace Pasadena's Rose Parade. A prairie schooner, covered with white poppies, disclosed a pair of corn kiddies; half-nude "Filipino savages" in cornhusk skirts were mimed by "the colored population"; fifty bright-eyed children sailed in a corn-covered ship; fifty more peeped from windows in a monster ear constructed from thirty-six bushels of corn. Not to mention marching bands, a giant cigar-store Indian of corn, an Old Glory of red, white and blue husks, minstrels dancing on street stages, boys holding red ears of corn like mistletoe above the heads of kissable girls and merrymakers buying bags of shelled corn to hurl like confetti until "the streets became veritable mills for the grinding of the corn." Welcoming all to the main carnival street was the Burlington Railroad's elephantine erection of an ear of corn four stories high, a fit symbol for the rising of King Corn in an age innocent of Freud.

King Corn had already informed the world of the grandeur of his territorial ambitions in 1893, at the World's Columbian Exposition in Chicago, America's first world fair. Here states competed side by side in erecting fantasies, deliriums, phantasmagories of corn. Photographs show corn draperies, columns, obelisks, pyramids, corbels, caryatids, Roman arches, Gothic arches, buttresses, arabesques. Every possible architectural motif that could be purloined from Victoria's Crystal Palace at the 1851 Great Exhibition in London was emulated by the new prairie metropolis that sought to conquer the world with corn. If the Crystal Palace was "a revelation of what could be done with steel and glass," America's state palaces were a revelation of what could be donewith corn.

At the Fair, Iowa's palace was built "Pompeian style," with a grapevine frieze of purple popcorn. The soil that produced such wonders was so sacred they exhibited it in long glass columns. Iowa's palace was rivaled only by Illinois's Corn Kitchen, ruled by the formidable Sarah T. Rorer of the Philadelphia Cooking School, there to tell the world of the superiority of corn over other grains as a cheap and nourishing staple and as an elegant dessert. When St. Louis staged its world fair in 1904 in the Louisiana Purchase Exposition, it was now Missouri that garnered kudos with its palace of corn. In addition to two corn towers, each thirty-eight feet high, flanking the Louisiana Purchase Monument and Emblem rendered in husks, the central Russo-Byzantine cupola stood sixty-five feet high and forty feet in diameter, giving rest to the weary in lounge chairs.

A partial view of Missouri's corn towers, pyramids, curtains and cupolas at the Louisiana Purchase Exposition of 1904. Courtesy Missouri Historical Society

If the medium was the message, the message outlined was "The Agricultural Conquest of the Earth." Illinois, Iowa and Missouri were doing for corn what Pittsburgh was doing for steel, noted a magazine called *The World's Work.* The fair as a whole provided a history of corn in America, beginning with the single-rowed small ears grown "by blind choice" before Columbus came. "But the new corn, the ideal corn," the magazine said, was developed by "initiative and brains" and the kind of farmer-boy contests that created "a new kind of Olympic game!" From such exhibits the rest of the world would not only learn how to grow corn scientifically, but would also discover all the products that grow from corn. The familiar litany of hidden corn products was recited in 1904, two decades before Henry Wallace extended King Corn's empire around the world:

All the products of the corn plant are there—oils, paper, pith (that is used in battleships to stop shot-holes below the waterline), whisky. There are three kinds of sugar and two each of syrup and molasses. There are many food elements—different kinds of cellulose, vicose, proxylene, and anyloid. There are many products useful in the arts: celluloid, collodion, sizing, varnishes, films, filaments for incandescent lights, artificial silk, guncotton, smokeless powder, and fine charcoal. There are many varieties of starch and of glucose, several kinds of gum, grape-sugar, corn-rubber (used in buffers on railway cars), corn-oilcake and meal, malt, beer, wines, alcohol, and fusel-oil. One may see also shuck-mats and shuck-mattresses. How many products of corn there are nobody knows, for new products are evolved every year.

Iowa, at the State Fair in Des Moines in 1908, exhibits itself in unconscious parody as a Corn Belt Coatlicue, both the mother and maw of corn. Lake County Museum, Curt Teich Postcard Collection, Wauconda, Illinois

Corn's conquest of the world happened within my lifetime. An unexpected form of bio-botanical destiny linked the personal roots of my family tree to the history of my country and its place in the world. For me, the glory of Mitchell's Corn Palace was that it put together what had so long been kept apart, and I saw conjunctions in the most commonplace artifacts and events. In the auditorium, a triple-header girls' basketball tourney was framed by scenes of Indian life. In the lobby, a mounted buffalo head marked the main exit and an elk head guarded the popcorn machine. On the wall, a 1930s snapshot of Jean Darling, of Our Gang, showed her in Indian regalia, being initiated into a Sioux tribe. The chief designer and architect of the palace's exterior for twenty-five years was a Sioux from Crow Creek Indian Reservation, Oscar Howe, a Second World War veteran who became professor of fine arts at the University of South Dakota and "South Dakota's Artist Laureate." When I learned that "Mitchell of the Palace grand" stands on the site of an eleventh-century Sioux village, I realized that there might be a way to speak two tongues at once. There might be a way to connect in heart as well as mind the vision of a Chicago prairie poet, "The wind blows, the corn leans" with a proverb of the Sioux, "A people without history is like wind on the buffalo grass."

SIX

*Closing
the
Circle*

The Night of Shalako

THERE IS NO WIND and no grass where I stand now, alone in the cold on this darkest night of the year, at winter solstice, New Year's Eve for the Zuni and the night of Shalako. My boots make a sucking sound as I lift first one and then the other to keep from sinking deeper in the mud. My poncho stinks of wet wool while I stand and drip in the unrelenting rain. Other strangers have been waiting in the rain for a good eight hours outside one or another of the seven host houses on the Zuni mesa, but I am as much a stranger to them as to the people we stare at through large glass windows set in the house walls, shut out from the lit interior where men and women sit in chairs, laughing, smoking, dozing, greeting friends who come and go, catching up a loose child, waiting—like us—for the dancing to begin.

The dancing place is a long rectangle dug out of the earth floor, like the womb of the Underworld from which the Zuni emerged. The room is hung with fabrics, with dancers' white kilts embroidered in red and black, Spanish shawls dripping fringe, hand-loomed Navaho rugs and factory-made nylon rugs next to animal skins and deer heads. These are backdrops for the family jewels displayed like a trader's dream of Indian goods—huge silver belts looped over clotheslines, thick strands of coral and shell tossed casually over a shawl. Even the deer heads are looped with turquoise neckpieces, their antlers chained with silver. At the west end of the dancing place are the totems and signs of the wooden altar, painted white, red, blue, yellow, many-hued and black, which center the cardinal points, the up and down and the here and now of this place, this house. In front of the altar the prayer plumes and baskets of seed and corn have already been blessed by a sprinkling of sacred meal. And in the corner the giant mask of the god Shalako stands guard, stands empty, while his young impersonator, stripped to the waist, drinks coffee and smokes a cigarette before the ritual begins.

Once begun, the chanting, drumming and dancing will not stop until all that is has been *named*—all the rivers, mesas, rocks, skies and seas, all the animals, plants, peoples and seeds, all the stars, planets and constellations, every kind of bean, every kind of corn, every kind of god. The drumming will not stop until the chanters have sung and the dancers danced the entire story of how the Zuni were created and emerged, how they wandered in search of the Middle Place and how they finally found their home in the middle of the Middle Place, here on this mesa on the border of Arizona and the Navaho Reservation, south of Gallup and Four Corners, west of the Continental Divide and north of Bat Cave, where a young Harvard graduate student struck popcorn.

Once begun, the chanting and dancing will go on until dawn, but now it's two hours past midnight and still we wait, outside, inside. I have come by bus from Santa Fe, climbing though piñon and juniper forests and red rock canyons to a flat scrubland of sage and the four-winged saltbush, and now not even scrub—just a brown sea of mud from which a few low, flat, ugly houses emerge around the cardinal points of Western Auto, White's Laundry, Pat's Chili Parlor and Chu-Chu's Pizzeria. In the parking lot outside Thriftway, teenagers hang out in pickup trucks blaring the Kinks; inside the Zuni Tribal Building, tourists hang out over counters of silver, hoping to hammer down prices.

I came before sunset, to see through the drizzle the procession of gods and priests come down from Greasy Hill and cross the Zuni River, no more than a creek, by the footpath near White's Laundry. Tonight the gods are masked as Shalakos, the monster kachinas, led by one that is ten feet tall, twice the height of the dancer who walks inside him. The chief Shalako wears horns for ears, popped orbs for eyes in a turquoise face that explodes in feathers—a ruff of black crow feathers at the neck, an arc of eagle feathers on top and, behind, a sunburst of parrot feathers loosening a cascade of black hair that falls to the hem of his stiff white skirt. His bright blue beak with three white teeth goes snicker-snack—four times, like a vaudeville clapper—before the four shrill hoots of his cry send chills down the spine. As a stranger, I don't know whether to laugh or run.

Shalako is but one of the strange gods that have been summoned by the tribe. There is Longhorn, the Rain Priest of the kiva of the north, who appears in a black and white mask with eye-slits like a Sherman tank and a yard-long horn instead of an ear. There is Hututu, his deputy from the kiva of the south, who ululates the high-pitched sound for which he is named. There are the twins

"What is usually termed the 'Discovery of America' must be seen to entail the problem of who is entering whose history."
—GORDON BROTHERSTON, *Image of the New World* (1979)

Yamuhakto of the east and west, who dangle feathered cottonwood sticks from their bright blue heads. And running in every direction are those "without place," the ten Mudheads, their mouths, eyes and ears like fat doughnuts in knobbed and lumpy adobe-colored heads, the progeny of incest. These are the clowns, like the *koshari* of the Corn Dances, who scare little boys and tweak dozing matrons awake. All are spirits of rain, summoned at this turn of the year, to renew the life of the crops and the life of the tribe.

Their chosen hosts have spent an entire year building new houses in which to receive them, and some of the rooms are only half-finished, but the dancing and eating rooms are done, for this is a feast. Strangers as well as visiting gods are welcomed at trestle tables of posole and cheese Danish and pots of hot coffee. The coffee tastes good after hours of wandering in the pitch black, looking for houses that are sometimes two or three miles apart. Skirting ditches but blinded by headlights along Highway 53, which runs straight through the village, I've fallen flat in the mud to emerge like a mud-volleyball player at Hog Days in Kewanee, or a Mudman in Zuniland. I've been lost many times this dark night because there are no maps for this territory and this is not my place.

"A curious thing about the Spirit of Place," D. H. Lawrence wrote, "is the fact that no place exerts its full influence upon a new-comer until the old inhabitant is dead or absorbed." On this night, this is a place where the old inhabitants are so long dead and so fully absorbed that it is a place chiefly of Spirit, not only of the masked gods, but of the animals sacred to each surrounding

The Shalakos in 1897 descended as they do today from Greasy Hill to cross the Zuni River at nightfall. Photo Ben Wittick, courtesy Museum of New Mexico, Neg. No. 16443

place—Mountain Lion in the Home of Barren Regions; Bear, old Clumsy Foot, in the Home of Waters; Badger, Black-Masked Face, in the Home of Beautiful Red Sunrise; Wolf, Hang Tail, in the Home of Day; Eagle in the Home on High; Mole in the Home of Low. Here there is a place for everything, as my Grandmother Harper would have said, and everything in the created world, if properly named, is in its place.

But in this place I am an intruder shut out by language. When the impersonators put out their cigarettes to put on their masks, to jangle the bells at their ankles, to clack their beaks to bird hoots and drumbeats, to begin the dance at last, I am outside, separated by a thick layer of glass. I am outside the blessing of cornmeal that the villagers have sprinkled on houses, walls, altars, hosts and dancing gods, and although I can follow the cornmeal path of the Shalakos to and from "the place whence they came" on Greasy Hill, I am but a spectator dripping mud.

Only once do I make connection, this in the darkness of the village itself, lit fitfully by the blue glare of television screens through open windows and doors. As I turn a corner by the little graveyard of the church, a window in front of me frames a face and a blanket-wrapped body in frozen profile. As I step close to make out what it is, the face turns slowly in my direction, a face so ancient I can see the skull beneath the skin, two holes for eyes, another one for mouth. My heart thumps and I turn and run. It is not the skull that unnerves me, it is the face. The face is my Grandmother Harper's the night she died.

THE BLESSINGWAY

"I AM THE WAY, the Truth, and the Life." My Grandmother Harper pointed to each word when she first taught me to read from the little Bible cards she kept in her family Bible. The image on the card was of a pale bearded man in a white nightshirt knocking at a door, his head backlit like a picture of Lillian Gish. When I grew up I departed from the Way of my ancestors, blazing my own path through urban wildernesses as remote as possible from the rebel stock of which I was a rebel child. I was a traveler of other times and places and had circled most of the world before that first reunion of cousins in Lincoln, Nebraska, planted my feet for the first time on the prairie sod of my grandparents. Like other Americans dispersed to the four corners of the continent, I had trouble putting the coasts together with the prairies. It was easy to miss the center when there was so much ground to cover.

"Yet one day the demons of America must be placated, the ghosts must be appeased, the Spirit of Place atoned for."
—D. H. LAWRENCE, "The Spirit of Place" (1923)

So many paths, so many dispersals, so many stranger tongues. Even in my own language, the King James Bible talk of my Grandmother Harper clashed with the *Origin of Species* talk of her son. It was hard to fit her Garden into his Galapagos; each place jostled for room on the family map, crowding the Brazilian jungles where Uncle Roy and Aunt Evelyn warbled "Bringing in the Sheaves." My parents were the last generation to inherit the American earth, at the turn of the century when America was still locked in land and clocked by the seasons. They were a generation glad to chuck annual blizzards in Kansas for an occasional earthquake in California, just as they were glad to chuck fresh corn for canned, horses for cars, trains for planes, full speed ahead.

From that first Culver in New England to the Harpers in Kansas and the Kennedys in Nebraska, they never looked back and they never looked down, down to the earth their ancestors had taken by God-given right. God's Will was theirs and their Way straight as Kansas corn rows and Union Pacific railroad tracks east to west. Their Way was lit by Old World myths which mutated into New World science, and the word of science was square. What wouldn't fit their straight and narrow was excised— the blood in the Garden, the transplanted mother who sickened and died in an orange grove.

But who had entered whose history? My people were deaf to ancient drumbeats and blind to the spirals, mazes and circles of people for whom the Word was round. Where my people shot into space, these people went underground, every year, year after year, for millennia. It was difficult to square space shuttles and earth navels, just as it was difficult to square the vision of those for whom only the future was real with the vision of those for whom the past was eternally present. This is the past Ivan Illich called "the lost time of the blue tortilla," made from the blue corn which shaped religious rites, hearthstones and lives long before Columbus or Cortez. The time of the blue tortilla my people had obliterated almost overnight with Reid Dent and the McCormick Reaper. The seeds of life preserved here for hundreds of generations my people had killed in the name of permanent shelf life. Different languages, different worlds. "Only by re-entering the

This drypainting, The Blessingway of the Black Cornfield, *was made by Maude Oakes, with singer Wilito Wilson, at Mariano Lake, New Mexico, in 1947. Pollen footprints lead up the cornstalk to Cornbeetle Girl and Pollen Boy, flanking a corn tassel beneath the double-arched rainbow. The painting is used "for anyone who has been harmed in a cornfield."* Repainted by M. S. Hurford. Courtesy Wheelwright Museum of the American Indian, PI-#21

present moment with knowledge of the lost time of the blue tortilla," Illich said, "will it be possible to establish a new way of seeing and a new set of terms."

Looking for just that, a new way of seeing, a new set of terms, a way to connect my people and this land, I stumbled onto the hidden corn road and the lost time of the blue tortilla. In a bowl of buttered popcorn, I found both Grandfather Harper and my son, microwaves and movie theaters connected to Peruvian corn poppers and Machu Picchu. Crunching into the sweet milk of fresh corn-on-the-cob, I found both Grandmother Kennedy and my daughter connected to the Green Corn Busks of the Moundbuilders in Ohio and Illinois. In a plate of mush I found Grandmother Harper connected to both Buffalo Bird Woman in her garden and Frank Cushing in his mesa. In a dried corncake I found the Empress of China and an African in Soweto. In Miss Sweet Corn in Hoopeston I found Xilonen of the Aztec, and in a bottle of Arkansas moonshine the *chicha*-drinking Inca. In the Corn Palace of South Dakota I found Palenque and Copán. In a box of Kellogg's corn flakes I found Grandma Helen Sekaquaptewa in Hopiland. Whether we knew it or not,

Wide Cornfield, *drawn by Maude Oakes, with singers Hosteen Yazzi and Tom Ration, at Smith Lake, New Mexico, in 1946/47, shows the pollen footsteps of Changing Woman and "the path of her mind" as she proceeds up the cornfield to stand between white male corn and yellow female corn under a rainbow arc. The ceremony is done in the cornfield to ensure good crops.* Courtesy Wheelwright Museum of the American Indian, PIA-#21

America was corn land, and both native and immigrant Americans were corn people.

But what immigrant wheat people fractured, atomized and smashed, corn natives kept whole. For the Navaho that wholeness was inscribed in song and ritual as the Blessingway. The task of a Blessingway singer was to remember by heart each part of the ceremony that blesses the center of the hogan, the homeplace, the center of harmony to which we give the names holy and blessed. The Blessingway enacts in song and ceremony what Navaho sand paintings depict in images. Both are forms of speaking pictures which First Man gave to Changing Woman when he gave her the magic corn bundle containing four jewels of four colors —white shell, turquoise, abalone and jet—each carved into a perfect ear of corn, each wrapped in a translucent sheet of its own gem, each containing the power to create the inner forms of outward shapes, the forms of the natural world.

When Changing Woman buried the medicine bundle in the heart of the earth, she created

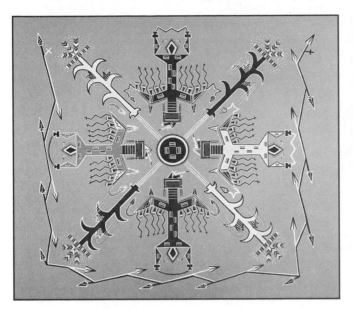

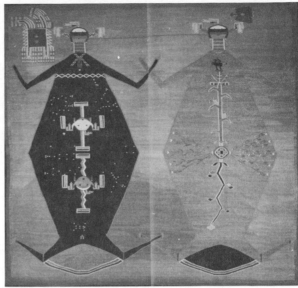

Thought in the shape of Long Life Boy and Speech in the shape of Happiness Girl. Together they made That Which Continues, the progeny of White Corn Boy and Yellow Corn Girl, Pollen Boy and Cornbeetle Girl. Changing Woman created the Mating Songs of these first corn couples by painting their forms, running her hand over the corn ears of shell, turquoise, abalone and jet, drawing their root legs and tasseled faces. She then told the Corn People that through them the earth would be permanent, the sky, flowers, dew, pollen, young men, young women, death, birth and rain, and "by that we will go about in blessing." When she gave them speech, she gave them pollen words. "You will speak for us with pollen words. You will talk for us with pollen words." The word of Blessingway is pollen.

Pollen is the way, the truth and the light of the Navaho ceremony in which the singer draws two pictures in the cornfield and two in the hogan and links them by pollen trails. In these paintings, the path of life is a cornstalk, hung with twelve ears of corn as a ladder for footprints to climb to a cloud terrace where a bluebird sings. The person who seeks blessing places his feet in the pollen path to circle each painting clockwise, before he steps inside the circle and enters the painting, for the painting is "the place where the gods come and go." Here the One Sung Over is absorbed into the image and here the pollen placed in his mouth becomes the fine carved jewel in the mouth of the god, immutable as the inner forms of the holy. These are the images that, in the beginning in the worlds below this one, the gods painted on sheets of sky and

(left) A figure from the Male Shooting Chant depicts "The Home of the Thunder Gods," painted here by Hosteen Bezody. Because the number four to the Navaho represents order, wholeness and harmony, four thunder gods are paired with four cornstalks, radiating toward the four cardinal points. The encircling guardian line (here zigzag arrows of lightning) opens always to the east. Photo by Gene Balzer, courtesy Museum of Northern Arizona

(right) The paired figures are "Blue Mother Earth and Black Father Sky," painted on the fourth day of Male Shootingway. Father Sky contains the sun, moon and Milky Way, Mother Earth the four sacred plants and a bird atop the cornstalk. The guardian figures at their heads are "Corn Bugs." A sandpainting rug woven by Atlnabah. Courtesy Museum of Northern Arizona

gave to the Navaho to copy on the floors of their hogans. When the singing is done, the singer erases the painting, but the image stays, like the songs and stories in the memory of the singer. "It is like the plants, like corn and beans," says a Blessingway singer of this century. "When they mature you pick the seed and you plant it again."

In the New World, pollen words were sung by the warrior-hero who, planted as a stalk of corn in the fecund earth, was transformed into a poet. For the Toltec singer, it was the hunter Nacxitl who gave language and art to the snake-columned city of Tula before he climbed the sky to fall as quetzal plumes and clear water. This earlier skywalker, like Quetzalcoatl, was born again as the blossoming flower of the earth's flesh and as the emblem of perfection, an ear of corn:

> *As white and yellow maize I am born,*
> *The many-coloured flower of living flesh rises up*
> *and opens its glistening seeds before the face of our mother . . .*
> *My song is heard and flourishes.*
> *My implanted word is sprouting,*
> *our flowers stand up in the rain.*

"Corn is the Navaho staff of life, and pollen is its essence."

—DONALD SANDNER, *Navaho Symbols of Healing* (1979)

Star-trek returned to earth-surge bends the arrow and completes the circle. In that round, the oppositions between Sky Father and Earth Mother, between movement through time and growth within, between the way up and the way down, are equidistant from the center, fixed and immutable. "In the song, corn grows upward, rain comes downward," says the Blessingway singer. "If there's no down, there's no up."

The migration of my ancestors was across continents, up and away from the earth navel of fallen man. My own journey had been down, down into the muddy cornfield and below, into the darkness of seeds and roots to find my dead mother and her mothers, my father and his fathers, in the womb not of Eden but of Mother Earth, the living flesh of the land where I was born. Here corn was not silent or invisible, but sprouted in many-colored songs and images of my own language, if I would but open my ears to hear and my eyes to see. Singers of my own time and tradition had found pollen words, had sung of pollen paths, in their own searchings for connection. In the dreaming eyes of "A Girl in the Library," the poet Randall Jarrell saw "The Corn King beckoning to his Spring Queen." All this time the poets and myth makers of America were implanting words and images to make them blossom in new soil. In a mythical cornfield in Gatlin, Nebraska, the storyteller Stephen King planted seeds of bloody horror in the *Children of the Corn,* invoking primal gods who slaughtered the innocent on crosses made of corn. In a real cornfield in Iowa, W. P. Kinsella saw his novel *Shoeless Joe* transformed into a cinematic *Field of Dreams,* where the mythic spirit of the cornfield summoned lost baseball gods and fathers. "Is this heaven?" asked the father. "No—it's Iowa," said the son.

In the language of art was a new way of seeing, a new set of terms, though long drowned in our country by the language of science and industry, by the clamorous shouts for a future devoid of the past. While we plotted the cosmos and crunched it into numbers for our mental machines, while we lost touch with the earth, lost living connection to the fecundity of the land, still our singers could make us feel corn songs we'd never heard. Even if we'd never once stood in a cornfield, Jim Harrison in *Farmer* could give us a glimpse of the ancient sexiness of corn: "They walked a few rows into the tall corn with a slight breeze rasping the stalks and leaves together. They giggled and drank their beer and began kissing kneeling there. She raised her dress some, not wanting to get it dirty. He had his hands on her thighs. She took his thing out saying she had never touched one before. Even in the cornfield the music was loud and people were shouting with the light of the bonfire against the stalk tips."

Corn was the center that contained and ordered the oppositions of male and female, youth and age, sex and death, even in our own polyglot anarchy of races and kinds. The land bound us all together, the land to which we came as strangers and the earth from which we were increasingly estranged. Corn was the connection between the tyranny of slavers and the innocence of the enslaved, between the violence of rape and the tenderness of love. Just as the Blessingway singer had to hold the implanted word in his memory that the pollen path might be walked again and again, so Sethe, the runaway slave in Toni Morrison's *Beloved,* had to hold in her memory the silk and juice of new corn to feel again her husband's heart:

Looking at Paul D's back, she remembered that some of the corn stalks broke, folded down over Halle's back, and among the things her fingers clutched were husk and cornsilk hair.

How loose the silk. How jailed down the juice.

The jealous admiration of the watching men melted with the feast of new corn they allowed themselves that night. Plucked from the broken stalks that Mr. Garner could not doubt was the fault of the raccoon. Paul F wanted his roasted; Paul A wanted his boiled and now Paul D couldn't remember how finally they'd cooked those ears too young to eat. What he

"Always with one thought
We shall live.
This is all.
Thus with plain words
We have passed you on your roads.

This is our father's waters,
His seeds,
His riches,
His power,
His strong strong spirit,
All this good fortune whatsoever,
We shall give to you
To the end, my fathers
My children,
Verily, so long as we enjoy this light of day,
We shall greet one another as kindred.
Verily, we shall pray that our roads may be fulfilled.
To where your sun father's road comes out
May your roads reach.
May your roads be fulfilled."
——Night Chant of Hekiapawa Shalako

did remember was parting the hair to get to the tip, the edge of his fingernail just under, so as not to graze a single kernel.

The pulling down of the tight sheath, the ripping sound always convinced her it hurt.

As soon as one strip of husk was down, the rest obeyed and the ear yielded up to him its shy rows, exposed at last. How loose the silk. How quick the jailed-up flavor ran free.

We too had our feasts of new corn, our rites of fertility shaped by husking the corn, loosening the silk, letting the juice run free. These were not just echoes of ceremonies indigenous to the first people of corn. We had formed rites in common, given them the accent of our multiple tongues, our Babel of voices. We made of the cornfields our own heaven and hell, transformed our own pollen words into immutable turquoise and jade, the inner forms "where the gods come and go." Corn was our mutual birthright, corn was our home. "Corn is the connection between my bottom and the chair," Wright Morris wrote of his own *Home Place* in Nebraska. "It's the cane seat Grandmother Osborn stretched between the long, long ago, and what she knew to be the never-never land. The figure in the carpet, if there is a carpet, is corn. Corn, I guess is the grass that grows wherever the land is—as Whitman put it—and sometimes it grows whether the water is there or not. No, it isn't the carpet. It's under the carpet. Corn is the floor."

What is the corn? The floor, earth, grass, leaves, the bluebird on top of the stalk, the evening and morning star, the man who tends it with his blood and the woman who grinds it with her sweat into meal, the lost time of the blue tortilla. "When the twelve Holy People came up to this world they made twelve paintings, each one his own, and they sang eleven songs, each one missing the hogan song except Changing Woman, who remembered it and sang it. While they sang during the night corn grew." So sang the singers of Blessingway, who remembered the pollen words. So sang the singer of Walden Pond, who believed in the forest, and in the meadow, and in the night in which the corn grows. So sang the singer of the Oglala Sioux, Black Elk, remembering his vision quest: "The life of man is a circle from childhood to childhood, and so it is in everything where power moves." Rain comes down, corn grows up, the way up is the way down, when the circle is complete. For a moment I completed my circle, where the way forward was the way back, where Coatlicue and the Couac monster were at my center as well as the Blessed Saints. "The old myths, the old gods, the old heroes have never died," said Stanley Kunitz, the singer of "Seedcorn and Windfall." "They are only sleeping at the bottom of our minds, waiting for our call." Waiting for the dancing to begin.

Photo by Gregory Thorp

SELECTED BIBLIOGRAPHY

In a library catalogue there is always too much and too little under the subject heading "Corn." In the New York Public Library alone there are more than a thousand entries, but 99 percent of them are technical works on agriculture, industry or economics. Corn as a generative force in works of mythology, anthropology, ethnology, geopolitics, poetry and fiction, folk ritual and art, food and drink is a hidden subject that must be hunted out. The works cited here are but a skeletal guide for the general reader who wants to explore some area of this vast territory without going corn mad. For a condensed map, take along Gordon Brotherston's *Image of the New World: The American Continent Portrayed in Native Texts* (London: Thames & Hudson, 1979). (Where relevant, I've cited a title in its original language with the probable date of composition or first publication, followed by its publication in English.)

GENERAL

Corn in the Development of the Civilization of the Americas. Eds. L. O. Bercaw *et al.* Washington, D.C.: Bureau of Agricultural Economics, 1940.

Corn of the Southwestern U.S. and Northwestern Mexico. Eds. G. Habhan *et al.* Tucson, Arizona: Native Seeds/SEARCH [n.d.].

Maíz: Bibliografía de las Publicaciónes Que Se Encuentran en la Biblioteca. Eds. A. Martinez and C. N. James. (Editorial material in Spanish and English.) 2 vols. and supplement. Turrialba, Costa Rica: Instituto Interamericano de Ciencias Agricolas, 1960–64.

Maize bibliography for the years 1917–1936; 1888–1916; 1937–1945. Ames, Iowa: Iowa State College, 1941, 1948, 1951.

Aliki. *Corn Is Maize: The Gift of the Indians.* New York: Harper & Row, 1976.

El maíz, fundamento de la cultura popular mexicana. Mexico City: Museo de culturas populares, 1982.

Giles, D. *Singing Valleys.* New York: Random House, 1940.

Hardeman, N. P. *Shucks, Shocks, and Hominy Blocks.* Baton Rouge and London: Louisiana State University Press, 1981.

Historical Atlas of the United States. Eds. W. E. Garrett *et al.* Washington, D.C.: National Geographic Society, 1988.

Longone, J. B. *Mother Maize and King Corn: The Persistence of Corn in the American Ethos.* Ann Arbor, Michigan: William L. Clements Library, 1986.

Nuestro Maíz. 2 vols. Mexico City: Museo national de culturas populares, 1982–83.

Visser, M. *Much Depends on Dinner.* New York: Grove Press, 1986.

MIDDLE AND SOUTH AMERICA

Acosta, J. de. (*Historia natural y moral de las Indias,* 1590.) *The Natural and Moral History of the Indies.* Trans. E. Grimston. New York: Burt Franklin [n.d.].

Bankes, G. *Peru Before Pizarro.* Oxford: Phaidon Press, 1977.

The Book of Counsel: The Popol Vuh of the Quiché Maya of Guatemala. Trans. M. Edmonson. New Orleans: Middle American Research Institute, Tulane University, Pub. 35, 1971.

Bray, W. *Everyday Life of the Aztecs.* London: Batsford, 1968.

Brundage, B. C. *Lords of Cuzco: A History and Description of the Inca People in Their Final Days.* Norman: University of Oklahoma Press, 1967.

Campbell, J. *Historical Atlas of World Mythology: Volume I: The Way of the Animal Powers.* London: Alfred van der Marck, 1983.

———. *Volume II: The Way of the Seeded Earth.*

Part 1: The Sacrifice. New York: Harper & Row, 1988.

Part 2: Mythologies of the Primitive Planters: The Northern Americas. New York: Harper & Row, 1989.

Part 3: The Middle and Southern Americas. New York: Harper & Row, 1989.

Cobo, B. (*Historia del nuevo mundo,* 1653.) *History of the Inca Empire.* Trans. R. Hamilton. Austin: University of Texas Press, 1979.

Coe, M. E. *Mexico.* London: Thames & Hudson, 1962.

———. *The Maya.* London: Thames & Hudson, 1966.

Davies, N. *The Aztecs.* London: Macmillan, 1973.

———. *The Toltecs.* Norman: University of Oklahoma Press, 1977.

Díaz del Castillo, B. (*Historia de la conquista de la Nueva España,* 1568.) *The Conquest of New Spain.* Trans. J. M. Cohen. Harmondsworth, England: Penguin, 1963.

Durán, D. (*Historia de las Indias de Nueva España e islas de la tierra firme,* 1574–81.) *The Aztecs: The History of the Indies of New Spain.* Trans. D. Heyden and F. Horcasitas. New York: Orion Press, 1964.

Frazer, Sir J. G. *The Golden Bough: A Study in Magic and Religion.* 3rd ed. London: Macmillan, 1966.

Galeano, E. *Memory of Fire: I. Genesis. II. Faces and Masks.* Trans. C. Belfrage. New York: Pantheon Books, 1985–87.

Garcilaso de la Vega. (*Los commentarios reales de los Incas,* 1609–17.) *Royal Commentaries of the Incas and General History of Peru.* Trans. H. V. Livermore. Austin: University of Texas Press, 1966.

Gauman Poma de Ayala, F. *Nueva corónica y buen gobierno* (1580–1620). Paris: Institut d'Ethnologie, 1936.

Handbook of South American Indians. Ed. J. H. Steward. 7 vols. Washington, D.C.: Government Printing Office, 1946–59.

Hemming, J. *The Conquest of the Incas.* New York: Harcourt Brace Jovanovich, 1970.

Joralemon, D. "A Study of Olmec Iconography." *Studies in Pre-Columbian Art and Archaeology No. 7.* Washington, D. C.: Dumbarton Oakes, 1971.

——— . "Ritual Blood-Sacrifice Among the Ancient Maya: Part I." *Primera Mesa Redonda de Palenque, Part II.* Ed. M. G. Robertson. Pebble Beach, California: Robert Louis Stevenson School, 1974.

Kubler, G. *The Art and Architecture of Ancient America: The Mexican, Maya and Andean Peoples.* 2nd ed. Harmondsworth, England: Penguin, 1975.

Landa, D. de. (*Relación de las cosas de Yucatán, 1566.*) *Yucatan Before and After the Conquest.* Trans. W. Gates. New York: Dover Publications, 1978.

Las Casas, B. de. (*Historia de las Indias, 1552–61.*) *Obras Escogidas de Fray Bartolomé de Las Casas* (1550). 5 vols. Madrid: Ediciones Atlas, 1957–58.

Laughlin, R. M. *The People of the Bat.* Washington, D.C.: Smithsonian Institution Press, 1988.

Lumbreras, L. G. *The Peoples and Cultures of Ancient Peru.* Trans. B. J. Meggers. Washington, D.C.: Smithsonian Institution Press, 1974.

MacNeish, R. S. "Ancient Mesoamerican Civilization." *Science* 143 (7 February 1964).

Maya Iconography. Eds. E. Benson and G. Griffin. Princeton, New Jersey: Princeton University Press, 1988.

Moctezuma, E. M. *The Great Temple of the Aztecs: Treasures of Tenochtitlán.* London: Thames & Hudson, 1988.

Neumann, E. *The Great Mother: An Analysis of the Archetype.* 2nd ed. Trans. R. Mannheim. Princeton, New Jersey: Princeton University Press, 1963.

Oviedo y Valdés, G. F. de. *Historia general y natural de las Indias* (1535–57). 4 vols. Madrid: Ediciónes Atlas, 1959.

Popol Vuh: The Definitive Edition of the Maya Book of the Dawn of Life and the Glories of God and Kings. Trans. D. Tedlock. New York: Simon & Schuster, 1985.

Robiscek, F., and D. Hales. *The Maya Book of the Dead: The Ceramic Codex.* Charlottesville: University of Virginia Museum, 1981.

Roys, R. L. *The ethno-botany of the Maya.* Middle American Research Series Publication No. 2, New Orleans: Tulane University, 1931.

———. *The book of Chalam Balam of Chumayel.* Carnegie Institution of Washington Publication 438, 1933.

Sahagún, B. de. (*Historia general de las cosas de Nueva España, 1558–69.*) *Florentine Codex: A General History of the Things of New Spain.* Trans. A. J. O. Anderson and C. E. Dibble. 12 vols. Santa Fe, N.M.: School of American Research and University of Utah, 1950–69.

Schele, L., and D. Freidel. *A Forest of Kings: The Untold Story of the Ancient Maya.* New York: William Morrow and Co., 1990.

Schele, L., and M. E. Miller. *The Blood of Kings: Dynasty and Ritual in Maya Art.* Fort Worth, Texas: Kimbell Art Museum, 1986.

Spinden, H. J. "A Study of Maya Art, Its Subject Matter and Historical Development." *Memoirs of the Peabody Museum of American Archaeology and Ethnology, Harvard University, VI.* Cambridge: Harvard University Press, 1913.

Stephens, J. L., and F. Catherwood. *Incidents of Travels in Central America, Chiapas, and Yucatan* (1841). Repr. 2 vols. New York: Dover Publications, 1969.

Taube, K. "The Classic Maya Maize God: A Reappraisal." *Fifth Palenque Round Table,*

1983, Vol. VII. Eds. M. G. Robertson and V. M. Fields. San Francisco: The Pre-Columbian Art Research Institute, 1985.

———. "The Maize Tamale in Classic Maya Diet, Epigraphy, and Art." *American Antiquity* 54 (1), 1989.

Thompson, J. E. S. *Maya Hieroglyphic Writing.* Norman: University of Oklahoma Press, 1960.

———. *A Catalogue of Maya Hieroglyphics.* Norman: University of Oklahoma Press, 1971.

———. *A Commentary on the Dresden Codex.* Philadelphia: American Philosophical Society, 1972.

Torquemada, J. de. *Monarquía Indiana (Los veinte i un libros rituales i monarchía indiana,* 1613). Mexico City: Editorial Chavez Hayhoe, 1943–44.

NORTH AMERICA

Adair, J. *History of the American Indians* (1775). Ed. S. C. Williams. Johnson City, Tennessee: Watauga Press, 1930.

Bahti, T. *Southwestern Indian Ceremonials* (1968). Rev. ed. Las Vegas, Nevada: KC Publications, 1982.

Barlow, J. *The Hasty Pudding* (1793). Ed. D. J. Browne (*with A Memoir on Maize or Indian Corn*). New York: W. H. Graham, 1867.

Bartram W. *Travels of William Bartram* (1791). Ed. M. Van Doren. Reprint of 1928 edition. New York: Dover Publications, 1955.

Benedict, R. *Tales of the Cochiti Indians* (1931). Albuquerque: University of New Mexico Press, 1981.

Berkhofer, R. F., Jr. *The White Man's Indian: Images of the American Indian from Columbus to the Present.* New York: Alfred A. Knopf, 1978.

Beverley, R. *The history and present state of Virginia.* London: 1705.

Billington, R. A. *Land of Savagery, Land of Promise: The European Image of the American Frontier in the Nineteenth Century.* New York: W. W. Norton & Company, 1981.

Boorstin, D. J. *The Americans: The Colonial Experience.* New York: Random House, 1958.

———. *The National Experience.* New York: Random House, 1965.

Bradford, W. *Of Plimouth Plantation, from the Original Manuscript.* Boston: Wright and Potter, 1898.

Catlin, G. *Letters and Notes on the Manners, Customs, and Condition of the North American Indians.* 2 vols. 3rd ed. New York: Wiley and Putnam, 1844.

Champlain, S. de. (*Les voyages de la nouvelle France occidentale, dicte Canada,* 1632.) *Voyages.* Trans. C. P. Otis. 3 vols. Boston: Prince Society, 1878–82.

Cronon, W. *Changes in the Land: Indians, Colonists, and the Ecology of New England.* New York: Hill and Wang, 1983.

Crosby, A. W. *The Columbian Exchange: Biological and Cultural Consequences of 1492.* Westport, Connecticut: Greenwood Press, 1972.

Cushing, F. H. *Zuni: Selected Writings of Frank Hamilton Cushing.* Ed. J. Green. Lincoln and London: University of Nebraska Press, 1979.

Debo, A. *The Rise and Fall of the Choctaw Republic.* Norman and London: University of Oklahoma Press, 1934–61.

Driver, H. E. *Indians of North America.* Chicago: University of Chicago Press, 1961.

Dutton, B. P. *American Indians of the Southwest* (1975). Rev. ed. Albuquerque: University of New Mexico Press, 1983.

Earle, E., and E. A. Kennard. *Hopi Kachinas.* New York: Museum of the American Indian, Heye Foundation, 1971.

Fergusson, E. *Dancing Gods: Indian Ceremonials of New Mexico and Arizona.* Albuquerque: University of New Mexico Press, 1966.

Hariot, T. *A briefe and true report of the new found land of Virginia* (1588). Reprint of 1590 ed. New York: Dover Publications, 1972.

The Jesuit Relations and Allied Documents, Travels and Explorations of the Jesuit Missionaries in New France, 1610–1791. Ed. R. G. Thwaites. 73 vols. Cleveland, Ohio: Burrows Bros., 1896–1901.

Josephy, A. M., Jr. *The Indian Heritage of America.* New York: Alfred A. Knopf, 1968.

Kalm, P. *Peter Kalm's Travels in North America* (1753). Ed. A. B. Benson. Reprint of 1937 ed. New York: Dover Publications, 1987.

Kopper, P. *The Smithsonian Book of North American Indians: Before the Coming of the Europeans.* Washington, D.C.: Smithsonian Books, 1986.

Moulard, B. L. *Within the Underworld Sky: Mimbres Ceramic Art in Context.* Pasadena, California: Twelvetrees Press, 1981.

Mullett, G. N. *Spider Woman Stories: Legends of the Hopi Indians.* Tuscon: University of Arizona Press, 1979.

Navajo Blessingway Singer: The Autobiography of Frank Mitchell 1881–1967. Eds. C. J. Frisbie and C. P. McAllester. Tucson: University of Arizona Press, 1978.

Neihardt, J. G. *Black Elk Speaks: Being the Life Story of a Holy Man of the Oglala Sioux.* Reprint of 1932 ed. Lincoln and London: University of Nebraska Press, 1979.

Newcomb, F. J., and G. A. Reichard. *Sandpaintings of the Navajo Shooting Chant.* Reprint of 1937 ed. New York: Dover Publications, 1975.

Ortiz, A. *The Tewa World: Space, Time, Being and Becoming in a Pueblo Society.* Chicago and London: University of Chicago Press, 1969.

Parker, A. C. *Iroquois Uses of Maize and Other Plants.* Albany: University of the State of New York, 1910.

Powell, J. W. *The Hopi Villages: The Ancient Province of Tusayan* (1875). Palmer Lake, Colorado: Filter Press, 1972.

Robbins, W. W., J. P. Harrington and B. Freire-Marreco. *Ethnobotany of the Tewa Indians.* Bureau of American Ethnology, Bulletin 55 (1916).

Rothenberg, J. *Shaking the Pumpkin.* Garden City, New York: Doubleday & Company, 1972.

Russell, H. S. *Indian New England Before the Mayflower.* Hanover, New Hampshire, and London: University Press of New England, 1980.

Sander, D. *Navajo Symbols of Healing.* New York and London: Harcourt Brace Jovanovich, 1979.

Schoolcraft, H. R. *The Indian Tribes of the United States.* Ed. F. S. Drake. 2 vols. Philadelphia: J. B. Lippincott, 1884.

Sekaquaptewa, H. *Me and Mine: The Life Story of Helen Sekaquaptewa.* Tuscon: University of Arizona Press, 1969.

Simpson, R. D. *The Hope Indians.* Los Angeles: Southwest Museum, 1953.

Stratton, J. L. *Pioneer Women: Voices from the Kansas Frontier.* New York: Simon & Schuster, 1981.

Strickland, R. *The Indians in Oklahoma*. Norman: University of Oklahoma Press, 1980.

Swanton, J. R. *The Indians of the Southeastern United States*. Reprint of Bureau of American Ethnology, Bulletin 137 (1946).

Underhill, R. *Red Man's America*. Chicago: University of Chicago Press, 1953.

———. *Singing for Power*. Berkeley: University of California Press, 1938.

Viola, H. J. *After Columbus: The Smithsonian Chronicle of the North American Indians*. Washington, D.C.: Smithsonian Books, 1990.

Waters, F. *Masked Gods: Navaho and Pueblo Ceremonialism* (1950). Athens, Ohio: Swallow Press reprint, 1984.

———. *Book of the Hopi*. New York: Viking Press, 1963.

Will, G. F., and G. E. Hyde. *Corn Among the Indians of the Upper Missouri*. St. Louis: W. H. Miner, 1917.

Williams, R. "A Key into the Language of the Natives in That Part of America Called New England (1643)," *Complete Writings*. 7 vols. New York: Russell and Russell, 1963.

Wilson, G. L. *Buffalo Bird Woman's Garden: Agriculture of the Hidatsa Indians* (1917). St. Paul: Minnesota Historical Society Press, 1987.

Winthrop, J. *The History of New England from 1630 to 1649*. Ed. J. Savage. 2 vols. Boston: Phelps and Farnham, 1825.

Witthoft, J. *Green Corn Ceremonialism in the Eastern Woodlands*. Ann Arbor: University of Michigan Press, 1949.

The World of the American Indian (1974). Eds. J. B. Billard *et al*. Rev. ed. Washington, D.C.: National Geographic Society, 1989.

Wyman, L. C. *Blessingway*. Tucson: University of Arizona Press, 1970.

AGRICULTURE

Ainsworth, W. T., and R. M. Ainsworth. *Practical Corn Culture*. Mason City, Illinois: W. T. Ainsworth & Sons, 1914.

Alternative Agriculture. National Research Council Committee on the Role of Alternative Farming Methods in Modern Production Agriculture. Washington, D.C.: National Academy Press, 1989.

Anderson, E. *Corn Before Columbus*. Des Moines, Iowa: Pioneer Hi-Bred International, 1945.

———. *Plants, Man and Life*. Berkeley: University of California Press, 1967.

Beadle, G. W. "The Mystery of Maize." Field Museum of Natural History, Bulletin 42 (1972).

———. "Teosinte and the Origin of Maize," *Maize Breeding and Genetics*. Ed. D. B. Walden. New York: John Wiley & Sons, 1978.

Benz, B. F. "Racial Systematics and the Evolution of Mexican Maize," *Studies in the Neolithic and Urban Revolutions*. Ed. L. Manzanilla. BAR International Series 349 (1987).

Bird, R. M. "Systematics of *Zea* and the Selection of Experimental Material," *Maize for Biological Research*. Ed. W. F. Sheridan. Charlottesville, Virginia: Plant Molecular Biology Association, 1982.

Bonafous, M. *Histoire naturelle, agricole et économique du maïs*. Paris: 1836.

Bruce, R. V. *The Launching of Modern American Science 1846–1876*. Ithaca, N.Y.: Cornell University Press, 1987.

Candolle, A. de. *Origin of Cultivated Plants.* New York: D. Appleton and Company, 1887.

The Canned Food Reference Manual. Eds. R. W. Pilcher *et al.* New York: American Can Company, 1947.

Corn and Corn Improvement. Ed. G. F. Sprague. Madison, Wisconsin: American Society of Agronomy, 1977.

Crabb, A. R. *The Hybrid-Corn Makers.* New Brunswick, N.J.: Rutgers University Press, 1947.

Dick, E. *Conquering the Great American Desert: Nebraska.* Lincoln: Nebraska State Historical Society, 1975.

Emrich, D. *Folklore on the American Land.* Boston and Toronto: Little, Brown and Company, 1972.

Enfield, E. *Indian Corn: Its Value, Culture, and Uses.* New York: D. Appleton and Company, 1866.

Field, J. *The Corn Lady: The Story of a Country Teacher's Work.* Chicago: A. Flanagan Company, 1911.

Galinat, W. "The Origin of Sweet Corn," Massachusetts Agricultural Experiment Station, *Research Bulletin* 591 (1971).

———. "The Origin of Corn," *Corn and Corn Improvement.* Ed. G. F. Sprague. Madison, Wisc.: American Society of Agronomy, 1977.

———. "Domestication and Diffusion of Maize," *Prehistoric Food Production in North America.* Ed. R. Ford. Anthropological Paper 75. Ann Arbor: Museum of Anthropology, University of Michigan, 1985.

Gerarde, J. *The Herball, or Generall Historie of Plantes.* London: 1597. (2nd ed. 1633.)

Goodman, M. M. "The History and Evolution of Maize," *CRC Critical Reviews in Plant Sciences* 7, no. 3 (1988).

Grobman, A., W. Salhauna and R. Sevilla, with P. C. Mangelsdorf. *Races of Maize in Peru: Their Origins, Evolution, and Classification.* Washington, D.C.: National Research Council, 1961.

Hurt, R. D. *American Farm Tools: From Hand-Power to Steam-Power.* Manhattan, Kan.: Sunflower University Press for *Journal of the West,* 1982.

Iltis, H. H. "Maize Evolution and Agricultural Origins," *Grass Systematics and Evolution.* Eds. T. R. Soderstrom *et al.* Washington, D.C.: Smithsonian Institution Press, 1987.

———. "From Teosinte to Maize: The Catastrophic Sexual Transmutation," *Science* 222, no. 4626 (25 November 1983).

Jacobs, L. J. *Battle of the Bangboards.* Des Moines, Iowa: Wallace Homestead Co., 1975.

Jefferson, T. *Thomas Jefferson's Farm Book.* Ed. E. M. Betts. Princeton, N.J.: Princeton University Press, 1953.

The Journal of Gastronomy 5, no. 2. Ed. N. H. Jenkins. San Francisco: American Institute of Wine & Food, 1989.

Kahn, E. J., Jr. *The Staffs of Life.* Boston: Little, Brown and Company, 1985.

Maize. Ed. E. Hafliger. Basel, Switzerland: CIBA-GEIGY Ltd., 1979.

Mangelsdorf, P. C. *Corn: Its Origin, Evolution, and Improvement.* Cambridge, Mass.: Harvard University Press, 1974.

Mangelsdorf, P. C., and R. G. Reeves. *The Origin of Indian Corn and Its Relatives.* Texas Agricultural Experiment Station, *Bulletin* 574 (1939).

McClintock, B. "Significance of chromosome constitutions in tracing origin and migration of races of maize in the Americas," *Maize Breeding and Genetics.* Ed. D. B. Walden. New York: John Wiley & Sons, 1978.

Nabhan, G. P. *Enduring Seeds: Native American Agriculture and Wild Plant Conservation.* San Francisco: North Point Press, 1989.

Report of the Commissioner of Patents for the Year 1860. Washington, D.C.: Government Printing Office, 1861.

Roe, K. E. *Corncribs: History, Folklife, and Architecture.* Ames: Iowa State University Press, 1988.

Sargent, F. L. *Corn Plants: Their Uses and Ways of Life.* Boston: Houghton Mifflin, 1901.

Sauer, C. *Agricultural Origins and Dispersals.* New York: American Geographical Society, 1952.

Sturtevant, E. L. *Indian Corn.* Albany: Charles Van Benthuysen and Sons, 1880.

Usborne, J. *Corn on the Cob.* London: Rupert Hart-Davis, 1956.

Walden, H. T. *Native Inheritance.* New York and London: Harper & Row, 1966.

Wallace, H. A., and E. N. Bressman. *Corn and Corn Growing* (1923). Rev. 5th ed. New York: John Wiley & Sons, 1949.

Wallace, H. A., and W. L. Brown. *Corn and Its Early Fathers.* East Lansing: Michigan State University Press, 1956.

Weatherwax, P. *Indian Corn in Old America.* New York: Macmillan, 1954.

Wellhausen, E. J., L. M. Roberts and X. E. Hernandez, with P. C. Mangelsdorf. *Races of Maize in Mexico.* Cambridge, Massachusetts: Bussey Institute, Harvard University, 1952.

Whealy, K., and A. Adelmann. *Seed Savers Exchange: The First Ten Years.* Decorah, Iowa: Seed Saver Publications, 1986.

Wilkes, G. "Maize: domestication, racial evolution, and spread," *Foraging and Farming.* Eds. D. R. Harris and G. C. Hillman. London: Institute of Archaeology, University College, 1989.

Yearbook of the United States Department of Agriculture 1915. Washington, D.C.: Government Printing Office, 1916.

FOOD AND DRINK

Anderson, E. A. *The Food of China.* New Haven: Yale University Press, 1988.

Bayless, R. D., and D. G. Bayless. *Authentic Mexican: Regional Cooking from the Heart of Mexico.* New York: William Morrow and Company, 1987.

Benitez, A. M. de. *Cocina prehispánica.* Mexico City: Ediciónes Euroamericas, 1976.

Benjamin Franklin on the Art of Eating. Princeton, N.J.: Printed for the American Philosophical Society by Princeton University Press, 1958.

Carnacina. *La Polenta.* Milan: Fratelli Fabbri Editori, 1973.

Carson, G. *Cornflake Crusade: From the Pulpit to the Breakfast Table.* New York and Toronto: Rinehart & Co., 1957.

———. *The Social History of Bourbon.* New York: Dodd, Mead and Company, 1963.

Cobbett, W. *A Treatise on Indian Corn.* London: W. Cobbett, 1828.

Coe, S. "Aztec Cuisine," *Pétits Propos Culinaires* 19–21. London: Prospect Books, 1985.

Crellin, J. K. *Plain Southern Eating: From the Reminiscences of A. L. Tommie Bass, Herbalist.* Durham, N.C.: Duke University Press, 1988.

Cushing, C. H., and B. Gray. *The Kansas Home Cook-Book.* Reprint of 1886 ed. Ed. L. Szathmary. New York: Arno Press, 1973.

Cushing, F. H. *Zuni Breadstuff.* Reprint of 1920 ed. New York: Museum of the American Indian Heye Foundation, 1974.

Dabney, J. E. *Mountain Spirits: A Chronicle of Corn Whiskey from King James' Ulster Plantation to America's Appalachians and the Moonshine Life.* New York: Charles Scribner's Sons, 1974.

Eddy, F. W. *Metates and Manos: The Basic Corn Grinding Tools of the Southwest* (1964). Rev. ed. Santa Fe: Museum of New Mexico Press, 1979.

Egerton, J. *Southern Food.* New York: Alfred A. Knopf, 1987.

Eustis, C. *Cooking in Old Creole Days* (1904). Reprint ed. New York: Arno Press, 1973.
———. *Fifty valuable and delicious recipes made with corn meal.* Aiken, S.C.: 1917.

Furnas, C. C., and S. M. Furnas. *Man, Bread and Destiny.* Baltimore: Williams and Wilkins Company, 1937.

Getz, O. *Whiskey.* New York: David McKay Company, 1978.

Hazard, T. R. *The Jonny-Cake Papers of 'Shepherd Tom.'* Reprint of 1915 ed. New York: Johnson Reprint Corp., 1968.

Hesse, Z. G. *Southwestern Indian Recipe Book.* Palmer Lake, Col.: Filter Press, 1973.

Humphrey, R. V. *Corn: The American Grain.* Kingston, Massachusetts: Teaparty Books, 1985.

Jaramillo, C. M. *The Genuine New Mexico Tasty Recipes* (1939). Reprint ed. Santa Fe, N.M.: Seton Village Press, 1942.

Kamman, M. *Madeleine Kamman's Savoie.* New York: Atheneum, 1989.

Katz, S. H. "Food and Biocultural Evolution: A Model for the Investigation of Modern Nutritional Problems," *Nutritional Anthropology.* Ed. F. E. Johnston. New York: Alan R. Liss, 1987.

Katz, S. H., M. L. Hediger and L. A. Valleroy. "Traditional Maize Processing Techniques in the New World," *Science* 184 (17 May 1974).

Kavena, J. T. *Hopi Cookery.* Tucson: University of Arizona Press, 1987.

Keegan, M. *Southwest Indian Cookbook.* Weehawken, New Jersey: Clear Light Publications, 1987.

Kennedy, D. *The Art of Mexican Cooking.* New York: Bantam Books, 1989.

Kimball, Y., and J. Anderson. *The Art of American Indian Cooking.* Garden City, N.Y.: Doubleday & Company, 1965.

Lang, G. *The Cuisine of Hungary.* New York: Bonanza Books, 1971.

Leslie, E. *The Indian Meal Book.* Philadelphia: Carey and Hart, 1847.
———. *New Receipts for Cooking.* Philadelphia: T. B. Peterson and Brothers, 1852.

Maize on the Menu. Pretoria, South Africa: published by Muller and Retief for the Maize Board, 1971.

McGee, H. *On Food and Cooking.* New York: Charles Scribner's Sons, 1984.

Miller, J. *True Grits.* New York: Workman Press, 1990.

Murphy, C. J. *American Indian Corn.* New York and London: G. P. Putnam's Sons, 1917.

Myrick, H. *The Book of Corn.* New York: Orange Judd, 1904.

Navajo Curriculum Center Cookbook. Rough Rock, Arizona: Rough Rock Demonstration School, 1986.

Niethammer, C. *American Indian Food and Lore.* New York and London: Collier Macmillan Publishers, 1974.

Ortiz, E. L. *The Book of Latin American Cooking.* New York: Alfred A. Knopf, 1979.

Page, L. G., and E. Wigginton. *The Foxfire Book of Appalachian Cookery.* New York: E. P. Dutton, 1984.

Peruvian Dishes: Platos Peruanos. Lima: Brenuil, 1980.

The Picayune's Creole Cook Book. New Orleans: The Picayune, 1901.

The Pocumtuc Housewife. Deerfield, Mass.: Deerfield Parish Guild, 1805.

Powell, H. B. *The Original Has This Signature—W. K. Kellogg.* Englewood Cliffs, New Jersey: Prentice-Hall, 1956.

Quintana, P. *The Taste of Mexico.* New York: Stewart, Tabori and Chang, 1986.

Randolph, V. *The Virginia House-wife* (1824). Facsimile ed. Ed. Karen Hess. Columbia: University of South Carolina Press, 1984.

Rivieccio, M. Z. *Polenta, Piatto da Re.* Milan: Idealibri, 1986.

Rose, P. *The Sensible Cook: Dutch Foodways in the Old and the New World.* Syracuse, New York: Syracuse University Press, 1989.

Root, W. *The Food of Italy.* New York: Atheneum, 1971.

Rutledge, S. *The Carolina Housewife* (1847). Facsimile ed. Ed. A. W. Rutledge. Columbia: University of South Carolina Press, 1979.

Simmons, A. *American Cookery* (1796). Facsimile ed. Ed. M. T. Wilson. New York: Oxford University Press, 1958.

Southwestern Cookery: Indian and Spanish Influences. Facsimile ed. Ed. L. Szathmary. New York: Arno Press, 1973.

Spitler, S., and N. Hauser. *The Popcorn Lover's Book.* Chicago: Contemporary Books, 1983.

Super, J. C. *Food, Conquest, and Colonization in Sixteenth-Century Spanish America.* Albuquerque: University of New Mexico Press, 1988.

Traditional Navajo Foods and Cooking. Pine Hill, N.M.: Tsa'Aszi Graphics Center, 1983.

Wade, M. L. *The Book of Corn Cookery.* Glenwood, Ill.: Meyerbooks, 1919.

Wigginton, E. *Foxfire.* New York: Doubleday & Company, 1972.

Wilkinson, A. *Moonshine: A Life in Pursuit of White Liquor.* New York: Alfred A. Knopf, 1985.

Woodier, O. *Corn: Meals & More.* Pownal, Vt.: Storey Communications, 1987.

Index

Page numbers in italics refer to illustrations and their captions.

Acoma Pueblo, *120*
Acton, Eliza, 182
Adena, 287
Adventures of Huckleberry Finn, The (Twain), 224
advertising for corn products, 172–4
African cornmeal cooking, 236–7, *237*
agribusiness, 145, 159–60, 161
Agricultural Adjustment Act (1933), 158
Allis Chalmers combine, 153
Alternative Agriculture (National Research Council), 162
American Agricultural Movement, 139–40
American Farm Tools (Hurt), 146
American Glucose Company, 269
American Indian Corn (Murphy), 185, 191, 218
American Lady's Cookery Book, The (Crowen), 179
American Philosophical Society, 64
Anasazi civilization, 47–9, 54–8, *55, 56*
Anderson, Edgar, 79, 85, 87
Anholt, Joe, 310, 311
animal fodder, 7, 76, 149, 152–3, 271
Anti, 297
Apache, 49
Arapaho, 23–4
Arawak, 16, 17
Archer Daniels Midland, 275
Arrowhead Mills, 219
Art of Mexican Cooking, The (Kennedy), 208–9
ashcakes, 224
ash cooking, 171, 175–6, 195–209; by Indians, 200–2, 204–5; products, 205–9, *205*

ash powders, 226–7
Association of Living Historical Farms and Agricultural Museums (ALHFAM), 156
Atchison, Kan., *316*, 317–18, *317*
Athore, Chief, *18*
Atkins, Josiah, 20
atole, 195
Au, M. Sing, 276
Authentic Mexican (Bayless), 199
"Aztec Cuisine" (Coe), 201
"Aztec Flyers" (dance troupe), 47
Aztec civilization, 16, 38, 108, 194, 212; agricultural methods, 108–13, *112*; calendar of, 113, *293*; food of, 200–2, 206; human sacrifices, 291–6, *291, 292, 295*; popcorn and, 10–11

Baker, Gerard, 131
Baker, Raymond "Bake," 74
baking powders (potash, pearlash, saleratus), 226–8
baking soda, 226
Banerjee, Umesh, 79, 83
Barlow, Joel, 231
Barnes, Carl and Karen, 95
Barstow, John, 187
Bartram, John, 184
Bartram, William, 186, 288–9
Bass, A. L. Tommie, 180, 198–9, 242
Bat Cave, 14, 79–80
Bayless, Rick and Deann, 199
Beadle, George Wells, 77–8, 85–6, 91
Beal, William James, 71
Beals, Fred, 219
Beard, James, 181, 195

Bedford, Edward Thomas, 275
Beecher, Catharine, 232
beer from corn, 249–53
Beeton, Isabella, 182
Beeton, Sam, 183
Beheaded God cache vessel, 35, *36*
Behr, Edward, 218–19
Bellon, Chavez, 251
Beloved (Morrison), 331–2
Benedict, Ruth, 240
Benz, Bruce, 88
Benzoni, Girolamo, *205*, 249–50
Berry, Wendell, 162
beverages from corn, 194–5, 249–63
Beverley, Robert, 177, 288
Biggs, William, 199
Billings, Josh, 243–4
binders, 149–50
Bingham, Hiram, 100
Bird, Junius, 79
Bitchknee, Paul, 24
Black Elk, 332–3
Blessingway ritual, 328–30, 332
Bliss, Mrs., 228
Blood of Kings, The (Schele and Miller), 32, 33
blue-corn cuisine, 131, 204
Blue-eyes, George, 120–1
Bonafous, Matthieu, 92, *92*, 240, *240*
Booge, James E., 315
Book of Corn, The, 210, 214
Book of Household Management, The (Beeton), 182, 183
Boorstin, Daniel, 25, 312
Bordley, J. B., 226
Borlaug, Norman, 93
Boston Cooking-School Cook Book, The (Fannie Farmer), 187
Bouma, Dirk and Shabnum, 22
bourbon, 257, 258–9
Bourke, Capt. John G., 283–4
Boyd, Benjamin, 222–3
Bracklin, Vera, 132
Bradbury, John, 124, 132, 178
Bradley, Richard, 252–3
Bradstreet, Anne, 21
Bray, William, 110

breads, 5, 7, 186, 200, 205–6, *205*; *see also* cornmeal cooking
Bressman, Earl N., 61
Bronz, Ruth, 10
broom corn, 64
Brown, George W., 146, 147
Brown, William, 74–5
Brown, William Hoover, 13
Browne, D. J., 12, 185
Buckeye Cookbook, The, 186
Buckner, Carroll, 245
Buescher Corn Cob Pipe Company, 245
Buffalo Bird Woman, 12, 65, 121, 124–31, *125*, 132, 174, 177, 178–9, 186, 189–90, 328
Buffalo Bird Woman's Garden (Wilson), 125
Buffalo Cow Woman, 122
Burpee Company, 90
burrstones, 216–17
Burr-Leaming double-cross, 73
Butler, James, 133, 134, 142–3

Cahokia (temple complex), 24
Calendar Round, 289
Camerarius, 60
Campbell, Joseph, 41, 42
Candolle, Alphonse de, 18
canned corn, 188–9, 191, 192, 304
cannibalism, 38–9, *38*, 41–2, 201, 297
canning industry, 144, 302, 303–4
Carnegie Institution Station for Experimental Evolution, 73
carnivals of corn, *316*, 317–18, *317*
Carolina Housewife, The (Rutledge), 187, 200, 223
Carson, Gerald, 257
Carver, George Washington, 70
Castañeda, Pedro de, 115
Caswell, James, 223
Catastrophic Sexual Transmutation Theory (CSTT), 84–5, *84*
Cates, J. S., 157
Catlin, George, 290
ceramic corn art, 87
ceremonies of corn, 282–300, *282, 283, 284, 287, 291, 295*

Chaco Canyon, *see* Anasazi civilization
Champlain, Samuel de, 123, 206
Chan-Bahlum, Lord, 33, 35
Chapalote (pod popcorn), 15, 88
Charleston Receipts, 200
Cherokee, 186, 242–3
Cheyenne, 23, 122
chicha (corn beer), 249–52, 254, 297, 299,
    300, *300*, 328
Chickasaw, 174
chicos, 190–1
Child, Lydia, 221–2, 232
Childhood, A (Crews), 245
Children of the Corn (King), 330–1
China, 7, 18, 75, 162–4, *163*
chinampas (floating gardens), 108–9
Chinese Cook Book (Au), 276
Chippewa, 16
Choat, Duane and Addis, 134–40
chowchow (relish), 192
Christensen, David, 72
cleansers from corn, 246
Clinton Corn Processing Co., 273
clouded yarn, 244
Coatlicue (goddess), 292–3, *292*
Cobbett, William, 178, 182
Cobo, Bernabé, 101, 104–6, 298, 299
Coe, Michael, 31
Coe, Sophie, 201
coffee from corn, 242
Coffin, Robert Tristram, 134
Colgate & Company, 144, 266–7
Colony, Okla., 22–4
Colquitt, Harriett Ross, 223
Columbus, Christopher, 17
combines, 135–6, *136*, 138, 153
Comentarios reales de los Incas
    (Garcilaso), 102, 103, 104
Committee on the Preservation of
    Indigenous Strains of Maize, 93
commodification of corn, 265–78;
    methodology of, 266
Complete Cook Book, The (Harland), 181
Condon, Richard and Wendy, 195
Consumer Reports, 11
Cooper, Joseph, 65
Copán (Maya ruin), 31–2

Cope family, 189
corn: anatomy of, 60–1, ancient draw-
    ings of, *19, 20, 60;* breeding of, 64–7,
    70, 78; classification of, 86–8, *86, 88;*
    evangelists of, 69–71, 92; germ in, as
    related to oil, 212, 216; growing pro-
    cess, 26, 61–4; *62;* man, relation to,
    20–1; mass production of, 68, 70;
    nutrients and protein value of, 60–1,
    204–6; origin theories of, 77–82, 84–5;
    prehistoric, 15, 79–81, *79,* 100–1;
    processing of 174–8, 202, 204–7, 216,
    266–75; seed preservation, 93–6; *see
    also specific names and products*
*Corn: Its Origin, Evolution, and Improve-
    ment* (Mangelsdorf), 79
Corn, Okla., 21–2, 23–4
"corn," Old English meaning of, 19
*Corn Among the Indians of the Upper
    Missouri* (Will and Hyde), 66, 128
Corn and Corn Growing (Wallace and
    Bressman), 61
Corn and Its Early Fathers (Wallace and
    Brown), 68
cornballs, 132
corncobs, uses for, 245–6, *246*
corn costumes, 318, *318*
corncribs, 151–2, *152*
*Corncribs: In History, Folklife, and
    Architecture* (Roe), 152
corn dance ceremonies, 282–6, *282, 283,
    284*
cornflakes, 167, 168, 172–3, *173*
cornhusking, 6, 305–11, *305, 306*
cornhusks, uses for, 243–4
Corn Items Collectors, 310
Corn Kitchen, 319
corn-leaf blight, 93
corn madness, 8–9, 78
cornmeal, cleaning with, 246
cornmeal cooking, named dishes:
    blue-corn, 172–3, 206; dodgers, 223,
    226; hoe-cakes, 20, 222–4; hush pup-
    pies, 226; johnnycakes, 217–8, 222–5;
    pone, 21, 222–3, 226–7, 229; spoon-
    bread, 230; *see also under individual
    countries*

cornmeal mush, 229–38, *230, 235, 237;*
  mamaliga, 15, 229, 236; *sofki* (or
  posho), 196–7, 236–7, *237*
corn network, 7–8
corn oil, 277
corn-on-the-cob, 178–84, *178, 179, 180,*
  184
corn palaces: Mitchell (S.D.), 8, 311–12,
  315–17, *315,* 320; North Dakota, *314;*
  Sioux City, 313–15, *313*
Corn Plants (Sargent), 62, 68
corn pollen, 25, 40, 62–3, 66, 69, 71, 95,
  117, 329–30
Corn Products Company International,
  269
Corn Products Cook Book, 275–6, *275*
Corn Products Refining Company,
  275–7
corn races, 15–16, 86–96, *88;* classification system, 60, 87; elimination of,
  86–7, 93; inventory of, 87–8; preservation of, 93–6
Corn Refiners Association, 265
corn sleds, 149
corn smut (*huitlacoche*), 117, 238–41, *240*
cornstarch, 7–8, 11, 276; extraction of,
  266–8, *268;* sugar, conversion to,
  269–74
corn syrup, 8, 269–74
Corn War, 76–86
Corson, Juliet, 276
costumes of corn, 318, *318*
Country Gentleman (sweet corn), 90
Country Housewife and Lady's Director,
  The (Bradley), 253
Courtland, Kan., 308–9
Cracker Jack, 10, 13, *13*
Craig, Rev. Elijah, 257
creamed corn, 186, 187, 188–9
creation stories, Indian, 49–54
Creek, 288–9
Creoles, 196
Cretors, Charles C., 12–13, *12*
Crews, Harry, 245
criollo cooking, 191
crooning feast, 213
Crowen, Mrs. T. J., 179

Cuisine of Hungary, The (Lang), 236
cultivators, 137, 147–8
Curley, Louis, 306
Curley, Slim, 95
Cushing, Frank H., 49, 115–16, *116,* 117,
  118, 119, 167, 171, 174, 184, 186, 192,
  193, 194, 204–5, 206, 213, 233, 239, 328
Cutler, Hugh, 87
Cutler, Jonathan, 151

Dabney, Joseph Earl, 260, 262
Daniel, Jack Newton, 260
Darwin, Charles, 71, 100, 143
Davis, Wade, 247–8
Deere, John, 145, 147
Delavan, Ill., 305
Della Rocco, Michael, 14
De Natura Stirpium (Ruel), 18
desert farming, 48, 113–21, *117, 118, 120*
detasseling, 74
dextrin, 7, 8, 271–2
dextrose, 269, 272
Díaz del Castillo, Bernal, 201
Dick, Herbert, 14–15
Dill, Agnes, 212
Dinner Year-Book, The (Harland), 181,
  188, 233
diploperennis, 81–2, *81*
Directions for Cookery in Its Various
  Branches (Leslie), 181
diseases of corn, 93
Doerper, John, 180
dolls, cornhusk, *244*
Doolin, Elmer, 209
Dovetec Corn Heater, 245
Drees, Mahina, 95
dried corn, 189–91
Drumm, Paul, 216
Dudley, Paul, 64
Duryea, Wright, 267
Dust Bowl, 134
Dutch immigrants, 221
Duvick, Donald, 74
dyes from corn, 244

Eagle, Chief, 310
East, Edward Murray, 73

Eating Well (Doerper), 180

Eberhart, Steve, 94

Eckstein, Henry, 13

Effects of Cross and Self Fertilization . . .
    (Darwin), 71

electroporation, 92

Enciso, Paul, 212

Enduring Seeds (Nabhan), 115

Enfield, Edward, 67–8, 144, 239

ensilage, 152–3

Erickson, Vernon, 308, 311

Escoffier, Auguste, 183, 187

*Essays and Notes on Husbandry and Rural
    Affairs* (Bordley), 226

Esterly, George, 147

ethanol, 274

Etiquette (Post), 184

Eustis, Célestine, 200, 228, 233

"Ever-Normal Granary Plan," 158

Everyday Life of the Aztecs (Bray), 110

fairs, exhibits at, 318–20, *319, 320*

Falls Mill, *216, 218*

Farmer (magazine), 307, 331

Farmer, Fannie, 187, 232–3

farming of corn, 133–64; changes in
    farming, impact of, 140–3; computer-
    ization of, 139; consulting businesses,
    140–1; corporate takeover of land,
    139–40; costs of, 136, 138; excess
    capacity problem, 135–6, 139, 153,
    158–9, *159*; exhaustion of land, 155–6,
    156; family farming, decline of,
    157–8; government's role, 139, 156,
    158, 160–1; harvesting, 138; Indian
    methods, 48, 101–5, 108–31, *112, 117,
    118, 120, 122, 130*; planting year,
    136–8; reform proposals for, 161–2,
    164; *see also* mechanization of
    farming

farming revolution, 3–4, 67–76; effi-
    ciency of production, 68–9; genetic
    engineering, 74–5, 91–3; hybrids, 5,
    69–74, *72*, 75–6; purebred strains and,
    72–3; religious aspects, 69; results of,
    74, 75; visionaries of, 67–70

farm museums, 155–6

farm population, 5

Faulkner, William, 246

feedstock, corn as, 275

Ferguson, Bob, 309, 310, 311

fermented corn, 206

fertilizers, 137, 158, 159

festivals of corn, 72, 301–11, 314–18, *316,
    317*

Field of Dreams (film), 331

Fifty Valuable and Delicious Recipes . . .
    (Eustis), 220, 228

fishing/hunting with corn, 247

Fletcher, Horace, 172

floating gardens (*chinampas*), 108–9

Florentine Codex, 38, 110–11, *112, 291,
    295*

Foliated Maize God, 33–4, *34*

food from corn, 167–76; European atti-
    tudes toward, 174, 175; green-corn
    cooking, 176–95; Indian cooking
    methods, 168–72, *170,* 174–6; *see also*
    ash cooking; cornmeal cooking; *and
    specific foods*

Food of Italy, The (Root), 234, 235

Food of the Western World, The, 174

Ford, Frank, 219

Ford, Henry, 153, 168

Ford, Richard, 247

Franklin, Benjamin, 11, 64, 177, 192–3,
    220–1, 230–1, 241

Frazer, Sir James, 292

French cornmeal cooking, 228, 235–6

fried corn, 187

Frito company, 209

fritters, 187–8

Frobenius, Leo, 42

fructose, 269, 270, 273

Frugal Housewife, The (Child), 221–2

Furnas, C. C. and S. M., 202, 203

Galinat, Walton C., 15, *61,* 67, 82–4, *83,*
    85, 88, 89

Garcilaso de la Vega, 99, 102, 103, 104,
    206, 297, 298, 299

Garden Book (Jefferson), 90

Garrard, Rev. James, 257

Garreau, Joel, *4, 6, 141,* 144

Garst, Roswell, 164
Garst & Thomas Hybrid Corn
    Company, 95–6
gasohol, 274
Gates, Merrill E., 154
Gay, Lettie, 199
Genesee Farmer, The, 186, 227
genetic engineering, 74–5, 91–3
Gentlewoman's Housewifery, Ye, 231
Gerarde, John, 18–19
German immigrant farmers, 32, 144
Germplasm Resources Information
    Network (GRIN), 94
Gibson, Chief, 186
"Girl in the Library, A" (Jarrell), 330
Glasse, Hannah, 231
glucose, 269, 270
glues from corn, 244–5
gluten, 271
Goldberger, Joseph, 202, 203
Golden Bantam (sweet corn), 89, 90
Goodbird, Edward, 125, *125*, 128, 131
Good Form (Stokes), 183
Goodman, Major, 85
"Good Morning, America" (Sandburg),
    141–2
grain elevators, *138*
Grand Voyage du Pays des Hurons
    (Sagard), 190
graters, 186–7
Gray's Gristmill, 215
Green Corn Busk, 286–91, *287*
green-corn cooking, 176–95
Grider, Rufus A., 229, 230
Grinding Corn as I Have Seen It
    (Jeffries), 267–8
gristmills, 215–18, *216, 217*, 219
grits, 196, 197–8, 199
Guide culinaire (Escoffier), 183
Guzman, Rafael, 81–2, *81*

Hadley, A. N., 150
Hammond, John, 253
Hanson, Charles, 131
Hardeman, Nicholas, 149, 213
Hariot, Thomas, 122–3, *184*, 252
Harland, Marion, 181, 187, 188, 232, 233

Harris, Joel Chandler, 221
Harris, John, 19
Harrison, Jim, 331
harrows, 145
Hart, Charles, 153
harvesters, 74, 148–9, *148*, 303–4
hasty pudding, 231
Haury, Mary, 23–4
Hazard, Thomas Robinson, 198, 216–17
health food, 172
heating with corncobs, 245
Heckewelder, J. G. E., 194
Heidecker, Ernest, 310
Hellmann's mayonnaise, 277–8, *278*
Henry, Alexander, 124
herbicides and insecticides, 135, 137,
    141, 158, 159
Hernández, Francisco, 202, 206, 207
Hiawatha story, 53–4
Hidatsa culture, 12, 121–2, 124, 127, 177,
    189–90; farming methods, 124–31, *130*
High Fructose Corn Syrup (HFCS),
    270, 273–4
Hinz, A. C., 25–6
Histoire de la Louisiane, L' (Pratz), 193
Histoire naturelle . . . (Bonafous), 92, 240
Historia del Mundo Nuevo, La
    (Benzoni), 249–50
Historia del Nuevo Mundo, La (Cobo),
    104–6
*Historia general de las cosas de Nueva
    España, La* (Sahagún), 110–13, 201,
    202
*History, manners, and customs of the
    Indian nations, The* (Heckewelder),
    194
hoe-cakes, 220, 221, 222, 223
hog and hominy, 199, *199*
Hog Days festival, 304, 305
Hohokam culture, 113–15
Holden, P. G., 71–2, 73
Home Country (Pyle), 134
Home Place (Morris), 332
Homestead National Monument, 155
hominy, 176, 195–200, *196*
Hoopeston, Ill., 301–4
Hopewell, 24, 122

Hopi Cookery (Kavena), 190
Hopi culture, 6, 92, 96, 204, 205–6, *213*, 239; food from corn, 167, 168–72, *170*
Household Discoveries (Morse), 191, 246
Howard, Josefina, 240–1
Howe, Oscar, 320
Huayna Picchu, 107, *107*
Hughes, Phyllis, 195
Huie, William Bradford, 224–5
huitlacoche (corn smut), 238–41, *240*
human sacrifice, 291–6, *291, 292, 295*, 298–9
Hungary, 236
Huron, 38, 123
Hurt, R. Douglas, 146
huskers, 150–1
hybrids, 5, 69–74, *72*, 75–6, 82, 92–6
Hyde, George E., 66, 211–12
hydrol (corn sugar molasses), 272

Illich, Ivan, 327–8
Iltis, Hugh H., 79, 81, 82, 84–5
Inca civilization, 16, 99–108, 249–50, 296–300; farming methods, 101–5; sacredness of corn, 99–100; solstice festival, 107–8
Incidents of Travel in Central America, Chiapas & Yucatan (Stephens), 207
Indian Colony, 23–4
Indian Cookbook, The (Lawson), 190
Indian Cook Book, 186
Indian Corn (Enfield), 67, 144, 239
Indian Meal Book, The (Leslie), 222
Indian pudding (hasty pudding), 230–3
Indians, *see individual peoples and cultures*
International Harvester, 148, 153
International Maize and Wheat Improvement Center (CIMMYT), 93
Inter-Tribal Ceremonial, 47
Inti Raymi, feast of, 296–300
Iowa State Fair (1908), 320
Iroquois culture, 16, 38, 52–3, *52*, 123, 124, 179, 194, 197, 212, 232, 241
irrigation systems, 113–14, 137–8, 141
Isomerose, 273–4
Italian cornmeal cooking, 233–5

Jack Daniel Distillery, 259–60
Jackson, John, 310
Jackson, Wes, 161
Jaramillo, Cleofas, 190–1, 194–5
Jarrell, Randall, 330
Jefferson, Thomas, 90, 146, 263
Jeffries, Frederick, 267–8
johnnycakes, 222–3
Johnny's Selected Seeds, 91
Johnson, Edward, 21
Johnson, Everett, 153–4
Johnston, Robert L., 91
Jones, Orlando, 144, 266
Jonny-Cake Papers of 'Shepherd Tom,' The (Hazard), 216–17, 230
Josselyn, John, 241–2
Journal of Gastronomy, The, 161–2
Joy of Cooking (Rombauer), 179, 187–9
"jumping genes," 92

Kansas Home Cook-Book, The (Thompson), 242
Kan sign, 34
Karo syrup, 276–7
Katz, Solomon, 175, 176, 204
Kauger, John and Alice, 22–3, 24
Kavasch, Barrie, 233
Kavena, Juanita Tiger, 180, 190, 204
Kellogg, John H., 168, *168*, 172, 173–4
Kellogg, William K., 168, 172, 173–4
Kennedy, Diana, 208–9
Kewanee, Ill., 305–6
Key into the Language of America, A (Williams), 174
Kickapoo, 199
King, Stephen, 330–1
Kingsford, Thomas, 267
Kingsford's Corn Starch, 275–7, *275*
Kinsella, W. P., 331
Kitchen Companion, The (Parloa), 218
kivas, 47–8, 54, 56, 57–8
Knife River Indian Villages, 131
Krug, George, 75
Kunitz, Stanley, 333
Kupferschmid, Gary, 310–11

lactic acid, 272–3
Lane, Ralph, 241
Lang, George, 236
Lappé, Frances Moore, 162
Laughman, J. R., 90
Lawrence, D. H., 283, 325
Lawson, Roberta, 190
Leah and Rachel . . . (Hammond), 253
Leslie, Eliza, 181, 187, 198, 222, 226–7, 232
Lewis, Edna, 198
Lincoln, Mary, 242
Linnaeus, Carolus, 60
liquor from corn, 253–64, *258*; factory production, 258–60; moonshining, 253–8, *258*, 260; Prohibition, 264; Whiskey Rebellion, 263
Loft, W. E., 314
Logan, James, 64
Longfellow, Henry Wadsworth, 53
Lorain, John, 64
Lord Chan-Balum (Maya king), *31*, 33
Louisiana Purchase Exposition, 319, *319*
Lumbra, Puffer, 245
lye hominy, 197, 198–9
Lyte, Henry, 59

Machu Picchu, 106–7, *107*, 108
MacNeish, Richard, 40, 80, 101
Ma Cuisine (Escoffier), 187
Madison cake, 226–7
Madrid Codex, 34
Maize (Corn) Gods: Beheaded, 35, *36*; Cinteotl, 113, 292; Changing Woman, 328–9, *328*, 332; Coatlicue (Chicome Couatl), 113, 292–3, *292*, 320; Earth and Corn Mothers, 50–4, 103, 113, 281, 283, 286, 312–13, 330; Foliated (Tonsured), 33–4, *34*; God K, 30, 37; Mondawin, 16, 53, 290; Olmec, 35; One Hunahpu, 16, 43–4, *44*; Quetzalcoatl, 16, 30, 291, 293–6, *294*, 330; Xilonen, 16, 113, 291–2, 318; Xochipilli, 113, 293; Young Lord, 30–2, *32*, 33
maltose, 270
Man, Bread and Destiny (Furnas), 202

Mandan, 124, 131–2
Mangelsdorf, Paul C., *10*, 15, 40, 59, 77, 78–81, *78*, 82, 87, 101
Mansur, Alvah, 147
Marion Harland's Complete Cook Book, 181
Marks, Copeland, 237
Martineau, Harriet, 174
masa, 175, 206, *208*
Massasoit, Chief, 287
Mather, Cotton, 63–4
mattresses, cornshuck, 243–4
Maya civilization, 16, 45, 293; language of corn, 37; mythology of, 29–35, *31, 32, 34, 36*
mayonnaise, 277–8, *278*
Mazola corn oil, 275–7, *275*
McClintock, Barbara, 73, 91–2
McElhany, John and Clyde, 302
McGee, Harold, 203, 270
McTague, Tim, 215
mechanization of farming, 74, 143–54; beginnings of, 143–5; consequences of, 153–4; lawsuits on patents, 147
medicinal uses of corn, 241–3
Memoir on Maize . . . (Browne), 12, 185
Mencken, H. L., 223
Mennonites, 22
metates, 211, *212, 213*
Mexico, 40, 81–2, 87
M'Gillivray, Anthony Alexander, 288
milking the cob, 185–6
Miller, Mary Ellen, 33
milling, 210–20, *215*; Indian techniques, 169, 211–13, *211, 213, 214*, 218; roller milling, 173, 214, 218–19; traditional methods, return to, 215–16, 219
Mills, George, 70, 157, 158
Minataree, 290
Miss Parloa's Kitchen Companion, 218
Mitchell, S. D., 8, 311–12, 315–17, 320
Mobile Corporation, 275
Moche pottery, *11, 102*
Moctezuma, 38, 200–1, 293
Modern Cookery (Acton), 182
Monks Mound, 288
Montigny, Dumont de, 174, 194

moonshine, 253–8, *255, 258*, 260–1
Moore, Milan, 140–1, 161
Morris, Wright, 332
Morrison, Toni, 331–2
Morse, Sidney, 191, 246
Motlow, Lemuel, 260
Moundbuilders, 24, 122
Mountain Spirits (Dabney), 260, 262
Mrs. Curtis's Cookbook, 222
Mrs. Rorer's New Cook Book, 183–4
Mud on the Stars (Huie), 224–5
Mundus Novus (Vespucci), 38
Murphy, Charles, 185, 191, 218, 242
mush, *see* cornmeal mush
Musk, Pat, 302
mythology of corn, 29–36, *31, 32, 34, 35, 36,* 41, 42–6, 49–52, 283–4, 286, 289–99

Nabhan, Gary, 95, 115, 131–2
Narragansett, 174, 184, 193
Natchez, 193, 194
National Corn Growers Association, 139, 265
National Research Council, 162
National Seed Storage Laboratory, 93–5
Native Harvests (Kavasch), 233
Native Inheritance (Walden), 8
Native Seeds/SEARCH, 95
Nature and Reason Harmonized in the Practice of Husbandry (Lorain), 64
Navaho cakes, 225
Navaho, 49, 225, 328–30, 332
Nebraska, 3–6, 133–4, 143
Nebraska (Butler), 133
Nebraska Farmer, The, 230
Neumann, Erich, 113, 295
New Cook Book (Rorer), 183–4
New Herbal, A (Lyte), 59
New Receipts (Leslie), 198, 226, 232
Nine Nations of America (Garreau), 4, 141
nixtamal, 195, 200 ff., 208
nokehick, 193
North Central Plant Introduction Station, 94
"Notes on the Florida Seminole" (Skinner), 197

Oglala aquifer, 133
Ojibway, 53–4
Oklahoma, 21–4, 25–6
Old World perception of corn, 16–20
!Ole Mole! (Condon), 195
Olmec culture, 35, *35,* 41
Omaha people, 142
On Food and Cooking (McGee), 203, 270
origin of corn, 76–86, *79, 84, 85*; place of origin, 18, 40, 101
Oroke, Ray, 311
Ortiz, Alfonso, 56, 120, 283, 285
Oswald, George, 217

Paleo-Indians, 39–42
Pampa de las Llamas-Moxeke, 100–1
parched corn, 11, 13, 192–4; *atole* and *pinole*, 194–5; sagamite, 146, 196
Parkinson, John, 18
Parloa, Maria, 195, 218
Parr, Charles, 153
Pawnee, 16, 291
Peck, A. S., 149
pellagra, 176, 202–3
Peterson, Orville, 309
Pfister Hybrid Corn, 75–6
Philipsburg Manor gristmill, 219
Picayune's Creole Cook Book, 187, 192
pickled corn, 191–2
Pierce, Samuel S., 144, 304
Pieri, Paul, 215
piki, 167–72, *170,* 176
Pioneer Cookery Around Oklahoma, 192
Pioneer Hi-Bred International, 69–70, 96, 139
Pioneer Women (Stratton), 244
pits for winter storage, 130, *130*
Plambeck, Herb, 309–10, 311
planters, 145–7, *144, 146*
plastics, cornstarch degradable, 274–5
plows, 145, *145*
Pocumtuc Housewife, The, 231, 233
pod corn, 78, 79–81, *79*
polenta, 229, 233–6, *235*
Polich, Tony, 310, 311
pollen, *see* corn pollen
Poma de Ayala, Guaman, *100,* 101–2

pone, 224–5, 229
popcorn, 9–15, *10, 11, 12, 13, 14,* 79, 82, 88, 192–4
population, relationship to corn, 162–3
Popul Vuh (Quiché Maya text), 33, 42–6
Post, C. W., 172, 173
Post, Emily, 184
powdered ("journeying") corn, 194
Powell, Maj. John Wesley, 205–6, 294
pozole (posole), 196, 199–200
Practical Cook Book, The (Bliss), 228
Practical Housekeeping, 231–2
Prairie Farmer, The, 90, 245
prairie grass, 155
Pratz, Antoine Le Page Du, 193
primer nueva corónica y buen govierno, El (Poma de Ayala), 101–2, *100, 101*
processed food products, 275–8
Prohibition, 264
protein value of corn, 202–4
puddings, custards and sauces, 186–9
Pueblo Bonito, 47–8
Pueblo culture, 48, 49–52, 171, 212, 282–6
Pueblo Indian Cookbook, The (Hughes), 195
puliszka, 235, 236
Pyle, Ernie, 134

Quetzalcoatl (god), 108, 293–6, *294*
Quiché Maya, 42–6

Races of Maize in Mexico, The, 87
rain prayers, 116–17
raised-field agriculture, 41, 108–11, *111*
Ramsay, John, 147
Randolph, Virginia, 226, 233
Redenbacher, Orville, 14, *14*
Reeves, Robert G., *10,* 79
Reid, Robert, 72, 305
Reid's Yellow Dent hybrid, 72, *72,* 73, 75
Relación de la jornada de Cíbola (Castañeda), 115
Relación de las cosas de Yucatán (Landa), 32–3
relishes, 192
Ren Rui, 163
Resneder, James, 22, 23

Rhodes, Richard, 161
Richardson, Flo Ratzluff, 21, 22
Robbins, Martin, 146
Roe, Keith, 152
Rojas-Lombardo, Felipe, 251
Romanian cornmeal cooking, 236
Rombauer, Irma, 179, 187–9
Root, Waverley, 234, 235
Rorer, Sarah T., 183–4, 319
Rose, Bill, 305–6, 307–8, 311
Rueckheim, Frederick and Louis, 13
Ruel, Jean, 18
Rutledge, Sarah, 187, 200, 223
Rutter, J. F., 304

Sacsahuaman fortress, *56, 57,* 299, *299*
sagamite, 194, 196
Sagard, Gabriel, 123, 190, 206
Sahagún, Bernardino de, 10–11, 38, 110–13, 201, 202, 206, 292, 294
St. Lawrence Starch Company, 274
Salt River Indian Community, 95
samp, 197, 198
Sanctuary (Faulkner), 246
sancu, 295–6
Sandburg, Carl, 141–2
San Domingo pueblo, 282–3
Sargent, Frederick, 62, 63, 68
Saunsoci, Frank and Alice, 142
Savannah Cook Book, The (Colquitt), 223
scarecrows, 83, *83,* 118, *118,* 127
Schele, Linda, 33
Schubeline, Peter, 174
Scotti family, 234
scrapple, 237–8
Sears, Paul B., 80
"Seedcorn and Windfall" (Kunitz), 333
Seed Savers Exchange (SSE), 72, 94–5
Seger, John, 23
Seiler, Carl, 310
Sekaquaptewa, Helen, 167, 169–71, 176, 328
Sekatau, Ella Thomas, 233
Sellars, Jim, 317
Sells, William, 304
Seminole, 197
Seneca, 52–3, 126, 186, 193

Serpent and Rainbow (Davis), 247–8
serpent effigy (Ohio), 287–8
Seventy-Five Receipts (Leslie), 232
Shalako ritual, 323–6, *325*
shaman-priests, 289–91
shellers, 151
shifting-field farming (*milpa*), 41
shocking of corn, 150, *150*
Shucks, Shocks and Hominy Blocks
(Hardeman), 149
Shull, George Harrison, 73
silos, 152–3
Simmons, Amelia, 221, 226, 232, 245
Simon, Henry, 214
Singing for Power (Underhill), 114–15
singing plow, 145, *145*
Sioux City, Iowa, 312–15, *313*
Skinner, Alanson, 197
Sky Woman (Smith), 52
slave ships, *236*
Slosser, Alwin, 304
Smith, C. B., 156, 159
Smith, Earle, 14–15
Smith, Ernest, 52
Smith, Capt. John, 39, 154
smoked corn, 142, 194
Social History of Bourbon, The (Carson),
257
Society for the Preservation of Old
Mills, 219
Society for the Propagation of the Jon-
nycake Tradition, 215–16, 222
Sofaer, Anna, 54, 55
sofki, 196–7, 237
Song of Hiawatha, The (Longfellow), 53
songs: for grinding corn, 212–13; for
planting corn, 1, 14–15, 126, 127
sorbitol, 273
Soth, Lauren, 158, 159, 160
soups and chowders, 188
soybeans, 164
Speck, Frank, 212
spiritual/demonic uses of corn, 247–8
spoonbread, 228
Sprague, Welcome, 304
Spruce, Elizabeth Woolsey, 243
Squanto, 24–5

Staley, Gene, 269
Staley Manufacturing firm, 269, 275
Stedman, Ebenezer Hiram, 259
Stephens, John, 207
stills, *255*, 256–60, *258*
Stokes, Frederick, 183
storage technology, 138–9, *138*, 151–3
Strachey, William, 197
Stratton, Joanna, 244
Stuhr Museum of the Prairie Pioneer,
155
succotash, 184–5, *184*
sugar from corn, 8, 241, 269–74
Sun Dagger calendar, 54–6, *55*
Sunday, Billy, 264
sweet corn, evolution of, 89–91
sweet corn, specific names of : Chullpi,
89; Country Gentleman, 90; Early, 90;
Golden Bantam, 90; Old Colony, 90;
Papoon, 89–90; Stowell's Evergreen,
90
sweet-corn festivals, 301–4

Tales of the Cochiti Indians (Benedict), 240
tamales, 37, 175, 200, 202, 206, 209
Taylor, Col. Edmund Hayes, Jr., 259
Taylor, Edward, 21
tea from corn silks, 242
Tedlock, Dennis, 45
Tehuacán, caves of, 40, 80–1
Tello, Julio C., 101
temperance movement, 264
Temple of the Foliated Cross, 30, *31*, 33
Temple of the Sun (Cuzco), 99, 297–8
teosinte, 77–8, 79, 81–2, *81*, 84–5
Tewa culture, 56, 171, 174, 204, 212,
242; corn dance, 282–6, *282*, *283*, *284*
Theatrum Botanicum (Parkinson), 18
Thompson, John, 147
Thompson, Libbie, 242
Thoreau, Henry David, 11, 146, 224,
289
Thorpe, Capt. George, 252
Tibbe, Henrick, 245
time measurement with corn, 248
Timmons, Clyde, 303, 304
Torquemada, Juan de, 108–9, 292

tortillas, 175, 200, 202, 202–3, 205, 206–9, *208*
tractors, 153
Traditional Navajo Foods . . . , 225
Treatise on Indian Corn, A (Cobbett), 178, 182
"tripartite theory" of corn's origin, 78–81, *79*
Tripsacum (grass), 79, 82
Trollope, Frances, 174, 200
True History of the Conquest of Mexico, The (Díaz), 201
Tschirky, Oscar, 182–3
Tupinambas, 38
Turkish wheat, 18, 19, *19*
Twain, Mark, 164, 224
200 Years of Charleston Cooking (Gay), 199

Udall, Louise, 176
Ulster Presbyterians, 262–3
Underhill, Ruth, 114–15
Union Sugar Company, 269
Utatlán, citadel of, 45–6

Van der Post, Laurens, 237
Vespucci, Amerigo, 38
Vilcanota Valley, 106
"virgin hash," 206
Virginia Housewife, The (Randolph), 226, 233
Voyages & Travels (Harris), 19

wafer breads, 171–2
waffle gardens, 49, 117, 120
Waheenee: An Indian Girl's Story, 121
Wakefield, Robert, 215
Walbot, Virginia, 92
Walden (Thoreau), 224
Walden, Howard, 8
Walker, Gene, 308–9
Walker, Herb, 200
Wallace, Henry A., 61, 67, 68, 69, 70, *70*, 73–4, 75, 86–7, 92, 157–8, 306, 307
Wallace, James, 6, 69, 70, 73–4
War Eagle Mill, 219
Water Chief, 128

Waters, Alice, 162
Waugh, F. W., 179, 185, 197, 206, 212, 232
Weatherwax, Paul, 79, 82, 103, 239–40
Weddington, Frank and Paralee, 254–5, 256, 261–2
Weinstein, Jeff, 167, 172
wet-milling, 7, 144, 266
Whealy, Kent and Diane, 94, 95
Whiskey Rebellion, 263
White, Jasper, 181
White Cap flint corn, 215
White House Cook Book, The, 181–2
white lightning, 257
Wilcox, Neb., 4, 155
Wilkes, Garrison, 21, 82, 85, 93, 94
Will, George F., 66–7, 128, 129, 211–12, 291
Williams, Roger, 174, 184, 193, 242
Wilson, Gilbert L., 124, 125, 129
Winslow, Capt. Isaac, 144, 304
Winthrop, John, 11, 24
Wolfert, Paula, 235
Wonder Working Providence of Sions Savior in New England (Johnson), 21
Woodland people, 289–92
World's Columbian Exposition, 72, 318–19
World's Work, The (magazine), 319
Wyandot Popcorn company, 13

Xilonen (Aztec goddess), 328

Yax-Pac, Maya king, 31–2
Ya Ya Ceremony, 57–8
Youhoena, Frieda, 170, 171
Young Maize God, 31–2, *32*, 44, *44*
Yuchi, 212

Zapotec funeral urn, *109*
Zea mays (corn), 15, 60
Zinacantec, 29–30
Zuni Breadstuff (Cushing), 171
Zuni culture, 49, 51–2, 167, 171, 174, 184, 186, 193–4, 204–5, 213, 233, 239, 248; corn farming, 115–21, *117, 118, 120*; Shalako ritual, 323–6, *325*, 332

As you read questions in the left-hand column, use this handy bookmark to cover the right-hand column until you've chosen each answer.

Test-taking strategies

Try these tried-and-true test-taking tips! (Wow, betcha can't say that three times fast!)

• Read each question and all options carefully before making your selection.

• Pay special attention to such words as *best, most, first,* and *not* when reading the stem of a question. These words usually provide clues to the correct response.

• Try to predict the answer as you read the stem. If your prediction is among the four options, it's probably the correct response.

• If a question focuses on an unfamiliar topic, determine the nursing principles involved. Then eliminate options until you can select an answer.

Other *Incredibly Easy* books

Anatomy & Physiology Made Incredibly Easy
Assessment Made Incredibly Easy, 2nd edition
Charting Made Incredibly Easy, 2nd edition
Clinical Pharmacology Made Incredibly Easy
Diagnostic Tests Made Incredibly Easy
Dosage Calculations Made Incredibly Easy, 2nd edition
ECG Interpretation Made Incredibly Easy, 2nd edition
Fluids & Electrolytes Made Incredibly Easy, 2nd edition
I.V. Therapy Made Incredibly Easy, 2nd edition
Medical Spanish Made Incredibly Easy
Medical Terminology Made Incredibly Easy
Medication Administration Made Incredibly Easy
NCLEX-PN Review Made Incredibly Easy
Nursing Procedures Made Incredibly Easy
Nutrition Made Incredibly Easy
Pathophysiology Made Incredibly Easy, 2nd edition
Patient Teaching Made Incredibly Easy
Studying & Test Taking Made Incredibly Easy

For more information about *Incredibly Easy* books or other Lippincott Williams & Wilkins products, visit **www.lww.com** on the Internet.

Use this side of the card to jot notes to yourself about topics you need to review more.

Here's to NCLEX success!

SPRINGHOUSE
LIPPINCOTT WILLIAMS & WILKINS

NCLEX-PN Questions & Answers

made

Incredibly Easy!®

3,000 + Questions!

LIPPINCOTT WILLIAMS & WILKINS
A **Wolters Kluwer** Company

Philadelphia • Baltimore • New York • London
Buenos Aires • Hong Kong • Sydney • Tokyo

Staff

Publisher
Judith A. Schilling McCann, RN, MSN

Editorial Director
David Moreau

Clinical Director
Joan M. Robinson, RN, MSN

Senior Art Director
Arlene Putterman

Art Director
Mary Ludwicki

Clinical Editors
Beverly Ann Tscheschlog, RN, BS, and Denise D. Hayes, RN, MSN, CRNP (clinical project managers); Joanne Bartelmo, RN, MSN; Maryann Foley, RN, BSN; Pamela Kovach, RN, BSN; Barbara Stiebeling, RN; Dorothy P. Terry, RN; Patricia Fischer, RN, BSN

Editors
Julie Munden (senior editor), Ty Eggenberger, Brenna H. Mayer, Carol H. Munson

Copy Editors
Kimberly Bilotta (supervisor), Heather Ditch, Amy Furman, Shana Harrington, Elizabeth Mooney, Judith Orioli, Pamela Wingrod

Designer
Lynn Foulk

Illustrator
Bot Roda

Digital Composition Services
Diane Paluba (manager), Joyce Rossi Biletz, (senior desktop assistant)

Manufacturing
Patricia K. Dorshaw (senior manager), Beth Janae Orr

Editorial Assistants
Danielle J. Barsky, Beverly Lane, Linda Ruhf

Indexer
Karen C. Comerford

IEQAPN – D N O S
05 04   10 9 8 7 6 5 4 3

Library of Congress Cataloging-in-Publication Data

NCLEX-PN questions & answers made incredibly easy / Judith A. Schilling McCann ... [et al.].
    p. ; cm.
    Includes index.
    1. Practical nursing—Examinations, questions, etc.
    [DNLM: 1. Nursing, Practical—United States—Examination Questions. WY 18.2 N3365 2003] I. Title: NCLEX-PN questions and answers made incredibly easy. II. McCann, Judith A. Schilling.
RT62 .N377 2003
610.73'076—dc21
ISBN 1-58255-240-1 (pbk. : alk. paper)   2002154012

Contents

Contributors and consultants   *v*

Foreword   *vii*

Part I Surviving the NCLEX

1	Preparing for the NCLEX	3
2	Passing the NCLEX	19

Part II Care of the adult

3	Cardiovascular disorders	33
4	Hematologic & immune disorders	63
5	Respiratory disorders	83
6	Neurosensory disorders	113
7	Musculoskeletal disorders	143
8	Gastrointestinal disorders	165
9	Endocrine disorders	187
10	Genitourinary disorders	211
11	Integumentary disorders	235

Part III Care of the psychiatric client

12	Essentials of psychiatric care	249
13	Somatoform & sleep disorders	251
14	Anxiety & mood disorders	263
15	Cognitive disorders	285
16	Personality disorders	297
17	Schizophrenic & delusional disorders	317
18	Substance abuse disorders	333
19	Dissociative disorders	351
20	Sexual & gender identity disorders	363
21	Eating disorders	375

Part IV Maternal-neonatal care

22	Antepartum care	389
23	Intrapartum care	403
24	Postpartum care	417
25	Neonatal care	431

Part V Care of the child

26	Growth & development	451
27	Cardiovascular disorders	453
28	Hematologic & immune disorders	483
29	Respiratory disorders	509
30	Neurosensory disorders	539
31	Musculoskeletal disorders	559
32	Gastrointestinal disorders	579
33	Endocrine disorders	603
34	Genitourinary disorders	627
35	Integumentary disorders	653

Part VI Coordinated care

36	Concepts of management & supervision	679
37	Ethical & legal issues	683

Appendices and index

Comprehensive test 1	691
Comprehensive test 2	712
Comprehensive test 3	733
Comprehensive test 4	755
Index	776

Contributors and consultants

Bertha L. Almendarez, RN, MSN
Chairperson, Department of Vocational
  Nurse Education
Del Mar College
Corpus Christi, Tex.

Melody C. Antoon, RN, MSN
Instructor
Lamar University
Beaumont, Tex.

Laura Aromando, MSN, ARNP
Department Chair, Nursing Programs
Seminole Community College
Sanford, Fla.

Adrianne E. Avillion, RN, DEd
President
AEA Consulting
York, Pa.

Peggy D. Baikie, RN, MS, CNNP, CPN,
Senior Instructor
University of Colorado School of Nursing
Nurse Practitioner
Gilliam Youth Center
Denver

Pennie Sessler Branden, CNM, BSN, MS
Consultant
Maternal Instinct Corporation
Woodbridge, Conn.

Barbara S. Broome, RN, PhD, CNS
Chair, Community and Mental Health
University of South Alabama
Mobile

Garry Brydges, RN, MSN, ACNP
Medical Consultant
Houston

Michelle Byrne, RN, MS, PhD, CNOR
Instruction of Nursing and Independent
  Nurse Consultant
North Georgia College and State University
Dahlonega

Nan S. Carey, RN, MSN
Assistant Professor
Baptist College of Health Sciences
Memphis

Rebecca Sue Chamberlain, RN, MSN, CRNP
Pediatric Cardiology/Cardiovascular
  Surgery Advanced Practice Nurse
Children's National Medical Center
Washington, D.C.

Jacqueline Clarke, RN
PN Nursing Instructor
Camden County Technical School
Sicklerville, N.J.

Linda Carman Copel, RN, PhD, CS, DAPA
Associate Professor
Villanova (Pa.) University College of
  Nursing

Arlene Coughlin, RN, MSN
Nursing Faculty
Holy Name Hospital School of Nursing
Teaneck, N.J.

Sheryl L. Currie, RN, MN
Lead Instructor
Butler County Community College
El Dorado, Kans.

Louise Diehl-Oplinger, RN, MSN, APRN, BC, CCRN, CLWC
Staff Development Instructor
Warren Hospital
Phillipsburg, N.J.

Jennifer Elizabeth DiMedio, RN, MSN, CRNP
Family Nurse Practitioner
West Chester (Pa.) Family Practice
(University of Pennsylvania affiliate)

Shelba Durston, RN, MSN, CCRN
Adjunct Faculty
San Joaquin Delta College
Stockton, Calif.
Staff Nurse
San Joaquin General Hospital
French Camp, Calif.

Emilie M. Fedorov, RN, MSN, CS
Clinical Nurse Manager, Department of
  Neurology
The Cleveland Clinic Foundation

Linda C. Finch, RN, PhD
Assistant Professor of Nursing
Baptist College of Health Sciences
Memphis

Frances D. Green, RN, MSN
Independent Consultant
Charlotte, N.C.

Patricia S. Greer, RN
Instructor Practical Nursing
Tennessee Technology Center
Paris

Maureen A. Heaton, RN, MSN
Associate Professor
Northern Michigan University
Marquette

Ruth Howell, RN, BSN, MEd
Director, Practical Nursing Program
Tri County Technology Center
Bartlesville, Okla.

Bobbie L. Hunter, RN, MSN, CFNP
Nursing Instructor, Practical Nursing
Program Manager
Columbus (Ga.) Technical College

Dianne Jensen, RN, BA
PN Faculty
Anoka-Hennepin Technical College
Anoka, Minn.

Michelle M. Johnson, RN, MSN
Assistant Professor
Northern Michigan University
Marquette

Karla Jones, RN, MS
Nursing Faculty
Treasure Valley Community College
Ontario, Ore.

Gwendolyn L. Jordan, RN, MSN
Faculty
Carolinas College of Health Sciences
Charlotte, N.C.

Rhonda M. Kearns, RN, BSN
Instructor
Orleans-Niagara BOCES
Medina, N.Y.

Jacqueline M. Lamb, RN, PhD
Instructor
University of Pittsburgh School of Nursing

Annie M. Lewis, RN, BSN, MS
Practical Nursing Instructor
Columbus (Ga.) Technical College

Sandra Liming, RN, MN, PhC
PN Program Coordinator and Instructor
North Seattle Community College

Patricia B. Lisk, RN, BSN
Acting Department Chairperson and
    Instructor
Augusta (Ga.) Technical College

Jennifer McWha, RN, MSN
Nursing Instructor
Del Mar College
Corpus Christi, Tex.

Marilynn V. Mettler, RN, BSN, MS, PhD
Nursing Faculty
Pueblo (Colo.) Community College

Robyn M. Nelson, RN, DNSc
Chair and Professor
Division of Nursing
California State University, Sacramento

Karin K. Roberts, RN, PhD
Associate Professor
Research College of Nursing
Kansas City, Mo.

Judith A. Robertson, RN, MN
Instructor
Seminole Community College
Sanford, Fla.

Kristi Robinia, RN, MSN
Assistant Professor of Practical Nursing
Northern Michigan University
Marquette

Bruce Austin Scott, MSN, APRN, BC
Nursing Instructor
San Joaquin Delta College
Stockton, Calif.
Staff Nurse
University of California, Davis, Medical
    Center
Sacramento

Catherine Shields, RN, BSN
Teacher of Practical Nursing
Career and Technical School of Practical
    Nursing
Lakehurst, N.J.

**Debra Siela, RN, DNSc, CCNS, CCRN,
CNS, CS, CCRN, RRT**
Assistant Professor of Nursing
Ball State University School of Nursing
ICU Clinical Nurse Specialist
Ball Memorial Hospital
Muncie, Ind.

Yolanda M. Williams, MSN
Practical Nursing Instructor
Columbus (Ga.) Technical College

Linda Wood, RN, MSN
Director of Practical Nursing
Massanutten Technical Center
Harrisonburg, Va.

Foreword

Congratulations! You've just completed your PN program and are ready to begin preparing for the NCLEX examination. As a new graduate you are, no doubt, excited. However, this is also a time when you may be experiencing a lot of anxiety as you try to prepare for the unknown: What's the NCLEX all about? What kind of questions will it include? What's the best way to prepare for it? And, finally, will I pass it?

Welcome to *NCLEX-PN Questions & Answers Made Incredibly Easy* — your answer to those questions and more! The purpose of this book is to help you become familiar with NCLEX-style questions, analyze your knowledge base, and help you practice, practice, practice! To enhance your preparation process, the questions appear on the left side of the page while the answers and other information are on the right. This allows you, in one quick glance, to review the answers, rationales, and client needs categories for each question. A convenient card in the front of the book can be detached and used to cover the answers while you read the questions on the left.

Fun cartoon characters throughout each chapter provide helpful tips and words of encouragement to help you through the review process. These characters help promote relaxation and encourage active participation as well as jog your memory about certain key points and important elements when preparing for the all-important NCLEX.

NCLEX-PN Questions & Answers Made Incredibly Easy includes six parts that address all the major topics included on the NCLEX. Part I addresses facts about the actual NCLEX-PN examination, including studying and test-taking strategies, test formatting, and computer adaptive testing (CAT). This information is provided to help you thoroughly prepare for the NCLEX-PN.

Parts II through V provide many questions related to each of four main areas included on the NCLEX: Care of adult, psychiatric, and maternal-neonatal clients and care of the child. Each part is carefully divided into chapters covering specific disorders. For example, Part II (Care of the adult) is divided into nine chapters: cardiovascular disorders, hematologic & immune disorders, respiratory disorders, neurosensory disorders, musculoskeletal disorders, gastrointestinal disorders, endocrine disorders, genitourinary disorders, and integumentary disorders. Part VI (Coordinated care) covers questions related to concepts of leadership and management as well as ethical and legal issues.

After reviewing and practicing with each chapter, you'll find four 85-question comprehensive tests at the back of the book. These tests are designed to help you further improve your test-taking skills and assess areas of weakness so you can focus your remaining study time.

I feel certain that using *NCLEX-PN Questions & Answers Made Incredibly Easy* will increase your confidence when taking the NCLEX-PN examination. Know that I and many other nurses — as well as Nurse Joy and the other characters in the book — are rooting for you! Remember all that you've accomplished and what you've learned from this book and others: Their words of wisdom will help through the process!

Sandra Liming, RN, MN, PhC
PN Program Coordinator and Instructor
North Seattle Community College

Part I   Surviving the NCLEX

1   Preparing for the NCLEX   3

2   Passing the NCLEX   19

Understanding the NCLEX goals and structure is an important first step in proper preparation for the test. This chapter explains how best to prepare for this important examination.

Chapter 1
Preparing for the NCLEX

Just the facts

In this chapter, you'll learn:

♦ about the NCLEX examination and why you must take it

♦ what you need to know about taking the NCLEX by computer

♦ strategies to use when answering NCLEX questions

♦ how to avoid common mistakes when taking an NCLEX examination.

All about the NCLEX

Passing the National Council Licensure Examination (NCLEX) is a vital step in your career as a nurse. The first step on your way to passing the NCLEX is to understand what the NCLEX is and how it's administered.

NCLEX structure

The NCLEX is a multiple-choice test written by nurses who, like most of your nursing instructors, have master's degrees and clinical expertise in particular areas. Only one small difference distinguishes nurses who write NCLEX questions: They're trained to write questions in a style particular to the NCLEX.

If you've completed an accredited nursing program, you've already taken numerous tests written by nurses with backgrounds and experiences similar to those who write for the NCLEX. The test-taking experience you've gained will help you pass the NCLEX. So your NCLEX review should be *just* that—a review.

The point of it all

The NCLEX is designed for one purpose — to determine whether it's appropriate for you to receive a license to practice as a nurse. By passing the NCLEX, you demonstrate that you possess the minimum level of knowledge necessary to practice nursing safely.

An integrated exam

In nursing school, you probably took courses organized by the medical model. Courses were separated into such subjects as medical-surgical, pediatrics, and psychiatric nursing. By contrast, the NCLEX is integrated, meaning that different subjects are mixed together.

As you answer NCLEX questions, you may encounter patients at any stage of life, from neonatal to geriatric. These patients — clients in NCLEX lingo — may be of any background, may be completely well or extremely ill, and may have various disorders.

Believe me, you're more ready for the NCLEX than you probably think.

What you need to know about client needs

The NCLEX draws questions from four categories of client needs that were developed by the National Council of State Boards of Nursing, the organization that sponsors and manages the NCLEX. Client needs categories ensure that many topics appear on every NCLEX.

The National Council of State Boards of Nursing developed client needs categories after conducting a work-study analysis of new nurses. All aspects of nursing care observed in the study were broken down into categories. Categories were broken down further into subcategories. (See *Client needs categories*.)

The categories and subcategories are used to develop the NCLEX test plan, the content guidelines for the distribution of test questions. Question writers and the people who put the NCLEX together use the test plan and client needs categories to make sure that a full spectrum of nursing activities are covered. Client needs categories appear in most NCLEX review and question-and-answer books, including this one. The truth is, however, that as a test-taker you don't have to concern yourself with client needs categories. You'll see those categories for each question and answer in this book, but they'll be invisible on the actual NCLEX.

Testing by computer

The NCLEX, like many standardized tests today, is administered by computer. That means you won't be filling in empty circles, sharpening pencils, or erasing frantically. It also means that you must become familiar with computer tests if you aren't already.

Now I get it!

Client needs categories

The NCLEX assigns each question a certain category based on client needs. This chart lists client needs categories and subcategories and the percentages of each type of question that appear on the NCLEX.

Category	Subcategories	Percentage of NCLEX questions
Safe, effective care environment	Coordinated care	6% to 12%
	Safety and infection control	7% to 13%
Health promotion and maintenance	Growth and development through the life span	4% to 10%
	Prevention and early detection of disease	4% to 10%
Psychosocial integrity	Coping and adaptation	6% to 12%
	Psychosocial adaptation	4% to 10%
Physiological integrity	Basic care and comfort	10% to 16%
	Pharmacological therapies	5% to 11%
	Reduction of risk potential	11% to 17%
	Physiological adaptation	13% to 19%

Fortunately, the skills required to take the NCLEX on a computer are simple enough to allow you to focus on the questions, not the keyboard.

Eek, a mouse!

You'll use a mouse to select your answer to each question. A tutorial will tell you how to choose and register your answer. You'll have a chance to practice using the mouse on three sample questions before you begin the test. (See *A sample NCLEX question,* page 6.)

Onscreen calculations

During the test you may use an onscreen calculator to calculate medication dosages. You'll be shown how to use the calculator during the tutorial before the test begins.

There's no going back

When you take the NCLEX, the computer will show you only one question at a time on the screen. Take a reasonable amount of time to answer each question. After your answer is recorded, you

---

A sample NCLEX question

The NCLEX, given on a computer, provides a question and four options. Many questions present a clinical scenario. Here's an example of a typical NCLEX question.

29. Which of the following is a characteristic sign of systemic lupus erythematosus (SLE)?

1. Butterfly rash
2. Watery eyes
3. Diplopia
4. Ptosis

(Correct answer: 1. Butterfly rash)

---

can't go back to review the question or to change your answer. You can't skip a question either— they must all be answered.

> I react to you!

Computer-adaptive testing

The NCLEX is a computer-adaptive test, meaning that the computer reacts to the answers you give, supplying more difficult questions if you answer correctly and slightly easier questions if you answer incorrectly. Each test is thus uniquely adapted to the individual test-taker.

The goal of computer-adaptive testing is to find the point at which each student answers 50% of the questions correctly. Imagine the questions all lined up, from easiest to most difficult. At some point you would be answering more questions incorrectly than correctly. This is where you reach your 50% mark or competency level. When you've answered the minimum number of questions, the computer compares your competency level to the standard required for passing. If your level is above the passing standard, you pass the test.

A matter of time

You have a great deal of flexibility with the time you can spend on individual questions. The examination lasts a maximum of 5 hours, however, so don't waste time. If you fail to answer a set number of questions within 5 hours, the computer may determine that you lack minimum competency.

The 5-hour period includes the mouse tutorial, three sample test questions, and rest periods. There's a mandatory 10-minute rest period at the end of 2 hours and an optional 10-minute rest period after 3½ hours.

Most students have plenty of time to complete the test, so take as much time as you need to get the question right without wasting time. Keep moving at a decent pace to help you maintain concentration. A good rule-of-thumb is to answer one question per minute.

Difficult items = Good news

If you find as you progress through the test that the questions seem to be more difficult, it's a good sign. The more questions you answer correctly, the more difficult the questions become.

Some students, however — knowing that questions get progressively harder — focus on the degree of difficulty of subsequent questions to try to figure out if they're answering questions correctly. Avoid the temptation to do this. Stay focused on selecting the best answer for each question put before you.

Finished

The computer test finishes when one of the these events occurs:
• You demonstrate minimum competency, according to the computer program.
• You demonstrate a lack of minimum competency, according to the computer program.
• You've answered the maximum number of questions (265 total questions).
• You've used the maximum time allowed (5 hours).

I'm unique, and so is my NCLEX test.

Answering NCLEX questions

NCLEX questions are commonly fairly long. As a result, it's easy to become overloaded with information. (See *Anatomy of an NCLEX question,* page 8.) To focus on the question and avoid becoming overwhelmed, apply proven strategies for answering NCLEX questions, including:
• determining what the question asks
• determining relevant facts about the client
• rephrasing the question
• choosing the best option.

Can't figure out what the question is asking? Knock it down into smaller pieces.

What's the question asking?

Read the question twice. If the answer isn't apparent, rephrase the question in simpler, more personal terms. Breaking down the question into easier, less intimidating phrases may help you to focus more accurately on the correct answer.

Now I get it!

Anatomy of an NCLEX question

When taking the NCLEX, you may note that the structure and tone of questions is repetitive. That's no accident. NCLEX questions are constructed according to strict standards. As shown below, each question has a stem and four options: a key (correct answer) and three distractors (incorrect answers). A brief case study may precede the question.

Case study

A client with Alzheimer's disease has a nursing diagnosis of *Risk for injury related to memory loss, wandering, and disorientation.*

Stem	Options
Which nursing intervention should appear in the client's care plan to prevent injury?	1. Remove hazards from the environment. (key)
	2. Provide many detailed instructions to the client. (distractor)
	3. Keep the client sedated whenever possible. (distractor)
	4. Use restraints at all times. (distractor)

Sample question

For example, a question might be, "A 20-year-old female with cystic fibrosis has a small bowel obstruction. She's admitted to the medical-surgical unit for treatment, which involves placement of an intestinal tube connected to intermittent suction. Which nursing intervention would be most effective for this client?"

The options for this question — each numbered from 1 to 4 — might include:
1. Record intake and output accurately.
2. Turn the client from side to side, as prescribed.
3. Give the client sips of water to facilitate passage of the tube through the bowel.
4. Add antacids to the intestinal tube to reduce bowel reaction.

Hocus, focus on the question

Read the question again, ignoring all details except what's being asked. Focus on the last line of the question. It asks you to select the *most* effective nursing intervention for this client.

What facts about the client are relevant?

Next, sort out the relevant client information. Start by asking whether the information provided about the client is *not* relevant. For instance, do you need to know that the client has been admitted to the medical-surgical unit? Probably not; her care plan won't be affected by her location in the hospital.

Determine what you *do* know about the client. In the example, you know that:
- she's a 20-year-old female
- she has a small bowel obstruction
- she has cystic fibrosis (a fact that may be relevant).

Rephrase the question

After you've determined relevant information about the client and the question being asked, consider rephrasing the question to make it more clear. Eliminate jargon, and put the question in simpler, more personal terms. Here's how you might rephrase the question in the example: "My client has a small bowel obstruction. She requires placement of an intestinal tube, which will be connected to intermittent suction. She's 20 years old and has a history of cystic fibrosis. Which nursing intervention would be most effective for this client?"

Choose the best option

Armed with all the information you now have, it's time to select an option. You know that the client will have an intestinal tube placed and connected to intermittent suction. You know that while the tube is being advanced into the intestine, the client will lie quietly on her right side for about 2 hours to promote the tube's passage, eliminating option 2. You also know that nothing but normal saline solution should be instilled into the intestinal tube, a fact that eliminates option 4. In addition, you know it isn't likely that the client will be permitted anything by mouth, eliminating option 3. By process of elimination, option 1, *Record intake and output accurately,* is the best option because monitoring fluid balance is the *most* effective nursing intervention for this client.

Hmmm, what happens when two options seem correct?

When more than one option seems correct

What if you encounter a question for which two or more options appear correct? You're most likely facing a common NCLEX question, one for which there may be two or more appropriate options. NCLEX questions commonly include phrases, such as:
- most appropriate

- best
- first
- last
- next
- most helpful
- most suitable.

   Such questions ask you to determine priority. Determining priority means deciding which answer is best, most appropriate, or should be implemented first.

Key strategies

Regardless of the type of question or the number of seemingly correct options, four key strategies will help you determine the correct answer for each question. (See *Strategies for success.*) These strategies are:
- considering the nursing process
- referring to Maslow's hierarchy of needs
- reviewing patient safety
- reflecting on principles of therapeutic communication.

Nursing process

One of the ways to determine which answer takes priority is to apply the nursing process. Steps in the nursing process include:
- data collection
- planning
- implementation
- evaluation.

First things first

The nursing process may provide insights to help you analyze a question and eliminate incorrect options. According to the nursing process, data collection comes before planning, which comes before implementation, which comes before evaluation.

   You're halfway to the correct answer when you encounter a question that asks you to assess the situation and then provides two assessment options and two implementation options. You can immediately eliminate the implementation options, which then gives you — at worst — a 50-50 chance of selecting the correct answer. Use the following sample question to apply the nursing process:

A client returns from an endoscopic procedure during which he was sedated. Before offering the client food, which of the following actions should the nurse take?

Put that ol' nursing process to work for you to answer my questions.

Advice from the experts

Strategies for success

Keeping a few main strategies in mind as you answer each NCLEX question can help ensure greater success. These four strategies are critical for answering NCLEX questions correctly:
• If the question asks what you should do in a situation, use the nursing process to determine which step in the process would be next.
• If the question asks what the client needs, use Maslow's hierarchy to determine which need to address first.
• If the question indicates that the client doesn't have an urgent physiologic need, focus on the patient's safety.
• If the question involves communicating with a patient, use the principles of therapeutic communication.

1. Monitor the client's respiratory status.
2. Check the client's gag reflex.
3. Place the client in a side-lying position.
4. Have the client drink a few sips of water.

Assess before intervening

According to the nursing process, the nurse must collect data from a client before performing an intervention. Does the question indicate that you have enough data regarding the client? No, it doesn't. Therefore, you can eliminate options 3 and 4 because they're both interventions.

That leaves options 1 and 2, both of which involve collecting data. Your nursing knowledge should tell you the correct answer— in this case, option 2. The sedation required for an endoscopic procedure may impair the client's gag reflex, so you should assess the gag reflex before giving food to the client to reduce the risk of aspiration and airway obstruction.

Final elimination

Why not select option 1, monitor the client's respiratory status? You might select this option, but the question is specifically asking about offering the client food, an action that wouldn't be taken if the client's respiratory status was at all compromised. In this case, you're making a judgment based on the phrase, "Before offering the client food." If the question was trying to test your knowledge of respiratory depression following an endoscopic procedure, it probably wouldn't mention a function — such as giving food to a client — that so clearly occurs only after the client's respiratory status has been stabilized.

Maslow's hierarchy

Knowledge of Maslow's hierarchy of needs can be a vital tool for establishing priorities on the NCLEX. Maslow's theory states that physiologic needs are the most basic human needs of all. Only after physiologic needs have been met can safety concerns be addressed. Only after safety concerns are met can concerns involving love and belonging be addressed, and so forth. (See *Maslow's hierarchy of needs.*) Apply the principles of Maslow's hierarchy of needs to the following sample question:

A client complains of severe pain 2 days after surgery. Which of the following actions should the nurse perform first?
1. Offer reassurance to the client that he will feel less pain tomorrow.
2. Allow the client time to verbalize his feelings.
3. Check the client's vital signs.
4. Administer an analgesic.

Maslow? Who's this Maslow? Oh, you mean that Maslow.

Phys before psych

In this example, two of the options — 3 and 4 — address physiologic needs. The other two options — 1 and 2 — address psychosocial concerns. According to Maslow, physiological needs must be met before psychosocial needs, so you can eliminate options 1 and 2.

Final elimination

Now use your nursing knowledge to choose the best answer from the two remaining options. In this case, 3 is correct because the client's vital signs should be checked before administering an analgesic (data collection before intervention). When prioritizing according to Maslow's hierarchy, remember your ABCs — airway, breathing, circulation — to help you further prioritize. Check for a patent airway before addressing breathing. Check breathing before checking the health of the cardiovascular system.

One caveat...

Just because an option appears on the NCLEX doesn't mean it's a viable choice for the client referred to in the question. Always examine your choice in light of your knowledge and experience. Ask yourself, "Does this choice make sense for this client?" Allow yourself to eliminate choices — even ones that might normally take priority — if they don't make sense for a particular client's situation.

Advice from the experts

Maslow's hierarchy of needs

Maslow's hierarchy of needs is a vital tool for establishing priorities on the NCLEX. These illustrations show Maslow's hierarchy of needs and a definition for each stage in the hierarchy. The stages, from most basic to most complex, are physiologic needs, safety and security, love and belonging, self-esteem, and self-actualization.

Self-actualization
Recognition and realization of one's
potential growth, health, and autonomy

Self-esteem
Sense of self-worth, self-respect, independence,
dignity, privacy, and self-reliance

Love and belonging
Affiliation, affection, intimacy, support,
and reassurance

Safety and security
Safety from physiologic and psychological threat,
protection, continuity, stability, and lack of danger

Physiologic needs
Oxygen, food, elimination, temperature
control, sex, movement, rest, and comfort

Client safety

As you might expect, client safety takes high priority on the NCLEX. You'll encounter many questions that can be answered by asking yourself, "Which answer will best ensure the safety of this client?" Use client safety criteria for situations involving laboratory values, drug administration, or nursing care procedures.

> Remember, client safety takes high priority on the NCLEX.

Client 1st, equipment 2nd

You may encounter a question in which some options address the client and others address the equipment. When in doubt, select an option relating to the client; never place equipment before a client.

For instance, suppose a question asks what the nurse should do first when entering a client's room where an infusion pump alarm is sounding. If two options deal with the infusion pump, one with the infusion tubing, and another with the client's catheter insertion site, select the one relating to the client's catheter insertion site. Always check the client first; the equipment can wait.

Therapeutic communication

Some NCLEX questions focus on the nurse's ability to communicate effectively with the client. Therapeutic communication incorporates verbal or nonverbal responses and involves:
- listening to the client
- understanding the client's needs
- promoting clarification and insight about the client's condition.

Poor therapeutic communication

Like other NCLEX questions, those dealing with therapeutic communication commonly require choosing the best response from among the four options. First, eliminate options that indicate the use of poor therapeutic communication techniques, such as those in which the nurse:
- tells the client what to do without regard to the client's feelings or desires (the "do this" response)
- asks a question that can be answered "yes" or "no" or with another one-syllable response
- seeks reasons for the client's behavior
- implies disapproval of the client's behavior
- offers false reassurances

- attempts to interpret the client's behavior rather than allowing the client to express his feelings
- offers a response that focuses on the nurse, not the client.

Good therapeutic communication

When answering NCLEX questions, look for responses that:
- allow the client time to think and reflect
- encourage the client to talk
- encourage the client to describe a particular experience
- reflect that the nurse has listened to the client, such as through paraphrasing the client's response.

The best therapeutic communication responds to the client's feelings and provides accurate information.

Avoiding pitfalls

Even the most knowledgeable students can get tripped up on certain NCLEX questions. (See *Handling further teaching questions,* page 16.) Students commonly cite two areas that can be difficult for unwary test takers:

 knowing the difference between the NCLEX and the "real world"

knowing laboratory values.

NCLEX versus the real world

Some students who take the NCLEX have extensive practical experience in health care. For example, many test takers have worked as nurse's assistants. In one of those capacities, test takers might have been exposed to less than optimum clinical practice and may carry those experiences over to the NCLEX.

However, the NCLEX is a textbook examination — not a test of clinical skills. Take the NCLEX with the understanding that what happens in the real world may differ from what the NCLEX and your nursing school say *should* happen.

Remember, this is the real world. The NCLEX may not always reflect what happens in it.

Don't take shortcuts

If you've had practical experience in health care, you may know a quicker way to perform a procedure or tricks to get by when you don't have the right equipment. Situations such as staff shortages may force you to improvise. On the NCLEX, such scenarios can lead to trouble. Always check your practical experiences against textbook nursing care, taking care to select the response that follows the textbook.

Advice from the experts

Handling further teaching questions

The NCLEX occasionally asks a particular kind of question called the "further teaching" question, which involves client-teaching situations. These questions can be tricky. You'll have to choose the response that suggests that the client has *not* learned the correct information. Here's an example:

37. During a routine prenatal visit, the nurse teaches the client how to relieve her constipation. Which statement by the client indicates that she requires further teaching?
  1. "I'll decrease my intake of green leafy vegetables."
  2. "I'll limit daily fluid intake to four 8-oz glasses."
  3. "I'll increase my intake of unrefined grains."
  4. "I'll take iron supplements regularly."

The answer you should choose is 2 because it indicates the client has a poor understanding of what she needs to include or increase in her diet and what to avoid. *Remember:* If you see the phrase *further teaching* or *further instruction,* you're looking for a wrong answer by the client.

Calling in reinforcements

Deciding when to notify a physician, social worker, or other hospital staff comprises an important element of nursing care. On the NCLEX, however, choices that involve notifying the physician are usually incorrect. Remember that the NCLEX wants to see you, the nurse, at work.

If you're sure the correct answer is to notify the physician, however, make sure the client's safety has been addressed before notifying a physician or other staff member. On the NCLEX, the client's safety has a higher priority than notifying other health care providers.

Knowing laboratory values

Some NCLEX questions supply laboratory results without indicating normal levels. As a result, answering questions involving laboratory values requires you to have the normal range of the most common laboratory values memorized to make an informed decision.

Quick quiz

1. To register your answer for an NCLEX question, you'll press:
- A. the *ENTER* key.
- B. touch the appropriate answer on the screen.
- C. click the mouse.

Answer: C. After deciding the option you want and double-checking to make certain that option is the one you want, click the mouse to register the answer.

2. Because the NCLEX is a computer-adaptive test, the computer provides questions according to:
- A. the number of questions you answer correctly.
- B. a random selection of questions in specific client needs categories.
- C. a strict pattern established after studying the responses of new nurses.

Answer: A. The NCLEX is a computer-adaptive test, meaning that the computer reacts to the answers you give, supplying more difficult questions if you answer correctly and slightly easier questions if you answer incorrectly.

3. When answering an NCLEX question that focuses on the nurse's ability to communicate with the client, which type of communication is involved?
- A. Verbal
- B. Therapeutic
- C. Nontherapeutic

Answer: B. Therapeutic communication incorporates verbal and nonverbal responses and involves listening to the client.

4. When answering an NCLEX question, which client priority should the nurse take care of first?
- A. Notifying the physician
- B. Addressing the client's safety
- C. Calling the social worker

Answer: B. Make sure the client's safety has been addressed before notifying the physician or another staff member. On the NCLEX, the client's safety has a higher priority than notifying other health care providers.

Scoring

☆☆☆   If you answered all four questions correctly, wowsa, wowsa, *wowsa!* You're ready for the NCLEX!

☆☆   If you answered two to three questions correctly, super! You've got a handle on the NCLEX!

☆   If you answered fewer than two questions correctly, take it easy, kiddo! By the time you finish working with this book, you'll be so ready for the NCLEX that you won't be able to *stand* it. *That's* how ready you'll be. Promise.

Chapter 2
Passing the NCLEX

Just the facts

In this chapter, you'll learn:

♦ how to properly prepare for the NCLEX

♦ how to concentrate during difficult study times

♦ ways to make more effective use of your time

♦ why creative studying strategies can enhance learning

♦ how to get the most out of NCLEX practice tests.

Study preparations

If you're like most people preparing to take a test, you're probably feeling nervous, anxious, or concerned. Keep in mind that most test-takers pass the NCLEX the first time around.

Passing the test won't happen by accident, though; you'll need to prepare carefully and efficiently. To help jump-start your preparations:

• determine your strengths and weaknesses
• create a study schedule
• set realistic goals
• find an effective study space
• think positively
• start studying sooner rather than later.

Most students pass the NCLEX on the first try. I know you can pass it too!

Strengths and weaknesses

Most students recognize that, even at the end of their nursing studies, they know more about some topics than others. Because the NCLEX covers a broad range of material, you should make some decisions about how intensively you'll review each topic.

Make a list

Base those decisions on a list. Divide a sheet of paper in half vertically. On one side, list topics you think you know well. On the other side, list topics you feel less secure about. Pay no attention if one side is longer than the other. When you're done studying, you'll feel strong in every area.

Where the list comes from

To make sure your list reflects a comprehensive view of all the areas you studied in school, look at the contents page in the front of this book. For each topic listed, place it in the "know well" column or "needs review" column. Separating content areas this way shows immediately which topics need less study time and which need more time.

> Because I'm strong in certain areas, I'll spend less time studying them.

Scheduling study time

Study when you're most alert. Most people can identify a period of the day when they feel most alert. If you feel most alert and energized in the morning, for example, set aside sections of time in the morning for topics that need a lot of review. Then you can use the evening, a time of lesser alertness, for topics that need some refreshing. The opposite is true as well; if you're more alert in the evening, study difficult topics at that time.

> Because I'm more alert in the morning, that's when I'll cover difficult topics.

What you'll do, when

Set up a basic schedule for studying. Using a calendar or organizer, determine how much time remains before you'll take the NCLEX. (See *2 to 3 months before the NCLEX*.) Fill in the remaining days with specific times and topics to be studied. For example, you might schedule the respiratory system on a Tuesday morning and the GI system that afternoon. Remember to schedule difficult topics during your most alert times.

Keep in mind that you shouldn't fill each day with studying. Be realistic and set aside time for normal activities. Try to create ample study time before the NCLEX and then stick to the schedule.

Set goals you can meet

Part of creating a schedule means setting goals you can accomplish. You no doubt studied a great deal in nursing school, and by now you have a sense of your own capabilities. Ask yourself, "How much can I cover in a day?" Set that amount of time aside and then stay on task. You'll feel better about yourself — and your chances of passing the NCLEX — when you meet your goals regularly.

To-do list

2 to 3 months before the NCLEX

With 2 to 3 months remaining before you plan to take the examination, take these steps:
• Establish a study schedule. Set aside ample time to study, but also leave time for social activities, exercise, family or personal responsibilities, and other matters.
• Become knowledgeable about the NCLEX, its content, the types of questions it asks, and the testing format.
• Begin studying your notes, textbooks, and other study materials.
• Take some NCLEX practice questions to help you discover strengths and weaknesses as well as to become familiar with NCLEX-style questions.

I know how much I can study in one day, so I've made up my schedule taking that into account.

Study space

Find a space conducive to effective learning and then study there. Whatever you do, don't study with a television on in the room. Instead, find a quiet, inviting study space that:
• is located in a convenient place, away from normal traffic patterns
• contains a solid chair that encourages good posture (Avoid studying in bed; you'll be more likely to fall asleep and not accomplish your goals.)
• uses comfortable, soft lighting with which you can see clearly without eye strain
• has a temperature between 65° and 70° F
• contains flowers or green plants, familiar photos or paintings, and easy access to soft, instrumental background music.

Accentuate the positive

Consider taping positive messages around your study space. Make signs with words of encouragement, such as, "You can do it!" "Keep studying!" and "Remember the goal!" These upbeat messages can help keep you going when your attention begins to waver.

Feel free to hang this upbeat message in your study area.

Maintaining concentration

When you're faced with reviewing the amount of information covered by the NCLEX, it's easy to become distracted and lose your concentration. When you lose concentration, you make less effective use of valuable study time. To help stay focused, keep these tips in mind:

• Alternate the order of the subjects you study during the day to add variety. Try alternating between topics you find most interesting and those you find least interesting.

• Approach your studying with enthusiasm, sincerity, and determination.

• After you've decided to study, begin immediately. Don't let anything interfere with your thought processes after you've begun.

• Concentrate on accomplishing one task at a time, to the exclusion of everything else.

• Don't try to do two things at once, such as studying and watching television or conversing with friends.

• Work continuously without interruption for a while, but don't study for such a long period that the whole experience becomes grueling or boring.

• Allow time for periodic breaks to give yourself a change of pace. Use these breaks to ease your transition into studying a new topic.

• When studying in the evening, wind down from your studies slowly. Don't progress directly from studying to sleeping.

A short jog now will help me to concentrate later.

Taking care of yourself

Never neglect your physical and mental well-being in favor of longer study hours. Maintaining physical and mental health are critical for success in taking the NCLEX. (See *4 to 6 weeks before the NCLEX,* page 24.)

A few simple rules

You can increase your likelihood of passing the test by following these simple health rules:

• Get plenty of rest. You can't think deeply or concentrate for long periods when you're tired.

• Eat nutritious meals. Maintaining your energy level is impossible when you're undernourished.

• Exercise regularly. Regular exercise helps you work harder and think more clearly. As a result, you'll study more efficiently and increase the likelihood of success on the all-important NCLEX.

Memory powers, activate!

If you're having trouble concentrating, but would rather push through than take a break, try making your studying more active by reading out loud. Active studying can renew your powers of concentration. By reading review material out loud to yourself, you're engaging your ears as well as your eyes — and making your studying a more active process. Hearing the material out loud also fosters memory and subsequent recall.

You can also rewrite in your own words a few of the more difficult concepts you're reviewing. Explaining these concepts in writing forces you to think through the material and can jumpstart your memory.

Study schedule

When you were creating your schedule, you might have asked yourself, "How long should I study? One hour at a stretch? Two hours? Three?" To make the best use of your study time, you'll need to answer those questions.

Optimum study time

Experts are divided about the optimum length of study time. Some say you should study no more than an hour at a time several times a day. Their reasoning: You remember the material you study at the beginning and end of a session best and tend to remember less material studied in the middle of the session.

Other experts say you should hold longer study sessions because you lose time in the beginning, when you're just getting warmed up, and again at the end, when you're cooling down.

Therefore, say those experts, a long, concentrated study period will allow you to cover more material.

To thine own self be true

So what's the answer? It doesn't matter as long as you determine what's best for *you*. At the beginning of your NCLEX study schedule, try study periods of varying lengths. Pay close attention to those that seem more successful.

Remember that you're a trained nurse who's competent at assessment. Think of yourself as a client, and assess your own progress. Then implement the strategy that works best for you.

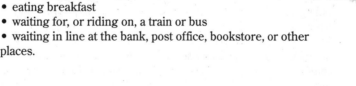

I've found that hour-and-a-half sessions work best for me.

Finding time to study

So does that mean that short sections of time are useless? Not at all. We all have spaces in our day that might otherwise be dead time. (See *1 week before the NCLEX*.) These are perfect times to review for the NCLEX, but not to cover new material because by the time you get deep into new material, your time will be over. Always keep some flash cards or a small notebook handy for situations when you have a few extra minutes.

You'll be amazed how many short sessions you can find in a day and how much reviewing you can do in 5 minutes. The following places offer short stretches of time you can use:
• eating breakfast
• waiting for, or riding on, a train or bus
• waiting in line at the bank, post office, bookstore, or other places.

To-do list

1 week before the NCLEX

With 1 week remaining before the NCLEX, take these steps:
• Take a review test to measure your progress.
• Record key ideas and principles on flash cards or audiotapes.
• Rest, eat well, and avoid thinking about the examination during nonstudy times.
• Treat yourself to one special event. You've been working hard, and you deserve it!

Creative studying

Even when you study in a perfect study space and concentrate better than ever, studying for the NCLEX can get a little, well, dull. Even people with terrific study habits occasionally feel bored or sluggish. That's why it's important to have some creative tricks in your study bag to liven up your studying during those down times.

Creative studying doesn't have to be hard work. It involves making efforts to alter your study habits a bit. Some techniques that might help include studying with a partner or group and creating flash cards or other audiovisual study tools.

Study partners

Studying with a partner or group of students can be an excellent way to energize your studying. Working with a partner allows you to test each other on the material you've reviewed. Your partner can give you encouragement and motivation. Perhaps most important, working with a partner can provide a welcome break from solitary studying.

I'll take the one in the red hat

Exercise some care when choosing a study partner or assembling a study group. A partner who doesn't fit your needs won't help you make the most of your study time. Look for a partner who:

• possesses similar goals to yours. For example, someone taking the NCLEX at approximately the same date who feels the same sense of urgency as you do might make an excellent partner.
• possesses about the same level of knowledge as you. Tutoring someone can sometimes help you learn, but partnering should be give-and-take so both partners can gain knowledge.
• can study without excess chatting or interruptions. Socializing is an important part of creative study, but remember, you've still got to pass the NCLEX — so stay serious!

Find a partner with similar goals and strengths. But remember to stay focused.

Audiovisual tools

Using flash cards and other audiovisual tools foster retention and make learning and reviewing fun.

Flash Gordon? No, it's Flash Card!

Flash cards can provide you with an excellent study tool. The process of writing material on a flash card will help you remem-

To-do list

The day before the NCLEX

With 1 day before the NCLEX, take these steps:
• Drive to the test site, review traffic patterns, and find out where to park. If your route to the test site occurs during heavy traffic or if you're expecting bad weather, set aside extra time to ensure prompt arrival.
• Do something relaxing during the day.
• Avoid concentrating on the test.
• Rest, eat well, and avoid dwelling on the NCLEX during nonstudy periods.
• Call a supportive friend or relative for some last-minute words of encouragement.

ber it. In addition, flash cards are small and easily portable, perfect for those 5-minute slivers of time that show up during the day.

Creating a flash card should be fun. Use magic markers, highlighters, and other colorful tools to make them visually stimulating. The more effort you put into creating your flash cards, the better you'll remember the material contained on the cards.

Other visual tools

Flowcharts, drawings, diagrams, and other image-oriented study aids can also help you learn material more effectively. Substituting images for text can be a great way to give your eyes a break and recharge your brain. Remember to use vivid colors to make your creations visually engaging.

Hear's the thing

If you learn more effectively when you hear information rather than see it, consider recording key ideas using a handheld tape recorder. Recording information helps promote memory because you say the information aloud when taping and then listen to it when playing it back. Like flash cards, tapes are portable and perfect for those short study periods during the day. (See *The day before the NCLEX*.)

Charts, drawings, and diagrams help me learn more efficiently.

Practice tests

Practice questions should comprise an important part of your NCLEX study strategy. Practice questions can improve your studying by helping you review material and familiarizing you with the exact style of questions you'll encounter on the NCLEX.

To-do list

The day of the NCLEX

On the day of the NCLEX, take these steps:
- Get up early.
- Wear comfortable clothes, preferably with layers you can adjust to the room temperature.
- Leave your house early.
- Arrive at the test site early.
- Avoid looking at your notes as you wait for your test computer.
- Listen carefully to the instructions given before entering the test room.
- Succeed, succeed, *succeed!*

Because I've taken lots of practice tests, I understand how the questions work.

Practice at the beginning

Consider working through some practice questions as soon as you begin studying for the NCLEX. For example, you might try a half-dozen questions from each chapter in this book.

If you score well, you probably know the material contained in that chapter fairly well and can spend less time reviewing that particular topic. If you have trouble with the questions, spend extra study time on that topic.

I'm getting there

Practice questions can also provide an excellent means of marking your progress. Don't worry if you have trouble answering the first few practice questions you take; you'll need time to adjust to the way the questions are asked. Eventually you'll become accustomed to the question format and begin to focus more on the questions themselves.

If you make practice questions a regular part of your study regimen, you'll be able to notice areas in which you're improving. You can then adjust your study time accordingly.

Practice makes perfect

As you near the examination date, continue to do practice questions, but also set aside time to take an entire NCLEX practice test. (We've included four at the back of this book.) That way, you'll know exactly what to expect. (See *The day of the NCLEX.*) The more you know ahead of time, the better you're likely to do on the NCLEX.

Taking an entire practice test is also a way to gauge your progress. When you find yourself answer-

The test is getting awfully close. I'm not nervous. I look nervous? I'm NOT!

ing questions correctly, it will give you the confidence you need to conquer the NCLEX for real.

Quick quiz

1. The best time to study is:
   A. in the morning.
   B. early in the evening.
   C. when you feel most alert.

Answer: C. Study when you're most alert. If you feel most alert and energized in the morning, for example, set aside sections of time in the morning for topics that need a lot of review.

2. The temperature of the ideal study area should be between:
   A. 60° and 65° F.
   B. 65° and 70° F.
   C. 70° and 75° F.

Answer: B. The ideal study area has a temperature between 65° and 70° F.

3. To help you maintain concentration during long study periods, recommended study strategies include:
   A. studying the topics you find most interesting first, followed by the topics you find least interesting.
   B. studying the topics you find least interesting first, followed by the topics you find most interesting.
   C. alternating the order of the subjects you study during the day.

Answer: C. Alternating the order of the subjects you study during the day adds variety to your study, helps you remain focused, and makes the most of your study time.

4. When selecting a study partner, choose one who:
   A. possesses similar goals as you.
   B. is highly social and will keep you entertained.
   C. isn't as knowledgeable as you so you can tutor him.

Answer: A. A partner who doesn't fit your needs won't help you make the most of your study time. Look for a partner who possesses similar goals to yours, possesses about the same level of knowledge as you, and won't spend too much time socializing.

Scoring

☆☆☆   If you answered all four questions correctly, that's *it!* We're calling the NCLEX testing center right now. You're *ready!*

☆☆   If you answered three questions correctly, outstanding! All you need now is a snack before the test, and you'll be rarin' to go!

☆   If you answered fewer than three questions correctly, fear not. You've got *NCLEX success* written all over your future!

Part II Care of the adult

3 Cardiovascular disorders   33

4 Hematologic & immune disorders   63

5 Respiratory disorders   83

6 Neurosensory disorders   113

7 Musculoskeletal disorders   143

8 Gastrointestinal disorders   165

9 Endocrine disorders   187

10 Genitourinary disorders   211

11 Integumentary disorders   235

Chapter 3
Cardiovascular disorders

1. Which of the following arteries primarily feeds the anterior wall of the heart?
1. Circumflex artery
2. Internal mammary artery
3. Left anterior descending artery
4. Right coronary artery

Different arteries supply my different sections with the blood I need to stay healthy.

2. When do coronary arteries primarily receive blood flow?
1. During cardiac standstill
2. During diastole
3. During expiration
4. During systole

3. Which of the following illnesses is the leading cause of death in the United States?
1. Cancer
2. Coronary artery disease
3. Liver failure
4. Renal failure

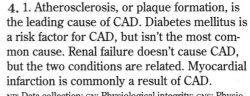

Stop and think! Although all of the answers relate to CAD, only one answer is correct.

4. Which of the following conditions most commonly results in coronary artery disease (CAD)?
1. Atherosclerosis
2. Diabetes mellitus
3. Myocardial infarction
4. Renal failure

5. Atherosclerosis impedes coronary blood flow by which of the following mechanisms?
1. Plaques obstruct the vein.
2. Plaques obstruct the artery.
3. Blood clots form outside the vessel wall.
4. Hardened vessels dilate to allow blood to flow through.

1. 3. The left anterior descending artery is the primary source of blood for the anterior wall of the heart. The circumflex artery supplies the lateral wall, the internal mammary artery supplies the mammary, and the right coronary artery supplies the inferior wall of the heart.
NP: Evaluation; CN: Physiological integrity; CNS: Physiological adaptation; CL: Knowledge

2. 2. Although the coronary arteries may receive a minute portion of blood during systole, most of the blood flow to coronary arteries is supplied during diastole. Blood doesn't flow during cardiac standstill. Breathing patterns are irrelevant to blood flow.
NP: Data collection; CN: Physiological integrity; CNS: Physiological adaptation; CL: Knowledge

3. 2. Coronary artery disease accounts for over 50% of all deaths in the United States. Cancer accounts for approximately 20%. Liver failure and renal failure account for less than 10% of all deaths in the United States.
NP: Data collection; CN: Health promotion and maintenance; CNS: Prevention and early detection of disease; CL: Knowledge

4. 1. Atherosclerosis, or plaque formation, is the leading cause of CAD. Diabetes mellitus is a risk factor for CAD, but isn't the most common cause. Renal failure doesn't cause CAD, but the two conditions are related. Myocardial infarction is commonly a result of CAD.
NP: Data collection; CN: Physiological integrity; CNS: Physiological adaptation; CL: Analysis

5. 2. Arteries, not veins, provide coronary blood flow. Atherosclerosis is a direct result of plaque formation in the artery. Hardened vessels can't dilate properly and, therefore, constrict blood flow.
NP: Data collection; CN: Physiological integrity; CNS: Physiological adaptation; CL: Knowledge

NP: Nursing process   CN: Client needs category   CNS: Client needs subcategory   CL: Cognitive level

6. Which of the following risk factors for coronary artery disease cannot be corrected?
  1. Cigarette smoking
  2. Diabetes mellitus
  3. Heredity
  4. Hypertension

Exercise does wonders for me!

6. 3. Because "heredity" refers to our genetic makeup, it can't be changed. Cigarette smoking cessation is a lifestyle change that involves behavior modification. Diabetes mellitus is a risk factor that can be controlled with medication. Altering one's diet, exercise, and medication can correct hypertension.

NP: Evaluation; CN: Physiological integrity; CNS: Reduction of risk potential; CL: Analysis

7. Exceeding which of the following serum cholesterol levels significantly increases the risk of coronary artery disease?
  1. 100 mg/dl
  2. 150 mg/dl
  3. 175 mg/dl
  4. 200 mg/dl

7. 4. Cholesterol levels above 200 mg/dl are considered excessive. They require dietary restriction and, perhaps, medication. The other levels listed are all below the nationally accepted levels for cholesterol and carry a lesser risk for coronary artery disease.

NP: Data collection; CN: Physiological integrity; CNS: Reduction of risk potential; CL: Comprehension

8. Which of the following actions is the first priority of care for a client exhibiting signs and symptoms of coronary artery disease?
  1. Decrease anxiety.
  2. Enhance myocardial oxygenation.
  3. Administer sublingual nitroglycerin.
  4. Educate the client about his symptoms.

Careful! This question is asking you to prioritize!

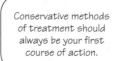

8. 2. Enhancing myocardial oxygenation is always the first priority when a client exhibits signs or symptoms of cardiac compromise. Without adequate oxygen, the myocardium suffers damage. Sublingual nitroglycerin dilates the coronary vessels to increase blood flow, but its administration isn't the first priority. Although educating the client and decreasing anxiety are important in care delivery, neither are priorities when a client is compromised.

NP: Implementation; CN: Physiological integrity; CNS: Physiological adaptation; CL: Application

9. Medical treatment of coronary artery disease includes which of the following procedures?
  1. Cardiac catheterization
  2. Coronary artery bypass surgery
  3. Oral medication administration
  4. Percutaneous transluminal coronary angioplasty

Conservative methods of treatment should always be your first course of action.

9. 3. Oral medication administration is a noninvasive, medical treatment for coronary artery disease. Cardiac catheterization isn't a treatment, but a diagnostic tool. Coronary artery bypass surgery and percutaneous transluminal coronary angioplasty are invasive, surgical treatments.

NP: Evaluation; CN: Physiological integrity; CNS: Physiological adaptation; CL: Analysis

10. Prolonged occlusion of the right coronary artery produces an infarction in which of the following areas of the heart?
  1. Anterior
  2. Apical
  3. Inferior
  4. Lateral

10. 3. The right coronary artery supplies the right ventricle, or the inferior portion of the heart. Therefore, prolonged occlusion could produce an infarction in that area. The right coronary artery doesn't supply the anterior portion (left ventricle), lateral portion (some of the left ventricle and the left atrium), or the apical portion (left ventricle) of the heart.

NP: Knowledge; CN: Physiological integrity; CNS: Physiological adaptation; CL: Knowledge

NP: Nursing process   CN: Client needs category   CNS: Client needs subcategory   CL: Cognitive level

11. Which of the following is the most common symptom of myocardial infarction (MI)?
1. Chest pain
2. Dyspnea
3. Edema
4. Palpitations

I may experience many symptoms during a heart attack, but which one are you most likely to see?

12. Which of the following landmarks is the correct one for obtaining an apical pulse?
1. Left fifth intercostal space, midaxillary line
2. Left fifth intercostal space, midclavicular line
3. Left second intercostal space, midclavicular line
4. Left seventh intercostal space, midclavicular line

13. Which of the following systems is the most likely origin of pain the client describes as knifelike chest pain that increases in intensity with inspiration?
1. Cardiac
2. Gastrointestinal
3. Musculoskeletal
4. Pulmonary

What should be the most appropriate statement here?

14. A male client is hospitalized to rule out an acute myocardial infarction (MI); he has a normal lactate dehydrogenase level and an elevated creatine kinase (CK-MB) level. The nurse enters the client's room and finds him pacing the floor. Which statement by the nurse would be most appropriate in this situation?
1. "You've had a heart attack. Get back in bed."
2. "You seem upset. Why don't you get into bed and, if you wish, we can talk for a while."
3. "You sure you have a lot of energy; do you want to play cards?"
4. "Your physician doesn't want you up. Would you please get back into your bed?"

11. 1. The most common symptom of an MI is chest pain, resulting from deprivation of oxygen to the heart. Dyspnea is the second most common symptom, related to an increase in the metabolic needs of the body during an MI. Edema is a later sign of heart failure, commonly seen after an MI. Palpitations may result from reduced cardiac output, producing arrhythmias.
NP: Data collection; CN: Safe, effective care environment; CNS: Management of care; CL: Analysis

12. 2. The correct landmark for obtaining an apical pulse is the left fifth intercostal space in the midclavicular line. This is the point of maximum impulse and the location of the left ventricle. The left second intercostal space in the midclavicular line is where pulmonic sounds are auscultated. Normally, heart sounds aren't heard in the midaxillary line or the seventh intercostal space in the midclavicular line.
NP: Data collection; CN: Physiological integrity; CNS: Basic care and comfort; CL: Application

13. 4. Pulmonary pain is generally described by these symptoms. Musculoskeletal pain only increases with movement. Cardiac and GI pains don't change with respiration.
NP: Evaluation; CN: Physiological integrity; CNS: Physiological adaptation; CL: Analysis

14. 2. Given the laboratory data, especially the elevated CK-MB level, the nurse should realize that the client probably had an MI and that he needs to lie down and rest his heart. However, the nurse also should realize the need to respond to the client's emotional distress by acknowledging his feelings and offering to discuss the situation. Telling the client that he had a heart attack would be giving a medical diagnosis that hasn't yet been made and would also be practicing outside the scope of nursing. A comment about his energy level acknowledges the client's pacing, but not his underlying concerns. Stating the physician's preferences attempts to impose authority to control the client's behavior. It doesn't acknowledge the client's distress.
NP: Implementing care; CN: Psychosocial integrity; CNS: Coping and adaptation; CL: Application

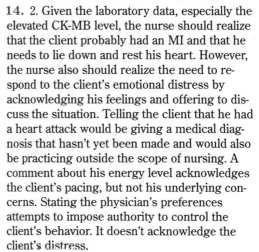

15. Which of the following blood tests is most indicative of cardiac damage?
1. Arterial blood gas (ABG) levels
2. Complete blood count (CBC)
3. Complete chemistry
4. Creatine kinase isoenzymes (CK-MB)

Think of what happens when cardiac tissue is damaged.

16. What's the primary reason for administering morphine to a client with a myocardial infarction?
1. To sedate the client
2. To decrease the client's pain
3. To decrease the client's anxiety
4. To decrease oxygen demand on the client's heart

This question is looking for the most common cause of myocardial infarction.

17. Which of the following conditions is most commonly responsible for myocardial infarction (MI)?
1. Aneurysm
2. Heart failure
3. Coronary artery thrombosis
4. Renal failure

18. What supplemental medication is most frequently ordered in conjunction with furosemide (Lasix)?
1. Chloride
2. Digoxin
3. Potassium
4. Sodium

It's important to know what physiologic changes occur in your client after a heart attack.

19. After a myocardial infarction (MI), serum glucose levels and free fatty acid production increase. What type of physiologic changes are these?
1. Electrophysiologic
2. Hematologic
3. Mechanical
4. Metabolic

15. 4. CK-MB enzymes are present in the blood after a myocardial infarction. These enzymes spill into the plasma when cardiac tissue is damaged. ABG levels are obtained to review respiratory function, a CBC is obtained to review blood counts, and a complete chemistry is obtained to review electrolytes.
NP: Data collection; CN: Health promotion and maintenance; CNS: Prevention and early detection of disease; CL: Analysis

16. 4. Morphine is administered because it decreases myocardial oxygen demand. Morphine will also decrease pain and anxiety while causing sedation, but it isn't primarily given for those reasons.
NP: Implementation; CN: Physiological integrity; CNS: Pharmacological therapies; CL: Application

17. 3. Coronary artery thrombosis causes an occlusion of the artery, leading to myocardial death. An aneurysm is an outpouching of a vessel and doesn't cause an MI. Renal failure can be associated with MI, but isn't a direct cause. Heart failure is usually the result of an MI.
NP: Data collection; CN: Physiological integrity; CNS: Physiological adaptation; CL: Knowledge

18. 3. Supplemental potassium is given with furosemide because of the potassium loss that occurs as a result of this diuretic. Chloride and sodium aren't lost during diuresis. Digoxin acts to increase contractility, but isn't given routinely with furosemide.
NP: Data collection; CN: Physiological integrity; CNS: Pharmacological therapies; CL: Knowledge

19. 4. Glucose and fatty acids are metabolites whose levels increase after an MI. Mechanical changes are those that affect the pumping action of the heart, and electrophysiologic changes affect conduction. Hematologic changes would affect the blood.
NP: Data collection; CN: Physiological integrity; CNS: Physiological adaptation; CL: Knowledge

20. Which of the following complications is indicated by a third heart sound (S_3)?
1. Ventricular dilation
2. Systemic hypertension
3. Aortic valve malfunction
4. Increased atrial contractions

21. After an anterior wall myocardial infarction (MI), which of the following problems is indicated by auscultation of crackles in the lungs?
1. Left-sided heart failure
2. Pulmonic valve malfunction
3. Right-sided heart failure
4. Tricuspid valve malfunction

22. Which of the following diagnostic tools is most commonly used to determine myocardial damage?
1. Cardiac catheterization
2. Cardiac enzymes
3. Echocardiogram
4. Electrocardiogram (ECG)

23. What is the first intervention for a client experiencing myocardial infarction (MI)?
1. Administer morphine.
2. Administer oxygen.
3. Administer sublingual nitroglycerin.
4. Obtain an electrocardiogram (ECG).

Twenty questions done! Good job!

The most common tool is usually the most accurate.

Prioritize!

20. 1. Rapid filling of the ventricle causes vasodilation, which is auscultated as S_3. Increased atrial contraction or systemic hypertension can result in a fourth heart sound. Aortic valve malfunction is heard as a murmur.
NP: Data collection; CN: Health promotion and maintenance; CNS: Prevention and early detection of disease; CL: Analysis

21. 1. The left ventricle is responsible for most of the cardiac output. An anterior wall MI may result in a decrease in left ventricular function. When the left ventricle doesn't function properly, resulting in left-sided heart failure, fluid accumulates in the interstitial and alveolar spaces in the lungs and causes crackles. Pulmonic and tricuspid valve malfunction causes right-sided heart failure.
NP: Data collection; CN: Physiological integrity; CNS: Physiological adaptation; CL: Analysis

22. 4. The ECG is the quickest, most accurate, and most widely used tool to diagnose myocardial infarction (MI). Cardiac enzymes also are used to diagnose MI, but the results can't be obtained as quickly and generally aren't as accurate as an ECG. An echocardiogram is used most widely to view myocardial wall function after an MI has been diagnosed. Cardiac catheterization is an invasive study for determining coronary artery disease that can also be used to treat the disease.
NP: Data collection; CN: Safe, effective care environment; CNS: Management of care; CL: Knowledge

23. 2. Administering supplemental oxygen to the client is the first priority of care. The myocardium is deprived of oxygen during an infarction, so additional oxygen is administered to assist in oxygenation and prevent further damage. Morphine and sublingual nitroglycerin are also used to treat MI, but they're more commonly administered after the oxygen. An ECG is the most common diagnostic tool used to evaluate MI.
NP: Application; CN: Physiological integrity; CNS: Pharmacological therapies; CL: Application

24. What's the most appropriate nursing response to a client with a myocardial infarction (MI) who's fearful of dying?
    1. "Tell me about your feelings right now."
    2. "When the doctor arrives, everything will be fine."
    3. "This is a bad situation, but you'll feel better soon."
    4. "Please be assured we're doing everything we can to make you feel better."

Be sensitive to your client's feelings during an emergency.

24. 1. Validation of a client's feelings is the most appropriate response. It gives the client a feeling of comfort and safety. The other three responses give the client false hope. No one can determine if a client experiencing an MI will feel or get better and, therefore, these responses are inappropriate.
NP: Data collection; CN: Psychosocial integrity; CNS: Coping and adaptation; CL: Comprehension

25. Which of the following classes of medications protects the ischemic myocardium by decreasing catecholamines and sympathetic nerve stimulation?
    1. Beta-adrenergic blockers
    2. Calcium channel blockers
    3. Narcotics
    4. Nitrates

25. 1. Beta-adrenergic blockers work by decreasing catecholamines and sympathetic nerve stimulation. They protect the myocardium, helping to reduce the risk of another infarction by decreasing the heart's workload. Calcium channel blockers reduce workload by decreasing the heart rate. Narcotics reduce myocardial oxygen demand, promote vasodilation, and decrease anxiety. Nitrates reduce myocardial oxygen consumption and decrease the risk of ventricular arrhythmias.
NP: Data collection; CN: Physiological integrity; CNS: Pharmacological therapies; CL: Application

26. What's the most common complication of a myocardial infarction (MI)?
    1. Cardiogenic shock
    2. Heart failure
    3. Arrhythmias
    4. Pericarditis

26. 3. Arrhythmias, caused by oxygen deprivation to the myocardium, are the most common complication of an MI. Cardiogenic shock, another complication of MI, is defined as the end stage of left ventricular dysfunction. The condition occurs in approximately 15% of clients with MI. Because the pumping function of the heart is compromised by an MI, heart failure is the second most common complication. Pericarditis most commonly results from a bacterial or viral infection.
NP: Data collection; CN: Physiological integrity; CNS: Physiological adaptation; CL: Knowledge

This question wasn't asked in vein. (Hah! Get it? "Vein?" I kill myself!)

27. With which of the following disorders is jugular vein distention most prominent?
    1. Abdominal aortic aneurysm
    2. Heart failure
    3. Myocardial infarction (MI)
    4. Pneumothorax

27. 2. Elevated venous pressure, exhibited as jugular vein distention, indicates the heart's failure to pump. This symptom isn't a symptom of abdominal aortic aneurysm or pneumothorax. An MI, if severe enough, can progress to heart failure; however, in and of itself, an MI doesn't cause jugular vein distention.
NP: Data collection; CN: Physiological integrity; CNS: Physiological adaptation; CL: Analysis

28. Which classification of drugs dilate cardiac arteries and decrease afterload?
   1. Diuretics
   2. Antibiotics
   3. Beta-adrenergic blockers
   4. Calcium channel blockers

Keep in mind that digoxin strengthens myocardial contraction.

28. 4. Calcium channel blockers inhibit calcium influx through the coronary arteries and decrease afterload; they are used to treat effort-induced angina. Diuretics are used to treat heart failure. They promote removal of excess water by the kidneys and therefore reduce edema and make breathing easier. Antibiotics are used to fight infection and don't have an effect on cardiac output. Beta-adrenergic blockers, such as propranolol (Inderal), have a negative inotropic action. One of the adverse effects of this action is heart failure.
NP: Data collection; CN: Physiological integrity; CNS: Pharmacological therapies; CL: Comprehension

29. Which of the following parameters should be checked before administering digoxin?
   1. Apical pulse
   2. Blood pressure
   3. Radial pulse
   4. Respiratory rate

29. 1. An apical pulse is essential for accurately assessing the client's heart rate before administering digoxin. The apical pulse is the most accurate pulse point in the body. Blood pressure is usually only affected if the heart rate is too low, in which case the nurse would withhold digoxin. The radial pulse can be affected by cardiac and vascular disease and, therefore, won't always accurately depict the heart rate. Digoxin has no effect on respiratory function.
NP: Data collection; CN: Physiological integrity; CNS: Pharmacological therapies; CL: Knowledge

30. Toxicity from which of the following medications may cause a client to see a green halo around lights?
   1. Digoxin
   2. Furosemide (Lasix)
   3. Metoprolol (Lopressor)
   4. Enalapril (Vasotec)

30. 1. One of the most common signs of digoxin toxicity is the visual disturbance known as the green halo sign. The other medications aren't associated with such an effect.
NP: Evaluation; CN: Physiological integrity; CNS: Pharmacological therapies; CL: Knowledge

31. Which of the following symptoms is most commonly associated with left-sided heart failure?
   1. Crackles
   2. Arrhythmias
   3. Hepatic engorgement
   4. Hypotension

You're doing great! Keep going!

31. 1. Crackles in the lungs are a classic sign of left-sided heart failure. These sounds are caused by fluid backing up into the pulmonary system. Arrhythmias can be associated with right- and left-sided heart failure. Hepatic engorgement is associated with right-sided heart failure. Left-sided heart failure causes hypertension secondary to an increased workload on the system.
NP: Data collection; CN: Physiological integrity; CNS: Physiological adaptation; CL: Knowledge

32. In which of the following disorders would the nurse expect to assess sacral edema in a bedridden client?
1. Diabetes mellitus
2. Pulmonary emboli
3. Renal failure
4. Right-sided heart failure

32. 4. The most accurate area on the body to assess dependent edema in a bedridden client is the sacral area. Sacral, or dependent, edema is secondary to right-sided heart failure. Diabetes mellitus, pulmonary emboli, and renal disease aren't directly linked to sacral edema.
NP: Data collection; CN: Physiological integrity; CNS: Physiological adaptation; CL: Analysis

33. Which of the following symptoms may a client with right-sided heart failure exhibit?
1. Adequate urine output
2. Polyuria
3. Oliguria
4. Polydipsia

33. 3. Inadequate deactivation of aldosterone by the liver after right-sided heart failure leads to fluid retention, which causes oliguria. Adequate urine output, polyuria, and polydipsia aren't associated with right-sided heart failure.
NP: Data collection; CN: Physiological integrity; CNS: Physiological adaptation; CL: Application

34. Which of the following classes of medications maximizes cardiac performance in clients with heart failure by increasing ventricular contractility?
1. Beta-adrenergic blockers
2. Calcium channel blockers
3. Diuretics
4. Inotropic agents

Make sure you understand how these different drugs work to benefit the heart.

34. 4. Inotropic agents are administered to increase the force of the heart's contractions, thereby increasing ventricular contractility and ultimately increasing cardiac output. Beta-adrenergic blockers and calcium channel blockers decrease the heart rate and ultimately decrease the workload of the heart. Diuretics are administered to decrease the overall vascular volume, also decreasing workload.
NP: Data collection; CN: Physiological integrity; CNS: Pharmacological therapies; CL: Comprehension

35. Stimulation of the sympathetic nervous system produces which of the following responses?
1. Bradycardia
2. Tachycardia
3. Hypotension
4. Decreased myocardial contractility

35. 2. Stimulation of the sympathetic nervous system causes tachycardia, or an increase in heart rate. This response causes an increase in contractility, which compensates for the response. The other symptoms listed are related to the parasympathetic nervous system, which is responsible for slowing the heart rate.
NP: Data collection; CN: Physiological integrity; CNS: Physiological adaptation; CL: Knowledge

---

NP: Nursing process   CN: Client needs category   CNS: Client needs subcategory   CL: Cognitive level

36. Which of the following conditions is most closely associated with weight gain, nausea, and a decrease in urine output?
1. Angina pectoris
2. Cardiomyopathy
3. Left-sided heart failure
4. Right-sided heart failure

This question asks you to distinguish between symptoms of left- and right-sided heart failure.

37. What's the most common cause of an abdominal aortic aneurysm?
1. Atherosclerosis
2. Diabetes mellitus
3. Hypertension
4. Syphilis

38. In which of the following areas is an abdominal aortic aneurysm most commonly located?
1. Distal to the iliac arteries
2. Distal to the renal arteries
3. Adjacent to the aortic arch
4. Proximal to the renal arteries

39. A client with pulmonary edema is given digoxin. What is digoxin's most direct and beneficial effect on myocardial contraction in the failing heart?
1. Decreases cardiac output
2. Decreases ventricular emptying capacity
3. Increases circulating blood volume
4. Slows conduction of impulses through the atrioventricular (AV) node

Looking good! Keep at it!

40. What is the most common symptom in a client with an abdominal aortic aneurysm?
1. Abdominal pain
2. Diaphoresis
3. Headache
4. Upper back pain

36. 4. Weight gain, nausea, and a decrease in urine output are secondary effects of right-sided heart failure. Cardiomyopathy is usually identified as a symptom of left-sided heart failure. Left-sided heart failure causes primarily pulmonary symptoms rather than systemic ones. Angina pectoris doesn't cause weight gain, nausea, or a decrease in urine output.
NP: Evaluation; CN: Physiological integrity; CNS: Physiological adaptation; CL: Application

37. 1. Atherosclerosis accounts for 75% of all abdominal aortic aneurysms. Plaques build up on the wall of the vessel and weaken it, causing an aneurysm. Although the other conditions are related to the development of aneurysm, none is a direct cause.
NP: Data collection; CN: Health promotion and maintenance; CNS: Prevention and early detection of disease; CL: Knowledge

38. 2. The portion of the aorta distal to the renal arteries is more prone to an aneurysm because the vessel isn't surrounded by stable structures, unlike the proximal portion of the aorta. Distal to the iliac arteries, the vessel is again surrounded by stable vasculature, making this an uncommon site for an aneurysm. There's no area adjacent to the aortic arch, which bends into the thoracic (descending) aorta.
NP: Data collection; CN: Physiological integrity; CNS: Physiological adaptation; CL: Knowledge

39. 4. Digoxin's physiologic effect on the heart slows impulse conduction through the AV node. Digoxin increases cardiac output and ventricular emptying capacity. Digoxin also promotes diuresis, thereby decreasing the circulating blood volume.
NP: Evaluation; CN: Physiological integrity; CNS: Pharmacological therapies; CL: Analysis

40. 1. Abdominal pain in a client with an abdominal aortic aneurysm results from the disruption of normal circulation in the abdominal region. Lower back pain, not upper, is a common symptom, usually signifying expansion and impending rupture of the aneurysm. Headache and diaphoresis aren't associated with abdominal aortic aneurysm.
NP: Data collection; CN: Physiological integrity; CNS: Basic care and comfort; CL: Comprehension

41. Which of the following symptoms usually signifies rapid expansion and impending rupture of an abdominal aortic aneurysm?
1. Abdominal pain
2. Absent pedal pulses
3. Angina
4. Lower back pain

Knowing which symptom relates to which complication will help you to make a more accurate diagnosis.

41. 4. Lower back pain results from expansion of the aneurysm. The expansion applies pressure in the abdominal cavity, and the pain is referred to the lower back. Abdominal pain is the most common symptom resulting from impaired circulation. Absent pedal pulses are a sign of no circulation and would occur after a ruptured aneurysm or in peripheral vascular disease. Angina is associated with atherosclerosis of the coronary arteries.
NP: Data collection; CN: Physiological integrity; CNS: Basic care and comfort; CL: Comprehension

42. What's the definitive test used to diagnose an abdominal aortic aneurysm?
1. Abdominal X-ray
2. Arteriogram
3. Computed tomography (CT) scan
4. Ultrasound

42. 2. An arteriogram accurately and directly depicts the vasculature; therefore, it clearly delineates the vessels and any abnormalities. An abdominal aneurysm would only be visible on an X-ray if it were calcified. CT scan and ultrasound don't give a direct view of the vessels and don't yield as accurate a diagnosis as the arteriogram.
NP: Data collection; CN: Health promotion and maintenance; CNS: Prevention and early detection of disease; CL: Knowledge

This question is asking you to prioritize, once again.

43. Which of the following complications is of greatest concern when caring for a preoperative abdominal aortic aneurysm client?
1. Hypertension
2. Aneurysm rupture
3. Cardiac arrhythmias
4. Diminished pedal pulses

43. 2. Rupture of the aneurysm is a life-threatening emergency and is of the greatest concern for the nurse caring for this type of client. Hypertension should be avoided and controlled because it can cause the weakened vessel to rupture. Diminished pedal pulses, a sign of poor circulation to the lower extremities, are associated with an aneurysm but aren't life-threatening. Cardiac arrhythmias aren't directly linked to an aneurysm.
NP: Evaluation; CN: Physiological integrity; CNS: Basic care and comfort; CL: Analysis

44. Which of the following blood vessel layers may be damaged in a client with an aneurysm?
1. Externa
2. Interna
3. Media
4. Interna and media

44. 3. The factor common to all types of aneurysms is a damaged media. The media has more smooth muscle and less elastic fibers, so it's more capable of vasoconstriction and vasodilation. The interna and externa are generally not damaged in an aneurysm.
NP: Data collection; CN: Physiological integrity; CNS: Physiological adaptation; CL: Knowledge

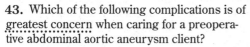

NP: Nursing process   CN: Client needs category   CNS: Client needs subcategory   CL: Cognitive level

45. Which precaution should the nurse take when caring for a client with a myocardial infarction who has received a thrombolytic agent?

1. Avoid puncture wounds.
2. Monitor the potassium level.
3. Maintain a supine position.
4. Force fluids.

45. 1. Thrombolytic agents are declotting agents that place the client at risk for hemorrhage from puncture wounds. All unnecessary needle sticks and invasive procedures should be avoided. The potassium level should be monitored in all cardiac clients, not just those receiving a thrombolytic agent. Although no specific position is required, most cardiac clients seem more comfortable in semi-Fowler's position. The client's fluid balance must be carefully monitored, so it may be inappropriate to force fluids at this time.

NP: Implementation; CN: Physiological integrity; CNS: Reduction of risk potential; CL: Application

46. Which of the following conditions is linked to more than 50% of clients with an abdominal aortic aneurysm?

1. Diabetes mellitus
2. Hypertension
3. Peripheral vascular disease
4. Syphilis

I'm suddenly feeling a little tense, a little hyper. (Get it?)

46. 2. Continuous pressure on the vessel walls from hypertension causes the walls to weaken and an aneurysm to occur. Atherosclerotic changes can occur with peripheral vascular diseases and are linked to aneurysms, but the link isn't as strong as it is with hypertension. Only 1% of clients with syphilis experience an aneurysm. Diabetes mellitus isn't directly linked to aneurysm.

NP: Data collection; CN: Health promotion and maintenance; CNS: Prevention and early detection of disease; CL: Comprehension

47. Which of the following sounds is distinctly heard on auscultation over the abdominal region of a client with an abdominal aortic aneurysm?

1. Bruit
2. Crackles
3. Dullness
4. Friction rubs

This is a real bruit of a question.

47. 1. A bruit, a vascular sound resembling heart murmur, suggests partial arterial occlusion. Crackles are indicative of fluid in the lungs. Dullness is heard over solid organs such as the liver. Friction rubs indicate inflammation of the peritoneal surface.

NP: Evaluation; CN: Physiological integrity; CNS: Basic care and comfort; CL: Comprehension

48. Which of the following groups of symptoms indicates a ruptured abdominal aortic aneurysm?

1. Lower back pain, increased blood pressure, decreased red blood cell (RBC) count, increased white blood cell (WBC) count
2. Severe lower back pain, decreased blood pressure, decreased RBC count, increased WBC count
3. Severe lower back pain, decreased blood pressure, decreased RBC count, decreased WBC count
4. Intermittent lower back pain, decreased blood pressure, decreased RBC count, increased WBC count

48. 2. Severe lower back pain indicates an aneurysm rupture, secondary to pressure being applied within the abdominal cavity. When rupture occurs, the pain is constant because it can't be alleviated until the aneurysm is repaired. Blood pressure decreases due to the loss of blood. After the aneurysm ruptures, the vasculature is interrupted and blood volume is lost, so blood pressure wouldn't increase. For the same reason, the RBC count is decreased — not increased. The WBC count increases as cells migrate to the site of injury.

NP: Data collection; CN: Physiological integrity; CNS: Physiological adaptation; CL: Knowledge

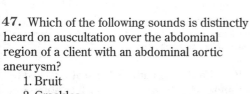

49. Which of the following complications of an abdominal aortic repair is indicated by detection of a hematoma in the perineal area?
1. Hernia
2. Stage 1 pressure ulcer
3. Retroperitoneal rupture at the repair site
4. Rapid expansion of the aneurysm

49. 3. Blood collects in the retroperitoneal space and is exhibited as a hematoma in the perineal area. This rupture is most commonly caused by leakage at the repair site. A hernia doesn't cause vascular disturbances, nor does a pressure ulcer. Because no bleeding occurs with rapid expansion of the aneurysm, a hematoma won't form.
NP: Evaluation; CN: Physiological integrity; CNS: Physiological adaptation; CL: Comprehension

You're finished 50 questions! Good job!

50. Which hereditary disease is most closely linked to aneurysm?
1. Cystic fibrosis
2. Lupus erythematosus
3. Marfan syndrome
4. Myocardial infarction (MI)

50. 3. Marfan syndrome results in the degeneration of the elastic fibers of the aortic media. Therefore, clients with the syndrome are more likely to develop an aneurysm. Although cystic fibrosis is hereditary, it hasn't been linked to aneurysms. Lupus erythematosus isn't hereditary. MI is neither hereditary nor a disease.
NP: Data collection; CN: Health promotion and maintenance; CNS: Prevention and early detection of disease; CL: Knowledge

51. Which of the following treatments is the definitive one for a ruptured aneurysm?
1. Antihypertensive medication administration
2. Aortogram
3. Beta-adrenergic blocker administration
4. Surgical intervention

This question is asking for a treatment, not a preventive measure.

51. 4. When the vessel ruptures, surgery is the only intervention that can repair it. Administration of antihypertensive medications and beta-adrenergic blockers can help control hypertension, reducing the risk of rupture. An aortogram is a diagnostic tool used to detect an aneurysm.
NP: Intervention; CN: Physiological integrity; CNS: Basic care and comfort; CL: Application

52. Which of the following heart muscle diseases is unrelated to other cardiovascular disease?
1. Cardiomyopathy
2. Coronary artery disease (CAD)
3. Myocardial infarction (MI)
4. Pericardial effusion

52. 1. Cardiomyopathy isn't usually related to an underlying heart disease such as atherosclerosis. The etiology in most cases is unknown. CAD and MI are directly related to atherosclerosis. Pericardial effusion is the escape of fluid into the pericardial sac, a condition associated with pericarditis and advanced heart failure.
NP: Data collection; CN: Physiological integrity; CNS: Physiological adaptation; CL: Knowledge

NP: Nursing process CN: Client needs category CNS: Client needs subcategory CL: Cognitive level

53. Which of the following types of cardiomy-opathy can be associated with childbirth?
1. Dilated
2. Hypertrophic
3. Myocarditis
4. Restrictive

53. 1. Although the cause isn't entirely known, cardiac dilation and heart failure may develop during the last month of pregnancy or the first few months after childbirth. The condition may result from a preexisting cardiomyopathy not apparent before pregnancy. Hypertrophic cardiomyopathy is an abnormal symmetry of the ventricles that has an unknown etiology, but a strong familial tendency. Myocarditis isn't specifically associated with childbirth. Restrictive cardiomyopathy indicates constrictive pericarditis; the underlying cause is usually myocardial.

NP: Data collection; CN: Physiological integrity; CNS: Physiological adaptation; CL: Knowledge

54. Septal involvement occurs in which type of cardiomyopathy?
1. Congestive
2. Dilated
3. Hypertrophic
4. Restrictive

54. 3. In hypertrophic cardiomyopathy, hypertrophy of the ventricular septum — not the ventricle chambers — is apparent. This abnormality isn't seen in other types of cardiomyopathy.

NP: Data collection; CN: Physiological integrity; CNS: Physiological adaptation; CL: Knowledge

This is a tough one! Keep plugging along!

55. Which of the following recurring conditions most commonly occurs in clients with cardiomyopathy?
1. Heart failure
2. Diabetes mellitus
3. Myocardial infarction (MI)
4. Pericardial effusion

55. 1. Because the structure and function of the heart muscle is affected, heart failure most commonly occurs in clients with cardiomyopathy. MI results from atherosclerosis. Pericardial effusion is most predominant in clients with pericarditis. Diabetes mellitus is unrelated to cardiomyopathy.

NP: Data collection; CN: Physiological integrity; CNS: Physiological adaptation; CL: Comprehension

56. What's the term used to describe an enlargement of the heart muscle?
1. Cardiomegaly
2. Cardiomyopathy
3. Myocarditis
4. Pericarditis

56. 1. Cardiomegaly denotes an enlarged heart muscle. Cardiomyopathy is a heart muscle disease of unknown origin. Myocarditis refers to inflammation of heart muscle, rather than enlargement of the muscle itself. Pericarditis is an inflammation of the pericardium, the sac surrounding the heart, and isn't associated with heart muscle enlargement.

NP: Data collection; CN: Physiological integrity; CNS: Physiological adaptation; CL: Knowledge

I don't know about you, but I'm getting kinda fried.

57. Dyspnea, cough, expectoration, weakness, and edema are classic signs and symptoms of which of the following conditions?
1. Pericarditis
2. Hypertension
3. Myocardial infarction (MI)
4. Heart failure

This question is a classic one, if you catch my drift.

57. 4. These are the classic signs and symptoms of heart failure, the most common problem related to cardiomyopathy. Pericarditis is exhibited by a feeling of fullness in the chest and auscultation of a pericardial friction rub. Hypertension is usually exhibited by headaches, visual disturbances, and a flushed face. MI causes heart failure but isn't related to these symptoms.
NP: Evaluation; CN: Physiological integrity; CNS: Reduction of risk potential; CL: Knowledge

58. The nurse is planning care for a client with heart failure. Which nursing diagnosis should receive priority?
1. Ineffective tissue perfusion (cardiopulmonary, renal) related to sympathetic response to heart failure
2. Imbalanced nutrition: Less than body requirements related to rapid tiring while feeding
3. Anxiety related to unknown nature of illness
4. Decreased cardiac output related to decreased pumping ability

58. 4. The primary nursing diagnosis for a client with heart failure is *Decreased cardiac output related to pumping ability*. *Ineffective tissue perfusion, imbalanced nutrition,* and *anxiety* don't take priority over decreased cardiac output. The client's heart must produce cardiac output sufficient to meet the body's metabolic demands.
NP: Planning; CN: Physiological integrity; CNS: Physiological adaptation; CL: Application

59. The nurse is caring for a 59-year-old male client diagnosed with myocardial infarction (MI). The purpose of giving nitrates to a client who has had an MI is to:
1. relieve pain.
2. dilate coronary arteries.
3. relieve headaches caused by other medications.
4. calm and relax the client.

It's important to know what nitrates do.

59. 2. Nitrates dilate the arteries, allowing oxygen to continue flowing to the heart. Nitrates can cause headaches but don't relieve pain and don't calm or relax the client.
NP: Implementation; CN: Physiological integrity; CNS: Pharmacological therapies; CL: Application

60. Which of the following classes of drugs is most widely used in the treatment of cardiomyopathy?
1. Antihypertensives
2. Beta-adrenergic blockers
3. Calcium channel blockers
4. Nitrates

60. 2. By decreasing the heart rate and contractility, beta-adrenergic blockers improve myocardial filling and cardiac output, which are primary goals in the treatment of cardiomyopathy. Antihypertensives aren't usually indicated because they would decrease cardiac output in clients who are often already hypotensive. Calcium channel blockers are sometimes used for the same reasons as beta-adrenergic blockers; however, they aren't as effective as beta-adrenergic blockers and cause increased hypotension. Nitrates aren't used because of their dilating effects, which would further compromise the myocardium.
NP: Data collection; CN: Physiological integrity; CNS: Pharmacological therapies; CL: Comprehension

61. If medical treatments fail, which of the following invasive procedures is necessary for treating cardiomyopathy?
1. Cardiac catheterization
2. Coronary artery bypass graft (CABG)
3. Heart transplantation
4. Intra-aortic balloon pump (IABP)

61. 3. The only definitive treatment for cardiomyopathy that can't be controlled medically is a heart transplant because the damage to the heart muscle is irreversible. Cardiac catheterization is an invasive diagnostic procedure for coronary artery disease. CABG is a surgical intervention used for atherosclerotic vessels. An IABP is an invasive treatment that assists the failing heart; however, it's only a temporary solution because it can't be used for an extended time.

NP: Data collection; CN: Physiological integrity; CNS: Physiological adaptation; CL: Comprehension

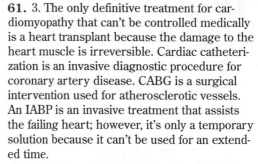

Stress can be a real pain in the heart.

62. Which of the following conditions is associated with a predictable level of pain that occurs as a result of physical or emotional stress?
1. Anxiety
2. Stable angina
3. Unstable angina
4. Variant angina

62. 2. The pain of stable angina is predictable in nature, builds gradually, and quickly reaches maximum intensity. Anxiety generally isn't described as painful. Unstable angina doesn't always need a trigger, is more intense, and lasts longer than stable angina. Variant angina usually occurs at rest — not as a result of exertion or stress.

NP: Evaluation; CN: Physiological integrity; CNS: Physiological adaptation; CL: Analysis

63. After undergoing a cardiac catheterization, the client has a large puddle of blood under his buttocks. Which of the following steps should the nurse take first?
1. Call for help.
2. Obtain vital signs.
3. Ask the client to "lift up."
4. Apply gloves and assess the groin site.

Prioritize!

63. 4. Observing standard precautions is the first priority when dealing with body fluids. Assessment of the groin site is the second priority. This establishes where the blood is coming from and determines how much blood has been lost. The goal in this situation is to stop the bleeding. The nurse would call for help if it were warranted after assessing the situation. After determining the extent of the bleeding, vital signs assessment is important. The nurse should never move the client in case a clot has formed. Moving can disturb the clot and cause rebleeding.

NP: Evaluation; CN: Safe, effective care environment; CNS: Management of care; CL: Comprehension

64. Which of the following types of pain is most characteristic of angina?
1. Knifelike
2. Sharp
3. Shooting
4. Tightness

64. 4. The pain of angina usually ranges from a vague feeling of tightness to heavy, intense pain. Pain impulses originate in the most visceral muscles and may move to such areas as the chest, neck, and arms. Pain described as knifelike, sharp, or shooting is more characteristic of pulmonary or pleuritic pain.

NP: Data collection; CN: Physiological integrity; CNS: Basic care and comfort; CL: Knowledge

65. Which of the following types of angina is most closely associated with an impending myocardial infarction (MI)?
1. Angina decubitus
2. Chronic stable angina
3. Nocturnal angina
4. Unstable angina

66. Which of the following medications is the drug of choice for angina pectoris?
1. Aspirin
2. Furosemide (Lasix)
3. Nitroglycerin
4. Nifedipine (Procardia)

You're over halfway finished! Keep going!

67. Which of the following conditions is the predominant cause of angina?
1. Increased preload
2. Decreased afterload
3. Coronary artery spasm
4. Inadequate oxygen supply to the myocardium

68. Which of the following tests is used most often to diagnose angina?
1. Chest X-ray
2. Echocardiogram
3. Cardiac catheterization
4. 12-lead electrocardiogram (ECG)

This question is trying to determine if you know when to use each diagnostic tool.

65. 4. Unstable angina progressively increases in frequency, intensity, and duration and is related to an increased risk of MI within 3 to 18 months. Angina decubitus, chronic stable angina, and nocturnal angina aren't associated with an increased risk of MI.
NP: Data collection; CN: Physiological integrity; CNS: Physiological adaptation; CL: Knowledge

66. 3. Nitroglycerin is administered to reduce myocardial demand, which decreases ischemia and relieves pain. In addition, nitroglycerin dilates the vasculature, thereby reducing preload. Aspirin is administered to reduce clot formation in clients having a myocardial infarction. Furosemide is a loop diuretic that won't directly reduce pain. Nifedipine is a calcium channel blocker primarily used to decrease coronary artery spasm, as in variant angina.
NP: Data collection; CN: Physiological integrity; CNS: Pharmacological therapies; CL: Knowledge

67. 4. Inadequate oxygen supply to the myocardium is responsible for the pain accompanying angina. Increased preload would be responsible for right-sided heart failure. Decreased afterload causes low cardiac output. Coronary artery spasm is responsible for variant angina.
NP: Data collection; CN: Physiological integrity; CNS: Physiological adaptation; CL: Knowledge

68. 4. The 12-lead ECG will indicate ischemia, showing T-wave inversion. In addition, with variant angina, the ECG shows ST-segment elevation. A chest X-ray will show heart enlargement or signs of heart failure, but isn't used to diagnose angina. An echocardiogram is used to detect wall function and valvular function and is most accurate in diagnosing myocardial infarction. Cardiac catheterization is used to diagnosis coronary artery disease, which can cause angina.
NP: Data collection; CN: Health promotion and maintenance; CNS: Prevention and early detection of disease; CL: Knowledge

69. Which of the following results is the primary treatment goal for angina?
1. Reversal of ischemia
2. Reversal of infarction
3. Reduction of stress and anxiety
4. Reduction of associated risk factors

69. 1. Reversal of ischemia is the primary goal, achieved by reducing oxygen consumption and increasing oxygen supply. An infarction is permanent and can't be reversed. Reduction of associated risk factors, such as stress and anxiety, is a progressive, long-term treatment goal that has cumulative effects. Reduction of these factors will decrease the risk for angina, but this usually isn't an immediate goal.
NP: Data collection; CN: Physiological integrity; CNS: Physiological adaptation; CL: Comprehension

70. Which of the following treatments is a suitable surgical intervention for unstable angina?
1. Cardiac catheterization
2. Echocardiogram
3. Nitroglycerin
4. Percutaneous transluminal coronary angioplasty (PTCA)

70. 4. PTCA can alleviate the blockage and restore blood flow and oxygenation. An echocardiogram is a noninvasive diagnostic test. Nitroglycerin is an oral medication. Cardiac catheterization is a diagnostic tool, not a treatment.
NP: Data collection; CN: Physiological integrity; CNS: Physiological adaptation; CL: Knowledge

Prioritize, prioritize, prioritize.

71. Which of the following interventions should be the first priority when treating a client experiencing chest pain while walking?
1. Sit the client down.
2. Get the client back to bed.
3. Obtain an electrocardiogram (ECG).
4. Administer sublingual nitroglycerin.

71. 1. The initial priority is to decrease the oxygen consumption; this would be achieved by sitting the client down. An ECG can be obtained after the client is sitting down. After the ECG, sublingual nitroglycerin would be administered. When the client's condition is stabilized, he can be returned to bed.
NP: Implementation; CN: Physiological integrity; CNS: Basic care and comfort; CL: Application

72. Which of the following terms is used to describe reduced cardiac output and perfusion impairment due to ineffective pumping of the heart?
1. Anaphylactic shock
2. Cardiogenic shock
3. Distributive shock
4. Myocardial infarction (MI)

Pay attention. This info could really shock you!

72. 2. Cardiogenic shock is shock related to ineffective pumping of the heart. Anaphylactic shock results from an allergic reaction. Distributive shock results from changes in the intravascular volume distribution and is usually associated with increased cardiac output. MI isn't a shock state; however, a severe MI can lead to shock.
NP: Data collection; CN: Physiological integrity; CNS: Physiological adaptation; CL: Knowledge

73. Which of the following conditions most commonly causes cardiogenic shock?
  1. Acute myocardial infarction (MI)
  2. Coronary artery disease (CAD)
  3. Decreased hemoglobin level
  4. Hypotension

This question is asking you to distinguish between the causes and the symptoms of shock.

73. 1. Of all clients with an acute MI, 15% suffer cardiogenic shock secondary to the myocardial damage and decreased function. CAD causes MI. Hypotension is the result of a reduced cardiac output produced by the shock state. A decreased hemoglobin level is a result of bleeding.
NP: Data collection; CN: Physiological integrity; CNS: Reduction of risk potential; CL: Knowledge

74. The nurse is caring for a client admitted with a possible myocardial infarction (MI). Which diagnostic test is used to rule out an MI?
  1. Complete blood count
  2. Cardiac biopsy
  3. Arterial blood gas analysis
  4. Cardiac enzyme analysis

74. 4. Certain isoenzymes come only from myocardial cells and are released when the cells are damaged. Creatine kinase (CK) and its isoenzyme CK-MB are the least specific enzymes analyzed in acute MI. These cardiac enzymes should be tested in relation to the time of onset of chest discomfort or other symptoms to rule out MI. Blood counts, blood gases, and cardiac biopsies don't play large roles in ruling out an MI.
NP: Evaluation; CN: Physiological adaptation; CNS: Reduction of risk potential; CL: Comprehension

75. The nurse is taking a 49-year-old woman's health history when the client mentions that her heart sometimes seems to race. Which condition is a life-threatening cardiac rhythm?
  1. Ventricular fibrillation (VF)
  2. Ventricular tachycardia (VT)
  3. Premature ventricular contractions (PVCs)
  4. Premature atrial contractions (PACs)

75. 1. VF is a life-threatening arrhythmia. It occurs when the ventricle fibrillates, failing to fully contract and pump blood through the heart. VT, PVCs, and PACs aren't life-threatening.
NP: Evaluation; CN: Health promotion and maintenance; CNS: Prevention and early detection of disease; CL: Knowledge

76. Which of the following factors would be most useful in detecting a client's risk of developing cardiogenic shock?
  1. Decreased heart rate
  2. Decreased cardiac index
  3. Decreased blood pressure
  4. Decreased cerebral blood flow

Pay attention. This question involves assessing the order of symptoms.

76. 2. The cardiac index, a figure derived by dividing the cardiac output by the client's body surface area, is used to identify whether the cardiac output is meeting a client's needs. Decreased cerebral blood flow, blood pressure, and heart rate are less useful in detecting the risk of cardiogenic shock.
NP: Data collection; CN: Physiological integrity; CNS: Physiological adaptation; CL: Analysis

77. Which of the following symptoms is one of the earliest signs of cardiogenic shock?
  1. Tachycardia
  2. Decreased urine output
  3. Presence of a fourth heart sound (S_4)
  4. Altered level of consciousness

77. 4. Initially, the decrease in cardiac output results in a decrease in cerebral blood flow that causes restlessness, agitation, or confusion. Tachycardia, decreased urine output, and presence of an S_4 heart sound are all later signs of shock.
NP: Evaluation; CN: Physiological integrity; CNS: Basic care and comfort; CL: Application

78. Which of the following diagnostic studies can determine when cellular metabolism becomes anaerobic and when pH decreases?
 1. Arterial blood gas (ABG) levels
 2. Complete blood count (CBC)
 3. Electrocardiogram (ECG)
 4. Lung scan

79. Which of the following is the initial treatment goal for cardiogenic shock?
 1. Correct hypoxia
 2. Prevent infarction
 3. Correct metabolic acidosis
 4. Increase myocardial oxygen supply

You're getting there! Keep it up!

80. Which of the following drugs is most commonly used to treat cardiogenic shock?
 1. Dopamine (Intropin)
 2. Enalapril (Vasotec)
 3. Furosemide (Lasix)
 4. Metoprolol (Lopressor)

When answering questions, look for phrases such as most commonly. These are typically hints.

81. Which of the following instruments is used as a diagnostic and monitoring tool for determining the severity of a shock state?
 1. Arterial line
 2. Indwelling urinary catheter
 3. Intra-aortic balloon pump (IABP)
 4. Pulmonary artery (PA) catheter

78. 1. ABG levels reflect cellular metabolism and indicate hypoxia. A CBC is performed to determine various constituents of venous blood. An ECG shows the electrical activity of the heart. A lung scan is performed to view the lungs' function.
NP: Data collection; CN: Health promotion and maintenance; CNS: Prevention and early detection of disease; CL: Analysis

79. 4. A balance must be maintained between oxygen supply and demand. In a shock state, the myocardium requires more oxygen. If it can't get more oxygen, the shock worsens. Increasing the oxygen will also help correct metabolic acidosis and hypoxia. Infarction typically causes the shock state, so prevention isn't an appropriate goal for this condition.
NP: Implementation; CN: Physiological integrity; CN: Physiological adaptation; CL: Comprehension

80. 1. Dopamine, a sympathomimetic drug, improves myocardial contractility and blood flow through vital organs by increasing perfusion pressure. Enalapril is an angiotensin-converting enzyme inhibitor that directly lowers blood pressure. Furosemide is a diuretic and doesn't have a direct effect on contractility or tissue perfusion. Metoprolol is a beta-adrenergic blocker that slows the heart rate and blood pressure, neither of which is a desired effect in the treatment of cardiogenic shock.
NP: Data collection; CN: Physiological integrity; CNS: Pharmacological therapies; CL: Application

81. 4. A PA catheter is used to give accurate pressure measurements within the heart, which aids in determining the course of treatment. An arterial line is used to directly assess blood pressure continuously. An indwelling urinary catheter is used to drain the bladder. An IABP is an assistive device used to rest the damaged heart.
NP: Data collection; CN: Physiological integrity; CNS: Pharmacological therapies; CL: Knowledge

82. Which of the following parameters represents the World Health Organization's definition of hypertension?
1. Systolic blood pressure of 160 mm Hg or higher, or diastolic blood pressure of 95 mm Hg or higher
2. Systolic blood pressure of 160 mm Hg or higher, or diastolic blood pressure of 95 mm Hg or lower
3. Systolic blood pressure below 160 mm Hg, or diastolic blood pressure of 95 mm Hg or higher
4. Systolic blood pressure below 160 mm Hg, or diastolic blood pressure of 95 mm Hg or lower

83. Which of the following sounds will be heard during the first phase of Korotkoff's sounds?
1. Disappearance of sounds
2. Faint, clear tapping sounds
3. A murmur or swishing sound
4. Soft, muffling sounds

84. Which of the following parameters is the major determinant of diastolic blood pressure?
1. Baroreceptors
2. Cardiac output
3. Renal function
4. Vascular resistance

85. Which of the following factors can cause blood pressure to drop to normal levels?
1. Kidneys' excretion of sodium only
2. Kidneys' retention of sodium and water
3. Kidneys' excretion of sodium and water
4. Kidneys' retention of sodium and excretion of water

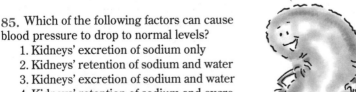

How would you respond to a rise in blood pressure?

82. 1. These values are compared to a normal adult with a systolic blood pressure of 140 mm Hg or lower, and a diastolic blood pressure of 90 mm Hg or lower. The other values are outside the acceptable standards put forth by the World Health Organization.
NP: Evaluation; CN: Health promotion and maintenance; CNS: Prevention and early detection of disease; CL: Knowledge

83. 2. In phase I, auscultation produces a faint, clear tapping sound that gradually increases in intensity. Phase II produces a murmur sound, and precedes Phase III, the phase marked by an increased intensity of sound. Phase IV produces a muffling sound that gives a soft blowing noise. Phase V, the final phase, is marked by the disappearance of sounds.
NP: Evaluation; CN: Physiological integrity; CNS: Basic care and comfort; CL: Application

84. 4. Vascular resistance is the impedance of blood flow by the arterioles that most predominantly affects the diastolic pressure. Baroreceptors are nerve endings that are embedded in the blood vessels and respond to the stretching of vessel walls. They don't directly affect diastolic blood pressure. Cardiac output determines systolic blood pressure. Renal function helps control blood volume and indirectly affects diastolic blood pressure.
NP: Data collection; CN: Physiological integrity; CNS: Physiological adaptation; CL: Knowledge

85. 3. The kidneys respond to a rise in blood pressure by excreting sodium and excess water. This response ultimately affects systolic blood pressure by regulating blood volume. The retention of either sodium or water would only further increase blood pressure. Sodium and water travel together across the membrane in the kidneys; one can't travel without the other.
NP: Data collection; CN: Physiological integrity; CNS: Physiological adaptation; CL: Comprehension

NP: Nursing process   CN: Client needs category   CNS: Client needs subcategory   CL: Cognitive level

86. Chemoreceptors in the carotid artery walls, aorta, and medulla respond to which of the following conditions?
  1. Decreased blood pressure
  2. Increased blood pressure
  3. Decreased pulse
  4. Increased pulse

87. Which of the following hormones is responsible for raising arterial pressure and promoting venous return?
  1. Angiotensin I
  2. Angiotensin II
  3. Epinephrine
  4. Renin

Don't get too tense to answer this question. (Oh, I am just two, er, too clever for words.)

88. Which of the following terms is used to describe persistently elevated blood pressure with an unknown cause that accounts for approximately 90% of hypertension cases?
  1. Accelerated hypertension
  2. Malignant hypertension
  3. Primary hypertension
  4. Secondary hypertension

Note the words most common in this question. That's a hint.

89. Which of the following symptoms of hypertension is most common?
  1. Blurred vision
  2. Epistaxis
  3. Headache
  4. Peripheral edema

86. 1. Chemoreceptors respond to a decrease in blood pressure by stimulating sympathetic nervous system activity. The receptors don't respond to the other conditions.
NP: Data collection; CN: Physiological integrity; CNS: Physiological adaptation; CL: Knowledge

87. 2. Angiotensin II, triggered by angiotensin I, is responsible for vasoconstriction, thereby increasing arterial blood pressure. Angiotensin I is the hormone that causes angiotensin II to respond. Epinephrine is a direct sympathetic nervous system cardiovascular stimulant that increases the heart rate. Renin produces angiotensin I when triggered by reduced blood flow.
NP: Data collection; CN: Physiological integrity; CNS: Physiological adaptation; CL: Knowledge

88. 3. Characterized by a progressive, usually asymptomatic blood pressure increase over several years, primary hypertension is the most common type. Malignant hypertension, also known as accelerated hypertension, is rapidly progressive, uncontrollable, and causes a rapid onset of complications. Secondary hypertension occurs secondary to a known, correctable cause.
NP: Evaluation; CN: Physiological integrity; CNS: Reduction of risk potential; CL: Analysis

89. 3. An occipital headache is typical of hypertension secondary to continued increased pressure on the cerebral vasculature. Epistaxis (nosebleed) occurs far less frequently than a headache, but can also be a diagnostic sign of hypertension. Blurred vision can result from hypertension due to the arteriolar changes in the eye. Peripheral edema can also occur from an increase in sodium and water retention, but is usually a latent sign.
NP: Evaluation; CN: Health promotion and maintenance; CNS: Prevention and early detection of disease; CL: Comprehension

90. The diaphragm of the stethoscope is most commonly placed over which of the following arteries to obtain a blood pressure measurement?
1. Brachial
2. Brachiocephalic
3. Radial
4. Ulnar

91. Which of the following statements explains why furosemide (Lasix) is administered to treat hypertension?
1. It dilates peripheral blood vessels.
2. It decreases sympathetic cardioacceleration.
3. It inhibits the angiotensin-converting enzyme.
4. It inhibits reabsorption of sodium and water in the loop of Henle.

92. The hypothalamus responds to a decrease in blood pressure by secreting which of the following substances?
1. Angiotensin
2. Antidiuretic hormone (ADH)
3. Epinephrine
4. Renin

93. Which of the following parts of the eye is examined to see arterial changes caused by hypertension?
1. Cornea
2. Fovea
3. Retina
4. Sclera

This question is asking for the mechanism of action of furosemide.

Wow! You're almost at 100. Cool!

90. 1. The brachial artery is most commonly used because of its easy accessibility and location. The brachiocephalic artery isn't accessible for blood pressure measurement. The radial and ulnar arteries can be used in extraordinary circumstances, but the measurement may not be as accurate.
NP: Data collection; CN: Physiological integrity; CNS: Basic care and comfort; CL: Application

91. 4. Furosemide is a loop diuretic that inhibits sodium and water reabsorption in the loop of Henle, thereby causing a decrease in blood pressure. Vasodilators cause dilation of peripheral blood vessels, directly relaxing vascular smooth muscle and decreasing blood pressure. Adrenergic blockers decrease sympathetic cardioacceleration and decrease blood pressure. Angiotensin-converting enzyme inhibitors decrease blood pressure due to their action on angiotensin.
NP: Data collection; CN: Physiological integrity; CNS: Pharmacological therapies; CL: Comprehension

92. 2. ADH acts on the renal tubules to promote water retention, which increases blood pressure. Angiotensin, epinephrine, and renin aren't stored in the hypothalamus, but they all help to increase blood pressure.
NP: Data collection; CN: Physiological integrity; CNS: Physiological adaptation; CL: Knowledge

93. 3. The retina is the only site in the body where arteries can be seen without invasive techniques. Changes in the retinal arteries signal similar damage to vessels elsewhere. The cornea is the nonvascular, transparent fibrous coat where the iris can be seen. The fovea is the point of central vision. The sclera is the fibrous tissue that forms the outer protective covering over the eyeball.
NP: Data collection; CN: Health promotion and maintenance; CNS: Prevention and early detection of disease; CL: Knowledge

94. Which of the following conditions causes varicose veins?
1. Tunica media tear
2. Intraluminal occlusion
3. Intraluminal valvular compression
4. Intraluminal valvular incompetence

95. Which of the following factors causes primary varicose veins?
1. Hypertension
2. Pregnancy
3. Thrombosis
4. Trauma

Oooh, I love babies. But not the varicose veins that go with them!

96. Which of the following symptoms commonly occur in a client with varicose veins?
1. Fatigue and pressure
2. Fatigue and cool feet
3. Sharp pain and fatigue
4. Sharp pain and cool feet

97. In which of the following veins do varicose veins most commonly occur?
1. Brachial
2. Femoral
3. Renal
4. Saphenous

98. Which of the following conditions is caused by increased hydrostatic pressure and chronic venous stasis?
1. Venous occlusion
2. Cool extremities
3. Nocturnal calf muscle cramps
4. Diminished blood supply to the feet

94. 4. Varicose veins, dilated tortuous surface veins engorged with blood, result from intraluminal valvular incompetence. A tunica media tear would result in a hematoma, not a varicosity. An intraluminal occlusion would result from plaque or thrombosis. Varicose veins aren't a result of incompetence in the tunica adventicia or the fibrous elastic covering of the veins but involve the valves.
NP: Data collection; CN: Physiological integrity; CNS: Physiological adaptation; CL: Knowledge

95. 2. Primary varicose veins have a gradual onset and progressively worsen. In pregnancy, the expanding uterus and increased vascular volume impede blood return to the heart. The pressure places increased stress on the veins. Hypertension has no role in varicose vein formation. Thrombosis and trauma cause valvular incompetence and so are secondary causes of varicosities, not primary.
NP: Data collection; CN: Health promotion and maintenance; CNS: Prevention and early detection of disease; CL: Comprehension

96. 1. Fatigue and pressure are classic signs of varicose veins, secondary to increased blood volume and edema. Sharp pain and cool feet are symptoms of alteration in arterial blood flow.
NP: Evaluation; CN: Physiological integrity; CNS: Physiological adaptation; CL: Comprehension

97. 4. Varicose veins occur most commonly in the saphenous veins of the lower extremities. They don't develop in the brachial, femoral, or renal veins.
NP: Data collection; CN: Physiological integrity; CNS: Physiological adaptation; CL: Knowledge

98. 3. Calf muscle cramps result from increased pressure and venous stasis secondary to varicose veins. An occlusion is a blockage of blood flow. Cool extremities and diminished blood supply to the feet are symptoms of arterial blood flow changes.
NP: Data collection; CN: Health promotion and maintenance; CNS: Prevention and early detection of disease; CL: Analysis

99. Which of the following activities should a client with varicose veins avoid?
1. Exercise
2. Leg elevations
3. Prolonged lying
4. Wearing tight clothing

Be careful! Using the word *avoid* makes this almost a negative question.

100. Which of the following tests demonstrates the backward flow of blood through incompetent valves of superficial veins?
1. Trendelenburg's test
2. Manual compression test
3. Perthes' test
4. Plethysmography

Congratulations! You've finished 100 questions. You're almost there!

101. Which of the following signs and symptoms are produced by secondary varicose veins?
1. Pallor and severe pain
2. Severe pain and edema
3. Edema and pigmentation
4. Absent hair growth and pigmentation

102. Which of the following treatments can be used to eliminate varicose veins?
1. Ablation therapy
2. Cold therapy
3. Ligation and stripping
4. Radiation

103. Which of the following treatments is recommended for postoperative management of a client who has undergone ligation and stripping?
1. Sitting
2. Bed rest
3. Ice packs
4. Elastic leg compression

99. 4. Tight clothing, especially below the waist, increases vascular volume and impedes blood return to the heart. Exercise, leg elevations, and lying down usually relieve symptoms of varicose veins.
NP: Data collection; CN: Health promotion and maintenance; CNS: Prevention and early detection of disease; CL: Comprehension

100. 1. Trendelenburg's test is the most accurate tool used to determine retrograde venous filling. The manual compression test is a quick, easy test done by palpation and usually isn't diagnostic of the backward flow of blood. Perthes' test easily indicates whether the deeper venous system and communicating veins are competent. Plethysmography allows measurement of changes in venous blood volume.
NP: Data collection; CN: Health promotion and maintenance; CNS: Prevention and early detection of disease; CL: Application

101. 3. Secondary varicose veins result from an obstruction of the deep veins. Incompetent valves lead to impaired blood flow, and edema and pigmentation result from venous stasis. Severe pain, pallor, and absent hair growth are symptoms of an altered arterial blood flow.
NP: Data collection; CN: Physiological integrity; CNS: Physiological adaptation; CL: Comprehension

102. 3. Ligation and stripping of the vein can rid the vein of varicosity. This invasive procedure will take care of current varicose veins only; it won't prevent others from forming. The other procedures aren't used for varicose veins.
NP: Data collection; CN: Physiological integrity; CNS: Pharmacological therapies; CL: Knowledge

103. 4. Elastic leg compression helps venous return to the heart, thereby decreasing venous stasis. Sitting and bed rest are contraindicated because both promote decreased blood return to the heart and venous stasis. Although ice packs would help reduce edema, they would also cause vasoconstriction and impede blood flow.
NP: Implementation; CN: Physiological integrity; CNS: Basic care and comfort; CL: Application

104. Which of the following factors usually causes deep vein thrombosis (DVT)?
1. Aerobic exercise
2. Inactivity
3. Pregnancy
4. Tight clothing

I'll answer this question after I've had my nap.

104. 2. A thrombus lodged in a vein can cause venous occlusion as a result of venous stasis. Inactivity can cause venous stasis, leading to DVT. Aerobic exercise helps to prevent venous stasis. Pregnancy and tight clothing can cause varicose veins, which can lead to venous stasis and eventually DVT, but these aren't primary causes.

NP: Data collection; CN: Health promotion and maintenance; CNS: Prevention and early detection of disease; CL: Knowledge

105. Which of the following terms is used to describe a thrombus lodged in the lungs?
1. Hemothorax
2. Pneumothorax
3. Pulmonary embolism
4. Pulmonary hypertension

I was never any good at PE when I was in school. Were you? (That's a clue, don'tcha know.)

105. 3. A pulmonary embolism is a blood clot lodged in the pulmonary vasculature. A hemothorax refers to blood in the pleural space. A pneumothorax is caused by an opening in the pleura. Pulmonary hypertension is an increase in pulmonary artery pressure, which increases the workload of the right ventricle.

NP: Data collection; CN: Physiological integrity; CNS: Physiological adaptation; CL: Knowledge

106. Which of the following terms refers to the condition of blood coagulating faster than normal, causing thrombin and other clotting factors to multiply?
1. Embolus
2. Hypercoagulability
3. Venous stasis
4. Venous wall injury

This question is asking you to characterize the type of pain experienced during deep vein thrombosis.

106. 2. Hypercoagulability is the condition of blood coagulating faster than normal, causing thrombin and other clotting factors to multiply. This condition, along with venous stasis and venous wall injury, accounts for the formation of deep vein thrombosis. An embolus is a blood clot or fatty globule that formed in one area and is carried through the bloodstream to another area.

NP: Data collection; CN: Physiological integrity; CNS: Physiological adaptation; CL: Comprehension

107. Which of the following characteristics is typical of the pain associated with deep vein thrombosis (DVT)?
1. Dull ache
2. No pain
3. Sudden onset
4. Tingling

107. 3. DVT is associated with deep leg pain of sudden onset, which occurs secondary to the occlusion. A dull ache is more commonly associated with varicose veins. A tingling sensation is associated with an alteration in arterial blood flow. If the thrombus is large enough, it will cause pain.

NP: Data collection; CN: Physiological integrity; CNS: Basic care and comfort; CL: Analysis

108. Which of the following treatments should be included in the discharge treatment plan for a client with deep vein thrombosis (DVT)?
1. Application of heat
2. Bed rest
3. Exercise
4. Leg elevation

108. 4. Leg elevation alleviates the pressure caused by thrombosis and occlusion by assisting venous return. The application of heat would dilate the vessels and pool blood in the area of the thrombus, increasing the risk of further thrombus formation. Bed rest adds to venous stasis, thereby increasing the risk of thrombosis formation. When DVT is diagnosed, exercise isn't recommended until the clot has dissolved.
NP: Data collection; CN: Physiological integrity; CNS: Basic care and comfort; CL: Comprehension

109. Which of the following terms best describes the findings on cautious palpation of the vein in typical superficial thrombophlebitis?
1. Dilated
2. Knotty
3. Smooth
4. Tortuous

109. 2. The knotty feeling is secondary to the emboli adhering to the vein wall. Varicose veins may be described as dilated and tortuous. Normal veins feel smooth.
NP: Evaluation; CN: Physiological integrity; CNS: Physiological adaptation; CL: Application

110. Which of the following terms is used to describe pain in the calf due to sharp dorsiflexion of the foot?
1. Dyskinesia
2. Eversion
3. Positive Babinski's reflex
4. Positive Homans' sign

I have a pain in my foot. I'm going to go Homan rest it. (Hah!)

110. 4. A positive Homans' sign (elicited by quickly dorsiflexing the foot), when accompanied by other findings, is diagnostic of deep vein thrombosis (DVT). Alone, however, Homans' sign can't be used to diagnose DVT because other conditions of the calf can produce a positive Homans' sign. Dyskinesia is the inability to perform voluntary movement. Eversion is the outward movement of the transverse tarsal joint. A positive Babinski's reflex is an extensor plantar response.
NP: Evaluation; CN: Health promotion and maintenance; CNS: Prevention and early detection of disease; CL: Comprehension

111. Which of the following conditions causes intermittent claudication (cramplike pains in the calves)?
1. Inadequate blood supply
2. Elevated leg position
3. Dependent leg position
4. Inadequate muscle oxygenation

111. 4. When a muscle is starved of oxygen, it produces pain much like that of angina. Inadequate blood supply would cause necrosis. Leg position either alleviates or aggravates the condition.
NP: Data collection; CN: Physiological integrity; CNS: Physiological adaptation; CL: Knowledge

112. Which of the following medical treatments should be administered to treat intermittent claudication?
1. Analgesics
2. Warfarin (Coumadin)
3. Heparin
4. Pentoxifylline (Trental)

112. 4. Pentoxifylline decreases blood viscosity, increases red blood cell flexibility, and improves flow through small vessels. Analgesics are administered for pain relief. Warfarin and heparin are anticoagulants.
NP: Data collection; CN: Physiological integrity; CNS: Pharmacological therapies; CL: Knowledge

NP: Nursing process   CN: Client needs category   CNS: Client needs subcategory   CL: Cognitive level

113. Which of the following oral medications is administered to prevent further thrombus formation?
1. Warfarin (Coumadin)
2. Heparin
3. Furosemide (Lasix)
4. Metoprolol (Lopressor)

This question requires you to know how certain drugs are administered.

113. 1. Warfarin prevents vitamin K from synthesizing certain clotting factors. This oral anticoagulant can be given long-term. Heparin is a parenteral anticoagulant that interferes with coagulation by readily combining with antithrombin; it can't be given by mouth. Neither furosemide nor metoprolol affect anticoagulation.

NP: Data collection; CN: Physiological integrity; CNS: Pharmacological therapies; CL: Application

114. Which of the following positions would best aid breathing for a client with acute pulmonary edema?
1. Lying flat in bed
2. Left side-lying
3. In high Fowler's position
4. In semi-Fowler's position

114. 3. A high Fowler's position facilitates breathing by reducing venous return. Lying flat and side-lying positions worsen breathing and increase the heart's workload. Semi-Fowler's position won't reduce the workload of the heart as well as high Fowler's position will.

NP: Implementation; CN: Physiological integrity; CNS: Basic care and comfort; CL: Comprehension

115. Which of the following blood gas abnormalities is initially most suggestive of pulmonary edema?
1. Anoxia
2. Hypercapnia
3. Hyperoxygenation
4. Hypocapnia

115. 4. In an attempt to compensate for the increased work of breathing due to hyperventilation, CO_2 decreases, causing hypocapnia. If the condition persists, CO_2 retention occurs and hypercapnia results. Although oxygenation is relatively low, the client isn't anoxic. Hyperoxygenation would result if the client was given oxygen in excess. However, secondary to fluid build-up, the client would have a low oxygenation level.

NP: Data collection; CN: Physiological integrity; CNS: Physiological adaptation; CL: Analysis

116. Which of the following responses does the body initially experience when cardiac output falls?
1. Decreased blood pressure
2. Alteration in level of consciousness (LOC)
3. Decreased blood pressure and diuresis
4. Increased blood pressure and fluid volume

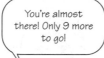

You're almost there! Only 9 more to go!

116. 4. The body compensates for a decrease in cardiac output with a rise in blood pressure due to the stimulation of the sympathetic nervous system, and an increase in fluid volume as the kidneys retain sodium and water. Blood pressure doesn't initially drop in response to the compensatory mechanism of the body. Alteration in LOC will occur only if decreased cardiac output persists.

NP: Data collection; CN: Physiological integrity; CNS: Physiological adaptation; CL: Knowledge

117. Which of the following actions is the appropriate initial response to a client coughing up pink, frothy sputum?
1. Call for help.
2. Call the physician.
3. Start an I.V. line.
4. Suction the client.

Again, you need to prioritize.

117. 1. Production of pink, frothy sputum is a classic sign of acute pulmonary edema. Because the client is at high risk for decompensation, the nurse should call for help, but not leave the room. The other three interventions would immediately follow.

NP: Evaluation; CN: Physiological integrity; CNS: Physiological adaptation; CL: Comprehension

118. Which of the following precautions should a client be instructed to take after an episode of acute pulmonary edema?
1. Limit calorie intake.
2. Restrict carbohydrates.
3. Measure weight twice each day.
4. Call the physician if there's weight gain of more than 3 lb in 1 day.

Don't weight to answer this question (Get it? Tee-hee!)

118. 4. Gaining 3 lb in 1 day is indicative of fluid retention that would increase the heart's workload, thereby putting the client at risk for acute pulmonary edema. Restricting carbohydrates wouldn't affect fluid status. The body needs carbohydrates for energy and healing. Limiting calorie intake doesn't influence fluid status. The client must be weighed in the morning after the first urination. If the client is weighed later in the day, the finding wouldn't be accurate because of fluid intake during the day.

NP: Data collection; CN: Physiological integrity; CNS: Reduction of risk potential; CL: Knowledge

119. Which of the following terms describes the force against which the ventricle must expel blood?
1. Afterload
2. Cardiac output
3. Overload
4. Preload

119. 1. Afterload refers to the resistance normally maintained by the aortic and pulmonic valves, the condition and tone of the aorta, and the resistance offered by the systemic and pulmonary arterioles. Cardiac output is the amount of blood expelled from the heart per minute. Overload refers to an abundance of circulating volume. Preload is the volume of blood in the ventricle at the end of diastole.

NP: Data collection; CN: Physiological integrity; CNS: Physiological adaptation; CL: Knowledge

120. After recovery from an episode of acute pulmonary edema, why would an angiotensin-converting enzyme (ACE) inhibitor be administered?
1. To promote diuresis
2. To increase contractility
3. To decrease contractility
4. To reduce blood pressure

This is another question that tests your knowledge of what certain drugs do.

120. 4. ACE inhibitors are given to reduce blood pressure by decreasing the heart's workload. Diuretics are given to promote diuresis. Inotropic agents increase contractility. Negative inotropic agents decrease contractility.

NP: Data collection; CN: Physiological integrity; CNS: Pharmacological therapies; CL: Knowledge

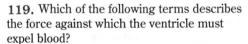

NP: Nursing process   CN: Client needs category   CNS: Client needs subcategory   CL: Cognitive level

121. A 71-year-old male client is told that he needs to increase his potassium intake. Which food should the nurse recommend?
1. Bananas
2. Oatmeal
3. Carrots
4. Corn chips

121. 1. Bananas are high in potassium. Oatmeal is high in fiber. Carrots are high in beta-carotene. Corn chips aren't high in potassium.
NP: Implementation; CN: Health promotion and maintenance; CNS: Prevention and early detection of disease; CL: Comprehension

122. How quickly can an episode of acute pulmonary edema develop?
1. In minutes
2. In ½ hour
3. In 1 hour
4. In 3 hours

How fast did this happen?

122. 1. Pulmonary edema can develop in minutes, secondary to a sudden fluid shift from the pulmonary vasculature to the lung's interstitial alveoli.
NP: Evaluation; CN: Physiological integrity; CNS: Physiological adaptation; CL: Comprehension

123. A 56-year-old woman is having a consultation about possible circulation problems. Which information should the nurse provide about varicosities?
1. They're located only in the calves of the lower extremities.
2. They become stronger as the walls of the vessels hold the blood.
3. They're caused by incompetent arteries.
4. They're veins that have become dilated and lost their elasticity.

123. 4. Varicosities are veins that become dilated and lose their elasticity. They can be located in the lower extremities, esophagus, or rectal area. They're incompetent valves and veins, not arteries.
NP: Evaluation; CN: Physiological integrity; CNS: Reduction of risk potential; CL: Application

124. Which of the following actions should a nurse take when administering a new blood pressure medication to a client?
1. Administer the medication to the client without explanation.
2. Inform the client of the new drug only if he asks about it.
3. Inform the client of the new medication, its name, use, and the reason for the change.
4. Administer the medication and inform the client that the physician will later explain the medication.

Which answer best promotes compliance?

124. 3. Informing the client of the medication, its use, and the reason for the change is important to the care of the client. Teaching the client about his treatment regimen promotes compliance. The other responses are inappropriate.
NP: Implementation; CN: Safe, effective care environment; CNS: Management of care; CL: Application

125. Antihypertensives should be used cautiously in clients taking which of the following drugs?

1. Ibuprofen (Advil)
2. Diphenhydramine (Benadryl)
3. Thioridazine (Mellaril)
4. Vitamins

125. 3. Thioridazine affects the neurotransmitter norepinephrine, which causes hypotension and other cardiovascular effects. Administering an antihypertensive to a client who already has hypotension could have serious adverse effects. Ibuprofen is an anti-inflammatory that doesn't interfere with the cardiovascular system. Although diphenhydramine does have histaminic effects, such as sedation, it isn't known to decrease blood pressure. Vitamins aren't drugs and don't interfere with cardiovascular function.

NP: Data collection; CN: Physiological integrity; CNS: Pharmacological therapies; CL: Knowledge

This challenging chapter covers HIV infection, AIDS, rheumatoid arthritis, ITP, and lots of other complex disorders. You can handle it, though; I know you can. Go for it!

Chapter 4
Hematologic & immune disorders

1. For a client to be diagnosed with acquired immunodeficiency syndrome (AIDS), which of the following conditions must be present?

1. Infection with human immunodeficiency virus (HIV), tuberculosis, and cytomegalovirus infection
2. Infection with HIV, an alternative lifestyle, and a T-cell count above 200 cells/µl
3. Infection with HIV, $CD4^+$ count below 200 cells/µl, and a T-cell count above 400 cells/µl
4. Infection with HIV, a history of acute HIV infection, and a $CD4^+$ T-cell count below 200 cells/µl

2. Which of the following body fluids <u>most easily</u> transmits human immunodeficiency virus (HIV)?

1. Feces and saliva
2. Blood and semen
3. Breast milk and tears
4. Vaginal secretions and urine

3. Immediately after giving an injection, a nurse is accidentally stuck with the needle when a client becomes agitated. When is the best time for the employer to test the nurse for human immunodeficiency virus (HIV) antibodies to determine if she became infected as a result of the needle stick?

1. Immediately and then again in 6 weeks
2. Immediately and then again in 3 months
3. In 2 weeks and then again in 6 months
4. In 2 weeks and then again in 1 year

More than one answer may be correct. Be careful to choose the best answer.

Time is commonly a crucial element in testing for HIV.

1. 4. Three criteria must be met for a client to be diagnosed with AIDS. He must be HIV-positive, have a $CD4^+$ T-cell count below 200 cells/µl, and the presence of one or more specific conditions that include acute infection with HIV. Because HIV attaches to the $CD4^+$ receptor sites of the T cell, a T-cell value alone is incorrect.

NP: Data collection; CN: Safe, effective care environment; CNS: Management of care; CL: Knowledge

2. 2. HIV is most easily transmitted in blood, semen, and vaginal secretions. However, it has also been found in urine, feces, saliva, tears, and breast milk.

NP: Data collection; CN: Health promotion and maintenance; CNS: Prevention and early detection of disease; CL: Knowledge

3. 2. The employer will want to test the nurse immediately to determine whether a preexisting infection is present, and then again in 3 months to detect seroconversion as a result of the needle stick. Waiting 2 weeks to perform the first test is too late to detect preexisting infection. Testing sooner than 3 months may yield false-negative results.

NP: Implementation; CN: Health promotion and maintenance; CNS: Prevention and early detection of disease; CL: Application

NP: Nursing process   CN: Client needs category   CNS: Client needs subcategory   CL: Cognitive level

4. Which of the following blood tests is used first to identify a response to human immunodeficiency virus (HIV) infection?
1. Western blot
2. CD4$^+$ T-cell count
3. Erythrocyte sedimentation rate
4. Enzyme-linked immunosorbent assay (ELISA)

Question 4 is asking you to prioritize.

4. 4. The ELISA is the first screening test for HIV. A Western blot test confirms a positive ELISA test. Other blood tests that support the diagnosis of HIV include CD4$^+$ and CD8$^+$ counts, complete blood cell counts, immunoglobulin levels, p24 antigen assay, and quantitative ribonucleic acid assays.

NP: Data collection; CN: Health promotion and maintenance; CNS: Prevention and early detection of disease; CL: Knowledge

5. The nurse would instruct a client with human immunodeficiency virus who has frequent bouts of diarrhea to avoid eating or drinking:
1. milk.
2. red licorice.
3. chicken soup.
4. broiled meat.

Hmmm, what shall I have?

5. 1. Clients with chronic diarrhea may develop intolerance to lactose, which may worsen the diarrhea. Although red licorice may be eaten, black licorice should be avoided. Other foods that the client should avoid include fatty foods, other lactose-containing foods, caffeine, and sugar. Chicken soup and broiled meat may be consumed.

NP: Implementation; CN: Physiological integrity; CNS: Reduction of risk potential; CL: Application

6. Which of the following dietary recommendations may help the client with rheumatoid arthritis reduce inflammation?
1. Fish oil
2. Vitamin D
3. Iron-rich foods
4. Calcium carbonate

6. 1. The therapeutic effect of fish oil suppresses inflammatory mediator production (such as prostaglandins); how it works is unknown. Iron-rich foods are recommended to decrease the anemia associated with rheumatoid arthritis. Calcium and vitamin D supplements may help reduce bone resorption.

NP: Planning; CN: Physiological integrity; CNS: Physiological adaptation; CL: Knowledge

7. Which of the following nonsteroidal anti-inflammatory drugs (NSAIDs) is most often used to treat rheumatoid arthritis?
1. Furosemide
2. Haloperidol
3. Ibuprofen
4. Methotrexate

More than one answer may seem right. Read the key words carefully.

7. 3. Ibuprofen, fenoprofen, naproxen, piroxicam, and indomethacin are NSAIDs used for clients with rheumatoid arthritis. Furosemide is a loop diuretic, and haloperidol is an antipsychotic agent, neither of which is used to treat rheumatoid arthritis. Methotrexate is an immunosuppressant used in the *early* treatment of rheumatoid arthritis.

NP: Implementation; CN: Physiological integrity; CNS: Pharmacological therapies; CL: Knowledge

8. Which of the following methods of transmission has the <u>most</u> risk for exposure to human immunodeficiency virus (HIV)?
1. Routine teeth cleaning
2. Intercourse with your spouse
3. Unprotected, noninsertive relations
4. Intercourse with a new partner without a condom

The key word in question 8 is most.

8. 4. Having intercourse with a new partner is risky because of the unknown I.V. drug use and sexual history. Use of a condom may increase the protection against HIV exposure. Safer sex practices include autosexual activities, abstinence, relations in a monogamous uninfected couple, and noninsertive relations. Having your teeth cleaned isn't a risk factor if the dental office properly sterilizes the equipment.
NP: Planning; CN: Health promotion and maintenance; CNS: Prevention and early detection of disease; CL: Comprehension

9. Which of the following clients is most at risk for developing rheumatoid arthritis?
1. A 25-year-old woman
2. A 40-year-old man
3. A 65-year-old woman
4. A 70-year-old man

9. 3. Rheumatoid arthritis affects women three times more often than men. The average age of onset is age 55.
NP: Data collection; CN: Health promotion and maintenance; CNS: Prevention and early detection of disease; CL: Analysis

10. Which instruction would be appropriate when teaching a client with human immunodeficiency virus who's at high risk for altered oral mucous membranes?
1. "Brush your teeth frequently with a firm toothbrush."
2. "Use mouthwash that contains an astringent agent."
3. "Be sure to heat all your food."
4. "Lubricate your lips."

10. 4. Lubricating the lips will keep them moist and prevent cracking. A firm toothbrush would damage already sensitive gums. An astringent would be painful, as would foods that are too hot.
NP: Implementation; CN: Physiological integrity; CNS: Reduction of risk potential; CL: Application

11. A pregnant woman has just been diagnosed with human immunodeficiency virus (HIV). Which of the following methods or actions does not put the neonate at risk for infection? •••
1. Vaginal birth
2. Vertical (in utero)
3. Breast-feeding after birth
4. Changing diapers after birth

Watch out! This is a negative question.

11. 4. Changing diapers doesn't put the baby at risk for HIV. HIV infection can occur vertically during pregnancy or a vaginal birth and through breast-feeding.
NP: Implementation; CN: Safe, effective care environment; CNS: Safety and infection control; CL: Application

12. Which of the following groups or factors is linked to higher morbidity and mortality in human immunodeficiency virus (HIV)–infected clients?
1. Homosexual men
2. Lower socioeconomic levels
3. Treatment in a large teaching hospital
4. Treatment by a physician who specializes in HIV infection

12. 2. Morbidity and mortality have been associated with lower socioeconomic status, receiving care in a community hospital or by a physician without much experience with HIV infection, or lack of access to adequate health care.
NP: Data collection; CN: Physiological integrity; CNS: Physiological adaptation; CL: Analysis

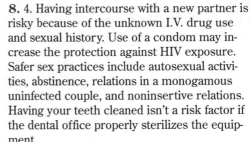

13. After teaching a client about rheumatoid arthritis, which of the following statements indicates the client understands the disease process?

1. "It will get better and worse again."
2. "Once it clears up, it will never come back."
3. "I'll definitely have to have surgery for this."
4. "It will never get any better than it is right now."

14. Which of the following medications would be prescribed <u>first</u> for a client with rheumatoid arthritis?

1. Aspirin
2. Cytoxan
3. Ferrous sulfate
4. Prednisone

The word first is a clue!

15. Human immunodeficiency virus (HIV) primarily attacks the immune system. HIV does <u>not</u> cause which of the following illnesses?

1. Anemia
2. Arthritis
3. Cardiomyopathy
4. Glaucoma

Watch out! This is another negative question.

16. Which of the following blood components is decreased in anemia?

1. Erythrocytes
2. Granulocytes
3. Leukocytes
4. Platelets

This means prioritize!

17. A middle-aged client arrives at the emergency department complaining of chest and stomach pain. He also reports passing black stools for a month. Which of the following interventions is done <u>first</u>?

1. Give nasal oxygen.
2. Take his vital signs.
3. Begin cardiac monitoring.
4. Draw blood for laboratory analysis.

13. 1. The client with rheumatoid arthritis needs to understand it's a somewhat unpredictable disease characterized by periods of exacerbation and remission. There's no cure, but symptoms can be managed at times. Surgery may be indicated in some cases, but not always.

NP: Evaluation; CN: Psychosocial integrity; CNS: Coping and adaptation; CL: Application

14. 1. Nonsteroidal anti-inflammatory drugs (NSAIDs), such as aspirin, are considered first-line therapy by some physicians. Cytoxan may be used in cases of severe synovitis, rather than as first-line therapy. Ferrous sulfate isn't used to treat rheumatoid arthritis. Prednisone may be used to control inflammation when NSAIDs aren't tolerated.

NP: Implementation; CN: Physiological integrity; CNS: Pharmacological therapies; CL: Application

15. 4. Glaucoma hasn't been associated with HIV infection. HIV has been shown to cause neuropathy, cardiomyopathy, psoriasis, arthritis, cervicitis, uveitis, pneumonia, malabsorption of the small bowel, nephritis, gonad dysfunction, anemia, thrombocytopenia, granulocytopenia, and adrenalitis.

NP: Evaluation; CN: Physiological integrity; CNS: Reduction in risk potential; CL: Knowledge

16. 1. Anemia is defined as a decreased number of erythrocytes. Leukopenia is a decreased number of leukocytes. Thrombocytopenia is a decreased number of platelets. Granulocytopenia is a decreased number of granulocytes.

NP: Data collection; CN: Health promotion and maintenance; CNS: Prevention and early detection of disease; CL: Knowledge

17. 2. His vital signs will show hemodynamic stability, and monitoring his heart rhythm may be indicated based on your findings. Giving nasal oxygen and drawing blood require a physician's order and wouldn't be part of a screening evaluation.

NP: Data collection; CN: Physiological integrity; CNS: Physiological adaptation; CL: Application

NP: Nursing process   CN: Client needs category   CNS: Client needs subcategory   CL: Cognitive level

18. A physician evaluates a client who arrived at the emergency department with chest and stomach pain and a report of black, tarry stools for several months. Which of the following orders is appropriate?

1. Cardiac monitoring, oxygen, creatine kinase (CK) and lactate dehydrogenase (LDH) levels
2. Prothrombin time (PT), partial thromboplastin time (PTT), fibrinogen and fibrin split product values
3. Electrocardiogram (ECG), complete blood count, testing for occult blood, comprehensive serum metabolic panel
4. EEG, alkaline phosphatase and aspartate aminotransferase (AST) levels, basic serum metabolic panel

19. Which of the following goals for medications prescribed to treat rheumatoid arthritis is accurate?

1. To cure the disease
2. To prevent osteoporosis
3. To control inflammation
4. To encourage bone regeneration

20. A client with anemia may be tired due to a tissue deficiency of which of the following substances?

1. Carbon dioxide
2. Factor VIII
3. Oxygen
4. T-cell antibodies

21. The nurse in a family health clinic is caring for a female client with anemia. The nurse recognizes that she:

1. needs to restrict activity as much as possible.
2. should be encouraged to eat foods high in calcium.
3. needs to have activities spaced to allow for rest periods.
4. should be supervised when ambulating.

Make sure you know the action of the drug before you administer it.

So, where do you stem from?

18. 3. An ECG evaluates the complaint of chest pain, laboratory tests determine anemia, and the stool test for occult blood determines blood in the stool. Cardiac monitoring, oxygen, and CK and LDH levels are appropriate for a primary cardiac problem. A basic metabolic panel and alkaline phosphatase and AST levels assess liver function. PT, PTT, fibrinogen, and fibrin split products are measured to verify bleeding dyscrasias. An EEG evaluates the brain's electrical activity.

NP: Evaluation; CN: Physiological integrity; CNS: Reduction of risk potential; CL: Analysis

19. 3. The goal of medications in the treatment of rheumatoid arthritis is to control inflammation. There's no cure for rheumatoid arthritis. Rheumatoid arthritis causes bone erosion at the joints, not osteoporosis. Medications aren't available to replace bone lost through erosion.

NP: Implementation; CN: Physiological integrity; CNS: Pharmacological therapies; CL: Comprehension

20. 3. Anemia stems from a decreased number of red blood cells and the resulting deficiency in oxygen in body tissues. Clotting factors, such as factor VIII, relate to the body's ability to form blood clots and aren't related to anemia, nor is carbon dioxide or T-cell antibodies.

NP: Data collection; CN: Physiological integrity; CNS: Physiological adaptation; CL: Application

21. 3. Clients with anemia fatigue easily and need rest between activities to conserve energy. Activities don't need to be severely restricted for clients with anemia. The client needs to eat food that is high in iron, such as lean red meat and fortified breakfast cereal, not calcium. The client doesn't need close supervision when walking.

NP: Planning; CN: Physiological integrity; CNS: Physiological adaptation; CL: Application

22. Which of the following conditions is not a cause of blood disorders such as anemia?
1. Chronic disease
2. Malnutrition
3. Medications
4. Rhinovirus infection

Careful. Not makes this a negative question.

22. 4. Trauma, chronic illness, surgery, malnutrition, medication, toxins, radiation, genetic and congenital disorders, and sepsis can cause anemia. It isn't caused by a rhinovirus infection, which causes the common cold.
NP: Evaluation; CN: Health promotion and maintenance; CNS: Prevention and early detection of disease; CL: Comprehension

23. A client is admitted with acute lymphoblastic leukemia. He's scheduled to have a bone marrow transplant in which he'll receive his own harvested bone marrow. What type of transplant does this describe?
1. Allogeneic transplant
2. Autologous transplant
3. Solid organ transplant
4. Syngeneic transplant

23. 2. An autologous transplant is the use of the client's own bone marrow. An allogeneic bone marrow transplant is from another donor. A solid organ transplant is the transplant of an organ, such as the heart, lung, or kidney. A syngeneic transplant is a transplant between twins.
NP: Implementation; CN: Physiological integrity; CNS: Physiological adaptation; CL: Knowledge

24. Which of the following symptoms is expected with a hemoglobin level of 5 to 7 g/dl?
1. None
2. Pallor
3. Palpitations
4. Shortness of breath

24. 1. Mild anemia usually has no clinical signs. Palpitations, shortness of breath, and pallor are associated with *severe* anemia.
NP: Data collection; CN: Physiological adaptation; CNS: Reduction of risk potential; CL: Application

25. Which of the following is not a common cause of anemia?
1. Lack of dietary iron
2. Vitamin C deficiency
3. GI bleeding
4. Hereditary disorders of the red blood cells

The word most is a hint!

25. 2. Anemia can be caused by a lack of vitamin B_{12}, iron, and folic acid, not by a lack of vitamin C. It can also be caused by bleeding from any organ or system, hereditary disorders, and hematopoietic disorders.
NP: Data collection; CN: Physiological integrity; CNS: Physiological adaptation; CL: Comprehension

26. Which of the following age-groups is most at risk for developing anemia?
1. Younger than 24 months
2. 18 to 30 years
3. 30 to 65 years
4. 65 years and older

26. 4. Elderly clients are most at risk for anemia, usually due to financial concerns affecting protein intake or poor dentition that interferes with chewing meat.
NP: Data collection; CN: Health promotion and maintenance; CNS: Prevention and early detection of disease; CL: Comprehension

NP: Nursing process   CN: Client needs category   CNS: Client needs subcategory   CL: Cognitive level

27. The nurse is caring for a 41-year-old male client with rheumatoid arthritis. The nurse plans to ambulate the client:
  1. when the client first awakens in the morning.
  2. after returning from physical therapy.
  3. after the client has a bath.
  4. just before the noontime meal.

You've almost finished one-third of the questions. Way to go!

28. For which of the following conditions is a client who has just had an appendectomy most at risk?
  1. Anemia
  2. Polycythemia
  3. Purpura
  4. Thrombocytopenia

29. Which of the following terms describes a decreased number of platelets?
  1. Thrombectomy
  2. Thrombocytopenia
  3. Thrombocytopathy
  4. Thrombocytosis

This question is a classic. Get it?

30. Which of the following signs and symptoms are classic for thrombocytopenia?
  1. Weakness and fatigue
  2. Dizziness and vomiting
  3. Bruising and petechiae
  4. Light-headedness and nausea

31. Which explanation about the action of heparin should the nurse provide to a client who has just started taking the drug?
  1. It slows the time it takes the blood to clot.
  2. It stops the blood from clotting.
  3. It thins the blood.
  4. It dissolves clots in the arteries of the heart.

27. 3. Warmth and the movement of the extremities during a bath eases the stiffness and pain of rheumatoid arthritis. Ambulation when the client first awakens is the worst time because pain and stiffness are greatest after long periods of immobility. The client may be too tired to walk soon after returning from therapy. There's no relationship between eating and ease of ambulation in rheumatoid arthritis.
NP: Planning; CN: Physiological integrity; CNS: Basic care and comfort; CL: Comprehension

28. 1. Surgery is a risk factor for anemia. Polycythemia can occur from severe hypoxia due to congenital heart and pulmonary disease. Purpura and thrombocytopenia may result from decreased bone marrow production of platelets and doesn't result from surgery.
NP: Data collection; CN: Physiological integrity; CNS: Reduction of risk potential; CL: Application

29. 2. Thrombocytopenia is a decreased number of platelets. Thrombocytosis is an excess number of platelets, and thrombocytopathy is platelet dysfunction. Thrombectomy is the surgical removal of a thrombus.
NP: Data collection; CN: Physiological integrity; CNS: Physiological adaptation; CL: Knowledge

30. 3. Platelets are necessary for clot formation, so petechiae and bruising are signs of a decreased number of platelets. Weakness and fatigue are signs of anemia. Light-headedness, nausea, dizziness, and vomiting are *not* usual signs of thrombocytopenia.
NP: Data collection; CN: Health promotion and maintenance; CNS: Prevention and early detection of disease; CL: Analysis

31. 1. Heparin prolongs the time needed for blood to clot; however, it doesn't thin the blood. If given in large doses, heparin may stop the blood from clotting; however, this isn't why heparin is usually given. Heparin doesn't dissolve clots.
NP: Implementation; CN: Physiological integrity; CNS: Pharmacological therapies CL: Application

32. Disseminated intravascular coagulation (DIC) typically results in complications <u>initially</u> associated with which of the following organs?
1. Brain
2. Kidney
3. Lung
4. Stomach

Pay close attention to the word *initially* before choosing an answer.

32. 2. DIC usually affects the kidneys and extremities, but if it isn't treated, it will affect the lungs, brain, stomach, and adrenal and pituitary glands.

NP: Data collection; CN: Physiological integrity; CNS: Reduction of risk potential; CL: Comprehension

33. A 70-year-old female client with cancer and iron deficiency anemia is on iron replacement therapy. Which of the following statements indicates the client has a good understanding of her medication?
1. "I'll always take the iron supplement on an empty stomach."
2. "I'll take the iron supplement with orange juice."
3. "When I take liquid iron, I should swish it in my mouth before swallowing."
4. "Iron supplements might change my stools to a lighter color."

33. 2. Taking iron with orange juice or ascorbic acid enhances absorption. Taking iron after meals or a snack, rather than on an empty stomach, decreases GI upset. Liquid iron can permanently stain the teeth. The client should use a straw to swallow the liquid. Iron will turn the client's stools dark and tarry, not lighter.

NP: Planning; CN: Physiological integrity; CNS: Pharmacological therapies; CL: Knowledge

34. A pregnant woman arrives at the emergency department with abruptio placentae at 34 weeks' gestation. She's at risk for which of the following blood dyscrasias?
1. Thrombocytopenia
2. Idiopathic thrombocytopenic purpura (ITP)
3. Disseminated intravascular coagulation (DIC)
4. Heparin-associated thrombosis and thrombocytopenia (HATT)

34. 3. Abruptio placentae is a cause of DIC because of activation of the clotting cascade after hemorrhage. Thrombocytopenia results from decreased bone marrow production. ITP can result in DIC, but not because of abruptio placentae. A client with abruptio placentae wouldn't receive heparin and, as a result, wouldn't be at risk for HATT.

NP: Data collection; CN: Physiological integrity; CNS: Reduction of risk potential; CL: Application

35. Which of the following conditions is <u>not</u> caused by disseminated intravascular coagulation (DIC)?
1. Organ tissue damage
2. Depletion of circulating clotting factors
3. Thrombus formation in the large vessels
4. Activation of the clotting-dissolving process

Another negative question!

35. 3. DIC occurs in response to a primary problem that initiates the clotting cascade, resulting in hypercoagulability. This state produces clots that can block the microcirculation. As the clotting factors are depleted and fibrinolysis occurs, then a hypocoagulable state results. Clotting and bleeding in the microcirculation can cause tissue damage.

NP: Data collection; CN: Physiological integrity; CNS: Physiological adaptation; CL: Comprehension

36. A client with thrombocytopenia, secondary to leukemia, develops epistaxis. The nurse should instruct the client to:
1. lie supine with his neck extended.
2. sit upright, leaning slightly forward.
3. blow his nose and then put lateral pressure on it.
4. hold his nose while bending forward at the waist.

36. 2. The upright position, leaning slightly forward, avoids increasing the vascular pressure in the nose and helps the client avoid aspirating blood. Lying supine won't prevent aspiration of blood. Nose blowing can dislodge any clotting that has occurred. Bending at the waist increases vascular pressure and promotes bleeding rather than stopping it.
NP: Implementation; CN: Physiological integrity CNS: Physiological adaptation; CL: Application

So many tests! Which is the best?

37. Which of the following laboratory tests, besides a platelet count, is best to confirm the diagnosis of essential thrombocytopenia?
1. Bleeding time
2. Complete blood count (CBC)
3. Immunoglobulin (Ig) G level
4. Prothrombin time (PT) and International Normalized Ratio (INR)

37. 1. After a platelet count, the best test to determine thrombocytopenia is a bleeding time. The platelet count is decreased and bleeding time is prolonged. IgG assays are nonspecific, but may help determine the diagnosis. A CBC shows the hemoglobin levels, hematocrit levels, and white blood cell values. PT and INR evaluate the effect of warfarin therapy.
NP: Data collection; CN: Physiological integrity; CNS: Physiological adaptation; CL: Application

38. The nurse is documenting care for a client with iron deficiency anemia. Which of the following nursing diagnoses is most appropriate?
1. Impaired gas exchange
2. Deficient fluid volume
3. Ineffective airway clearance
4. Ineffective breathing pattern

38. 1. Iron is necessary for hemoglobin synthesis. Hemoglobin is responsible for oxygen transport in the body. Iron deficiency anemia causes subnormal hemoglobin levels, which impair tissue oxygenation and bring about a nursing diagnosis of impaired gas exchange. Iron deficiency anemia doesn't cause a deficient fluid volume and is less directly related to ineffective airway clearance and ineffective breathing pattern than it is to ineffective gas exchange.
NP: Evaluation; CN: Physiological integrity; CNS: Physiological adaptation; CL: Analysis

Don't quit now! You're getting there!

39. Which of the following organs are part of the immune system?
1. Adenoids and tonsils
2. Adrenals and kidneys
3. Lymph nodes and thymus
4. Pancreas and liver

39. 3. The immune system includes the lymph nodes, thymus, and spleen. Adenoids and tonsils are part of the respiratory system. The adrenals are endocrine organs. Kidneys belong to the genitourinary system. The liver and pancreas are part of the GI system.
NP: Data collection; CN: Physiological integrity; CNS: Physiological adaptation; CL: Knowledge

40. Which of the following definitions best explains the function of the thymus gland in the immune system?
1. The thymus gland is a reservoir for blood.
2. The thymus gland stores blood cells until they mature.
3. The thymus gland protects the body against ingested pathogens.
4. The thymus gland removes bacteria and toxins from the circulatory system.

41. T cells are involved in which of the following types of immunity?
1. Humoral immunity
2. Cell-mediated immunity
3. Antigen-mediated immunity
4. Immunoglobulin-mediated immunity

42. A client is scheduled to receive a heart valve replacement with a porcine valve. Which of the following types of transplant is this?
1. Allogeneic
2. Autologous
3. Syngeneic
4. Xenogeneic

43. Which of the following clients is most at risk for developing malignant lymphoma?
1. A 22-year-old man with a history of mononucleosis
2. A 25-year-old man who smokes a pack of cigarettes a day
3. A 33-year-old man whose sister has Hodgkin's disease
4. A 40-year-old woman with a history of human immunodeficiency virus (HIV) infection

44. What is the life span for normal platelets?
1. 1 to 3 days
2. 3 to 5 days
3. 7 to 10 days
4. 3 to 4 months

Let's see, what kind of immunity do I have a hankering for today?

40. 2. Bone marrow produces immature blood cells (stem cells). Those that become lymphocytes migrate to the bone marrow for maturation (to B lymphocytes) or to the thymus for maturation (to T lymphocytes). These lymphocytes are responsible for cell-mediated immunity. The spleen is a reservoir for blood cells. The tonsils shield against airborne and ingested pathogens, and the lymph nodes remove bacteria and toxins from the bloodstream.
NP: Evaluation; CN: Physiological integrity; CNS: Physiological adaptation; CL: Knowledge

41. 2. T cells are responsible for cell-mediated immunity, in which the T cells respond directly to the antigen. B cells are responsible for humoral or immunoglobulin-mediated immunity. There's no antigen-mediated immunity.
NP: Data collection; CN: Physiological integrity; CNS: Physiological adaptation; CL: Knowledge

42. 4. An xenogeneic transplant is between humans and another species. A syngeneic transplant is between identical twins, allogeneic transplant is between two humans, and autologous is a transplant from the same individual.
NP: Data collection; CN: Physiological integrity; CNS: Physiological adaptation; CL: Comprehension

43. 1. Malignant lymphoma has a peak incidence between ages 20 and 30 and after age 50. It's more common in men than in women and is associated with a history of Epstein-Barr virus (which causes mononucleosis). There's also an increased incidence of the disease among siblings. There's no reported association between malignant lymphoma and smoking, I.V. drug use, or HIV infection.
NP: Data collection; CN: Health promotion and maintenance; CNS: Prevention and early detection of disease; CL: Comprehension

44. 3. The life span of a normal platelet is 7 to 10 days. However, in idiopathic thrombocytopenia, the platelet life span is reduced to 1 to 3 days.
NP: Planning; CN: Physiological integrity; CNS: Physiological adaptation; CL: Knowledge

NP: Nursing process   CN: Client needs category   CNS: Client needs subcategory   CL: Cognitive level

45. What's the normal life span for healthy red blood cells (RBCs)?
1. 60 days
2. 90 days
3. 120 days
4. 240 days

46. Which of the following statements indicates that a client with thrombocytopenia understands the function of platelets in her body?
1. "Platelets regulate acid-base balance."
2. "Platelets regulate the immune response."
3. "Platelets protect the body from infection."
4. "Platelets stop bleeding when arteries and veins are injured."

47. A 43-year-old woman is undergoing treatment for colon cancer. The physician documents thrombocytopenia on the client's diagnosis list. What observations can the nurse expect?
1. Diarrhea
2. Thin, brittle hair
3. Bruises on the skin
4. Urinary urgency

48. A client involved in a motor vehicle accident arrives in the emergency department unconscious and severely hypotensive. He's suspected to have several fractures (pelvis and legs). Which of the following parenteral fluids is the best choice for his current condition?
1. Whole blood
2. Normal saline solution
3. Lactated Ringer's solution
4. Packed red blood cells (RBCs)

Congratulations! You're more than halfway finished!

What observations can the nurse expect with this question?

45. 3. A healthy RBC lives for about 120 days.
NP: Data collection; CN: Physiological integrity; CNS: Physiological adaptation; CL: Knowledge

46. 4. Platelets clump together to plug small breaks in blood vessels. They also initiate the clotting cascade by releasing thromboplastin, which (in the presence of calcium) converts prothrombin into thrombin. Platelets don't perform the other functions.
NP: Evaluation; CN: Physiological integrity; CNS: Reduction of risk potential; CL: Application

47. 3. With thrombocytopenia, there's an abnormal decrease in the number of blood platelets, which can result in bruises and bleeding. The client may have constipation, but usually not diarrhea. Thin, brittle hair isn't a sign of thrombocytopenia, but could be a sign of hypothyroidism. Urinary urgency could be a sign of urinary tract infection, but not thrombocytopenia.
NP: Data collection; CN: Physiological integrity; CNS: Reduction of risk potential; CL: Application

48. 4. In a trauma situation, the first blood product given is unmatched (O negative) packed RBCs. Fresh frozen plasma is commonly used to replace clotting factors. Normal saline or lactated Ringer's solution is used to increase volume and blood pressure, but too much colloid will hemodilute the blood and won't improve oxygen-carrying capacity, whereas RBCs would.
NP: Planning; CN: Physiological integrity; CNS: Physiological adaptation; CL: Application

49. A nurse is monitoring a 17-year-old male client who's receiving a blood transfusion for volume replacement. The client complains of itching about 20 minutes after the infusion begins. The nurse should:
1. stop the infusion immediately.
2. call the physician immediately.
3. give the client diphenhydramine (Benadryl) and continue the infusion.
4. do nothing because the time has passed when a reaction could have occurred.

Stop! Think carefully about this question.

49. 1. Itching is a sign of an adverse reaction, so the nurse should stop the infusion immediately. The physician should be called, but only after the infusion has been stopped and the client is assessed. No medications should be administered without a physician's order. Reactions can occur up to several hours after the infusion.

NP: Implementation; CN: Physiological integrity; CNS: Reduction of risk potential; CL: Application

50. Which of the following clients is at the highest risk for systemic lupus erythematosus (SLE)?
1. A 20-year-old White man
2. A 25-year-old Black woman
3. A 45-year-old Hispanic man
4. A 65-year-old Black woman

50. 2. SLE affects women eight times more often than men and usually strikes during childbearing age. It's three times more common in Black women than in White women.

NP: Data collection; CN: Health promotion and maintenance; CNS: Prevention and early detection of disease; CL: Comprehension

Pay close attention to the key word *primarily* here.

51. Systemic lupus erythematosus (SLE) primarily attacks which of the following tissues?
1. Connective
2. Heart
3. Lung
4. Nerve

51. 1. SLE is a chronic, inflammatory, autoimmune disorder that primarily affects connective tissue. It also affects the skin and kidneys and may affect the pulmonary, cardiac, neural, and renal systems.

NP: Data collection; CN: Physiological integrity; CNS: Physiological adaptation; CL: Knowledge

52. Which of the following symptoms is most commonly an early indication of stage I Hodgkin's disease?
1. Pericarditis
2. Night sweats
3. Splenomegaly
4. Persistent hypothermia

52. 2. In stage I, symptoms include a single enlarged lymph node (usually), unexplained fever, night sweats, malaise, and generalized pruritus. Although splenomegaly may be present in some clients, night sweats are generally more prevalent. Pericarditis isn't associated with Hodgkin's disease. Persistent hypothermia is associated with Hodgkin's disease, but isn't an early sign.

NP: Data collection; CN: Health promotion and maintenance; CNS: Prevention and early detection of disease; CL: Knowledge

53. Which of the following statements shows that a client does not understand the cause of an exacerbation of systemic lupus erythematosus (SLE)?

1. "I need to stay away from sunlight."
2. "I don't have to worry if I get a strep throat."
3. "I need to work on managing stress in my life."
4. "I don't have to worry about changing my diet."

54. Which of the following complications of systemic lupus erythematosus (SLE) is most common and most serious?

1. Arthritis
2. Nephritis
3. Pericarditis
4. Pleural effusion

55. A 73-year-old female client is about to receive a blood transfusion to treat severe anemia. She asks the nurse how long the procedure will take. The nurse explains that the treatment takes:

1. 8 hours.
2. at least 12 hours.
3. at least 24 hours.
4. 4 hours.

56. A 61-year-old male client is receiving a blood transfusion to treat volume depletion that occurred during recent surgery. What solution can safely be given during a blood transfusion?

1. Normal saline solution
2. Any I.V. antibiotic
3. Total parenteral nutrition
4. Half-normal saline solution

57. Which of the following symptoms is a classic sign of systemic lupus erythematosus (SLE)?

1. Vomiting
2. Weight loss
3. Difficulty urinating
4. Superficial lesions over the cheeks and nose

Watch out for this negative question!

This hint is a doubleheader!

Always remember safety.

53. 2. Infection may cause an exacerbation of SLE. Other factors that can precipitate an exacerbation are immunizations, sunlight exposure, and stress. A client's diet doesn't exacerbate SLE.

NP: Evaluation; CN: Health promotion and maintenance; CNS: Prevention and early detection of disease; CL: Application

54. 2. About 50% of the clients with SLE have some type of nephritis, and kidney failure is the most common cause of death for clients with SLE. Pericarditis is the most common cardiovascular manifestation of SLE, but it isn't usually life-threatening. Arthritis is very common (95%), as are pleural effusions (50%), but neither is life-threatening.

NP: Data collection; CN: Physiological integrity; CNS: Physiological adaptation; CL: Analysis

55. 4. The American Association of Blood Banks recommends that blood or blood components should be transfused within 4 hours. If they aren't, they should be divided and stored appropriately in the blood bank. Any length of time over 4 hours would compromise the integrity of the transfusion components.

NP: Implementation; CN: Safe, effective care environment; CNS: Safety and infection control; CL: Application

56. 1. The only solution that can safely be given during a blood transfusion is normal saline solution. If given to a client during a blood transfusion, these other solutions could cause incompatibility reactions.

NP: Implementation; CN: Safe, effective care environment; CNS: Safety and infection control; CL: Application

57. 4. Although all these symptoms can be signs of SLE, the classic sign is the butterfly rash over the cheeks and nose.

NP: Data collection; CN: Physiological integrity; CNS: Physiological adaptation; CL: Knowledge

58. Which of the following laboratory test results supports the diagnosis of systemic lupus erythematosus (SLE)?
  1. Elevated serum complement level
  2. Thrombocytosis, elevated sedimentation rate
  3. Pancytopenia, elevated antinuclear antibody (ANA) titer
  4. Leukocytosis, elevated blood urea nitrogen (BUN) and creatinine levels

58. 3. Laboratory findings for clients with SLE usually show pancytopenia, elevated ANA titer, and decreased serum complement levels. Clients may have elevated BUN and creatinine levels from nephritis, but the increase does *not* indicate SLE. Thrombocytosis and elevated sedimentation rate usually indicate polyarteritis nodosa, not SLE.
NP: Data collection; CN: Physiological integrity; CNS: Physiological adaptation; CL: Application

59. Which of the following laboratory values is expected for a client just diagnosed with chronic lymphocytic leukemia?
  1. Elevated erythrocyte sedimentation rate (ESR)
  2. Uncontrolled proliferation of granulocytes
  3. Thrombocytopenia and increased lymphocytes
  4. Elevated aspartate aminotransferase (AST) and alanine aminotransferase (ALT) levels

59. 3. Chronic lymphocytic leukemia shows a proliferation of small abnormal mature B lymphocytes and decreased antibody response. Thrombocytopenia also is commonly present. Uncontrolled proliferation of granulocytes occurs in myelogenous leukemia. AST, ALT, and ESR values are *not* affected.
NP: Data collection; CN: Physiological integrity; CNS: Physiological adaptation; CL: Knowledge

60. At the time of diagnosis of Hodgkin's disease, which of the following areas is commonly involved?
  1. Back
  2. Chest
  3. Groin
  4. Neck

60. 4. At the time of diagnosis, a painless cervical lesion is commonly present. The back, chest, and groin areas aren't involved.
NP: Data collection; CN: Physiological integrity; CNS: Physiological adaptation; CL: Knowledge

61. According to a standard staging classification of Hodgkin's disease, which of the following criteria reflects stage II?
  1. Involvement of extralymphatic organs or tissues
  2. Involvement of a single lymph node region or structure
  3. Involvement of two or more lymph node regions or structures
  4. Involvement of lymph node regions or structures on both sides of the diaphragm

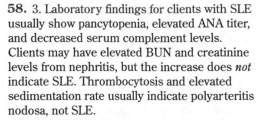
You're doing great! Don't quit now!

61. 3. Stage II involves two or more lymph node regions. Stage I involves only one lymph node region; stage III involves nodes on both sides of the diaphragm; and stage IV involves extralymphatic organs or tissues.
NP: Data collection; CN: Physiological integrity; CNS: Physiological adaptation; CL: Knowledge

NP: Nursing process   CN: Client needs category   CNS: Client needs subcategory   CL: Cognitive level

62. A client is about to start chemotherapy for acute lymphocytic leukemia. Which of the following statements shows he does not understand this phase of chemotherapy?

1. "I'll have treatments only once per month."
2. "I'll be getting high doses of chemotherapy."
3. "I won't get sick at this stage of the treatment."
4. "The purpose of these treatments is to induce a remission."

Warning: The word *not* makes this a negative question.

62. 1. The initial phase of chemotherapy is called the induction phase and is designed to put the client into remission by giving high doses of the drugs. Treatments will be closer together than once per month. Monthly treatments usually occur during the maintenance phase of chemotherapy.

NP: Evaluation; CN: Physiological integrity; CNS: Basic care and comfort; CL: Knowledge

63. When protective isolation is not indicated, which of the following activities is recommended for a client receiving chemotherapy?

1. Bed rest
2. Activity as tolerated
3. Walk to bathroom only
4. Out of bed for brief periods

63. 2. It's important that the client be able to engage in activities that are of interest and to maintain as much independence and autonomy as possible. Bed rest isn't necessary, nor is it necessary to limit the client's activity to only walks to the bathroom or out of bed for brief periods.

NP: Intervention; CN: Health promotion and maintenance; CNS: Prevention and early detection of disease; CL: Application

Hint! Hint! Hint!

64. The nurse is teaching a client with allergies how to prevent anaphylaxis. Which recommendation is most appropriate for this client?

1. Dry mop all hardwood floors.
2. Wear a medical identification bracelet or necklace.
3. Have carpet installed in every room of the house.
4. Advise family and friends not to visit during the winter.

64. 2. If the client becomes unconscious or couldn't report allergies, medical identification jewelry could provide that information and help prevent anaphylaxis caused by administration of the antigen. The client should wet mop hardwood floors because dry mopping scatters dust, which can trigger allergies. The client should minimize the amount of carpet in the home because carpet traps allergens, such as dust and dirt. Unless the client is ill, the nurse may encourage visits by family and friends to promote healthy social interaction.

NP: Implementation; CN: Health promotion and maintenance; CNS: Prevention and early detection of disease; CL: Comprehension

65. Which nursing intervention is most appropriate for a client with multiple myeloma?

1. Monitoring respiratory status
2. Balancing rest and activity
3. Restricting fluid intake
4. Preventing bone injury

65. 4. When caring for a client with multiple myeloma, the nurse should focus on relieving pain, preventing bone injury and infection, and maintaining hydration. Monitoring respiratory status and balancing rest and activity are appropriate interventions for any client. To prevent such complications as pyelonephritis and renal calculi, the nurse should keep the client well hydrated, not restrict the client's fluid intake.

NP: Implementation; CN: Safe, effective care environment; CNS: Safety and infection control; CL: Application

66. Which of the following foods should a client with leukemia avoid?
1. White bread
2. Carrot sticks
3. Stewed apples
4. Medium-rare steak

67. A client with leukemia has neutropenia. Which of the following functions must be frequently monitored?
1. Blood pressure
2. Bowel sounds
3. Heart sounds
4. Breath sounds

68. Which process removes excess white blood cells (WBCs) from the body?
1. Erythrapheresis
2. Granulapheresis
3. Leukapheresis
4. Plasmapheresis

69. Which of the following clients is at the highest risk for developing multiple myeloma?
1. A 20-year-old Asian woman
2. A 30-year-old White man
3. A 50-year-old Hispanic woman
4. A 60-year-old Black man

70. Which of the following substances has abnormal values in multiple myeloma?
1. Immunoglobulins
2. Platelets
3. Red blood cells (RBCs)
4. White blood cells (WBCs)

71. For which of the following conditions is a client with multiple myeloma monitored?
1. Hypercalcemia
2. Hyperkalemia
3. Hypernatremia
4. Hypermagnesemia

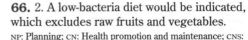

66. 2. A low-bacteria diet would be indicated, which excludes raw fruits and vegetables.
NP: Planning; CN: Health promotion and maintenance; CNS: Prevention and early detection of disease; CL: Application

67. 4. Pneumonia — viral and fungal — is a common cause of death in clients with neutropenia, so frequent assessment of respiratory rate and breath sounds is required. Although assessing blood pressure, bowel sounds, and heart sounds is important, it won't help detect pneumonia.
NP: Data collection; CN: Physiological integrity; CNS: Physiological adaptation; CL: Application

68. 3. Leukapheresis is the removal of excess WBCs. Plasmapheresis is the filtering of the plasma. There are no processes called granulapheresis or erythrapheresis.
NP: Data collection; CN: Physiological integrity; CNS: Basic care and comfort; CL: Knowledge

69. 4. Multiple myeloma is more common in middle-aged and older clients (the median age at diagnosis is age 60) and is twice as common in Blacks as Whites. It occurs most commonly in Black men.
NP: Data collection; CN: Health promotion and maintenance; CNS: Prevention and early detection of disease; CL: Comprehension

70. 1. Multiple myeloma is characterized by an overproduction of immunoglobulins and multiple tumors throughout the skeleton, which results in significant bone marrow loss and hard bone tissue. This produces pancytopenia and osteoporosis. Platelet counts and RBC and WBC levels aren't usually affected in clients with multiple myeloma.
NP: Data collection; CN: Health promotion and maintenance; CNS: Prevention and early detection of disease; CL: Knowledge

71. 1. Calcium is released when bone is destroyed. This causes an increase in serum calcium levels. Multiple myeloma doesn't affect potassium, sodium, or magnesium levels.
NP: Planning; CN: Physiological integrity; CNS: Physiological adaptation; CL: Application

NP: Nursing process   CN: Client needs category   CNS: Client needs subcategory   CL: Cognitive level

72. Which of the following symptoms commonly occurs with hypercalcemia?
1. Tremors
2. Headache
3. Confusion
4. Muscle weakness

73. Which of the following conditions or symptoms is typically a secondary complication of hypercalcemia in a client with multiple myeloma?
1. Pneumonia
2. Muscle spasms
3. Renal dysfunction
4. Myocardial irritability

74. The neurologic complications of multiple myeloma usually involve which of the following body systems?
1. Brain
2. Spinal column
3. Autonomic nervous system
4. Parasympathetic nervous system

75. Which of the following interventions is stressed in teaching about multiple myeloma?
1. Maintain bed rest.
2. Enforce fluid restriction.
3. Drink 3 L of fluid daily.
4. Keep the lower extremities elevated.

76. Although a client's physiologic response to a health crisis is important to the health outcome, which of the following nursing interventions also must be addressed?
1. Teach the family how to care for the client.
2. Help the client effectively cope with the crisis.
3. Maintain I.V. access, medications, and diet.
4. Teach the client basic information about the illness.

Don't get confused by this prefix! (Tee-hee!)

Wait a second! Here's a hint for you!

Here's a question that uses a clever little technique — must — to get you to prioritize.

72. 3. Signs of hypercalcemia include confusion, anorexia, nausea, vomiting, abdominal pain, ileus, constipation and, eventually, impaired renal function. Tremors, headache, and muscle weakness aren't common symptoms of hypercalcemia.
NP: Data collection; CN: Physiological integrity; CNS: Physiological adaptation; CL: Application

73. 3. Twenty percent of multiple myeloma clients with hypercalcemia and hyperuricemia develop renal insufficiency. Hypocalcemia causes muscle spasms, and hypokalemia causes myocardial irritability. Pneumonia doesn't result from hypercalcemia.
NP: Planning; CN: Health promotion and maintenance; CNS: Prevention and early detection of disease; CL: Comprehension

74. 2. Back pain or paresthesia in the lower extremities may indicate impending spinal cord compression from a spinal tumor. This should be recognized and treated promptly because progression of the tumor may result in paraplegia. The other options, which reflect parts of the nervous system, aren't usually affected by multiple myeloma.
NP: Data collection; CN: Physiological integrity; CNS: Physiological adaptation; CL: Application

75. 3. The client needs to drink 3 to 5 L of fluid each day to dilute calcium and uric acid to try to reduce the risk of renal dysfunction. Walking is encouraged to prevent further bone demineralization. The lower extremities don't need to be elevated.
NP: Planning; CN: Physiological integrity; CNS: Basic care and comfort; CL: Application

76. 2. Although all of the answers are important in the care of the client, if the individual isn't able to cope with the emotional, spiritual, and psychological aspects of his crisis, the other components of care may be ineffective as well.
NP: Implementation; CN: Psychosocial integrity; CNS: Coping and adaptation; CL: Application

77. To promote healing of a laceration, which of the following interventions is correct?
1. Elevate the body part.
2. Monitor blood pressure.
3. Apply a pressure dressing and heat.
4. Apply a pressure dressing and ice pack.

Only 11 more questions! You're a whiz at this!

77. 4. Pressure dressings help clotting by promoting the localization of microorganisms and the development of meshwork for repair and healing. Ice decreases blood flow to the site, slowing the bleeding. Heat increases blood flow to the site, increasing the bleeding. Monitoring blood pressure is important when the individual is bleeding, but does nothing to promote clotting. Elevating the body part helps reduce edema, but doesn't directly promote healing.

NP: Implementation; CN: Physiological integrity; CNS: Physiological adaptation; CL: Application

78. An elderly client has a wound that isn't healing normally. Interventions should be based on which of the following principles or test results?
1. Laboratory test results
2. Kidney function test results
3. Poor wound healing expected as part of the aging process
4. Diminished immune function interfering with ability to fight infection

78. 4. Immune function is important in the healing process, and diminished response may slow or prevent the healing process from taking place. Although immune function declines with age, there are healthy behaviors that will enhance the elderly individual's response to tissue trauma (nutrition, exercise). Kidney function and laboratory results are important, but are *not* solely responsible for health outcomes.

NP: Data collection; CN: Physiological integrity; CNS: Physiological adaptation; CL: Analysis

79. Which of the following conditions or symptoms is an appropriate immune response?
1. Allergies
2. Autoimmune disorder
3. Inflammation and increased temperature
4. Insufficient protection (immunodeficiency)

Is this question appropriate, or is it just me?

79. 3. Inflammation and increased temperature are normal immune responses to antigens (virus, bacteria, injury). Autoimmune disorders result when immune cells attack self cells and result in long-term diseases. Allergies are heightened responses to antigens to which the body was previously exposed. Insufficient protection results in immunodeficiency disorders that compromise the individual's ability to ward off infections.

NP: Data collection; CN: Physiological integrity; CNS: Reduction of risk potential; CL: Analysis

80. What is the average length of time from human immunodeficiency virus (HIV) infection to the development of acquired immunodeficiency syndrome (AIDS)?
1. Less than 5 years
2. 5 to 7 years
3. 10 years
4. More than 10 years

80. 3. Epidemiological studies show that the average time from initial contact with HIV to the development of AIDS is 10 years.

NP: Planning; CN: Health promotion and maintenance; CNS: Prevention and early detection of disease; CL: Knowledge

NP: Nursing process   CN: Client needs category   CNS: Client needs subcategory   CL: Cognitive level

81. Which of the following conditions or factors may cause an acquired immune deficiency?
1. Age
2. Genetics
3. Environment
4. Medical treatments

82. Which of the following statements best explains the reason for using stress management with clients?
1. Everyone is stressed.
2. It has become an accepted practice.
3. Eastern health practices have shown its effectiveness.
4. Prolonged psychological stress may contribute to the development of physical illness.

83. Corticosteroids are potent suppressors of the body's inflammatory response. Which of the following conditions or actions do they suppress?
1. Arthritis
2. Pain receptors
3. Immune response
4. Neural transmission

84. During the recovery phase of a surgical client's hospitalization, a nurse notes that the client's immune status appears to be altered. Although there's no obvious rationale for the immunocompromise, which of the following areas should be further investigated?
1. Nutrition
2. Acquired immune disorder
3. Family history of immune problems
4. Personal history of substance abuse or use

Almost at the finish line!

Five questions left! Keep going!

81. 4. Immune deficiencies may result from medical treatments, such as medications, radiation, or transplants. Immune function may decline with age, but it isn't considered the cause of acquired immune deficiency. Genetics and environment haven't been shown to be factors in acquired immune deficiency.
NP: Evaluation; CN: Physiological integrity; CNS: Physiological adaptation; CL: Analysis

82. 4. Psychological and emotional stress stimulate the central nervous system, increasing the levels of corticotropin and cortisol, which results in harmful effects on immune, cardiac, neural, and endocrine function. Although stress management may be a common therapy for stressed individuals and Eastern countries may promote its use, nursing interventions need to focus on research-based rationales for intervention. Many people report high levels of stress, but not everyone would claim to be stressed.
NP: Planning; CN: Psychosocial integrity; CNS: Coping and adaptation; CL: Knowledge

83. 3. Corticosteroids suppress eosinophils, lymphocytes, natural-killer cells, and other microorganisms, inhibiting the natural inflammatory process in an infected or injured part of the body. This promotes resolution of inflammation, stabilizes lysosomal membranes, decreases capillary permeability, and depresses phagocytosis of tissues by white blood cells, thus blocking the release of more inflammatory materials. Although corticosteroids help relieve pain from arthritis and other conditions, they act by suppressing the natural immune response.
NP: Data collection; CN: Physiological integrity; CNS: Pharmacological therapies; CL: Knowledge

84. 4. Substance abuse, including alcohol consumption and tobacco or marijuana use, influences immunocompetence and overall health status. Although nutrition is important for immunocompetence, it would be part of the client's daily assessment. A family history would have been assessed initially. Assessing the client for an acquired immune disorder would be a joint effort with the physician and wouldn't be conducted independently.
NP: Data collection; CN: Psychosocial integrity; CNS: Psychosocial adaptation; CL: Application

85. In community health and epidemiological studies, which of the following definitions of disease prevalence is correct?
1. The number of individuals affected by a particular disease at a specific time
2. The rate at which individuals without a specific disease develop that disease
3. The proportion of individuals affected by the disease who live for a particular period
4. The proportion of individuals without the disease who eventually develop the disease within a specific period

85. 1. Prevalence is the number of individuals affected by the disease at a specific time. Risk is the proportion of individuals without the disease who develop the disease within a particular period. Incidence rate is the rapidity with which individuals without the disease contract it. Survival is the proportion of individuals affected by the disease who live for a particular length of time.

NP: Data collection; CN: Health promotion and maintenance; CNS: Prevention and early detection of disease; CL: Analysis

86. Hepatitis B immunizations shouldn't be given to which of the following groups?
1. Neonates
2. Immigrants
3. Health care professionals
4. Individuals older than age 65

86. 1. The immune system of the neonate isn't mature enough to handle the hepatitis B vaccine. Elderly clients, health care professionals, and immigrants are all candidates for the vaccine and may need the protection.

NP: Data collection; CN: Physiological integrity; CNS: Pharmacological therapies; CL: Analysis

87. Epidemiological research has contributed to our knowledge of acquired immunodeficiency syndrome (AIDS) by which of the following actions?
1. Finding a cure
2. Developing vaccines for future use
3. Requiring the reporting of all diagnosed clients and their partners
4. Identifying risk factors that compare between affected and unaffected individuals

87. 4. Epidemiological research has been pivotal in identifying the behaviors that place individuals at risk for AIDS. No vaccine has been shown to be effective to date, but work is ongoing. Not all states require the reporting of sexual partners for clients with AIDS. No cure has been found.

NP: Data collection; CN: Health promotion and maintenance; CNS: Prevention and early detection of disease; CL: Knowledge

88. A 32-year-old client is admitted with a tentative diagnosis of acquired immunodeficiency syndrome (AIDS). The preliminary report of biopsies done on her facial lesions indicates Kaposi's sarcoma. Which of the following approaches would be most appropriate?
1. Tell the client that Kaposi's sarcoma is common in people with AIDS.
2. Pretend not to notice the lesions on the client's face.
3. Inform the client of the biopsy results and support her emotionally.
4. Explore the client's feelings about her facial disfigurement.

Success! You did it!

88. 4. Facial lesions can contribute to decreased self-esteem and a disturbed body image. Discussing AIDS with a client whose diagnosis isn't final may be inappropriate and doesn't provide emotional support. Pretending not to notice visible lesions ignores the client's concerns. The primary care provider — not the nurse — should inform the client of the biopsy results.

NP: Implementation; CN: Psychosocial integrity; CNS: Psychosocial adaptation; CL: Application

Looking for the latest information about respiratory disorders? Check out the American Association for Respiratory Care's Web site at **www.aarc.org.** It'll respire — er, inspire — you

Chapter 5
Respiratory disorders

1. Clients with chronic illnesses are <u>more likely</u> to get pneumonia when which of the following situations is present?
1. Dehydration
2. Group living
3. Malnutrition
4. Severe periodontal disease

In question 1, the word *chronic* is a hint for finding the correct answer.

2. Which of the following pathophysiological mechanisms that occur in the lung parenchyma allows pneumonia to develop?
1. Atelectasis
2. Bronchiectasis
3. Effusion
4. Inflammation

3. Which of the following organisms most commonly causes community-acquired pneumonia in adults?
1. *Haemophilus influenzae*
2. *Klebsiella pneumoniae*
3. *Streptococcus pneumoniae*
4. *Staphylococcus aureus*

Pssst. Yeah. I'm talking to you. You'd better strep lively — er, I mean, step lively. Got it?

4. An elderly client with pneumonia may appear with which of the following symptoms first?
1. Altered mental status and dehydration
2. Fever and chills
3. Hemoptysis and dyspnea
4. Pleuritic chest pain and cough

1. 2. Clients with chronic illness generally have poor immune systems. Often, residing in group living situations increases the chance of disease transmission. Adequate fluid intake, adequate nutrition, and proper oral hygiene help maintain normal defenses and can reduce the incidence of getting such diseases as pneumonia.
NP: Evaluation; CN: Physiological integrity; CNS: Physiological adaptation; CL: Comprehension

2. 4. The common feature of all types of pneumonia is an inflammatory pulmonary response to the offending organism or agent. Atelectasis and bronchiectasis indicate a collapse of a portion of the airway that doesn't occur in pneumonia. An effusion is an accumulation of excess pleural fluid in the pleural space, which may be a secondary response to pneumonia.
NP: Evaluation; CN: Physiological integrity; CNS: Physiological adaptation; CL: Knowledge

3. 3. Pneumococcal or streptococcal pneumonia, caused by *Streptococcus pneumoniae,* is the most common cause of community-acquired pneumonia. *Haemophilus influenzae* is the most common cause of infection in children. *Klebsiella* species is the most common gram-negative organism found in the hospital setting. *Staphylococcus aureus* is the most common cause of hospital-acquired pneumonia.
NP: Evaluation; CN: Physiological integrity; CNS: Physiological adaptation; CL: Knowledge

4. 1. Fever, chills, hemoptysis, dyspnea, cough, and pleuritic chest pain are the common symptoms of pneumonia, but elderly clients may first appear with only an altered mental status and dehydration due to a blunted immune response.
NP: Data collection; CN: Physiological integrity; CNS: Physiological adaptation; CL: Application

NP: Nursing process CN: Client needs category CNS: Client needs subcategory CL: Cognitive level

5. When auscultating the chest of a client with pneumonia, the nurse would expect to hear which of the following sounds over areas of consolidation?
   1. Bronchial
   2. Bronchovesicular
   3. Tubular
   4. Vesicular

For question 5, think of where you normally hear each type of breath sound.

5. 1. Chest auscultation reveals bronchial breath sounds over areas of consolidation. Bronchovesicular breath sounds are normal over midlobe lung regions, tubular sounds are commonly heard over large airways, and vesicular breath sounds are commonly heard in the bases of the lung fields.

NP: Data collection; CN: Physiological integrity; CNS: Physiological adaptation; CL: Application

6. A diagnosis of pneumonia is typically achieved by which of the following diagnostic tests?
   1. Arterial blood gas (ABG) analysis
   2. Chest X-ray
   3. Blood cultures
   4. Sputum culture and sensitivity

6. 4. Sputum culture and sensitivity is the best way to identify the organism causing the pneumonia. Chest X-ray will show the area of lung consolidation. ABG analysis will determine the extent of hypoxia present due to the pneumonia, and blood cultures will help determine if the infection is systemic.

NP: Implementation; CN: Physiological integrity; CNS: Physiological adaptation; CL: Application

7. A client with pneumonia has a nursing diagnosis of *Ineffective airway clearance related to increased secretions and ineffective cough.* Which intervention would facilitate effective coughing?
   1. Lying in semi-Fowler's position
   2. Sipping water, hot tea, or coffee
   3. Inhaling and exhaling from pursed lips
   4. Using thoracic breathing

7. 2. Sips of water, hot tea, or coffee may stimulate coughing. The best position is sitting in a chair with the knees flexed and the feet placed firmly on the floor. The client should inhale through the nose and exhale through pursed lips. Diaphragmatic, not thoracic, breathing helps to facilitate coughing.

NP: Planning; CN: Physiological integrity; CNS: Basic care and comfort; CL: Application

8. On entering the room of a client with chronic obstructive pulmonary disease (COPD), the nurse notices that the client is receiving oxygen at 4 L/minute by way of a nasal cannula. The nurse's actions should be based on which of the following statements?
   1. The flow rate is too high.
   2. The flow rate is too low.
   3. The flow rate is correct.
   4. The client shouldn't receive oxygen.

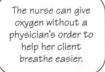

The nurse can give oxygen without a physician's order to help her client breathe easier.

8. 1. The administration of oxygen at 1 to 2 L/minute by way of a nasal cannula is recommended for clients with COPD: therefore, a rate of 4 L/minute is too high. The normal mechanism that stimulates breathing is a rise in blood carbon dioxide. Clients with COPD retain blood carbon dioxide, so their mechanism for stimulating breathing is a low blood oxygen level. High levels of oxygen may cause hypoventilation and apnea. Oxygen delivered at 1 to 2 L/minute should aid in oxygenation without causing hypoventilation. Oxygen therapy is the only therapy that has been demonstrated to be life-preserving for patients with COPD.

NP: Implementation; CN: Safe, effective care environment; CNS: Safety and infection control; CL: Application

9. A client has been treated with antibiotic therapy for right lower-lobe pneumonia for 10 days and will be discharged today. Which of the following physical findings would lead the nurse to believe it is appropriate to discharge this client?

1. Continued dyspnea
2. Fever of 102° F (38.9° C)
3. Respiratory rate of 32 breaths/minute
4. Vesicular breath sounds in right base

10. A 20-year-old client is being treated for pneumonia. He has a persistent cough and complains of severe pain on coughing. What type of instruction could be given to help the client reduce the discomfort he is having?

1. "Hold in your cough as much as possible."
2. "Place the head of your bed flat to help with coughing."
3. "Restrict fluids to help decrease the amount of sputum."
4. "Splint your chest wall with a pillow for comfort."

Make sure your answer responds to the question asked.

11. An elderly client with pneumonia has a nursing diagnosis of *Ineffective airway clearance*. Which intervention would be the most appropriate?

1. Monitor the need for suctioning every hour.
2. Suction every hour.
3. Suction once per shift.
4. Ask the physician for an order to suction.

12. A nurse is working in a walk-in clinic. She has been alerted that there is an outbreak of tuberculosis (TB). Which of the following clients entering the clinic today is most likely to have TB?

1. A 16-year-old female high school student
2. A 33-year-old day-care worker
3. A 43-year-old homeless man with a history of alcoholism
4. A 54-year-old businessman

Another test-taking hint! Choose the most likely response about today's population.

9. 4. If the client still has pneumonia, the breath sounds in the right base will be bronchial, not the normal vesicular breath sounds. If the client still has dyspnea, fever, and increased respiratory rate, the client should be examined by the physician before discharge because he may have another source of infection or still have pneumonia.
NP: Evaluation; CN: Physiological integrity; CNS: Physiological adaptation; CL: Analysis

10. 4. Showing this client how to splint his chest wall will help decrease discomfort when coughing. Holding in his coughs will only increase the amount of pain he has. Placing the head of the bed flat may increase the frequency of his cough and require more work; a 45-degree angle may help his cough more efficiently and with less pain. Increasing fluid intake will help thin his secretions, making it easier for him to clear them. Promoting fluid intake is appropriate in this situation.
NP: Implementation; CN: Physiological integrity; CNS: Physiological adaptation; CL: Application

11. 1. Suctioning should be performed only when necessary, based on the client's condition at the time of assessment. Suctioning is a nursing procedure and doesn't require a physician's order
NP: Planning; CN: Physiological integrity; CNS: Basic care and comfort; CL: Application

12. 3. Clients who are economically disadvantaged, malnourished, and have reduced immunity, such as a client with a history of alcoholism, are at extremely high risk for developing TB. A high school student, businessman, and day-care worker probably have a much lower risk of contracting TB.
NP: Evaluation; CN: Physiological integrity; CNS: Physiological adaptation; CL: Comprehension

13. Tuberculosis (TB) is a communicable disease transmitted by which of the following methods?
1. Sexual contact
2. Using dirty needles
3. Using an infected person's eating utensils
4. Inhaling droplets exhaled from an infected person

This droplet transmission stuff sure beats the subway! Wheeee!

13. 4. The TB bacillus is airborne and carried in droplets exhaled by an infected person who is coughing, sneezing, laughing, or singing. Sexual contact and dirty needles don't spread the TB bacillus, but may spread other communicable diseases. It's never advisable to use dirty utensils, but if cleaned normally, it isn't necessary to dispose of eating utensils used by someone infected with TB.
NP: Evaluation; CN: Physiological integrity; CNS: Physiological adaptation; CL: Knowledge

14. An adult client is being screened in the clinic today for tuberculosis. He reports having negative purified protein derivative (PPD) test results in the past. The nurse performs a PPD test on his right forearm today. When should he return to have the test read?
1. Immediately after performing the test
2. 24 hours after performing the test
3. 48 hours after performing the test
4. 1 week after performing the test

The timing is the clue!

14. 3. PPD tests should be read in 48 to 72 hours. If read too early or too late, the results won't be accurate.
NP: Data collection; CN: Health promotion and maintenance; CNS: Prevention and early detection of disease; CL: Knowledge

15. The right forearm of a client who had a purified protein derivative (PPD) test for tuberculosis (TB) is reddened and raised about 5 mm where the test was given. This PPD would be read as having which of the following results?
1. Indeterminate
2. Needs to be redone
3. Negative
4. Positive

15. 3. This test would be classed as negative. A 5-mm raised area would be a positive result if a client had recent close contact with someone diagnosed with or suspected of having infectious TB. Follow-up should be done with this client, and a chest X-ray should be ordered. The test can be redone in 6 months to see if the client's test results change. If the PPD test is reddened and raised 10 mm or more, it's considered positive according to the Centers for Disease Control and Prevention. *Indeterminate* isn't a term used to describe results of a PPD test.
NP: Evaluation; CN: Health promotion and maintenance; CNS: Prevention and early detection of disease; CL: Knowledge

16. A client with a primary tuberculosis (TB) infection can expect to develop which of the following conditions?
1. Active TB within 2 weeks
2. Active TB within 1 month
3. A fever and require hospitalization
4. A positive skin test

16. 4. A primary TB infection occurs when the bacillus has successfully invaded the entire body after entering through the lungs. At this point, the bacilli are walled off and skin tests read positive. However, all but infants and immunosuppressed people will remain asymptomatic. The general population has a 10% risk of developing active TB over their lifetime, often because of a break in the body's immune defenses. The active stage shows the classic symptoms of TB: fever, hemoptysis, and night sweats.
NP: Data collection; CN: Physiological integrity; CNS: Physiological adaptation; CL: Application

17. A client was infected with tuberculosis (TB) bacillus 10 years ago but never developed the disease. He's now being treated for cancer. The client begins to develop signs of TB. This is known as which of the following types of infection?

1. Active infection
2. Primary infection
3. Superinfection
4. Tertiary infection

18. A client has active tuberculosis (TB). Which of the following symptoms will he exhibit?

1. Chest and lower back pain
2. Chills, fever, night sweats, and hemoptysis
3. Fever of more than 104° F and nausea
4. Headache and photophobia

19. Which of the following diagnostic tests is definitive for tuberculosis?

1. Chest X-ray
2. Mantoux skin test
3. Sputum culture
4. Tuberculin skin test

20. A client with a positive Mantoux skin test result will be sent for a chest X-ray. For which of the following reasons is this done?

1. To confirm the diagnosis
2. To determine if a repeat skin test is needed
3. To determine the extent of lesions
4. To determine if this is a primary or secondary infection

21. A chest X-ray shows a client's lungs to be clear. His tuberculin Mantoux skin test is positive, with 10 mm of induration. His previous test was negative. These test results are possible because of which of the following reasons?

1. He had tuberculosis (TB) in the past and no longer has it.
2. He was successfully treated for TB but skin tests always stay positive.
3. He is a "seroconverter," meaning the TB has gotten to his blood stream.
4. He is a "tuberculin converter," which means he has been infected with TB since his last skin test.

Think: How does cancer affect the immune system?

You don't need to be a fortune teller! The right diagnostic test can give you the answer!

Remember what a positive skin test means.

17. 1. Some people carry dormant TB infections that may develop into active disease. In addition, primary sites of infection containing TB bacilli may remain latent for years and then activate when the client's resistance is lowered, as when a client is being treated for cancer. There's no such thing as tertiary infection, and superinfection doesn't apply in this case.
NP: Evaluation; CN: Physiological integrity; CNS: Physiological adaptation; CL: Application

18. 2. Typical signs and symptoms are chills, fever, night sweats, and hemoptysis. Clients with TB typically have low-grade fevers, not higher than 102° F. Chest pain may be present from coughing, but isn't usual. Nausea, headache, and photophobia aren't usual TB symptoms.
NP: Data collection; CN: Physiological integrity; CNS: Physiological adaptation; CL: Application

19. 3. Skin tests may be falsely positive or falsely negative. Lesions in the lung may not be big enough to be seen on X-ray. The sputum culture for *Mycobacterium tuberculosis* is the only method of confirming the diagnosis.
NP: Data collection; CN: Physiological integrity; CNS: Physiological adaptation; CL: Knowledge

20. 3. If the lesions are large enough, the chest X-ray will show their presence in the lungs. Sputum culture confirms the diagnosis. There can be false-positive and false-negative skin test results. A chest X-ray can't determine if this is a primary or secondary infection.
NP: Data collection; CN: Physiological integrity; CNS: Physiological adaptation; CL: Application

21. 4. A tuberculin converter's skin test will be positive, meaning he's been exposed to and infected with TB and now has a cell-mediated immune response to the skin test. The client's blood and X-ray results may stay negative. It doesn't mean the infection has advanced to the active stage. Because his X-ray is negative, he should be monitored every 6 months to see if he develops changes in his chest X-ray or pulmonary examination. Being a seroconverter doesn't mean the TB has gotten into his bloodstream, it means it can be detected by a blood test.
NP: Evaluation CN: Physiological integrity; CNS: Physiological adaptation; CL: Application

22. A client with a positive skin test for tuberculosis (TB) isn't showing signs of active disease. To help prevent the development of active TB, the client should be treated with isoniazid, 300 mg daily, for how long?
1. 10 to 14 days
2. 2 to 4 weeks
3. 3 to 6 months
4. 9 to 12 months

23. A client with a productive cough, chills, and night sweats, is suspected of having active tuberculosis (TB). The physician should take which of the following actions?
1. Admit him to the hospital in respiratory isolation.
2. Prescribe isoniazid and tell him to go home and rest.
3. Give a tuberculin skin test and tell him to come back in 48 hours to have it read.
4. Give a prescription for isoniazid, 300 mg daily for 2 weeks, and send him home.

24. A client is diagnosed with active tuberculosis and started on triple antibiotic therapy. What signs and symptoms would the client show if therapy is inadequate?
1. Decreased shortness of breath
2. Improved chest X-ray
3. Nonproductive cough
4. Positive acid-fast bacilli in a sputum sample after 2 months of treatment

25. Which of the following instructions should the nurse give a client about his active tuberculosis (TB)?
1. "It's okay to miss a dose every day or two."
2. "If side effects occur, stop taking the medication."
3. "Only take the medication until you feel better."
4. "You must comply with the medication regimen to treat TB."

Think about TB's resistant strains.

Read question 24 carefully! It's easy to miss the two little letters in front of "adequate."

Hmmm. Is it OK to miss a dose of an antitubercular drug?

22. 4. Because of the increasing incidence of resistant strains of TB, the disease must be treated for up to 24 months in some cases, but treatment typically lasts from 9 to 12 months. Isoniazid is the most common medication used for the treatment of TB, but other antibiotics are added to the regimen to obtain the best results.
NP: Implementation; CN: Physiological integrity; CNS: Pharmacological therapies; CL: Application

23. 1. This client is showing signs and symptoms of active TB and, because of the productive cough, is highly contagious. He should be admitted to the hospital, placed in respiratory isolation, and three sputum cultures should be obtained to confirm the diagnosis. He would most likely be given isoniazid and two or three other antitubercular antibiotics until the diagnosis is confirmed, and then isolation and treatment would continue if the cultures were positive for TB. After 7 to 10 days, three more consecutive sputum cultures will be obtained. If they're negative, he would be considered noncontagious and may be sent home, although he'll continue to take the antitubercular drugs for 9 to 12 months.
NP: Implementation; CN: Physiological integrity; CNS: Physiological adaptation; CL: Application

24. 4. Continuing to have acid-fast bacilli in the sputum after 2 months indicates continued infection. The other choices would all indicate improvement with therapy.
NP: Evaluation; CN: Physiological integrity; CNS: Physiological adaptation; CL: Application

25. 4. The regimen may last up to 24 months. It's essential that the client comply with therapy during that time or resistance will develop. At no time should he stop taking the medications before his physician tells him to.
NP: Evaluation; CN: Physiological integrity; CNS: Physiological adaptation; CL: Analysis

26. A client diagnosed with active tuberculosis (TB) would be hospitalized primarily for which of the following reasons?
1. To evaluate his condition
2. To determine his compliance
3. To prevent spread of the disease
4. To determine the need for antibiotic therapy

You're doing great! Keep it up!

26. 3. The client with active TB is highly contagious until three consecutive sputum cultures are negative, so he's put in respiratory isolation in the hospital. Neither assessment of physical condition, antibiotic therapy, nor determinations of compliance are primary reasons for hospitalization in this case.
NP: Implementation; CN: Physiological integrity; CNS: Physiological adaptation; CL: Application

27. A 7-year-old client is brought to the emergency department. He is tachypneic and afebrile and has a respiratory rate of 36 breaths/minute and a nonproductive cough. He recently had a cold. From this history, the client may have which of the following conditions?
1. Acute asthma
2. Bronchial pneumonia
3. Chronic obstructive pulmonary disease
4. Emphysema

27. 1. Based on the client's history and symptoms, acute asthma is the most likely diagnosis. He's too young to have developed chronic obstructive pulmonary disease and emphysema, and he's unlikely to have bronchial pneumonia without a productive cough and fever.
NP: Data collection; CN: Physiological integrity; CNS: Physiological adaptation; CL: Analysis

Listen to how we sound — any changes from normal are clues.

28. Which of the following assessment findings would help confirm a diagnosis of asthma in a child suspected of having the disorder?
1. Circumoral cyanosis
2. Increased forced expiratory volume
3. Inspiratory and expiratory wheezing
4. Normal breath sounds

28. 3. Inspiratory and expiratory wheezes are typical findings in asthma. Circumoral cyanosis may be present in extreme cases of respiratory distress. The nurse would expect the client to have a decreased forced expiratory volume because asthma is an obstructive pulmonary disease. Breath sounds will be "tight" sounding or markedly decreased; they won't be normal.
NP: Data collection; CN: Physiological integrity; CNS: Physiological adaptation; CL: Analysis

29. A 22-year-old female client is experiencing a new-onset asthmatic attack. Which position is best for this client?
1. High Fowler's
2. Left side-lying
3. Right side-lying
4. Supine with pillows under each arm

29. 1. The best position is high Fowler's, which helps lower the diaphragm and facilitates passive breathing and thereby improves air exchange. A side-lying position won't facilitate the client's breathing. A supine position increases the breathing difficulty of an asthmatic client.
NP: Implementation; CN: Physiological integrity; CNS: Physiological adaptation; CL: Application

Another hint! Question 30 asks you to set priorities.

30. A client with acute asthma showing inspiratory and expiratory wheezes and a decreased forced expiratory volume should be treated with which of the following classes of medication right away?
1. Beta-adrenergic blockers
2. Bronchodilators
3. Inhaled steroids
4. Oral steroids

30. 2. Bronchodilators are the first line of treatment for asthma because bronchoconstriction is the cause of reduced airflow. Inhaled or oral steroids may be given to reduce the inflammation but aren't used for emergency relief. Beta-adrenergic blockers aren't used to treat asthma and can cause bronchoconstriction.
NP: Implementation; CN: Physiological integrity; CNS: Pharmacological therapies; CL: Application

31. A 19-year-old client comes to the emergency department with acute asthma. His respiratory rate is 44 breaths/minute, and he appears in acute respiratory distress. Which of the following actions should be taken first?
1. Take a full medical history.
2. Give a bronchodilator by nebulizer.
3. Apply a cardiac monitor to the client.
4. Provide emotional support to the client.

Question 31 asks you to prioritize. Which action comes first?

32. A client is found to be allergic to Chinese food, which causes acute asthma. Which of the following instructions should the nurse give the client?
1. "Only eat Chinese food once a month."
2. "Use your inhalers before eating Chinese food."
3. "Avoid Chinese food because this is a trigger for you."
4. "Determine other causes, because Chinese food wouldn't cause such a violent reaction."

What would you do if you were allergic to Chinese food?

33. Which nursing activity requires greater caution when performed on a client with chronic obstructive pulmonary disease (COPD)?
1. Administering narcotics for pain relief
2. Increasing the client's fluid intake
3. Monitoring the client's cardiac rhythm
4. Assisting the client with coughing and deep breathing

31. 2. The client having an acute asthma attack needs to increase oxygen delivery to the lung and body. Nebulized bronchodilators open airways and increase the amount of oxygen delivered. It may not be necessary to place the client on a cardiac monitor because he's only 19 years old, unless he has a medical history of cardiac problems. First resolve the acute phase of the attack, then obtain a full medical history to determine the cause of the attack and how to prevent attacks in the future.
NP: Implementation; CN: Physiological integrity; CNS: Physiological adaptation; CL: Application

32. 3. If the trigger of an acute asthma attack is known, this trigger should be avoided at all times. Using an inhaler before eating wouldn't prevent the attack, and food is typically a trigger for an acute asthma attack.
NP: Implementation; CN: Physiological integrity; CNS: Reduction of risk potential; CL: Application

33. 1. Narcotics suppress the respiratory center in the medulla. Both COPD and pneumonia cause alterations in gas exchange; any further problems with oxygenation could result in respiratory failure and cardiac arrest. Increasing the fluid intake would help to thin the client's secretions. Although the nurse would need to monitor the intake and output and watch for signs of heart failure, this isn't as critical as administering narcotics. The cardiac rhythm provides an indication of the client's myocardial oxygenation; it should be a part of the nurse's regular assessment. Helping with coughing and deep breathing should be included in the plan of care. The only caution would be to assess for possible rupture of emphysematous alveolar sacs and pneumothorax.
NP: Implementation; CN: Physiological integrity; CNS: Physiological adaptation; CL: Application

34. The term "blue bloater" refers to which of the following conditions?
1. Acute respiratory distress syndrome (ARDS)
2. Asthma
3. Chronic obstructive bronchitis
4. Emphysema

Terms such as this can help you remember the symptoms of some diseases.

34. 3. Clients with chronic obstructive bronchitis appear bloated; they have large barrel chests and peripheral edema, cyanotic nail beds and, at times, circumoral cyanosis. Clients with asthma don't exhibit characteristics of chronic disease. Clients with emphysema appear pink and cachectic, and clients with ARDS are acutely short of breath and frequently need intubation for mechanical ventilation and large amounts of oxygen.

NP: Data collection; CN: Physiological integrity; CNS: Physiological adaptation; CL: Application

35. The term "pink puffer" refers to the client with which of the following conditions?
1. Acute respiratory distress syndrome (ARDS)
2. Asthma
3. Chronic obstructive bronchitis
4. Emphysema

Here's another term to help remember symptoms.

35. 4. Because of the large amount of energy it takes to breathe, clients with emphysema are usually cachectic. They're pink and usually breathe through pursed lips, hence the term "puffer." Clients with asthma don't have any particular characteristics. Clients with chronic obstructive bronchitis are bloated and cyanotic in appearance, and clients with ARDS are usually acutely short of breath.

NP: Data collection; CN: Physiological integrity; CNS: Physiological adaptation; CL: Application

36. A 66-year-old client has marked dyspnea at rest, is thin, and uses accessory muscles to breathe. He's tachypneic, with a prolonged expiratory phase. He has no cough. He leans forward with his arms braced on his knees to support his chest and shoulders for breathing. This client has symptoms of which of the following respiratory disorders?
1. Acute respiratory distress syndrome (ARDS)
2. Asthma
3. Chronic obstructive bronchitis
4. Emphysema

36. 4. These are classic signs and symptoms of a client with emphysema. Clients with asthma are acutely short of breath during an attack and appear very frightened. Clients with bronchitis are bloated and cyanotic in appearance, and clients with ARDS are acutely short of breath and require emergency care.

NP: Data collection; CN: Physiological integrity; CNS: Physiological adaptation; CL: Application

37. It's highly recommended that clients with asthma, chronic bronchitis, and emphysema have Pneumovax and flu vaccinations for which of the following reasons?
1. All clients are recommended to have these vaccines.
2. These vaccines produce bronchodilation and improve oxygenation.
3. These vaccines help reduce the tachypnea these clients experience.
4. Respiratory infections can cause severe hypoxia and possibly death in these clients.

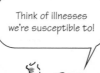

Think of illnesses we're susceptible to!

37. 4. It's highly recommended that clients with respiratory disorders be given vaccines to protect against respiratory infection. Infections can cause these clients to need intubation and mechanical ventilation, and it may be difficult to wean these clients from the ventilator. The vaccines have no effect on respiratory rate or bronchodilation.

NP: Implementation; CN: Health promotion and maintenance CNS: Prevention and early detection of disease; CL: Application

38. Exercise has which of the following effects on clients with asthma, chronic bronchitis, and emphysema?
1. It enhances cardiovascular fitness.
2. It improves respiratory muscle strength.
3. It reduces the number of acute attacks.
4. It worsens respiratory function and is discouraged.

38. 1. Exercise can improve cardiovascular fitness and help the client tolerate periods of hypoxia better, perhaps reducing the risk of heart attack. Most exercise has little effect on respiratory muscle strength, and these clients can't tolerate the type of exercise necessary to do this. Exercise won't reduce the number of acute attacks. In some instances, exercise may be contraindicated, and the client should check with his physician before starting any exercise program.
NP: Implementation; CN: Health promotion and maintenance; CNS: Prevention and early detection of disease; CL: Application

Hint! Hint! Hint!

39. Clients with chronic obstructive bronchitis are given diuretic therapy. Which of the following reasons best explains why?
1. Reducing fluid volume reduces oxygen demand.
2. Reducing fluid volume improves clients' mobility.
3. Reducing fluid volume reduces sputum production.
4. Reducing fluid volume improves respiratory function.

39. 1. Reducing fluid volume reduces the workload of the heart, which reduces oxygen demand and, in turn, reduces the respiratory rate. Sputum may get thicker and make it harder to clear airways. Reducing fluid volume won't improve respiratory function, but may improve oxygenation. Reducing fluid volume may reduce edema and improve mobility a little, but exercise tolerance will still be poor.
NP: Implementation; CN: Physiological integrity; CNS: Physiological adaptation; CL: Application

40. A 69-year-old client appears thin and cachectic. He's short of breath at rest and his dyspnea increases with the slightest exertion. His breath sounds are diminished even with deep inspiration. These signs and symptoms fit which of the following conditions?
1. Acute respiratory distress syndrome (ARDS)
2. Asthma
3. Chronic obstructive bronchitis
4. Emphysema

40. 4. In emphysema, the wall integrity of the individual air sacs is damaged, reducing the surface area available for gas exchange. Very little air movement occurs in the lungs because of bronchiole collapse, as well. In asthma and bronchitis, wheezing is prevalent. In ARDS, the client's condition is more acute and typically requires mechanical ventilation.
NP: Data collection; CN: Physiological integrity; CNS: Physiological adaptation; CL: Application

41. A client with emphysema should receive only 1 to 3 L/minute of oxygen, if needed, or he may lose his hypoxic drive. Which of the following statements is correct about hypoxic drive?
1. The client doesn't notice he needs to breathe.
2. The client breathes only when his oxygen levels climbs above a certain point.
3. The client breathes only when his oxygen levels dip below a certain point.
4. The client breathes only when his carbon dioxide level dips below a certain point.

We keep moving thanks to our hypoxic drive!

41. 3. Clients with emphysema breathe when their oxygen levels drop to a certain level; this is known as the hypoxic drive. Clients with emphysema and chronic obstructive pulmonary disease take a breath when they've reached this low oxygen level. They don't take a breath when their levels of carbon dioxide are higher than normal, as do those with healthy respiratory physiology. If too much oxygen is given, the client has little stimulus to take another breath. In the meantime, his carbon dioxide levels continue to climb, and he will pass out, leading to a respiratory arrest.
NP: Implementation; CN: Physiological integrity; CNS: Physiological adaptation; CL: Application

NP: Nursing process   CN: Client needs category   CNS: Client needs subcategory   CL: Cognitive level

42. Teaching for a client with chronic obstructive pulmonary disease should include which of the following topics?
1. How to listen to his own lungs
2. How to improve his oxygenation
3. How to treat respiratory infections without going to the physician
4. How to recognize the signs of an impending respiratory infection

42. 4. Respiratory infection in clients with a respiratory disorder can be fatal. It's important that the client understands how to recognize the signs and symptoms of an impending respiratory infection. It isn't appropriate to teach a client how to listen to his own lungs. If the client has signs and symptoms of an infection, he should contact his physician at once to obtain prompt treatment.
NP: Implementation; CN: Physiological integrity; CNS: Reduction of risk potential; CL: Application

43. Which of the following respiratory disorders is most common in the first 24 to 48 hours after surgery?
1. Atelectasis
2. Bronchitis
3. Pneumonia
4. Pneumothorax

You're one-third finished with this chapter. Way to go!

43. 1. Atelectasis develops when there's interference with the normal negative pressure that promotes lung expansion. Clients in the postoperative phase often splint their breathing because of pain and positioning, which causes hypoxia. It's uncommon for any of the other respiratory disorders to develop.
NP: Implementation; CN: Physiological integrity; CNS: Physiological adaptation; CL: Application

44. Which of the following measures can reduce or prevent the incidence of atelectasis in a postoperative client?
1. Chest physiotherapy
2. Mechanical ventilation
3. Reducing oxygen requirements
4. Use of an incentive spirometer

44. 4. Using an incentive spirometer requires the client to take deep breaths and promotes lung expansion. Chest physiotherapy helps mobilize secretions but won't prevent atelectasis. Reducing oxygen requirements or placing someone on mechanical ventilation doesn't affect the development of atelectasis.
NP: Implementation; CN: Physiological integrity; CNS: Basic care and comfort; CL: Application

45. Emergency treatment of a client in status asthmaticus includes which of the following medications?
1. Inhaled beta-adrenergic agents
2. Inhaled corticosteroids
3. I.V. beta-adrenergic agents
4. Oral corticosteroids

45. 1. Inhaled beta-adrenergic agents help promote bronchodilation, which improves oxygenation. I.V. beta-adrenergic agents can be used but have to be monitored because of their greater systemic effects. They're typically used when the inhaled beta-adrenergic agents don't work. Corticosteroids are slow acting, so their use won't reduce hypoxia in the acute phase.
NP: Evaluation; CN: Physiological integrity; CNS: Pharmacological therapies; CL: Application

For question 46, you need to choose the best out of several possible goals.

46. Which of the following treatment goals is best for the client with status asthmaticus?
1. Avoid intubation
2. Determine the cause of the attack
3. Improve exercise tolerance
4. Reduce secretions

46. 1. Inhaled beta-adrenergic agents, I.V. corticosteroids, and supplemental oxygen are used to reduce bronchospasm, improve oxygenation, and avoid intubation. Improvement in exercise tolerance and determining the trigger for the client's attack are later goals. Typically, secretions aren't a problem in status asthmaticus.
NP: Planning; CN: Physiological integrity; CNS: Physiological adaptation; CL: Application

47. A client was given morphine for pain. He's sleeping and his respiratory rate is 4 breaths/minute. If action isn't taken quickly, he might have which of the following reactions?
1. Asthma attack
2. Respiratory arrest
3. Seizure
4. Wake up on his own

47. 2. Narcotics suppress the respiratory center in the medulla and can cause respiratory arrest if given in large quantities. It's unlikely he'll have an asthma attack or a seizure or wake up on his own.
NP: Evaluation; CN: Physiological integrity; CNS: Pharmacological therapies CL: Application

48. Which of the following additional data should immediately be gathered to determine the status of a client with a respiratory rate of 4 breaths/minute?
1. Arterial blood gas (ABG) and breath sounds
2. Level of consciousness and a pulse oximetry value
3. Breath sounds and reflexes
4. Pulse oximetry value and heart sounds

Ask yourself. "Which collected data would help first, and which could wait?"

48. 2. First, the nurse should attempt to rouse the client, because this should increase the client's respiratory rate. If available, a spot pulse oximetry check should be done and breath sounds should be checked. The physician should be notified immediately of the findings. He'll probably order an ABG to determine specific carbon dioxide and oxygen levels, which will indicate the effectiveness of ventilation. Heart sounds and reflexes will be part of the more extensive examination done after these initial actions are completed.
NP: Data collection; CN: Physiological integrity; CNS: Physiological adaptation; CL: Comprehension

49. A client is in danger of respiratory arrest following the administration of a narcotic analgesic. An arterial blood gas is obtained. The nurse would expect the $Paco_2$ to be which of the following values?
1. 15 mm Hg
2. 30 mm Hg
3. 40 mm Hg
4. 80 mm Hg

49. 4. A client about to go into respiratory arrest will have inefficient ventilation and will be retaining carbon dioxide. The value expected would be around 80 mm Hg. All other values are lower than normal.
NP: Evaluation; CN: Physiological integrity; CNS: Physiological adaptation; CL: Analysis

50. A client's arterial blood gas (ABG) results are as follows: pH, 7.16; $Paco_2$, 80 mm Hg; Pao_2, 46 mm Hg; HCO_3^-, 24 mEq/L; Sao_2, 81%. This ABG result represents which of the following conditions?
1. Metabolic acidosis
2. Metabolic alkalosis
3. Respiratory acidosis
4. Respiratory alkalosis

Which condition would most likely cause us to fail?

50. 3. The pH is less than 7.35, acidemic, which eliminates metabolic and respiratory alkalosis as possibilities. Because the $Paco_2$ is high at 80 mm Hg and the metabolic measure, HCO_3^-, is normal, the client has a respiratory acidosis.
NP: Evaluation; CN: Physiological integrity; CNS: Physiological adaptation; CL: Analysis

51. Clients at high risk for respiratory failure include those with which of the following diagnoses?
1. Breast cancer
2. Cervical sprains
3. Fractured hip
4. Guillain-Barré syndrome

51. 4. Guillain-Barré syndrome is a progressive neuromuscular disorder that can affect the respiratory muscles and cause respiratory failure. The other conditions typically don't affect the respiratory system.
NP: Evaluation; CN: Physiological integrity; CNS: Physiological adaptation; CL: Analysis

NP: Nursing process   CN: Client needs category   CNS: Client needs subcategory   CL: Cognitive level

52. A client has started a new drug for hypertension. Thirty minutes after he takes the drug, he develops chest tightness and becomes short of breath and tachypneic. He has a decreased level of consciousness. These signs indicate which of the following conditions?
1. Asthma attack
2. Pulmonary embolism
3. Respiratory failure
4. Rheumatoid arthritis

53. Emergency treatment for a client with impending anaphylaxis secondary to hypersensitivity to a drug should include which of the following actions first?
1. Administer oxygen.
2. Insert an I.V. catheter.
3. Obtain a complete blood count (CBC).
4. Take vital signs.

54. Following the initial care of a client with asthma and impending anaphylaxis from hypersensitivity to a drug, the nurse should take which of the following steps next?
1. Administer beta-adrenergic blockers.
2. Administer bronchodilators.
3. Obtain serum electrolyte levels.
4. Have the client lie flat in the bed.

55. A 19-year-old client went to a party, took "some pills," and drank beer. He's brought to the emergency department because he won't wake up. When collecting data from him, the nurse would expect to find which of the following reactions?
1. Hyperreflexive reflexes
2. Muscle spasms
3. Shallow respirations
4. Tachypnea

56. Which of the following actions should be taken first in response to an initial assessment of probable drug overdose complicated with alcohol ingestion?
1. Administer I.V. fluids.
2. Administer I.V. naloxone (Narcan).
3. Continue close monitoring of vital signs.
4. Draw blood for a drug screen.

Set your priorities properly. What should be done first?

Next is another hint that means to prioritize.

In a drug overdose, you'll want to do only one of these first. Which one?

52. 3. These signs indicate a hypersensitivity to the new medication, leading to anaphylaxis and respiratory failure.
NP: Data collection; CN: Physiological integrity; CNS: Pharmacological therapies; CL: Analysis

53. 1. Giving oxygen would be the best first action in this case. Vital signs then should be checked and the physician immediately notified. If the client doesn't already have an I.V. catheter, one may be inserted now if anaphylactic shock is developing. Obtaining a CBC would not help the emergency situation.
NP: Planning; CN: Physiological integrity; CNS: Pharmacological therapies; CL: Application

54. 2. Bronchodilators would help open the client's airway and improve his oxygenation status. Beta-adrenergic blockers aren't indicated in the management of asthma because they may cause bronchospasm. Obtaining laboratory values wouldn't be done on an emergency basis, and having the client lie flat in bed could worsen the client's ability to breathe.
NP: Planning; CN: Physiological integrity; CNS: Physiological adaptation; CL: Application

55. 3. The client probably can't be roused from the combination of pills and alcohol he has taken. This has probably caused him to breathe shallowly, which, if not monitored closely, could lead to respiratory arrest. The nurse wouldn't expect to find tachypnea and doesn't have enough information about which drugs he took to expect muscle spasms or hyperreflexia.
NP: Data collection; CN: Physiological integrity; CNS: Physiological adaptation; CL: Application

56. 2. If the client took narcotics, giving naloxone could reverse the effects and awaken the client. I.V. fluids will most likely be administered, and he'll be closely monitored over a period of several hours to several days. A drug screen should be drawn in the emergency department, but results may not come back for several hours.
NP: Planning; CN: Physiological integrity; CNS: Physiological adaptation; CL: Application

57. An unconscious client who overdosed on a narcotic receives naloxone (Narcan) to reverse the overdose. After he awakens, which of the following actions by the nurse would be the best?

1. Feed the client.
2. Teach the client about the effects of taking pills and alcohol together.
3. Discharge the client from the hospital.
4. Admit the client to a psychiatric facility.

58. A 45-year-old male client is brought to the hospital with smoke inhalation due to a house fire. What is the nurse's first priority for this client?

1. Checking the oral mucous membranes
2. Checking for any burned areas
3. Obtaining a medical history
4. Ensuring a patent airway

59. In a client with smoke inhalation, the nurse would expect to hear which of the following breath sounds?

1. Crackles
2. Decreased breath sounds
3. Inspiratory and expiratory wheezing
4. Upper airway rhonchi

60. A client is receiving oxygen by way of a nasal cannula at a rate of 2 L/minute. How should the oxygen flow meter be set?

1. The bottom of the ball should sit on top of the line marked "2."
2. The top of the ball should sit below the line marked "2."
3. The line marked "2" should cut the ball in half.
4. Any part of the ball should touch the line marked "2."

61. Which of the following statements best describes what happens to the alveoli in acute respiratory distress syndrome (ARDS)?

1. Alveoli are overexpanded.
2. Alveoli increase perfusion.
3. Alveolar spaces are filled with fluid.
4. Alveoli improve gaseous exchange.

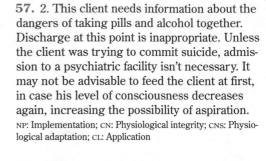

Crackle, wheeze, or rattle? Each sound is another clue.

Note the key word: best

57. 2. This client needs information about the dangers of taking pills and alcohol together. Discharge at this point is inappropriate. Unless the client was trying to commit suicide, admission to a psychiatric facility isn't necessary. It may not be advisable to feed the client at first, in case his level of consciousness decreases again, increasing the possibility of aspiration.
NP: Implementation; CN: Physiological integrity; CNS: Physiological adaptation; CL: Application

58. 4. The nurse's first priority is to make sure the airway is open and the client is breathing. Checking the mucous membranes and burned areas is important but not as vital as maintaining a patent airway. Obtaining a medical history can be pursued after ensuring a patent airway.
NP: Evaluation; CN: Physiological integrity; CNS: Physiological adaptation; CL: Application

59. 1. In acute respiratory distress syndrome, the most frequently heard sounds are crackles throughout the lung fields. Decreased breath sounds or inspiratory and expiratory wheezing are associated with asthma, and rhonchi are heard when there is sputum in the airways.
NP: Data collection; CN: Physiological integrity; CNS: Physiological adaptation; CL: Application

60. 3. The oxygen flow rate is set by centering the indicator on the line marked "2."
NP: Implementation; CN: Safe, effective care environment; CNS: Safety and infection control; CL: Knowledge

61. 3. In ARDS, the alveolar membranes are more permeable and the spaces are fluid-filled. The fluid interferes with gas exchange and reduces perfusion.
NP: Evaluation; CN: Physiological integrity; CNS: Physiological adaptation; CL: Knowledge

NP: Nursing process   CN: Client needs category   CNS: Client needs subcategory   CL: Cognitive level

62. A 69-year-old client develops acute shortness of breath and progressive hypoxia requiring mechanical ventilation after repair of a fractured right femur. The hypoxia was <u>probably</u> caused by which of the following conditions?
1. Asthma attack
2. Atelectasis
3. Bronchitis
4. Fat embolism

63. The nurse is caring for a client with a fracture of the right femur caused by a skiing accident. Which is an early sign of fat emboli?
1. Abdominal cramping
2. Fatty stools
3. Confusion
4. Numbness in the right foot

64. If a client with a fat embolism continues to be hypoxic following respiratory therapy, what can be done to reduce oxygen demand?
1. Give diuretics.
2. Give neuromuscular blockers.
3. Put the head of the bed flat.
4. Use bronchodilators.

65. Positive end-expiratory pressure (PEEP) therapy has which of the following effects on the heart?
1. Bradycardia
2. Tachycardia
3. Increased blood pressure
4. Reduced cardiac output

Several answers may seem correct, but the word probably points to the best one.

You're halfway there! Keep going!

62. 4. Long bone fractures are correlated with fat emboli, which cause shortness of breath and hypoxia. It's unlikely the client has developed asthma or bronchitis without a previous history. He could develop atelectasis, but it typically doesn't produce progressive hypoxia.
NP: Evaluation; CN: Physiological integrity; CNS: Physiological adaptation; CL: Application

63. 3. Irritation and confusion are signs of hypoxia, which is caused by the fat emboli traveling to the lungs and producing an inflammatory response to the lung tissue. Abdominal cramping may be a sign of abdominal distention and constipation caused by immobility. Fatty stools occur with pancreatitis. Numbness may be secondary to neurovascular impairment.
NP: Data collection; CN: Physiological integrity; CNS: Physiological adaptation; CL: Application

64. 2. Neuromuscular blockers cause skeletal muscle paralysis, reducing the amount of oxygen used by the restless skeletal muscles. This should improve oxygenation. Bronchodilators may be used, but they typically don't have enough of an effect to reduce the amount of hypoxia present. The head of the bed should be partially elevated to facilitate diaphragm movement, and diuretics can be administered to reduce pulmonary congestion. However, bronchodilators, diuretics, and head elevation would improve oxygen delivery, not reduce oxygen demand.
NP: Planning; CN: Physiological integrity; CNS: Physiological adaptation; CL: Application

65. 4. PEEP reduces cardiac output by increasing intrathoracic pressure and reducing the amount of blood delivered to the left side of the heart, thereby reducing cardiac output. It doesn't affect heart rate, but a decrease in cardiac output may reduce blood pressure, commonly causing a compensatory tachycardia.
NP: Implementation; CN: Physiological integrity; CNS: Physiological adaptation; CL: Application

66. Occasionally clients with acute respiratory distress syndrome (ARDS) are placed in the prone position. How does this position help the client?
1. It improves cardiac output.
2. It makes the client more comfortable.
3. It prevents skin breakdown.
4. It recruits more alveoli.

66. 4. Turning the client to the prone position may recruit new alveoli in the posterior region of the lung and improve oxygenation status. Cardiac output shouldn't be affected by the prone position. Skin breakdown can still occur over the new pressure points. Generally the client is obtunded when this measure is used. If not, he should be well-sedated.
NP: Implementation; CN: Physiological integrity; CNS: Physiological adaptation; CL: Application

67. Which of the following conditions could lead to acute respiratory distress syndrome (ARDS)?
1. Appendicitis
2. Massive trauma
3. Receiving conscious sedation
4. Right meniscus injury

It's important to know which values to monitor.

67. 2. The client with massive trauma will require multiple transfusions. Blood products are preserved with citrate, which causes increased permeability in the lungs, the defect that allows ARDS to develop. Appendicitis, unless it causes overwhelming sepsis, won't lead to ARDS. Injuries to the meniscus and conscious sedation don't lead to ARDS.
NP: Evaluation; CN: Physiological integrity; CNS: Physiological adaptation; CL: Application

68. Which of the following indicators would show if the condition of the client with acute respiratory distress syndrome (ARDS) is improving?
1. Arterial blood gas (ABG) values
2. Bronchoscopy results
3. Increased blood pressure
4. Sputum culture and sensitivity results

68. 1. Improved ABG results would indicate that the client's oxygenation status is improved. Hypoxia is the problem in ARDS, so bronchoscopy and sputum culture results may have no bearing on the improvement of ARDS. Increased blood pressure isn't relative to the client's respiratory condition.
NP: Evaluation; CN: Physiological integrity; CNS: Physiological adaptation; CL: Analysis

69. A high level of oxygen exerts which of the following effects on the lung?
1. Improves oxygen uptake
2. Increases carbon dioxide levels
3. Stabilizes carbon dioxide levels
4. Reduces amount of functional alveolar surface area

In question 70, break kyphoscoliosis down to find its meaning — and a clue.

69. 4. Oxygen toxicity causes direct pulmonary trauma, reducing the amount of alveolar surface area available for gaseous exchange, which results in increased carbon dioxide levels and decreased oxygen uptake.
NP: Implementation; CN: Physiological integrity; CNS: Physiological adaptation; CL: Application

70. Which of the following effects can thoracic kyphoscoliosis have on lung function?
1. Improves lung expansion
2. Obstructs lung deflation
3. Reduces alveolar compression during expiration
4. Restricts lung expansion

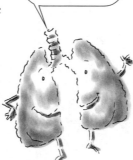

70. 4. Thoracic kyphoscoliosis causes lung compression, restricts lung expansion, and results in more rapid and shallow respiration. It doesn't cause obstruction or reduce alveolar compression during expiration. It also doesn't improve lung expansion because of the compression.
NP: Implementation; CN: Physiological integrity; CNS: Physiological adaptation; CL: Application

71. A 24-year-old client comes into the clinic complaining of right-sided chest pain and shortness of breath. He reports that it started suddenly. The data collection should include which of the following?
1. Auscultation of breath sounds
2. Chest X-ray
3. Echocardiogram
4. Electrocardiogram

72. A client with shortness of breath has decreased to absent breath sounds on the right side, from the apex to the base. Which of the following conditions would best explain this?
1. Acute asthma
2. Chronic bronchitis
3. Pneumonia
4. Spontaneous pneumothorax

73. Which of the following treatments would the nurse expect for a client with spontaneous pneumothorax?
1. Antibiotics
2. Bronchodilators
3. Chest tube placement
4. Hyperbaric chamber

74. A 60-year-old client was in a motor vehicle accident as an unrestrained driver. He's now in the emergency department complaining of difficulty breathing and chest pain. On auscultation of his lung fields, no breath sounds are present in the left upper lobe. This client may have which of the following conditions?
1. Bronchitis
2. Pneumonia
3. Pneumothorax
4. Tuberculosis (TB)

75. Which of the following methods is the best way to confirm the diagnosis of a pneumothorax?
1. Auscultate breath sounds.
2. Have the client use an incentive spirometer.
3. Take a chest X-ray.
4. Stick a needle in the area of the decreased breath sounds.

Listening to breath sounds sure helps diagnose a lot of conditions, doesn't it?

You're doing great! Keep going!

71. 1. Because he's short of breath, listening to breath sounds is a good idea. He may need a chest X-ray and an electrocardiogram, but a physician must order that. Unless a cardiac source for his pain is identified, he won't need an echocardiogram.
NP: Data collection; CN: Physiological integrity; CNS: Physiological adaptation; CL: Application

72. 4. A spontaneous pneumothorax occurs when the client's lung collapses, causing an acute decrease in the amount of functional lung used in oxygenation. The sudden collapse was the cause of his chest pain and shortness of breath. An asthma attack would show wheezing breath sounds, and bronchitis would have rhonchi. Pneumonia would have bronchial breath sounds over the area of consolidation.
NP: Evaluation; CN: Physiological integrity; CNS: Physiological adaptation; CL: Application

73. 3. The only way to reexpand the lung is to place a chest tube on the right side so the air in the pleural space can be removed and the lung reexpanded. Antibiotics and bronchodilators would have no effect on lung reexpansion, nor would the hyperbaric chamber.
NP: Implementation; CN: Physiological integrity; CNS: Physiological adaptation; CL: Application

74. 3. From the trauma the client experienced, it's unlikely he has bronchitis, pneumonia, or TB; rhonchi with bronchitis, bronchial breath sounds with pneumonia, and rhonchorous breath sounds with TB would be heard.
NP: Evaluation; CN: Physiological integrity; CNS: Physiological adaptation; CL: Application

75. 3. A chest X-ray will show the area of collapsed lung if a pneumothorax is present, as well as the volume of air in the pleural space. Listening to breath sounds won't confirm a diagnosis. The client wouldn't do well with an incentive spirometer at this time. A needle thoracostomy is done only in an emergency and only by someone trained to do it.
NP: Implementation; CN: Physiological integrity; CNS: Physiological adaptation; CL: Application

76. A client is about to have a chest tube inserted in the left upper chest. When the tube is inserted, it begins to drain a large amount of serosanguineous fluid. Which of the following explanations best describes what caused this?
　1. The chest tube was inserted improperly.
　2. This always happens when a chest tube is inserted.
　3. An artery was nicked when the chest tube was placed.
　4. The client had a hemothorax instead of a pneumothorax.

Always choose the best response.

76. 4. Because of the traumatic cause of injury, the client had a hemothorax, in which blood collection causes the collapse of the lung. The placement of the chest tube will drain the blood from the space and reexpand the lung. There is a very slight chance of nicking an intercostal artery during insertion, but it's fairly unlikely if the person placing the chest tube has been trained. The initial chest X-ray would help confirm whether there was blood in the pleural space or just air.

NP: Implementation; CN: Physiological integrity; CNS: Physiological adaptation; CL: Application

77. A hospitalized client needs a central I.V. catheter inserted. The physician places the catheter in the subclavian vein. Shortly afterward, the client develops shortness of breath and appears restless. Which of the following actions would the nurse do first?
　1. Administer a sedative.
　2. Advise the client to calm down.
　3. Auscultate breath sounds.
　4. Check to see if the client can have medication.

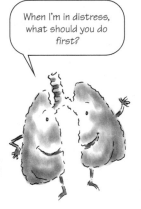

When I'm in distress, what should you do first?

77. 3. Because this is an acute episode, listen to the client's lungs to see if anything has changed. Don't give this client medication, especially sedatives, if he's having trouble breathing. Give the client emotional support and contact the physician who placed the central venous access.

NP: Planning; CN: Safe, effective care environment; CNS: Coordinated care; CL: Application

78. Which of the following measures would be ordered for a client who recently had a central venous catheter inserted and who now appears short of breath and anxious?
　1. Chest X-ray
　2. Electrocardiogram
　3. Laboratory tests
　4. Sedation

78. 1. Inserting an I.V. catheter in the subclavian vein can result in a pneumothorax, so a chest X-ray should be done. If it's negative, then other tests should be done but they aren't appropriate as the first intervention.

NP: Planning; CN: Health promotion and maintenance; CNS: Prevention and early detection of disease; CL: Knowledge

Anticipation is a necessary part of nursing.

79. A client needs to have a chest tube inserted in the right upper chest. Which of the following actions is part of the nurse's role?
　1. The nurse isn't needed.
　2. Prepare the chest tube drainage system.
　3. Bring the chest X-ray to the client's room.
　4. Insert the chest tube.

79. 2. The nurse must anticipate that a drainage system is required and set this up before insertion so the tube can be directly connected to the drainage system. The chest X-ray need not be brought to the client's room. A physician will insert the chest tube.

NP: Planning; CN: Physiological integrity; CNS: Physiological adaptation; CL: Application

80. When assessing the closed chest drainage system (Pleur-evac) of a client who has just returned from a lobectomy, the nurse must ensure which of the following?
1. The fluid in the water-seal chamber rises from inspiration and falls with expiration.
2. The tubing remains looped below the level of the bed.
3. The drainage chamber doesn't drain more than 100 ml in 8 hours.
4. The suction-control chamber bubbles vigorously when connected to suction.

81. Which of the following measures <u>best</u> determines that a chest tube is no longer needed for a client who had a pneumothorax?
1. The drainage from the chest tube is minimal.
2. Arterial blood gas (ABG) levels are obtained to ensure proper oxygenation.
3. It's removed and the client is assessed to see if he's breathing adequately.
4. No fluctuation in the water seal chamber occurs when no suction is applied.

82. Which of the following interventions should be done before a chest tube is removed?
1. Disconnect the drainage system from the tube.
2. Obtain a chest X-ray to document reexpansion.
3. Obtain an arterial blood gas to document oxygen status.
4. Sedate the client and the physician will slip the tube out without warning the client.

83. Which of the following actions or conditions is the <u>number one</u> cause of lung cancer?
1. Genetics
2. Occupational exposures
3. Smoking a pipe
4. Smoking cigarettes

Wait a minute. What breath do I expect to see after a chest tube insertion?

Go for the best!

Remember, choose the most common cause.

80. 1. Rise and fall of the water-seal chamber immediately after surgery indicates patency of the chest tube drainage system. The tubing should be coiled on the bed, without dependent loops, to promote drainage. Up to 500 ml of drainage can occur in the first 24 hours after surgery. Gentle bubbling is indicated after surgery to prevent excessive evaporation.
NP: Data collection; CN: Physiological integrity; CNS: Physiological adaptation; CL: Application

81. 4. The chest tube isn't removed until it's determined the client's lung has adequately reexpanded and will stay that way. One indication of reexpansion is the cessation of fluctuation in the water seal chamber when suction isn't applied. After the lung stays expanded, the chest tube is removed. Drainage should be minimal before the chest tube is removed. An ABG test may be done to ensure proper oxygenation but isn't necessary if clinical assessment criteria are met.
NP: Implementation; CN: Physiological integrity; CNS: Physiological adaptation; CL: Application

82. 2. A chest X-ray should be done to ensure and document the lung is reexpanded and has remained expanded since suction was discontinued. Client cooperation is desirable; if the client can hold his breath while the chest tube is removed, there's less chance that air will be drawn back into the pleural space during removal. The drainage system shouldn't be disconnected from the tube while still in the client because that could cause a pneumothorax to recur. A pulse oximetry measurement is sufficient to track oxygenation before the tube is removed.
NP: Implementation; CN: Physiological integrity; CNS: Physiological adaptation; CL: Application

83. 4. As many as 90% of clients with lung cancer smoke cigarettes. Cigarette smoke contains several organ-specific carcinogens. There may be a genetic predisposition for the development of cancer. Occupational hazards such as pollutants can cause cancer. Pipe smokers inhale less often and tend to develop cancers of the lip and mouth.
NP: Implementation; CN: Physiological integrity; CNS: Physiological adaptation; CL: Application

84. The client with which of the following types of lung cancer has the best prognosis?
1. Adenocarcinoma
2. Oat cell
3. Squamous cell
4. Small cell

85. Warning signs and symptoms of lung cancer include persistent cough, bloody sputum, dyspnea, and which of the following other symptoms?
1. Dizziness
2. Generalized weakness
3. Hypotension
4. Recurrent pleural effusions

That's no smoke screen in question 86. Think about the risks of smoking.

86. Which laboratory test value is elevated in clients who smoke and therefore can't be used as a general indicator of cancer?
1. Acid phosphatase level
2. Serum calcitonin level
3. Alkaline phosphatase level
4. Carcinoembryonic antigen level

87. A definitive diagnosis of lung cancer is obtained by which of the following evaluations?
1. Bronchoscopy
2. Chest X-ray
3. Computerized tomography of the chest
4. Surgical biopsy

Tell me. Am I being upstaged — or staged — here?

88. Staging of lung cancer is done for which of the following reasons?
1. To identify the type of cancer
2. To identify the best treatment
3. To identify if metastasis occurred
4. To identify the location of the lesion

84. 3. Squamous cell carcinoma is a slow-growing, rarely metastasizing type of cancer. Adenocarcinoma is the next best lung cancer to have in terms of prognosis. Oat cell and small cell carcinoma are the same. Small cell carcinoma grows rapidly and is quick to metastasize.
NP: Evaluation; CN: Physiological integrity; CNS: Physiological adaptation; CL: Knowledge

85. 4. Recurring episodes of pleural effusions can be caused by the tumor and should be investigated. Dizziness, hypotension, and generalized weakness are not typically considered warning signals, but may occur in advanced stages of cancer.
NP: Data collection; CN: Physiological integrity; CNS: Physiological adaptation; CL: Application

86. 4. Because the level of carcinoembryonic antigen is elevated in clients who smoke, it can't be used as a general indicator of cancer. However, the carcinoembryonic antigen level is helpful in monitoring cancer treatment because it usually falls to normal within 1 month if treatment is successful. An elevated acid phosphatase level may indicate prostate cancer. An elevated alkaline phosphatase level may reflect bone metastasis. An elevated serum calcitonin level usually signals thyroid cancer.
NP: Data collection; CN: Physiological integrity; CNS: Physiological adaptation; CL: Knowledge

87. 4. Only surgical biopsy and cytologic examination of the cells can give a definitive diagnosis of the type of cancer. Bronchoscopy gives positive results in only 30% of the cases. Chest X-ray and computerized tomography can identify location, but don't diagnose the type of cancer.
NP: Implementation; CN: Physiological integrity; CNS: Physiological adaptation; CL: Application

88. 2. Staging the cancer allows the physician to determine appropriate therapies and estimate the client's prognosis. The staging information may include whether the cancer has metastasized, but this information is obtained by a body scan, such as magnetic resonance imaging or computerized tomography. The location and type of cancer has already been identified before the staging is done.
NP: Implementation; CN: Physiological integrity; CNS: Physiological adaptation; CL: Application

89. Which of the following interventions is the key to increasing the survival rates of clients with lung cancer?
    1. Early bronchoscopy
    2. Early detection
    3. High-dose chemotherapy
    4. Smoking cessation

89. 2. Detecting cancer early when the cells may be premalignant and potentially curable would be most beneficial. However, a tumor must be 1 cm in diameter before it is detectable on a chest X-ray, so this is difficult. If the cancer is detected early, a bronchoscopy may help identify cell type. Smoking cessation won't reverse the process but may prevent further decompensation.
NP: Implementation; CN: Health promotion and maintenance; CNS: Prevention and early detection of disease; CL: Application

90. A client with a chest tube has accidentally removed it. What should be done first?
    1. Lie the client down on his left side.
    2. Lie the client down on his right side.
    3. Apply an occlusive dressing over the site.
    4. Reinsert the chest tube that fell out.

You're almost there! Keep up the good work!

90. 3. To prevent the client from sucking air into the pleural space and causing a pneumothorax, an occlusive dressing should be put over the hole where the tube came out. The physician should be called and the client checked for signs of respiratory distress. It isn't advisable for the physician to reinsert the old tube because it's no longer sterile. Positioning the client on either the left or right side won't make a difference.
NP: Implementation; CN: Physiological integrity; CNS: Physiological adaptation; CL: Application

91. A client has been diagnosed with lung cancer and requires a wedge resection. How much of the lung is removed?
    1. One entire lung
    2. A lobe of the lung
    3. A small, localized area near the surface of the lung
    4. A segment of the lung, including a bronchiole and its alveoli

91. 3. A very small area of tissue close to the surface of the lung is removed in a wedge resection. A segment of the lung is removed in a segmental resection, a lobe is removed in a lobectomy, and an entire lung is removed in a pneumonectomy.
NP: Implementation; CN: Physiological integrity; CNS: Physiological adaptation; CL: Application

92. When a client has a lobectomy, what fills the space where the lobe was?
    1. The space stays empty.
    2. The surgeon fills the space with a gel.
    3. The lung space fills up with serous fluid.
    4. The remaining lobe or lobes overexpand to fill the space.

Consider: what would keep everything in balance?

92. 4. The remaining lobe or lobes overexpand slightly to fill the space previously occupied by the removed tissue. The diaphragm is carried higher on the operative side to further reduce the empty space. The surgeon doesn't use gel to fill the space. Serous fluid overproduction would compress the remaining lobes and diminish their function, and also possibly cause a mediastinal shift. The space can't remain "empty," because truly empty would imply a vacuum, which would interfere with the intrathoracic pressure changes that allow breathing.
NP: Implementation; CN: Physiological integrity; CNS: Physiological adaptation; CL: Application

93. If a client requires a pneumonectomy, what fills the area of the thoracic cavity?
1. The space remains filled with air only.
2. The surgeon fills the space with a gel.
3. Serous fluid fills the space and consolidates the region.
4. The tissue from the other lung grows over to the other side.

94. During a pneumonectomy, the phrenic nerve on the surgical side is often cut to cause hemidiaphragm paralysis. Why is this done?
1. Paralyzing the diaphragm reduces oxygen demand.
2. Cutting the phrenic nerve is a mistake during surgery.
3. The client isn't using that lung to breathe any longer.
4. Paralyzing the diaphragm reduces the space left by the pneumonectomy.

95. Which of the following is the primary goal of surgical resection for lung cancer?
1. Remove the tumor and all surrounding tissue.
2. Remove the tumor and as little surrounding tissue as possible.
3. Remove all the tumor and any collapsed alveoli in the same region.
4. Remove as much of the tumor as possible, without removing any alveoli.

96. Which position is contraindicated when caring for a client who has had a pneumonectomy for lung cancer?
1. Semi-Fowler's
2. Lying on the nonoperative side
3. Reverse Trendelenburg's
4. Prone

What happens when there's only one of us?

Here's another test-taking hint for you!

93. 3. Serous fluid fills the space and eventually consolidates, preventing extensive mediastinal shift of the heart and remaining lung. There's no gel that can be placed in the pleural space. The tissue from the other lung can't cross the mediastinum, although a temporary mediastinal shift exists until the space is filled. Air can't be left in the space.
NP: Implementation; CN: Physiological integrity; CNS: Physiological adaptation; CL: Application

94. 4. Because the hemidiaphragm is a muscle that doesn't contract when paralyzed, an uncontracted hemidiaphragm remains in an "up" position, which reduces the space left by the pneumonectomy. Serous fluid has less space to fill, thus reducing the extent and duration of mediastinal shift after surgery. Although it's true that the client no longer needs the hemidiaphragm on the operative side to breathe, this alone wouldn't be sufficient justification for cutting the phrenic nerve. Paralyzing the hemidiaphragm doesn't significantly decrease total-body oxygen demand.
NP: Implementation; CN: Physiological integrity; CNS: Physiological adaptation; CL: Application

95. 2. The goal of surgical resection is to remove the lung tissue that has tumor in it while saving as much surrounding tissue as possible. It may be necessary to remove alveoli and bronchioles, but care is taken to make sure only what's absolutely necessary is removed.
NP: Implementation; CN: Physiological integrity; CNS: Physiological adaptation; CL: Application

96. 2. A client who has undergone a pneumonectomy doesn't have a chest tube in place. Therefore, the accumulation of any blood or fluid around the operative site could cause pressure on the operative side and compromise ventilation. Any accumulation may be avoided by lying the client on his nonoperative side.
NP: Implementation; CN: Physiological integrity; CNS: Reduction of risk potential; CL: Application

97. Preoperative teaching for the client having surgery should focus on which of the following areas?
1. Deciding if the client should have the surgery
2. Giving emotional support to the client and family
3. Giving minute details of the surgery to the client and family
4. Providing general information to reduce client and family anxiety

The nurse does a lot of teaching but in this case, what would be the main focus?

97. 4. The nurse's role is to provide general information about the surgery and what to expect before and after surgery, and to give emotional support during this time. The nurse's role isn't to decide if the client should have surgery or to give minute details of the surgery unless the client or family requests them, in which case the surgeon should answer the questions. Emotional support alone during this time isn't sufficient.
NP: Implementation; CN: Physiological integrity; CNS: Physiological adaptation; CL: Application

98. The client with a benign tumor is treated in which of the following ways?
1. The tumor is treated with radiation only.
2. The tumor is treated with chemotherapy only.
3. The tumor is left alone unless symptoms are present.
4. The tumor is removed, involving the least possible amount of tissue.

98. 4. The tumor is removed to prevent further compression of lung tissue as the tumor grows, which could lead to respiratory decompensation. If for some reason it can't be removed, then chemotherapy or radiation may be used to try to shrink the tumor.
NP: Implementation; CN: Physiological integrity; CNS: Physiological adaptation; CL: Application

99. In the client with terminal lung cancer, the focus of nursing care is on which of the following nursing interventions?
1. Provide emotional support.
2. Provide nutritional support.
3. Provide pain control.
4. Prepare the client's will.

Prioritize!

99. 3. The client with terminal lung cancer may have extreme pleuritic pain and should be treated to reduce his discomfort. Preparing the client and his family for the impending death is also important but shouldn't be the primary focus until pain is under control. Nutritional support may be provided, but as the terminal phase advances, the client's nutritional needs greatly decrease. Nursing care doesn't focus on helping the client prepare a will.
NP: Planning; CN: Physiological integrity; CNS: Physiological adaptation; CL: Application

100. Which of the following statements best defines a pulmonary embolism?
1. It's a blood clot that originates in the lung.
2. It's a blood clot that has occluded an alveolus.
3. It's a blood clot that has occluded a bronchiole.
4. It's a blood clot that has occluded a pulmonary blood vessel.

100. 4. A pulmonary embolism is a blood clot or some other material that, after traveling from some other place in the body through the bloodstream, becomes lodged in one of the pulmonary blood vessels. An embolism remains in the bloodstream and specifically isn't in the bronchial tree, the "air side" of the lung architecture; therefore, it isn't in the bronchiole or the alveoli.
NP: Implementation; CN: Physiological integrity; CNS: Physiological adaptation; CL: Application

101. Which of the following is the most common origin for a pulmonary embolism?
1. Amniotic fluid
2. Bone marrow
3. Septic thrombi
4. Venous thrombi

This clue asks you to select the most common, even when all answers are correct.

101. 4. Venous thrombi in the thigh and pelvis are the most common sources for pulmonary emboli. Clients who are immobile form clots from this source. When dislodged, the clots are carried through the bloodstream and lodge in the pulmonary vasculature. The other options are also sources, but not the most common.
NP: Implementation; CN: Physiological integrity; CNS: Physiological adaptation; CL: Application

102. Clients most at risk for pulmonary embolism are those with which of the following conditions?
1. Arthritis
2. Diabetes
3. Pregnancy
4. Trauma to the pelvis or lower extremities

Hint! Check out the word best!

102. 4. Trauma to the pelvis or lower extremities requires the client to be immobile for a long recovery period, which allows the blood to pool and clot in this region and ultimately become an embolus. Clients with the other conditions aren't at as much risk for pulmonary embolism.
NP: Implementation; CN: Physiological integrity; CNS: Physiological adaptation; CL: Application

103. Which of the following measures to prevent pulmonary embolism after lower extremity surgery is the best?
1. Early ambulation
2. Frequent chest X-rays to find a pulmonary embolism
3. Frequent lower extremity scans
4. Intubation of the client

103 1. Early ambulation helps reduce pooling of blood, which reduces the tendency of the blood to form a clot that could then dislodge. Frequent lower extremity scans or chest X-rays don't prevent pulmonary embolism. Intubation of the client will not prevent the occurrence of a pulmonary embolism.
NP: Implementation; CN: Physiological integrity; CNS: Physiological adaptation; CL: Application

104. At 8 a.m., the nurse assesses a client who is scheduled for surgery at 10 a.m. During the assessment, the nurse detects dyspnea, a nonproductive cough, and back pain. What should the nurse do next?
1. Check to see that the chest X-ray was done yesterday, as ordered.
2. Check the serum electrolyte levels and complete blood count (CBC).
3. Notify the physician immediately of these findings.
4. Sign the preoperative checklist for this client.

Another hint: Notice the word next!

104. 3. The nurse should notify the physician immediately because dyspnea, a nonproductive cough, and back pain may signal a change in the client's respiratory status. The nurse should check any ordered tests (such as chest X-ray, serum electrolyte levels, and CBC) after notifying the physician because they may help explain the change in the client's condition. The nurse should sign the preoperative checklist *after* notifying the physician of the client's condition and learning the physician's decision on whether to proceed with surgery.
NP: Implementation; CN: Safe, effective care environment; CNS: Coordinated care; CL: Application

105. When a client's ventilation is impaired, the body retains which substance?
1. Sodium bicarbonate
2. Carbon dioxide
3. Nitrous oxide
4. Oxygen

106. When a client has a pulmonary embolism, he develops chest pain caused by which of the following conditions?
1. Costochondritis
2. Myocardial infarction
3. Pleuritic pain
4. Referred pain from the pelvis to the chest

107. The client with a pulmonary embolism frequently feels apprehension or a sense of "impending doom" because of which of the following reasons?
1. Inflammatory reaction in the lung parenchyma
2. Loss of chest expansion
3. Loss of lung tissue
4. Sudden reduction in adequate oxygenation

108. Hemoptysis may be present in the client with a pulmonary embolism because of which of the following reasons?
1. Alveolar damage in the infarcted area
2. Involvement of major blood vessels in the occluded area
3. Loss of lung parenchyma
4. Loss of lung tissue

109. A client with a massive pulmonary embolism will have an arterial blood gas analysis performed to determine the extent of hypoxia. The acid-base disorder that may be present is which of the following conditions?
1. Metabolic acidosis
2. Metabolic alkalosis
3. Respiratory acidosis
4. Respiratory alkalosis

Here's a hint for question 106: think about what causes the condition.

Almost at the finish line. Nothing can stop you now!

105. 2. When ventilation is impaired, the body retains carbon dioxide because the carbonic acid level increased in the blood. Sodium bicarbonate is used to treat acidosis. Nitrous oxide, which has analgesic and anesthetic properties, commonly is administered before minor surgical procedures. When ventilation is impaired, the body doesn't retain oxygen. Instead, the tissues use oxygen, and carbon dioxide is the end result.
NP: Data collection; CN: Physiological integrity; CNS: Physiological adaptation; CL: Knowledge

106. 3. Pleuritic pain is caused by the inflammatory reaction of the lung parenchyma. The pain isn't associated with myocardial infarction, costochondritis, or referred pain from the pelvis to the chest.
NP: Implementation; CN: Physiological integrity; CNS: Physiological adaptation; CL: Application

107. 4. The client with a pulmonary embolism has less lung involved in oxygenation, causing the client to feel apprehensive. If the area involved is large, the apprehension can be great, giving the client the feeling of "impending doom." There's no actual loss of lung tissue, and chest expansion isn't affected.
NP: Implementation; CN: Physiological integrity; CNS: Physiological adaptation; CL: Application

108. 1. The infarcted area produces alveolar damage that can lead to the production of bloody sputum, sometimes in massive amounts. There's a loss of lung parenchyma and subsequent scar tissue formation.
NP: Implementation; CN: Physiological integrity; CNS: Physiological adaptation; CL: Application

109. 4. A client with a massive pulmonary embolism will have a large region of lung tissue unavailable for perfusion. This causes the client to hyperventilate and blow off large amounts of carbon dioxide, which crosses the unaffected alveolar-capillary membrane more readily than does oxygen and results in respiratory alkalosis.
NP: Implementation CN: Physiological integrity; CNS: Physiological adaptation; CL: Application

110. A ventilation-perfusion scan is frequently performed to diagnose a pulmonary embolism. This test provides what type of information?
1. Amount of perfusion present in the lung
2. Extent of the occlusion and amount of perfusion lost
3. Location of the pulmonary embolism
4. Location and size of the pulmonary embolism

111. Which of the following tests definitively diagnoses a pulmonary embolism?
1. Arterial blood gas (ABG) analysis
2. Computed tomography scan
3. Pulmonary angiogram
4. Ventilation-perfusion scan

OK, Mr. Ventilation-Perfusion Scan. Talk to me!

In question 111... watch that word, definitively.

112. Which of the following medications is prescribed after a pulmonary embolism is diagnosed?
1. Warfarin (Coumadin)
2. Heparin
3. Streptokinase
4. Urokinase

Knowing what a treatment should achieve will help you monitor the response.

113. I.V. heparin is given to clients with pulmonary embolism for which of the following reasons?
1. To dissolve the clot
2. To break up the pulmonary embolism
3. To slow the development of other clots
4. To prevent clots from breaking off and embolizing to the lung

110. 2. The ventilation-perfusion scan provides information on the extent of occlusion caused by the pulmonary embolism and the amount of lung tissue involved in the area not perfused.
NP: Data collection; CN: Physiological integrity; CNS: Physiological adaptation; CL: Application

111. 3. Pulmonary angiogram is used to definitively diagnose a pulmonary embolism. A catheter is passed through the circulation to the region of the occlusion; the region can be outlined with an injection of contrast medium and viewed by fluoroscopy. This shows the location of the clot, as well as the extent of the perfusion defect. Computed tomography scan can show the location of infarcted or ischemic tissue. ABG levels can define the amount of hypoxia present. The ventilation-perfusion scan can report whether there is a ventilation-perfusion mismatch present and define the amount of tissue involved.
NP: Data collection; CN: Physiological integrity; CNS: Physiological adaptation; CL: Application

112. 2. Heparin is started I.V. once a pulmonary embolism is diagnosed to reduce further clot formation. When a therapeutic level of heparin is established, warfarin is started. It can take up to 3 days before a therapeutic level of warfarin is achieved. Streptokinase and urokinase are both fibrinolytics, and their usefulness in the management of pulmonary embolism is unclear.
NP: Implementation; CN: Physiological integrity; CNS: Pharmacological therapies; CL: Application

113. 3. Heparin doesn't break up pulmonary embolisms or dissolve clots already formed, but it slows the development of other clots. Heparin doesn't stop clots from going to the lung.
NP: Implementation; CN: Physiological integrity; CNS: Pharmacological therapies; CL: Application

114. A client with a pulmonary embolism is discharged but will remain on warfarin (Coumadin) therapy for up to 6 months to accomplish which of the following actions?

1. Prevent further embolism formation
2. Minimize the growth of new or existing thrombi
3. Continue to reduce the size of the pulmonary embolism
4. Break up the existing pulmonary embolism until it's totally gone

115. The goal of oxygen therapy for a client with a pulmonary embolism is to obtain which of the following values?

1. $Paco_2$ above 40 mm Hg
2. $Paco_2$ below 40 mm Hg
3. Pao_2 above 60 mm Hg
4. Pao_2 below 60 mm Hg

In question 115, think about therapy goals.

116. A client is receiving oxygen via a nasal cannula at 2 L/minute. What percentage of oxygen concentration is coming through the cannula?

1. 23% to 30%
2. 30% to 40%
3. 40% to 60%
4. 50% to 75%

Watch for clues for what you're treating.

117. A client with a pulmonary embolism typically has chest pain and apprehension, which can be treated by which of the following methods?

1. Analgesics
2. Guided imagery
3. Positioning the client on the left side
4. Providing emotional support

114. 2. Warfarin minimizes the growth of new or existing thrombi. It's impossible to tell whether a client who developed a thrombus that became a pulmonary embolism will develop more thrombi. Therefore, current therapy is to treat clients who had a pulmonary embolism with warfarin for up to 6 months, as long as it isn't contraindicated. Warfarin doesn't reduce the size of existing pulmonary emboli or break them up.

NP: Implementation; CN: Physiological integrity; CNS: Pharmacological therapies; CL: Application

115. 3. The goal of oxygen therapy for a client with a pulmonary embolism is to have a Pao_2 greater than 60 mm Hg on an Fio_2 of 40% or less. The normal range of the $Paco_2$ is 35 to 45 mm Hg. In the absence of other pathologic states, it should reach normal levels before the Pao_2 does on room air because carbon dioxide crosses the alveolar-capillary membrane with greater ease.

NP: Implementation; CN: Physiological integrity; CNS: Physiological adaptation; CL: Application

116. 1. The percent of oxygen concentration as it passes out of the nasal cannula at 2 L/minute is 23% to 30%. The oxygen concentration via cannula at 3 to 5 L/minute is 30% to 40%. A simple mask at 6 to 8 L/minute delivers 40% to 60% of oxygen. A partial rebreather mask at 8 to 11 L/minutes delivers 50% to 75% of oxygen.

NP: Data collection; CN: Psychological integrity CNS: Reduction of risk potential; CL: Comprehension

117. 1. Once the pulmonary embolism has been diagnosed and the amount of hypoxia determined, apprehension can be treated with analgesics as long as respiratory status isn't compromised by analgesics. Guided imagery and providing emotional support can be used as alternatives. Positioning the client on the left side when a pulmonary embolism is suspected may prevent a clot that has extended through the capillaries and into the pulmonary veins from breaking off and traveling through the heart into the arterial circulation, leading to a massive stroke.

NP: Implementation; CN: Physiological integrity; CNS: Physiological adaptation; CL: Application

118. A client with a pulmonary embolism may have an umbrella filter placed in the vena cava for which of the following reasons?
   1. The filter prevents further clot formation.
   2. The filter collects clots so they don't go to the lung.
   3. The filter break up clots into insignificantly small pieces.
   4. The filter contains anticoagulants that are slowly released, dissolving any clots.

Always know the whys of a condition, especially the most common ones, like this one.

119. A client with a pulmonary embolism may need an embolectomy, which involves which of the following actions?
   1. Removal of an embolism in the lower extremity
   2. Sucking the embolism out of the lung by bronchoscopy
   3. Surgical removal of the embolism source in the pelvis
   4. Surgical removal of the embolism in the pulmonary vasculature

120. Nursing management of a client with a pulmonary embolism focuses on which of the following actions?
   1. Assessing oxygenation status
   2. Monitoring the oxygen delivery device
   3. Monitoring for other sources of clots
   4. Determining whether the client requires another ventilation-perfusion scan

You're close to the finish. Keep going!

121. A pulse oximeter gives what type of information about the client?
   1. Amount of carbon dioxide in the blood
   2. Amount of oxygen in the blood
   3. Percentage of hemoglobin carrying oxygen
   4. Respiratory rate

118. 3. The umbrella filter is placed in a client at high risk for the formation of more clots that could potentially become pulmonary emboli. The filter breaks the clots into small pieces that won't significantly occlude the pulmonary vasculature. The filter doesn't release anticoagulants and doesn't prevent further clot formation. The filter doesn't collect the clots, because if it did, it would have to be emptied periodically, causing the client to require surgery in the future.
NP: Implementation; CN: Physiological integrity; CNS: Physiological adaptation; CL: Application

119. 4. If the pulmonary embolism is large and doesn't respond to treatment, surgical removal may be necessary to restore perfusion to the area of the lung. This is rarely done because of the associated high mortality risk. It's impossible to remove a pulmonary embolism through bronchoscopy because the defect isn't in the bronchial tree. A thrombectomy can be performed at other sources of clot, but when a pulmonary embolism has already occurred, it would have little effect on oxygenation.
NP: Implementation; CN: Physiological integrity; CNS: Physiological adaptation; CL: Application

120. 1. Nursing management of a client with a pulmonary embolism focuses on assessing oxygenation status and ensuring treatment is adequate. If the client's status begins to deteriorate, it's the nurse's responsibility to contact the physician and attempt to improve oxygenation. Monitoring for other clot sources and ensuring the oxygen delivery device is working properly are other nursing responsibilities, but they aren't the focus of care.
NP: Implementation; CN: Physiological integrity; CNS: Physiological adaptation; CL: Application

121. 3. The pulse oximeter determines the percentage of hemoglobin carrying oxygen. This doesn't ensure that the oxygen being carried through the bloodstream is actually being taken up by the tissue. The pulse oximeter doesn't provide information about the amount of oxygen or carbon dioxide in the blood or the client's respiratory rate.
NP: Implementation; CN: Physiological integrity; CNS: Physiological adaptation; CL: Application

122. What effect does hemoglobin amount have on oxygenation status?
1. No effect
2. More hemoglobin reduces the client's respiratory rate
3. Low hemoglobin levels cause reduced oxygen-carrying capacity
4. Low hemoglobin levels cause increased oxygen-carrying capacity

To answer question 122, think about hemoglobin's role.

122. 3. Hemoglobin carries oxygen to all tissues in the body. If the hemoglobin level is low, the amount of oxygen-carrying capacity is also low. More hemoglobin will increase oxygen-carrying capacity and thus increase the total amount of oxygen available in the blood. If the client has been tachypneic during exertion, or even at rest, because oxygen demand is higher than the available oxygen content, then an increase in hemoglobin may decrease the respiratory rate to normal levels.
NP: Implementation; CN: Physiological integrity; CNS: Physiological adaptation; CL: Application

123. How does positive end-expiratory pressure (PEEP) improve oxygenation?
1. It provides more oxygen to the client.
2. It opens up bronchioles and allows oxygen to get in the lungs.
3. It opens up collapsed alveoli and helps keep them open.
4. It adds pressure to the lung tissue, which improves gaseous exchange.

123. 3. PEEP delivers positive pressure to the lung at the end of expiration. This helps open collapsed alveoli and helps them stay open so gas exchange can occur in these newly opened alveoli, improving oxygenation. The bronchioles don't participate in gas exchange except to act as a conduit for inspired and expired air. The walls are rigid enough they generally don't collapse. PEEP doesn't directly add pressure to the lung tissue or provide more oxygen to the client.
NP: Implementation; CN: Physiological integrity; CNS: Physiological adaptation; CL: Application

You're the best for catching these hints!

124. Which of the following statements best explains how opening up collapsed alveoli improves oxygenation?
1. Alveoli need oxygen to live.
2. Alveoli have no effect on oxygenation.
3. Collapsed alveoli increase oxygen demand.
4. Gaseous exchange occurs in the alveolar membrane.

124. 4. Gaseous exchange occurs in the alveolar membrane, so if the alveoli collapse, no exchange occurs. Collapsed alveoli receive oxygen, as well as other nutrients, from the bloodstream. Collapsed alveoli have no effect on oxygen demand, though by decreasing the surface area available for gas exchange, they decrease oxygenation of the blood.
NP: Implementation; CN: Physiological integrity; CNS: Physiological adaptation; CL: Application

125. Continuous positive airway pressure can be provided through an oxygen mask to improve oxygenation in hypoxic clients by which of the following methods?
1. The mask provides 100% oxygen to the client.
2. The mask provides continuous air that the client can breathe.
3. The mask provides pressurized oxygen so the client can breathe more easily.
4. The mask provides pressurized oxygen at the end of expiration to open collapsed alveoli.

125. 3. The mask provides pressurized oxygen continuously through both inspiration and expiration. By providing a client with pressurized oxygen, the client has less resistance to overcome in taking in his next breath, making it easier to breathe. The mask can be set to deliver any amount of oxygen needed. Pressurized oxygen delivered at the end of expiration is positive end-expiratory pressure, not continuous positive airway pressure.
NP: Implementation; CN: Physiological integrity; CNS: Physiological adaptation; CL: Application

126. Bilevel positive airway pressure (BiPAP) is delivered though a special oxygen mask that performs which of the following functions?
1. The mask provides 100% oxygen at both inspiration and expiration.
2. The mask provides pressurized oxygen so the client can breathe more easily.
3. The mask provides pressurized oxygen at the end of expiration to open collapsed alveoli.
4. The mask provides both continuous positive airway pressure (CPAP) and positive end-expiratory pressure (PEEP) to provide optimal oxygenation and ventilation.

127. Which of the following states best describes pleural effusion?
1. The collapse of alveoli
2. The collapse of a bronchiole
3. The fluid in the alveolar space
4. The accumulation of fluid between the linings of the pleural space

128. If a pleural effusion develops, which of the following actions best describes how the fluid can be removed from the pleural space and proper lung status restored?
1. Insert a chest tube
2. Perform thoracentesis
3. Perform paracentesis
4. Allow the pleural effusion to drain by itself

126. 4. BiPAP delivers both CPAP and PEEP. It provides the differing pressures throughout the respiratory cycle, attempting to optimize a client's oxygenation and ventilation. It's used in an effort to avoid intubation for mechanical ventilation. Inspiratory and expiratory pressures are set separately to optimize the client's ventilatory status, and F_{IO_2} is adjusted to optimize oxygenation. The second choice describes only the CPAP component of BiPAP, and the third choice describes the PEEP component.
NP: Implementation; CN: Physiological integrity; CNS: Physiological adaptation; CL: Application

127. 4. Pleural fluid normally seeps continually into the pleural space from the capillaries lining the parietal pleura and is reabsorbed by the visceral pleural capillaries and lymphatics. Any condition that interferes with either the secretion or drainage of this fluid will lead to a pleural effusion. The collapse of alveoli or a bronchiole has no particular name. Fluid within the alveolar space can be caused by heart failure or adult respiratory distress syndrome.
NP: Implementation; CN: Physiological integrity; CNS: Physiological adaptation; CL: Application

Hurray! You did it! Are you the best, or what!

128. 2. Thoracentesis is used to remove excess pleural fluid. The fluid is then analyzed to determine if it's transudative or exudative. Transudates are substances that have passed through a membrane and usually occur in low protein states. Exudates are substances that have escaped from blood vessels. They contain an accumulation of cells and have a high specific gravity and a high lactate dehydrogenase level. Exudates usually occur in response to a malignancy, infection, or inflammatory process. Paracentesis is the removal of fluid from the abdomen. A chest tube is rarely necessary because the amount of fluid is typically not large enough to warrant such a measure. Pleural effusions can't drain by themselves.
NP: Evaluation; CN: Physiological integrity; CNS: Physiological adaptation; CL: Analysis

NP: Nursing process   CN: Client needs category   CNS: Client needs subcategory   CL: Cognitive level

Stroke, subdural hematoma, laminectomy—they're all here in this comprehensive chapter on neurosensory disorders in adults. I've got a sixth sense you're going to do great!

1. An elderly client had a stroke and can only see the nasal visual field on one side and the temporal portion on the opposite side. Which of the following terms correctly describes this condition?

1. Astereognosis
2. Homonymous hemianopia
3. Oculogyric crisis
4. Receptive aphasia

A stroke can change my normal views.

1. 2. Homonymous hemianopia describes the loss of visual field on the nasal side and the opposite temporal side due to damage of the optic nerves. Receptive aphasia is the inability to understand words or word meaning. Oculogyric crisis, a fixed position of the eyeballs that can last for minutes or hours, occurs in response to antipsychotic medications. Astereognosis is the inability to identify common objects through touch.
NP: Data collection; CN: Physiological integrity; CNS: Physiological adaptation; CL: Application

2. A client had an embolic stroke. Which of the following conditions places a client at risk for thromboembolic stroke?

1. Atrial fibrillation
2. Bradycardia
3. Deep vein thrombosis (DVT)
4. History of myocardial infarction (MI)

Read question 2 carefully.

2. 1. Atrial fibrillation occurs with the irregular and rapid discharge from multiple ectopic atrial foci that causes quivering of the atria without atrial systole. This asynchronous atrial contraction predisposes to mural thrombi, which may embolize, leading to a stroke. Bradycardia, past MI, or DVT won't lead to arterial embolization.
NP: Planning; CN: Physiological integrity; CNS: Physiological adaptation; CL: Application

3. A heparin infusion at 1,500 units/hour is ordered for a 65-year-old client with a stroke in evolution. The infusion contains 25,000 units of heparin in 500 ml of saline solution. How many milliliters per hour should be given?

1. 15 ml/hour
2. 30 ml/hour
3. 45 ml/hour
4. 50 ml/hour

Practice makes infusion calculations easy.

3. 2. An infusion prepared with 25,000 units of heparin in 500 ml of saline solution yields 50 units of heparin per milliliter of solution. The equation is set up as 50 units times X (the unknown quantity) equals 1,500 units/hour; X equals 30 ml/hour.
NP: Implementation; CN: Physiological integrity; CNS: Pharmacological therapies; CL: Application

4. Which of the following medications may be prescribed to prevent a thromboembolic stroke?

1. Acetaminophen
2. Streptokinase
3. Ticlopidine
4. Methylprednisolone

4. 3. Ticlopidine inhibits platelet aggregation by interfering with adenosine diphosphate release in the coagulation cascade and is used to prevent thromboembolic stroke. Aspirin, not acetaminophen, interferes with platelet aggregation. Streptokinase is a medication used with evolving myocardial infarctions and dissolves clots. Methylprednisolone is a steroid with anticoagulant properties.
NP: Planning; CN: Physiological integrity; CNS: Pharmacological therapies; CL: Application

NP: Nursing process   CN: Client needs category   CNS: Client needs subcategory   CL: Cognitive level

5. To maintain airway patency during a stroke in evolution, which of the following nursing interventions is appropriate?
1. Thicken all dietary liquids.
2. Restrict dietary and parenteral fluids.
3. Place the client in the supine position.
4. Have tracheal suction available at all times.

6. For a client with a stroke, which of the following criteria must be fulfilled before the client is fed?
1. The gag reflex returns.
2. Speech returns to normal.
3. Cranial nerves III, IV, and VI are intact.
4. The client swallows small sips of water without coughing.

Think: How well can a dysphagic client chew and swallow?

7. Which of the following diets would be least likely to lead to aspiration in a client who had a stroke with residual dysphagia?
1. Clear liquid
2. Full liquid
3. Mechanical soft
4. Thickened liquid

8. A 77-year-old client had a thromboembolic right stroke; his left arm is swollen. Which of the following conditions may cause swelling after a stroke?
1. Elbow contracture secondary to spasticity
2. Loss of muscle contraction decreasing venous return
3. Deep vein thrombosis (DVT) due to immobility of the ipsilateral side
4. Hypoalbuminemia due to protein escaping from an inflamed glomerulus

Where is the hemiplegia in a right stroke?

9. After a brain stem infarction, a nurse would observe for which of the following conditions?
1. Aphasia
2. Bradypnea
3. Contralateral hemiplegia
4. Numbness and tingling to the face or arm

5. 4. Because of a potential loss of gag reflex and potential altered level of consciousness, the client should be kept in Fowler's or a semi-prone position with tracheal suction available at all times. Unless heart failure is present, restricting fluids isn't indicated. Thickening dietary liquids isn't done until the gag reflex returns or the stroke has evolved and the deficit can be assessed.
NP: Implementation; CN: Physiological integrity; CNS: Reduction of risk potential; CL: Application

6. 1. An intact gag reflex shows a properly functioning cranial nerve IX (glossopharyngeal). A nurse shouldn't offer food or fluids without assessing for intact gag reflexes. Cranial nerves III, IV, and VI evaluate eye movement and accommodation. Speech may be normal while the gag reflex is absent.
NP: Planning; CN: Physiological integrity; CNS: Reduction of risk potential; CL: Application

7. 4. Thickened liquids are easiest to form into a bolus and swallow. Clear and full liquids are amorphous and can't easily form a bolus. A mechanical soft diet may be too hard to chew and too dry to swallow when dysphagia is present.
NP: Planning; CN: Physiological integrity; CNS: Reduction of risk potential; CL: Application

8. 2. In clients with hemiplegia or hemiparesis, loss of muscle contraction decreases venous return and may cause swelling of the affected extremity. Stroke isn't linked to protein loss. DVT may develop in clients with a stroke but is more likely in the lower extremities. Contractures, or bony calcifications, may occur with stroke, but don't appear with swelling.
NP: Planning; CN: Physiological integrity; CNS: Physiological adaptation; CL: Analysis

9. 2. The brain stem contains the medulla and the vital cardiac, vasomotor, and respiratory centers. A brain stem infarction leads to vital sign changes such as bradypnea. Numbness, tingling in the face or arm, contralateral hemiplegia, and aphasia may occur with a stroke.
NP: Data collection; CN: Physiological integrity; CNS: Physiological adaptation; CL: Application

---

NP: Nursing process   CN: Client needs category   CNS: Client needs subcategory   CL: Cognitive level

10. Which of the following conditions is a risk factor for the development of cataracts in a 40-year-old client?
1. History of frequent streptococcal throat infections
2. Maternal exposure to rubella during pregnancy
3. Increased intraocular pressure
4. Prolonged use of steroidal anti-inflammatory agents

11. A client who had cataract surgery should be told to call his physician if he has which of the following conditions?
1. Blurred vision
2. Eye pain
3. Glare
4. Itching

12. Clear fluid is draining from the nose of a client who had a head trauma 3 hours ago. This may indicate which of the following conditions?
1. Basilar skull fracture
2. Cerebral concussion
3. Cerebral palsy
4. Sinus infection

Which symptom would spell danger?

13. A 19-year-old client with a mild concussion is discharged from the emergency department. Before discharge, he complains of a headache. When offered acetaminophen, his mother tells the nurse the headache is severe and she would like her son to have something stronger. Which response by the nurse is appropriate?
1. "Your son had a mild concussion, acetaminophen is strong enough."
2. "Aspirin is avoided because of the danger of Reye's syndrome in children or young adults."
3. "Narcotics are avoided after a head injury because they may hide a worsening condition."
4. "Stronger medications may lead to vomiting, which increases intracranial pressure (ICP)."

You're doing a great job!

10. 4. Age 50 and older, corticosteroids use, exposure to ultraviolet light or radiation, or diabetes are risk factors for cataracts.
NP: Data collection; CN: Health promotion and maintenance; CNS: Prevention and early detection of disease; CL: Application

11. 2. Pain shouldn't be present after cataract surgery. The client should be told the other symptoms might be present.
NP: Implementation; CN: Physiological integrity; CNS: Physiological adaptation; CL: Comprehension

12. 1. Clear fluid draining from the ear or nose of a client may mean a cerebrospinal fluid leak, which is common in basilar skull fractures. Concussion is associated with a brief loss of consciousness, sinus infection is associated with facial pain and pressure with or without nasal drainage, and cerebral palsy is associated with nonprogressive paralysis present since birth.
NP: Data collection; CN: Physiological integrity; CNS: Physiological adaptation; CL: Analysis

13. 3. Narcotics may mask changes in the level of consciousness that indicate increased ICP and shouldn't be given. Saying acetaminophen is strong enough ignores the mother's question and therefore isn't appropriate. Aspirin is contraindicated in conditions that may have bleeding, such as trauma, and for children or young adults with viral illnesses due to the danger of Reye's syndrome. Aspirin isn't a stronger analgesic than acetaminophen.
NP: Implementation; CN: Physiological integrity; CNS: Reduction of risk potential; CL: Application

14. A client admitted to the hospital with a subarachnoid hemorrhage (SAH) has complaints of severe headache, nuchal rigidity, and projectile vomiting. The nurse knows lumbar puncture (LP) would be contraindicated in this client in which of the following circumstances?
1. Vomiting continues.
2. Intracranial pressure (ICP) is increased.
3. The client needs mechanical ventilation.
4. Blood is anticipated in the cerebrospinal fluid.

14. 2. Sudden removal of cerebrospinal fluid results in pressures lower in the lumbar area than the brain and favors herniation of the brain; therefore, LP is contraindicated with increased ICP. Vomiting may be caused by reasons other than increased ICP; therefore, LP isn't strictly contraindicated. Blood in the cerebrospinal fluid is diagnostic for SAH and was obtained before signs and symptoms of increased ICP. An LP may be performed on clients needing mechanical ventilation.
NP: Planning; CN: Physiological integrity; CNS: Physiological adaptation; CL: Application

Hint, hint!

15. A client with head trauma develops a urine output of 300 ml/hour, dry skin, and dry mucous membranes. Which of the following nursing interventions is the most appropriate to perform immediately?
1. Evaluate urine specific gravity.
2. Anticipate treatment for renal failure.
3. Provide emollients to the skin to prevent breakdown.
4. Slow the I.V. fluids and notify the physician.

15. 1. Urine output of 300 ml/hour may indicate diabetes insipidus, which is failure of the pituitary to produce antidiuretic hormone. This may occur with increased intracranial pressure and head trauma; the nurse evaluates for low urine specific gravity, increased serum osmolarity, and dehydration. There's no evidence that the client is experiencing renal failure. Providing emollients to prevent skin breakdown is important, but doesn't need to be performed immediately. Slowing the rate of I.V. fluid would contribute to dehydration when polyuria is present.
NP: Evaluation; CN: Physiological integrity; CNS: Physiological adaptation; CL: Analysis

16. Stool softeners would be given to a client after a repair of a cerebral aneurysm for which of the following reasons?
1. To stimulate the bowel due to loss of nerve innervation
2. To prevent straining, which increases intracranial pressure (ICP)
3. To prevent the Valsalva maneuver, which may lead to bradycardia
4. To prevent constipation when osmotic diuretics are used

Knowing why you do something can help prevent errors.

16. 2. Straining when having a bowel movement, sneezing, coughing, and suctioning may lead to increased ICP and should be avoided when potential increased ICP exists. Although the Valsalva maneuver may lead to bradycardia and reflex tachycardia, this rationale doesn't apply to this client.
NP: Planning; CN: Physiological integrity; CNS: Reduction of risk potential; CL: Application

17. A client with a subdural hematoma becomes restless and confused, with dilation of the ipsilateral pupil. The physician orders mannitol for which of the following reasons?
1. To reduce intraocular pressure
2. To prevent acute tubular necrosis
3. To promote osmotic diuresis to decrease intracranial pressure (ICP)
4. To draw water into the vascular system to increase blood pressure

17. 3. Mannitol promotes osmotic diuresis by increasing the pressure gradient, drawing fluid from intracellular to intravascular spaces. Although mannitol is used for all the reasons described, the reduction of ICP in this client is of concern.
NP: Planning; CN: Physiological integrity; CNS: Pharmacological therapies; CL: Application

NP: Nursing process   CN: Client needs category   CNS: Client needs subcategory   CL: Cognitive level

18. A client with a subdural hematoma was given mannitol to decrease intracranial pressure (ICP). Which of the following results would best show the mannitol was effective?

1. Urine output increases.
2. Pupils are 8 mm and nonreactive.
3. Systolic blood pressure remains at 150 mm Hg.
4. Blood urea nitrogen (BUN) and creatinine levels return to normal.

19. When evaluating an arterial blood gas from a client with a subdural hematoma, the nurse notes the $Paco_2$ is 30 mm Hg. Which response best describes this result?

1. Appropriate; lowering carbon dioxide (CO_2) reduces intracranial pressure (ICP)
2. Emergent; the client is poorly oxygenated
3. Normal
4. Significant; the client has alveolar hypoventilation

20. Which of the following nursing interventions should be used to prevent footdrop and contractures in a client recovering from a subdural hematoma?

1. High-topped sneakers
2. Low-dose heparin therapy
3. Physical therapy consultation
4. Sequential compression device

21. A client who is diagnosed with a right subarachnoid hemorrhage should be placed in which position?

1. With the head of the bed elevated
2. On his right side
3. On his left side
4. Flat in bed

22. After a hypophysectomy, vasopressin is given I.M. for which of the following reasons?

1. To prevent GI bleeding
2. To prevent the syndrome of inappropriate antidiuretic hormone (SIADH)
3. To reduce cerebral edema and lower intracranial pressure
4. To replace antidiuretic hormone (ADH) normally secreted from the pituitary

Choose the best response.

Hmmm, does health insurance cover high-topped sneakers?

18. 1. Mannitol promotes osmotic diuresis by increasing the pressure gradient in the renal tubules. No information is given about abnormal BUN and creatinine levels or that mannitol is being given for renal dysfunction or blood pressure maintenance. Fixed and dilated pupils are symptoms of increased ICP or cranial nerve damage.

NP: Evaluation; CN: Physiological integrity; CNS: Physiological adaptation; CL: Application

19. 1. A normal $Paco_2$ value is 35 to 45 mm Hg. CO_2 has vasodilating properties; therefore, lowering $Paco_2$ through hyperventilation will lower ICP caused by dilated cerebral vessels. Alveolar hypoventilation would be reflected in an increased $Paco_2$. Oxygenation is evaluated through Pao_2 and oxygen saturation.

NP: Evaluation; CN: Physiological integrity; CNS: Physiological adaptation; CL: Analysis

20. 1. High-topped sneakers are used to prevent footdrop and contractures in neurologic clients. Low-dose heparin therapy and sequential compression boots will prevent deep-vein thrombosis. Although a consultation with physical therapy is important to prevent footdrop, a nurse may use high-topped sneakers independently.

NP: Implementation; CN: Physiological integrity; CNS: Reduction of risk potential; CL: Application

21. 2. Elevating the head of the bed enhances cerebral venous return and thereby decreases intracranial pressure (ICP). The other positions wouldn't decrease ICP.

NP: Implementation; CN: Safe, effective care environment; CNS: Safety and infection control; CL: Application

22. 4. After hypophysectomy, or removal of the pituitary gland, the body can't synthesize ADH. Although vasopressin tannate may be used I.V. to decrease portal pressure, it isn't correct in this instance. SIADH results from excessive ADH secretion. Mannitol or corticosteroids are used to decrease cerebral edema.

NP: Implementation; CN: Physiological integrity; CNS: Pharmacological therapies; CL: Application

23. Which of the following values is considered normal for intracranial pressure (ICP)?
1. 0 to 15 mm Hg
2. 25 mm Hg
3. 35 to 45 mm Hg
4. 120/80 mm Hg

24. A 33-year-old client undergoes an L4-L5 laminectomy. Which of the following methods would be best to prevent skin breakdown and hypostatic pneumonia?
1. Apply an alternating air mattress to the bed.
2. Use a foam wedge when propping the client on his side.
3. Logroll the client with assistance, if necessary, to keep his back in alignment.
4. Teach the client to reach for the side rail with his dominant hand to pull himself onto his side.

25. Voiding small frequent amounts of urine after a lumbar laminectomy may indicate which of the following conditions?
1. Diabetes insipidus
2. Diabetic ketoacidosis
3. Urine retention
4. Urinary tract infection (UTI)

26. A client with lower back pain and a herniated nucleus pulposus should be taught that strengthening which of the following muscles after laminectomy will prevent lower back pain?
1. Abdominal
2. Diaphragm
3. Gluteus
4. Rectus femoris

27. When preparing a client with suspected herniated nucleus pulposus (HNP) for myelography, which of the following nursing interventions should be done before the test?
1. Question the client about allergy to iodine.
2. Mark distal pulses on the foot in ink.
3. Assess and document pain along the sciatic nerve.
4. Tell the client he may be asked to cough or pant to clear the dye.

Be familiar with normal values to recognize an abnormal result.

Client teaching is a primary responsibility of the nurse.

23. 1. Normal ICP is 0 to 15 mm Hg.
NP: Data collection; CN: Physiological integrity; CNS: Physiological adaptation; CL: Comprehension

24. 3. Turning the client as a unit by logrolling is necessary to prevent stress on the intradiskal area. Having the client reach out will cause lordosis of the spine. An air mattress will help prevent skin breakdown, but the client must be turned to prevent stasis of pulmonary secretions. A foam wedge will maintain the client's position but not prevent breakdown.
NP: Implementation; CN: Physiological integrity; CNS: Reduction of risk potential; CL: Application

25. 3. Swelling or pressure on the peripheral nerves controlling micturition, anesthesia, or use of an indwelling urinary catheter may lead to urine retention with overflow of small frequent amounts of urine. UTI may be shown by dysuria and small frequent amounts of voided urine, but would be less likely in this situation. Diabetes insipidus and diabetic ketoacidosis are shown by polyuria.
NP: Evaluation; CN: Physiological integrity; CNS: Basic care and comfort; CL: Analysis

26. 1. Strengthening abdominal muscles will support the back, preventing lower back pain.
NP: Planning; CN: Physiological integrity; CNS: Reduction of risk potential; CL: Application

27. 1. A radiopaque dye, commonly iodine-based, is instilled into the spinal canal to outline structures during myelography. Pain may be expected along the sciatic nerve with HNP. During cardiac catheterization, a client is asked to cough or pant to clear the dye, and before cardiac catheterization or arteriogram, the nurse marks pedal pulses in ink.
NP: Planning; CN: Physiological integrity; CNS: Reduction of risk potential; CL: Application

NP: Nursing process   CN: Client needs category   CNS: Client needs subcategory   CL: Cognitive level

28. A client is scheduled for chemonucleolysis with chymopapain to relieve the pain of a herniated disk. Which of the following factors should be assessed before the procedure?
1. Allergy to meat tenderizers
2. Allergy to shellfish
3. Ability to lie flat during the procedure
4. Range of motion on the affected side

29. When prioritizing care, which of the following clients should the nurse assess *first*?
1. A 17-year-old client 24-hours postappendectomy
2. A 33-year-old client with a recent diagnosis of Guillain-Barré
3. A 50-year-old client 3 days post-myocardial infarction (MI)
4. A 50-year-old client with diverticulitis

30. A client is newly diagnosed with myasthenia gravis. Client teaching would include which of the following conditions as the cause of this disease?
1. A postviral illness characterized by ascending paralysis
2. Loss of the myelin sheath surrounding peripheral nerves
3. Inability of basal ganglia to produce sufficient dopamine
4. Destruction of acetylcholine receptors causing muscle weakness

31. Which of the following conditions is an early symptom commonly seen in myasthenia gravis?
1. Dysphagia
2. Fatigue improving at the end of the day
3. Ptosis
4. Respiratory distress

Here's a test-taking hint. Prioritize.

Keep your eyes open for clues.

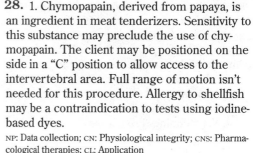

28. 1. Chymopapain, derived from papaya, is an ingredient in meat tenderizers. Sensitivity to this substance may preclude the use of chymopapain. The client may be positioned on the side in a "C" position to allow access to the intervertebral area. Full range of motion isn't needed for this procedure. Allergy to shellfish may be a contraindication to tests using iodine-based dyes.
NP: Data collection; CN: Physiological integrity; CNS: Pharmacological therapies; CL: Application

29. 2. Guillain-Barré syndrome is characterized by ascending paralysis and potential respiratory failure. The order of client assessment should follow client priorities, with disorders of airway, breathing, and then circulation. There's no information to suggest the post-MI client has an arrhythmia or other complication. There's no evidence to suggest hemorrhage or perforation for the remaining clients as a priority of care.
NP: Planning; CN: Safe, effective care environment; CNS: Coordinated care; CL: Analysis

30. 4. Myasthenia gravis, an autoimmune disorder, is caused by the destruction of acetylcholine receptors. Multiple sclerosis is caused by loss of the myelin sheath. Guillain-Barré syndrome is a postviral illness characterized by ascending paralysis, and Parkinson's disease is caused by the inability of basal ganglia to produce sufficient dopamine.
NP: Planning; CN: Health promotion and maintenance; CNS: Prevention and early detection of disease; CL: Comprehension

31. 3. Ptosis and diplopia are early signs of myasthenia gravis; respiratory distress and dysphagia occur later. Symptoms are typically milder in the morning and may be exacerbated by stress or lack of rest.
NP: Data collection; CN: Health promotion and maintenance; CNS: Prevention and early detection of disease; CL: Application

32. One hour after receiving pyridostigmine (Mestinon), a client reports difficulty swallowing and excessive respiratory secretions. The nurse notifies the physician and prepares to administer which of the following medications?
1. Additional pyridostigmine
2. Atropine
3. Edrophonium (Tensilon)
4. Neostigmine (Prostigmin)

32. 2. These symptoms suggest cholinergic crisis or excessive acetylcholinesterase medication, typically appearing 45 to 60 minutes after the last dose of acetylcholinesterase inhibitor. Atropine, an anticholinergic drug, is used to antagonize acetylcholinesterase inhibitors. The other drugs are acetylcholinesterase inhibitors. Tensilon is used for the diagnosis and Mestinon and Prostigmin are used for the treatment of myasthenia gravis and would worsen these symptoms.
NP: Data collection; CN: Physiological integrity; CNS: Pharmacological therapies; CL: Analysis

Watch out for words like *isn't*.

33. A client with suspected myasthenia gravis is to undergo a Tensilon test. Tensilon is used to diagnose — but not treat — myasthenia gravis. Why isn't it used for treatment?
1. It isn't available in an oral form.
2. With repeated use, immunosuppression may occur.
3. Dry mouth and abdominal cramps may be intolerable adverse effects.
4. The short half-life of Tensilon makes it impractical for long-term use.

33. 4. The duration of action of Tensilon is 1 to 2 minutes, making it impractical for the long-term management of myasthenia gravis. Immunosuppression with repeated use is an adverse effect of steroid administration, a medication used to treat myasthenia gravis. Dry mouth and abdominal cramps are adverse effects of increased acetylcholine in the parasympathetic nervous system.
NP: Planning; CN: Physiological integrity; CNS: Pharmacological therapies; CL: Application

34. A 20-year-old client with myasthenia gravis will undergo plasmapheresis. Which of the following actions describes the purpose of this procedure?
1. Prevent exacerbations during pregnancy.
2. Remove T and B lymphocytes that attack acetylcholine receptors.
3. Deliver acetylcholinesterase inhibitor directly into the bloodstream.
4. Separate and remove acetylcholine receptor antibodies from the blood.

34. 4. The purpose of plasmapheresis in myasthenia gravis is to separate and remove circulating acetylcholine receptor antibodies from the blood of clients refractory to the usual therapies or clients in crisis. Although stress, including pregnancy, may precipitate crisis, this isn't the purpose of the procedure. Plasmapheresis doesn't remove T and B lymphocytes, nor does it deliver acetylcholinesterase inhibitor directly into the bloodstream.
NP: Planning; CN: Physiological integrity; CNS: Pharmacological therapies; CL: Application

Way to go!

35. When assessing a client with glaucoma, a nurse expects which of the following findings?
1. Complaints of double vision
2. Complaints of halos around lights
3. Intraocular pressure of 15 mm Hg
4. Soft globe on palpation

35. 2. Glaucoma is largely asymptomatic. Symptoms of glaucoma can include loss of peripheral vision or blind spots, reddened sclera, firm globe, decreased accommodation, halos around lights, and occasional eye pain, but clients may be asymptomatic. Normal intraocular pressure is 10 to 21 mm Hg.
NP: Data collection; CN: Physiological integrity; CNS: Physiological adaptation; CL: Application

36. A client at the eye clinic is newly diagnosed with glaucoma. Client teaching includes the need to take his medication because noncompliance may lead to which of the following conditions?
1. Diplopia
2. Permanent vision loss
3. Progressive loss of peripheral vision
4. Pupillary constriction

36. 2. Without treatment, glaucoma may progress to irreversible blindness. Treatment won't restore visual damage, but will halt disease progression. Miotics, which constrict the pupil, are used in the treatment of glaucoma to permit outflow of the aqueous humor. Central vision loss is typical in glaucoma. Blurred or foggy vision, not diplopia, is typical in glaucoma.

NP: Implementation; CN: Physiological integrity; CNS: Pharmacological therapies; CL: Application

Don't get all discombobulated, now. Remember the abbreviations?

37. Pilocarpine, 2 gtt OU q.i.d., should be instilled according to which of the following procedures?
1. Two drops of the drug in both eyes four times daily
2. Two drops on the sclera of the left eye four times daily
3. Two drops over the lacrimal duct of the right eye four times daily
4. Two drops of the drug toward the nasal side of each conjunctival sac four times daily

37. 1. The abbreviation OU means both eyes, OS refers to the left eye, and OD to the right eye. Q.I.D. means four times daily and gtt means drops. Medications placed on the nasal side near the lacrimal duct will enter the nose and be ineffective.

NP: Implementation; CN: Physiological integrity; CNS: Pharmacological therapies; CL: Application

38. When evaluating the extent of Parkinson's disease, a nurse observes for which of the following conditions?
1. Bulging eyeballs
2. Diminished distal sensation
3. Increased dopamine levels
4. Muscle rigidity

38. 4. Parkinson's disease is characterized by the slowing of voluntary muscle movement, muscular rigidity, and resting tremor. Dopamine is deficient in this disorder. Diminished distal sensation doesn't occur in Parkinson's disease. Bulging eyeballs (exophthalmos) occur in Graves disease.

NP: Data collection; CN: Physiological integrity; CNS: Physiological adaptation; CL: Application

Check out the word best.

39. Which of the following statements best describes the cause of Parkinson's disease?
1. Loss of the myelin sheath surrounding peripheral nerves
2. Degeneration of the substantia nigra, depleting dopamine
3. Bleeding into the brain stem, resulting in motor dysfunction
4. An autoimmune disorder that destroys acetylcholine receptors

39. 2. Parkinson's disease is caused by degeneration of the substantia nigra in the basal ganglia of the brain, where dopamine is produced and stored. This results in motor dysfunction. Loss of the myelin sheath around the peripheral nerves describes multiple sclerosis. Myasthenia gravis is an autoimmune disorder that destroys acetylcholine receptors, and bleeding into the brain stem resulting in motor dysfunction describes a hemorrhagic stroke.

NP: Evaluation; CN: Physiological integrity; CNS: Physiological adaptation; CL: Comprehension

40. Which of the following clients would be most at risk for secondary Parkinson's disease caused by pharmacotherapy?
1. A 30-year-old client with schizophrenia taking chlorpromazine (Thorazine)
2. A 50-year-old client taking nitroglycerin tablets for angina
3. A 60-year-old client taking prednisone for chronic obstructive pulmonary disease
4. A 75-year-old client using naproxen for rheumatoid arthritis

41. Which of the following symptoms occurs initially in Parkinson's disease?
1. Akinesia
2. Aspiration of food
3. Dementia
4. Pill rolling movements of the hand

42. To evaluate the effectiveness of levodopa-carbidopa, a nurse would watch for which of the following results?
1. Improved visual acuity
2. Decreased dyskinesia
3. Reduction in short-term memory
4. Lessened rigidity and tremor

43. Two days after starting therapy with trihexyphenidyl (Artane), a client complains of a dry mouth. Which of the following nursing interventions would best relieve the client's dry mouth?
1. Offer the client ice chips and frequent sips of water.
2. Withhold the drug and notify the physician.
3. Change the client's diet to clear liquid until the symptoms subside.
4. Encourage the use of supplemental puddings and shakes to maintain weight.

44. Which of the following anti-Parkinson drugs can cause drug tolerance or toxicity if taken for too long at one time?
1. Amantadine (Symmetrel)
2. Levodopa-carbidopa (Sinemet)
3. Pergolide (Permax)
4. Selegiline (Eldepryl)

Read question 41 carefully to get the correct answer.

My mouth isn't dry anymore. But hmmm, that's odd. Now I'm transparent.

40. 1. Phenothiazines such as Thorazine deplete dopamine, which may lead to tremor and rigidity (extrapyramidal effects). The other clients aren't at a greater risk for developing Parkinson's disease.
NP: Evaluation; CN: Physiological integrity; CNS: Pharmacological therapies; CL: Application

41. 4. Early symptoms of Parkinson's disease include coarse resting tremors of the fingers and thumb. Akinesia and aspiration are late signs of Parkinson's disease. Dementia occurs in only 20% of the clients with Parkinson's disease.
NP: Data collection; CN: Health promotion and maintenance; CNS: Prevention and early detection of disease; CL: Application

42. 4. Levodopa-carbidopa increases the amount of dopamine in the central nervous system, allowing for more smooth, purposeful movements. The drug doesn't affect visual acuity and should improve dyskinesia and short-term memory.
NP: Evaluation; CN: Physiological integrity; CNS: Pharmacological therapies; CL: Application

43. 1. Trihexyphenidyl is an anticholinergic agent that causes blurred vision, dry mouth, constipation, and urine retention. There's no need to withhold the drug unless hypotension or tachyarrhythmia occurs. Weight loss may occur with Parkinson's disease; however, the question relates to effects of trihexyphenidyl. A clear liquid diet doesn't provide adequate nutrition and may be more difficult to swallow than thickened liquids if dysphagia is present; it isn't indicated at this time.
NP: Implementation; CN: Physiological integrity; CNS: Pharmacological therapies; CL: Analysis

44. 2. Long-term therapy with levodopa can result in drug tolerance or toxicity, shown by confusion, hallucinations, or decreased drug effectiveness. The other drugs don't require that the client stop using them for a short period.
NP: Evaluation; CN: Physiological integrity; CNS: Pharmacological therapies; CL: Application

NP: Nursing process   CN: Client needs category   CNS: Client needs subcategory   CL: Cognitive level

45. Which of the following clients would be most likely to develop multiple sclerosis (MS)?
1. A 20-year-old soccer player
2. A 35-year-old White female teacher
3. A 45-year-old type A male smoker
4. A 50-year-old Black female with hypertension

46. Which of the following pathophysiological processes are involved in multiple sclerosis (MS)?
1. Destruction of the brain stem and basal ganglia in the brain
2. Degeneration of the nucleus pulposus, causing pressure on the spinal cord
3. Chronic inflammation of rhizomes just outside the central nervous system
4. Development of demyelinization of the myelin sheath, interfering with nerve transmission

47. Which of the following symptoms frequently occurs early in multiple sclerosis (MS)?
1. Diplopia
2. Grief
3. Hemiparesis
4. Recent memory loss

48. Which of the following measures would be included in teaching for the client with multiple sclerosis (MS) to avoid exacerbation of the disease?
1. Patch the affected eye.
2. Sleep 8 hours each night.
3. Take hot baths for relaxation.
4. Drink 1,500 to 2,000 ml of fluid daily.

49. Which of the following conditions or activities may exacerbate multiple sclerosis (MS)?
1. Pregnancy
2. Range-of-motion (ROM) exercises
3. Swimming
4. Urine retention

Close in on the clues.

In question 47, early is another test-taking clue.

Client teaching is very important.

45. 2. MS is more common in White women ages 20 to 40 years, with a secondary onset between ages 40 to 60 years.
NP: Data collection; CN: Physiological integrity; CNS: Physiological adaptation; CL: Comprehension

46. 4. MS results from chronic progressive demyelination of the myelin sheath, interfering with nerve impulse transmission. The other processes don't describe MS.
NP: Data collection; CN: Physiological integrity; CNS: Physiological adaptation; CL: Comprehension

47. 1. Early symptoms of MS include slurred speech and diplopia. Paralysis is a late symptom. Although depression and a short attention span may occur, dementia is rarely associated with MS.
NP: Data collection; CN: Physiological integrity; CNS: Physiological adaptation; CL: Application

48. 2. MS is exacerbated by exposure to stress, fatigue, and heat. Clients should balance activity with rest. Patching the affected eye may result in improvement in vision and balance but won't prevent exacerbation of the disease. Adequate hydration will help prevent urinary tract infections secondary to a neurogenic bladder.
NP: Implementation; CN: Physiological integrity; CNS: Reduction of risk potential; CL: Application

49. 1. Pregnancy, stress, fatigue, and heat may exacerbate MS. Urine retention is common due to neurogenic bladder but doesn't lead to the exacerbation of symptoms. Exercise to maintain ROM is encouraged; swimming is particularly effective due to weightlessness and the cooling of nerves.
NP: Implementation; CN: Physiological integrity; CNS: Reduction of risk potential; CL: Application

50. A client with suspected multiple sclerosis (MS) undergoes a lumbar puncture. Which of the following abnormalities is typically found in the cerebrospinal fluid (CSF) of clients with MS?

1. Blood or increased red blood cells
2. Elevated white blood cells or pus
3. Increased glucose concentrations
4. Increased protein levels

51. Which of the following terms describes involuntary, jerking, rhythmic movements of the eyes?

1. Diplopia
2. Exophthalmos
3. Nystagmus
4. Oculogyric crisis

52. Which of the following nursing interventions takes priority for the client having a tonic-clonic seizure?

1. Maintain a patent airway.
2. Time the duration of the seizure.
3. Note the origin of seizure activity.
4. Insert a padded tongue blade to prevent the client from biting his tongue.

Question 52 screams out, "Prioritize!"

53. A client recalls smelling an unpleasant odor before his seizure. Which of the following terms describes to this?

1. Atonic seizure
2. Aura
3. Icterus
4. Postictal experience

Question 54 is "loaded" with clues. Focus on loading.

54. A client with new-onset seizures of unknown cause is started on phenytoin (Dilantin), 750 mg I.V. now and 100 mg P.O. t.i.d. Which of the following statements best describes the purpose of the loading dose?

1. To ensure that the drug reaches the cerebrospinal fluid
2. To prevent the need for surgical excision of the epileptic focus
3. To reduce secretions in case another seizure occurs
4. To more quickly attain therapeutic levels

50. 4. Elevated gamma globulin fraction in CSF without an elevated level in the blood occurs in MS. Blood may be found with trauma or subarachnoid hemorrhage. Increased glucose concentration is a nonspecific finding indicating infection or subarachnoid hemorrhage. White blood cells or pus indicate infection.
NP: Data collection; CN: Physiological integrity; CNS: Physiological adaptation; CL: Analysis

51. 3. Nystagmus refers to jerking movements of the eye. Oculogyric crisis involves deviation of the eyes. Exophthalmos refers to bulging eyeballs, seen in Graves disease. Diplopia means double vision.
NP: Data collection; CN: Health promotion and maintenance; CNS: Prevention and early detection of disease; CL: Application

52. 1. The priority during and after a seizure is to maintain a patent airway. Nothing should be placed in the client's mouth during a seizure because teeth may be dislodged or the tongue pushed back, further obstructing the airway. Noting the origin of motor dysfunction and timing the seizure activity are done, but not first.
NP: Implementation; CN: Physiological integrity; CNS: Reduction of risk potential; CL: Application

53. 2. An aura occurs in some clients as a warning before a seizure. The client may experience a certain smell, a vision such as flashing lights, or a sensation. Postictal experience occurs after a seizure, during which the client may be confused, somnolent, fatigued, and may need to sleep. Atonic seizure or drop attack refers to an abrupt loss of muscle tone. Icterus refers to jaundice.
NP: Implementation; CN: Physiological integrity; CNS: Physiological adaptation; CL: Application

54. 4. A loading dose of phenytoin and other drugs is given to reach therapeutic levels more quickly; maintenance dosing follows. A loading dose of phenytoin can be oral or parenteral. Surgical excision of an epileptic focus is considered when seizures aren't controlled with anticonvulsant therapy. Phenytoin doesn't reduce secretions.
NP: Implementation; CN: Physiological integrity; CNS: Pharmacological therapies; CL: Application

NP: Nursing process   CN: Client needs category   CNS: Client needs subcategory   CL: Cognitive level

55. Which of the following adverse effects may occur during phenytoin (Dilantin) therapy?
1. Dry mouth
2. Furry tongue
3. Somnolence
4. Tachycardia

56. Which of the following symptoms may occur with a phenytoin level of 32 mg/dl?
1. Ataxia and confusion
2. Sodium depletion
3. Tonic-clonic seizure
4. Urinary incontinence

Caution! Can you find the clues here?

57. Which of the following precautions must be taken when giving phenytoin (Dilantin) to a client with a nasogastric (NG) tube for feeding?
1. Check the phenytoin level after giving the drug to check for toxicity.
2. Elevate the head of the bed before giving phenytoin through the NG tube.
3. Give phenytoin 1 hour before or 2 hours after NG tube feedings to ensure absorption.
4. Verify proper placement of the NG tube by placing the end of the tube in a glass of water and observing for bubbles.

58. Clients taking phenytoin should avoid using alcohol for which of the following reasons?
1. Alcohol increases phenytoin activity.
2. Alcohol raises the seizure threshold.
3. Alcohol impairs judgment and coordination.
4. Alcohol decreases the effectiveness of phenytoin.

So, why don't I get along with phenytoin?

59. When obtaining vital signs in a client with a seizure disorder, which of the following measures is used?
1. Check for a pulse deficit.
2. Check for pulsus paradoxus.
3. Take an axillary instead of oral temperatures.
4. Check the blood pressure for an auscultatory gap.

55. 3. Adverse effects of phenytoin include sedation, drowsiness, gingival hyperplasia, blood dyscrasia, and toxicity. The other symptoms aren't adverse effects of phenytoin.
NP: Implementation; CN: Physiological integrity; CNS: Pharmacological therapies; CL: Application

56. 1. A therapeutic phenytoin level is 10 to 20 mg/dl. A level of 32 mg/dl indicates phenytoin toxicity. Symptoms of toxicity include confusion and ataxia. Phenytoin doesn't cause hyponatremia, seizure, or urinary incontinence. Incontinence may occur during or after a seizure.
NP: Evaluation; CN: Physiological integrity CNS: Pharmacological therapies; CL: Analysis

57. 3. Nutritional supplements and milk interfere with the absorption of phenytoin, decreasing its effectiveness. The nurse verifies NG tube placement by checking for stomach contents before giving drugs and feedings. The head of the bed is elevated when giving all drugs or solutions and isn't specific to phenytoin administration. Phenytoin levels are checked before giving the drug and the drug is withheld for elevated levels to avoid compounding toxicity.
NP: Implementation; CN: Physiological integrity; CNS: Pharmacological therapies; CL: Application

58. 4. Although alcohol impairs judgment and coordination, the larger concern is a lowered phenytoin level and lower threshold for seizures.
NP: Implementation; CN: Physiological integrity; CNS: Pharmacological therapies; CL: Application

59. 3. To reduce the risk for injury, the nurse should take an axillary temperature or use a metal thermometer when taking an oral temperature to prevent injury if a seizure occurs. An auscultatory gap occurs in hypertension. Pulse deficit occurs in an arrhythmia. Pulsus paradoxus may occur with cardiac tamponade.
NP: Implementation; CN: Physiological integrity; CNS: Reduction of risk potential; CL: Application

60. A client in status epilepticus arrives at the emergency department. The family is interviewed to determine the cause of this problem. Which of the following events may have predisposed the client to this condition?
1. Abruptly stopping anticonvulsant therapy
2. Airplane travel
3. Exposure to sunlight
4. Recent upper respiratory infection

60. 1. Status epilepticus (seizures not responsive to usual therapies) occurs with the abrupt cessation of anticonvulsant drugs or ethanol intake. The other options don't cause status epilepticus.
NP: Evaluation; CN: Physiological integrity; CNS: Physiological adaptation; CL: Analysis

The hint hunt is on! See it?

61. A client comes to the emergency department after hitting his head in a motor vehicle accident. He's alert and oriented. Which of the following nursing interventions should be done first?
1. Assess full range of motion to determine the extent of injuries.
2. Call for an immediate chest X-ray.
3. Immobilize the client's head and neck.
4. Open the airway with the head-tilt chin-lift maneuver.

61. 3. All clients with a head injury are treated as if a cervical spine injury is present until X-rays confirm their absence. Range of motion would be contraindicated at this time. There's no indication the client needs a chest X-ray. The airway doesn't need to be opened since the client appears alert and not in respiratory distress. In addition, the head-tilt chin-lift maneuver wouldn't be used until cervical spine injury is ruled out.
NP: Implementation; CN: Physiological integrity; CNS: Reduction of risk potential; CL: Application

62. A client with a C6 spinal injury would most likely have which of the following symptoms?
1. Aphasia
2. Hemiparesis
3. Paraplegia
4. Tetraplegia

62. 4. Tetraplegia (quadriplegia) occurs as a result of cervical spine injuries. Paraplegia occurs as a result of injury to the thoracic cord and below. Hemiparesis describes weakness of one side of the body. Aphasia refers to difficulty expressing or understanding spoken words.
NP: Data collection; CN: Physiological integrity; CNS: Physiological adaptation; CL: Application

Let's see, what do all three of these brain areas have in common?

63. A 22-year-old client has a spinal cord transection at the T4 level. The nurse can expect the client to have which of the following symptoms?
1. Paraplegia
2. Quadriplegia
3. Autonomic dysreflexia
4. No deficits

63. 2. Spinal cord injuries at the T4 level affect all motor and sensory nerves below the level of injury and result in dysfunction of both arms, legs, bowel, and bladder. Paraplegic injuries result from a lesion involving the thoracic, lumbar, or sacral region of the spinal cord. Autonomic dysreflexia occurs because of a massive sympathetic discharge of stimuli from the autonomic nervous system.
NP: Data collection; CN: Physiological integrity; CNS: Physiological adaptation; CL: Application

64. A 30-year-old client is admitted to the progressive care unit with a C5 fracture from a motorcycle accident. Which of the following assessments would take priority?
1. Bladder distention
2. Neurologic deficit
3. Pulse oximetry readings
4. The client's feelings about the injury

65. While in the emergency department, a client with C8 quadriplegia develops a blood pressure of 80/44 mm Hg, pulse of 48 beats/minute, and respiratory rate of 18 breaths/minute. The nurse suspects which of the following conditions?
1. Autonomic dysreflexia
2. Hemorrhagic shock
3. Neurogenic shock
4. Pulmonary embolism

66. A client is admitted with a spinal cord injury at the level of T12. He has limited movement of his upper extremities. Which of the following medications would be used to control edema of the spinal cord?
1. Acetazolamide (Diamox)
2. Furosemide (Lasix)
3. Methylprednisolone (Solu-Medrol)
4. Sodium bicarbonate

67. A 22-year-old client with quadriplegia is apprehensive and flushed, with a blood pressure of 210/100 mm Hg and heart rate of 50 beats/minute. Which of the following nursing interventions should be done first?
1. Place the client flat in bed.
2. Assess patency of the indwelling urinary catheter.
3. Give one sublingual nitroglycerin tablet.
4. Raise the head of the bed immediately to 90 degrees.

Hoppin' hints! Here they come again!

Hurray! Halfway done! And here's another hint!

64. 3. After a spinal cord injury, ascending cord edema may cause a higher level of injury. The diaphragm is innervated at the level of C4, so assessment of adequate oxygenation and ventilation is necessary. Although the other options would be necessary at a later time, observation for respiratory failure is the priority.
NP: Implementation; CN: Safe, effective care environment; CNS: Coordinated care; CL: Application

65. 3. Symptoms of neurogenic shock include hypotension, bradycardia, and warm, dry skin due to loss of adrenergic stimulation below the level of the lesion. Hypertension, bradycardia, flushing, and sweating of the skin are seen with autonomic dysreflexia. Hemorrhagic shock presents with anxiety, tachycardia, and hypotension; this wouldn't be suspected without an injury. Pulmonary embolism presents with chest pain, hypotension, hypoxemia, tachycardia, and hemoptysis; this may be a later complication of spinal cord injury due to immobility.
NP: Data collection; CN: Health promotion and maintenance; CNS: Prevention and early detection of disease; CL: Analysis

66. 3. High doses of methylprednisolone are used within 24 hours of spinal cord injury to reduce cord swelling and limit neurological deficit. The other drugs aren't indicated in this circumstance.
NP: Planning; CN: Physiological integrity; CNS: Pharmacological therapies; CL: Application

67. 4. Anxiety, flushing above the level of the lesion, piloerection, hypertension, and bradycardia are symptoms of autonomic dysreflexia, typically caused by such noxious stimuli as a full bladder, fecal impaction, or pressure ulcer. Putting the client flat will cause the blood pressure to increase more. Nitroglycerin is given to relieve chest pain and reduce preload; it isn't used for hypertension or dysreflexia. The indwelling urinary catheter should be assessed immediately after the head of the bed is raised.
NP: Implementation; CN: Physiological integrity; CNS: Physiological adaptation; CL: Analysis

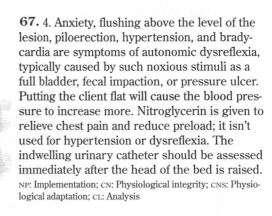

68. A client with paraplegia from a T10 injury is getting ready to transfer to a rehabilitation hospital. When a nurse offers to assist him, the client throws his suitcase on the floor and says, "You don't want to help me." Which of the following responses would be the most appropriate for the nurse to give?

1. "You know I want to help you, I offered."
2. "I'll pick these things up for you and come back later."
3. "You seem angry today. Going to rehab may be scary."
4. "When you get to rehab, they won't let you behave like a spoiled brat."

69. A client with a cervical spine injury has Gardner Wells tongs inserted for which of the following reasons?

1. Hasten wound healing
2. Immobilize the surgical spine
3. Prevent autonomic dysreflexia
4. Hold bony fragments of the skull together

70. When a client with a halo vest is discharged from the hospital, which of the following instructions should the nurse give the client and family?

1. Don't use the wheelchair while the halo vest is in place.
2. Clean the pin sites with povidone-iodine and apply water-soluble lubricant.
3. Keep the wrench that opens the vest attached to the client at all times.
4. Perform range-of-motion exercises to the neck and shoulders four times daily.

71. A 20-year-old with structural scoliosis has spinal fusion surgery. Which of the following positions would be best during the postoperative period?

1. Supine in bed
2. Side-lying
3. Semi-Fowler's
4. High Fowler's

Choose the most therapeutic response.

Teaching is inseparable from nursing.

68. 3. The nurse should always focus on the feelings underlying a particular action. Options 1 and 4 are confrontational. Offering to pick up the client's belongings doesn't deal with the situation and assumes he can't do it alone.
NP: Implementation; CN: Psychosocial integrity; CNS: Psychosocial adaptation; CL: Application

69. 2. Gardner-Wells, Vinke, and Crutchfield tongs immobilize the spine until surgical stabilization is accomplished. Tongs don't hasten wound healing, prevent autonomic dysreflexia, or hold bony fragments of the skull together.
NP: Implementation; CN: Physiological integrity; CNS: Physiological adaptation; CL: Comprehension

70. 3. The wrench must be attached at all times to remove the vest in case the client needs cardiopulmonary resuscitation. The pins are cleaned with peroxide and povidone-iodine. The purpose of the vest is to immobilize the neck; range-of-motion exercises to the neck are prohibited but should be performed to other areas. The vest is designed to improve mobility; the client may use a wheelchair.
NP: Implementation; CN: Physiological integrity; CNS: Reduction of risk potential; CL: Application

71. 1. After spinal fusion surgery, the client must remain flat in bed. The latch on a manual bed should be taped and electric beds should be unplugged to prevent the client from arising the head or foot of the bed. Other positions, such as side-lying, semi-Fowler, and high Fowler positions, could prove damaging because the spine must be maintained in a straight position.
NP: Implementation; CN: Physiological integrity; CNS: Reduction of risk potential; CL: Application

72. Which of the following interventions describes an appropriate bladder program for a client in rehabilitation for spinal cord injury?
1. Insert an indwelling urinary catheter.
2. Schedule intermittent catheterization every 2 to 4 hours.
3. Perform a straight catheterization every 8 hours while awake.
4. Perform Credé's maneuver to the lower abdomen before the client voids.

72. 2. Intermittent catheterization should begin every 2 to 4 hours early in treatment. When residual volume is less than 400 ml, the schedule may advance to every 4 to 6 hours. Indwelling catheters may predispose the client to infection and are removed as soon as possible. Credé's maneuver is applied after voiding to enhance bladder emptying.
NP: Implementation; CN: Physiological integrity; CNS: Basic care and comfort; CL: Application

73. A 46-year-old client with breast cancer complains of back pain and difficulty moving her legs. Which of the following nursing interventions is the most appropriate?
1. Notify the physician.
2. Position the client on her side, and prop her with a foam wedge.
3. Ask the physician for a physical therapy consultation.
4. Give acetaminophen, and reassure the client the pain will disappear soon.

Choose the best of the correct answers.

73. 1. Symptoms of back pain and neurologic deficits may be symptoms of metastasis. The physician should be notified. Repositioning the client, physical therapy, or acetaminophen may help the pain but may delay evaluation and treatment.
NP: Implementation; CN: Health promotion and maintenance; CNS: Prevention and early detection of disease; CL: Analysis

74. A client was admitted to the hospital because of a transient ischemic attack secondary to atrial fibrillation. He would be given which of the following medications to prevent further neurologic deficit?
1. Digoxin (Lanoxin)
2. Diltiazem (Cardizem)
3. Heparin
4. Quinidine gluconate (Quinaglute Dura-Tabs)

74. 3. Atrial fibrillation may lead to the formation of mural thrombi, which may embolize to the brain. Heparin will prevent further clot formation and prevent clot enlargement. The other drugs are used in the treatment and control of atrial fibrillation but won't affect clot formation.
NP: Implementation; CN: Physiological integrity; CNS: Pharmacological therapies; CL: Application

75. A client is diagnosed with Ménière's disease. Which of the following nursing diagnoses would take priority for this client?
1. Ineffective tissue perfusion: Cerebral
2. Imbalanced nutrition: More than body requirements
3. Impaired social interaction
4. Risk for injury

You're doing great! I knew you would!

75. 4. Ménière's disease results in dizziness so the client should be protected from falling. Although hearing loss may occur, causing impaired social interaction, this isn't a priority. Ménière's disease doesn't alter cerebral tissue perfusion or directly affect nutrition.
NP: Implementation; CN: Safe, effective care environment; CNS: Safety and infection control; CL: Application

76. Which of the following positions would be the most appropriate for a client who has undergone stapedectomy?
1. On the affected side
2. On the unaffected side
3. Prone
4. Sims'

76. 2. The client should be positioned with the operative ear up, on the unaffected side. Although Sims' position is a side-lying position, it doesn't consider which side is best for after-ear surgery.
NP: Implementation; CN: Physiological integrity; CNS: Basic care and comfort; CL: Application

77. Which of the following symptoms would the nurse expect to find when collecting data from a client with Ménière's disease?
1. Epistaxis
2. Facial pain
3. Ptosis
4. Tinnitus

77. 4. Tinnitus, dizziness, and vertigo occur in Ménière's disease. Facial pain may occur with trigeminal neuralgia. Ptosis occurs with a variety of conditions, including myasthenia gravis. Epistaxis may occur with a variety of blood dyscrasias or local lesions.
NP: Data collection; CN: Physiological integrity; CNS: Physiological adaptation; CL: Application

Choose the priority.

78. Which of the following nursing interventions has priority for an occupational nurse treating a client with a foreign body protruding from the eye?
1. Irrigate the eye with sterile saline.
2. Assess visual acuity with a Snellen chart.
3. Remove the foreign body with sterile forceps.
4. Patch both eyes until seen by the ophthalmologist.

78. 4. One or both eyes may be patched to prevent pain with extraocular movement or accommodation. Assessment of visual acuity isn't a priority, although it may be done after treatment. Chemicals or small foreign bodies may be irrigated. Protruding objects aren't removed by the nurse because the vitreous body may rupture.
NP: Implementation; CN: Safe, effective care environment; CNS: Safety and infection control; CL: Application

79. A client with severe eye pain requests a prescription for the topical anesthetic the ophthalmologist instilled. A nurse explains these drugs shouldn't be used on an ongoing basis for which of the following reasons?
1. They are a way for pathogens to enter.
2. They cause dependence and rebound pain.
3. Damage could occur to the cornea due to lack of sensation.
4. The resulting blurred vision from mydriasis makes activity hazardous.

79. 3. Corneal damage may occur with the prolonged use of topical anesthetics. Dependence and rebound don't occur from topical anesthetics. Anesthetics don't cause mydriasis. If the bottle isn't touched to the eye or lashes, the entry of pathogens should be limited.
NP: Implementation; CN: Physiological integrity; CNS: Reduction of risk potential; CL: Application

Talk so I can hear ya!

80. An 86-year-old client admitted to the hospital with chest pain is hard of hearing. Which of the following methods should be used when collecting data from a client?
1. Obtain an ear wick.
2. Shout into the better ear.
3. Lower the voice pitch while facing the client.
4. Ask the family to go home and get the client's hearing aid.

80. 3. Hearing loss in the elderly typically involves the upper ranges; lowering the pitch of the voice and facing the client is essential for the client to use others means of understanding, such as lip reading, mood, and so on. Shouting is typically in the upper ranges and could cause anxiety to an already anxious client. An ear wick is used to allow medications to enter the ear canal. Alternate means of communication such as writing may also be used to assess chest pain while waiting for the family to bring the hearing aid from home.
NP: Data collection; CN: Physiological integrity; CNS: Basic care and comfort; CL: Application

---

NP: Nursing process   CN: Client needs category   CNS: Client needs subcategory   CL: Cognitive level

81. A client is scheduled for magnetic resonance imaging (MRI) of the head. Which of the following areas is essential to assess before the procedure?

1. Food or drink intake within the past 8 hours
2. Metal fillings, prostheses, or a pacemaker
3. The presence of carotid artery disease
4. Voiding before the procedure

82. When giving hydrocortisone, 1 gtt AU t.i.d., which of the following methods would the nurse use?

1. One drop into each ear three times daily
2. One drop into each eye three times daily
3. One drop of gold salts into each eye three times daily
4. Three drops into the right eye once daily

83. To properly instill eardrops in a 28-year-old client with otitis externa, which of the following methods is correct?

1. Pull the pinna down and back.
2. Pull the pinna up and back.
3. Pull the tragus up and back.
4. Separate the palpebral fissures with a clean gauze pad.

84. Which of the following instructions given to a client after cataract surgery is inappropriate?

1. Avoid bending and straining.
2. Avoid high-sodium foods to reduce intraocular pressure.
3. Don't drive or sleep on the affected side.
4. Don't use makeup on the affected eye.

85. Which of the following symptoms of increased intracranial pressure (ICP) after head trauma would appear first?

1. Bradycardia
2. Large amounts of very dilute urine
3. Restlessness and confusion
4. Widened pulse pressure

Prevent errors. Know your abbreviations.

Lock in on the *in* before *appropriate*.

81. 2. Strong magnetic waves may dislodge metal in the client's body, causing tissue injury. Although the client may be told to restrict food for 8 hours, particularly if contrast is used, metal is an absolute contraindication for this procedure. Voiding beforehand would make the client more comfortable and better able to remain still during the procedure, but isn't essential for the test. Having carotid artery disease isn't a contraindication to having an MRI.

NP: Data collection; CN: Safe, effective care environment; CNS: Safety and infection control; CL: Application

82. 1. A is the abbreviation for ears; O, for eyes; D, for right or dexterous; and S, for left or sinestre. AU refers to both ears. Although Au is the symbol for gold, gold isn't placed in the eye and isn't a correct option here.

NP: Implementation; CN: Physiological integrity; CNS: Pharmacological therapies; CL: Comprehension

83. 2. To straighten the ear canal of an adult, the pinna is pulled up and back. The palpebral fissures are in the eye. The other options aren't appropriate methods for preparing the ear to receive eardrops.

NP: Implementation; CN: Physiological integrity; CNS: Pharmacological therapies; CL: Application

84. 2. After cataract surgery, there's no need to restrict sodium. Using makeup, bending, straining, lifting, vomiting, and sleeping on the affected side may increase intraocular pressure and put strain on the sutures.

NP: Data collection; CN: Physiological integrity; CNS: Reduction of risk potential; CL: Application

85. 3. The earliest symptom of increased ICP is a change in mental status. Bradycardia, widened pulse pressure, and bradypnea occur later. The client may void large amounts of very dilute urine if there's damage to the posterior pituitary.

NP: Data collection; CN: Physiological integrity; CNS: Physiological adaptation; CL: Application

86. A client admitted to the emergency department for head trauma is diagnosed with an epidural hematoma. The underlying cause of epidural hematoma is <u>usually</u> related to which of the following conditions?
   1. Laceration of the middle meningeal artery
   2. Rupture of the carotid artery
   3. Thromboembolism from a carotid artery
   4. Venous bleeding from the arachnoid space

87. A 23-year-old client has been hit on the head with a baseball bat. The nurse notes clear fluid draining from his ears and nose. Which of the following nursing interventions should be done first?
   1. Position the client flat in bed.
   2. Check the fluid for dextrose with a dipstick.
   3. Suction the nose to maintain airway patency.
   4. Insert nasal and ear packing with sterile gauze.

88. When discharging a client from the emergency department after a head trauma, the nurse teaches the guardian to observe for a lucid interval. Which of the following statements best describes a lucid interval?
   1. An interval when the client's speech is garbled
   2. An interval when the client is alert but can't recall recent events
   3. An interval when the client is oriented but then becomes somnolent
   4. An interval when the client has a "warning" symptom, such as an odor or visual disturbance

89. When teaching the family of a client with C4 quadriplegia how to suction his tracheostomy, the nurse includes which of the following instructions?
   1. Suction for 10 to 15 seconds at a time.
   2. Regulate the suction machine to −300 cm suction.
   3. Apply suction to the catheter during insertion only.
   4. Pass the suction catheter into the opening of the tracheostomy tube 2 to 3 cm.

The word usually helps you choose between more than one correct answer.

Hint: How do you determine whether the fluid is mucus or cerebral spinal fluid?

Teaching the family is as important as teaching the client.

86. 1. Epidural hematoma or extradural hematoma is usually caused by laceration of the middle meningeal artery. An embolic stroke is a thromboembolism from a carotid artery that ruptures. Venous bleeding from the arachnoid space is usually observed with subdural hematoma.
NP: Planning; CN: Physiological integrity; CNS: Physiological adaptation; CL: Comprehension

87. 2. Clear liquid from the nose (rhinorrhea) or ear (otorrhea) can be determined to be cerebral spinal fluid or mucus by the presence of dextrose. Placing the client flat in bed may increase intracranial pressure and promote pulmonary aspiration. Nothing is inserted into the ears or nose of a client with a skull fracture because of the risk for infection. The nose wouldn't be suctioned because of the risk for suctioning brain tissue through the sinuses.
NP: Data collection; CN: Physiological integrity; CNS: Physiological adaptation; CL: Analysis

88. 3. A lucid interval is described as a brief period of unconsciousness followed by alertness; after several hours, the client again loses consciousness. Garbled speech is known as dysarthria. An interval in which the client is alert but can't recall recent events is known as amnesia. Warning symptoms or auras typically occur before seizures.
NP: Implementation; CN: Health promotion and maintenance; CNS: Prevention and early detection of disease; CL: Comprehension

89. 1. Suction should be applied for 10 to 15 seconds at a time. When suctioning the trachea, the catheter is inserted 4 to 6 inches or until resistance is felt. Suction should be applied only during withdrawal of the catheter. Suction is regulated to 80 to 120 cm.
NP: Implementation; CN: Physiological integrity; CNS: Reduction of risk potential; CL: Application

90. Which of the following conditions is a risk factor for hemorrhagic stroke?
  1. Coronary artery disease
  2. Diabetes
  3. Hypertension
  4. Recent viral infection

91. An 86-year-old client with a stroke in evolution and a history of coronary artery disease is brought to the medical-surgical floor. His medications include heparin, isosorbide, and verapamil. Which of the following conditions should be avoided in a client with a stroke?
  1. Dehydration
  2. Hypocarbia
  3. Hypotension
  4. Tube feeding

92. Which of the following clients on the rehabilitation unit is most likely to develop autonomic dysreflexia?
  1. A client with brain injury
  2. A client with herniated nucleus pulposus
  3. A client with a high cervical spine injury
  4. A client with a stroke

93. Which of the following conditions indicates that spinal shock is resolving in a client with C7 quadriplegia?
  1. Absence of pain sensation in chest
  2. Spasticity
  3. Spontaneous respirations
  4. Urinary continence

94. When discharging a client from the hospital after a laminectomy, the nurse recognizes that the client needs further teaching when he makes which of the following statements?
  1. "I'll sleep on a firm mattress."
  2. "I won't drive for 2 to 4 weeks."
  3. "When I pick things up, I'll bend my knees."
  4. "I can't wait to toss my granddaughter up in the air."

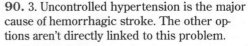

You're doing great!

To solve question 93, clue in on the word resolving.

90. 3. Uncontrolled hypertension is the major cause of hemorrhagic stroke. The other options aren't directly linked to this problem.
NP: Evaluation; CN: Physiological integrity; CNS: Reduction of risk potential; CL: Application

91. 3. Isosorbide and verapamil can cause hypotension, which reduces brain perfusion and should be avoided in a client with a stroke. Hypocarbia is used to reduce intracranial pressure through cerebral vasoconstriction. Nutrition may be delivered by tube when dysphagia exists. Dehydration would be inappropriate in this instance.
NP: Planning; CN: Physiological integrity; CNS: Physiological adaptation; CL: Application

92. 3. Autonomic dysreflexia refers to uninhibited sympathetic outflow in clients with spinal cord injuries above the level of T10. The other clients aren't prone to dysreflexia.
NP: Planning; CN: Physiological integrity; CNS: Physiological adaptation; CL: Application

93. 2. Spasticity, the return of reflexes, is a sign of resolving shock. Spinal or neurogenic shock is characterized by hypotension, bradycardia, dry skin, flaccid paralysis, or the absence of reflexes below the level of injury. Slight muscle contraction at the bulbocavernosus reflex occurs, but not enough for urinary continence. Spinal shock descends from the injury, and respiratory difficulties occur at C4 and above. The absence of pain sensation in the chest doesn't apply to spinal shock.
NP: Data collection; CN: Physiological integrity; CNS: Physiological adaptation; CL: Application

94. 4. Lifting more than 10 lb (4.5 kg) for several weeks after surgery is contraindicated. The other responses are appropriate.
NP: Evaluation; CN: Physiological integrity; CNS: Reduction of risk potential; CL: Analysis

95. When assessing a client with herniated nucleus pulposus (HNP) of L4-L5, the nurse would expect to find which of the following symptoms of spinal cord compression?
1. Low back pain
2. Pain radiating across the buttocks
3. Positive Kernig's sign
4. Urinary incontinence

96. A nurse assesses a client who has episodes of autonomic dysreflexia. Which of the following conditions can cause autonomic dysreflexia?
1. Headache
2. Lumbar spinal cord injury
3. Neurogenic shock
4. Noxious stimuli

97. During an episode of autonomic dysreflexia in which the client becomes hypertensive, the nurse should perform which of the following interventions?
1. Elevate the client's legs.
2. Put the client flat in bed.
3. Put the bed in Trendelenburg's position.
4. Put the client in high Fowler's position.

98. A client recovering from a spinal cord injury has a great deal of spasticity. Which of the following medications may be used to control spasticity?
1. Hydralazine (Apresoline)
2. Baclofen (Lioresal)
3. Lidocaine (Xylocaine)
4. Methylprednisolone (Medrol)

99. Client teaching for a client with Gardner-Wells tongs would include which of the following reasons for their use?
1. To reduce intracranial pressure (ICP)
2. To reduce dislocations and pain
3. To prevent deep vein thrombosis (DVT)
4. To improve neurologic outcome

Correct positioning will help relieve your client's symptoms.

Keep a-goin'!

95. 4. Progressive neurological deficits at L4-L5, including worsening muscle weakness, paresthesia, and loss of bowel and bladder control, are symptoms of spinal cord compression. The other symptoms usually occur in clients with HNP.
NP: Planning; CN: Physiological integrity; CNS: Reduction of risk potential; CL: Analysis

96. 4. Noxious stimuli, such as a full bladder, fecal impaction, or a pressure ulcer, may cause autonomic dysreflexia. Neurogenic shock isn't a cause of dysreflexia. Autonomic dysreflexia is most commonly seen with injuries at T10 or above. A headache is a symptom of autonomic dysreflexia, not a cause.
NP: Data collection; CN: Physiological integrity; CNS: Physiological adaptation; CL: Application

97. 4. Putting the client in high-Fowler's position will decrease cerebral blood flow, decreasing hypertension. Elevating the client's legs, putting the client flat in bed, and putting the bed in Trendelenburg's position place the client in positions that improve cerebral blood flow, worsening hypertension.
NP: Implementation; CN: Physiological integrity; CNS: Reduction of risk potential; CL: Application

98. 2. Baclofen is a skeletal muscle relaxant used to decrease spasms. Methylprednisolone, an anti-inflammatory drug, is used to decrease spinal cord edema. Hydralazine is an antihypertensive and afterload reducing agent. Lidocaine is an antiarrhythmic and a local anesthetic agent.
NP: Planning; CN: Physiological integrity; CNS: Pharmacological therapies; CL: Application

99. 2. Gardner-Wells tongs are used to reduce dislocations, subluxations, pain, and spasm in cervical spinal cord injuries. They aren't used to reduce ICP, prevent DVT, or improve neurologic outcome.
NP: Implementation; CN: Physiological integrity; CNS: Reduction of risk potential; CL: Application

NP: Nursing process   CN: Client needs category   CNS: Client needs subcategory   CL: Cognitive level

100. A client with a T1 spinal cord injury arrives at the emergency department with a blood pressure of 82/40 mm Hg; pulse, 34 beats/minute; dry skin, and flaccid paralysis of the lower extremities. Which of the following conditions would most likely be suspected?

1. Autonomic dysreflexia
2. Hypervolemia
3. Neurogenic shock
4. Sepsis

101. A client has a cervical spinal cord injury at the level of C5. Which of the following conditions would the nurse anticipate during the acute phase?

1. Absent corneal reflex
2. Decerebrate posturing
3. Movement of only the right or left half of the body
4. The need for mechanical ventilation

I'm not trying to be cute. There really is a clue in question 101.

102. When caring for a client with quadriplegia, which of the following nursing interventions takes priority?

1. Forcing fluids to prevent renal calculi
2. Maintaining skin integrity
3. Obtaining adaptive devices for more independence
4. Preventing atelectasis

103. A client with C7 quadriplegia is flushed and anxious and complains of a pounding headache. Which of the following symptoms would also be anticipated?

1. Decreased urine output or oliguria
2. Hypertension and bradycardia
3. Respiratory depression
4. Symptoms of shock

Keep alert! You're almost finished!

104. A client has a diagnosis of stroke versus transient ischemic attack (TIA). Which of the following statements shows the difference between a TIA and a stroke?

1. TIAs typically resolve in 24 hours.
2. TIAs may be hemorrhagic in origin.
3. TIAs may cause a permanent motor deficit.
4. TIAs may predispose the client to a myocardial infarction (MI).

100. 3. Loss of sympathetic control and unopposed vagal stimulation below the level of the injury typically cause hypotension, bradycardia, pallor, flaccid paralysis, and warm, dry skin in the client in neurogenic shock. Autonomic dysreflexia occurs after neurogenic shock abates. Signs of sepsis would include elevated temperature, increased heart rate, and increased respiratory rate.

NP: Data collection; CN: Physiological integrity; CNS: Physiological adaptation; CL: Analysis

101. 4. The diaphragm is stimulated by nerves at the level of C4. Initially, this client may need mechanical ventilation due to cord edema. This may resolve in time. Decerebrate posturing, hemiplegia, and absent corneal reflexes occur with brain injuries, not spinal cord injuries.

NP: Planning; CN: Physiological integrity; CNS: Physiological adaptation; CL: Application

102. 4. Clients with quadriplegia have paralysis or weakness of the diaphragm, abdominal, or intercostal muscles. Maintenance of airway and breathing take top priority. Although forcing fluids, maintaining skin integrity, and obtaining adaptive devices for more independence are all important interventions, preventing atelectasis has more priority.

NP: Planning; CN: Physiological integrity; CNS: Reduction of risk potential; CL: Application

103. 2. Hypertension, bradycardia, anxiety, blurred vision, and flushing above the lesion occur with autonomic dysreflexia due to uninhibited sympathetic nervous system discharge. The other options are incorrect.

NP: Data collection; CN: Physiological integrity; CNS: Physiological adaptation; CL: Analysis

104. 1. Symptoms of TIA result from a transient lack of oxygen to the brain and usually resolve within 24 hours. Hemorrhage into the brain has the worst neurological outcome and isn't associated with a TIA. Permanent motor deficits don't result from TIA. Unstable angina, not a TIA, may predispose the client to a future MI.

NP: Planning; CN: Physiological integrity; CNS: Physiological adaptation; CL: Comprehension

105. A client with a right stroke has a flaccid left side. Which of the following interventions would best prevent shoulder subluxation?
1. Splint the wrist.
2. Use an air splint.
3. Put the affected arm in a sling.
4. Perform range-of-motion exercises on the affected side.

106. A 40-year-old paraplegic must perform intermittent catheterization of the bladder. Which of the following instructions should be given?
1. Clean the meatus from back to front.
2. Measure the quantity of urine.
3. Gently rotate the catheter during removal.
4. Clean the meatus with soap and water.

Keep on teaching!

107. Which of the following methods should be used to assess pupil accommodation?
1. Assess for peripheral vision.
2. Touch the cornea lightly with a wisp of cotton.
3. Have the client follow an object upward, downward, obliquely, and horizontally.
4. Observe for pupil constriction and convergence while focusing on an object coming toward the client.

108. A client at the eye clinic reports difficulty seeing at night. This may result from which of the following nutritional deficiencies?
1. Vitamin A
2. Vitamin B_6
3. Vitamin C
4. Vitamin K

Which procedure or program should I plan for this client?

109. A client with a spinal cord injury has a neurogenic bladder. When planning for discharge, the nurse anticipates the client will need which of the following procedures or programs?
1. Intermittent catheterization program
2. Kock pouch
3. Transurethral prostatectomy
4. Ureterostomy

105. 3. Due to the weight of the flaccid extremity, the shoulder may disarticulate. A sling will support the extremity. The other options won't support the shoulder.
NP: Planning; CN: Physiological integrity; CNS: Basic care and comfort; CL: Application

106. 4. Intermittent catheterization may be performed chronically with clean technique, using soap and water to clean the urinary meatus. The meatus is always cleaned from front to back in a woman, or in expanding circles working outward from the meatus in a man. The catheter doesn't need to be rotated during removal. It isn't necessary to measure the urine.
NP: Planning; CN: Physiological integrity; CNS: Basic care and comfort; CL: Application

107. 4. Accommodation refers to convergence and constriction of the pupil while focusing on a nearing object. Touching the cornea lightly with a wisp of cotton describes assessment of the corneal reflex. Having the client follow an object upward, downward, obliquely, and horizontally refers to cardinal fields of gaze. Assessing for peripheral vision refers to visual fields.
NP: Data collection; CN: Physiological integrity; CNS: Physiological adaptation; CL: Application

108. 1. Night blindness (nyctalopia) may be caused from a vitamin A deficiency or dysfunctional rod receptors. None of the other deficiencies lead to nyctalopia.
NP: Data collection; CN: Physiological integrity; CNS: Physiological adaptation; CL: Application

109. 1. Intermittent catheterization, starting with 2-hour intervals and increasing to 4- to 6-hour intervals, is used to manage neurogenic bladder. A Kock pouch is a continent ileostomy. An ileostomy or ureterostomy isn't necessary. Transurethral prostatectomy is indicated for obstruction to urinary outflow by benign prostatic hyperplasia or for the treatment of cancer.
NP: Planning; CN: Physiological integrity; CNS: Basic care and comfort; CL: Application

NP: Nursing process   CN: Client needs category   CNS: Client needs subcategory   CL: Cognitive level

110. When using a Snellen alphabet chart, the nurse records the client's vision as 20/40. Which of the following statements best describes 20/40 vision?

1. The client has alterations in near vision and is legally blind.
2. The client can see at 20 feet what the person with normal vision sees at 40 feet.
3. The client can see at 40 feet what the person with normal vision sees at 20 feet.
4. The client has a 20% decrease in acuity in one eye, and 40% decrease in the other eye.

Keep those numbers straight!

110. 2. The numerator refers to the client's vision while comparing the normal vision in the denominator. Legal blindness refers to 20/150 or less. Alterations in near vision may be due to loss of accommodation caused by the aging process (presbyopia) or farsightedness.
NP: Data collection; CN: Physiological integrity; CNS: Physiological adaptation; CL: Analysis

111. Which of the following instruments is used to record intraocular pressure?

1. Goniometer
2. Ophthalmoscope
3. Slit lamp
4. Tonometer

111. 4. A tonometer is a device used in glaucoma screening to record intraocular pressure. A goniometer measures joint movement and angles. An ophthalmoscope examines the interior of the eye, especially the retina. A slit lamp evaluates structures in the anterior chamber of the eye.
NP: Data collection; CN: Physiological integrity; CNS: Reduction of risk potential; CL: Comprehension

112. After a nurse instills atropine drops into both eyes for a client undergoing an ophthalmic examination, which of the following instructions would be given to the client?

1. Be careful because the blink reflex is paralyzed.
2. Avoid wearing your regular glasses when driving.
3. Be aware that the pupils may be unusually small.
4. Wear dark glasses in bright light because the pupils are dilated.

112. 4. Atropine, an anticholinergic drug, has mydriatic effects causing pupil dilation. This allows more light onto the retina and causes photophobia and blurred vision. Atropine doesn't paralyze the blink reflex or cause miosis (pupil constriction). Driving may be contraindicated due to blurred vision.
NP: Implementation; CN: Physiological integrity; CNS: Reduction of risk potential; CL: Application

113. Which of the following procedures or assessments must the nurse perform when preparing a client for eye surgery?

1. Clip the client's eyelashes.
2. Verify the affected eye has been patched for 24 hours before surgery.
3. Verify the client has had nothing by mouth since midnight or at least 8 hours before surgery.
4. Obtain informed consent with the client's signature and place the forms in the chart.

Way to go!

113. 3. Maintaining nothing-by-mouth status for at least 8 hours before surgical procedures prevents vomiting and aspiration. The physician is responsible for obtaining informed consent; the nurse validates that the consent is obtained. There's no need to patch an eye before most surgeries or to clip the eyelashes unless specifically ordered by the physician.
NP: Planning; CN: Safe, effective care environment; CNS: Safety and infection control; CL: Application

114. Which of the following statements indicates that a client needs additional teaching after cataract surgery?
1. "I'll avoid eating until the nausea subsides."
2. "I can't wait to pick up my granddaughter."
3. "I'll avoid bending over to tie my shoelaces."
4. "I'll avoid touching the dropper to my eye when using my eyedrops."

115. Cataract surgery results in aphakia. Which of the following statements best describes this term?
1. Absence of the crystalline lens
2. A "keyhole" pupil
3. Loss of accommodation
4. Retinal detachment

116. When developing a teaching session on glaucoma for the community, which of the following statements would the nurse stress?
1. Glaucoma is easily corrected with eyeglasses.
2. White and Asian individuals are at the highest risk for glaucoma.
3. Yearly screening for people ages 20 to 40 is recommended.
4. Glaucoma can be painless and vision may be lost before the person is aware of a problem.

117. For a client having an episode of acute narrow-angle glaucoma, a nurse expects to give which of the following medications?
1. Acetazolamide (Diamox)
2. Atropine
3. Furosemide (Lasix)
4. Streptokinase (Streptase)

118. Which of the following symptoms would occur in a client with a detached retina?
1. Flashing lights and floaters
2. Homonymous hemianopia
3. Loss of central vision
4. Ptosis

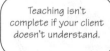

Teaching isn't complete if your client doesn't understand.

You don't need 20/20 vision to see how important this information is!

114. 2. Lifting, often involving the Valsalva maneuver, increases intraocular pressure and strain on the surgical site. Preventing nausea and subsequent vomiting will prevent increased intraocular pressure, as will avoiding bending or placing the head in a dependent position. Touching the eye dropper to the eye will contaminate the dropper and thus the entire bottle of medication.
NP: Evaluation; CN: Physiological integrity; CNS: Reduction of risk potential; CL: Analysis

115. 1. Aphakia means without lens. A keyhole pupil results from iridectomy. Loss of accommodation is a normal response to aging. A retinal detachment is usually associated with retinal holes created by vitreous traction.
NP: Planning; CN: Physiological integrity; CNS: Physiological adaptation; CL: Comprehension

116. 4. Open-angle glaucoma causes a painless increase in intraocular pressure with loss of peripheral vision. Individuals older than age 40 should be screened. Blacks have a threefold greater chance of developing glaucoma with an increased chance of blindness than other groups. A variety of miotics and agents to decrease intraocular pressure and occasionally surgery are used to treat glaucoma.
NP: Planning; CN: Health promotion and maintenance; CNS: Prevention and early detection of disease; CL: Application

117. 1. Acetazolamide, a carbonic anhydrase inhibitor, decreases intraocular pressure by decreasing the secretion of aqueous humor. Streptokinase is a thrombolytic agent, and furosemide is a loop diuretic; these aren't used in the treatment of glaucoma. Atropine dilates the pupil and decreases outflow of aqueous humor, causing a further increase in intraocular pressure.
NP: Planning; CN: Physiological integrity; CNS: Pharmacological therapies; CL: Application

118. 1. Signs and symptoms of retinal detachment include abrupt flashing lights, floaters, loss of peripheral vision, or a sudden shadow or curtain in the vision. Occasionally, vision loss is gradual.
NP: Planning; CN: Physiological integrity; CNS: Physiological adaptation; CL: Application

NP: Nursing process   CN: Client needs category   CNS: Client needs subcategory   CL: Cognitive level

119. A client underwent an enucleation of the right eye for a malignancy. Which of the following interventions will the nurse perform?
1. Instill miotics as ordered to the affected eye.
2. Teach the client to clean the prosthesis in soap and water.
3. Assess reactivity of the pupils to light and accommodation.
4. Teach the client to avoid straining at stool to prevent intraocular pressure.

If you're unsure about an order, ask.

120. A nurse would question an order to irrigate the ear canal in which of the following circumstances?
1. Ear pain
2. Hearing loss
3. Otitis externa
4. Perforated tympanic membrane

121. Which of the following interventions is essential when instilling Cortisporin suspension, 2 gtts AD?
1. Verify the proper client and route.
2. Warm the solution to prevent dizziness.
3. Hold an emesis basin under the client's ear.
4. Place the client in the semi-Fowler's position.

Giving drugs isn't a game of chance. Know the right way to do it.

122. When teaching the client with Ménière's disease, which of the following instructions would a nurse give about vertigo?
1. Report dizziness at once.
2. Drive in daylight hours only.
3. Get up slowly, turning the entire body.
4. Change your position using the logroll technique.

In question 123, look your best.

123. Which response by a client best indicates an understanding of the adverse effects of phenytoin (Dilantin)?
1. "I should take the medication with food to prevent nausea."
2. "I need to take this medication until my seizures stop."
3. "I need to see the dentist every 6 months."
4. "I need to report any drowsiness to the physician immediately."

119. 2. Enucleation of the eye refers to surgical removal of the entire eye; therefore, the client needs instructions about the prosthesis. There are no activity restrictions or need for eye drops; however, prophylactic antibiotics may be used in the immediate postoperative period.
NP: Planning; CN: Physiological integrity; CNS: Physiological adaptation; CL: Application

120. 4. Irrigation of the ear canal is contraindicated with perforation of the tympanic membrane because solution entering the inner ear may cause dizziness, nausea, vomiting, and infection. The other conditions aren't contraindications to irrigation of the ear canal.
NP: Planning; CN: Physiological integrity; CNS: Reduction of risk potential; CL: Application

121. 1. When giving medications, a nurse follows the five "R's" of medication administration: right client, right drug, right dose, right route, and right time. Put the client in the lateral position, not semi-Fowler's position, to prevent the drops from draining out for 5 minutes. The drops may be warmed to prevent pain or dizziness, but this action isn't essential. An emesis basin would be used for irrigation of the ear.
NP: Planning; CN: Physiological integrity; CNS: Pharmacological therapies; CL: Application

122. 3. Turning the entire body, not the head, will prevent vertigo. Turning the client in bed slowly and smoothly will be helpful; logrolling isn't needed. The client shouldn't drive as he may reflexively turn the wheel to correct for vertigo. Dizziness is expected but can be prevented with Ménière's disease.
NP: Implementation; CN: Physiological integrity; CNS: Reduction of risk potential; CL: Application

123. 3. Phenytoin can cause hypertrophy of the gums and gingivitis; therefore, regular dental checkups are essential. Phenytoin doesn't need to be taken with food and should never be discontinued unless ordered by a physician. Some drowsiness is expected initially; however, this usually decreases with continued use.
NP: Evaluation; CN: Physiological therapy; CNS: Pharmacological therapies; CL: Analysis

124. An 18-year-old client was hit in the head with a baseball during practice. When discharging him to the care of his mother, the nurse gives which of the following instructions?
1. Watch him for keyhole pupil for the next 24 hours.
2. Expect profuse vomiting for 24 hours after the injury.
3. Wake him every hour and assess orientation to person, time, and place.
4. Notify the physician immediately if he has a headache.

Instruct the client's mother about the care for her son's injury.

124. 3. Changes in level of consciousness may indicate expanding lesions such as subdural hematoma; orientation and level of consciousness are assessed frequently for 24 hours. Profuse or projectile vomiting is a symptom of increased intracranial pressure and should be reported immediately. A slight headache may last for several days after concussion; severe or worsening headaches should be reported. A keyhole pupil is found after iridectomy.
NP: Implementation; CN: Physiological integrity; CNS: Physiological adaptation; CL: Application

125. A client taking carbamazepine (Tegretol) should be monitored for which of the following potential complications?
1. Adult respiratory distress syndrome
2. Diplopia
3. Elevated levels of phenytoin (Dilantin)
4. Leukocytosis

125. 2. Carbamazepine is more likely to cause diplopia, dizziness, ataxia, and a rash. Carbamazepine causes agranulocytosis because of the reduction in leukocytes. Adult respiratory distress syndrome isn't a complication of carbamazepine. Carbamazepine decreases blood levels of phenytoin and oral contraceptives.
NP: Evaluation; CN: Physiological integrity; CNS: Pharmacological therapies; CL: Comprehension

126. When assessing the pupil's ability to constrict, which cranial nerve is being tested?
1. II
2. III
3. IV
4. V

126. 2. Cranial nerve (CN) III, the oculomotor nerve, controls pupil constriction. CN II is the optic nerve, which controls vision. CN IV is the trochlear nerve, which coordinates eye movement. CN V is the trigeminal nerve, which innervates the muscles of chewing.
NP: Data collection; CN: Physiological integrity; CNS: Physiological adaptation; CL: Application

127. Nursing care of a client with damage to the thalamus, hypothalamus, and pineal gland would be based on knowing the client has problems in which of the following areas?
1. Seizure control
2. Identifying foreign agents
3. Difficulty regulating emotions
4. Initiating movements and maintaining temperature control and the sleep-awake cycle

Different areas of the brain (Hey, that's me!) regulate different functions.

127. 4. The thalamus is the relay center of communication of sensory and motor information between the higher and lower regions of the brain. The hypothalamus regulates temperature control, and the pineal gland is important to the sleep-awake cycle. Seizure control is more related to neurotransmitter dysfunction. Identifying foreign agents is a function of the immune system. Regulating emotions is a function of the limbic system.
NP: Data collection; CN: Physiological integrity; CNS: Physiological adaptation; CL: Knowledge

128. Problems with memory and learning would relate to which of the following lobes?
1. Frontal
2. Occipital
3. Parietal
4. Temporal

Lost? Need a map?

128. 4. The temporal lobe functions to regulate memory and learning problems because of the integration of the hippocampus. The parietal lobe primarily functions with sensory function. The occipital lobe functions to regulate vision. The frontal lobe primarily functions to regulate thinking, planning, and judgment.
NP: Data collection; CN: Physiological integrity; CNS: Physiological adaptation; CL: Knowledge

129. While cooking, your client couldn't feel the temperature of a hot oven. Which lobe could be dysfunctional?
1. Frontal
2. Occipital
3. Parietal
4. Temporal

129. 3. The parietal lobe regulates sensory function, which would include the ability to sense hot or cold objects. The occipital lobe is primarily responsible for vision function. The temporal lobe regulates memory, and the frontal lobe regulates thinking, planning, and judgment.
NP: Data collection; CN: Safe, effective care environment; CNS: Safety and infection control; CL: Knowledge

130. Which neurotransmitter is responsible for many of the functions of the frontal lobe?
1. Dopamine
2. Gamma-aminobutyric acid (GABA)
3. Histamine
4. Norepinephrine

Stay the course, you're almost done!

130. 1. The frontal lobe primarily functions to regulate thinking, planning, and affect. Dopamine is known to circulate widely throughout this lobe, which is why it's such an important neurotransmitter in schizophrenia. GABA is widely circulated in the hippocampus and hypothalamus. Histamine is primarily found in the hypothalamus. Norepinephrine primarily functions with the hippocampal region not located in the frontal lobe.
NP: Data collection; CN: Physiological integrity; CNS: Physiological adaptation; CL: Knowledge

131. The nurse is discussing the purpose of an EEG with the family of a client with massive cerebral hemorrhage and loss of consciousness. It would be most accurate for the nurse to tell family members that the test measures which of the following conditions?
1. Extent of intracranial bleeding
2. Sites of brain injury
3. Activity of the brain
4. Percent of functional brain tissue

131. 3. An EEG measures the electrical activity of the brain. Extent of intracranial bleeding and location of the injury site would be determined by computerized tomography or magnetic resonance imaging. Percent of functional brain tissue would be determined by a series of tests.
NP: Implementation; CN: Physiological integrity; CNS: Physiological adaptation; CL: Comprehension

132. The nurse is teaching a client and his family about dietary practices related to Parkinson's disease. Which of the following signs and symptoms would be most important for the to nurse address?
1. Fluid overload and drooling
2. Aspiration and anorexia
3. Choking and diarrhea
4. Dysphagia and constipation

133. In some clients with multiple sclerosis (MS), plasmapheresis diminishes symptoms. Plasmapheresis achieves this effect by removing which of the following blood components?
1. Catecholamines
2. Antibodies
3. Plasma proteins
4. Lymphocytes

132. 4. The eating problems associated with Parkinson's disease include dysphagia, risk for choking, aspiration, and constipation. Fluid overload, anorexia, and diarrhea aren't problems specifically related to Parkinson's disease.
NP: Implementation; CN: Physiological integrity; CNS: Reduction of risk potential; CL: Analysis

133. 2. In plasmapheresis, antibodies are removed from the client's plasma. Antibodies attack the myelin sheath of the neuron causing the manifestations of MS. The treatment of MS with plasmapheresis isn't for the purpose of removing catecholamines, plasma proteins, or lymphocytes.
NP: Evaluation; CN: Physiological integrity; CNS: Physiological adaptation; CL: Comprehension

Here's a test that covers nursing care for clients with a disorder of the musculoskeletal system. So get moving! (Get it? Moving? Hmmm. I must be losing my touch.)

Chapter 7
Musculoskeletal disorders

1. Osteoblast activity is needed for which of the following functions?
1. Bone formation
2. Estrogen production
3. Hematopoiesis
4. Muscle formation

Your hint here is in the Latin root of the word *osteoblast*.

1. 1. Osteoblast activity is necessary for bone formation. Osteoblasts are bone-forming cells; they don't have a role in muscle formation. Estrogen is linked to calcium reuptake and building of bone tissue. Hematopoiesis is the production of red blood cells in the bone marrow.

NP: Data collection; CN: Physiological integrity; CNS: Physiological adaptation; CL: Knowledge

2. Which of the following conditions is the primary complication of osteoporosis?
1. Pain
2. Fractures
3. Hardening of the bones
4. Increased bone matrix and remineralization

Watch out! In question 2, *primary* means main or principle. In question 3, it refers to a type of disease.

2. 2. The primary complication of osteoporosis is fractures. Bones soften, and there's a decrease in bone matrix and remineralization. Pain may occur, but fractures can be life-threatening.

NP: Data collection; CN: Physiological integrity; CNS: Physiological adaptation; CL: Knowledge

3. Which of the following conditions is the cause of primary osteoporosis?
1. Alcoholism
2. Hormonal imbalance
3. Malnutrition
4. Osteogenesis imperfecta

3. 2. Hormonal imbalance, faulty metabolism, and poor dietary intake of calcium cause primary osteoporosis. Malnutrition, alcoholism, osteogenesis imperfecta, rheumatoid arthritis, liver disease, scurvy, lactose intolerance, hyperthyroidism, and trauma cause secondary osteoporosis.

NP: Data collection; CN: Physiological integrity; CNS: Physiological adaptation; CL: Knowledge

4. A 42-year-old woman just had a total hysterectomy. Is she at risk for osteoporosis?
1. No, because she still has her thyroid gland.
2. No, because she isn't at risk until she's older.
3. Yes, because she's still producing hormones.
4. Yes, because she's just had surgically induced menopause.

4. 4. Menopause at any age puts women at risk for osteoporosis because of the associated hormonal imbalance. This client's thyroid gland won't protect her from menopause. With her ovaries removed, she's no longer producing hormones.

NP: Data collection; CN: Physiological integrity; CNS: Physiological adaptation; CL: Comprehension

NP: Nursing process   CN: Client needs category   CNS: Client needs subcategory   CL: Cognitive level

5. Primary prevention of osteoporosis includes which of the following measures?
1. Place items within reach of the client.
2. Install bars in the bathroom to prevent falls.
3. Maintain the optimal calcium intake, and use estrogen replacement therapy.
4. Use a professional alert system in the home in case a fall occurs when the client is alone.

In question 5, *primary* refers to a level of prevention.

6. Which of the following mechanisms is believed to cause gout?
1. Overproduction of calcium
2. Underproduction of calcium
3. Overproduction of uric acid
4. Underproduction of uric acid

7. In advanced gout, urate crystal deposits develop on the hands, knees, feet, forearms, ear, and Achilles tendon. Which of the following terms refers to these deposits?
1. Arthralgia
2. Gout nodules
3. Pinna
4. Tophi

8. Which of the following phrases best explains the usual pattern of nonchronic gout?
1. Frequent painful attacks
2. Generally painful joints at all times
3. Painful attacks with pain-free periods
4. Painful attacks with less painful periods, but pain never subsides

You'll answer these questions in no time.

9. A client has been prescribed a diet that limits purine-rich foods. Which of the following foods would the nurse teach her to avoid eating?
1. Bananas and dried fruits
2. Milk, ice cream, and yogurt
3. Wine, cheese, preserved fruits, meats, and vegetables
4. Anchovies, sardines, kidneys, sweetbreads, and lentils

5. 3. Primary prevention of osteoporosis includes maintaining optimal calcium intake and using estrogen replacement therapy. Using a professional alert system in the home, installing bars in bathrooms to prevent falls, and placing items within reach of the client are all secondary and tertiary prevention methods.
NP: Implementation; CN: Health promotion and maintenance; CNS: Prevention and early detection of disease; CL: Application

6. 3. Although the exact cause of primary gout remains unknown, it seems linked to a genetic defect in purine metabolism that causes overproduction of uric acid, retention of uric acid, or both. Gout isn't related to calcium production.
NP: Data collection; CN: Physiological integrity; CNS: Physiological adaptation; CL: Knowledge

7. 4. Urate crystal deposits are known as tophi. The pinna is the outside of the ear. Gout nodule is an incorrect term. Arthralgia describes painful joints.
NP: Data collection; CN: Physiological integrity; CNS: Physiological adaptation; CL: Knowledge

8. 3. The usual pattern of gout involves painful attacks with pain-free periods. Chronic gout may lead to frequent attacks with persistently painful joints.
NP: Implementation; CN: Physiological integrity; CNS: Physiological adaptation; CL: Knowledge

9. 4. Anchovies, sardines, kidneys, sweetbreads, and lentils are high in purines. Bananas and dried fruits are high in potassium. Milk, ice cream, and yogurt are rich in calcium. Wine, cheese, preserved fruits, meats, and vegetables contain tyramine.
NP: Implementation CN: Health promotion and maintenance; CNS: Prevention and early detection of disease; CL: Application

NP: Nursing process CN: Client needs category CNS: Client needs subcategory CL: Cognitive level

10. A 61-year-old male client has had gout for a long time and has developed nodules. While teaching the client, the nurse explains that these nodules are referred to by which of the following terms?

1. Cysts
2. Stones
3. Tophi
4. Calculi

11. The nurse is collecting a health history from a 63-year-old man who may have gout. What joint is most often affected in the client with gout?

1. Great toe
2. Wrist
3. Ankle
4. Knee

12. A client has been diagnosed with gout and wants to know why colchicine is used in the treatment of gout. Which of the following actions of colchicine explains why it's effective for gout?

1. Replaces estrogen
2. Decreases infection
3. Decreases inflammation
4. Decreases bone demineralization

13. A client with gout is encouraged to increase fluid intake. Which of the following statements best explains why increased fluids are encouraged for gout?

1. Fluids decrease inflammation.
2. Fluids increase calcium absorption.
3. Fluids promote the excretion of uric acid.
4. Fluids provide a cushion for weakened bones.

14. Gout pain usually occurs in which of the following locations?

1. Joints
2. Tendons
3. Long bones
4. Areas of striated muscle

Let's get the joint jumping! It'd be easier if I didn't have gout in the joint of the answer to question 11!

In question 13, look for the best answer!

10. 3. Tophi are nodular deposits of sodium urate crystals. They can occur as a result of chronic gout. A cyst is a closed sac or pouch with a definite wall that contains fluid, semi-fluid, or solid material. Stones are abnormal concretions usually composed of mineral salts. Calculi is another name for stones formed by mineral salts.

NP: Evaluation; CN: Physiological integrity; CNS: Reduction of risk potential; CL: Comprehension

11. 1. The great toe is most commonly affected in clients with gout. Gout can affect other joints but most commonly affects the great toe.

NP: Implementation CN: Physiological integrity; CNS: Reduction of risk potential; CL: Knowledge

12. 3. The action of colchicine is to decrease inflammation by reducing the migration of leukocytes to synovial fluid. Colchicine doesn't decrease infection, replace estrogen, or decrease bone demineralization.

NP: Planning; CN: Physiological integrity; CNS: Pharmacological therapies; CL: Comprehension

13. 3. With gout, fluids promote the excretion of uric acid. Fluids don't provide a cushion for weakened bones, decrease inflammation, or increase calcium absorption.

NP: Planning; CN: Physiological integrity; CNS: Physiological adaptation; CL: Comprehension

14. 1. The pain of gout is usually found in joints. Gout isn't found in long bones, striated muscle, or tendons.

NP: Data collection; CN: Physiological integrity; CNS: Physiological adaptation; CL: Knowledge

15. A client with gout is receiving indomethacin for pain. Which of the following instructions should be given to a client taking nonsteroidal anti-inflammatory drugs (NSAIDs)?
  1. Bleeding isn't a problem with NSAIDs.
  2. Take NSAIDs with food to avoid an upset stomach.
  3. Take NSAIDs on an empty stomach to increase absorption.
  4. Don't take NSAIDs at bedtime because they may cause excitement.

Your client depends on you for information.

15. 2. Indomethacin, like other NSAIDs, should be taken with food because it can be irritating to the GI mucosa and lead to GI bleeding. It can cause drowsiness and potential bleeding complications.
NP: Planning; CN: Physiological integrity; CNS: Pharmacological therapies; CL: Application

16. A client asks for information about osteoarthritis. Which of the following statements about osteoarthritis is correct?
  1. Osteoarthritis is rarely debilitating.
  2. Osteoarthritis is a rare form of arthritis.
  3. Osteoarthritis is the most common form of arthritis.
  4. Osteoarthritis afflicts people over age 60.

16. 3. Osteoarthritis is the most common form of arthritis. It can afflict people of any age, although most are elderly. It can be extremely debilitating.
NP: Planning; CN: Physiological integrity; CNS: Physiological adaptation; CL: Knowledge

17. Which of the following conditions or actions can cause primary osteoarthritis?
  1. Overuse of joints, aging, obesity
  2. Obesity, diabetes mellitus, aging
  3. Congenital abnormality, aging, overuse of joints
  4. Diabetes mellitus, congenital abnormality, aging

Don't be fooled. What does primary mean in this question?

17. 1. Primary osteoarthritis may be caused by the overuse of joints, aging, or obesity. Congenital abnormalities and diabetes mellitus can cause secondary osteoarthritis.
NP: Data collection; CN: Physiological integrity; CNS: Physiological adaptation; CL: Knowledge

18. The nurse knows that a client with osteoarthritis of the knee understands the discharge instructions when the client makes which of the following statements?
  1. "I'll take my ibuprofen (Motrin) on an empty stomach."
  2. "I'll try taking a warm shower in the morning."
  3. "I'll wear my knee splint every night."
  4. "I'll jog at least a mile every evening."

18. 2. A client with osteoarthritis has joint stiffness that may be partially relieved with a warm shower on arising in the morning. Ibuprofen should be taken with food, as should all nonsteroidal anti-inflammatory medications. Splints are usually used by clients with rheumatoid arthritis. Because the problem is one of continued stress on the joint, the client may want to try to an exercise that puts less strain on the joint such as swimming.
NP: Evaluation; CN: Physiological integrity; CNS: Basic care and comfort; CL: Application

NP: Nursing process   CN: Client needs category   CNS: Client needs subcategory   CL: Cognitive level

19. The treatment for osteoarthritis commonly includes salicylates. Salicylates can be dangerous in older people because they can cause which of the following adverse effects?
1. Hearing loss
2. Increased pain in joints
3. Decreased calcium absorption
4. Increased bone demineralization

Way to go!

20. Clients with osteoarthritis may be on bed rest for prolonged periods. Which of the following nursing interventions would be appropriate for these clients?
1. Encourage coughing and deep breathing, and limit fluid intake.
2. Provide only passive range of motion (ROM), and decrease stimulation.
3. Have the client lie as still as possible, and give adequate pain medicine.
4. Turn the client every 2 hours, and encourage coughing and deep breathing.

It's important to know the difference between these two common diseases.

21. Which of the following statements explains the main difference between rheumatoid arthritis and osteoarthritis?
1. Osteoarthritis is gender-specific, rheumatoid arthritis isn't.
2. Osteoarthritis is a localized disease, rheumatoid arthritis is systemic.
3. Osteoarthritis is a systemic disease, rheumatoid arthritis is localized.
4. Osteoarthritis has dislocations and subluxations, rheumatoid arthritis doesn't.

22. During the health history of a client, signs of osteoarthritis are noted. Which of the following data findings would indicate osteoarthritis?
1. Elevated sedimentation rate
2. Multiple subcutaneous nodules
3. Asymmetrical joint involvement
4. Signs of inflammation, such as heat, fever, and malaise

19. 1. Many elderly people already have diminished hearing, and salicylate use can lead to further or total hearing loss. Salicylates can cause fluid retention and edema, which is worrisome in older clients, especially those with heart failure. Salicylates don't increase bone demineralization, decrease calcium absorption, or increase pain in joints.
NP: Data collection; CN: Physiological integrity; CNS: Pharmacological therapies; CL: Comprehension

20. 4. A bedridden client needs to be turned every 2 hours, have adequate nutrition, and cough and deep-breathe. Adequate pain medication, active and passive ROM, and hydration are also appropriate nursing measures. The client shouldn't lie as still as possible, to prevent contractures, or limit his fluid intake.
NP: Implementation; CN: Physiological integrity; CNS: Basic care and comfort; CL: Application

21. 2. Osteoarthritis is a localized disease, rheumatoid arthritis is systemic. Osteoarthritis isn't gender-specific, but rheumatoid arthritis is. Clients have dislocations and subluxations in both disorders.
NP: Data collection; CN: Physiological integrity; CNS: Physiological adaptation; CL: Knowledge

22. 3. Asymmetrical joint involvement is present in osteoarthritis. Multiple subcutaneous nodules, elevated sedimentation rate, and such signs of inflammation as heat, fever, and malaise are all present in rheumatoid arthritis.
NP: Data collection; CN: Physiological integrity; CNS: Physiological adaptation; CL: Comprehension

23. Which of the following instructions would be considered primary prevention of injury from osteoarthritis?
1. Stay on bed rest.
2. Avoid physical activity.
3. Perform only repetitive tasks.
4. Warm up before exercise and avoid repetitive tasks.

24. A client with osteoarthritis wants to know what it is. Client teaching would include which of the following descriptions for osteoarthritis?
1. A systemic inflammatory joint disease
2. A disease involving fusion of the joints in the hands
3. An inflammatory joint disease, with degeneration and loss of articular cartilage in synovial joints
4. A noninflammatory joint disease, with degeneration and loss of articular cartilage in synovial joints

25. Use of which of the following articles or types of clothing would help a client with osteoarthritis perform activities of daily living at home?
1. Zippered clothing
2. Tied shoes to promote stability
3. Velcro clothing, slip-on shoes, and rubber grippers
4. Buttoned clothing, slip-on shoes, and rubber grippers

26. A client with osteoarthritis is refusing to perform her own daily care. Which of the following approaches would be most appropriate to use with this client?
1. Perform the care for the client.
2. Explain that she needs to maintain complete independence.
3. Encourage her to perform as much care as her pain will allow.
4. Tell her that once she's completed her care, she'll receive her pain medication.

Read carefully to understand what *primary* means in this question.

You've finished 25 questions already!

What would be the best way to help the client?

23. 4. Primary prevention of injury from osteoarthritis includes warming up and avoiding repetitive tasks. Physical activity is important to remain fit and healthy and to maintain joint function. Bed rest would contribute to many other systemic complications.
NP: Implementation; CN: Health promotion and maintenance; CNS: Prevention and early detection of disease; CL: Application

24. 4. Osteoarthritis is a noninflammatory joint disease, with degeneration and loss of articular cartilage in synovial joints. Rheumatoid arthritis is a systemic inflammatory joint disease. Arthrodesis is fusion of the joints.
NP: Implementation; CN: Physiological integrity; CNS: Physiological adaptation; CL: Knowledge

25. 3. Velcro clothing, slip-on shoes, and rubber grippers make it easier for the client to dress and grip objects. Zippers, ties, and buttons may be difficult for the client to use.
NP: Implementation; CN: Physiological integrity; CNS: Basic care and comfort; CL: Application

26. 3. A client with osteoarthritis should be encouraged to perform as much of her care as she's able to. The nurse's goal should be to allow her to maintain her self-care abilities with help as needed. It's never appropriate to use pain medication as a bargaining tool.
NP: Implementation; CN: Psychosocial integrity; CNS: Coping and adaptation; CL: Application

NP: Nursing process   CN: Client needs category   CNS: Client needs subcategory   CL: Cognitive level

27. Clients in the late stages of osteoarthritis commonly use which of the following terms to describe joint pain?
1. Grating
2. Dull ache
3. Deep aching pain
4. Deep aching, relieved with rest

Ask your client about the pain; you don't need a crystal ball.

27. 1. In the late stages of osteoarthritis, the client typically describes joint pain as grating. As the disease progresses, the cartilage covering the ends of bones is destroyed and bones rub against each other. Osteophytes, or bone spurs, may also form on the ends of bones. A dull ache and deep aching pain with or without relief with rest is usually seen in the earlier stages of osteoarthritis.
NP: Data collection; CN: Physiological integrity; CNS: Physiological adaptation; CL: Analysis

28. A client uses a cane for assistance in walking. Which of the following statements is true about a cane or other assistive devices?
1. A walker is a better choice than a cane.
2. The cane should be used on the affected side.
3. The cane should be used on the unaffected side.
4. A client with osteoarthritis should be encouraged to ambulate without the cane.

28. 3. A cane should be used on the unaffected side. A client with osteoarthritis should be encouraged to ambulate with a cane, walker, or other assistive device as needed; their use takes weight and stress off joints.
NP: Implementation; CN: Physiological integrity; CNS: Physiological adaptation; CL: Application

29. Which of the following instructions about activity should be given to a client with osteoarthritis after he returns home?
1. Learn to pace activity.
2. Remain as sedentary as possible.
3. Return to a normal level of activity.
4. Include vigorous exercise in your daily routine.

Here are those key words again: *most appropriate*. Some interventions may fit, but only one is most appropriate.

29. 1. A client with osteoarthritis should pace his activities and avoid overexertion. Overexertion can increase degeneration and cause pain. The client shouldn't become sedentary because he'll have a high risk of pneumonia and contractures.
NP: Planning; CN: Physiological integrity; CNS: Physiological adaptation; CL: Application

30. A client was prescribed an anti-inflammatory drug for osteoarthritis 5 days ago. She says the pain has decreased some but not completely. Which of the following nursing interventions would be the most appropriate?
1. Continue the present dose and offer other pain measures.
2. Notify the physician and suggest increasing the dose.
3. Notify the physician and suggest stopping the medication.
4. Notify the physician and suggest adding another medication.

30. 1. Anti-inflammatory medications may take up to 2 to 3 weeks for full benefits to be appreciated. If the client can tolerate the pain, continue on the present medication and offer other pain measures, such as rest, massage, heat, or cold.
NP: Implementation; CN: Physiological integrity; CNS: Pharmacological therapies; CL: Comprehension

31. A client is diagnosed with a herniated nucleus pulposus (HNP), or herniated disk. Which of the following statements about a herniated disk is correct?
1. The disk slips out of alignment.
2. The disk shatters and fragments place pressure on nerve roots.
3. The nucleus tissue itself remains centralized, and the surrounding tissue is displaced.
4. The nucleus of the disk puts pressure on the anulus, causing pressure on the nerve root.

Keep up the good work!

31. 4. With an HNP, or herniated disk, the nucleus of the disk puts pressure on the anulus, causing pressure on the nerve root. The disk itself doesn't slip, rupture, or shatter. The nucleus tissue usually moves from the center of the disk.
NP: Implementation; CN: Physiological integrity; CNS: Physiological adaptation; CL: Knowledge

32. Which of the following symptoms would occur with a herniated nucleus pulposus (HNP)?
1. Pain, numbness, weakness, leg or foot pain
2. Shortness of breath, weakness, leg or foot pain
3. Inability to walk, shortness of breath, weakness, leg or foot pain
4. Unstable gait, disorientation, numbness, weakness, leg or foot pain

32. 1. Common symptoms of an HNP are pain, numbness, weakness, and leg or foot pain.
NP: Data collection; CN: Physiological integrity; CNS: Physiological adaptation; CL: Comprehension

33. Conservative treatment of a herniated nucleus pulposus (HNP) would include which of the following measures?
1. Surgery
2. Bone fusion
3. Bed rest, pain medication, physiotherapy
4. Strenuous exercise, pain medication, physiotherapy

The word conservative is a clue to the right answer.

33. 3. Conservative treatment of an HNP may include bed rest, pain medication, and physiotherapy. Aggressive treatment may include surgery, such as a bone fusion.
NP: Implementation; CN: Physiological integrity; CNS: Reduction of risk potential; CL: Application

34. Closed spine surgery is a new technique to fix a herniated disk. Which of the following statements is true about closed spine surgery?
1. There's a greater associated risk.
2. Intense physical therapy is needed.
3. An endoscope is used to perform the surgery.
4. Recovery time is longer than with open spine surgery.

34. 3. Closed spine surgery uses endoscopy to fix a herniated disk. It's less risky than open surgery and has a shorter recovery time; it's commonly done as a same-day surgical procedure. Physical therapy may be less intensive or not needed at all.
NP: Implementation; CN: Physiological integrity; CNS: Physiological adaptation; CL: Knowledge

35. Which of the following descriptions best identifies the position of an intervertebral disk?
1. Encloses the anulus fibrosus
2. Surrounds the nucleus pulposus
3. Located between the vertebrae and the spinal column
4. Located between spinal nerves in the vertebral column

36. Herniation of a vertebral disk can occur under which of the following conditions?
1. Major trauma or stress
2. Minor trauma or stress
3. With a history of back problems
4. Either major or minor trauma or stress

37. Which of the following areas of vertebral herniation is the most common?
1. L1-L2, L4-L5
2. L1-L2, L5-S1
3. L4-L5, L5-S1
4. L5-S1, S2-S3

38. Which of the following tests may be used to diagnose a herniated nucleus pulposus?
1. Chest X-ray, magnetic resonance imaging (MRI), computed tomography (CT) scan
2. Lumbar puncture, chest X-ray, MRI, CT scan
3. Lumbar puncture, chest X-ray, myelography
4. Myelography, MRI, CT scan

39. Which of the following instructions would be included when teaching a client how to protect his back?
1. Sleep on your side, and carry objects at arm's length.
2. Sleep on your back, and carry objects at arm's length.
3. Sleep on your side, and carry objects close to your body.
4. Sleep on your back, and carry objects close to your body.

Now where was that disk?

35. 3. Intervertebral disks are between the vertebrae and the spinal column. The other answers are incorrect descriptions of vertebral disks.
NP: Data collection; CN: Physiological integrity; CNS: Physiological adaptation; CL: Knowledge

36. 4. Herniation of a vertebral disk can occur with either major or minor trauma or stress. It can occur in anyone regardless of history.
NP: Data collection; CN: Physiological integrity; CNS: Physiological adaptation; CL: Knowledge

37. 3. The most common areas of herniation are L4-L5, L5-S1.
NP: Data collection; CN: Physiological integrity; CNS: Physiological adaptation; CL: Knowledge

38. 4. Tests used to diagnose a herniated nucleus pulposus include myelography, MRI, and CT scan. Chest X-ray and lumbar puncture aren't conclusive.
NP: Data collection; CN: Physiological integrity; CNS: Physiological adaptation; CL: Knowledge

Teach! Teach! Teach!

39. 3. By sleeping on the side and carrying objects close to the body, there's less strain on the back. Sleeping on the back and carrying objects at arm's length adds pressure to the back.
NP: Implementation; CN: Health promotion and maintenance; CNS: Prevention and early detection of disease; CL: Knowledge

40. Skeletal muscle relaxants may be used in the acute treatment of a herniated nucleus pulposus. Which of the following instructions should be included in client teaching?
1. Change your position quickly to avoid dizziness.
2. Double a missed dose to ensure proper muscle relaxation.
3. Cough and cold medications are appropriate to take if needed.
4. Avoid activities that require alertness; muscle relaxants can cause drowsiness.

41. A client with a recent fracture is suspected of having compartment syndrome. Data findings may include which of the following symptoms?
1. Body-wide decrease in bone mass
2. A growth in and around the bone tissue
3. Inability to perform active movement, pain with passive movement
4. Inability to perform passive movement, pain with active movement

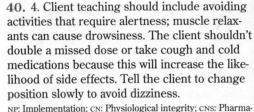

You're great at data collection!

42. Compartment syndrome occurs under which of the following conditions?
1. Increase in scar tissue
2. Increase in bone mass
3. Decrease in bone mass
4. Hemorrhage into the muscle

You've got it!

43. The hemorrhage that occurs in compartment syndrome causes which of the following symptoms?
1. Edema
2. Decreased venous pressure
3. Increased venous circulation
4. Increased arterial circulation

44. In compartment syndrome, how long would it take for tissue death to occur?
1. 2 to 4 hours
2. 6 to 8 hours
3. 24 hours
4. 72 hours

40. 4. Client teaching should include avoiding activities that require alertness; muscle relaxants can cause drowsiness. The client shouldn't double a missed dose or take cough and cold medications because this will increase the likelihood of side effects. Tell the client to change position slowly to avoid dizziness.
NP: Implementation; CN: Physiological integrity; CNS: Pharmacological therapies; CL: Comprehension

41. 3. With compartment syndrome, the client is unable to perform active movement, and pain occurs with passive movement. A bone tumor shows growth in and around the bone tissue. Osteoporosis has a body-wide decrease in bone mass.
NP: Data collection; CN: Physiological integrity; CNS: Physiological adaptation; CL: Application

42. 4. Compartment syndrome occurs when pressure within the muscle compartment, resulting from edema or bleeding, increases to the point of interfering with circulation. Crush injuries, burns, bites, and fractures requiring casts or dressings may cause this syndrome. It isn't a result of scar tissue or an increase or decrease in bone mass.
NP: Data collection; CN: Physiological integrity; CNS: Physiological adaptation; CL: Knowledge

43. 1. The hemorrhage in compartment syndrome causes edema, increased venous pressure, and decreased venous and arterial circulation.
NP: Data collection;; CN: Physiological integrity; CNS: Physiological adaptation; CL: Comprehension

44. 1. Tissue death can occur in 2 to 4 hours in compartment syndrome.
NP: Data collection; CN: Physiological integrity; CNS: Physiological adaptation; CL: Knowledge

45. Treatment of compartment syndrome includes which of the following measures?
1. Amputation
2. Casting
3. Fasciotomy
4. Observation, no treatment is necessary

Here's a question your client with a similar problem is bound to ask.

45. 3. Treatment of compartment syndrome includes fasciotomy. A fasciotomy involves cutting the fascia over the affected area to permit muscle expansion. Casting and amputation aren't treatments for compartment syndrome.
NP: Implementation; CN: Physiological integrity; CNS: Physiological adaptation; CL: Comprehension

46. A client is admitted to the emergency department with a foot fracture. Which of the following reasons explains why the foot is placed in a brace?
1. To act as a splint
2. To prevent infection
3. To allow for movement
4. To encourage direct contact

46. 1. The purpose of the brace is to act as a splint, prevent direct contact, and maintain immobility. A brace doesn't prevent infection.
NP: Implementation; CN: Physiological integrity; CNS: Reduction of risk potential; CL: Comprehension

47. After treatment of compartment syndrome, a client reports experiencing paresthesia. Which of the following symptoms would be seen with paresthesia?
1. Fever and chills
2. Change in range of motion (ROM)
3. Pain and blanching
4. Numbness and tingling

47. 4. Paresthesia is described as numbness and tingling. It isn't described as pain or blanching or associated with fever and chills or change in ROM.
NP: Data collection; CN: Physiological integrity; CNS: Physiological adaptation; CL: Comprehension

48. Which of the following characteristics of the fascia can cause it to develop compartment syndrome?
1. It's highly flexible.
2. It's fragile and weak.
3. It's unable to expand.
4. It's the only tissue within the compartment.

48. 3. Compartment syndrome occurs because the fascia can't expand. It isn't flexible or weak. The compartment contains blood vessels and nerves.
NP: Data collection; CN: Physiological integrity; CNS: Physiological adaptation; CL: Comprehension

49. Compartment syndrome can occur from which of the following conditions?
1. Internal pressure
2. External pressure
3. Increased blood pressure
4. Internal and external pressure

Hint! Hint! Hint!

49. 4. Compartment syndrome can occur from internal (bleeding) and external pressure (cast or dressing). Blood pressure isn't a cause of compartment syndrome.
NP: Data collection; CN: Physiological integrity; CNS: Physiological adaptation; CL: Knowledge

50. Which of the following symptoms is an early sign of compartment syndrome?
1. Heat
2. Paresthesia
3. Skin pallor
4. Swelling

50. 2. Paresthesia is the earliest sign of compartment syndrome. Pain, swelling, and heat are also signs but occur after paresthesia.
NP: Data collection; CN: Physiological integrity; CNS: Physiological adaptation; CL: Knowledge

51. Which of the following symptoms are considered signs of a fracture?
1. Tingling, coolness, loss of pulses
2. Loss of sensation, redness, warmth
3. Coolness, redness, new site of pain
4. Redness, warmth, pain at the site of injury

52. Which of the following areas would be included in a neurovascular assessment?
1. Orientation, movement, pulses, warmth
2. Capillary refill, movement, pulses, warmth
3. Orientation, pupillary response, temperature, pulses
4. Respiratory pattern, orientation, pulses, temperature

53. Which of the following methods is the correct way to measure limb circumference?
1. Use a measuring tape.
2. Visually compare limbs bilaterally.
3. Check the client's medical history.
4. Follow the standardized chart for limb circumference.

54. If pulses aren't palpable, which of the following interventions should be performed first?
1. Check again in 1 hour.
2. Alert the nurse in charge immediately.
3. Verify the findings with Doppler ultrasonography.
4. Alert the physician immediately.

55. A client describes a foul odor from his cast. Which of the following responses or interventions would be the most appropriate?
1. Assess further because this may be a sign of infection.
2. Teach him proper cast care, including hygiene measures.
3. This is normal, especially when a cast is in place for a few weeks.
4. Assess further because this may be a sign of neurovascular compromise.

Don't take a wrong turn in your data collection. Know the signs to look out for.

Question 54 asks you to prioritize your response.

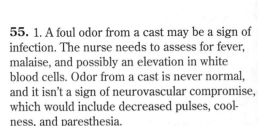

51. 4. Signs of a fracture may include redness, warmth, and new site of pain. Coolness, tingling, and loss of pulses are signs of a vascular problem.
NP: Data collection; CN: Physiological integrity; CNS: Physiological adaptation; CL: Application

52. 2. A correct neurovascular assessment should include capillary refill, movement, pulses, and warmth. Neurovascular assessment involves nerve and blood supply to an area. Respiratory pattern, orientation, temperature, and pupillary response aren't part of a neurovascular examination.
NP: Data collection; CN: Physiological integrity; CNS: Physiological adaptation; CL: Application

53. 1. The right way to assess limb circumference is to use a measuring tape. Visual inspection and checking the past history are unreliable. There isn't a standardized chart for limb circumference.
NP: Data collection; CN: Health promotion and maintenance; CNS: Prevention and early detection of disease; CL: Knowledge

54. 3. If pulses aren't palpable, verify the assessment with Doppler ultrasonography. If pulses can't be found with Doppler ultrasonography, immediately notify the physician.
NP: Implementation; CN: Physiological integrity; CNS: Physiological adaptation; CL: Application

55. 1. A foul odor from a cast may be a sign of infection. The nurse needs to assess for fever, malaise, and possibly an elevation in white blood cells. Odor from a cast is never normal, and it isn't a sign of neurovascular compromise, which would include decreased pulses, coolness, and paresthesia.
NP: Implementation; CN: Health promotion and maintenance; CNS: Prevention and early detection of disease; CL: Analysis

NP: Nursing process   CN: Client needs category   CNS: Client needs subcategory   CL: Cognitive level

56. To reduce the roughness of a cast, which of the following measures should be used?
1. Petal the edges.
2. Elevate the limb.
3. Break off the rough area.
4. Distribute pressure evenly.

57. Elevating a limb with a cast will prevent swelling. Which of the following actions best describes how this is done?
1. Place the limb with the cast close to the body.
2. Place the limb with the cast at the level of the heart.
3. Place the limb with the cast below the level of the heart.
4. Place the limb with the cast above the level of the heart.

58. A client asks why a cast can't get wet. Which of the following responses would be the most appropriate?
1. A wet cast can cause a foul odor.
2. A wet cast will weaken or be destroyed.
3. A wet cast is heavy and difficult to maneuver.
4. It's all right to get the cast wet, just use a hair dryer to dry it off.

59. A male client with a fractured femur is in Russell's traction. He asks the nurse to help him with back care. Which nursing action is most appropriate?
1. Tell the client that he can't have back care while he's in traction.
2. Remove the weight to give the client more slack to move.
3. Support the weight to give the client more slack to move.
4. Tell the client to use the trapeze to lift his back off the bed.

60. Which of the following fractures is classic for occurring from trauma?
1. Brachial and clavicle
2. Brachial and humerus
3. Humerus and clavicle
4. Occipital and humerus

You've finished more than half these questions! Hooray!

Your client depends on you to know the most appropriate answers.

The word *classic* clues you in to a commonly occurring sign, symptom, or event.

56. 1. To reduce the roughness of the cast, petal the edges. Elevating the limb will prevent swelling. Distributing pressure evenly will prevent pressure ulcers. Never break a rough area off the cast.
NP: Implementation; CN: Physiological integrity; CNS: Basic care and comfort; CL: Application

57. 4. To reduce swelling, place the limb with the cast above the level of the heart. Placing it below or at the level of the heart won't reduce swelling. To elevate a cast, the limb may need to be extended from the body.
NP: Implementation; CN: Physiological integrity; CNS: Physiological adaptation; CL: Application

58. 2. A wet cast will weaken or be destroyed. A foul odor is a sign of infection. It's never all right to get a cast wet.
NP: Implementation; CN: Physiological integrity; CNS: Physiological adaptation; CL: Knowledge

59. 4. The traction must not be disturbed to maintain correct alignment. Therefore, the client should use the trapeze to lift his back off of the bed. The client can have back care as long as he uses the trapeze and doesn't disturb the alignment. The weight shouldn't be moved without a physician's order. The weight should hang freely without touching anything.
NP: Implementation; CN: Physiological integrity; CNS: Reduction of risk potential; CL: Application

60. 3. Classic fractures that occur with trauma are those of the humerus and clavicle. The brachial and occipital bones aren't usually involved in a traumatic delivery.
NP: Data collection; CN: Physiological integrity; CNS: Physiological adaptation; CL: Knowledge

61. Which of the following characteristics applies to a closed fracture?
1. Extensive tissue damage
2. Increased risk of infection
3. Same as for a compound fracture
4. Intact skin over the fracture site

62. A client is put in traction before surgery. Which of the following reasons for the traction is correct?
1. To prevent skin breakdown
2. To aid in turning the client
3. To help the client become active
4. To prevent trauma and overcomes muscle spasms

63. When the fracture line is straight across the bone, the fracture is known as which of the following types?
1. Linear
2. Longitudinal
3. Oblique
4. Transverse

64. Which of the following fractures commonly occurs with such bone diseases as osteomalacia and Paget's disease?
1. Linear
2. Longitudinal
3. Oblique
4. Transverse

65. Which of the following fractures is commonly seen in the upper extremities and is related to physical abuse?
1. Longitudinal
2. Oblique
3. Spiral
4. Transverse

66. Which of the following mechanisms or conditions causes healing of a fracture?
1. Scar tissue
2. Displacement
3. Necrotic tissue formation
4. Formation of new bone tissue

You're doing great!

Think about cause and location.

61. 4. A closed fracture maintains intact skin over the fracture site. An open fracture has extensive tissue damage, an increased risk of infection, and is also known as a compound fracture
NP: Data collection; CN: Physiological integrity; CNS: Physiological adaptation; CL: Knowledge

62. 4. Traction prevents trauma and overcomes muscle spasms. Traction doesn't help in turning the client, prevent skin breakdown, or help the client become active.
NP: Implementation; CN: Physiological integrity; CNS: Basic care and comfort; CL: Knowledge

63. 4. A fracture line straight across the bone is called a transverse fracture. A linear fracture has an intact fracture line. A fracture line at a 45-degree angle to the shaft of the bone is an oblique fracture. A longitudinal fracture has a longitudinal fracture line.
NP: Data collection; CN: Physiological integrity; CNS: Physiological adaptation; CL: Knowledge

64. 4. A transverse fracture commonly occurs with such bone diseases as osteomalacia and Paget's disease. Linear, oblique, and longitudinal fractures generally occur with trauma.
NP: Data collection; CN: Physiological integrity; CNS: Physiological adaptation; CL: Knowledge

65. 3. Spiral fractures are commonly seen in the upper extremities and are related to physical abuse. Oblique and longitudinal fractures generally occur with trauma. A transverse fracture commonly occurs with such bone diseases as osteomalacia and Paget's disease.
NP: Data collection; CN: Physiological integrity; CNS: Physiological adaptation; CL: Knowledge

66. 4. Healing of a fracture occurs by the formation of new bone tissue. Bone doesn't heal by forming scar tissue or necrotic tissue or by displacement.
NP: Data collection; CN: Physiological integrity; CNS: Physiological adaptation; CL: Knowledge

67. A 20-year-old male client has just had a plaster cast applied to his right forearm following reduction of a closed radius fracture due to an in-line skating accident. It's most important for the nurse to check which of the following?

1. Whether the cast is completely dry
2. Sensation and movement of the fingers
3. Whether the client is having any pain
4. Whether the cast needs pedaling

The words most important are a clue that should help you select the answer.

67. 2. Neurovascular checks are most important because they're used to determine if any impairment exists after cast application and reduction of the fracture. Checking to see if the cast is completely dry isn't the nurse's highest priority. Checking to see if the client has pain is important but not the highest priority. Pedaling to smooth cast edge is done when the cast is completely dry.

NP: Data collection; CN: Physiological integrity; CNS: Reduction of risk potential; CL: Application

68. Which of the following serious complications can occur with long bone fractures?

1. Bone emboli
2. Fat emboli
3. Platelet emboli
4. Serous emboli

68. 2. A serious complication of long bone fractures is the development of fat emboli. Platelet or bone emboli are rare occurrences. There aren't emboli known as serous emboli.

NP: Data collection; CN: Physiological integrity; CNS: Physiological adaptation; CL: Knowledge

69. Which of the following signs and symptoms can occur with fat emboli?

1. Tachypnea, tachycardia, shortness of breath, paresthesia
2. Paresthesia, bradypnea, bradycardia, petechial rash on chest and neck
3. Bradypnea, bradycardia, shortness of breath, petechial rash on chest and neck
4. Tachypnea, tachycardia, shortness of breath, petechial rash on chest and neck

Concentrate. Lists are tricky!

69. 4. Signs and symptoms of fat emboli include tachypnea, tachycardia, shortness of breath, and a petechial rash on the chest and neck. The fat molecules enter the venous circulation and travel to the lung, obstructing pulmonary circulation. Bradycardia, bradypnea, and paresthesia aren't usual symptoms.

NP: Data collection CN: Health promotion and maintenance; CNS: Prevention and early detection of disease; CL: Knowledge

70. The community health nurse found an elderly female client lying in the snow, unable to move her right leg because of a fracture. What's the nurse's first priority?

1. Realign the fracture ends.
2. Reduce the fracture.
3. Immobilize the fracture in its present position.
4. Elevate the leg on whatever is available.

Why has a high-protein diet been ordered for the client in question 71?

70. 3. Initial treatment of obvious and suspected fractures includes immobilizing and splinting the limb. Any attempt to realign or rest the fracture at the site may cause further injury and complications. The leg may be elevated only after immobilization.

NP: Implementation; CN: Physiological integrity CNS: Physiological adaptation; CL: Application

71. A high-protein diet is ordered for a client recovering from a fracture. High protein is ordered for which of the following reasons?

1. Protein promotes gluconeogenesis.
2. Protein has anti-inflammatory properties.
3. Protein promotes cell growth and bone union.
4. Protein decreases pain medication requirements.

71. 3. High-protein intake promotes cell growth and bone union. Protein doesn't decrease pain medication requirements, exert anti-inflammatory properties, or promote gluconeogenesis.

NP: Implementation; CN: Physiological integrity; CNS: Basic care and comfort; CL: Application

72. The nurse is instructing a nursing assistant on the proper care of a client in Buck's extension traction following a fracture of his left fibula. Which observation indicates that the teaching was effective?
   1. The weights are allowed to hang freely over the end of the bed.
   2. The nursing assistant lifts the weights when assisting the client to move up in bed.
   3. The leg in traction is kept externally rotated.
   4. The nursing assistant instructs the client to perform ankle rotation exercises.

73. Which of the following types of traction is used for children who weigh under 35 lb (15.9 kg)?
   1. Bryant's traction
   2. Buck's traction
   3. Pelvic belt
   4. Russell traction

74. The nurse is instructing a nursing assistant on how to properly position a 45-year-old male client who underwent total hip replacement. The nurse explains that the client's hip needs to be in which of the following positions?
   1. Straight with the knee flexed
   2. In an abducted position
   3. In an adducted position
   4. Externally rotated

75. A 51-year-old client has undergone a total hip replacement on her right side. After surgery, the nurse should turn the client how often?
   1. Every 1 to 2 hours, from the unaffected side to the back
   2. Every 1 to 2 hours, from the affected side to the back
   3. Every 4 to 6 hours, from the unaffected side to the back
   4. Every 4 to 6 hours, from the affected side to the back

Wow! You've finished 72 questions! Outstanding!

They say location is everything in real estate. Well, position can be everything when caring for a client with a fracture.

72. 1. In Buck's extension traction, the weights should hang freely without touching the bed or floor. Lifting the weights would break the traction. The client should be moved up in bed, allowing the weights to move freely along with the client. The leg should be kept in straight alignment. Performing ankle rotation exercises could cause the leg to go out of alignment.

NP: Evaluation; CN: Physiological integrity; CNS: Basic care and comfort; CL: Knowledge

73. 1. Bryant's traction is used for children who weigh less than 35 lb. A pelvic belt is conservative treatment for low back pain. Russell's traction is used for leg traction. Buck's traction is used for leg or arm traction.

NP: Implementation; CN: Physiological integrity; CNS: Physiological adaptation; CL: Knowledge

74. 2. An abducted position keeps the new joint from becoming displaced out of the socket. The client can keep his hip straight with the knee flexed as long as an abductor pillow is kept in place. Keeping the hip adducted or externally rotated can dislocate the hip joint.

NP: Implementation; CN: Physiological integrity; CNS: Basic care and comfort; CL: Application

75. 1. The client should be turned at least every 2 hours and always from the unaffected side to the back. The client should never be placed on the affected side. Turning the client every 4 to 6 hours places her at greater risk for skin breakdown.

NP: Implementation; CN: Physiological integrity; CNS: Reduction of risk potential; CL: Application

76. Which of the following interventions would help prevent deep vein thrombosis (DVT) after hip surgery?
1. Bed rest
2. Egg crate mattress
3. Vigorous pulmonary care
4. Subcutaneous heparin and pneumatic compression boots

Wow! You've finished 76 questions! Outstanding!

76. 4. To prevent DVT after hip surgery, subcutaneous heparin and pneumatic compression boots are used. Egg crate mattresses and pulmonary care don't prevent DVT. Bed rest can cause DVT.
NP: Implementation; CN: Health promotion and maintenance; CNS: Prevention and early detection of disease; CL: Application

77. The nurse is caring for a client who is on complete bed rest following complete hip replacement. In an effort to reduce sensory deprivation, the nursing assistant should be instructed to do which of the following?
1. Provide mouth care before meals.
2. Monitor the client's urine output every 2 hours.
3. Check bilateral hand grasps every 4 hours.
4. Orient the client to date and time frequently.

77. 1. Cleaning the mouth before meals enhances sensual stimuli and taste bud function. Checking urine output doesn't affect sensory input. Checking hand grasps and orienting the client would help to assess and stimulate neurologic deficits.
NP: Implementation; CN: Safe, effective care environment; CNS: Coordinated care; CL: Application

78. A client's left leg is in skeletal traction with a Thomas leg splint and Pearson attachment. Which intervention should the nurse include in this client's plan of care?
1. Apply the traction straps snugly.
2. Assess the client's level of consciousness.
3. Remove the traction at least every 8 hours.
4. Teach the client how to prevent problems caused by immobility.

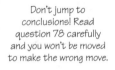

Don't jump to conclusions! Read question 78 carefully and you won't be moved to make the wrong move.

78. 4. By teaching the client about prevention measures, the nurse can help avoid problems caused by immobility, such as hypostatic pneumonia, muscle contracture, and atrophy. The nurse applies traction straps for skin traction, not skeletal traction. For a client in skeletal traction, the nurse should assess the affected limb, rather than assessing the level of consciousness. Removal of skeletal traction is the physician's responsibility, not the nurse's.
NP: Planning; CN: Physiological integrity; CNS: Reduction of risk potential; CL: Comprehension

79. Which of the following nursing interventions is appropriate for a client in traction?
1. Assess the pin sites every shift and as needed.
2. Add and remove weights as the client wants.
3. Make sure the knots in the rope catch on the pulley.
4. Give range of motion (ROM) to all joints, including those immediately proximal and distal to the fracture, every shift.

79. 1. Nursing care for a client in traction may include assessing pin sites every shift and as needed and making sure the knots in the rope don't catch on the pulley. Add and remove weights as the physician orders, and give ROM to all joints except those immediately proximal and distal to the fracture every shift.
NP: Implementation; CN: Physiological integrity; CNS: Basic care and comfort; CL: Application

80. A client with a right hip fracture is complaining of left-sided leg pain and edema and has a positive Homans' sign. Which of the following conditions would show those symptoms?

1. Deep vein thrombosis (DVT)
2. Fat emboli
3. Infection
4. Pulmonary embolism

80. 1. Unilateral leg pain and edema with a positive Homans' sign (not always present) might be symptoms of DVT. Tachycardia, chest pain, and shortness of breath may be symptoms of a pulmonary embolism. It's unlikely an infection would occur on the opposite side of the fracture without cause. Symptoms of fat emboli include restlessness, tachypnea, and tachycardia and are more common in long-bone injuries.

NP: Implementation; CN: Physiological integrity; CNS: Reduction of risk potential; CL: Knowledge

81. A female client who fell while washing her outside windows has a fractured right ankle and is being fitted with a cast. After assisting with the cast application, what instructions should the nurse give the client?

1. Go home and stay in bed for about 5 days.
2. Keep the cast covered with plastic until it feels dry.
3. Move the right toes for several minutes every hour.
4. Expect some swelling and blueness of the toes.

81. 3. Moving the toes is encouraged to facilitate circulation and prevent swelling. By moving the toes, the client will be aware of any numbness or swelling and can take appropriate action, such as elevating the extremity and reporting the findings to her physician. Usually, clients are instructed to remain in bed for 24 hours while the cast dries. Prolonged immobility creates problems for the client. While the cast is still damp, the ankle should be elevated on a pillow that is protected with plastic; the cast itself should be left open to the air. Swelling and a bluish color of the toes indicate compromised circulation and should be reported immediately.

NP: Implementation; CN: Physiological integrity CNS: Reduction of risk potential; CL: Application

82. Synthetic casts take approximately how long to set?

1. Immediately
2. 20 minutes
3. 45 minutes
4. 2 hours

So, when can I sign your cast?

82. 2. Synthetic casts take about 20 minutes to set.

NP: Implementation; CN: Physiological integrity; CNS: Reduction of risk potential; CL: Knowledge

83. As a cast is drying, a client complains of heat from the cast. Which of the following interventions is the most appropriate?

1. Remove the cast immediately.
2. Explain this is a normal sensation.
3. Notify the physician.
4. Assess the client for other signs of infection.

83. 2. Normally as the cast dries, a client may complain of heat from the cast. Offer reassurance. Don't notify the physician or remove the cast. Heat from the cast isn't a sign of infection.

NP: Implementation; Physiological integrity; CNS: Reduction of risk potential; CL: Comprehension

84. To prevent footdrop in a leg with a cast, which of the following interventions is appropriate?

1. Encourage bed rest.
2. Support the foot with 45 degrees of flexion.
3. Support the foot with 90 degrees of flexion.
4. Place a stocking on it to provide warmth.

85. The physician has just removed the cast from a 20-year-old male client's lower leg. During the removal, a small superficial abrasion occurred over the ankle. Which statement by the client indicates the need for additional client teaching?

1. "I can use a moisturizing lotion on the dry areas."
2. "The dry, peeling skin will go away by itself."
3. "I can wash the abrasion on my ankle with soap and water."
4. "I'll wait until the abrasion is healed before I go swim."

86. Which information is critical to include in the discharge plans for a client leaving the hospital in a leg cast?

1. Cast care procedures and devices to relieve itching
2. Skin care, mouth care, and cast removal procedures
3. Cast care, neurovascular checks, and hygiene measures
4. Cast removal procedures, neurovascular checks, and devices to relieve itching

The word *prevent* is a hint. Look for a preventive measure, not a treatment.

You're almost finished!

84. 3. To prevent footdrop in a leg with a cast, the foot should be supported with 90 degrees of flexion. Bed rest can cause footdrop. Keeping the extremity warm won't prevent footdrop.
NP: Implementation; CN: Health promotion and maintenance; CNS: Prevention and early detection of disease; CL: Comprehension

85. 1. The dry, peeling skin will heal in a few days with normal cleaning; therefore, lotions are unnecessary. Vigorous scrubbing isn't necessary. Washing the abrasion and delaying swimming until healing are correct procedures to follow after removal of a cast.
NP: Evaluation; CN: Physiological integrity; CNS: Reduction of risk potential; CL: Application

86. 3. Proper cast care procedures include observing the skin nearest the cast edges for signs of pressure ulcers, keeping the cast dry and intact, and avoiding the use of insertable devices (such as wire hangers or sticks) to relieve itching. Frequent neurovascular checks can reveal evidence of pressure or impaired circulation to the leg under the cast. This includes checking the toes frequently for discoloration, swelling, or lack of movement or sensation. Hygiene measures should focus on the client's normal elimination patterns and the importance of cleanliness after elimination as well as on the need to maintain skin integrity by taking sponge baths and caring for dry skin. Devices should never be inserted between the cast and the skin. Although mouth care and cast removal are important issues, they aren't priority discharge instructions in this case.
NP: Implementation; CN: Physiological integrity; CNS: Physiological adaptation; CL: Application

87. Which of the following forms new tissue after an injury instead of scar tissue?
1. Bone
2. Brain
3. Kidney
4. Liver

88. The nurse is caring for a client with a cast on his left arm. Which assessment finding is most significant for this client?
1. Normal capillary refill in the great toe
2. Presence of a normal popliteal pulse
3. Intact skin around the cast edges
4. Ability to move all toes

I'll be a good host of this quiz show and point your attention toward the words most significant.

89. Which of the following statements explains an open reduction of a fractured femur?
1. Traction will be used.
2. A cast will be applied.
3. Crutches will be used after surgery.
4. Some form of screw, plate, nail, or wire is usually used to maintain alignment.

Only seven more questions. You can do it!

90. A client has just returned from the post-anesthesia care unit after undergoing internal fixation of a fractured left femoral neck. The nurse should place the client in which position?
1. On the left side with the right knee bent
2. On the back with two pillows between the legs
3. On the right side with the left knee bent
4. Sitting at a 90-degree angle

87. 1. Bone tissue regenerates; all other tissues form scar tissue after injury.
NP: Data collection; CN: Physiological integrity; CNS: Physiological adaptation; CL: Knowledge

88. 3. Because a cast can irritate the skin, the nurse should inspect for this complication. Usually, the skin remains intact around the cast edges. Normally capillary refill in the left thumb and fingers is more significant than in the great toe because an arm cast can impair circulation in the affected arm. Similarly, the presence of a normal radial pulse is more noteworthy than popliteal pulse because a cast on the left arm could affect circulation in that limb. Movement of this client's fingers is more critical than movement of the toes because a left arm cast could compress a nerve, preventing movement of fingers in the left arm.
NP: Data collection; CN: Physiological integrity; CNS: Physiological adaptation; CL: Application

89. 4. Open reduction means that the tissue must be surgically opened and the fractured bones realigned. To maintain proper alignment, a screw, plate, nail, or wire is inserted to prevent the bones from separating. Although traction may have been used before surgery, it won't be needed any longer once the fracture is reduced. Crutches or a cast may be used after surgery, but the question asks specifically about the surgical procedure.
NP: Implementation; CN: Physiological integrity; CNS: Physiological adaptation; CL: Application

90. 2. The operative leg must be kept abducted to prevent dislocation of the hip. Placing the client on the left or right side with knee bent doesn't promote abduction. Acute flexion of the operated hip may cause dislocation. The head of the bed may be raised 35 to 40 degrees.
NP: Implementation; CN: Physiological integrity; CNS: Reduction of risk potential; CL: Application

91. A 20-year-old client developed osteo-myelitis 2 weeks after a fishhook was removed from his foot. Which of the following rationales best explains the expected long-term antibiotic therapy needed?
1. Bone has poor circulation.
2. Tissue trauma requires antibiotics.
3. Feet are normally more difficult to treat.
4. Fishhook injuries are highly contaminat-ed.

92. Degenerative joint disease, also commonly known as osteoarthritis, is a term describing which of the following conditions?
1. Noninflammatory joint disease
2. Immune-mediated joint disease
3. Joint inflammation after a viral infection
4. Joint inflammation related to systemic in-fections

93. Client education about gout includes which of the following information?
1. Good foot care will reduce complications.
2. Increased dietary intake of purine is needed.
3. Production of uric acid in the kidney af-fects joints.
4. Uric acid crystals cause inflammatory de-struction of the joint.

94. If I.V. antibiotics don't eliminate osteomyelitis, which of the following treat-ments is most commonly used next?
1. Bone grafts
2. Hyperbaric oxygen therapy
3. Amputation of the extremity
4. Debridement of necrotic tissue

The words most commonly in question 94 are clues to the answer. Am I right, or am I right?

91. 1. Bone has poor blood circulation, making it difficult to treat an infection in the bone. This requires the long-term use of I.V. antibiotics to make sure the infection is cleared. Tissue trau-ma doesn't always require antibiotics, at least not long term. Fishhooks may not be any more contaminated than another instrument that caused an injury. Feet aren't more difficult to treat than other parts of the body unless the client has a circulatory problem or diabetes mellitus.
NP: Implementation; CN: Physiological integrity; CNS: Pharma-cological therapies; CL: Application

92. 1. Degenerative joint disease is joint dis-ease due to the wear and tear on joints and is often seen in athletes. It isn't inflammatory, immune-mediated, or caused by systemic infec-tions.
NP: Data collection; CN: Physiological integrity; CNS: Physio-logical adaptation; CL: Comprehension

93. 4. The client needs to know that uric acid crystals collect in the joint of the great toe and cause inflammation. The kidney excretes uric acid, an end product of metabolism. A diet low in purines would be indicated. Good foot care doesn't affect the development of complica-tions, but increasing water intake may help pre-vent urinary stone formation.
NP: Implementation; CN: Physiological integrity; CNS: Reduc-tion of risk potential; CL: Application

94. 4. The tissues may need to be debrided to eliminate necrotic tissue and allow new tissue to form. Amputation isn't indicated in the treat-ment of acute osteomyelitis. A bone graft would be done after debridement. Hyperbaric oxygen therapy is a new treatment modality that has been used in the successful treatment of osteomyelitis, but it isn't universally available.
NP: Implementation; CN: Physiological integrity; CNS: Physio-logical adaptation; CL: Comprehension

95. Nursing interventions to treat a musculoskeletal injury may include cold or heat therapy. Cold therapy decreases pain by which of the following actions?
1. Promotes analgesia and circulation
2. Numbs the nerves and dilates the vessels
3. Promotes circulation and reduces muscle spasms
4. Causes local vasoconstriction and prevents edema or muscle spasm

Only two more questions. You're terrific!

95. 4. Cold causes the blood vessels to constrict, which reduces the leakage of fluid into the tissues and prevents swelling and muscle spasms. Cold therapy may reduce pain by numbing the nerves and tissues. Heat therapy promotes circulation, enhances flexibility, reduces muscle spasms, and also provides analgesia.
NP: Implementation; CN: Physiological integrity; CNS: Basic care and comfort; CL: Comprehension

96. A client has been fitted for crutches, and the nurse is evaluating the outcome of patient-teaching sessions for crutch walking. Which outcome is desired?
1. The client's weight is being borne by the axilla.
2. The client is wearing slippers.
3. The elbows are bent at a slight angle.
4. When descending stairs, the unaffected leg moves first.

96. 3. The elbows shouldn't be locked open but rather bent at a slight angle if the crutches are fitted correctly. Weight shouldn't be borne on the axilla; this could cause nerve damage. It should be borne by the heel pad of the hand. Crutches shouldn't be used in stocking feet, slippers, or high heels. Only sturdy, low-heeled shoes should be worn to prevent injuries. For proper support, the client leads off with the unaffected foot when ascending stairs and with the affected foot and crutches when descending. The client should remember "Down with the bad and up with the good."
NP: Evaluation; CN: Safe, effective, care environment; CNS: Safety and infection control; CL: Analysis

97. A 55-year-old male client has been hospitalized for 1 week due to injuries sustained in a motor vehicle accident. He's weak and continues to require some assistance with activities of daily living. The nurse helps the client to maintain muscle tone through which of the following nursing actions?
1. Getting the client out of bed even if he's on bed rest
2. Providing range-of-motion (ROM) exercises if the client is on bed rest
3. Limiting the client's activity
4. Encouraging the client to turn himself in bed

You did it! Congratulations!

97. 2. ROM exercises are used to move an extremity through its range without putting stress on the joints. If the client is on bed rest, never get him out of bed without an order; this could cause undue harm to the client. Limiting the client's activity isn't a good choice unless there are specific orders to do so from the physician. Encouraging the client to turn himself may prevent skin breakdown but doesn't help to maintain muscle tone. Always check the physician's order before you initiate any care.
NP: Planning; CN: Physiological integrity; CNS: Reduction of risk potential; CL: Application

From hiatal hernia to diverticulitis to pancreatitis, this chapter covers all the GI disorders you could ask for, in one handy package. Gotta love it!

Chapter 8
Gastrointestinal disorders

1. Which of the following conditions can cause a hiatal hernia?
1. Increased intrathoracic pressure
2. Weakness of the esophageal muscle
3. Increased esophageal muscle pressure
4. Weakness of the diaphragmatic muscle

2. Which of the following interventions can the nurse teach a client with a sliding hiatal hernia to improve the client's comfort?
1. To drink carbonated cola beverages with meals
2. To lie down immediately after eating
3. To eat three large, high-carbohydrate meals each day
4. To sleep with his head elevated 30 degrees

Read #1 carefully. This question is looking for a cause.

3. Which of the following symptoms is common with a hiatal hernia?
1. Left arm pain
2. Lower back pain
3. Esophageal reflux
4. Abdominal cramping

1. 4. A hiatal hernia is caused by weakness of the diaphragmatic muscle and increased intra-abdominal — not intrathoracic — pressure. This weakness allows the stomach to slide into the esophagus. The esophageal supports weaken, but esophageal muscle weakness or increased esophageal muscle pressure isn't a factor in hiatal hernia.
NP: Data collection; CN: Physiological integrity; CNS: Physiological adaptation; CL: Knowledge

2. 4. With a hiatal hernia, sleeping with the head of the bed elevated 30 degrees (about 3″ to 4″ [7.5 to 10 cm]) prevents stomach acids from refluxing into the esophagus. Carbonated beverages would create gas and possibly irritate the herniated area as the client begins to tolerate bland foods. Lying down immediately after eating would facilitate the reflux of stomach acids, causing irritation and possible aspiration. Small meals are recommended for clients with hiatal hernia.
NP: Implementation; CN: Physiological integrity; CNS: Physiological adaptation; CL: Comprehension

3. 3. Esophageal reflux is a common symptom of hiatal hernia. This seems to be associated with chronic exposure of the lower esophageal sphincter to the lower pressure of the thorax, making it less effective. Left arm pain is a common symptom of heart attack. Lower back pain can be caused by lumbar strain. Abdominal cramping can be caused by intestinal infection.
NP: Data collection; CN: Physiological integrity; CNS: Physiological adaptation; CL: Comprehension

4. Which of the following tests can be performed to <u>diagnose</u> a hiatal hernia?
1. Colonoscopy
2. Lower GI series
3. Barium swallow
4. Abdominal X-ray series

Which test allows the radiologist to see the stomach in relation to the diaphragm?

4. 3. A barium swallow with fluoroscopy shows the position of the stomach in relation to the diaphragm. A colonoscopy and a lower GI series show disorders of the intestine. An abdominal X-ray series will show structural defects but not necessarily a hiatal hernia, unless it's sliding or rolling at the time of the X-ray.

NP: Implementation; CN: Health promotion and maintenance; CNS: Prevention and early detection of disease; CL: Comprehension

5. When is the <u>best</u> time for the nurse to teach a client scheduled for an appendectomy about incision splinting and leg exercises?
1. On the client's admission to the recovery room
2. When the client returns from the recovery room
3. During the intraoperative period
4. Before the surgical procedure

This question is looking for the best.

5. 4. Teaching is most effective when done before the surgical procedure. During this time, the nurse should tell the client what to expect in the immediate postoperative period. On admission to the recovery room, the client is usually very drowsy, making this an inopportune time for teaching. On the client's return from the recovery room, the client may remain drowsy. During the intraoperative period, anesthesia alters the client's mental status, rendering teaching ineffective.

NP: Planning; CN: Physiological integrity; CNS: Reduction of risk potential; CL: Application

6. Which of the following terms best describes the pain associated with appendicitis?
1. Aching
2. Fleeting
3. Intermittent
4. Steady

6. 4. The pain begins in the epigastrium or periumbilical region, then shifts to the right lower quadrant and becomes steady. The pain may be moderate to severe.

NP: Data collection; CN: Physiological integrity; CNS: Physiological adaptation; CL: Knowledge

7. Which of the following positions should a client with appendicitis assume to help relieve the pain?
1. Prone
2. Sitting
3. Supine, stretched out
4. Lying with legs drawn up

The last two items are similar. Know the difference before answering this one.

7. 4. Lying still with the legs drawn up toward the chest helps relieve tension on the abdominal muscles, which helps to reduce the amount of discomfort felt. Lying flat or sitting may increase the amount of pain experienced.

NP: Data collection; CN: Physiological integrity; CNS: Physiological adaptation; CL: Comprehension

NP: Nursing process   CN: Client needs category   CNS: Client needs subcategory   CL: Cognitive level

8. Which of the following nursing interventions should be implemented when caring for a client with appendicitis?
1. Monitoring for pain
2. Encouraging oral intake of clear fluids
3. Providing discharge teaching
4. Monitoring for symptoms of peritonitis

8. 4. The focus of care is to monitor for peritonitis, or inflammation of the peritoneal cavity. Peritonitis is most commonly caused by appendix rupture and invasion of bacteria, which could be lethal. The client with appendicitis will have pain that should be controlled with analgesia. The nurse should discourage oral intake in preparation for surgery. Discharge teaching is important; however, in the acute phase, management should focus on minimizing preoperative complications and recognizing when they may be occurring.
NP: Implementation; CN: Physiological integrity; CNS: Reduction of risk potential; CL: Application

9. Which of the following definitions best describes gastritis?
1. Erosion of the gastric mucosa
2. Inflammation of a diverticulum
3. Inflammation of the gastric mucosa
4. Reflux of stomach acid into the esophagus

9. 3. Gastritis is an inflammation of the gastric mucosa that may be acute (often resulting from exposure to local irritants) or chronic (associated with autoimmune infections or atrophic disorders of the stomach). Erosion of the mucosa results in ulceration. Inflammation of a diverticulum is called diverticulitis; reflux of stomach acid is known as gastroesophageal reflux disease.
NP: Data collection; CN: Physiological integrity; CNS: Physiological adaptation; CL: Knowledge

Congratulations! You've finished the first 10 questions!

10. Which of the following substances is most likely to cause gastritis?
1. Milk
2. Bicarbonate of soda, or baking soda
3. Enteric-coated aspirin
4. Nonsteroidal anti-inflammatory drugs (NSAIDs)

10. 4. NSAIDs are a common cause of gastritis because they inhibit prostaglandin synthesis. Milk, once thought to help reduce gastritis, has little effect on the stomach mucosa. Bicarbonate of soda, or baking soda, may be used to neutralize stomach acid, but it should be used cautiously because it may lead to metabolic acidosis. Aspirin with enteric coating shouldn't contribute significantly to gastritis because the coating limits the aspirin's effect on the gastric mucosa.
NP: Data collection; CN: Physiological integrity; CNS: Reduction of risk potential; CL: Knowledge

Know what signs to look for in your client with a gastric resection.

11. Which of the following tasks should be included in the immediate postoperative care of a client who has undergone gastric resection?
1. Monitoring gastric pH to detect complications
2. Assessing for bowel sounds
3. Providing nutritional support
4. Monitoring for symptoms of hemorrhage

11. 4. The client should be monitored closely for signs and symptoms of hemorrhage, such as bright red blood in the nasogastric tube suction, tachycardia, or a drop in blood pressure. Gastric pH may be monitored to evaluate the need for histamine-2 receptor antagonists. Bowel sounds may not return for up to 72 hours postoperatively. Nutritional needs should be addressed soon after surgery.
NP: Data collection; CN: Physiological integrity; CNS: Reduction of risk potential; CL: Comprehension

12. Which of the following treatments should be included in the immediate care of a client with acute gastritis?
1. Reducing work stress
2. Completing gastric resection
3. Treating the underlying cause
4. Administering enteral tube feedings

13. Which of the following risk factors can lead to chronic gastritis?
1. Young age
2. Antibiotic usage
3. Gallbladder disease
4. *Helicobacter pylori* infection

14. Which of the following factors associates chronic gastritis with pernicious anemia?
1. Increased absorption of vitamin B_{12}
2. Inability to absorb vitamin B_{12}
3. Overproduction of stomach acid
4. Overproduction of vitamin B_{12}

15. Which of the following definitions best describes diverticulosis?
1. An inflamed outpouching of the intestine
2. A noninflamed outpouching of the intestine
3. The partial impairment of the forward flow of intestinal contents
4. An abnormal protrusion of an organ through the structure that usually holds it

16. Which of the following types of diets is implicated in the development of diverticulosis?
1. Low-fiber diet
2. High-fiber diet
3. High-protein diet
4. Low-carbohydrate diet

Read the first two options carefully. They're very similar.

12. 3. Discovering and treating the cause of gastritis is the most beneficial approach. Reducing or eliminating oral intake until the symptoms are gone and reducing the amount of stress are important in the recovery phase. A gastric resection is an option only when serious erosion has occurred.
NP: Data collection; CN: Safe, effective care environment; CNS: Coordinated care; CL: Comprehension

13. 4. *H. pylori* infection can lead to chronic atrophic gastritis. Chronic gastritis can occur at any age but is more common in older adults. It may be caused by conditions that allow reflux of bile acids into the stomach. Drugs such as nonsteroidal anti-inflammatory agents, not antibiotics, may cause gastritis. Chronic gastritis isn't related to gallbladder disease.
NP: Data collection; CN: Physiological integrity; CNS: Physiological adaptation; CL: Comprehension

14. 2. With gastritis, the stomach lining becomes thin and atrophic, decreasing stomach acid secretion (the source of intrinsic factor). This causes a reduction in the absorption of vitamin B_{12}, leading to pernicious anemia.
NP: Data collection; CN: Physiological integrity; CNS: Physiological adaptation; CL: Comprehension

15. 2. Diverticulosis involves a noninflamed outpouching of the intestine. Diverticulitis involves an inflamed outpouching. The partial impairment of forward flow of the intestine is an obstruction; abnormal protrusion of an organ is a hernia.
NP: Data collection; CN: Physiological integrity; CNS: Physiological adaptation; CL: Comprehension

16. 1. Low-fiber diets have been implicated in the development of diverticula because these diets decrease the bulk in the stool and predispose the person to the development of constipation. A high-fiber diet is recommended to help prevent diverticulosis. A high-protein or low-carbohydrate diet has no effect on the development of diverticulosis.
NP: Data collection; CN: Health promotion and maintenance; CNS: Prevention and early detection of disease; CL: Comprehension

17. Which of the following foods should be avoided by the client with diverticulosis to reduce the risk of complications?
1. Bran cereal and lettuce
2. Dried apricots and raisins
3. Cucumbers and tomatoes
4. Brown rice and fresh peeled apples

17. 3. A client with diverticulosis should avoid cucumbers and fresh tomatoes because their seeds can irritate the bowel and contribute to diverticular obstruction. To promote regular bowel movements, the client should consume foods that are high in undigestible fiber, such as bran cereal, lettuce, apricots, raisins, brown rice, and fresh peeled apples.

NP: Planning; CN: Health promotion and maintenance; CNS: Prevention and early detection of disease; CL: Knowledge

18. Which of the following symptoms indicates diverticulosis?
1. No symptoms exist
2. Change in bowel habits
3. Anorexia and low-grade fever
4. Episodic, dull or steady midabdominal pain

Be careful. This isn't a negative question, but the answer is tricky.

18. 1. Diverticulosis is an asymptomatic condition. The other choices are signs and symptoms of diverticulitis.

NP: Data collection; CN: Physiological integrity; CNS: Physiological adaptation; CL: Comprehension

19. Which of the following tests should be administered to a client suspected of having diverticulosis?
1. Abdominal ultrasound
2. Barium enema
3. Barium swallow
4. Gastroscopy

19. 2. A barium enema will cause diverticula to fill with barium and be easily seen on an X-ray. An abdominal ultrasound can tell more about such structures as the gallbladder, liver, and spleen than the intestine. A barium swallow and gastroscopy view upper GI structures.

NP: Planning; CN: Health promotion and maintenance; CNS: Prevention and early detection of disease; CL: Comprehension

20. Medical management of the client with acute diverticulitis should include which of the following treatments?
1. Reduced fluid intake
2. Increased fiber in diet
3. Administration of antibiotics
4. Exercises to increase intra-abdominal pressure

20. 3. Antibiotics are used to reduce inflammation. The client with acute diverticulitis typically isn't allowed anything orally until the acute episode subsides. Parenteral fluids are given until the client feels better; then it's recommended that the client drink eight 8-oz (237-ml) glasses of water per day and gradually increase fiber in the diet to improve intestinal motility. During the acute phase, activities that increase intra-abdominal pressure should be avoided in order to decrease pain and the chance of intestinal obstruction.

NP: Data collection; CN: Safe, effective care environment; CNS: Safety and infection control; CL: Comprehension

You're really pumping up now!

21. The nurse has taught a teenage client and his parents about Crohn's disease and the dietary changes needed to manage it. Which statement by the parents indicates an accurate understanding of their child's dietary needs?
1. "We'll need to include plenty of calories in the diet."
2. "We'll need to make certain that all food is gluten-free."
3. "We'll be sure to use only low-sodium foods."
4. "We'll provide high-fiber foods."

21. 1. Crohn's disease is an inflammatory bowel disease that causes diarrhea with subsequent weight loss and malnutrition. A high-calorie, low-protein diet helps replenish nutrients that are lost through the affected bowel. A gluten-free diet is appropriate for a client with celiac disease, not Crohn's disease. A client with Crohn's disease doesn't need to restrict dietary sodium but should avoid high-fiber foods because they can contribute to bowel irritation.
NP: Evaluation; CN: Physiological integrity; CNS: Basic care and comfort; CL: Application

22. Which area of the alimentary canal is the most common location of Crohn's disease?
1. Ascending colon
2. Descending colon
3. Sigmoid colon
4. Terminal ileum

22. 4. Studies have shown that the terminal ileum is the most common site for recurrence in clients with Crohn's disease. The other areas may be involved but aren't as common.
NP: Data collection; CN: Physiological integrity; CNS: Physiological adaptation; CL: Knowledge

23. Which of the following factors is believed to be linked to Crohn's disease?
1. Constipation
2. Diet
3. Heredity
4. Lack of exercise

So tell me, Gene, what's your answer? Eh? Gene? (Hint)

23. 3. Although the definitive cause of Crohn's disease is unknown, it's thought to be associated with infectious, immune, or psychological factors. Because it has a higher incidence in siblings, it may have a genetic cause. Constipation isn't linked to Crohn's disease. On the contrary, Crohn's disease causes bouts of diarrhea. Diet may contribute to exacerbations of Crohn's disease but isn't considered a cause. A lack of exercise isn't considered a cause of Crohn's disease.
NP: Data collection; CN: Health promotion and maintenance; CNS: Prevention and early detection of disease; CL: Comprehension

24. Which of the following factors is believed to cause ulcerative colitis?
1. Acidic diet
2. Altered immunity
3. Chronic constipation
4. Emotional stress

24. 2. Several theories exist regarding the cause of ulcerative colitis. One suggests altered immunity as the cause based on the extraintestinal characteristics of the disease, such as peripheral arthritis and cholangitis. Diet and constipation have no effect on the development of ulcerative colitis. Emotional stress may exacerbate the attacks but isn't believed to be the primary cause.
NP: Data collection; CN: Health promotion and maintenance; CNS: Prevention and early detection of disease; CL: Comprehension

25. Fistulas are most common in which of the following bowel disorders?
1. Crohn's disease
2. Diverticulitis
3. Diverticulosis
4. Ulcerative colitis

26. Which of the following areas is the most common site of fistulas in clients with Crohn's disease?
1. Anorectal
2. Ileum
3. Rectovaginal
4. Transverse colon

27. A client who reports a number of GI-related symptoms is diagnosed with ulcerative colitis. The nurse explains that ulcerative colitis is:
1. an inflammatory disease of the colon and rectum.
2. an inflammatory disease of the GI tract from the mouth to the anus.
3. a disease that forms outpouches in the large colon.
4. an opening between two or more body structures or spaces.

28. Which of the following associated disorders may the client with Crohn's disease exhibit?
1. Ankylosing spondylitis
2. Colon cancer
3. Malabsorption
4. Pernicious anemia

29. Which of the following symptoms may be exhibited by a client with Crohn's disease?
1. Bloody diarrhea
2. Cramping abdominal pain
3. Nausea and vomiting
4. Excessive amounts of fats in the feces

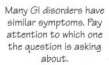

NCLEX questions sure can be a fistula of trouble!

Many GI disorders have similar symptoms. Pay attention to which one the question is asking about.

25. 1. The lesions of Crohn's disease are *transmural;* that is, they involve all thicknesses of the bowel. These lesions may perforate the bowel wall, forming fistulas in adjacent structures. Fistulas don't develop in diverticulitis or diverticulosis. The ulcers that occur in the submucosal and mucosal layers of the intestine in ulcerative colitis usually don't progress to fistula formation as in Crohn's disease.
NP: Data collection; CN: Physiological integrity; CNS: Physiological adaptation; CL: Comprehension

26. 1. Fistulas occur in all these areas, but the anorectal area is most common because of the relative thinness of the intestinal wall in this area.
NP: Data collection; CN: Physiological integrity; CNS: Physiological adaptation; CL: Comprehension

27. 1. Ulcerative colitis is an inflammatory disease that affects the mucosal lining of the colon and rectum. Crohn's disease commonly affects the entire GI tract, from the mouth to the anus. Diverticulitis causes outpouching in the large colon. A fistula is an opening between two or more body structures.
NP: Implementation; CN: Physiological integrity; CNS: Physiological adaptation; CL: Application

28. 3. Because of the transmural nature of Crohn's disease lesions, malabsorption may occur. Ankylosing spondylitis and colon cancer are more commonly associated with ulcerative colitis. Pernicious anemia is associated with vitamin B_{12} deficiency.
NP: Data collection; CN: Physiological integrity; CNS: Physiological adaptation; CL: Comprehension

29. 4. Excessive amounts of fats in the feces due to malabsorption can occur with Crohn's disease. Cramping abdominal pain before defecation, nausea, vomiting, and bloody diarrhea are symptoms of ulcerative colitis.
NP: Data collection; CN: Health promotion and maintenance; CNS: Prevention and early detection of disease; CL: Comprehension

30. Which of the following symptoms is associated with ulcerative colitis?
1. Nutritional deficit
2. Rectal bleeding
3. Soft stools
4. Weight loss

30. 2. In ulcerative colitis, rectal bleeding is the predominant symptom. Soft stools, nutritional deficit, and weight loss are more commonly associated with Crohn's disease, in which malabsorption is more of a problem.
NP: Data collection; CN: Health promotion and maintenance; CNS: Prevention and early detection of disease; CL: Knowledge

31. If a client had irritable bowel syndrome, which of the following diagnostic tests would determine whether the diagnosis is Crohn's disease or ulcerative colitis?
1. Abdominal computed tomography (CT) scan
2. Abdominal X-ray
3. Barium swallow
4. Colonoscopy with biopsy

One of these tests can help you distinguish between two GI disorders.

31. 4. A colonoscopy with biopsy can be performed to determine the state of the colon's mucosal layers, presence of ulcerations, and level of cytologic involvement. An abdominal X-ray or CT scan wouldn't provide the cytologic information necessary to diagnose which disease it is. A barium swallow doesn't involve the intestine.
NP: Data collection; CN: Physiological integrity; CNS: Physiological adaptation; CL: Comprehension

32. Which of the following interventions should be included in the medical management of Crohn's disease?
1. Increasing oral intake of fiber
2. Administering laxatives
3. Using long-term steroid therapy
4. Increasing physical activity

32. 3. Management of Crohn's disease may include long-term steroid therapy to reduce the extensive inflammation associated with the deeper layers of the bowel wall. Other management focuses on bowel rest (not increasing oral intake) and reducing diarrhea with medications (not giving laxatives). The pain associated with Crohn's disease may require bed rest, not an increase in physical activity.
NP: Implementation; CN: Physiological integrity; CNS: Basic care and comfort; CL: Application

33. Which of the following actions would not be a desired effect of antibiotic therapy in a client with Crohn's disease?
1. Decrease in bleeding
2. Decrease in temperature
3. Decrease in body weight
4. Decrease in number of stools

Careful! This is a negative question.

33. 3. A decrease in body weight may occur during therapy due to inadequate dietary intake but it isn't related to antibiotic therapy. Effective antibiotic therapy will be noted by a decrease in temperature, number of stools, and bleeding.
NP: Evaluation; CN: Physiological integrity; CNS: Pharmacological therapies; CL: Comprehension

34. Surgery may be performed to treat which of the following complications of ulcerative colitis?
1. Gastritis
2. Bowel herniation
3. Bowel outpouching
4. Bowel perforation

34. 4. Perforation, obstruction, hemorrhage, and toxic megacolon are common complications of ulcerative colitis that may require surgery. Herniation and gastritis aren't associated with irritable bowel diseases, and outpouching of the bowel wall is diverticulosis.
NP: Implementation; CN: Safe, effective care environment; CNS: Safety and infection control; CL: Comprehension

---

NP: Nursing process   CN: Client needs category   CNS: Client needs subcategory   CL: Cognitive level

35. Which of the following medications is most effective for treating the underlying cause of pain associated with irritable bowel disease?
  1. Acetaminophen
  2. Opiates
  3. Steroids
  4. Stool softeners

This question is asking you to prioritize.

36. During the first few days of recovery from ostomy surgery for ulcerative colitis, which of the following aspects should be the first priority of client care?
  1. Body image
  2. Ostomy care
  3. Sexual concerns
  4. Skin care

37. Colon cancer is most closely associated with which of the following conditions?
  1. Appendicitis
  2. Hemorrhoids
  3. Hiatal hernia
  4. Ulcerative colitis

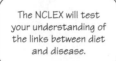

The NCLEX will test your understanding of the links between diet and disease.

38. Which of the following diets is most commonly associated with colon cancer?
  1. Low fiber, high fat
  2. Low fat, high fiber
  3. Low protein, high carbohydrate
  4. Low carbohydrate, high protein

39. Which of the following diagnostic tests should be performed annually after age 40 to screen for colon cancer?
  1. Abdominal computed tomography (CT) scan
  2. Abdominal X-ray
  3. Colonoscopy
  4. Digital rectal examination

35. 3. The pain of irritable bowel disease is caused by inflammation, which steroids can reduce. Stool softeners aren't necessary. Acetaminophen has little effect on the pain, and opiate narcotics won't treat its underlying cause.
NP: Data collection; CN: Physiological integrity; CNS: Pharmacological therapies; CL: Comprehension

36. 2. Although all of these are concerns the nurse should address, being able to safely manage the ostomy is crucial for the client before discharge.
NP: Implementation; CN: Physiological integrity; CNS: Basic care and comfort; CL: Application

37. 4. Chronic ulcerative colitis, granulomas, and familial polyposis seem to increase a person's chance of developing colon cancer. The other conditions listed have no known effect on colon cancer risk.
NP: Data collection; CN: Health promotion and maintenance; CNS: Prevention and early detection of disease; CL: Comprehension

38. 1. A low-fiber, high-fat diet reduces motility and increases the chance of constipation. The metabolic end products of this type of diet are carcinogenic. A low-fat, high-fiber diet is recommended to help avoid colon cancer. Carbohydrates and protein aren't necessarily associated with colon cancer.
NP: Data collection; CN: Health promotion and maintenance; CNS: Prevention and early detection of disease; CL: Knowledge

39. 4. One-third of malignant tumors of the distal colon and rectum can be palpated with the examining finger, and so a digital rectal examination is recommended annually after age 40. Abdominal X-ray and CT scan can help establish tumor size and metastasis. A colonoscopy can help locate a tumor as well as polyps, which can be removed before they become malignant.
NP: Data collection; CN: Health promotion and maintenance; CNS: Prevention and early detection of disease; CL: Knowledge

40. Radiation therapy is used to treat colon cancer before surgery for which of the following reasons?
  1. Reducing the size of the tumor
  2. Eliminating the malignant cells
  3. Curing the cancer
  4. Helping heal the bowel after surgery

Another 10 down! Good job!

41. Which of the following symptoms is a client with colon cancer most likely to exhibit?
  1. A change in appetite
  2. A change in bowel habits
  3. An increase in body weight
  4. An increase in body temperature

Again, make sure you read the question carefully and know what disorder is being addressed.

42. A client has just had surgery for colon cancer. The nurse should observe this client for symptoms that might indicate the development of which of the following complications?
  1. Peritonitis
  2. Diverticulosis
  3. Partial bowel obstruction
  4. Complete bowel obstruction

43. A client with gastric cancer may exhibit which of the following symptoms?
  1. Abdominal cramping
  2. Constant hunger
  3. Feeling of fullness
  4. Weight gain

44. Which of the following diagnostic tests may be performed to determine whether a client has gastric cancer?
  1. Barium enema
  2. Colonoscopy
  3. Gastroscopy
  4. Serum chemistry levels

40. 1. Radiation therapy is used to treat colon cancer before surgery to reduce the size of the tumor, making it easier to resect. Radiation therapy isn't curative, can't eliminate the malignant cells (though it helps to define tumor margins), and could slow postoperative healing.
NP: Implementation; CN: Physiological integrity; CNS: Reduction of risk potential; CL: Comprehension

41. 2. The most common complaint of the client with colon cancer is a change in bowel habits. The client may have anorexia, secondary abdominal distention, or weight loss. Fever isn't related to colon cancer.
NP: Data collection; CN: Physiological integrity; CNS: Basic care and comfort; CL: Application

42. 1. Bowel spillage could occur during surgery, resulting in peritonitis. Complete or partial intestinal obstruction may occur *before* bowel resection. Diverticulosis doesn't result from surgery for colon cancer.
NP: Evaluation; CN: Physiological integrity; CNS: Physiological adaptation; CL: Application

43. 3. The client with gastric cancer may report a feeling of fullness in the stomach, but not enough to cause him to seek medical care. Abdominal cramping isn't associated with gastric cancer. Anorexia and weight loss (not increased hunger or weight gain) are common symptoms of gastric cancer.
NP: Data collection; CN: Physiological integrity; CNS: Physiological adaptation; CL: Comprehension

44. 3. A gastroscopy will allow direct visualization of the tumor. A colonoscopy or barium enema would help to diagnose *colon* cancer. Serum chemistry levels don't contribute data useful to the assessment of gastric cancer.
NP: Data collection; CN: Health promotion and maintenance; CNS: Prevention and early detection of disease; CL: Comprehension

45. A client with gastric cancer can expect to have surgery for resection. Which of the following should be the nursing care priority for the preoperative client with gastric cancer?
1. Discharge planning
2. Correction of nutritional deficits
3. Prevention of deep vein thrombosis
4. Instruction regarding radiation treatment

Prioritize!

45. 2. Clients with gastric cancer often have nutritional deficits and may be cachectic. Discharge planning before surgery is important, but correcting the nutritional deficit is a higher priority. At present, radiation therapy hasn't been proven effective for gastric cancer, and teaching about it preoperatively wouldn't be appropriate. Prevention of deep vein thrombosis also isn't a high priority prior to surgery, though it assumes greater importance after surgery.
NP: Planning; CN: Physiological integrity; CNS: Basic care and comfort; CL: Application

46. Care of the postoperative client after gastric resection should focus on which of the following problems?
1. Body image
2. Nutritional needs
3. Skin care
4. Spiritual needs

What's the key to the treatment of GI disorders in general?

46. 2. After gastric resection, a client may require total parenteral nutrition or jejunostomy tube feedings to maintain adequate nutritional status. Body image isn't much of a problem for this client because clothing can cover the incision site. Wound care of the incision site is necessary to prevent infection; otherwise, the skin shouldn't be affected. Spiritual needs may be a concern, depending on the client, and should be addressed as the client demonstrates readiness to share concerns.
NP: Planning; CN: Physiological integrity; CNS: Basic care and comfort; CL: Comprehension

47. The nurse should teach the client to watch for which of the following complications of gastric resection?
1. Constipation
2. Dumping syndrome
3. Gastric spasm
4. Intestinal spasms

Knowing what to teach your clients can help avoid problems down the road.

47. 2. Dumping syndrome is a problem that occurs postprandially after gastric resection because ingested food rapidly enters the jejunum without proper mixing and without the normal duodenal digestive processing. Diarrhea, not constipation, may also be a symptom. Gastric or intestinal spasms don't occur, but antispasmodics may be given to slow gastric emptying.
NP: Planning; CN: Health promotion and maintenance; CNS: Prevention and early detection of disease; CL: Application

48. A client with rectal cancer may exhibit which of the following symptoms?
1. Abdominal fullness
2. Gastric fullness
3. Rectal bleeding
4. Right upper quadrant pain

48. 3. Rectal bleeding is a common symptom of rectal cancer. Rectal cancer may be missed because other conditions such as hemorrhoids can cause rectal bleeding. Abdominal fullness may occur with colon cancer, gastric fullness may occur with gastric cancer, and right upper quadrant pain may occur with liver cancer.
NP: Data collection; CN: Physiological integrity; CNS: Physiological adaptation; CL: Comprehension

49. A client with which of the following conditions may be <u>likely</u> to develop rectal cancer?
1. Adenomatous polyps
2. Diverticulitis
3. Hemorrhoids
4. Peptic ulcer disease

Note the word *likely*. It's a clue to the answer.

49. 1. A client with adenomatous polyps has a higher risk of developing rectal cancer than others do. Clients with diverticulitis are more likely to develop colon cancer. Hemorrhoids don't increase the chance of any type of cancer. Clients with peptic ulcer disease have a higher incidence of gastric cancer.

NP: Data collection; CN: Health promotion and maintenance; CNS: Prevention and early detection of disease; CL: Analysis

50. A client with peptic ulcer disease hasn't responded to traditional medical management. The physician is planning to do a vagotomy. The nurse should explain to the client that this is:
1. removal of the antrum of the stomach.
2. removal of 60% to 80% of the stomach.
3. severing of the vagus nerve.
4. enlargement of the pyloric sphincter.

50. 3. The vagus nerve receives impulses from the brain to secrete hydrochloric acid. A vagotomy severs the vagus nerve, which decreases gastric acid by diminishing cholinergic stimulation to the parietal cells, making them less responsive to gastrin. Removal of the antrum is an antrectomy. Removal of 60% to 80% of the stomach is a gastrectomy. Surgically enlarging the pyloric sphincter is a pyloroplasty.

NP: Evaluation; CN: Physiological integrity; CNS: Physiological adaptation; CL: Knowledge

51. Which of the following conditions may cause hemorrhoids?
1. Diarrhea
2. Diverticulosis
3. Portal hypertension
4. Rectal bleeding

51. 3. Portal hypertension and other conditions associated with persistently high intra-abdominal pressure, such as pregnancy, can aggravate hemorrhoids. The passing of hard stool, not diarrhea, can lead to hemorrhoids. Diverticulosis has no relationship to hemorrhoids. Rectal bleeding can be a symptom of hemorrhoids.

NP: Data collection; CN: Physiological integrity; CNS: Physiological adaptation; CL: Analysis

52. Which of the following assessments would you expect the physician to make when forming a diagnosis of hemorrhoids?
1. Abdominal assessment
2. Diet history
3. Digital rectal examination
4. Sexual history

This question is looking for the most important assessment for diagnosis.

52. 3. Digital rectal examination is important to assess for internal hemorrhoids and to determine if other causes of the pain and bleeding are present. Abdominal assessment isn't necessary for hemorrhoids. Diet history is relevant because constipation can worsen hemorrhoids but isn't as important to diagnosis as a digital rectal examination. Sexual history may also be relevant, but again, the history isn't as important as a digital rectal examination.

NP: Data collection; CN: Physiological integrity; CNS: Physiological adaptation; CL: Comprehension

53. Medical management of hemorrhoids includes which of the following treatments?
  1. Recommending a high-fiber diet
  2. Applying cold to reduce swelling
  3. Using astringent lotions to reduce swelling
  4. Elevating the buttocks to reduce engorgement

54. Which of the following responses should the nurse offer to a client who asks why he's having a vagotomy to treat his ulcer?
  1. To repair a hole in the stomach
  2. To reduce the ability of the stomach to produce acid
  3. To prevent the stomach from sliding into the chest
  4. To remove a potentially malignant lesion in the stomach

55. What's the best postoperative position for a client who had a ruptured appendix and who now has peritonitis?
  1. Dorsal decubitus
  2. Lateral decubitus
  3. Flat
  4. Semi-Fowler's

56. A client is admitted with absent bowel sounds, abdominal distention, rebound tenderness, and muscle rigidity. Which diagnosis would the nurse expect to receive?
  1. Peritonitis
  2. Diverticulitis
  3. Dumping syndrome
  4. Normal symptoms of diverticulosis

This question is looking for the appropriate treatment — not just symptom alleviation.

Client education is an important tool for obtaining compliance.

You're doing an excellent job!

53. 1. A high-fiber diet will add bulk to the stool and ease its passage through the rectum. Application of cold isn't recommended because it can cause injury to the tissue. Astringent lotions can be used to reduce pain, but they aren't a treatment. The buttocks should be elevated only when prolapsed hemorrhoids are present.
NP: Data collection; CN: Physiological integrity; CNS: Physiological adaptation; CL: Comprehension

54. 2. A vagotomy is performed to eliminate the acid-secreting stimulus to gastric cells. A perforation would be repaired with a gastric resection. Repair of hiatal hernia (fundoplication) prevents the stomach from sliding through the diaphragm. Removal of a potentially malignant tumor wouldn't reduce the entire acid-producing mechanism.
NP: Implementation; CN: Physiological integrity; CNS: Reduction of risk potential; CL: Application

55. 4. Clients with peritonitis are maintained in semi-Fowler's position to keep abdominal contents below the diaphragm to promote lung expansion. The other positions would be inappropriate because the client's head should be elevated to maintain comfort and prevent complications.
NP: Implementation; CN: Physiological integrity; CNS: Physiological adaptation; CL: Application

56. 1. Absent bowel sounds, abdominal distention, rebound tenderness, and muscle rigidity are symptoms of peritonitis. Diverticulitis is multiple diverticula of the colon. The major signs of diverticulitis are intervals of diarrhea, abrupt onset of cramping pain in the lower left quadrant of the abdomen, and a low-grade fever. Dumping syndrome is rapid emptying of the stomach contents into the small intestine. It produces sweating and weakness after eating. Diverticulosis is an asymptomatic condition.
NP: Evaluation; CN: Health promotion and maintenance; CNS: Prevention and early detection of disease; CL: Comprehension

57. Which of the following laboratory results would be expected in a client with peritonitis?
1. Partial thromboplastin time above 100 seconds
2. Hemoglobin level below 10 mg/dl
3. Potassium level above 5.5 mEq/L
4. White blood cell (WBC) count above 15,000/μl

58. Which of the following therapies is not included in the medical management of a client with peritonitis?
1. Broad-spectrum antibiotics
2. Electrolyte replacement
3. I.V. fluids
4. Regular diet

Watch it! This is another negative question.

59. A client had a gastroscopy while under local anesthesia. Before resuming the client's oral fluid intake, what should the nurse do first?
1. Listen for bowel sounds.
2. Determine whether the client can talk.
3. Check for a gag reflex.
4. Determine the client's mental status.

60. Which of the following factors is most commonly associated with the development of pancreatitis?
1. Alcohol abuse
2. Hypercalcemia
3. Hyperlipidemia
4. Pancreatic duct obstruction

61. Which of the following actions of pancreatic enzymes can cause pancreatic damage?
1. Utilization by the intestine
2. Autodigestion of the pancreas
3. Reflux into the pancreas
4. Clogging of the pancreatic duct

Read question 61 carefully before you answer. Make sure you know what it's asking.

57. 4. Because of infection, the client's WBC count will be elevated. A hemoglobin level below 10 mg/dl may occur from hemorrhage. A partial thromboplastin time longer than 100 seconds may suggest disseminated intravascular coagulation, a serious complication of septic shock. A potassium level above 5.5 mEq/L may suggest renal failure.
NP: Data collection; CN: Physiological integrity; CNS: Physiological adaptation; CL: Application

58. 4. The client with peritonitis commonly isn't allowed anything orally until the source of the peritonitis is confirmed and treated. The client also requires broad-spectrum antibiotics to combat the infection. I.V. fluids are given to maintain hydration and hemodynamic stability and to replace electrolytes.
NP: Implementation; CN: Safe, effective care environment; CNS: Coordinated care; CL: Knowledge

59. 3. After a gastroscopy, the nurse should check for the presence of a gag reflex before giving oral fluids. This is essential to prevent aspiration. The presence of bowel sounds, the ability to speak, and a clear mental status wouldn't ensure the presence of a gag reflex.
NP: Implementation; CN: Physiological integrity; CNS: Reduction of risk potential; CL: Application

60. 1. Alcohol abuse is the major cause of acute pancreatitis in males, although gallbladder disease is more often implicated in women. Hypercalcemia, hyperlipidemia, and pancreatic duct obstruction are also causes of pancreatitis, but they occur less frequently.
NP: Data collection; CN: Physiological integrity; CNS: Reduction of risk potential; CL: Comprehension

61. 2. In pancreatitis, pancreatic enzymes become activated and begin to autodigest the pancreas. The enzymes are activated but aren't used properly by the intestine. Reflux of bile into the pancreatic duct and clogging of the pancreatic duct may occur before autodigestion of the pancreas occurs.
NP: Data collection; CN: Physiological integrity; CNS: Physiological adaptation; CL: Knowledge

NP: Nursing process   CN: Client needs category   CNS: Client needs subcategory   CL: Cognitive level

62. Which of the following laboratory tests is used to diagnose pancreatitis?
1. Lipase level
2. Hemoglobin level
3. Blood glucose level
4. White blood cell (WBC) count

Looking good! Keep going!

63. A preschooler is admitted to the hospital with a diagnosis of acute abdominal pain. What's the most appropriate measure the nurse can take to provide pain relief?
1. Administer clear liquids by mouth.
2. Apply moist heat to the abdomen.
3. Administer a saline enema.
4. Allow the child to assume a comfortable position.

64. Which of the following symptoms observed in a client with acute pancreatitis should be of primary concern to the nurse?
1. Increased appetite
2. Vomiting
3. Hypoglycemia
4. Pain

Prioritize!

65. What's the most important nursing goal for a client who has had a barium enema?
1. To prevent fecal incontinence
2. To monitor for bleeding
3. To prevent constipation
4. To limit fluid intake

62. 1. Lipase is an enzyme secreted by the pancreas; when elevated, it's useful in diagnosing pancreatitis. Hemoglobin level can be low in pancreatitis, but there are other causes for this. The blood glucose level may be elevated with pancreatitis, but this factor isn't diagnostic. The WBC count may also be elevated in pancreatitis, but this symptom can be caused by infection.

NP: Data collection; CN: Health promotion and maintenance; CNS: Prevention and early detection of disease; CL: Analysis

63. 4. The child should assume the position he finds most comfortable. Acute abdominal pain may signal acute appendicitis, so the nurse shouldn't apply heat or administer an enema because these measures stimulate the GI tract, which increases the risk of perforation in appendicitis. The child should also receive nothing by mouth until it's determined whether or not he needs surgery.

NP: Implementation; CN: Physiological integrity; CNS: Physiological adaptation; CL: Application

64. 2. Acute pancreatitis is commonly associated with fluid isolation and accumulation in the bowel secondary to ileus or peripancreatic edema. Fluid and electrolyte loss from vomiting is the primary concern. Although pain is an important concern, it's less significant than vomiting. A client with acute pancreatitis may have increased pain on eating and is unlikely to demonstrate an increased appetite. A client with acute pancreatitis is at risk for hyperglycemia, not hypoglycemia.

NP: Implementation; CN: Physiological integrity; CNS: Physiological adaptation; CL: Comprehension

65. 3. Barium should be promptly eliminated from the client's system after it has been introduced into the colon to prevent mass formation and possible bowel obstruction. Therefore, laxatives or enemas are commonly given after a barium enema to prevent the client from becoming constipated. Fecal incontinence isn't an issue because the passage of stool is desired. Bleeding isn't commonly anticipated after a barium enema. The client should be encouraged to increase, not decrease, fluid intake.

NP: Planning; CN: Physiological integrity; CNS: Reduction of risk potential; CL: Application

66. If a gastric ulcer perforates, which of the following actions should <u>not</u> be included in the immediate management of the client?
1. Blood replacement
2. Antacid administration
3. Nasogastric (NG) tube suction
4. Fluid and electrolyte replacement

Another negative question?

66. 2. Antacids aren't helpful in perforation. The client should be treated with antibiotics as well as fluid, electrolyte, and blood replacement. NG tube suction also should be performed to prevent further spillage of stomach contents into the peritoneal cavity.
NP: Planning; CN: Physiological integrity; CNS: Physiological adaptation; CL: Application

67. Which of the following symptoms would be unusual for the client with chronic pancreatitis to exhibit?
1. Abdominal pain
2. Diabetes mellitus
3. Blood in the stool
4. Weight loss

67. 3. Blood in the stool (hematochezia) is unusual for a client with pancreatitis. The client is more likely to exhibit abdominal pain, diabetes mellitus, and weight loss.
NP: Data collection; CN: Physiological integrity; CNS: Physiological adaptation; CL: Knowledge

68. In alcohol-related pancreatitis, which of the following interventions is the best way to reduce the exacerbation of pain?
1. Lying supine
2. Taking aspirin
3. Eating a low-fat diet
4. Abstaining from alcohol

68. 4. Abstaining from alcohol is imperative to reduce injury to the pancreas; in fact, it may be enough to completely control pain. Lying supine usually aggravates the pain because it stretches the abdominal muscles. Taking aspirin can cause bleeding in hemorrhagic pancreatitis. During an attack of acute pancreatitis, the client usually isn't allowed to ingest anything orally.
NP: Planning; CN: Physiological integrity; CNS: Reduction of risk potential; CL: Application

Just because something is common worldwide doesn't mean it's common in the United States.

69. Which of the following types of cirrhosis is most common in the United States?
1. Biliary
2. Cardiac
3. Laënnec's
4. Postnecrotic

69. 3. Sixty-five percent of clients with cirrhosis in the United States have Laënnec's, or alcohol-related, cirrhosis. Postnecrotic cirrhosis is the most common form worldwide.
NP: Data collection; CN: Physiological integrity; CNS: Physiological adaptation; CL: Knowledge

70. Which of the following factors causes biliary cirrhosis?
1. Acute viral hepatitis
2. Alcohol hepatotoxicity
3. Chronic biliary inflammation or obstruction
4. Heart failure with prolonged venous hepatic congestion

70. 3. Chronic biliary inflammation or obstruction causes biliary cirrhosis. Acute viral hepatitis can cause postnecrotic cirrhosis. Alcohol hepatotoxicity is Laënnec's cirrhosis. Heart failure with prolonged venous hepatic congestion will cause cardiac cirrhosis.
NP: Data collection; CN: Physiological integrity; CNS: Physiological adaptation; CL: Comprehension

71. Which of the following findings would <u>strongly</u> indicate the possibility of cirrhosis?
1. Dry skin
2. Hepatomegaly
3. Peripheral edema
4. Pruritus

72. Which of the following diagnostic tests helps determine a definitive diagnosis for cirrhosis?
1. Albumin level
2. Bromsulphalein dye excretion
3. Liver biopsy
4. Liver enzyme levels

73. A client with cirrhosis may have alterations in which of the following laboratory values?
1. Carbon dioxide level
2. pH
3. Prothrombin time
4. White blood cell (WBC) count

74. Which of the following measures should the nurse focus on for the client with esophageal varices?
1. Recognizing hemorrhage
2. Controlling blood pressure
3. Encouraging nutritional intake
4. Teaching the client about varices

75. Which of the following conditions is most likely to cause hepatitis?
1. Bacterial infection
2. Biliary dysfunction
3. Metastasis
4. Viral infection

One of these findings is a red flag for cirrhosis.

Give yourself a pat on the back! You've completed 75 questions!

71. 2. The client with cirrhosis has a liver that's enlarged (hepatomegaly), fibrotic, and nodular, which makes it palpable. The client may develop dry skin, pruritus, and peripheral edema, but these symptoms may have other causes.
NP: Data collection; CN: Physiological integrity; CNS: Physiological adaptation; CL: Comprehension

72. 3. A liver biopsy can reveal the exact cause of the hepatomegaly. The albumin level will be low, but that can be caused by poor nutritional status. Liver enzymes may be elevated, but other liver conditions may cause these elevations. Bromsulphalein dye excretion may be reduced, but other hepatocirculatory disorders could also cause this.
NP: Data collection; CN: Physiological integrity; CNS: Physiological adaptation; CL: Comprehension

73. 3. Clotting factors may not be produced normally when a client has cirrhosis, increasing the potential for bleeding. There's no associated change in carbon dioxide level or pH unless the client is developing other comorbidities such as metabolic alkalosis. The WBC count can be elevated in acute cirrhosis but isn't always altered.
NP: Data collection; CN: Physiological integrity; CNS: Physiological adaptation; CL: Comprehension

74. 1. Recognizing the rupture of esophageal varices, or hemorrhage, is the focus of nursing care because the client could succumb to this quickly. Controlling blood pressure is also important because it helps reduce the risk of variceal rupture. It's also important to teach the client what varices are and what foods he should avoid such as spicy foods.
NP: Planning; CN: Physiological integrity; CNS: Physiological adaptation; CL: Application

75. 4. The most common types of hepatitis are those caused by viruses—hepatitis A, B, C, D, and E. Other types of hepatitis are alcoholic, toxic, and chronic. Bacterial infections rarely cause hepatitis, and biliary dysfunction isn't a cause. Liver metastasis from colon cancer causes liver cancer, not hepatitis.
NP: Data collection; CN: Health promotion and maintenance; CNS: Prevention and early detection of disease; CL: Knowledge

76. Which of the following factors can cause hepatitis A?
1. Blood contact
2. Blood transfusion
3. Contaminated shellfish
4. Sexual contact

Remember, hepatitis A is highly contagious.

76. 3. Hepatitis A can be caused by infected water, milk, or food, especially shellfish from contaminated waters. Hepatitis B is caused by blood contact and sexual contact. Hepatitis C is usually caused by contact with infected blood, including blood transfusions.
NP: Data collection; CN: Health promotion and maintenance; CNS: Prevention and early detection of disease; CL: Knowledge

77. The nurse is developing a care plan for a client with hepatitis A. What's the main route of transmission of this virus?
1. Sputum
2. Feces
3. Blood
4. Urine

77. 2. The hepatitis A virus is transmitted by the fecal-oral route, primarily through ingestion of contaminated food or liquids. It isn't transmitted via sputum, blood, or urine. However, the hepatitis B virus is transmitted primarily through contact with contaminated blood, human secretions, and feces.
NP: Planning; CN: Safe, effective care environment; CNS: Safety and infection control; CL: Knowledge

78. A client with viral hepatitis may exhibit which of the following symptoms?
1. Arthralgia
2. Excitability
3. Headache
4. Polyphagia

78. 1. Arthralgia is very common in clients with viral hepatitis. Other symptoms of viral hepatitis include lethargy, flulike symptoms, anorexia, nausea and vomiting, abdominal pain, diarrhea, constipation, and fever. Excitability, headache, and polyphagia are *not* symptoms of viral hepatitis.
NP: Data collection; CN: Physiological integrity; CNS: Physiological adaptation; CL: Knowledge

79. To prevent infection transmission when caring for a client with hepatitis A, what should the nurse do?
1. Wear a mask for all client care.
2. Keep the client's room door closed.
3. Put on gloves to empty the emesis basin.
4. Wear a gown to deliver the meal tray.

Aren't you just itching to get to the next question? Oops, did I give it away? (Tee-hee!)

79. 3. The nurse should wear gloves to empty the emesis basin because hepatitis A is transmitted by the fecal-oral route, which makes body fluids potentially infectious. The nurse doesn't need to wear a mask or keep the client's door closed because hepatitis A isn't transmitted by airborne droplets. Generally, the nurse should wear a gown if soiling is likely; however, delivery of a meal tray poses a low risk of soiling.
NP: Implementation; CN: Safe, effective care environment; CNS: Safety and infection control; CL: Knowledge

80. Nursing interventions for the client with toxic hepatitis include which of the following actions?
1. Not allowing oral ingestion
2. Administering corticosteroids as ordered
3. Encouraging increased activity
4. Treating adverse effects such as itching

80. 4. Along with removal of the causative agent, alleviation of adverse effects such as itching is paramount to care of the client with toxic hepatitis. Other interventions include providing a high-calorie diet and promoting rest. Corticosteroids are rarely used in the care of the client with hepatitis.
NP: Data collection; CN: Safe, effective care environment; CNS: Coordinated care; CL: Comprehension

NP: Nursing process   CN: Client needs category   CNS: Client needs subcategory   CL: Cognitive level

81. Which of the following tests would be the most accurate for diagnosing liver cancer?
1. Abdominal ultrasound
2. Abdominal flat plate X-ray
3. Cholangiogram
4. Computed tomography (CT) scan

82. Immediately after a liver biopsy, which of the following complications should the client be closely monitored for?
1. Abdominal cramping
2. Hemorrhage
3. Nausea and vomiting
4. Potential infection

83. When a client who has a liver disorder is having an invasive procedure, the nurse helps ensure safety by assessing the results of which of the following tests?
1. Coagulation studies
2. Liver enzyme levels
3. Serum chemistries
4. White blood cell count

84. The nurse is giving preoperative and postoperative instructions to a client who will undergo a liver biopsy the following morning. In this situation, patient-teaching information for which problem is most critical?
1. Paralytic ileus
2. Hemorrhage
3. Renal shutdown
4. Constipation

Coagulations! Oops! I meant congratulations! You're getting near the end!

85. An alcoholic client is hospitalized with cirrhosis of the liver. In this client, which findings may be early signs of alcohol withdrawal?
1. Hypertension, hallucinations, and seizures
2. Muscle spasms, cyanosis, and hypocalcemia
3. Hand tremor, irritability, and anxiety
4. Chest pain, weakness, and vertigo

Careful, this question is asking you to identify early signs.

81. 4. A client with suspected liver cancer will likely undergo CT imaging to identify tumors. The results of a CT scan are much more definitive than the findings of an X-ray or ultrasound. A cholangiogram evaluates the gallbladder, not the liver.
NP: Data collection; CN: Health promotion and maintenance; CNS: Prevention and early detection of disease; CL: Knowledge

82. 2. The liver is very vascular, and taking a biopsy could cause the client to hemorrhage. The client may experience some discomfort but typically not cramping. Nausea and vomiting may be present and infection may occur, but not immediately after the procedure.
NP: Data collection; CN: Physiological integrity; CNS: Reduction of risk potential; CL: Application

83. 1. The liver produces coagulation factors. If the liver is affected negatively, production of these factors may be altered, placing the client at risk for hemorrhage. The other laboratory tests should be monitored as well, but the results may not necessarily relate to the safety of the procedure.
NP: Planning; CN: Physiological integrity; CNS: Reduction of risk potential; CL: Analysis

84. 2. Because the most common adverse effect of a liver biopsy is bleeding, the nurse should provide relevant information regarding the potential for hemorrhage. There's no reason to provide the client with information about a paralytic ileus. Renal shutdown isn't an expected complication after a liver biopsy. The nurse would have no reason to suspect that the client will have a problem with constipation after a liver biopsy.
NP: Planning; CN: Physiological integrity; CNS: Reduction of risk potential; CL: Application

85. 3. Early signs of alcohol withdrawal include hand tremor, irritability, anxiety, nausea, and slight sweating. Later signs may include hypertension, hallucinations, seizures, vomiting, tachycardia, and marked confusion. Muscle spasms, cyanosis, and hypocalcemia are signs of parathyroid hormone deficiency. Chest pain, weakness, and vertigo may signal pacemaker failure.
NP: Data collection; CN: Physiological integrity; CNS: Reduction of risk potential; CL: Knowledge

86. When counseling a client in the ways to prevent cholecystitis, which of the following guidelines is most important?

1. Eat a low-protein diet.
2. Eat a high-fat, high-cholesterol diet.
3. Limit exercise to 10 minutes a day.
4. Keep weight proportional to height.

It takes a lot of gall to ask a question like this. (Oh, I just know I'll hate myself in the morning!)

86. 4. Obesity is a known cause of gallstones, and maintaining a recommended weight will help to protect against cholecystitis. Excessive dietary intake of cholesterol is associated with the development of gallstones in many people. Dietary protein isn't implicated in cholecystitis. Liquid protein and low-calorie diets (with rapid weight loss of more than 5 lb [2.3 kg] per week) *are* implicated as the cause of some cases of cholecystitis. Regular exercise (30 minutes/three times a week) may help to reduce weight and improve fat metabolism.

NP: Implementation; CN: Health promotion and maintenance; CNS: Prevention and early detection of disease; CL: Application

87. The anesthesiologist orders atropine sulfate for a client who will undergo cholecystectomy. Why is this drug administered preoperatively?

1. To relieve pain and anxiety
2. To ease anesthesia induction
3. To reduce respiratory secretions
4. To prevent infection caused by the surgery

87. 3. Atropine, an anticholinergic agent, is commonly prescribed preoperatively to reduce respiratory tract secretions and prevent the reflex slowing of the heart that occurs during anesthesia. Anxiolytic agents are used to relieve anxiety. Narcotic analgesics are used to relieve pain and ease anesthesia induction. Antimicrobials are used to prevent infection resulting from contamination during surgery.

NP: Implementation; CN: Physiological integrity; CNS: Pharmacological therapies; CL: Comprehension

88. Which of the following tests is used most commonly to diagnose cholecystitis?

1. Abdominal computed tomography (CT) scan
2. Abdominal ultrasound
3. Barium swallow
4. Endoscopy

88. 2. An abdominal ultrasound can show if the gallbladder is enlarged, if gallstones are present, if the gallbladder wall is thickened, or if distention of the gallbladder lumen is present. An abdominal CT scan can be used to diagnose cholecystitis, but it usually isn't necessary. A barium swallow looks at the stomach and the duodenum. Endoscopy looks at the esophagus, stomach, and duodenum.

NP: Data collection; CN: Health promotion and maintenance; CNS: Prevention and early detection of disease; CL: Comprehension

89. A client being treated for chronic cholecystitis should be given which of the following instructions?

1. Increase rest.
2. Avoid antacids.
3. Increase fat in diet.
4. Use anticholinergics as prescribed.

89. 4. Conservative therapy for chronic cholecystitis includes weight reduction by increasing physical activity, a low-fat diet, antacid use to treat dyspepsia, and anticholinergic use to relax smooth muscles and reduce ductal tone and spasm, thereby reducing pain.

NP: Implementation; CN: Physiological integrity; CNS: Pharmacological therapies; CL: Comprehension

90. While a client is being prepared for discharge, the nasogastric (NG) feeding tube becomes clogged. To remedy this problem and teach the client's family how to deal with it at home, what should the nurse do?

 1. Irrigate the tube with cola.
 2. Advance the tube into the colon.
 3. Apply intermittent suction to the tube.
 4. Withdraw the clog with a 30-ml syringe.

Think: What's the conservative approach?

90. 1. The nurse should irrigate the tube with cola because its effervescence and acidity are suited to the purpose, it's inexpensive, and it's readily available in most homes. Advancing the NG tube is inappropriate because the tube is designed to stay in the stomach and isn't long enough to reach the intestines. Applying intermittent suction or using a syringe for aspiration is unlikely to dislodge the clog and may create excess pressure. Intermittent suction may even collapse the tube.

NP: Implementation; CN: Physiological integrity; CNS: Basic care and comfort; CL: Application

91. The client with a duodenal ulcer may exhibit which of the following signs and symptoms?

 1. Hematemesis
 2. Malnourishment
 3. Melena
 4. Pain with eating

91. 3. The client with a duodenal ulcer may have bleeding at the ulcer site, which shows up as melena. The other findings are consistent with a gastric ulcer.

NP: Data collection; CN: Physiological integrity; CNS: Physiological adaptation; CL: Comprehension

92. Which of the following characteristics is associated with most stress ulcers?

 1. A single, well-defined lesion
 2. Increased gastric acid production
 3. Decreased gastric mucosal blood flow
 4. Increased blood flow to gastric mucosa

Concentrate. This is a tough one.

92. 3. Contrary to popular belief, overproduction of gastric acid is rarely the cause of stress ulcers. Rather, stress states decrease gastric mucosal blood flow, resulting in mucosal breakdown. Stress ulcers are usually more shallow and diffuse than peptic ulcers, which tend to be singular and well-defined.

NP: Data collection; CN: Physiological integrity; CNS: Physiological adaptation; CL: Comprehension

93. The nurse is caring for a client with suspected upper GI bleeding. The nurse should monitor this client for:

 1. hemoptysis.
 2. hematuria.
 3. passage of bright red blood in the stool.
 4. black, tarry stools.

93. 4. As blood from the GI tract passes through the intestines, bacterial action causes it to become black. Hemoptysis involves coughing up blood from the lungs. Hematuria is blood in the urine. Bright red blood in the stools indicates bleeding from the lower GI tract.

NP: Data collection; CN: Physiological integrity; CNS: Physiological adaptation; CL: Application

94. Which of the following tests is the major diagnostic test for ulcers?

 1. Abdominal X-ray
 2. Barium swallow
 3. Computed tomography (CT) scan
 4. Esophagogastroduodenoscopy (EGD)

94. 4. The EGD can visualize the entire upper GI tract as well as allow for tissue specimens and electrocautery if needed. The barium swallow could locate a gastric ulcer. A CT scan and an abdominal X-ray aren't useful in the diagnosis of an ulcer.

NP: Data collection; CN: Health promotion and maintenance; CNS: Prevention and early detection of disease; CL: Knowledge

95. Which of the following best describes the method of action of medications such as ranitidine (Zantac), which are used in the treatment of peptic ulcer disease?
1. Neutralize acid
2. Reduce acid secretions
3. Stimulate gastrin release
4. Protect the mucosal barrier

The answer to question 95 is no secret, if you get my drift!

95. 2. Ranitidine is a histamine-2 receptor antagonist that reduces acid secretion by inhibiting gastrin secretion. Antacids neutralize acid, and mucosal barrier fortifiers protect the mucosal barrier.

NP: Data collection; CN: Physiological integrity; CNS: Pharmacological therapies; CL: Knowledge

96. A client with a peptic ulcer is receiving Maalox and oral cimetidine (Tagamet). The nurse administers the drugs correctly by giving:
1. Cimetidine with meals and Maalox 1 hour after meals.
2. Maalox and cimetidine together with meals.
3. Maalox and cimetidine together between meals.
4. Maalox and cimetidine together with other scheduled medications.

96. 1. Maalox should be given 1 hour after meals and at bedtime, and cimetidine should be given with meals and at bedtime. These drugs should be given at least 1 hour apart. Maalox and cimetidine shouldn't be given together.

NP: Implementation; CN: Physiological integrity; CNS: Pharmacological therapies; CL: Application

97. Which of the following instructions should the nurse give a client with pancreatitis during discharge teaching?
1. Consume high-fat meals.
2. Consume low-calorie meals.
3. Limit daily intake of alcohol.
4. Avoid beverages that contain caffeine.

97. 4. A client with pancreatitis must avoid foods or beverages that can cause a relapse of the disease. Caffeine must be avoided because it's a stimulant that will further irritate the pancreas. The client with pancreatitis must avoid all alcohol because chronic alcohol use is one of the causes of pancreatitis. The diet should be high in calories, especially carbohydrates, and low in fats.

NP: Implementation; CN: Physiological integrity; CNS: Reduction of risk potential; CL: Application

98. A client with irritable bowel syndrome is being prepared for discharge. Which of the following meal plans should the nurse give the client?
1. Low-fiber, low-fat
2. High-fiber, low-fat
3. Low-fiber, high-fat
4. High-fiber, high-fat

Congratulations! You finished!

98. 2. The client with irritable bowel syndrome needs to be on a diet that contains at least 25 g of fiber per day. Fatty foods are to be avoided because they may precipitate symptoms.

NP: Implementation; CN: Physiological integrity; CNS: Reduction of risk potential; CL: Application

NP: Nursing process   CN: Client needs category   CNS: Client needs subcategory   CL: Cognitive level

Chapter 9
Endocrine disorders

Make sure you don't confuse hyper and hypo!

1. Which of the following signs and symptoms indicate hyperglycemia?
1. Polydipsia, polyuria, and weight loss
2. Weight gain, tiredness, and bradycardia
3. Irritability, diaphoresis, and tachycardia
4. Diarrhea, abdominal pain, and weight loss

2. A client presents with diaphoresis, palpitations, jitters, and tachycardia approximately 1½ hours after taking his regular morning insulin. Which of the following treatments is appropriate for this client?
1. Check blood glucose level, and administer carbohydrates.
2. Give nitroglycerin, and perform an electrocardiogram (ECG).
3. Check pulse oximetry, and administer oxygen therapy.
4. Restrict salt, administer diuretics, and perform paracentesis.

3. Which of the following medication orders would be appropriate on the morning of surgery for a client with type 1 diabetes mellitus?
1. The client should take half of his usual daily insulin.
2. The client should receive an oral antidiabetic agent.
3. The client should receive an I.V. insulin infusion.
4. The client should take his full daily insulin dose with no dextrose infusion.

1. 1. Symptoms of hyperglycemia include polydipsia, polyuria, and weight loss. Weight gain, tiredness, and bradycardia are symptoms of hypothyroidism. Irritability, diaphoresis, and tachycardia are symptoms of hypoglycemia. Symptoms of Crohn's disease include diarrhea, abdominal pain, and weight loss.
NP: Data collection; CN: Physiological integrity; CNS: Reduction of risk potential; CL: Knowledge

2. 1. The client is experiencing symptoms of hypoglycemia. Checking the blood glucose level and administering carbohydrates will elevate blood glucose. Giving insulin will further lower blood glucose. ECG and nitroglycerin are treatments for myocardial infarction. Administering oxygen won't help correct the low blood glucose level. Restricting salt, administering diuretics, and performing paracentesis are treatments for ascites.
NP: Data collection; CN: Physiological integrity; CNS: Physiological adaptation; CL: Comprehension

3. 1. If the client takes his full daily dose of insulin when he isn't allowed anything orally before surgery, he'll become hypoglycemic. Half the insulin dose will provide all that's needed. Clients with type 1 diabetes don't take oral antidiabetic agents. I.V. insulin infusions aren't standard for routine surgery; they're used in the management of clients undergoing stressful procedures, such as transplants or coronary artery bypass surgery.
NP: Planning; CN: Physiological integrity; CNS: Physiological adaptation; CL: Comprehension

4. Which of the following types of diabetes is controlled primarily through diet, exercise, and oral antidiabetic agents?

    1. Diabetes insipidus
    2. Diabetic ketoacidosis
    3. Type 1 diabetes mellitus
    4. Type 2 diabetes mellitus

Reread this question carefully. Controlled is the key word.

4. 4. Type 2 diabetes mellitus is controlled primarily through diet, exercise, and oral antidiabetic agents. Diet and exercise are important in type 1 diabetes mellitus, but blood glucose levels are controlled by insulin injections in that disorder. Desmopressin acetate, a long-acting vasopressin given intranasally, is the treatment of choice for diabetes insipidus. Treatment for diabetic ketoacidosis includes restoration of fluid volume, electrolyte management, reversal of acidosis, and control of blood glucose.

NP: Data collection; CN: Physiological integrity; CNS: Reduction of risk potential; CL: Knowledge

5. Which of the following nursing interventions should be taken for a client who complains of nausea and vomits 1 hour after taking his morning glyburide (DiaBeta)?

    1. Give glyburide again.
    2. Give subcutaneous insulin, and monitor blood glucose.
    3. Monitor blood glucose closely, and look for signs of hypoglycemia.
    4. Monitor blood glucose, and assess for symptoms of hyperglycemia.

5. 3. When a client who has taken an oral antidiabetic agent vomits, the nurse would monitor glucose and assess him frequently for signs of hypoglycemia. Most of the medication has probably been absorbed. Therefore, repeating the dose would further lower glucose levels later in the day. Giving insulin will also lower glucose levels, causing hypoglycemia. The client wouldn't have hyperglycemia if the glyburide were absorbed.

NP: Data collection; CN: Physiological integrity; CNS: Pharmacological therapies; CL: Analysis

Nutritional education is crucial to your diabetic client's recovery.

6. When teaching a newly diagnosed diabetic client about diet and exercise, it's important to include which of the following aspects?

    1. Use of fiber laxatives and bulk-forming agents
    2. Management of fluid, protein, and electrolytes
    3. Reduction of calorie intake before exercising
    4. Caloric goals, food consistency, and physical activity

6. 4. Diabetic clients must be taught the relationship among caloric goals, consistency of food composition, and physical activity. Management of fluids, proteins, and electrolytes is important for a client with acute renal failure. Fiber laxatives and bulk-forming agents are treatments for constipation. The diabetic client may need to intake additional calories before exercising.

NP: Planning; CN: Health promotion and maintenance; CNS: Prevention and early detection of disease; CL: Comprehension

7. Which of the following chronic complications is associated with diabetes mellitus?

    1. Dizziness, dyspnea on exertion, and angina
    2. Retinopathy, neuropathy, and coronary artery disease
    3. Leg ulcers, cerebral ischemic events, and pulmonary infarcts
    4. Fatigue, nausea, vomiting, muscle weakness, and cardiac arrhythmias

7. 2. Retinopathy, neuropathy, and coronary artery disease are all chronic complications of diabetes mellitus. Dizziness, dyspnea on exertion, and angina are symptoms of aortic valve stenosis. Leg ulcers, cerebral ischemic events, and pulmonary infarcts are complications of sickle cell anemia. Hyperparathyroidism causes fatigue, nausea, vomiting, muscle weakness, and cardiac arrhythmias.

NP: Data collection; CN: Physiological integrity; CNS: Reduction of risk potential; CL: Knowledge

8. Rotating injection sites when administering insulin prevents which of the following complications?

1. Insulin edema
2. Insulin lipodystrophy
3. Insulin resistance
4. Systemic allergic reactions

Rotating injection sites is important for preventing complications.

8. 2. Insulin lipodystrophy produces fatty masses at the injection sites, causing unpredictable absorption of insulin injected into these sites. Insulin edema is generalized retention of fluid, sometimes seen after normal blood glucose levels are established in a client with prolonged hyperglycemia. Insulin resistance occurs mostly in overweight clients and is due to insulin binding with antibodies, decreasing the amount of absorption. Systemic allergic reactions range from hives to anaphylaxis; rotating injection sites won't prevent these.

NP: Planning; CN: Health promotion and maintenance; CNS: Prevention and early detection of disease; CL: Comprehension

9. Which of the following tests allows a rapid measurement of glucose in whole blood?

1. Capillary blood glucose test
2. Serum ketone test
3. Serum T_4 test
4. Urine glucose test

This question is asking for the best test for a specific indication.

9. 1. A capillary blood glucose test is a rapid test used to show blood glucose levels. A urine glucose test monitors glucose levels in urine and is influenced by both glucose and water excretion. Therefore, results correlate poorly with blood glucose levels. A serum T_4 test is used to diagnosis thyroid disorders. A serum ketone test is used to document diabetic ketoacidosis by titration and may allow determination of serum ketone concentration. Most of the time, however, neither serum ketone levels nor T_4 levels are useful in determining blood glucose levels.

NP: Evaluation; CN: Physiological integrity; CNS: Reduction of risk potential; CL: Comprehension

10. How long does the peak effect last for Novolin NPH, an intermediate-acting insulin?

1. 15 minutes to 1 hour
2. 2 to 6 hours
3. 6 to 16 hours
4. 14 to 26 hours

You've finished 10 questions already! Good job!

10. 3. Novolin NPH has a peak effect of 6 to 16 hours. The peak effect of rapid-acting insulin is 2 to 6 hours. Long-acting insulin has a peak effect of 14 to 26 hours. The onset of rapid-acting insulin is 15 minutes to 1 hour.

NP: Data collection; CN: Physiological integrity; CNS: Pharmacological therapies; CL: Knowledge

11. The hormones thyroxine (T_4) and triiodothyronine (T_3) affect which of the following body processes?

1. Blood glucose level and glycogenolysis
2. Growth and development as well as metabolic rate
3. Growth of bones, muscles, and other organs
4. Bone resorption, calcium absorption, and blood calcium levels

11. 2. T_3 and T_4 are thyroid hormones that affect growth and development as well as metabolic rate. Bone resorption and increased calcium absorption are the principal effects of parathyroid hormone. Glucagon raises blood glucose levels and stimulates glycogenolysis. The growth hormone (somatotrophin) affects the growth of bones, muscles, and other organs.

NP: Data collection; CN: Physiological integrity; CNS: Physiological adaptation; CL: Knowledge

12. The nurse should anticipate administration of which of the following medications to a client with hypothyroidism?

1. Dexamethasone
2. Lactulose
3. Levothyroxine
4. Lidocaine

12. 3. Levothyroxine, a synthetic form of the thyroid hormone T_4, is the medication of choice for treating hypothyroidism. Dexamethasone is a steroid and an antithyroid medication. Lactulose is used to produce an osmotic diarrhea, resulting in an acidic diarrhea. Lidocaine is used to treat ventricular arrhythmias.
NP: Implementation; CN: Physiological integrity; CNS: Pharmacological therapies; CL: Knowledge

Client-teaching skills are important for all nurses!

13. The nurse and a client have just discussed the client's recent diagnosis of hypothyroidism and its causes and effects. Which statement indicates that the client needs further instruction?

1. "Now I see. My clumsiness is caused by a hormone problem."
2. "I just eat too much. That's why I'm depressed and overweight."
3. "No wonder I'm constipated. I'm predisposed to it no matter what I eat."
4. "I'm not cold all the time because I'm getting older. I'm cold because of a metabolic problem."

13. 2. Hypothyroidism causes inadequate secretion of thyroid hormones, which slows metabolic processes and can cause depression and weight gain. A client with hypothyroidism who insists that overeating has caused depression and obesity indicates a need for further instruction about the effects of the disease. The other statements reflect an accurate understanding that hypothyroidism can cause clumsiness, constipation, and a feeling of coldness.
NP: Evaluation; CN: Physiological integrity; CNS: Reduction of risk potential; CL: Analysis

Don't get myxed up on this one! (That's not a typo; it's a hint!)

14. A client with hypothyroidism who experiences trauma, emergency surgery, or severe infection is at risk for developing which of the following conditions?

1. Hepatitis B
2. Malignant hyperthermia
3. Myxedema coma
4. Thyroid storm

14. 3. Myxedema coma represents the most severe form of hypothyroidism. The client develops severe hypothermia and hypoglycemia and becomes comatose. Myxedema coma can be precipitated by narcotics, stress (such as surgery), trauma, and infections. The client would be hypothermic, not hyperthermic. Thyroid storm is a complication of hyperthyroidism. Hepatitis B is a virus and isn't caused by thyroid disorders.
NP: Data collection; CN: Physiological integrity; CNS: Pharmacological therapies; CL: Application

15. Which of the following potentially serious complications may occur when a client is treated for hypothyroidism?

1. Acute hemolytic reaction
2. Angina or cardiac arrhythmia
3. Retinopathy
4. Thrombocytopenia

15. 2. Precipitation of angina or cardiac arrhythmia is a potentially serious complication of hypothyroidism treatment, especially for elderly clients or those with underlying heart disease. Acute hemolytic reaction is a complication of blood transfusions. Retinopathy is usually a complication of diabetes mellitus. Thrombocytopenia is defined as a platelet count of less than 150,000/µl and doesn't result from treating hypothyroidism.
NP: Evaluation; CN: Physiological integrity; CNS: Reduction of risk potential; CL: Comprehension

16. What tests should the nurse expect to be ordered if hypothyroidism is suspected?
1. Liver function tests
2. Hemoglobin A_{1C}
3. T_4 and thyroid-stimulating hormone (TSH)
4. 24-hour urine free cortisol measurement

Know which tests the physician will order to ensure a proper diagnosis.

16. 3. Levels of TSH and T_4 should be measured if hypothyroidism is suspected. Hemoglobin A_{1C} measurement is used to assess hyperglycemia. Liver function tests are used to determine liver disease. As part of the screening process for Cushing's syndrome, a 24-hour urinary free cortisol measurement is completed.
NP: Planning; CN: Physiological integrity; CNS: Physiological adaptation; CL: Knowledge

17. The nurse can expect to see which of the following symptoms in a client who has hypothyroidism?
1. Polyuria, polydipsia, and weight loss
2. Heat intolerance, nervousness, weight loss, and hair loss
3. Coarsening of facial features and extremity enlargement
4. Tiredness, cold intolerance, weight gain, and constipation

17. 4. Tiredness, cold intolerance, weight gain, and constipation are symptoms of hypothyroidism, secondary to a decrease in cellular metabolism. Polyuria, polydipsia, and weight loss are symptoms of type 1 diabetes mellitus. Hyperthyroidism has symptoms of heat intolerance, nervousness, weight loss, and hair loss. Coarsening of facial features and extremity enlargement are symptoms of acromegaly.
NP: Data collection; CN: Physiological integrity; CNS: Physiological adaptation; CL: Application

18. Which of the following groups of hormones are released by the medulla of the adrenal gland?
1. Epinephrine and norepinephrine
2. Glucocorticoids, mineralocorticoids, and androgens
3. Thyroxine, triiodothyronine, and calcitonin
4. Insulin, glucagon, and somatostatin

18. 1. The medulla of the adrenal gland causes the release of epinephrine and norepinephrine. Glucocorticoids, mineralocorticoids, and androgens are released from the adrenal cortex. Thyroxine, triiodothyronine, and calcitonin are secreted by the thyroid gland. The islet cells of the pancreas secrete insulin, glucagon, and somatostatin.
NP: Data collection; CN: Physiological integrity; CNS: Physiological adaptation; CL: Knowledge

19. When a client is scheduled for a thyroid test, the nurse must determine whether the client has taken any medication containing iodine, which would alter the test results. Which of the following medications contain iodine?
1. Acetaminophen and aspirin
2. Estrogen and amphetamines
3. Insulin and oral antidiabetic agents
4. Contrast media, topical antiseptics, and multivitamins

19. 4. Contrast media, topical antiseptics, and multivitamins contain iodine and can alter thyroid function test results. Estrogen and amphetamines don't contain iodine but may alter thyroid function test results. Insulin, oral antidiabetic agents, acetaminophen, and aspirin won't affect a thyroid test.
NP: Planning; CN: Physiological integrity; CNS: Pharmacological therapies; CL: Knowledge

20. While monitoring a client with hypothyroidism, the nurse would expect to observe:
1. hypoactive bowel sounds.
2. hypertension.
3. photophobia.
4. flushed skin.

Looking good! Keep at it!

20. 1. Hypothyroidism is associated with a general slowing of the body systems, as indicated by hypoactive bowel sounds. The nurse would expect to find the client hypotensive, not hypertensive. Photophobia and flushed skin are symptoms associated with hyperthyroidism.
NP: Data collection; CN: Physiological integrity; CNS: Physiological adaptation; CL: Application

21. A client with Cushing's syndrome is admitted to the medical-surgical unit. During the admission assessment, the nurse notes that the client is agitated and irritable, has poor memory, reports loss of appetite, and appears disheveled. These findings are consistent with which problem?
1. Depression
2. Neuropathy
3. Hypoglycemia
4. Hyperthyroidism

21. 1. Agitation, irritability, poor memory, loss of appetite, and neglect of one's appearance may signal depression, which is common in clients with Cushing's syndrome. Neuropathy affects clients with diabetes mellitus, not Cushing's syndrome. Although hypoglycemia can cause irritability, it also produces increased appetite, rather than loss of appetite. Hyperthyroidism typically causes such signs as goiter, nervousness, heat intolerance, and weight loss despite increased appetite.
NP: Data collection; CN: Psychosocial integrity; CNS: Psychosocial adaptation; CL: Comprehension

22. Which of the following conditions is caused by excessive secretion of vasopressin?
1. Thyrotoxic crisis
2. Diabetes insipidus
3. Primary adrenocortical insufficiency
4. Syndrome of inappropriate antidiuretic hormone secretion (SIADH)

Add another point to your score if you get this one right.

22. 4. SIADH occurs as a result of excessive vasopressin. Diabetes insipidus is a deficiency of vasopressin. Adrenocortical insufficiency (Addison's disease) is caused by a deficiency of cortical hormones. Thyrotoxic crisis occurs with severe hyperthyroidism.
NP: Data collection; CN: Physiological integrity; CNS: Reduction of risk potential; CL: Knowledge

23. A client with muscle weakness, anorexia, dark pigmentation of the skin, and laboratory findings of low serum sodium and high potassium levels may be presenting with which of the following conditions?
1. Addison's disease
2. Cushing's syndrome
3. Diabetes insipidus
4. Thyrotoxic crisis

23. 1. The clinical picture of Addison's disease includes muscle weakness, anorexia, darkening of the skin's pigmentation, low sodium level, and high potassium level. Cushing's syndrome presents with obesity, "buffalo hump," "moonface," and thin extremities. Symptoms of diabetes insipidus include excretion of large volumes of dilute urine, leading to hypernatremia and dehydration. Thyrotoxic crisis can occur with severe hyperthyroidism.
NP: Data collection; CN: Physiological integrity; CNS: Physiological adaptation; CL: Comprehension

NP: Nursing process   CN: Client needs category   CNS: Client needs subcategory   CL: Cognitive level

24. Head trauma, brain tumor, or surgical ablation of the pituitary gland can lead to which of the following conditions?
1. Addison's disease
2. Cushing's syndrome
3. Diabetes insipidus
4. Hypothyroidism

25. If fluid intake is limited in a client with diabetes insipidus, which of the following complications will he be at risk for developing?
1. Hypertension and bradycardia
2. Glucosuria and weight gain
3. Peripheral edema and hyperglycemia
4. Severe dehydration and hypernatremia

Does anyone have a glass of water?

26. Adequate fluid replacement, vasopressin replacement, and correction of underlying intracranial pathology are goals in the medical management of which of the following disease processes?
1. Diabetes mellitus
2. Diabetes insipidus
3. Diabetic ketoacidosis
4. Syndrome of inappropriate antidiuretic hormone secretion (SIADH)

27. Which of the following disorders is suggested by polydipsia and large amounts of waterlike urine with a specific gravity of 1.003?
1. Diabetes mellitus
2. Diabetes insipidus
3. Diabetic ketoacidosis
4. Syndrome of inappropriate antidiuretic hormone secretion (SIADH)

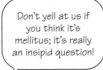

Don't yell at us if you think it's mellitus; it's really an insipid question!

24. 3. The cause of diabetes insipidus is unknown, but it may be secondary to head trauma, brain tumors, or surgical ablation of the pituitary gland. Addison's disease is caused by a deficiency of cortical hormones, whereas Cushing's syndrome is an excess of cortical hormones. Hypothyroidism occurs when the thyroid gland secretes low levels of thyroid hormone.
NP: Data collection; CN: Physiological integrity; CNS: Physiological adaptation; CL: Comprehension

25. 4. A client with diabetes insipidus has high volumes of urine, even without fluid replacement. Therefore, limiting fluid intake will cause severe dehydration and hypernatremia. A client undergoing a fluid deprivation test may experience tachycardia and hypotension. A client with diabetes insipidus will usually experience weight loss, and his urine won't contain glucose. Diabetes insipidus has no effect on blood glucose; therefore, the client wouldn't suffer from hyperglycemia. Peripheral edema isn't a symptom of diabetes insipidus.
NP: Data collection; CN: Physiological integrity; CNS: Physiological adaptation; CL: Comprehension

26. 2. Maintaining adequate fluid and replacing vasopressin are the main objectives in treating diabetes insipidus. An excess of vasopressin leads to SIADH, causing the client to retain fluid. Diabetic ketoacidosis is a result of severe insulin deficiency. Diabetes mellitus doesn't involve vasopressin or an intracranial pathology but rather a disturbance in the production or use of insulin.
NP: Implementation; CN: Physiological integrity; CNS: Pharmacological therapies; CL: Comprehension

27. 2. Diabetes insipidus is characterized by a great thirst (polydipsia) and large amounts of waterlike urine, which has a specific gravity of 1.001 to 1.005. Diabetes mellitus presents with polydipsia, polyuria, and polyphagia, but the client also has hyperglycemia. Diabetic ketoacidosis presents with weight loss, polyuria, and polydipsia, and the client has a severe acidosis. A client with SIADH can't excrete a dilute urine; he retains fluid and develops a sodium deficiency.
NP: Data collection; CN: Physiological integrity; CNS: Physiological adaptation; CL: Analysis

28. Which of the following medications would the nurse expect the physician to order when treating a client with diabetes insipidus?
1. Desmopressin acetate
2. Glucocorticoids
3. Insulin
4. Oral antidiabetic agents

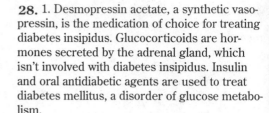

Stay calm. You can pick out the right answer!

28. 1. Desmopressin acetate, a synthetic vasopressin, is the medication of choice for treating diabetes insipidus. Glucocorticoids are hormones secreted by the adrenal gland, which isn't involved with diabetes insipidus. Insulin and oral antidiabetic agents are used to treat diabetes mellitus, a disorder of glucose metabolism.
NP: Implementation; CN: Physiological integrity; CNS: Pharmacological therapies; CL: Knowledge

29. Diabetes insipidus is a disorder of which of the following glands?
1. Adrenal gland
2. Parathyroid gland
3. Pituitary gland
4. Thyroid gland

29. 3. Diabetes insipidus is a disorder of the posterior pituitary gland. The adrenal, thyroid, and parathyroid glands aren't involved.
NP: Data collection; CN: Physiological integrity; CNS: Reduction of risk potential; CL: Knowledge

30. A deficiency of which of the following hormones causes diabetes insipidus?
1. Androgen
2. Epinephrine
3. Norepinephrine
4. Vasopressin

30. 4. Clients with diabetes insipidus have a deficiency of vasopressin, the antidiuretic hormone. Epinephrine, norepinephrine, and androgen are hormones secreted by the adrenal gland and aren't related to diabetes insipidus.
NP: Data collection; CN: Physiological integrity; CNS: Physiological adaptation; CL: Knowledge

31. Which of the following tests would the nurse expect to see ordered for a client suspected of having diabetes insipidus?
1. Capillary blood glucose test
2. Fluid deprivation test
3. Serum ketone test
4. Urine glucose test

You're doing great! Press on!

31. 2. The fluid deprivation test involves withholding water for 4 to 18 hours and checking urine osmolarity periodically. Plasma osmolarity is also checked. A client with diabetes insipidus will have an increased serum osmolarity (of less than 300 mOsm/kg). Urine osmolarity won't increase. The capillary blood glucose test allows a rapid measurement of glucose in whole blood. The serum ketone test documents diabetic ketoacidosis. The urine glucose test monitors glucose levels in urine, but diabetes insipidus doesn't affect urine glucose levels.
NP: Evaluation; CN: Physiological integrity; CNS: Reduction of risk potential; CL: Analysis

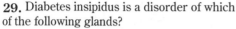

32. Which of the following medical emergencies may occur when the client with Addison's disease develops acute hypotension, secondary to hypoadrenocorticism?
1. Addisonian crisis
2. Diabetic ketoacidosis
3. Myxedema
4. Thyrotoxic crisis

33. Which of the following diseases is caused by a deficiency of cortical hormones?
1. Addison's disease
2. Cushing's syndrome
3. Diabetes mellitus
4. Diabetic ketoacidosis

34. Laboratory findings indicating excessive levels of adrenocortical hormones would correlate with which disease?
1. Addison's disease
2. Cushing's syndrome
3. Diabetes mellitus
4. Hypothyroidism

35. A client presenting with high levels of triiodothyronine (T_3) and thyroxine (T_4) in the blood or urine may have which of the following conditions?
1. Addison's disease
2. Cushing's syndrome
3. Hyperthyroidism
4. Hypopituitarism

36. The nurse would expect the physician to order I.V. hydrocortisone for which of the following diseases?
1. Addison's disease
2. Cushing's syndrome
3. Hyperthyroidism
4. Hypoparathyroidism

There's a hint built into the question on this one.

Read question 35 carefully. It says high levels.

32. 1. As Addison's disease progresses, the client moves into an addisonian crisis, a medical emergency marked by cyanosis, fever, and signs of shock. Diabetic ketoacidosis is a form of hyperglycemia. Myxedema is a form of severe hypothyroidism. Thyrotoxic crisis is a form of severe hyperthyroidism.
NP: Data collection; CN: Physiological integrity; CNS: Physiological adaptation; CL: Comprehension

33. 1. Addison's disease is caused by a deficiency of cortical hormones. Cushing's syndrome is the opposite of Addison's disease and includes excessive adrenocortical activity. Diabetes mellitus is a deficiency of insulin. Diabetic ketoacidosis is a state of severe hyperglycemia, causing acidosis.
NP: Data collection; CN: Physiological integrity; CNS: Physiological adaptation; CL: Knowledge

34. 2. Cushing's syndrome is indicated by excessive levels of adrenocortical hormones. Low levels of glucose and sodium, along with high levels of potassium and white blood cells, are diagnostic of Addison's disease. Diabetes mellitus would have increased blood glucose levels. Hypothyroidism would have low levels of thyroid hormone.
NP: Data collection; CN: Physiological integrity; CNS: Physiological adaptation; CL: Comprehension

35. 3. Hyperthyroidism has high levels of T_3 and T_4. A definitive diagnosis of Addison's disease must have low levels of adrenocortical hormones. Cushing's syndrome would have excessive amounts of adrenocortical hormones. Lower pituitary hormone secretion levels are consistent with hypopituitarism.
NP: Data collection; CN: Physiological integrity; CNS: Physiological adaptation; CL: Comprehension

36. 1. I.V. hydrocortisone is the proper treatment for Addison's disease because it replaces the glucocorticoid deficiency. Cushing's syndrome has excessive amounts of glucocorticoids. Hyperthyroidism and hypoparathyroidism aren't treated with hydrocortisone.
NP: Implementation; CN: Physiological integrity; CNS: Pharmacological therapies; CL: Knowledge

37. An appropriate nursing diagnosis for a client with Addison's disease would include which of the following assessments?
1. Risk for injury
2. Excessive fluid volume
3. Ineffective thermoregulation
4. Impaired gas exchange

Don't risk not reading this question again. (Aren't I clever?)

37. 1. Clients with Addison's disease have inadequate production of adrenal hormones. Therefore, they have inadequate responses to stress. Abdominal and muscle pain are often experienced. Clients with Addison's disease experience fluid volume deficit, secondary to decreased mineralocorticoid secretion. Heat intolerance is a symptom of hyperthyroidism. The respiratory system isn't directly affected, and gas exchange shouldn't be affected.
NP: Data collection; CN: Health promotion and maintenance; CNS: Prevention and early detection of disease; CL: Analysis

38. A 35-year-old client is admitted with a diagnosis of Addison's disease. Which nursing intervention is most appropriate?
1. Provide frequent rest periods.
2. Administer diuretics.
3. Encourage a high-potassium diet.
4. Maintain fluid restrictions.

38. 1. A client with Addison's disease is dehydrated, hypotensive, and very weak. Frequent rest periods are needed to prevent exhausting the client. Diuretics would cause further dehydration and are contraindicated in a client with Addison's disease. Potassium levels are usually elevated in Addison's disease because aldosterone secretion is decreased, resulting in decreased sodium and increased potassium. Fluid intake would be encouraged, not restricted, in a dehydrated client.
NP: Implementation; CN: Physiological integrity; CNS: Physiological adaptation; CL: Analysis

39. A client had a subtotal thyroidectomy in the early morning. During evening rounds, the nurse assesses the client, who now has nausea, a temperature of 105° F (40.6° C), tachycardia, and extreme restlessness. What's the most likely cause of these signs?
1. Diabetic ketoacidosis
2. Thyroid crisis
3. Hypoglycemia
4. Tetany

Wow! You've finished 40 questions already! That's excellent!

39. 2. Thyroid crisis usually occurs in the first 12 hours after thyroidectomy and causes exaggerated signs of hyperthyroidism, such as high fever, tachycardia, and extreme restlessness. Diabetic ketoacidosis is more likely to produce polyuria, polydipsia, and polyphagia. Hypoglycemia typically produces weakness, tremors, profuse perspiration, and hunger. Tetany typically causes uncontrollable muscle spasms, stridor, cyanosis and, possibly, asphyxia.
NP: Data collection; CN: Physiological integrity; CNS: Basic care and comfort; CL: Application

40. Overproduction of which of the following hormones leads to growth arrest, obesity, and musculoskeletal changes?
1. Adrenocortical hormone
2. Follicle-stimulating hormone (FSH)
3. Parathyroid hormone
4. Thyroid hormone

40. 1. Overproduction of adrenocortical hormone leads to growth arrest, obesity, and musculoskeletal changes. FSH is responsible for development of follicles and seminiferous tubules. Parathyroid hormone is responsible mainly for bone resorption. Overproduction of thyroid hormone would lead to weight loss, heat intolerance, and other symptoms of hyperthyroidism.
NP: Data collection; CN: Health promotion and maintenance; CNS: Prevention and early detection of disease; CL: Knowledge

41. A client has thin extremities but an obese truncal area and a "buffalo hump" at the shoulder area. The client also complains of weakness and disturbed sleep. Which of the following disorders is the most likely diagnosis?
1. Addison's disease
2. Cushing's syndrome
3. Graves' disease
4. Hyperparathyroidism

This is a pretty cushy job I have. (I'm kidding. I'm a kidder.)

41. 2. Clients with Cushing's syndrome have truncal obesity with thin extremities and a fatty "buffalo hump" at the back of the neck. Clients with Addison's disease show signs of weakness, anorexia, and dark pigmentation of the skin. Clients with Graves' disease (hyperthyroidism) have symptoms of heat intolerance, irritability, and bulging eyes. Hyperparathyroidism is characterized by osteopenia and renal calculi.

NP: Data collection; CN: Physiological integrity; CNS: Physiological adaptation; CL: Knowledge

42. Sodium and water retention in a client with Cushing's syndrome contributes to which of the following commonly seen disorders?
1. Hypoglycemia and dehydration
2. Hypotension and hyperglycemia
3. Pulmonary edema and dehydration
4. Hypertension and heart failure

42. 4. Increased mineralocorticoid activity in a client with Cushing's syndrome commonly contributes to hypertension and heart failure. Hypoglycemia and dehydration are uncommon in a client with Cushing's syndrome. Diabetes mellitus may develop, but hypotension isn't part of the disease process. Pulmonary edema and dehydration also aren't complications of Cushing's syndrome.

NP: Data collection; CN: Physiological integrity; CNS: Physiological adaptation; CL: Analysis

43. Which roommate is most appropriate for a client with Cushing's syndrome who has a nursing diagnosis of *Risk for infection?*
1. Postmastectomy client
2. Client with a tracheostomy
3. Client with a pressure ulcer
4. Client with diarrhea

43. 1. The postmastectomy client has the least chance of having an infection that could be transferred to the client with Cushing's syndrome. The clients in the other nursing diagnoses are all at higher risk for infection.

NP: Implementation; CN: Safe, effective care environment; CNS: Safety and infection control; CL: Application

44. Which of the following tests is used to diagnose Cushing's syndrome?
1. Fluid deprivation test
2. Glucose tolerance test
3. Low-dose dexamethasone suppression test
4. Thallium stress test

Only one of these tests is used to diagnose Cushing's syndrome. Or so I'm told.

44. 3. A low-dose dexamethasone suppression test is used to detect changes in plasma cortisol levels. A fluid deprivation test is used to diagnosis diabetes insipidus. The glucose tolerance test is used to determine gestational diabetes in pregnant women. A thallium stress test is used to monitor heart function under stress.

NP: Evaluation; CN: Physiological integrity; CNS: Physiological adaptation; CL: Knowledge

45. Treatment for Cushing's syndrome may involve removal of one of the adrenal glands, which could cause a temporary state of which of the following conditions?
1. Hyperkalemia
2. Adrenal insufficiency
3. Excessive levels of adrenal hormone
4. Syndrome of inappropriate antidiuretic hormone secretion (SIADH)

Watch the word temporary.

45. 2. Removing a major source of adrenal hormones may cause a state of temporary adrenal insufficiency, requiring short-term replacement therapy. When both adrenal glands are removed, the client requires lifelong hormone replacement. A client with Cushing's syndrome would have a low—not high—potassium level. The client wouldn't have excessive levels of adrenal hormone if all or part of the adrenal glands were removed. SIADH secretion doesn't involve the adrenal gland; it involves the pituitary gland.
NP: Implementation; CN: Physiological integrity; CNS: Physiological adaptation; CL: Comprehension

46. Which of the following nursing diagnoses is appropriate for a client with Addison's disease?
1. Risk for injury
2. Deficient fluid volume
3. Acute pain with movement
4. Functional urinary incontinence

46. 2. Fluid volume deficit, related to inadequate adrenal hormones, is common in clients with Addison's disease. Clients with Cushing's syndrome have an increased susceptibility to injury or infection, secondary to the immunosuppression caused by excessive cortisol. Pain and functional incontinence aren't common in Cushing's syndrome.
NP: Planning; CN: Physiological integrity; CNS: Reduction of risk potential; CL: Application

47. Which of the following nursing interventions should be performed for a client with Cushing's syndrome?
1. Suggest clothing or bedding that's cool and comfortable.
2. Suggest consumption of high-carbohydrate and low-protein foods.
3. Explain that physical changes are a result of excessive corticosteroids.
4. Explain the rationale for increasing salt and fluid intake in times of illness, increased stress, and very hot weather.

Explaining physical changes that may result from corticosteroid use can help ease your client's mind.

47. 3. Clients with Cushing's syndrome have physical changes related to excessive corticosteroids. Clients with hyperthyroidism are heat intolerant and must have comfortable, cool clothing and bedding. Clients with Cushing's syndrome should have a high-protein, not a low-protein, diet. Clients with Addison's disease must increase sodium intake and fluid intake in times of stress to prevent hypotension.
NP: Data collection; CN: Physiological integrity; CNS: Physiological adaptation; CL: Application

48. Which of the following disease processes is caused by an absence of insulin or an inadequate amount of insulin, resulting in hyperglycemia and leading to a series of biochemical disorders?
1. Diabetes insipidus
2. Hyperaldosteronism
3. Diabetic ketoacidosis
4. Hyperosmolar hyperglycemic nonketotic syndrome (HHNS)

48. 3. Diabetic ketoacidosis is caused by inadequate amounts of insulin or absence of insulin and leads to a series of biochemical disorders. Diabetes insipidus is caused by a deficiency of vasopressin. Hyperaldosteronism is an excess in aldosterone production, causing sodium and fluid excesses and hypertension. HHNS is a coma state in which hyperglycemia and hyperosmolarity dominate.
NP: Data collection; CN: Physiological integrity; CNS: Physiological adaptation; CL: Knowledge

NP: Nursing process   CN: Client needs category   CNS: Client needs subcategory   CL: Cognitive level

49. An insulin-dependent client who fails to take insulin regularly is at risk for which of the following complications?
1. Diabetic ketoacidosis
2. Hypoglycemia
3. Pancreatitis
4. Respiratory failure

49. 1. A client who fails to regularly take his insulin is at risk for hyperglycemia, which could lead to diabetic ketoacidosis. Hypoglycemia wouldn't occur because the lack of insulin would lead to increased levels of sugar in the blood. A client with chronic pancreatitis may develop diabetes (secondary to the pancreatitis), but insulin-dependent diabetes mellitus doesn't lead to pancreatitis. Respiratory failure isn't related to insulin levels.

NP: Evaluation; CN: Physiological integrity; CNS: Physiological adaptation; CL: Application

50. A diabetic client presents with polyphagia, polydipsia, and oliguria; he also complains of headache, malaise, and some visual changes. Assessment shows signs of dehydration. Which of the following diagnoses could be made?
1. Diabetes insipidus
2. Diabetic ketoacidosis
3. Hypoglycemia
4. Syndrome of inappropriate antidiuretic hormone secretion (SIADH)

Good going! You're over halfway finished!

50. 2. Early manifestations of diabetic ketoacidosis include polydipsia, polyphagia, and polyuria. As the client dehydrates and loses electrolytes, this condition often leads to oliguria, malaise, and visual changes. Diabetes insipidus may result in dehydration but not polyphagia and polydipsia. Symptoms of hypoglycemia include diaphoresis, tachycardia, and nervousness. A client with SIADH is unable to excrete a dilute urine, causing hypernatremia.

NP: Data collection; CN: Physiological integrity; CNS: Physiological adaptation; CL: Application

51. The thyroid gland is located in which of the following areas of the body?
1. Upper abdomen
2. Inferior aspect of the brain
3. Upper portion of the kidney
4. Lower neck, anterior to the trachea

51. 4. The thyroid gland is in the lower neck, anterior to the trachea. The pancreas is in the upper abdomen. The pituitary gland is located in the inferior aspect of the brain. The adrenal glands are attached to the upper portion of the kidneys.

NP: Data collection; CN: Health promotion and maintenance; CNS: Prevention and early detection of disease; CL: Knowledge

Do you know which glands produce which hormones? Here's a chance to test your knowledge.

52. The thyroid gland produces which of the following groups of hormones?
1. Amylase, lipase, and trypsin
2. Thyroxine (T_4), triiodothyronine (T_3), and calcitonin
3. Glucocorticoids, mineralocorticoids, and androgens
4. Vasopressin, oxytocin, and thyroid-stimulating hormone (TSH)

52. 2. T_4, T_3, and calcitonin are all secreted by the thyroid gland. Amylase, lipase, and trypsin are enzymes produced by the pancreas that aid in digestion. Glucocorticoids, mineralocorticoids, and androgens are produced by the adrenal gland. The pituitary gland secretes vasopressin, oxytocin, and TSH.

NP: Data collection; CN: Health promotion and maintenance; CNS: Prevention and early detection of disease; CL: Knowledge

53. Secretion of thyroid-stimulating hormone (TSH) by which of the following glands controls the rate at which thyroid hormone is released?
1. Adrenal gland
2. Parathyroid gland
3. Pituitary gland
4. Thyroid gland

TSH is a trophic hormone with a target gland.

53. 3. By secreting TSH, the pituitary gland controls the rate of thyroid hormone released. The adrenal gland isn't involved with the release of thyroid hormone. The parathyroid gland secretes parathyroid hormones, depending on the levels of calcium and phosphorus in the blood. The thyroid gland secretes thyroid hormone but doesn't control how much is released.
NP: Data collection; CN: Physiological integrity; CNS: Physiological adaptation; CL: Knowledge

54. Which of the following treatments can be used for hyperthyroidism?
1. Cholelithotomy
2. Irradiation of the thyroid
3. Administration of oral thyroid hormones
4. Whipple procedure

54. 2. Irradiation, involving administration of ^{131}I, destroys the thyroid gland and thereby treats hyperthyroidism. Oral thyroid hormones are the treatment for hypothyroidism. Cholelithotomy is used to treat gallstones. The Whipple procedure is a surgical treatment for pancreatic cancer.
NP: Data collection; CN: Physiological integrity; CNS: Pharmacological therapies; CL: Knowledge

Elderly clients are different from other clients. Be aware of age-related symptoms.

55. Which of the following group of symptoms of hyperthyroidism is most commonly found in elderly clients?
1. Depression, apathy, and weight loss
2. Palpitations, irritability, and heat intolerance
3. Cold intolerance, weight gain, and thinning hair
4. Numbness, tingling, and cramping of extremities

55. 1. Most elderly clients present with depression, apathy, and weight loss, which are typical signs and symptoms of hyperthyroidism. Palpitations, irritability, and heat intolerance can be present with hyperthyroidism, but these aren't typical symptoms in elderly clients. Cold intolerance, weight gain, and thinning hair are some of the signs of hypothyroidism. Numbness, tingling, and cramping of extremities are symptoms of hypocalcemia, which may be a symptom of hypoparathyroidism.
NP: Data collection; CN: Physiological integrity; CNS: Physiological adaptation; CL: Application

56. Which of the following forms of severe hyperthyroidism is life-threatening and produces high fever, extreme tachycardia, and altered mental status?
1. Hepatic coma
2. Thyroid storm
3. Myxedema coma
4. Hyperosmolar hyperglycemic nonketotic syndrome (HHNS)

56. 2. Thyroid storm is a form of severe hyperthyroidism that can be precipitated by stress, injury, or infection. Hepatic coma occurs in clients with profound liver failure. Myxedema coma is a rare disorder characterized by hypoventilation, hypotension, hypoglycemia, and hypothyroidism. HHNS occurs in clients with type 2 diabetes mellitus who are dehydrated and have severe hyperglycemia.
NP: Data collection; CN: Physiological integrity; CNS: Physiological adaptation; CL: Knowledge

57. Hyperthyroidism is commonly known as which of the following disorders?
1. Addison's disease
2. Buerger's disease
3. Cushing's syndrome
4. Graves' disease

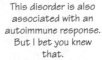

This disorder is also associated with an autoimmune response. But I bet you knew that.

57. 4. Hyperthyroidism is known as Graves' disease. Addison's disease is a deficiency of cortical hormones. Cushing's syndrome is an excess of cortical hormones. Buerger's disease is a recurring inflammation of blood vessels in the upper and lower extremities.
NP: Data collection; CN: Physiological integrity; CNS: Physiological adaptation; CL: Knowledge

58. Excessive output of thyroid hormone from abnormal stimulation of the thyroid gland is the etiology of which condition?
1. Hyperparathyroidism
2. Hypoparathyroidism
3. Hyperthyroidism
4. Hypothyroidism

58. 3. Excessive output of thyroid hormone is the etiology of hyperthyroidism. Hyperparathyroidism is overproduction of parathyroid hormone, characterized by bone calcification or kidney stones. Hypoparathyroidism is inadequate secretion of parathyroid hormone that occurs after interruption of the blood supply or surgical removal of the parathyroid. Hypothyroidism is slow production of thyroid hormone.
NP: Data collection; CN: Physiological integrity; CNS: Physiological adaptation; CL: Knowledge

59. A client presents with flushed skin, bulging eyes, and perspiration, and states that he has been "irritable" and having palpitations. This client is presenting with symptoms of which disorder?
1. Hyperthyroidism
2. Myocardial infarction
3. Pancreatitis
4. Type 1 diabetes mellitus

59. 1. Signs and symptoms of hyperthyroidism include nervousness, palpitations, irritability, bulging eyes, heat intolerance, weight loss, and weakness. Myocardial infarction usually presents with chest pain that may radiate to arms, back, or neck and shortness of breath. Pancreatitis presents with severe abdominal pain and back tenderness. Type 1 diabetes mellitus presents with polyuria, polydipsia, and weight loss.
NP: Data collection; CN: Physiological integrity; CNS: Physiological adaptation; CL: Comprehension

60. What can the nurse do to prevent lipodystrophy when administering insulin to a diabetic client?
1. Grasp the skin tightly.
2. Rotate injection sites.
3. Massage the site vigorously.
4. Inject into the deltoid muscle.

Another 10 down! Great job!

60. 2. The nurse should rotate insulin injection sites systematically to minimize tissue damage, promote absorption, avoid discomfort, and prevent lipodystrophy. Grasping the skin tightly can traumatize the skin; therefore, the nurse should hold it gently but firmly when giving an injection. Massaging the site vigorously is contraindicated because it hastens insulin absorption, which isn't recommended. Usually, insulin is administered subcutaneously, rather than into a muscle, because muscular activity increases insulin absorption.
NP: Implementation; CN: Physiological integrity; CNS: Pharmacological therapies; CL: Application

61. A client is brought into the emergency department with a brain stem contusion. Two days after admission, the client has a large amount of urine and a serum sodium level of 155 mEq/dl. Which of the following conditions may be developing?
1. Myxedema coma
2. Diabetes insipidus
3. Type 1 diabetes mellitus
4. Syndrome of inappropriate antidiuretic hormone secretion (SIADH)

61. 2. Two leading causes of diabetes insipidus are hypothalamic or pituitary tumors and closed head injuries. Myxedema coma is a form of hypothyroidism. Type 1 diabetes mellitus isn't caused by a brain injury. A client with SIADH would have hyponatremia; this client's sodium level was 155 mEq/dl, which is above normal levels of 135 to 145 mEq/dl.

NP: Data collection; CN: Physiological integrity; CNS: Physiological adaptation; CL: Application

62. Which of the following conditions could be diagnosed in a client with serum ketones and a serum glucose level above 300 mg/dl?
1. Diabetes insipidus
2. Diabetic ketoacidosis
3. Hypoglycemia
4. Somogyi phenomenon

62. 2. Clients with serum ketones and serum glucose levels above 300 mg/dl could be diagnosed with diabetic ketoacidosis. Diabetes insipidus is an overproduction of antidiuretic hormone and doesn't create ketones in the blood. Hypoglycemia causes low blood glucose levels. The Somogyi phenomenon is rebound hyperglycemia following an episode of hypoglycemia.

NP: Data collection; CN: Physiological integrity; CNS: Physiological adaptation; CL: Knowledge

Before you answer, make sure you know which disorder the question is asking you to treat.

63. Objectives for treating diabetic ketoacidosis include administration of which of the following treatments?
1. Glucagon
2. Blood products
3. Glucocorticoids
4. Insulin and I.V. fluids

63. 4. A client with diabetic ketoacidosis would receive insulin to lower glucose and I.V. fluids to correct hypotension. Blood products aren't needed to correct diabetic ketoacidosis. Glucagon is given to treat hypoglycemia; diabetic ketoacidosis involves hyperglycemia. Glucocorticoids aren't needed because the adrenal glands aren't involved.

NP: Implementation; CN: Physiological integrity; CNS: Pharmacological therapies; CL: Application

64. Which of the following methods of insulin administration would the nurse expect to be used in the initial treatment of hyperglycemia in a client with diabetic ketoacidosis?
1. Subcutaneous
2. Intramuscular
3. I.V. bolus only
4. I.V. bolus, followed by continuous infusion

64. 4. An I.V. bolus of insulin is given initially to control the hyperglycemia, followed by a continuous infusion, titrated to control blood glucose. After the client is stabilized, subcutaneous insulin is given. Insulin is never given intramuscularly.

NP: Implementation; CN: Physiological integrity; CNS: Pharmacological therapies; CL: Application

Isn't life a bolus of cherries? (They told me to say that!)

65. Insulin forces which of the following electrolytes out of the plasma and into the cells?
1. Calcium
2. Magnesium
3. Phosphorus
4. Potassium

66. Which of the following combinations of adverse effects must be carefully monitored when administering I.V. insulin to a client diagnosed with diabetic ketoacidosis?
1. Hypokalemia and hypoglycemia
2. Hypocalcemia and hyperkalemia
3. Hyperkalemia and hyperglycemia
4. Hypernatremia and hypercalcemia

The correct answers may appear in more than one choice. Make sure you get the right combo!

67. A diabetic client suddenly develops hypoglycemia. What should the nurse do first?
1. Give the client a glass of orange juice to drink.
2 Administer 5 more units of regular insulin.
3. Check the client's blood glucose level.
4. Call the client's physician.

68. For a diabetic patient with a foot ulcer, the physician orders bed rest, a wet-to-dry dressing change every shift, and blood glucose monitoring before meals and at bedtime. Why are wet-to-dry dressings used for this client?
1. They contain exudate and provide a moist wound environment.
2. They protect the wound from mechanical trauma and promote healing.
3. They debride the wound and promote healing by secondary intention.
4. They prevent the entrance of microorganisms and minimize wound discomfort.

A right answer here shows you've been doing your homework.

65. 4. Insulin forces potassium out of the plasma, back into the cells, causing hypokalemia. Potassium is needed to help transport glucose and insulin into the cells. Calcium, magnesium, and phosphorus aren't affected by insulin.
NP: Data collection; CN: Physiological integrity; CNS: Physiological adaptation; CL: Knowledge

66. 1. Blood glucose needs to be monitored because there's a chance for hypokalemia or hypoglycemia. Hypokalemia might occur because I.V. insulin forces potassium into cells, thereby lowering the plasma levels of potassium. Hypoglycemia might occur if too much insulin is administered. The client wouldn't have hyperkalemia. Calcium and sodium levels aren't affected.
NP: Data collection; CN: Physiological integrity; CNS: Pharmacological therapies; CL: Application

67. 1. Drinking a glass of orange juice should raise the client's blood glucose level, thus correcting hypoglycemia. Receiving additional insulin would lower the client's blood glucose level even further, causing hypoglycemia to worsen. The nurse shouldn't take the time to check the client's blood glucose level or call the physician at this time because the client needs immediate attention to prevent loss of consciousness.
NP: Implementation; CN: Physiological integrity; CNS: Physiological adaptation; CL: Application

68. 3. For this client, wet-to-dry dressings are most appropriate because they clean the foot ulcer by debriding exudate and necrotic tissue, thus promoting healing by secondary intention. Moist, transparent dressings contain exudate and provide a moist wound environment. Hydrocolloid dressings prevent the entrance of microorganisms and minimize wound discomfort. Dry sterile dressings protect the wound from mechanical trauma and promote healing.
NP: Implementation; CN: Physiological integrity; CNS: Physiological adaptation; CL: Knowledge

69. A diabetic client who had a stroke has right-sided paralysis and incontinence and is in the rehabilitation center. Which action should be the nurse's <u>priority</u> in caring for the client?

1. Apply body powder every 4 hours to keep the client dry.
2. To conserve energy, maintain bed rest when the client isn't in therapy.
3. Insert an indwelling urinary catheter to keep the client continent.
4. Wash the client's skin with soap and water, gently patting it dry.

This question is asking you to prioritize!

69. 4. The skin of a diabetic client should be kept dry to prevent breakdown and infection. The nurse should avoid excessive use of powders, which can cake with perspiration and cause irritation. Clients undergoing rehabilitation should be upright in a chair, except for short rest periods during the day, to promote optimal recovery. Diabetic clients are especially prone to infections. Urinary tract infections are commonly caused by the use of indwelling catheters. Other methods should be used to encourage continence.
NP: Planning; CN: Physiological integrity; CNS: Basic care and comfort; CL: Application

70. An 86-year-old female client who has been on long-term steroid therapy now has drug-induced Cushing's syndrome. She's residing in an extended-care facility because of her multiple chronic health problems. Which condition is closely related to chronic use of steroids?

1. Periods of hypoglycemia
2. Muscle wasting in the abdominal area
3. Thin, easily damaged skin
4. Weight loss

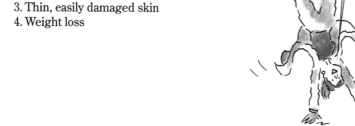

Don't stop now! You're on a roll!

70. 3. Clients taking steroids on a long-term basis lose subcutaneous fat under the skin and are especially vulnerable to skin breakdown and bruising. Such clients should take great care when performing tasks that may injure the skin and should anticipate delayed healing when injuries occur. Clients taking long-term steroids are likely to have hyperglycemia. Typically, clients with muscle wasting in the abdominal area experience weight gain in the abdomen and thinning of the extremities while on steroids. Clients who experience weight loss should be monitored for weight gain and edema.
NP: Data collection; CN: Physiological integrity; CNS: Pharmacological therapies; CL: Application

71. The nurse is assessing a 7-year-old child with possible hyperaldosteronism. Which factor is the nurse's <u>top priority</u> in assessing a child with hyperaldosteronism?

1. Blood glucose level
2. Breath sounds
3. Arterial blood gas (ABG) levels
4. Weekly weights

71. 2. Aldosterone is secreted by the adrenal cortex. One of its major functions is causing the kidneys to retain saline in the body. A child with hyperaldosteronism should be observed for signs of respiratory distress, crackles, and the use of accessory muscles. Blood glucose and ABG levels aren't immediate priorities when assessing a child with hyperaldosteronism. Rapid weight gain can indicate fluid volume excess, but the child should be weighed daily, not weekly.
NP: Data collection; CN: Physiological integrity; CNS: Physiological adaptation; CL: Comprehension

72. Which of the following conditions is characterized by osteopenia and renal calculi?

1. Hyperparathyroidism
2. Hypoparathyroidism
3. Hypopituitarism
4. Hypothyroidism

Again, make sure you don't confuse hypo and hyper.

72. 1. Hyperparathyroidism is characterized by osteopenia and renal calculi, secondary to overproduction of parathyroid hormone. Hypoparathyroidism's main symptoms include tetany from hypocalcemia. Hypopituitarism presents extreme weight loss and atrophy of all endocrine glands. Symptoms of hypothyroidism include hair loss, weight gain, and cold intolerance.

NP: Data collection; CN: Physiological integrity; CNS: Physiological adaptation; CL: Comprehension

73. Renal calculi formation in clients with hyperparathyroidism is caused by an excess of which of the following electrolytes?

1. Calcium and magnesium
2. Calcium and phosphorus
3. Potassium and magnesium
4. Potassium and phosphorus

Be careful! Look for the correct combo!

73. 2. Renal calculi usually consist of calcium and phosphorus. In hyperparathyroidism, serum calcium levels are high, leading to kidney stone formation. Potassium and magnesium don't form renal calculi, and levels of these aren't high in clients with hyperparathyroidism.

NP: Data collection; CN: Physiological integrity; CNS: Physiological adaptation; CL: Comprehension

74. Which of the following laboratory results supports a diagnosis of primary hyperparathyroidism?

1. High parathyroid hormone and high calcium levels
2. High magnesium and high thyroid hormone levels
3. Low parathyroid hormone and low potassium levels
4. Low thyroid-stimulating hormone (TSH) and high phosphorus levels

74. 1. A diagnosis of primary hyperparathyroidism is established based on increased serum calcium levels and elevated parathyroid hormone levels. Potassium, magnesium, TSH, and thyroid hormone levels aren't used to diagnose hyperparathyroidism.

NP: Data collection; CN: Physiological integrity; CNS: Physiological adaptation; CL: Application

75. Which of the following types of medication is contraindicated in the treatment of clients with hyperparathyroidism?

1. Acetaminophen
2. Aspirin
3. Potassium-wasting diuretics
4. Thiazide diuretics

Caution: The word contraindicated makes this almost a negative question.

75. 4. Thiazide diuretics shouldn't be used because they decrease renal excretion of calcium, thereby raising serum calcium levels even higher. There are no contraindications to aspirin or acetaminophen for clients with hyperparathyroidism. Potassium loss isn't an issue for clients with hyperparathyroidism.

NP: Planning; CN: Physiological integrity; CNS: Pharmacological therapies; CL: Application

76. In clients with hyperparathyroidism, which of the following calcium levels would be considered an acute hypercalcemic crisis?

1. 2 mg/dl
2. 4 mg/dl
3. 10.5 mg/dl
4. 15 mg/dl

76. 4. Normal calcium levels are 8.5 to 10.5 mg/dl, so a level of 15 mg/dl is dangerously high.

NP: Evaluation; CN: Physiological integrity; CNS: Physiological adaptation; CL: Knowledge

77. Hyperphosphatemia and hypocalcemia are indicative of which of the following disorders?
1. Cushing's syndrome
2. Graves' disease
3. Hypoparathyroidism
4. Hypothyroidism

Choose carefully. This question is asking for the chief sign.

77. 3. Symptoms of hypoparathyroidism include hyperphosphatemia and hypocalcemia. Excessive adrenocortical activity indicates Cushing's syndrome. Excessive thyroid hormone levels indicate Graves' disease (hyperthyroidism). Low thyroid hormone levels indicate hypothyroidism.

NP: Data collection; CN: Physiological integrity; CNS: Physiological adaptation; CL: Knowledge

78. Which of the following symptoms is the chief sign of hypoparathyroidism?
1. Chest pain
2. Exophthalmos
3. Shortness of breath
4. Tetany

78. 4. Tetany is the chief symptom of hypoparathyroidism. Shortness of breath and chest pain aren't usually symptoms of hypoparathyroidism. Exophthalmos, or bulging eyes, is a common symptom of hyperthyroidism.

NP: Data collection; CN: Physiological integrity; CNS: Physiological adaptation; CL: Knowledge

79. Which of the following vitamins is used to enhance absorption of calcium from the GI tract?
1. Vitamin A
2. Vitamin C
3. Vitamin D
4. Vitamin E

Question 79 asks about enhancing absorption.

79. 3. Variable doses of vitamin D preparations enhance the absorption of calcium from the GI tract. Vitamins A, C, and E aren't involved with this process.

NP: Implementation; CN: Physiological integrity; CNS: Pharmacological therapies; CL: Knowledge

80. The nurse is caring for a 30-year-old female client who had a subtotal thyroidectomy to treat hyperthyroidism. Which sign or symptom could indicate bleeding at the incision site?
1. Hoarseness
2. Severe stridor
3. Complaints of pressure at the incision site
4. Difficulty swallowing

80. 3. Complaints of pressure at the incision site indicate postoperative bleeding. Hoarseness or severe stridor indicates damage to the laryngeal nerve. Difficulty swallowing doesn't indicate postoperative bleeding.

NP: Data collection; CN: Physiological integrity; CNS: Physiological adaptation; CL: Application

Congratulations! You've finished 81 questions!

81. Hypersecretion of which of the following hormones would cause pituitary gigantism?
1. Follicle-stimulating hormone (FSH)
2. Growth hormone
3. Parathyroid hormone
4. Thyroid-stimulating hormone (TSH)

81. 2. Hypersecretion of growth hormone causes pituitary gigantism. FSH is involved with the development of ovaries and sperm. Hypersecretion of parathyroid hormone would cause hyperparathyroidism. Hypersecretion of TSH would cause hyperthyroidism.

NP: Data collection; CN: Physiological integrity; CNS: Physiological adaptation; CL: Knowledge

NP: Nursing process   CN: Client needs category   CNS: Client needs subcategory   CL: Cognitive level

82. A female client has been diagnosed with hyperthyroidism. In planning her care, the nurse should give priority to which goal?
1. Keeping the client warm
2. Reducing the client's calorie intake
3. Providing adequate rest and sleep
4. Forcing fluids and roughage

Here's a case for knowing your priorities.

83. The nurse prepares to administer regular insulin to a client based on the physician's order for "Regular insulin 6 units U100." Which statement regarding the insulin is true?
1. The insulin is cloudy.
2. The insulin should be rolled and gently rotated before drawing the medication into the syringe.
3. U100 means that there are 100 units in each milliliter of the insulin, and 6 units are to be administered.
4. The insulin should be drawn up in a tuberculin syringe.

You can call me Nancy, but don't call me Tubercu-Lynn!

84. After undergoing a thyroidectomy, a client develops hypocalcemia and tetany. Which of the following medications should the nurse anticipate administering?
1. Calcium gluconate
2. Potassium chloride
3. Sodium bicarbonate
4. Sodium phosphorus

85. Which of the following disorders may cause acute pancreatitis?
1. Gallstones
2. Crohn's disease
3. High gastric acid levels
4. Low thyroid hormone level

I can't believe we had the gall to ask you this question! (Oooh, that one hurt.)

82. 3. Clients with hyperthyroidism are typically anxious, diaphoretic, nervous, and fatigued. They need a calm, restful environment to relax and get adequate rest and sleep. Clients with hyperthyroidism are usually warm and diaphoretic and need a cool environment. Clients with hyperthyroidism require a high-calorie diet and should eat four to six small meals per day. Hyperthyroidism causes diarrhea from hypermotility of the GI tract. Excessive fluids and foods high in roughage should be avoided.
NP: Planning; CN: Physiological integrity; CNS: Basic care and comfort; CL: Knowledge

83. 3. There are 100 units of insulin in each milliliter of U100 insulin and it should be drawn up in a U100 syringe (orange needle cap). Regular insulin is clear. Cloudy insulin has a zinc precipitate that must be evenly distributed in the solution. Insulin syringes are the only type of syringe used for drawing up insulin.
NP: Implementation; CN: Safe, effective care environment; CNS: Safety and infection control; CL: Knowledge

84. 1. Immediate treatment for a client who develops hypocalcemia and tetany after thyroidectomy is calcium gluconate. Potassium chloride and sodium bicarbonate aren't indicated. Sodium phosphorus wouldn't be given because phosphorus levels are already elevated.
NP: Planning; CN: Physiological integrity; CNS: Pharmacological therapies; CL: Application

85. 1. Gallstones may cause obstruction and swelling at the ampulla of Vater, preventing flow of pancreatic juices into the duodenum and leading to pancreatitis. Gallbladder obstruction and alcoholism are the major causes of acute pancreatitis. Crohn's disease usually involves the terminal ileum and wouldn't affect the pancreas. Low thyroid hormone levels and high gastric acid levels aren't related to an acute pancreatitis attack.
NP: Data collection; CN: Physiological integrity; CNS: Physiological adaptation; CL: Comprehension

86. Severe abdominal pain in the midepigastric region, back tenderness, nausea, and vomiting are symptoms of which of the following conditions?
  1. Acute pancreatitis
  2. Crohn's disease
  3. Hypophysectomy
  4. Pheochromocytoma

87. Which of the following organs can become inflamed from autodigestion by the enzymes it produces, principally trypsin?
  1. Adrenal gland
  2. Appendix
  3. Kidney
  4. Pancreas

88. An acute pancreatitis attack may be more severe when combined with which of the following factors?
  1. Bowel movement
  2. Cold weather
  3. A heavy meal
  4. Hot weather

Alcohol doesn't mix with pancreatitis.

89. If a pancreatitis attack has been brought on by gallstones or gallbladder disease, a client may require reinforcement about the need to follow which of the following types of diets?
  1. High-calorie, high-protein diet
  2. High-fiber diet, encouraging fluid intake
  3. Low-fat diet, avoiding heavy meals
  4. Diet high in protein, calcium, and vitamin D

86. 1. Severe abdominal pain in the midepigastric region, back tenderness, nausea, and vomiting are due caused by irritation of the pancreas. A client with Crohn's disease would have abdominal pain, but the pain would be in the lower quadrant; the client would also have diarrhea. Hypophysectomy is removal of the pituitary gland. Pheochromocytoma is a benign tumor of the adrenal medulla.
NP: Data collection; CN: Physiological integrity; CNS: Physiological adaptation; CL: Knowledge

87. 4. Acute pancreatitis stems from the digestion of the pancreas by the very enzymes it produces, especially trypsin. The kidney, adrenal gland, and appendix don't produce enzymes.
NP: Data collection; CN: Physiological integrity; CNS: Physiological adaptation; CL: Knowledge

88. 3. An acute pancreatitis attack may be more severe after a heavy meal or alcohol ingestion. Eating a heavy meal triggers the need for digestive enzymes, which are trapped in the pancreas. Bowel movements aren't related to the pancreas. Cold and heat intolerance are related to thyroid disorders.
NP: Data collection; CN: Physiological integrity; CNS: Physiological adaptation; CL: Knowledge

89. 3. A client who survives an acute pancreatitis attack caused by gallstones or gallbladder disease requires reinforcement to maintain a low-fat diet and to avoid heavy meals. A high-calorie, high-protein diet is appropriate for clients with hyperthyroidism. A diet high in fiber, encouraging fluid intake, is recommended for constipation. A client with Cushing's syndrome should follow a diet high in protein, calcium, and vitamin D.
NP: Implementation; CN: Physiological integrity; CNS: Reduction of risk potential; CL: Application

90. Which of the following disorders is an inflammatory disease characterized by progressive anatomic and functional destruction of the pancreas?
1. Acute pancreatitis
2. Addison's disease
3. Chronic pancreatitis
4. Graves' disease

You're getting there! Only 6 more to go!

90. 3. Chronic pancreatitis is an inflammatory disease characterized by progressive anatomic and functional destruction of the pancreas. Acute pancreatitis is a digestion of the pancreas by the enzymes it produces. Addison's disease is a deficiency of adrenocortical hormones. Graves' disease is another name for hyperthyroidism.
NP: Data collection; CN: Physiological integrity; CNS: Physiological adaptation; CL: Knowledge

91. A client has diabetic ketoacidosis, secondary to infection. As the condition progresses, which of the following symptoms might the nurse see?
1. Kussmaul's respirations and a fruity odor on the breath
2. Shallow respirations and severe abdominal pain
3. Decreased respirations and increased urine output
4. Cheyne-Stokes respirations and foul-smelling urine

91. 1. Coma and severe acidosis are ushered in with Kussmaul's respirations (very deep but not labored respirations) and a fruity odor on the breath (acidemia). Shallow respirations and severe abdominal pain may be symptoms of pancreatitis. Decreased respirations and increased urine output aren't symptoms related to acidemia. Cheyne-Stokes respirations and foul-smelling urine don't result from diabetic ketoacidosis.
NP: Data collection; CN: Physiological integrity; CNS: Physiological adaptation; CL: Application

92. A client with diabetes mellitus and a hearing impairment is admitted to the medical-surgical unit. For this client, the nursing team is developing a care plan that includes daily self-administration of insulin. Which intervention should the team include in this plan?
1. Use facial expressions as needed.
2. Chew gum while giving instructions.
3. Shine a light on the client's face.
4. Raise an arm or hand to get attention.

The most effective client teaching depends on making sure the client can understand the instructions.

92. 4. To get the client's attention, the nurse should raise an arm or hand before presenting instructions. The nurse should avoid relying on facial expressions because they may be misinterpreted as signs of annoyance or other feelings by a client who depends on visual cue. The nurse shouldn't chew gum when teaching a client with limited hearing because it can distort the sounds that the client can hear. If needed, a light may be shone on the nurse's face — not the client's face — to help the client lip read.
NP: Planning; CN: Safe, effective care environment; CNS: Coordinated care; CL: Application

93. Clients with insulin-dependent diabetes mellitus may require which of the following changes to their daily routine during periods of infection?
1. No changes
2. Less insulin
3. More insulin
4. Oral antidiabetic agents

93. 3. During periods of infection or illness, insulin-dependent clients may need even more insulin, rather than reducing the levels or not making any changes in their daily insulin routines, to compensate for increased blood glucose levels. Clients usually aren't switched from injectable insulin to oral antidiabetic agents during periods of infection.
NP: Data collection; CN: Physiological integrity; CNS: Pharmacological therapies; CL: Application

94. The nurse has just taught a client how to perform blood glucose monitoring and is observing as the client practices the technique. Which client action indicates an accurate understanding of the technique?

    1. The client places the reagent strip on a dry paper towel.

    2. The client smears the blood across the reagent strip.

    3. The client selects a puncture site beside a bone.

    4. The client uses the first drop of blood.

You have a greater chance of getting this one correct if you read all of the choices carefully.

94. 1. The client should place the reagent strip on a clean, dry paper towel; a damp surface could change the strip, thereby altering the test results. The client should avoid smearing blood across the reagent strip because it can cause an inaccurate reading. The client should select a highly vascular puncture site, rather than a site beside a bone (which isn't a vascular area). The client should wipe away the first drop of blood and use the second drop because the first drop usually contains a greater proportion of serous fluid, which can alter test results.

NP: Evaluation; CN: Physiological integrity; CNS: Reduction of risk potential; CL: Application

95. A 55-year-old client with type 2 diabetes is obese and hasn't been successful at controlling the condition by diet alone. The physician has prescribed glipizide (Glucotrol). The nurse knows that glipizide is commonly used for type 2 diabetes because it:

    1. is an oral form of insulin.

    2. stimulates the pancreas to secrete more insulin.

    3. promotes weight reduction, which lowers the blood glucose level.

    4. is a sulfa drug that treats pancreatic infections.

95. 2. Oral antidiabetic drugs work by stimulating the beta cells of the pancreas to secrete more insulin and decrease insulin resistance at the receptor sites. Oral insulin would be destroyed in the GI tract. Weight reduction would help to reduce the blood glucose level; however, glipizide is classified as an oral antidiabetic drug, not an appetite suppressant. Oral antidiabetics are sulfa-based drugs; however, they aren't indicated in diabetes for their anti-infective actions.

NP: Evaluation; CN: Physiological integrity; CNS: Pharmacological therapies; CL: Application

96. A boy with diabetes has a blood glucose level of 58 mg/dl before breakfast. The nurse is to administer regular and NPH insulin as prescribed for his routine morning insulin dose. Which nursing action is appropriate?

    1. Provide extra servings of food for breakfast.

    2. Notify the physician, and clarify the morning insulin order.

    3. Wait about 15 minutes after the insulin injection before serving his breakfast.

    4. Wait at least 45 minutes after the insulin injection before serving his breakfast.

Congratulations! You finished! Good work!

96. 3. A diabetic client should normally wait about 30 minutes after insulin administration before eating. However, if the blood glucose level is less than 60 mg/dl, the client should eat within about 15 minutes after insulin injection. If the blood glucose level is greater than 240 mg/dl, the client should wait 45 to 60 minutes after insulin injection before eating. Regular insulin begins to act within 30 to 60 minutes after injection. The blood glucose level isn't low enough to warrant extra food or notification of the physician. The blood glucose level isn't higher than 240 mg/dl, so waiting for breakfast wouldn't be appropriate.

NP: Planning; CN: Physiological integrity; CNS: Pharmacological therapies; CL: Application

For more information about genitourinary system disorders, visit the Internet site of the National Institute of Diabetes and Digestive and Kidney Diseases at **www.niddk.nih.gov/**.

Chapter 10
Genitourinary disorders

1. To treat cervical cancer, a client has had an applicator of radioactive material placed in her vagina. Which observation by the nurse indicates a radiation hazard?
1. The client is on strict bed rest.
2. The head of the bed is set at a 30-degree angle.
3. The client receives a complete bed bath each morning.
4. The nurse checks the applicator's position every 4 hours.

2. A physician tells a client to return 1 week after treatment to have a repeat culture done to verify the cure. This order would be appropriate for a woman with which of the following conditions?
1. Genital warts
2. Genital herpes
3. Gonorrhea
4. Syphilis

3. When assessing a client with gonorrhea, the nurse should expect to discover which sign or symptom?
1. Burning on urination
2. Dry, hacking cough
3. Diffuse skin rash
4. Painless chancre

Ask yourself: Which observation indicates a radiation hazard?

Choose the correct sign or symptom here.

1. 3. The client shouldn't receive a complete bed bath while the applicator is in place. In fact, she shouldn't be bathed below the waist because of the risk of radiation exposure to the nurse. During this treatment, the client should remain on strict bed rest, but the head of her bed may be raised to a 30- to 45-degree angle. The nurse should check the applicator's position every 4 hours to ensure that it remains in the proper place.
NP: Data collection; CN: Safe, effective care environment; CNS: Safety and infection control; CL: Application

2. 3. Gonococcal infections can be completely eliminated by drug therapy. This is documented by a negative culture 4 to 7 days after therapy is finished. Genital warts aren't curable and are identified by appearance, not culture. The diagnosis of syphilis is by darkfield microscopy or serologic tests. Genital herpes isn't curable and is identified by the appearance of the lesions or cytologic studies.
NP: Planning; CN: Physiological integrity; CNS: Physiological adaptation; CL: Application

3. 1. Burning on urination may be a symptom of gonorrhea or urinary tract infection. A dry, hacking cough is a sign of a respiratory infection, not gonorrhea. A diffuse rash may indicate secondary stage syphilis. A painless chancre is the hallmark of primary syphilis. It appears wherever the organisms enter the body, such as on the genitalia, anus, or lips.
NP: Data collection; CN: Physiological integrity; CNS: Physiological adaptation; CL: Knowledge

NP: Nursing process   CN: Client needs category   CNS: Client needs subcategory   CL: Cognitive level

4. Which of the following statements made by a client with a chlamydial infection indicates understanding of the potential complications?
1. "I'm glad I'm not pregnant; I'd hate to have a malformed baby from this disease."
2. "I hope this medicine works before this disease gets into my urine and destroys my kidneys."
3. "If I had known a diaphragm would put me at risk for this, I would have taken birth control pills."
4. "I need to treat this infection so it doesn't spread into my pelvis because I want to have children some day."

5. Which of the following comfort measures can be recommended to a client with genital herpes?
1. Wear loose cotton underwear.
2. Apply a water-based lubricant to the lesions.
3. Rub rather than scratch in response to an itch.
4. Pour hydrogen peroxide and water over the lesions.

6. The nurse is checking the laboratory values of a client with chronic renal failure (CRF) who has just begun hemodialysis. The nurse should expect to find an improvement in which value?
1. Complete blood count (CBC)
2. White blood cell (WBC) count
3. Calcium level
4. Blood urea nitrogen (BUN) level

You've finished five questions already and are doing great. Keep it up.

4. 4. Chlamydia is a common cause of pelvic inflammatory disease and infertility. It doesn't affect the kidneys or cause birth defects. It can cause conjunctivitis and respiratory infection in neonates exposed to infected cervicovaginal secretions during delivery. Use of a diaphragm isn't a risk factor.
NP: Evaluation; CN: Physiological integrity; CNS: Reduction of risk potential; CL: Application

5. 1. Wearing loose cotton underwear promotes drying and helps avoid irritation of the lesions. The use of hydrogen peroxide and water on lesions isn't recommended. The use of lubricants is contraindicated because they can prolong healing time and increase the risk of secondary infection. Lesions shouldn't be rubbed or scratched because of the risk of tissue damage and additional infection. Cool, wet compresses can be used to soothe the itch.
NP: Implementation; CN: Physiological integrity; CNS: Basic care and comfort; CL: Comprehension

6. 4. The BUN level reflects the amount of urea and nitrogenous waste products in the blood. Dialysis removes excess amounts of these elements from the blood, which is reflected in a lower BUN level. Hemodialysis doesn't affect the WBC count or the CBC. Calcium levels are usually low in clients with CRF. These levels are corrected by giving calcium supplements, not through hemodialysis.
NP: Evaluation; CN: Physiological integrity; CNS: Physiological adaptation; CL: Comprehension

NP: Nursing process   CN: Client needs category   CNS: Client needs subcategory   CL: Cognitive level

7. A client with nephritis is taking the diuretic furosemide (Lasix) as prescribed. To avoid potassium depletion, the nurse teaches the client prevention techniques. Which client statement indicates an accurate understanding of these techniques?

1. "I'll avoid consuming magnesium-rich foods."
2. "I'll watch for and report signs of hypercalcemia."
3. "I'll eat such foods as apricots, dates, and citrus fruits."
4. "I'll take furosemide with the usual dose of my antihypertensive drug."

Alas, poor Yorick! Er, Yorange? (Okay, that's a hint!)

7. 3. Because furosemide is a potassium-wasting diuretic, the client should eat potassium-rich foods, such as apricots, dates, and citrus fruits, to prevent potassium depletion. The other client statements have no relationship to potassium balance. The client may consume magnesium-rich foods as desired. The client should watch for signs of adverse reactions to furosemide such as hypocalcemia — not hypercalcemia. The client should take furosemide with an antihypertensive drug only if prescribed; the combination may produce hypotension but doesn't cause potassium depletion.

NP: Evaluation; CN: Health promotion and maintenance; CNS: Prevention and early detection of disease; CL: Application

8. The nurse has taught a client with genital herpes how to prevent the spread of the herpes simplex virus. Which client behavior demonstrates an accurate understanding of transmission prevention technique?

1. The client keeps the area moist.
2. The client keeps fingernails long.
3. The client wears tight-fitting blue jeans.
4. The client washes hands before and after touching lesions.

Nurses teach about prevention as well as treatment.

8. 4. Because hand-to-body contact is a common method of transmitting the herpes simplex virus, the client should wash his hands before and after touching the lesions to prevent the spread of the disease. To promote lesion drying and client comfort, the client should keep the affected area dry. To prevent scratching of the lesions, the client should keep his fingernails short, instead of long. Because tight-fitting clothes help retain heat and moisture, which can delay healing and cause discomfort, the client should wear loose-fitting garments.

NP: Evaluation; CN: Safe, effective care environment; CNS: Safety and infection control; CL: Comprehension

9. In which of the following groups is it most important for the client to understand the importance of an annual Papanicolaou test?

1. Clients with a history of recurrent candidiasis
2. Clients with a pregnancy before age 20
3. Clients infected with the human papillomavirus (HPV)
4. Clients with a long history of oral contraceptive use

An annual Pap test may be important for all of these groups, but for which group is it most important?

9. 3. HPV causes genital warts, which are associated with an increased incidence of cervical cancer. Recurrent candidiasis, use of oral contraceptives, and pregnancy before age 20 don't increase the risk of cervical cancer.

NP: Planning; CN: Health promotion and maintenance; CNS: Prevention and early detection of disease; CL: Application

10. Which of the following factors in a client's history indicates she's at risk for candidiasis?
1. Nulliparity
2. Menopause
3. Use of corticosteroids
4. Use of spermicidal jelly

10. 3. Small numbers of the fungus *Candida albicans* are commonly in the vagina. Because corticosteroids decrease host defense, they increase the risk of candidiasis. Candidiasis is rare before menarche and after menopause. The use of oral contraceptives, not spermicidal jelly, increases the risk of candidiasis. Pregnancy, not nulliparity, increases the risk of candidiasis.
NP: Data collection; CN: Health promotion and maintenance; CNS: Prevention and early detection of disease; CL: Application

You've already finished 11 questions and are well on your way toward the finish line.

11. Copious amounts of frothy, greenish vaginal discharge would be a symptom of which of the following infections?
1. Candidiasis
2. *Gardnerella vaginalis* vaginitis
3. Gonorrhea
4. Trichomoniasis

11. 4. The discharge associated with infection caused by *Trichomonas* organisms is homogenous, greenish gray, watery, and frothy or purulent. The discharge associated with infection due to *G. vaginalis* is thin and grayish white, with a marked fishy odor, while that associated with candidiasis is thick, white, and resembles cottage cheese in appearance. With gonorrhea, vaginal discharge is purulent when present but, in many women, gonorrhea is asymptomatic.
NP: Data collection; CN: Physiological integrity; CNS: Physiological adaptation; CL: Application

12. A 19-year-old woman reports an intermittent milky vaginal discharge. She isn't sexually active and doesn't report itching or burning. Which of the following factors is the most likely cause of the milky discharge?
1. Inadequate cleaning of the perineal area
2. Sensitivity to a feminine hygiene product
3. Normal fluctuation in estrogen and progesterone levels
4. Reaction to heat and moisture from wearing tight clothing

Any of these factors might be the cause, but this item is asking you to choose the *most* likely cause.

12. 3. Vaginal fluid is clear, milky, or cloudy, depending on the fluctuating levels of estrogen and progesterone. A milky vaginal discharge is normal and isn't associated with sensitivity, reaction to heat or moisture, or inadequate cleaning.
NP: Data collection; CN: Health promotion and maintenance; CNS: Growth and development through the life span; CL: Application

13. A 59-year-old client who underwent mastectomy of the left breast the day before has an I.V. catheter in her lower right arm. To check the client's blood pressure, what should the nurse do?
1. Use the right arm above the I.V. site.
2. Use the left arm, and pump the cuff only as necessary.
3. Use the leg.
4. Stop the I.V. infusion, and take the blood pressure in the right arm.

13. 1. To obtain a blood pressure reading, the nurse should always use the arm opposite the side of the mastectomy. Never take a blood pressure reading in the arm on the affected side. Blood pressure readings may be taken in the leg in some cases, but this isn't necessary in this case. It isn't necessary to stop the I.V. fluid to take a blood pressure reading.
NP: Implementation; CN: Health promotion and maintenance; CNS: Prevention and early detection of disease; CL: Knowledge

NP: Nursing process   CN: Client needs category   CNS: Client needs subcategory   CL: Cognitive level

14. The nurse must administer a douche to a client scheduled for a vaginal hysterectomy. How should she perform the procedure?
1. Separate the labia, clean the vaginal orifice, insert the douche nozzle 2″ (5.1 cm), and administer room temperature solution well above the client's hip level.
2. Separate the labia, clean the vaginal orifice, insert the douche nozzle 2″, and administer cool solution well above the client's hip level.
3. Separate the labia, clean the vaginal orifice, insert the douche nozzle 2″, and administer 100° F (37.8° C) solution 2′ (61 cm) above the client's hip level.
4. Separate the labia, clean the vaginal orifice, insert the douche nozzle 2″, and administer 100° F solution 3′ (91 cm) above the client's hip level.

14. 3. The correct procedure for administering a douche is to separate the labia, clean the vaginal orifice, insert the douche nozzle 2″, and administer 100° F solution 2′ above the client's hip level.

NP: Implementation; CN: Safe, effective care environment; CNS: Safety and infection control; CL: Application

Which action could harm your client?

15. A nurse enters the room of a client who had a left modified mastectomy 8 hours earlier. Which of the following observations indicates that the nursing assistant assigned to the client needs further instruction and guidance?
1. The client is squeezing a ball in her left hand.
2. The client is wearing a robe with elastic cuffs.
3. The client's affected arm is elevated on a pillow.
4. A blood pressure cuff is on the client's right arm.

15. 2. Elastic cuffs can contribute to the development of lymphedema and should be avoided. Blood pressure measurements in the affected arm should also be avoided. Simple exercises such as squeezing a ball help promote circulation and should be started as soon as possible after surgery. Elevation of the affected arm promotes venous and lymphatic return from the extremity.

NP: Evaluation; CN: Safe, effective care environment; CNS: Coordinated care; CL: Analysis

16. For which of the following symptoms should a client at risk for evisceration be monitored after an abdominal hysterectomy?
1. Tachycardia accompanied by a weak, thready pulse
2. Hypotension with a decreased level of consciousness
3. Shallow, rapid respirations and increasing vaginal drainage
4. Low-grade fever with increasing serosanguineous incisional drainage

16. 4. Signs of impending evisceration are low-grade fever and increasing serosanguineous drainage. Tachycardia; weak, thready pulse; shallow, rapid respirations; vaginal drainage; hypotension; and decreased level of consciousness after abdominal hysterectomy are all unrelated to impeding evisceration, although they may be associated with other serious problems such as shock.

NP: Planning; CN: Physiological integrity; CNS: Reduction of risk potential; CL: Comprehension

17. Which of the following findings indicates that oxycodone (Percodan) given to a client with breast cancer metastasized to the bone is exerting the desired effect?
1. Bone density is increased.
2. Pain is 0 to 2 on a 10-point scale.
3. Alphafetoprotein level is decreased.
4. Serum calcium level is within normal range.

For what purpose are drugs that end with "one" given?

17. 2. Oxycodone is an opioid analgesic used for alleviating severe pain, especially in terminal illness. If a client's pain has decreased to 0 to 2 on a 10-point scale (0 is no pain and 10 is the worst pain), the medication is working as desired. The drug doesn't directly affect bone density, serum calcium level, or alphafetoprotein level.

NP: Evaluation; CN: Physiological integrity; CNS: Pharmacological therapies; CL: Analysis

18. Which of the following instructions should be given to a client with prostatitis who's receiving co-trimoxazole double strength (Bactrim DS)?
1. Don't expect improvement of symptoms for 7 to 10 days.
2. Drink six to eight glasses of fluid daily while taking this medication.
3. If a sore mouth or throat develops, take the medication with milk or an antacid.
4. Use a sunscreen of at least SPF-15 with PABA to protect against drug-induced photosensitivity.

18. 2. Six to eight glasses of fluid daily are needed to prevent renal problems, such as crystalluria and calculi formation. The symptoms should improve in a few days if the drug is effective. Sore throat and sore mouth are adverse effects that should be reported right away. The drug causes photosensitivity, but a PABA-free sunscreen should be used because PABA can interfere with the drug's action.

NP: Planning; CN: Physiological integrity; CNS: Pharmacological therapies; CL: Application

You're doing great! Keep going!

19. Which of the following factors should be checked when evaluating the effectiveness of an alpha-adrenergic blocker given to a client with benign prostatic hyperplasia (BPH)?
1. Voiding pattern
2. Size of the prostate
3. Creatinine clearance
4. Serum testosterone level

19. 1. Alpha-adrenergic blockers relax the smooth muscle of the bladder neck and prostate, so the urinary symptoms (frequency, urgency, hesitancy) of BPH are reduced in many clients. These drugs don't affect the size of the prostate, production or metabolism of testosterone, or renal function.

NP: Evaluation; CN: Physiological integrity; CNS: Pharmacological therapies; CL: Application

20. A nursing diagnosis addressing *Risk for impaired tissue integrity* would be most appropriate for which of the following clients?
1. A client with endometriosis
2. A client taking oral contraceptives
3. A client with a vaginal packing in place
4. A client having reconstructive breast surgery

The appropriate nursing diagnosis helps everyone work toward the same goal.

20. 4. Reconstructive breast surgery places the client at risk for insufficient blood supply to the muscle graft and skin, which can lead to tissue necrosis. Endometriosis or oral contraceptives aren't generally associated with altered tissue perfusion. Pressure from vaginal packing can sometimes put pressure on the bladder neck and interfere with voiding, not tissue perfusion.

NP: Planning; CN: Physiological integrity; CNS: Reduction of risk potential; CL: Analysis

21. The physician has ordered a condom catheter for a male client. While cleaning the client's perineal area, the nurse observes irritation, excoriation, and swelling of the penis. What should the nurse do next?

   1. Twist the condom after application
   2. Apply the condom with adhesive tape.
   3. Inform the charge nurse of these findings.
   4. Roll the condom down securely over the tip of the penis.

22. Which of the following instructions is the most important to give women to decrease the risk of toxic shock syndrome?

   1. Avoid douching.
   2. Wear loose cotton underwear.
   3. Use pads, not tampons, overnight.
   4. Avoid sexual intercourse during menses.

The words most important offer a hint!

23. Which of the following responses is the most appropriate when a client asks what activity limitations are necessary after a dilatation and curettage procedure?

   1. Tampons may be used during exercise.
   2. Avoid strenuous work and sexual intercourse for at least 2 weeks.
   3. Stay on bed rest for 3 days; then gradually resume normal activity.
   4. Engage in activity as tolerated, and take a soaking tub bath each day to promote relaxation.

24. Which of the following assessment findings is abnormal in a 72-year-old man?

   1. Decreased sperm count
   2. Small firm testes on palpation
   3. History of slowed sexual response
   4. Decreased plasma testosterone level

Read carefully. The word *abnormal* can be tricky.

21. 3. The nurse should inform the charge nurse of the baseline data because a condom catheter shouldn't be used on a client with penile irritation, excoriation, swelling, or discoloration. A condom catheter shouldn't be twisted because twisting can obstruct urine flow. It should be secured with elastic tape or Velcro rather than adhesive tape, which is inflexible and can stop blood flow. It should have a 1″ (2.5-cm) gap between the tip of the penis and the connecting tube to prevent penile irritation and allow complete urine drainage.

NP: Data collection; CN: Safe, effective care environment; CNS: Coordinated care; CL: Comprehension

22. 3. The cause of toxic shock syndrome is a toxin produced by *Staphylococcus aureus* bacteria. It occurs most often in menstruating women using tampons. Tampons, particularly when left in place for more than 8 hours (such as overnight), are believed to provide a good environment for growth of the bacteria, which then enter the bloodstream through breaks in the vaginal mucosa. Douching, use of loose cotton underwear, and sexual intercourse during menstruation have no direct association with toxic shock syndrome.

NP: Planning; CN: Health promotion and maintenance; CNS: Prevention and early detection of disease; CL: Comprehension

23. 2. Strenuous work, which can result in increased bleeding, should be avoided for 2 weeks to allow time for healing. Sexual intercourse should also be avoided for 2 weeks to allow healing and decrease the risk of infection. Overall activity should be gradually resumed, reaching preoperative levels in the 2-week period, but bed rest isn't necessary. Tampons and tub baths should be avoided for 1 week. No other restrictions are routinely necessary.

NP: Implementation; CN: Physiological integrity; CNS: Reduction of risk potential; CL: Application

24. 1. Normal sperm production continues despite the age-related degenerative changes that occur in the male reproductive system. Among the normal age-related changes are decreased size and increased firmness of the testes, decreased production of testosterone and progesterone, and a decrease in sexual potency.

NP: Data collection; CN: Physiological integrity; CNS: Physiological adaptation; CL: Application

25. After having transurethral resection of the prostate (TURP), a client returns to the unit with a three-way indwelling urinary catheter and continuous closed bladder irrigation. Which finding suggests that the client's catheter is occluded?
1. The urine in the drainage bag appears red to pink.
2. The client reports bladder spasms and the urge to void.
3. The normal saline irrigation is infusing at the rate of 50 gtt/minute.
4. About 1,000 ml of irrigant have been instilled, and 1,200 ml of drainage have been returned.

25. 2. Reports of bladder spasms and the urge to void suggest that a blood clot may be occluding the catheter. After TURP, urine normally appears red to pink, and normal saline irrigant usually is infused at a rate of 40 to 60 gtt/minute or according to facility protocol. The amount of returned fluid (1,200 ml) should correspond to the amount of instilled fluid, plus the client's urine output (1,000 ml + 200 ml), which reflects catheter patency.
NP: Evaluation; CN: Physiological integrity; CNS: Basic care and comfort; CL: Application

26. Which of the following comments made by a client being treated for chronic prostatitis indicates that self-care instructions need to be clarified?
1. "I miss not being able to have sex."
2. "I enjoy frequent soaking in a hot tub of water."
3. "Cutting down on coffee hasn't been as hard as I expected."
4. "I'm used to getting up and moving, not just sitting for long times."

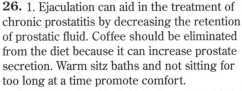

Client teaching includes making sure your client understands your instructions.

26. 1. Ejaculation can aid in the treatment of chronic prostatitis by decreasing the retention of prostatic fluid. Coffee should be eliminated from the diet because it can increase prostate secretion. Warm sitz baths and not sitting for too long at a time promote comfort.
NP: Evaluation; CN: Physiological integrity; CNS: Physiological adaptation; CL: Application

27. Perineal pain in the absence of any observable cause is suggestive of which of the following conditions?
1. Endometriosis
2. Internal hemorrhoids
3. Prostatitis
4. Renal calculi

27. 3. Prostatitis can cause prostate pain, which is felt as perineal discomfort. Renal calculi typically produce flank pain. Hemorrhoids cause rectal pain and pressure. Endometriosis can cause pain low in the abdomen, deep in the pelvis, or in the rectal or sacrococcygeal area, depending on the location of the ectopic tissue.
NP: Data collection; CN: Health promotion and maintenance; CNS: Prevention and early detection of disease; CL: Comprehension

28. Which of the following findings observed in a client taking finasteride (Proscar) would be a cause for alarm?
1. Azotemia
2. Breast enlargement
3. Decreased prostate size
4. Flushing

Good job!

28. 1. Azotemia, a buildup of nitrogenous waste products in the blood, indicates impaired renal function. Proscar is prescribed for chronic urinary retention with large residual volumes secondary to benign prostatic hypertrophy. Azotemia in a client on finasteride therapy can indicate the drug isn't effective in relieving the urinary symptoms associated with benign prostatic hypertrophy or that an unrelated renal problem has occurred. Flushing, breast enlargement, and decrease in prostate size are expected effects of finasteride, an antiandrogenic agent.
NP: Data collection; CN: Physiological integrity; CNS: Pharmacological therapies; CL: Application

NP: Nursing process   CN: Client needs category   CNS: Client needs subcategory   CL: Cognitive level

29. Which of the following treatments is appropriate for a client with cervical polyps who has been treated with cryosurgery?
1. Daily douche
2. Oral antibiotics
3. Intravaginal antibiotic cream
4. Use of tampons for 72 hours

30. Which of the following interventions would <u>not</u> be correct for a woman having intracavitary radiation for cancer of the cervix?
1. Low-residue diet
2. Fowler's position when in bed
3. Indwelling urinary catheter to gravity drainage
4. Diphenoxylate hydrochloride and atropine (Lomotil) 2 mg four times daily

Careful! This question asks what is *not* correct!

31. Which of the following conditions of the female reproductive system generally requires the identification and treatment of sexual partners?
1. Bartholinitis
2. Candidiasis
3. *Chlamydia trachomatis* infection
4. Endometriosis

32. Which of the following pieces of information should be given to a client taking metronidazole (Flagyl)?
1. Breathlessness and cough are common adverse effects.
2. Urine may develop a greenish tinge while the client is taking this drug.
3. Mixing this drug with alcohol causes severe nausea and vomiting.
4. Heart palpitations may occur and should be immediately reported.

29. 3. Intravaginal antibiotic cream is often used to aid healing and prevent infection. Oral antibiotics are used for clients with acute cervicitis or perimetritis. Douching is generally avoided for 2 weeks, as is the use of tampons.
NP: Planning; CN: Physiological integrity; CNS: Reduction of risk potential; CL: Application

30. 2. Clients having intracavitary radiation therapy are on strict bed rest, with the head of the bed elevated no more than 10 to 15 degrees to avoid displacing the radiation source. An order for Fowler's position when in bed is incorrect. An indwelling urinary catheter is used to prevent urine from distending the bladder and changing the position of tissues relative to the radiation source. A low-residue diet and diphenoxylate hydrochloride and atropine are used to prevent diarrhea during treatment.
NP: Planning; CN: Physiological integrity; CNS: Reduction of risk potential; CL: Analysis

31. 3. Chlamydia is a common sexually transmitted disease requiring the treatment of all current sexual partners to prevent reinfection. Bartholinitis results from obstruction of a duct. Endometriosis occurs when endometrial cells are seeded throughout the pelvis and isn't a sexually transmitted disease. Candidiasis is a yeast infection that often occurs as a result of antibiotic use. Sexual partners may become infected, although men can usually be treated with over-the-counter products.
NP: Planning; CN: Health promotion and maintenance; CNS: Prevention and early detection of disease; CL: Application

32. 3. When mixed with alcohol, metronidazole causes a disulfiram-like effect involving nausea, vomiting, and other unpleasant symptoms. Urine may turn reddish brown, not greenish, from the drug. Cardiovascular or respiratory effects aren't associated with use of this drug.
NP: Planning; CN: Physiological integrity; CNS: Pharmacological therapies; CL: Application

33. Which of the following symptoms is an adverse reaction of hydrocodone with acetaminophen (Percocet) that a client with metastatic prostate cancer should report to the physician?

1. Blurred vision
2. Diarrhea
3. Unusual dreams
4. Vomiting

Here's the key to this question: adverse reaction.

33. 4. Vomiting is an adverse reaction to the drug that should be reported because it impairs the client's quality of life and places the client at risk for dehydration. Taking the medication with food may prevent vomiting. If not, other opiate analgesics may be better tolerated. Blurred vision and diarrhea aren't associated with the use of hydrocodone with acetaminophen. Unusual dreams are a common adverse effect but don't need to be reported unless bothersome to the client.

NP: Implementation; CN: Physiological integrity; CNS: Pharmacological therapies; CL: Application

34. Which of the following interventions is appropriate for a client having hysterosalpingography?

1. Give the client a perineal pad to wear after the procedure.
2. Give the client nothing by mouth after midnight the night before the procedure.
3. Position the client in the knee-chest position during the procedure.
4. Keep the client in a dorsal recumbent position for 4 hours after the procedure.

Keep going! You're doing an outstanding job!

34. 1. A perineal pad is needed after hysterosalpingography because the contrast medium may leak from the vagina for several hours and stain the clothing. The bowel needs to be cleansed before the procedure, but the client doesn't have to refrain from having anything by mouth after midnight. The procedure is performed with the client in the lithotomy position, and no special positioning is required after the procedure.

NP: Planning; CN: Physiological integrity; CNS: Basic care and comfort; CL: Application

35. Which of the following instructions apply to a vaginal irrigation?

1. Insert the nozzle about 3″ (7.5 cm) into the vagina.
2. Direct the tip of the nozzle toward the sacrum.
3. Instill the solution in a constant flow over 5 to 10 minutes.
4. Raise the solution at least 24″ (61 cm) above the client's hip level.

35. 2. The normal position of the vagina slants up and back toward the sacrum. Directing the tip of the nozzle toward the sacrum allows it to follow the normal slant of the vagina and minimizes tissue trauma. The nozzle should be inserted about 2″ (5 cm). The fluid can be instilled intermittently and, for best therapeutic results, over 20 to 30 minutes. The container should be no higher than 24″ above the client's hip level to avoid forcing fluid and bacteria through the cervical os into the uterus.

NP: Implementation; CN: Safe, effective care environment; CNS: Safety and infection control; CL: Comprehension

36. A 36-year-old man who has never had mumps reports that he was just notified that an 8-year-old child of a family with whom he stayed recently has been diagnosed with mumps. Which of the following treatments should the man receive?

1. I.V. antibiotics
2. Ice packs to the scrotum
3. Application of a scrotal support
4. Administration of gamma globulin

36. 4. Gamma globulin provides passive immunity to mumps. Antibiotic therapy is used in the treatment of bacterial orchitis. Ice and the use of a scrotal support are used as comfort measures in the treatment of orchitis.

NP: Planning; CN: Health promotion and maintenance; CNS: Prevention and early detection of disease; CL: Application

NP: Nursing process   CN: Client needs category   CNS: Client needs subcategory   CL: Cognitive level

37. Which of the following statements by a man scheduled for a vasectomy indicates he needs further teaching about the procedure?
1. "If I decide I want a child, I'll just get a reversal."
2. "Amazing! I can make sperm but classify as sterile."
3. "I'm sure glad I made some deposits in the sperm bank."
4. "I can't believe I still have to worry about contraception after this surgery."

Look for signs that your client doesn't understand the information.

38. Which of the following areas of client teaching should be stressed when the goal is preventing the development of phimosis in a 20-year-old uncircumcised man?
1. Proper cleaning of the prepuce
2. Importance of regular ejaculation
3. Technique of testicular self-examination
4. Proper hand washing before touching the genitals

39. Which of the following statements should be included when teaching a client newly diagnosed with testicular cancer?
1. Testicular cancer isn't responsive to chemotherapy, but it's highly curative with surgery.
2. Radiation therapy is never used, so the unaffected testicle remains healthy.
3. Testicular self-examination is still important because there's increased risk of a second tumor.
4. Taking testosterone after orchiectomy prevents changes in appearance and sexual function.

You can ease your client's worries through effective teaching efforts.

40. Which of the following statements shows the significance of a persistent elevation in alphafetoprotein (AFP) level after orchiectomy for testicular cancer?
1. Fertility is maintained.
2. The cancer has recurred.
3. There's metastatic disease.
4. Testosterone levels are low.

37. 1. Vasectomy procedures can be reversed but with varying degrees of success. Because of the variable success, a client can't be sure of reversibility and needs to consider vasectomy a permanent sterilization procedure when deciding to have it done. After vasectomy, the client remains fertile until sperm stored distal to the severed vas are evacuated. Once this occurs, sperm are still produced, but they don't enter the ejaculate and are absorbed by the body.
NP: Evaluation; CN: Physiological integrity; CNS: Physiological adaptation; CL: Application

38. 1. Proper cleaning of the preputial area to remove secretions is critical to the prevention of noncongenital phimosis. Hand washing is important in preventing the spread of infection, and testicular self-examination is important in the early detection and treatment of testicular cancer. Regular ejaculation can decrease the symptoms of chronic prostatitis, but it has no effect on the development of phimosis.
NP: Implementation; CN: Health promotion and maintenance; CNS: Prevention and early detection of disease; CL: Application

39. 3. A history of a testicular malignancy puts the client at increased risk for a second tumor. Testicular self-examination allows for early detection and treatment and is critical. Radiation therapy is used on the retroperitoneal lymph nodes. Chemotherapy is added for clients who have evidence of metastasis after irradiation. Testosterone usually isn't needed because the unaffected testis usually produces sufficient hormone.
NP: Implementation; CN: Physiological integrity; CNS: Reduction of risk potential; CL: Application

40. 3. AFP is a tumor marker elevated in non-seminomatous malignancies of the testicle. After the tumor is removed, the level should decrease. A persistent elevation after orchiectomy indicates tumor is present someplace outside the testicle that was removed. A recurrence of the cancer is indicated by a postsurgical decrease in AFP level followed by an elevation as a new tumor starts to grow. The level of AFP isn't related to fertility or testosterone level.
NP: Data collection; CN: Physiological integrity; CNS: Physiological adaptation; CL: Analysis

41. Which of the following discharge instructions should be given to a client after a prostatectomy?

1. Avoid straining at stool.
2. Report clots in the urine right away.
3. Soak in a warm tub daily for comfort.
4. Return to your usual activities in 3 weeks.

41. 1. Straining at stool after prostatectomy can cause bleeding. Small blood clots or pieces of tissue commonly are passed in the urine for up to 2 weeks postoperatively. Tub baths are prohibited because they cause dilation of pelvic blood vessels. Other activities are resumed based on the guidance of the physician. Sexual intercourse and driving are usually prohibited for about 3 weeks. Exercising and returning to work are usually prohibited for about 6 weeks.
NP: Implementation; CN: Physiological integrity; CNS: Reduction of risk potential; CL: Comprehension

42. After a biopsy of the prostate, which of the following symptoms should be reported?

1. Pain on ejaculation
2. Blood in the semen
3. Difficulty urinating
4. Temperature of more than 99°F (37.2° C)

42. 3. Difficulty urinating suggests urethral obstruction. Temperature of more than 101°F (38.3° C) should be reported because it suggests infection. Blood in the semen is an expected finding for months, and discomfort on ejaculation is expected for weeks.
NP: Implementation; CN: Physiological integrity; CNS: Reduction of risk potential; CL: Comprehension

Good client teaching is an important responsibility for any nurse.

43. Two days after a transrectal biopsy of the prostate, a client calls the clinic to report his stool is streaked with blood. Which of the following responses is appropriate?

1. Tell the client to take a laxative.
2. Tell the client to come in for examination.
3. Reassure the client that this is an expected occurrence.
4. Ask the client to collect a stool specimen for testing.

43. 3. After a transrectal prostatic biopsy, blood in the stool is expected for a number of days. Because blood in the stool is expected, testing the stool or examining the client isn't necessary. Stool softeners are prescribed if the client complains of constipation; straining at stool can precipitate bleeding, but laxatives generally aren't necessary.
NP: Implementation; CN: Physiological integrity; CNS: Reduction of risk potential; CL: Application

44. When assessing a client with a history of genital herpes, which of the following symptoms would indicate that an outbreak of lesions is imminent?

1. Headache and fever
2. Vaginal and urethral discharge
3. Dysuria and lymphadenopathy
4. Genital pruritus and paresthesia

Teach your client the prodromal signs of an outbreak of genital herpes.

44. 4. Pruritus and paresthesia as well as redness of the genital area are prodromal symptoms of recurrent herpes infection. These symptoms occur 30 minutes to 48 hours before the lesions appear. Headache and fever are symptoms of viremia associated with the primary infection. Dysuria and lymphadenopathy are local symptoms of primary infection that may also occur with recurrent infection. Vaginal and urethral discharge is also a local sign of primary infection.
NP: Data collection; CN: Physiological integrity; CNS: Physiological adaptation; CL: Comprehension

45. Which of the following instructions should be given to a woman newly diagnosed with genital herpes?
1. Obtain a Papanicolaou (Pap) test every year.
2. Have your partner use a condom when lesions are present.
3. Use a water-soluble lubricant for relief of pruritus.
4. Limit stress and emotional upset as much as possible.

You're at the halfway point! Great job!

45. 4. Stress, anxiety, and emotional upset seem to predispose a client to recurrent outbreaks of genital herpes. During an outbreak, creams and lubricants should be avoided because they may prolong healing. Sexual intercourse should be avoided during outbreaks, and a condom should be used between outbreaks; it isn't known whether the virus can be transmitted at this time. Because a relationship has been found between genital herpes and cervical cancer, a Pap test is recommended every 6 months.
NP: Implementation; CN: Physiological integrity; CNS: Physiological adaptation; CL: Application

46. Which of the following symptoms is consistent with primary syphilis?
1. A painless genital ulcer that appears about 3 weeks after unprotected sex
2. Copper-colored macules on the palms and soles after a brief fever
3. Patchy hair loss in red, broken skin involving the scalp, eyebrows, and beard area
4. One or more flat, wartlike papules in the genital area that are sensitive to the touch

46. 1. A painless genital ulcer is a symptom of primary syphilis. Macules on the palms and soles after fever are indicative of secondary syphilis, as is patchy hair loss. Wartlike papules are indicative of genital warts.
NP: Data collection; CN: Physiological integrity; CNS: Physiological adaptation; CL: Knowledge

47. During a routine physical examination, a firm mass is palpated in the right breast of a 35-year-old woman. Which of the following findings or client history would suggest cancer of the breast as opposed to fibrocystic disease?
1. History of early menarche
2. Cyclic change in mass size
3. History of anovulatory cycles
4. Increased vascularity of the breast

47. 4. Increase in breast size or vascularity is consistent with breast cancer. Early menarche as well as late menopause or a history of anovulatory cycles is associated with fibrocystic disease. Masses associated with fibrocystic disease of the breast are firm, are most often located in the upper outer quadrant of the breast, and increase in size prior to menstruation. They may be bilateral in a mirror image and are typically well demarcated and freely moveable.
NP: Data collection; CN: Health promotion and maintenance; CNS: Prevention and early detection of disease; CL: Application

48. After which of the following procedures is the use of aseptic technique in the provision of client hygiene most critical?
1. Radical prostatectomy
2. Perineal prostatectomy
3. Suprapubic prostatectomy
4. Transurethral prostatectomy (TURP)

48. 2. The incision in a perineal prostatectomy is close to the rectum, which normally contains gram-negative organisms that can cause infection if introduced into other areas of the body. The use of proper aseptic technique, including washing front to back, in providing hygiene is more critical to these clients than those with either no external incision, as with TURP, or abdominal incisions, as with suprapubic or radical prostatectomy.
NP: Implementation; CN: Physiological integrity; CNS: Reduction of risk potential; CL: Application

49. Which of the following conditions is a common cause of prerenal acute renal failure (ARF)?
1. Atherosclerosis
2. Decreased cardiac output
3. Prostatic hypertrophy
4. Rhabdomyolysis

Hints galore!

49. 2. Prerenal ARF refers to renal failure due to an interference with renal perfusion. Decreased cardiac output causes a decrease in renal perfusion, which leads to a lower glomerular filtration rate. Atherosclerosis and rhabdomyolysis are renal causes of ARF. Prostatic hypertrophy would be an example of a postrenal cause of ARF.
NP: Data collection; CN: Physiological integrity; CNS: Physiological adaptation; CL: Knowledge

50. A client admitted for acute pyelonephritis is about to start antibiotic therapy. Which of the following symptoms would be expected in this client?
1. Hypertension
2. Flank pain on the affected side
3. Pain that radiates toward the unaffected side
4. No tenderness with deep palpation over the costovertebral angle

50. 2. The client may complain of pain on the affected side because the kidney is enlarged and might have formed an abscess. Hypertension is associated with chronic pyelonephritis. The client would have tenderness with deep palpation over the costovertebral angle. Pain may radiate down the ureters or to the epigastrium.
NP: Data collection; CN: Physiological integrity; CNS: Physiological adaptation; CL: Knowledge

51. Discharge instructions for a client treated for acute pyelonephritis should include which of the following statements?
1. Avoid taking any dairy products.
2. Return for follow-up urine cultures.
3. Stop taking the prescribed antibiotics when the symptoms subside.
4. Recurrence is unlikely because you've been treated with antibiotics.

Here's another question about client teaching, an essential topic on NCLEX examinations.

51. 2. The client needs to return for follow-up urine cultures because bacteriuria may be present but asymptomatic. Pyelonephritis often recurs as a relapse or new infection and frequently recurs within 2 weeks of completing therapy. Intake of dairy products won't contribute to pyelonephritis. Antibiotics need to be taken for the full course of therapy regardless of symptoms.
NP: Planning; CN: Health promotion and maintenance; CNS: Prevention and early detection of disease; CL: Comprehension

52. A client is complaining of severe flank and abdominal pain. A flat plate of the abdomen shows urolithiasis. Which of the following interventions is important?
1. Strain all urine.
2. Limit fluid intake.
3. Enforce strict bed rest.
4. Encourage a high-calcium diet.

52. 1. Urine should be strained for calculi and sent to the laboratory for analysis. Fluid intake of 3 to 4 L/day is encouraged to flush the urinary tract and prevent further calculi formation. A low-calcium diet is recommended to help prevent the formation of calcium calculi. Ambulation is encouraged to help pass the calculi through gravity.
NP: Implementation; CN: Physiological integrity; CNS: Reduction of risk potential; CL: Application

53. A client is receiving a radiation implant for the treatment of bladder cancer. Which of the following interventions is appropriate?

1. Flush all urine down the toilet.
2. Restrict the client's fluid intake.
3. Place the client in a semiprivate room.
4. Monitor the client for signs and symptoms of cystitis.

54. A client has undergone a radical cystectomy and has an ileal conduit for the treatment of bladder cancer. Which of the following postoperative assessment findings must be reported to the physician immediately?

1. A red, moist stoma
2. A dusky-colored stoma
3. Urine output more than 30 ml/hour
4. Slight bleeding from the stoma when changing the appliance

55. Which of the following instructions should be given to a client with an ileal conduit about skin care at the stoma site?

1. Change the appliance at bedtime.
2. Leave the stoma open to air while changing the appliance.
3. Clean the skin around the stoma with mild soap and water and dry it thoroughly.
4. Cut the faceplate or wafer of the appliance no more than 4 mm larger than the stoma.

56. A client is diagnosed with cystitis. Client teaching aimed at preventing a recurrence should include which of the following instructions?

1. Bathe in a tub.
2. Wear cotton underpants.
3. Use a feminine hygiene spray.
4. Limit your intake of cranberry juice.

Question 54 asks you to prioritize responses according to which is most urgent.

Stay awake now. You're closer to home!

53. 4. Cystitis is the most common adverse reaction of clients undergoing radiation therapy; symptoms include dysuria, frequency, urgency, and nocturia. Clients with radiation implants require a private room. Urine of clients with radiation implants for bladder cancer should be sent to the radioisotopes laboratory for monitoring. It's recommended that fluid intake be increased.

NP: Data collection; CN: Physiological integrity; CNS: Reduction of risk potential; CL: Knowledge

54. 2. The stoma should be red and moist, indicating adequate blood flow. A dusky or cyanotic stoma indicates insufficient blood supply and is an emergency needing prompt intervention. Urine output less than 30 ml/hour or no urine output for more than 15 minutes should be reported. Slight bleeding from the stoma when changing the appliance may occur because the intestinal mucosa is very fragile.

NP: Data collection; CN: Physiological integrity; CNS: Reduction of risk potential; CL: Comprehension

55. 3. Cleaning the skin around the stoma with mild soap and water and drying it thoroughly helps keep the area clean from urine, which can irritate the skin. Change the appliance in the early morning when urine output is less to decrease the amount of urine in contact with the skin. The faceplate or wafer of the appliance shouldn't be more than 3 mm larger than the stoma to reduce the skin area in contact with urine. The stoma should be covered with a gauze pad when changing the appliance to prevent seepage of urine onto the skin.

NP: Implementation; CN: Physiological integrity; CNS: Basic care and comfort; CL: Knowledge

56. 2. Cotton underpants prevent infection because they allow for air to flow to the perineum. Women should shower instead of taking a tub bath to prevent infection. Feminine hygiene spray can act as an irritant. Cranberry juice helps prevent cystitis because it increases urine acidity; alkaline urine supports bacterial growth.

NP: Planning; CN: Health promotion and maintenance; CNS: Prevention and early detection of disease; CL: Application

57. Care for an indwelling urinary catheter should include which of the following interventions?

1. Insert the catheter using clean technique.
2. Keep the drainage bag on the bed with the client.
3. Clean around the catheter at the meatus with soap and water.
4. Lay the drainage bag on the floor to allow for maximum drainage through gravity.

58. Which of the following methods should be used to collect a specimen for urine culture?

1. Have the client void in a clean container.
2. Clean the foreskin of the penis of uncircumcised men before specimen collection.
3. Have the client void into a urinal and then pour the urine into the specimen container.
4. Have the client begin the stream of urine in the toilet and catch the urine in a sterile container midstream.

59. A client with a history of chronic renal failure is admitted to the unit with pulmonary edema after missing her dialysis treatment yesterday. Blood is drawn and sent for a chemistry analysis. Which of the following results is expected?

1. Alkalemia
2. Hyperkalemia
3. Hypernatremia
4. Hypokalemia

60. Dialysis allows for the exchange of particles across a semipermeable membrane by which of the following actions?

1. Osmosis and diffusion
2. Passage of fluid toward a solution with a lower solute concentration
3. Allowing the passage of blood cells and protein molecules through it
4. Passage of solute particles toward a solution with a higher concentration

Here's a question straight from Nursing 101. Remember?

If I fail and can't do my job properly, what will happen?

57. 3. It's important to clean around the meatus at the catheter site to decrease the chance of infection. The drainage bag shouldn't be placed on the floor because of the increased risk of infection due to microorganisms. It should hang on the bed in a dependent position. The catheter should be inserted using sterile technique. Keeping the drainage bag in the bed with the client causes backflow of urine into the urethra, increasing the chance of infection.
NP: Implementation; CN: Physiological integrity; CNS: Reduction of risk potential; CL: Comprehension

58. 4. Catching urine midstream reduces the amount of contamination by microorganisms at the meatus. Voiding in a urinal doesn't allow for an uncontaminated specimen because the urinal isn't sterile. When cleaning an uncircumcised male, the foreskin should be retracted, and the glans penis should be cleaned to prevent specimen contamination. Voiding in a clean container is done for a random specimen, not a clean-catch specimen for urine culture.
NP: Implementation; CN: Physiological integrity; CNS: Reduction of risk potential; CL: Knowledge

59. 2. The kidneys are responsible for excreting potassium. In renal failure, the kidneys are no longer able to excrete potassium, resulting in hyperkalemia. Hypokalemia is generally seen in clients undergoing diuresis. The kidneys are responsible for regulating the acid-base balance; in renal failure, acidemia, not alkalemia, would be seen. Generally, hyponatremia, not hypernatremia, would be seen due to the dilutional effect of water retention.
NP: Data collection; CN: Physiological integrity; CNS: Physiological adaptation; CL: Comprehension

60. 1. Osmosis allows for the removal of fluid from the blood by allowing it to pass through the semipermeable membrane to an area of high concentration (dialysate), and diffusion allows for passage of particles (electrolytes, urea, and creatinine) from an area of higher concentration to an area of lower concentration. Fluid passes to an area with a higher solute concentration. The pores of the semipermeable membrane are small, thus preventing the flow of blood cells and protein molecules through it.
NP: Planning; CN: Physiological integrity; CNS: Physiological adaptation; CL: Knowledge

61. A client has just received a renal transplant and has started cyclosporine therapy to prevent graft rejection. Which of the following conditions is a major complication of this drug therapy?

1. Depression
2. Hemorrhage
3. Infection
4. Peptic ulcer disease

62. A client received a kidney transplant 2 months ago. He's admitted to the hospital with the diagnosis of acute rejection. Which of the following assessment findings would be expected?

1. Hypotension
2. Normal body temperature
3. Decreased white blood cell (WBC) counts
4. Elevated blood urea nitrogen (BUN) and creatinine levels

63. A client is undergoing peritoneal dialysis. The dialysate dwell time is completed, and the clamp is opened to allow the dialysate to drain. The nurse notes that drainage has stopped and only 500 ml has drained; the amount of dialysate instilled was 1,500 ml. Which of the following interventions would be done first?

1. Change the client's position.
2. Call the physician.
3. Check the catheter for kinks or obstruction.
4. Clamp the catheter and instill more dialysate at the next exchange time.

64. A client receiving hemodialysis treatment arrives at the hospital with a blood pressure of 200/100 mm Hg, a heart rate of 110 beats/minute, and a respiratory rate of 36 breaths/minute. Oxygen saturation in room air is 89%. He complains of shortness of breath, and +2 pedal edema is noted. His last hemodialysis treatment was yesterday. Which of the following interventions should be done first?

1. Administer oxygen.
2. Elevate the foot of the bed.
3. Restrict the client's fluids.
4. Prepare the client for hemodialysis.

Remember, cyclosporine is an immune system suppressant.

First means prioritize!

There's a lot of information given here. Look closely for the common patterns.

61. 3. Infection is the major complication to watch for in clients on cyclosporine therapy because it's an immunosuppressive drug. Depression may occur posttransplantation, but not because of cyclosporine. Hemorrhage is a complication associated with anticoagulant therapy. Peptic ulcer disease is a complication of steroid therapy.
NP: Data collection; CN: Physiological integrity; CNS: Pharmacological therapies; CL: Knowledge

62. 4. In a client with acute renal graft rejection, evidence of deteriorating renal function, including elevated BUN and creatinine levels, is expected. The nurse would see elevated, not decreased, WBC counts as well as fever because the body is recognizing the graft as foreign and is attempting to fight it. The client would most likely have acute hypertension.
NP: Data collection; CN: Physiological integrity; CNS: Reduction of risk potential; CL: Analysis

63. 3. The first intervention should be to check for kinks and obstructions because that could be preventing drainage. After checking for kinks, have the client change position to promote drainage. Don't give the next scheduled exchange until the dialysate is drained because abdominal distention will occur, unless the output is within the parameters set by the physician. If unable to get more output despite checking for kinks and changing the client's position, the nurse should then call the physician to determine the proper intervention.
NP: Implementation; CN: Physiological integrity; CNS: Reduction of risk potential; CL: Analysis

64. 1. Airway and oxygenation are always the first priority. Because the client is complaining of shortness of breath and his oxygen saturation is only 89%, the nurse needs to try to increase the partial pressure of arterial oxygen by administering oxygen. The client is in pulmonary edema from fluid overload and will need to be dialyzed and have his fluids restricted, but the first intervention should be aimed at the immediate treatment of hypoxia. The foot of the bed may be elevated to reduce edema, but this isn't a priority.
NP: Implementation; CN: Physiological integrity; CNS: Physiological adaptation; CL: Analysis

65. A client with renal insufficiency is admitted with a diagnosis of pneumonia. He's being treated with I.V. antibiotics and has had episodes of hypotension. Which of the following laboratory values would be monitored closely?
1. Blood urea nitrogen (BUN) and creatinine levels
2. Arterial blood gas (ABG) levels
3. Platelet count
4. Potassium level

66. A client had transurethral prostatectomy for benign prostatic hypertrophy. He's currently being treated with a continuous bladder irrigation and is complaining of an increase in severity of bladder spasms. Which of the following interventions should be done first?
1. Administer an oral analgesic.
2. Stop the irrigation and call the physician.
3. Administer a belladonna and opium suppository as ordered by the physician.
4. Check for the presence of clots, and make sure the catheter is draining properly.

67. A client with bladder cancer has had his bladder removed and an ileal conduit created for urine diversion. While changing this client's pouch, the nurse observes that the area around the stoma is red, weeping, and painful. What should the nurse conclude?
1. The skin wasn't lubricated before the pouch was applied.
2. The pouch faceplate doesn't fit the stoma.
3. A skin barrier was applied properly.
4. Stoma dilation wasn't performed.

68. A client has an indwelling urinary catheter, and urine is leaking from a hole in the collection bag. Which of the following nursing interventions would be most appropriate?
1. Cover the hole with tape.
2. Remove the catheter, and insert a new one using sterile technique.
3. Disconnect the drainage bag from the catheter, and replace it with a new bag.
4. Place a towel under the bag to prevent spillage of urine on the floor, which could cause the client to slip and fall.

Lots to do, but what's *most* important?

Here are those words most appropriate again. Don't you just love 'em?

65. 1. BUN and creatinine levels are used to monitor renal function. Because the client is receiving I.V. antibiotics, which can be nephrotoxic, these tests would be used to closely monitor renal function. The client is also hypotensive, which is a prerenal cause of acute renal failure. ABG determinations are inappropriate for this situation. Platelets and potassium levels should be monitored according to routine.
NP: Data collection; CN: Physiological integrity; CNS: Reduction of risk potential; CL: Analysis

66. 4. Blood clots and blocked outflow of the urine can increase spasms. The irrigation shouldn't be stopped as long as the catheter is draining because clots will form. A belladonna and opium suppository should be given to relieve spasms but only *after* assessment of the drainage. Oral analgesics should be given if the spasms are unrelieved by the belladonna and opium suppository.
NP: Implementation; CN: Physiological integrity; CNS: Physiological adaptation; CL: Comprehension

67. 2. If the pouch faceplate doesn't fit the stoma properly, the skin around the stoma will be exposed to continuous urine flow from the stoma, causing excoriation and red, weeping, painful skin. A lubricant shouldn't be used because it would prevent the pouch from adhering to the skin. When properly applied, a skin barrier prevents skin excoriation. Stoma dilation isn't performed with an ileal conduit, although it may be done with a colostomy, if ordered.
NP: Data collection; CN: Physiological integrity; CNS: Basic care and comfort; CL: Analysis

68. 2. The system is no longer a closed system, and bacteria might have been introduced into the system, so a new sterile catheter should be inserted. Placing a towel under the bag and taping up the hole leave the system open, which increases the risk of infection. Replacing the drainage bag by disconnecting the old one from the catheter opens up the entire system and isn't recommended because of the increased risk of infection.
NP: Implementation; CN: Safe, effective care environment; CNS: Safety and infection control; CL: Analysis

NP: Nursing process   CN: Client needs category   CNS: Client needs subcategory   CL: Cognitive level

69. During a health history, which statement by the client indicates a risk of renal calculi?
  1. "I've been drinking a lot of cola soft drinks lately."
  2. "I've been jogging more than usual."
  3. "I've had more stress since we adopted a child last year."
  4. "I'm a vegetarian and eat cheese two to three times each day."

69. 4. Renal calculi are commonly composed of calcium. Diets high in calcium may predispose a person to renal calculi. Milk and milk products are high in calcium. Cola soft drinks don't contain ingredients that would increase the risk for renal calculi. Jogging and increased stress aren't considered risk factors for renal calculi formation.

NP: Data collection; CN: Health promotion and maintenance; CNS: Prevention and early detection of disease; CL: Analysis

Almost finished. Don't quit now!

70. A client is admitted with severe nausea, vomiting, and diarrhea and is hypotensive. She's noted to have severe oliguria with elevated blood urea nitrogen (BUN) and creatinine levels. The physician will most likely write an order for which of the following treatments?
  1. Force oral fluids.
  2. Give furosemide 20 mg I.V.
  3. Start hemodialysis after a temporary access is obtained.
  4. Start I.V. fluid of normal saline solution bolus followed by a maintenance dose.

70. 4. The client is prerenal secondary to hypovolemia. I.V. fluids should be given to rehydrate the client, urine output should increase, and the BUN and creatinine levels will normalize. The client wouldn't be able to tolerate oral fluids because of the nausea, vomiting, and diarrhea. The client isn't fluid overloaded, and her urine output won't increase with furosemide. The client won't need dialysis because the oliguria and increased BUN and creatinine levels are due to dehydration.

NP: Planning; CN: Physiological integrity; CNS: Physiological adaptation; CL: Analysis

71. A woman who reports painful urination during or after voiding might have a problem in which of the following locations?
  1. Bladder
  2. Kidneys
  3. Ureters
  4. Urethra

71. 1. Pain during or after voiding indicates a bladder problem, usually infection. Kidney and ureter pain would be in the flank area, and problems of the urethra would cause pain at the external orifice that's often felt at the start of voiding.

NP: Data collection; CN: Health promotion and maintenance; CNS: Prevention and early detection of disease; CL: Knowledge

What time is best for the client?

72. What's the best time of day to give diuretics?
  1. Anytime
  2. Bedtime
  3. Morning
  4. Noon

72. 3. A diuretic given in the morning has time to work throughout the day. Diuretics given at nighttime will cause the client to get up to go to the bathroom frequently, interrupting sleep.

NP: Implementation; CN: Physiological integrity; CNS: Pharmacological therapies; CL: Application

73. Which of the following interventions would be inappropriate to help a client with postoperative urinary retention?

1. Give a diuretic.
2. Pour warm water over the perineum.
3. Consider inserting a bladder catheter.
4. Place the client in a sitting or semi-Fowler position.

Watch out for the word inappropriate. It can be tricky.

73. 1. Urinary retention reflects bladder distention from urine. A diuretic isn't necessary. Sitting upright and pouring water over the perineum may help the client void. If these measures aren't successful, the nurse should consider inserting a bladder catheter to drain the bladder, which requires an order from the physician.
NP: Planning; CN: Physiological integrity; CNS: Basic care and comfort; CL: Application

74. Which of the following factors may place a surgical client at risk for urinary retention?

1. Dehydration
2. History of smoking
3. Duration of surgery
4. Anticholinergic medication before surgery

74. 4. Anticholinergic medications, such as atropine and scopolamine, may cause urinary retention, particularly for the client who has surgery in the pelvic area (inguinal hernia, hysterectomy). Dehydration, smoking, and duration of surgery aren't risk factors for urine retention.
NP: Planning; CN: Physiological integrity; CNS: Reduction of risk potential; CL: Comprehension

75. Which type of catheter is generally used for the client with urine retention?

1. Coudé
2. Indwelling urinary
3. Straight
4. Three-way

All these catheters aren't the same.

75. 3. Urine retention is usually a temporary problem. The other catheters are used for longer-term bladder problems. The three-way catheter is used for clients who need bladder irrigation such as after a prostate resection. A catheter coudé is used only when it's difficult to insert a standard catheter, usually because of an enlarged prostate.
NP: Implementation; CN: Physiological integrity; CNS: Basic care and comfort; CL: Comprehension

76. An 80-year-old man reports urine retention. Which of the following factors may contribute to this client's problem?

1. Benign prostatic hyperplasia
2. Diabetes
3. Diet
4. Hypertension

76. 1. An enlarged prostate gland is common among elderly men and often results in urine retention, frequency, dribbling, and difficulty starting the urine stream. Diet, hypertension, and diabetes usually don't affect urine retention. Diabetes can cause renal failure.
NP: Data collection; CN: Physiological integrity; CNS: Reduction of risk potential; CL: Application

NP: Nursing process   CN: Client needs category   CNS: Client needs subcategory   CL: Cognitive level

77. A client with chronic renal failure (CRF) is admitted to the urology unit. Which diagnostic test results are consistent with CRF?

1. Increased pH with decreased hydrogen ions
2. Increased serum levels
3. Blood urea nitrogen (BUN) level of 100 mg/dl and serum creatinine level of 6.5 mg/dl
4. Uric acid level of 3.5 mg/dl and phenol-sulfonphthalein (PSP) excretion of 75%

Almost finished! Keep pluggin' away!

77. 3. The normal BUN level ranges from 8 to 23 mg/dl, and the normal serum creatinine level ranges from 0.7 to 1.5 mg/dl. BUN level of 100 mg/dl and serum creatinine level of 6.5 mg/dl are abnormally elevated, reflecting CRF and the kidneys' decreased ability to remove nonprotein nitrogen waste from the blood. CRF causes decreased pH and increased hydrogen ions, not vice versa. CRF also increases serum levels of potassium, magnesium, and phosphatase and decreases serum levels of calcium. A uric acid level of 3.5 mg/dl falls within the normal range of 2.7 to 7.7. mg/dl; PSP excretion of 75% also falls within the normal range of 60% to 75%.

NP: Data collection; CN: Physiological integrity; CNS: Physiological adaptation; CL: Knowledge

78. Which steps should the nurse follow to insert a straight urinary catheter?

1. Create a sterile field, drape the client, clean the meatus, and insert the catheter only 6″ (15.2 cm).
2. Put on gloves, prepare the equipment, create a sterile field, expose the urinary meatus, and insert the catheter 6″.
3. Prepare the client and equipment, create a sterile field, put on gloves, clean the urinary meatus, and insert the catheter until the urine flows.
4. Prepare the client, prepare the equipment, create a sterile field, test the catheter balloon, clean the meatus, and insert the catheter until urine flows.

78. 3. Preparing the client and equipment, creating a sterile field, putting on gloves, cleaning the urinary meatus, and inserting the catheter until the urine flows are the vital steps for inserting a straight catheter. The nurse must prepare the client and equipment *before* creating a sterile field. The nurse shouldn't put on gloves before creating a sterile field or performing the other tasks. The nurse would test a catheter balloon when inserting a retention catheter, rather than a straight catheter.

NP: Implementation; CN: Safe, effective care environment; CNS: Safety and infection control; CL: Knowledge

79. A client is injected with radiographic contrast medium and immediately shows signs of dyspnea, flushing, and pruritus. Which of the following interventions should take priority?

1. Check vital signs.
2. Make sure the airway is patent.
3. Apply a cold pack to the I.V. site.
4. Call the physician.

This word — priority — is your biggest hint.

79. 2. The client is showing symptoms of an allergy to the iodine in the contrast medium. The first action is to make sure the client's airway is patent. If compromised, call a cardiac arrest code. Checking vital signs and calling for the physician are important nursing actions but should follow making sure the airway is patent. A cold pack isn't indicated.

NP: Implementation; CN: Physiological integrity; CNS: Physiological adaptation; CL: Application

80. An 80-year-old man is admitted for a cystoscopy with biopsy of the bladder. After the physician obtains a history, the surgery is postponed. Which of the following reasons would n͟o͟t be a reason postpone this client's surgery?
1. The client is on an anticoagulant.
2. The client has a urinary tract infection (UTI).
3. The client might have carcinoma of the bladder.
4. The client reports chest pain at rest for the past 3 days.

Watch out! This question asks what *not* to do.

80. 3. Suspected bladder carcinoma is probably the reason for the planned biopsy. Bladder biopsies shouldn't be done when an active UTI is present because sepsis may result. Chest pain at rest may indicate myocardial ischemia and would be an indication to postpone the biopsy until it can be further investigated. Anticoagulants should be discontinued for 3 to 5 days before the procedure.
NP: Implementation; CN: Physiological integrity; CNS: Reduction of risk potential; CL: Analysis

81. Unless there are postoperative complications, a cystoscopy client is discharged to home within 24 hours. Which of the following instructions is given at discharge?
1. Expect bloody urine for about a week.
2. Drink 8 to 10 glasses of water every 8 hours.
3. Try to urinate frequently, and measure your output.
4. Check the color, consistency, and amount of urine in the indwelling urinary catheter bag every 4 to 8 hours.

What would your client do without your professional, comprehensive teaching?

81. 3. The bladder needs to be emptied frequently, and output should be measured to make sure the bladder is emptying. Large amounts of fluids help flush microorganisms out of the body, but 8 to 10 glasses every 8 hours may not be reasonable. Also, families don't tend to think in time periods, so instructions should be given per day. The client may not have an indwelling urinary catheter. Blood in the urine isn't normal except for small amounts during the first 24 hours after the procedure.
NP: Implementation; CN: Physiological integrity; CNS: Reduction of risk potential; CL: Application

82. Before a renal biopsy, which of the following pieces of information is most important to tell the physician?
1. The client signed a consent.
2. The client understands the procedure.
3. The client has normal urinary elimination.
4. The client regularly takes aspirin or nonsteroidal anti-inflammatory drugs (NSAIDs).

82. 4. Aspirin and NSAIDs cause increased bleeding times and often result in hemorrhaging when biopsies are performed. It's the physician's responsibility to make sure the client understands the procedure, which is needed for informed consent. It isn't necessary to report that the client has normal urinary elimination.
NP: Implementation; CN: Physiological integrity; CNS: Reduction of risk potential; CL: Application

83. Kegel exercises are used to gain control of bladder function in women with stress incontinence and some men after prostate surgery. Which of the following instructions would help the client perform these exercises?
1. Completely empty the bladder.
2. Do the exercise 200 times a day.
3. Sit or stand with your legs together.
4. Drink small amounts of fluid frequently.

Kegel exercises can help you gain bladder control.

83. 2. Exercises begin with tightening and relaxing the vagina, rectum, and urethra four or five times during each session and gradually increasing to 25 times for each session. The client stops the flow of urine during urination to practice holding the flow. Standing or sitting with the legs apart facilitates the exercise. Clients should drink plenty of fluids to prevent urinary problems.
NP: Evaluation; CN: Physiological integrity; CNS: Physiological adaptation; CL: Application

NP: Nursing process   CN: Client needs category   CNS: Client needs subcategory   CL: Cognitive level

84. Which of the following instructions is given to clients with chronic pyelonephritis?
1. Stay on bed rest for up to 2 weeks.
2. Use analgesia on a regular basis for up to 6 months.
3. Have a urine culture every 2 weeks for up to 6 months.
4. You may need antibiotic treatment for several weeks or months.

85. Which of the following factors causes the nausea associated with renal failure?
1. Oliguria
2. Gastric ulcers
3. Electrolyte imbalance
4. Accumulation of metabolic wastes

86. Which of the following clients is at greatest risk for developing acute renal failure?
1. A dialysis client who gets influenza
2. A teenager who has an appendectomy
3. A pregnant woman who has a fractured femur
4. A client with diabetes who has a heart catheterization

87. Which of the following interventions would be done for a client with urinary calculus?
1. Save any stone larger than 0.25 cm.
2. Strain the urine, limit oral fluids, and give pain medications.
3. Encourage fluid intake, strain the urine, and give pain medications.
4. Insert an indwelling urinary catheter, check intake and output, and give pain medications.

Remember, nausea is commonly caused by a systemic condition.

Only a few more questions to go!

84. 4. Chronic pyelonephritis can be a long-term condition and requires close monitoring to prevent permanent damage to the kidneys. Analgesia and bed rest may be used during the acute stage but usually aren't required long term. A urine culture is done 2 weeks after stopping antibiotics to make sure the infection has been eradicated.
NP: Implementation; CN: Physiological integrity; CNS: Reduction of risk potential; CL: Application

85. 4. Although a client with renal failure can develop stress ulcers, nausea is usually related to the poisons of metabolic wastes that accumulate when the kidneys are unable to eliminate them. Although a client may have electrolyte imbalances and oliguria, these conditions don't directly cause nausea.
NP: Data collection; CN: Physiological integrity; CNS: Physiological adaptation; CL: Comprehension

86. 4. Clients with diabetes are prone to renal insufficiency and renal failure. The contrast used for heart catheterization must be eliminated by the kidneys, which further stresses them and may produce acute renal failure. A teenager who has an appendectomy and a pregnant woman who fractures a femur aren't at increased risk for renal failure. A dialysis client already has end-stage renal disease and wouldn't develop acute renal failure.
NP: Data collection; CN: Health promotion and maintenance; CNS: Prevention and early detection of disease; CL: Analysis

87. 3. Force fluids on the client and strain all urine, saving any and all stones, including "flecks." Give pain medications because kidney stones are extremely painful. Indwelling urinary catheters usually aren't needed.
NP: Planning; CN: Physiological integrity; CNS: Basic care and comfort; CL: Application

88. Which of the following interventions is <u>inappropriate</u> for a client on hemodialysis?
 1. Palpate for a thrill on the arm with the fistula.
 2. Auscultate for a bruit on the arm with the fistula.
 3. Report the absence of a thrill or bruit on the arm with the fistula.
 4. Take blood pressure or start an I.V. on the arm with the fistula.

89. A client has passed a kidney stone. The nurse sends the specimen to the laboratory so it can be analyzed for which of the following factors?
 1. Antibodies
 2. Type of infection
 3. Composition of stone
 4. Size and number of stones

90. A young school-age child who recently underwent kidney transplantation is taking prednisone (Deltasone) to prevent organ rejection. During this client's long-term steroid therapy, which parameter should the nurse monitor closely?
 1. Red blood cell (RBC) count
 2. Growth and development
 3. Platelet count
 4. Hearing

91. Adverse reactions of prednisone therapy include which of the following conditions?
 1. Acne and bleeding gums
 2. Sodium retention and constipation
 3. Mood swings and increased temperature
 4. Increased blood sugar levels and decreased wound healing

Careful. The two little letters before *appropriate* can be easily missed.

Congratulations! You did it! I'm proud of you!

88. 4. The nurse would palpate for a thrill and auscultate for a bruit on the arm with a fistula, but no procedures (I.V. access, blood pressure, or blood draw) should be done on the arm with a fistula because it could damage the fistula. The absence of a thrill or bruit should be reported promptly to the physician because it indicates an occlusion.
NP: Implementation; CN: Physiological integrity; CNS: Reduction of risk potential; CL: Application

89. 3. The stone should be analyzed for composition to determine appropriate interventions such as dietary restrictions. Stones don't result from infections. The size and number of the stones aren't relevant, and they don't contain antibodies.
NP: Evaluation; CN: Physiological integrity; CNS: Reduction of risk potential; CL: Application

90. 2. For a child receiving steroids, the nurse should monitor growth and development closely because prolonged steroid use interferes with pituitary gland function, resulting in linear growth retardation. The nurse doesn't need to monitor the RBC or platelet count or the client's hearing because they aren't affected by steroids.
NP: Evaluation; CN: Health promotion and maintenance; CNS: Growth and development through the life span; CL: Application

91. 4. Steroid use tends to increase blood sugar levels, particularly in clients with diabetes and borderline diabetes. Steroids also contribute to poor wound healing and may cause acne, mood swings, and sodium and water retention. Steroids don't affect thermoregulation, bleeding tendencies, or constipation.
NP: Data collection; CN: Physiological integrity; CNS: Pharmacological therapies; CL: Analysis

So, you think the care of the patient with a skin disorder isn't your strong suit, eh? Maybe you'd like to check out this Internet site before taking on this chapter: **www.aad.org/.** Enjoy!

1. Adequate nutrition is essential for a burn client. Which of the following statements is correct about the nutritional needs of a burn client?
　1. The client needs 100 cal/kg throughout hospitalization.
　2. The hypermetabolic state after a burn injury leads to poor healing.
　3. Controlling the temperature of the environment decreases caloric demands.
　4. Maintaining a hypermetabolic rate decreases the client's risk of infection.

2. Which of the following characteristics is correct for a deep partial-thickness burn?
　1. Pain and redness
　2. Minimal damage to the epidermis
　3. Necrotic tissue through all layers of skin
　4. Necrotic tissue through most of the dermis

3. Which of the following characteristics is a classic sign of a bite from a brown recluse spider?
　1. Bull's-eye rash
　2. Painful rash around a necrotic lesion
　3. Herald patch of oval lesions
　4. Line of papules and vesicles that appear 1 to 3 days after exposure

You know these answers. It won't be hard!

You're looking not just for a sign but a classic sign.

1. 2. A burn injury causes a hypermetabolic state resulting in protein and lipid catabolism that affects wound healing. Calories need to be 1½ to 2 times the basal metabolic rate, with at least 1.5 to 2 g/kg of body weight of protein daily. High metabolic rates increase the risk of infection. An environmental temperature within normal range lets the body function efficiently and devote caloric expenditure to healing and normal physiologic processes. If the temperature is too warm or too cold, the body gives energy to warming or cooling, which takes away from energy used for tissue repair.

NP: Data collection; CN: Physiological integrity; CNS: Basic care and comfort; CL: Knowledge

2. 4. A deep partial-thickness burn causes necrosis of the epidermal and dermal layers. Necrosis through all skin layers is seen with full-thickness injuries. Redness and pain are characteristics of a superficial injury. Superficial burns cause slight epidermal damage. With deep burns, the nerve fibers are destroyed and the client doesn't feel pain in the affected area.

NP: Data collection; CN: Physiological integrity; CNS: Physiological adaptation; CL: Knowledge

3. 2. Necrotic, painful rashes are associated with the bite of a brown recluse spider. A bull's-eye rash located primarily at the site of the bite is a classic sign of Lyme disease. A herald patch—a slightly raised, oval lesion about 2 to 6 cm in diameter and appearing anywhere on the body—is indicative of pityriasis rosea. A linear, papular, vesicular rash is characteristic of exposure to poison ivy.

NP: Data collection; CN: Physiological integrity; CNS: Physiological adaptation; CL: Knowledge

NP: Nursing process　CN: Client needs category　CNS: Client needs subcategory　CL: Cognitive level

4. Which of the following instructions should be observed when administering a Mantoux test?
1. Use the deltoid muscle.
2. Rub the site to help absorption.
3. Read the results within 72 hours.
4. Read the results by checking for a rash.

5. A woman is worried she might have lice. Which of the following findings is associated with this infestation?
1. Diffuse pruritic wheals
2. Oval, white dots stuck to the hair shafts
3. Pain, redness, and edema with an embedded stinger
4. Pruritic papules, pustules, and linear burrows of the finger and toe webs

6. A client is brought to the emergency department with second- and third-degree burns on his left arm, left anterior leg, and anterior trunk. Using the Rule of Nines, what percentage of the total body surface area has been burned?
1. 18%
2. 27%
3. 30%
4. 36%

7. A 19-year-old woman comes to the clinic with dark red lesions on her hands, wrist, and waistline. She has scratched several of the lesions so they're open and bleeding. The nurse instructs the client to try pressing on the itchy lesions. What is the rationale for this intervention?
1. Pressing the skin spreads beneficial microorganisms.
2. Pressing is suggested before scratching.
3. Pressing the skin promotes breaks in the skin.
4. Pressing the skin stimulates nerve endings.

You're doing great! Keep up the good work!

4. 3. The test results should be read 48 to 72 hours after placement by measuring the diameter of the induration that develops at the site. The purified protein derivative test is injected intradermally on the volar surface of the forearm, not I.M. Rubbing the site could cause leakage from the injection site.
NP: Data collection; CN: Physiological integrity; CNS: Reduction of risk potential; CL: Knowledge

5. 2. Nits, the eggs of lice, are seen as white, oval dots. Diffuse itching wheals are associated with an allergic reaction. Bites from honeybees are associated with a stinger, pain, and redness. Pruritic papules, vesicles, and linear burrows are diagnostic for scabies.
NP: Data collection; CN: Physiological integrity; CNS: Physiological adaptation; CL: Knowledge

6. 4. The Rule of Nines divides body surface into percentages that, when totaled, equal 100%. According to the Rule of Nines, the arms account for 9% each, the anterior legs account for 9% each, and the anterior trunk accounts for 18%. Therefore, this client's burns cover 36% of his body surface area.
NP: Data collection; CN: Physiological integrity; CNS: Physiological adaptation; CL: Knowledge

7. 4. Pressing the skin stimulates nerve endings and can reduce the sensation of itching. Scratching (not pressing) the skin spreads microorganisms and opens portals of entry for bacteria. Scratching isn't recommended at all. Pressing the skin doesn't promote breaks in the skin.
NP: Evaluation; CN: Physiological integrity; CNS: Physiological adaptation; CL: Application

---

NP: Nursing process   CN: Client needs category   CNS: Client needs subcategory   CL: Cognitive level

8. A man arrives at the office of his physician complaining of several palpable elevated masses. Which of the following terms accurately describes these masses?
1. Erosions
2. Macules
3. Papules
4. Vesicles

9. Which of the following adverse reactions may be caused by the use of isotretinoin (Accutane)?
1. Birth defects
2. Nausea and vomiting
3. Vaginal yeast infection
4. Gram-negative folliculitis

10. The nurse is caring for a 12-year-old child with a diagnosis of eczema. Which nursing interventions are appropriate for a child with eczema?
1. Administer antibiotics as prescribed.
2. Administer antifungals as ordered.
3. Administer tepid baths, and use moisturizers immediately after the bath.
4. Administer hot baths, and pat dry or air-dry the affected areas.

11. Which of the following instructions is given to a client taking nystatin oral solution?
1. Take the drug right after meals.
2. Take the drug right before meals.
3. Mix the drug with small amounts of food.
4. Take half the dose before and half after meals.

12. A client is examined and found to have pinpoint, pink-to-purple, nonblanching macular lesions 1 to 3 mm in diameter. Which of the following terms best describes these lesions?
1. Ecchymosis
2. Hematoma
3. Petechiae
4. Purpura

Which of these terms describes palpable, elevated masses? It's that simple!

Questions about adverse reactions are common on NCLEX examinations. Better bone up on 'em!

You are doing petechiae-ly well so far. (Ha! I'm too funny!)

8. 3. Papules are elevated up to 0.5 cm, and nodules and tumors are masses elevated more than 0.5 cm. Erosions are characterized as loss of the epidermal layer. Macules and patches are nonpalpable, flat changes in skin color. Fluid-filled lesions are vesicles and pustules.
NP: Data collection; CN: Health promotion and maintenance; CNS: Prevention and early detection of disease; CL: Knowledge

9. 1. Even small amounts of Accutane are associated with severe birth defects. Most female clients are also prescribed oral contraceptives. Clindamycin phosphate (Cleocin T Gel), another medicine used in the treatment of acne, can cause both diarrhea and gram-negative folliculitis. Tetracycline is associated with yeast infections.
NP: Data collection; CN: Health promotion and maintenance; CNS: Growth and development through the life span; CL: Knowledge

10. 3. Tepid baths and moisturizers are indicated to keep the infected areas clean and minimize itching. Antibiotics are given only when there's superimposed infection. Antifungals aren't usually administered in the treatment of eczema. Hot baths can exacerbate the condition and increase itching.
NP: Implementation; CN: Physiological integrity; CNS: Physiological adaptation; CL: Application

11. 1. Nystatin oral solution should be swished around the mouth after eating for the best contact with mucous membranes. Taking the drug before or with meals doesn't allow for the best contact with the mucous membranes.
NP: Implementation; CN: Physiological integrity; CNS: Pharmacological therapies; CL: Application

12. 3. Petechiae are small macular lesions 1 to 3 mm in diameter. Ecchymosis is a purple-to-brown bruise, macular or papular, and varied in size. A hematoma is a collection of blood from ruptured blood vessels that's more than 1 cm in diameter. Purpura are purple macular lesions larger than 1 cm.
NP: Data collection; CN: Physiological integrity; CNS: Physiological adaptation; CL: Knowledge

13. A client has a rash consisting of scattered lesions on various parts of the body. Which of the following types of rash is this?
1. Annular
2. Confluent
3. Diffuse
4. Linear

14. Which of the following terms best describes hair loss in small round areas on the scalp?
1. Alopecia
2. Amblyopia
3. Exotropia
4. Seborrhea

Concentrate:
What's the meaning
of diffuse?

15. When helping to plan nursing care to maintain skin integrity for an adult client, the nurse should remember which general guideline?
1. Apply a pleasantly scented dusting powder to the axillae and groin, beneath the breasts, and between the toes.
2. Remember to apply a deodorant or antiperspirant immediately after shaving under the arms.
3. Try to keep skin intact because healthy skin is the body's first line of defense.
4. Always use alcohol for back rubs instead of lotion.

16. A client has rough papules on the soles of his feet that are sometimes painful when he walks. Which of the following terms best describes this condition?
1. Filiform wart
2. Flat wart
3. Plantar wart
4. Venereal wart

Don't quit
now!

13. 3. A diffuse rash usually has widely distributed scattered lesions. An annular rash is ring shaped. Confluent lesions are touching or adjacent to each other. Linear rashes are lesions arranged in a line.

NP: Data collection; CN: Physiological integrity; CNS: Physiological adaptation; CL: Knowledge

14. 1. Alopecia is the correct term for thinning hair loss. Exotropia and amblyopia are eye disorders. Seborrhea is a chronic inflammatory dermatitis that occurs in infants.

NP: Data collection; CN: Physiological integrity; CNS: Physiological adaptation; CL: Knowledge

15. 3. Because healthy skin is the body's first line of defense, a key nursing goal is to keep the skin intact. To reduce moisture, the nurse can apply a nonirritating dusting powder, such as cornstarch, to the client's axillae and groin, beneath the breasts, and between the toes after those areas are dry. However, scented powder shouldn't be used because it can irritate the skin. Deodorants and antiperspirants shouldn't be applied to the skin immediately after shaving because they may cause irritation. The nurse should use lotion for back rubs because alcohol dries the skin and can irritate it.

NP: Planning; CN: Physiological integrity; CNS: Basic care and comfort; CL: Knowledge

16. 3. Plantar warts are rough papules commonly found on the soles of the feet. Filiform warts are long spiny projections from the skin surface. Flat warts are flat-topped, smooth-surfaced lesions. Venereal warts appear on the genital mucosa and are confluent papules with rough surfaces.

NP: Data collection; CN: Physiological integrity; CNS: Physiological adaptation; CL: Knowledge

---

NP: Nursing process   CN: Client needs category   CNS: Client needs subcategory   CL: Cognitive level

17. A client reports circular lesions on his neck. Which of the following conditions is the most likely cause?
 1. Candidiasis
 2. Molluscum contagiosum
 3. Tinea corporis
 4. Tinea pedis

Look carefully for this answer.

17. 3. Tinea corporis, or ringworm, is a flat scaling papular lesion with raised borders. Candidiasis is a fungal infection of the skin or mucous membranes commonly found in the oral, vaginal, and intestinal mucosal tissue. Molluscum contagiosum is a viral skin infection with small, red, papular lesions. Tinea pedis is a superficial fungal infection on the feet, commonly called athletes' foot, that causes itching and sweating and a foul odor.

NP: Data collection; CN: Physiological integrity; CNS: Physiological adaptation; CL: Knowledge

18. A client has thick, discolored nails with splintered hemorrhages, easily separated from the nail bed. There are also "ice pick" pits and ridges. Which of the following terms best describes these symptoms?
 1. Paronychia
 2. Psoriasis
 3. Seborrhea
 4. Scabies

18. 2. Psoriasis, a chronic skin disorder with an unknown cause, shows these characteristic skin changes. A paronychia is a bacterial infection of the nail bed. Seborrhea is a chronic inflammatory dermatitis known as cradle cap. Scabies are mites that burrow under the skin, generally between the webbing of the fingers and toes.

NP: Data collection; CN: Physiological integrity; CNS: Physiological adaptation; CL: Knowledge

19. A client found unconscious at home is brought to the emergency department. Physical examination shows cherry-red mucous membranes, nail beds, and skin. Which of the following factors is the most likely cause of his condition?
 1. Spider bite
 2. Aspirin ingestion
 3. Hydrocarbon ingestion
 4. Carbon monoxide poisoning

Color can be a classic sign of some conditions.

19. 4. Cherry-red skin indicates exposure to high levels of carbon monoxide. Spider bite reactions are usually localized to the area of the bite. Hydrocarbon or petroleum ingestion usually causes respiratory symptoms and tachycardia. Nausea and vomiting and pale skin are symptoms of aspirin ingestion.

NP: Data collection; CN: Physiological integrity; CNS: Physiological adaptation; CL: Knowledge

20. Which of the following terms describes a fungal infection of the scalp?
 1. Tinea capitis
 2. Tinea corporis
 3. Tinea cruris
 4. Tinea pedis

20. 1. Tinea capitis is a fungal infection of the scalp. Tinea corporis describes fungal infections of the body. Tinea cruris describes fungal infections of the inner thigh and inguinal creases, and tinea pedis is the term for fungal infections of the foot.

NP: Data collection; CN: Physiological integrity; CNS: Physiological adaptation; CL: Application

21. In the client with burns on his legs, which nursing intervention helps prevent contractures?
1. Applying knee splints
2. Elevating the foot of the bed
3. Hyperextending the client's palms
4. Performing shoulder range-of-motion (ROM) exercises

21. 1. Applying knee splints prevents leg contractures by holding the joints in a position of function. Elevating the foot of the bed can't prevent contractures because this action doesn't hold the joints in a position of function. Hyperextending a body part for an extended time is inappropriate because it can cause contractures. Performing shoulder ROM exercises can prevent contractures in the shoulders, but not in the legs.
NP: Implementation; CN: Physiological integrity; CNS: Reduction of risk potential; CL: Application

22. The nurse is planning care for a client with burns on the upper torso. Which nursing diagnosis should take the highest priority?
1. Ineffective airway clearance related to edema of the respiratory passages
2. Impaired physical mobility related to the disease process
3. Disturbed sleep pattern related to facility environment
4. Risk for infection related to breaks in the skin

Prioritize!

22. 1. When caring for a client with upper torso burns, the nurse's primary goal is to maintain respiratory integrity. Therefore, a diagnosis of ineffective airway clearance related to edema of the respiratory passages should take the highest priority. The second diagnosis isn't appropriate because burns aren't a disease. Although the third and fourth diagnoses may be appropriate, they don't command a higher priority than the first because they don't reflect immediate life-threatening problems.
NP: Planning; CN: Physiological integrity; CNS: Physiological adaptation; CL: Application

23. A client has just arrived at the emergency department after sustaining a major burn injury. Which of the following metabolic alterations is expected during the first 8 hours postburn?
1. Hyponatremia and hypokalemia
2. Hyponatremia and hyperkalemia
3. Hypernatremia and hypokalemia
4. Hypernatremia and hyperkalemia

Which way did the electrolytes go?

23. 2. During the first 48 hours after a burn, capillary permeability increases, allowing fluids to shift from the plasma to the interstitial spaces. This fluid is high in sodium, causing a decrease in serum sodium levels. Potassium also leaks from the cells into the plasma, causing hyperkalemia.
NP: Data collection; CN: Physiological integrity; CNS: Physiological adaptation; CL: Analysis

24. In a client who has been burned, which medication should the nurse expect to use to prevent infection?
1. Lindane (G-well)
2. Diazepam (Valium)
3. Mafenide (Sulfamylon)
4. Meperidine (Demerol)

24. 3. The topical antibiotic mafenide is prescribed to prevent infection in clients with second- and third-degree burns. Lindane is a pediculicide used to treat lice infestation. Diazepam is an antianxiety agent that may be administered to clients with burns, but not to prevent infection. The narcotic analgesic meperidine is used to help control pain in clients with burns.
NP: Planning; CN: Physiological integrity; CNS: Pharmacological therapies; CL: Comprehension

25. Which of the following methods is best for making sure a client receives an ordered dressing change each shift?
1. Write the order in the client's care Kardex.
2. Put a sign above the head of the client's bed.
3. Tell the nurse about the treatment in the report.
4. Document the dressing change in the narrative note.

26. Which of the following nursing diagnoses is correct for a client with a reddened sacrum unrelieved by position change?
1. Risk for pressure ulcer
2. Risk for impaired skin integrity
3. Impaired skin integrity related to infrequent turning and positioning
4. Impaired skin integrity related to the effects of pressure and shearing force

27. A surgical client has just been admitted to a unit from the recovery room. When should the nurse change the dressing for the first time?
1. Two hours after admission
2. When it becomes saturated
3. Based on written orders for dressing changes
4. The surgeon changes the first dressing; then follow written orders

28. Which of the following techniques is incorrect when giving a client a bath?
1. Close the bed curtain to provide privacy.
2. Change the bath water when it becomes tepid.
3. Wash the female perineum from the pubis toward the anus.
4. Wash the client without gloves to improve the nurse-client relationship.

Congrats! You're halfway finished!

Careful! This is a negative term!

25. 1. Writing the order in the client's care Kardex notifies everyone of the treatment. Verbally reporting to the nurse on the upcoming shift doesn't ensure the dressing change will be done. Although the intervention should be documented in the narrative note, this doesn't guarantee that the next nurse will do the treatment. Posting a sign above the head of the bed is a good reminder but doesn't ensure that the treatment will be performed.
NP: Implementation; CN: Safe, effective care environment; CNS: Coordinated care; CL: Application

26. 4. The client's impaired skin integrity is the result of pressure and shearing forces. The first option isn't an approved nursing diagnosis. This client has an actual—not potential—skin impairment. The third option is an improperly written nursing diagnosis. The use of infrequent turning and positioning noted in the statement implies the nurse hasn't done the required nursing care.
NP: Planning; CN: Physiological integrity; CNS: Basic care and comfort; CL: Application

27. 4. The surgeon should always perform the first postoperative dressing change. Generally, the surgeon will change the dressing and assess the wound the next morning during rounds. The dressing shouldn't need to be changed 2 hours after the procedure. If the first dressing becomes saturated, it may be secured with additional tape or bandages. If the nurse notes hemorrhage or excessive amount of drainage, the surgeon should be notified.
NP: Implementation; CN: Physiological integrity; CNS: Basic care and comfort; CL: Application

28. 4. Wearing gloves when providing personal care is part of standard infection control precautions. Latex gloves should be worn to wash a client to prevent transmission of microorganisms from the client to the nurse and from the nurse to the client. Providing privacy, preventing chills, and preventing organisms from the anus from entering the vagina and the urethra are essential to giving a proper bath.
NP: Implementation; CN: Safe, effective care environment; CNS: Coordinated care; CL: Comprehension

29. Which of the following techniques is correct for obtaining a wound culture from a surgical site?

1. Thoroughly irrigate the wound before collecting the culture.
2. Use a sterile swab, and wipe the crusty area around the outside of the wound.
3. Gently roll a sterile swab from the center of the wound outward to collect drainage.
4. Use one sterile swab to collect drainage from several possible infected sites along the incision.

30. Which of the following characteristics of an abdominal incision would indicate a potential for delayed wound healing?

1. Sutures dry and intact
2. Wound edges in close approximation
3. Purulent drainage on soiled wound dressing
4. Sanguineous drainage in wound collection drainage bag

31. A nursing assistant will be changing the soiled bed linens of a client with a draining pressure ulcer. Which of the following protective equipment should the nursing assistant wear?

1. Mask
2. Clean gloves
3. Sterile gloves
4. Shoe protectors

32. Which of the following interventions is most appropriate for preventing pressure ulcers in a bedridden elderly client?

1. Slide instead of lift the client when turning.
2. Turn and reposition the client at least every 8 hours.
3. Apply lotion after bathing the client, and vigorously massage the skin.
4. Post a turning schedule at the client's bedside, and adapt position changes to the client's situation.

Question 30 is looking for something that would delay healing. It's almost a negative question.

Preventive care is a focus of many NCLEX questions.

29. 3. Rolling a swab from the center outward is the right way to culture a wound. Irrigating the wound washes away drainage, debris, and many of the microorganisms colonizing or infecting the wound. All sources of drainage in an incision or surgical wound may not be infected or may be infected with different microorganisms, so each swab should be used on only one site. The outside of the wound may be colonized with microorganisms from this wound, another wound, or the normal microorganisms on the client's skin. These may grow in culture and confuse the interpretation of results.

NP: Implementation; CN: Safe, effective care environment; CNS: Safety and infection control; CL: Knowledge

30. 3. Purulent drainage contains white blood cells, which fight infection. Wound edges can't approximate with an infection in the wound. The sutures from a wound draining purulent secretions would pull away with an infection. Sanguineous drainage indicates bleeding, not infection.

NP: Data collection; CN: Physiological integrity; CNS: Physiological adaptation; CL: Comprehension

31. 2. Clean gloves protect the hands and wrists from microorganisms in the linens. Sterile gloves allow one to touch a sterile object or area without contaminating it. A mask protects the wearer and client from droplet nuclei and large particle aerosols. Shoe protectors prevent static and microorganism transmission from the floor of one room to another.

NP: Planning; CN: Safe, effective care environment; CNS: Safety and infection control; CL: Analysis

32. 4. A turning schedule with a signing sheet will ensure that the client gets turned. A client in bed for prolonged periods should be turned every 1 to 2 hours. Apply lotion to keep the skin moist, but refrain from vigorous massage to avoid damaging capillaries. When moving a client, lift, rather than slide, the client to avoid shearing.

NP: Planning; CN: Safe, effective care environment; CNS: Safety and infection control; CL: Analysis

33. Which of the following instructions is the most important to give a client who has recently had a skin graft?
1. Continue physical therapy.
2. Protect the graft from direct sunlight.
3. Use cosmetic camouflage techniques.
4. Apply lubricating lotion to the graft site.

34. Which of the following factors is the most common cause of basal cell epithelioma?
1. Burns
2. Exposure to sun
3. Immunosuppression
4. Exposure to radiation

35. Which of the following statements is correct regarding skin turgor?
1. Overhydration causes the skin to tent.
2. Normal skin turgor is moist and boggy.
3. Inelastic skin turgor is a normal part of aging.
4. Dehydration causes the skin to appear edematous and spongy.

36. The nurse is caring for a geriatric client with a pressure ulcer on the sacrum. When teaching the client about dietary intake, which foods should the nurse plan to emphasize?
1. Legumes and cheese
2. Whole-grain products
3. Fruits and vegetables
4. Lean meats and low-fat milk

37. Which of the following water temperatures is correct for a bed bath?
1. 97° to 100° F (36.1° to 37.8° C)
2. 100° to 110° F (37.8° to 43.3° C)
3. 110° to 115° F (43.3° to 46.1° C)
4. 115° to 120° F (46.1° to 48.9° C)

Question 33 is asking you to prioritize.

You're really moving! Keep it up!

And the correct temperature is...

33. 2. To avoid burning and sloughing, the client must protect the graft from direct sunlight. The other three interventions are all helpful to the client and his recovery.
NP: Implementation; CN: Physiological integrity; CNS: Physiological adaptation; CL: Analysis

34. 2. The sun is the best-known and most common cause of basal cell epithelioma. Immunosuppression, radiation, and burns are less common causes.
NP: Data collection; CN: Physiological integrity; CNS: Reduction of risk potential; CL: Knowledge

35. 3. Inelastic skin turgor is a normal part of aging. Overhydration causes the skin to appear edematous and spongy. Normal skin turgor is dry and firm. Dehydration causes inelastic skin with tenting.
NP: Data collection; CN: Physiological integrity; CNS: Basic care and comfort; CL: Knowledge

36. 4. Although the client should eat a balanced diet with foods from all food groups, the diet should emphasize foods that supply complete protein, such as lean meats and low-fat milk. Protein helps build and repair body tissue, which promotes healing. Legumes provide incomplete protein. Cheese contains complete protein but also fat, which should be limited to 30% or less of caloric intake. Whole-grain products supply incomplete proteins and carbohydrates. Fruits and vegetables provide mainly carbohydrates.
NP: Planning; CN: Physiological integrity; CNS: Basic care and comfort; CL: Application

37. 3. The client should be protected from becoming chilled. The water temperature should be 110° to 115° F to compensate for evaporative body cooling during and after the bath. Water temperature the same as body temperature, slightly cooler, or slightly warmer will eventually cool, which could cause discomfort and increase evaporative body cooling. Water of 115° to 120° F would be too hot and put the client at risk for burns or discomfort. Water should always be tested with a bath thermometer.
NP: Planning; CN: Physiological integrity; CNS: Basic care and comfort; CL: Knowledge

38. A client received burns to his entire back and left arm. Using the Rule of Nines, the nurse calculates that he has sustained burns to which of the following percentages of his body?
1. 9%
2. 18%
3. 27%
4. 36%

One back and one arm equals...

39. Which of the following techniques will maintain surgical asepsis?
1. Change the sterile field after sterile water is spilled on it.
2. Put on sterile gloves; then open a container of sterile saline.
3. Place a sterile dressing ½″ (1.25 cm) from the edge of the sterile field.
4. Clean the wound with a circular motion, moving from outer circles toward the center.

40. Which of the following information is not critical during an assessment of skin integrity?
1. Family history of pressure ulcers
2. Presence of existing pressure ulcers
3. Overall risk of developing pressure ulcers
4. Potential areas of pressure ulcer development

This is another negative question!

41. Which of the following pieces of equipment for the wheelchair- or bed-bound client impedes circulation to the area it's meant to protect?
1. Waterbed
2. Ring or donut
3. Gel flotation pad
4. Polyurethane foam mattress

38. 3. According to the Rule of Nines, the posterior trunk, anterior trunk, and legs are each 18% of the total body surface. The head, neck, and arms are each 9% of total body surface, and the perineum is 1%. In this case, the client received burns to his back (18%) and one arm (9%), totaling 27% of his body.
NP: Data collection; CN: Physiological integrity; CNS: Reduction of risk potential; CL: Application

39. 1. A sterile field is considered contaminated when it becomes wet. Moisture can act as a wick, allowing microorganisms to contaminate the field. The outside of containers such as sterile saline bottles aren't sterile. The containers should be opened before sterile gloves are put on, and the solution poured over the sterile dressings placed in a sterile basin. Wounds should be cleaned from the most contaminated area to the least contaminated area, for example, from the center outward. The outer inch of a sterile field isn't considered sterile.
NP: Implementation; CN: Safe, effective care environment; CNS: Safety and infection control; CL: Application

40. 1. Family history isn't important for the assessment of skin integrity. Areas already broken down need immediate treatment. Plans should be developed to avoid skin breakdown in areas at high risk. Certain clients are at high risk because of internal factors, such as malnutrition, cachexia, obesity, and diabetes.
NP: Data collection; CN: Physiological integrity; CNS: Reduction of risk potential; CL: Knowledge

41. 2. Rings or donuts shouldn't be used because they restrict circulation. Foam mattresses evenly distribute pressure. Gel pads give with weight. The waterbed also distributes pressure over the entire surface.
NP: Implementation; CN: Physiological integrity; CNS: Reduction of risk potential; CL: Application

42. Which of the following interventions is inappropriate for the treatment of a postoperative wound evisceration?

1. Give prophylactic antibiotics as ordered.
2. Have the client drink as much fluid as possible.
3. Explain to the client what's happening and give support.
4. Cover the protruding internal organs with sterile gauze moistened with sterile saline.

Inappropriate is the key word to focus on here.

42. 2. Evisceration requires emergency surgery. The nurse should place the client on nothing-by-mouth status immediately. Evisceration is a frightening situation for any client. While the nurse works quickly to get the client treated, providing support that will reduce the client's anxiety is important. Covering the wound with moistened gauze will prevent the organs from drying. Both the gauze and the saline must be sterile to reduce the risk of infection. Antibiotics will usually be ordered immediately and should be started as soon as possible.

NP: Implementation; CN: Physiological integrity; CNS: Physiological adaptation; CL: Application

43. Which of the following interventions is performed first when changing a dressing or giving wound care?

1. Put on gloves.
2. Wash hands thoroughly.
3. Slowly remove the soiled dressing.
4. Observe the dressing for the amount, type, and odor of drainage.

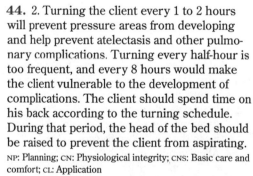

Prioritize!

43. 2. The first thing the nurse must do is wash her hands. Putting on gloves, removing the dressing, and observing the drainage are all parts of the dressing change procedure.

NP: Implementation; CN: Physiological integrity; CNS: Basic care and comfort; CL: Application

44. When caring for a client who spends all or most of the time in bed, a turning schedule prevents the development of complications. Which of the following schedules is best for most clients?

1. Turn every half-hour.
2. Turn every 1 to 2 hours.
3. Turn once every 8 hours.
4. Keep the client on his back as much as possible.

44. 2. Turning the client every 1 to 2 hours will prevent pressure areas from developing and help prevent atelectasis and other pulmonary complications. Turning every half-hour is too frequent, and every 8 hours would make the client vulnerable to the development of complications. The client should spend time on his back according to the turning schedule. During that period, the head of the bed should be raised to prevent the client from aspirating.

NP: Planning; CN: Physiological integrity; CNS: Basic care and comfort; CL: Application

45. Which of the following instructions is most important when teaching a client about hypersensitivity skin test results?

1. Wash the sites daily with a mild soap.
2. Have the sites read on the correct date.
3. Keep the skin test areas moist with a mild lotion.
4. Stay out of direct sunlight until the tests are read.

Here's another prioritizing question.

45. 2. An important facet of evaluating skin tests is to read the skin test results at the proper time. Evaluating the skin test too late or too early will give inaccurate and unreliable results. The sites should be kept dry. There's no need to wash the sites with soap, and direct sunlight isn't prohibited.

NP: Implementation; CN: Health promotion and maintenance; CNS: Prevention and early detection of disease; CL: Application

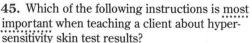

46. A client with disseminated herpes zoster is given I.V. hydrocortisone (Solu-Cortef). Which of the following serum chemistry values may be elevated as a result of this therapy?
1. Calcium
2. Glucose
3. Magnesium
4. Potassium

46. 2. Corticosteroids increase blood sugar and tend to lower serum potassium and calcium levels. Their effect on magnesium isn't substantial.
NP: Data collection; CN: Physiological integrity; CNS: Pharmacological therapies; CL: Comprehension

47. A client is admitted to the burn unit with extensive full-thickness burns. Which of the following considerations has priority?
1. Fluid status
2. Body image
3. Level of pain
4. Risk of infection

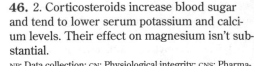

You know what this means! Prioritize!

47. 1. In early burn care, the client's greatest need is fluid resuscitation because of large-volume fluid loss through the damaged skin. Infection, body image, and pain are important concerns in the nursing care of a burn client but don't take precedence over fluid management in the early phase of burn care.
NP: Data collection; CN: Physiological integrity; CNS: Physiological adaptation; CL: Analysis

48. Which of the following characteristics is seen with deep partial-thickness burn wounds?
1. Blanching
2. Erythema
3. Eschar
4. Fluid-filled vesicles

48. 4. Fluid-filled vesicles are characteristic of deep partial-thickness skin destruction (also referred to as second-degree burns). Erythema and blanching on pressure are characteristic of partial-thickness skin destruction. Eschar is a manifestation of full-thickness skin destruction.
NP: Data collection; CN: Physiological integrity; CNS: Physiological adaptation; CL: Knowledge

49. A client with extensive burns has a new donor site. Which of the following considerations is important in positioning the client?
1. Make the site dependent.
2. Avoid pressure on the site.
3. Keep the site fully covered.
4. Allow ventilation of the site.

You did it! Now take a break, have a nutritious snack, and then forge ahead. Nothing can stop you now!

49. 2. A universal concern in the care of donor sites for burn care is to keep the site away from sources of pressure. Ventilation of the site and keeping the site fully covered are practices in some institutions but aren't hallmarks of donor site care. Placing the site in a position of dependence isn't a justified aspect of donor site care.
NP: Implementation; CN: Physiological integrity; CNS: Physiological adaptation; CL: Application

50. Prevention and early treatment of Lyme disease are crucial because which of the following complications can occur late in the disease?
1. Arthritis
2. Lung abscess
3. Renal failure
4. Sterility

50. 1. If Lyme disease goes untreated, arthritis, neurologic problems, and cardiac abnormalities may arise as late complications. Lung abscess, renal failure, and sterility aren't complications of Lyme disease.
NP: Data collection; CN: Physiological integrity; CNS: Reduction of risk potential; CL: Knowledge

NP: Nursing process   CN: Client needs category   CNS: Client needs subcategory   CL: Cognitive level

Part III Care of the psychiatric client

12 Essentials of psychiatric care   249

13 Somatoform & sleep disorders   251

14 Anxiety & mood disorders   263

15 Cognitive disorders   285

16 Personality disorders   297

17 Schizophrenic & delusional disorders   317

18 Substance abuse disorders   333

19 Dissociative disorders   351

20 Sexual & gender identity disorders   363

21 Eating disorders   375

Before you take the tests related to psychiatric care, take this test relating to tests (and treatments, too) in psychiatric care.

Chapter 12
Essentials of psychiatric care

1. A 50-year-old client is scheduled for electro-convulsive therapy (ECT). The nurse knows that ECT is most commonly prescribed for which of the following conditions?
1. Major depression
2. Antisocial personality disorder
3. Chronic schizophrenia
4. Somatoform disorder

All of these options might be acceptable but which is best?

1. 1. ECT is most commonly used for the treatment of major depression in clients who haven't responded to antidepressants or who have medical problems that contraindicate the use of antidepressants. ECT isn't commonly used for treatment of personality disorders. ECT doesn't appear to be of value to individuals with chronic schizophrenia and isn't the treatment of choice for clients with dissociative disorders.

NP: Data collection; CN: Safe, effective care environment; CNS: Coordinated care; CL: Knowledge

2. A client with bipolar disorder becomes verbally aggressive in a group therapy session. Which of the following responses by the nurse would be best?
1. "You're behaving in an unacceptable manner and you need to control yourself."
2. "If you continue to talk like that, no one will want to be around you."
3. "You're frightening everyone in the group. Leave the room immediately."
4. "Other people are disturbed by your profanity. I'll walk with you down the hall to help release some of that energy."

2. 4. This response informs the client that, although the behavior is unacceptable, the client is still worthy of help. The other responses indicate that the client is in control of the behavior.

NP: Implementation; CN: Safe, effective care environment; CNS: Coordinated care; CL: Application

3. The nurse is caring for a female client who had been extremely suicidal but now appears very serene. She says she has finally gotten things worked out. What would be the best way to respond to this client?
1. "I'm glad things are working out for you."
2. "How have you managed to work things out so quickly?"
3. "I'm not clear about what has changed for you; tell me more.
4. "You look so much better; you must be feeling relieved."

3. 3. Asking the client to tell her more allows the nurse to seek further clarification without placing words in the client's mouth. Choices 1 and 4 block further exploration and indicate the nurse's lack of insight. Choice 2 allows the client to give an answer that hides underlying intent (for example, the client may respond, "I prayed about it and believe that God is now in charge of my life").

NP: Implementation; CN: Psychosocial integrity; CNS: Coping and adaptation; CL: Application

NP: Nursing process   CN: Client needs category   CNS: Client needs subcategory   CL: Cognitive level

4. A depressed client states that she prefers to remain in her room and not socialize with other clients on the unit. Which response by the nurse is most appropriate?
 1. Allow the client to remain in her room.
 2. Insist that the client interact with others throughout the day.
 3. Sit with the client in her room for brief intervals to establish a trusting relationship.
 4. Have other clients go to the client's room to keep her company.

4. 3. Depressed people shouldn't be allowed to isolate themselves, as isolation only further lowers their self-esteem. Brief periods of interaction would accomplish two goals: To prevent the client's withdrawal, and to begin establishing trust and demonstrating genuine interest in the client. Allowing her to remain alone in her room wouldn't be therapeutic. Insisting that she interact with others or having other clients go to her room would be too threatening to the client at this time.
NP: Implementation; CN: Psychosocial integrity; CNS: Psychosocial adaptation; CL: Application

5. Two 16-year-old clients are being treated in an adolescent unit. During a recreational activity they begin a physical fight. How should the nurse intervene?
 1. Remove the teenagers to separate areas and set limits.
 2. Remind the teenagers of the unit rules.
 3. Obtain an order to place the teenagers in seclusion.
 4. Obtain an order to place the teenagers in restraints.

5. 1. Setting limits and removing the clients from the situation is the best way to handle aggression. Reminders of appropriate behaviors aren't likely to be effective at this time, and seclusion and restraints are reserved for more serious situations.
NP: Implementation; CN: Physiological integrity; CNS: Physiological adaptation; CL: Application

6. Which of the following statements best describes the key advantage of using groups in psychotherapy?
 1. Decreases the focus on the individual
 2. Fosters the physician-client relationship
 3. Confronts individuals with their shortcomings
 4. Fosters a new learning environment

6. 4. In a group, the individual has the opportunity to learn that others have the same problems and needs. The group can also provide an arena where new methods of relating to others can be tried. Decreasing focus on the individual isn't a key advantage (and sometimes isn't an advantage at all). Groups don't, by themselves, foster the physician-client relationship, and they aren't always used to confront individuals.
NP: Evaluation; CN: Physiological integrity; CNS: Physiological adaptation; CL: Analysis

Wow! That was easy. Way to go!

7. Which of the following therapies has been most strongly advocated for the treatment of posttraumatic stress disorder?
 1. Electroconvulsive therapy (ECT)
 2. Group therapy
 3. Hypnotherapy
 4. Individual therapy

7. 2. Group therapy has been especially effective with survivors of combat trauma. ECT, hypnotherapy, and individual therapy may be useful to the client but these therapies aren't as strongly advocated as group therapy.
NP: Planning; CN: Safe, effective care environment; CNS: Coordinated care; CL: Knowledge

NP: Nursing process   CN: Client needs category   CNS: Client needs subcategory   CL: Cognitive level

Chapter 13
Somatoform & sleep disorders

1. Which of the following statements is correct about clients who have somatoform disorders?
   1. They usually seek medical attention.
   2. They have organic pathologic disorders.
   3. They regularly attend psychotherapy sessions without encouragement.
   4. They're eager to discover the true reasons for their physical symptoms.

Not to worry. You'll finish these questions before you know it!

2. Which of the following statements is correct about the diagnosis of somatoform disorders?
   1. The somatic complaints are limited to one organ system.
   2. The event preceding the physical illness occurred recently.
   3. They're physical conditions with organic pathologic causes.
   4. They're disorders that occur in the absence of organic findings.

3. Which of the following actions best accounts for the physical symptoms in a client with a somatoform disorder?
   1. To cope with delusional thinking
   2. To provide attention for the individual
   3. To prevent or relieve symptoms of anxiety
   4. To protect the client from family conflict

1. 1. A client with a somatoform disorder usually seeks medical attention. These clients have a history of multiple physiological complaints without associated demonstrable organic pathologic causes. The expected behavior for this type of disorder is to seek treatment from several medical physicians for somatic complaints, not psychiatric evaluation.
NP: Data collection; CN: Health promotion and maintenance; CNS: Prevention and early detection of disease; CL: Knowledge

2. 4. The essential feature of somatoform disorders is a physical or somatic complaint without any demonstrable organic findings to account for the complaint. There are no known physiological mechanisms to explain the findings. Somatic complaints aren't limited to one organ system. The diagnostic criteria for somatoform disorders state that the client has a history of many physical complaints beginning before age 30 that occur over several years.
NP: Data collection; CN: Psychosocial integrity; CNS: Psychosocial adaptation; CL: Analysis

3. 3. Anxiety and depression commonly occur in somatoform disorders. The client prevents or relieves symptoms of anxiety by focusing on physical symptoms. Somatic delusions occur in schizophrenia. Somatization in dysfunctional families shifts the open conflict to the client's illness, thus providing some stability for the family, not the client. The symptoms allow the client to avoid unpleasant activity, not to seek individual attention.
NP: Data collection; CN: Psychosocial integrity; CNS: Coping and adaptation; CL: Analysis

4. Which of the following ego defense mechanisms describes the underlying dynamics of somatization disorder?

1. Repression of anger
2. Suppression of grief
3. Denial of depression
4. Preoccupation with pain

Remember you're looking for a defense mechanism.

4. 1. One psychodynamic theory states that somatization is the transformation of aggressive and hostile wishes toward others into physical complaints. Repressed anger originating from past disappointments and unfilled needs for nurturing and caring are expressed by soliciting other people's concern and rejecting them as ineffective. Denial, suppression, and preoccupation aren't the defense mechanisms underlying the dynamics of somatization disorder.

NP: Data collection; CN: Psychosocial integrity; CNS: Coping and adaptation; CL: Knowledge

5. For the past 6 months, an 86-year-old client in an extended care facility has been complaining about breast pain and is convinced she has cancer despite medical diagnostic tests indicating that she doesn't. The symptoms—feelings of pain and a belief that she has a serious disease—may indicate which disorder?

1. Conversion disorder
2. Hypochondriasis
3. Severe anxiety
4. Sublimation

5. 2. Hypochondriasis is defined as a morbid fear or belief that one has a serious disease, though none exists. No neurologic symptoms are apparent as with conversion disorders. There are no apparent symptoms related to severe anxiety. There's no evidence of channeling maladaptive feelings or impulses into socially acceptable behavior that would be evident in a client experiencing sublimation.

NP: Data collection; CN: Physiological integrity; CN: Physiological adaptation; CL: Analysis

6. A client reports severe pain in the back and joints. On reviewing the client's history, the nurse notes a diagnosis of depression and frequent hospitalizations for psychosomatic illness. What should the nurse encourage this client to do?

1. Tell the physician about the pain so that its cause can be determined.
2. Remember all the previous health problems that weren't real.
3. Try to get more rest and use relaxation techniques.
4. Ignore the pain and focus on happy things.

6. 1. Initially, the nurse should treat all symptoms as indicators of possible pathology because a history of psychosomatic illness doesn't rule out a physical illness as a cause of the client's current symptoms. The other options assume that the client has a psychosomatic illness, which could lead to ignoring a physical illness or condition. A client with a psychosomatic illness experiences real physical symptoms that are triggered or exacerbated by psychological issues. The physical symptoms aren't imagined or fabricated by the client.

NP: Data collection; CN: Physiological integrity; CNS: Reduction of risk potential; CL: Application

7. A college student frequently visited the health center during the past year with multiple vague complaints of GI symptoms before course examinations. Although physical causes have been eliminated, the student continues to express her belief that she has a serious illness. These symptoms are typical of which of the following disorders?
1. Conversion disorder
2. Depersonalization
3. Hypochondriasis
4. Somatization disorder

8. A nursing goal for a client diagnosed with hypochondriasis should focus on which of the following areas?
1. Determining the cause of a sleep disturbance
2. Relieving the fear of serious illness
3. Recovering the lost or altered function
4. Giving positive reinforcement for accomplishments related to physical appearance

9. Which of the following nursing interventions is appropriate for a client diagnosed with hypochondriasis?
1. Teach the client adaptive coping strategies.
2. Help the client eliminate the stress in her life.
3. Confront the client with the statement, "It's all in your head."
4. Encourage the client to focus on identification of physical symptoms.

Nursing goals target emotional outcomes as well as physical ones.

See what word is missing from question 9? There's no "*most*" preceding "*appropriate.*" That means, in this case, there's only one right answer!

7. 3. Hypochondriasis in this case is shown by the client's belief she has a serious illness, although pathologic causes have been eliminated. The disturbance usually lasts at least 6 months, and the GI system is commonly affected. Exacerbations are usually associated with identifiable life stressors that, in this case, can be related to the client's examinations. Conversion disorders are characterized by one or more neurologic symptoms. Depersonalization refers to persistent, recurrent episodes of feeling detached from one's self or body. Somatoform disorders generally have a chronic course with few remissions.

NP: Data collection; CN: Psychosocial integrity; CNS: Psychosocial adaptation; CL: Analysis

8. 2. The nursing goal for hypochondriasis is relief of fear. For insomnia, the goal is focused on determining the cause of the sleep disturbance. An appropriate goal for body dysmorphic disorder focuses on positive reinforcement for accomplishments related to physical appearance. The nursing goal for a conversion disorder focuses on the recovery of the lost or altered function.

NP: Planning; CN: Psychosocial integrity; CNS: Coping and adaptation; CL: Application

9. 1. Because of weak ego strength, a client with hypochondriasis is unable to use coping mechanisms effectively. The nursing focus is to teach adaptive coping mechanisms. It isn't realistic to eliminate all stress. A client should never be confronted with the statement, "It's all in your head," because this wouldn't facilitate a long-term therapeutic relationship, which is necessary to offer reassurance that no physical disease is present.

NP: Implementation; CN: Psychosocial integrity; CNS: Coping and adaptation; CL: Application

10. A client is exhibiting anxiety, which is evidenced by muscle tension, distractibility, and increased vital signs. Which nursing intervention has top priority?
1. Remain with the client and use a soft voice and reassuring approach.
2. Assist the client to identify factors that contribute to anxiety.
3. Teach relaxation techniques such as deep breathing and muscle relaxation.
4. Administer antianxiety medications as appropriate.

10. 1. The priority nursing intervention is to remain with the client and use a soft voice and reassuring approach. Remaining with the client provides for his safety and a soft voice is calming and reassuring, which will add to his feelings of safety and protection. Interventions, such as identifying factors that contribute to anxiety, teaching relaxation techniques, and administering antianxiety medications, are included in the client's care plan but should be addressed later.
NP: Implementation; CN: Psychosocial integrity; CNS: Psychosocial adaptation; CL: Application

11. Which of the following therapeutic strategies can a nurse use to reduce anxiety in a client diagnosed with hypochondriasis?
1. Suicide precautions
2. Relaxation exercises
3. Electroconvulsive therapy
4. Pharmacologic intervention

11. 2. A nurse can initiate relaxation exercises to decrease anxiety without an order from the physician. Medical intervention would include electroconvulsive therapy and pharmacologic intervention. In a hypochondriasis disorder, no threat of suicide exists.
NP: Implementation; CN: Physiological integrity; CNS: Physiological adaptation; CL: Application

12. Which of the following pharmacologic agents is a sedative-hypnotic medication used to induce sleep for clients experiencing a sleep disorder?
1. Triazolam (Halcion)
2. Paroxetine (Paxil)
3. Fluoxetine (Prozac)
4. Risperidone (Risperdal)

12. 1. Triazolam is one of a group of sedative-hypnotic medications that can be used only for a limited time because of the risk of dependence. Paroxetine is a serotonin-specific reuptake inhibitor used for treatment of depression, panic disorder, and obsessive-compulsive disorder. Fluoxetine is a serotonin-specific reuptake inhibitor used for depressive disorders and obsessive-compulsive disorders. Risperidone is indicated for psychotic disorders.
NP: Implementation; CN: Physiological integrity; CNS: Pharmacological therapies; CL: Knowledge

13. Which of the following measures should be included when teaching a client strategies to help promote sleep?
1. Keep the room warm.
2. Eat a large meal before bedtime.
3. Schedule bedtime when you feel tired.
4. Avoid caffeine, excessive fluid intake, alcohol, and stimulating drugs before bedtime.

There's no caffeine in this meal!

13. 4. Caffeine, excessive fluid intake, alcohol, and stimulating drugs act as stimulants; avoiding them should promote sleep. Excessive fullness or hunger may interfere with sleep. Setting a regular bedtime and wake-up time facilitates physiological patterns. Maintaining a cool temperature in the room will better facilitate sleeping.
NP: Implementation; CN: Health promotion and maintenance; CNS: Prevention and early detection of disease; CL: Application

NP: Nursing process   CN: Client needs category   CNS: Client needs subcategory   CL: Cognitive level

14. Which of the following sleep disorders indicates that an individual can't prevent falling asleep, even in the middle of a task or sentence?
   1. Hypersomnia
   2. Insomnia
   3. Narcolepsy
   4. Parasomnia

14. 3. Narcolepsy is also known as sleep attacks. Insomnia is a sleep disorder in which an individual has difficulty initiating or maintaining sleep. Hypersomnia, or somnolence, refers to excessive sleepiness or seeking excessive amounts of sleep. Parasomnia refers to unusual or undesired behavior that occurs during sleep, such as nightmares and sleepwalking.
NP: Data collection; CN: Physiological integrity; CNS: Physiological adaptation; CL: Knowledge

15. Treatments for sleep disorders include which of the following methods?
   1. Behavior therapy
   2. Biofeedback
   3. Group therapy
   4. Insight-oriented psychotherapy

15. 2. Biofeedback, relaxation therapy, and psychopharmacology are appropriate treatments for sleep disorders. Behavior therapy, insight-oriented psychotherapy, and group psychotherapy are treatments related to somatoform disorders.
NP: Implementation; CN: Physiological integrity; CNS: Basic care and comfort; CL: Application

Give question 16 the most you have to offer!

16. Which nursing intervention would be most appropriate for a depressed client with a nursing diagnosis of *Disturbed sleep pattern related to external factors?*
   1. Consult the physician about prescribing a bedtime sleep medication.
   2. Allow the client to sit at the nurses' station for comfort.
   3. Allow the client to watch television until he's sleepy.
   4. Encourage the client to take a warm bath before retiring.

16. 4. Sleep-inducing activities, such as a warm bath, help promote relaxation and sleep. Although consulting a physician about prescribing a bedtime sleep medication is possible, it wouldn't be the best nursing intervention for this client. Encouraging the client to watch television or sit at the nurses' station wouldn't necessarily promote sleep. In fact, they may provide too much stimulation, further preventing sleep.
NP: Implementation; CN: Physiological integrity; CNS: Basic care and comfort; CL: Application

17. The nurse is reviewing a nursing care plan for a 45-year-old male client with a psychophysiological disorder. Which intervention should be included?
   1. Don't include any physical symptoms that aren't life-threatening.
   2. Consider only the physical symptoms that are distressing the client.
   3. Include physical, psychosocial, and spiritual problems.
   4. Consider only psychosocial symptoms.

17. 3. Physical, psychosocial, and spiritual problems are thoroughly and continuously assessed with each client. The nurse needs to include all symptoms, even those that aren't life-threatening. Consider all physical symptoms, even those the client doesn't find distressing. Psychosocial symptoms should be considered but all three areas must be assessed to provide a thorough care plan.
NP: Planning; CN: Psychosocial integrity; CNS: Psychosocial adaptation; CL: Comprehension

18. The nurse is instructing a 38-year-old male client undergoing treatment for anxiety and insomnia. The physician has prescribed lorazepam (Ativan) 1 mg by mouth three times per day. The nurse should instruct the client to do which of the following?

1. Avoid caffeine.
2. Avoid aged cheese.
3. Avoid sunlight.
4. Maintain an adequate salt intake.

As a nurse, you'll always be a prime time player!

18. 1. Lorazepam is a benzodiazepine used to treat various forms of anxiety and insomnia. Caffeine is contraindicated because it's a stimulant and increases anxiety. A client on a monoamine oxidase inhibitor should avoid aged cheeses. Clients taking certain antipsychotic medications should avoid sunlight. Salt intake has no effect on lorazepam.

NP: Implementation; CN: Physiological integrity; CNS: Pharmacological therapies; CL: Knowledge

19. A home health nurse is caring for a client diagnosed with a conversion disorder manifested by paralysis in the left arm. An organic cause for the deficit has been ruled out. Which nursing intervention is most appropriate for this client?

1. Perform all physical tasks for the client to foster dependence.
2. Allot an hour each day to discuss the paralysis and its cause.
3. Identify primary or secondary gains that the physical symptom provides.
4. Allow the client to withdraw from all physical activities.

19. 3. Primary or secondary gains should be identified because these are etiological factors that can be used in problem resolution. The nurse should encourage the client to be as independent as possible and intervene only when the client requires assistance. The nurse shouldn't focus on the disability. The nurse should encourage the client to perform physical activities to the greatest extent possible.

NP: Implementation; CN: Psychosocial integrity; CNS: Psychosocial adaptation; CL: Application

20. A 35-year-old female client is diagnosed with conversion disorder with paralysis of the legs. What is the best nursing intervention for the nurse to use?

1. Discuss with the client ways to live with the paralysis.
2. Focus interactions on her results of medical tests.
3. Encourage the client to move her legs as much as possible.
4. Avoid focusing on the client's physical limitations.

Don't lose sleep over question 21 — focus on the most appropriate answer.

20. 4. The paralysis is used as an unhealthy way of expressing unmet psychological needs. The nurse should avoid speaking about the paralysis to shift the client's attention and focus on the mental aspect of the disorder. The other options focus too much on the paralysis, which doesn't allow recognition of the underlying psychological motivations.

NP: Implementation; CN: Psychosocial integrity; CNS: Psychosocial adaptation; CL: Application

21. Which outcome criteria is most appropriate for a teenager who hasn't slept well in 6 months, is irritable, and has dropped out of social activities?

1. The client will sleep well at night.
2. The parents will stop worrying about the client.
3. The client will obtain appropriate mental health services.
4. The parents will impose strict behavior guidelines for the client to follow.

21. 3. Mental health services can protect the client and offer the best means of regaining mental health. The client could reestablish a healthy sleeping pattern without addressing underlying issues. The parents' worrying is unrelated to the child's immediate need for help. The child's behavior suggests the need for professional service, not disciplinary measures.

NP: Planning; CN: Psychosocial integrity; CNS: Coping and adaptation; CL: Application

NP: Nursing process   CN: Client needs category   CNS: Client needs subcategory   CL: Cognitive level

22. The mother of a client approaches the nurses' station in tears. She says that she's upset about her daughter's diagnosis of conversion disorder. Which response is best?
1. "What is it that upsets you the most?"
2. "Are you afraid your daughter will never get well?"
3. "Her behavior is typical of someone with conversion disorder."
4. "Let me give you some information about her illness."

You're given more hints in questions 22 and 23.

22. 1. Asking the mother an open-ended question permits the nurse to collect data reported in the mother's own words. The other options narrow the data collection process prematurely. Also, it's important to hear what the mother has to say without planting suggestions about her daughter's condition.

NP: Data collection; CN: Psychosocial integrity; CNS: Coping and adaptation; CL: Application

23. Which of the following interventions would most likely be found in a teaching plan for an anxious client who reports difficulty settling down for sleep?
1. Teaching time management skills
2. Teaching conflict resolution skills
3. Teaching progressive muscle relaxation
4. Teaching adverse effects of antipsychotic medication

You've passed the halfway mark!

23. 3. Progressive muscle relaxation is a systematic tensing and relaxing of separate muscle groups. As the technique is mastered, relaxation results. Time management skills and conflict resolution skills are helpful in an overall effort to reduce stress and anxiety but don't provide immediate relief in an effort to sleep. Antipsychotic medications aren't usually used to treat anxiety or difficulty with sleep.

NP: Planning; CN: Psychosocial integrity; CNS: Coping and adaptation; CL: Application

24. Which of the following statements is correct about conversion disorders?
1. The symptoms can be controlled.
2. The psychological conflict is repressed.
3. The client is aware of the psychological conflict.
4. The client shouldn't be made aware of the conflicts underlying the symptoms.

24. 2. In conversion disorders, the client isn't conscious of intentionally producing symptoms that can't be self-controlled. The symptoms are characterized by one or more neurologic symptoms. Understanding the principles and conflicts behind the symptoms can prove helpful during a client's therapy.

NP: Data collection; CN: Psychosocial integrity; CNS: Psychosocial adaptation; CL: Comprehension

25. A client is admitted for abrupt onset of paralysis in her left arm. Although no physiological cause has been found, the symptoms are exacerbated when she speaks of losing custody of her children in a recent divorce. These findings are characteristic of which of the following disorders?
1. Body dysmorphic disorder
2. Conversion disorder
3. Delusional disorder
4. Malingering

You're terrific at answering these questions!

25. 2. Conversion disorders are characterized by one or more neurologic symptoms associated with psychological conflict. The client doesn't have a delusion; this is the sole manifestation of a delusional disorder. Body dysmorphic disorder is an imagined belief that there's a defect in the appearance of all or part of the body. Malingering is the intentional production of symptoms to avoid obligations or obtain rewards.

NP: Data collection; CN: Psychosocial integrity; CNS: Psychosocial adaptation; CL: Analysis

26. A client has been diagnosed with conversion-disorder blindness. The client shows "la belle indifference." Which of the following statements best describes this term?
1. The client is suppressing her true feelings.
2. The client's anxiety has been relieved through her physical symptoms.
3. The client is acting indifferent because she doesn't want to show her actual fear.
4. The client's needs are being met, so she doesn't need to be anxious.

26. 2. Conversion accomplishes anxiety reduction through the production of a physical symptom symbolically linked to an underlying conflict. The client isn't aware of the internal conflict. Hospitalization doesn't remove the source of the conflict.
NP: Data collection; CN: Psychosocial integrity; CNS: Psychosocial adaptation; CL: Analysis

> Take pains to choose the most appropriate answer.

27. A client with hypochondriasis complains of pain in her right side that she hasn't had before. Which response is the most appropriate?
1. "It's time for group therapy now."
2. "Tell me about this new pain you're having. You'll miss group therapy today."
3. "I'll report this pain to your physician. In the meantime, group therapy starts in 5 minutes. You must leave now to be on time."
4. "I'll call your physician and see if he'll order a new pain medication. Why don't you get some rest for now?"

27. 3. The amount of time focused on discussing physical symptoms should be decreased. Lack of positive reinforcement may help stop the maladaptive behavior. All physical complaints need to be evaluated for physiological causes by the physician. Avoiding the statement demeans the client and doesn't address the underlying problem. Allowing verbalization of the new onset of pain emphasizes physical symptoms and prevents the client from attending group therapy.
NP: Implementation; CN: Psychosocial integrity; CNS: Coping and adaptation; CL: Application

28. Which of the following interventions would help a client with conversion-disorder blindness to eat?
1. Direct the client to independently locate items on the tray and feed himself.
2. See to the needs of the other clients in the dining room, then feed this client last.
3. Establish a "buddy" system with other clients who can feed the client at each meal.
4. Expect the client to feed himself after explaining the location of food on the tray.

28. 4. The client is expected to maintain some level of independence by feeding himself, while at the same time the nurse is supportive in a matter-of-fact way. Feeding the client leads to dependence.
NP: Implementation; CN: Psychosocial integrity; CNS: Coping and adaptation; CL: Application

> Nursing goals provide a means by which staff can work together.

29. A client diagnosed with conversion disorder who is experiencing left-sided paralysis tells the nurse she's received a lot of attention in the hospital and it's unfortunate that others outside the hospital don't find her interesting. Which of the following nursing diagnoses is appropriate for this client?
1. Interrupted family processes
2. Ineffective health maintenance
3. Ineffective coping
4. Social isolation

29. 3. The client can't express her internal conflicts in appropriate ways. Emphasis on physical symptoms can become frustrating for family members. There are no defining characteristics to support the other nursing diagnoses.
NP: Planning; CN: Psychosocial integrity; CNS: Psychosocial adaptation; CL: Application

30. To help a client with conversion disorder increase self-esteem, which of the following nursing interventions is appropriate?
1. Set large goals so the client can see positive gains.
2. Focus attention on the client as a person rather than on the symptom.
3. Discuss the client's childhood to link present behaviors with past traumas.
4. Encourage the client to use avoidant-interactional patterns rather than assertive patterns.

30. 2. Focusing on the client directs attention away from the symptom. This approach eventually reduces the client's need to gain attention through physical symptoms. Discussion of childhood has no correlation with self-esteem. Avoiding interactional situations doesn't foster self-esteem. Small goals ensure success and reinforce self-esteem.
NP: Implementation; CN: Psychosocial integrity; CNS: Coping and adaptation; CL: Application

31. According to diagnostic criteria, which of the following conditions occurs with conversion disorders?
1. Delusions
2. Feelings of depression or euphoria
3. A feeling of dread accompanied by somatic signs
4. One or more neurologic symptoms associated with psychological conflict or need

31. 4. Symptoms of conversion disorders are neurologic in nature (paralysis, blindness). Anxiety is characterized by a feeling of dread. Mood disorders are characterized by abnormal feelings of depression or euphoria. Delusional disorders are characterized by delusions.
NP: Data collection; CN: Health promotion and maintenance; CNS: Prevention and early detection of disease; CL: Knowledge

32. Which of the following nursing diagnoses is appropriate for a client with conversion disorder who has little energy to expend on activities or interactions with friends?
1. Powerlessness
2. Hopelessness
3. Impaired social interaction
4. Compromised family coping

32. 3. When clients focus their mental and physical energy on somatic symptoms, they have little energy to expend on social or diversional activities. Such a client needs nursing assistance to become involved in social interactions. Although the other diagnoses are common for a client with conversion disorder, the information given doesn't support them.
NP: Planning; CN: Psychosocial integrity; CNS: Coping and adaptation; CL: Application

33. A client diagnosed with conversion disorder has a nursing diagnosis of *Interrupted family processes related to the client's disability.* Which of the following goals is appropriate for this client?
1. The client will resume former roles and tasks.
2. The client will take over roles of other family members.
3. The client will rely on family members to meet all client needs.
4. The client will focus energy on problems occurring in the family.

Way to go!

33. 1. The client who uses somatization has typically adopted a sick role in the family, characterized by dependence. Increasing independence and resumption of former roles are necessary to change this pattern. The client shouldn't be expected to take on the roles or responsibilities of other family members.
NP: Planning; CN: Psychosocial integrity; CNS: Psychosocial adaptation; CL: Application

34. A new client admitted to a psychiatric unit is diagnosed with conversion disorder. The client shows a lack of concern for her sudden paralysis, though her athletic abilities have always been a source of pride to her. This manifestation is known as which of the following conditions?
  1. Acute dystonia
  2. La belle indifference
  3. Malingering
  4. Secondary gain

34. 2. La belle indifference is a lack of concern about the present illness in some clients. Acute dystonia refers to muscle spasms. Secondary gain refers to the benefits of illness. Malingering is voluntary production of symptoms.
NP: Data collection; CN: Psychosocial integrity; CNS: Coping and adaptation; CL: Knowledge

35. Which of the following nursing interventions is the most appropriate for a client who had pseudoseizures and is diagnosed with conversion disorder?
  1. Explain that the pseudoseizures are imaginary.
  2. Promote dependence so that unfilled dependency needs are met.
  3. Encourage the client to discuss his feelings about the pseudoseizures.
  4. Promote independence and withdraw attention from the pseudoseizures

> The phrase most appropriate is a hint! Don't miss it!

35. 4. Successful performance of independent activities enhances self-esteem. Positive reinforcement encourages the continual use of the maladaptive responses. Confronting the client with a statement that the symptoms are imaginary may jeopardize a long-term relationship with the health care professional. Focus shouldn't be on the disability because it may provide positive gains for the client.
NP: Implementation; CN: Physiological integrity; CNS: Physiological adaptation; CL: Application

36. Which of the following therapeutic approaches would enable a client to cope effectively with life stress without using conversion?
  1. Focus on the symptoms.
  2. Ask for clarification of the symptoms.
  3. Listen to the client's symptoms in a matter-of-fact manner.
  4. Point out that the client's symptoms are an escape from dealing with conflict.

36. 3. Listening in a matter-of-fact manner doesn't focus on the client's symptoms. All other interventions focus on the client's symptoms, which draw attention to the physical symptoms, not the underlying cause.
NP: Implementation; CN: Health promotion and maintenance; CNS: Prevention and early detection of disease; CL: Application

37. Which of the following statements made by a client shows the nurse that the goal of stress management was attained?
  1. "My arm hurts."
  2. "I enjoy being dependent on others."
  3. "I don't really understand why I'm here."
  4. "My muscles feel relaxed after that progressive relaxation exercise."

> Read between the lines when the client speaks.

37. 4. The client is experiencing positive results from the relaxation exercise. All other responses alert the nurse that the client needs further interventions.
NP: Evaluation; CN: Physiological integrity; CNS: Physiological adaptation; CL: Analysis

38. Pain disorders, such as headaches and musculoskeletal pain, are most frequently associated with which of the following groups?
1. Men
2. Women
3. Less educated population
4. Low socioeconomic population

39. Which of the following descriptions best defines a somatoform pain disorder?
1. A preoccupation with pain in the absence of physical disease
2. A physical or somatic complaint without any demonstrable organic findings
3. A morbid fear or belief that one has a serious disease where none exists
4. One or more neurologic symptoms associated with psychological conflict or need

40. Which of the following features would not be found in malingering?
1. A collection of symptoms inconsistent with any one medical profile
2. An exaggeration of symptoms
3. A medical condition confirmed by diagnostic testing
4. A group of symptoms not believable

41. Which of the following statements best defines malingering?
1. It's a preoccupation with pain in the absence of physical disease.
2. It's a voluntary production of a physical symptom for a secondary gain.
3. It's a morbid fear or belief that one has a serious disease where none exists.
4. It's associated with a psychological need or conflict in which the client shows one or more neurologic symptoms.

42. Which of the following nursing diagnoses is appropriate for a client with somatoform pain disorder?
1. Interrupted family processes
2. Disturbed thought processes
3. Ineffective denial
4. Ineffective coping

You're moving right along!

Remember, diagnostic criteria for psychological problems come from the latest edition of the *Diagnostic & Statistical Manual of Mental Disorders*, currently known as *DSM-IV*.

38. 2. Pain disorders are more common in women than men. Pain disorders can occur at any age, especially during the 30s and 40s. There isn't a high correlation of pain disorders with less educated individuals or those from a low socioeconomic population.
NP: Data collection; CN: Physiological integrity; CNS: Physiological adaptation; CL: Knowledge

39. 1. Somatoform pain disorder is a preoccupation with pain in the absence of physical disease. A physical or somatic complaint refers to somatoform disorders in general. Neurologic symptoms are associated with conversion disorders. A morbid fear of serious illness is hypochondriasis.
NP: Data collection; CN: Psychosocial integrity; CNS: Coping and adaptation; CL: Knowledge

40. 3. Although there's no specific method to determine whether an individual is a malingerer, all of the above features except confirmation of a medical condition are seen in this condition. Psychological tests may also help to detect this condition.
NP: Data collection; CN: Psychosocial integrity; CNS: Psychosocial adaptation; CL: Knowledge

41. 2. Malingering is defined as a voluntary production of physical or psychological symptoms to accomplish a specific goal or secondary gain (avoidance of a specific situation such as a jail term or to obtain money). The other statements are associated with conversion, hypochondriasis, and pain disorders.
NP: Data collection; CN: Health promotion and maintenance; CNS: Prevention and early detection of disease; CL: Comprehension

42. 4. A somatoform pain disorder is closely associated with the client's inability to handle stress and conflict. Disturbed thought processes aren't directly correlated with this disorder. While interrupted family processes and ineffective denial may be present, they aren't the primary focus.
NP: Planning; CN: Psychosocial integrity; CNS: Psychosocial adaptation; CL: Application

43. Based on a nursing diagnosis of *Ineffective coping* for a client with somatoform pain disorder, which nursing goal is most realistic?
 1. The client will be free from injury.
 2. The client will recognize sensory impairment.
 3. The client will discuss beliefs about spiritual issues.
 4. The client will verbalize absence or significant reduction of physical symptoms.

44. Which of the following statements made by a client best meets the diagnostic criteria for pain disorder?
 1. "I can't move my right leg."
 2. "I'm having severe stomach and leg pain."
 3. "I'm so afraid I might have human immunodeficiency virus."
 4. "I'm having chest pain and pain radiating down my left arm that began more than 1 hour ago."

This word *initial* is a hint for the right answer.

45. Which of the following initial therapeutic interventions is the most appropriate for a client diagnosed with *Ineffective coping related to a pain disorder?*
 1. Make an accurate assessment.
 2. Promote expressions of feelings.
 3. Promote insight into the disorder.
 4. Help the client develop alternative coping strategies.

46. Which of the following would most likely be considered an appropriate intervention for a client with a somatoform pain disorder?
 1. Analgesics as needed
 2. Exploring environmental stressors
 3. Exploring pain symptoms with the client
 4. Encouraging the client to discuss pain symptoms with her primary care physician

Hooray for you! You're the best!

43. 4. Expression of feelings enables the client to ventilate emotions, which decreases anxiety and draws attention away from the physical symptoms. The client isn't experiencing a safety issue. Spiritual issues are related to spiritual distress, and no evidence exists to support that the client is having spiritual distress. There's also no apparent correlation with any sensory-perceptual alterations.
NP: Planning; CN: Physiological integrity; CNS: Basic care and comfort; CL: Analysis

44. 2. The diagnostic criteria for pain disorders state that pain in one or more anatomic sites is the predominant focus of the clinical presentation and is of sufficient severity to warrant clinical attention. Hypochondriasis is a morbid fear or belief that one has a serious disease where none exists. A client with a conversion disorder can experience a motor neurologic symptom such as paralysis. Unremitting chest pain with radiation of pain down the left arm is symptomatic of a myocardial infarction.
NP: Data collection; CN: Psychosocial integrity; CNS: Psychosocial adaptation; CL: Application

45. 1. It's essential to accurately assess the client first before any interventions. Promoting insight, expression of feelings, and helping the client develop coping strategies are appropriate interventions that can be implemented after the initial assessment.
NP: Implementation; CN: Psychosocial integrity; CNS: Psychosocial adaptation; CL: Application

46. 2. Somatoform pain disorders are related to psychological conflicts that are often made worse by environmental stressors. Analgesics usually don't help because the source of pain isn't physical. A discussion of pain with the nurse or physician shouldn't be the focus of a therapeutic intervention for this condition.
NP: Implementation; CN: Psychosocial integrity; CNS: Coping and adaptation; CL: Application

Chapter 14
Anxiety & mood disorders

1. Which of the following statements is typical of a client who experiences periodic panic attacks while sleeping?
1. "Yesterday, I sat up in bed and just felt so scared."
2. "I have difficulty sleeping because I'm so anxious."
3. "Sometimes I have the most wild and vivid dreams."
4. "When I drink beer, I fall asleep without any problems."

The key word in question 1 is typical.

2. During the admission assessment, a client with a panic disorder begins to hyperventilate and says, "I'm going to die if I don't get out of here right now!" What is the nurse's best response?
1. "Just calm down. You're getting overly anxious."
2. "What do you think is causing your panic attack."
3. "You can rest alone in your room until you feel better."
4. "You're having a panic attack. I'll stay here with you."

3. A client with a history of panic attacks who says, "I felt so trapped," right after an attack most likely has which of the following fears?
1. Loss of control
2. Loss of identity
3. Loss of memory
4. Loss of maturity

Question 3 doesn't ask for all possible answers, just the most likely one.

1. 1. A person who suffers a panic attack while sleeping experiences an abrupt awakening and feelings of fear. People with severe anxiety often have symptoms related to a sleep disorder; they wouldn't typically experience a sleep panic attack. A panic attack while sleeping often causes an inability to remember dreams. Intake of alcohol initially produces a drowsy feeling, but after a short period of time alcohol causes restless, fragmented sleep and strange dreams.
NP: Data collection; CN: Psychosocial integrity; CNS: Psychosocial adaptation; CL: Knowledge

2. 4. During a panic attack, the nurse's best approach is to orient the client to what's happening and provide reassurance that the client won't be left alone. The client's anxiety level is likely to increase—and the panic attack is likely to continue—if the client is told to calm down, asked the reasons for the attack, or left alone.
NP: Implementation; CN: Psychosocial integrity; CNS: Coping and adaptation; CL: Application

3. 1. People who fear loss of control during a panic attack often make statements about feeling trapped, getting hurt, or having little to no personal control over their situations. People who experience panic attacks don't tend to have memory impairment or loss of identity. People who have panic attacks also don't regress or become immature.
NP: Data collection; CN: Psychosocial integrity; CNS: Coping and adaptation; CL: Application

NP: Nursing process   CN: Client needs category   CNS: Client needs subcategory   CL: Cognitive level

4. Which of the following should the nurse be most concerned about when caring for a client taking an antianxiety medication?
1. Physical and psychological dependence
2. Transient hypertension
3. Abrupt withdrawal
4. Constipation

4. 3. Abrupt discontinuation of an antianxiety drug can lead to withdrawal symptoms. Antianxiety medications usually are prescribed for short periods. However, if used over a prolonged period, such drugs may produce psychological or physical dependence. Transient hypertension and constipation aren't associated with antianxiety drugs.
NP: Planning; CN: Physiological integrity; CNS: Pharmacological therapies; CL: Application

Careful. Question 5 is asking you to prioritize.

5. Which of the following nursing interventions is given priority in a care plan for a person having a panic disorder?
1. Tell the client to take deep breaths.
2. Have the client talk about the anxiety.
3. Encourage the client to verbalize feelings.
4. Ask the client about the cause of the attack.

5. 1. During a panic attack, the nurse should remain with the client and direct what's said toward changing the physiological response, such as taking deep breaths. During an attack, the client is unable to talk about anxious situations and isn't able to address feelings, especially uncomfortable feelings and frustrations. While having a panic attack, the client is also unable to focus on anything other than the symptoms, so the client won't be able to discuss the cause of the attack.
NP: Planning; CN: Psychosocial integrity; CNS: Psychosocial adaptation; CL: Knowledge

6. Which of the following instructions should the nurse include in a teaching session about panic disorder for clients and their family?
1. Identifying when anxiety is escalating
2. Determining how to stop a panic attack
3. Addressing strategies to reduce physical pain
4. Preventing the client from depending on others

6. 1. By identifying the presence of anxiety, it's possible to take steps to prevent its escalation. A panic attack can't be stopped. The nurse can take steps to assist the client safely through the attack. Later, the nurse can assist the client to alleviate the precipitating stressors. Clients who experience panic disorder don't tend to be in physical pain. The client experiencing a panic disorder may need to periodically depend on other people when having a panic attack.
NP: Planning; CN: Psychosocial integrity; CNS: Coping and adaptation; CL: Comprehension

For the NCLEX, you should memorize the definitions of common fears such as agoraphobia.

7. Which of the following questions should a nurse ask to determine how agoraphobia affects the life of a client who has panic attacks with agoraphobia?
1. How realistic are your goals?
2. Are you able to go shopping?
3. Do you struggle with impulse control?
4. Who else in your family has panic disorder?

7. 2. The client with agoraphobia often restricts himself to home and is unable to carry out normal socializing and life-sustaining activities. Clients with panic disorder are able to set realistic goals and tend to be cautious and reclusive rather than impulsive. Although there is a familial tendency toward panic disorder, information about client needs must be obtained to determine how agoraphobia affects the client's life.
NP: Planning; CN: Psychosocial integrity; CNS: Coping and adaptation; CL: Comprehension

NP: Nursing process   CN: Client needs category   CNS: Client needs subcategory   CL: Cognitive level

8. Which of the following facts about the relationship between substance abuse and panic attacks would be helpful to a client considering lifestyle changes as part of a behavior modification program to treat his panic attacks?

1. Cigarettes can trigger panic episodes.
2. Fermented foods can cause panic attacks.
3. Hormonal therapy can induce panic attacks.
4. Tryptophan can predispose a person to panic attacks.

8. 1. Cigarettes are considered stimulants and can trigger panic attacks. Fermented foods, hormonal therapy, or tryptophan don't cause panic attacks.

NP: Implementation; CN: Psychosocial integrity; CNS: Psychosocial adaptation; CL: Comprehension

9. The nurse is caring for a 27-year-old male client who's in the panic level of anxiety. Which of the following is the nurse's highest priority?

1. Encourage the client to discuss his feelings.
2. Provide for the client's safety needs.
3. Decrease the environmental stimuli.
4. Respect the client's personal space.

9. 2. A client in panic level of anxiety doesn't comprehend and can't follow instructions or care for his own basic needs. The client is unable to express feelings due to the level of anxiety. Decreased environmental stimulus is needed but only after the client's safety needs and other basic needs are met. The nurse needs to enter the client's personal space to provide personal care because a client in panic is unable to do this for himself.

NP: Planning; CN: Psychosocial integrity; CNS: Coping and adaptation; CL: Knowledge

10. Which of the following short-term client outcomes is appropriate for a client with panic disorder?

1. Identify childhood trauma.
2. Monitor nutritional intake.
3. Institute suicide precautions.
4. Decrease episodes of disorientation.

Congratulations! You've finished 10 questions!

10. 3. Clients with panic disorder are at risk for suicide. Childhood trauma is associated with posttraumatic stress disorder, *not* panic disorder. Nutritional problems don't typically accompany panic disorder. Clients aren't typically disoriented; they may have a temporary altered sense of reality, but that lasts only for the duration of the attack.

NP: Planning; CN: Psychosocial integrity; CNS: Psychosocial adaptation; CL: Comprehension

11. A client diagnosed with panic disorder with agoraphobia is talking with the nurse about the progress made in treatment. Which of the following statements indicates a positive client response?

1. "I went to the mall with my friend last Saturday."
2. "I'm hyperventilating only when I have a panic attack."
3. "Today I decided that I can stop taking my medication."
4. "Last night I decided to eat more than a bowl of cereal."

11. 1. Clients with panic disorder tend to be socially withdrawn. Going to the mall is a sign of working on avoidance behaviors. Hyperventilation is a key symptom of panic disorder. Teaching breathing control is a major intervention for clients with panic disorder. The client taking medications for panic disorder, such as tricyclic antidepressants and benzodiazepines, must be weaned off these drugs. Most clients with panic disorder with agoraphobia don't have nutritional problems.

NP: Evaluation; CN: Psychosocial integrity; CNS: Psychosocial adaptation; CL: Analysis

12. Which of the following group therapy interventions would be of primary importance to a client with panic disorder?
1. Explore how secondary gains are derived from the disorder.
2. Discuss new ways of thinking and feeling about panic attacks.
3. Work to eliminate manipulative behavior used for meeting needs.
4. Learn the risk factors and other demographics associated with panic disorder.

Question 12 wants you to focus on the intervention of *primary* importance.

12. 2. Discussion of new ways of thinking and feeling about panic attacks can enable others to learn and benefit from a variety of intervention strategies. There are usually no secondary gains obtained from having a panic disorder. People with panic disorder aren't using the disorder as a way to manipulate others. Learning the risk factors could be accomplished in another format, such as a psychoeducational program.

NP: Implementation; CN: Psychosocial integrity; CNS: Coping and adaptation; CL: Analysis

13. A client with social phobia should be observed for which of the following symptoms?
1. Self-harm
2. Poor self-esteem
3. Compulsive behavior
4. Avoidance of social situations

All of the answers indicate common fears, but only one is associated with social phobia.

13. 4. Clients with social phobia avoid social situations for fear of being humiliated or embarrassed. They generally don't tend to be at risk for self-harm and usually don't demonstrate compulsive behavior. Not all individuals with social anxiety have low self-esteem.

NP: Data collection; CN: Psychosocial integrity; CNS: Coping and adaptation; CL: Knowledge

14. Which of the following statements is typical of a client with social phobia?
1. "Without people around, I just feel so lost."
2. "There's nothing wrong with my behavior."
3. "I always feel like I'm the center of attention."
4. "I know I can't accept that award for my brother."

14. 4. People who have a social phobia usually undervalue themselves and their talents. They fear social gatherings and dislike being the center of attention. They tend to stay away from situations in which they may feel humiliated and embarrassed. They don't like to be in feared social situations or around many people. They're very critical of themselves and believe that others also will be critical.

NP: Data collection; CN: Psychosocial integrity; CNS: Coping and adaptation; CL: Application

15. Clients with a social phobia would most likely fear which of the following situations?
1. Dental procedures
2. Meeting strangers
3. Being bitten by a dog
4. Having a car accident

Allow me to interject. Could I be the cause of blood-injection-injury phobia?

15. 2. Fear of meeting strangers is a common example of social phobia. Fears of having a dental procedure, being bitten by a dog, or having an accident *aren't* social phobias.

NP: Data collection; CN: Psychosocial integrity; CNS: Coping and adaptation; CL: Knowledge

16. Which of the following factors would the nurse find most helpful in assessing a client for a blood-injection-injury phobia?
1. Episodes of fainting
2. Gregarious personality
3. Difficulty managing anger
4. Dramatic, overreactive personality

16. 1. Many people with a history of blood-injection-injury phobia report frequently fainting when exposed to this type of situation. All personality styles can develop phobias, so personality type doesn't provide information for assessing phobias. Information about a client's difficulty managing anger isn't related to a specific phobic disorder. Individuals with blood-injection-injury phobias aren't being dramatic or overreactive.

NP: Data collection; CN: Psychosocial integrity; CNS: Coping and adaptation; CL: Comprehension

NP: Nursing process   CN: Client needs category   CNS: Client needs subcategory   CL: Cognitive level

17. A client on the sixth floor of a psychiatric unit has a morbid fear of elevators. She's scheduled to attend occupational therapy, which is located on the ground floor of the hospital. The client refuses to take the elevator, insisting that the stairs are safer. Which nursing action would be best given the client's refusal to use the elevator?

1. Insist that she take the elevator.
2. Offer a special reward if she rides the elevator.
3. Withhold her occupational therapy privileges until she's able to ride the elevator.
4. Allow her to use the stairs.

18. Which of the following behavior modification techniques is useful in the treatment of phobias?

1. Aversion therapy
2. Imitation or modeling
3. Positive reinforcement
4. Systematic desensitization

19. Which of the following statements would be useful when teaching the client and family about phobias and the need for a strong support system?

1. The use of a family support system is only temporary.
2. The need to be assertive can be reinforced by the family.
3. The family needs to set limits on inappropriate behaviors.
4. The family plays a role in promoting client independence.

20. Which nursing diagnosis would the nurse expect to find on the care plan of a client with a phobia about elevators?

1. Social isolation related to a lack of social skills
2. Disturbed thought processes related to a fear of elevators
3. Ineffective coping related to poor coping skills
4. Anxiety related to fear of elevators

Along with the data collection, you also need to familiarize yourself with common treatments.

17. 4. This client has a phobia and must not be forced to ride the elevator because of the risk of panic-level anxiety, which can occur if she's forced to contact the phobic object. This client can't control her fear; therefore, stating that she must take the elevator or promising a reward won't work. Occupational therapy is a treatment the client needs, not a reward.
NP: Implementation; CN: Psychosocial integrity; CNS: Psychosocial adaptation; CL: Application

18. 4. Systematic desensitization is a common behavior modification technique successfully used to help treat phobias. Aversion therapy and positive reinforcement *aren't* behavior modification techniques used with treatment of phobias. The techniques of imitation or modeling are social learning techniques, not behavior modification techniques.
NP: Implementation; CN: Psychosocial integrity; CNS: Psychosocial adaptation; CL: Application

19. 4. The family plays a vital role in supporting the client in treatment and preventing the client from using the phobia to obtain secondary gains. Family support must be ongoing, not temporary. The family can be more helpful focusing on effective handling of anxiety, rather than focusing energy on developing assertiveness skills. People with phobias are already restrictive in their behavior; more restrictions aren't necessary.
NP: Implementation; CN: Psychosocial integrity; CNS: Psychosocial adaptation; CL: Analysis

20. 3. Poor coping skills can cause ineffective coping. Such a client isn't relegated to social isolation and lack of social skills has nothing to do with phobia. Fear of elevators is a manifestation, not the cause, of altered thoughts and anxiety.
NP: Planning; CN: Psychosocial integrity; CNS: Psychosocial adaptation; CL: Analysis

21. A client suspected of having posttraumatic stress disorder should be assessed for which of the following problems?
  1. Eating disorder
  2. Schizophrenia
  3. Suicide
  4. "Sundown" syndrome

Whoopee! Another 10 questions down! Good job!

21. 3. Clients who experience posttraumatic stress disorder are at high risk for suicide and other forms of violent behaviors. Eating disorders are possible, but not a common complication of posttraumatic stress disorder. Clients with posttraumatic stress disorder don't usually have their extreme anxiety manifest itself as schizophrenia. "Sundown" syndrome is an increase in agitation accompanied by confusion. It's commonly seen in clients with dementia, not clients with posttraumatic stress disorder.
NP: Data collection; CN: Psychosocial integrity; CNS: Coping and adaptation; CL: Comprehension

22. Which of the following actions explains why tricyclic antidepressant medication is given to a client who has severe posttraumatic stress disorder?
  1. It prevents hyperactivity and purposeless movements.
  2. It increases the client's ability to concentrate.
  3. It helps prevent experiencing the trauma again.
  4. It facilitates the grieving process.

22. 3. Tricyclic antidepressant medication will decrease the frequency of reenactment of the trauma for the client. It will help memory problems and sleeping difficulties and will decrease numbing. The medication won't prevent hyperactivity and purposeless movements nor increase the client's concentration. No medication will facilitate the grieving process.
NP: Implementation; CN: Physiological integrity; CNS: Pharmacological therapies; CL: Application

23. Which of the following nursing actions would be included in a care plan for a client with posttraumatic stress disorder who states that the experience was "bad luck"?
  1. Encourage the client to verbalize the experience.
  2. Assist the client in defining the experience as a trauma.
  3. Work with the client to take steps to move on with life.
  4. Help the client accept positive and negative feelings.

A client describing a trauma as "bad luck" may be in denial.

23. 2. The client must define the experience as traumatic to realize the situation wasn't under personal control. Encouraging the client to verbalize the experience without first addressing the denial isn't a useful strategy. The client can only move on with life after acknowledging the trauma and processing the experience. Acknowledgment of the actual trauma and verbalization of the event should come *before* the acceptance of feelings.
NP: Planning; CN: Psychosocial integrity; CNS: Psychosocial adaptation; CL: Analysis

24. One year ago, a client was in a plane crash and several people were killed. The client is now experiencing nightmares, insomnia, headaches, loss of appetite, and fatigue. Which of the following is the client most likely experiencing?
  1. Panic disorder
  2. Posttraumatic stress disorder
  3. Bipolar disorder
  4. Conversion disorder

24. 2. The client is reliving the crash because it's close to the first anniversary of the accident. Panic disorder isn't correct because the client is still able to function. Bipolar disorder isn't appropriate because the client isn't exhibiting mania and depression. Conversion disorder isn't correct because the client isn't exhibiting any physical symptoms, such as paralysis, that aren't supported by medical diagnosis.
NP: Data collection; CN: Psychosocial integrity; CNS: Coping and adaptation CL: Application

NP: Nursing process   CN: Client needs category   CNS: Client needs subcategory   CL: Cognitive level

25. Which of the following approaches should the nurse use with the family when a posttraumatic stress disorder client states, "My family doesn't believe anything about posttraumatic stress disorder"?
1. Provide the family with information.
2. Teach the family about problem solving.
3. Discuss the family's view of the problem.
4. Assess for the presence of family violence.

Caring for clients often means working with their families — a skill you can expect the NCLEX to test.

25. 1. If the family can understand posttraumatic stress disorder, they can more readily participate in the client's care and be supportive. Learning problem-solving skills doesn't help clarify posttraumatic stress disorder. After giving the family information about posttraumatic stress disorder, the family can then ask questions and present its views. The family must first have information about posttraumatic stress disorder; then, the discussion about violence to self or others can be addressed.
NP: Implementation; CN: Safe, effective care environment; CNS: Coordinated care; CL: Comprehension

26. While caring for a client with posttraumatic stress disorder, the family notices that loud noises cause a serious anxiety response. Which of the following explanations would help the family understand the client's response?
1. Environmental triggers can cause the client to react emotionally.
2. Clients often experience extreme fear about normal environmental stimuli.
3. After a trauma, the client can't respond to stimuli in an appropriate manner.
4. The response indicates another emotional problem needs investigation.

26. 1. Repeated exposure to environmental triggers can cause the client to experience a hyperarousal state because there's a loss of physiological control of incoming stimuli. After experiencing a trauma, the client may have strong reactions to stimuli similar to those that occurred during the traumatic event. However, not *all* stimuli will cause an anxiety response. The client's anxiety response is typically seen after a traumatic experience and doesn't indicate the presence of another problem.
NP: Planning; CN: Psychosocial integrity; CNS: Coping and adaptation; CL: Application

27. Which of the following psychological symptoms would the nurse expect to find in a hospitalized client who's the only survivor of a train accident?
1. Denial
2. Indifference
3. Perfectionism
4. Trust

27. 1. Denial can act as a protective response. The client tends to be overwhelmed and disorganized by the trauma, not indifferent to it. Perfectionism is more commonly seen in clients with eating disorders, not in clients with posttraumatic stress disorder. Clients who have had a severe trauma often experience an inability to trust others.
NP: Data collection; CN: Psychosocial integrity; CNS: Coping and adaptation; CL: Comprehension

28. If a client suffering from posttraumatic stress disorder says, "I've decided to just avoid everything and everyone," the nurse might suspect the client is at greatest risk for which of the following behaviors?
1. Becoming homeless
2. Exhausting finances
3. Terminating employment
4. Using substances

All of the answers may be possible risks, but you've got to choose the greatest risk.

28. 4. The use of substances is a way for the client to deny problems and self-medicate distress. There are few homeless people with posttraumatic stress disorder as the cause of their homelessness. Most clients with posttraumatic stress disorder can manage money and maintain employment.
NP: Data collection; CN: Psychosocial integrity; CNS: Psychosocial adaptation; CL: Application

29. Which action would be most appropriate when speaking with a client with posttraumatic stress disorder about the trauma?
1. Obtain validation of what the client says from another party.
2. Request that the client write down what's being said.
3. Ask questions to convey an interest in the details.
4. Listen attentively.

30. Which of the following client statements indicates an understanding of survivor guilt?
1. "I think I can see the purpose of my survival."
2. "I can't help but feel that everything is their fault."
3. "I now understand why I'm not able to forgive myself."
4. "I wish I could stop sabotaging my family relationships."

Thirty questions! You're well into the race now.

31. Which of the following nursing interventions would best help the client with posttraumatic stress disorder and his family handle interpersonal conflict at home?
1. Have the family teach the client to identify defensive behaviors.
2. Have the family discuss how to change dysfunctional family patterns.
3. Have the family agree not to tell the client what to do about problems.
4. Have the family arrange for the client to participate in social activities.

32. The effectiveness of monoamine oxidase (MAO) inhibitor drug therapy in a client with posttraumatic stress disorder can be demonstrated by which of the following client self-reports?
1. "I'm sleeping better and don't have nightmares."
2. "I'm not losing my temper as much."
3. "I've lost my craving for alcohol."
4. "I've lost my phobia of water."

29. 4. An effective communication strategy for a nurse to use with a posttraumatic stress disorder client is listening attentively and staying with the client. There's no need to obtain validation about what the client says by asking for information from another party, asking them to write what's being said, or distracting them by asking questions.
NP: Implementation; CN: Psychosocial integrity; CNS: Psychosocial adaptation; CL: Application

30. 3. Survivor guilt occurs when the person has almost constant thoughts about the other people who perished in the event. They don't understand why they survived when their friend or loved one didn't. Blaming self, rather than blaming others, is a component of survivor guilt. Survivor guilt and impaired interpersonal relationships are two different categories of responses to trauma and don't indicate that the person understands survivor guilt.
NP: Evaluation; CN: Psychosocial integrity; CNS: Coping and adaptation; CL: Analysis

31. 2. Discussion of dysfunctional family patterns allows the family to determine why and how these patterns are maintained. Having family members point out the defensive behaviors of the client may inadvertently produce more defensive behavior. Families can be a source of support and assistance. Therefore, inflexible rules aren't useful to either the client or the family. The family shouldn't be encouraged to arrange social activities for the client. Social activities outside of the home don't assist the family to handle conflict within the home.
NP: Implementation; CN: Psychosocial integrity; CNS: Psychosocial adaptation; CL: Application

32. 1. MAO inhibitors are used to treat sleep problems, nightmares, and intrusive daytime thoughts in individuals with posttraumatic stress disorder. MAO inhibitors aren't used to help control flashbacks or phobias or to decrease the craving for alcohol.
NP: Evaluation; CN: Physiological integrity; CNS: Pharmacological therapies; CL: Analysis

33. The nurse is caring for a Vietnam veteran with a history of explosive anger, unemployment, and depression since being discharged from the service. The client reports feeling ashamed of being "weak" and of letting past experiences control his thoughts and actions in the present. What is the nurse's best response?
   1. "Many people who have been in your situation experience similar emotions and behaviors."
   2. "No one can predict how he'll react in a traumatic situation."
   3. "It's not too late for you to make changes in your life."
   4. "Weak people don't want to make changes in their lives."

34. Which of the following results is the major purpose for providing group therapy to a group of adolescents who witnessed the violent death of a peer?
   1. To learn violence prevention strategies
   2. To talk about appropriate expression of anger
   3. To discuss the effect of the trauma on their lives
   4. To develop trusting relationships among their peers

35. The nurse is caring for a 53-year-old male client with posttraumatic stress disorder who's experiencing a frightening flashback. The nurse can best offer reassurance of safety and security through which of the following nursing actions?
   1. Encouraging the client to talk about the traumatic event
   2. Assessing for maladaptive and coping strategies
   3. Staying with the client
   4. Acknowledging feelings of guilt or self-blame

The term *major purpose* indicates that there may be more than one right answer. You need to choose the best answer.

33. 1. By providing that extreme anger and other reactions are normal responses to trauma, the nurse assists the client to deal with his shame over a perceived lack of control over his feelings and to gain confidence in his ability to alter behaviors. The other responses are clichés and don't address the client's feelings.
NP: Implementation; CN: Psychosocial integrity; CNS: Coping and adaptation; CL: Application

34. 3. By discussing the effect of the trauma on their lives, the adolescents can grieve and develop effective coping strategies. Learning violence prevention strategies isn't the most immediate concern after a trauma occurs nor is working on developing healthy relationships. It's appropriate to talk about how to express anger appropriately after the trauma is addressed.
NP: Implementation; CN: Psychosocial integrity; CNS: Coping and adaptation; CL: Application

35. 3. The nurse should stay with the client during periods of flashbacks and nightmares, offer reassurance of safety and security, and assure the client that these symptoms aren't uncommon following a severe trauma. Encouraging him to talk about the traumatic event, observing for maladaptive and coping strategies, and acknowledging feelings of guilt or self-blame may be carried out in the future. The nurse's top priority during the flashback is to stay with the client.
NP: Implementation; CN: Psychosocial integrity; CNS: Psychosocial adaptation CL: Application

36. Which of the following nursing behaviors would demonstrate caring to clients with a diagnosis of anxiety disorder?
1. Verbalize concern about the client.
2. Arrange group activities for clients.
3. Have the client sign the treatment plan.
4. Hold psychoeducational groups on medications.

36. 1. The nurse who verbally expresses concern about a client's well being is acting in a caring and supportive manner. Arranging for group activities may be an action where the nurse has no direct client contact and is therefore unable to have interpersonal contact with clients. Having a client sign the treatment plan may not be viewed as a sign of caring. Having a psychoeducational group on medications may be viewed by clients as a teaching experience and not interpersonal contact because the nurse may have limited interactions with them.
NP: Implementation; CN: Health promotion and maintenance; CNS: Prevention and early detection of disease; CL: Application

Conquer your own anxieties! Feel confident about your answers!

37. Which of the following findings should the nurse expect when talking about school to a child diagnosed with a generalized anxiety disorder?
1. The child has been fighting with peers for the past month.
2. The child can't stop lying to parents and teachers.
3. The child has gained 15 lb (6.8 kg) in the past month.
4. The child expresses concerns about grades.

37. 4. Children with generalized anxiety disorder will worry about how well they're performing in school. Children with generalized anxiety disorder don't tend to be involved in conflict. They're more oriented toward good behavior. Children with generalized anxiety disorder don't tend to lie to others. They would want to do their best and try to please others. A weight gain of 15 lb isn't a typical characteristic of a child with anxiety disorder.
NP: Data collection; CN: Psychosocial integrity; CNS: Coping and adaptation; CL: Analysis

38. A client with a generalized anxiety disorder also may have which of the following concurrent diagnoses?
1. Bipolar disorder
2. Gender identity disorder
3. Panic disorder
4. Schizoaffective disorder

38. 3. Approximately 75% of clients with generalized anxiety disorder also may have a diagnosis of phobia, panic disorder, or substance abuse. Clients with generalized anxiety disorder don't tend to have a coexisting diagnosis of gender identity disorder, bipolar disorder, or schizoaffective disorder.
NP: Data collection; CN: Psychosocial integrity; CNS: Coping and adaptation; CL: Application

What do you know about the relationship between anxiety and caffeine?

39. Which of the following statements indicates a positive response from a generalized anxiety disorder client to a nurse's teaching about nutrition?
1. "I've stopped drinking so much diet cola."
2. "I've reduced my intake of carbohydrates."
3. "I now eat less at dinner and before bedtime."
4. "I've cut back on my use of dairy products."

39. 1. Clients with generalized anxiety disorder can decrease anxiety by eliminating caffeine from their diets. It isn't necessary for clients with generalized anxiety to decrease their carbohydrate intake, eat less at dinner or before bedtime (unless there are other compelling health reasons), or cut back on their use of dairy products.
NP: Evaluation; CN: Health promotion and maintenance; CNS: Prevention and early detection of disease; CL: Application

40. A client with generalized anxiety disorder is prescribed a benzodiazepine, but the client doesn't want to take the medication. Which of the following explanations for this behavior would most likely be correct?
1. "I don't think the psychiatrist likes me."
2. "I want to solve my problems on my own."
3. "I'll wait several weeks to see if I need to take it."
4. "I think my family gains by keeping me medicated."

40. 2. Many clients don't want to take medications because they believe that using a medication is a sign of personal weakness and that they can't solve their problems by themselves. Thinking that the psychiatrist dislikes them reflects paranoid thinking that isn't usually seen in clients with generalized anxiety disorder. By waiting several weeks to take the medication, the client could be denying that the medication is necessary or beneficial. By focusing on the negative motives of family members, the client could be avoiding talking about himself.

NP: Data collection; CN: Psychosocial integrity; CNS: Coping and adaptation; CL: Analysis

41. A female client describes her unpredictable episodes of acute anxiety as "just awful." She says that she feels like she's about to die and can hardly breathe. These symptoms are most characteristic of which of the following disorders?
1. Agoraphobia
2. Dissociative disorder
3. Posttraumatic stress disorder
4. Panic disorder

41. 4. This client is describing the characteristics of someone with panic disorder. Agoraphobia is characterized by fear of public places; dissociative disorder, by lost periods of time; and posttraumatic stress disorder, by hypervigilance and sleep disturbance.

NP: Data collection; CN: Psychosocial integrity; CNS: Coping and adaptation; CL: Knowledge

42. Which of the following statements by a 44-year-old client with a diagnosis of generalized anxiety disorder would convince the nurse that anxiety has been a long-standing problem?
1. "I was, and still am, an impulsive person."
2. "I've always been hyperactive, but not in useful ways."
3. "When I was in college, I never thought I would finish."
4. "All my life I've had intrusive dreams and scary nightmares."

42. 3. For many people who have a generalized anxiety disorder, the age of onset is during young adulthood. The symptoms of impulsiveness and hyperactivity aren't commonly associated with a diagnosis of generalized anxiety disorder. The symptom of intrusive dreams and nightmares is associated with posttraumatic stress disorder rather than generalized anxiety disorder.

NP: Data collection; CN: Psychosocial integrity; CNS: Coping and adaptation; CL: Comprehension

43. Which of the following symptoms would the client with generalized anxiety disorder most likely display when assessed for muscle tension?
1. Difficulty sleeping
2. Restlessness
3. Strong startle response
4. Tachycardia

43. 2. Restlessness is a symptom associated with muscle tension. Difficulty sleeping and a strong startle response are considered symptoms of vigilance and scanning of the environment, not muscle tension. Tachycardia is classified as a symptom of autonomic hyperactivity, not muscle tension.

NP: Data collection; CN: Physiological integrity; CNS: Physiological adaptation; CL: Knowledge

> Certain psychological disorders tend to appear at particular ages — that's information to know for the NCLEX.

44. When planning the care of a client with general anxiety disorder, which intervention is most important to include?
1. Encourage the client to engage in activities that increase feelings of power and self-esteem.
2. Promote the client's interaction and socialization with others.
3. Assist the client to make plans for regular periods of leisure time.
4. Encourage the client to use a diary to record when anxiety occurred, its cause, and which interventions may have helped.

45. Which of the following instructions helps the nurse deal with escalating client anxiety?
1. Explore feelings about current life stressors.
2. Discuss the need to flee from painful situations.
3. Have the client develop a realistic view of self.
4. Provide appropriate phone numbers for hotlines and clinics.

46. The nurse is teaching a 53-year-old male client with the nursing diagnosis *Anxiety related to lack of knowledge about an impending surgical procedure.* Which of the following is an appropriate nursing intervention?
1. Reassure the client that there are many treatments for the problem.
2. Calmly ask the patient to describe the procedure that is to be done.
3. Instruct the client that the nursing staff will help in any way they can.
4. Tell the client that he shouldn't keep his feelings to himself.

47. A client with a diagnosis of generalized anxiety disorder wants to stop taking his lorazepam (Ativan). Which of the following important facts should the nurse discuss with the client about discontinuing the medication?
1. Stopping the drug may cause depression.
2. Stopping the drug increases cognitive abilities.
3. Stopping the drug decreases sleeping difficulties.
4. Stopping the drug can cause withdrawal symptoms.

Thorough teaching plans usually include ways to contact support organizations.

You should be familiar with the symptoms that discontinuing key medications can cause.

44. 4. One of the nurse's goals is to help the client associate symptoms with an event, thereby beginning to learn appropriate ways to eliminate or reduce distress. A diary can be a beneficial tool for this purpose. Although encouraging the client to engage in activities that increase feelings of power and self-esteem, promoting interaction and socialization with others, and assisting the client to make plans for regular periods of leisure time may be appropriate, they aren't the priority.
NP: Planning; CN: Psychosocial integrity; CNS: Coping and adaptation; CL: Application

45. 4. By having information on hotlines and clinics, the client can pursue help when the anxiety is escalating. Discussion about current life stressors isn't useful when focusing on how best to handle the client's escalating anxiety. Fleeing from painful situations and discussing views of oneself aren't the best strategies; neither allows for problem solving.
NP: Implementation; CN: Psychosocial integrity; CNS: Psychosocial adaptation; CL: Knowledge

46. 2. An appropriate short-term goal in this case is, "The client will repeat to the nurse the major points of the procedure." By asking the client to describe the procedure, the nurse can assess his level of understanding and address his anxiety by providing necessary teaching. Reassuring him that there are many treatments, instructing him that the nursing staff will help, and telling him that he shouldn't keep his feelings to himself don't address the client's anxiety or lack of knowledge about the procedure.
NP: Evaluation; CN: Psychosocial integrity; CNS: Coping and adaptation; CL: Comprehension

47. 4. Stopping antianxiety drugs such as benzodiazepines can cause the client to have withdrawal symptoms. Stopping a benzodiazepine doesn't tend to cause depression, increase cognitive abilities, or decrease sleeping difficulties.
NP: Implementation; CN: Physiological integrity; CNS: Pharmacological therapies; CL: Application

NP: Nursing process   CN: Client needs category   CNS: Client needs subcategory   CL: Cognitive level

48. Five days after running out of medication, a client taking clonazepam (Klonopin) says to the nurse, "I know I shouldn't have just stopped the drug like that, but I'm okay." Which of the following responses would be best?
1. "Let's monitor you for problems, in case something else happens."
2. "You could go through withdrawal symptoms for up to 2 weeks."
3. "You have handled your anxiety and you now know how to cope with stress."
4. "If you're fine now, chances are you won't experience withdrawal symptoms."

48. 2. Withdrawal syndrome symptoms can appear after 1 or 2 weeks because the benzodiazepine has a long half-life. Looking for another problem unrelated to withdrawal isn't the nurse's best strategy. The act of discontinuing an antianxiety medication doesn't indicate that a client has learned to cope with stress. Every client taking medication needs to be monitored for withdrawal symptoms when the medication is stopped abruptly.
NP: Implementation; CN: Physiological integrity; CNS: Pharmacological therapies; CL: Analysis

49. A client taking alprazolam (Xanax) reports light-headedness and nausea every day while getting out of bed. Which of the following actions should the nurse take to objectively validate this client's problem?
1. Take the client's blood pressure.
2. Monitor body temperature.
3. Teach the Valsalva maneuver.
4. Obtain a blood chemical profile.

49. 1. The nurse should take a blood pressure reading to validate orthostatic hypotension. A body temperature reading or chemistry profile won't yield useful information about hypotension. The Valsalva maneuver is performed to lower the heart rate and isn't an appropriate intervention.
NP: Implementation; CN: Physiological integrity; CNS: Reduction of risk potential; CL: Knowledge

50. A client with generalized anxiety disorder complains to the nurse about several other minor health problems. Which of the following concerns must the nurse keep in mind when preparing instructions for this client?
1. Clients may have a variety of somatic symptoms.
2. Clients undergo an alteration in their self-care skills.
3. Clients are prone to unhealthy binge eating episodes.
4. Clients will experience secondary gains from mental illness.

50. 1. Clients with anxiety disorders often experience somatic symptoms. They don't usually experience problems with self-care. Eating problems aren't a typical part of the diagnostic criteria for anxiety disorders. Not all clients obtain secondary gains from mental illness.
NP: Planning; CN: Psychosocial integrity; CNS: Psychosocial adaptation; CL: Comprehension

51. Which of the following communication guidelines should the nurse use when talking with a client experiencing mania?
1. Address the client in a light and joking manner.
2. Focus and redirect the conversation as necessary.
3. Allow the client to talk about several different topics.
4. Ask only open-ended questions to facilitate conversation.

Stay focused. You're making solid progress.

51. 2. To decrease stimulation, the nurse should attempt to redirect and focus the client's communication, not allow the client to talk about different topics. By addressing the client in a light and joking manner, the conversation may contribute to the client feeling out of control. For a manic client, it's best to ask closed questions because open-ended questions may enable the client to talk endlessly, again possibly contributing to the client feeling out of control.
NP: Planning; CN: Psychosocial integrity; CNS: Psychosocial adaptation; CL: Comprehension

52. Which of the following adverse effects would the nurse explain to the client with bipolar disorder and the client's family before administering electroconvulsive therapy (ECT)?

1. Cholestatic jaundice
2. Hypertensive crisis
3. Mouth ulcers
4. Respiratory arrest

53. A client who has just had electroconvulsive therapy (ECT) asks for a drink. Which of the following observations is a priority when meeting the client's request?

1. Take the client's blood pressure.
2. Monitor the gag reflex.
3. Obtain a body temperature.
4. Determine the level of consciousness.

54. A client who's in the manic phase of bipolar disorder constantly belittles other clients and demands special favors from the nurses. Which nursing intervention would be most appropriate for this client?

1. Ask other clients and staff members to ignore the client's behavior.
2. Set limits with consequences for belittling or demanding behavior.
3. Offer the client an antianxiety agent when this behavior occurs.
4. Offer the client a variety of stimulating activities.

55. A client with bipolar disorder who complains of headache, agitation, and indigestion is most likely experiencing which of the following problems?

1. Depression
2. Cyclothymia
3. Hypomania
4. Mania

You can probably have a drink, but I need to check something first.

The words most likely can help you focus on the answer.

52. 4. Respiratory arrest may occur as a complication of the anesthesia used with ECT. Cholestatic jaundice, hypertensive crisis, and mouth ulcers don't occur during or as a result of ECT.

NP: Implementation; CN: Physiological integrity; CNS: Physiological adaptation; CL: Knowledge

53. 2. The nurse must check the client's gag reflex before allowing the client to have a drink after an ECT procedure. Blood pressure and body temperature don't influence whether the client may have a drink after the procedure. The client would obviously be conscious if he's requesting a glass of water.

NP: Data collection; CN: Safe, effective care environment; CNS: Safety and infection control; CL: Knowledge

54. 2. By setting limits with consequences for noncompliance, the nurse can protect others from a client who exhibits belittling and demanding behaviors. Asking others to ignore the client is likely to increase those behaviors. Offering the client an antianxiety agent or stimulating activities provides no motivation for the client to change the problematic behaviors.

NP: Planning; CN: Psychosocial integrity; CNS: Coping and adaptation; CL: Application

55. 4. Headache, agitation, and indigestion are symptoms suggestive of mania in a client with a history of bipolar disorder. These symptoms *aren't* suggestive of depression, cyclothymia, or hypomania.

NP: Data collection; CN: Physiological integrity; CNS: Physiological adaptation; CL: Knowledge

56. Which of the following high-risk behaviors should the nurse assess for in a client with bipolar disorder who has abruptly stopped taking the prescribed medication and experiences a manic episode?
1. Binge eating
2. Relationship avoidance
3. Sudden relocation
4. Thoughtless spending

Know the high-risk behaviors associated with each disorder.

56. 4. Thoughtless or reckless spending is a common symptom of a manic episode. Binge eating isn't a behavior that's characteristic of a client during a manic episode. Relationship avoidance doesn't occur in a client experiencing a manic episode. During episodes of mania, a client may in fact interact with many people and participate in unsafe sexual behavior. Sudden relocation isn't a characteristic of impulsive behavior demonstrated by a client with bipolar disorder.
NP: Data collection; CN: Psychosocial integrity; CNS: Coping and adaptation; CL: Knowledge

57. Which of the following interventions would assist the client with bipolar disorder to maintain adequate nutrition?
1. Determine the client's metabolic rate.
2. Make the client sit down for each meal and snack.
3. Give the client foods to be eaten while he's active.
4. Have the client interact with a dietitian twice a week.

57. 3. By giving the client high-calorie foods that can be eaten while he's active, the nurse facilitates the client's nutritional intake. Determining the client's metabolic rate isn't useful information when the client is experiencing mania. During a manic episode, the client can't be still or focused long enough to interact with a dietitian or sit still long enough to eat.
NP: Implementation; CN: Psychosocial integrity; CNS: Psychosocial adaptation; CL: Knowledge

58. Which of the following adverse reactions does the client with bipolar disorder taking lithium need to report?
1. Black tongue
2. Increased lacrimation
3. Periods of disorientation
4. Persistent GI upset

58. 4. Persistent GI upset indicates a mild-to-moderate toxic reaction. Black tongue is an adverse reaction of mirtazapine (Remeron), not lithium. Increased lacrimation isn't an adverse effect of lithium. Periods of disorientation don't tend to occur with the use of lithium.
NP: Implementation; CN: Physiological integrity; CNS: Pharmacological therapies; CL: Knowledge

59. When providing discharge instructions for a client with bipolar disorder who's receiving lithium carbonate, the nurse should stress which of the following?
1. The client should take the medication with food.
2. The client should have blood serum levels checked regularly.
3. The client should take the medication on an empty stomach.
4. The client should avoid operating heavy machinery.

59. 2. Regularly checking serum lithium levels is a critical part of medication management. The client is at high risk for toxicity with this drug, so regular monitoring is key to the success of treatment. Taking lithium carbonate with food or on an empty stomach and avoiding operating heavy machinery aren't necessary when taking this medication.
NP: Implementation; CN: Physiological integrity; CNS: Reduction of risk potential; CL: Application

60. Which activity is appropriate for a client with a diagnosis of bipolar disorder, manic phase?
1. Playing a card game
2. Playing a vigorous basketball game
3. Playing a board game
4. Painting

61. Which of the following short-term goals is most appropriate for a client with bipolar disorder who's having difficulty sleeping?
1. Obtain medication for sleep.
2. Work on solving a problem.
3. Exercise before bedtime.
4. Develop a sleep ritual.

Remember, you're looking for the *most* appropriate sleep-related goal.

62. Which of the following topics should the nurse discuss with the family of a client with bipolar disorder if the family is distressed about the client's episodes of manic behavior?
1. Ways to protect themselves from the client's behavior
2. How to proceed with an involuntary commitment
3. Where to confront the client about the reckless behavior
4. When to safely increase medication during manic periods

63. Which of the following statements should the nurse explain to a client newly diagnosed with bipolar disorder who doesn't understand why frequent blood work is necessary while he's taking lithium?
1. Frequent lithium levels will help the primary care provider spot liver and renal damage early.
2. Frequent lithium levels will demonstrate whether the client is taking a high enough dosage.
3. Frequent lithium levels will indicate whether the drug passes through the blood-brain barrier.
4. Frequent lithium levels are unnecessary if the client takes the drug as ordered.

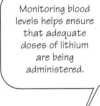

Monitoring blood levels helps ensure that adequate doses of lithium are being administered.

60. 4. An activity, such as painting, that promotes minimal stimulation is the best choice. Activities, such as playing cards, basketball, or a board game, may escalate hyperactivity and should be avoided.
NP: Implementation; CN: Psychosocial integrity; CNS: Psychosocial adaptation; CL: Application

61. 4. A sleep ritual or nighttime routine helps the client to relax and prepare for sleep. Obtaining sleep medication is a temporary solution. Working on problem solving may excite the client rather than tire him. Exercise before retiring is inappropriate.
NP: Planning; CN: Physiological integrity; CNS: Reduction of risk potential; CL: Knowledge

62. 1. Family members need to assess their needs and develop ways to protect themselves. Clients who have symptoms of impulsive or reckless behavior might not be candidates for hospitalization. Confronting the client during a manic episode may escalate the behavior. The family must never increase the dosage of prescribed medication without first consulting the primary health care provider.
NP: Implementation; CN: Safe, effective care environment; CNS: Safety and infection control; CL: Application

63. 2. Lithium levels determine whether an effective dose of lithium is being given to maintain a therapeutic level of the drug. The drug is contraindicated for clients with renal, cardiac, or liver disease. Lithium levels aren't drawn for the purpose of determining if the drug passes through the blood-brain barrier. Taking the drug as ordered doesn't eliminate the need for blood work.
NP: Implementation; CN: Physiological integrity; CNS: Pharmacological therapies; CL: Comprehension

NP: Nursing process   CN: Client needs category   CNS: Client needs subcategory   CL: Cognitive level

64. A client with a diagnosis of bipolar disorder, manic phase, is at risk for exhaustion and inadequate food intake. How can the nurse best meet the client's nutritional needs?

1. Establish a set time to eat meals.
2. Order high-protein milk shakes between meals.
3. Allow family members to bring the client's favorite foods to the hospital.
4. Provide finger foods that can be eaten on the go.

Anxiety and mood disorders can affect the nutritional status of your client. A chef's hat is yet another hat the nurse may have to wear!

64. 4. Finger foods that can be eaten quickly while the client is in a highly energetic state are best. This client can't sit still long enough to eat. Favorite foods and high-protein shakes may help with the client's intake but don't necessarily provide balanced nutrition.

NP: Implementation; CN: Physiological integrity; CNS: Basic care and comfort; CL: Application

65. Which of the following statements made by a client with bipolar disorder indicates that the nurse's teaching on coping strategies was effective?

1. "I can decide what to do to prevent family conflict."
2. "I can handle problems without asking for any help."
3. "I can stay away from my friends when I feel distressed."
4. "I can ignore things that go wrong instead of getting upset."

65. 1. The client should be focusing on his strengths and abilities to prevent family conflict. Not being able to ask for help is problematic and not a good coping strategy. Avoiding problems also isn't a good coping strategy. It's better to identify and handle problems as they arise. Ignoring situations that cause discomfort won't facilitate solutions or allow the client to demonstrate effective coping skills.

NP: Evaluation; CN: Psychosocial integrity; CNS: Coping and adaptation; CL: Analysis

66. Which of the following short-term goals would be given high priority for a client with depression admitted to the inpatient unit because of attempted suicide?

1. The client will seek out the nurse when feeling self-destructive.
2. The client will identify and discuss actual and perceived losses.
3. The client will learn strategies to promote relaxation and self-care.
4. The client will establish healthy and mutually caring relationships.

Question 66 is asking you to prioritize. Read it carefully!

66. 1. By seeking out the nurse when feeling self-destructive, the client can feel safe and begin to see that there are coping skills to assist in dealing with self-destructive tendencies. Discussion of losses also is important when dealing with feelings of depression, but the priority intervention is still to promote immediate client safety. Although relationship building and learning strategies to promote relaxation and self-care are important goals, safety is the priority intervention.

NP: Planning; CN: Safe, effective care environment; CNS: Safety and infection control; CL: Comprehension

67. In conferring with the treatment team, the nurse should make which of the following recommendations for a client who tells the nurse that everyday thoughts of suicide are present?

1. A no-suicide contract
2. Weekly outpatient therapy
3. A second psychiatric opinion
4. Intensive inpatient treatment

67. 4. For a client thinking about suicide on a daily basis, inpatient care would be the best intervention. Although a no-suicide contract is an important strategy, this client needs additional care. The client needs a more intensive level of care than weekly outpatient therapy. Immediate intervention is paramount, not a second psychiatric opinion.

NP: Planning; CN: Safe, effective care environment; CNS: Safety and infection control; CL: Comprehension

68. Which of the following short-term goals should the nurse focus on for a client who makes statements about not deserving things?

1. Identify distorted thoughts.
2. Describe self-care patterns.
3. Discuss family relationships.
4. Explore communications skills.

Note that question 68 asks about short-term goals, rather than long-term goals.

68. 1. It's important to identify distorted thinking because self-deprecating thoughts lead to depression. Self-care patterns don't necessarily reflect distorted thinking. Family relationships might not influence distorted thinking patterns. A form of communication called negative self-talk would be explored only after distorted thinking patterns were identified.

NP: Planning; CN: Psychosocial integrity; CNS: Coping and adaptation; CL: Application

69. Which of the following interventions should be of primary importance to the nurse working with a client to modify the client's negative expectations?

1. Encourage the client to discuss spiritual matters.
2. Assist the client to learn how to problem solve.
3. Help the client explore issues related to loss.
4. Have the client identify positive aspects of self.

69. 4. An important intervention used to counter negative expectations is to focus on the positive and have the client explore positive aspects of himself. Discussion of spiritual matters doesn't address the need to change negative expectations. Learning how to problem solve won't modify the client's negative expectations. If the client dwells on the negative and focuses on loss, it will be natural to have negative expectations.

NP: Implementation; CN: Psychosocial integrity; CNS: Psychosocial adaptation; CL: Application

70. Which of the following intervention strategies is most appropriate to use with a client with depression, who's suspected of being suicidal?

1. Speak to family members to ascertain if the client is suicidal.
2. Talk to the client to determine if the client is an attention seeker.
3. Arrange for the client to be placed on immediate suicidal precautions.
4. Ask a direct question such as, "Do you ever think about killing yourself?"

70. 4. The best approach to determining if a client is suicidal is to ask about thoughts of suicide in a direct and caring manner. Assessing for attention-seeking behaviors doesn't deal directly with the problem. The client should be assessed directly, not through family members. Assessment must be performed before determining if suicide precautions are necessary.

NP: Implementation; CN: Safe, effective care environment; CNS: Safety and infection control; CL: Knowledge

You may have misconceptions about depression.

71. Which of the following instructions should the nurse include when teaching the family of a client with major depression?

1. Address how depression is a lifelong illness.
2. Explain that depression is an illness and can be treated.
3. Describe how depression masks a person's true feelings.
4. Teach how depression causes frequent disorganized thinking.

71. 2. The nurse must help the family understand depression, its impact on the family, and recommended treatments. Depression doesn't need to be a lifelong illness. It's important to help families understand that depression can be successfully treated and that, in some situations, depression can reoccur during the life cycle. The feelings expressed by the client are genuine; they reflect cognitive distortions and disillusionment. Disorganized thinking is more commonly associated with schizophrenia rather than depression.

NP: Planning; CN: Psychosocial integrity; CNS: Psychosocial adaptation; CL: Application

NP: Nursing process   CN: Client needs category   CNS: Client needs subcategory   CL: Cognitive level

72. A client with major depression asks why he is taking mirtazapine (Remeron) instead of imipramine (Tofranil). Which of the following explanations is most accurate?

1. The newer serotonin reuptake inhibitor drugs are better-tested drugs.
2. The serotonin reuptake inhibitors have few adverse effects.
3. The serotonin reuptake inhibitors require a low dose of antidepressant drug.
4. The serotonin reuptake inhibitors are as good as other antidepressant drugs.

Roses are red. Violets are blue. Just 14 more questions and you'll be through!

72. 2. The serotonin reuptake inhibitors are drugs with few adverse effects and are unlikely to be toxic in an overdose. All drugs must be tested through a government-specified protocol. Comparison of two different types of antidepressant medications isn't useful. The final statement doesn't give the client helpful information.

NP: Implementation; CN: Physiological integrity; CNS: Pharmacological therapies; CL: Analysis

73. Which of the following interventions is most likely to promote a positive sense of self in a client with depression whose goal is enhancing self-esteem?

1. Playing cards
2. Praying daily
3. Taking medication
4. Writing poetry

73. 4. Writing poetry or engaging in some other creative outlet will enhance self-esteem. Playing cards and praying don't necessarily promote self-esteem. Taking medication will decrease symptoms of depression after a blood level is established but it won't, by itself, promote self-esteem.

NP: Implementation; CN: Psychosocial integrity; CNS: Psychosocial adaptation; CL: Comprehension

74. An adolescent who's depressed and reported by his parents as having difficulty in school is brought to the community mental health center to be evaluated. Which of the following other health problems would the nurse suspect?

1. Anxiety disorder
2. Behavioral difficulties
3. Cognitive impairment
4. Labile moods

These teens may not be subtle or quiet.

74. 2. Adolescents tend to demonstrate severe irritability and behavioral problems rather than simply a depressed mood. Anxiety disorder is more commonly associated with small children rather than with adolescents. Cognitive impairment is typically associated with delirium or dementia. Labile mood is more characteristic of a client with cognitive impairment or bipolar disorder.

NP: Data collection; CN: Psychosocial integrity; CNS: Coping and adaptation; CL: Analysis

75. Which of the following nursing interventions is most effective in lowering a client's risk for suicide?

1. Using a caring approach
2. Developing a strong relationship with the client
3. Establishing a suicide contract to ensure his safety
4. Encouraging avoidance of overstimulating activities

75. 3. Establishing a suicide contract with the client demonstrates that the nurse's concern for his safety is a priority and that his life is of value. When a client agrees to a suicide contract, it decreases his risk for a successful attempt. Caring alone ignores the underlying mechanism of the client's wish to commit suicide. Merely developing a strong relationship with the client isn't addressing the potential the client has for harming himself. Encouraging the client to stay away from activities could cause isolation, which would be detrimental to the client's well-being.

NP: Planning; CN: Safe, effective care environment; CNS: Safety and infection control; CL: Application

76. The nurse is instructing a 53-year-old female client about using the antianxiety medication lorazepam (Ativan). Which of the following statements by the client indicates a need for further teaching?
1. "I should get up slowly from a sitting or lying position."
2. "I shouldn't stop taking this medicine abruptly."
3. "I usually drink a beer every night to help me sleep."
4. "If I have a sore throat, I should report it to the physician."

77. Which of the following is typically the etiology for posttraumatic stress disorder?
1. Ineffective coping
2. A life-threatening or catastrophic event
3. Posttraumatic phobia and uncontrolled anxiety
4. Altered role performance

78. A client has been receiving treatment for depression for 3 weeks. Which behavior suggests that the client is less depressed?
1. Talking about the difficulties of returning to college after discharge
2. Spending most of the day sitting alone in the corner of the room
3. Wearing a hospital gown instead of street clothes
4. Showing no emotion when visitors leave

79. A lack of dietary salt intake can have which of the following effects on lithium levels?
1. Decrease
2. Increase
3. Increase then decrease
4. No effect at all

Your client shouldn't imbibe while taking this drug.

I bet you're feeling less depressed knowing there are only a few more questions to go!

76. 3. The client shouldn't consume alcohol or any other central nervous system depressant while taking this drug. All of the other statements indicate that the client understands the nurse's instructions.
NP: Data collection; CN: Physiological integrity; CNS: Pharmacological therapies; CL: Application

77. 2. Posttraumatic stress disorder is usually caused by some life-threatening or catastrophic event, such as a war, rape, or natural disaster. Ineffective coping, posttraumatic phobia and uncontrolled anxiety, and altered role performance are all manifestations, not causes, of this disorder.
NP: Data collection; CN: Psychosocial integrity; CNS: Psychosocial adaptation; CL: Knowledge

78. 1. By talking about returning to college, the client is demonstrating an interest in making plans for the future, which is a sign of beginning recovery from depression. Decreased socialization, lack of interest in personal appearance, and lack of emotion are all symptoms of depression.
NP: Evaluation; CN: Physiological integrity; CN: Reduction of risk potential; CL: Application

79. 2. There's a direct relationship between the amount of salt and the plasma levels of lithium. Lithium plasma levels increase when there is a decrease in dietary salt. An increase in dietary salt causes the opposite effect of decreasing lithium plasma levels. It's important that the nurse monitors adequate dietary sodium.
NP: Data collection; CN: Physiological integrity; CNS: Pharmacological therapies; CL: Knowledge

80. A client with bipolar disorder has been taking lithium (Lithotabs), as prescribed, for the past 3 years. Today, family members brought this client to the hospital because the client hadn't slept, bathed, or changed clothes for 4 days; had lost 10 lb (4.5 kg) in the past month; and woke the entire family at 4 a.m. with plans to fly them to Hawaii for a vacation. Based on this information, what may the nurse assume?

1. The family isn't supportive of the client.
2. The client had stopped taking the pre-scribed medication.
3. The client hasn't accepted the diagnosis of bipolar disorder.
4. The lithium level should be measured be-fore the client receives the next lithium dose.

81. A client taking antidepressants for major depression for about 3 weeks is expressing that he's feeling better. Which of the following com-plications should he now be assessed for?

1. Manic depression
2. Potential for violence
3. Substance abuse
4. Suicidal ideation

82. A 40-year-old client has been brought to the hospital by her husband because she has refused to get out of bed for 2 days. She won't eat, she's been neglecting household responsi-bilities, and she's tired all the time. Her diagno-sis on admission is major depression. Which of the following questions is most appropriate for the admitting nurse to ask at this point?

1. "What has been troubling you?"
2. "Why do you dislike yourself?"
3. "How do you feel about your life?"
4. "What can we do to help?"

83. A client on the psychiatric unit is receiving lithium therapy and has a lithium level of 1 mEq/L. The nurse notes that the client has fine tremors of the hands. What should the nurse do?

1. Hold the client's next lithium dose.
2. Notify the physician immediately.
3. Have the lithium level repeated.
4. Realize that a fine tremor is expected.

Timing is critical here. Read the question again if you have any doubts about what is being asked.

80. 4. Measuring the lithium level is the best way to evaluate the effectiveness of lithium therapy and begin to assess the client's current status. The client's unsupportive family, stop-ping of the medication, and not accepting the diagnosis may contribute to his manic episode, but the nurse can't assume these to be true until after assessing the client and family more fully.

NP: Data collection; CN: Psychosocial integrity; CNS: Reduc-tion of risk potential; CL: Application

81. 4. After a client has been on antidepres-sants and is feeling better, he often then has the energy to harm himself. Manic depression isn't treated with antidepressants. Nothing in the client's history suggests a potential for vio-lence. There are no signs or symptoms sug-gesting substance abuse.

NP: Data collection; CN: Safe, effective care environment; CNS: Safety and infection control; CL: Analysis

82. 3. The nurse must develop nursing inter-ventions based on the client's perceived prob-lems and feelings. Asking the client to draw a conclusion may be difficult for her at this time. *Why* questions can place the client in a defen-sive position. Requiring the client to find possi-ble solutions is beyond the scope of her pre-sent abilities.

NP: Implementation; CN: Psychosocial integrity; CNS: Coping and adaptation; CL: Analysis

83. 4. Fine tremors of the hands are consid-ered normal with lithium therapy. The lithium level is within normal limits so there's no need to hold a dose, notify the physician, or repeat the blood work.

NP: Implementation; CN: Psychosocial integrity; CNS: Psy-chosocial adaptation; CL: Application

84. A 23-year-old client has been on the psychiatric unit for 2 days. She has a history of bipolar disorder and came to the hospital in the manic phase. She stopped taking lithium (Eskalith) 2 weeks ago. Which of the following findings would the nurse be least likely to see?
1. Flight of ideas
2. Delusions of grandeur
3. Increased appetite
4. Restlessness

85. An acutely manic client kisses the nurse on the lips and asks her to marry him. The nurse is taken by surprise. Which of the following responses would be best?
1. Seclude the client for his inappropriate behavior.
2. Ask the client what he's trying to prove by his behavior.
3. Have the client help her fold some laundry.
4. Tell the client that his behavior is offensive.

86. Which of the following discharge instructions is most important for a client taking lithium (Eskalith)?
1. Limit fluids to 1,500 ml daily.
2. Maintain a high fluid intake.
3. Take advantage of the warm weather by exercising outside whenever possible.
4. When feeling a cold coming, it's okay to take over-the-counter (OTC) cold remedies.

Did you catch the phrase least likely?

84. 3. The manic client is usually unwilling or unable to slow down enough to eat. Flight of ideas, delusions of grandeur, and restlessness are associated with the manic phase.
NP: Data collection; CN: Psychosocial integrity; CNS: Psychosocial adaptation; CL: Application

85. 3. Having the client help with laundry rechannels his energy in a positive activity. The client needs direction and structure, not seclusion. Asking the client what he's trying to prove ignores his impaired judgment and poor impulse control. Telling the client that his behavior is offensive doesn't assist him in controlling his behavior.
NP: Implementation; CN: Psychosocial integrity; CNS: Psychosocial adaptation; CL: Application

86. 2. Clients taking lithium need to maintain a high fluid intake. Fluids shouldn't be limited when taking lithium. Exercising outside may not be safe; photosensitivity occurs with lithium use. In addition, activity in warm weather could increase sodium loss, which would predispose the client to a toxic reaction to lithium. The client shouldn't take OTC drugs without the physician's approval.
NP: Implementation; CN: Physiological integrity; CNS: Pharmacological therapies; CL: Application

Congratulations! You did it! Good job!

NP: Nursing process   CN: Client needs category   CNS: Client needs subcategory   CL: Cognitive level

This chapter covers a host of cognitive disorders. Are your own cognitive powers ready? Ok, let's go!

Chapter 15
Cognitive disorders

1. Which of the following disorders is related to dementia?

1. Alcohol withdrawal
2. Alzheimer's disease
3. Obsessive-compulsive disorder
4. Postpartum depression

1. 2. Dementia occurs in Alzheimer's disease and is generally progressive and deteriorating. The symptoms related to alcohol withdrawal result from alcohol intoxication. Effects of alcohol on the central nervous system include loss of memory, concentration, insight, and motor control. Obsessive-compulsive disorders are recurrent ideas, impulses, thoughts, or patterns of behavior that produce anxiety if resisted. Postpartum depression doesn't lead to dementia.

NP: Data collection; CN: Physiological integrity; CNS: Physiological adaptation; CL: Comprehension

2. A physician diagnoses a client with dementia of the Alzheimer's type. Which of the following statements about possible causes of this disorder is most accurate?

1. Alzheimer's disease is most commonly caused by cerebral abscess.
2. Chronic alcohol abuse plays a significant role in Alzheimer's disease.
3. Multiple small brain infarctions typically lead to Alzheimer's disease.
4. The cause of Alzheimer's disease is currently unknown.

2. 4. Several hypotheses suggest genetic factors, trauma, accumulation of aluminum, alterations in the immune system, or alterations in acetylcholine as contributing to the development of Alzheimer's disease, but the exact cause of Alzheimer's disease is unknown.

NP: Data collection; CN: Health promotion and maintenance; CNS: Prevention and early detection of disease; CL: Application

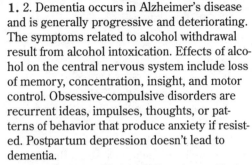

This word — reversible — clarifies the answer.

3. Of the following conditions that can cause a dementia similar to Alzheimer's disease, which is reversible?

1. Multiple sclerosis
2. Electrolyte imbalance
3. Multiple small brain infarctions
4. Human immunodeficiency virus infection

3. 2. Electrolyte imbalance is a correctable metabolic abnormality. The other conditions are irreversible.

NP: Data collection; CN: Physiological integrity; CNS: Physiological adaptation; CL: Application

NP: Nursing process   CN: Client needs category   CNS: Client needs subcategory   CL: Cognitive level

4. In the early stages of Alzheimer's disease, which of the following symptoms is expected?
1. Dilated pupils
2. Rambling speech
3. Elevated blood pressure
4. Significant recent memory impairment

Another hint!

5. Which of the following pathophysiological changes in the brain causes the symptoms of Alzheimer's disease?
1. Glucose inadequacy
2. Atrophy of the frontal lobe
3. Degeneration of the cholinergic system
4. Intracranial bleeding in the limbic system

Data collection is a skill of critical importance!

6. Which of the following data collection findings shows impairment in abstract thinking and reasoning?
1. The client can't repeat a sentence.
2. The client has problems calculating simple problems.
3. The client doesn't know the president of the United States.
4. The client can't find similarities and differences between related words or objects.

7. In addition to disturbances in cognition and orientation, a client with Alzheimer's disease may also show changes in which of the following areas?
1. Appetite
2. Energy levels
3. Hearing
4. Personality

Two more questions and you're at number 10!

8. Which of the following interventions would help a client diagnosed with Alzheimer's disease perform activities of daily living?
1. Have the client perform all basic care without help.
2. Tell the client morning care must be done by 9 a.m.
3. Give the client a written list of activities he's expected to do.
4. Encourage the client, and give ample time to complete basic tasks.

4. 4. Significant recent memory impairment, indicated by the inability to verbalize remembrances after several minutes to an hour, can be assessed in the early stages of Alzheimer's disease. Dilated pupils, increased blood pressure, and rambling speech are expected symptoms of delirium.
NP: Data collection; CN: Physiological integrity; CNS: Physiological adaptation; CL: Application

5. 3. Research related to Alzheimer's disease indicates the enzyme needed to produce acetylcholine is dramatically reduced. The other pathophysiological changes don't cause the symptoms of Alzheimer's disease.
NP: Data collection; CN: Physiological integrity; CNS: Physiological adaptation; CL: Analysis

6. 4. Abstract thinking is assessed by noting similarities and differences between related words or objects. Not knowing the president of the United States is a deficiency in general knowledge. Not being able to do a simple calculation shows a client's inability to concentrate and focus on thoughts.
NP: Data collection; CN: Physiological integrity; CNS: Physiological adaptation; CL: Analysis

7. 4. Personality change is common in dementia. There shouldn't be a remarkable change in appetite, energy level, or hearing.
NP: Data collection; CN: Physiological integrity; CNS: Physiological adaptation; CL: Application

8. 4. Clients with Alzheimer's disease respond to the affect of those around them. A gentle calm approach is comforting and nonthreatening, and a tense, hurried approach may agitate the client. The client has problems performing independently. These expectations may lead to frustration.
NP: Implementation; CN: Safe, effective care environment; CNS: Coordinated care; CL: Application

NP: Nursing process   CN: Client needs category   CNS: Client needs subcategory   CL: Cognitive level

9. Which of the following medications for Alzheimer's disease, approved by the Food and Drug Administration, is a moderately long-acting inhibitor of cholinesterase?
   1. Bupropion (Wellbutrin)
   2. Haloperidol (Haldol)
   3. Tacrine (Cognex)
   4. Triazolam (Halcion)

10. A nurse places an object in the hand of a client with Alzheimer's disease and asks the client to identify the object. Which of the following terms represents the client's inability to name the object?
   1. Agnosia
   2. Aphasia
   3. Apraxia
   4. Perseveration

I know that term is in here somewhere!

11. Which of the following nursing interventions will help a client with progressive memory deficit function in his environment?
   1. Help the client do simple tasks by giving step-by-step directions.
   2. Avoid frustrating the client by performing basic care routines for the client.
   3. Stimulate the client's intellectual functioning by bringing new topics to the client's attention.
   4. Promote the use of the client's sense of humor by telling jokes or riddles and discussing cartoons.

12. Which of the following interventions is an important part of providing care to a client diagnosed with Alzheimer's disease?
   1. Avoid physical contact.
   2. Apply wrist and ankle restraints.
   3. Provide a high level of sensory stimulation.
   4. Monitor the client carefully.

Look for the answer that makes the most safety sense.

9. 3. Tacrine is used to improve cognition and functional autonomy in mild to moderate dementia of the Alzheimer's type. Wellbutrin is used for depression. Haldol is used for agitation, aggression, hallucinations, thought disturbances, and wandering. Halcion is used for sleep disturbances.
NP: Implementation; CN: Physiological integrity; CNS: Pharmacological therapies; CL: Knowledge

10. 1. Agnosia is the inability to recognize familiar objects. Aphasia is characterized by an impaired ability to speak. Apraxia refers to the client's inability to use objects properly. All three impairments usually occur in stage 3 of Alzheimer's disease. Perseveration is continued repetition of a meaningless word or phrase that occurs in stage 2 of Alzheimer's disease.
NP: Data collection; CN: Health promotion and maintenance; CNS: Prevention and early detection of disease; CL: Application

11. 1. Clients with cognitive impairment should do all the tasks they can. By giving simple directions in a step-by-step fashion, the client can better process information and perform tasks. Clients with cognitive impairment may not be able to understand the joke or riddle, and cartoons may add to their confusion. Stimulation of intellect can be accomplished by discussing familiar topics with them; changes in topics may add to their confusion.
NP: Implementation; CN: Psychosocial integrity; CNS: Psychosocial adaptation; CL: Application

12. 4. Whenever client safety is at risk, careful observation and supervision are of ultimate importance in avoiding injury. Applying restraints may cause agitation and combativeness. Physical contact is implemented during basic care. A high level of sensory stimulation may be too stimulating and distracting.
NP: Implementation; CN: Safe, effective care environment; CNS: Coordinated care; CL: Application

13. Which of the following nursing interventions is the most important in caring for a client diagnosed with Alzheimer's disease?
  1. Make sure the environment is safe to prevent injury.
  2. Make sure the client receives food she likes to prevent hunger.
  3. Make sure the client meets other clients to prevent social isolation.
  4. Make sure the client takes care of her daily physical care to prevent dependence.

14. Which of the following medications is used to decrease the agitation, violence, and bizarre thoughts associated with dementia?
  1. Diazepam (Valium)
  2. Ergoloid (Hydergine)
  3. Haloperidol (Haldol)
  4. Tacrine (Cognex)

15. A client diagnosed with Alzheimer's disease tells the nurse that today she has a luncheon date with her daughter, who isn't visiting that day. Which of the following responses by the nurse would be most appropriate for this situation?
  1. "Where are you planning on having your lunch?"
  2. "You're confused and don't know what you're saying."
  3. "I think you need some more medication, and I'll bring it to you."
  4. "Today is Monday, March 8, and we'll be eating lunch in the dining room."

16. Which of the following features is characteristic of cognitive disorders?
  1. Catatonia
  2. Depression
  3. Feeling of dread
  4. Deficit in cognition or memory

Question 13 asks you to prioritize.

The phrases underlined in question 15 add key points in your evaluation.

13. 1. Providing client safety is the number one priority when caring for any client but particularly when a client is already compromised and at greater risk for injury.
NP: Implementation; CN: Safe, effective care environment; CNS: Safety and infection control; CL: Application

14. 3. Haloperidol is an antipsychotic that decreases the symptoms of agitation, violence, and bizarre thoughts. Diazepam is used for anxiety and muscle relaxation. Ergoloid is an adrenergic blocker used to block vascular headaches. Tacrine is used for improvement of cognition.
NP: Implementation; CN: Physiological integrity; CNS: Reduction of risk potential; CL: Application

15. 4. The best nursing response is to reorient the client to the date and environment. Medication won't provide immediate relief for memory impairment. Confrontation can provoke an outburst.
NP: Implementation; CN: Psychosocial integrity; CNS: Psychosocial adaptation; CL: Application

16. 4. Cognitive disorders represent a significant change in cognition or memory from a previous level of functioning. Catatonia is a type of schizophrenia characterized by periods of physical rigidity, negativism, excitement, and stupor. Depression is a feeling of sadness and apathy and is part of major depressive and other mood disorders. A feeling of dread is characteristic of an anxiety disorder.
NP: Data collection; CN: Physiological integrity; CNS: Physiological adaptation; CL: Application

NP: Nursing process   CN: Client needs category   CNS: Client needs subcategory   CL: Cognitive level

17. Which of the following disorders is a degenerative disorder of cognition primarily associated with age or metabolic deterioration?
1. Delirium
2. Dementia
3. Neurosis
4. Psychosis

Another hint. Only one of the disorders is degenerative.

18. Which of the following characteristics best defines dementia?
1. Personal neglect in self-care
2. Poor judgment, especially in social situations
3. Memory loss occurring as a natural consequence of aging
4. Loss of intellectual abilities sufficient to impair the ability to perform basic care

19. Asking a client with a suspected dementia disorder to recall what she ate for breakfast would assess which of the following areas?
1. Food preferences
2. Recent memory
3. Remote memory
4. Speech

Great job!

20. It's important to collect data for a definitive diagnosis of a dementia disorder to determine which of the following factors?
1. Prognosis
2. Genetic information
3. Degree of impairment
4. Implications for treatment

21. Which of the following factors is considered a cause of vascular dementia?
1. Head trauma
2. Genetic factors
3. Acetylcholine alteration
4. Interruption of blood flow to the brain

17. 2. Dementia is progressive and often associated with aging or underlying metabolic or organic deterioration. Delirium is characterized by abrupt, spontaneous cognitive dysfunction with an underling organic mental disorder. Neurosis and psychosis are psychological diagnoses.
NP: Data collection; CN: Physiological integrity; CNS: Physiological adaptation; CL: Knowledge

18. 4. The ability to perform self-care is an important measure of the progression of dementia. Memory loss reflects underlying physical, metabolic, and pathologic processes. Personal neglect and poor judgment typically occur in dementia but aren't considered defining characteristics.
NP: Data collection; CN: Physiological integrity; CNS: Physiological adaptation; CL: Knowledge

19. 2. Persons with dementia have difficulty in recent memory or learning, which may be a key to early detection. Assessing food preferences may be helpful in determining what the client likes to eat, but this assessment has no direct correlation in assessing dementia. Speech difficulties, such as rambling, irrelevance, and incoherence, may be related to delirium.
NP: Implementation; CN: Health promotion and maintenance; CNS: Prevention and early detection of disease; CL: Application

20. 4. The reversibility of a dementia is a function of the underlying pathologic states, so it's important to collect data and treat the underlying cause. The degree of impairment is necessary information for developing a care plan. Genetic information isn't relevant. Prognosis isn't the most important factor when making a diagnosis.
NP: Data collection; CN: Health promotion and maintenance; CNS: Prevention and early detection of disease; CL: Comprehension

21. 4. The cause of vascular dementia is directly related to an interruption of blood flow to the brain. Head trauma, genetic factors, and acetylcholine alteration are causative factors related to dementia of the Alzheimer's type.
NP: Data collection; CN: Physiological integrity; CNS: Physiological adaptation; CL: Comprehension

22. Which of the following factors is considered the <u>most significant</u> cause of vascular dementia?
1. Arterial hypertension
2. Hypoxia
3. Infection
4. Toxins

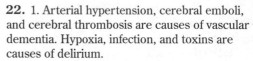

This question asks you to select the most significant cause.

23. Which of the following assessment findings is expected for a client with vascular dementia?
1. Hypersomnolence
2. Insomnia
3. Restlessness
4. Small-stepped gait

24. In which of the following ways is vascular dementia different from Alzheimer's disease?
1. Vascular dementia has a more abrupt onset.
2. The duration of vascular dementia is usually brief.
3. Personality change is common in vascular dementia.
4. The inability to perform motor activities is common in vascular dementia.

25. A progression of symptoms that occurs in steps rather than a gradual deterioration indicates which type of dementia?
1. Alzheimer's dementia
2. Parkinson's dementia
3. Substance-induced dementia
4. Vascular dementia

Hey, you're half finished. Super! Keep going!

22. 1. Arterial hypertension, cerebral emboli, and cerebral thrombosis are causes of vascular dementia. Hypoxia, infection, and toxins are causes of delirium.
NP: Data collection; CN: Physiological integrity; CNS: Reduction of risk potential; CL: Knowledge

23. 4. Focal neurologic signs commonly seen with vascular dementia include weakness of the limbs, small-stepped gait, and difficulty with speech. Insomnia, hypersomnolence, and restlessness are symptoms related to delirium.
NP: Data collection; CN: Physiological integrity; CNS: Physiological adaptation; CL: Application

24. 1. Vascular dementia differs from Alzheimer's disease in that it has a more abrupt onset and runs a highly variable course. Personality change is common in dementia. The duration of delirium is usually brief. The inability to carry out motor activities is common in dementia.
NP: Data collection; CN: Health promotion and maintenance; CNS: Prevention and early detection of disease; CL: Analysis

25. 4. Vascular dementia differs from Alzheimer's disease in that vascular dementia has a more abrupt onset and progresses in steps. At times the dementia seems to clear up, and the individual shows fairly lucid thinking. Dementia of the Alzheimer's type has a slow onset with a progressive and deteriorating course. Dementia of Parkinson's sometimes resembles the dementia of Alzheimer's disease. Substance-induced dementia is related to the persisting effects of use of a substance.
NP: Data collection; CN: Physiological integrity; CNS: Physiological adaptation; CL: Analysis

26. Which of the following disorders is characterized by a general impairment in intellectual functioning?
    1. Alzheimer's disease
    2. Amnesia
    3. Delirium
    4. Dementia

26. 4. Dementia is characterized by general impairment in intellectual functioning and occurs in a progressive, irreversible course. One of the distinguishing characteristics of delirium is the development of symptoms over a short time, usually hours to days. Alzheimer's disease is a progressive dementia in which all known reversible causes are eliminated. Amnesia refers to recent short-term and long-term memory loss.
NP: Data collection; CN: Physiological integrity; CNS: Physiological adaptation; CL: Knowledge

27. An elderly client has experienced memory and attention deficits that developed over a 3-day period. These symptoms are characteristic of which of the following disorders?
    1. Alzheimer's disease
    2. Amnesia
    3. Delirium
    4. Dementia

27. 3. Delirium is characterized by an abrupt onset of fluctuating levels of awareness, clouded consciousness, perceptual disturbances, and disturbed memory and orientation. Alzheimer's disease is a progressive dementia in which all known reversible causes are eliminated. Amnesia refers to recent short-term and long-term memory loss. Dementia is characterized by general impairment in intellectual functioning and occurs in a progressive, irreversible course.
NP: Data collection; CN: Physiological integrity; CNS: Physiological adaptation; CL: Application

28. Which of the following age groups is at high risk for developing a state of delirium?
    1. Adolescent
    2. Elderly
    3. Middle-aged
    4. School-aged

Use the hint *high risk* to help you choose the answer.

28. 2. The elderly population, because of normal physiological changes, is highly susceptible to delirium. Adolescent, middle-aged, and school-aged groups are incorrect.
NP: Data collection; CN: Health promotion and maintenance; CNS: Prevention and early detection of disease; CL: Application

29. Which of the following nursing diagnoses is best for an elderly client experiencing visual and auditory hallucinations?
    1. Interrupted family processes
    2. Ineffective role performance
    3. Impaired verbal communication
    4. Disturbed sensory perception

29. 4. The client is experiencing visual and auditory hallucinations related to a sensory alteration. Interrupted family processes, ineffective role performance, and impaired verbal communication don't address the hallucinations the client is experiencing.
NP: Planning; CN: Health promotion and maintenance; CNS: Prevention and early detection of disease; CL: Application

30. Which of the following interventions is the most appropriate for clients with cognitive disorders?
1. Promote socialization.
2. Maintain optimal physical health.
3. Provide frequent changes in personnel.
4. Provide an overstimulating environment.

Which is the best response?

31. A newly admitted client diagnosed with delirium has a history of hypertension and anxiety. The client had been taking digoxin, furosemide (Lasix), and diazepam (Valium) for anxiety. This client's impairment may be related to which of the following conditions?
1. Infection
2. Metabolic acidosis
3. Drug intoxication
4. Hepatic encephalopathy

32. Which of the following environments is the most appropriate for a client experiencing sensory-perceptual alterations?
1. A softly lit room around the clock
2. A brightly lit room around the clock
3. Sitting by the nurses' desk while out of bed
4. A quiet, well-lit room without glare during the day and a darkened room for sleeping

Your clients depend on you for meeting their environmental needs as well as their physical ones.

33. As a nurse enters a client's room, the client says, "They're crawling on my sheets! Get them off my bed!" Which of the following assessments is the most accurate?
1. The client is experiencing aphasia.
2. The client is experiencing dysarthria.
3. The client is experiencing a flight of ideas.
4. The client is experiencing visual hallucinations.

30. 2. A client's cognitive impairment may hinder self-care abilities. An overstimulating environment, frequent changes in staff members, and more socialization would only increase anxiety and confusion.
NP: Implementation; CN: Health promotion and maintenance; CNS: Prevention and early detection of disease; CL: Application

31. 3. This client was taking several medications that have a propensity for producing delirium: digoxin (a cardiac glycoside), furosemide (a thiazide diuretic), and diazepam (a benzodiazepine). Sufficient supporting data don't exist to suspect infection, metabolic acidosis, or hepatic encephalopathy as causes.
NP: Data collection; CN: Physiological integrity; CNS: Physiological adaptation; CL: Analysis

32. 4. A quiet, shadow-free environment produces the fewest sensory-perceptual distortions for a client with cognitive impairment associated with delirium.
NP: Implementation; CN: Safe, effective care environment; CNS: Safety and infection control; CL: Application

33. 4. The presence of a sensory stimulus correlates with the definition of a hallucination, which is a false sensory perception. Aphasia refers to a communications problem. Dysarthria is difficulty in speech production. Flight of ideas is rapid shifting from one topic to another.
NP: Data collection; CN: Health promotion and maintenance; CNS: Prevention and early detection of disease; CL: Application

34. A delirious client is shouting for someone to get the bugs off her. Which of the following responses is the most appropriate?
 1. "Don't worry, I'll stay here and brush away the bugs for you."
 2. "Try to relax. The crawling sensation will go away sooner if you can relax."
 3. "There are no bugs on your legs. It's just your imagination playing tricks on you."
 4. "I know you're frightened. I don't see bugs crawling on your legs, but I'll stay here with you."

34. 4. Never argue about hallucinations with a client. Instead, promote an environment of trust and safety by acknowledging the client's perceptions.
NP: Implementation; CN: Physiological integrity; CNS: Basic care and comfort; CL: Application

Keep up the super work!

35. Which of the following descriptions of a client's experience and behavior can be assessed as an illusion?
 1. The client tries to hit the nurse when vital signs must be taken.
 2. The client says, "I keep hearing a voice telling me to run away."
 3. The client becomes anxious whenever the nurse leaves the bedside.
 4. The client looks at the shadows on a wall and tells the nurse she sees frightening faces on the wall.

35. 4. An illusion is an inaccurate perception or false response to a sensory stimulus. Auditory hallucinations are associated with sound and are more common in schizophrenia. Anxiety and agitation can be secondary to illusions.
NP: Data collection; CN: Physiological integrity; CNS: Physiological adaptation; CL: Analysis

36. Which of the following neurologic changes is an <u>expected</u> characteristic of aging?
 1. Widening of the sulci
 2. Depletion of neurotransmitters
 3. Neurofibrillary tangles and plaques
 4. Degeneration of the frontal and temporal lobes

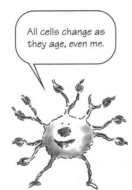
All cells change as they age, even me.

36. 3. Aging isn't necessarily associated with significant decline, but neurofibrillary tangles and plaques are expected changes. These normal occurrences are sometimes referred to as benign senescent forgetfulness of age-associated memory impairment.
NP: Data collection; CN: Physiological integrity; CNS: Physiological adaptation; CL: Analysis

37. A major consideration in assessing memory impairment in an elderly individual includes which of the following factors?
 1. Allergies
 2. Past surgery
 3. Age at onset of symptoms
 4. Social and occupational lifestyle

37. 4. Minor memory problems are distinguished from dementia by their minor severity and their lack of significant interference with the client's social or occupational lifestyle. Allergies, past surgery, or age at onset of symptoms would be included in the history data but don't directly correlate with the client's memory impairment.
NP: Data collection; CN: Psychosocial integrity; CNS: Coping and adaptation; CL: Analysis

38. The nurse is caring for an 88-year-old female client in a nursing home. The client is confused and thinks she's in a train station. Which response to this behavior is the nurse's top priority?
  1. Correct errors in the client's perception of reality in a matter-of-fact manner.
  2. Have a conversation with the client when this behavior occurs.
  3. Observe the client and know her whereabouts at all times.
  4. Refer to the date, time and place during interactions with client.

39. A nursing assistant tells a nurse, "The client with amnesia looks fine but responds to questions in a vague, distant manner. What should I be doing for her?" Which of the following responses is the most appropriate?
  1. "Give her lots of space to test her independence."
  2. "Keep her busy and make sure she doesn't take naps during the day."
  3. "Whenever you think she needs direction, use short simple sentences."
  4. "Spend as much time with her as you can, and ask questions about her recent life."

40. With amnesic disorders, in which of the following cognitive areas is a change expected?
  1. Speech
  2. Concentration
  3. Intellectual function
  4. Recent short-term and long-term memory

41. Which of the following nursing actions is the best way to help a client with mild Alzheimer's disease to remain functional?
  1. Obtain a physician's order for a mild anxiolytic to control behavior.
  2. Call attention to all mistakes so they can be quickly corrected.
  3. Advise the client to move into a retirement center.
  4. Maintain a stable, predictable environment and daily routine.

Teaching nursing assistants helps to improve the quality of care they provide.

Question 41 asks you to focus on the best and ignore the rest!

38. 3. The nurse's top priority is to maintain safety and security for the client by observing her and knowing her whereabouts at all times, otherwise, the client may wander off and endanger herself. Correcting errors in the client's perception, conversing with her, and orienting her to date, time, and place are important but don't have the highest priority in this case.
NP: Implementation; CN: Psychosocial integrity; CNS: Psychosocial adaptation; CL: Application

39. 3. Disruptions in the ability to perform basic care, confusion, and anxiety are often apparent in clients with amnesia. Offering simple directions to promote daily functions and reduce confusion helps increase feelings of safety and security. Giving this client lots of space may make her feel insecure. Asking her many questions that she won't be able to answer just intensifies her anxiety level. There is no significant rationale for keeping her busy all day with no rest periods; the client may become more tired and less functional at other basic tasks.
NP: Implementation; CN: Safe, effective care environment; CNS: Coordinated care; CL: Application

40. 4. The primary area affected in amnesia is memory; all other areas of cognition are normal.
NP: Data collection; CN: Physiological integrity; CNS: Physiological adaptation; CL: Application

41. 4. Clients in the early stages of Alzheimer's disease remain fairly functional with familiar surroundings and a predictable routine. They become easily disoriented with surprises and social overstimulation. Anxiolytics can impair memory and worsen the problem. Calling attention to all the client's mistakes is nonproductive and serves to lower the client's self-esteem. Moving to an unfamiliar environment will heighten the client's agitation and confusion.
NP: Planning; CN: Psychosocial integrity; CNS: Psychosocial adaptation; CL: Application

42. The nurse is providing nursing care to a client with Alzheimer-type dementia. Which nursing intervention takes top priority?
1. Establish a routine that supports former habits.
2. Maintain physical surroundings that are cheerful and pleasant.
3. Maintain an exact routine from day to day.
4. Control the environment by providing structure, boundaries, and safety.

43. The nurse finds a 78-year-old client with Alzheimer's type dementia wandering in the hall at 3 a.m. The client has removed his clothing and says to the nurse, "I'm just taking a stroll through the park." What's the best approach to this behavior?
1. Immediately help the client back to his room and into some clothing.
2. Tell the client that such behavior won't be tolerated.
3. Tell the client it's too early in the morning to be taking a stroll.
4. Ask the client if he would like to go back to his room.

44. Which of the following nursing diagnoses is appropriate for a client diagnosed with an amnesic disorder?
1. Anticipatory grieving related to loss of functional ability
2. Ineffective denial
3. Ineffective coping
4. Risk for injury related to impaired cognition

45. The nurse is planning care for a client who was admitted with dementia due to Alzheimer's disease. The family reports that the client has to be watched closely for wandering behavior at night. Which nursing diagnosis is the nurse's top priority for this client?
1. Disturbed sleep pattern
2. Activity intolerance
3. Disturbed sensory perception
4. Risk for injury

You're at question 42 already? You're the tops! Get it? That's meant as both a boost to your self-esteem and a clue!

Five more questions! Now's the time to kick into high gear!

42. 4. By controlling the environment and providing structure and boundaries, the nurse is helping to keep the client safe and secure, which is a top-priority nursing measure. Establishing a routine that supports former habits and maintaining cheerful and pleasant surroundings and an exact routine foster a supportive environment but keeping the client safe and secure takes priority.
NP: Implementation; CN: Psychosocial integrity; CNS: Psychosocial adaptation; CL: Application

43. 1. The nurse shouldn't allow the client to embarrass himself in front of others. Intervene as soon as the behavior is observed. Scolding the client isn't helpful because it isn't something the client can understand. Don't engage in social chatter. The interaction with this client should be concrete and specific. Don't ask the client to choose unnecessarily. The client may not be able to make appropriate choices.
NP: Implementation; CN: Psychosocial integrity; CNS: Psychosocial adaptation; CL: Application

44. 4. Changes in cognitive ability place a client at high risk for injury. The client isn't aware of a loss and therefore doesn't grieve for it or experience ineffective denial. The client isn't aware of a need to cope.
NP: Data collection; CN: Safe, effective care environment; CNS: Safety and infection control; CL: Analysis

45. 4. Providing a safe, effective care environment takes priority in this case. Disturbed sleep pattern, activity intolerance, and disturbed sensory perception fall under physical integrity and, although important, they aren't as important as providing a safe, effective care environment.
NP: Planning; CN: Safe, effective care environment; CNS: Safety and infection control; CL: Application

46. Which of the following medical conditions may be associated with an amnesic disorder?
1. Drug overdose
2. Cerebral anoxia
3. Medications (anticonvulsants)
4. Lead, mercury, and carbon dioxide toxins

46. 2. A variety of medical conditions are related to amnesic disorders, such as head trauma, stroke, cerebral neoplastic disease, herpes simplex, encephalitis, poorly controlled insulin-dependent diabetes, and cerebral anoxia. Drug overdose, medications, and toxins are substance induced.
NP: Data collection; CN: Physiological integrity; CNS: Physiological adaptation; CL: Application

47. Which of the following laboratory evaluations is an expected part of the initial workup for an amnesic disorder?
1. Angiography
2. Cardiac catheterization
3. Electrocardiography
4. Metabolic and endocrine tests

Check out the hints!

47. 4. An amnesic disorder is characterized by impairment in memory that's either due to direct physiological effects of a medical condition or effects of a substance, medication, or toxin, which is why metabolic and endocrine tests should be done. Angiography, cardiac catheterization, and electrocardiography are diagnostic tests related to the cardiovascular system.
NP: Data collection; CN: Health promotion and maintenance; CNS: Prevention and early detection of disease; CL: Application

48. Which of the following nursing interventions is appropriate for a client with memory impairment?
1. Speak to the client in a high-pitched tone.
2. Offer low-dose sedative/hypnotic medications.
3. Ask a series of questions when obtaining information.
4. Identify yourself and look directly into the client's eyes.

48. 4. Clients with memory impairments need to have the nurse's identification constantly reestablished. Asking a series of questions may create confusion; ask only one question at a time. A low-dose sedative/hypnotic medication may be used to promote rapid-eye-movement sleep. High-pitched tones may create more anxiety.
NP: Implementation; CN: Safe, effective care environment; CNS: Coordinated care; CL: Application

49. Transient global amnesia is generally associated with which of the following disorders?
1. Adjustment disorders
2. Cardiac anomalies
3. Cerebrovascular disease
4. Sleep disorders

Success! My, how far you've come in such a short time. Stupendous!

49. 3. Transient global amnesia is usually associated with cerebrovascular disease that involves transient impairment in blood flow through the vertebrobasilar arteries. No correlation exists with adjustment or sleep disorders or cardiac anomalies.
NP: Data collection; CN: Physiological integrity; CNS: Physiological adaptation; CL: Analysis

50. During conversation with a client, the nurse observes that he shifts from one topic to the next on a regular basis. Which of the following terms describes this disorder?
1. Flight of ideas
2 Concrete thinking
3 Ideas of reference
4. Loose associations

50. 4. Loose associations are by definition conversations that reflect a constant shift in topic. Concrete thinking implies highly definitive thought processes. A concrete thinker doesn't typically shift topics in conversation. Flight of ideas is characterized by the onset of the conversation being disorganized. Loose associations don't necessarily start in a disorganized way; the conversations can begin cogently and then become loose.
NP: Data collection; CN: Psychosocial integrity; CNS: Psychosocial adaptation; CL: Knowledge

NP: Nursing process   CN: Client needs category   CNS: Client needs subcategory   CL: Cognitive level

No, this chapter doesn't cover quirks of the rich and famous. It's all about mental disorders affecting the personality. Have a blast!

Chapter 16
Personality disorders

1. A client tells the nurse that her coworkers are sabotaging her computer. When the nurse asks questions, the client becomes argumentative. This behavior shows personality traits associated with which of the following personality disorders?
   1. Antisocial
   2. Histrionic
   3. Paranoid
   4. Schizotypal

Lots of questions to go, but you're good! Get to it!

1. 3. Because of their suspiciousness, paranoid personalities ascribe malevolent activities to others and tend to be defensive, becoming quarrelsome and argumentative. Clients with antisocial personality disorder can also be antagonistic and argumentative, but are less suspicious than paranoid personalities. Clients with a histrionic personality disorder are dramatic, not suspicious and argumentative. Clients with schizoid personality disorder are usually detached from others and tend to have eccentric behavior.

NP: Data collection; CN: Psychosocial integrity; CNS: Coping and adaptation; CL: Comprehension

2. A nurse notices a client is mistrustful and shows hostile behavior. Which of the following types of personality disorder is associated with these characteristics?
   1. Antisocial
   2. Avoidant
   3. Borderline
   4. Paranoid

2. 4. Paranoid individuals have a need to constantly scan the environment for signs of betrayal, deception, and ridicule, appearing mistrustful and hostile. They expect to be tricked or deceived by others. The extreme suspiciousness is lacking in antisocial personalities, who tend to be more arrogant and self-assured despite their vigilance and mistrust. Individuals with avoidant personality disorders are guarded, fearing interpersonal rejection and humiliation. Clients with borderline personality disorders behave impulsively and tend to manipulate others.

NP: Data collection; CN: Psychosocial integrity; CNS: Coping and adaptation; CL: Knowledge

Understanding a client's traits will help you deal with him effectively.

3. Which of the following traits is expected from a client with paranoid personality disorder?
   1. The client can't follow limits set on behavior.
   2. The client is afraid another person will inflict harm.
   3. The client avoids responsibility for health care actions.
   4. The client depends on others to make important decisions.

3. 2. A client with paranoid personality disorder is afraid others will inflict harm. An individual with antisocial personality disorder won't be able to follow the limits set on behavior. An individual with an avoidant personality might avoid responsibility for health care because he tends to scan the environment for threatening things. A client with dependent personality disorder is likely to want others to make important decisions for him.

NP: Data collection; CN: Psychosocial integrity; CNS: Coping and adaptation; CL: Comprehension

NP: Nursing process   CN: Client needs category   CNS: Client needs subcategory   CL: Cognitive level

4. Which of the following statements is typical of a client diagnosed with paranoid personality disorder?
    1. "I understand you're to blame."
    2. "I must be seen first; it's not negotiable."
    3. "I see nothing humorous in this situation."
    4. "I wish someone would select the outfit for me."

5. The nurse is caring for a 33-year-old male client diagnosed with borderline personality disorder. The nurse tells the client that they'll be meeting for 1 hour every week on Monday at 1 p.m. Which of the following statements best describes the rationale for setting limits for a client with personality disorder?
    1. It helps the client clarify limits.
    2. It encourages the client to be manipulative.
    3. It provides the nurse with leverage against unacceptable behavior.
    4. It provides an opportunity for the client to assess the situation.

You're already finished 5 questions! Keep going!

6. A nurse suspects a client has paranoid personality disorder. Which of the following findings confirms the nurse's suspicion?
    1. Exhibitionism
    2. Impulsiveness
    3. Secretiveness
    4. Self-destructiveness

7. Which of the following types of behavior is expected from a client diagnosed with paranoid personality disorder?
    1. Eccentric
    2. Exploitative
    3. Hypersensitive
    4. Seductive

4. 3. Clients with paranoid personality disorder tend to be extremely serious and lack a sense of humor. Clients with borderline personality disorder tend to blame others for their problems. Clients with narcissistic personality disorders have a sense of self-importance and entitlement. Clients with dependent personality disorder want others to make their decisions.
NP: Data collection; CN: Psychosocial integrity; CNS: Coping and adaptation; CL: Analysis

5. 1. Clarifying limits and making clear what may be unclear to the client helps the client establish boundaries himself. This fosters a therapeutic, trusting relationship between the nurse and the client. The nurse should never encourage manipulation. The nurse should never attempt to gather leverage against the client; that would be unprofessional. The client needs to understand his behavior patterns before he can start assessing the situation.
NP: Implementation; CN: Psychosocial integrity; CNS: Psychosocial adaptation; CL: Application

6. 3. Clients with paranoid personality disorder tend to be secretive. Clients with histrionic personality disorder tend to be exhibitionists, and those with borderline personality disorder tend to be impulsive and self-destructive.
NP: Data collection; CN: Psychosocial integrity; CNS: Coping and adaptation; CL: Comprehension

7. 3. People with paranoid personality disorders are hypersensitive to perceived threats. Schizotypal personalities appear eccentric and engage in activities others find perplexing. Clients with narcissistic personality disorder are interpersonally exploitative to enhance themselves or indulge their own desires. A client with histrionic personality disorder can be extremely seductive when in search of stimulation and approval.
NP: Data collection; CN: Psychosocial integrity; CNS: Coping and adaptation; CL: Analysis

8. A client with paranoid personality disorder is discussing current problems with a nurse. Which of the following nursing interventions has priority in the care plan?
1. Have the client look at sources of frustration.
2. Have the client focus on ways to interact with others.
3. Have the client discuss the use of defense mechanisms.
4. Have the client clarify thoughts and beliefs about an event.

Prioritize!

9. A client with a paranoid personality disorder makes an inappropriate and unreasonable report to a nurse. Which of the following principles of good communication skills is important to use?
1. Use logic to address the client's concern.
2. Confront the client about the stated misperception.
3. Use nonverbal communication to address the issue.
4. Tell the client matter-of-factly that you don't share his interpretation.

Identify and understand the key words before answering this question!

10. Which of the following short-term goals is most appropriate for the client with paranoid personality disorder who has impaired social skills?
1. Obtain feedback from other people.
2. Discuss anxiety-provoking situations.
3. Address positive and negative feelings about self.
4. Identify personal feelings that hinder social interaction.

8. 4. Clarifying thoughts and beliefs helps the client avoid misinterpretations. Clients with a paranoid personality disorder tend to mistrust people and don't see interacting with others as a way to handle problems. They tend to be aggressive and argumentative rather than frustrated. The client's priority must be to interpret his thoughts and beliefs realistically, rather than discuss defensive mechanisms. A paranoid client will focus on defending himself rather than acknowledging the use of defense mechanisms.
NP: Planning; CN: Psychosocial integrity; CNS: Psychosocial adaptation; CL: Analysis

9. 4. Telling the client you don't share his interpretation helps the client differentiate between realistic and emotional thoughts and conclusions. When the nurse uses logic to respond to a client's inappropriate statement, the nurse risks creating a power struggle with the client. The use of nonverbal communication will probably be misinterpreted and arouse the client's suspicion. It's unwise to confront a client with a paranoid personality disorder as the client will immediately become defensive.
NP: Implementation; CN: Psychosocial integrity; CNS: Psychosocial adaptation; CL: Analysis

10. 4. The client must address the feelings that impede social interactions before developing ways to address impaired social skills. Feedback can only be obtained after action is taken to improve or change the situation. Discussion of anxiety-provoking situations is important but doesn't help the client with impaired social skills. Addressing the client's positive and negative feelings about himself won't directly influence impaired social skills.
NP: Planning; CN: Psychosocial integrity; CNS: Psychosocial adaptation; CL: Application

11. The nurse is caring for a client diagnosed with histrionic personality disorder. The client is observed tearing pages out of the books in the unit library and putting them into the ventilation system. Which of the following is the best initial nursing intervention?
1. Place the client in a safe, secluded environment.
2. Help the client develop more acceptable methods of seeking attention.
3. Withdraw attention from the client at this time.
4. Identify inappropriate behaviors to the client in a matter-of-fact manner.

11. 1. If the client begins destroying property or presenting potential harm to himself or others, it may be necessary to immediately place him in a safe, secluded environment. When the client regains control and ceases the behavior, then attempt to talk to him to explore more acceptable ways of handling frustration and expressing feelings. Lack of attention from the nurse wouldn't reduce the client's attention-seeking behaviors. When the client regains control and ceases the behavior, the nurse must make it clear which behaviors are inappropriate.
NP: Implementation; CN: Psychosocial integrity; CNS: Psychosocial adaptation; CL: Analysis

12. Which of the following approaches should be used with a client with paranoid personality disorder who misinterprets many things the health care team says?
1. Limit interaction to activities of daily living.
2. Address only problems and causes of distress.
3. Explore anxious situations and offer reassurance.
4. Speak in simple messages without details.

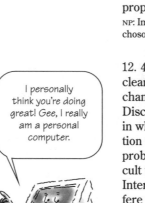

I personally think you're doing great! Gee, I really am a personal computer.

12. 4. If the nurse speaks to the client with clear and simple messages, there's less of a chance that information will be misinterpreted. Discussing complex topics creates a situation in which the client will have additional information to misinterpret. If the nurse addresses only problems and specific stressors, it will be difficult to establish a trusting relationship. Interaction can't be limited because it will interfere with working on identified treatment goals.
NP: Implementation; CN: Psychosocial integrity; CNS: Psychosocial adaptation; CL: Application

13. A client with paranoid personality disorder responds aggressively during a psychoeducational group to something another client said about him. Which of the following explanations is the most likely?
1. The client doesn't want to participate in the group.
2. The client took the statement as a personal criticism.
3. The client is impulsive and was acting out frustrations.
4. The client was attempting to handle emotional distress.

13. 2. Clients with paranoid personality disorder tend to be hypersensitive and take what other people say as a personal attack on their character. The client is driven by the suspicion that others will inflict harm. Group participation would be minimal because the client is directing energy toward emotional self-protection. Clients with a paranoid personality disorder tend to be rigid and guarded rather than expressive and acting out. The client with a paranoid personality disorder is acting to defend himself, not handle emotional distress.
NP: Data collection; CN: Psychosocial integrity; CNS: Psychosocial adaptation; CL: Analysis

14. A client with a paranoid personality disorder tells a nurse of his decision to stop talking to his wife. Which of the following areas should be assessed?
1. The client's doubts about the partner's loyalty
2. The client's need to be alone and have time for self
3. The client's decision to separate from the marital partner
4. The client's fears about becoming too much like the partner

15. Which of the following characteristics of a client with a paranoid personality disorder makes it difficult for a nurse to establish an interpersonal relationship?
1. Dysphoria
2. Hypervigilance
3. Indifference
4. Promiscuity

16. The wife of a client diagnosed with paranoid personality disorder tells the client she wants a divorce. When discussing this situation with the couple, which of the following factors would help the nurse form a care plan for this couple?
1. Denied grief
2. Intense jealousy
3. Exploitation of others
4. Self-destructive tendencies

17. A client with a paranoid personality disorder tells a nurse that another nurse is out to get him. Which of the following actions by the nurse may cause this paranoid client distress?
1. Giving as-needed medication to another client
2. Taking the clients outside the unit for exercise
3. Checking vital signs of each person on the unit
4. Talking to another client in the corner of the lounge

Try to establish a therapeutic relationship with the client.

Keep going!

14. 1. Clients with paranoid personality disorder are preoccupied with the loyalty or trustworthiness of people, especially family and friends. People commonly withdraw from a client with paranoid personality disorder due to the difficulty in maintaining a healthy relationship. The client's need to be alone and have time for self isn't related to the decision to stop talking to a partner. These clients focus on the belief that others will harm them, not that they may become like a marital partner.
NP: Implementation; CN: Psychosocial integrity; CNS: Coping and adaptation; CL: Application

15. 2. Clients with paranoid personality disorder think others will harm, deceive, or exploit them in some way, and they're often guarded and ready to defend themselves from actual or perceived attacks. They don't tend to be dysphoric, indifferent, or promiscuous.
NP: Planning; CN: Psychosocial integrity; CNS: Coping and adaptation; CL: Knowledge

16. 2. Clients with paranoid personality disorder are often extremely suspicious and jealous and make frequent accusations of partners and family members. Clients with paranoid personality disorder don't tend to struggle with the denial of grief. Clients with narcissistic personality disorder tend to exploit other people. Clients with borderline personality disorder have self-destructive tendencies.
NP: Planning; CN: Psychosocial integrity; CNS: Psychosocial adaptation; CL: Comprehension

17. 4. Clients with paranoid personality disorder tend to interpret any discussion that doesn't include them as evidence of a plot against them. Giving medication to another client wouldn't alarm the client. Checking vital signs on each client on the unit or taking the clients outside for exercise probably wouldn't be seen as a threat to the client's well-being.
NP: Implementation; CN: Psychosocial integrity; CNS: Psychosocial adaptation; CL: Application

18. Which of the following statements made by a client with paranoid personality disorder shows teaching about social relationships is effective?

1. "As long as I live, I won't abide by social rules."
2. "Sometimes I can see what causes relationship problems."
3. "I'll find out what problems others have so I won't repeat them."
4. "I don't have problems in social relationships; I never really did."

19. Which of the following long-term goals is appropriate for a client with paranoid personality disorder who's trying to improve peer relationships?

1. The client will verbalize a realistic view of self.
2. The client will take steps to address disorganized thinking.
3. The client will become appropriately interdependent on others.
4. The client will become involved in activities that foster social relationships.

20. A family of a client with paranoid personality disorder is trying to understand the client's behavior. Which of the following interventions would help the family?

1. Help the family find ways to handle stress.
2. Explore the possibility of finding respite care.
3. Help the family manage the client's eccentric action.
4. Encourage the family to focus on the client's strength.

The word *long-term* is a clue to the correct choice.

18. 2. Progress is shown when the client addresses behaviors that negatively impact relationships. Clients with paranoid personality disorder tend to have impaired social relationships and are very uncomfortable in social settings. Not recognizing the problem indicates the client is in denial. Knowing other people's problems isn't useful; the client must focus on his own issues. Clients with paranoid personality disorder struggle to understand and express their feelings about social rules.

NP: Evaluation; CN: Psychosocial integrity; CNS: Psychosocial adaptation; CL: Application

19. 4. An appropriate long-term goal is for the client to increase interactions, social skills, and make the commitment to become involved with others on a long-term basis. A client with paranoid personality disorder won't allow himself to be interdependent on others. To verbalize a realistic view of self is a short-term goal. The client with a paranoid personality disorder doesn't tend to have disorganized thinking.

NP: Planning; CN: Psychosocial integrity; CNS: Psychosocial adaptation; CL: Analysis

20. 3. The family needs to know how to handle the client's symptoms and eccentric behaviors. Focusing on the client's strengths is a positive action, but the family in this situation must learn how to manage the client's behavior. There's no need to find respite care for a client with a paranoid personality disorder. All people need to learn strategies for handling stress, but the focus must be on helping the family learn how to handle symptoms.

NP: Implementation; CN: Psychosocial integrity; CNS: Psychosocial adaptation; CL: Application

21. A client with antisocial personality disorder is trying to convince a nurse that he deserves special privileges and that an exception to the rules should be made for him. Which of the following responses is the most appropriate?

1. "I believe we need to sit down and talk about this."
2. "Don't you know better than to try to bend the rules?'
3. "What you're asking me to do for you is unacceptable."
4. "Why don't you bring this request to the community meeting?"

22. A client with antisocial personality disorder tells a nurse, "Life has been full of problems since childhood." Which of the following situations or conditions would the nurse explore in the assessment?

1. Birth defects
2. Distracted easily
3. Hypoactive behavior
4. Substance abuse

23. Which of the following behaviors by a client with antisocial personality disorder alerts a nurse to the need for teaching related to interaction skills?

1. Frequently crying
2. Having panic attacks
3. Avoiding social activities
4. Failing to follow social norms

24. Which of the following behaviors is suspected when a client with antisocial personality disorder shows blatant disregard for the law?

1. Labeling self as bad
2. Resisting spiritual healing
3. Engaging in criminal behaviors
4. Successful manipulation of the legal system

Therapeutic interventions must be appropriate for the client.

This is almost one of those *further teaching* questions. It asks you to identify a behavior that requires more teaching.

21. 3. These clients often try to manipulate the nurse to get special privileges or make exceptions to the rules on their behalf. By informing the client directly when actions are inappropriate, the nurse helps the client learn to control unacceptable behaviors by setting limits. By sitting down to talk about the request, the nurse is telling the client there's room for negotiation when there's none. By implying that the client wants to bend the rules humiliates him. The client's behavior is unacceptable and shouldn't be brought to a community meeting.
NP: Implementation; CN: Psychosocial integrity; CNS: Psychosocial adaptation; CL: Application

22. 4. Clients with antisocial personality disorder often engage in substance abuse during childhood. They don't have a higher incidence of birth defects than other people. They tend to be hyperactive, not hypoactive. Clients with antisocial personality disorder are often manipulative and are no more distracted from issues than others.
NP: Data collection; CN: Psychosocial integrity; CNS: Coping and adaptation; CL: Comprehension

23. 4. Failure to abide by social norms influences the client's ability to interact in a healthy manner with peers. Clients with antisocial personality disorders don't have frequent crying episodes or panic attacks. Avoiding social activities is more likely observed in avoidance personality style.
NP: Data collection; CN: Psychosocial integrity; CNS: Coping and adaptation; CL: Analysis

24. 3. Clients with antisocial personality disorder tend to engage in a variety of criminal activities. They don't tend to label themselves as bad, they have no interest in spiritual healing, no sense of remorse, and no inner need for forgiveness. They don't tend to successfully manipulate the legal system, but many of these clients are referred for treatment by the criminal justice system.
NP: Data collection; CN: Psychosocial integrity; CNS: Coping and adaptation; CL: Comprehension

25. Which of the following interventions should be done first for a client who has an antisocial personality disorder and a history of polysubstance abuse?
1. Human immunodeficiency virus (HIV) testing
2. Electrolyte profile
3. Anxiety screening
4. Psychological testing

This question is asking you to prioritize.

25. 1. A client who engages in high-risk behaviors such as polysubstance abuse should undergo HIV testing. This client would benefit from an entire chemistry profile as part of a complete medical examination, rather than a single test for electrolytes. An anxiety screen isn't needed for a client with antisocial personality disorder. Information from psychological testing is valuable when developing a treatment but isn't an immediate concern.
NP: Planning; CN: Physiological integrity; CNS: Reduction of risk potential; CL: Application

26. Which of the following short-term goals is appropriate for a client with an antisocial personality disorder who acts out when distressed?
1. Develop goals for personal improvement.
2. Identify situations that are out of the client's control.
3. Encourage the client to identify traumatic life events.
4. Learn to express feelings in a nondestructive manner.

26. 4. By working on appropriate expression of feelings, the client learns how to talk about what's stressful, rather than hurt himself or others. The most pressing need is to learn to cope and talk about problems rather than act out. Developing goals for personal improvement is a long-term goal, not a short-term one. Although it's important to differentiate what is and isn't under the client's control, the most important goal for handling distress is to talk about feelings appropriately. The identification of traumatic life events will occur only after the client begins to express feelings appropriately.
NP: Planning; CN: Safe, effective care environment; CNS: Safety and infection control; CL: Application

Read this question carefully. It seems to be asking you for a positive response, but it isn't.

27. A nurse notices other clients on the unit avoiding a client diagnosed with antisocial personality disorder. When discussing appropriate behavior in group therapy, which of the following comments is expected about this client by his peers?
1. Lack of honesty
2. Belief in superstitions
3. Show of temper tantrums
4. Constant need for attention

27. 1. Clients with antisocial personality disorder tend to engage in acts of dishonesty, shown by lying. Clients with schizotypal personality disorder tend to be superstitious. Clients with histrionic personality disorders tend to overreact to frustrations and disappointments, have temper tantrums, and seek attention.
NP: Data collection; CN: Psychosocial integrity; CNS: Coping and adaptation; CL: Application

NP: Nursing process   CN: Client needs category   CNS: Client needs subcategory   CL: Cognitive level

28. During a family meeting for a client with an antisocial personality disorder, which of the following statements is expected from an exasperated family member?
1. "Today I'm the enemy, but tomorrow I'll be a saint to him."
2. "When he's wrong, he never apologizes or even acts sorry."
3. "Sometimes I can't believe how he exaggerates about everything."
4. "There are times when his compulsive behavior is too much to handle."

29. Which of the following goals is most appropriate for a client with antisocial personality disorder with a high risk for violence directed at others?
1. The client will discuss the desire to hurt others rather than act.
2. The client will be given something to destroy to displace the anger.
3. The client will develop a list of resources to use when anger escalates.
4. The client will understand the difference between anger and physical symptoms.

30. A client with antisocial personality disorder says, "I always want to blow things off." Which of the following responses is the most appropriate?
1. "Try to focus on what needs to be done and just do it."
2. "Let's work on considering some options and strategies."
3. "Procrastinating is a part of your illness that we'll work on."
4. "The best thing to do is decide on some useful goals to accomplish."

31. Which of the following goals for the family of a client with antisocial disorder should the nurse stress in her teaching?
1. The family must assist the client to decrease ritualistic behavior.
2. The family must learn to live with the client's impulsive behavior.
3. The family must stop reinforcing inappropriate negative behavior.
4. The family must start to use negative reinforcement of the client's behavior.

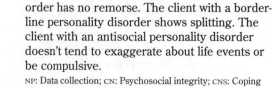
Time to prioritize.

Part of the job is teaching the family.

28. 2. The client with antisocial personality disorder has no remorse. The client with a borderline personality disorder shows splitting. The client with an antisocial personality disorder doesn't tend to exaggerate about life events or be compulsive.
NP: Data collection; CN: Psychosocial integrity; CNS: Coping and adaptation; CL: Analysis

29. 1. By discussing the desire to be violent toward others, the nurse can help the client get in touch with the pain associated with the angry feelings. It isn't helpful to have the client destroy something. The client needs to talk about strong feelings in a nonviolent manner, not refer to a list of crisis references. Helping the client understand the relationship between feelings and physical symptoms can be done after discussing the desire to hurt others.
NP: Planning; CN: Psychosocial integrity; CNS: Coping and adaptation; CL: Analysis

30. 2. By considering options or strategies, the client gains skills to overcome ineffective behaviors. The client tends to be irresponsible and needs guidance on what specifically to focus on to change behavior. Clients with an antisocial personality disorder don't tend to struggle with procrastination; instead, they show reckless and irresponsible behaviors. It's premature to decide on goals when the client needs to address the mental mindset and work to change the irresponsible behavior.
NP: Implementation; CN: Psychosocial integrity; CNS: Psychosocial adaptation; CL: Analysis

31. 3. The family needs help learning how to stop reinforcing inappropriate client behavior. Negative reinforcement is an inappropriate strategy for the family to use to support the client. The family can set limits and reinforce consequences when the client shows short-sightedness and poor planning. Clients with antisocial personality disorder don't show ritualistic behaviors.
NP: Implementation; CN: Psychosocial integrity; CNS: Psychosocial adaptation; CL: Analysis

32. Which of the following nursing interventions has priority in the care plan for a client with antisocial personality disorder who shows defensive behaviors?
1. Help the client accept responsibility for his own decisions and behaviors.
2. Work with the client to feel better about himself by taking care of basic needs.
3. Teach the client to identify the defense mechanisms used to cope with distress.
4. Confront the client about the disregard of social rules and the feelings of others.

33. A client with antisocial personality disorder is trying to manipulate the health care team. Which of the following strategies is important for the staff to use?
1. Focus on how to teach the client more effective behaviors for meeting basic needs.
2. Help the client verbalize underlying feelings of hopelessness and learn coping skills.
3. Remain calm and not emotionally respond to the manipulative actions of the client.
4. Help the client eliminate the intense desire to have everything in life turn out perfectly.

34. A client with dependent personality disorder is working to increase self-esteem. Which of the following statements by the client shows teaching was successful?
1. "I'm not just going to look at the negative things about myself."
2. "I'm most concerned about my level of competence and progress."
3. "I'm not as envious of the things other people have as I used to be."
4. "I find I can't stop myself from taking over things others should be doing."

35. Which of the following characteristics or client histories substantiates a diagnosis of antisocial personality disorder?
1. Delusional thinking
2. Feelings of inferiority
3. Disorganized thinking
4. Multiple criminal charges

Staff must work together to help the client.

You've finished 35 questions! Good job!

32. 1. Clients with antisocial personality disorder tend to blame other people for their behaviors and need to be taught how to take responsibility for their actions. Clients with antisocial personality disorder don't tend to have problems with self-care habits or meeting their own basic needs. Clients with antisocial personality disorder will deny they're defensive or distressed. Most often, these clients feel justified with retaliatory behavior. To confront the client would only cause him to become even more defensive.
NP: Implementation; CN: Psychosocial integrity; CNS: Psychosocial adaptation; CL: Analysis

33. 3. The best strategy to use with a client trying to manipulate staff is to stay calm and refrain from responding emotionally. Negative reinforcement of inappropriate behavior increases the chance it will be repeated. Later, it may be possible to address how to meet the client's basic needs. Clients with antisocial personality disorder don't tend to experience feelings of hopelessness or the desire for life events to turn out perfectly. In most cases, these clients negate responsibility for their behavior.
NP: Implementation; CN: Psychosocial integrity; CNS: Psychosocial adaptation; CL: Analysis

34. 1. As the client makes progress on improving self-esteem, self-blame and negative self-evaluations will decrease. A client with dependent personality disorder tends to feel fragile and inadequate and would be extremely unlikely to discuss their level of competence and progress. These clients focus on self and aren't envious or jealous. Individuals with dependent personality disorders don't take over situations because they see themselves as inept and inadequate.
NP: Evaluation; CN: Psychosocial integrity; CNS: Psychosocial adaptation; CL: Application

35. 4. Clients with antisocial personality disorder are often sent for treatment by the court after multiple crimes or for the use of illegal substances. Clients with antisocial personality disorder don't tend to have feelings of inferiority, delusional thinking, or disorganized thinking.
NP: Data collection; CN: Psychosocial integrity; CNS: Coping and adaptation; CL: Comprehension

NP: Nursing process   CN: Client needs category   CNS: Client needs subcategory   CL: Cognitive level

36. A nurse on the psychiatric unit is caring for a client with an antisocial personality disorder. Which behavior is the nurse most likely to observe?
1. Manipulation, shallowness, and the need for immediate gratification
2. Tendency to profit from mistakes or learn from past experiences
3. Expression of guilt and anxiety regarding behavior
4. Acceptance of authority and discipline

37. Which of the following findings is consistent with a diagnosis of antisocial personality disorder?
1. Problematic work history
2. Struggle with severe anxiety
3. Severe physical health conditions
4. Being critical of positive feedback

38. A client with antisocial personality disorder is beginning to practice several socially acceptable behaviors in the group setting. Which of the following outcomes will result from this change?
1. Fewer panic attacks
2. Acceptance of reality
3. Improved self-esteem
4. Decreased physical symptoms

39. A client with borderline personality disorder is admitted to the unit after slashing his wrist. Which of the following goals is most important after promoting safety?
1. Establish a therapeutic relationship with the client.
2. Identify if splitting is present in the client's thoughts.
3. Talk about his acting out and self-destructive tendencies.
4. Encourage the client to understand why he blames others.

Notice the most likely hint.

Prioritize!

36. 1. Due to the client's lack of scruples and underlying powerlessness, the nurse expects to see manipulation, shallowness, impulsivity, and self-centered behavior. This client doesn't profit from mistakes and learn from past experiences, lacks anxiety and guilt, and is unable to accept authority and discipline.
NP: Data collection; CN: Psychosocial integrity; CNS: Psychosocial adaptation; CL: Application

37. 1. Clients with a diagnosis of antisocial personality disorder tend to have problems in their job roles and a poor work history. They don't have severe anxiety disorders or severe physical health problems and are able to accept positive feedback from others.
NP: Data collection; CN: Psychosocial integrity; CNS: Coping and adaptation; CL: Comprehension

38. 3. When clients with antisocial personality disorder begin to practice socially acceptable behaviors, they also frequently experience a more positive sense of self. Clients with antisocial personality disorder don't tend to have panic attacks, somatic manifestations of their illness, or withdrawal or alteration in their perception of reality.
NP: Data collection; CN: Psychosocial integrity; CNS: Coping and adaptation; CL: Comprehension

39. 1. After promoting client safety, the nurse establishes a rapport with the client to facilitate appropriate expression of feelings. At this time, the client isn't ready to address unhealthy behavior. A therapeutic relationship must be established before the nurse can effectively work with the client on self-destructive tendencies. A therapeutic relationship must be established before working on the issue of splitting.
NP: Planning; CN: Psychosocial integrity; CNS: Psychosocial adaptation; CL: Comprehension

40. Which of the following nursing interventions is most appropriate in helping a client with a borderline personality disorder identify appropriate behaviors?

1. Schedule a family meeting.
2. Place the client in seclusion.
3. Formulate a behavioral contract.
4. Perform a mental status assessment.

41. Which of the following statements is typical of a client with borderline personality disorder who has recurrent suicidal thoughts?

1. "I can't believe how everyone has suddenly stopped believing in me."
2. "I don't care what other people say, I know how badly I looked to them."
3. "I might as well check out since my boyfriend doesn't want me anymore."
4. "I won't stop until I've gotten revenge on all those people who blamed me."

42. Which of the following findings is expected when taking a health history from a client with borderline personality disorder?

1. A negative sense of self
2. A tendency to be compulsive
3. A problem with communication
4. An inclination to be philosophical

43. Which of the following characteristics or situations is indicated when a client with borderline personality disorder has a crisis?

1. Antisocial behavior
2. Suspicious behavior
3. Relationship problems
4. Auditory hallucinations

44. Which of the following assessment findings is seen in a client diagnosed with borderline personality disorder?

1. Abrasions in various healing stages
2. Intermittent episodes of hypertension
3. Alternating tachycardia and bradycardia
4. Mild state of euphoria with disorientation

When I typed out question 41, I paused at the word "typical." So should you!

You're moving through these questions quite nicely. Keep moving along!

40. 3. The use of a behavioral contract establishes a framework for healthier functioning and places responsibility for actions back on the client. Seclusion will reinforce the fear of abandonment found in clients with borderline personality. Performing a mental status assessment or scheduling a family meeting won't help the client identify appropriate behaviors.
NP: Implementation; CN: Psychosocial integrity; CNS: Psychosocial adaptation; CL: Application

41. 3. This statement is typical for the borderline personality disorder client who is suicidal and reflects the tendency toward all-or-nothing thinking. The first statement indicates the client has experienced a credibility problem, the second statement indicates the client is extremely embarrassed, and the last statement indicates the client has an antisocial personality disorder. None of these is a characteristic of a borderline personality disorder.
NP: Data collection; CN: Safe, effective care environment; CNS: Safety and infection control; CL: Application

42. 1. Clients with a borderline personality disorder have low self-esteem and a negative sense of self. They have little or no problem expressing themselves and communicating with others, and although they have a tendency to be impulsive, they aren't usually compulsive or philosophical.
NP: Data collection; CN: Psychosocial integrity; CNS: Coping and adaptation; CL: Comprehension

43. 3. Relationship problems can precipitate a crisis because they bring up issues of abandonment. Clients with borderline personality disorder aren't usually suspicious; they're more likely to be depressed or highly anxious. They don't have symptoms of antisocial behavior or auditory hallucinations.
NP: Data collection; CN: Psychosocial integrity; CNS: Coping and adaptation; CL: Analysis

44. 1. Clients with borderline personality disorder tend to self-mutilate and have abrasions in various stages of healing. Intermittent episodes of hypertension, alternating tachycardia and bradycardia, or a mild state of euphoria with disorientation don't tend to occur with this disorder.
NP: Data collection; CN: Psychosocial integrity; CNS: Coping and adaptation; CL: Application

45. Which of the following short-term goals is appropriate for a client with borderline personality disorder with low self-esteem?

1. Write in a journal daily.
2. Express fears and feelings.
3. Stop obsessive-compulsive behaviors.
4. Decrease dysfunctional family conflicts.

46. Which of the following interventions is important to include in a teaching plan for a family with a member diagnosed with borderline personality disorder?

1. Teach the family methods for handling the client's anxiety.
2. Explore how the family reinforces the sick role with the client.
3. Encourage the family to have the client express intense emotions.
4. Help the family put pressure on the client to improve current behavior.

47. In planning care for a client with borderline personality disorder, a nurse must be aware that this client is prone to develop which of the following conditions?

1. Binge eating
2. Memory loss
3. Cult membership
4. Delusional thinking

48. Which of the following statements is expected from a client with borderline personality disorder with a history of dysfunctional relationships?

1. "I won't get involved in another relationship."
2. "I'm determined to look for the perfect partner."
3. "I've decided to learn better communication skills."
4. "I'm going to be an equal partner in a relationship."

The key word clarifies the answer. Don't miss it!

Way to go!

45. 2. Acknowledging fears and feelings can help the client identify parts of himself that are uncomfortable, and he can begin to work on developing a positive sense of self. Writing in a daily journal isn't a short-term goal to enhance self-esteem. A client with borderline personality disorder doesn't struggle with obsessive-compulsive behaviors. Decreasing dysfunctional family conflicts is a long-term goal.
NP: Planning; CN: Psychosocial integrity; CNS: Coping and adaptation; CL: Analysis

46. 1. The family needs to learn how to handle the client's intense stress and low tolerance for frustration. Family members don't want to reinforce the sick role; they're more concerned with preventing anxiety from escalating. Clients with borderline personality disorder already maintain intense emotions and it's not safe to encourage further expression of them. The family doesn't need to put pressure on the client to change behavior; this approach will only cause inappropriate behavior to escalate.
NP: Planning; CN: Health promotion and maintenance; CNS: Prevention and early detection of disease; CL: Application

47. 1. Clients with borderline personality disorder are likely to develop dysfunctional coping and act out in self-destructive ways, such as binge eating. They aren't prone to develop memory loss or delusional thinking. Becoming involved in cults may be seen in some clients with antisocial personality disorder.
NP: Planning; CN: Psychosocial integrity; CNS: Coping and adaptation; CL: Analysis

48. 2. Clients with borderline personality disorder would decide to look for a perfect partner. This characteristic is a result of the dichotomous manner in which these clients view the world. They go from relationship to relationship without taking responsibility for their behavior. It's unlikely an unsuccessful relationship will cause clients to make a change. They tend to be demanding and impulsive in relationships. There is no thought given to what one wants or needs from a relationship. Because they tend to blame others for problems, it's unlikely they would express a desire to learn communication skills.
NP: Data collection; CN: Psychosocial integrity; CNS: Psychosocial adaptation; CL: Analysis

49. Which of the following nursing interventions is the most appropriate for a client with borderline personality disorder working on developing healthy relationships?
1. Have the client assess current behaviors.
2. Work with the client to develop outgoing behavior.
3. Limit the client's interactions to family members only.
4. Encourage the client to approach others for interactions.

Appropriate care considers the client's best interests.

49. 1. Self-assessment of behavior enables the client to look at himself and identify social behaviors that need to be changed. Clients with borderline personality disorder don't tend to have difficulty approaching and interacting with other people. It's unrealistic to have clients with borderline personality disorder limit their interactions to family members only. Clients with borderline personality disorder tend to be demanding and the center of attention. It's not useful to work on developing outgoing behavior.
NP: Implementation; CN: Psychosocial integrity; CNS: Psychosocial adaptation; CL: Analysis

50. Which of the following defense mechanisms is most likely to be seen in a client with borderline personality disorder?
1. Compensation
2. Displacement
3. Identification
4. Projection

You're way more than halfway finished! Keep it up!

50. 4. Clients with borderline personality disorder tend to blame and project their feelings and inadequacies onto others. They don't model themselves after other people or tend to use compensation to handle distress. Clients with borderline personality disorder are impulsive and tend to react immediately. It's unlikely they would displace their feelings onto others.
NP: Data collection; CN: Psychosocial integrity; CNS: Coping and adaptation; CL: Analysis

51. Which of the following conditions is likely to coexist in clients with a diagnosis of borderline personality disorder?
1. Avoidance
2. Delirium
3. Depression
4. Disorientation

51. 3. Chronic feelings of emptiness and sadness predispose a client to depression. About 40% of the clients with borderline personality disorder struggle with depression. Clients with borderline personality disorder don't tend to develop delirium or become disoriented. This is only a possibility if the client becomes intoxicated. Clients with borderline personality disorder tend to disregard boundaries and limits. Avoidance isn't an issue with these clients.
NP: Data collection; CN: Psychosocial integrity; CNS: Coping and adaptation; CL: Comprehension

Prioritizing is a crucial part of nursing!

52. Which of the following nursing interventions has priority for a client with borderline personality disorder?
1. Maintain consistent and realistic limits.
2. Give instructions for meeting basic self-care needs.
3. Engage in daytime activities to stimulate wakefulness.
4. Have the client attend group therapy on a daily basis.

52. 1. Clients with borderline personality disorder who are needy, dependent, and manipulative will benefit greatly from maintaining consistent and realistic limits. They don't tend to have difficulty meeting their self-care needs. They enjoy attending group therapy because they often attempt to use the opportunity to become the center of attention. They don't tend to have sleeping difficulties.
NP: Implementation; CN: Psychosocial integrity; CNS: Psychosocial adaptation; CL: Comprehension

53. Which of the following outcomes indicates individual therapy has been effective for a client with borderline personality disorder?
1. The client accepts that medication isn't a treatment of choice.
2. The client agrees to undergo hypnosis for suppression of memories.
3. The client understands the organic basis for the problematic behavior.
4. The client verbalizes awareness of the consequences for unacceptable behaviors.

54. Which of the following actions by a client with borderline personality disorder indicates adequate learning about personal behavior?
1. The client talks about intense anger.
2. The client smiles while making demands.
3. The client decides never to engage in conflict.
4. The client stops the family from controlling finances.

Evaluate if learning has occurred.

55. A nurse is planning care for a client with borderline personality disorder who has been agitated. Which of the following instructions is included for the client and family?
1. Encourage the rebuilding of family relationships.
2. Help the client handle anxiety before it escalates.
3. Have the client participate in a weekly support group.
4. Discuss the client's bad habits that need to be changed.

When treatment isn't working, reevaluation may be necessary.

56. A client with a borderline personality disorder isn't making progress on the identified goals. Which of the following client factors should be reevaluated?
1. Memory
2. Motivation
3. Orientation
4. Perception

53. 4. An indication of effective individual therapy for this client is his expressed awareness of consequences for unacceptable behaviors. Medications can control symptoms. However, monitoring for reckless use or abuse of drugs must be done for the client with borderline personality disorder. Hypnosis isn't a treatment used with a client with borderline personality disorder. There is no organic basis for the development of this disorder.
NP: Evaluation; CN: Psychosocial integrity; CNS: Psychosocial adaptation; CL: Application

54. 1. Learning has occurred when anger is discussed rather than acted out in unhealthy ways. The behavior to change would be the demands placed on others. Smiling while making these demands shows manipulative behavior. Not engaging in conflict is unrealistic. It's important to help this client slowly develop financial responsibility rather than just stopping the family from monitoring the client's overspending.
NP: Evaluation; CN: Psychosocial integrity; CNS: Psychosocial adaptation; CL: Application

55. 2. The client needs help handling anxiety because escalating anxiety can trigger self-destructive behaviors in clients with borderline personality disorder. When a client with borderline personality disorder is agitated, it's difficult to communicate, let alone rebuild family relationships. Participation in a weekly support group won't be enough to help the client handle agitation. When a client is agitated, it isn't appropriate to discuss bad habits that need to be changed. This action may further agitate the client.
NP: Planning; CN: Psychosocial integrity; CNS: Psychosocial adaptation; CL: Application

56. 2. Clients with borderline personality disorders tend to be poorly motivated for treatment. They don't tend to have perception problems such as hallucinations or illusions, problems in orientation, or memory problems.
NP: Evaluation; CN: Psychosocial integrity; CNS: Psychosocial adaptation; CL: Analysis

57. A nurse is assessing a client diagnosed with dependent personality disorder. Which of the following characteristics is a major component of this disorder?
1. Abrasive to others
2. Indifferent to others
3. Manipulative of others
4. Overreliance on others

58. A client with dependent personality disorder is working on goals for self-care. Which of the following short-term goals is <u>most important</u> to the client's everyday activities of daily living?
1. Do all self-care activities independently.
2. Write a daily schedule for each day of the week.
3. Do self-care activities in a minimal amount of time.
4. Determine activities that can be performed without help.

59. Which of the following information must be included for the family of a person diagnosed with dependent personality disorder?
1. Address coping skills.
2. Explore panic attacks.
3. Promote exercise programs.
4. Decrease aggressive outbursts.

60. Which of the following strategies is appropriate for a client with dependent personality disorder?
1. Orient the client to current surroundings.
2. Reassure the client about personal safety.
3. Ask questions to help the client recall problems.
4. Differentiate between positive and negative feedback.

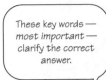

These key words — most important — clarify the correct answer.

Only 16 more questions!

57. 4. Clients with dependent personality disorder are extremely overreliant on other people; they're not abrasive or assertive. They're clinging and demanding of others; they don't manipulate. People with dependent personality disorder rely on others and want to be taken care of. They aren't indifferent.
NP: Data collection; CN: Psychosocial integrity; CNS: Coping and adaptation; CL: Knowledge

58. 4. By determining activities that can be performed without assistance, the client can then begin to practice them independently. Writing a daily schedule doesn't help the client focus on what needs to be done to promote self-care. If the nurse only encourages a client to perform self-care activities independently, nothing may change. The amount of time needed to perform self-care activities isn't important. If time pressure is put on the client, there may be more reluctance to perform self-care activities.
NP: Planning; CN: Psychosocial integrity; CNS: Psychosocial adaptation; CL: Comprehension

59. 1. The family needs information about coping skills to help the client learn to handle stress. Clients with dependent personality disorder don't have aggressive outbursts; they tend to be passive and submit to others. They don't tend to have panic attacks. Exercise is a health promotion activity for all clients. Clients with dependent personality disorder wouldn't need exercise promoted more than other people.
NP: Implementation; CN: Safe, effective care environment; CNS: Coordinated care; CL: Comprehension

60. 4. Clients with dependent personality disorder tend to view all feedback as criticism; they frequently misinterpret another's remarks. Clients with dependent personality disorder don't need orientation to their surroundings. Memory problems aren't associated with this disorder, so asking questions to stimulate one's memory isn't necessary. Personal safety isn't an issue because a person with dependent personality disorder typically isn't self-destructive.
NP: Implementation; CN: Psychosocial integrity; CNS: Psychosocial adaptation; CL: Application

61. A client with dependent personality disorder is crying after a family meeting. Which of the following statements by a family member is most likely the cause of upset to this client?
  1. "You take advantage of people, especially the people in our family."
  2. "You act like you love me one minute, but hate me the next minute."
  3. "You feel like you deserve everything, whether you work for it or not."
  4. "You always agree to everything, but deep down inside you feel differently."

62. A 31-year-old male client with borderline personality is instructed to stay in the game room. He asks the nurse to "bend the rules a little, just this one time." Which response is most appropriate?
  1. "No."
  2. "OK, but just this once."
  3. Ignore the client.
  4. "Let me explain the rules to you."

63. A client with borderline personality disorder confides in one nurse that she's disappointed in the other nurses because of their lack of sensitivity to her. What's the best action for the nurse to take?
  1. Explain the other nurses' behaviors to the client.
  2. Ask for more information about the client's disappointment.
  3. Call a team meeting to help others understand the client better.
  4. Advise the client to seek out the nurses and tell them directly.

64. In planning care for a client with borderline personality disorder, the nurse must account for which of the following behavioral traits?
  1. An inability to make decisions independently
  2. A propensity to act out when feeling afraid, alone, or devalued
  3. A belief she deserves special privileges not accorded to others
  4. A display of inappropriately seductive appearance and behavior

You know the expression, "That's a likely story?" Well, the story here is when answering question 61, you're looking for the most likely cause.

61. 4. The client was confronted by a family member about behavior that doesn't represent the client's true feelings. Clients are afraid they won't be taken care of if they disagree. Clients with a dependent personality disorder don't have a sense of entitlement and don't take advantage of other people, but they subordinate their needs to others. They don't show the defense mechanism of splitting, where a person is valued and then devalued.
NP: Data collection; CN: Psychosocial integrity; CNS: Psychosocial adaptation; CL: Analysis

62. 4. This allows the nurse to be firm but not confrontational with the client. A simple "no" is too harsh by itself. The nurse's response needs more of an explanation. By agreeing to bend the rules, the nurse would allow the client to manipulate the caregiver. Ignoring the client would do nothing but frustrate him.
NP: Implementation; CN: Psychosocial integrity; CNS: Psychosocial adaptation; CL: Application

63. 4. One communication strategy when working with a client with borderline personality disorder is to never engage in third-party conversations. Explaining the other nurses' behaviors to the client, asking for more information about his disappointment, or calling a team meeting encourage splitting behaviors and are nonproductive.
NP: Implementation; CN: Psychosocial integrity; CNS: Psychosocial adaptation; CL: Application

64. 2. Those with borderline personality disorder have an intense fear of abandonment. These clients are able to make decisions independently. Feeling deserving of special privileges is characteristic of a person with narcissistic personality disorder. Inappropriate seductive appearance and behavior is characteristic of someone with histrionic personality disorder.
NP: Planning; CN: Psychosocial integrity; CNS: Psychosocial adaptation; CL: Application

65. Which of the following information is expected in a history of a client with dependent personality disorder?
1. Lack of relationships
2. Lack of self-confidence
3. Preoccupation with rules
4. Tendency to overvalue self

66. Which of the following short-term goals is appropriate for a client with dependent personality disorder experiencing excessive dependency needs?
1. Verbalize self-confidence in own abilities.
2. Decide relationships don't take energy to sustain.
3. Discuss feelings related to frequent mood swings.
4. Stop obsessive thinking that impedes daily social functioning.

67. A client with dependent personality disorder is thinking about getting a part-time job. Which of the following nursing interventions will help this client when employment is obtained?
1. Help the client develop strategies to control impulses.
2. Explain there are consequences for inappropriate behaviors.
3. Have the client work to sustain healthy interpersonal relationships.
4. Help the client decrease the use of regression as a defense mechanism.

68. A client with dependent personality disorder has difficulty expressing personal concerns. Which of the following communication techniques is best to teach the client?
1. Questioning
2. Reflection
3. Silence
4. Touch

Question 66 asks for short-term goals, not long-term ones. Keep that in mind.

This is one way to communicate. Is it the best?

65. 2. Clients with a dependent personality disorder lack self-confidence and have low self-esteem. They tend to undervalue, not overvalue self. They have unhealthy relationships, allowing others to take over their lives. They aren't preoccupied with rules. They focus on their need to be taken care of by others.
NP: Data collection; CN: Psychosocial integrity; CNS: Coping and adaptation; CL: Knowledge

66. 1. Individuals with dependent personalities believe they must depend on others to be competent for them. They need to gain more self-confidence in their own abilities. The client must realize relationships take energy to develop and sustain. Clients with dependent personality disorder usually don't have obsessive thinking or mood swings to interfere with their socialization.
NP: Planning; CN: Psychosocial integrity; CNS: Psychosocial adaptation; CL: Application

67. 3. Sustaining healthy relationships will help the client be comfortable with peers in the job setting. Clients with dependent personality disorder don't usually use regression as a defense mechanism. It's common to see denial and introjection used. They don't usually have trouble with impulse control or offensive behavior that would lead to negative consequences.
NP: Implementation; CN: Psychosocial integrity; CNS: Psychosocial adaptation; CL: Analysis

68. 1. Questioning is a way to learn to identify feelings and express self. The use of reflection isn't a communication technique that will help the client express personal feelings and concerns. The use of touch to express feelings and personal concerns must be used very judiciously. Using silence won't help the client identify and discuss personal concerns.
NP: Implementation; CN: Psychosocial integrity; CNS: Psychosocial adaptation; CL: Application

NP: Nursing process   CN: Client needs category   CNS: Client needs subcategory   CL: Cognitive level

69. A nurse is evaluating the effectiveness of an assertiveness group that a client with dependent personality disorder attended. Which of the following client statements indicates the group had therapeutic value?
1. "I can't seem to do the things other people do."
2. "I wish I could be more organized like other people."
3. "I want to talk about something that's bothering me."
4. "I just don't want people in my family to fight any more."

70. After a family visit, a client with dependent personality disorder becomes anxious. Which of the following situations is a possible cause of the anxiety?
1. Sensitivity to criticism
2. Discussion of family rules
3. Being asked personal questions
4. Identification of eccentric behavior

The finish line is just 6 questions away!

71. A 46-year-old male client undergoing treatment for paranoia refuses to take his fluphenazine (Prolixin) because he says he thinks it's poisoned. Which of the following responses is best?
1. Omit the dose and notify the physician.
2. Tell him that he'll receive an injection if he refuses the oral medication.
3. Put the medication in his juice without informing him.
4. Allow him to examine the medication to see that it isn't poisoned.

72. Which of the following emotional health problems may potentially coexist in a client with dependent personality disorder?
1. Psychotic disorder
2. Acute stress disorder
3. Alcohol-related disorder
4. Posttraumatic stress disorder

69. 3. By asking to talk about a bothersome situation, the client has taken the first step toward assertive behavior. To smooth over or minimize troubling events isn't an assertive position. The first statement reflects a lack of self-confidence; it isn't an assertive statement. Statements that express the client's wishes aren't assertive statements.
NP: Evaluation; CN: Psychosocial integrity; CNS: Psychosocial adaptation; CL: Application

70. 1. Clients with dependent personality disorder are extremely sensitive to criticism and can become very anxious when they feel interpersonal conflict or tension. When they're asked personal questions, they don't necessarily become anxious. When they have discussions about family rules, they try to become submissive and please others rather than become anxious. Clients with dependent personality disorder don't tend to show eccentric behavior that causes them anxiety.
NP: Data collection; CN: Psychosocial integrity; CNS: Coping and adaptation; CL: Application

71. 1. The nurse's best response is to omit the dose and notify the physician because insisting that the client take the medication will only increase his paranoia and agitation. Forcing injections and tricking clients into taking medication is illegal. Don't put the medication in his juice without informing him. A rational approach (such as allowing the client to examine the medication) to irrational ideas seldom works.
NP: Data collection; CN: Psychosocial integrity; CNS: Coping and adaptation; CL: Application

72. 2. Because they have placed their own needs in the hands of others, clients with dependent personalities are extremely vulnerable to anxiety disorder. They don't tend to have coexisting problems of posttraumatic stress disorder, psychotic disorder, or alcohol-related disorder.
NP: Data collection; CN: Psychosocial integrity; CNS: Coping and adaptation; CL: Analysis

73. A nurse notices that a client with dependent personality disorder is depressed. Which of the following factors is assessed as contributing to depression?
1. Unmet needs
2. Sense of smothering
3. Messy, unkempt appearance
4. Difficulty delaying gratification

74. A client with dependent personality disorder makes the following statement, "I'll never be able to take care of myself." Which of the following responses is best?
1. "How can you say that? You can function."
2. "Let's talk about what's making you feel so fearful."
3. "I think we need to work on identifying your strengths."
4. "Can we talk about this tomorrow at the family meeting?"

75. A client on your unit says the Mafia has a contract out on him. He refuses to leave his semiprivate room and insists on frisking his roommate before allowing him to enter. Which of the following actions should you take first?
1. Have the client transferred to a private room.
2. Acknowledge the client's fear when he refuses to leave his room or wants to frisk his roommate.
3. Transfer the roommate to another room.
4. Lock the client out of his room for a short while each day so that he can see that he's safe.

Keep in mind the type of personality this client has.

73. 1. Having many unmet needs is a precursor to depression. Clients with dependent personality disorders don't tend to have panic disorders. Clients who experience panic disorders often have a racing pulse and the sensation of smothering. Clients with problems delaying gratification tend to have anxiety problems, not problems with depression. Poor hygiene is often a *manifestation* of depression, *not* a cause of it.
NP: Data collection; CN: Psychosocial integrity; CNS: Coping and adaptation; CL: Analysis

74. 2. The client with dependent personality disorder is afraid of abandonment, rejection, and being unable to care for himself. Talking about the client's fears is a useful strategy. The first response is inappropriate because the nurse is failing to recognize the client's feelings. When the client makes a desperate statement like this, the nurse must respond to the client's feelings, rather than insert her opinion. Working on identifying a client's strengths will only add to his feelings of not being strong enough to take care of himself. Waiting to talk about a client concern until the family meeting minimizes its importance.
NP: Implementation; CN: Psychosocial integrity; CNS: Coping and adaptation; CL: Analysis

One more time: Prioritize! Hooray! You're done.

75. 2. Acknowledging underlying feelings may help defuse the client's anxiety without promoting his delusional thinking. This, in turn, may help the client distinguish between his emotional state and external reality. Transferring either client to another room would validate the client's delusional thinking. Locking the client out of his room may further escalate the client's anxiety and stimulate aggressive acting-out behavior.
NP: Implementation; CN: Psychosocial integrity; CNS: Coping and adaptation; CL: Application

So I'm talking to Queen Elizabeth the other day, and she said you would do spectacularly well on this chapter. (I'm kidding. What do I look, delusional?) Personally, I think you'll do even better than that. Go for it!

Chapter 17
Schizophrenic & delusional disorders

Therapeutic intervention is the key.

1. A schizophrenic client tells his primary nurse that he's scheduled to meet the King of Samoa at a special time, making it impossible for the client to leave his room for dinner. Which of the following responses by the nurse is most appropriate?
1. "It's meal time. Let's go so you can eat."
2. "The King of Samoa told me to take you to dinner."
3. "Your physician expects you to follow the unit's schedule."
4. "People who don't eat on this unit aren't being cooperative."

2. While looking out the window, a client with schizophrenia remarks, "That school across the street has creatures in it that are waiting for me." Which of the following terms best describes what the creatures represent?
1. Anxiety attack
2. Projection
3. Hallucination
4. Delusion

3. Which nursing intervention is most important for a client with schizophrenia?
1. Teach the client about the illness.
2. Initiate a behavioral contract with the client.
3. Require the client to attend all unit functions.
4. Provide a consistent predictable environment.

1. 1. A delusional client is so wrapped up in his false beliefs that he tends to disregard activities of daily living, such as nutrition and hydration. He needs clear, concise, firm directions from a caring nurse to meet his needs. The second response belittles and tricks the client, possibly evoking mistrust on the part of the client. The third response evades the issue of meeting his basic needs. The last response is demeaning and doesn't address the delusion.
NP: Implementation; CN: Health promotion and maintenance; CNS: Prevention and early detection of disease; CL: Application

2. 4. A delusion is a false belief based on a misrepresentation of a real event or experience. Hallucinations are perceptual disorders of the five senses and are part of most psychoses. The person sees, tastes, feels, smells, and hears things in the absence of external stimulation. Although anxiety can increase delusional responses, it isn't considered the primary symptom.
NP: Data collection; CN: Physiological integrity; CNS: Physiological adaptation; CL: Knowledge

3. 4. A consistent, predictable environment helps the client remain as functional as possible and prevents sensory overload. Teaching the client about his illness is important but not a priority. A behavioral contract and required attendance at all functions aren't particularly effective with schizophrenia.
NP: Implementation; CN: Psychosocial integrity; CNS: Psychosocial adaptation; CL: Application

NP: Nursing process   CN: Client needs category   CNS: Client needs subcategory   CL: Cognitive level

4. A 22-year-old schizophrenic client was admitted to the psychiatric unit during the night. The next morning, he began to misidentify the nurse and call her by his sister's name. Which of the following interventions is best?
1. Assess the client for potential violence.
2. Take the client to his room, where he'll feel safer.
3. Assume the misidentification makes the client feel more comfortable.
4. Correct the misidentification and orient the client to the unit and staff.

5. Which of the following terms describes an effect of isolation?
1. Delusions
2. Hallucinations
3. Lack of volition
4. Waxy flexibility

6. A client diagnosed with schizophrenia several years ago tells the nurse that he feels "very sad." The nurse observes that he's smiling when he says it. Which of the following terms best describes the nurse's observation?
1. Inappropriate affect
2. Extrapyramidal
3. Insight
4. Inappropriate mood

7. Which of the following conditions or characteristics is related to the cluster of symptoms associated with disorganized schizophrenia?
1. Odd beliefs
2. Flat affect
3. Waxy flexibility
4. Systematized delusions

You're really moving through these questions!

The word disorganized refers to a type of schizophrenia.

4. 4. Misidentification can contribute to anxiety, fear, aggression, and hostility. Orienting a new client to the hospital unit, staff, and other clients, along with establishing a nurse-client relationship, can decrease these feelings and help the client feel in control. Assessing for potential violence is an important nursing function for any psychiatric client, but a perceived supportive environment reduces the risk of violence. Withdrawing to his room, unless interpersonal relationships have become nontherapeutic for him, encourages the client to remain in his fantasy world.
NP: Implementation; CN: Psychosocial integrity; CNS: Coping and adaptation; CL: Application

5. 2. Prolonged isolation can produce sensory deprivation, manifested by hallucinations. A delusion is a false perception caused by misinterpretation of a real object. Lack of volition is a symptom associated with type I negative symptoms of schizophrenia. Waxy flexibility is a motor disturbance that's a predominant feature of catatonic schizophrenia.
NP: Data collection; CN: Psychosocial integrity; CNS: Psychosocial adaptation; CL: Application

6. 1. Affect refers to behaviors such as facial expression that can be observed when a person is expressing and experiencing feelings. If the client's affect doesn't reflect the emotional content of the statement, the affect is considered inappropriate. Mood is an extensive and sustained feeling tone. Insight is a component of the mental status examination and is the ability to perceive oneself realistically and understand if a problem exists. Extrapyramidal symptoms are adverse effects of some categories of medication.
NP: Data collection; CN: Psychosocial integrity; CNS: Psychosocial adaptation; CL: Comprehension

7. 2. Flat affect, the lack of facial or behavioral manifestations of emotion, is related to disorganized schizophrenia. Waxy flexibility occurs in catatonic schizophrenia. Systematized delusions occur most commonly in paranoid residual type schizophrenia, characterized by odd beliefs or unusual perceptions rather than prominent delusions or hallucinations.
NP: Data collection; CN: Psychosocial integrity; CNS: Psychosocial adaptation; CL: Knowledge

NP: Nursing process   CN: Client needs category   CNS: Client needs subcategory   CL: Cognitive level

8. A client on the psychiatric unit is copying and imitating the movements of his primary nurse. During recovery, he said, "I thought the nurse was my mirror. I felt connected only when I saw my nurse." This behavior is known by which of the following terms?
1. Modeling
2. Echopraxia
3. Ego-syntonicity
4. Ritualism

8. 2. Echopraxia is the copying of another's behaviors and is the result of the loss of ego boundaries. Modeling is the conscious copying of someone's behaviors. Ego-syntonicity refers to behaviors that correspond with the individual's sense of self. Ritualistic behaviors are repetitive and compulsive.
NP: Data collection; CN: Psychosocial integrity; CNS: Coping and adaptation; CL: Application

9. The teenage son of a father with schizophrenia is worried that he might have schizophrenia as well. Which of the following behaviors would be an indication that he should be evaluated for signs of the disorder?
1. Moodiness
2. Preoccupation with his body
3. Spending more time away from home
4. Changes in sleep patterns

9. 4. In conjunction with other signs, changes in sleep patterns are distinctive initial signs of schizophrenia. Other signs include changes in personal care habits and social isolation. Moodiness, preoccupation with the body, and spending more time away from home are normal adolescent behaviors.
NP: Data collection; CN: Health promotion and maintenance; CNS: Prevention and early detection of disease; CL: Application

10. Which of the following symptoms is usually responsive to <u>traditional</u> antipsychotic drugs?
1. Apathy
2. Delusions
3. Social withdrawal
4. Attention impairment

Pay attention to the key word in questions 10 and 11.

10. 2. Positive symptoms, such as delusions, hallucinations, thought disorder, and disorganized speech, respond to traditional antipsychotic drugs. Apathy, social withdrawal, and attention impairment are part of the category of negative symptoms, including affective flattening, restricted thought and speech, apathy, anhedonia, asociality, and attention impairment, and are more responsive to the new atypical antipsychotics, such as clozapine (Clozaril), risperidone (Risperdal), and olanzapine (Zyprexa).
NP: Data collection; CN: Physiological integrity CNS: Pharmacological therapies; CL: Comprehension

11. A client was hospitalized after his son filed a petition for involuntary hospitalization for safety reasons. The son seeks out the nurse because his father is angry and refuses to talk with him. He's frustrated and feeling guilty about his decision. Which of the following responses to the son is the most <u>empathic</u>?
1. "Your father is here because he needs help."
2. "He'll feel differently about you as he gets better."
3. "It sounds like you're feeling guilty about leaving your father here."
4. "This is a stressful time for you, but you'll feel better as he gets well."

11. 3. This response focuses on the son and helps him discuss and deal with his feelings. Unresolved feelings of guilt, shame, isolation, and loss of hope impact on the family's ability to manage the crisis and be supportive to the client. The other responses offer premature reassurance and cut off the opportunity for the son to discuss his feelings.
NP: Implementation; CN: Psychosocial integrity; CNS: Coping and adaptation; CL: Application

12. A client is about to be discharged with a prescription for the antipsychotic agent haloperidol (Haldol), 10 mg P.O. b.i.d. During a discharge teaching session, the nurse should provide which instruction to the patient?

1. Take the medication 1 hour before a meal.
2. Decrease the dosage if signs of illness decrease.
3. Apply a sunscreen before being exposed to the sun.
4. Increase the dosage up to 50 mg twice a day if signs of illness don't decrease.

13. Which of the following signs indicate tardive dyskinesia?

1. Involuntary movements
2. Blurred vision
3. Restlessness
4. Sudden fever

Signs and symptoms are big on NCLEX examinations. Better know them cold.

14. A client approaches a nurse and tells her that he hears a voice telling him that he's evil and deserves to die. Which of the following terms describes the client's perception?

1. Delusion
2. Disorganized speech
3. Hallucination
4. Idea of reference

15. A client is referred to a self-help group after discharge from the hospital. Which of the following types of individuals leads these groups?

1. Individuals concerned about coping with mutual concerns
2. Social workers familiar with the problems of the mentally ill
3. Psychiatrists trained to help with stressors in the community
4. Psychiatric nurse specialists able to provide help with monitoring medication

Self-help groups may be beneficial for your client.

12. 3. Because haloperidol can cause photosensitivity and precipitate severe sunburn, the nurse should instruct the client to apply a sunscreen before exposure to the sun. The nurse should teach the client to take haloperidol with meals to prevent gastric upset or irritation, rather than 1 hour before, and should instruct the patient not to decrease or increase the dosage unless the physician orders it.
NP: Implementation; CN: Physiological integrity; CNS: Pharmacological therapies; CL: Application

13. 1. Symptoms of tardive dyskinesia include tongue protrusion, lip smacking, chewing, blinking, grimacing, choreiform movements of limbs and trunk, and foot tapping. Blurred vision is a common adverse reaction of antipsychotic drugs and usually disappears after a few weeks of therapy. Restlessness is associated with akathisia. Sudden fever is a symptom of a malignant neurologic disorder.
NP: Data collection; CN: Physiological integrity; CNS: Reduction of risk potential; CL: Application

14. 3. Hallucinations are sensory experiences that are misrepresentations of reality or have no basis in reality. Delusions are beliefs not based in reality. Disorganized speech is characterized by jumping from one topic to the next or using unrelated words. An idea of reference is a belief that an unrelated situation holds special meaning for the client.
NP: Data collection; CN: Psychosocial integrity; CNS: Coping and adaptation; CL: Knowledge

15. 1. Self-help groups are organized and led by consumers of mental health services who come together to share and offer support and advice to each other about a specific problem. These groups aren't led by such professionals as nurses, social workers, and psychiatrists.
NP: Implementation; CN: Psychosocial integrity; CNS: Coping and adaptation; CL: Knowledge

NP: Nursing process   CN: Client needs category   CNS: Client needs subcategory   CL: Cognitive level

16. Which of the following numbers of members in a therapy group is ideal?
1. 1 to 4
2. 4 to 7
3. 7 to 10
4. 10 to 15

17. A client admitted to an inpatient unit approaches a nursing student saying he descended from a long line of people of a "superrace." Which of the following actions is correct?
1. Smile and walk into the nurse's station.
2. Challenge the client's false belief.
3. Listen for hidden messages in themes of delusion, indicating unmet needs.
4. Introduce yourself, shake hands, and sit down with the client in the day room.

Always be sensitive to the client's needs.

18. The nurse assesses a schizophrenic client for auditory hallucinations. Which of the following is most suggestive of this symptom?
1. Speaking loudly when engaged in conversation
2. Ignoring comments by the nurse
3. Responding only to the same person
4. Tilting the head to one side

Don't boast you've got the right answer until you've considered the word most!

19. A nurse is assisting with morning care when a client suddenly throws off the covers and starts shouting, "My body is changing and disintegrating because I'm not of this world." Which of the following terms best describes this behavior?
1. Depersonalization
2. Ideas of reference
3. Looseness of association
4. Paranoid ideation

16. 3. The ideal number of members in an inpatient group is 7 to 10. Having fewer than 7 members provides inadequate interaction and material for successful group process. Having more than 10 members doesn't allow for adequate time for individual participation.
NP: Implementation; CN: Psychosocial integrity; CNS: Coping and adaptation; CL: Application

17. 4. The first goal is to establish a relationship with the client, which includes creating psychological space for the creation of trust. The student should sit and make herself available, reflecting concern and interest. Walking into the nurse's station would indicate disinterest and lack of concern about the client's feelings. After establishing a relationship and lessening the client's anxiety, the student can orient the client to reality, listen to his concerns and fears, and try to understand the feelings reflected in the delusions. Delusions are firmly maintained false beliefs, and attempts to dismiss them don't work.
NP: Implementation; CN: Safe, effective care environment; CNS: Coordinated care; CL: Application

18. 4. A client who's having auditory hallucinations may tilt his head to one side, as if listening to someone or something. Speaking loudly, ignoring comments, and responding only to one person are indicative of hearing deficit, anxiety, and paranoid behavior, respectively.
NP: Data collection; CN: Psychosocial integrity; CNS: Psychosocial adaptation; CL: Analysis

19. 1. Depersonalization is a state in which the client feels unreal or believes parts of the body are being distorted. Ideas of reference are beliefs unrelated to situations and hold special meaning for the individual. The term loose associations refers to sentences that have vague connections to each other. Paranoid ideations are beliefs that others intend to harm the client in some way.
NP: Data collection; CN: Psychosocial integrity; CNS: Coping and adaptation; CL: Analysis

20. Many clients with schizophrenia simultaneously have opposing emotions. Which of the following terms describes this phenomenon?
  1. Double bind
  2. Ambivalence
  3. Loose associations
  4. Inappropriate affect

Question 20 already? You're really good!

20. 2. Ambivalence, one of the symptoms associated with schizophrenia, immobilizes the person from acting. A double bind presents two conflicting messages—for example, saying that you trust someone but then not allowing the person into your room. Loose association involves rapid shifts of ideas from one subject to another in an unrelated manner. Inappropriate affect refers to an observable expression of emotions incongruent with the emotion felt.
NP: Data collection; CN: Psychosocial integrity; CNS: Coping and adaptation; CL: Comprehension

21. A client is admitted to a psychiatric unit with a diagnosis of undifferentiated schizophrenia. Which of the following defense mechanisms is probably used?
  1. Projection
  2. Rationalization
  3. Regression
  4. Repression

21. 3. Regression, a return to earlier behavior to reduce anxiety, is the basic defense mechanism in schizophrenia. Repression is the basic defense mechanism in the neuroses. Rationalization is a defense mechanism used to justify one's actions. Projection is a defense mechanism in which one blames others and attempts to justify actions; it's used primarily by people with paranoid schizophrenia and delusional disorder.
NP: Data collection; CN: Psychosocial integrity; CNS: Coping and adaptation; CL: Comprehension

Appropriate means therapeutic for the client.

22. A nurse on a psychiatric unit observes a client in the corner of the room moving his lips as if he were talking to himself. Which of the following actions is the most appropriate?
  1. Ask him why he's talking to himself.
  2. Leave him alone until he stops talking.
  3. Tell him it's not good for him to talk to himself.
  4. Invite him to join in a card game with the nurse.

22. 4. Being with the nurse and playing a game provide stimulation that competes with the hallucinations. Being alone keeps the client in his fantasy world when the nurse expects the client to take the first step. Telling him it's no good talking to himself fails to understand how real his fantasy world and hallucinations are.
NP: Implementation; CN: Psychosocial integrity; CNS: Coping and adaptation; CL: Application

23. A client makes vague statements with no logical connections. He asks whether the nurse understands. Which of the following responses is best?
  1. "Why don't we wait until later to talk about it?"
  2. "You're not making sense, so I won't talk about this topic."
  3. "Yes, I understand the overall sense of the logical connections from the idea."
  4. "I want to understand what you're saying, but I'm having difficulty following you."

23. 4. The nurse needs to communicate that she wants to understand without blaming the client for the lack of understanding. Telling the client that he isn't making sense is judgmental and could impair the therapeutic relationship. Asking the client to wait because he's too confused cuts off an attempt to communicate and asks the client to do what he can't at present. Pretending to understand is a violation of trust and can damage the therapeutic relationship.
NP: Implementation; CN: Psychosocial integrity; CNS: Coping and adaptation; CL: Application

24. A client asks a nurse if she hears the voice of the nonexistent man speaking to him. Which of the following responses is best?

1. "No one is in your room except you."
2. "Yes, I hear him, but I won't listen to him."
3. "What has he told you? Is it helpful advice?"
4. "No, I don't hear him, but I know you do. What's he saying?"

Think therapeutic.

25. Which of the following instructions is correct for a client taking chlorpromazine (Thorazine)?

1. Reduce the dosage if you feel better.
2. Occasional social drinking isn't harmful.
3. Stop taking the drug immediately if adverse reactions develop.
4. Schedule routine medication checks.

26. A 34-year-old woman is referred to a mental health clinic by the court. The client harassed a couple next-door to her with charges that the husband was in love with her. She wrote love notes and called him on the telephone throughout the night. The client is employed and has had no problems in her job. Which of the following disorders is suspected?

1. Major depression
2. Paranoid schizophrenia
3. Delusional disorder
4. Bipolar affective disorder

You're doing great! Keep going!

27. A client with a diagnosis of paranoid type schizophrenia is receiving an antipsychotic medication. His physician has just prescribed benztropine (Cogentin). The nurse realizes that this medication was most likely prescribed in response to which of the following possible adverse reactions?

1. Tardive dyskinesia
2. Hypertensive crisis
3. Acute dystonia
4. Orthostatic hypotension

24. 4. This response points out reality and shows concern and support. Attempting to argue the client out of the belief might entrench him more firmly in his belief, making him feel more out of control because of the negative and fearful nature of hallucinations. The other two responses violate the trust of the therapeutic relationship.
NP: Implementation; CN: Safe, effective care environment; CNS: Coordinated care; CL: Application

25. 4. Ongoing assessment by a primary health care provider is important to assess for adverse reactions and continued therapeutic effectiveness. Adverse reactions should be reported immediately to determine if the drug should be discontinued. Alcoholic beverages are contraindicated while taking an antipsychotic drug. The dosage should be cut only after checking with the primary care provider.
NP: Implementation; CN: Physiological integrity; CNS: Pharmacological therapies; CL: Application

26. 3. The client has a delusional disorder with erotomanic delusions as her primary symptom and believes she is loved intensely by a married person showing no interest in her. No symptoms of major depression exist. The client doesn't believe someone is trying to harm her, the hallmark characteristic of paranoia. Bipolar affective disorder is characterized by cycles of extreme emotional highs (mania) and lows (depression).
NP: Data collection; CN: Psychosocial integrity; CNS: Coping and adaptation; CL: Application

27. 3. Benztropine is used as adjunctive therapy in parkinsonism and for all conditions and medications that produce extrapyradimal symptoms, except tardive dyskenia. Its anticholinergic effect reduces the extrapyramidal effects associated with antipsychotic drugs. Hypertensive crisis and orthostatic hypotension aren't associated with extrapyramidal symptoms.
NP: Evaluation; CN: Physiological integrity; CNS: Pharmacological therapies; CL: Analysis

28. A 20-year-old client is admitted to the hospital with a diagnosis of schizophrenia. During the initial assessment, he points to the nurse's stethoscope and says it's a snake. Which of the following terms describes this phenomenon?
1. Abstraction
2. Delusion
3. Hallucination
4. Illusion

29. When asking a family about the onset of a schizophrenic client's illusions, which of the following periods is the most likely answer?
1. Adolescence
2. Early childhood
3. Late adulthood
4. Late childhood

30. A client can't eat because he believes his bowels have turned against him. Which of the following terms describes this phenomenon?
1. Conversion hysteria
2. Depersonalization
3. Hypochondriasis
4. Somatic delusion

It's important to know the correct terminology for making accurate diagnoses.

31. Which of the following actions by the nurse is an appropriate therapeutic intervention for a client experiencing hallucinations?
1. Confine him in his room until he feels better.
2. Provide a competing stimulus that distracts from the hallucinations.
3. Discourage attempts to understand what precipitates his hallucinations.
4. Support perceptual distortions until he gives them up of his own accord.

28. 4. An illusion is a misinterpretation of an actual sensory stimulation. An abstraction is an idea or concept such as love or a belief that can't be represented by a concrete object. A hallucination is a false sensory perception without a stimulus, and a delusion is a fixed belief.
NP: Data collection; CN: Psychosocial integrity; CNS: Coping and adaptation; CL: Comprehension

29. 1. The usual onset of schizophrenia is between adolescence and early adulthood.
NP: Implementation; CN: Psychosocial integrity; CNS: Psychosocial adaptation; CL: Knowledge

30. 4. A somatic delusion is a fixed false belief pertaining to the body and body parts. Hypochondriasis is somatic overconcern with a morbid attention to details of body functioning. Depersonalization is a feeling of unreality concerning self and a loss of self-identity, with things around the person seeming different, strange, or unreal. Conversion hysteria is a somatoform disorder in which there are symptoms of some physical illness without any underlying organic cause.
NP: Data collection; CN: Psychosocial integrity; CNS: Coping and adaptation; CL: Comprehension

31. 2. Providing competing stimuli acknowledges the presence of the hallucination and teaches ways to decrease the frequency of hallucinations. The other nursing actions support and maintain hallucinations or deny their existence.
NP: Implementation; CN: Psychosocial integrity; CNS: Coping and adaptation; CL: Application

32. A client with schizophrenia reports that her hallucinations have decreased in frequency. Which of the following interventions would be appropriate to begin addressing the client's problem with social isolation?
 1. Have the client join in a group game.
 2. Name the client as the leader of the client support group.
 3. Have the client play solitaire.
 4. Ask the client to participate in a group sing-along.

The answer is in the cards! Or is it?

32. 4. Having the client participate in a non-competitive group activity that doesn't require individual participation won't present a threat to the client. Games can become competitive and lead to anxiety or hostility. The client probably lacks sufficient social skills to lead a group at this time. Playing solitaire doesn't encourage socialization.
NP: Implementation; CN: Psychosocial integrity; CNS: Coping and adaptation; CL: Application

33. A single 24-year-old client is admitted with acute schizophrenic reaction. Which of the following methods is appropriate therapy for this type of schizophrenia?
 1. Counseling to produce insight into behavior
 2. Biofeedback to reduce agitation associated with schizophrenia
 3. Drug therapy to reduce symptoms associated with acute schizophrenia
 4. Electroconvulsive therapy to treat the mood component of schizophrenia

33. 3. Drug therapy is usually successful in normalizing behavior and reducing or eliminating hallucinations, delusions, thought disorder, affect flattening, apathy, avolition, and asociality. Electroconvulsive therapy might be considered for schizoaffective disorder, which has a mood component, and is a treatment of choice for clinical depression. Biofeedback reduces anxiety and modifies behavioral responses but isn't the major component in the treatment of schizophrenia.
NP: Implementation; CN: Psychosocial integrity; CNS: Coping and adaptation; CL: Comprehension

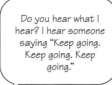

Do you hear what I hear? I hear someone saying "Keep going. Keep going. Keep going."

34. A client tells a nurse that voices are telling him to do "terrible things." Which of the following actions is part of the initial therapy?
 1. Find out what the voices are telling him.
 2. Let him go to his room to decrease his anxiety.
 3. Begin talking to the client about an unrelated topic.
 4. Tell the client the voices aren't real.

34. 1. For safety purposes, the nurse must find out whether the voices are directing the client to harm himself or others. Further assessment can help identify appropriate therapeutic interventions. Changing the topic indicates that the nurse isn't concerned about the client's fears. Dismissing the voices shuts down communication between the client and the nurse. Isolating a person during this intense sensory confusion often reinforces the psychosis.
NP: Implementation; CN: Safe, effective care environment; CNS: Safety and infection control; CL: Application

35. A newly admitted client is diagnosed with schizophrenia. The client tells the nurse that the police are looking for him and will kill him if they find him. The nurse recognizes this as delusion of which type?
 1. Paranoid
 2 Religious
 3. Grandiose
 4. Somatic

35. 1. This client is exhibiting paranoid delusions, which are excessive or irrational suspicions or distrust of others. A religious delusion is the belief that one is favored by a higher being or is an instrument of a higher being. A grandiose delusion is the belief that one possesses greatness or special powers. A somatic delusion is the belief that one's body or parts of one's body are distorted or diseased.
NP: Data collection; CN: Psychosocial integrity; CNS: Coping and adaptation; CL: Knowledge

36. A client has started taking haloperidol (Haldol). Which of the following instructions are <u>most appropriate</u> for a client taking haloperidol?
1. You should report feelings of restlessness or agitation at once.
2. You can take your herbal supplements safely with this drug.
3. Be aware you'll feel increased energy taking this drug.
4. This drug will indirectly control essential hypertension.

Don't get anxious now, but check out the words most appropriate.

37. The nurse is having an interaction with a 38-year-old male delusional client. Which is the <u>best</u> nursing action?
1. Tell the client the delusions aren't real.
2. Explain the delusion to the client.
3. Encourage the client to remain delusional.
4. Begin to develop a trusting relationship with the client.

38. Which of the following cluster of symptoms would indicate schizophrenia?
1. Persistent, intrusive thoughts leading to repetitive, ritualistic behaviors
2. Feelings of helplessness and hopelessness
3. Unstable moods and delusions of grandeur
4. Hallucinations or delusions and decreased ability to function in society

You're doing a great job!

39. The nurse is caring for a 58-year-old male client diagnosed with paranoid schizophrenia. When the client says, "the earth and the roof of the house rule the political structure with particles of rain," the nurse recognizes this as which of following types of expression?
1. Tangentiality
2. Perseveration
3. Loose associations
4. Thought blocking

36. 1. Agitation and restlessness are adverse effects of haloperidol and can be treated with anticholinergic drugs. Haloperidol isn't likely to cause essential hypertension. Using herbal supplements while taking haloperidol may interfere with the drug's effectiveness. Although the client may experience increased concentration and activity, these effects are due to a decrease in symptoms, not the drug itself.
NP: Implementation; CN: Physiological integrity; CNS: Pharmacological therapies; CL: Application

37. 4. Developing a trusting relationship gives the nurse more therapeutic time with the client. Never argue with or try to talk the client out of a delusion. The delusions are very real to him. Explaining the delusions helps the nurse, not the client. Encouraging the client to remain delusional isn't therapeutic.
NP: Implementation; CN: Psychosocial integrity; CNS: Psychosocial adaptation; CL: Application

38. 4. Schizophrenia is a brain disease characterized by a variety of symptoms, including hallucinations, delusions, and asociality. Clients with obsessive-compulsive disorder experience intrusive thoughts and ritualistic behaviors. Feelings of helplessness and hopelessness are pivotal symptoms of clinical depression. Unstable moods and delusions of grandeur are characteristics of bipolar affective disorder.
NP: Data collection; CN: Psychosocial integrity; CNS: Coping and adaptation; CL: Comprehension

39. 3. Loose association refers to changing ideas from one unrelated theme to another. Tangentiality is the wandering from topic to topic. Perseveration is involuntary repetition of the answer to a question in response to a new question. Thought blocking is having difficulty articulating a response or stopping midsentence.
NP: Data collection; CN: Psychosocial integrity; CNS: Psychosocial adaptation; CL: Application

40. Which of the following terms describes a marked decrease in the variation or intensity of emotional expression?
1. Anhedonia
2. Depression
3. Flat affect
4. Incongruence

Some illnesses can leave you feeling flat.

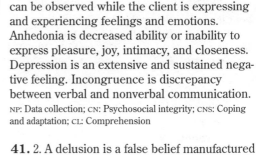

40. 3. Flat affect indicates that no expression can be observed while the client is expressing and experiencing feelings and emotions. Anhedonia is decreased ability or inability to express pleasure, joy, intimacy, and closeness. Depression is an extensive and sustained negative feeling. Incongruence is discrepancy between verbal and nonverbal communication.
NP: Data collection; CN: Psychosocial integrity; CNS: Coping and adaptation; CL: Comprehension

41. A client is admitted after being found on a highway, hitting at cars and yelling at motorists. When approached by the nurse, the client shouts, "You're the one who stole my husband from me." Which of the following conditions describes the client's condition?
1. Hallucinatory experience
2. Delusional experience
3. Disorientation to the environment
4. Asking for limit setting from the staff

41. 2. A delusion is a false belief manufactured without appropriate or sufficient evidence to support it. The client's statements don't represent hallucinations because they aren't perceptual disorders. No information in the question addresses orientation. Although limit setting is integral to a safe environment, the client's statement reflects issues about self-esteem.
NP: Data collection; CN: Psychosocial integrity; CNS: Coping and adaptation; CL: Application

42. When teaching the family of a client with schizophrenia, the nurse should provide which information?
1. Relapse can be prevented if the client takes medication.
2. Support is available to help family members meet their own needs.
3. Improvement should occur if the client has a stimulating environment.
4. Stressful situations in the family can precipitate a relapse in the client.

Don't surf through these questions. Pick up on words like most!

42. 2. Because family members of a client with schizophrenia face difficult situations and great stress, the nurse should inform them of support services that can help them cope with such problems. The nurse also should teach them that medication can't prevent relapses and that environmental stimuli may precipitate symptoms. Although stress can trigger symptoms, the nurse shouldn't make the family feel responsible for relapses.
NP: Implementation; CN: Health promotion and maintenance; CNS: Growth and development through the life span; CL: Application

43. A nurse is caring for a client hospitalized on an inpatient psychiatric unit. The client repeats the nurse's phrases and shows motor immobility with prominent grimacing. Which of the following medical diagnoses is most likely?
1. Catatonic schizophrenia
2. Disorganized schizophrenia
3. Residual schizophrenia
4. Undifferentiated schizophrenia

43. 1. Motor immobility, parroting phrases, and grimacing are characteristics of catatonic schizophrenia. Symptoms of disorganized schizophrenia are disorganized speech and behaviors with flat or inappropriate affect. Symptoms of residual schizophrenia include odd beliefs, eccentric behavior, and illogical thinking. Undifferentiated schizophrenia is characterized by delusions or hallucinations and other behaviors common to other types of schizophrenia.
NP: Data collection; CN: Psychosocial integrity; CNS: Coping and adaptation; CL: Comprehension

44. A nurse on an inpatient unit is having a discussion with a client with schizophrenia about his schedule for the day. The client comments that he was highly active at home, and then explains the volunteer job he held. Which of the following terms describes the client's thinking?
1. Circumstantiality
2. Loose associations
3. Referential
4. Tangentiality

45. While talking to a client with schizophrenia, a nurse notes that the client frequently uses unrecognizable words with no common meaning. Which of the following terms describes this?
1. Echolalia
2. Clang association
3. Neologisms
4. Word salad

46. While caring for a hospitalized client diagnosed with schizophrenia, a nurse observes the client watching television. The client tells the nurse the television is speaking directly to him. Which of the following terms describes this belief?
1. Autistic thinking
2. Concrete thinking
3. Paranoid thinking
4. Referential thinking

Do you understand what question 46 refers to?

47. A nurse is talking with a family of a client diagnosed with schizophrenia. The mother asks, "What causes this disease?" Which of the following explanations for the disorder is the most widely accepted?
1. Prenatal or postpartum central nervous system damage
2. Bacterial infections of the mother during pregnancy or delivery
3. A biological predisposition exacerbated by environmental stressors
4. Lack of bonding and attachment during infancy, which leads to depression in later life

Pay attention to the key words here.

44. 4. Tangentiality describes thought patterns loosely connected but not directly related to the topic. In circumstantiality, the person digresses with unnecessary details. Loose associations are rapid shifts in the expression of ideas from one subject to another in an unrelated manner. Referential thinking is when an individual incorrectly interprets neutral incidents and external events as having a particular or special meaning for him.
NP: Data collection; CN: Psychosocial integrity; CNS: Coping and adaptation; CL: Application

45. 3. Neologisms are newly coined words with personal meanings to the client with schizophrenia. Word salads are stringing words in sequence that have no connection to one another. Echolalia is parrotlike echoing of spoken words or sounds. Clanging is the association of words by sound rather than meaning.
NP: Data collection; CN: Psychosocial integrity; CNS: Psychosocial adaptation; CL: Application

46. 4. Referential, or primary process thinking, is a belief that incidents and events in the environment have special meaning for the client. Autistic thinking is a disturbance in thought due to the intrusion of a private fantasy world, internally stimulated, resulting in abnormal responses to people. Concrete thinking is the literal interpretation of words and symbols. Paranoid thinking is the belief that others are trying to harm you.
NP: Data collection; CN: Psychosocial integrity; CNS: Coping and adaptation; CL: Comprehension

47. 3. The holistic theory, currently the most widely accepted theory of its type, states that an interaction between biological predisposition and environmental stressors is the cause of schizophrenia. The psychoanalytic perspective involves the belief that the mother-infant bond is the source of the schizophrenia. The biological explanation states that schizophrenia is caused by a brain disease, a bacterial infection in utero, or early brain damage.
NP: Implementation; CN: Physiological integrity; CNS: Physiological adaptation; CL: Application

48. A 31-year-old female client with a diagnosis of schizophrenia walks with the nurse to the dayroom but refuses to speak. What's the most therapeutic nursing intervention?
1. Ignore the refusal to speak and talk about something such as the weather.
2. Tell the client that the refusal to speak is making others uncomfortable.
3. Plan to spend time with the client, even in silence.
4. Make the client attend therapy with others so the other clients can encourage talking.

49. A client is taking chlorpromazine (Thorazine) for the treatment of schizophrenia. This drug blocks the transmission of which of the following substances?
1. Dopamine
2. Epinephrine
3. Norepinephrine
4. Thyroxine

50. In preparation for discharge, a client diagnosed with schizophrenia was taught self-symptom management as part of a relapse prevention program. Which of the following statements indicates the client understands symptom monitoring?
1. "When I hear voices, I become afraid I'll relapse."
2. "My parents aren't involved enough to be aware if I begin to relapse."
3. "My family is more protected from stress if I keep them out of my illness process."
4. "When I'm feeling stressed, I go to a quiet room of the house by myself and do imagery."

51. A client diagnosed with schizophrenia has been taking haloperidol (Haldol) for 1 week when a nurse observes that the client's eyeball is fixed on the ceiling. Which of the following specific conditions is the client exhibiting?
1. Akathisia
2. Neuroleptic malignant syndrome
3. Oculogyric crisis
4. Tardive dyskinesia

Feedback alerts you to a lack of understanding.

You'll need to read the key word carefully on this one.

48. 3. Spending time with the client lets her know that the nurse cares. Although sitting in silence can be awkward, it's therapeutic. Ignoring the behavior and using small talk encourages the silence. Telling the client about making others uncomfortable usually has no effect on the behavior. Coercing the client to attend therapy with others would increase the silence.
NP: Implementation; CN: Psychosocial integrity; CNS: Coping and adaptation; CL: Application

49. 1. Most antipsychotic agents block the transmission of dopamine to the brain. Other transmitters linked to schizophrenia include serotonin, acetylcholine, and norepinephrine. Epinephrine, norepinephrine, and thyroxine aren't blocked by chlorpromazine.
NP: Implementation; CN: Physiological integrity; CNS: Pharmacological therapies; CL: Knowledge

50. 4. This statement indicates the client has learned a technique for coping with stress with the use of imagery technique. The other statements don't show an understanding of self-symptom monitoring and may result in symptom intensification and possible relapse.
NP: Evaluation; CN: Psychosocial integrity; CNS: Coping and adaptation; CL: Application

51. 3. An oculogyric crisis involves a fixed positioning of the eyes, typically in an upward gaze. The condition is uncomfortable but not life-threatening. Akathisia is a restlessness that can cause pacing and tapping of the fingers or feet. Stereotyped involuntary movements (tongue protrusion, lip smacking, chewing, blinking, and grimacing) characterize tardive dyskinesia.
NP: Data collection; CN: Physiological integrity; CNS: Pharmacological therapies; CL: Application

52. A client taking antipsychotic medications shows dystonic reactions, including torticollis and oculogyric crisis. Which of the following medications is given?
1. Benztropine (Cogentin)
2. Chlordiazepoxide (Librium)
3. Diazepam (Valium)
4. Fluoxetine (Prozac)

A frequent mistake on a test like this is to miss words like *most* frequent.

52. 1. Benztropine and trihexyphenidyl (Artane) are anticholinergic drugs used to counteract the dystonic reactions and adverse reactions of antipsychotic drugs. The antihistamine diphenhydramine (Benadryl) is also effective in treating extrapyramidal symptoms. Fluoxetine is an antidepressant, and diazepam and chlordiazepoxide are minor tranquilizers or antianxiety agents.

NP: Implementation; CN: Physiological integrity; CNS: Pharmacological therapies; CL: Application

53. A client diagnosed with schizophrenia is having hallucinations. The most frequent type of hallucination involves which of the following senses?
1. Hearing
2. Smell
3. Touch
4. Vision

53. 1. Hallucinations are sensory impressions or experiences without external stimuli. Auditory hallucinations are the most frequent symptoms in clients with schizophrenia. Olfactory, visual, and tactile hallucinations are seen in a number of psychiatric disorders, such as substance abuse disorders, delirium, and dementia, but not as often in schizophrenia.

NP: Data collection; CN: Physiological integrity; CNS: Physiological adaptation; CL: Comprehension

54. A 22 year-old woman was brought to the hospital by her parents because of her bizarre behavior. The parents reported that she stayed in her room, refused to eat meals with the family, and talked to herself almost constantly. She lost interest in her job and had been fired because of her inability to perform. During the intake process, the nurse notes that the client is unkempt, has body odor, exhibits illogical thought patterns, and appears to be listening to someone no one else can see or hear. Which nursing goal is the priority for this client?
1. Maintain safety.
2. Ensure adequate nutrition.
3. Orient the client.
4. Provide hygiene measures.

Prioritize!

54. 1. Whenever a client is hallucinating or otherwise out of touch with reality, safety is always the primary concern. Although ensuring adequate nutrition, orienting the client, and providing hygiene measures are valid concerns, they're all secondary to ensuring a safe environment.

NP: Planning; CN: Safe, effective care environment; CNS: Safety and infection control; CL: Application

55. As a nurse approaches the nurses' station, a client with the diagnosis of delusional disorder raises his voice and says, "You're following me. What do you want?" To prevent escalating fear and anger, the nurse takes a nonthreatening posture and makes which of the following responses in a calm voice?
1. "Are you frightened?"
2. "You know I'm not following you."
3. "You'll have to go into seclusion if you continue to threaten me."
4. "I'm sorry if I frightened you. I was returning to the nurses' station after going out for lunch."

56. Which of the following actions by a client with stable schizophrenia is most important for preventing relapse?
1. Attending group therapy sessions
2. Participating in family support meetings
3. Attending social skills training sessions
4. Consistently taking prescribed medications

57. A client approaches the nurse and points at the sky, showing her where the men would be coming from to get him. Which of the following responses is most therapeutic?
1. "Why do you think the men are coming here?"
2. "You're safe here, we won't let them harm you."
3. "It seems like the world is pretty scary for you, but you're safe here."
4. "There are no bad men in the sky because no one lives that close to earth."

58. A client is brought to the crisis response center by his family. During evaluation, he reports being depressed for the last month and complains about voices constantly whispering to him. Which of the following diagnoses is the most likely?
1. Catatonic schizophrenia
2. Disorganized schizophrenia
3. Paranoid schizophrenia
4. Schizoaffective disorder

The words most important in question 56 indicate the need to prioritize. As a matter of fact, there's lots of prioritizing on this page.

55. 4. Being clear in communication, remaining calm, and showing concern increases the chance the client will cooperate, lessening the potential for violence. The first response tries to identify the client's feelings but doesn't convey warmth and concern. The second response isn't empathic and shows no indication of trying to reach the client at a level beyond content of communication. The third response may increase the client's anxiety, fear, and mistrust when the nurse engages in a power struggle and triggers competitiveness within the client.
NP: Implementation; CN: Psychosocial integrity; CNS: Coping and adaptation; CL: Application

56. 4. Although all of the choices are important for preventing relapse, compliance with the medication regimen is central to the treatment of schizophrenia, a brain disease.
NP: Evaluation; CN: Safe, effective care environment; CNS: Coordinated care; CL: Application

57. 3. This response acknowledges the client's fears, listens to his feelings, and offers a sense of security as the nurse tries to understand the concerns behind the symbolism. She reflects these concerns to the client, along with reassurance of safety. The first response validates the delusion, not the feelings and fears, and doesn't orient the client to reality. The second response gives false reassurance. Because the nurse isn't sure of the symbolism, she can't make this promise. The last response rejects the client's feelings and doesn't address the client's fears.
NP: Implementation; CN: Safe, effective care environment; CNS: Coordinated care; CL: Application

58. 4. A client with a major depressive episode who begins to hear voices and at times thinks someone is after him is most likely schizoaffective. The client who repeats phrases and shows waxy flexibility or stupor with prominent grimaces is most likely catatonic. The client who expresses thoughts of people spying on him, attributes ulterior motives to others, and has a flat affect is most likely paranoid schizophrenic. The client with disorganized speech and behavior and a flat or inappropriate affect most likely has disorganized schizophrenia.
NP: Data collection; CN: Psychosocial integrity; CNS: Coping and adaptation; CL: Analysis

59. Which of the following nursing interventions is most appropriate for use with a client with paranoid schizophrenia?
1. Defend yourself when the client is verbally hostile toward you.
2. Provide a warm approach by touching the client.
3. Explain everything you're doing before you do it.
4. Clarify the content of the client's delusions.

60. Since admission 4 days ago, a client has refused to take a shower, stating, "There are poison crystals hidden in the shower head. They'll kill me if I take a shower." Which nursing action is most appropriate?
1. Dismantle the shower head and show the client that there's nothing in it.
2. Explain that other clients are complaining about the client's body odor.
3. Ask a security guard to assist you in giving the client a shower.
4. Accept these fears, and allow the client to take a sponge bath.

61. A client who's diagnosed with paranoid schizophrenia tells the nurse that a computer chip placed in his brain has informed him that his wife is cheating. What's the nurse's best response?
1. Attempt to disprove the delusion.
2. Ignore the client's statement and engage him in a game of pool.
3. Allow the client to write a letter to his wife expressing his thoughts.
4. Convey acceptance of the client's need for false belief but that you don't share this belief.

62. Low-potency psychotropic agents are likely to cause which of the following adverse reactions?
1. Akathisia
2. Dystonia
3. Sedation
4. Tardive dyskinesia

You're almost done!

You finished chapter 17! Congratulations!

59. 3. Explaining everything you do will prevent misinterpretation of your actions. A nondefensive stance provides an atmosphere in which the client's angry feelings can be explored. Touching the paranoid client should be avoided because it can be interpreted as threatening. The content of delusions shouldn't be the focus of your care because the content is illogical.
NP: Implementation; CN: Psychosocial integrity; CNS: Psychosocial adaptation; CL: Analysis

60. 4. By acknowledging the client's fears, the nurse can arrange to meet his hygiene needs in another way. Because these fears are real to the client, providing a demonstration of reality wouldn't be effective at this time. Telling the client that other's are complaining of his body odor or having a security guard help you force him take a shower would violate the client's rights by shaming or embarrassing him.
NP: Implementation; CN: Psychosocial integrity; CNS: Psychosocial adaptation; CL: Application

61. 4. By conveying acceptance of the client's need for false belief, the nurse promotes trust with the client. The client must understand that the nurse doesn't view the idea as real. Attempting to disprove the delusion may provoke agitation and doesn't foster trust or enhance the client's sense of safety and reality. Provide noncompetitive activities and avoid aggressive games requiring close physical contact. Allowing the client to write a letter to his wife expressing irrational thoughts would lead the client to think that the nurse shares the false belief and wouldn't be therapeutic.
NP: Implementation; CN: Psychosocial integrity; CNS: Psychosocial adaptation; CL: Application

62. 3. Low-potency agents usually have a weak affinity for dopamine-2 receptors and cause sedation. Tardive dyskinesia, dystonia, and akathisia are more likely to be caused by high-potency psychotropic agents because more of the drug is needed for therapeutic effects, increasing the likelihood of these extrapyramidal symptoms.
NP: Data collection; CN: Physiological integrity; CNS: Pharmacological therapies; CL: Knowledge

---

NP: Nursing process   CN: Client needs category   CNS: Client needs subcategory   CL: Cognitive level

About the only substance of abuse this chapter doesn't cover is my personal weakness — chocolate mousse! Think of me as you work through this chapter. I'll be the one with chocolate smudges on her fingers. Tee-hee!

Chapter 18
Substance abuse disorders

1. Family members of a client who abuses alcohol asks a nurse to help them intervene. Which of the following actions is essential for a successful intervention?
1. All family members must tell the client they're powerless.
2. All family members must describe how the addiction affects them.
3. All family members must come up with their share of financial support.
4. All family members must become caregivers during the detoxification period.

2. A client who abuses alcohol tells a nurse, "I'm sure I can become a social drinker." Which of the following responses is <u>most appropriate</u>?
1. "When do you think you can become a social drinker?"
2. "What makes you think you'll learn to drink normally?"
3. "Does your alcohol use cause major problems in your life?"
4. "How many alcoholic beverages can a social drinker consume?"

3. A client asks a nurse not to tell his parents about his alcohol problem. Which of the following responses is <u>most appropriate</u>?
1. "How can you not tell them? Is that being honest?"
2. "Don't you think you'll need to tell them someday?"
3. "Do alcohol problems run in either side of your family?"
4. "What do you think will happen if you tell your parents?"

You can do it!

The words most appropriate help clarify the correct answer.

1. 2. After the family is taught about addiction, they must write down examples of how the addiction has affected each of them and use this information during the intervention. It isn't necessary to tell the client the family is powerless. The family is empowered through this intervention experience. In many cases, a third-party payer will help with treatment costs. Doing an intervention doesn't make family members responsible for financial support or providing care and support during the detoxification period.
NP: Implementation; CN: Psychosocial integrity; CNS: Psychosocial adaptation; CL: Analysis

2. 3. This question may help the client recall the problematic results of using alcohol and the reasons the client began treatment. Asking how many alcoholic beverages a social drinker can consume and why the client thinks he can drink normally will encourage the addicted person to defend himself and deny the problem. Asking when he believes he can become a social drinker will only encourage the addicted person to deny the problem and develop an unrealistic, self-defeating goal.
NP: Implementation; CN: Psychosocial integrity; CNS: Psychosocial adaptation; CL: Application

3. 4. Clients who struggle with addiction problems often believe people will be judgmental, rejecting, and uncaring if they're told that the client is recovering from alcohol abuse. The first response challenges the client and will put him on the defensive. The second response will make the client defensive and construct rationalizations as to why his parents don't need to know. The third response is a good assessment question, but it isn't an appropriate question to ask a client who's afraid to tell others about his addiction.
NP: Implementation; CN: Psychosocial integrity; CNS: Psychosocial adaptation; CL: Analysis

NP: Nursing process   CN: Client needs category   CNS: Client needs subcategory   CL: Cognitive level

4. A nurse assesses a client for signs of alcohol withdrawal. During the period of early withdrawal, which of the following findings are expected?
 1. Depression
 2. Hyperactivity
 3. Insomnia
 4. Nausea

A disease can have many stages, each with its own symptoms.

5. Which of the following health findings is expected in a client who chronically abuses alcohol?
 1. Enlarged liver
 2. Nasal irritation
 3. Muscle wasting
 4. Limb paresthesia

6. Within 8 hours of her last drink, an alcoholic client experiences tremors, loss of appetite, disordered thinking, and insomnia. The nurse believes this client is in second-stage withdrawal. What should the nurse do next?
 1. Give disulfiram (Antabuse) as prescribed.
 2. Obtain a physician's order for lorazepam (Ativan).
 3. Help the client to engage in progressive muscle-relaxation techniques.
 4. Provide the client with constant one-on-one monitoring.

Accuracy is a skill that serves a nurse well. Aim for it in question 7!

7. A client who abuses alcohol tells a nurse, "Alcohol helps me sleep." Which of the following information about alcohol use affecting sleep is most accurate?
 1. Alcohol doesn't help promote sleep.
 2. Continued alcohol use causes insomnia.
 3. One glass of alcohol at dinnertime can induce sleep.
 4. Sometimes alcohol can make one drowsy enough to fall asleep.

4. 4. Nausea and, later, vomiting are early signs of alcohol withdrawal. Insomnia, hyperactivity, and depression aren't associated with early alcohol withdrawal.
NP: Data collection; CN: Psychosocial integrity; CNS: Coping and adaptation; CL: Knowledge

5. 1. A major effect of alcohol on the body is liver impairment, and an enlarged liver is a common physical finding. Muscle wasting and limb paresthesia don't tend to occur with clients who abuse alcohol. Nasal irritation is commonly seen with clients who snort cocaine.
NP: Data collection; CN: Psychosocial integrity; CNS: Coping and adaptation; CL: Knowledge

6. 2. A client in second-stage withdrawal should be medicated with a benzodiazepine, such as lorazepam, to prevent progression of symptoms to delirium tremors, a life-threatening withdrawal syndrome. Disulfiram is used during early recovery, not during detoxification. Progressive muscle relaxation isn't particularly effective during withdrawal. Close monitoring during withdrawal is appropriate after the client has been medicated for withdrawal symptoms.
NP: Implementation; CN: Psychosocial integrity ; CNS: Psychosocial adaptation; CL: Application

7. 2. Alcohol use may initially promote sleep, but with continued use, it causes insomnia. Evidence shows that alcohol doesn't facilitate sleep. One glass of alcohol at dinnertime won't induce sleep. The last statement that alcohol can make one drowsy enough to fall asleep doesn't give information about how alcohol affects sleep. It makes the client think alcohol use to induce sleep is an appropriate strategy to try.
NP: Data collection; CN: Psychosocial integrity; CNS: Psychosocial adaptation; CL: Analysis

8. A family expresses concern when a family member withdrawing from alcohol is given lorazepam (Ativan). Which of the following information should be given to the family about the medication?

1. The medication promotes a sense of well-being during the client's difficult withdrawal period.
2. The medication is given for a short time to help the client complete the withdrawal process.
3. The medication will help the client forget about the physical sensations that go with alcohol withdrawal.
4. The medication helps in the treatment of coexisting diseases, such as cardiac problems and hypertension.

It's important to teach the family and the client

8. 2. Lorazepam is a short-acting benzodiazepine usually given for 1 week to help the client in alcohol withdrawal. The medication isn't given to help forget the experience; it lessens the symptoms of withdrawal. It isn't used to treat coexisting cardiovascular problems or promote a sense of well-being.

NP: Implementation; CN: Physiological integrity; CNS: Pharmacological therapies; CL: Comprehension

9. A client who abuses alcohol tells a nurse everyone in his family has an alcohol problem and nothing can be done about it. Which of the following responses is the most appropriate?

1. "You're right; it's much harder to become a recovering person."
2. "This is just an excuse for you so you don't have to work on becoming sober."
3. "Sometimes nothing can be done, but you may be the exception in this family."
4. "Alcohol problems can occur in families, but you can decide to take the steps to become and stay sober."

9. 4. This statement challenges the client to become proactive and take the steps necessary to maintain a sober lifestyle. The first response agrees with the client's denial and isn't a useful response. The second response confronts the client and may make him more adamant in defense of this position. The third response agrees with the client's denial and isn't a useful response.

NP: Implementation; CN: Psychosocial integrity; CNS: Psychosocial adaptation; CL: Application

10. Which of the following major cardiovascular problems may occur in a client with chronic alcoholism?

1. Arteriosclerosis
2. Heart failure
3. Heart valve damage
4. Pericarditis

Alcohol can really take a toll on me!

10. 2. Heart failure is a severe cardiac consequence associated with long-term alcohol use. Heart valve damage, pericarditis, and arteriosclerosis aren't medical consequences of alcoholism.

NP: Data collection; CN: Physiological integrity; CNS: Reduction of risk potential; CL: Knowledge

11. Which of the following assessment findings is commonly associated with the abuse of alcohol in a young, depressed adult woman?

1. Defiant responses
2. Infertility
3. Memory loss
4. Sexual abuse

11. 4. Many women diagnosed with substance abuse problems also have a history of physical or sexual abuse. Alcohol abuse isn't a common finding in a young woman showing defiant behavior or experiencing infertility. Memory loss isn't a common finding in a young woman experiencing alcohol abuse.

NP: Data collection; CN: Psychosocial integrity; CNS: Coping and adaptation; CL: Analysis

12. A nurse determines that a client who abused alcohol has nutritional problems. Which of the following strategies is best for addressing the client's nutritional needs?

1. Encourage the client to eat a diet high in calories.
2. Help the client recognize and follow a balanced diet.
3. Have the client drink liquid protein supplements daily.
4. Have the client monitor the calories consumed each day.

Hmmmm, you said the best?

12. 2. Clients who abuse alcohol are often malnourished and need help to follow a balanced diet. Increasing calories may cause the client to just eat empty calories. The client must be involved in the decision to supplement the daily dietary intake. The nurse can't force the client to drink liquid protein supplements. Having the client monitor calorie intake could be done only after the client recognizes the need to maintain a balanced diet. Calorie counts usually aren't needed in most recovering clients who begin to eat from the basic food groups.

NP: Implementation; CN: Physiological integrity; CNS: Reduction of risk potential; CL: Comprehension

13. A client with a history of alcohol abuse refuses to take vitamins. Which of the following statements is most appropriate for explaining why vitamins are important?

1. "It's important to take vitamins to stop your craving."
2. "Prolonged use of alcohol can cause vitamin depletion."
3. "For every vitamin you take, you'll help your liver heal."
4. "By taking vitamins, you don't need to worry about your diet."

13. 2. Chronic alcoholism interferes with the metabolism of many vitamins. Vitamin supplements can prevent deficiencies from occurring. Taking vitamins won't stop a person from craving alcohol or help a damaged liver heal. A balanced diet is *essential* in addition to taking multivitamins.

NP: Implementation; CN: Physiological integrity; CNS: Reduction of risk potential; CL: Application

14. Which of the following behaviors in a client who abuses alcohol indicates a knowledge deficit in nutrition?

1. Avoiding foods high in fat
2. Eating only one adequate meal each day
3. Taking vitamin and mineral supplements
4. Eating large portions of foods containing fiber

Careful! The word deficit makes this a negative question.

14. 2. If the client eats only one adequate meal each day, there will be a deficit of essential nutrients. It's appropriate for the client to take vitamin and mineral supplements to prevent deficiency in these nutrients. Avoiding foods high in fat content and consuming large portions of foods containing fiber indicate the client has good knowledge about nutrition.

NP: Data collection; CN: Physiological integrity; CNS: Reduction of risk potential; CL: Knowledge

15. The nurse is caring for a client who typically consumes 15 to 20 beers per week and is extremely defensive about his alcohol intake. He admits to experiencing blackouts and has had three alcohol-related accidents. What is the best action for this client to take?

1. Monitor his alcohol intake.
2. Switch to low-alcohol beer or wine cooler.
3. Limit his intake to no more than three beers per drinking occasion.
4. Abstain from alcohol all together.

15. 4. This client demonstrates behaviors consistent with middle-stage addiction. Once addicted, the only way to control intake is to abstain altogether—monitoring or limiting intake, or switching to low-alcoholic drinks, will not work for this client.

NP: Planning; CN: Psychosocial integrity; CNS: Psychosocial adaptation; CL: Comprehension

16. The nurse is caring for a client who is undergoing treatment for acute alcohol dependence. The client tells the nurse, "I don't have a problem. My wife made me come here." What defense mechanism is the client using?
1. Projection and suppression
2. Denial and rationalization
3. Rationalization and repression
4. Suppression and denial

16. 2. The client is using denial and rationalization. Denial is the unconscious disclaimer of unacceptable thoughts, feelings, needs, or certain external factors. Rationalization is the unconscious effort to justify intolerable feelings, behaviors, and motives. The client isn't using projection, suppression, or repression. Emotions, behavior, and motives, which are consciously intolerable, are denied and then attributed to others in projection. Suppression is a conscious effort to control and conceal unacceptable ideas and impulses into the unconscious. Repression is the unconscious placement of unacceptable feelings into the unconscious mind.

NP: Data collection; CN: Physiological integrity; CNS: Physiological adaptation; CL: Application

17. Which of the following short-term goals should be a priority for a client with a knowledge deficit about the effects of alcohol on the body?
1. Test blood chemistries daily.
2. Verbalize the results of substance use.
3. Talk to a pharmacist about the substance.
4. Attend a weekly aerobic exercise program.

Prioritize to find this answer!

17. 2. It's important for the client to talk about the health consequences of the continued use of alcohol. Testing blood chemistries daily gives the client minimal knowledge about the effects of alcohol on the body and isn't the most useful information in a teaching plan. A pharmacist isn't the appropriate health care professional to educate the client about the effects of alcohol use on the body. Although exercise is an important goal of self-care, it doesn't address the client's knowledge deficit about the effects of alcohol on the body.

NP: Planning; CN: Physiological integrity; CNS: Reduction of risk potential; CL: Application

18. The nurse is assigned to care for a recently admitted client who has attempted suicide. What should the nurse be sure to do?
1. Search the client's belongings and room carefully for items that could be used to attempt suicide.
2. Express trust that the client won't cause self-harm while in the facility.
3. Respect the client's privacy and don't search any belongings.
4. Remind all staff members to check on the client frequently.

18. 1. Because a client who has attempted suicide could try again, the nurse should search the client's belongings and room to remove any items that could be used in another suicide attempt. Expressing trust that the client won't cause self-harm may increase guilt and pain if the client can't live up to that trust. The nurse should search the client's belongings because the need to maintain a safe environment supersedes the client's right to privacy. Although frequent checks by staff members are helpful, they aren't enough because the client may attempt suicide between checks.

NP: Planning; CN: Safe, effective care environment; CNS: Safety and infection control; CL: Application

19. A client withdrawing from alcohol says he's worried about periodic hallucinations. Which of the following interventions is best for this client's problem?
  1. Point out that the sensation doesn't exist.
  2. Allow the client to talk about the experience.
  3. Encourage the client to wash the body areas well.
  4. Determine if the client has a cognitive impairment.

19. 2. The client needs to talk about the periodic hallucinations to prevent them from becoming triggers to acting out behaviors and possible self-injury. Determining if the client has a cognitive impairment and encouraging the client to wash the body areas well don't address the problem of periodic hallucinations. The client's experience of sensory-perceptual alterations must be acknowledged.
NP: Implementation; CN: Psychosocial integrity; CNS: Psychosocial adaptation; CL: Application

Alcohol can make me more susceptible to infections.

20. A client who has been drinking alcohol for 30 years asks a nurse if permanent damage has occurred to his immune system. Which of the following responses is the best?
  1. "There is often less resistance to infections."
  2. "Sometimes the body's metabolism will increase."
  3. "Put your energies into maintaining sobriety for now."
  4. "Drinking puts you at high risk for disease later in life."

20. 1. Chronic alcohol use depresses the immune system and causes increased susceptibility to infections. A nutritionally well-balanced diet that includes foods high in protein and B vitamins will help develop a strong immune system. The potential damage to the immune system doesn't increase the body's metabolism. The third response negates the client's concern and isn't an appropriate or caring response. Drinking alcohol may put the client at risk for immune system problems at any time in life.
NP: Implementation; CN: Physiological integrity; CNS: Physiological adaptation; CL: Analysis

21. A client experiencing alcohol withdrawal is upset about going through detoxification. Which of the following goals is a priority?
  1. The client will commit to a drug-free lifestyle.
  2. The client will work with the nurse to remain safe.
  3. The client will drink plenty of fluids on a daily basis.
  4. The client will make a personal inventory of strengths.

Question 21 is asking you to prioritize!

21. 2. The priority goal is for client safety. Although drinking enough fluids, identifying personal strengths, and committing to a drug-free lifestyle are important goals, the nurse's first priority must be to promote client safety.
NP: Planning; CN: Psychosocial integrity; CNS: Psychosocial adaptation; CL: Comprehension

22. A client recovering from alcohol abuse needs to develop effective coping skills to handle daily stressors. Which of the following interventions is most useful to the client?
  1. Determine the client's verbal skills.
  2. Help the client avoid conflict.
  3. Discuss examples of successful coping behavior.
  4. Teach the client to accept uncomfortable situations.

22. 3. The client needs help identifying a successful coping behavior and developing ways to incorporate that behavior into daily functioning. There are many skills for coping with stress, and determining the client's level of verbal skills may not be important. Encouraging the client to avoid conflict or to accept uncomfortable situation prevents him from learning skills to handle daily stressors.
NP: Implementation; CN: Psychosocial integrity; CNS: Psychosocial adaptation; CL: Analysis

NP: Nursing process   CN: Client needs category   CNS: Client needs subcategory   CL: Cognitive level

23. A client is struggling with alcohol dependence. Which of the following communication strategies would be most effective?

1. Speak briefly and directly.
2. Avoid blaming or preaching to the client.
3. Confront feelings and examples of perfectionism.
4. Determine if nonverbal communication will be more effective.

24. A 30-year-old woman signed herself into an alcohol treatment program. During the first visit with the nurse, she vehemently maintains that she has no problem with alcohol. She states that she's in this program only because her husband issued an ultimatum. Which response is best?

1. "I wonder why your husband would issue such an ultimatum?"
2. "Because you came voluntarily, you're free to leave anytime you wish."
3. "From your point of view, what is most important for me to know about you?"
4. "You sound pretty definite about not having a problem with alcohol."

25. A client recovering from alcohol addiction has limited coping skills. Which of the following characteristics would indicate relationship problems?

1. The client is prone to panic attacks.
2. The client doesn't pay attention to details.
3. The client has poor problem-solving skills.
4. The client ignores the need to relax and rest.

26. A nurse suggests to a client struggling with alcohol addiction that keeping a journal may be helpful. Which of the following reasons best explains this?

1. The client can identify stressors and responses to them.
2. The client will be better able to understand the diagnosis.
3. The client can help others by reading the journal to them.
4. The client will develop an emergency plan for use in a crisis.

Effective communication helps in all areas of recovery.

When you want an explanation from someone, you expect the best one, right? Well, NCLEX expects the same!

23. 2. Blaming or preaching to the client causes negativity and prevents the client from hearing what the nurse has to say. Speaking briefly to the client may not allow time for adequate communication. Perfectionism doesn't tend to be an issue. Determining if nonverbal communication will be more effective is better suited for a client with cognitive impairment.
NP: Implementation; CN: Psychosocial integrity; CNS: Psychosocial adaptation; CL: Analysis

24. 3. The third response allows the nurse to collect more information. The first response focuses on the husband, who isn't the client. The second response is abrasive and blocks communication. The fourth response doesn't allow for further exploration.
NP: Implementation; CN: Psychosocial integrity; CNS: Psychosocial adaptation; CL: Application

25. 3. To have satisfying relationships, a person must be able to communicate and problem solve. Relationship problems don't predispose people to panic attacks more than other psychosocial stressors. Paying attention to details isn't a major concern when addressing the client's relationship difficulties. Although ignoring the need for rest and relaxation is unhealthy, it shouldn't pose a major relationship problem.
NP: Data collection; CN: Psychosocial integrity; CNS: Coping and adaptation; CL: Analysis

26. 1. Keeping a journal enables the client to identify problems and patterns of coping. From this information, the difficulties the client faces can be addressed. A journal isn't necessarily kept to promote better understanding of the client's illness, but it helps the client understand himself better. Journals aren't read to other people unless the client wants to share a particular part. Journals aren't typically used for identifying an emergency plan for use in a crisis.
NP: Implementation; CN: Psychosocial integrity; CNS: Coping and adaptation; CL: Application

27. Which of the following information is most important to use in a teaching plan for a client who abused alcohol?
1. Personal needs
2. Illness exacerbation
3. Cognitive distortions
4. Communication skills

Prioritize!

27. 4. Addicted clients often have difficulty communicating their needs in an appropriate way. Learning appropriate communication skills is a major goal of treatment. Next, behavior that focuses on the self and meeting personal needs will be addressed. Teaching about illness exacerbation isn't a skill, but it is essential for relaying information about relapse. The identification of cognitive distortions would be difficult if the client has poor communication skills.
NP: Planning; CN: Psychosocial integrity; CNS: Psychosocial adaptation; CL: Analysis

28. Which of the following assessments must be done before starting a teaching session with a client who abuses alcohol?
1. Sleep patterns
2. Decision making
3. Note-taking skills
4. Readiness to learn

28. 4. It's important to know if the client's current situation helps or hinders the potential to learn. Decision making and sleep patterns aren't factors that must be assessed before teaching about addiction. Note-taking skills aren't a factor in determining whether the client will be receptive to teaching.
NP: Data collection; CN: Psychosocial integrity; CNS: Coping and adaptation; CL: Knowledge

29. A nurse is developing strategies to prevent relapse with a client who abuses alcohol. Which of the following client interventions is important?
1. Avoid taking over-the-counter medications.
2. Limit monthly contact with the family of origin.
3. Refrain from becoming involved in group activities.
4. Avoid people, places, and activities from the former lifestyle.

29. 4. Changing the client's old habits is essential for sustaining a sober lifestyle. Certain over-the-counter medications that don't contain alcohol will probably need to be used by the client at certain times. It's unrealistic to have the client abstain from all such medications. Contact with the client's family of origin may not be a trigger to relapse, so limiting contact wouldn't be useful. Refraining from group activities isn't a good strategy to prevent relapse. Going to Alcoholics Anonymous and other support groups will help prevent relapse.
NP: Implementation; CN: Psychosocial integrity; CNS: Psychosocial adaptation; CL: Analysis

All answers may seem right, but choose the most appropriate one.

30. A client asks a nurse, "Why does it matter if I talk to my peers in group therapy?" Which of the following responses is most appropriate?
1. "Group therapy lets you to see what you're doing wrong in your life."
2. "Group therapy acts as a defense against your disorganized behavior."
3. "Group therapy provides a way to ask for support as well as to support others."
4. "In group therapy, you can vent your frustrations and others will listen."

30. 3. The best response addresses how group therapy provides opportunities to communicate, learn, and give and get support. Group members will give a client feedback, not just point out what a client is doing wrong. Group therapy isn't a defense against disorganized behavior. People can express all kinds of feelings and discuss a variety of topics in group therapy. Interactions are goal oriented and not just vehicles to vent one's frustrations.
NP: Implementation; CN: Psychosocial integrity; CNS: Psychosocial adaptation; CL: Application

31. A family meeting is held with a client who abuses alcohol. While listening to the family, which of the following unhealthy communication patterns might be identified?
1. Use of descriptive jargon
2. Disapproval of behaviors
3. Avoidance of conflicting issues
4. Unlimited expression of nonverbal communication

Again, you have been asked to prioritize!

32. A client addicted to alcohol begins individual therapy with a nurse. Which of the following interventions should be a priority?
1. Learn to express feelings.
2. Establish new roles in the family.
3. Determine strategies for socializing.
4. Decrease preoccupation with physical health.

33. A client recovering from alcohol addiction asks a nurse how to talk to his children about the impact of his addiction on them. Which of the following responses is most appropriate?
1. "Try to limit references to the addiction and focus on the present."
2. "Talk about all the hardships you've had in working to remain sober."
3. "Tell them you're sorry and emphasize that you're doing so much better now."
4. "Talk to them by acknowledging the difficulties and pain your drinking caused."

34. A client with alcoholism has just completed a residential treatment program. What can this client reasonably expect?
1. Her family will no longer be dysfunctional.
2. She'll need ongoing support to remain abstinent.
3. She doesn't need to be concerned about abusing alcohol in the future.
4. She can learn to consume alcohol without problems.

31. 3. The interaction pattern of a family with a member who abuses alcohol often revolves around denying the problem, avoiding conflict, or rationalizing the addiction. Health care providers are more likely to use jargon. The family might have problems setting limits and expressing disapproval of the client's behavior. Nonverbal communication often gives the nurse insight into family dynamics.
NP: Data collection; CN: Psychosocial integrity; CNS: Psychosocial adaptation; CL: Analysis

32. 1. The client must address issues, learn ways to cope effectively with life stressors, and express his needs appropriately. After the client establishes sobriety, the possibility of taking on new roles can become a reality. Determining strategies for socializing isn't the priority intervention for an addicted client. Usually, these clients need to change former socializing habits. Clients addicted to alcohol don't tend to be preoccupied with physical health problems.
NP: Implementation; CN: Psychosocial integrity; CNS: Psychosocial adaptation; CL: Comprehension

33. 4. Part of the healing process for the family is to acknowledge the pain, embarrassment, and overall difficulties the client's drinking problem caused family members. The first response facilitates the client's ability to deny the problem. The second prevents the client from acknowledging the difficulties the children endured. The third leads the client to believe only a simple apology is needed. The addiction must be addressed, and the children's pain acknowledged.
NP: Implementation; CN: Psychosocial integrity; CNS: Psychosocial adaptation; CL: Comprehension

34. 2. Addiction is a relapsing illness. Support is helpful to most people in maintaining an abstinent lifestyle. The family dynamics probably will change as a result of the client's abstinence; however, there is no way to predict whether these changes will be healthy. An alcoholic client always remains at risk for abusing alcohol. Most addicted people can't consume alcohol in moderation.
NP: Planning; CN: Psychosocial integrity; CNS: Coping and adaptation; CL: Application

35. A client who abused alcohol for more than 20 years is diagnosed with cirrhosis of the liver. Which of the following statements by the client shows that teaching has been effective?
1. "If I decide to stop drinking, I won't kill myself."
2. "If I watch my blood pressure, I should be okay."
3. "If I take vitamins, I can undo some liver damage."
4. "If I use nutritional supplements, I won't have problems."

36. A client tells a nurse, "I'm not going to have problems from smoking marijuana." Which of the following responses is most accurate?
1. "Evidence shows it can cause major health problems."
2. "Marijuana can cause reproductive problems later in life."
3. "Smoking marijuana isn't as dangerous as smoking cigarettes."
4. "Some people have minor or no reactions to smoking marijuana."

37. During an assessment of a client with a history of polysubstance abuse, which of the following information should be obtained after the names of the drugs?
1. Age at last use
2. Route of administration
3. How the drug was obtained
4. The place the drug was used

38. A client says, "I started using cocaine as a recreational drug, but now I can't seem to control the use." That statement identifies which of the following drug behaviors?
1. Toxic dose
2. Dual diagnosis
3. Cross-tolerance
4. Compulsive use

Your client teaching stems in part from the severity of disease.

Take aim at accuracy again!

35. 1. This statement reflects the client's perception of the severity of the condition and the life-threatening complications that can result from continued use of alcohol. Aggressive treatment is required, not merely watching one's blood pressure. At this point in the illness, there is little likelihood that liver damage from cirrhosis can be altered. The fourth statement denies the severity of the problem and negates the life-threatening complications common with a diagnosis of cirrhosis.
NP: Evaluation; CN: Physiological integrity; CNS: Reduction of risk potential; CL: Comprehension

36. 2. Marijuana causes cardiac, respiratory, immune, and reproductive health problems. Most people who smoke marijuana don't have major health problems. The residues from marijuana are more toxic than those from cigarettes. All people who smoke marijuana have symptoms of intoxication.
NP: Implementation; CN: Physiological integrity; CNS: Reduction of risk potential; CL: Comprehension

37. 2. The route of administration gives information about the effects of the drug and what immediate treatment is necessary. How the drug was obtained, place it was used, and age at last use aren't essential information for treatment.
NP: Data collection; CN: Physiological integrity; CNS: Reduction of risk potential; CL: Knowledge

38. 4. Compulsive drug use involves taking a substance for a period of time significantly longer than intended. A toxic dose is the amount of a drug that causes a poisonous effect. Dual diagnosis is the coexistence of a drug problem and a mental health problem. Cross-tolerance occurs when the effects of a drug are decreased and the client takes larger amounts to achieve the desired drug effect.
NP: Data collection; CN: Psychosocial integrity; CNS: Coping and adaptation; CL: Comprehension

39. A client says he used amphetamines to be productive at work. Which of the following symptoms commonly occurs when the drug is abruptly discontinued?
1. Severe anxiety
2. Increased yawning
3. Altered perceptions
4. Amotivational syndrome

You should know which drugs cause which adverse effects!

39. 1. When amphetamines are abruptly discontinued, the client may experience severe anxiety or agitation. Increased yawning is a symptom of opioid withdrawal. Altered perceptions occur when a client is withdrawing from hallucinogens. Amotivational syndrome is seen with clients using marijuana.
NP: Data collection; CN: Psychosocial integrity; CNS: Coping and adaptation; CL: Application

40. The use of which of the following drugs since early adolescence could lead to bone marrow depression?
1. Amphetamines
2. Cocaine
3. Inhalants
4. Marijuana

40. 3. Inhalants cause severe bone marrow depression. Marijuana, cocaine, and amphetamines do not cause bone marrow depression.
NP: Data collection; CN: Physiological integrity; CNS: Reduction of risk potential; CL: Knowledge

41. Which of the following reasons best explains why it's important to monitor behavior in a client who has stopped using phencyclidine (PCP)?
1. Fatigue can cause feelings of being overwhelmed.
2. Agitation and mood swings can occur during withdrawal.
3. Bizarre behavior can be a precursor to a psychotic episode.
4. Memory loss and forgetfulness can cause unsafe conditions.

You're doing a great job!

41. 3. Bizarre behavior and speech are associated with PCP withdrawal and can indicate psychosis. Fatigue isn't necessarily a problem when a client stops using PCP. Agitation, mood swings, memory loss, and forgetfulness don't tend to occur when a client has stopped using PCP.
NP: Data collection; CN: Psychosocial integrity; CNS: Psychosocial adaptation; CL: Analysis

42. A client is seeking help to stop using amphetamines. Which of the following symptoms indicates the client is experiencing withdrawal?
1. Disturbed sleep
2. Increased yawning
3. Psychomotor agitation
4. Inability to concentrate

More oxygen!

42. 1. It's common for a person withdrawing from amphetamines to experience disturbed sleep and unpleasant dreams. Increased yawning is seen with clients withdrawing from opioids. The inability to concentrate is seen in caffeine withdrawal, and psychomotor agitation is seen in cocaine withdrawal.
NP: Data collection; CN: Psychosocial integrity; CNS: Coping and adaptation; CL: Knowledge

43. Which of the following conditions can occur in a client who has just used cocaine?
1. Increased heart rate
2. Elevated temperature
3. Increased neck distention
4. Decreased respiratory rate

43. 1. An increase in heart rate is common because cocaine increases the heart's demand for oxygen. Cocaine doesn't decrease the client's respiratory rate, increase body temperature, or cause increased neck distention.
NP: Data collection; CN: Physiological integrity; CNS: Physiological adaptation; CL: Application

44. Which of the following information is most important in teaching a client who abuses prescription drugs?
1. Herbal substitutes are safer to use.
2. Medication should be used only for the reason prescribed.
3. The client should consult a physician before using a drug.
4. Consider if family members influence the client to use drugs.

44. 2. People usually take prescribed drugs for reasons other than those intended, primarily to self-medicate or experience a sense of euphoria. The safety and efficacy of most herbal remedies hasn't been established. Sometimes over-the-counter medications are necessary for minor problems. There may be a family history of substance abuse, but it isn't a priority when planning nursing care.
NP: Implementation; CN: Psychosocial integrity; CNS: Psychosocial adaptation; CL: Application

Choose the best answer!

45. The family of an adolescent who smokes marijuana asks a nurse if the use of marijuana leads to abuse of other drugs. Which of the following responses is best?
1. "Use of marijuana is a stage your child will go through."
2. "Many people use marijuana and don't use other street drugs."
3. "Use of marijuana can lead to abuse of more potent substances."
4. "It's difficult to answer that question as I don't know your child."

45. 3. Marijuana is considered a "gateway drug" because it tends to lead to the abuse of more potent drugs. People who use marijuana tend to use or at least experiment with more potent substances. Marijuana isn't a part of a developmental stage that adolescents go through. It isn't important that the nurse knows the child.
NP: Implementation; CN: Psychosocial integrity; CNS: Psychosocial adaptation; CL: Application

46. A pregnant client is thinking about stopping cocaine use. Which of the following statements by the client indicates effective teaching about pregnancy and drug use?
1. "Right after birth, I'll give the baby up for adoption."
2. "I'll help the baby get through the withdrawal period."
3. "I don't want the baby to have withdrawal symptoms."
4. "It's scary to think the baby may have Down syndrome."

46. 3. Neonates born to mothers addicted to cocaine have withdrawal symptoms at birth. If the client says she'll give the baby up for adoption after birth or help the baby get through the withdrawal period, the teaching was ineffective because the mother doesn't see the impact of her drug use on the child. Use of cocaine during pregnancy doesn't contribute to the baby having Down syndrome.
NP: Evaluation; CN: Physiological integrity; CNS: Reduction of risk potential; CL: Analysis

Congratulations! You've finished 47 questions!

47. A client with a history of cocaine abuse returns from a pass to an inpatient drug and alcohol facility showing behavior changes. Which of the following tests shows the presence of cocaine in the body?
1. Antibody screen
2. Glucose screen
3. Hepatic screen
4. Urine screen

47. 4. A urine toxicology screen would show the presence of cocaine in the body. Glucose, hepatic, or antibody screening wouldn't show the presence of cocaine in the body.
NP: Data collection; CN: Psychosocial integrity; CNS: Psychosocial adaptation; CL: Knowledge

NP: Nursing process   CN: Client needs category   CNS: Client needs subcategory   CL: Cognitive level

48. The nurse is assessing a client who has just been admitted to the emergency department. Which signs would suggest an overdose with an antianxiety agent?
1. Combativeness, sweating, and confusion
2. Agitation, hyperactivity, and grandiose ideation
3. Emotional lability, euphoria, and impaired memory
4. Suspiciousness, dilated pupils, and increased blood pressure

49. Which of the following interventions is most important in planning care for a client recovering from cocaine use?
1. Skin care
2. Suicide precautions
3. Frequent orientation
4. Nutrition consultation

Prioritize!

50. Which of the following clinical conditions is commonly seen with substance abuse clients who repeatedly use cocaine?
1. Panic attacks
2. Bipolar cycling
3. Attention deficits
4. Expressive aphasia

51. A client who uses cocaine finally admits other drugs were also abused to equalize the effect of cocaine. Which of the following substances might be included in the client's pattern of polysubstance abuse?
1. Alcohol
2. Amphetamines
3. Caffeine
4. Phencyclidine

What would have the opposite effect?

52. A mother asks the nurse what she should do about her son's behavior, which has been erratic for 6 months. Which response is best?
1. Tell her how to set daily goals for her son.
2. Reassure her that her child is going through a phase that will pass.
3. Discuss the child's specific behaviors, and suggest possible actions to take.
4. Explain that the child seems fragile and that she needs to be patient with him.

48. 3. Signs of antianxiety agent overdose include emotional lability, euphoria, and impaired memory. Phencyclidine overdose can cause combativeness, sweating, and confusion; amphetamine overdose can result in agitation, hyperactivity, and grandiose ideation; hallucinogen overdose can produce suspiciousness, dilated pupils, and increased blood pressure.
NP: Data collection; CN: Physiological integrity; CNS: Reduction of risk potential; CL: Knowledge

49. 2. Clients recovering from cocaine use are prone to "postcoke depression" and have a likelihood of becoming suicidal if they can't take the drug. Frequent orientation and skin care are routine nursing interventions but aren't the most immediate considerations for this client. Nutrition consultation isn't the most pressing intervention for this client.
NP: Planning; CN: Psychosocial integrity; CNS: Psychosocial adaptation; CL: Comprehension

50. 2. Clients who frequently use cocaine will experience the rapid cycling effect of excitement and then severe depression. They don't tend to experience panic attacks, expressive aphasia, or attention deficits.
NP: Data collection; CN: Psychosocial integrity; CNS: Coping and adaptation; CL: Analysis

51. 1. A cocaine addict will usually use alcohol to decrease or equalize the stimulating effects of cocaine. Caffeine, phencyclidine, and amphetamines aren't used to equalize the stimulating effects of cocaine.
NP: Data collection; CN: Psychosocial integrity; CNS: Coping and adaptation; CL: Application

52. 3. This parent requires guidance and direction to consider possible alternatives. The child needs to set his own goals. Erratic behavior isn't typical of a passing phase. More than patience is needed to deal with erratic behavior that persists for 6 months.
NP: Implementation; CN: Psychosocial integrity; CNS: Coping and adaptation; CL: Application

53. Which of the following statements by a client indicates teaching about cocaine use has been effective?
  1. "I wasn't using cocaine to feel better about myself."
  2. "I started using cocaine more and more until I couldn't stop."
  3. "I'm not addicted to cocaine because I don't use it every day."
  4. "I'm not going to be a chronic user, I only use it on holidays."

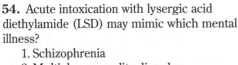

Can you tell which one is the real me? As a nurse you'll need to determine the real problem when symptoms point you in different directions.

54. Acute intoxication with lysergic acid diethylamide (LSD) may mimic which mental illness?
  1. Schizophrenia
  2. Multiple personality disorder
  3. Generalized anxiety disorder
  4. Acute mania

55. Which of the following psychiatric or medical emergencies is most likely to occur when a client is using phencyclidine (PCP)?
  1. Cardiac arrest
  2. Seizure disorder
  3. Violent behavior
  4. Delirium reaction

This client has amotivational syndrome.

56. A client who smoked marijuana daily for 10 years tells a nurse, "I don't have any goals, and I just don't know what to do." Which of the following communication techniques is the most useful when talking to this client?
  1. Focus the interaction.
  2. Use nonverbal methods.
  3. Use reflection techniques.
  4. Use open-ended questions.

53. 2. This statement reflects the trajectory or common pattern of cocaine use and indicates successful teaching. The first statement reflects the client's denial. People gravitate to the drug and continue its use because it gives them a sense of well-being, competency, and power. Cocaine abusers tend to be binge users and can be drug-free for days or weeks between use, but they still have a drug problem. The fourth statement indicates the client is in denial about the drug's potential to become a habit. Effective teaching didn't occur.
NP: Evaluation; CN: Psychosocial integrity; CNS: Psychosocial adaptation; CL: Analysis

54. 1. Visual distortions and gross distortion of reality are found in both LSD intoxication and schizophrenia. LSD intoxication doesn't manifest in switching personalities; total anxiety and tenseness; or hyperactivity, euphoria, or emotional liability.
NP: Data collection; CN: Physiological integrity; CNS: Physiological adaptation; CL: Knowledge

55. 3. When a client is using PCP, an acute psychotic reaction can occur. The client is capable of sudden, explosive, violent behavior. PCP doesn't tend to cause cardiac arrest or a seizure disorder. Delirium is associated with inhalant intoxication.
NP: Data collection; CN: Psychosocial integrity; CNS: Psychosocial adaptation; CL: Knowledge

56. 1. A client with amotivational syndrome from chronic use of marijuana tends to talk in tangents and needs the nurse to focus the conversation. Nonverbal communication or reflection techniques wouldn't be useful as this client must focus and learn to identify and accomplish goals. Using only open-ended questions won't allow the client to focus and establish specific goals.
NP: Implementation; CN: Psychosocial integrity; CNS: Psychosocial adaptation; CL: Comprehension

NP: Nursing process CN: Client needs category CNS: Client needs subcategory CL: Cognitive level

57. A nurse is assessing a client who uses heroin to determine if there are physical health problems. Which of the following medical consequences of heroin use frequently occurs?
1. Hepatitis
2. Peptic ulcers
3. Hypertension
4. Chronic pharyngitis

58. The family of a client withdrawing from heroin asks a nurse why the client is receiving naltrexone (ReVia). Which of the following responses is correct?
1. To help reverse withdrawal symptoms
2. To keep the client sedated during withdrawal
3. To take the place of detoxification with methadone
4. To decrease the client's memory of the withdrawal experience

Maybe we should be called priority engineers instead of nurses.

59. Which of the following nursing interventions has priority in a care plan for a client recovering from cocaine addiction?
1. Help the client find ways to be happy and competent.
2. Foster the creative use of self in community activities.
3. Teach the client to handle stresses in the work setting.
4. Help the client acknowledge the current level of dependency.

Don't quit now!

60. A client tells a nurse, "I've been clean from drugs for the past 5 years, but my life really hasn't changed." Which of the following concepts should be explored with this client?
1. Further education
2. Conflict resolution
3. Career development
4. Personal development

57. 1. Hepatitis is the most frequent medical complication of heroin abuse. Peptic ulcers are more likely to be a complication of caffeine use, hypertension is a complication of amphetamine use, and chronic pharyngitis is a complication of marijuana use.
NP: Data collection; CN: Physiological integrity; CNS: Reduction of risk potential; CL: Application

58. 1. Naltrexone is an opioid antagonist and helps reverse the symptoms of opioid withdrawal. Keeping the client sedated during withdrawal isn't the reason for giving this drug. The drug doesn't decrease the client's memory of the withdrawal experience and isn't used in place of detoxification with methadone.
NP: Implementation; CN: Physiological integrity; CNS: Reduction of risk potential; CL: Application

59. 1. The major component of a treatment program for a client with cocaine addiction is to have the client feel happy and competent. Cocaine addiction is difficult to treat because the drug actions reinforce its use. There are often perceived positive effects. Clients often credit the drug with giving them creative energy instead of looking within themselves. The second nursing intervention may inadvertently reinforce the client's drug use. The third nursing intervention is appropriate but isn't the most immediate nursing action. Examining the client's level of dependency isn't the immediate choice as the client needs to work on remaining drug-free.
NP: Implementation; CN: Psychosocial integrity; CNS: Psychosocial adaptation; CL: Application

60. 4. True recovery involves changing the client's distorted thinking and working on personal and emotional development. Before the client pursues further education, career development, or conflict resolution skills, it's imperative the client devotes energy to emotional and personal development.
NP: Implementation; CN: Psychosocial integrity; CNS: Psychosocial adaptation; CL: Analysis

61. A client discusses how drug addiction has made life unmanageable. Which of the following information does the client need to start coping with the drug problem?
1. How peers have committed to sobriety
2. How to accomplish family of origin work
3. The addiction process and tools for recovery
4. How environmental stimuli serve as drug triggers

Even the longest of journeys start with a first step.

61. 3. When the client admits life has become unmanageable, the best strategy is to teach about the addiction, how to obtain support, and how to develop new coping skills. Information about how peers committed to sobriety would be shared with the client as the treatment process begins. Family of origin work would be a later part of the treatment process. Initially, the client must commit to sobriety and learn skills for recovery. Identification of how environmental stimuli serve as drug triggers would be a later part of the treatment process.
NP: Implementation; CN: Psychosocial integrity; CNS: Psychosocial adaptation; CL: Analysis

62. Which of the following physical health findings is expected during an assessment of a client with a history of cocaine abuse?
1. Glossitis
2. Pharyngitis
3. Bilateral ear infections
4. Perforated nasal septum

62. 4. When cocaine is snorted frequently, the client commonly develops a perforated nasal septum. Bilateral ear infections, pharyngitis, and glossitis aren't common physical findings for a client with a history of cocaine abuse.
NP: Data collection; CN: Physiological integrity; CNS: Physiological adaptation; CL: Knowledge

63. A client recovering from cocaine abuse is participating in group therapy. Which of the following statements by the client indicates the client has benefited from the group?
1. "I think the laws about drug possession are too strict in this country."
2. "I'll be more careful about talking about my drug use to my children."
3. "I finally realize the short high from cocaine isn't worth the depression."
4. "I can't understand how I could get all these problems that we talked about in group."

63. 3. This is a realistic appraisal of a client's experience with cocaine and how harmful the experience is. The first statement indicates the client was distracting self from personal issues and isn't working on goals in the group setting. Talking about drugs to children must be reinforced with nonverbal behavior, and not talking about drugs may give children the wrong message about drug use. The fourth statement indicates the client is in denial about the consequences of cocaine use.
NP: Implementation; CN: Psychosocial integrity; CNS: Psychosocial adaptation; CL: Analysis

Only a mere handful or so more questions. Badda bing, badda boom!

64. A family expresses concern that a member who stopped using amphetamines 3 months ago is acting paranoid. Which of the following explanations is the best?
1. A person gets symptoms of paranoia with polysubstance abuse.
2. When a person uses amphetamines, paranoid tendencies may continue for months.
3. Sometimes family dynamics and a high suspicion of continued drug use make a person paranoid.
4. Amphetamine abusers may have severe anxiety and paranoid thinking.

64. 2. After a client uses amphetamines, there may be long-term effects that exist for months after use. Two common effects are paranoia and ideas of reference. Even with polysubstance abuse, the paranoia comes from the chronic use of amphetamines. The third explanation blames the family when the paranoia comes from the drug use. Severe anxiety isn't typically manifested in paranoid thinking.
NP: Implementation; CN: Psychosocial integrity; CNS: Psychosocial adaptation; CL: Analysis

65. A nurse is trying to determine if a client who abuses heroin has any drug-related legal problems. Which of the following assessment questions is the best to ask the client?
1. When did your spouse become aware of your use of heroin?
2. Do you have a probation officer that you report to periodically?
3. Have you experienced any legal violations while being intoxicated?
4. Do you have a history of frequent visits with the employee assistance program manager?

Choose the best answer!

65. 3. This question focuses on obtaining direct information about drug-related legal problems. When a spouse becomes aware of a partner's substance abuse, the first action isn't necessarily to institute legal action. Even if the client reports to a probation officer, the offense isn't necessarily a drug-related problem. Asking if the client has a history of frequent visits with the employee assistance program manager isn't useful. It assumes any visit to the employee assistance program manager is related to drug issues.
NP: Data collection; CN: Psychosocial integrity; CNS: Coping and adaptation; CL: Analysis

66. A male client returns to the psychiatric unit after being on a 6-hour pass. The nurse observes that the client is agitated and ataxic and that he exhibits nystagmus and general muscle hypertonicity. The nurse suspects that the client was using drugs while away from the unit. His symptoms are most indicative of intoxication with which of the following drugs?
1. Phencyclidine (PCP)
2. Crack cocaine
3. Heroin
4. Cannabis

66. 1. The client's behavior suggests the use of PCP. Crack intoxication is characterized by euphoria, grandiosity, aggressiveness, paranoia, and depression. Heroin intoxication is characterized by euphoria followed by sleepiness. Cannabis intoxication is characterized by panic state and visual hallucinations.
NP: Data collection; CN: Psychosocial integrity; CNS: Psychosocial adaptation; CL: Knowledge

67. A client who uses cocaine denies that drug use is a problem. Which of the following intervention strategies would be best to confront the client's denial?
1. State ways to cope with stress.
2. Repeat the drug facts as needed.
3. Identify the client's ambivalence.
4. Use open-ended, factual questions.

Think therapeutic.

67. 4. The use of open-ended, factual questions will help the client acknowledge that a drug problem is present. Repeating drug facts won't be effective, as the client will perceive it as preaching or nagging. Stating ways to cope with stress and identifying the client's ambivalence won't be effective for breaking through a client's denial.
NP: Implementation; CN: Psychosocial integrity; CNS: Psychosocial adaptation; CL: Application

68. The nurse is instructing a client who is to receive disulfiram (Antabuse). Which of the following substances would be safe for the client?
1. Aftershave lotion
2 Mouthwash
3 Backrub preparations
4. Antacids

68. 4. Antacids don't interact with disulfiram. The client should avoid anything containing alcohol, including aftershave lotion, cough medicines, back rub preparations, and some mouthwashes.
NP: Planning; CN: Psychosocial integrity; CNS: Psychosocial adaptation; CL: Comprehension

69. Which of the following conditions may be present when a client addicted to cocaine becomes pessimistic about treatment?
   1. Depression
   2. Estrangement
   3. Fatigue
   4. Impulsiveness

69. 1. Clients withdrawing from drugs, such as cocaine, commonly experience depression. It's common for drug-addicted clients to experience fatigue without becoming pessimistic. Being impulsive or having feelings of estrangement aren't necessarily related to a client becoming pessimistic about treatment.

NP: Data collection; CN: Psychosocial integrity; CNS: Coping and adaptation; CL: Knowledge

Finished!
Congratulations!

For more information on dissociative disorders, check this independent Web site: www.mentalhealth.com/.

Chapter 19
Dissociative disorders

1. Which of the following factors is associated with a client with dissociative identity disorder (DID)?

1. An absent father
2. An inflated sense of self-esteem
3. Vivid memories of childhood trauma
4. A parent who was alternately loving and abusive

You'll do great answering these questions!

2. Which of the following characteristics is expected for a person with dissociative identity disorder (DID)?

1. Ritualistic behavior
2. Out-of-body experiences
3. History of severe childhood abuse
4. Ability to give a thorough personal history

Nursing diagnoses standardize client care.

3. Which of the following nursing diagnoses would you expect to find for a client with dissociative identity disorder (DID)?

1. Disturbed thought processes related to delusional ideations
2. Risk for self-directed violence related to suicidal ideations or gestures
3. Deficient diversional activity: Self-directed related to lack of environmental stimulation
4. Disturbed sensory perception: Visual hallucinations related to altered sensory reception of visual stimulation

1. 4. Repeated exposure to a childhood environment that alternates between highly stressful and then loving and supportive can be a factor in the development of DID. Many children grow up in a household without a father but don't develop DID. Because of dissociation from the trauma, a client with DID usually can't recall childhood traumatic events. Clients with DID commonly have low self-esteem.
NP: Data collection; CN: Psychosocial integrity; CNS: Psychosocial adaptation; CL: Knowledge

2. 3. DID is theorized to develop as a protective response to such traumatic experiences as severe child abuse. Ritualistic behavior is seen with obsessive-compulsive disorders. Because of the dissociative response to personal experiences, people with DID are usually unable to give a thorough personal history. Out-of-body experiences are more commonly associated with depersonalization disorder.
NP: Data collection; CN: Psychosocial integrity; CNS: Coping and adaptation; CL: Knowledge

3. 2. A common reason clients with DID are admitted to a psychiatric facility is because one of the alter personalities is trying to kill another personality. Hallucinations and delusions are commonly associated with schizophrenic disorders. Because of the assortment of alter personalities controlling the client with DID, diversional activity deficit is rarely a problem.
NP: Planning; CN: Safe, effective care environment; CNS: Coordinated care; CL: Application

---

NP: Nursing process   CN: Client needs category   CNS: Client needs subcategory   CL: Cognitive level

4. Which of the following nursing interventions is important for a client with dissociative identity disorder (DID)?
1. Give antipsychotic medications as prescribed.
2. Maintain consistency when interacting with the client.
3. Confront the client about the use of alter personalities.
4. Prevent the client from interacting with others when one of the alter personalities is in control.

5. The nurse is caring for a 25-year-old female who is exhibiting signs of low self-esteem. Which client outcome is desirable?
1. The client will participate in new activities.
2. The client will sleep without interruption at night.
3. The client will be encouraged to confront fear of failure.
4. The nurse will spend more time with the client.

6. Which of the following goals would be realistic for a client with dissociative identity disorder (DID)?
1. Confronts the abuser
2. Attends the unit's milieu meetings
3. Prevents alter personalities from emerging
4. Reports no longer having feelings of anger about childhood traumas

7. Which of the following expectations is most appropriate for a client with dissociative identity disorder (DID)?
1. To complain of physical health problems with no organic basis
2. To be unable to account for certain times on a day-to-day basis
3. To participate in discussions about abusive incidents that occurred in the past
4. To be able to form a therapeutic relationship with the nurse after meeting with her twice

There's a trick to apply to question 5. Distinguishing between client outcomes and nursing interventions will help you to the right answer!

4. 2. Establishing trust and support is important when interacting with a client with DID. Many of these clients have had few healthy relationships. Medication hasn't proven effective in the treatment of DID. Confronting the client about the alter personalities would be ineffective because the client has little, if any, knowledge of the presence of these other personalities. Isolating the client wouldn't be therapeutically beneficial.
NP: Implementation; CN: Safe, effective care environment; CNS: Coordinated care; CL: Analysis

5. 1. A desired client outcome for a person with low self-esteem is participation in new activities without exhibiting fear of failure. It hasn't been established that this client has difficulty sleeping. Allowing the client to confront fear of failure and spending more time with the client are nursing interventions.
NP: Implementation CN: Psychosocial integrity; CNS: Psychosocial adaptation; CL: Application

6. 2. Attending milieu meetings decreases feelings of isolation and shows the client has begun to trust the nurse. The client is often unaware of an alter personality and thus can't prevent these alter personalities from emerging. Often the abuser was a part of the client's childhood, and confrontation in adulthood may not be possible or therapeutic. Clients with DID have dissociated from painful experiences, so the host personality often doesn't have negative feelings about such experiences.
NP: Planning; CN: Psychosocial integrity; CNS: Psychosocial adaptation; CL: Analysis

7. 2. When alter personalities are in control, periods of amnesia are common for clients with DID. Complaining of physical health problems with no organic basis is more descriptive of clients with somatoform disorder. The client doesn't have memories of abusive episodes, making it difficult to discuss past incidents. These clients are typically slow in forming trusting relationships because many past relationships have been hurtful.
NP: Data collection; CN: Psychosocial integrity; CNS: Coping and adaptation; CL: Comprehension

NP: Nursing process   CN: Client needs category   CNS: Client needs subcategory   CL: Cognitive level

8. Which of the following findings is expected for a client with dissociative identity disorder (DID)?
 1. A close relationship with her mother
 2. A history of performing poorly in school
 3. Inability to recall certain events or experiences
 4. Consistency in the performance of certain tasks or skills

You're almost at question 10!

9. A hospitalized client with dissociative identity disorder (DID) reports hearing voices. Which of the following nursing interventions is most appropriate?
 1. Tell the client to lie down and rest.
 2. Give an as-needed dose of haloperidol (Haldol).
 3. Encourage the client to continue with his daily activities.
 4. Notify the physician the client is having a psychotic episode.

Which way is *most appropriate?*

10. The initial reason clients with dissociative identity disorder (DID) seek assistance from a health care professional is usually to achieve which of the following goals?
 1. Learn how to control periods of mania.
 2. Learn how to integrate all the alternate personalities.
 3. Develop coping strategies to deal with the traumatic childhood.
 4. Determine what is causing them to feel they have periods of "lost time."

Think therapeutic here.

11. A client with dissociative identity disorder reports hearing voices and asks the nurse if that means she's "crazy." Which of the following responses would be the most appropriate?
 1. "What do the voices tell you?"
 2. "Why would you think you're crazy?"
 3. "Clients with DID commonly report hearing voices."
 4. "Hearing voices is typically a symptom of schizophrenia."

8. 3. Clients with DID often experience bouts of amnesia when alter personalities are in control. A client with DID has learned in childhood how to live in two separate worlds: one in the daytime, in which she's able to perform well in school and have friendships, and one at nighttime, when the abuse occurs. The alter personalities may vary in the ability and type of skills and tasks performed.
NP: Data collection; CN: Psychosocial integrity; CNS: Coping and adaptation; CL: Comprehension

9. 3. Because many clients with DID hear voices, it's appropriate to have the client continue with daily activities. The voices the client hears are probably alter personalities communicating. This doesn't indicate a psychotic episode, so the physician wouldn't be notified to prescribe such antipsychotic medication as haloperidol. Having the client lie down and rest would have no therapeutic value.
NP: Implementation; CN: Safe, effective care environment; CNS: Coordinated care; CL: Application

10. 4. The initial symptom many clients with DID experience is the sensation of lost time. These are times the alter personalities are in control. Before therapeutic interventions, clients with DID may not even be aware of childhood trauma because of dissociation from the event. Initially, the client with DID isn't aware of the presence of alternate personalities. Depression, not mania, may be another early symptom of clients with DID.
NP: Data collection; CN: Health promotion and maintenance; CNS: Prevention and early detection of disease; CL: Comprehension

11. 3. The most therapeutic answer is to give correct information. Asking what the voices tell the client would be changing the topic without answering the question. Asking "why" questions can put the client on the defensive. Schizophrenia isn't the only cause of hearing voices, and this response suggests the client may be schizophrenic.
NP: Implementation; CN: Psychosocial integrity; CNS: Coping and adaptation; CL: Analysis

12. Which of the following interventions would most likely appear in the care plan of a client with dissociative identity disorder (DID)?
1. Arrange to have staff check on the client every 15 to 30 minutes.
2. Prevent all family from visiting until the third day of hospitalization.
3. Make sure the staff understands the client will be on seizure precautions.
4. Place the client in a quiet room away from the noise of the nurse's station.

12. 1. A common reason for clients with DID to be hospitalized is for suicidal ideations or gestures. For the client's safety, frequent checks should be done. Family interactions might be therapeutic for the client, and the family may be able to provide a more thorough history because of the client's dissociation from traumatic events. Seizure activity isn't an expected symptom of DID. Because of the possibility of suicide, the client's room should be close to the nurse's station.
NP: Planning; CN: Safe, effective care environment; CNS: Safety and infection control; CL: Application

It helps to know which symptoms to watch for.

13. A client is being treated at a community mental health clinic. A nurse has been instructed to observe for any behaviors indicating dissociative identity disorder (DID). Which of the following behaviors would be included?
1. Delusions of grandeur
2. Reports of fatigue
3. Changes in dress, mannerisms, and voice
4. Refusal to make a follow-up appointment

13. 3. When alter personalities are in control, the person will have complete personality changes. Delusions of grandeur are more frequently associated with disorders such as manic states and schizophrenia. Complaints of fatigue aren't a main symptom of DID. The refusal to make a follow-up appointment could indicate many problems, including noncompliance.
NP: Data collection; CN: Psychosocial integrity; CNS: Coping and adaptation; CL: Application

14. Which of the following statements is correct for clients with a dissociative identity disorder (DID)?
1. They rarely marry.
2. They rarely improve with treatment.
3. They can be treated with antianxiety drugs.
4. They often improve with long-term therapy.

14. 4. Most clients with DID can be successfully treated with long-term therapy. For many of the conditions, pharmacologic therapy has little effect. Many clients with DID marry.
NP: Data collection; CN: Psychosocial integrity; CNS: Psychosocial adaptation; CL: Knowledge

You can't act as a casting director for a client's alter personalities. Stay focused on therapeutic interventions.

15. When interacting with a client with a dissociative identity disorder, a nurse observes that one of the alter personalities is in control. Which of the following interventions is the most appropriate?
1. Give recognition to the alter personality.
2. Notify the physician.
3. Immediately stop interacting with the client.
4. Ignore the alter personality, and ask to speak to the host personality.

15. 1. By giving recognition to the alter personalities, the nurse conveys to the client that she believes the alter personalities exist. Asking to speak to the host personality or immediately stopping interaction with the client won't stop the client from being controlled by alter personalities. The physician doesn't need to be notified because this is an expected occurrence.
NP: Implementation; CN: Psychosocial integrity; CNS: Psychosocial adaptation; CL: Analysis

16. The nurse on the psychiatric unit is caring for a 51-year-old male client who is suicidal. Which nursing intervention takes priority?
1. Discourage sleep except at bedtime.
2. Make a verbal contract with the client to notify the staff of suicidal thoughts.
3. Limit time spent alone by encouraging the client to participate in group activities.
4. Create a safe physical and interpersonal environment.

17. A family member of a client with dissociative identity disorder (DID) asks a nurse if hypnotic therapy might help the client. Which of the following responses would be most appropriate?
1. "What would make you think that?"
2. "No, hypnosis is rarely used in the treatment of psychiatric conditions."
3. "Yes, but this treatment is used only after other types of therapy have failed."
4. "Yes, often the client doesn't have conscious awareness of alter personalities."

18. Which of the following interventions is appropriate when caring for a client with dissociative identity disorder?
1. Remind the alter personalities they're part of the host personality.
2. Interact with the client only when the host personality is in control.
3. Establish an empathetic relationship with each emerging personality.
4. Provide positive reinforcement to the client when calm alter personalities are present instead of angry ones.

Hints, hints, hints!

16. 4. Creating a safe environment is the nurse's highest priority. This includes removing obvious hazards, recognizing nonobvious and emergency hazards, maintaining close observation, serving as a client advocate in interpersonal situations, and communicating concern to the client in both verbal and nonverbal ways. Other interventions, such as discouraging sleep except at bedtime, making a verbal contract, and encouraging participation in group activities should be included in the client's plan, but these don't have top priority.
NP: Implementation; CN: Psychosocial integrity; CNS: Psychosocial adaptation; CL: Application

17. 4. Because of dissociation from painful events, hypnosis is usually effective in the treatment of clients with DID. It may be under hypnosis that alter personalities start to emerge. The first response could place the family member on the defensive. Hypnosis is used in a variety of psychiatric conditions. Hypnosis is commonly a first-line treatment for the client with DID.
NP: Implementation; CN: Psychosocial integrity; CNS: Psychosocial adaptation; CL: Application

18. 3. Establishing an empathetic relationship with each emerging personality provides a therapeutic environment to care for the client. Interacting with the client only when the host personality is in control would be useless because the client has limited, if any, control or awareness when alter personalities are in control.
NP: Implementation; CN: Psychosocial integrity; CNS: Psychosocial adaptation; CL: Application

19. While interacting with a client with dissociative identity disorder (DID), a nurse observes one of the alter personalities take over. The client goes from being calm to angry and shouting. Which of the following responses would be most appropriate?

1. "Is one of you upset?"
2. "Why have you become angry?"
3. "Tell me what you're feeling right now."
4. "Let me speak to someone who isn't angry."

19. 3. This response encourages integration and discourages dissociation. When interacting with clients with DID, the nurse always wants to remind the client that the alter personalities are a component of one person. Responses reinforcing interaction with only one alter personality instead of trying to interact with the individual as a single person aren't appropriate. Asking "why" questions can put the client on the defensive and impede further communication.

NP: Implementation; CN: Psychosocial integrity; CNS: Psychosocial adaptation; CL: Application

20. A client with dissociative identity disorder has been in therapy for 2 years and just learned her father passed away. Her father sexually abused her throughout her childhood. Which of the following interventions would be most appropriate?

1. Have the client seek inpatient therapy.
2. Encourage the client's verbalization of feelings of anger and guilt.
3. Encourage the client's alter personalities to emerge during this stressful time.
4. Stress to the client that the death of the abuser should be helpful in her healing process.

20. 2. The death of the abuser may cause the client to experience feelings of anger and guilt. Encouraging the client's alter personalities to emerge could result in further dissociation. Unless the client becomes suicidal or rapidly deteriorates, inpatient treatment won't be necessary. The death of the abuser can be a stressful event and can leave the client with unresolved feelings.

NP: Implementation; CN: Health promotion and maintenance; CNS: Prevention and early detection of disease; CL: Analysis

21. The nurse finds a suicidal client trying to hang himself in his room. In order to preserve the client's self-esteem and safety, what should the nurse do?

1. Place the client in seclusion with checks every 15 minutes.
2. Assign a nursing staff member to remain with the client at all times.
3. Make the client stay with the group at all times.
4. Refuse to let the client in his room.

Keep going. You're doing great!

21. 2. Implementing a one-to-one staff-to-client ratio is the nurse's highest priority. This allows the client to maintain his self-esteem and keeps him safe. Seclusion would damage the client's self-esteem. Forcing the client to stay with the group and refusing to let him in his room doesn't guarantee his safety.

NP: Implementation; CN: Psychosocial integrity; CNS: Psychosocial adaptation; CL: Application

22. The nurse observes that the alter personality of the client with a dissociative identity disorder is in control. The client is sitting in the dayroom, interacting with others. Which of the following actions would be most appropriate?
1. Allow the client to continue interacting with clients in the dayroom.
2. Ask to speak to one of the adult alter personalities of the host personality.
3. Remove the client from the dayroom, and allow the client to play with toys.
4. Remove the client from the dayroom, and reorient her that she's in a safe place.

23. Which of the following feelings or background histories is most commonly reported by clients with dissociative identity disorder (DID) who seek help?
1. Feelings of loneliness
2. Supportive family system
3. Feelings of profound sadness
4. An almost uncontrollable urge to kill the abuser

Here's a hint: *most commonly.*

24. Which of the following characteristics best applies to dissociative disorders?
1. A group of disorders with the common symptom of hallucinations
2. A group of disorders with a rapid disruption of the client's memory
3. A group of disorders with impairment of memory or identity due to the development of organic changes in the brain
4. A group of disorders with impairment of memory or identity due to an unconscious attempt to protect the person from emotional pain or traumatic experiences

25. When asked about behaviors of a client with dissociative identity disorder (DID), family members might include which of the following observations?
1. Statements by the client that he "feels like a robot"
2. A desire to sleep for long periods
3. Unpredictable and sometimes bizarre changes in behavior
4. A need to continually verbalize feelings about the abuse he suffered as a child

22. 4. Removing the client at this time may protect her from future embarrassment. Reorienting the client discourages dissociation and encourages integration. Asking to speak to an alter personality encourages dissociation. Allowing the client to play with toys also reinforces this behavior and encourages dissociation.
NP: Implementation; CN: Safe, effective care environment; CNS: Safety and infection control; CL: Analysis

23. 3. Many clients with DID initially experience problems with sadness. Usually, the memories of abuse are repressed; the client may not have a conscious awareness of being angry with the abuser. Although clients with DID may feel lonely, this is secondary to sadness. Typically, the family is dysfunctional and unlikely to be supportive.
NP: Data collection; CN: Psychosocial integrity; CNS: Psychosocial adaptation; CL: Knowledge

24. 4. A group of disorders in which there's impairment of memory or identity due to an unconscious attempt to protect the client from emotional pain or traumatic experiences describes dissociative disorders. The onset of dissociative disorders may be gradual, sudden, or chronic. There's no known organic cause for dissociative disorders. Hallucinations are associated with schizophrenic disorders.
NP: Data collection; CN: Psychosocial integrity; CNS: Psychosocial adaptation; CL: Knowledge

25. 3. Changes in behavior are one of the cardinal findings of DID. This occurs as the alter personalities take control. Feeling like a robot is more likely to occur in depersonalization disorder. The desire to sleep for long periods is more descriptive of depression. Usually, the client with DID isn't aware of the occurrence of child abuse.
NP: Data collection; CN: Psychosocial integrity; CNS: Psychosocial adaptation; CL: Knowledge

26. A client with dissociative identity disorder (DID) is admitted to an inpatient psychiatric unit. A nurse-manager asked all staff to attend a meeting. Which of the following reasons for the meeting is the most likely?
1. To review the restraint protocol with the staff
2. To inform the staff that no one should refuse to work with the client
3. To warn the staff that this client may be difficult and challenging to work with
4. To allow staff members to discuss concerns about working with a client with DID

These words — most likely — can help you select the correct answer.

26. 4. Allowing all staff members to meet together may prevent the staff from splitting into groups of those who believe the validity of this diagnosis and those who don't. Unless this client shows behaviors harmful to himself or others, restraints aren't needed. Telling the staff that no one should refuse to work with the client or that this client will probably be difficult and challenging sets a negative tone as staff plan and provide care for the client.
NP: Planning; CN: Safe, effective care environment; CNS: Coordinated care; CL: Application

27. A 26-year-old man is reported missing after being the victim of a violent crime. Two months later, a family member finds him working in a city 100 miles from his home. The man doesn't recognize the family member or recall being the victim of a crime. He most likely has which of the following conditions?
1. Depersonalization disorder
2. Dissociative amnesia
3. Dissociative fugue
4. Dissociative identity disorder (DID)

27. 3. Dissociative fugue is sudden flight after a traumatic event. During the episode, the person may assume a new identity and not recognize people from his past. Depersonalization disorder is the sudden loss of the sense of one's own reality. Dissociative amnesia doesn't involve flight from work or home. DID is the coexistence of two or more personalities in one person.
NP: Data collection; CN: Psychosocial integrity; CNS: Psychosocial adaptation; CL: Knowledge

28. Which of the following nursing interventions is most appropriate for a client who has just had an episode of dissociative fugue?
1. Let the client verbalize the fear and anxiety he feels.
2. Encourage the client to share his experiences during the episode.
3. Have the client sign a contract stating he won't leave the premises again.
4. Tell the client he won't resolve his problems by running away from them.

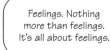

Feelings. Nothing more than feelings. It's all about feelings.

28. 1. An episode of dissociative fugue can be a frightening experience; encouraging the client to discuss his fears will help establish a plan for coping with them. The client rarely remembers the events during the episode, and this can increase anxiety. Signing a contract would have little effect because a dissociative fugue episode isn't something the client consciously wanted to do. Because the client isn't conscious of "running away," the fourth response isn't helpful.
NP: Implementation; CN: Health promotion and maintenance; CNS: Growth and development through the life span; CL: Analysis

29. Which statement best describes the cause of dissociative disorders?
1. They occur as a result of incest.
2. They occur as a result of substance abuse.
3. They occur in more than 40% of all people.
4. They occur as result of the brain trying to protect the person from severe stress.

29. 4. This answer best describes the cause of a dissociative disorder. Incest is only one of many reasons dissociative disorders occur. Typically, substance abuse isn't a cause (but may be an effect) of a dissociative disorder. Dissociative disorders are actually rare.
NP: Data collection; CN: Psychosocial integrity; CNS: Psychosocial adaptation; CL: Knowledge

NP: Nursing process   CN: Client needs category   CNS: Client needs subcategory   CL: Cognitive level

30. Which of the following nursing interventions would be most appropriate when working with a client who had a recent episode of dissociative fugue?

1. Place the client on elopement precautions.
2. Help the client identify resources to deal with stressful situations.
3. Allow the client to share his experiences about the dissociative fugue episode.
4. Confront the client about his running away from problems instead of dealing with them.

Hint! Hint! Hint!

31. A 32-year-old client lost her home in a flood last month. When questioned about her feelings about the loss, she doesn't remember being in a flood or owning a home. This client most likely has which of the following disorders?

1. Depersonalization disorder
2. Dissociative amnesia
3. Dissociative fugue
4. Dissociative identity disorder (DID)

32. Dissociative amnesia is most likely to occur as a result of which of the following circumstances?

1. Binge drinking
2. A hostage situation
3. A closed-head injury
4. A fight with a family member

Perhaps all of these circumstances can cause dissociative amnesia, but which is most likely?

33. A client was the driver in an automobile accident in which a 3-year-old boy was killed. The client now has dissociative amnesia. He verbalizes an understanding of his treatment plan when he makes which of the following statements?

1. "I won't drive a car again for at least 1 year."
2. "I'll take my Ativan anytime I feel upset about this situation."
3. "I'll visit the child's grave as soon as I'm released from the hospital."
4. "I'll attend my hypnotic therapy sessions prescribed by my psychiatrist."

30. 2. Dissociative fugue is precipitated by stressful situations. Helping the client identify resources could prevent recurrences. Once the dissociative fugue episode is over, the client returns to normal functioning; he wouldn't be an elopement risk. The client often has amnesia about the events during the dissociative fugue episode; therefore, asking him to share or remember his experience can increase his anxiety.
NP: Implementation; CN: Psychosocial integrity; CNS: Psychosocial adaptation; CL: Analysis

31. 2. Dissociative amnesia often occurs after a person has been in a traumatic event. Depersonalization disorder is characterized by recurrent sensations of loss of one's own reality. Dissociative fugue is the sudden departure from one's home or work. DID is the coexistence of two or more personalities within the same individual.
NP: Data collection; CN: Psychosocial integrity; CNS: Psychosocial adaptation; CL: Knowledge

32. 2. Dissociative amnesia typically occurs after the person has experienced a stressful, traumatic situation. A closed-head injury could result in physiologic but not dissociative amnesia. Having a fight with a family member typically wouldn't be stressful enough to cause dissociative amnesia. Binge drinking doesn't cause dissociative amnesia.
NP: Data collection; CN: Psychosocial integrity; CNS: Coping and adaptation; CL: Knowledge

33. 4. Hypnosis can be beneficial to this client because it allows repressed feelings and memories to surface. Visiting the child's grave on release from the hospital may be too traumatic and encourage continuation of the amnesia. The client needs to learn other coping mechanisms besides taking a highly addictive drug such as lorazepam (Ativan). The client may be ready to drive again, and circumstances may dictate that he drives again before 1 year has passed.
NP: Evaluation; CN: Psychosocial integrity; CNS: Coping and adaptation; CL: Application

34. A client with dissociative amnesia shows understanding of her condition when she makes which of the following statements?
1. "I'll probably never be able to regain my memories of the fire."
2. "I have problems with my memory due to my abuse of tranquilizers."
3. "If I concentrate hard enough, I'll be able to bring up memories of the car accident."
4. "To protect my mental well-being, my brain has temporarily hidden my memories of the rape from me."

35. Which of the following indicates the highest risk of suicide?
1. Suicide plan, handy means of carrying out plan, and history of previous attempt
2. Preoccupation with morbid thoughts and limited support system
3. Suicidal ideation, active suicide planning, and family history of suicide
4. Threats of suicide, recent job loss, and intact support system

36. A client with dissociative amnesia indicates understanding about the use of amobarbital (Amytal) in her treatment when she makes which of the following statements?
1. "This medication helps me sleep."
2. "This medication helps me control my anxiety."
3. "I must take this drug once a day after discharge if the drug is to be therapeutically beneficial."
4. "I'm given this medication during therapy sessions to increase my ability to remember forgotten events."

37. A client with dissociative amnesia says, "You must think I'm really stupid because I have no recollection of the accident." Which of the following responses would be most appropriate?
1. "Why would I think you're stupid?"
2. "Have I acted like I think you're stupid?"
3. "What kind of grades did you get in school?"
4. "As a protective measure, the brain sometimes doesn't let us remember traumatic events."

You're being asked to consider the highest — think top — risk for suicide.

Which of these statements can you almost hear one of your brightest clients saying?

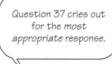

Question 37 cries out for the most appropriate response.

34. 4. One of the cardinal features of dissociative amnesia is that the person has loss of memory of a traumatic event. With this disorder, the loss of memory is a protective function performed by the brain and isn't within the person's conscious control. This type of amnesia isn't related to substance abuse. With therapy and time, the person will probably be able to recall the traumatic events.
NP: Data collection; CN: Psychosocial integrity; CNS: Psychosocial adaptation; CL: Knowledge

35. 1. A lethal plan with a handy means of carrying it out poses the highest risk and requires immediate intervention. Although all the remaining risk factors can lead to suicide, they aren't considered as high a risk as a formulated lethal plan and the means at hand. However, a client exhibiting any of these risk factors should be taken seriously and considered at risk for suicide.
NP: Data collection; CN: Psychosocial integrity; CNS: Psychosocial adaptation; CL: Application

36. 4. This drug is given to the client with dissociative amnesia to help her remember forgotten events. It isn't prescribed as a sleep aid or antianxiety agent. Because the drug is given during therapy to recall forgotten events, there would be no therapeutic benefit to taking this drug at home.
NP: Evaluation; CN: Psychosocial integrity; CNS: Coping and adaptation; CL: Analysis

37. 4. This provides a simple explanation for the client. The use of "why" can put someone on the defensive. The second choice takes the focus off the client. The third choice changes the topic.
NP: Implementation; CN: Psychosocial integrity; CNS: Psychosocial adaptation; CL: Application

38. Which of the following nursing interventions is important in caring for the client with a dissociative disorder?
1. Encourage the client to participate in unit activities and meetings.
2. Question the client about the events triggering the dissociative disorder.
3. Allow the client to remain in his room anytime he's experiencing feelings of dissociation.
4. Encourage the client to form friendships with other clients in his therapy groups to decrease his feelings of isolation.

39. Which of the following characteristics applies to depersonalization disorder?
1. Disorientation to time, place, and person
2. Sensation of detachment from body or mind
3. Unexpected and sudden travel to another location
4. A feeling that one's environment will never change

40. A client with depersonalization disorder verbalizes understanding of the ways to decrease his symptoms when he makes which of the following statements?
1. "I'll avoid any stressful situation."
2. "Meditation will help control my symptoms."
3. "I'll need to practice relaxation exercises regularly."
4. "I may need to remain on antipsychotic medication for the rest of my life."

41. A client with depersonalization disorder spends much of her day in a dreamlike state during which she ignores personal care needs. Which nursing diagnosis would you expect to find identified for this client?
1. Disturbed thought processes related to organic brain damage
2. Impaired memory related to frequently being in a dreamlike state
3. Dressing or grooming self-care deficit related to perceptual impairment
4. Deficient knowledge related to performance of personal care needs due to lack of information

Almost finished!

How can you tell when your client understands your instructions?

Making the right nursing diagnosis is critical for effective nursing care.

38. 1. Attending unit activities and meetings helps decrease the client's sense of isolation. Typically, the client can't recall the events that triggered the dissociative disorder. The client would need to be isolated from others only if he can't interact appropriately. A client with a dissociative disorder has typically had few healthy relationships. Forming friendships with others in therapy could be setting the client up to continue in unhealthy relationships.
NP: Implementation; CN: Safe, effective care environment; CNS: Coordinated care; CL: Application

39. 2. In depersonalization disorder, the person feels detached from his body and mental processes. The person is usually oriented to time, place, and person. Unexpected and sudden travel to another location is one of the characteristics of dissociative fugue. Clients with depersonalization disorder often feel the outside world has changed.
NP: Data collection; CN: Psychosocial integrity; CNS: Psychosocial adaptation; CL: Comprehension

40. 3. Relaxation can lead to a decrease in maladaptive responses. Although stress can be a predisposing factor in depersonalization disorder, it's impossible to avoid all stressful situations. This isn't a psychotic disorder, so antipsychotic medication wouldn't be therapeutic or beneficial. Meditation is the voluntary induction of the sensation of depersonalization.
NP: Evaluation; CN: Psychosocial integrity; CNS: Coping and adaptation; CL: Analysis

41. 3. Because of time spent in a dreamlike state, many clients with depersonalization disorder ignore self-care needs. There's no known organic brain damage with this disorder. Memory impairment is more of a problem with other dissociative disorders, such as dissociative identity disorder and dissociative amnesia. The dreamlike state can lead to problems meeting personal care needs, not a knowledge deficit.
NP: Planning; CN: Safe, effective care environment; CNS: Safety and infection control; CL: Application

42. A client reports frequently feeling that he's floating above his body. During these times, he says he's aware of who he is and where he's located. The client is describing which of the following types of dissociative disorder?
1. Depersonalization disorder
2. Dissociative amnesia
3. Dissociative identity disorder (DID)
4. Dissociative fugue

42. 1. One of the cardinal symptoms of depersonalization disorder is feeling detached from one's body or mental processes. During the feelings of detachment, the person doesn't become disoriented. Dissociative amnesia is defined as one or more episodes of being unable to recall important information. DID is the existence of two or more personalities that take control of the person's behavior. In a dissociative fugue, the person has no memory of his life before the flight.
NP: Data collection; CN: Psychosocial integrity; CNS: Psychosocial adaptation; CL: Knowledge

43. A client with depersonalization disorder tells the nurse, "I feel like such a freak when I have an out-of-body experience." Which response would be most appropriate?
1. "How often do you have these feelings?"
2. "I'm not sure what you mean by a freak."
3. "Tell me more about these out-of-body experiences."
4. "How does your husband feel about you having these experiences?"

Which kinds of questions tend to encourage discussion?

43. 3. This open-ended response allows the client to focus and expand on this topic. Asking how often the experiences occur is a closed-ended question that doesn't encourage discussion of the experience. The second response could cause the client to focus too narrowly on only one aspect of the topic. Asking how the client's husband feels makes it appear as if the nurse wants to change the topic.
NP: Implementation; CN: Psychosocial integrity; CNS: Psychosocial adaptation; CL: Analysis

44. In which of the following settings does treatment for clients with depersonalization disorder typically occur?
1. Inpatient psychiatric hospital
2. Community mental health clinic
3. Family practice physician's office
4. Support group for clients with depersonalization disorder

44. 2. Most clients with depersonalization disorder can be treated successfully on an outpatient basis. These clients only need to be hospitalized if they become suicidal or have severe depression or anxiety. Because no organic basis for the disorder usually exists, these clients aren't treated in a family practice physician's office. The disorder is rare, so few support groups have only clients with this disorder.
NP: Data collection; CN: Psychosocial integrity; CNS: Psychosocial adaptation; CL: Knowledge

You did it! Yippee!

45. A client with depersonalization disorder tells the nurse, "I feel like my arm isn't attached to my body." Which of the following responses would be most appropriate?
1. "Do you know where you are?"
2. "What makes you feel that way?"
3. "Don't worry because I can see your arm is attached to your body."
4. "This disorder causes people to feel that body parts may be unattached to the rest of the body."

45. 4. Reinforcing that what the client feels is an expected result of the disease process would be most appropriate. Asking if the client knows where he is changes the topic. Asking why he feels that way could put the client on the defensive. Stating that his arm is attached to his body belittles the client's feelings.
NP: Implementation; CN: Psychosocial integrity; CNS: Psychosocial adaptation; CL: Comprehension

NP: Nursing process   CN: Client needs category   CNS: Client needs subcategory   CL: Cognitive level

This chapter will test your knowledge of disorders of a highly sensitive nature. Remain professional at all times, and you'll do great. Good luck!

1. A male client has undergone surgery for the repair of an abdominal aortic aneurysm. Which of the following responses is most appropriate to the client's wife when she asks if her husband will be impotent?

1. "Don't worry; he will be all right."
2. "He has other problems to worry about."
3. "We will cross that bridge when we come to it."
4. "There's a chance of impotence after repair of an abdominal aortic aneurysm."

2. Which of the following discharge instructions would be most accurate to provide to a female client who has suffered a spinal cord injury at the C4 level?

1. After a spinal cord injury, women usually remain fertile; therefore, you may consider contraception if you don't want to become pregnant.
2. After a spinal cord injury, women are usually unable to conceive a child.
3. Sexual intercourse shouldn't be different for you.
4. After a spinal cord injury, menstruation usually stops.

3. Which of the following permanent complications might the nurse expect to see in a client who has just undergone a perineal prostatectomy?

1. Bleeding
2. Erectile dysfunction
3. Infection
4. Pneumonia

Therapeutic communication involves demonstrating sensitivity to your client's and his family's concerns.

Note the word permanent in question 3 before you answer.

1. 4. Impotence and retrograde ejaculation are sexual dysfunctions commonly experienced by male clients after abdominal aortic aneurysm. Telling a family member that the client will be all right is offering false assurance. Stating that he has other problems isn't therapeutic and doesn't address the wife's concern. Telling the client's wife to "cross that bridge when we come to it" ignores her concerns and isn't therapeutic.

NP: Implementation; CN: Psychosocial integrity; CNS: Coping and adaptation; CL: Application

2. 1. After a spinal cord injury, women remain fertile and can conceive and deliver a child. If a woman doesn't want to become pregnant, she *must* use contraception. Menstruation isn't affected by a spinal cord injury, but sexual functioning may be different.

NP: Data collection; CN: Physiological integrity; CNS: Physiological adaptation; CL: Application

3. 2. After a perineal prostatectomy, a major complication is erectile dysfunction. As with any surgery, the client also is at risk for bleeding, pneumonia, and infection, but these aren't permanent conditions.

NP: Data collection; CN: Physiological integrity; CNS: Reduction of risk potential; CL: Comprehension

4. A client with chronic obstructive pulmonary disease (COPD) tells the nurse, "I no longer have enough energy to make love to my husband." Which of the following nursing interventions would be most appropriate?
1. Refer the couple to a sex therapist.
2. Advise the woman to seek a gynecologic consult.
3. Suggest methods and measures that facilitate sexual activity.
4. Tell the client, "If you talk this over with your husband, he'll understand."

5. A client with an ileostomy tells the nurse he can't have an erection. Which of the following pertinent information should the nurse know?
1. The client will never regain functioning.
2. The client needs an abdominal X-ray.
3. The client has no problem with self-control.
4. Impotence is uncommon following an ileostomy.

6. A recently divorced 40-year-old male who has undergone radiation therapy for testicular cancer tells the nurse he can't achieve an erection. Which of the following nursing diagnoses is most appropriate?
1. Ineffective coping related to radiation therapy
2. Sexual dysfunction related to the effects of radiation therapy
3. Disturbed body image related to the effects of radiation therapy
4. Imbalanced nutrition: Less than body requirements related to radiation therapy

7. Which of the following actions should the nurse include in the teaching plan of a newly married female client with a cervical spinal cord injury who doesn't wish to become pregnant at this time?
1. Provide the client with brochures on sexual practice.
2. Provide the client's husband with material on vasectomy.
3. Instruct the client on the rhythm method of contraception.
4. Instruct the client's husband on inserting a diaphragm with contraceptive jelly.

In question 4 you've got to chose the most appropriate answer, not just a possible one.

There's that phrase most appropriate again.

4. 3. Sexual dysfunction in COPD clients is the direct result of dyspnea and reduced energy levels. Measures to reduce physical exertion, enhance oxygenation, and accommodate decreased energy levels may aid sexual activity. If the problem persists, a consult with a therapist might be necessary. A gynecologic consult isn't necessary. Discussing this with her husband may not resolve the problem.
NP: Implementation; CN: Physiological integrity; CNS: Reduction of risk potential; CL: Application

5. 4. Sexual dysfunction is uncommon after an ileostomy. Psychological causes of impotence should be explored. An abdominal X-ray isn't indicated for sexual dysfunction. An ileostomy can change a person's self-control, making sexual functioning difficult.
NP: Data collection; CN: Psychosocial integrity; CNS: Psychosocial adaptation; CL: Analysis

6. 2. Radiation or chemotherapy may cause sexual dysfunction. Libido may only be temporarily affected, and the client should be provided with emotional support. The client hasn't verbalized fear or concern related to the cancer. The client may experience alopecia or skin changes as well as weight loss, but the client isn't verbalizing concern in this area. Nutrition hasn't been mentioned.
NP: Planning; CN: Psychosocial integrity; CNS: Coping and adaptation; CL: Analysis

7. 4. Because the client experienced a cervical spinal cord injury, she won't be able to insert any form of contraception protection; therefore, it's vital to provide her husband with instruction on insertion of a diaphragm. Providing the couple with literature on sexual practice doesn't address the client's concerns. During this time of crisis, the couple doesn't wish to have children but they may reconsider; so providing information on vasectomy isn't appropriate. The rhythm method isn't the most effective way to prevent pregnancy.
NP: Implementation; CN: Psychosocial integrity; CNS: Coping and adaptation; CL: Application

NP: Nursing process   CN: Client needs category   CNS: Client needs subcategory   CL: Cognitive level

8. A client tells the nurse she's having her menstrual period every 2 weeks, and it lasts for 1 week. Which of the following conditions is best defined by this menstrual pattern?
1. Amenorrhea
2. Dyspareunia
3. Menorrhagia
4. Metrorrhagia

9. The nurse is caring for a 39-year-old male client who recently underwent surgery and is having difficulty accepting changes in his body image. Which nursing intervention is appropriate?
1. Allow the client to express both positive and negative feelings about his body image.
2. Restrict the client's opportunity to view the incision and dressing because it's upsetting.
3. Assist the client to focus on future plans for recovery.
4. Assist the client to repress anger while discussing the body image alteration.

10. A 38-year-old woman must undergo a hysterectomy for uterine cancer. The nurse planning her care should include which of the following actions to meet the woman's body image changes?
1. Ask her if she is having pain.
2. Refer her to a psychotherapist.
3. Don't discuss the subject with her.
4. Encourage her to verbalize her feelings.

11. A 50-year-old male who had a myocardial infarction 8 weeks ago tells a nurse, "My wife wants to make love, but I don't think I can. I'm worried that it might kill me." Which of the following responses from the nurse would be most appropriate?
1. "Tell me about your feelings."
2. "Let's increase your rehabilitation schedule."
3. "Let me call the primary health care provider for you."
4. "Tell your wife when you're able you'll make love."

What's the difference between menorrhagia and metrorrhagia? You have to know to answer question 8.

10 done! Keep going!

Let's put together a care plan that serves both your needs.

8. 3. Menorrhagia is an excessive menstrual period. Amenorrhea is lack of menstruation. Dyspareunia is painful intercourse. Metrorrhagia is uterine bleeding from another cause other than menstruation.
NP: Data collection; CN: Physiological integrity; CNS: Reduction of risk potential; CL: Knowledge

9. 1. The nurse needs to observe for any indication that the client is ready to address his body image change. The client should be allowed to look at the incision and dressing if he wants to do so. It's too soon to focus on the future with this client. The nurse should allow the client to express his feelings and not repress them, because repression prolongs recovery.
NP: Planning; CN: Psychosocial integrity; CNS: Coping and adaptation; CL: Application

10. 4. Encourage the client to verbalize her feelings because loss of one's reproductive organs may bring on feelings of loss of sexuality. Pain is a concern after surgery, but it has no bearing on body image. Referring her to a psychotherapist may be premature; the client should be given time to work through her feelings. Avoidance of the subject isn't a therapeutic nursing intervention.
NP: Implementation; CN: Psychosocial integrity; CNS: Coping and adaptation; CL: Application

11. 1. The nurse should address the client's concerns. Asking the client to verbalize his feelings will permit the nurse to gain insight into the problem. The rehabilitation schedule shouldn't be increased until the nurse assesses the situation and is sure no harm will come to the client. Calling the primary health care provider before a complete assessment is made is inappropriate. Telling the wife that eventually the client will make love may place strain on the marriage.
NP: Implementation; CN: Psychosocial integrity; CNS: Coping and adaptation; CL: Application

12. A 55-year-old female client who is in cardiac rehabilitation tells a nurse that she is unable to make love to her husband because she often feels fatigued and has a sense of doom. Which of the following nursing interventions is most appropriate?

1. Instruct her not to have intercourse until she is ready.
2. Instruct her to take a nitroglycerin tablet prior to intercourse.
3. Encourage her to learn additional methods to use for sexual intercourse.
4. Encourage her to verbalize her feelings while you perform a physical examination on her.

13. A 33-year-old female tells the nurse she has never had an orgasm. She tells the nurse that her partner is upset that he can't meet her needs. Which of the following nursing interventions is most appropriate?

1. Ask the client if she desires intercourse.
2. Assess the couple's perception of the problem.
3. Tell the client that most women don't reach orgasm.
4. Refer the client to a therapist because she has sexual aversion disorder.

14. A 20-year-old female client is in the emergency department after being sexually assaulted by a stranger. Which nursing intervention has the highest priority?

1. Assist her in identifying which of her behaviors placed her at risk for the attack.
2. Make an appointment for her in 6 weeks at a local sexual assault crisis center.
3. Encourage discussion of her early childhood experiences.
4. Assist her in identifying family or friends who could provide immediate support for her.

Just as sensitive client issues demand the most appropriate nursing behavior, tricky NCLEX questions demand the most appropriate answers. Catch my meaning?

Priority alert! Priority alert! Sirens should be flashing in your head!

12. 4. Because the client has a complaint of fatigue, she should be examined, and her feelings should be explored. Instructing her not to have intercourse doesn't address her concerns. She shouldn't take nitroglycerin before intercourse until her fatigue is evaluated. Before recommending alternative methods for intercourse, the client should be assessed physically and psychologically.

NP: Implementation; CN: Psychosocial integrity; CNS: Coping and adaptation; CL: Application

13. 2. Assessing the couple's perception of the problem will define the problem and assist the couple and the nurse in understanding it. A nurse can't make a medical diagnosis such as sexual aversion disorder. Most individuals can be taught to reach orgasm if there's no underlying medical condition. When assessing the client, the nurse should be professional and matter-of-fact and shouldn't make the client feel inadequate or defensive.

NP: Implementation; CN: Psychosocial integrity; CNS: Coping and adaptation; CL: Application

14. 4. The client needs a lot of support to help her through this ordeal. Assisting the client in identifying behaviors that placed her at risk for the attack places the blame on the client. Waiting 6 weeks to make an appointment is incorrect—the local crisis center needs to be called immediately. Some psychiatric disorders are related to early childhood experiences, but rape isn't.

NP: Implementation; CN: Psychosocial integrity; CNS: Coping and adaptation; CL: Comprehension

15. A 50-year-old male who is taking antihypertensive medication tells the office nurse who is monitoring his blood pressure that he can't have sexual intercourse with his wife anymore. Which of the following problems is most likely the cause?
   1. His advancing age
   2. His blood pressure
   3. His stressful lifestyle
   4. His blood pressure medication

16. Research performed on sexual disorders has shown that victims of sexual abuse have a tendency to experience which of the following results?
   1. Have higher hormonal levels
   2. Remain celibate throughout life
   3. Become sex offenders themselves
   4. Have normal sexual experiences throughout life

Good work! Keep it up!

You can tie yourself up in knots trying to answer question 17 if you miss the *not*.

17. Which of the following conditions or habits is <u>not</u> associated with female infertility?
   1. Sexually transmitted diseases
   2. Smoking
   3. Hormone imbalances
   4. Osteoporosis

18. A male client reports to the nurse that he has a strong desire to live and be treated as a woman. He confesses that he's uncomfortable with his assigned sex. Which of the following best describes these symptoms?
   1. Delusions
   2. Gender identity issues
   3. Homosexuality
   4. Hormonal imbalances

19. Which of the following treatments might be used for a patient with gender identity disorder?
   1. Group therapy
   2. Surgical sexual reassignment
   3. Relaxation techniques
   4. Antipsychotic agent

15. 4. Antihypertensive medication may cause impotence in men. Men are usually able to have an erection throughout their life. Blood pressure itself doesn't cause impotence, but its treatment does. Stress may cause erectile dysfunction, but there's no evidence that the client is under stress.
NP: Data collection; CN: Psychosocial integrity; CNS: Coping and adaptation; CL: Knowledge

16. 3. Children who have been sexually abused have a predisposition for becoming a sex offender. Research doesn't show that victims of sexual abuse have higher hormonal levels. They probably won't have normal sexual experiences throughout life without intense therapy. Research also shows that victims of sexual abuse are at risk for paraphiliac disorders.
NP: Data collection; CN: Psychosocial integrity; CNS: Psychosocial adaptation; CL: Analysis

17. 4. Osteoporosis is a condition in which the risk increases after menopause. All other conditions or habits have been associated with infertility.
NP: Data collection; CN: Health promotion and maintenance; CNS: Prevention and early detection of disease; CL: Knowledge

18. 2. A persistent cross-gender identification and dissatisfaction with one's assigned sex are major characteristics of gender identity disorders. Delusions are firmly held beliefs not substantiated in reality. Homosexuality is an attraction to members of the same sex. Hormonal imbalances aren't relevant to the diagnosis of gender identity issues.
NP: Data collection; CN: Psychosocial integrity; CNS: Psychosocial adaptation; CL: Knowledge

19. 2. A surgical sexual reassignment operation is the only effective treatment for gender identity disorder. This is undertaken only after careful clinical interviews and evaluations by specialists in this field. Group therapy and relaxation techniques may be helpful to this client, but aren't specific to the treatment of this disorder. Antipsychotic drugs aren't appropriate for this condition.
NP: Planning; CN: Psychosocial integrity; CNS: Psychosocial adaptation; CL: Application

20. A married couple in their early thirties has unsuccessfully tried to conceive for 1 year. The male partner will begin an infertility workup. Which of the following is not a part of this process?
1. Physical examination
2. Reproductive history
3. Semen analysis
4. Psychological testing

Prioritizing correctly is extremely important for question 21.

21. A 38-year-old woman was returning home from the store late one evening and was sexually assaulted. When she's brought to the emergency department, she's crying. Which of the following concerns for this client should be the nurse's first priority?
1. Filing a police report
2. Calling the client's family
3. Encouraging the client to enroll in a self-defense class
4. Remaining with the client and assisting her through the crisis

22. A 20-year-old single college female has recently been diagnosed with human papilloma virus (HPV). She has researched this condition and comes to the health clinic with many fears and concerns. Which of the following would be the best initial intervention by the nurse?
1. Ask the client to discuss her concerns.
2. Provide her with reliable information about this condition.
3. Refer her to her gynecologist.
4. Discuss the dangers of multiple sex partners.

Clients with sexual disorders may be ashamed and unwilling to discuss the problem.

23. A client is admitted to the hospital for treatment of pedophilia and tells the nurse that he doesn't want to talk to her about his sexual behaviors. Which of the following responses from the nurse is most appropriate?
1. "I need to ask you the questions on the database."
2. "It is your right not to answer my questions."
3. "OK, I'll just write 'no comment.'"
4. "I know this must be difficult for you."

20. 4. Psychological testing isn't a typical component of an infertility workup in males. All other examinations would be included in a thorough examination to determine if there's a problem, whether it's reversible, or whether it's life-threatening.
NP: Implementation; CN: Physiological integrity; CNS: Reduction of risk potential; CL: Knowledge

21. 4. Sexual assault is treated as a medical emergency, and the client requires constant attention and assistance during the crisis. Filing a police report wouldn't take precedence over a medical emergency. Comforting the client by contacting her family should be carried out after her injuries are treated. Encouraging the client to enroll in a self-defense class isn't appropriate during crisis.
NP: Implementation; CN: Psychosocial integrity; CNS: Psychosocial adaptation; CL: Application

22. 1. Encouraging the client to discuss her concerns establishes a nonjudgmental, therapeutic relationship and would be the best initial response. Other interventions might be appropriate at some point. The nurse should take care to discuss the dangers of multiple sex partners in a nonjudgmental manner.
NP: Implementation; CN: Psychosocial integrity; CNS: Psychosocial adaptation; CL: Knowledge

23. 4. Stating "I know this must be difficult for you" acknowledges the client's feelings and opens communications. Insisting that the form needs to be completed doesn't open up communications or acknowledge the client's feelings. Clients have rights, but data collection is necessary so that help with the problem can be offered. Writing "no comment" alone would be inappropriate.
NP: Implementation; CN: Psychosocial integrity; CNS: Psychosocial adaptation; CL: Application

24. A single woman with ulcerative colitis has recently had a colostomy. She's distraught about how this will affect future relationships. What would be an appropriate initial intervention for this client?

1. Refer her to a support group.
2. Allow her to express her concerns to the nurse.
3. Offer to research statistics on this topic.
4. Explore the positive aspects of living single.

25. When treating a client admitted to the psychiatric unit for transvestic fetishism, the nurse should develop a care plan based on which of the following nursing diagnoses?

1. Ineffective health maintenance
2. Ineffective sexuality patterns
3. Dysfunctional grieving
4. Self-care deficit

26. When working with a client with pedophilia, which of the following goals is appropriate for the client?

1. To attend all meetings on the unit
2. To use triggers to initiate sexual behaviors
3. To inform his employer of the reason for his hospitalization
4. To verbalize appropriate methods to meet sexual needs upon discharge

27. A client admitted to the hospital with a diagnosis of pedophilia tells his roommate about his problems. His roommate runs down the hall yelling at the nurse, "I don't want to be in here with a child molester." Which of the following responses from the nurse is most appropriate?

1. "Stop acting out."
2. "Calm down, and go back to your room."
3. "Your roommate isn't a child molester."
4. "I can see you're upset. Sit down and we will talk."

Don't stop now! You're more than half finished!

Effective care may involve managing interactions between clients.

24. 2. Allowing the client to express her concerns is a therapeutic first step. Referring her to a support group is premature. Offering to research statistics or to explore aspects of living single negate the emotional aspect of this problem, and the client might conclude that it isn't acceptable to discuss her feelings.
NP: Implementation; CN: Physiological integrity; CNS: Reduction of risk potential; CL: Knowledge

25. 2. Ineffective sexuality patterns would be appropriate because transvestic fetishism refers to intense sexual arousal with cross-dressing. Ineffective health maintenance is an appropriate diagnosis for someone experiencing a health problem. Dysfunctional grieving refers to the inability to recover from a loss. Self-care deficit is a diagnosis for the inability to meet self-care needs. The client hasn't exhibited any problems with health, self-care, or loss.
NP: Planning; CN: Psychosocial integrity; CNS: Coping and adaptation; CL: Application

26. 4. Upon discharge, the client should verbalize an alternative appropriate method to meet his sexual needs and effective strategies to prevent relapse. It isn't imperative that the client attend all meetings on the unit, but it's important that the client attend the prescribed group sessions. A client with pedophilia should recognize triggers that initiate inappropriate sexual behavior and learn ways to direct his impulses. The client may wish to discuss the disorder with his spouse but not necessarily with his employer.
NP: Planning; CN: Psychosocial integrity; CNS: Psychosocial adaptation; CL: Analysis

27. 4. Acknowledging that the client is upset and sitting down and talking with him allows the client to verbalize his feelings. Stating that the pedophile isn't a child molester doesn't acknowledge the client's feelings. If a client were agitated or anxious over his roommate, it wouldn't be therapeutic or safe to keep those clients together without intervention. Telling the client to stop acting out isn't a therapeutic response.
NP: Implementation; CN: Psychosocial integrity; CNS: Psychosocial adaptation; CL: Application

28. The nurse is obtaining a health history from a client when he states that he has been diagnosed with voyeurism. The nurse knows which of the following actions is characteristic of a voyeur?
1. Observing others while they disrobe
2. Wearing clothing of the opposite sex
3. Rubbing against a nonconsenting person
4. Using rubber sheeting for sexual arousal

28. 1. Voyeurism is sexual arousal from secretly observing someone who is disrobing. Rubbing against someone who is nonconsenting is frottage. Using objects for sexual arousal is fetishism. A transvestic fetishism describes someone who enjoys cross-dressing.
NP: Data collection; CN: Psychosocial integrity; CNS: Psychosocial adaptation; CL: Comprehension

29. A female being treated for infertility confides to the nurse that she hasn't told her partner she had been treated for a sexually transmitted disease in the past. What would be the most therapeutic response for the nurse to give?
1. "Do you think withholding this information is the basis for a trusting relationship?"
2. "Don't you think your partner deserves to know?"
3. "What concerns do you have about sharing this information?"
4. "I can understand why you would want to keep this information from him."

29. 3. This response encourages the client to verbalize her concerns in a safe environment and begin to choose a course of action for how to deal with this issue now. Telling the client that she's withholding information that may cause distrust in her relationship and that her partner deserves to know convey negative judgments towards the client. The fourth response doesn't encourage discussion and problem solving.
NP: Implementation; CN: Psychosocial integrity; CNS: Coping and adaptation; CL: Application

30. After learning that his gay roommate has tested positive for the human immunodeficiency virus, a client asks the nurse about moving to another room on the psychiatric unit because the client doesn't feel "safe" now. What should the nurse do first?
1. Move the client to another room.
2. Ask the client to describe any fears.
3. Move the client's roommate to a private room.
4. Explain that such a move wouldn't be therapeutic for the client or his roommate.

30. 2. To intervene effectively, the nurse must first understand the client's fears. After exploring the client's fears, the nurse may move the client or his roommate or explain why such a move wouldn't be therapeutic as needed.
NP: Data collection; CN: Psychosocial integrity; CNS: Coping and adaptation; CL: Application

You're doing extremely well, and I'm very proud!

31. A rape victim comes to the emergency department for treatment. Which nursing diagnosis would most likely be identified to address the psychological component of this crisis?
1. Sexual dysfunction
2. Risk for infection
3. Acute pain
4. Rape-trauma syndrome

31. 4. There's no evidence that this client has a sexual dysfunction. Risk for infection and acute pain may be appropriate diagnoses related to the physical aspect of this trauma, but they don't address emotional issues.
NP: Planning; CN: Psychosocial integrity; CNS: Coping and adaptation; CL: Comprehension

32. A nurse lecturing on paraphilias informs her audience that recidivism is high for paraphilias. Which of the following definitions best describes recidivism?
 1. Insight into treatment
 2. Aggressive sexual assault
 3. Behaviors associated with sexual deviation
 4. Continued inappropriate behavior after treatment

32. 4. Recidivism is defined as continuing in an unacceptable behavior after completing treatment to correct that behavior. Sexually deviant behaviors are known as paraphilias. Aggressive sexual assault is a type of paraphilia. High level of insight isn't connected with any specific disorder.
NP: Data collection; CN: Psychosocial integrity; CNS: Coping and adaptation; CL: Analysis

The NCLEX often tests your ability to educate accurately.

33. A married client with a month-long history of erectile dysfunction is visiting his physician for an initial evaluation. Prior to the examination, the nurse interviews the client. Which response wouldn't lead to useful information related to this problem?
 1. "Tell me about your smoking habits."
 2. "What are your current stressors?"
 3. "Tell me about your nutritional habits."
 4. "How would you describe your relationship with your wife?"

33. 3. There are medical and psychological causes for erectile dysfunction. Smoking habits, drug and alcohol consumption, and many medical conditions can contribute to this problem. Psychological stressors, depression, and interpersonal relationships can contribute as well. There are no known correlations between erectile dysfunction and nutritional habits.
NP: Data collection; CN: Health promotion and maintenance; CNS: Prevention and early detection of disease; CL: Knowledge

34. The nurse meets with an infertile couple to assist with coping strategies. Which of the following emotional experiences is not typical for those facing this problem?
 1. Guilt
 2. Loss of control
 3. Frustration
 4. Indifference

34. 4. Infertility usually results in a crisis for most couples who experience it. Indifference would be uncharacteristic. Guilt over past behaviors, a loss of control over reaching a lifelong goal, and frustration over not being able to conceive may play a major role in the emotions of any infertile couple.
NP: Data collection; CN: Psychosocial integrity; CNS: Coping and adaptation; CL: Comprehension

So, when do gender differences start?

35. The nurse knows that gender is part of one's identity. Which of the following events signifies when gender is first ascribed?
 1. A neonate is born.
 2. A child attends school.
 3. A child receives sex-specific toys.
 4. A child receives sex-specific clothing.

35. 1. As soon as a neonate is born, gender is ascribed. In the hospital, a neonate is given either a pink or blue name band, card, or blanket. Gender identification is perpetuated throughout life with sex-specific clothing and toys. Sexual identity is reaffirmed throughout the school years.
NP: Data collection; CN: Psychosocial integrity; CNS: Coping and adaptation; CL: Comprehension

36. A mother brings her 14-year-old son to the psychiatric crisis room. The client's mother states, "He's always dressing in female clothing. There must be something wrong with him." Which of the following responses from the nurse would be most appropriate?

1. "Your son will be evaluated shortly."
2. "I'll tell your son that this isn't appropriate."
3. "I know you're upset. Would you like to talk?"
4. "I wouldn't want my son to dress in girl's clothing."

37. A woman taking antidepressant medication complains that she has a decreased desire for sex which is causing significant marital stress. She wonders if she should discontinue her medication. What would be the best initial response from the nurse?

1. "You should not stop your medication."
2. "What are your thoughts on how you should handle this?"
3. "Doesn't your husband understand the importance of your medication?"
4. "Have you discussed this with your physician?"

38. Which of the following reasons explains the rationale for estrogen therapy for a male client who wishes to undergo sexual reassignment surgery?

1. To develop breasts
2. To cause menstruation
3. To assist with cross-dressing
4. To develop body hair and lack of menstruation

39. The nurse is developing a teaching plan on rape prevention. All of the following guidelines would be present on the plan except for which one?

1. Take responsible action. Avoid putting yourself in a vulnerable position such as attending a party alone and drinking heavily.
2. Don't walk alone at night.
3. Learn ways to defend yourself.
4. Always take the shortest driving route home.

You've finished 38 questions already. Only 7 more to go!

36. 3. Acknowledging the mother's feelings and offering her an opportunity to verbalize her concerns provides a forum for open communication. The nurse shouldn't offer an opinion by stating she wouldn't want her son dressing in female clothing. Telling the client's mother that he will be evaluated shortly doesn't address her concerns. Telling the client that this behavior isn't appropriate doesn't assess his feelings, nor does it analyze the behavior.
NP: Implementation; CN: Psychosocial integrity; CNS: Coping and adaptation; CL: Application

37. 2. Encouraging the client to verbalize her thoughts will help the client to problem solve and identify feelings related to different choices. The first response is too directive and doesn't encourage exploration on the part of the client. The third response conveys negative judgment. The fourth response might be appropriate, but it may also give the impression that the nurse doesn't want to discuss this issue with the client.
NP: Implementation; CN: Psychosocial integrity; CNS: Coping and adaptation; CL: Comprehension

38. 1. A male who receives long-term estrogen therapy will develop female secondary sexual characteristics such as breasts. A male on estrogen won't menstruate as he doesn't have a uterus. Estrogen has no bearing on cross-dressing. Androgens would be taken by a female to develop body hair and stop menstruation.
NP: Implementation; CN: Psychosocial integrity; CNS: Psychosocial adaptation; CL: Analysis

39. 4. The shortest driving route might take you through a high crime neighborhood. Learn alternative routes so you have options if your safety seems compromised. All other guidelines are prudent in rape prevention.
NP: Planning; CN: Health promotion and maintenance; CNS: Prevention and early detection of disease; CL: Comprehension

NP: Nursing process   CN: Client needs category   CNS: Client needs subcategory   CL: Cognitive level

40. A nurse working in a correctional facility is assigned to a unit treating sexual offenders. Which of the following disorders would she <u>not</u> encounter on this unit?
1. Pedophilia
2. Voyeurism
3. Transvestitism
4. Exhibitionism

40. 3. All of the above disorders have legal consequences except for transvestitism.
NP: Data collection; CN: Psychosocial integrity; CNS: Psychosocial adaptation; CL: Knowledge

Let me be most helpful by pointing out that you don't want to choose just any response for question 41 but the most helpful!

41. A group of female college students were walking back to their dorm at night. Suddenly a man ran out from the bushes and exposed himself to them. All of the women were initially startled, but all but one began to laugh. This one female went to the health clinic because she was so upset. Which response by the nurse would be <u>most helpful</u> psychologically to this student?
1. "Can you imagine a flasher on campus!"
2. "I will call security right away."
3. "I can see you're upset. Tell me more about this."
4. "Please describe this person to me."

41. 3. Acknowledging the client's emotions is the best initial step in helping her talk about her concerns. The first response is not therapeutic. The second and fourth responses are appropriate interventions but don't help the student deal with her emotional reaction.
NP: Implementation; CN: Psychosocial integrity; CNS: Growth and development through the life span; CL: Analysis

42. According to Erikson, an adolescent who is suffering from gender identity disorder can't progress through which of the following developmental tasks?
1. Initiative versus guilt
2. Intimacy versus isolation
3. Industry versus inferiority
4. Identity versus role confusion

42. 4. According to developmentalist Erik Erikson, adolescence is a time when role identity is found as a result of independence and sexual maturity; role confusion would result from the inability to integrate all experiences. Initiative versus guilt is when a child begins to conceptualize and interpersonalize relationships. Intimacy versus isolation is a stage in which the adult meets other adults and establishes relationships. Industry versus inferiority is when a child incorporates and acquires social skills.
NP: Data collection; CN: Health promotion and maintenance; CNS: Growth and development through the life span; CL: Analysis

In question 43, notice the words most appropriate.

43. A 35-year-old male who has been married for 10 years arrives at the psychiatric clinic stating, "I can't live this lie any more. I wish I were a woman. I don't want my wife. I need a man." Which of the following initial actions would be <u>most appropriate</u> from the nurse?
1. Call the primary health care provider.
2. Encourage the client to speak to his wife.
3. Have the client admitted.
4. Sit down with the client, and talk about his feelings.

43. 4. Sitting down with the client and exploring his feelings will allow the nurse to assess him. The primary health care provider shouldn't be notified until an assessment is made. An assessment of the client should be made *before* admitting the client to the unit. He shouldn't speak to his wife until he has processed his feelings.
NP: Implementation; CN: Psychosocial integrity; CNS: Psychosocial adaptation; CL: Application

44. A 14-year-old female client admits to having transsexual feelings and states, "I would rather die than live in this body." Which of the following <u>initial</u> actions would be most appropriate for the nurse to take?

1. Explain to her that she is too young to have these feelings.
2. Call her parents, and let them know about her feelings.
3. Encourage her to verbalize her feelings.
4. Ask her if she plans to kill herself.

44. 4. Whenever a client verbalizes feelings of preferring death to life, the nurse should always make sure that the client doesn't have a plan. Transsexual tendencies usually arise during the adolescent years, so it is appropriate for the client to have these feelings. Calling her parents wouldn't be a priority until after a psychological safety assessment is completed. Encouraging her to verbalize her feelings isn't an initial action for the nurse.

NP: Implementation; CN: Psychosocial integrity; CNS: Psychosocial adaptation; CL: Application

45. A client with panic disorder seeks out the nurse to discuss her concerns about another individual on the unit whom she describes as a "man wearing women's clothing." Which of the following responses by the nurse would be most helpful?

1. "Tell me more about your concerns."
2. "Don't worry about Joe; he's harmless!"
3. "Have you ever encountered a cross-dresser before?"
4. "I can't discuss other clients on the unit."

45. 1. Eliciting the client's concerns will help to build a trusting relationship so the client can work through troubling issues. The other responses are not therapeutic.

NP: Implementation; CN: Psychosocial integrity; CNS: Coping and adaptation; CL: Comprehension

Congratulations! You're finished!

New information about eating disorders is released continuously. For the latest about disorders of critical importance for young people, check the Internet site of the American Anorexia Bulimia Association at **www.edap.org/**.

Chapter 21
Eating disorders

Choose the best answer!

1. A parent with a daughter with bulimia nervosa asks a nurse, "How can my child have an eating disorder when she isn't underweight?" Which of the following responses is best?
1. "A person with bulimia nervosa can maintain a normal weight."
2. "It's hard to face this type of problem in a person you love."
3. "At first there is no weight loss; it comes later in the disease."
4. "This is a serious problem even though there is no weight loss."

2. A nurse is collecting data on an adolescent girl with symptoms of bulimia nervosa. Which of the following findings is expected based on laboratory test results?
1. Hypocalcemia
2. Hypoglycemia
3. Hypokalemia
4. Hypophosphatemia

3. Which of the following statements about the binge-purge cycle that occurs with bulimia nervosa is correct?
1. There are emotional triggers connected to bingeing.
2. Over time, people usually grow out of bingeing behaviors.
3. Bingeing isn't the problem; purging is the issue to address.
4. When a person gets too hungry, there's a tendency to binge.

Help your client understand her behavior.

1. 1. A client with bulimia nervosa may be of normal weight, overweight, or underweight. Weight loss isn't a clinical criterion for bulimia nervosa. The second response doesn't address the need for information about the relationship between weight change and bulimia nervosa. The fourth response doesn't address the issue of weight change in a client with bulimia nervosa.
NP: Implementation; CN: Psychosocial integrity; CNS: Psychosocial adaptation; CL: Knowledge

2. 3. Clients who are bulimic will have hypokalemia, or decreased potassium levels, due to purging behaviors. Hypophosphatemia, hypoglycemia, and hypocalcemia don't tend to occur in clients with bulimia nervosa.
NP: Data collection; CN: Psychosocial integrity; CNS: Coping and adaptation; CL: Knowledge

3. 1. It's important for the client to understand the emotional triggers to bingeing, such as disappointment, depression, and anxiety. People don't outgrow eating behaviors. This leads a person to believe binge eating is a normal part of growth and development when it definitely isn't. The third statement negates the seriousness of bingeing and leads the client to believe only vomiting is a problem, not overeating. Physiologic hunger doesn't predispose a client to binge behaviors.
NP: Data collection; CN: Psychosocial integrity; CNS: Psychosocial adaptation; CL: Comprehension

NP: Nursing process   CN: Client needs category   CNS: Client needs subcategory   CL: Cognitive level

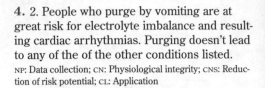

4. A client with bulimia and a history of purging by vomiting is hospitalized for further observation because she's at risk for which of the following?
1. Diabetes mellitus
2. Cardiac arrhythmias and electrolyte imbalances
3. GI obstruction or paralytic ileus
4. Septicemia from a low white blood cell count

5. A client with a diagnosis of bulimia nervosa is working on relationship issues. Which of the following nursing interventions is the most important?
1. Have the client work on developing social skills.
2. Focus on how relationships cause bulimic behavior.
3. Help the client identify feelings about relationships.
4. Discuss how to prevent getting overinvolved in relationships.

6. A young woman with bulimia nervosa wants to lessen her feelings of powerlessness. Which of the following short-term goals is most important initially?
1. Learn problem-solving skills.
2. Decrease symptoms of anxiety.
3. Perform self-care activities daily.
4. Verbalize how to set limits with others.

7. A client with bulimia nervosa tells a nurse her parents don't know about her eating disorder. Which of the following goals is appropriate for this client and her family?
1. Decrease the chaos in the family unit.
2. Learn effective communication skills.
3. Spend time together in social situations.
4. Discuss the client's need to be responsible.

Question 6 is asking you to prioritize.

4. 2. People who purge by vomiting are at great risk for electrolyte imbalance and resulting cardiac arrhythmias. Purging doesn't lead to any of the of the other conditions listed.
NP: Data collection; CN: Physiological integrity; CNS: Reduction of risk potential; CL: Application

5. 3. The client needs to address personal feelings, especially uncomfortable ones because they may trigger bingeing behavior. The client isn't necessarily overinvolved in relationships; the issue may be the lack of satisfying relationships in the person's life. Social skills are important to a client's well-being, but they aren't typically a major problem for the client with bulimia nervosa. Relationships *don't cause* bulimic behaviors. It's the inability to handle stress or conflict that arises from interactions that causes the client to be distressed.
NP: Implementation; CN: Psychosocial integrity; CNS: Psychosocial adaptation; CL: Application

6. 1. If the client can learn effective problem-solving skills, she'll gain a sense of control and power over her life. Anxiety is commonly caused by feelings of powerlessness. Performing daily self-care activities won't reduce one's sense of powerlessness. Verbalizing how to set limits and protect self from the intrusive behavior of others is a necessary life skill, but problem-solving skills take priority.
NP: Planning; CN: Psychosocial integrity; CNS: Psychosocial adaptation; CL: Comprehension

7. 2. A major goal for the client and her family is to learn to communicate directly and honestly. To change the chaotic environment, the family must first learn to communicate effectively. Families with a member who has an eating disorder are often enmeshed and don't need to spend more time together. Before discussing the client's level of responsibility, the family needs to establish effective ways of communicating with each other.
NP: Planning; CN: Psychosocial integrity; CNS: Psychosocial adaptation; CL: Comprehension

8. When discussing self-esteem with a client with bulimia nervosa, which of the following areas is the most important?
1. Personal fears
2. Family strengths
3. Negative thinking
4. Environmental stimuli

9. Which of the following complications of bulimia nervosa is life-threatening?
1. Amenorrhea
2. Bradycardia
3. Electrolyte imbalance
4. Yellow skin

10. A nurse is talking to a client with bulimia nervosa about the complications of laxative abuse. Which of the following complications should be included?
1. Loss of taste
2. Swollen glands
3. Dental problems
4. Malabsorption of nutrients

11. Which nursing diagnosis should the nurse expect to find in the nursing plan of care of a client with bulimia?
1. Decreased cardiac output related to muscle spasms in the hands and feet
2. Risk for infection related to enlargement of salivary and parotid glands
3. Disturbed thought processes related to personal identify disturbance
4. Ineffective denial related to underlying need for acceptance

12. A client is talking with a nurse about her binge-purge cycle. Which of the following questions should the nurse ask about the cycle?
1. "Do you know how to stop the binge-purge cycle?"
2. "Does the binge-purge cycle help you lose weight?"
3. "Can the binge-purge cycle take away your anxiety?"
4. "How often do you go through the binge-purge cycle?"

Again, prioritize

Focus on the correct answer.

8. 3. Clients with bulimia nervosa need to work on identifying and changing their negative thinking and distortion of reality. Environmental stimuli don't cause bulimic behaviors. Personal fears are related to negative thinking. Exploring family strengths isn't a priority; it's more appropriate to explore the client's strengths.
NP: Planning; CN: Psychosocial integrity; CNS: Psychosocial adaptation; CL: Application

9. 3. Electrolyte imbalance can be a life-threatening complication of bulimia nervosa due to purging behaviors. Amenorrhea, yellow skin, and bradycardia are complications of bulimia nervosa that don't tend to be life-threatening.
NP: Data collection; CN: Physiological integrity; CNS: Reduction of risk potential; CL: Knowledge

10. 4. A serious complication of laxative abuse is malabsorption of nutrients, such as proteins, fats, and calcium. Dental problems are a complication from purging. Swollen glands are a complication associated with purging. Laxative abuse doesn't tend to affect the client's sense of taste.
NP: Implementation; CN: Physiological integrity; CNS: Reduction of risk potential; CL: Knowledge

11. 4. An unmet need for acceptance can cause denial about self-destructive behaviors and is congruent with the underlying dynamics of eating disorders. Decreased cardiac output isn't caused by muscle spasms in the hands and feet. Enlarged parotid and salivary glands don't cause infection. Disturbed thought processes aren't caused by personal identity disturbance.
NP: Planning; CN: Psychosocial integrity; CNS: Psychosocial adaptation; CL: Application

12. 4. This is an important question because there's usually a range of frequencies, such as from a once-a-week pattern to multiple times each day. The binge-purge cycle doesn't decrease anxiety; it tends to generate overall negative feelings about self. It's common for clients with bulimia nervosa to experience daily fluctuations in their weight. Some clients report weight variations of up to 10 lb (4.5 kg). The first question isn't an appropriate question because it will generate feelings of self-blame and shame.
NP: Data collection; CN: Psychosocial integrity; CNS: Coping and adaptation; CL: Application

13. A nurse is collecting data on a client with bulimia nervosa for possible substance abuse. Which of the following questions is best to obtain information about this possible problem?

1. "Have you ever used diet pills?"
2. "Where would you go to buy drugs?"
3. "At what age did you start drinking?"
4. "Do your peers ever offer you drugs?"

14. A client with bulimia nervosa is discussing her abnormal eating behaviors. Which of the following statements by the client indicates she's beginning to understand this eating disorder?

1. "When my loneliness gets to me, I start to binge."
2. "I know that when my life gets better I'll eat right."
3. "I know I waste food and waste my money on food."
4. "After my parents divorce, I'll talk about bingeing and purging."

15. A nurse is collecting data on a client to determine the distress experienced after binge eating. Which of the following symptoms are typical after bingeing?

1. Ageusia
2. Headache
3. Pain
4. Sore throat

16. A mother of a client with bulimia nervosa asks a nurse if bulimia nervosa will stop her daughter from menstruating. Which of the following responses is best?

1. "All women with anorexia nervosa or bulimia nervosa will have amenorrhea."
2. "When your daughter is bingeing and purging, she won't have normal periods."
3. "The eating disorder must be ongoing for your daughter's menstrual cycle to change."
4. "Women with bulimia nervosa may have a normal or abnormal menstrual cycle, depending on the severity of the problem."

Choose the best answer!

You're doing great!

13. 1. Some clients with bulimia nervosa have a history of or actively use amphetamines to control weight. The use of alcohol and street drugs is also common. The second and fourth questions could be answered by the client without revealing drug use. The age the client started drinking may not show current substance use.

NP: Data collection; CN: Psychosocial integrity; CNS: Coping and adaptation; CL: Comprehension

14. 1. Binge eating is a way to handle the uncomfortable feelings of frustration, loneliness, anger, and fear. The second statement indicates the client is experiencing denial of the eating disorder. The third statement addresses the client's guilt feelings; it doesn't reflect knowledge of her eating disorder. The fourth statement shows the client isn't ready to discuss her eating disorder.

NP: Evaluation; CN: Psychosocial integrity; CNS: Psychosocial adaptation; CL: Analysis

15. 3. After a binge episode, the client commonly has abdominal distention and stomach pain. A sore throat is associated with vomiting. Headache or ageusia (loss of taste) aren't associated with binge eating.

NP: Data collection; CN: Physiological integrity; CNS: Physiological adaptation; CL: Knowledge

16. 4. Women with bulimia nervosa may have a normal or abnormal menstrual cycle, depending on the severity of the eating disorder. Not all women with eating disorders have amenorrhea. The eating disorder can disrupt the menstrual cycle at any point in the illness.

NP: Implementation; CN: Physiological integrity; CNS: Physiological adaptation; CL: Analysis

NP: Nursing process   CN: Client needs category   CNS: Client needs subcategory   CL: Cognitive level

17. Which of the following difficulties are commonly found in families with a member who has bulimia nervosa?
1. Mental illness
2. Multiple losses
3. Chronic anxiety
4. Substance abuse

18. A client with bulimia nervosa tells a nurse her major problem is eating too much food in a short period of time and then vomiting. Which of the following short-term goals is the most important?
1. Help the client understand every person has a satiety level.
2. Encourage the client to verbalize fears and concerns about food.
3. Determine the amount of food the client will eat without purging.
4. Obtain a therapy appointment to look at the emotional causes of bulimia nervosa.

The key words in question 18 mean prioritize.

19. Which of the following statements indicates a client with bulimia nervosa is making progress in interrupting the binge-purge cycle?
1. "I called my friend the last two times I got upset."
2. "I know I'll have this problem with eating forever."
3. "I started asking my mother or sister to watch me eat each meal."
4. "I can have my boyfriend bring me home from parties if I want to purge."

You're getting multiple answers correct!

20. A client with bulimia nervosa asks a nurse, "How can I ask for help from my family?" Which of the following responses is the most appropriate?
1. "When you ask for help, make sure you really need it."
2. "Have you ever asked for help before?"
3. "Ask family members to spend time with you at mealtime."
4. "Think about how you can handle this situation without help."

17. 2. Families with a member who has bulimia nervosa usually struggle with multiple losses. Mental illness, substance abuse, or chronic anxiety doesn't tend to be a theme in the family background of the client with bulimia nervosa.
NP: Data collection; CN: Psychosocial integrity; CNS: Coping and adaptation; CL: Analysis

18. 3. The client must meet her nutritional needs to prevent further complications, so she must identify the amount of food she can eat without purging as her first short-term goal. Therapy is an important part of dealing with this disorder, but it's not the first step. Obtaining knowledge or verbalizing her fears and feelings about food are *not* priority goals for this client. All clients must take steps to meet their nutritional needs.
NP: Planning; CN: Physiological integrity; CNS: Reduction of risk potential; CL: Application

19. 1. A sign of progress is when the client begins to verbalize feelings and interact with people instead of going to food for comfort. The second statement indicates the client needs more information on how to handle the disorder. Having another person watch the client eat isn't a helpful strategy as the client will depend on others to help control food intake. The last statement indicates the client is in denial about the severity of the problem.
NP: Evaluation; CN: Psychosocial integrity; CNS: Psychosocial adaptation; CL: Application

20. 2. Determine whether the client has ever been successful in asking for help. Previous experiences affect the client's ability to ask for help now. Developing a support system is imperative for this client. Having other people around at mealtime isn't the only way to ask for help. The client needs to ask for help anytime without analyzing the level of need.
NP: Implementation; CN: Psychosocial integrity; CNS: Psychosocial adaptation; CL: Analysis

21. A client with bulimia nervosa tells a nurse that she doesn't eat during the day, but after 5 p.m. she begins to binge and vomit. Which of the following interventions would be the <u>most</u> useful to this client?

1. Help the client stop eating the foods on which she binges.
2. Discuss the effects of fasting on the client's pattern of eating.
3. Encourage the client to become involved in food preparation.
4. Teach the client to eat earlier in the day and decrease intake at night.

Don't quit now!

21. 2. If a person fasts for most of the day, it's common to become extremely hungry, overeat by bingeing, and then feel the need to purge. Restricting food intake can actually trigger the binge-purge cycle. In treatment, the client is taught to identify foods that trigger eating, discuss the feelings associated with these foods, and work to eat them in normal amounts. Involvement in food preparation won't promote changes in the client's behaviors. The last intervention doesn't address how fasting can trigger the binge-purge cycle.
NP: Implementation; CN: Psychosocial integrity; CNS: Psychosocial adaptation; CL: Application

22. A client with bulimia nervosa tells a nurse she was doing well until last week, when she had a fight with her father. Which nursing intervention would be most helpful?

1. Examine the relationship between feelings and eating.
2. Discuss the importance of therapy for the entire family.
3. Encourage the client to avoid certain family members.
4. Identify daily stressors and learn stress management skills.

22. 1. The client needs to understand her feelings and develop healthy coping skills to handle unpleasant situations. Family therapy may be indicated but shouldn't be an immediate intervention. Avoidance isn't a useful coping strategy; eventually the underlying issues need to be explored. All clients can benefit from stress management skills but, for this client, care must focus on the relationship between feelings and eating behaviors.
NP: Implementation; CN: Psychosocial integrity; CNS: Psychosocial adaptation; CL: Application

23. Which of the following statements from a bulimic client shows that she understands the concept of relapse?

1. "If I can't maintain control over things, I'll have problems."
2. "If I have problems, then that says I haven't learned much."
3. "If this illness becomes chronic, I won't be able to handle it."
4. "If I have problems, I can start over again and not feel hopeless."

Pick yourself up, dust yourself off, and, um, that's enough of a hint.

23. 4. This statement indicates that the client knows a relapse is just a slip, and positive gains made from treatment haven't been lost. Negative self-statements can lead to relapse. Control issues relate to powerlessness, which contribute to relapse.
NP: Evaluation; CN: Psychosocial integrity; CNS: Psychosocial adaptation; CL: Comprehension

24. Which of the following medical conditions is commonly found in clients with bulimia nervosa?

1. Allergies
2. Cancer
3. Diabetes
4. Hepatitis A

24. 3. Diabetes, heart disease, and hypertension are medical complications frequently seen in clients with bulimia nervosa. Cancer, hepatitis A, and allergies are *not* medical complications commonly associated with bulimia nervosa.
NP: Data collection; CN: Health promotion and maintenance; CNS: Prevention and early detection of disease; CL: Application

25. Which of the following is the treatment team's priority in planning the care of a client with an eating disorder?
1. Preventing the client from performing any muscle-building exercises
2. Keeping the client on bed rest until she attains a specified weight
3. Meeting daily to discuss countertransferances and splitting behaviors
4. Monitoring the client's weight and vital signs daily

Congratulations! You've passed the halfway point.

25. 3. Clients with eating disorders commonly use manipulative ploys to resist weight gain if they restrict food intake or to maintain purging practices if they are bulimic. They frequently play staff members against one another and hone in on their caretaker's vulnerabilities. Muscle building is acceptable because compared with aerobic exercise it burns relatively few calories. Keeping the patient on best rest until she reaches a specified weight can result in unnecessary power struggles and prevents staff from focusing on more pertinent issues and problems. Monitoring the client's weight and vital signs is important but not vital on a daily basis unless the client's physiologic condition warrants such close scrutiny.

NP: Planning; CN: Psychosocial integrity; CNS: Psychosocial adaptation; CL: Application

26. A client with anorexia nervosa attended psychoeducational sessions on principles of adequate nutrition. Which of the following statements by the client indicates the teaching was effective?
1. "I eat while I'm doing things to distract myself."
2. "I eat all my food at night right before I go to bed."
3. "I eat small amounts of food slowly at every meal."
4. "I eat only when I'm with my family and trying to be social."

26. 3. Slowly eating small amounts of food facilitates adequate digestion and prevents distention. Healthy eating is best accomplished when a person isn't doing other things while eating. Eating right before bedtime isn't a healthy eating habit. If a client eats only when the family is present or when trying to be social, eating is tied to social or emotional cues rather than nutritional needs.

NP: Evaluation; CN: Physiological integrity; CNS: Reduction of risk potential; CL: Comprehension

27. A client with anorexia nervosa tells a nurse, "I'll never have the slender body I want." Which of the following interventions is best to use to handle this problem?
1. Call a family meeting to get help from the parents.
2. Help the client work on developing a realistic body image.
3. Make an appointment to see the dietitian on a weekly basis.
4. Develop an exercise program the client can do twice a week.

All options may be good, but choose the best one.

27. 2. With anorexia nervosa, the client pursues thinness and has a distorted view of self. Clients with anorexia nervosa typically exercise excessively. Although meeting with a dietitian might be helpful, it isn't a priority. A family meeting may not help the client develop a more realistic view of the body.

NP: Implementation; CN: Psychosocial integrity; CNS: Psychosocial adaptation; CL: Application

28. For a client with anorexia nervosa, which goal takes the highest priority?
1. The client will establish adequate daily nutritional intake.
2. The client will make a contract with the nurse that sets a target weight.
3. The client will identify self-perceptions about body size as unrealistic.
4. The client will verbalize the possible physiological consequences of self–starvation.

29. Which of the following communication strategies is best to use with a client with anorexia nervosa who is having problems with peer relationships?
1. Use concrete language and maintain a focus on reality.
2. Direct the client to talk about what's causing the anxiety.
3. Teach the client to communicate feelings and express self appropriately.
4. Confront the client about being depressed and self-absorbed.

30. The nurse plans to include the parents of a client with anorexia nervosa in therapy sessions along with the client. What fact should the nurse remember about the parents of clients with anorexia?
1. They tend to overprotect their children.
2. They usually have a history of substance abuse.
3. They maintain emotional distance from their children.
4. They alternate between loving and rejecting their children.

31. A nurse is talking to a family of a client with anorexia nervosa. Which of the following family behaviors is most likely to be seen during the family's interaction?
1. Sibling rivalry
2. Rage reactions
3. Parental disagreement
4. Excessive independence

Stick with the best and forget the rest! Now we're pulling up the right answers!

28. 1. According to Maslow's hierarchy of needs, all humans need to meet basic physiological needs first. Because a client with anorexia nervosa eats little or nothing, the nurse must plan to help the client meet this basic, immediate physiological need first. The nurse may give lesser priority to goals that address long-term plans, self-perception, and potential complications.
NP: Planning; CN: Physiological integrity; CNS: Basic care and comfort; CL: Application

29. 3. Clients with anorexia nervosa commonly communicate on a superficial level and avoid expressing feelings. Identifying feelings and learning to express them are initial steps in decreasing isolation. Clients with anorexia nervosa are usually able to discuss abstract and concrete issues. Confrontation usually isn't an effective communication strategy as it may cause the client to withdraw and become more depressed.
NP: Implementation; CN: Psychosocial integrity; CNS: Psychosocial adaptation; CL: Application

30. 1. Clients with anorexia nervosa typically come from a family with parents who are controlling and overprotective. These clients use eating to gain control of an aspect of their lives. Having a history of substance abuse, maintaining an emotional distance, and alternating between love and rejection aren't typical characteristics of parents of children with anorexia.
NP: Planning; CN: Psychosocial integrity; CNS: Psychosocial adaptation; CL: Application

31. 3. In many families with a member with anorexia nervosa, there's marital conflict and parental disagreement. Sibling rivalry is a common occurrence and not specific to a family with a member with anorexia nervosa. In these families, the members tend to be enmeshed and dependent on each other. The family with an anorexic member is usually one that looks good to the outside observer. Emotions are overcontrolled, and there's difficulty appropriately expressing negative feelings.
NP: Data collection; CN: Psychosocial integrity; CNS: Coping and adaptation; CL: Comprehension

NP: Nursing process   CN: Client needs category   CNS: Client needs subcategory   CL: Cognitive level

32. A nurse is working with a client with anorexia nervosa who has skin problems in her extremities. Which of the following short-term goals is the most important for the client?

1. Do daily range-of-motion exercises.
2. Eat some fatty foods daily.
3. Check neurologic reflexes.
4. Promote adequate circulation.

Hint: Fatty foods are almost never the right answer.

33. A woman with anorexia nervosa is discharged from the hospital after gaining 12 lb. Which of the following statements indicates the client still has a lack of knowledge about her condition?

1. "I plan to eat six small meals a day."
2. "I feel this is scary, and I need to write in my journal about it."
3. "I have to diet because I've gained 12 pounds."
4. "I'll need to attend therapy for support to stay healthy."

Careful! Here's one of those cleverly disguised further teaching questions.

34. A client with anorexia nervosa tells a nurse she always feels fat. Which of the following interventions is the best for this client?

1. Talk about how important the client is.
2. Encourage her to look at herself in a mirror.
3. Address the dynamics of the disorder.
4. Talk about how she's different from her peers.

You're heading down to the home stretch. Keep going!

35. The grandparents of a client with anorexia nervosa want to support the client but aren't sure what they should do. Which of the following interventions is best?

1. Promote positive expressions of affection.
2. Encourage behaviors that enhance socialization.
3. Discuss how eating disorders create powerlessness.
4. Discuss the meaning of hunger and body sensations.

32. 4. Circulation changes will cause extremities to be cold, numb, and have dry and flaky skin. Exercise may help prevent contractures and muscle atrophy, but it may have only a limited secondary effect on promoting circulation. Intake of fatty foods won't have an impact on the client's skin problems. Checking neurologic reflexes won't necessarily assist with handling skin problems.
NP: Planning; CN: Physiological integrity; CNS: Reduction of risk potential; CL: Knowledge

33. 3. The client is still relying on a mental image of herself as being fat. Eating small meals each day is a realistic plan for meeting nutritional needs without feeling bloated. Feeling insecure when leaving a controlled environment is a common response to discharge. Planning to attend therapy after discharge shows an understanding of the need for continued therapy.
NP: Evaluation; CN: Psychosocial integrity; CNS: Psychosocial adaptation; CL: Analysis

34. 3. The client can benefit from understanding the underlying dynamics of the eating disorder. The client with anorexia nervosa has low self-esteem and won't believe the positive statements. Although the client may look at herself in the mirror, in her mind she'll still see herself as fat. Pointing out differences will only diminish her already low self-esteem.
NP: Implementation; CN: Psychosocial integrity; CNS: Psychosocial adaptation; CL: Application

35. 1. Clients with eating disorders need emotional support and expressions of affection from family members. Talking about hunger and other sensations won't give the grandparents useful strategies. Although clients with eating disorders feel powerless, it's better to have the grandparents focus on something positive. It wouldn't be an appropriate strategy to have the grandparents promote socialization.
NP: Implementation; CN: Psychosocial integrity; CNS: Psychosocial adaptation; CL: Application

36. A 15-year-old teenager is brought to the clinic by her parents who are concerned because she has lost a significant amount of weight in the past 4 months. If anorexia nervosa is the cause of the weight loss, the nurse is most likely to find which of the following when assessing this client?
1. Hypertension
2. Amenorrhea
3. Hyperthermia
4. Diarrhea

37. An adolescent client with anorexia nervosa tells a nurse about her outstanding academic achievements and her thoughts about suicide. Which of the following factors must the nurse consider when contributing to the care plan for this client?
1. Self-esteem
2. Physical illnesses
3. Paranoid delusions
4. Relationship avoidance

38. In contributing to the care plan for a family with a member who has anorexia nervosa, which of the following information should be included?
1. Coping mechanisms used in the past
2. Concerns about changes in lifestyle and daily activities
3. Rejection of feedback from family and significant others
4. Appropriate eating habits and social behaviors centering on eating

39. Which goal is best to help a client with anorexia nervosa recognize self-distortions?
1. Identify the client's misperceptions of self.
2. Acknowledge immature and childlike behaviors.
3. Determine the consequences of a faulty support system.
4. Recognize the age-appropriate tasks to be accomplished.

Care plans encourage staff to work toward the same goals.

Only 8 more questions to go!

36. 2. Anorexia nervosa is characterized by profound weight loss caused by severe restriction of food intake by the client. It commonly causes amenorrhea in females, along with decreased body temperature and hypotension—not hypertension. It usually doesn't produce diarrhea, but it may produce constipation because decreased oral intake leads to decreased GI motility.
NP: Data collection; CN: Physiological integrity; CNS: Reduction of risk potential; CL: Knowledge

37. 1. The client lacks self-esteem, which contributes to her level of depression and feelings of personal ineffectiveness, which in turn may lead to suicidal thoughts. Physical illnesses are common with clients with anorexia nervosa, but they don't relate to this situation. Paranoid delusions refer to false ideas that others want to harm you. No evidence exists that this client is socially isolated.
NP: Planning; CN: Psychosocial integrity; CNS: Psychosocial adaptation; CL: Analysis

38. 1. Examination of positive and negative coping mechanisms used by the family allows the nurse to build a care plan specific to the family's strengths and weaknesses. The way the family copes with concerns is more important than the concerns themselves. Feedback from the family and significant others is vital when building a plan of care. Eating habits and behaviors are symptoms of the way people cope with problems.
NP: Planning; CN: Psychosocial integrity; CNS: Psychosocial adaptation; CL: Application

39. 1. Questioning the client's misperceptions and distortions will create doubt about how the client views himself. Recognizing the age-appropriate tasks to be accomplished won't help the client recognize distortions. Acknowledging immature behaviors or determining the consequences of a faulty support system won't promote recognition of self-distortions.
NP: Planning; CN: Psychosocial integrity; CNS: Psychosocial adaptation; CL: Analysis

40. Parents of a client with anorexia nervosa ask about the risk factors for this disorder. Which of the following definitions is the most accurate?

1. Inability to be still and emotional lability
2. High level of anxiety and disorganized behavior
3. Low self-esteem and problems with family relationships
4. Lack of life experiences and no opportunities to learn skills

41. A client with anorexia nervosa has started taking fluoxetine (Prozac). Which of the following adverse reactions complicates the treatment of this eating disorder?

1. Drowsiness
2. Dry mouth
3. Light-headedness
4. Nausea

42. A client with anorexia nervosa is worried about rectal bleeding. Which question will help obtain more information about the problem?

1. "How often do you use laxatives?"
2. "How many days ago did you stop vomiting?"
3. "Are you eating anything that causes irritation?"
4. "Do you bleed before or after exercise?"

43. The nurse is conducting a discharge evaluation on a 20-year-old female client with anorexia nervosa. Which outcome shows that the client's self-concept has improved?

1. The client reaches an appropriate body weight.
2. The client states viewing her body as fat was a misperception.
3. The client consumes adequate calories.
4. The client understands her use of maladaptive eating behaviors.

Knowing adverse reactions to key drugs is important.

40. 3. There are several risk factors for eating disorders, including low self-esteem, history of depression, substance abuse, and dysfunctional family relationships. A lack of life experiences and an absence of opportunities to learn life skills may be a result of anorexia. Restlessness and emotional lability are symptoms of manic depressive illness. Anxiety and disorganized behavior could be signs of a psychotic disorder.
NP: Data collection; CN: Psychosocial integrity; CNS: Psychosocial adaptation; CL: Knowledge

41. 4. Nausea is an adverse reaction to the drug that compounds the eating disorder problem, and the client must be closely monitored. Although the adverse reactions of drowsiness, light-headedness, or dry mouth may occur, they aren't likely to interfere with treatment.
NP: Data collection; CN: Physiological integrity; CNS: Pharmacological therapies; CL: Knowledge

42. 1. Excessive use of laxatives will cause GI irritation and rectal bleeding. If the client stopped vomiting but is still using laxatives, rectal bleeding can occur. Clients who are anorexic eat very little, and what they eat won't cause rectal bleeding. Exercise doesn't cause rectal bleeding.
NP: Data collection; CN: Health promotion and maintenance; CNS: Prevention and early detection of disease; CL: Application

43. 2. When the client's self-concept has improved, she will state that viewing her body as fat was a misperception. Reaching an appropriate body weight, consuming adequate calories, and understanding her use of maladaptive eating behaviors show that other goals of therapy have been met.
NP: Data collection; CN: Psychosocial integrity; CNS: Psychosocial adaptation; CL: Application

44. A client with anorexia nervosa is talking with a nurse about her group therapy. Which of the following statements shows the group experience has helped the client?
1. "I feel I'm different and I don't need a lot of friends."
2. "I'll tell my parents it isn't just me who has problems."
3. "I can see how to do things better and become the best."
4. "I think I have some unrealistic expectations of myself."

Question 44 is looking for something that proved helpful.

44. 4. A goal of group therapy is to provide methods to assess whether personal expectations are unrealistic. Other goals are to learn to handle problems; not to blame parents or others; decrease perfectionist tendencies; and decrease isolation and learn to have healthy peer relationships.
NP: Evaluation; CN: Psychosocial integrity; CNS: Psychosocial adaptation; CL: Application

45. A nurse and her client who has anorexia nervosa are working on the goal of developing social relationships. Which of the following actions by the client is an indication the client is meeting her goal?
1. The client talks about the value of peer relationships.
2. The client decides to talk to her parents about her friends.
3. The client expresses the need to establish trust relationships.
4. The client attends an activity without prompting from others.

The client must agree to the goal or it won't work.

45. 4. When a client with anorexia nervosa attends an activity without prompting from others, it's a positive sign that the client is working toward developing social relationships. Expressing the need to establish trust relationships is a first step, but an indication of success would be actually initiating such a relationship. Talking to parents about friends is a start but doesn't necessarily indicate that the client can establish relationships. Talking about the value of relationships is also beneficial but is only the first step in establishing them.
NP: Evaluation; CN: Psychosocial integrity; CNS: Psychosocial adaptation; CL: Application

46. A client with anorexia nervosa tells a nurse, "I feel so awful and inadequate." Which of the following responses is the best?
1. "You're being too hard on yourself."
2. "Someday you'll feel better about things."
3. "Tell me something you like about yourself."
4. "Maybe relaxing by yourself will help you feel better."

46. 3. This statement redirects the client to talk about positive aspects of self. The other responses minimize her feelings or don't address the client's concerns or encourage the client to change her self-image.
NP: Implementation; CN: Psychosocial integrity; CNS: Psychosocial adaptation; CL: Application

47. Which of the following physiologic complications of anorexia nervosa can quickly become life-threatening?
1. Cardiac arrhythmias
2. Decreased blood pressure
3. Decreased metabolic rate
4. Altered thyroid functioning

Hurray! I knew you could do it!

47. 1. With starvation, the heart muscle becomes irritable and less efficient, causing sudden, severe arrhythmias and, possibly, death. Decreased blood pressure, decreased metabolic rate, and altered thyroid functioning do not tend to be life-threatening.
NP: Data collection; CN: Health promotion and maintenance; CNS: Prevention and early detection of disease; CL: Knowledge

NP: Nursing process   CN: Client needs category   CNS: Client needs subcategory   CL: Cognitive level

Part IV Maternal-neonatal care

22 Antepartum care   389

23 Intrapartum care   403

24 Postpartum care   417

25 Neonatal care   431

Chapter 22
Antepartum care

1. During an examination, a client who is 32 weeks' pregnant becomes dizzy, light-headed, and pale while supine. What should the nurse do first?
  1. Listen to fetal heart tones.
  2. Take the client's blood pressure.
  3. Ask the client to breathe deeply.
  4. Turn the client on her left side.

Think about the development of the fetus at 32 weeks to answer question 1.

2. The nurse has just taught a client about the signs of true and false labor. Which client statement indicates an accurate understanding of this information?
  1. "False labor contractions are regular."
  2. "False labor contractions intensify with walking."
  3. "False labor contractions usually occur in the abdomen."
  4. "False labor contractions move from the back to the front of the abdomen."

Careful! Question 3 asks about the recipient twin and not the donor.

3. In twin-to-twin transfusion syndrome, the arterial circulation of one twin is in communication with the venous circulation of the other twin. One fetus is considered the "donor" twin, and one becomes the "recipient" twin. Observation of the recipient twin would most likely show which of the following conditions?
  1. Anemia
  2. Oligohydramnios
  3. Polycythemia
  4. Small fetus

1. 4. As the enlarging uterus increases pressure on the inferior vena cava, it compromises venous return, which can cause dizziness, light-headedness, and pallor when the client is supine. The nurse can relieve these symptoms by turning the client on her left side, which relieves pressure on the vena cava and restores venous return. Although they're valuable assessments, fetal heart tone and maternal blood pressure measurements don't correct the problem. Because deep breathing has no effect on venous return, it can't relieve the client's symptoms.
NP: Implementation; CN: Physiological integrity; CNS: Reduction of risk potential; CL: Application

2. 3. False labor contractions are usually felt in the abdomen, are irregular, and are typically relieved by walking. True labor contractions move from the back to the front of the abdomen, are regular, and aren't relieved with walking.
NP: Evaluation; CN: Health promotion and maintenance; CNS: Growth and development through the life span; CL: Comprehension

3. 3. The recipient twin in twin-twin transfusion syndrome is transfused by the other twin. The recipient twin then becomes polycythemic and often has heart failure due to circulatory overload. The donor twin becomes anemic. The recipient twin has polyhydramnios, not oligohydramnios. The recipient twin is usually large, whereas the donor twin is usually small.
NP: Data collection; CN: Physiological integrity; CNS: Physiological adaptation; CL: Analysis

NP: Nursing process   CN: Client needs category   CNS: Client needs subcategory   CL: Cognitive level

4. A pregnant client who reports painless vaginal bleeding at 28 weeks' gestation is diagnosed with placenta previa. The placental edge reaches the internal os. This type of placenta previa is known as which of the following types?

1. Low-lying placenta previa
2. Marginal placenta previa
3. Partial placenta previa
4. Total placenta previa

5. Expectant management of the client with placenta previa includes which of the following procedures or treatments?

1. Stat culture and sensitivity
2. Antenatal steroids after 34 weeks' gestation
3. Ultrasound examination every 2 to 3 weeks
4. Scheduled delivery of the fetus before fetal maturity in a hemodynamically stable mother

6. A client with painless vaginal bleeding has just been diagnosed as having placenta previa. Which of the following procedures is usually performed to diagnose placenta previa?

1. Amniocentesis
2. Digital or speculum examination
3. External fetal monitoring
4. Ultrasound

7. A client is being admitted to the antepartum unit for hypovolemia secondary to hyperemesis gravidarum. Which factor predisposes the development of this condition?

1. Trophoblastic disease
2. Maternal age older than 35 years
3. Malnourished or underweight clients
4. Low levels of human chorionic gonadotropin (HCG)

The word expectant is a clue to answering question 5.

What do these procedures evaluate?

This word — predisposes — says there's more than one factor to consider.

4. 2. A marginal placenta previa is characterized by implantation of the placenta in the margin of the cervical os, not covering the os. A low-lying placenta is implanted in the lower uterine segment but doesn't reach the cervical os. A partial placenta previa is the partial occlusion of the cervical os by the placenta. The internal cervical os is completely covered by the placenta in a total placenta previa.
NP: Data collection; CN: Physiological integrity; CNS: Physiological adaptation; CL: Analysis

5. 3. Fetal surveillance through ultrasound examination every 2 to 3 weeks is indicated to evaluate fetal growth, amniotic fluid, and placental location in clients with placenta previa being expectantly managed. A stat culture and sensitivity would be done for severe bleeding or maternal or fetal distress and isn't part of expectant management. Antenatal steroids may be given to clients between 26 and 32 weeks' gestation to enhance fetal lung maturity. In a hemodynamically stable mother, delivery of the fetus should be delayed until fetal lung maturity is attained.
NP: Planning; CN: Physiological integrity; CNS: Reduction of risk potential; CL: Knowledge

6. 4. Once the mother and fetus are stabilized, ultrasound evaluation of the placenta should be done to determine the cause of the bleeding. Amniocentesis is contraindicated in placenta previa. A digital or speculum examination shouldn't be done as this may lead to severe bleeding or hemorrhage. External fetal monitoring won't detect a placenta previa, although it will detect fetal distress, which may result from blood loss or placental separation.
NP: Implementation; CN: Physiological integrity; CNS: Reduction of risk potential; CL: Knowledge

7. 1. Trophoblastic disease is associated with hyperemesis gravidarum. Obesity and maternal age younger than 20 years are risk factors for developing hyperemesis gravidarum. High levels of estrogen and HCG have been associated with hyperemesis.
NP: Data collection; CN: Physiological integrity; CNS: Reduction of risk potential; CL: Knowledge

NP: Nursing process   CN: Client needs category   CNS: Client needs subcategory   CL: Cognitive level

8. The nurse is assisting in developing a teaching plan for a client who is about to enter the third trimester of pregnancy. The teaching plan should include identification of which danger sign that must be reported immediately?

1 Hemorrhoids
2. Blurred vision
3. Dyspnea on exertion
4. Increased vaginal mucus

Sometimes it's hard to keep your symptoms straight, isn't it?

9. Which of the following symptoms occurs with a hydatidiform mole?

1. Heavy, bright red bleeding every 21 days
2. Fetal cardiac motion after 6 weeks' gestation
3. Benign tumors found in the smooth muscle of the uterus
4. "Snowstorm" pattern on ultrasound with no fetus or gestational sac

10. A 21-year-old client has just been diagnosed with having a hydatidiform mole. Which of the following is considered a risk factor for developing a hydatidiform mole?

1. Age in 20s or 30s
2. High socioeconomic status
3. Primigravida
4. Prior molar gestation

Now you're getting up to speed. Way to go!

11. A 21-year-old client arrives at the emergency department with complaints of cramping abdominal pain and mild vaginal bleeding. Pelvic examination shows a left adnexal mass that is tender when palpated. Culdocentesis shows blood in the cul-de-sac. This client probably has which of the following conditions?

1. Abruptio placentae
2. Ectopic pregnancy
3. Hydatidiform mole
4. Pelvic inflammatory disease

8. 2. During pregnancy, blurred vision may be a danger sign of preeclampsia or eclampsia, complications that require immediate attention because they can cause severe maternal and fetal consequences. Although hemorrhoids may occur during pregnancy, they don't require immediate attention. Dyspnea on exertion and increased vaginal mucus are common discomforts caused by the physiological changes of pregnancy.
NP: Planning; CN: Physiological integrity; CNS: Reduction of risk potential; CL: Application

9. 4. Ultrasound is the technique of choice in diagnosing a hydatidiform mole. The chorionic villi of a molar pregnancy resemble a snowstorm pattern on ultrasound. Bleeding with a hydatidiform mole is usually dark brown and may occur erratically for weeks or months. There's no cardiac activity because there's no fetus. Benign tumors found in the smooth muscle of the uterus are leiomyomas or fibroids.
NP: Data collection; CN: Physiological integrity; CNS: Reduction of risk potential; CL: Analysis

10. 4. A previous molar gestation increases a woman's risk for developing a subsequent molar gestation by 4 to 5 times. Adolescents and women ages 40 years and older are at increased risk for molar pregnancies. Multigravidas, especially women with a prior pregnancy loss, and women with lower socioeconomic status are at an increased risk for this problem.
NP: Data collection; CN: Physiological integrity; CNS: Physiological adaptation; CL: Analysis

11. 2. Most ectopic pregnancies don't appear as obvious life-threatening medical emergencies. Ectopic pregnancies must be considered in any woman of childbearing age who complains of menstrual irregularity, cramping abdominal pain, and mild vaginal bleeding. Pelvic inflammatory disease, abruptio placentae, and hydatidiform moles won't show blood in the cul-de-sac.
NP: Data collection; CN: Physiological integrity; CNS: Reduction of risk potential; CL: Analysis

12. A client, 34 weeks' pregnant, arrives at the emergency department with severe abdominal pain, uterine tenderness, and an increased uterine tone. The client denies vaginal bleeding. The external fetal monitor shows fetal distress with severe, variable decelerations. The client most likely has which of the following conditions?
1. Abruptio placentae
2. Ectopic pregnancy
3. Molar pregnancy
4. Placenta previa

13. During a routine visit to the clinic, a client tells the nurse that she thinks she may be pregnant. The physician orders a pregnancy test. The nurse should know the purpose of this test is to determine which change in the client's hormone level?
1. Increase in human chorionic gonadotropin (HCG)
2. Decrease in HCG
3. Increase in luteinizing hormone (LH)
4. Decrease in LH

14. Which of the following changes in respiratory functioning during pregnancy is considered normal?
1. Increased tidal volume
2. Increased expiratory volume
3. Increased inspiratory capacity
4. Decreased oxygen consumption

15. A 23-year-old client who is 27 weeks' pregnant, arrives at her physician's office with complaints of fever, nausea, vomiting, malaise, unilateral flank pain, and costovertebral angle tenderness. Which of the following would the nurse most likely suspect?
1. Asymptomatic bacteriuria
2. Bacterial vaginosis
3. Pyelonephritis
4. Urinary tract infection (UTI)

So many hormones to learn; so little time!

Be careful with question 14. It asks what's normal during pregnancy.

12. 1. A client with severe abruptio placentae will often have severe abdominal pain. The uterus will have increased tone with little to no return to resting tone between contractions. The fetus will start to show signs of distress, with decelerations in the heart rate or even fetal death with a large placental separation. A molar pregnancy generally would be detected before 34 weeks' gestation. Placenta previa usually involves painless vaginal bleeding without uterine contractions. An ectopic pregnancy, which usually occurs in the fallopian tubes, would rupture well before 34 weeks.
NP: Data collection; CN: Physiological integrity; CNS: Reduction of risk potential; CL: Analysis

13. 1. HCG increases in a woman's blood and urine to fairly large concentrations until the 15th week of pregnancy. The other hormone values aren't indicative of pregnancy.
NP: Data collection; CN: Health promotion and maintenance; CNS: Prevention and early detection of disease; CL: Comprehension

14. 1. A pregnant client breathes deeper, which increases the tidal volume of gas moved in and out of the respiratory tract with each breath. The expiratory volume and residual volume decrease as the pregnancy progresses. The inspiratory capacity increases during pregnancy. The increased oxygen consumption in the pregnant client is 15% to 20% greater than in the nonpregnant state.
NP: Data collection; CN: Health promotion and maintenance; CNS: Growth and development through the life span; CL: Knowledge

15. 3. The symptoms indicate acute pyelonephritis, a serious condition in a pregnant client. UTI symptoms include dysuria, urgency, frequency, and suprapubic tenderness. Asymptomatic bacteriuria doesn't cause symptoms. Bacterial vaginosis causes milky white vaginal discharge but no systemic symptoms.
NP: Data collection; CN: Physiological integrity; CNS: Reduction of risk potential; CL: Analysis

NP: Nursing process   CN: Client needs category   CNS: Client needs subcategory   CL: Cognitive level

16. Which of the following terms applies to the tiny, blanched, slightly raised end arterioles found on the face, neck, arms, and chest during pregnancy?
  1. Epulis
  2. Linea nigra
  3. Striae gravidarum
  4. Telangiectasias

16. 4. The dilated arterioles that occur during pregnancy are due to the elevated level of circulating estrogen. An epulis is a red raised nodule on the gums that may develop at the end of the first trimester and continue to grow as the pregnancy progresses. The linea nigra is a pigmented line extending from the symphysis pubis to the top of the fundus during pregnancy. Striae gravidarum, or stretch marks, are slightly depressed streaks that commonly occur over the abdomen, breast, and thighs during the second half of pregnancy.
NP: Implementation; CN: Health promotion and maintenance; CNS: Growth and development through the life span; CL: Knowledge

Watch out for these true-false kinds of questions. Read carefully!

17. Which of the following statements is true about dizygotic twins?
  1. They occur most frequently in Asian women.
  2. There is a decreased risk with increased parity.
  3. There is an increased risk with increased maternal age.
  4. Use of fertility drugs poses no additional risk.

17. 3. Dizygotic twinning is influenced by race (most common in Black women and least common in Asian women), age (increased risk with increased maternal age), parity (increased risk with increased parity), and fertility drugs (increased risk with the use of fertility drugs, especially ovulation-inducing drugs).
NP: Data collection; CN: Health promotion and maintenance; CNS: Growth and development through the life span; CL: Knowledge

18. Which of the following conditions is common in pregnant clients in the second trimester of pregnancy?
  1. Mastitis
  2. Metabolic alkalosis
  3. Physiologic anemia
  4. Respiratory acidosis

18. 3. Hemoglobin level and hematocrit decrease during pregnancy as the increase in plasma volume exceeds the increase in red blood cell production. Alterations in acid-base balance during pregnancy result in a state of respiratory alkalosis, compensated by mild metabolic acidosis. Mastitis is an infection in the breast characterized by a swollen tender breast and flulike symptoms. This condition is most commonly seen in breast-feeding clients.
NP: Data collection; CN: Health promotion and maintenance; CNS: Growth and development through the life span; CL: Knowledge

Take a second glance at the word — second.

19. A 21-year-old client, 6 weeks' pregnant, is diagnosed with hyperemesis gravidarum. The nurse should be alert for which of the following conditions?
   1. Bowel perforation
   2. Electrolyte imbalance
   3. Miscarriage
   4. Pregnancy-induced hypertension (PIH)

19. 2. Excessive vomiting in clients with hyperemesis gravidarum often causes weight loss and fluid, electrolyte, and acid-base imbalances. PIH and bowel perforation aren't related to hyperemesis. The effects of hyperemesis on the fetus depend on the severity of the disorder. Clients with severe hyperemesis may have a low-birth-weight infant, but the disorder isn't generally life-threatening.
NP: Planning; CN: Physiological integrity; CNS: Reduction of risk potential; CL: Analysis

20. Clients with gestational diabetes are usually managed by which of the following therapies?
   1. Diet
   2. Long-acting insulin
   3. Oral hypoglycemic drugs
   4. Glucagon

20. 1. Clients with gestational diabetes are usually managed by diet alone to control their glucose intolerance. Long-acting insulin usually isn't needed for blood glucose control in the client with gestational diabetes. Oral hypoglycemic drugs are contraindicated in pregnancy. Glucagon raises blood glucose and is used to treat hypoglycemic reactions.
NP: Implementation; CN: Health promotion and maintenance; CNS: Growth and development through the life span; CL: Knowledge

21. Magnesium sulfate is given to pregnant clients with preeclampsia to prevent which of the following conditions?
   1. Hemorrhage
   2. Hypertension
   3. Hypomagnesemia
   4. Seizures

Here's an item asking you to identify a preventive measure.

21. 4. The anticonvulsant mechanism of magnesium is believed to depress seizure foci in the brain and peripheral neuromuscular blockade. Hypomagnesemia isn't a complication of preeclampsia. Antihypertensive drugs other than magnesium are preferred for sustained hypertension. Magnesium doesn't help prevent hemorrhage in preeclamptic clients.
NP: Implementation; CN: Physiological integrity; CNS: Pharmacological therapies; CL: Analysis

22. A pregnant client has a negative contraction stress test (CST). Which of the following statements describes negative CST results?
   1. Persistent late decelerations in fetal heartbeat occurred, with at least three contractions in a 10-minute window.
   2. Accelerations of fetal heartbeat occurred, with at least 15 beats/minute, lasting 15 to 30 seconds in a 20-minute period.
   3. Accelerations of fetal heartbeat were absent or didn't increase by 15 beats/minute for 15 to 30 seconds in a 20-minute period.
   4. There was good fetal heart rate variability and no decelerations from contraction in a 10-minute period in which there were three contractions.

22. 4. A CST measures the fetal response to uterine contractions. A client must have three contractions in a 10-minute period. A negative CST shows good fetal heart rate variability with no decelerations from uterine contractions. No accelerations in the heartbeat of at least 15 beats/minute for 15 to 30 seconds in a 20-minute period indicates a nonreactive nonstress test (NST). Reactive NSTs have accelerations in the fetal heartbeat of at least 15 beats/minute lasting 15 to 30 seconds in a 20-minute period. Persistent late decelerations with contractions is a positive CST.
NP: Data collection; CN: Health promotion and maintenance; CNS: Growth and development through the life span; CL: Analysis

23. A pregnant client with sickle cell anemia is at an increased risk for having a sickle cell crisis during pregnancy. Aggressive management of a sickle cell crisis includes which of the following measures?

1. Antihypertensive agents
2. Diuretic agents
3. I.V. fluids
4. Acetaminophen (Tylenol) for pain

24. Which of the following cardiac conditions is normal during pregnancy?

1. Cardiac tamponade
2. Heart failure
3. Endocarditis
4. Systolic murmur

Hey sickle cells! Want to fight?

25. While assessing a client in her 24th week of pregnancy, the nurse learns that the client has been experiencing signs and symptoms of pregnancy-induced hypertension, or preeclampsia. Which sign or symptom helps differentiate preeclampsia from eclampsia?

1. Seizures
2. Headaches
3. Blurred vision
4. Weight gain

26. Magnesium sulfate is given to clients with pregnancy-induced hypertension (PIH) to prevent seizure activity. Which of the following magnesium levels is therapeutic for clients with preeclampsia?

1. 4 to 7 mEq/L
2. 8 to 10 mEq/L
3. 10 to 12 mEq/L
4. Greater than 15 mEq/L

Keep those numbers and letters straight, now.

23. 3. A sickle cell crisis during pregnancy is usually managed by exchange transfusion, oxygen, and I.V. fluids. The client usually needs a stronger analgesic than acetaminophen to control the pain of a crisis. Antihypertensive drugs usually aren't necessary. Diuretics wouldn't be used unless fluid overload resulted.
NP: Planning; CN: Physiological integrity; CNS: Reduction of risk potential; CL: Analysis

24. 4. Systolic murmurs are heard in up to 90% of pregnant clients, and the murmur disappears soon after the delivery. Cardiac tamponade, which causes effusion of fluid into the pericardial sac, isn't normal during pregnancy. Despite the increases in intravascular volume and work load of the heart associated with pregnancy, heart failure isn't normal in pregnancy. Endocarditis is most commonly associated with I.V. drug use and isn't a normal finding in pregnancy.
NP: Data collection; CN: Health promotion and maintenance; CNS: Growth and development through the life span; CL: Knowledge

25. 1. The primary difference between preeclampsia and eclampsia is the occurrence of seizures, which occur when the client becomes eclamptic. Headaches, blurred vision, weight gain, increased blood pressure, and edema of the hands and feet are all indicative of preeclampsia.
NP: Data collection; CN: Physiological integrity; CNS: Physiological adaptation; CL: Comprehension

26. 1. The therapeutic level of magnesium for clients with PIH is 4 to 7 mEq/L. A serum level of 8 to 10 mEq/L may cause the absence of reflexes in the client. Serum levels of 10 to 12 mEq/L may cause respiratory depression, and a serum level of magnesium greater than 15 mEq/L may result in respiratory paralysis.
NP: Data collection; CN: Physiological integrity; CNS: Pharmacological therapies; CL: Knowledge

27. A client is receiving I.V. magnesium sulfate for severe preeclampsia. Which of the following adverse effects is associated with magnesium sulfate?
1. Anemia
2. Decreased urine output
3. Hyperreflexia
4. Increased respiratory rate

With the right drug, I can get rid of that extra magnesium.

27. 2. Decreased urine output may occur in clients receiving I.V. magnesium and should be monitored closely to keep urine output at greater than 30 ml/hour because magnesium is excreted through the kidneys and can easily accumulate to toxic levels. Anemia isn't associated with magnesium therapy. The client should be monitored for respiratory depression and paralysis when serum magnesium levels reach approximately 15 mEq/L. Magnesium infusions may cause depression of deep tendon reflexes.
NP: Data collection; CN: Physiological integrity; CNS: Pharmacological therapies; CL: Analysis

28. The antagonist for magnesium sulfate should be readily available to any client receiving I.V. magnesium. Which of the following drugs is the antidote for magnesium toxicity?
1. Calcium gluconate (Kalcinate)
2. Hydralazine (Apresoline)
3. Naloxone (Narcan)
4. Rh$_o$(D) immune globulin (RhoGAM)

Now I've got it!

28. 1. Calcium gluconate is the antidote for magnesium toxicity. Ten milliliters of 10% calcium gluconate is given by I.V. push over 3 to 5 minutes. Hydralazine is given for sustained elevated blood pressures in preeclamptic clients. Naloxone is used to correct narcotic toxicity. Rh$_o$(D) immune globulin is given to women with Rh-negative blood to prevent antibody formation from Rh-positive conceptions.
NP: Planning; CN: Physiological integrity; CNS: Pharmacological therapies; CL: Analysis

29. A pregnant client is screened for tuberculosis during her first prenatal visit. An intradermal injection of purified protein derivative (PPD) of the tuberculin bacilli is given. The client is considered to have a positive test for which of the following results?
1. An indurated wheal under 10 mm in diameter appears in 6 to 12 hours.
2. An indurated wheal over 10 mm in diameter appears in 48 to 72 hours.
3. A flat circumcised area under 10 mm in diameter appears in 6 to 12 hours.
4. A flat circumcised area over 10 mm in diameter appears in 48 to 72 hours.

Not that kind of trial!

29. 2. A positive PPD result would be an indurated wheal over 10 mm in diameter that appears in 48 to 72 hours. The area must be a raised wheal, not a flat circumcised area.
NP: Data collection; CN: Health promotion and maintenance; CNS: Prevention and early detection of disease; CL: Knowledge

30. Clients with which of the following conditions would be appropriate for a trial of labor after a prior cesarean delivery?
1. Complete placenta previa
2. Invasive cervical cancer
3. Premature rupture of membranes
4. Prior classical cesarean delivery

30. 3. Clients with premature rupture of membranes are permitted a trial of labor after a previous cesarean delivery. A client with invasive cervical cancer should be scheduled for a cesarean delivery. Clients with placenta previa or a prior classical cesarean delivery shouldn't be given a trial of labor due to the risk of uterine rupture or severe bleeding.
NP: Data collection; CN: Physiological integrity; CNS: Physiological adaptation; CL: Analysis

NP: Nursing process   CN: Client needs category   CNS: Client needs subcategory   CL: Cognitive level

31. Rh isoimmunization in a pregnant client develops during which of the following conditions?
1. Rh-positive maternal blood crosses into fetal blood, stimulating fetal antibodies.
2. Rh-positive fetal blood crosses into maternal blood, stimulating maternal antibodies.
3. Rh-negative fetal blood crosses into maternal blood, stimulating maternal antibodies.
4. Rh-negative maternal blood crosses into fetal blood, stimulating fetal antibodies.

You're doing a great job!

31. 2. Rh isoimmunization occurs when Rh-positive fetal blood cells cross into the maternal circulation and stimulate maternal antibody production. In subsequent pregnancies with Rh-positive fetuses, maternal antibodies may cross back into the fetal circulation and destroy the fetal blood cells.
NP: Data collection; CN: Physiological integrity; CNS: Reduction of risk potential; CL: Knowledge

32. A client who is into week 32 of her pregnancy is having contractions. The physician prescribes terbutaline (Brethine). This drug helps to do which of the following?
1. Stabilize blood pressure.
2. Relax smooth muscle.
3. Control bleeding.
4. Stimulate contractions.

32. 2. Terbutaline, a selective beta$_2$-adrenergic agonist, relaxes smooth muscles of the bronchi and uterus. It's commonly used for bronchospasm and premature labor. This drug doesn't stabilize blood pressure and has no effect on bleeding. Terbutaline is given to slow contractions, not stimulate them.
NP: Implementation; CN: Physiological integrity; CNS: Pharmacological therapies; CL: Comprehension

33. A client hospitalized for premature labor tells the nurse she's having occasional contractions. Which of the following nursing interventions would be the most appropriate?
1. Teach the client the possible complications of premature birth.
2. Tell the client to walk to see if she can get rid of the contractions.
3. Encourage her to empty her bladder and drink plenty of fluids, and give I.V. fluids.
4. Notify anesthesia for immediate epidural placement to relieve the pain associated with contractions.

Another hint.

33. 3. An empty bladder and adequate hydration may help decrease or stop labor contractions. Walking may encourage contractions to become stronger. Teaching the client potential complications is likely to increase her anxiety rather than help her relax. It would be inappropriate to call anesthesia and have an epidural placed because further assessment of the contractions is necessary.
NP: Implementation; CN: Physiological integrity; CNS: Reduction of risk potential; CL: Application

34. The phrase gravida 4, para 2 indicates which of the following prenatal histories?
1. A client has been pregnant four times and had two miscarriages.
2. A client has been pregnant four times and had two live-born children.
3. A client has been pregnant four times and had two cesarean deliveries.
4. A client has been pregnant four times and had two spontaneous abortions.

34. 2. *Gravida* refers to the number of times a client has been pregnant; *para* refers to the number of viable children born. Therefore, the client who's gravida 4, para 2 has been pregnant four times and had two live-born children.
NP: Data collection; CN: Health promotion and maintenance; CNS: Growth and development through the life span; CL: Knowledge

35. Which of the following factors would contribute to a high-risk pregnancy?
1. Blood type O positive
2. First pregnancy at age 33
3. History of allergy to honey bee pollen
4. Type 1 diabetes

35. 4. A woman with a history of diabetes has an increased risk for perinatal complications, including hypertension, preeclampsia, and neonatal hypoglycemia. The age of 33 without other risk factors doesn't increase risk, nor does type O positive blood or environmental allergens.
NP: Data collection; CN: Health promotion and maintenance; CNS: Prevention and early detection of disease; CL: Comprehension

The answer to question 36 has to be correct in two ways.

36. Which of the following complications can be potentially life-threatening and can occur in a client receiving a tocolytic agent?
1. Diabetic ketoacidosis
2. Hyperemesis gravidarum
3. Pulmonary edema
4. Sickle cell anemia

36. 3. Tocolytics are used to stop labor contractions. The most common adverse effect associated with the use of these drugs is pulmonary edema. Clients who don't have diabetes don't need to be observed for diabetic ketoacidosis. Hyperemesis gravidarum doesn't result from tocolytic use. Sickle cell anemia is an inherited genetic condition and doesn't develop spontaneously.
NP: Data collection; CN: Physiological integrity; CNS: Pharmacological therapies; CL: Knowledge

37. Which of the following hormones would be administered for the stimulation of uterine contractions?
1. Estrogen
2. Fetal cortisol
3. Oxytocin
4. Progesterone

37. 3. Oxytocin is the hormone responsible for stimulating uterine contractions. Pitocin, the synthetic form, may be given to clients who are past their due date. Progesterone has a relaxing effect on the uterus. Fetal cortisol is believed to slow the production of progesterone by the placenta. Although estrogen has a role in uterine contractions, it's not given in a synthetic form to help uterine contractility.
NP: Planning; CN: Physiological integrity; CNS: Pharmacological therapies; CL: Knowledge

Do your best with question 38! (Best, get it? It's a clue!)

38. Which of the following answers best describes the stage of pregnancy in which maternal and fetal blood are exchanged?
1. Conception
2. 9 weeks' gestation, when the fetal heart is well developed
3. 32 to 34 weeks' gestation (third trimester)
4. Maternal and fetal blood are never exchanged

38. 4. Only nutrients and waste products are transferred across the placenta. Blood exchange never occurs. Complications and some medical procedures can cause an exchange to occur accidentally.
NP: Data collection; CN: Physiological integrity; CNS: Physiological adaptation; CL: Comprehension

39. Which of the following rationales best explains why a pregnant client is urged to lie on her side when resting or sleeping in the later stages of pregnancy?
1. To facilitate digestion
2. To facilitate bladder emptying
3. To prevent compression of the vena cava
4. To avoid the development of fetal anomalies

40. A pregnant client is concerned about lack of fetal movement. What instructions would the nurse give that might offer reassurance?
1. Start taking two prenatal vitamins.
2. Take a warm bath to facilitate fetal movement.
3. Eat foods that contain a high sugar content to enhance fetal movement.
4. Lie down once a day and count the number of fetal movements for 15 to 30 minutes.

Client teaching is an important role for the nurse.

More than one answer to question 41 may be right, but which is the *most appropriate*?

41. What would be the most appropriate recommendation to a pregnant client who complains of swelling in her feet and ankles?
1. Limit fluid intake.
2. Buy walking shoes.
3. Sit and elevate the feet twice daily.
4. Start taking a diuretic as needed daily.

42. Which of the following interventions would the nurse recommend to a client having severe heartburn during her pregnancy?
1. Eat several small meals daily.
2. Eat crackers on waking every morning.
3. Drink a preparation of salt and vinegar.
4. Drink orange juice frequently during the day.

39. 3. The weight of the pregnant uterus is sufficiently heavy to compress the vena cava, which could impair blood flow to the uterus, possibly supplying insufficient oxygen to the fetus. The side-lying position hasn't been shown to prevent fetal anomalies, nor does it facilitate bladder emptying or digestion.
NP: Planning; CN: Physiological integrity; CNS: Reduction of risk potential; CL: Analysis

40. 4. Having the client lie down once during the day will allow her to concentrate on detecting fetal movement, which can be reassuring. Additionally, when the mother is up and actively walking around, it tends to be soothing to the fetus, resulting in sleep promotion. Lying down will make it easier for the client to detect movement. Eating additional sugary foods isn't recommended as some pregnant clients are more susceptible to cavities. Taking a warm bath is again likely to be soothing to the fetus. There's also a risk for hyperthermia if the water is too warm or the client is immersed too long. Instructing her to take additional prenatal vitamins isn't recommended as vitamins can be toxic.
NP: Implementation; CN: Psychosocial integrity; CNS: Coping and adaptation; CL: Application

41. 3. Sitting down and putting up her feet at least once daily will promote venous return and therefore decrease edema. Limiting fluid intake isn't recommended unless there are additional medical complications such as heart failure. Buying walking shoes won't necessarily decrease edema. Diuretics aren't recommended during pregnancy because it's important to maintain an adequate circulatory volume.
NP: Planning; CN: Physiological integrity; CNS: Basic care and comfort; CL: Application

42. 1. Eating small frequent meals will place less pressure on the esophageal sphincter, reducing the likelihood of the regurgitation of stomach contents into the lower esophagus. None of the other suggestions have been shown to decrease heartburn.
NP: Planning; CN: Physiological integrity; CNS: Basic care and comfort; CL: Application

43. Which of the following maternal complications is associated with obesity in pregnancy?
1. Mastitis
2. Placenta previa
3. Preeclampsia
4. Rh isoimmunization

You're almost finished! Hang in there!

43. 3. The incidence of preeclampsia in obese clients is about seven times more than that in a nonobese pregnant client. Placenta previa, mastitis, and Rh isoimmunization aren't associated with increased incidence in obese pregnant clients.

NP: Data collection; CN: Physiological integrity; CNS: Reduction of risk potential; CL: Analysis

44. Because uteroplacental circulation is compromised in clients with preeclampsia, a nonstress test (NST) is performed to detect which of the following conditions?
1. Anemia
2. Fetal well-being
3. Intrauterine growth retardation (IUGR)
4. Oligohydramnios

44. 2. An NST is based on the theory that a healthy fetus will have transient fetal heart rate accelerations with fetal movement. A fetus with compromised uteroplacental circulation usually won't have these accelerations, which indicate a nonreactive NST. Serial ultrasounds will detect IUGR and oligohydramnios in a fetus. An NST can't detect anemia in a fetus.

NP: Data collection; CN: Health promotion and maintenance; CNS: Growth and development through the life span; CL: Analysis

45. A client is 33 weeks' pregnant and has had diabetes since she was 21. When checking her fasting blood glucose level, which of the following values would indicate the client's disease was controlled?
1. 45 mg/dl
2. 85 mg/dl
3. 120 mg/dl
4. 136 mg/dl

45. 2. Recommended fasting blood glucose levels in pregnant clients with diabetes are 60 to 90 mg/dl. A fasting blood glucose level of 45 g/dl is low and may result in symptoms of hypoglycemia. A blood glucose level below 120 mg/dl is recommended for 1-hour postprandial values. A blood glucose level above 136 mg/dl in a pregnant client indicates hyperglycemia.

NP: Data collection; CN: Health promotion and maintenance; CNS: Growth and development through the life span; CL: Analysis

These hints are everywhere!

46. Which of the following techniques is best to monitor the fetus of a client with diabetes in her third trimester?
1. Ultrasound examination weekly
2. Nonstress test (NST) twice weekly
3. Daily contraction stress test (CST) at 32 weeks' gestation
4. Monitoring of fetal activity by client weekly

46. 2. The NST is the preferred antepartum heart rate screening test for pregnant clients with diabetes. NSTs should be done at least twice weekly, starting at 32 weeks' gestation because fetal deaths in clients with diabetes have been noted within 1 week of a reactive NST. Ultrasounds should be done every 4 to 6 weeks to monitor fetal growth. CST wouldn't be initiated at 32 weeks' gestation. Maternal fetal activity monitoring should be done daily.

NP: Planning; CN: Physiological integrity; CNS: Reduction of risk potential; CL: Knowledge

47. A client is diagnosed with preterm labor at 28 weeks' gestation. She comes to the emergency department saying, "I think I'm in labor." The nurse would expect her physical examination to show which of the following conditions?
  1. Painful contractions with no cervical dilation
  2. Regular uterine contractions with cervical dilation
  3. Irregular uterine contractions with no cervical dilation
  4. Irregular uterine contractions with cervical effacement

48. Which of the following conditions is the most common cause of anemia in pregnancy?
  1. Alpha thalassemia
  2. Beta thalassemia
  3. Iron deficiency anemia
  4. Sickle cell anemia

49. Which of the following tests should be ordered to confirm a diagnosis of beta thalassemia?
  1. Complete blood count (CBC)
  2. Hemoglobin A_{1c}
  3. Hemoglobin electrophoresis
  4. Iron level

50. A pregnant adolescent client is at risk for which of the following complications?
  1. Gestational diabetes
  2. Low-birth-weight neonate
  3. Macrosomic neonate
  4. Placenta previa

Two more hints, is this fun, or what?

47. 2. Regular uterine contractions (every 10 minutes or more) along with cervical dilation change before 36 weeks is considered preterm labor. No cervical change with uterine contractions isn't considered preterm labor.
NP: Data collection; CN: Health promotion and maintenance; CNS: Growth and development through the life span; CL: Knowledge

48. 3. Iron deficiency anemia accounts for approximately 95% of anemia in pregnancy. Sickle cell anemia is an inherited chronic disease that results from abnormal hemoglobin synthesis. Thalassemias are the most common genetic disorders of the blood. These anemias cause a reduction or absence of the alpha or beta hemoglobin chain.
NP: Data collection; CN: Health promotion and maintenance; CNS: Growth and development through the life span; CL: Knowledge

49. 3. Diagnosis of the specific type of thalassemia is achieved by hemoglobin electrophoresis. This test detects high levels of hemoglobin A_2 or F. The CBC includes white blood cell, hemoglobin, hematocrit, and platelet values. Hemoglobin A_{1c} values show the client's blood glucose levels over the past 120 days. A direct iron level can't be tested. Iron status is indirectly assessed through hemoglobin, hematocrit, mean corpuscular volume, and other such values. A client history and diet record can also help determine iron intake.
NP: Data collection; CN: Physiological integrity; CNS: Reduction of risk potential; CL: Knowledge

50. 2. Nutritional counseling should be included as part of prenatal care for adolescent clients. Adolescent clients are at risk for delivering low-birth-weight neonates, not macrosomic neonates. Adolescents aren't at increased risk for developing gestational diabetes or placenta previa.
NP: Data collection; CN: Health promotion and maintenance; CNS: Growth and development through the life span; CL: Knowledge

51. With which of the following methods would the nurse expect the client to use effleurage as a technique to displace pain?
1. Bradley method
2. Hydrotherapy
3. Lamaze method
4. Nubain method

51. 3. The Lamaze method incorporates effleurage, or light abdominal massage, to reduce pain during labor. With the Bradley method, the client reduces pain through abdominal breathing, ambulation, and use of an internal focus point as a disassociation technique. Hydrotherapy is the use of water during labor, including showers and tubs. Nubain, a narcotic analgesic, is commonly used during labor.

NP: Implementation; CN: Physiological integrity; CNS: Pharmacological therapies; CL: Knowledge

52. Which drug would the nurse choose to utilize as an antagonist for magnesium sulfate?
1. Oxytocin (Pitocin)
2. Terbutaline (Brethine)
3. Calcium gluconate
4. Naloxone (Narcan)

52. 3. Calcium gluconate should be kept at the bedside while a client is receiving a magnesium infusion. If magnesium toxicity occurs, calcium gluconate is administered as an antidote. Oxytocin (Pitocin) is the synthetic form of the naturally occurring pituitary hormone used to initiate or augment uterine contractions. Terbutaline is a beta$_2$-adrenergic agonist that may be used to relax the smooth muscle of the uterus, especially for preterm labor and uterine hyperstimulation. Naloxone is an opiate antagonist administered to reverse the respiratory depression that may follow administration of opiates.

NP: Implementation; CN: Physiological integrity; CNS: Pharmacological therapies; CL: Analysis

53. When auscultating the heart sounds of a client who is 34 weeks pregnant, the nurse detects a murmur. What should the nurse do?
1. Consult with a cardiologist.
2. Contact the primary care provider.
3. Consider the murmur a normal finding.
4. Ask the physician to order appropriate laboratory tests.

53. 3. During pregnancy, increased blood volume and cardiac output and changes in heart size and position normally may lead to murmurs and other heart sound changes. Such murmurs don't require treatment unless the pregnant client has a history of preexisting heart disease. Therefore, the nurse should record this finding and check for it at the next visit. No laboratory tests are required, and the services of a cardiologist or primary care provider aren't needed.

NP: Implementation; CN: Physiological integrity; CNS: Reduction of risk potential; CL: Application

Three cheers! You did it!

The antepartum and postpartum periods are important to know about. However, the intrapartum period — that's where the action is! This chapter covers the intrapartum period, perhaps the most critical of the three.

Chapter 23
Intrapartum care

1. A woman with a term, uncomplicated pregnancy comes into the labor-and-delivery unit in early labor saying that she thinks her water has broken. Which of the following actions by the nurse would be most appropriate?
1. Prepare the woman for delivery.
2. Note the color, amount, and odor of the fluid.
3. Immediately contact the physician.
4. Collect a sample of the fluid for microbial analysis.

2. A woman who is 36 weeks pregnant comes into the labor-and-delivery unit with mild contractions. Which of the following complications should the nurse watch for when the client informs her that she has placenta previa?
1. Sudden rupture of membranes
2. Vaginal bleeding
3. Emesis
4. Fever

3. A client's labor doesn't progress. After ruling out cephalopelvic disproportion, the physician orders I.V. administration of 1,000 ml normal saline solution with oxytocin (Pitocin) 10 U to run at 2 milliunits/minute. Two milliunits/minute is equivalent to how many ml/minute?
1. 0.002
2. 0.02
3. 0.2
4. 2.0

Don't miss the words most appropriate in question 1. They're hints to the answer.

That's three down, 47 to go. Keep at it!

1. 2. Noting the color, amount, and odor of the fluid will help guide the nurse in her next action. There's no need to call the client's physician immediately or prepare the client for delivery if the fluid is clear and delivery isn't imminent. Rupture of membranes isn't unusual in the early stages of labor. Fluid collection for microbial analysis isn't routine if there's no concern for infection (maternal fever).

NP: Implementation; CN: Physiological integrity; CNS: Reduction of risk potential; CL: Application

2. 2. Contractions may disrupt the microvascular network in the placenta of a client with placenta previa and result in bleeding. If the separation of the placenta occurs at the margin of the placenta, the blood will escape vaginally. Sudden rupture of the membranes isn't related to placenta previa. Fever would indicate an infectious process, and emesis isn't related to placenta previa.

NP: Data collection; CN: Physiological integrity; CNS: Reduction of risk potential; CL: Application

3. 3. The answer is found by setting up a ratio and following through with the calculations, shown below. Each unit of oxytocin contains 1,000 milliunits. Therefore, 1,000 ml of I.V. fluid contains 10,000 milliunits (10 units) of Pitocin. All other options are incorrect.

$$\frac{10,000}{1,000} = \frac{2}{X}$$

$$10,000X = 2,000$$

$$X = \frac{2,000}{10,000}$$

$$X = 0.2 \text{ ml}$$

NP: Implementation; CN: Physiological integrity; CNS: Pharmacological therapies; CL: Analysis

NP: Nursing process   CN: Client needs category   CNS: Client needs subcategory   CL: Cognitive level

4. A client in labor has been receiving oxytocin (Pitocin) to aid her progress. The nurse caring for her notes that a contraction has remained strong for 60 seconds. Which of the following actions should the nurse take first?
   1. Stop the oxytocin infusion.
   2. Notify the physician.
   3. Monitor fetal heart tones as usual.
   4. Turn the client on her left side.

Question 4 wants you to prioritize.

5. A client at term arrives in the labor unit experiencing contractions every 4 minutes. After a brief assessment, she's admitted and an electric fetal monitor is applied. Which of the following observations would alert the nurse to an increased potential for fetal distress?
   1. Total weight gain of 30 lb (13.6 kg)
   2. Maternal age of 32 years
   3. Blood pressure of 146/90 mm Hg
   4. Treatment for syphilis at 15-weeks' gestation

6. Cervical effacement and dilation aren't progressing in a client in labor. The physician orders intravenous administration of oxytocin (Pitocin). During oxytocin administration, why must the nurse monitor the client's fluid intake and output closely?
   1. Oxytocin causes water intoxication.
   2. Oxytocin causes excessive thirst.
   3. Oxytocin is toxic to the kidneys.
   4. Oxytocin has a diuretic effect.

7. After an amniotomy, which client goal should take the highest priority?
   1. The client will express increased knowledge about amniotomy.
   2. The fetus will maintain adequate tissue perfusion.
   3. The fetus will display no signs of infection.
   4. The client will report relief of pain.

4. 1. A contraction that remains strong for 60 seconds with no sign of letting up signals approaching tetany and could cause rupture of the uterus. Oxytocin stimulates contractions and should be stopped. The nurse should monitor the fetal heart tones and notify the physician but only after stopping the oxytocin.
NP: Implementation; CN: Physiological integrity; CNS: Reduction of risk potential; CL: Knowledge

5. 3. A blood pressure of 146/90 mm Hg indicates pregnancy-induced hypertension (PIH). Over time, PIH reduces blood flow to the placenta and can cause intrauterine growth retardation and other problems that make the fetus less able to tolerate the stress of labor. A weight gain of 30 lb is within expected parameters for a healthy pregnancy. A woman over age 30 doesn't have a greater risk of complications if her general condition is healthy before pregnancy. Syphilis that has been treated doesn't pose an additional risk.
NP: Data collection; CN: Physiological integrity; CNS: Reduction of risk potential; CL: Application

6. 1. The nurse should monitor fluid intake and output because prolonged oxytocin infusion may cause severe water intoxication, leading to seizure, coma, and death. Excessive thirst results from the work of labor and limited oral fluid intake, not oxytocin. Oxytocin has no nephrotoxic or diuretic effects; in fact, it produces an antidiuretic effect.
NP: Implementation; CN: Physiological integrity; CNS: Pharmacological therapies; CL: Knowledge

7. 2. Amniotomy increases the risk of umbilical cord prolapse, which would impair the fetal blood supply and tissue perfusion. Because the fetus's life depends on the oxygen carried by that blood, maintaining fetal tissue perfusion takes priority over goals related to increased knowledge, infection prevention, and pain relief.
NP: Planning; CN: Safe, effective care environment; CNS: Safety and infection control; CL: Application

8. A client at 42-weeks' gestation is 3 cm dilated, 30% effaced, with membranes intact and the fetus at +2 station. Fetal heart rate (FHR) is 140 to 150 beats/minute. After 2 hours, the nurse notes on the external fetal monitor that for the past 10 minutes the FHR ranged from 160 to 190 beats/minute. The client states that her baby has been extremely active. Uterine contractions are strong, occurring every 3 to 4 minutes and lasting 40 to 60 seconds. Which of the following findings would indicate fetal hypoxia?
   1. Abnormally long uterine contractions
   2. Abnormally strong uterine intensity
   3. Excessively frequent contractions, with rapid fetal movement
   4. Excessive fetal activity and fetal tachycardia

Remember that every piece of information provided may not be necessary to answer the question.

8. 4. Fetal tachycardia and excessive fetal activity are the first signs of fetal hypoxia. The duration of uterine contractions is within normal limits. Uterine intensity can be mild to strong and still be within normal limits. The frequency of contractions is within the normal limits for the active phase of labor.
NP: Evaluation; CN: Physiological integrity; CNS: Reduction of risk potential; CL: Analysis

9. A client at 33 weeks' gestation and leaking amniotic fluid is placed on an external fetal monitor. The monitor indicates uterine irritability, and contractions are occurring every 4 to 6 minutes. The physician orders terbutaline (Brethine). Which of the following teaching statements is appropriate for this client?
   1. "This medicine will make you breathe better."
   2. "You may feel a fluttering or tight sensation in your chest."
   3. "This will dry your mouth and make you feel thirsty."
   4. "You'll need to replace the potassium lost by this drug."

9. 2. A fluttering or tight sensation in the chest is a common adverse reaction to terbutaline. Although terbutaline relieves bronchospasm, this client is receiving it to reduce uterine motility. Mouth dryness and thirst occur with the inhaled form of terbutaline but are unlikely with the subcutaneous form. Hypokalemia is a potential adverse reaction following large doses of terbutaline but not at doses of 0.25 mg.
NP: Implementation; CN: Health promotion and maintenance; CNS: Prevention and early detection of disease; CL: Application

10. A 17-year-old primigravida with severe pregnancy-induced hypertension has been receiving magnesium sulfate I.V. for 3 hours. The latest assessment reveals deep tendon reflexes (DTR) of +1, blood pressure of 150/100 mm Hg, a pulse of 92 beats/minute, a respiratory rate of 10 breaths/minute, and urine output of 20 ml/hour. Which of the following actions would be most appropriate?
   1. Continue monitoring per standards of care.
   2. Stop the magnesium sulfate infusion.
   3. Increase the infusion rate by 5 gtt/minute.
   4. Decrease the infusion rate by 5 gtt/minute.

This question isn't a dice roll. Here are those key words again.

10. 2. Magnesium sulfate should be withheld if the client's respiratory rate or urine output falls or if reflexes are diminished or absent, all of which are true for this client. The client also shows other signs of impending toxicity, such as flushing and feeling warm. Inaction won't resolve the client's suppressed DTRs and low respiratory rate and urine output. The client is already showing central nervous system depression because of excessive magnesium sulfate, so increasing the infusion rate is inappropriate. Impending toxicity indicates that the infusion should be stopped rather than just slowed down.
NP: Implementation; CN: Physiological integrity; CNS: Pharmacological therapies; CL: Application

11. During a vaginal examination of a client in labor, the nurse palpates the fetus's larger, diamond-shaped fontanel toward the anterior portion of the client's pelvis. Which of the following statements best describes this situation?
 1. The client can expect a brief and intense labor, with potential for lacerations.
 2. The client is at risk for uterine rupture and needs constant monitoring.
 3. The client may need interventions to ease her back labor and change the fetal position.
 4. The client needs to be told that the fetus will be delivered using forceps or a vacuum extractor.

12. The cervix of a 26-year-old primigravida in labor is 5 cm dilated and 75% effaced, and the fetus is at 0 station. The physician prescribes an epidural regional block. Into which of the following positions should the nurse place the client when the epidural is administered?
 1. Lithotomy
 2. Supine
 3. Prone
 4. Lateral

13. Which of the following terms would the nurse use to describe the thinning and shortening of the cervix that occurs just before and during labor?
 1. Ballottement
 2. Dilation
 3. Effacement
 4. Multiparous

14. Which of the following fetal positions is best for birth?
 1. Vertex position
 2. Transverse position
 3. Frank breech position
 4. Posterior position of the fetal head

Be careful of the words will be in statement 4. They indicate an absolute, a rarity in health care.

I can tell you love to face a challenge! (Tee-hee!)

11. 3. The fetal position is occiput posterior, a position that commonly produces intense back pain during labor. Most of the time, the fetus rotates during labor to occiput anterior position. Positioning the client on her side can facilitate this rotation. An occiput posterior position would most likely result in prolonged labor. Occiput posterior alone doesn't create a risk of uterine rupture. The fetus would be delivered with forceps or vacuum extractor only if it doesn't rotate spontaneously.
NP: Planning; CN: Safe, effective care environment; CNS: Coordinated care; CL: Analysis

12. 4. The client should be placed on her left side or sitting upright, with her shoulders parallel and legs slightly flexed. Her back shouldn't be flexed because this position increases the possibility that the dura may be punctured and the anesthetic will accidentally be given as spinal, not epidural, anesthesia. None of the other positions allows proper access to the epidural space.
NP: Data collection; CN: Safe, effective care environment; CNS: Safety and infection control; CL: Application

13. 3. Effacement is cervical shortening and thinning while dilation is widening of the cervix; both facilitate opening the cervix in preparation for delivery. Ballottement is the ability of another individual to move the fetus by externally manipulating the maternal abdomen. A ballotable fetus hasn't yet engaged in the maternal pelvis. Multiparous refers to the number of live births a woman has had.
NP: Implementation; CN: Physiological integrity; CNS: Physiological adaptation; CL: Comprehension

14. 1. Vertex position (flexion of the fetal head) is the optimal position for passage through the birth canal. Transverse positioning generally results in poor labor contractions and an unacceptable fetal position for birth. Frank breech positioning, in which the buttocks present first, is a difficult delivery. Posterior positioning of the fetal head makes it difficult for the fetal head to pass under the maternal symphysis pubis bone.
NP: Data collection; CN: Physiological integrity; CNS: Reduction of risk potential; CL: Analysis

15. During which of the following stages would the nurse expect birth to occur?
1. First stage of labor
2. Second stage of labor
3. Third stage of labor
4. Fourth stage of labor

15. 2. The second stage of labor begins with complete dilation (10 cm) and ends with the expulsion of the fetus. The first stage of labor is the stage of dilation, which is divided into three distinct phases: latent, active, and transition. The third stage of labor begins with the birth of the neonate and ends with the expulsion of the placenta. The fourth stage of labor is the first 4 hours after placental expulsion, in which the client's body begins the recovery process.
NP: Planning; CN: Physiological integrity; CNS: Basic care and comfort; CL: Knowledge

To answer question 16 correctly, it's critical that you understand you're being asked to consider only those laboratory values that are critical.

16. Which of the following laboratory values would be critical for a client admitted to the labor-and-delivery unit?
1. Blood type
2. Calcium
3. Iron
4. Oxygen saturation

16. 1. Blood type would be a critical value to have because the risk of blood loss is always a potential complication during the labor-and-delivery process. Approximately 40% of a woman's cardiac output is delivered to the uterus, therefore, blood loss can occur quite rapidly in the event of uncontrolled bleeding. Calcium and iron aren't critical values and oxygen saturation isn't a laboratory value.
NP: Planning; CN: Physiological integrity; CNS: Reduction of risk potential; CL: Analysis

17. Which of the following fetal heart rates would be expected in the fetus of a laboring woman who is full-term?
1. 80 to 100 beats/minute
2. 100 to 120 beats/minute
3. 120 to 160 beats/minute
4. 160 to 180 beats/minute

17. 3. A rate of 120 to 160 beats/minute in the fetal heart is appropriate for filling the heart with blood and pumping it out to the system. Faster or slower rates don't accomplish perfusion adequately.
NP: Data collection; CN: Health promotion and maintenance; CNS: Growth and development through the life span; CL: Knowledge

18. A laboring client has external electronic fetal monitoring in place. Which of the following data can be determined by examining the fetal heart rate strip produced by the external electronic fetal monitor?
1. Gender of the fetus
2. Fetal position
3. Labor progress
4. Oxygenation

You're doing great! Keep going!

18. 4. Oxygenation of the fetus may be indirectly determined through fetal monitoring by closely examining the fetal heart rate strip. Accelerations in the fetal heart rate indicate good oxygenation, whereas decelerations in the fetal heart rate sometimes indicate poor fetal oxygenation. The fetal heart rate strip can't determine the gender of the fetus or fetal position. Labor progress can be directly monitored only through cervical examination; although, commonly, the woman's body language can give some indication of the progression of labor.
NP: Data collection; CN: Physiological integrity; CNS: Reduction of risk potential; CL: Comprehension

19. Which of the following nursing actions is required before a client in labor receives an epidural?
1. Giving a fluid bolus of 500 ml
2. Checking for maternal pupil dilation
3. Testing maternal reflexes
4. Observing maternal gait

20. Which of the following complications is possible with an episiotomy?
1. Blood loss
2. Uterine disfigurement
3. Prolonged dyspareunia
4. Hormonal fluctuation postpartum

Don't prolong this!
(Oops, did I give it
away? Oh, darn!)

21. A client in early labor states that she has been able to express a clear, milky discharge from both of her breasts. Which of the following actions by the nurse would be most appropriate?
1. Tell her that her milk is starting to come in because she's in labor.
2. Complete a thorough breast examination, and document the results in the chart.
3. Perform a culture on the discharge, and inform the client that she might have mastitis.
4. Inform the client that the discharge is colostrum, normally present after the 4th month of pregnancy.

22. When participating in care planning for a client in labor, the nurse expects to monitor the blood pressure frequently for which of the following reasons?
1. Decreased blood pressure is a sign of maternal pain.
2. Alterations in cardiovascular function affect the fetus.
3. Blood pressure decreases at the acme of each contraction.
4. Decreased blood pressure is the first sign of preeclampsia.

19. 1. One of the major adverse effects of epidural administration is hypotension. Therefore, a 500-ml fluid bolus is usually administered to help prevent hypotension in the client who wishes to receive an epidural for pain relief. Checking maternal reflexes, pupil response, and gait aren't necessary.
NP: Planning; CN: Physiological integrity; CNS: Reduction of risk potential; CL: Analysis

20. 3. Prolonged dyspareunia (painful intercourse) may result when complications such as infection interfere with wound healing. Minimal blood loss occurs when an episiotomy is performed. The uterus isn't affected by episiotomy because it's the perineum that's cut to accommodate the fetus. Hormonal fluctuations that occur during the postpartum period aren't the result of an episiotomy.
NP: Data collection; CN: Physiological integrity; CNS: Basic care and comfort; CL: Analysis

21. 4. After the 4th month, colostrum may be expressed. The breasts normally produce colostrum for the first few days after delivery. Milk production begins 1 to 3 days postpartum. A clinical breast examination isn't usually indicated in the intrapartum setting. Although a culture may be indicated, it requires advanced assessment as well as a medical order.
NP: Implementation; CN: Health promotion and maintenance; CNS: Growth and development through the life span; CL: Knowledge

22. 2. During contractions, blood pressure increases and blood flow to the intervillous spaces changes, compromising the fetal blood supply. Therefore, the nurse should assess the client's blood pressure frequently to determine if it returns to precontraction levels and allows adequate fetal blood flow again. During pain and contractions, the maternal blood pressure usually increases rather than decreases. Preeclampsia causes the blood pressure to increase, not decrease.
NP: Planning; CN: Physiological integrity; CNS: Reduction of risk potential; CL: Comprehension

NP: Nursing process   CN: Client needs category   CNS: Client needs subcategory   CL: Cognitive level

23. A client in early labor is concerned about the pinkish "stretch marks" on her abdomen. Which of the following observations by the nurse shows an accurate understanding of the marks on the client's abdomen?
 1. Striae are common in pregnancy and will fade away completely after the uterus contracts to its prepregnant state.
 2. Striae are common in pregnancy, and will fade after delivery, but don't disappear.
 3. Striae are common in pregnancy and will fade away after application of an emollient cream.
 4. This is a sign of a separation of the rectus muscle and will require further assessment by the physician.

Stretch marks are yet another reminder of the joy of giving birth.

23. 2. Striae are wavy, depressed streaks that may occur over the abdomen, breasts, or thighs as pregnancy progresses. They fade with time to a silvery color but won't disappear. Creams may soften the skin but won't remove the striae. Separation of the rectus muscle, diastasis, is a condition of pregnancy whereby the abdominal wall has difficulty stretching enough to accommodate the growing fetus, causing the muscle to separate.

NP: Implementation; CN: Health promotion and maintenance; CNS: Growth and development through the life span; CL: Knowledge

24. Which of the following positions increases cardiac output and stroke volume of a client in labor?
 1. Supine
 2. Sitting
 3. Side-lying
 4. Semi-Fowler's

24. 3. In the side-lying position, cardiac output increases, stroke volume increases, and the pulse rate decreases. In the supine position, the blood pressure can drop severely, due to the pressure of the fetus on the vena cava, resulting in supine hypotensive syndrome or vena caval syndrome. Neither the sitting nor semi-Fowler's position increase cardiac output or stroke volume.

NP: Data collection; CN: Health promotion and maintenance; CNS: Prevention and early detection of disease; CL: Application

25. When caring for a client in the first stage of labor, the nurse documents cervical dilation of 9 cm and intense contractions that last 45 to 60 seconds and occur about every 2 minutes. Based on these findings, the nurse should recognize that the client is in which phase of labor?
 1. Active phase
 2. Latent phase
 3. Descent phase
 4. Transitional phase

Congratulations! You're halfway there! Keep at it!

25. 4. In the transitional phase, the cervix dilates from 8 to 10 cm, and intense contractions occur every 1½ to 2 minutes and last for 45 to 90 seconds. In the active phase, the cervix dilates from 5 to 7 cm, and moderate contractions progress to strong contractions that last 60 seconds. In the latent phase, the cervix dilates 3 to 4 cm, and contractions are short, irregular, and mild. No descent phase exists. (Fetal descent may begin several weeks before labor, but usually doesn't occur until the second stage of labor.)

NP: Data collection; CN: Physiological integrity; CNS: Physiological adaptation; CL: Comprehension

26. When assessing a client several minutes after vaginal delivery, the nurse notes blood gushing from the vagina, umbilical cord lengthening, and a globe-shaped uterus. What do these findings suggest?
1. Uterine involution
2. Cervical laceration
3. Placental separation
4. Postpartum hemorrhage

26. 3. Placental separation causes a sudden gush or trickle of blood from the vagina, rise of the fundus in the abdomen, increased umbilical cord length at the introitus, and a globe-shaped uterus. Uterine involution causes a firmly contracted uterus, which can't occur until the placenta is delivered. Cervical lacerations produce a steady flow of bright red blood in a client with a firmly contracted uterus. Postpartum hemorrhage results in excessive vaginal bleeding and signs of shock, such as pallor and a rapid, thready pulse.

NP: Data collection; CN: Health promotion and maintenance; CNS: Growth and development through the life span; CL: Comprehension

Think about which part is the presenting part for question 27.

27. During labor involving a breech presentation, the nurse should anticipate which of the following?
1. A greater amount of bloody show
2. Slower than normal labor progress
3. More intense labor contractions
4. Precipitous delivery

27. 2. Because the presenting part doesn't fit the cervix snugly in a breech presentation, dysfunctional labor may result. A breech presentation wouldn't account for a greater amount of bloody show. Fetal positioning wouldn't affect uterine contractions. The delivery of a neonate in a breech presentation is more likely to be prolonged, not swift.

NP: Planning; CN: Physiological integrity; CNS: Physiological adaptation; CL: Comprehension

28. Which of the following descriptions best fits Braxton Hicks contractions?
1. Contractions beginning irregularly, becoming regular and predictable
2. Contractions causing cervical effacement and dilation
3. Contractions felt initially in the lower back and radiating to the abdomen in a wavelike motion
4. Contractions that begin and remain irregular

Check the word best in question 28. It's a hint to the answer.

28. 4. Braxton Hicks contractions begin and remain irregular. They're felt in the abdomen and remain confined to the abdomen and groin. They often disappear with ambulation and don't dilate the cervix. True contractions begin irregularly but become regular and predictable, causing cervical effacement and dilatation. True contractions are felt initially in the lower back and radiate to the abdomen in a wavelike motion.

NP: Data collection; CN: Physiological integrity; CNS: Physiological adaptation; CL: Knowledge

29. Which of the following descriptions best fits the term effacement?
1. Enlargement of the cervical canal
2. Expulsion of the mucus plug
3. Shortening and thinning of the cervical canal
4. Downward movement of the fetal head

29. 3. With effacement, the cervical canal shortens and thins due to longitudinal traction from the contracting uterine fundus. Dilation is the enlargement of the cervical canal to approximately 10 cm. Show is the expulsion of the mucus plug, followed by a seepage of cervical capillary blood. Descent is a mechanism of labor whereby the biparietal diameter of the fetal head descends into the pelvic inlet.

NP: Data collection; CN: Health promotion and maintenance; CNS: Growth and development through the life span; CL: Knowledge

NP: Nursing process   CN: Client needs category   CNS: Client needs subcategory   CL: Cognitive level

30. A laboring client is in the first stage of labor and has progressed from 4 to 7 cm in cervical dilation. In which of the following phases of the first stage does cervical dilation occur most rapidly?
1. Preparatory phase
2. Latent phase
3. Active phase
4. Transition phase

The question asks about rapid dilation.

31. During which of the following stages of labor would the nurse expect crowning to occur?
1. First
2. Second
3. Third
4. Fourth

32. The nurse suspects that the laboring client may have been physically abused by her male partner. Which of the following interventions by the nurse would be most appropriate?
1. Confront the male partner.
2. Question the woman in front of her partner.
3. Contact hospital security.
4. Collaborate with the physician to make a referral to social services.

Did you notice the words most appropriate in question 32? Another hint!

30. 3. Cervical dilation occurs more rapidly during the active phase than any of the previous phases. The active phase is characterized by cervical dilation that progresses from 4 to 7 cm. The preparatory, or latent, phase begins with the onset of regular uterine contractions and ends when rapid cervical dilation begins. Transition is defined as cervical dilation beginning at 8 cm and lasting until 10 cm or complete dilation.
NP: Data collection; CN: Health promotion and maintenance; CNS: Growth and development through the life span; CL: Comprehension

31. 2. The second stage of labor begins at full cervical dilation (10 cm) and ends when the infant is born. Crowning is present during this stage as the fetal head, pushed against the perineum, causes the vaginal introitus to open, allowing the fetal scalp to be visible. The first stage begins with true labor contractions and ends with complete cervical dilation. The third stage is from the time the infant is born until the delivery of the placenta. The fourth stage is the first 1 to 4 hours following delivery of the placenta.
NP: Planning; CN: Health promotion and maintenance; CNS: Growth and development through the life span; CL: Comprehension

32. 4. Collaborating with the physician to make a referral to social services will aid the client by creating a plan and providing support. Additionally, by law, the nurse or nursing supervisor must report the suspected abuse to the police and follow up with a written report. Although confrontation can be used therapeutically, this action will most likely provoke anger in the suspected abuser. Questioning the woman in front of her partner doesn't allow her the privacy required to address this issue and may place her in greater danger. If the woman isn't in imminent danger, there's no need to call hospital security.
NP: Implementation; CN: Physiological integrity; CNS: Reduction of risk potential; CL: Analysis

33. For a client in active labor, the nurse-midwife plans to use an internal electronic fetal monitoring (EFM) device. What must occur before the internal EFM can be applied?

1. The membranes must rupture.
2. The fetus must be at 0 station.
3. The cervix must be dilated fully.
4. The client must receive anesthesia.

34. The nurse has just admitted a client in the labor-and-delivery unit who has been diagnosed by her physician as having diabetes mellitus. Which of the following measures would be most appropriate for this situation?

1. Ask the client about her most recent blood glucose levels.
2. Prepare oral hypoglycemic medications for administration during labor.
3. Notify the neonatal intensive care unit that you'll be admitting a client with diabetes.
4. Prepare the client for cesarean delivery.

35. During a vaginal examination of a client in labor, the obstetrician determines that the biparietal diameter of the fetal head has reached the level of the ischial spines. How should the nurse document fetal station?

1. –1
2. 0
3. +1
4. +2

36. A 30-year-old multiparous client admitted to the labor-and-delivery unit hasn't received prenatal care for this pregnancy. Which of the following data is most relevant to the nursing assessment?

1. Date of last menstrual period (LMP)
2. Family history of sexually transmitted diseases (STDs)
3. Name of insurance provider
4. Number of siblings

Be careful and remember where the device is being placed.

Ah, the sweet sound of most appropriate

33. 1. Internal EFM can be applied only after the client's membranes have ruptured, when the fetus is at least at the –1 station, and when the cervix is dilated at least 2 cm. Although the client may receive anesthesia, it isn't required before application of an internal EFM device.
NP: Planning; CN: Physiological integrity; CNS: Basic care and comfort; CL: Comprehension

34. 1. As part of the history, asking about the client's most recent blood glucose levels will indicate how well her diabetes has been controlled. Oral hypoglycemic drugs are never used during pregnancy because they cross the placental barrier, stimulate fetal insulin production, and are potentially teratogenic. Plans to admit the neonate to the neonatal intensive care unit are premature. Cesarean delivery is no longer the preferred delivery for clients with diabetes. Vaginal birth is preferred and presents a lower risk to the mother and fetus.
NP: Data collection; CN: Physiological integrity; CNS: Reduction of risk potential; CL: Application

35. 2. When the largest diameter of the presenting part (typically the biparietal diameter of the fetal head) is level with the ischial spines, the fetus is at station 0. At –4 station, the fetal head is at the pelvic outlet. A station of –1 indicates that the fetal head is 1 cm above the ischial spines; of +1, that it is 1 cm below the ischial spines; of +2, that it is 2 cm below the ischial spines.
NP: Data collection; CN: Health promotion and maintenance; CNS: Growth and development through the life span; CL: Comprehension

36. 1. The date of the LMP is essential to estimate the date of delivery. The nursing history would also include subjective information, such as personal (but not necessarily family) history of STDs, gravidity, and parity. Although beneficial to the hospital for financial reimbursement, the insurance provider has no bearing on the nursing history. Likewise, the number of siblings isn't pertinent to the assessment.
NP: Data collection; CN: Health promotion and maintenance; CNS: Prevention and early detection of disease; CL: Analysis

37. Which of the following is the nurse least likely to observe in laboring clients with pregnancy-induced hypertension (PIH)?
1. Elevated blood pressure
2. Polyuria
3. Facial and hand edema
4. Epigastric discomfort

Careful. The question is looking for the least likely answer.

37. 2. Renal plasma flow and glomerular filtration are decreased in PIH, so increasing oliguria indicates a worsening condition. Blood pressure increases as a result of increased peripheral resistance. Facial and hand edema is due to protein loss, sodium retention, and a lowered glomerular filtration rate, moving fluid from intravascular to extravascular spaces. Epigastric discomfort may be due to abdominal edema or pancreatic or hepatic ischemia.
NP: Data collection; CN: Health promotion and maintenance; CNS: Prevention and early detection of disease; CL: Application

38. While in the first stage of labor, a client with active genital herpes is admitted to the labor and delivery area. Which type of birth should the nurse anticipate for this client?
1. Mid-forceps
2. Low forceps
3. Induction
4. Cesarean

38. 4. For a client with active genital herpes, cesarean birth helps avoid infection transmission to the neonate, which could occur during a vaginal birth. Mid-forceps and low forceps are types of vaginal births that could transmit the herpes infection to the neonate. Induction is used only during vaginal birth; therefore, it's inappropriate for this client.
NP: Planning; CN: Physiological integrity; CNS: Reduction of risk potential; CL: Application

39. A client is admitted to the labor-and-delivery unit in labor, with blood flowing down her legs. Which of the following nursing interventions would be most appropriate?
1. Placing an indwelling catheter
2. Monitoring fetal heart tones
3. Performing a vaginal examination
4. Preparing the client for cesarean delivery

Question 39 is asking you to prioritize!

39. 2. Monitoring fetal heart tones would be the first step because it's necessary to establish fetal well-being due to a possible placenta previa or abruptio placentae. Although an indwelling catheter may be placed, it isn't an early intervention. Performing a vaginal examination would be contraindicated, as any agitation of the cervix with a previa can result in hemorrhage and death for the mother or fetus. Preparing the client for a cesarean delivery may not be indicated. A sonogram will need to be performed to determine the cause of bleeding. If the diagnosis is a partial placenta previa, the client may still be able to deliver vaginally.
NP: Implementation; CN: Physiological integrity; CNS: Reduction of risk potential; CL: Comprehension

40. During labor, a primigravid client receives epidural anesthesia, and the nurse assists in monitoring maternal and fetal status. Which finding suggests an adverse reaction to the anesthesia?

1. Increased variability
2. Maternal hypotension
3. Fetal tachycardia
4. Anuria

41. Which of the following drugs would the nurse choose to use as an antagonist for magnesium sulfate?

1. Oxytocin (Pitocin)
2. Terbutaline (Brethine)
3. Calcium gluconate
4. Naloxone (Narcan)

Only 10 more to go.

42. The physician has ordered an I.V. of 5% dextrose in lactated Ringer's solution at 125 ml/hour. The I.V. tubing delivers 10 drops per ml. How many drops per minute should fall into the drip chamber?

1. 10 to 11
2. 12 to 13
3. 20 to 21
4. 22 to 24

Your math skills are being tested on this one!

43. The nurse receives an order to start an infusion for a client who's hemorrhaging from placenta previa. Which of the following supplies will be needed?

1. Y tubing, normal saline solution, and an 18G catheter
2. Y tubing, normal saline solution, and a 16G catheter
3. Y tubing, lactated Ringer's solution, and an 18G catheter
4. Y tubing, lactated Ringer's solution, and a 16G catheter

40. 2. As the epidural anesthetic agent spreads through the spinal canal, it may produce hypotensive crisis, which is characterized by maternal hypotension, decreased beat-to-beat variability, and fetal bradycardia. Although the client may experience some postpartum urine retention, anuria isn't associated with epidural anesthesia.

NP: Data collection; CN: Health promotion and maintenance; CNS: Prevention and early detection of disease; CL: Comprehension

41. 3. Calcium gluconate should be kept at the bedside while a client is receiving a magnesium infusion. If magnesium toxicity occurs, administer calcium gluconate as an antidote. Oxytocin is the synthetic form of the naturally occurring pituitary hormone used to initiate or augment uterine contractions. Terbutaline is a smooth muscle relaxant used to relax the uterus, especially for preterm labor and uterine hyperstimulation. Naloxone is an opiate antagonist administered to reverse the respiratory depression that sometimes follows doses of opiates.

NP: Implementation; CN: Physiological integrity; CNS: Pharmacological therapies; CL: Analysis

42. 3. Multiply the number of milliliters to be infused (125) by the drop factor (10); $125 \times 10 = 1,250$. Then divide the answer by the number of minutes to run the infusion (60); $1,250 \div 60 = 20.83$, or 20 to 21 gtt/minute.

NP: Implementation; CN: Physiological integrity; CNS: Pharmacological therapies; CL: Knowledge

43. 2. Blood transfusions require Y tubing, normal saline solution to mix with the blood product, and a 16G catheter to avoid lysing (breaking) the red blood cells.

NP: Data collection; CN: Physiological integrity; CNS: Pharmacological therapies; CL: Comprehension

44. Which of the following drugs is preferred for treating pregnancy-induced hypertension (PIH)?
1. Terbutaline (Brethine)
2. Oxytocin (Pitocin)
3. Magnesium sulfate
4. Calcium gluconate (Kalcinate)

44. 3. Magnesium sulfate is the drug of choice to treat PIH because it reduces edema by causing a shift from the extracellular spaces into the intestines. It also depresses the central nervous system, which decreases the incidence of seizures. Terbutaline is a smooth-muscle relaxant used to relax the uterus. Oxytocin is the synthetic form of the pituitary hormone used to stimulate uterine contractions. Calcium gluconate is the antagonist for magnesium toxicity.
NP: Planning; CN: Physiological integrity; CNS: Pharmacological therapies; CL: Knowledge

45. Which of the following characteristics best describes variable decelerations?
1. Predictable
2. Indicators of fetal well-being
3. Indicative of cord compression
4. Periodic decreases in fetal heart rate resulting from pressure on the fetal head

45. 3. Variable decelerations are commonly seen in labor when the membranes are ruptured, decreasing protection to the cord, especially as the fetus descends the birth canal. Variable decelerations are unpredictable. Atypical variable decelerations are suggestive of fetal hypoxia. Decreases in the fetal heart rate are due to decreased protection on the cord, not the fetal head. *Early* decelerations are related to pressure on the fetal head.
NP: Data collection; CN: Physiological integrity; CNS: Physiological adaptation; CL: Knowledge

46. Which of the following characteristics best describes early decelerations?
1. Reassuring, beginning at the onset of the contraction and ending with the end of the contraction
2. Decelerations initiated 30 to 40 seconds after the onset of the contraction
3. Decreased blood flow to the fetus due to uteroplacental insufficiency
4. Loss of variability with fetal hypoxia

Here are your last 5 questions. You're almost home!

46. 1. Early decelerations are reassuring and are due to pressure on the fetal head as it progresses down the birth canal. They have a uniform waveform that resembles an inverted mirror of the corresponding contraction. They begin at the onset of the contraction and end with the end of the contraction. Decelerations initiated 30 to 40 seconds after the onset of the contraction are termed late decelerations and are due to uteroplacental insufficiency from decreased blood flow during uterine contractions. Early decelerations aren't associated with loss of variability or fetal hypoxia.
NP: Data collection; CN: Physiological integrity; CNS: Physiological adaptation; CL: Knowledge

47. Which of the following characteristics best describes spontaneous (periodic) accelerations?

1. Occur with contractions
2. Reflect the shape of contractions
3. Are associated with decelerations
4. Indicate fetal well-being and aren't associated with contractions.

47. 4. Spontaneous accelerations are symmetrical, uniform increases in fetal heart rate. They indicate fetal well-being, represent an intact central nervous system response to fetal movement or stimulation, and aren't necessarily associated with contractions or deceleration. Uniform accelerations are symmetric, occur with contractions, and reflect the shape of the contraction. Decelerations are periodic decreases in the fetal heart rate from the baseline heart rate.

NP: Data collection; CN: Physiological integrity; CNS: Physiological adaptation; CL: Knowledge

Another math question! You can do it!

48. A client admitted in labor reports that the first day of her last menstrual period (LMP) was October 10. Using Nägele's rule, what is the estimated date of delivery?

1. July 10
2. July 17
3. August 10
4. August 17

48. 2. After determining the 1st day of the LMP, the nurse would subtract 3 months and add 7 days. If the client's LMP was October 10, subtracting 3 months is July 10, and adding 7 days brings the date to July 17.

NP: Data collection; CN: Health promotion and maintenance; CNS: Growth and development through the life span; CL: Analysis

49. At 1 minute of life, a neonate is crying vigorously, has a heart rate of 98, is active with normal reflexes, and has a pink body and blue extremities. Which of the following Apgar scores would be correct for this infant?

1. 6
2. 7
3. 8
4. 9

49. 3. Five signs are used to determine the Apgar score: heart rate, respiratory effort, muscle tone, reflex irritability, and color. Each of the signs is assigned a score of 0, 1, or 2. The highest possible score is 10. This infant lost 1 point for a slower-than-normal heart rate and 1 point for its acrocyanosis, a common finding in which the trunk is pink but the extremities are bluish.

NP: Data collection; CN: Physiological integrity; CNS: Reduction of risk potential; CL: Analysis

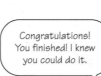

Congratulations! You finished! I knew you could do it.

50. A client at 30 weeks' gestation enters the labor-and-delivery unit with ruptured membranes. Which of the following drugs is administered from the 24th to 32nd week of pregnancy to promote fetal lung maturity?

1. Betamethasone (Celestone)
2. Magnesium sulfate
3. Oxytocin (Pitocin)
4. Terbutaline (Brethine)

50. 1. Betamethasone is a glucocorticoid capable of inducing pulmonary maturation in the fetus and decreasing the incidence of respiratory distress syndrome in preterm infants when administered to the mother prior to delivery. Magnesium sulfate is also a beta-adrenergic used for tocolysis. Oxytocin is the synthetic form of the naturally occurring pituitary hormone used to initiate or augment uterine contractions. Terbutaline is a beta-adrenergic used to relax the smooth muscle of the uterus, especially for preterm labor and uterine hyperstimulation.

NP: Planning; CN: Physiological integrity; CNS: Reduction of risk potential; CL: Knowledge

Before taking off through the chapter, why not spend a few minutes browsing the birthing stories at **www.birthstories.com/**? It will get you in just the right mood to tackle care of the postpartum client. Enjoy!

Chapter 24
Postpartum care

Be careful! This question is asking you to prioritize!

1. When completing the morning postpartum data collection, the nurse notices the client's perineal pad is completely saturated. Which of the following actions should be the nurse's **first** response?

    1. Vigorously massage the fundus.

    2. Immediately call the primary care provider.

    3. Have the charge nurse review the assessment.

    4. Ask the client when she last changed her perineal pad.

1. 4. If the morning assessment is done relatively early, it's possible that the client hasn't yet been to the bathroom, in which case her perineal pad may have been in place all night. Secondly, her lochia may have pooled during the night, resulting in a heavy flow in the morning. Vigorous massage of the fundus wouldn't be recommended as a *first* response until the client had gone to the bathroom, changed her perineal pad, and emptied her bladder. The nurse wouldn't want to call the primary care provider unnecessarily. If the nurse were uncertain, it would be appropriate to have another qualified individual check the client but only after a complete assessment of the client's status.

NP: Planning; CN: Physiological integrity; CNS: Physiological adaptation; CL: Analysis

2. Which of the following factors might result in a decreased supply of breast milk in a postpartum client?

    1. Supplemental feedings with formula

    2. Maternal diet high in vitamin C

    3. An alcoholic drink

    4. Frequent feedings

2. 1. Routine formula supplementation may interfere with establishing an adequate milk volume because decreased stimulation to the client's nipples affects hormonal levels and milk production. Vitamin C levels haven't been shown to influence milk volume. One drink containing alcohol generally tends to relax the client, facilitating letdown. Excessive consumption of alcohol may block letdown of milk to the infant, though supply isn't necessarily affected. Frequent feedings are likely to increase milk production.

NP: Data collection; CN: Physiological integrity; CNS: Physiological adaptation; CL: Application

NP: Nursing process   CN: Client needs category   CNS: Client needs subcategory   CL: Cognitive level

3. Which of the following interventions would be helpful to a breast-feeding client who is experiencing hard or engorged breasts?
  1. Applying ice
  2. Applying a breast binder
  3. Applying warm compresses
  4. Administering bromocriptine (Parlodel)

Which of these options would promote comfort best?

3. 1. Ice promotes comfort by decreasing blood flow (vasoconstriction), numbing the area, and discouraging further letdown of milk. Breast binders aren't effective in relieving the discomforts of engorgement. Warm compresses will promote blood flow and hence, milk production, worsening the problem of engorgement. Bromocriptine has been removed from the market for lactation suppression.
NP: Implementation; CN: Physiological integrity; CNS: Basic care and comfort; CL: Application

4. Which of the following tests should be performed routinely in the postpartum client?
  1. Antibody screen
  2. Babinski's reflex
  3. Homans' sign
  4. Patellar reflex

4. 3. Homans' sign, or pain on dorsiflexion of the foot, may indicate deep vein thrombosis (DVT). Postpartum clients are at increased risk for DVT because of changes in clotting mechanisms to control bleeding at delivery. An antibody screen, Babinski's reflex, and the patellar reflex aren't routinely assessed in the postpartum client.
NP: Planning; CN: Health promotion and maintenance; CNS: Prevention and early detection of disease; CL: Analysis

5. Which of the following reasons explains why women should be encouraged to perform Kegel exercises after they deliver a child?
  1. They assist with lochia removal.
  2. They promote the return of normal bowel function.
  3. They promote blood flow, allowing for healing and muscle strengthening.
  4. They assist the woman in burning calories for rapid postpartum weight loss.

If you know what Kegel exercises are, you should get question 5 correct easily.

5. 3. Exercising the pubococcygeal muscle increases blood flow to the area. The increased blood flow brings oxygen and other nutrients to the perineal area to aid in healing. Additionally, these exercises help to strengthen the musculature, thereby decreasing the risk of future complications, such as incontinence and uterine prolapse. Performing Kegel exercises may assist with lochia removal, but that isn't their main purpose. Bowel function isn't influenced by Kegel exercises. Kegel exercises don't generate sufficient energy expenditure to burn many calories.
NP: Implementation; CN: Health promotion and maintenance; CNS: Prevention and early detection of disease; CL: Analysis

6. When monitoring a postpartum client, the nurse is aware that what percentage of women experience "postpartum blues"?
  1. 25%
  2. 50%
  3. 75%
  4. 100%

6. 3. Postpartum blues, or mild depression during the first 10 days after giving birth, affects 75% to 80% of women who give birth. More intense depression during this period is referred to as postpartum depression, which affects approximately 10% to 15% of postpartum clients. Postpartum depression can be severe with implications for maternal and neonatal well-being.
NP: Planning; CN: Psychosocial integrity; CNS: Coping and adaptation; CL: Knowledge

NP: Nursing process   CN: Client needs category   CNS: Client needs subcategory   CL: Cognitive level

7. Which of the following practices would the nurse recommend to a client who has had a cesarean delivery?
1. Frequent douching after she's discharged
2. Coughing and deep-breathing exercises
3. Sit-ups for 2 weeks postoperatively
4. Side-rolling exercises

Remember to be sensitive to your postpartum client's needs.

8. Which of the following reasons explains why a client might express disappointment after having a cesarean delivery instead of a vaginal delivery?
1. Cesarean deliveries cost more.
2. Depression is more common after a cesarean delivery.
3. The client is usually more fatigued after cesarean delivery.
4. The client may feel a loss for not having experienced a "normal" birth.

9. Which of the following findings is normal during the postpartum period of a client who has experienced a vaginal birth?
1. Redness or swelling in the calves
2. A palpable uterine fundus beyond 6 weeks
3. Vaginal dryness after the lochial flow has ended
4. Dark red lochia for approximately 6 weeks after the birth

10. On completing fundal palpation, the nurse notes that the fundus is situated in the client's left abdomen. Which of the following actions is appropriate?
1. Ask the client to empty her bladder.
2. Straight catheterize the client immediately.
3. Call the client's primary health care provider for direction.
4. Straight catheterize the client for half of her urine volume.

7. 2. As for any postoperative client, coughing and deep-breathing exercises should be taught to keep the alveoli open and prevent infection. Frequent douching isn't recommended and is contraindicated in clients who have just given birth. Sit-ups at 2 weeks postpartum could potentially damage the healing of the incision. Side-rolling exercises aren't an accepted medical practice.
NP: Implementation; CN: Physiological integrity; CNS: Reduction of risk potential; CL: Application

8. 4. Clients occasionally feel a loss after a cesarean delivery. They may feel they're inadequate because they couldn't deliver their infant vaginally. The cost of cesarean delivery doesn't generally apply because the client isn't directly responsible for payment. No conclusive studies support the theory that depression is more common after cesarean delivery when compared to vaginal delivery. Although clients are usually more fatigued after a cesarean delivery, fatigue hasn't been shown to cause feelings of disappointment over the method of delivery.
NP: Data collection; CN: Psychosocial integrity; CNS: Coping and adaptation; CL: Analysis

9. 3. Vaginal dryness is a normal finding during the postpartum period due to hormonal changes. Redness or swelling in the calves may indicate thrombophlebitis. The fundus shouldn't be palpable beyond 6 weeks. Dark red lochia (indicating fresh bleeding) should only last 2 to 3 days postpartum.
NP: Data collection; CN: Physiological integrity; CNS: Physiological adaptation; CL: Knowledge

10. 1. A full bladder may displace the uterine fundus to the left or right side of the abdomen. A straight catheterization is unnecessarily invasive if the client can urinate on her own. Nursing interventions should be completed before notifying the primary health care provider in a nonemergency situation.
NP: Data collection; CN: Physiological integrity; CNS: Physiological adaptation; CL: Application

11. The nurse should inform the client with mastitis that the disorder is <u>most commonly</u> caused by which of the following organisms?
1. *Escherichia coli*
2. Group beta-hemolytic streptococci (GBS)
3. *Staphylococcus aureus*
4. *Streptococcus pyogenes*

12. The nurse is observing a client who gave birth yesterday. Where should the nurse expect to find the top of the client's fundus?
1. One fingerbreadth above the umbilicus
2. One fingerbreadth below the umbilicus
3. At the level of the umbilicus
4. Below the symphysis pubis

13. The nurse should be alert for which of the following complications in a client with type 1 diabetes mellitus whose delivery was complicated by polyhydramnios and macrosomia?
1. Postpartum mastitis
2. Increased insulin needs
3. Postpartum hemorrhage
4. Pregnancy-induced hypertension

14. Infections in diabetic clients tend to be more severe and can quickly lead to which of the following complications?
1. Anemia
2. Ketoacidosis
3. Respiratory acidosis
4. Respiratory alkalosis

Bacteria can strike when you least expect us!

When observing postpartum fundus height, timing is everything!

11. 3. The most common cause of mastitis is *S. aureus*, transmitted from the neonate's mouth. Mastitis isn't harmful to the neonate. *E. coli*, GBS, and *S. pyogenes* aren't associated with mastitis. GBS infection *is* associated with neonatal sepsis and death.
NP: Implementation; CN: Health promotion and maintenance; CNS: Growth and development through the life span; CL: Knowledge

12. 2. After a client gives birth, the height of her fundus should decrease by about one fingerbreadth (about 1 cm) each day. So by the end of the first postpartum day, the fundus should be one fingerbreadth below the umbilicus. Immediately after birth, the fundus may be above the umbilicus; 6 to 12 hours after birth, it should be at the level of the umbilicus; 10 days after birth, it should be below the symphysis pubis.
NP: Data collection; CN: Physiological integrity; CNS: Physiological adaptation; CL: Knowledge

13. 3. The client is at risk for a postpartum hemorrhage from the overdistention of the uterus because of the extra amniotic fluid and the large neonate. The uterus may not be able to contract as well as it would normally. The diabetic client usually has decreased insulin needs for the first few days postpartum. Neither polyhydramnios nor macrosomia would increase the client's risk of pregnancy-induced hypertension or mastitis.
NP: Data collection; CN: Physiological integrity; CNS: Reduction of risk potential; CL: Knowledge

14. 2. Diabetic clients who become pregnant tend to become sicker and develop illnesses more quickly than pregnant clients without diabetes. Severe infections in diabetes can lead to diabetic ketoacidosis. Anemia, respiratory acidosis, and respiratory alkalosis aren't generally associated with infections in diabetic clients.
NP: Data collection; CN: Physiological integrity; CNS: Reduction of risk potential; CL: Knowledge

15. Which of the following statements regarding mastitis is correct?
1. The most common pathogen is group A beta-hemolytic streptococci.
2. A breast abscess is a common complication of mastitis.
3. Mastitis usually develops in both breasts of a breast-feeding client.
4. Symptoms include fever, chills, malaise, and localized breast tenderness.

16. Which of the following measurements best describes delayed postpartum hemorrhage?
1. Blood loss in excess of 300 ml, occurring 24 hours to 6 weeks after delivery
2. Blood loss in excess of 500 ml, occurring 24 hours to 6 weeks after delivery
3. Blood loss in excess of 800 ml, occurring 24 hours to 6 weeks after delivery
4. Blood loss in excess of 1,000 ml, occurring 24 hours to 6 weeks after delivery

17. From which of the following items would the nurse gather data for a client in the immediate postpartum period (first 2 hours)?
1. Blood glucose level
2. Electrocardiogram (ECG)
3. Height of fundus
4. Stool test for occult blood

18. When monitoring a postpartum client 2 hours after delivery, the nurse notices heavy bleeding with large clots. Which of the following responses is most appropriate initially?
1. Massaging the fundus firmly
2. Performing bimanual compressions
3. Administering ergonovine (Ergotrate)
4. Notifying the primary health care provider

Is it getting chilly in here, or is it just me? (Hey, that's a hint!)

If you catch on fast enough, question 17 can be a lot of fun-dus!

Be careful. The word *initially* is a clue in question 18.

15. 4. Mastitis is an infection of the breast characterized by flulike symptoms, along with redness and tenderness in the breast. The most common causative agent is *Staphylococcus aureus*. Breast abscess is rarely a complication of mastitis if the client continues to empty the affected breast. Mastitis usually occurs in one breast, not bilaterally.
NP: Data collection; CN: Physiological integrity; CNS: Physiological adaptation; CL: Knowledge

16. 2. Postpartum hemorrhage involves blood loss in excess of 500 ml. Most delayed postpartum hemorrhages occur between the fourth and ninth days postpartum. The most frequent causes of a delayed postpartum hemorrhage include retained placental fragments, intrauterine infection, and fibroids.
NP: Data collection; CN: Physiological integrity; CNS: Reduction of risk potential; CL: Knowledge

17. 3. A complete physical examination should be performed every 15 minutes for the first 1 to 2 hours postpartum, including determination of the fundus, lochia, perineum, blood pressure, pulse, and bladder function. A blood glucose level must be obtained only if the client has risk factors for an unstable blood glucose level or if she has symptoms of an altered blood glucose level. An ECG would be necessary only if the client is at risk for cardiac difficulty. A stool test for occult blood generally wouldn't be valid during the immediate postpartum period because it's difficult to sort out lochial bleeding from rectal bleeding.
NP: Data collection; CN: Health maintenance and promotion; CNS: Growth and development through the life span; CL: Knowledge

18. 1. Initial management of excessive postpartum bleeding is firm massage of the fundus and administration of oxytocin (Pitocin). Bimanual compression is performed by a primary health care provider. Ergotrate should be used only if the bleeding doesn't respond to massage and oxytocin. The primary health care provider should be notified if the client doesn't respond to fundal massage, but other measures can be taken in the meantime.
NP: Implementation; CN: Physiological integrity; CNS: Physiological adaptation; CL: Analysis

19. The nurse is about to give a client with class B diabetes (age of onset greater than 20 years, duration less than 10 years) insulin before breakfast on her first day postpartum. Which of the following answers best describes insulin requirements immediately postpartum?
1. Lower than during her pregnancy
2. Higher than during her pregnancy
3. Lower than before she became pregnant
4. Higher than before she became pregnant

You're hustling! Keep up the good work!

20. Which of the following findings would be expected when monitoring the postpartum client?
1. Fundus 1 cm above the umbilicus 1 hour postpartum
2. Fundus 1 cm above the umbilicus on post-partum day 3
3. Fundus palpable in the abdomen at 2 weeks postpartum
4. Fundus slightly to right; 2 cm above um-bilicus on postpartum day 2

21. A client is complaining of painful contrac-tions, or afterpains, on postpartum day 2. Which of the following conditions could increase the severity of afterpains?
1. Bottle-feeding
2. Diabetes
3. Multiple gestation
4. Primiparity

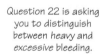

Question 22 is asking you to distinguish between heavy and excessive bleeding.

22. When giving a postpartum client self-care instructions, the nurse instructs her to report heavy or excessive bleeding. Which of the fol-lowing would indicate heavy bleeding?
1. Saturating 1 pad in 15 minutes
2. Saturating 1 pad in 1 hour
3. Saturating 1 pad in 4 to 6 hours
4. Saturating 1 pad in 8 hours

19. 3. Postpartum insulin requirements are usually significantly lower than prepregnancy requirements. Occasionally, clients may require little to no insulin during the first 24 to 48 hours postpartum.

NP: Planning; CN: Physiological integrity; CNS: Reduction of risk potential; CL: Knowledge

20. 1. Within the first 12 hours postpartum, the fundus usually is approximately 1 cm above the umbilicus. The fundus should be below the umbilicus by postpartum day 3. The fundus shouldn't be palpated in the abdomen after day 10. A uterus that isn't midline or is above the umbilicus on postpartum day 3 might be caused by a full, distended bladder or a uterine infection.

NP: Data collection; CN: Health promotion and maintenance; CNS: Growth and development through the life span ; CL: Knowledge

21. 3. Multiple gestation, breast-feeding, multi-parity, and conditions that cause overdistention of the uterus will increase the intensity of after-pains. Bottle-feeding and diabetes aren't direct-ly associated with increasing severity of after-pains, unless the client has delivered a macro-somic neonate.

NP: Data collection; CN: Health promotion and maintenance; CNS: Growth and development through the life span; CL: Knowledge

22. 2. Bleeding is considered heavy when a woman saturates 1 sanitary pad in 1 hour. Excessive bleeding occurs when a postpartum client saturates 1 pad in 15 minutes. Moderate bleeding occurs when the bleeding saturates less than 6″ (15 cm) of 1 pad in 1 hour.

NP: Evaluation; CN: Health promotion and maintenance; CNS: Growth and development through the life span; CL: Knowl-edge

23. On which of the following postpartum days can the client expect lochia serosa (old blood, serum, leukocytes, and tissue debris)?
1. Days 1 to 3 postpartum
2. Days 3 to 10 postpartum
3. Days 10 to 14 postpartum
4. Days 14 to 42 postpartum

Before you answer, make sure you know what type of lochia this question is asking for.

23. 2. On about the third but usually the fourth postpartum day, the lochia becomes a pale pink or brown, called lochia serosa, and contains old blood, serum, leukocytes, and tissue debris. This type of lochia usually lasts until postpartum day 10. Lochia rubra lasts for 1 to 3 days postpartum and consists of blood, decidua, and trophoblastic debris. Lochia alba, which contains leukocytes, decidua, epithelial cells, mucus, and bacteria, may continue for 2 to 6 weeks postpartum.

NP: Data collection; CN: Health promotion and maintenance; CNS: Growth and development through the life span; CL: Knowledge

24. A client and her neonate have a blood incompatibility, and the neonate has had a positive direct Coombs' test. Which of the following nursing interventions is appropriate?
1. Because the client has been sensitized, give $Rh_0(D)$ immune globulin (RhoGAM).
2. Because the client hasn't been sensitized, give RhoGAM.
3. Because the client has been sensitized, don't give RhoGAM.
4. Because the client hasn't been sensitized, don't give RhoGAM.

24. 3. A positive Coombs' test means that the Rh-negative client is now producing antibodies to the Rh-positive blood of the neonate. RhoGAM shouldn't be given to a sensitized client because it won't be able to prevent antibody formation.

NP: Implementation; CN: Physiological integrity; CNS: Reduction of risk potential; CL: Analysis

25. Which of the following definitions best describes a puerperal infection?
1. An infection in the uterus of a postpartum client
2. An infection in the bladder of a postpartum client
3. An infection in the perineum of a postpartum client
4. An infection in the genital tract of a postpartum client

25. 4. A puerperal infection is an infection of the genital tract after delivery through the first 6 weeks postpartum. Endometritis is an infection of the mucous membrane or endometrium of the uterus. Cystitis is an infection of the bladder. Infection of the perineum or episiotomy site usually results in localized pain, low-grade fever, and redness and swelling at the wound edges.

NP: Data collection; CN: Physiological integrity; CNS: Reduction of risk potential; CL: Knowledge

Don't pass on this question. It's important! (Now that's a subtle hint!)

26. Which of the following behaviors characterizes the postpartum client in the taking-in phase?
1. Passive and dependent
2. Striving for independence and autonomy
3. Curious and interested in care of the neonate
4. Exhibiting maximum readiness for new learning

26. 1. During the taking-in phase, which usually lasts 2 to 3 days, the client is passive and dependent and expresses her own needs rather than the neonate's needs. The taking-hold phase usually lasts from days 3 to 10 postpartum. During this stage, the client strives for independence and autonomy; she also becomes curious and interested in the care of the neonate and is most ready to learn.

NP: Data collection; CN: Psychosocial integrity; CNS: Coping and adaptation; CL: Knowledge

27. Which of the following verbalizations should concern the nurse treating a postpartum client within a few days of delivery?

    1. The client is nervous about taking the baby home.

    2. The client feels empty since she delivered the neonate.

    3. The client would like to watch the nurse give the baby her first bath.

    4. The client would like the nurse to take her baby to the nursery so she can sleep.

28. Which of the following complications may be indicated by continuous seepage of blood from the vagina of a postpartum client when palpation of the uterus reveals a firm uterus 1 cm below the umbilicus?

    1. Retained placental fragments

    2. Urinary tract infection

    3. Cervical laceration

    4. Uterine atony

29. Discharge teaching of the postpartum client who is receiving anticoagulant therapy for deep venous thrombophlebitis includes which of the following instructions?

    1. Avoid iron replacement therapy.

    2. Avoid over-the-counter (OTC) salicylates.

    3. Wear girdles and knee-high stockings when possible.

    4. Shortness of breath is a common adverse effect of the medication.

30. TORCH is an acronym for maternal infections associated with congenital malformations and disorders. Which of the following disorders does the *H* represent?

    1. Hemophilia

    2. Hepatitis B virus

    3. Herpes simplex virus

    4. Human immunodeficiency virus

Choose carefully. The word *empty* is a clue to the client's feelings.

CAUTION!!

Another 10 down! Keep at it!

27. 2. A client experiencing postpartum blues may say she feels empty now that the infant is no longer in her uterus. She may also verbalize that she feels unprotected now. Many first-time mothers are nervous about caring for their neonates by themselves after discharge. New mothers may want a demonstration before doing a task themselves. A client may want to get some uninterrupted sleep, so she may ask that the neonate be taken to the nursery.
NP: Data collection; CN: Psychosocial integrity; CNS: Psychosocial adaptation; CL: Knowledge

28. 3. Continuous seepage of blood may be due to cervical or vaginal lacerations if the uterus is firm and contracting. Retained placental fragments and uterine atony may cause subinvolution of the uterus, making it soft, boggy, and larger than expected. Urinary tract infection won't cause vaginal bleeding, although hematuria may be present.
NP: Data collection; CN: Physiological integrity; CNS: Reduction of risk potential; CL: Knowledge

29. 2. Discharge teaching should include informing the client to avoid OTC salicylates, which may potentiate the effects of anticoagulant therapy. Iron won't affect anticoagulation therapy. Restrictive clothing should be avoided to prevent the recurrence of thrombophlebitis. Shortness of breath should be reported immediately because it may be a symptom of pulmonary embolism.
NP: Planning; CN: Physiological integrity; CNS: Reduction of risk potential; CL: Knowledge

30. 3. TORCH represents the following maternal infections: Toxoplasmosis; Others, such as gonorrhea, syphilis, varicella, hepatitis, and human immunodeficiency virus; Rubella; Cytomegalovirus; and Herpes simplex virus. Hemophilia is a clotting disorder in which factors VII and X are deficient.
NP: Data collection; CN: Physiological integrity; CNS: Reduction of risk potential; CL: Knowledge

---

NP: Nursing process   CN: Client needs category   CNS: Client needs subcategory   CL: Cognitive level

31. Which of the following signs of grieving is dysfunctional in a client 3 days after a perinatal loss?
1. Lack of appetite
2. Denial of the death
3. Blaming herself
4. Frequent crying spells

32. Which of the following conditions in a postpartum client may cause fever not caused by infection?
1. Breast engorgement
2. Endometritis
3. Mastitis
4. Uterine involution

33. An RH-positive client vaginally delivers a 6-lb, 10-oz neonate after 17 hours of labor. Which of the following conditions puts this client at risk for infection?
1. Length of labor
2. Maternal Rh status
3. Method of delivery
4. Size of the neonate

34. Which of the management strategies should be implemented regarding breast-feeding after cesarean delivery?
1. Delay breast-feeding until 24 hours after delivery.
2. Breast-feed frequently during the day and every 4 to 6 hours at night.
3. Use the cradle hold position to avoid incisional discomfort.
4. Use the football hold position to avoid incisional discomfort.

35. What type of milk is present in the breasts 7 to 10 days postpartum?
1. Colostrum
2. Hind milk
3. Mature milk
4. Transitional milk

Use of the word *dysfunctional* in this question almost makes it a negative question.

Check out the options in question 34. It's like there are two pairs of options. Hmmmmm.

31. 2. Denial of the perinatal loss is dysfunctional grieving in the client. Lack of appetite, blaming oneself, and frequent crying spells are part of a normal grieving process.
NP: Data collection; CN: Psychosocial integrity; CNS: Psychosocial adaptation; CL: Knowledge

32. 1. Breast engorgement and dehydration are noninfectious causes of postpartum fevers. Mastitis and endometritis are both postpartum infections. Involution of the uterus won't cause temperature elevations.
NP: Data collection; CN: Health promotion and maintenance; CNS: Growth and development through the life span; CL: Knowledge

33. 1. A prolonged length of labor places the mother at increased risk for developing an infection. The average size of the neonate, vaginal delivery, and Rh status of the client don't place the mother at increased risk.
NP: Data collection; CN: Physiological integrity; CNS: Physiological adaptation; CL: Analysis

34. 4. When breast-feeding after a cesarean delivery, the client should be encouraged to use the football hold to avoid incisional discomfort. Breast-feeding should be initiated as soon after birth as possible. The client should be encouraged to breast-feed her neonate every 2 to 4 hours throughout the night as well as during the day to increase the milk supply.
NP: Planning; CN: Physiological integrity; CNS: Basic care and comfort; CL: Analysis

35. 4. Transitional milk is present 7 to 10 days postpartum and usually lasts until 2 weeks postpartum. Colostrum is a thin, milky fluid released by the breasts before and up to a few days after parturition. Hind milk, which satisfies the neonate's hunger and promotes weight gain, arrives approximately 10 minutes after each feeding starts. Mature milk is white and thinner than transitional milk and is present after 2 weeks postpartum.
NP: Data collection; CN: Health promotion and maintenance; CNS: Growth and development through the life span; CL: Knowledge

36. Which of the following recommendations should be given to a client with mastitis who is concerned about breast-feeding her neonate?
  1. She should stop breast-feeding until completing the antibiotic.
  2. She should supplement feeding with formula until the infection resolves.
  3. She shouldn't use analgesics because they aren't compatible with breast-feeding.
  4. She should continue to breast-feed; mastitis won't infect the neonate.

37. Which of the following terms is used to describe maladaptation to the stress and conflicts of the postpartum period, characterized by disabling feelings of inadequacy and an inability to cope?
  1. Postpartum blues
  2. Postpartum depression
  3. Postpartum neurosis
  4. Postpartum psychosis

Mastitis shouldn't interfere with breast-feeding.

36. 4. The client with mastitis should be encouraged to continue breast-feeding while taking antibiotics for the infection. No supplemental feedings are necessary because breast-feeding doesn't need to be altered and actually encourages resolution of the infection. Analgesics are safe and should be administered as needed.
NP: Planning; CN: Health promotion and maintenance; CNS: Growth and development through the life span; CL: Analysis

37. 2. Postpartum depression occurs in approximately 10% to 15% of all postpartum clients. This depression is characterized by disabling feelings of inadequacy and an inability to cope that can last up to 3 years. The client is often tearful and despondent. The client with postpartum blues experiences crying and sadness, generally between 3 to 5 days postpartum, but this condition resolves itself quickly. Postpartum neurosis includes neurotic behavior during the initial 6 weeks after birth. Postpartum psychosis includes hallucinations, delusions, and phobias.
NP: Data collection; CN: Psychosocial integrity; CNS: Psychosocial adaptation; CL: Knowledge

38. In which of the following time periods would the nurse most likely expect a client who has delivered twins to experience late postpartum hemorrhage?
  1. 24 to 48 hours after delivery
  2. 24 hours to 6 weeks after delivery
  3. 6 weeks to 3 months after delivery
  4. 6 weeks to 6 months after delivery

39. Which of the following complications is most likely responsible for a delayed postpartum hemorrhage?
  1. Cervical laceration
  2. Clotting deficiency
  3. Perineal laceration
  4. Uterine subinvolution

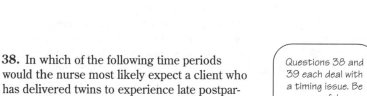

Questions 38 and 39 each deal with a timing issue. Be careful.

38. 2. Late or secondary postpartum hemorrhages occur more than 24 hours but less than 6 weeks postpartum. Early or primary postpartum hemorrhages occur within 24 hours of delivery.
NP: Planning; CN: Physiological integrity; CNS: Reduction of risk potential; CL: Knowledge

39. 4. Late postpartum bleeding is usually the result of subinvolution of the uterus. Retained products of conception or infection often cause subinvolution. Cervical or perineal lacerations can cause an immediate postpartum hemorrhage. A client with a clotting deficiency may also have an immediate postpartum hemorrhage, if the deficiency isn't corrected at the time of delivery.
NP: Data collection; CN: Physiological integrity; CNS: Physiological adaptation; CL: Knowledge

NP: Nursing process   CN: Client needs category   CNS: Client needs subcategory   CL: Cognitive level

40. A client needs to void 3 hours after a vaginal delivery. Which of the following risk factors necessitates assisting her out of bed?
1. Chest pain
2. Breast engorgement
3. Orthostatic hypotension
4. Separation of episiotomy incision

Excellent work! You're really flying through this test!

41. Before giving a postpartum client the rubella vaccine, which of the following facts should the nurse include in client teaching?
1. The vaccine is safe in clients with egg allergies.
2. Breast-feeding isn't compatible with the vaccine.
3. Transient arthralgia and rash are uncommon adverse effects.
4. The client should avoid getting pregnant for 3 months after the vaccination because the vaccine has teratogenic effects.

Don't lose your focus. Pay attention to the words *most commonly* in question 42.

42. Which of the following medications is most commonly used to treat preeclampsia during the prenatal and postpartum periods?
1. Diazepam (Valium)
2. Hydralazine (Apresoline)
3. Magnesium sulfate
4. Nifedipine (Procardia)

43. For which of the following complications would a nurse monitor a client who is receiving magnesium sulfate therapy?
1. Hypotension
2. Postpartum depression
3. Postpartum hemorrhage
4. Uterine infection

40. 3. The rapid decrease in intra-abdominal pressure occurring after birth causes splanchnic engorgement. The client is at risk for orthostatic hypotension when standing due to the blood pooling in this area. Breast engorgement is caused by vascular congestion in the breast before true lactation. The client shouldn't experience separation of the episiotomy incision or chest pain when standing.
NP: Implementation; CN: Health promotion and maintenance; CNS: Growth and development through the life span; CL: Knowledge

41. 4. The client must understand that she must not become pregnant for 2 to 3 months after the vaccination because of its potential teratogenic effects. The rubella vaccine is made from duck eggs, so an allergic reaction may occur in clients with egg allergies. The virus isn't transmitted via the breast milk, so clients may continue to breast-feed after vaccination. Transient arthralgia and rash are common adverse effects of the vaccine.
NP: Planning; CN: Health promotion and maintenance; CNS: Growth and development through the life span; CL: Knowledge

42. 3. Magnesium sulfate is commonly used in the treatment of preeclampsia to prevent seizures. It also produces a smooth-muscle depression effect, which can lower blood pressure. Diazepam may also be given for seizure activity. Nifedipine and hydralazine are used for clients with severe hypertensive preeclampsia.
NP: Implementation; CN: Physiological integrity; CNS: Pharmacological therapies; CL: Knowledge

43. 3. Because magnesium sulfate produces a smooth-muscle depressive effect, the uterus should be monitored for uterine atony, which would increase the risk of postpartum hemorrhage. Uterine infection and postpartum depression aren't associated with magnesium sulfate therapy. Magnesium sulfate is considered more of an anticonvulsant than an antihypertensive.
NP: Data collection; CN: Physiological integrity; CNS: Pharmacological therapies; CL: Comprehension

44. Which of the following changes best describes the <u>insulin needs</u> of a client with type 1 diabetes mellitus who has just delivered an infant vaginally without complications?
  1. Increased
  2. Decreased
  3. The same as before pregnancy
  4. The same as during pregnancy

To answer question 44, think about how insulin works in the body. Take it one step at a time.

44. 2. The placenta produces the hormone human placental lactogen, an insulin antagonist. After birth, the placenta, the major source of insulin resistance, is gone. Insulin needs decrease, and women with type 1 diabetes may need only one-half to two-thirds of the prenatal insulin dose during the first few postpartum days. Blood glucose levels should be monitored and insulin dosages adjusted as needed. The client should be encouraged to maintain appropriate dietary schedules, even if their infants are feeding on demand.
NP: Evaluation; CN: Physiological integrity; CNS: Physiological adaptation; CL: Comprehension

45. Which of the following responses is most appropriate for a client with diabetes who wants to breast-feed but is concerned about the effects of breast-feeding on her health?
  1. Diabetic clients who breast-feed have a hard time controlling their insulin needs.
  2. Diabetic clients shouldn't breast-feed because of potential complications.
  3. Diabetic clients shouldn't breast-feed; insulin requirements are doubled.
  4. Diabetic clients may breast-feed; insulin requirements may decrease from breast-feeding.

To breast-feed or not to breast-feed, that is the question for mothers today.

45. 4. Breast-feeding has an antidiabetic effect. Insulin needs are decreased because carbohydrates are used in milk production. Breast-feeding clients are at a higher risk for hypoglycemia in the first postpartum days after birth because glucose levels are lower. Diabetic clients should be encouraged to breast-feed.
NP: Implementation; CN: Physiological integrity; CNS: Pharmacological therapies; CL: Comprehension

46. A multiparous client vaginally delivered a neonate at 38 weeks without complications. She has three other children at home, two of whom were full-term and one a preterm. Using the GTPAL formula, which of the following classifications would accurately describe this client?
  1. Gravida 3 Para 4104
  2. Gravida 3 Para 3113
  3. Gravida 4 Para 2103
  4. Gravida 4 Para 3104

46. 4. Gravida is the number of times a client has been pregnant. Parity is the number of pregnancies that have reached viability (whether the fetus was born alive or was still-born). The T is term births; the P is preterm births; the A is the number of spontaneous or induced abortions; and the L is living children. This client is a Gravida 4 (counting this pregnancy and delivery of a full-term neonate). In her obstetrical history, she delivered two other full-term neonates and one preterm neonate; therefore, she is Para 3104.
NP: Data collection; CN: Health promotion and maintenance; CNS: Growth and development through the life span; CL: Knowledge

47. Which of the following factors puts a multiparous client on her first postpartum day at risk for developing hemorrhage?
    1. Hemoglobin level of 12 g/dl
    2. Uterine atony
    3. Thrombophlebitis
    4. Moderate amount of lochia rubra

47. 2. Multiparous women typically experience a loss of uterine tone due to frequent distentions of the uterus from previous pregnancies. As a result, this client is also at higher risk for hemorrhage. Thrombophlebitis doesn't increase the risk of hemorrhage during the postpartum period. The hemoglobin level and lochia flow are within acceptable limits.
NP: Data collection; CN: Health promotion and maintenance; CNS: Growth and development through the life span; CL: Analysis

Remember, when you're dealing with a postpartum client, you've got two clients to think about.

48. On the first postpartum night, a client requests that her neonate be sent back to the nursery so she can get some sleep. The client is most likely in which of the following phases?
    1. Depression phase
    2. Letting-go phase
    3. Taking-hold phase
    4. Taking-in phase

48. 4. The taking-in phase occurs in the first 24 hours after birth. The client is concerned with her own needs and requires support from staff and relatives. The taking-hold phase occurs when the client is ready to take responsibility for her care as well as her neonate's care. The letting-go phase begins several weeks later, when the client incorporates the new infant into the family unit. The depression phase isn't an appropriate answer.
NP: Data collection; CN: Health promotion and maintenance; CNS: Growth and development through the life span; CL: Analysis

49. The nurse provides information about various contraceptives to a postpartum client. The nurse knows that the client needs further instruction if she makes which statement?
    1. "To best protect me from herpes, my husband should use a latex condom."
    2. "I can expect some breast tenderness if I decide to use the pill."
    3. "If I choose the rhythm method, I must first determine when I ovulate."
    4. "It's okay for me to remove my diaphragm 1 hour after intercourse."

49. 4. A diaphragm should be left in place for at least 8 hours after intercourse. The other statements are all true.
NP: Evaluation; CN: Health promotion and maintenance; CNS: Prevention and early detection of disease; CL: Application

Think about what would make the client most comfortable.

50. The nurse is performing a postpartum check on a 40-year-old client. Which nursing measure is appropriate?
    1. Place the client in a supine position with her arms overhead for the examination of her breasts and fundus.
    2. Instruct the client to empty her bladder before the examination.
    3. Wear sterile gloves when assessing the pad and perineum.
    4. Perform the examination as quickly as possible.

50. 2. An empty bladder facilitates the examination of the fundus. The client should be in a supine position with her arms at her sides and her knees bent. The arms-overhead position is unnecessary. Clean gloves should be used when assessing the perineum. Sterile gloves aren't necessary. The postpartum examination shouldn't be done quickly. The nurse can take this time to teach the client about the changes in her body after delivery.
NP: Implementation; CN: Health promotion and maintenance; CNS: Growth and development through the life span; CL: Application

51. A client has delivered twins. Which of the following interventions would be most important for the nurse to perform?
1. Monitoring fundal tone and lochia flow
2. Applying a cold pack to the perineal area
3. Administering analgesics as ordered
4. Encouraging voiding by offering the bedpan

52. Which of the following physiological responses is considered normal in the early postpartum period?
1. Urinary urgency and dysuria
2. Rapid diuresis
3. Decrease in blood pressure
4. Increased motility of the GI system

53. During the third postpartum day, which of the following observations about the client would the nurse be most likely to make?
1. The client appears interested in learning more about neonate care.
2. The client wants to talk about her pregnancy.
3. The client wants to be waited on by the nurse.
4. The client requests help in choosing a name for the neonate.

54. Which of the following circumstances is most likely to cause uterine atony and lead to postpartum hemorrhage?
1. Hypertension
2. Cervical tears
3. Urine retention
4. Endometritis

Watch it! Question 52 is looking for a normal response.

You did it! Congratulations!

51. 1. Clients who deliver twins are at a higher risk for postpartum hemorrhage due to overdistention of the uterus, which causes uterine atony. Monitoring fundal tone and lochia flow helps to determine risks of hemorrhage. Applying cold packs to the perineum, administering analgesics as ordered, and offering the bedpan are all significant nursing interventions but aren't as important as preventing postpartum hemorrhage.
NP: Implementation; CN: Health promotion and maintenance; CNS: Growth and development through the life span; CL: Comprehension

52. 2. In the early postpartum period, there's an increase in the glomerular filtration rate and a drop in progesterone levels, which result in rapid diuresis. There should be no urinary urgency, although a woman may feel anxious about voiding. There is minimal change in blood pressure, and a residual decrease in GI motility following childbirth.
NP: Data collection; CN: Physiological integrity; CNS: Physiological adaptation; CL: Knowledge

53. 1. The third to tenth days of postpartum care are the taking-hold phase, in which the new mother strives for independence and is eager for her neonate. Talking about her pregnancy and wanting to be waited on describe the taking-in phase. Choosing a name describes the letting-go phase, as the client gives up the fantasized image of her neonate and accepts him.
NP: Evaluation; CN: Health promotion and maintenance; CNS: Growth and development through the life span; CL: Analysis

54. 3. Urinary retention causes a distended bladder to displace the uterus above the umbilicus and to the side, which prevents the uterus from contracting. The uterus needs to remain contracted if bleeding is to stay within normal limits. Cervical and vaginal tears can cause postpartum hemorrhage but are less common occurrences in the postpartum period. Endometritis and maternal hypertension don't cause postpartum hemorrhage.
NP: Implementation; CN: Health promotion and maintenance; CNS: Growth and development through the life span; CL: Knowledge

NP: Nursing process   CN: Client needs category   CNS: Client needs subcategory   CL: Cognitive level

Neonates depend on you for everything. Let's show 'em you've got what it takes for neonatal care!

Chapter 25
Neonatal care

1. The use of breast milk for premature neonates helps prevent which of the following conditions?

　　1. Down syndrome
　　2. Hyaline membrane disease
　　3. Necrotizing enterocolitis
　　4. Turner's syndrome

2. A client asks the nurse what surfactant is. Which of the following explanations would the nurse give as the <u>main role</u> of surfactant in the neonate?

　　1. Assists with ciliary body maturation in the upper airways
　　2. Helps maintain a rhythmic breathing pattern
　　3. Promotes clearing of mucus from the respiratory tract
　　4. Helps the lungs remain expanded after the initiation of breathing

3. The nurse observes that a 2-hour-old neonate has acrocyanosis. Which of the following nursing actions should be performed <u>immediately</u>?

　　1. Activate the code blue or emergency system.
　　2. Do nothing because acrocyanosis is normal in a neonate.
　　3. Immediately take the neonate's temperature according to facility policy.
　　4. Notify the physician of the need for genetic counseling.

Your baby is being given a drug to help him breathe better.

Immediately means to set priorities right away!

1. 3. Components specific to breast milk have been shown to lower the incidence of necrotizing enterocolitis in premature neonates. Hyaline membrane disease isn't directly influenced by breast milk or breast-feeding. Down syndrome and Turner's syndrome are genetic defects and not influenced by breast milk.
NP: Planning; CN: Health promotion and maintenance; CNS: Prevention and early detection of disease; CL: Application

2. 4. Surfactant works by reducing surface tension in the lung, which allows the lung to remain slightly expanded, decreasing the amount of work required for inspiration. Surfactant hasn't been shown to influence ciliary body maturation, clearing of the respiratory tract, or regulation of the neonate's breathing pattern.
NP: Implementation; CN: Health promotion and maintenance; CNS: Growth and development through the life span; CL: Knowledge

3. 2. Acrocyanosis, or bluish discoloration of the hands and feet in the neonate (also called peripheral cyanosis), is a normal finding and shouldn't last more than 24 hours after birth. The other choices are inappropriate.
NP: Implementation; CN: Physiological integrity; CNS: Physiological adaptation; CL: Application

---

NP: Nursing process　　CN: Client needs category　　CNS: Client needs subcategory　　CL: Cognitive level

4. When caring for a neonate, what is the most important step the nurse can take to prevent and control infection?
  1. Check frequently for signs of infection.
  2. Use sterile technique for all caregiving.
  3. Practice meticulous hand washing.
  4. Wear gloves at all times.

Knowing your risk factors can help guide your data collection.

4. 3. To prevent and control infection, the nurse should practice meticulous hand washing, scrubbing for 3 minutes before entering the nursery, washing frequently during caregiving activities, and scrubbing for 1 minute after providing care. Checking for signs of infection can detect, not *prevent*, infection. The nurse should use sterile technique for invasive procedures, not all caregiving. The nurse should wear gloves whenever contact with blood or body fluids is possible.
NP: Implementation; CN: Safe, effective care environment; CNS: Safety and infection control; CL: Knowledge

5. A woman with diabetes has just given birth. While caring for this neonate, the nurse is aware that he's at risk for which of the following complications?
  1. Anemia
  2. Hypoglycemia
  3. Nitrogen loss
  4. Thrombosis

5. 2. Neonates of mothers with diabetes are at risk for hypoglycemia due to increased insulin levels. During gestation, an increased amount of glucose is transferred to the fetus through the placenta. The neonate's liver can't initially adjust to the changing glucose levels after birth. This may result in an overabundance of insulin in the neonate, resulting in hypoglycemia. Neonates of mothers with diabetes aren't at increased risk for anemia, nitrogen loss, or thrombosis.
NP: Data collection; CN: Physiological integrity; CNS: Reduction of risk potential; CL: Knowledge

6. The nurse should be alert for which of the following complications in neonates who receive prolonged mechanical ventilation at birth?
  1. Bronchopulmonary dysplasia
  2. Esophageal atresia
  3. Hydrocephalus
  4. Renal failure

Knowing normal signs of development is also important.

6. 1. Bronchopulmonary dysplasia commonly results from the high pressures that must sometimes be used to maintain adequate oxygenation. Esophageal atresia, a structural defect in which the esophagus and trachea communicate with each other, doesn't relate to mechanical ventilation. Hydrocephalus and renal failure don't typically occur in these clients.
NP: Data collection; CN: Physiological integrity; CNS: Reduction of risk potential; CL: Analysis

7. Which of the following findings indicates adequate hydration in a neonate?
  1. Soft, smooth skin
  2. A sunken fontanel
  3. Frequent spitting up
  4. No urine output in the first 24 hours of life

7. 1. Soft, smooth skin is a sign of adequate hydration. A sunken fontanel and no urine output in the first 24 hours of life are signs of poor hydration. In the case of no urine output, kidney dysfunction would also be a concern. Frequent spitting up is normal in neonates. Excessive spitting up, however, may result in poor hydration.
NP: Data collection; CN: Physiological integrity; CNS: Physiological adaptation; CL: Analysis

NP: Nursing process   CN: Client needs category   CNS: Client needs subcategory   CL: Cognitive level

8. Which of the following signs is considered normal in a neonate?
1. Doll eyes
2. "Sunset" eyes
3. Positive Babinski's sign
4. Pupils that don't react to light

Question 8 asks what is *normal*. It could just as easily ask what's abnormal. Always double-check this kind of term.

9. Which technique is best when bathing a 5-hour-old girl?
1. Place her on a table covered with blankets, and give her a sponge bath.
2. Bathe her in a tub of warm water.
3. Keep her under a radiant warmer, and give her a sponge bath.
4. Wash only her hands and head because her condition isn't stable enough for her to have a complete bath.

Watch out! This question is asking for the best technique in this situation!

10. A client's mother asks the nurse why her newborn grandson is getting an injection. The nurse answers that the neonate is receiving an injection of vitamin K. Which of the following actions of vitamin K explains why the drug is given to neonates?
1. Assists with coagulation
2. Assists the gut to mature
3. Initiates the immunization process
4. Protects the brain from excess fluid production

8. 3. A positive Babinski's sign is present in infants until approximately age 1 and is normal in neonates, though abnormal in adults. The appearance of "sunset" eyes, in which the sclera is visible above the iris, results from cranial nerve palsies and may indicate increased intracranial pressure. Doll eyes is also a neurologic response but is noted in adults. A neonate's pupils normally react to light as an adult's would.

NP: Data collection; CN: Physiological integrity; CNS: Physiological adaptation; CL: Analysis

9. 3. During the first several hours after delivery, a neonate's thermal regulatory system is adapting to extrauterine life. When bathing a neonate under a radiant warmer, the external heat decreases the chances for cold stress by decreasing the number of internal mechanisms the neonate must use to stay warm. Bathing a neonate on a table, where she's exposed to air drafts and cooler air currents, can set her up for cold stress. Bathing the neonate in a tub and then removing her increases her heat loss and metabolism. Washing only the hands and head would chill the neonate and reduce thermoregulation because most heat is lost through the head.

NP: Implementation ; CN: Physiological integrity; CNS: Physiological adaptation; CL: Application

10. 1. Vitamin K, deficient in the neonate, is needed to activate clotting factors II, VII, IX, and X. In the event of trauma, the neonate would be at risk for excessive bleeding. Vitamin K doesn't assist the gut to mature, but the gut produces vitamin K after maturity is achieved. Vitamin K doesn't influence fluid production in the brain or the immunization process.

NP: Planning; CN: Physiological integrity; CNS: Reduction of risk potential; CL: Application

11. Neonates born to women infected with hepatitis B should receive which of the following treatments?
1. Hepatitis B vaccine at birth and 1 month
2. Hepatitis B immune globulin at birth; no hepatitis B vaccine
3. Hepatitis B immune globulin within 48 hours of birth and hepatitis B vaccine at 1 month
4. Hepatitis B immune globulin within 12 hours of birth and hepatitis B vaccine at birth, 1 month, and 6 months

11. 4. Hepatitis B immune globulin should be given as soon as possible after birth but within 12 hours. Neonates should also receive hepatitis B vaccine at regularly scheduled intervals. This sequence of care has been determined as superior to the other options provided.

NP: Planning; CN: Health promotion and maintenance; CNS: Prevention and early detection of disease; CL: Knowledge

12. The nurse is caring for a full-term neonate. Which finding is considered abnormal?
1. Respiratory rate of 52 breaths/minute
2. Small pinpoint white dots on the nose
3. Strabismus
4. Two veins and one artery in the umbilical cord

Be careful. Question 12 asks for an *abnormal* finding.

12. 4. The umbilical cord normally contains two arteries and one vein. A normal respiratory rate for a neonate is between 30 and 80 breaths/minute. Small white dots on the nose, which are known as milia, are normal. Strabismus is normal in the first month of life.

NP: Data collection; CN: Health promotion and maintenance; CNS: Growth and development through the life span; CL: Comprehension

13. Erythromycin ointment is administered to the neonate's eyes shortly after birth to prevent which of the following disorders?
1. Cataracts
2. Diabetic retinopathy
3. Ophthalmia neonatorum
4. Strabismus

13. 3. Eye prophylaxis is administered to the neonate immediately or soon after birth to prevent ophthalmia neonatorum. Strabismus is neuromuscular incoordination of the eye alignment. Cataracts are opacity of the lens of the eye associated with children with congenital rubella, galactosemia, and cortisone therapy. Diabetic retinopathy occurs in clients with diabetes when the retina bleeds into the vitreous humor causing scarring, after which neovascularization occurs.

NP: Planning; CN: Health promotion and maintenance; CNS: Prevention and early detection of disease; CL: Knowledge

14. A client with group AB blood whose husband has group O blood has just given birth. The major sign of ABO blood incompatibility in the neonate is which of the following complications or test results?
1. Negative Coombs' test
2. Bleeding from the nose or ear
3. Jaundice after the first 24 hours of life
4. Jaundice within the first 24 hours of life

This question is looking for the major sign.

14. 4. The neonate with an ABO blood incompatibility with its mother will have jaundice within the first 24 hours of life. The neonate would have a positive Coombs' test result. Jaundice after the first 24 hours of life is physiologic jaundice. Bleeding from the nose and ear should be investigated for possible causes but probably isn't related to ABO incompatibility.

NP: Data collection; CN: Health promotion and maintenance; CNS: Growth and development through the life span; CL: Knowledge

NP: Nursing process   CN: Client needs category   CNS: Client needs subcategory   CL: Cognitive level

15. Which of the following conditions of delivery would predispose a neonate to respiratory distress syndrome (RDS)?
1. Premature birth
2. Vaginal delivery
3. First born of twins
4. Postdate pregnancy

16. When caring for a male neonate on the day after he was circumcised, the nurse notices yellow-white exudate around the procedure site. Which of the following interventions should be done first?
1. Try to remove the exudate with a warm washcloth.
2. Take the neonate's temperature to check for infection.
3. Leave the area alone because this is part of the normal healing process.
4. Call the physician and notify him of the infection.

The color of an exudate helps determine its cause.

17. A client has just given birth at 42 weeks' gestation. The nurse should expect which of the following findings?
1. A sleepy, lethargic neonate
2. Lanugo covering the neonate's body
3. Desquamation of the neonate's epidermis
4. Vernix caseosa covering the neonate's body

18. The small-for-gestation neonate is at increased risk during the transitional period for which of the following complications?
1. Anemia probably due to chronic fetal hypoxia
2. Hyperthermia due to decreased glycogen stores
3. Hyperglycemia due to decreased glycogen stores
4. Polycythemia probably due to chronic fetal hypoxia

The test-taking expertise you're gaining from answering these questions will be well worth your effort. Keep at it!

15. 1. Prematurity is the single most important risk factor for developing RDS. The second born of twins and neonates born by cesarean delivery are also at increased risk for RDS. Surfactant deficiency, which frequently results in RDS, isn't a problem for postdate neonates.
NP: Data collection; CN: Physiological integrity; CNS: Reduction of risk potential; CL: Analysis

16. 3. The yellow-white exudate is part of the granulating process. This isn't a sign of infection and shouldn't be removed. Explain to the parents that this exudate will disappear as the site heals.
NP: Implementation; CN: Health promotion and maintenance; CNS: Growth and development through the life span; CL: Knowledge

17. 3. Postdate neonates lose the vernix caseosa, and the epidermis may become desquamated. A neonate at 42 weeks' gestation is usually very alert and missing lanugo.
NP: Data collection; CN: Health promotion and maintenance; CNS: Growth and development through the life span; CL: Knowledge

18. 4. The small-for-gestation neonate is at risk for developing polycythemia during the transitional period in an attempt to decrease hypoxia. This neonate is also at increased risk for developing hypoglycemia and hypothermia due to decreased glycogen stores.
NP: Data collection; CN: Health promotion and maintenance; CNS: Growth and development through the life span; CL: Analysis

19. Which of the following findings might be seen in a neonate suspected of having an infection?

    1. Flushed cheeks
    2. Increased temperature
    3. Decreased temperature
    4. Increased activity level

19. 3. Temperature instability, especially when it results in a low temperature in the neonate, may be a sign of infection. The neonate's color often changes with an infection process but generally becomes ashen or mottled. The neonate with an infection will usually show a decrease in activity level or lethargy.

NP: Data collection; CN: Health promotion and maintenance; CNS: Growth and development through the life span; CL: Analysis

20. Which of the following symptoms would indicate the neonate was adapting normally to extrauterine life without difficulty?

    1. Nasal flaring
    2. Light audible grunting
    3. Respiratory rate of 40 to 60 breaths/ minute
    4. Respiratory rate of 60 to 80 breaths/minute

Which answer indicates a normal response?

20. 3. A respiratory rate of 40 to 60 breaths/ minute is normal for a neonate during the transitional period. Nasal flaring, respiratory rate of more than 60 breaths/minute, and audible grunting are signs of respiratory distress.

NP: Data collection; CN: Health promotion and maintenance; CNS: Growth and development through the life span; CL: Knowledge

21. Which of the following drugs given to the mother is capable of causing respiratory depression in the neonate?

    1. Hydralazine (Apresoline)
    2. Magnesium sulfate
    3. Naloxone (Narcan)
    4. Penicillin

21. 2. An adverse reaction to magnesium sulfate causes respiratory depression in the neonate. Hydralazine, given to the mother for elevated blood pressure, isn't associated with respiratory depression in the neonate. Naloxone is given to reverse the respiratory depressive effects of narcotics. Penicillin, commonly given during labor for beta-hemolytic streptococcus organisms, isn't associated with respiratory depression.

NP: Data collection; CN: Health promotion and maintenance; CNS: Growth and development through the life span; CL: Knowledge

22. Which of the following interventions is helpful for the neonate experiencing drug withdrawal?

    1. Place the Isolette in a quiet area of the nursery.
    2. Withhold all medication to help the liver metabolize drugs.
    3. Dress the neonate in loose clothing so he won't feel restricted.
    4. Place the Isolette near the nurses' station for frequent contact with health care workers.

Some neonates need special interventions.

22. 1. Neonates experiencing drug withdrawal often have sleep disturbance. The neonate should be moved to a quiet area of the nursery to minimize environmental stimuli. The neonate should be swaddled to prevent him from flailing and stimulating himself. Medications, such as phenobarbital and paregoric, should be given as needed.

NP: Implementation; CN: Psychosocial integrity; CNS: Psychosocial adaptation; CL: Analysis

23. Neonates of diabetic mothers are at risk for which of the following complications?
1. Atelectasis
2. Microcephaly
3. Pneumothorax
4. Macrosomia

24. Neonates are given which of the following medications to prevent hemorrhagic disease?
1. Vitamin K
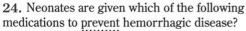
2. Heparin
3. Iron
4. Warfarin

Which measure is preventive, rather than curative or restorative?

25. During the transition period, a neonate can lose heat in many different ways. A neonate who isn't completely dried immediately after birth or a bath loses heat through which of the following methods?
1. Conduction
2. Convection
3. Evaporation
4. Radiation

26. By keeping the nursery temperature warm and wrapping the neonate in blankets, the nurse is preventing which of the following types of heat loss?
1. Conduction
2. Convection
3. Evaporation
4. Radiation

23. 4. Neonates of diabetic mothers are at increased risk for macrosomia (excessive fetal growth) due to the increased supply of maternal glucose combined with an increase in fetal insulin. Along with macrosomia, neonates of diabetic mothers are at risk for respiratory distress syndrome, hypoglycemia, hypocalcemia, hyperbilirubinemia, and congenital anomalies. They aren't at greater risk for atelectasis or pneumothorax. Microcephaly is usually the result of cytomegalovirus or rubella virus infection.

NP: Data collection; CN: Health promotion and maintenance; CNS: Growth and development through the life span; CL: Knowledge

24. 1. Neonates have coagulation deficiencies because of a lack of vitamin K in the intestines, which helps the liver synthesize clotting factors II, VII, IX, and X. Heparin and warfarin are given as anticoagulant therapy, not to prevent hemorrhagic disease in the neonate. Iron is stored in the fetal liver; hemoglobin binds to iron and carries oxygen.

NP: Implementation; CN: Health promotion and maintenance; CNS: Growth and development through the life span; CL: Knowledge

25. 3. Evaporation is the loss of heat that occurs when a liquid is converted to a vapor. In the neonate, heat loss by evaporation occurs as a result of vaporization of moisture from the skin. Convection is the flow of heat from the body surface to cooler air. Conduction is the loss of heat from the body surface to cooler surfaces in direct contact. Radiation is the loss of heat to a cooler surface that isn't in direct contact with the neonate.

NP: Implementation; CN: Health promotion and maintenance; CNS: Growth and development through the life span; CL: Knowledge

26. 2. Convection heat loss is the flow of heat from the body surface to cooler air. Evaporation is the loss of heat that occurs when a liquid is converted to a vapor. Conduction is the loss of heat from the body surface to cooler surfaces in direct contact. Radiation is the loss of heat from the body surface to cooler solid surfaces not in direct contact but in relative proximity.

NP: Implementation; CN: Health promotion and maintenance; CNS: Growth and development through the life span; CL: Knowledge

27. Which of the following statements is correct for physiologic hyperbilirubinemia?
1. The neonate usually also has a medical problem.
2. In term neonates, it usually appears after 24 hours.
3. It results in unusually elevated conjugated bilirubin levels.
4. It's usually progressive from the neonate's feet to his head.

28. A neonate has been diagnosed with caput succedaneum. Which of the following statements is correct about caput succedaneum?
1. It usually resolves in 3 to 6 weeks.
2. It doesn't cross the cranial suture line.
3. It's a collection of blood between the skull and the periosteum.
4. It involves swelling of the tissue over the presenting part of the fetal scalp.

29. The nurse teaches a postpartum client that her neonate's first stool will be meconium, which consists of intestinal secretions and cells. Which of the following colors and consistencies best describes the typical appearance of meconium?
1. Soft, pale yellow
2. Hard, pale brown
3. Sticky, greenish black
4. Loose, golden yellow

30. A 3-day-old neonate needs phototherapy for hyperbilirubinemia. Nursery care of a neonate receiving phototherapy includes which of the following treatments?
1. Tube feedings
2. Mask over the mouth
3. Eye patches to prevent retinal damage
4. Temperature monitored every 6 hours during phototherapy

Be careful with this prefix. *Hyper* is almost identical to *hypo*, but its meaning, of course, is vastly different.

When you're teaching a new mom, it helps to know what to expect at each stage in a neonate's development!

27. 2. Physiologic hyperbilirubinemia, or jaundice, in term neonates first appears after 24 hours. Jaundice usually appears in a cephalocaudal progression from head to feet. Neonates are otherwise healthy and have no medical problems. Hyperbilirubinemia is caused almost exclusively by unconjugated bilirubin.
NP: Data collection; CN: Physiological integrity; CNS: Reduction of risk potential; CL: Knowledge

28. 4. Caput succedaneum is the swelling of tissue over the presenting part of the fetal scalp due to sustained pressure. This boggy edematous swelling is present at birth, crosses the suture line, and most commonly occurs in the occipital area. A cephalhematoma is a collection of blood between the skull and periosteum, doesn't cross cranial suture lines, and resolves in 3 to 6 weeks. Caput succedaneum resolves within 3 to 4 days.
NP: Data collection; CN: Physiological integrity; CNS: Reduction of risk potential; CL: Knowledge

29. 3. Meconium collects in the GI tract during gestation and is initially sterile. Meconium is greenish black because of occult blood and is viscous. The stools of breast-fed neonates are loose and golden yellow after the transition to extrauterine life. The stools of formula-fed babies are typically soft and pale yellow after feeding is well established.
NP: Data collection; CN: Health promotion and maintenance; CNS: Growth and development through the life span; CL: Knowledge

30. 3. The neonate's eyes must be covered with eye patches to prevent damage. The mouth of the neonate doesn't need to be covered during phototherapy. The neonate can be removed from the lights and held for feeding. The neonate's temperature should be monitored at least every 2 to 4 hours because of the risk of hyperthermia with phototherapy.
NP: Implementation; CN: Physiological integrity; CNS: Physiological adaptation; CL: Analysis

NP: Nursing process   CN: Client needs category   CNS: Client needs subcategory   CL: Cognitive level

31. Which of the following neonates would be likely to develop hyperbilirubinemia?
1. Black neonate
2. Neonate of an Rh-positive mother
3. Neonate with ABO incompatibility
4. Neonate with Apgar scores 9 and 10 at 1 and 5 minutes

You're going strong. Keep at it!

31. 3. The mother's blood type, which is different from the neonate's, has an impact on the neonate's bilirubin level due to the antigen-antibody reaction. Black neonates tend to have lower mean levels of bilirubin. Chinese, Japanese, Korean, and Greek neonates tend to have higher incidences of hyperbilirubinemia. Neonates of Rh-negative, not Rh-positive, mothers tend to have hyperbilirubinemia. Low Apgar scores may indicate a risk for hyperbilirubinemia.
NP: Data collection; CN: Physiological integrity; CNS: Physiological adaptation; CL: Knowledge

32. Gram-positive cocci are responsible for causing 15% to 25% of major neonatal infections. Which of the following types of bacteria is gram-positive?
1. *Escherichia coli*
2. Group B streptococci
3. *Klebsiella* species
4. *Pseudomonas aeruginosa*

32. 2. Group B streptococci are gram-positive cocci that the neonate is exposed to if these bacteria are colonized in the vaginal tract. *E. coli, P. aeruginosa,* and *Klebsiella* species are gram-negative rods that produce 78% to 85% of the bacterial infection in neonates.
NP: Data collection; CN: Physiological integrity; CNS: Physiological adaptation; CL: Knowledge

33. The most common neonatal sepsis and meningitis infections seen within 24 hours after birth are caused by which of the following organisms?
1. *Candida albicans*
2. *Chlamydia trachomatis*
3. *Escherichia coli*
4. Group B beta-hemolytic streptococci

Here's an important clue to the correct answer: most common.

33. 4. Transmission of group B beta-hemolytic streptococci to the fetus results in respiratory distress that can rapidly lead to septic shock. *E. coli* is the second most common cause. *C. trachomatis* infection causes neonatal conjunctivitis and pneumonia. Candidiasis may be acquired from the birth canal.
NP: Data collection; CN: Physiological integrity; CNS: Physiological adaptation; CL: Knowledge

34. The nurse observes a neonate delivered at 28 weeks' gestation. Which finding would the nurse expect to see?
1. The skin is pale, and no vessels show through it.
2. Creases appear on the interior two-thirds of the sole.
3. The pinna of the ear is soft and flat and stays folded.
4. The neonate has 7 to 10 mm of breast tissue.

34. 3. The ear has a soft pinna that's flat and stays folded. Pale skin with no vessels showing through and 7 to 10 mm of breast tissue are characteristic of a neonate at 40 weeks' gestation. Creases on the anterior two-thirds of the sole are characteristic of a neonate at 36 weeks' gestation.
NP: Data collection; CN: Health promotion and maintenance; CNS: Growth and development through the life span; CL: Knowledge

35. When attempting to interact with a neonate experiencing drug withdrawal, which of the following behaviors would indicate that the neonate is willing to interact?
1. Gaze aversion
2. Hiccups
3. Quiet alert state
4. Yawning

35. 3. When caring for neonates experiencing drug withdrawal, the nurse needs to be alert for distress signals from the neonate. Stimuli should be introduced one at a time when the neonate is in a quiet alert state. Gaze aversion, yawning, sneezing, hiccups, and body arching are distress signals that the neonate can't handle stimuli at that time.

NP: Data collection; CN: Psychosocial integrity; CNS: Psychosocial adaptation; CL: Knowledge

36. When providing discharge instructions for umbilical cord care, which of the following information is given?
1. The stump should fall off 1 to 2 days after birth.
2. The stump should fall off 3 to 4 days after birth.
3. The stump should fall off 7 to 10 days after birth.
4. The stump should fall off 15 to 30 days after birth.

36. 3. The umbilical stump deteriorates over the first 7 to 10 days postpartum due to dry gangrene and usually falls off by day 10. Wiping the base of the stump with alcohol helps dry the area, facilitating separation.

NP: Implementation; CN: Health promotion and maintenance; CNS: Growth and development through the life span; CL: Knowledge

Watch this prefix. It's awfully close to hyper but means something quite different.

37. A client with diabetes has recently given birth to a boy. Which of the following findings is expected in a hypoglycemic neonate?
1. Jitteriness
2. Rooting reflex
3. Blood glucose level of 60 mg/dl
4. Blood glucose level of 82 mg/dl

37. 1. Evidence of hypoglycemia in the neonate includes jitteriness, lethargy, and the inability to maintain a normal body temperature. A rooting reflex is normal in a neonate. Neither a blood glucose level of 60 mg/dl nor one of 82 mg/dl indicates hypoglycemia.

NP: Data collection; CN: Physiological integrity; CNS: Physiological adaptation; CL: Knowledge

38. A neonate is born at 38 weeks' gestation. The mother asks what the thick, white, cheesy coating is on his skin. Which of the following terms is correct?
1. Lanugo
2. Milia
3. Nevus flammeus
4. Vernix

38. 4. Vernix is a white, cheesy material present on the neonate's skin at birth. Lanugo is the fine body hair on a neonate at birth. Milia are small white papules on the skin. Nevus flammeus is a reddish discoloration of an area of skin.

NP: Data collection; CN: Health promotion and maintenance; CNS: Growth and development through the life span; CL: Knowledge

Here's a big hint, the word routinely.

39. A nurse is giving care to a neonate. Which of the following drugs is routinely given to the neonate within 1 hour of birth?
1. Erythromycin ophthalmic ointment
2. Gentamycin
3. Nystatin
4. Vitamin A

39. 1. Erythromycin ophthalmic ointment is given for prophylactic treatment of ophthalmic neonatorum. Vitamin K, not vitamin A, is given. Gentamycin is an antibiotic used in the treatment of an infection in the neonate. Nystatin is used for treatment of thrush.

NP: Implementation; CN: Physiological integrity; CNS: Pharmacological therapies; CL: Analysis

40. Which of the following conditions or treatments is one of the most accurate indicators of lung maturity?
1. Meconium in the amniotic fluid
2. Glucocorticoid treatment just before delivery
3. Lecithin to sphingomyelin ratio of more than 2:1
4. Absence of phosphatidylglycerol in amniotic fluid

41. Which of the following factors is an unlikely risk factor for respiratory distress syndrome (RDS)?
1. Second born of twins
2. Neonate born at 34 weeks
3. Neonate of a diabetic mother
4. Chronic maternal hypertension

Be careful. This is almost a negative question.

42. Which of the following findings is considered common in the healthy neonate?
1. Simian crease
2. Oral moniliasis
3. Cystic hygroma
4. Bulging fontanel

43. When performing nursing care for a neonate after a birth, which of the following interventions should be done immediately?
1. Obtain a Dextrostix.
2. Give the initial bath.
3. Give the vitamin K injection.
4. Cover the neonate's head with a cap.

You should perform only one of these actions right away. Which one?

40. 3. Lecithin and sphingomyelin are phospholipids that help compose surfactant in the lungs; lecithin peaks at 36 weeks, and sphingomyelin concentrations remain stable. The presence of phosphatidylglycerol indicates lung maturity. Glucocorticoids must be given at least 48 hours before delivery. Meconium is released due to fetal stress before delivery, but it's chronic fetal stress that matures lungs.
NP: Data collection; CN: Physiological integrity; CNS: Physiological adaptation; CL: Knowledge

41. 4. Chronic maternal hypertension is an unlikely factor because chronic fetal stress tends to increase lung maturity. Premature neonates younger than 35 weeks are associated with RDS. Even with a mature lecithin to sphingomyelin ratio, neonates of diabetic mothers may still develop respiratory distress. Second twins may be prone to greater risk of asphyxia.
NP: Data collection; CN: Physiological integrity; CNS: Physiological adaptation; CL: Analysis

42. 2. Also known as thrush, oral moniliasis is a common finding in neonates, usually acquired from the mother during delivery. Bulging fontanels are a sign of intracranial pressure. Simian creases are present in 40% of neonates with trisomy 21. Cystic hygroma is a neck mass that can affect the airway.
NP: Data collection; CN: Health promotion and maintenance; CNS: Growth and development through the life span; CL: Analysis

43. 4. Covering the neonate's head with a cap helps prevent cold stress due to excessive evaporative heat loss from a neonate's wet head. Initial baths aren't given until the neonate's temperature is stable. Dextrostix, appropriate for neonates with risk factors, are obtained at 30 minutes to 1 hour of age. Vitamin K can be given within 4 hours after birth.
NP: Implementation; CN: Health promotion and maintenance; CNS: Growth and development through the life span; CL: Analysis

44. The nurse observes small white papules surrounded by erythematous dermatitis on a neonate's skin. This finding is characteristic of which of the following conditions?
1. Cutis marmorata
2. Epstein's pearls
3. Erythema toxicum
4. Mongolian spots

44. 3. Erythema toxicum has lesions that come and go on the face, trunk, and limbs. They're small white or yellow papules or vesicles with erythematous dermatitis. Cutis marmorata is bluish mottling of the skin. Mongolian spots are large macules or patches that are gray or blue green. Epstein's pearls, found in the mouth, are similar to facial milia.

NP: Data collection; CN: Health promotion and maintenance; CNS: Growth and development through the life span; CL: Knowledge

45. Which of the following nursing considerations is most important when giving a neonate his initial bath?
1. Giving a tub bath
2. Using water and mild soap
3. Giving the bath right after delivery
4. Using hexachlorophene soap

Question 45 asks you to prioritize according to importance.

45. 2. Use only water and mild soap on a neonate to prevent drying out the skin. The initial bath is given when the neonate's temperature is stable. Hexachlorophene soaps should be avoided; they're neurotoxic and may be absorbed through a neonate's skin. Tub baths are delayed until the umbilical cord falls off.

NP: Implementation; CN: Health promotion and maintenance; CNS: Growth and development through the life span; CL: Analysis

46. The pediatric nurse is caring for neonates in a busy nursery. When weighing a neonate, which action should the nurse take?
1. Leave the diaper on for comfort.
2. Place a sterile scale paper on the scale for infection control.
3. Keep a hand on the neonate's abdomen for safety.
4. Weigh the neonate at the same time each day for accuracy.

46. 4. A neonate of any age should be weighed at the same time each day, using the same technique and wearing the same clothes. A neonate should be weighed while undressed. Clean scale paper should be used when weighing the neonate. Sterile scale paper is unnecessary. The nurse should keep a hand above, not on, the abdomen when weighing the neonate.

NP: Data collection; CN: Health promotion and maintenance; CNS: Growth and development through the life span; CL: Application

47. A male neonate has just been circumcised. Which of the following interventions is part of the initial care of a circumcised neonate?
1. Apply alcohol to the site.
2. Change the diaper as needed.
3. Keep the neonate in the supine position.
4. Apply petroleum gauze to the site for 24 hours.

You're doing spectacularly!

47. 4. Petroleum gauze is applied to the site for the first 24 hours to prevent the skin edges from sticking to the diaper. Neonates are initially kept in the prone position. Diapers are changed more frequently to inspect the site. Alcohol is contraindicated for circumcision care.

NP: Implementation; CN: Health promotion and maintenance; CNS: Growth and development through the life span; CL: Application

48. Which of the following findings suggests that a neonate is hypothermic?
1. Bradycardia
2. Hyperglycemia
3. Metabolic alkalosis
4. Shivering

48. 1. Hypothermic neonates become bradycardic proportional to the degree of core temperature. Hypoglycemia is seen in hypothermic neonates. Shivering is rarely observed in neonates. Metabolic acidosis, not alkalosis, is seen due to slowed respirations.
NP: Data collection; CN: Health promotion and maintenance; CNS: Growth and development through the life span; CL: Analysis

An ounce of prevention is worth, oh, you know the rest.

49. Which of the following interventions helps prevent evaporative heat loss in the neonate after birth?
1. Administering warm oxygen
2. Controlling the drafts in the room
3. Immediately drying the neonate
4. Placing the neonate on a warm, dry towel

49. 3. Immediately drying the neonate decreases evaporative heat loss from his moist body from birth. Placing the neonate on a warm, dry towel decreases conductive losses. Controlling the drafts in the room and administering warmed oxygen help reduce convective loss.
NP: Implementation; CN: Health promotion and maintenance; CNS: Growth and development through the life span; CL: Analysis

50. Which of the following findings in a neonate would indicate a metabolic response to cold stress?
1. Arrhythmias
2. Hypoglycemia
3. Increase in liver function
4. Increase in blood pressure

50. 2. Hypoglycemia occurs as the consumption of glucose increases with the increase in metabolic rate. Arrhythmias and increases in blood pressure occur due to cardiorespiratory manifestations. Liver function declines in cold stress.
NP: Data collection; CN: Health promotion and maintenance; CNS: Growth and development through the life span; CL: Knowledge

The word first is a clue to the answer!

51. Which of the following goals should be placed first in regulating the temperature of a neonate?
1. Supply extra heat sources to the neonate.
2. Keep the ambient room temperature less than 100° F (37.8° C).
3. Minimize the energy needed for the neonate to produce heat.
4. Block radiant, convective, conductive, and evaporative losses.

51. 4. Prevention of heat loss is always the first goal in thermoregulation to avoid hypothermia. The second goal is to minimize the energy necessary for neonates to produce heat. Adding extra heat sources is a means of correcting hypothermia. The ambient room temperature should be kept at approximately 100° F.
NP: Planning; CN: Health promotion and maintenance; CNS: Growth and development through the life span; CL: Application

52. A neonate undergoing phototherapy treatment needs to be monitored for which of the following adverse effects?
1. Hyperglycemia
2. Increased insensible water loss
3. Severe decrease in platelet count
4. Increased GI transit time

52. 2. Increased insensible water loss is due to absorbed photon energy from the lights. GI transit time may decrease with use of phototherapy. Hyperglycemia isn't a characteristic effect of phototherapy treatment. Phototherapy may cause a mild decrease in platelet count.
NP: Data collection; CN: Physiological integrity; CNS: Reduction of risk potential; CL: Analysis

53. When maintaining thermoregulation, which of the following neonatal characteristics affects the establishment of a thermal neutral zone?
1. Flexed posture
2. Blood vessels that aren't close to the skin
3. Decreased subcutaneous fat and a thin epidermis
4. Increased subcutaneous fat and a thick epidermis

54. Which of the following clinical findings suggests physiologic hyperbilirubinemia?
1. Clinical jaundice before age 36 hours
2. Clinical jaundice lasting beyond 14 days
3. Bilirubin levels of 12 mg/dl by the third day of life
4. Serum bilirubin level increasing by more than 5 mg/dl/day

55. When caring for a neonate receiving phototherapy, the nurse should remember to:
1. decrease the amount of formula.
2. dress the neonate warmly.
3. massage the neonate's skin with lotion.
4. reposition the neonate frequently.

Did you remember to double-check that puzzling prefix? Excellent!

56. A 2-day-old boy is scheduled for circumcision without anesthesia. Which measure should be included in the infant's postcircumcision care plan?
1. Chart the time of the infant's voiding.
2. Keep the infant's penis exposed to air.
3. Feed the infant only clear fluids for the first 12 hours.
4. Place a small ice cap on the infant's penis.

53. 3. Decreased subcutaneous fat and a thin epidermis affect the establishment of a thermal neutral zone. The more insulated a neonate is, the greater the ability to cope with lower environmental temperatures. Blood vessels in neonates are close to the skin, so circulating blood is influenced by temperature changes. The flexed posture decreases the surface area exposed and decreases heat loss.
NP: Data collection; CN: Health promotion and maintenance; CNS: Growth and development through the life span; CL: Knowledge

54. 3. Increased bilirubin levels in the liver usually cause bilirubin levels of 12 mg/dl by the third day of life. This results from the impaired conjugation and excretion of bilirubin and difficulty clearing bilirubin from plasma. The other answers suggest nonphysiologic jaundice.
NP: Data collection; CN: Health promotion and maintenance; CNS: Growth and development through the life span; CL: Analysis

55. 4. Phototherapy works by the chemical interaction between a light source and the bilirubin in the neonate's skin. Therefore, the larger the skin area exposed to light, the more effective the treatment. Changing the neonate's position frequently ensures maximum exposure. Because the neonate will lose water through the skin as a result of evaporation, the amount of formula or water may need to be increased. The neonate is typically undressed to ensure maximum skin exposure. The eyes are covered to protect them from light, and an abbreviated diaper is used to prevent soiling. The skin should be clean and patted dry. Use of lotions would interfere with phototherapy.
NP: Implementation; CN: Physiological integrity CNS: Physiological adaptation; CL: Application

56. 1. After a circumcision, urine retention may occur. A petroleum dressing is commonly applied to the penis; then the infant is diapered. Because no anesthetic was given, feeding restrictions are unnecessary. Although the penis should be inspected for swelling and bleeding, further care is unnecessary.
NP: Planning; CN: Safe, effective care environment; CNS: Safety and infection control; CL: Application

NP: Nursing process   CN: Client needs category   CNS: Client needs subcategory   CL: Cognitive level

57. Which of the following findings might be seen in a neonate suspected of having breast-milk jaundice?
1. History of being a poor breast-feeder
2. Decreased bilirubin level around day 3 of life
3. Clinical jaundice evident after 96 hours
4. Interruption of breast-feeding, resulting in increased bilirubin levels between 24 to 72 hours

57. 3. Breast milk jaundice is an elevation of indirect bilirubin in a breast-fed neonate that develops following the first 4 to 7 days of life. History of being a poor breast-feeder and interruption of breast feeding are indicative of breast-feeding jaundice, which occurs before the first 4 to 7 days of life and is caused by insufficient production or intake of breast milk. Jaundice is an elevation, not a decrease, in bilirubin.

NP: Data collection; CN: Health promotion and maintenance; CNS: Growth and development through the life span; CL: Analysis

Early is the clue.

58. Which of the following signs appears early in a neonate with respiratory distress syndrome?
1. Bilateral crackles
2. Pale gray skin color
3. Tachypnea more than 60 breaths/minute
4. Poor capillary filling time (3 to 4 seconds)

58. 3. Tachypnea and expiratory grunting occur early in respiratory distress syndrome to help improve oxygenation. Poor capillary filling time, a later manifestation, occurs if signs and symptoms aren't treated. Crackles occur as the respiratory distress progressively worsens. A pale gray skin color obscures earlier cyanosis as respiratory distress symptoms persist and worsen.

NP: Data collection; CN: Health promotion and maintenance; CNS: Growth and development through the life span; CL: Analysis

59. Which of the following disorders would the nurse expect as a nonrespiratory cause of respiratory distress?
1. Choanal atresia
2. Meconium aspiration
3. Pulmonary hemorrhage
4. Retained lung fluid syndrome

59. 1. Choanal atresia is caused by protrusion of bone or membrane into nasal passages, causing blockage or narrowing. Meconium aspiration is meconium aspirated into the lungs during birth. Retained lung fluid syndrome is caused by a delay in removing excessive amounts of lung fluid. Pulmonary hemorrhage is bleeding into the alveoli.

NP: Data collection; CN: Health promotion and maintenance; CNS: Growth and development through the life span; CL: Knowledge

Not many more questions left! Outstanding!

60. Which of the following respiratory disorders in a neonate is usually mild and runs a self-limited course?
1. Pneumonia
2. Meconium aspiration syndrome
3. Transient tachypnea
4. Persistent pulmonary hypertension

60. 3. Transient tachypnea has an invariably favorable outcome after several hours to several days. The outcome of pneumonia depends on the causative agent involved and may have complications. Meconium aspiration, depending on severity, may have long-term adverse effects. In persistent pulmonary hypertension, the mortality rate is more than 50%.

NP: Planning; CN: Physiological integrity; CNS: Physiological adaptation; CL: Analysis

61. A 10-hour-old neonate appears exceptionally irritable; he cries easily and startles when touched. The physician orders a drug screen test, which indicates that the neonate is positive for cocaine. Which nursing action would best help to soothe this neonate?
1. Leaving the light beside the bassinet on at night
2. Wrapping the neonate snugly in a blanket
3. Providing multisensory stimulation while the neonate is awake
4. Giving the neonate a warm bath

This question gives you a lot of information. Be sure to pay attention to the most important facts to help you choose the correct answer.

61. 2. The practice of tightly wrapping or swaddling a cocaine-addicted neonate provides a safe, secure environment and maintains body warmth, both of which are soothing. A cocaine-addicted neonate typically experiences withdrawal 8 to 10 hours after birth; signs and symptoms include constant crying, jitteriness, poor feeding, emesis, respiratory distress, and seizures. To minimize or prevent these signs and symptoms, sensory stimulation is kept to a minimum and the neonate is typically kept in a quiet, dimly lit environment. A bath would necessitate the removal of clothing and exposure to changes in temperature, both of which are too stimulating.
NP: Implementation; CN: Physiological integrity; CNS: Basic care and comfort; CL: Application

62. Which of the following immunoglobulins (Ig) provides immunity against bacterial and viral pathogens through passive immunity?
1. IgA
2. IgE
3. IgG
4. IgM

62. 3. IgG is a major Ig of serum and interstitial fluid that crosses the placenta. IgE has a major role in allergic reactions. IgM and IgA don't cross the placenta.
NP: Data collection; CN: Health promotion and maintenance; CNS: Growth and development through the life span; CL: Knowledge

63. Which of the following facial changes is characteristic in a neonate with fetal alcohol syndrome (FAS)?
1. Macrocephaly
2. Microcephaly
3. Wide, palpebral fissures
4. Well-developed philtrum

63. 2. FAS infants are usually born with microcephaly. Their facial features include short, palpebral fissures and a thin upper lip.
NP: Data collection; CN: Physiological integrity; CNS: Physiological adaptation; CL: Knowledge

64. A 36-week neonate born weighing 1,800 g has microcephaly and microophthalmia. Based on these findings, which of the following risk factors might be expected in the maternal history?
1. Use of alcohol
2. Use of marijuana
3. Gestational diabetes
4. Positive group B streptococci

Read closely. Question 64 asks about maternal history, not the neonate's current condition.

64. 1. The most common sign of the effects of alcohol on fetal development is retarded growth in weight, length, and head circumference. Gestational diabetes usually produces large-for-gestational-age neonates. Intrauterine growth retardation isn't characteristic of marijuana use. Positive group B streptococcus isn't a relevant risk factor.
NP: Data collection; CN: Health promotion and maintenance; CNS: Growth and development through the life span; CL: Analysis

NP: Nursing process   CN: Client needs category   CNS: Client needs subcategory   CL: Cognitive level

65. Of the following cognitive defects, which is associated with fetal alcohol syndrome (FAS)?

    1. Hypoactivity

    2. Low birth weight

    3. Poor wake and sleep patterns

    4. High threshold of stimulation

66. Which of the following GI disorders is seen exclusively in neonates with cystic fibrosis?

    1. Duodenal obstruction

    2. Jejunal atresia

    3. Malrotation

    4. Meconium ileus

The biggest clue here is exclusively.

67. A diabetic client delivers a full-term neonate who weights 10 lb, 1 oz (4.6 kg). While caring for this large-for-gestational age (LGA) neonate, the nurse palpates the clavicles for which reason?

    1. Neonates of diabetic mothers have brittle bones.

    2. Clavicles are commonly absent in neonates of diabetic mothers.

    3. One of the neonate's clavicles may have been broken during delivery.

    4. LGA neonates have glucose deposits on their clavicles.

68. A heelstick hematocrit test on a 4-hour-old male neonate is 63%. This finding indicates which of the following?

    1. Serious anemia

    2. Nothing because this is normal for a neonate

    3. Possible risk of hyperbilirubinemia

    4. Need to repeat the test using venous blood

The finish line is just ahead!

65. 3. Altered sleep patterns are caused by disturbances in the central nervous system from alcohol exposure in utero. Hyperactivity is a characteristic deficit generally noted. Low birth weight is a physical defect seen in neonates with FAS. Neonates with FAS generally have a low threshold for stimulation.

NP: Data collection; CN: Physiological integrity; CNS: Physiological adaptation; CL: Knowledge

66. 4. Meconium ileus is a luminal obstruction of the distal small intestine by abnormal meconium seen in neonates with cystic fibrosis. Duodenal obstruction, malrotation, and jejunal atresia aren't characteristic findings in neonates with cystic fibrosis.

NP: Data collection; CN: Physiological integrity; CNS: Physiological adaptation; CL: Knowledge

67. 3. Because of the neonate's large size, clavicular fractures are common during delivery. The nurse should assess all LGA neonates for this occurrence. None of the other options are true.

NP: Implementation; CN: Physiological integrity; CNS: Physiological adaptation; CL: Application

68. 2. A hematocrit of 45% to 65% is normal in a neonate because of increased blood supply during intrauterine life. A hematocrit of 63% doesn't indicate serious anemia because the value is within the normal hematocrit range for a neonate. Although all neonates have the potential to develop hyperbilirubinemia, a hematocrit of 63% is inconclusive for this finding. If the heelstick blood test shows a hematocrit greater than 65%, a venous blood sample is obtained for testing because hemolysis of the heelstick sample can show a false reading. A hematocrit greater than 65% requires treatment after the physician has been notified.

NP: Data collection; CN: Health promotion and maintenance; CNS: Prevention and early detection of disease; CL: Comprehension

69. An initial inspection of a female neonate shows pink-streaked vaginal discharge. Which of the following factors is the probable cause?
1. Cystitis
2. Birth trauma
3. Neonatal candidiasis
4. Withdrawal of maternal hormones

69. 4. Withdrawal of maternal estrogen can produce pseudomenstruation. Cystitis or a urinary tract infection in a neonate would show generalized signs of sepsis. Birth trauma may cause surface abrasions but not vaginal discharge. Neonates with candidal infections usually have oral lesions (thrush) or monilial diaper rash.

NP: Data collection; CN: Health promotion and maintenance; CNS: Prevention and early detection of disease; CL: Knowledge

70. Which of the following symptoms is seen <u>first</u> with tracheoesophageal atresia?
1. Torticollis
2. Nasal stuffiness
3. Oligohydramnios
4. Excessive oral secretions

Question 70 asks about early signs of tracheoesophageal atresia.

70. 4. Accumulated secretions are copious because the neonate can't swallow. Torticollis would be present only if there were a defect of muscle or bone. Nasal stuffiness is common in neonates and doesn't indicate esophageal abnormalities. Atresia will produce polyhydramnios because the fetus can't swallow the amniotic fluid.

NP: Data collection; CN: Physiological integrity; CNS: Physiological adaptation; CL: Analysis

71. The neonate is vulnerable to heat loss because of which of the following anatomic characteristics?
1. Immature liver
2. Immature brain
3. Large skin surface area to body weight ratio
4. More brown fat (adipose tissue) than an adult

71. 3. The neonate has proportionally more surface area through which heat can dissipate. The liver isn't involved in thermoregulation. Brown fat is used to produce body heat and doesn't predispose the neonate to heat loss. The part of the brain that regulates temperature is mature at birth.

NP: Data collection; CN: Physiological integrity; CNS: Reduction of risk potential; CL: Knowledge

72. Maintaining thermoregulation in the neonate is an important nursing intervention because cold stress in the neonate can lead to which of the following conditions?
1. Anemia
2. Hyperglycemia
3. Metabolic alkalosis
4. Increased oxygen consumption

Success! We're super-de-dooper proud of you!

72. 4. The neonate's metabolic rate increases as a result of cold stress, which leads to an increased oxygen requirement. Cold stress doesn't increase erythrocyte destruction. Cold stress leads to anaerobic glycolysis, which results in metabolic acidosis. The increased metabolic rate leads to the use of glycogen stores and produces hypoglycemia.

NP: Planning; CN: Physiological integrity; CNS: Reduction of risk potential; CL: Analysis

NP: Nursing process   CN: Client needs category   CNS: Client needs subcategory   CL: Cognitive level

Part V Care of the child

26	Growth & development	451
27	Cardiovascular disorders	453
28	Hematologic & immune disorders	483
29	Respiratory disorders	509
30	Neurosensory disorders	539
31	Musculoskeletal disorders	559
32	Gastrointestinal disorders	579
33	Endocrine disorders	603
34	Genitourinary disorders	627
35	Integumentary disorders	653

Here's a short but important chapter that covers the growth and development of children. Enjoy!

Chapter 26
Growth & development

1. A client tells the nurse that her 22-month-old child says "no" to everything. When scolded, the toddler becomes angry and starts crying loudly but then immediately wants to be held. What is the best interpretation of this behavior?

1. The toddler isn't effectively coping with the stress.
2. The toddler's need for affection isn't being met.
3. This is normal behavior for a 2-year-old child.
4. This behavior suggests the need for counseling.

2. The mother of a 12-month-old infant expresses concern about the effect of frequent thumb sucking on her child's teeth. After the nurse teaches her about this matter, which response by the client indicates that the teaching has been effective?

1. "Thumb sucking should be discouraged at 12 months."
2. "I'll give the baby a pacifier instead."
3. "Sucking is important to the baby."
4. "I'll wrap the thumb in a bandage."

3. An adolescent client has just had surgery and has a dressing on the abdomen. Which of the following questions would the nurse expect the client to ask?

1. "Did the surgery go OK?"
2. "Will I have a large scar?"
3. "What complications can I expect?"
4. "When can I return to school?"

Hint! Hint! Hint!

Ditto! Ditto! Ditto!

1. 3. Toddlers are confronted with the conflict of achieving autonomy yet relinquishing the much-enjoyed dependence on—and affection of—others. As a result, their negativism is a necessary part of their growth and development. Nothing about this behavior indicates that the child is under stress, isn't receiving sufficient affection, or requires counseling.
NP: Data collection; CN: Health promotion and maintenance; CNS: Growth and development through the life span; CL: Comprehension

2. 3. Sucking is the infant's chief pleasure. However, thumb sucking can cause malocclusion if it persists after age 4. Many fetuses begin sucking their fingers in utero and, as infants, refuse a pacifier as a substitute. A young child is likely to chew on a bandage, which could lead to airway obstruction
NP: Evaluation; CN: Physiological integrity; CNS: Basic care and comfort; CL: Analysis

3. 2. Adolescents are deeply concerned about their body image and how they appear to others. An adolescent wouldn't ask how the surgery went or what complications to expect, although an adult probably would. Although an adolescent may be curious as to when he can return to school, it probably wouldn't be his primary concern.
NP: Planning; CN: Psychosocial integrity; CNS: Psychosocial adaptation; CL: Application

NP: Nursing process   CN: Client needs category   CNS: Client needs subcategory   CL: Cognitive level

4. For an 8-month-old infant, the nurse should plan to provide which of the following toys to promote the child's cognitive development?
1. Finger paints
2. Jack-in-the-box
3. Small rubber ball
4. Play gym strung across the crib

5. Before a routine checkup in the pediatrician's office, an 8-month-old infant is sitting contentedly on his mother's lap, chewing on a toy. When preparing to examine this infant, which of the following actions should the nurse do <u>first</u>?
1. Measure the head circumference.
2. Auscultate heart and lung sounds.
3. Elicit pupillary reaction.
4. Obtain body weight.

6. Which comment by a 7-year-old boy to his friend best typifies his developmental stage?
1. "Girls are so yucky."
2. "My mommy and I are always together."
3. "I can't decide if I like Amy or Heather better."
4. "I can turn into Batman when I come out of my closet."

7. The nurse should expect a 3-year-old child to be able to perform which of the following actions?
1. Ride a tricycle
2. Tie shoelaces
3. Roller skate
4. Jump rope

Prioritize!

Super job! You finished this chapter in record time! Way to go!

4. 2. According to Piaget's theory of cognitive development, an 8-month-old child will look for an object after it disappears from sight to develop the cognitive skill of object permanence. Finger paints and small balls are inappropriate because infants frequently put their fingers or objects into their mouth. Anything strung across an infant's crib is a safety hazard, especially to a child who may use it to pull to a standing position.
NP: Planning; CN: Health promotion and maintenance; CNS: Growth and development through the life span; CL: Knowledge

5. 2. Heart and lung auscultation shouldn't distress the infant, so it should be done early in the assessment. Placing a tape measure on the infant's head, shining a light in his eyes, or undressing the infant before weighing may cause distress, making the rest of the examination more difficult.
NP: Planning; CN: Health promotion and maintenance; CNS: Growth and development through life span; CL: Comprehension

6. 1. During the school-age years, the most important social interactions typically are those with peers. Peer-to-peer interactions lead to the formation of intimate friendships between same-sex children. Friendships with opposite-sex children are uncommon. At this age, children socialize more frequently with friends than with parents. Interest in peers of the opposite sex generally doesn't begin until ages 10 to 12. Magical thinking and fantasy play are more characteristic during the preschool years.
NP: Data collection; CN: Health promotion and maintenance; CNS: Growth and development through the life span; CL: Knowledge

7. 1. At age 3, gross motor development and refinement in hand-eye coordination enable a child to ride a tricycle. The fine motor skills required to tie shoelaces and the gross motor skills required for roller-skating and jumping rope develop around age 5.
NP: Data collection; CN: Health promotion and maintenance; CNS: Growth and development through the life span; CL: Knowledge

NP: Nursing process   CN: Client needs category   CNS: Client needs subcategory   CL: Cognitive level

Ah, there you are. You've reached our test on cardiovascular disorders in children. Before taking this comprehensive test, why not bolster yourself with a heart-healthy snack of celery and low-fat cream cheese? Yum!

Chapter 27
Cardiovascular disorders

1. A nurse is auscultating heart sounds on a 2-year-old child. The first heart sound can best be heard at which of the following locations?
1. Third or fourth intercostal space
2. The apex with the stethoscope bell
3. Second intercostal space, midclavicular line
4. Fifth intercostal space, left midclavicular line

2. When auscultating the heart, which of the following characteristics or statements best describes the first heart sound?
1. Heard late in diastole
2. Heard early in diastole
3. Closure of the mitral and tricuspid valves
4. Closure of the aortic and pulmonic valves

3. Which of the following characteristics best describes a grade 1 heart murmur?
1. Equal to the heart sounds
2. Softer than the heart sounds
3. Can be heard with the naked ear
4. Associated with a precordial thrill

You might have a lot of questions ahead of you, but I know you can do it!

1. 4. The first heart sound can best be heard at the fifth intercostal space, left midclavicular line. The second heart sound is heard at the second intercostal space. The third heart sound is heard with the stethoscope bell at the apex of the heart. The fourth heart sound can be heard at the third or fourth intercostal space.
NP: Data collection; CN: Health promotion and maintenance; CNS: Prevention and early detection of disease; CL: Knowledge

2. 3. The first heart sound occurs during systole with closure of the mitral and tricuspid valves. The second heart sound occurs during diastole with closure of the aortic and pulmonic valves. The third heart sound is heard early in diastole. The fourth heart sound is heard late in diastole and may be a normal finding in children.
NP: Data collection; CN: Health promotion and maintenance; CNS: Prevention and early detection of disease; CL: Knowledge

3. 2. A grade 1 heart murmur is usually difficult to hear and softer than the heart sounds. A grade 2 murmur is usually equal to the heart sounds. A grade 4 murmur can be associated with a precordial thrill. A thrill is a palpable manifestation associated with a loud murmur. A grade 6 murmur can be heard with the naked ear or with the stethoscope off the chest.
NP: Data collection; CN: Health promotion and maintenance; CNS: Prevention and early detection of disease; CL: Knowledge

NP: Nursing process   CN: Client needs category   CNS: Client needs subcategory   CL: Cognitive level

4. Which of the following phrases best defines the term *stroke volume?*
1. Volume of blood returning to the heart
2. Ability of the cardiac muscle to act as an efficient pump
3. Resistance against which the ventricles pump when ejecting blood
4. Amount of blood ejected by the heart in any one contraction

5. Which of the following statements best defines the term *cardiogenic shock?*
1. Decreased cardiac output
2. A reduction in circulating blood volume
3. Overwhelming sepsis and circulating bacterial toxins
4. Inflow or outflow obstruction of the main bloodstream

6. Which of the following signs is considered a late sign of shock in children?
1. Tachycardia
2. Hypotension
3. Delayed capillary refill
4. Pale, cool, mottled skin

7. Which of the following factors indicating a cardiac defect might be found when observing a 1-month-old infant?
1. Weight gain
2. Hyperactivity
3. Poor nutritional intake
4. Pink mucous membranes

8. A 2-year-old child is showing signs of shock. A 10-mg/kg bolus of normal saline solution is ordered. The child weighs 20 kg. How many milliliters should be administered?
1. 20 ml
2. 100 ml
3. 200 ml
4. 2,000 ml

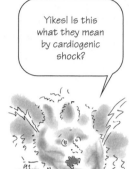

Yikes! Is this what they mean by cardiogenic shock?

Here's a hint for ya: late.

Yes, this is a math question. Stay cool, and you'll do great!

4. 4. Stroke volume is the amount of blood ejected by the heart in any one contraction. It's influenced by preload, afterload, and contractility. Preload is the amount of blood returning to the heart. Afterload is the resistance against which the ventricles pump when ejecting blood. Contractility is the ability of the cardiac muscle to act as an efficient pump.
NP: Evaluation; CN: Physiological integrity; CNS: Physiological adaptation; CL: Knowledge

5. 1. Cardiogenic shock occurs when cardiac output is decreased and tissue oxygen needs aren't adequately met. Hypovolemic shock describes a reduction in circulating blood volume. Septic shock describes overwhelming sepsis and circulating bacterial toxins. Obstructive shock is seen with an inflow or outflow obstruction of the main bloodstream.
NP: Evaluation; CN: Physiological integrity; CNS: Physiological adaptation; CL: Knowledge

6. 2. Hypotension is considered a late sign of shock in children. This represents a decompensated state and impending cardiopulmonary arrest. Tachycardia; pale, cool, mottled skin; and delayed capillary refill are earlier indicators of shock that may show compensation.
NP: Data collection; CN: Physiological integrity; CNS: Physiological adaptation; CL: Analysis

7. 3. Children with heart defects tend to have poor nutritional intake and weight loss, indicating poor cardiac output or hypoxemia. Pink, moist mucous membranes are normal. Gray, pale, or mottled skin may indicate hypoxia or poor cardiac output. Hypoxemia causes lethargy and fatigue, not hyperactivity.
NP: Data collection; CN: Health promotion and maintenance; CNS: Prevention and early detection of disease; CL: Analysis

8. 3. The correct formula for this calculation is 10 ml/kg × 20 kg. The correct answer is 200 ml. The other options are incorrect.
NP: Implementation; CN: Physiological integrity; CNS: Pharmacological therapies; CL: Analysis

---

NP: Nursing process   CN: Client needs category   CNS: Client needs subcategory   CL: Cognitive level

9. Which of the following arrhythmias is commonly found in neonates and infants?
1. Atrial fibrillation
2. Bradyarrhythmias
3. Premature atrial contractions
4. Premature ventricular contractions

Pick an arrhythmia, any correct arrhythmia.

9. 3. Premature atrial contractions are common in fetuses, neonates, and infants. They occur from increased automaticity of an atrial cell anywhere except the sinoatrial node. Premature ventricular contractions are more common in adolescents. Bradyarrhythmias are usually congenital, surgically acquired, or caused by infection. Atrial fibrillation is an uncommon arrhythmia in children occurring from a disorganized state of electrical activity in the atria.

NP: Evaluation; CN: Physiological integrity; CNS: Physiological adaptation; CL: Knowledge

10. Which part of an electrocardiogram reflects ventricular depolarization and contraction?
1. P wave
2. PR interval
3. QRS complex
4. T wave

Noninvasive is a clue. Narrow this question down by eliminating the invasive evaluations, and then choose the best answer from the remaining options.

10. 3. The QRS complex reflects ventricular depolarization and contraction. The P wave represents atrial depolarization and contraction. The PR interval represents the time it takes an impulse to trace from the atrioventricular node to the bundle of His. The T wave represents repolarization of the ventricles.

NP: Evaluation; CN: Physiological integrity; CNS: Physiological adaptation; CL: Knowledge

11. Which of the following evaluations of cardiovascular status is noninvasive?
1. Echocardiogram
2. Cardiac enzyme levels
3. Cardiac catheterization
4. Transesophageal pacing

11. 1. An echocardiogram is a noninvasive procedure to visualize the anatomy of the heart. Blood testing determines cardiac enzyme levels. Transesophageal pacing requires a probe to be placed in the esophagus for high-frequency ultrasound. Cardiac catheterization involves passing a catheter into the chambers of the heart for direct visualization of the heart and great vessels.

NP: Evaluation; CN: Physiological integrity; CNS: Physiological adaptation; CL: Analysis

12. A boy with a patent ductus arteriosus was delivered 6 hours earlier and is being held by his mother. As the nurse enters the room to assess the neonate's vital signs, the mother says, "The physician says that my baby has a heart murmur. Does that mean he has a bad heart?" Which response by the nurse would be most appropriate?
1. "He'll need more tests to determine his heart condition."
2. "He'll require oxygen therapy at home for awhile."
3. "He'll be fine. Don't worry about him."
4. "The murmur is caused by the natural opening, which can take a day or two to close. It's a normal part of your baby's transition."

12. 4. Although the nurse may want to tell the client not to worry, the most appropriate response would be to explain the neonate's present condition, to relieve her, and to acknowledge an awareness of the condition. A neonate's vascular system changes with birth; certain factors help to reverse the flow of blood through the ductus and ultimately favor its closure. This closure typically begins within the first 24 hours after birth and ends within a few days after birth. The other responses don't adequately address the client's question.

NP: Implementation; CN: Health promotion and maintenance; CNS: Growth and development through the life span; CL: Analysis

13. Before a cardiac catheterization, which of the following interventions is <u>most appropriate</u> for a child and his parents?
1. Supplying a map of the hospital
2. Limiting visitors to parents only
3. Offering a guided tour of the hospital and catheterization laboratory
4. Explaining that the child can't eat or drink for 1 to 2 days postoperatively

Your hint for question 13 is the phrase *most appropriate.*

13. 3. A guided tour will help minimize fears and allay anxieties for the child and parents. It gives the opportunity for questions and teaching. A map of the hospital is helpful, but a tour provides the family with more information. The child will be able to start clear liquids and advance as tolerated after the procedure is completed and the child is fully awake. Visitors should include all significant others and siblings as part of the preoperative teaching.
NP: Planning; CN: Physiological integrity; CNS: Physiological adaptation; CL: Analysis

14. Which of the following statements about cardiac catheterization is correct?
1. It's a noninvasive procedure.
2. General anesthesia is required.
3. High-frequency sound waves produce an image of the heart in motion.
4. It provides visualization of the heart and great vessels with radiopaque dye.

14. 4. Cardiac catheterization provides visualization of the heart and great vessels. High-frequency sound waves describe ultrasound and echocardiography. Conscious sedation is usually given before cardiac catheterization. General anesthesia may be used for more complex catheterizations or procedures that place the child at greater risk. It's an invasive procedure in which a thin catheter is passed into the chambers of the heart through a peripheral vein or artery.
NP: Planning; CN: Physiological integrity; CNS: Physiological adaptation; CL: Knowledge

15. Which of the following nursing interventions is <u>most appropriate</u> when caring for a child in the immediate postcatherization phase?
1. Elevate the head of the bed 45 degrees.
2. Encourage the child to remain flat.
3. Assess vital signs every 2 to 4 hours.
4. Replace a bloody groin dressing with a new dressing.

You've already answered 15 questions. See how time flies when you're taking a test?

15. 2. During recovery, the child should remain flat in bed, keeping the punctured leg straight for the prescribed time. The child should avoid raising the head, sitting, straining the abdomen, or coughing. Vital signs are taken every 15 minutes until the child is awake and stable, then every half hour, then hourly as ordered. If bleeding occurs at the insertion site, the nurse should reinforce the dressing and monitor for changes.
NP: Implementation; CN: Physiological integrity; CNS: Physiological adaptation; CL: Analysis

16. Which of the following home care instructions is included for a child postcatheterization?
1. Encourage fluids and regular diet.
2. Encourage physical activities.
3. The child can routinely bathe after returning home.
4. The child may return to school the next day.

16. 1. A regular diet and increased fluids are encouraged postcatheterization. Increased fluids may flush the injected dyes out of the system. Prolonged bathing can be resumed in about 2 days. A sponge bath is encouraged until then. The child may return to school in 2 to 3 days after discharge. Normal activities may be resumed, but physical activities or sports should be avoided for about 3 days.
NP: Planning; CN: Physiological integrity; CNS: Physiological adaptation; CL: Analysis

---

NP: Nursing process   CN: Client needs category   CNS: Client needs subcategory   CL: Cognitive level

17. A 2-year-old child is being monitored after cardiac surgery. Which of the following signs represents a decrease in cardiac output?
1. Hypertension
2. Increased urine output
3. Weak peripheral pulses
4. Capillary refill less than 2 seconds

18. A 3-year-old child is experiencing distress after having cardiac surgery. Which of the following signs indicates cardiac tamponade?
1. Hypertension
2. Muffled heart sounds
3. Widened pulse pressures
4. Increased chest tube drainage

19. A nurse is monitoring fluid and electrolyte balance in a child postoperatively. Which of the following findings is expected?
1. Increased urine output
2. Increased sodium level
3. Decreased sodium level
4. Increased potassium level

20. A nurse is teaching wound care to parents after cardiac surgery. Which of the following statements is most appropriate?
1. Lotions and powders are acceptable.
2. Your child can take a bath tomorrow.
3. Tingling, itching, and numbness are normal sensations at the wound site.
4. If the sterile adhesive strips over the incision fall off, call the physician.

21. Parents ask a nurse about a child's activity level after cardiac surgery. Which of the following responses would be best?
1. There are no exercise limitations.
2. The child may resume school in 3 days.
3. Encourage a balance of rest and exercise.
4. Climbing and contact sports are restricted for 1 week.

All of these signs may appear, but only one indicates cardiac tamponade. Which one?

Your hint? Why, the words most appropriate, of course.

17. 3. Signs of decreased cardiac output include weak peripheral pulses, low urine output, delayed capillary refill, hypotension, and cool extremities. Poor cardiac function leads to a decrease in cardiac output, so vasoactive and inotropic agents are needed to maintain adequate cardiac output.
NP: Data collection; CN: Physiological integrity; CNS: Physiological adaptation; CL: Analysis

18. 2. Symptoms of cardiac tamponade include muffled heart sounds, hypotension, sudden cessation of chest tube drainage, and a narrowing pulse pressure. Cardiac tamponade occurs when a large volume of fluid interferes with ventricular filling and pumping and collects in the pericardial sac, decreasing cardiac output.
NP: Data collection; CN: Physiological integrity; CNS: Physiological adaptation; CL: Analysis

19. 2. In response to surgery and cardiopulmonary bypass, the body secretes aldosterone and antidiuretic hormone. This in turn increases sodium levels, decreases potassium levels, and increases water retention.
NP: Data collection; CN: Physiological integrity; CNS: Physiological adaptation; CL: Analysis

20. 3. As the area heals, tingling, itching, and numbness are normal sensations and will eventually go away. A complete bath should be delayed for the first week. Lotions and powders should be avoided during the first 2 weeks after surgery. Steri-Strips may loosen or fall off on their own. This is a common and normal occurrence.
NP: Evaluation; CN: Physiological integrity; CNS: Physiological adaptation; CL: Analysis

21. 3. Activity should be increased gradually each day, allowing for a sensible balance of rest and exercise. School and large crowds should be avoided for at least 2 weeks to prevent exposure to people with active infections. Sports and contact activities should be restricted for about 6 weeks, giving the sternum enough time to heal.
NP: Evaluation; CN: Physiological integrity; CNS: Physiological adaptation; CL: Analysis

22. Which of the following home care instructions is most appropriate for a child after cardiac surgery?
1. Don't stop giving the child the prescribed drugs until the physician says so.
2. Maintain a sodium-restricted diet.
3. Routine dental care can be resumed.
4. Immunizations are delayed indefinitely.

23. A 4-year-old client with a chest tube is placed on water seal. Which of the following statements is correct?
1. The water level rises with inhalation.
2. Bubbling is seen in the suction chamber.
3. Bubbling is seen in the water seal chamber.
4. Water seal is obtained by clamping the tube.

24. Which of the following interventions is most appropriate when a chest tube falls out or becomes dislodged?
1. Place a dry gauze dressing.
2. Place a petroleum gauze dressing.
3. Wipe the tube with alcohol and reinsert it.
4. Call the physician immediately.

25. When observing a child with heart failure, which of the following findings would the nurse expect to find?
1. Bradycardia
2. Weight gain
3. Gallop murmur
4. Strong, bounding pulses

OK. Let's talk about what you need to know when you take Timmy home from the hospital.

Question 25 is a milestone. Way to go!

22. 1. Drugs, such as digoxin and furosemide, shouldn't be stopped abruptly. Immunizations may be delayed 6 to 8 weeks after surgery. Routine dental care is usually delayed 4 to 5 months after surgery. There are no diet restrictions. The child may resume his regular diet.
NP: Evaluation; CN: Physiological integrity; CNS: Physiological adaptation; CL: Analysis

23. 1. The water seal chamber is functioning appropriately when the water level rises in the chamber with inhalation and falls with expiration. This shows that negative pressure required in the lung is being maintained. Bubbling in the water seal chamber generally indicates the presence of an air leak. Bubbling in the suction chamber should only be seen when suction is being used. The chest tube should never be clamped; a tension pneumothorax may occur. Water seal is activated when the suction is disconnected.
NP: Data collection; CN: Physiological integrity; CNS: Physiological adaptation; CL: Application

24. 2. Petroleum gauze should be placed immediately to prevent a pneumothorax. The physician should be notified after this step. The tube is only reinserted by a physician using a sterile thoracotomy tray. A dry gauze dressing will allow air to escape, leading to a pneumothorax.
NP: Planning; CN: Physiological integrity; CNS: Physiological adaptation; CL: Application

25. 3. When the heart stretches beyond efficiency, an extra heart sound or S_3 gallop murmur may be audible. This is related to excessive preload and ventricular dilation. Tachycardia occurs as a compensatory mechanism to the decrease in cardiac output. It also attempts to increase the force and rate of myocardial contraction and increase oxygen consumption of the heart. Pulses are usually weak and thready. Children with heart failure tend to have difficulty feeding and tire easily. They're commonly diagnosed as failure to thrive and are in the lower percentiles of their growth charts for weight.
NP: Data collection; CN: Physiological integrity; CNS: Physiological adaptation; CL: Analysis

26. Which of the following symptoms is seen with pulmonary venous congestion or left-sided heart failure?
1. Weight gain
2. Peripheral edema
3. Neck vein distention
4. Tachypnea and dyspnea

What happens when my left side isn't working properly?

27. Which of the following interventions is most appropriate when caring for an infant with heart failure?
1. Limit fluid intake.
2. Avoid using infant seats.
3. Cluster nursing activities.
4. Place the infant prone or supine.

What's the diet plan?

28. Which of the following diet plans is recommended for an infant with heart failure?
1. Restrict fluids.
2. Weigh once a week.
3. Use low-sodium formula.
4. Increase caloric content per ounce.

29. Which of the following statements is true about giving oxygen to a client with heart failure?
1. Oxygen is contraindicated in this situation.
2. Oxygen is given at high levels only.
3. Oxygen is a pulmonary bed constrictor.
4. Oxygen decreases the work of breathing.

26. 4. Respiratory symptoms, such as tachypnea and dyspnea, are seen due to pulmonary congestion. Peripheral edema, neck vein distention, and weight gain are seen with systemic venous congestion or right-sided failure. Fluid accumulates in the interstitial spaces due to blood pooling in the venous circulation.
NP: Data collection; CN: Physiological integrity; CNS: Physiological adaptation; CL: Knowledge

27. 3. Energy expenditures need to be limited to reduce metabolic and oxygen needs. Nursing care should be clustered, followed by long periods of undisturbed rest. Infants should be placed in the semi-Fowler or upright position. Infant seats help maintain an upright position. This facilitates lung expansion, provides less restrictive movement of the diaphragm, relieves pressure from abdominal organs, and decreases pulmonary congestion. Fluid may be restricted in older children, but infants' nutritional requirements depend on fluid needs.
NP: Implementation; CN: Physiological integrity; CNS: Physiological adaptation; CL: Analysis

28. 4. Formulas with increased caloric content are given to meet the greater caloric requirements from the overworked heart and labored breathing. Daily weights at the same time of the day on the same scale before feedings are recommended to follow trends in nutritional stability and diuresis. Fluid restriction and low-sodium formulas aren't recommended. An infant's nutritional needs depend on fluid. Low-sodium formulas may cause hyponatremia.
NP: Implementation; CN: Physiological integrity; CNS: Physiological adaptation; CL: Application

29. 4. Oxygen decreases the work of breathing and increases arterial oxygen levels. Oxygen is a pulmonary bed dilator, not constrictor, and can exacerbate the condition in which the lungs are overloaded. Oxygen usually is administered at low levels with humidification.
NP: Implementation; CN: Physiological integrity; CNS: Physiological adaptation; CL: Analysis

30. A teenage client with heart failure is prescribed digoxin. To which of the following classifications does this drug belong?
1. Angiotensin-converting enzyme (ACE) inhibitor
2. Cardiac glycoside
3. Diuretic
4. Vasodilator

31. Which of the following toxic adverse reactions can be seen in a child taking digoxin?
1. Weight gain
2. Tachycardia
3. Nausea and vomiting
4. Purple tint around objects or halos

32. An 11-month-old infant with heart failure weighs 10 kg. Digoxin is prescribed as 0.01 mg/kg in divided doses every 12 hours. How much is given per dose?
1. 0.001 mg/dose
2. 0.05 mg/dose
3. 0.1 mg/dose
4. 0.5 mg/dose

33. A client with heart failure is given captopril (Capoten), an angiotensin-converting enzyme (ACE) inhibitor. Which of the following actions occurs with this type of drug?
1. Vasoconstriction
2. Increased sodium excretion
3. Decreased sodium excretion
4. Increased vascular resistance

34. Which of the following statements about patent ductus arteriosus is correct?
1. Heart failure is uncommon.
2. The ductus normally closes completely by 6 weeks of age.
3. An open ductus arteriosus causes decreased blood flow to the lungs.
4. It represents a cyanotic defect with decreased pulmonary blood flow.

Knowing the classification of a drug can help you remember its actions.

How are you figuring that dosage?

30. 2. Digoxin is a cardiac glycoside. It decreases the workload of the heart and improves myocardial function. Diuretics help remove excess fluid. Vasodilators enhance cardiac output by decreasing afterload. ACE inhibitors cause vasodilation and increase sodium excretion.
NP: Evaluation; CN: Physiological integrity; CNS: Pharmacological therapies; CL: Knowledge

31. 3. Digoxin toxicity in infants and children may present with nausea, vomiting, anorexia, or a slow, irregular apical heart rate. Visual disturbances are described as a green or yellow tint to objects or green or yellow halos around objects or bright lights.
NP: Evaluation; CN: Physiological integrity; CNS: Pharmacological therapies; CL: Analysis

32. 2. 10 kg × 0.01 mg/kg = 0.1 mg
24 hours/12 hours/dose = 2 doses
0.1 mg/2 doses = 0.05 mg/dose
NP: Implementation; CN: Physiological integrity; CNS: Pharmacological therapies; CL: Analysis

33. 2. ACE inhibitors block the conversion of angiotensin I to angiotensin II in the kidney. This causes decreased aldosterone, vasodilatation, and increased sodium excretion. As a vasodilator, it also acts to reduce vascular resistance by the manipulation of afterload.
NP: Evaluation; CN: Physiological integrity; CNS: Pharmacological therapies; CL: Knowledge

34. 2. At birth, oxygenated blood normally causes the ductus to constrict, and the vessel closes completely by 6 weeks of age. This defect is considered an acyanotic defect with increased pulmonary blood flow. The open ductus arteriosus can cause an excessive blood flow to the lungs because of the high pressure in the aorta. Heart failure is common in premature infants with a patent ductus arteriosus.
NP: Data collection; CN: Physiological integrity; CNS: Physiological adaptation; CL: Knowledge

35. During observation of a child who has undergone cardiac catheterization, the nurse notes bleeding from the percutaneous femoral catheterization site. Which action should be taken first?
1. Apply direct, continuous pressure.
2. Assess the pulse and blood pressure.
3. Seek the assistance of a registered nurse.
4. Check the pedal pulse in the affected leg.

36. Which intervention would be appropriate for an infant after cardiac catheterization?
1. Keep the leg on the operative site flexed to reduce bleeding.
2. Change the catheterization dressing immediately to reduce the risk of infection.
3. Apply pressure if oozing or bleeding is noted.
4. Keep the infant's temperature below normal to promote vasoconstriction and decrease bleeding.

37. Which of the following cardiovascular disorders is considered acyanotic?
1. Patent ductus arteriosus
2. Tetralogy of Fallot
3. Tricuspid atresia
4. Truncus arteriosus

38. Which of the following findings is <u>not</u> expected during an observation of a child with an acyanotic heart defect?
1. Weight gain
2. Bradycardia
3. Hepatomegaly
4. Decreased respiratory rate

Pay attention to question 35. All of these actions might be necessary, but which should come first?

Don't be fooled by question 38. The word *not* asks what is *not* expected. Be wary of questions phrased as a negative.

35. 1. Bleeding from a major vessel must be stopped immediately to prevent massive hemorrhage. Vital signs would be taken after bleeding control measures are instituted. Calling for help is important, but pressure on the site must be applied and maintained while help is found. Pedal pulses would be checked after bleeding is controlled.
NP: Implementation; CN: Physiological integrity; CNS: Reduction of risk potential; CL: Application

36. 3. Applying pressure to the site is appropriate if bleeding is noted. The leg should be kept straight and immobile to prevent trauma and bleeding. The pressure dressing shouldn't be changed, but it may be reinforced if bleeding occurs. Hypothermia causes stress in infants and should be avoided.
NP: Implementation; CN: Physiological integrity; CNS: Reduction of risk potential; CL: Application

37. 1. A patent ductus arteriosus represents an acyanotic heart defect with increased pulmonary blood flow. Tricuspid atresia and tetralogy of Fallot are cyanotic defects with decreased pulmonary blood flow. Truncus arteriosus is a cyanotic defect with increased pulmonary blood flow.
NP: Data collection; CN: Physiological integrity; CNS: Physiological adaptation; CL: Knowledge

38. 3. Hepatomegaly may result from blood backing up into the liver due to difficulty entering the right side of the heart. The increase in blood flow to the lungs may cause tachycardia and increased respiratory rates to compensate. Poor growth and development may be seen due to the increased energy required for breathing.
NP: Data collection; CN: Physiological integrity; CNS: Physiological adaptation; CL: Analysis

39. A pediatric client is scheduled for echocardiography. The nurse is providing teaching to the client's mother. Which statement about echocardiography indicates the need for further teaching?

1. "I'm glad my child won't have an I.V. catheter inserted for this procedure."
2. "I'm glad my child won't need to have dye injected before the procedure."
3. "How am I ever going to explain to my son that he can't have anything to eat before the test?"
4. "I know my child may need to lie on his left side and breathe in and out slowly during the procedure."

40. Which of the following signs may be seen in a child with ventricular septal defect?

1. Clubbing of the fingers
2. Above average height on growth chart
3. Above average weight gain on growth chart
4. Pink nailbeds with capillary refill less than 2 seconds

41. Which of the following conditions best describes ventricular septal defect?

1. Narrowing of the aortic arch
2. Failure of a septum to develop completely between the atria
3. Narrowing of the valves at the entrance of the pulmonary artery
4. Failure of a septum to develop completely between the ventricles

42. A 6-month-old infant with uncorrected tetralogy of Fallot suddenly becomes increasingly cyanotic and diaphoretic, with weak peripheral pulses and an increased respiratory rate. What should the nurse do immediately?

1. Administer oxygen.
2. Administer morphine sulfate.
3. Place the infant in a knee-chest position
4. Place the infant in Fowler's position.

You can see these signs without a magnifying glass!

Question 42 already? Wow! You're really moving!

39. 3. Echocardiography is a noninvasive procedure used to evaluate the size, shape, and motion of various cardiac structures. Therefore, it isn't necessary for the client to have an I.V. catheter, dye injected, or restrictions such as nothing by mouth, as would be the case with a cardiac catheterization. The child may need to lie on his left side and inhale and exhale slowly during the procedure.

NP: Evaluation; CN: Physiological integrity; CNS: Reduction of risk potential; CL: Analysis

40. 1. Clubbing and cyanotic nailbeds can be seen when shunting shifts right to left. These children usually show symptoms of heart failure, poor growth and development, and failure to thrive.

NP: Data collection; CN: Physiological integrity; CNS: Physiological adaptation; CL: Analysis

41. 4. Failure of a septum to develop between the ventricles results in a left-to-right shunt, which is noted as a ventricular septal defect. When the septum fails to develop between the atria, it's considered an atrial septal defect. The narrowing of the aortic arch describes coarctation of the aorta. Narrowing of the valves at the pulmonary artery describes pulmonary stenosis.

NP: Data collection; CN: Physiological integrity; CNS: Physiological adaptation; CL: Knowledge

42. 3. The knee-chest position reduces the workload of the heart by increasing the blood return to the heart and keeping the blood flow more centralized. Oxygen should be administered quickly but only after placing the infant in the knee-chest position. Morphine should be administered after positioning and oxygen administration are completed. Fowler's position wouldn't improve the situation.

NP: Implementation; CN: Physiological integrity; CNS: Physiological adaptation; CL: Application

NP: Nursing process   CN: Client needs category   CNS: Client needs subcategory   CL: Cognitive level

43. A client with a ventricular septal repair is receiving dopamine (Intropin) postoperatively. Which of the following responses is expected?
1. Decreased heart rate
2. Decreased urine output
3. Increased cardiac output
4. Decreased cardiac contractility

44. A pediatric client returns to his room after a cardiac catheterization. Which of the following nursing interventions is most appropriate?
1. Allow the client to sit in a chair with the affected extremity immobilized.
2. Maintain the client on bed rest with no further activity restrictions.
3. Maintain the client on bed rest with the affected extremity immobilized.
4. Allow the client to get out of bed to go to the bathroom, if necessary.

45. A nanny is taught to administer digoxin (Lanoxin) to a 6-month-old infant at home. Which statement by the nanny indicates the need for additional teaching?
1. "I'll count the baby's pulse before every dose."
2. "I'll make sure the pulse is regular before every dose."
3. "I'll measure the dose carefully."
4. "I'll withhold the medication if the pulse is below 60."

46. Which finding would concern the nurse who is caring for an infant after a right femoral cardiac catheterization?
1. Weak right dorsalis pedis pulse
2. Elevated temperature
3. Decreased urine output
4. Slight bloody drainage around catheterization site dressing

You know more about drugs than you realize!

Don't forget that this is a negative question!

43. 3. Dopamine stimulates beta$_1$ and beta$_2$ receptors. It's a selective cardiac stimulant that will increase cardiac output, heart rate, and cardiac contractility. Urine output increases in response to dilation of the blood vessels to the mesentery and kidneys.
NP: Evaluation; CN: Physiological integrity; CNS: Pharmacological therapies; CL: Knowledge

44. 3. The pediatric client should be maintained on bed rest with the affected extremity immobilized after cardiac catheterization to prevent hemorrhage. Allowing the client to sit in a chair with the affected extremity immobilized, moving the affected extremity while on bed rest, or allowing the client bathroom privileges places the client at risk for hemorrhage.
NP: Implementation; CN: Physiological integrity; CNS: Reduction of risk potential; CL: Knowledge

45. 4. A pulse rate under 60 beats/minute is an indication for withholding digoxin from an adult. Withholding digoxin from an infant is appropriate if the infant's pulse is under 90 beats/minute. The pulse rate must be counted before each dose of digoxin is given to an infant. An irregular pulse may be a sign of digoxin toxicity; if this occurs, the physician should be consulted before the drug is given. The dose must be measured carefully to decrease the risk of toxicity.
NP: Evaluation; CN: Physiological integrity; CNS: Pharmacological therapies; CL: Application

46. 1. The pulse below the catheterization site should be strong and equal to the unaffected extremity. A weakened pulse may indicate vessel obstruction or perfusion problems. Elevated temperature and decreased urine output are relatively normal findings after catheterization and may be the result of decreased oral fluids. A small amount of bloody drainage is normal; however, the site must be assessed frequently for increased bleeding.
NP: Data collection; CN: Physiological integrity; CNS: Reduction of risk potential; CL: Application

47. Which of the following cardiac anomalies produces a left-to-right shunt?
1. Atrial septal defect
2. Pulmonic stenosis
3. Tetralogy of Fallot
4. Total anomalous pulmonary venous return

My shunt is left to right. How about yours?

48. A child with an atrial septal repair is entering postoperative day 3. Which of the following interventions would be most appropriate?
1. Give the child nothing by mouth.
2. Maintain strict bed rest.
3. Take vital signs every 8 hours.
4. Administer an analgesic as needed.

All of the interventions in question 48 may be appropriate at some point, but which is most appropriate for this situation?

49. A 3-year-old child is on postoperative day 5 for an atrial septal repair. Which of the following nursing diagnoses would be most appropriate?
1. Activity intolerance
2. Chronic pain
3. Social isolation
4. Risk for imbalanced fluid volume

50. Which of the following conditions best describes coarctation of the aorta?
1. Absent tricuspid valve
2. Narrowing in the area of the aortic valve
3. Localized constriction or narrowing of the aortic wall
4. Narrowing at some location along the right ventricular outflow tract

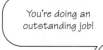

You're doing an outstanding job!

51. A client is diagnosed with coarctation of the aorta. Which finding would the nurse expect when observing this client?
1. Normal blood pressure
2. Increased blood pressure in the upper extremities
3. Decreased blood pressure in the upper extremities
4. Decreased or absent pulses in the upper extremities

47. 1. Atrial septal defects shunt from left to right because pressures are greater on the left side of the heart. Pulmonary stenosis, tetralogy of Fallot, and total anomalous pulmonary venous return will show a right-to-left shunting of blood.
NP: Evaluation; CN: Health promotion and maintenance; CNS: Prevention and early detection of disease; CL: Analysis

48. 4. Pain management is always a priority and should be given on an as-needed basis. By day 3, the child should be advancing to a regular diet. Vital signs should be performed routinely every 2 to 4 hours. Activity should be allowed as able in the step-down unit, with coughing and deep-breathing exercises.
NP: Planning; CN: Physiological integrity; CNS: Physiological adaptation; CL: Application

49. 4. Diuretics are still being used as needed at this point. By day 5, the child's activity level will be normal, and there will be little to no pain. The child isn't isolated from anyone and can usually attend the playroom on day 3 or 4.
NP: Evaluation; CN: Physiological integrity; CNS: Physiological adaptation; CL: Analysis

50. 3. Coarctation of the aorta consists of a localized constriction or narrowing of the aortic wall. Aortic stenosis is a narrowing in the area of the aortic valve. Tricuspid atresia is characterized by an absent tricuspid valve. Pulmonary stenosis consists of a narrowing along the right ventricular outflow tract.
NP: Data collection; CN: Health promotion and maintenance; CNS: Prevention and early detection of disease; CL: Knowledge

51. 2. As blood is pumped from the left ventricle to the aorta, some blood flows to the head and upper extremities while the rest meets obstruction and jets through the constricted area. Pressures and pulses are greater in the upper extremities. Decreased or absent pulses are found in the lower extremities.
NP: Data collection; CN: Physiological integrity; CNS: Physiological adaptation; CL: Analysis

52. Which of the following procedures is recommended to treat coarctation of the aorta?
1. Drug therapy
2. Balloon angioplasty
3. No surgical intervention required
4. Surgery with cardiopulmonary bypass

53. Which of the following factors is an important part of observing a child with a possible cardiac anomaly?
1. Heart rate
2. Temperature
3. Blood pressure
4. Blood pressure in four extremities

54. Which of the following interventions is recommended postoperatively for a client with repair of a coarctation of the aorta?
1. Give a vasoconstrictor.
2. Maintain hypothermia.
3. Maintain a low blood pressure.
4. Give a bolus of I.V. fluids.

55. Which of the following observations is expected in a child with tetralogy of Fallot?
1. Machinelike murmur
2. Eisenmenger's complex
3. Increasing cyanosis with crying or activity
4. Higher pressure in the upper extremities than in the lower extremities

56. A child with tetralogy of Fallot has clubbing of the fingers and toes, a finding related to which of the following conditions?
1. Polycythemia
2. Chronic hypoxia
3. Pansystolic murmur
4. Abnormal growth and development

You can use observation skills everywhere!

When it comes to observations, practice makes perfect!

52. 2. Balloon angioplasty in the catheterization laboratory is an option for many types of lesions. Surgery is usually performed without cardiopulmonary bypass. Drugs won't help this type of defect, and surgical intervention is almost always required.
NP: Planning; CN: Physiological integrity; CNS: Physiological adaptation; CL: Knowledge

53. 4. Measuring blood pressure in all four extremities is necessary to document hypertension and the blood pressure gradient between the upper and lower extremities. Temperature and heart rate are also important observations.
NP: Data collection; CN: Physiological integrity; CNS: Physiological adaptation; CL: Application

54. 3. Blood pressure is tightly managed and kept low so there's no excessive pressure on the fresh suture lines. Normal body temperature is maintained, and diuretics may be given to decrease fluid volume.
NP: Implementation; CN: Physiological integrity; CNS: Physiological adaptation; CL: Analysis

55. 3. A child with tetralogy of Fallot will be mildly cyanotic at rest and have increasing cyanosis with crying, activity, or straining, as with a bowel movement. A machinelike murmur is a characteristic of patent ductus arteriosus. Higher pressures in the upper extremities are characteristic of coarctation of the aorta. Eisenmenger's complex is a complication of ventricular resistance exceeding systemic pressure.
NP: Data collection; CN: Physiological integrity; CNS: Physiological adaptation; CL: Application

56. 2. Chronic hypoxia causes clubbing of the fingers and toes when untreated. Hypoxia varies with the degree of pulmonary stenosis. Growth and development may appear normal. Polycythemia is an increased number of red blood cells as a result of chronic hypoxemia. A pansystolic murmur is heard at the middle to lower left sternal border but has no impact on clubbing.
NP: Data collection; CN: Physiological integrity; CNS: Physiological adaptation; CL: Knowledge

57. A child with tetralogy of Fallot may assume which position of comfort during exercise?

    1. Prone
    2. Semi-Fowler
    3. Side-lying
    4. Squat

I bet you already know the answer!

57. 4. A child may squat or assume a knee-chest position to reduce venous blood flow from the lower extremities and to increase systemic vascular resistance, which diverts more blood flow into the pulmonary artery. Prone, semi-Fowler, and side-lying positions won't produce this effect.

NP: Evaluation; CN: Physiological integrity; CNS: Physiological adaptation; CL: Analysis

58. Which of the following statements is correct for children with tetralogy of Fallot?

    1. The condition is commonly referred to as "blue tets."
    2. They experience hypercyanotic, or "tet," spells.
    3. They experience frequent respiratory infections.
    4. They experience decreased or absent pulses in the lower extremities.

58. 2. Hypercyanotic, or "tet," spells may occur due to increasing obstruction of right ventricular outflow, resulting in decreased pulmonary blood flow and increased right-to-left shunting. Infants with mild obstruction to blood flow have little or no right-to-left shunting and appear pink, or "pink tets." Frequent respiratory infections are seen in defects with increased pulmonary blood flow such as a patent ductus arteriosus. Decreased or absent pulses in the lower extremities are a sign of coarctation of the aorta.

NP: Data collection; CN: Physiological integrity; CNS: Physiological adaptation; CL: Knowledge

59. Which of the following tests would show the direction and amount of shunting in a child with tetralogy of Fallot?

    1. Chest radiography
    2. Echocardiography
    3. Electrocardiography
    4. Cardiac catheterization

Which test is it?

59. 4. Cardiac catheterization provides specific information about the direction and amount of shunting, coronary anatomy, and each portion of the heart defect. Electrocardiograms show right ventricular hypertrophy with tall R waves. Echocardiogram scans define such defects as large ventricular septal defects, pulmonary stenosis, and malposition of the aorta. Chest radiographs will show right ventricular hypertrophy pushing the heart apex upward, resulting in a boot-shaped silhouette.

NP: Data collection; CN: Physiological integrity; CNS: Physiological adaptation; CL: Knowledge

60. A nurse is teaching parents about tricuspid atresia. Which of the following statements indicates the parents understand?

    1. "There's a narrowing at the aortic outflow tract."
    2. "The pulmonary veins don't return to the left atrium."
    3. "There's a narrowing at the entrance of the pulmonary artery."
    4. "There's no communication between the right atrium and the right ventricle."

60. 4. Tricuspid atresia is failure of the tricuspid valve to develop, leaving no communication between the right atrium and the right ventricle. Narrowing at the entrance of the pulmonary artery represents pulmonary stenosis. Narrowing at the aortic outflow tract is aortic stenosis. Total anomalous pulmonary venous return is a defect in which the pulmonary veins don't return to the left atrium but abnormally return to the right side of the heart.

NP: Evaluation; CN: Physiological integrity; CNS: Physiological adaptation; CL: Analysis

61. Which of the following characteristics can be noted when observing in a child with tricuspid atresia?
1. Cyanosis
2. Machinelike murmur
3. Decreased respiratory rate
4. Capillary refill more than 2 seconds

You don't need binoculars for this observation!

61. 1. Cyanosis is the most consistent clinical sign of tricuspid atresia. Tachypnea and dyspnea are typically present because of the pulmonary blood flow and right-to-left shunting. Decreased oxygenation would increase capillary refill time. Tricuspid atresia doesn't have a characteristic murmur. A machinelike murmur is characteristic of a patent ductus arteriosus.
NP: Data collection; CN: Physiological integrity; CNS: Physiological adaptation; CL: Analysis

62. Which of the following laboratory findings is expected in a child with tricuspid atresia?
1. Acidosis
2. Alkalosis
3. Normal red blood cell (RBC) count
4. Normal arterial oxygen saturation

62. 1. In tricuspid atresia, the tricuspid valve is completely closed so that no blood flows from the right atrium to the right ventricle. Therefore, no oxygenation of blood occurs. The child has chronic hypoxemia and acidosis, not alkalosis, due to decreased atrial oxygenation. This chronic hypoxemia leads to polycythemia, so a normal RBC wouldn't result. There's no normal arterial oxygenation as the blood bypasses the lungs and the step of oxygenation.
NP: Data collection; CN: Physiological integrity; CNS: Physiological adaptation; CL: Analysis

63. Which guideline should the nurse follow when administering digoxin (Lanoxin) to an infant?
1. Mix the digoxin with the infant's food.
2. Double the subsequent dose if a dose is missed.
3. Give the digoxin with antacids when possible.
4. Withhold the dose if the apical pulse rate is less than 90 beats/minute.

63. 4. Digoxin is used to decrease heart rate; however, the apical pulse must be carefully monitored to detect a severe reduction. Administering digoxin to an infant with a heart rate of under 90 beats/minute could further reduce the rate and compromise cardiac output. Mixing digoxin with other food may interfere with accurate dosing. Double-dosing should never be done. Antacids may decrease drug absorption.
NP: Implementation; CN: Physiological integrity; CNS: Pharmacological therapies; CL: Application

Read question 64 carefully to make sure you know what's being asked.

64. Which of the following conditions best describes total anomalous pulmonary venous return?
1. Pulmonary veins that don't return to the left atrium
2. A cyanotic defect with decreased pulmonary blood flow
3. An acyanotic defect with increased pulmonary blood flow
4. A single large vessel that arises from both ventricles astride a large ventricular septal defect

64. 1. Total anomalous pulmonary venous return is a condition in which the pulmonary veins don't return to the left atrium; instead, they abnormally return to the right side of the heart. This defect is classified as a cyanotic defect with increased pulmonary blood flow. A single large vessel arising from both ventricles astride a large ventricular septal defect describes truncus arteriosus.
NP: Data collection; CN: Health promotion and maintenance; CNS: Prevention and early detection of disease; CL: Knowledge

65. There are four different ways the pulmonary veins may be abnormally routed with a total anomalous pulmonary venous return defect. Which of the following types is the most common?

1. Cardiac
2. Infracardiac
3. Supracardiac
4. Mixed combination of supracardiac, cardiac, and infracardiac

Question 65 asks about what occurs most commonly.

65. 3. Supracardiac, the most common type, is characterized by the pulmonary veins draining directly into the superior vena cava. With the cardiac type, the pulmonary veins drain into the coronary sinuses or directly flow into the right atrium. The infracardiac type shows the common pulmonary vein running below the diaphragm into the portal system. The fourth type is a mixed combination of less common types.

NP: Evaluation; CN: Health promotion and maintenance; CNS: Prevention and early detection of disease; CL: Knowledge

66. Which of the following findings is common during an assessment of a child with a total anomalous pulmonary venous return defect?

1. Hypertension
2. Frequent respiratory infections
3. Normal growth and development
4. Above average weight gain on the growth chart

66. 2. Children with total anomalous pulmonary venous return defects are prone to repeated respiratory infections due to increased pulmonary blood flow. Poor feeding and failure to thrive are also signs. Infants look thin and malnourished. Hypertension usually occurs with coarctation of the aorta, an acyanotic defect with obstructive flow.

NP: Data collection; CN: Physiological integrity; CNS: Physiological adaptation; CL: Application

67. Which of the following complications may result after the repair of total anomalous pulmonary venous return?

1. Hypotension
2. Hypertension
3. Ventricular arrhythmias
4. Pulmonary vein dilatation

Watch the term after. It clues you in to the answer.

67. 2. Pulmonary hypertension, atrial arrhythmias, and pulmonary vein obstruction are complications that may result postoperatively. The left atrium is small and sensitive to fluid volume loading. An increase in the pressure in the right atria is required to ensure left atrial filling.

NP: Evaluation; CN: Physiological integrity; CNS: Physiological adaptation; CL: Knowledge

68. Which of the following statements about truncus arteriosus is correct?

1. It's classified as an acyanotic defect.
2. Systemic and pulmonary blood mix.
3. It can't be diagnosed until after birth.
4. There are two types of truncus arteriosus.

68. 2. Blood ejects from the left and right ventricles and enters a common trunk, mixing pulmonary and systemic blood. Diagnosis can be made in utero using echocardiography. Truncus arteriosus can be divided into four types or categories. It's classified as a cyanotic defect with increased pulmonary blood flow.

NP: Evaluation; CN: Health promotion and maintenance; CNS: Prevention and early detection of disease; CL: Knowledge

---

NP: Nursing process   CN: Client needs category   CNS: Client needs subcategory   CL: Cognitive level

69. Which of the following findings commonly occurs during observation of a child with truncus arteriosus?
 1. Weak, thready pulses
 2. Narrowed pulse pressure
 3. Pink, moist mucous membranes
 4. Harsh systolic ejection murmur

70. Treatment for truncus arteriosus includes digoxin and diuretics. Which of the following techniques would be best for giving these drugs to an infant?
 1. Use a measuring spoon.
 2. Use a graduated dropper.
 3. Mix the drug with baby food.
 4. Mix the drug in a bottle with juice or milk.

71. Which of the following statements best describes transposition of the great arteries?
 1. The body receives only saturated blood.
 2. It's classified as an acyanotic defect with increased pulmonary blood flow.
 3. The pulmonary artery leaves the left ventricle, and the aorta exits from the right ventricle.
 4. The right atrium and the left atrium empty into one ventricular chamber.

72. Which of the following statements about transposition of the great arteries is correct?
 1. Diagnosis is made at birth.
 2. Diagnosis can be made in utero.
 3. Chest X-ray can show an accurate view of the defect.
 4. Heart failure isn't a related complication.

There may be many techniques for administering these drugs, but question 70 is asking you to choose the best.

You're doing a great job! Way to go!

69. 4. As a result of the ventricular septal defect, a harsh systolic ejection murmur is heard along the left sternal border and is usually accompanied by a thrill. Increasing pulmonary blood flow causes bounding pulses and a widened pulse pressure. Pulmonary stenosis leads to mild or moderate cyanosis, so mucous membranes may appear dull or gray.
NP: Data collection; CN: Physiological integrity; CNS: Physiological adaptation; CL: Knowledge

70. 2. Using a graduated dropper allows the exact dosage to be given. Mixing drugs with juice, milk, or food may cause a problem because, if the child doesn't completely finish the meal, you can't determine the exact amount ingested. In addition, this may prevent the child from drinking or eating for fear of tasting the drug. A teaspoon isn't as exact as a graduated dropper.
NP: Evaluation; CN: Physiological integrity; CNS: Pharmacological therapies; CL: Application

71. 3. Transposition of the great arteries is a condition in which the pulmonary artery leaves the left ventricle and the aorta exits from the right ventricle. This type of circulation gives the body only desaturated blood. The mixing of blood characterizes this defect as a cyanotic defect with variable pulmonary blood flow or as a mixed defect. A single-ventricle defect is a condition in which the right and left atria empty into one ventricular chamber.
NP: Evaluation; CN: Health promotion and maintenance; CNS: Prevention and early detection of disease; CL: Knowledge

72. 2. Echocardiography done by a fetal cardiologist can diagnose transposition of the great arteries in utero. Other associated defects include a patent foramen ovale and a ventricular septal defect, which contribute to heart failure. Chest X-ray can show cardiomegaly and pulmonary vascular markings only. Echocardiography or cardiac catheterization may be required preoperatively to show the coronary artery anatomy before surgical repair.
NP: Data collection; CN: Physiological integrity; CNS: Physiological adaptation; CL: Analysis

73. Which of the following associated defects is most common with transposition of the great arteries?

1. Mitral atresia
2. Atrial septal defect
3. Patent foramen ovale
4. Hypoplasia of the left ventricle

73. 3. A patent foramen ovale, patent ductus arteriosus, and ventricular septal defect are associated defects related to transposition of the great arteries. A patent foramen ovale is the most common and is necessary to provide adequate mixing of blood between the two circulations. An atrial septal defect is common in association with total anomalous pulmonary venous return. Hypoplasia of the left ventricle and mitral atresia are two defects associated with hypoplastic left heart syndrome.

NP: Evaluation; CN: Physiological integrity; CNS: Physiological adaptation; CL: Knowledge

Don't get anxious about this question — hint, hint!

74. Which change would the nurse expect after administering oxygen to an infant with uncorrected tetralogy of Fallot?

1. Disappearance of the murmur
2. No evidence of cyanosis
3. Improvement of finger clubbing
4. Less agitation

74. 4. Supplemental oxygen will help the infant breathe more easily and feel less anxious or agitated. None of the other findings would occur as the result of supplemental oxygen administration.

NP: Evaluation; CN: Physiological integrity; CNS: Basic care and comfort; CL: Application

75. A nurse on the pediatric unit is reviewing the signs and symptoms of fluid volume deficit and excess. What are the signs and symptoms of extracellular fluid volume deficit?

1. Full, bounding pulse
2. Distended neck veins
3. Thready, rapid pulse
4. Decreased hemoglobin level and hematocrit

75. 3. Cardiac reflex responses to decreased vascular volume result in a thready, rapid pulse. A full, bounding pulse, distended neck veins, and decreased hemoglobin level and hematocrit are observed in fluid volume excess.

NP: Data collection; CN: Physiological integrity; CNS: Physiological adaptation; CL: Comprehension

76. Which of the following statements best describes a characteristic of pulmonic stenosis?

1. The valve is abnormal.
2. The right ventricle is hypoplastic.
3. Left ventricular hypertrophy develops.
4. Divisions between the cusps are fused.

76. 4. Blood flow through the valve is restricted by fusion of the divisions between the cusps. The valve may be normal or malformed. Right ventricular hypertrophy develops due to resistance to blood flow.

NP: Data collection; CN: Physiological integrity; CNS: Physiological adaptation; CL: Knowledge

77. The nurse is planning care for a 9-year-old male child with heart failure. Which nursing diagnosis should receive priority?
1. Ineffective tissue perfusion (cardiopulmonary, renal) related to sympathetic response to heart failure
2. Imbalanced nutrition: Less than body requirements related to rapid tiring while feeding
3. Anxiety (parent) related to unknown nature of child's illness
4. Decreased cardiac output related to cardiac defect

Question 77 is asking you to prioritize!

77. 4. The primary nursing diagnosis for a child with heart failure is *Decreased cardiac output related to cardiac defect*. The most common cause of heart failure in children is congenital heart defects. Some defects result from the blood being pumped from the left side of the heart to the right side of the heart. The heart is unable to manage the extra volume, resulting in the pulmonary system becoming overloaded. Ineffective tissue perfusion, imbalanced nutrition, and anxiety don't take priority over decreased cardiac output. The child's heart must produce cardiac output sufficient to meet the body's metabolic demands.

NP: Planning; CN: Physiological integrity; CNS: Physiological adaptation; CL: Application

78. Which of the following findings is seen during cardiac catheterization of a child with pulmonic stenosis?
1. Right-to-left shunting
2. Left-to-right shunting
3. Decreased pressure in the right side of the heart
4. Increased oxygenation in the left side of the heart

78. 1. Right-to-left shunting develops through a patent foramen ovale due to right ventricular failure and an increase in pressure in the right side of the heart. Decreased oxygenation in the left side of the heart is noted due to the right-to-left shunt.

NP: Data collection; CN: Physiological integrity; CNS: Physiological adaptation; CL: Analysis

79. Which of the following findings is associated with aortic stenosis?
1. Hypotension
2. Right ventricular failure
3. Increased cardiac output
4. Loud systolic murmur with a thrill

79. 1. Older children with aortic stenosis are at risk for hypotension, tachycardia, angina, syncope, left ventricular failure, dyspnea, fatigue, and palpitations. Poor left ventricular ejection leads to decreased cardiac output. Loud systolic murmurs are heard with ventricular septal defects.

NP: Data collection; CN: Physiological integrity; CNS: Physiological adaptation; CL: Knowledge

I'm impressed with your progress!

80. Which of the following statements about aortic stenosis is correct?
1. It can result from rheumatic fever.
2. It accounts for 25% of all congenital defects.
3. There are two subcategories: valvular and subvalvular.
4. It's classified as an acyanotic defect with increased pulmonary blood flow.

80. 1. Aortic stenosis can result from rheumatic fever, which damages the valve due to lesions on the leaflets and scarring that develop over time. It's classified as an acyanotic defect with obstructed flow from the ventricles. It accounts for about 5% of all congenital defects. There are three subcategories: valvular, subvalvular, and supravalvular.

NP: Data collection; CN: Physiological integrity; CNS: Physiological adaptation; CL: Knowledge

81. Which of the following instructions would be most appropriate for a child with aortic stenosis?

1. Restrict exercise.
2. Avoid prostaglandin E$_1$.
3. Avoid digoxin and diuretics.
4. Allow the child to exercise freely.

Yo! Is it OK if I run around? I need a break!

82. Which of the following statements is correct for hypoplastic left heart syndrome?

1. It can be diagnosed only at birth.
2. It includes a group of related anomalies.
3. It's classified as 25% of all congenital defects.
4. Surgical intervention is the only option for treatment.

83. Which of the following nursing diagnoses is the most appropriate when caring for an infant with hypoplastic left heart syndrome?

1. Anticipatory grieving
2. Delayed growth and development
3. Deficient diversional activity
4. Risk for activity intolerance

There's that most appropriate phrase again.

84. A 3-year-old client has a high red blood cell count and polycythemia. In planning care, the nurse would anticipate which goal to help prevent blood clot formation?

1. The child won't have signs of dehydration.
2. The child won't have signs of dyspnea.
3. The child will be pain-free.
4. The child will attain the 40th percentile of weight for his age.

81. 1. Exercise should be restricted due to low cardiac output and left ventricular failure. Strenuous activity has been reported to result in sudden death from the development of myocardial ischemia. Digoxin and diuretics may be required for critically ill infants experiencing heart failure as a result of severe aortic stenosis. Prostaglandin E$_1$ is recommended to maintain the patency of the ductus arteriosus in the neonates. This allows for improved systemic blood flow.

NP: Implementation; CN: Physiological integrity; CNS: Physiological adaptation; CL: Application

82. 2. Hypoplastic left heart syndrome includes a group of related anomalies, such as hypoplasia of the left ventricle, mitral atresia, aortic atresia, and hypoplasia of the ascending aorta and aortic arch. Treatment options include palliative surgery, cardiac transplantation, or no intervention, in which case death would occur. It can be diagnosed prenatally with a level 2 ultrasound examination. It's classified as about 1% to 2% of all congenital defects.

NP: Evaluation; CN: Health promotion and maintenance; CNS: Prevention and early detection of disease; CL: Knowledge

83. 1. Without intervention, death usually occurs within the first few days of life due to progressive hypoxia, acidosis, and shock as the ductus closes and systemic perfusion diminishes. If the parents choose cardiac transplantation, the child may die waiting for a donor heart. For those who choose surgery, the child may not survive the three stages of the surgery. The other three choices don't apply to this type of defect due to the low survival rates.

NP: Planning; CN: Physiological integrity; CNS: Physiological adaptation; CL: Analysis

84. 1. When dehydration occurs, blood is thicker and more prone to clotting. Dyspnea would be a sign of hypoxia. Pain would be an indicator of nutritional status, not the risk of embolism.

NP: Planning; CN: Physiological integrity; CNS: Reduction of risk potential; CL: Application

NP: Nursing process   CN: Client needs category   CNS: Client needs subcategory   CL: Cognitive level

85. A child receives prednisone after a heart transplant. Prednisone is in which of the following drug classifications?
1. Antibiotic
2. Cardiac glycoside
3. Corticosteroid
4. Diuretic

86. A child is given 0.5 mg/kg/day of prednisone divided into two doses. The child weighs 10 kg. How much is given in each dose?
1. 2.5 mg
2. 5 mg
3. 10 mg
4. 1.5 mg

87. Which of the following adverse reactions of prednisone is expected in a child after having a heart transplant?
1. Anorexia
2. Hypolipidemia
3. Decreased appetite
4. Poor wound healing

88. Which of the following characteristics best describes bacterial/infective endocarditis?
1. Bacteria invading only tissues of the heart
2. Infection of the valves and inner lining of the heart
3. Inappropriate fusion of the endocardial cushions in fetal life
4. Caused by alterations in cardiac preload, afterload, contractility, or heart rate

89. A child with suspected bacterial endocarditis arrives at the emergency department. Which of the following findings is expected during data collection?
1. Weight gain
2. Bradycardia
3. Low-grade fever
4. Increased hemoglobin level

Math problems get you stumped? Just break each problem into manageable pieces, one by one.

Hey, am I to blame for this or what?

85. 3. Prednisone is a corticosteroid used frequently during posttransplantation as part of antirejection therapy. The drug influences the immune system through its strong anti-inflammatory action and immunologic effect. The other three classes of drugs aren't indicated in this situation.
NP: Evaluation; CN: Physiological integrity; CNS: Pharmacological therapies; CL: Knowledge

86. 1. 0.5 mg/kg × 10 kg = 5 mg
5 mg/2 doses = 2.5 mg/dose
NP: Implementation; CN: Physiological integrity; CNS: Pharmacological therapies; CL: Analysis

87. 4. Common adverse reactions of prednisone include poor wound healing, increased appetite, weight gain, hyperlipidemia, delayed sexual maturation, growth impairment, and a cushingoid appearance. The school-age child is usually overweight and has a moon-shaped face.
NP: Evaluation; CN: Physiological integrity; CNS: Pharmacological therapies; CL: Knowledge

88. 2. Bacterial/infective endocarditis is an infection of the valves and inner lining of the heart. It's usually caused by the bacteria *Streptococcus viridans* and frequently affects children with acquired or congenital anomalies of the heart or great vessels. Endocardial cushion defects represent inappropriate fusion of the endocardial cushions in fetal life. Bacteria may grow into adjacent tissues and may break off and embolize elsewhere, such as the spleen, kidney, lung, skin, and central nervous system. Alterations in preload, afterload, contractility, or heart rate refer to heart failure.
NP: Evaluation; CN: Physiological integrity; CNS: Physiological adaptation; CL: Knowledge

89. 3. Symptoms may include a low-grade intermittent fever, decrease in hemoglobin level, tachycardia, anorexia, weight loss, and decreased activity level.
NP: Data collection; CN: Physiological integrity; CNS: Physiological adaptation; CL: Application

90. Which of the following factors may lead to bacterial endocarditis in a child with underlying heart disease?

1. History of a cold for 3 days
2. Dental work pretreated with antibiotics
3. Peripheral I.V. catheter in place for 1 day
4. Indwelling urinary catheter for 2 days leading to a urinary tract infection

Wow! You've finished question 90! You're in the home stretch!

90. 4. Bacterial organisms can enter the bloodstream from any site of infection such as a urinary tract infection. Gram-negative bacilli are common causative agents. A peripheral I.V. catheter is an entry site but only if signs and symptoms of infection are present. Long-term indwelling catheters pose a higher risk of infection. Dental work is a common portal of entry if not pretreated with antibiotics. Colds are usually viral, not bacterial. Heart surgery is also a common cause of endocarditis, especially if synthetic material is used.

NP: Data collection; CN: Physiological integrity; CNS: Physiological adaptation; CL: Analysis

91. Erythromycin is given to a 6-year-old child before dental work to prevent endocarditis. The child weighs 44 lb. The order is for 20 mg/kg by mouth 2 hours before the procedure. How many milligrams would that be?

1. 200 mg
2. 400 mg
3. 440 mg
4. 880 mg

91. 2. 44 lb/2.2 kg = 20 kg
20 mg/kg × 20 kg = 400 mg

NP: Implementation; CN: Physiological integrity; CNS: Pharmacological therapies; CL: Analysis

92. Which of the following adverse reactions may occur with the administration of erythromycin?

1. Weight gain
2. Constipation
3. Increased appetite
4. Nausea and vomiting

92. 4. Erythromycin is an antibacterial antibiotic. Common adverse effects include nausea, vomiting, diarrhea, abdominal pain, and anorexia. It should be given with a full glass of water and after meals or with food to lessen GI symptoms.

NP: Evaluation; CN: Physiological integrity; CNS: Pharmacological therapies; CL: Knowledge

93. A child is hospitalized with bacterial endocarditis. Which of the following nursing diagnoses is most appropriate?

1. Constipation
2. Excess fluid volume
3. Deficient diversional activity
4. Imbalanced nutrition: More than body requirements

Nursing diagnoses: Share the knowledge!

93. 3. Treatment for bacterial endocarditis requires long-term hospitalization or home care for I.V. antibiotics. Children may be bored and depressed, needing age-appropriate activities. *Deficient fluid volume, Diarrhea,* and *Imbalanced nutrition: Less than body requirements* may be possible nursing diagnoses related to such adverse reactions of antibiotics as GI upset.

NP: Evaluation; CN: Physiological integrity; CNS: Physiological adaptation; CL: Analysis

94. When assessing a child with suspected Kawasaki disease, which of the following symptoms is common?
  1. Low-grade fever
  2. "Strawberry" tongue
  3. Pink, moist mucous membranes
  4. Bilateral conjunctival infection with yellow exudate

95. Which of the following statements best describes Kawasaki disease?
  1. It mostly occurs in the summer and fall.
  2. Diagnosis can be determined by laboratory testing.
  3. It's an acute systemic vasculitis of unknown cause.
  4. It manifests in two different stages: acute and subacute.

96. Which of the following characteristics indicates a child with Kawasaki disease has entered the subacute phase?
  1. Polymorphous rash
  2. Normal blood values
  3. Cervical lymphadenopathy
  4. Desquamation of the hands and feet

97. A nurse is caring for a child with Kawasaki disease. Which of the following symptoms concerns the nurse the most?
  1. Mild diarrhea
  2. Pain in the joints
  3. Abdominal pain with vomiting
  4. Increased erythrocyte sedimentation rate

There's that million-dollar word again: *best.*

Clever clue, that subacute, eh?

94. 2. Inflammation of the pharynx and oral mucosa develops, causing red, cracked lips and a "strawberry" tongue, in which the normal coating of the tongue sloughs off. A high fever of 5 or more days unresponsive to antibiotics and antipyretics is also part of the diagnostic criteria. The eyes are generally dry without exudation.
NP: Data collection; CN: Physiological integrity; CNS: Physiological adaptation; CL: Knowledge

95. 3. Kawasaki disease can best be described as an acute systemic vasculitis of unknown cause. Diagnosis is based on clinical findings of five of the six diagnostic criteria and associated laboratory results. There is no specific laboratory test for diagnosis. There are three stages: acute, subacute, and convalescent. Most cases are geographic and seasonal, with most occurring in the late winter and early spring.
NP: Evaluation; CN: Physiological integrity; CNS: Physiological adaptation; CL: Knowledge

96. 4. The subacute phase shows characteristic desquamation of the hands and feet. Blood values return to normal at the end of the convalescent phase. Cervical lymphadenopathy and a polymorphous rash can be seen in the acute phase because of the onset of inflammation and fever.
NP: Data collection; CN: Physiological integrity; CNS: Physiological adaptation; CL: Analysis

97. 3. The most serious complication of this disease is cardiac involvement. Abdominal pain, vomiting, and restlessness are the main symptoms of an acute myocardial infarction in children. Pain in the joints is an expected sign of arthritis that usually occurs in the subacute phase. Mild diarrhea can be treated with oral fluids. An increased erythrocyte sedimentation rate is a reflection of the inflammatory process and may be seen for 2 to 4 weeks after the onset of symptoms.
NP: Data collection; CN: Physiological integrity; CNS: Physiological adaptation; CL: Analysis

98. The nurse is caring for a 2-year-old toddler diagnosed with Kawasaki disease. The nurse must assess for which signs and symptoms in a child with Kawasaki disease?
1. Chest pain, dyspnea, fever, and headache
2. Fever, headache, and erythema marginatum, a rash characterized by pink macules and blanching in the middle of the lesions
3. Bilateral conjunctivitis
4. Weight loss, abdominal pain, and cramping

Two more questions to number 100!

98. 3. Bilateral conjunctivitis is typically observed early in the illness. Chest pain and dyspnea are sometimes observed in endocarditis. Erythema marginatum is a rash observed in rheumatic fever. Weight loss, abdominal pain, and cramping are characteristic of ulcerative colitis.
NP: Data collection; CN: Physiological integrity; CNS: Physiological adaptation; CL: Comprehension

99. Therapy for Kawasaki disease includes I.V. gamma globulin, prescribed at 400 mg/kg/day for 4 days. The child weighs 10 kg. How much is given per dose?
1. 200 mg
2. 400 mg
3. 2,000 mg
4. 4,000 mg

99. 4. 400 mg/kg × 10 kg = 4,000 mg, or 4 g
NP: Implementation; CN: Physiological integrity; CNS: Pharmacological therapies; CL: Analysis

100. A 2-month-old infant arrives in the emergency department with a heart rate of 180 beats/minute and a temperature of 103.1° F (39.5° C) rectally. Which of the following interventions is most appropriate?
1. Give acetaminophen (Tylenol).
2. Encourage fluid intake.
3. Apply carotid massage.
4. Place the infant's hands in cold water.

100. 1. Acetaminophen should be given first to decrease the temperature. A heart rate of 180 beats/minute is normal in an infant with a fever. A tepid sponge bath may be given to help decrease the temperature and calm the infant. Carotid massage is an attempt to decrease the heart rate as a vagal maneuver. This won't work in this infant because the source of the increased heart rate is fever. Fluid intake is encouraged after the acetaminophen is given to help replace insensible fluid losses.
NP: Planning; CN: Physiological integrity; CNS: Physiological adaptation; CL: Application

101. A child is prescribed aspirin as part of the therapy for Kawasaki disease. The order is for 80 mg/kg/day orally in four divided doses until the child is afebrile. The child weighs 15 kg. How much is given in one dose?
1. 60 mg
2. 300 mg
3. 320 mg
4. 1,200 mg

Keep practicing that math. You'll get it!

101. 2. 80 mg/kg × 15 kg = 1,200 mg
1,200 mg/4 doses = 300 mg/dose
NP: Implementation; CN: Physiological integrity; CNS: Pharmacological therapies; CL: Analysis

---

NP: Nursing process   CN: Client needs category   CNS: Client needs subcategory   CL: Cognitive level

102. A nurse is giving discharge instructions to the parents of a child with Kawasaki disease. Which of the following statements shows an understanding of the treatment plan?
1. "A regular diet can be resumed at home."
2. "Black, tarry stools are considered normal."
3. "My child should use a soft-bristled toothbrush."
4. "My child can return to playing football next week."

103. A nurse is preparing the family of a client with Kawasaki disease for discharge. Which of the following instructions is most appropriate?
1. Stop the aspirin when you return home.
2. Immunizations can be given in 2 weeks.
3. The child may return to school in 1 week.
4. Frequent echocardiography will be needed.

104. Which of the following descriptions of acute rheumatic fever is correct?
1. A progressive inflammation of the small vessels
2. A mucocutaneous lymph node syndrome
3. A serious infection of the endocardial surface of the heart
4. A sequela of group A beta-hemolytic streptococcal infections

Listen for feedback to find out if your instructions were understood.

Question 104 asks about acute rheumatic fever, not chronic.

102. 3. Because of the anticoagulant effects of aspirin therapy, a soft-bristled toothbrush will prevent bleeding of the gums. Contact sports should be avoided because of the cardiac involvement and excessive bruising that may occur due to aspirin therapy. Black, tarry stools are abnormal and are signs of bleeding that should be reported to the physician immediately. A low-cholesterol diet should be followed until coronary artery involvement resolves, usually within 6 to 8 weeks.
NP: Evaluation; CN: Physiological integrity; CNS: Physiological adaptation; CL: Analysis

103. 4. Because of the risk of coronary artery involvement and possible aneurysm development, repeat echocardiography and electrocardiography will be required the first few weeks and at 6 months. Aspirin therapy may be continued for 2 weeks after the onset of symptoms. If signs of coronary artery involvement are present, aspirin therapy may be continued indefinitely. School should be avoided until cleared by the physician. Live-virus vaccines should be avoided for at least 5 months after gamma globulin therapy because of an increased risk of a cross-sensitivity reaction to the antibodies found in the dose given.
NP: Evaluation; CN: Physiological integrity; CNS: Physiological adaptation; CL: Analysis

104. 4. Acute rheumatic fever is a multisystem disorder caused by group A beta-hemolytic streptococcal infections. It may involve the heart, joints, central nervous system, and skin. Kawasaki disease is a mucocutaneous lymph node syndrome characterized by a progressive inflammation of the small vessels. Endocarditis describes a serious infection of the endocardial surface of the heart.
NP: Evaluation; CN: Health promotion and maintenance; CNS: Prevention and early detection of disease; CL: Knowledge

105. Which finding is expected in a child with acute rheumatic fever?
1. Leukocytosis
2. Normal electrocardiogram
3. High fever for 5 or more days
4. Normal erythrocyte sedimentation rate

Concentrate. You know the answer.

105. 1. Leukocytosis can be seen as an immune response triggered by colonization of the pharynx with group A streptococci. The electrocardiogram will show a prolonged PR interval as a result of carditis. The inflammatory response will cause an elevated erythrocyte sedimentation rate. A low-grade fever is a minor manifestation. A high fever of 5 or more days may represent Kawasaki disease.
NP: Data collection; CN: Physiological integrity; CNS: Physiological adaptation; CL: Knowledge

106. Which of the following criteria is required to establish a diagnosis of acute rheumatic fever?
1. Laboratory tests
2. Fever and four diagnostic criteria
3. Positive blood cultures for *Staphylococcus* organisms
4. Use of Jones criteria and presence of a streptococcal infection

106. 4. Two major or one major and two minor manifestations from Jones criteria and the presence of a streptococcal infection justify the diagnosis of rheumatic fever. There is no single laboratory test for diagnosis. Fever and four diagnostic criteria are required to diagnose Kawasaki disease. Blood cultures would be positive for *Streptococcus*, not *Staphylococcus*, organisms.
NP: Data collection; CN: Physiological integrity; CNS: Physiological adaptation; CL: Analysis

107. Which of the following manifestations is considered major for Jones criteria for acute rheumatic fever?
1. Carditis
2. Polyarthralgia
3. Low-grade fever
4. Previous heart disease

107. 1. Carditis is the most serious major manifestation of acute rheumatic fever. It's the only manifestation that can lead to death or long-term sequelae. Polyarthralgia, low-grade fever, and previous heart disease are considered minor manifestations for Jones criteria.
NP: Data collection; CN: Physiological integrity; CNS: Physiological adaptation; CL: Knowledge

108. A nurse is caring for a child with acute rheumatic fever. Which of the following symptoms can be recognized as Sydenham's chorea, a major manifestation of acute rheumatic fever?
1. Cardiomegaly
2. Regurgitant murmur
3. Pericardial friction rubs
4. Involuntary muscle movements

Almost finished! Keep up the good work!

108. 4. Sydenham's chorea is an involvement of the central nervous system by the rheumatic process. This is seen as muscular incoordination; purposeless, involuntary movements; and emotional liability. A regurgitant murmur, cardiomegaly, and a pericardial friction rub are clinical signs of rheumatic carditis.
NP: Data collection; CN: Physiological integrity; CNS: Physiological adaptation; CL: Knowledge

NP: Nursing process   CN: Client needs category   CNS: Client needs subcategory   CL: Cognitive level

109. Criteria for rheumatic fever are being discussed with parents. The nurse realizes that the parents understand chorea when they make which of the following statements?
1. "My child may not be able to walk."
2. "Long movies may help for relaxation."
3. "My child might have difficulty in school."
4. "Many activities and visitors are recommended."

110. A 3-year-old child has a positive culture for *Streptococcus* organisms. Which of the following interventions is most appropriate?
1. Give aspirin.
2. Give antibiotics.
3. Give corticosteroids.
4. Encourage fluid intake.

111. A nurse is preparing a child for discharge after being diagnosed with rheumatic fever. Which of the following interventions is recommended?
1. Give aspirin for signs of chorea.
2. Give penicillin for 1 month total.
3. Only give penicillin for dental work.
4. Give penicillin until age 20.

112. Which of the following nursing diagnoses is most appropriate for a child with rheumatic fever?
1. Imbalanced nutrition: More than body requirements
2. Risk for injury
3. Delayed growth and development
4. Impaired gas exchange

Can you guess what the hint is in this question? I knew you could!

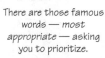

You're nearly finished! You can do it!

There are those famous words — *most appropriate* — asking you to prioritize.

109. 3. Chorea may last 1 to 6 months. Central nervous system involvement contributes to a shortened attention span, so children might have difficulty learning in school. A quiet environment is required for treatment. Muscle incoordination may cause the child to be more clumsy than usual when walking.
NP: Evaluation; CN: Physiological integrity; CNS: Physiological adaptation; CL: Analysis

110. 2. Infection caused by *Streptococcus* organisms is treated with antibiotics, mainly penicillin. Antipyretics, such as acetaminophen, may be given for fever. Aspirin isn't recommended. Fluid intake is encouraged to prevent dehydration from decreased oral intake due to a sore throat or to replace fluids lost due to possible diarrhea from the antibiotics. Corticosteroids have no implication.
NP: Planning; CN: Physiological integrity; CNS: Physiological adaptation; CL: Analysis

111. 4. Penicillin may be given preventively until age 20 to prevent additional damage from future attacks. Penicillin may be used for dental work or other invasive procedures to prevent bacterial endocarditis. Aspirin is administered for arthritis. Chorea is treated with sedatives and a quiet, relaxing environment.
NP: Planning; CN: Physiological integrity; CNS: Pharmacological therapies; CL: Application

112. 2. Due to symptoms of chorea, safety measures should be taken to prevent falls or injury. There may be an imbalance in nutrition of less than body requirements because of a sore throat and dysphagia. Growth and development usually aren't delayed. Gas exchange usually isn't an issue unless the condition worsens with carditis and heart failure is present.
NP: Planning; CN: Physiological integrity; CNS: Physiological adaptation; CL: Analysis

113. Sinus bradycardia can best be described as which of the following conditions?
1. Heart rate less than normal for age
2. Heart rate greater than normal for age
3. A variation of the normal cardiac rhythm
4. Increase in sinus node impulse formation

113. 1. Sinus bradycardia can best be described as a heart rate less than normal for age. Sinus tachycardia refers to a heart rate greater than normal for age or an increase in sinus node impulse formation. A sinus arrhythmia is a variation of the normal cardiac rhythm. NP: Data collection; CN: Physiological integrity; CNS: Physiological adaptation; CL: Knowledge

114. In which of the following conditions or groups is sinus bradycardia a normal finding?
1. Hypoxia
2. Hypothermia
3. Growth-delayed adolescent
4. Physically conditioned adolescent

Sometimes abnormal is normal and vice versa!

114. 4. A physically conditioned adolescent might have a lower than normal heart rate, which is of no significance. Hypoxia and hypothermia are pathologic states in which a slow heart rate may produce a compromised hemodynamic state. Growth-delayed adolescents won't have bradycardia as a normal finding. NP: Data collection; CN: Physiological integrity; CNS: Physiological adaptation; CL: Analysis

115. Treatment for a child with sinus bradycardia includes atropine 0.02 mg/kg. If the child weighs 20 kg, how much is given per dose?
1. 0.02 mg
2. 0.04 mg
3. 0.2 mg
4. 0.4 mg

115. 4. $0.02 \text{ mg/kg} \times 20 \text{ kg} = 0.4 \text{ mg}$ NP: Implementation; CN: Physiological integrity; CNS: Pharmacological therapies; CL: Analysis

116. Atropine is given to a child for sinus bradycardia. In which of the following classifications does atropine belong?
1. Anticholinergic
2. Beta-adrenergic agonist
3. Bronchodilator
4. Sympathomimetic

You're looking for an adverse reaction here, not a therapeutic action.

116. 1. Atropine can best be classified as an anticholinergic drug. It blocks vagal impulses to the myocardium and stimulates the cardioinhibitory center in the medulla, thereby increasing heart rate and cardiac output. NP: Evaluation; CN: Physiological integrity; CNS: Pharmacological therapies; CL: Knowledge

117. A nurse has given atropine to treat sinus bradycardia in an 11-month-old infant. Which of the following adverse reactions is noted?
1. Lethargy
2. Diarrhea
3. No tears when crying
4. Increased urine output

117. 3. Atropine dries up secretions and also lessens the response of ciliary and iris sphincter muscles in the eye, causing mydriasis. It usually causes paradoxical excitement in children. Constipation and urinary retention can be seen because of a decrease in smooth muscle contractions of the GI and genitourinary tracts. NP: Evaluation; CN: Physiological integrity; CNS: Pharmacological therapies; CL: Application

NP: Nursing process   CN: Client needs category   CNS: Client needs subcategory   CL: Cognitive level

118. Which of the following conditions could cause sinus tachycardia?
1. Fever
2. Hypothermia
3. Hypothyroidism
4. Hypoxia

119. Which of the following arrhythmias commonly seen in children involves heart rate changes related to respirations?
1. Sinus arrhythmia
2. Sinus block
3. Sinus bradycardia
4. Sinus tachycardia

Relax. Stressing out won't help you think!

120. Which of the following conditions may lead to sinus arrest or sinus pause in a child?
1. Hypokalemia
2. Hyperthermia
3. Valsalva's maneuver
4. Decreased intracranial pressure

121. For temporary treatment of sinus arrest, isoproterenol is given to increase the heart rate. The order is for 0.05 mcg/kg/minute. The child weighs 10 kg. How much should be given?
1. 0.05 mcg
2. 0.15 mcg
3. 0.25 mcg
4. 0.5 mcg

I knew you could get the hang of math!

118. 1. Sinus tachycardia can frequently be seen in children with a fever. It's frequently a result of a noncardiac cause. Hypothermia, hypothyroidism, and hypoxia will result in sinus bradycardia.
NP: Data collection; CN: Physiological integrity; CNS: Physiological adaptation; CL: Analysis

119. 1. With respirations, intrathoracic pressures decrease on inhalation, which decreases vagal stimulation, and increase on expiration, which increases vagal stimulation. The result is the perfectly normal arrhythmia called sinus arrhythmia, a common occurrence in childhood and adolescence. Sinus block, sinus bradycardia, and sinus tachycardia are respiration-independent arrhythmias.
NP: Data collection; CN: Physiological integrity; CNS: Physiological adaptation; CL: Analysis

120. 3. Sinus arrest may occur in children when vagal tone is increased, such as during Valsalva's maneuver, vomiting, gagging, or straining during a bowel movement. This condition involves a failure of the sinoatrial node to generate an impulse. A straight line or pause occurs, indicating the absence of electrical activity. After the pause, another impulse will be generated and a cardiac complex will appear. Hyperkalemia, not hypokalemia; hypothermia, not hyperthermia; and increased rather than decreased intracranial pressure are pathologic conditions that may also produce sinus pause.
NP: Data collection; CN: Physiological integrity; CNS: Physiological adaptation; CL: Analysis

121. 4. 0.05 mcg/kg × 10 kg = 0.5 mcg
NP: Implementation; CN: Physiological integrity; CNS: Pharmacological therapies; CL: Analysis

122. To which of the following classifications does isoproterenol belong?
1. Adrenergic agonist
2. Anticholinergic
3. Beta-adrenergic blocker
4. Vasopressor

123. Which of the following findings may be seen in a 1-year-old child with supraventricular tachycardia?
1. Heart rate of 100 beats/minute
2. Heart rate of 200 beats/minute
3. Heart rate less than 80 beats/minute
4. Heart rate more than 240 beats/minute

124. A 2-year-old child is experiencing supraventricular tachycardia. Which of the following interventions should be attempted first?
1. Administration of digoxin
2. Administration of verapamil
3. Synchronized cardioversion
4. Immersion of the child's hands in cold water

122. 3. Isoproterenol acts as a beta-adrenergic blocker to reduce peripheral resistance and increase the force of cardiac contraction without producing vasoconstriction. It also acts as a bronchodilator, relaxing bronchial smooth muscle and creating peripheral vasodilation.
NP: Data collection; CN: Physiological integrity; CNS: Pharmacological therapies; CL: Analysis

123. 4. Supraventricular tachycardia may be related to increased automaticity of an atrial cell other than the sinoatrial node, or as a reentry mechanism. The rhythm is regular and can occur at rates of 240 beats/minute or more. A heart rate less than 80 beats/minute can be characterized as sinus bradycardia. A heart rate around 200 beats/minute may represent sinus tachycardia. A heart rate of 100 beats/minute is a normal finding for a 1-year-old child.
NP: Data collection; CN: Physiological integrity; CNS: Physiological adaptation; CL: Knowledge

124. 4. Vagal maneuvers, such as immersion of the hands in cold water, are commonly tried first as a mechanism to decrease the heart rate. Other vagal maneuvers include breath-holding, carotid massage, gagging, and placing the head lower than the rest of the body. Synchronized cardioversion may be required if vagal maneuvers and drugs are ineffective. If a child has low cardiac output, cardioversion may be used instead of drugs. Verapamil isn't recommended. Digoxin is one of the most common drugs given to help decrease heart rate by increasing myocardial contractility and automaticity and reducing excitability.
NP: Planning; CN: Physiological integrity; CNS: Pharmacological therapies; CL: Analysis

This chapter covers sickle cell disease, varicella, Rocky Mountain spotted fever, leukemia, and many other blood and immune system disorders in kids. It's a whopper of a chapter on a critical area. If you're ready, let's begin!

Chapter 28
Hematologic & immune disorders

1. A child comes to the emergency department feeling feverish and lethargic. Which of the following assessment findings suggests Reye's syndrome?
1. Fever, profoundly impaired consciousness, and hepatomegaly
2. Fever, enlarged spleen, and hyperactive reflexes
3. Afebrile, intractable vomiting, and rhinorrhea
4. Malaise, cough, and sore throat

1. 1. Reye's syndrome is defined as toxic encephalopathy, characterized by fever, profoundly impaired consciousness, and disordered hepatic function. Intractable vomiting occurs during the first stage of Reye's syndrome, but rhinorrhea usually precedes the onset of the illness. Reye's syndrome doesn't affect the spleen but causes fatty degeneration of the liver. Hyperactive reflexes occur with central nervous system involvement. Malaise, cough, and sore throat are viral symptoms that often precede the illness.
NP: Data collection; CN: Physiological integrity; CNS: Physiological adaptation; CL: Knowledge

Early diagnosis is critical to treating Reye's syndrome.

2. Which of the following aspects is <u>most</u> <u>important</u> for successful management of the child with Reye's syndrome?
1. Early diagnosis
2. Initiation of antibiotics
3. Isolation of the child
4. Staging of the illness

2. 1. Early diagnosis and therapy are essential because of the rapid clinical course of the disease and its high mortality. Isolation isn't necessary because the disease isn't communicable. Reye's syndrome is associated with a viral illness, and antibiotic therapy isn't crucial to preventing the initial progression of the illness. Staging, although important to therapy, occurs after a differential diagnosis is made.
NP: Data collection; CN: Physiological integrity; CNS: Reduction of risk potential; CL: Knowledge

3. A child with Reye's syndrome is in stage I of the illness. Which of the following measures can be taken to prevent further progression of the illness?
1. Invasive monitoring
2. Endotracheal intubation
3. Hypertonic glucose solution
4. Pancuronium (Pavulon)

3. 3. For children in stage I of Reye's syndrome, treatment is primarily supportive and directed toward restoring blood glucose levels and correcting acid-base imbalances. I.V. administration of dextrose solutions with added insulin helps to replace glycogen stores. Noninvasive monitoring is adequate to assess status at this stage. Endotracheal intubation may be necessary later. Pancuronium is used as an adjunct to endotracheal intubation and wouldn't be used in this stage of Reye's syndrome.
NP: Planning; CN: Physiological integrity; CNS: Reduction of risk potential; CL: Application

NP: Nursing process   CN: Client needs category   CNS: Client needs subcategory   CL: Cognitive level

4. Which of the following groups of laboratory results, along with the clinical manifestations, establishes a diagnosis of Reye's syndrome?

   1. Elevated liver enzymes and prolonged prothrombin and partial thromboplastin times

   2. Increased serum glucose and insulin levels

   3. Increased bilirubin and alkaline phosphatase levels

   4. Decreased serum glucose and ammonia levels

The word *establishes* is a big 'ol hint. Catch it!

4. 1. Reye's syndrome causes fatty degeneration of the liver, altering results of liver function studies. Serum bilirubin and alkaline phosphatase usually aren't affected. Decreased serum glucose levels, with reduced insulin levels, occur secondary to dehydration caused by intractable vomiting.

NP: Data collection; CN: Physiological integrity; CNS: Physiological adaptation; CL: Analysis

5. The nurse is caring for a 9-month-old boy with Reye's syndrome. It's most important that the nurse plan to:

   1. check the skin for signs of breakdown every shift.

   2. perform range-of-motion (ROM) exercises every 4 hours.

   3. monitor the client's intake and output.

   4. place the client in protective isolation.

5. 3. Monitoring intake and output alerts the nurse to the development of dehydration and cerebral edema, complications of Reye's syndrome. Although checking the skin for signs of breakdown is important because the child may not be as active as normal, it isn't as critical as monitoring the client's intake and output. Active ROM exercises may not be needed and aren't as important as monitoring the client's intake and output. Placing the client in protective isolation isn't necessary.

NP: Planning; CN: Physiological integrity; CNS: Reduction of risk potential; CL: Application

6. Which change would indicate increased intracranial pressure (ICP) in a child acutely ill with Reye's syndrome?

   1. Irritability and quick pupil response

   2. Increased blood pressure and decreased heart rate

   3. Decreased blood pressure and increased heart rate

   4. Sluggish pupil response and decreased blood pressure

6. 2. A marked increase in ICP will trigger the pressure response; increased ICP produces an elevation in blood pressure with a reflex slowing of the heart rate. Irritability is often an early sign, but pupillary response becomes more sluggish in response to increased ICP.

NP: Data collection; CN: Physiological integrity; CNS: Physiological adaptation; CL: Application

Warning: The word *contraindicated* makes this almost a negative question.

7. A client with Reye's syndrome is exhibiting signs of increased intracranial pressure (ICP). Which of the following nursing interventions would be contraindicated for this client?

   1. Lung vibration

   2. Scheduled cluster care

   3. Suctioning and chest physiotherapy

   4. Positioning to avoid neck vein compression

7. 3. Nursing procedures tend to cause reactive pressure waves in many clients. Suctioning and percussion are poorly tolerated; thus, they're contraindicated. If concurrent respiratory problems exist, lung vibration provides good results without increasing ICP. Care should be provided at an optimal time for the client, clustering care to prevent overstimulation. Neck vein compression may further increase ICP by interfering with venous return.

NP: Implementation; CN: Physiological integrity; CNS: Physiological adaptation; CL: Application

---

NP: Nursing process   CN: Client needs category   CNS: Client needs subcategory   CL: Cognitive level

8. The goal of nursing care for a client with Reye's syndrome is to minimize intracranial pressure (ICP). Which of the following nursing interventions help to meet this goal?
1. Keeping the head of the bed flat
2. Frequent position changes
3. Positioning to avoid neck flexion
4. Suctioning and chest physiotherapy

Stay calm when encountering NCLEX questions you're unsure about. That's what I do.

9. Which of the following nursing interventions would be included in the care of an unconscious client with Reye's syndrome?
1. Providing mouth care only as needed
2. Placing the client on a sheepskin
3. Avoiding the use of lotions on skin
4. Placing the client in a supine position

Good job! You've finished 10 questions already!

10. Which of the following medications has been connected with the development of Reye's syndrome?
1. Acetaminophen (Tylenol)
2. Aspirin
3. Ibuprofen (Motrin)
4. Guaifenesin (Robitussin)

11. Which of the following therapeutic effects does pancuronium (Pavulon) have in a client with Reye's syndrome?
1. Decreased serum glucose levels
2. Elevated blood osmolality, and decreasing edema
3. Sedative effect, placing the child in a medication-induced coma
4. Paralysis, preventing activity that may increase intracranial pressure

8. 3. Neck vein compression can increase ICP by interfering with venous return. The head of the bed should be elevated to help promote venous return. Nursing procedures such as frequent positioning tend to cause overstimulation; therefore, care should be taken to avoid such procedures to prevent increased ICP. Suctioning and percussion are poorly tolerated and are contraindicated, unless concurrent respiratory problems are present.
NP: Implementation; CN: Physiological integrity; CNS: Physiological adaptation; CL: Application

9. 2. Placing the client on a sheepskin helps to prevent pressure on prominent areas of the body. Rubbing the extremities with lotion stimulates circulation and helps prevent drying of the skin. Mouth care should be performed at least twice daily because the mouth tends to become dry or coated with mucus. Placing the client supine would be contraindicated because of the risk of aspiration and increasing intracranial pressure. The supine position puts undue pressure on the sacral and occipital areas.
NP: Implementation; CN: Physiological integrity; CNS: Physiological adaptation; CL: Application

10. 2. There's speculation regarding the relationship between aspirin administration and the development of Reye's syndrome. It isn't uncommon for a child to receive large amounts of salicylates for fever associated with prior viral infection, such as chickenpox or influenza. Acetaminophen, ibuprofen, and guaifenesin haven't been associated with the development of Reye's syndrome. In fact, there has been a decreased incidence of Reye's syndrome with the increased use of acetaminophen and ibuprofen for management of fevers in children.
NP: Evaluation; CN: Physiological integrity; CNS: Pharmacological therapies; CL: Knowledge

11. 4. Pancuronium is a paralyzing agent and doesn't have a sedative effect or any effect on increasing blood glucose levels. Barbiturates are used for sedation to suppress sensory input. Mannitol (Osmitrol) elevates blood osmolality and is used to prevent cerebral edema.
NP: Planning; CN: Physiological integrity; CNS: Pharmacological therapies; CL: Knowledge

12. Which of the following nursing interventions would be included in the care of a client with Reye's syndrome who is receiving pancuronium (Pavulon)?
1. Applying artificial tears as needed
2. Providing regular tactile stimulation
3. Performing active range-of-motion (ROM) exercises
4. Placing the client in a supine position

13. Which of the following goals would be achieved by performing a craniectomy on a client with Reye's syndrome?
1. Decrease carbon dioxide levels
2. Determine the extent of brain injury
3. Reduce pressure from an edematous brain
4. Allow continuous monitoring of intracranial pressure (ICP)

14. Parents of a child with Reye's syndrome need a great deal of emotional support. Which of the following nursing interventions should be made to reduce stress and alleviate fears?
1. Not accepting aggressive behavior from parents
2. Not allowing the expression of feelings and concerns
3. Letting parents interpret the child's behaviors and responses
4. Explaining therapies and clarifying or reinforcing the information given

15. Which of the following clinical manifestations would you expect to see in a client in stage V of Reye's syndrome?
1. Vomiting, lethargy, and drowsiness
2. Seizures, flaccidity, and respiratory arrest
3. Hyperventilation and coma
4. Disorientation, aggressiveness, and combativeness

What are adverse effects of some paralytic agents?

Be sensitive not only to your client's needs but to the family's needs as well.

12. 1. Pancuronium suppresses the corneal reflex, making the eyes prone to irritation. Artificial tears prevent drying. Tactile stimulation isn't appropriate because it may elicit a pressure response. Active ROM exercises may cause an increase in pressure. The head of the bed should be elevated slightly, with the paralyzed client in a side-lying or semi-prone position to prevent aspiration and minimize intracranial pressure.
NP: Implementation; CN: Physiological integrity; CNS: Pharmacological therapies; CL: Application

13. 3. In severe cases of cerebral edema, creating bilateral bone flaps (craniectomy) is most effective in decreasing ICP. Carbon dioxide levels can be decreased through mechanical ventilation. Most clients with Reye's syndrome recover without any resulting brain injury. Continuous monitoring of ICP is implemented through central venous pressure lines.
NP: Evaluation; CN: Physiological integrity; CNS: Reduction of risk potential; CL: Knowledge

14. 4. Explaining treatments and therapies will help alleviate undue stress in the parents. An awareness of the potential for aggressive behaviors provides nurses with the understanding that helps them support the parents in their grief. Parents must express feelings and concerns so that the nurse can help them through the grief. Parents may need help interpreting their child's behavior to avoid assigning erroneous meanings to the many signs their child exhibits.
NP: Implementation; CN: Physiological integrity; CNS: Coping and adaptation; CL: Application

15. 2. Staging criteria were developed to help evaluate the client's progress and to evaluate the efficacy of therapies. The clinical manifestations of stage V include seizures, loss of deep tendon reflexes, flaccidity, and respiratory arrest. Vomiting, lethargy, and drowsiness occur in stage I. Hyperventilation and coma occur in stage III. Disorientation and aggressive behavior occur in stage II.
NP: Data collection; CN: Physiological integrity; CNS: Physiological adaptation; CL: Knowledge

16. Vaccinating a child against preventable diseases represents which of the following types of immunity?

1. Acquired immunity
2. Active immunity
3. Natural immunity
4. Passive immunity

Activ-ate your mind on this one! (That's a hint!)

16. 2. Active immunity occurs when the individual forms immune bodies against certain diseases, either by having the disease or by the introduction of a vaccine into the individual. Acquired immunity results from exposure to the bacteria, virus, or toxins. Natural immunity is resistance to infection or toxicity. Passive immunity is a temporary immunity caused by transfusion of immune plasma proteins.

NP: Evaluation; CN: Health promotion and maintenance; CNS: Prevention and early detection of disease; CL: Knowledge

17. What is the recommended age for beginning primary immunizations of normal infants?

1. Birth
2. 1 month
3. 6 months
4. 1 year

17. 1. According to the American Academy of Pediatrics, birth is the recommended age for beginning primary immunizations, starting with hepatitis B vaccine.

NP: Evaluation; CN: Health promotion and maintenance; CNS: Prevention and early detection of disease; CL: Knowledge

18. What is the recommended age for receiving the measles vaccine?

1. 6 months
2. 12 months
3. 15 months
4. 18 months

18. 2. Because of the presence of maternal antibodies, the measles vaccine should be delayed until age 12 months for infants who live in communities where the disease isn't prevalent. Only during outbreaks should the vaccine be given before age 12 months.

NP: Evaluation; CN: Health promotion and maintenance; CNS: Prevention and early detection of disease; CL: Knowledge

Whew! That's a lot of immunizations!

19. Which of the following immunizations should a healthy 4-month-old client receive?

1. Measles and inactivated poliovirus vaccine (IPV)
2. Measles, mumps, and rubella (MMR)
3. Diphtheria, tetanus, and pertussis (DTP)
4. DTP and IPV

19. 4. At age 4 months, DTP and OPV are the recommended immunizations. DTP is given again alone at age 6 months. MMR is given at age 12 months.

NP: Evaluation; CN: Health promotion and maintenance; CNS: Prevention and early detection of disease; CL: Knowledge

20. An infant with an immune deficiency should receive which type of polio vaccine?

1. Oral poliovirus vaccine
2. Inactivated poliovirus vaccine (IPV)
3. Either form
4. Neither form

20. 2. For all infants and children, an all-IPV schedule is recommended by the American Academy of Pediatrics for routine childhood polio vaccination in the United States. The four doses should be given at ages 2 months, 4 months, 6 to 18 months, and 4 to 6 years. The oral form is no longer given.

NP: Evaluation; CN: Health promotion and maintenance; CNS: Prevention and early detection of disease; CL: Knowledge

21. Which of the following conditions is the general contraindication for giving all immunizations?

1. Cold
2. Cough
3. Febrile illness
4. Poor appetite

22. Which of the following schedules is recommended for immunization of normal infants and children in the first year of life?

1. Birth, 2 months, 4 months, 6 months, 12 months
2. 1 month, 3 months, 5 months, 9 months
3. 2 months, 6 months, 9 months, 12 months
4. 2 months, 4 months, 6 months, 12 to 15 months

23. Which of the following directions is most important when administering immunizations?

1. Properly store the vaccine, and follow the recommended procedure for injection.
2. Monitor clients for approximately 1 hour after administration for adverse reactions.
3. Take the vaccine out of refrigeration 1 hour before administration.
4. Inject multiple vaccines at the same injection site.

24. Which of the following symptoms is the most common manifestation of severe combined immunodeficiency disease (SCID)?

1. Bruising
2. Failure to thrive
3. Prolonged bleeding
4. Susceptibility to infection

Remember, the word *contraindication* typically signals a seemingly negative question.

Stop and think back: Immunizations commonly come as suspensions, don't they?

21. 3. A child who is already ill should avoid the risk of additional adverse effects of the vaccine. Symptoms of the disease could be mistakenly identified as adverse effects of the vaccine. The presence of minor illnesses, such as a common cold, cough, or poor appetite, wouldn't be a contraindication.

NP: Evaluation; CN: Health promotion and maintenance; CNS: Prevention and early detection of disease; CL: Knowledge

22. 1. The nurse needs to be aware of the schedule for immunizations as well as the latest recommendations for their use. According to the American Academy of Pediatrics, the recommended age for beginning primary immunizations of normal infants is at birth.

NP: Planning; CN: Health promotion and maintenance; CNS: Prevention and early detection of disease; CL: Knowledge

23. 1. Vaccines must be properly stored to ensure their potency. The nurse must be familiar with the manufacturer's directions for storage and reconstitution of the vaccine. Faulty refrigeration is a major cause of primary vaccine failure. It isn't necessary to monitor the clients, but the nurse should teach parents to call the primary care provider and report any adverse effects. Taking the vaccine out of refrigeration too early can affect its potency. If more than one vaccine is to be administered, different injection sites should be used. The nurse should note which vaccine is given and at what site in case of a local reaction caused by the injection.

NP: Implementation; CN: Health promotion and maintenance; CNS: Prevention and early detection of disease; CL: Knowledge

24. 4. SCID is characterized by absence of both humoral and cell-mediated immunity. The most common manifestation is susceptibility to infection early in life, most often by age 3 months. SCID is characterized by chronic infection, failure to completely recover from an infection, and frequent reinfection. The history reveals no logical source for infection. Failure to thrive is a consequence of persistent illnesses. Prolonged bleeding and bruising aren't manifestations of SCID.

NP: Data collection; CN: Physiological integrity; CNS: Physiological adaptation; CL: Knowledge

NP: Nursing process   CN: Client needs category   CNS: Client needs subcategory   CL: Cognitive level

25. A child is admitted to the hospital for an asthma exacerbation. The nursing history reveals this client was exposed to chickenpox 1 week ago. When would this client require isolation if he were to remain hospitalized?
 1. Isolation isn't required.
 2. Immediate isolation is required.
 3. Isolation would be required 10 days after exposure.
 4. Isolation would be required 12 days after exposure.

26. On observation of a child's skin, the nurse notes a papular pruritic rash with some vesicles. The rash is profuse on the trunk and sparse on the distal limbs. Based on this observation, which of the following illnesses does the client have?
 1. Measles
 2. Mumps
 3. Roseola
 4. Chickenpox

27. Which of the following responses would be appropriate to a parent inquiring about when her child with chickenpox can return to school?
 1. When the client is afebrile
 2. When all vesicles have dried
 3. When vesicles begin to crust over
 4. When lesions and vesicles are gone

28. Which of the following symptoms are clinical manifestations associated with roseola?
 1. Apparent sickness, fever, and rash
 2. Fever for 3 to 4 days, followed by rash
 3. Rash, without history of fever or illness
 4. Rash for 3 to 4 days, followed by high fevers

Don't be chicken! (Wink, wink.) You're doing great!

Chickenpox is highly contagious. Teach parents how to determine when it's safe to send their children back to school.

25. 2. The incubation period for chickenpox is 2 to 3 weeks, commonly 13 to 17 days. A client is commonly isolated 1 week after exposure to avoid the risk of an earlier breakout. A person is infectious from 1 day before eruption of lesions to 6 days after the vesicles have formed crusts.
NP: Implementation; CN: Safe, effective care environment; CNS: Safety and infection control; CL: Application

26. 4. Chickenpox rash is highly pruritic. The rash begins as a macule, rapidly progresses to a papule, then becomes a vesicle. All three stages are present in varying degrees at one time. Measles begin as an erythematous maculopapular eruption on the face; the eruption gradually spreads downward. Mumps isn't associated with a skin rash. Roseola rash is nonpruritic and is described as discrete rose pink macules, appearing first on the trunk and then spreading to the neck, face, and extremities.
NP: Data collection; CN: Health promotion and maintenance; CNS: Prevention and early detection of disease; CL: Knowledge

27. 2. Chickenpox is contagious. It's transmitted through direct contact, droplet spread, and contact with contaminated objects. Vesicles break open; therefore, a person is potentially contagious until all vesicles have dried. It isn't necessary to wait until dried lesions have disappeared. Some vesicles may be crusted over, and new ones may have formed. Macules, papules, vesicles, and crusting are present in varying degrees at one time. A child may be fever-free but continue to have vesicles. Isolation is usually necessary only for about 1 week after the onset of the disease.
NP: Implementation; CN: Safe, effective care environment; CNS: Safety and infection control; CL: Knowledge

28. 2. Roseola is manifested by persistent high fever for 3 to 4 days in a client who appears well. Fever precedes the rash. When the rash appears, a precipitous drop in fever occurs and the temperature returns to normal.
NP: Data collection; CN: Health promotion and maintenance; CNS: Prevention and early detection of disease; CL: Knowledge

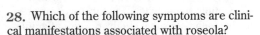

29. Which finding is consistent with a roseola rash?

1. Maculopapular red spots
2. Macular and pruritic, with papules and vesicles
3. Rose pink macules that fade on pressure
4. Red maculopapular eruption, beginning on the face

Everything's coming up roseola! Yuk, yuk, yuk.

29. 3. Roseola rashes are discrete, rose pink macules or maculopapules that fade on pressure and usually last 1 to 2 days. Chickenpox rash is macular, with papules and vesicles. Roseola isn't pruritic. Maculopapular red spots may be indicative of fifth disease. Measles begin as a maculopapular eruption on the face.
NP: Data collection; CN: Health promotion and maintenance; CNS: Prevention and early detection of disease; CL: Knowledge

30. Which of the following complications can be caused by a client with chickenpox scratching open and severely irritating the vesicles on his abdomen?

1. Myocarditis
2. Neuritis
3. Obstructive laryngitis
4. Secondary bacterial infection

30. 4. Secondary bacterial infections can occur as a complication of chickenpox. Irritation of skin lesions can lead to cellulitis or even an abscess. Myocarditis isn't considered a complication of chickenpox but has been noted as a complication of mumps. Neuritis has been associated with diphtheria. Obstructive laryngitis occurs as a complication of measles.
NP: Evaluation; CN: Physiological integrity; CNS: Reduction of risk potential; CL: Knowledge

31. Which of the following characteristics best describes the cough of an infant who was admitted to the hospital with suspected pertussis?

1. Nonexistent
2. Loose, nonproductive coughing episodes
3. Occurring more frequently during the day
4. Harsh, associated with a high-pitched crowing sound

31. 4. The cough associated with pertussis is a harsh series of short, rapid coughs, followed by a sudden inspiration and a high-pitched crowing sound. Cheeks become flushed or cyanotic, eyes bulge, and the tongue protrudes. Paroxysm may continue until a thick mucus plug is dislodged. This cough occurs most commonly at night.
NP: Data collection; CN: Physiological integrity; CNS: Physiological adaptation; CL: Knowledge

Think before you respond: Can a fetus contract scarlet fever?

32. Which of the following communicable diseases requires isolating an infected child from pregnant women?

1. Pertussis
2. Roseola
3. Rubella
4. Scarlet fever

32. 3. Rubella (German measles) has a teratogenic effect on the fetus. An infected child must be isolated from pregnant women. Pertussis, roseola, and scarlet fever don't have any teratogenic effects on a fetus.
NP: Implementation; CN: Safe, effective care environment; CNS: Safety and infection control; CL: Knowledge

33. Which of the following medications is the treatment of choice for scarlet fever?

1. Acyclovir (Zovirax)
2. Amphotericin B
3. Ibuprofen (Motrin)
4. Penicillin

33. 4. The causative agent of scarlet fever is group A beta-hemolytic streptococci, which is susceptible to penicillin. Erythromycin is used for penicillin-sensitive children. Anti-inflammatory drugs, such as ibuprofen, aren't indicated for these clients. Acyclovir is used in the treatment of herpes infections. Amphotericin B is used to treat fungal infections.
NP: Implementation; CN: Physiological integrity; CNS: Pharmacological therapies; CL: Knowledge

NP: Nursing process   CN: Client needs category   CNS: Client needs subcategory   CL: Cognitive level

34. Which of the following periods of isolation is indicated for a client with scarlet fever?
1. Until the associated rash disappears
2. Until completion of antibiotic therapy
3. Until the client is fever-free for 24 hours
4. Until 24 hours after initiation of treatment

35. Which of the following instructions should be included in the teaching about care of a client with chickenpox?
1. Administer penicillin or erythromycin as ordered.
2. Administer local or systemic antipruritics as ordered.
3. Offer periods of interaction with other children to provide distraction.
4. Avoid administering varicella-zoster immune globulin to clients receiving long-term salicylate therapy.

36. A child is being treated with penicillin for scarlet fever. Which of the following types of infection does scarlet fever follow?
1. Roseola
2. Staphylococcal parotitis
3. Streptococcal pharyngitis
4. Chickenpox

37. A client infected with human immunodeficiency virus (HIV) inquires about the possibility of breast-feeding her neonate. Which of the following would be a correct response?
1. Breast-feeding isn't an option.
2. Breast-feeding would be best for your neonate.
3. Breast-feeding is only an option if the client is taking zidovudine (Retrovir).
4. Breast-feeding is an option if milk is expressed and fed by a bottle.

Congratulations! You've finished 35 questions! Keep up the good work!

In which body fluids has HIV been isolated?

34. 4. A client requires respiratory isolation until 24 hours after initiation of treatment. Rash may persist for 3 weeks. It isn't necessary to wait until the end of treatment. Fever usually breaks 24 hours after therapy has begun. It isn't necessary to maintain isolation for an additional 24 hours.
NP: Implementation; CN: Safe, effective care environment; CNS: Safety and infection control; CL: Knowledge

35. 2. Chickenpox is highly pruritic. Preventing the child from scratching is necessary to prevent scarring and secondary infection caused by irritation of lesions. Penicillin and erythromycin aren't usually used in the treatment of chickenpox. Interaction with other children would be contraindicated due to the risk of communication, unless the other children previously have had chickenpox or been immunized. Varicella-zoster immune globulin *should* be administered to exposed children who are on long-term aspirin therapy because of the possible risk of Reye's syndrome.
NP: Implementation; CN: Physiological integrity; CNS: Pharmacological therapies; CL: Knowledge

36. 3. The causative agent of scarlet fever is group A beta-hemolytic streptococci; therefore, scarlet fever may follow a strep throat infection. Roseola, chickenpox, and parotitis aren't strep infections and don't contribute to scarlet fever.
NP: Data collection; CN: Physiological integrity; CNS: Reduction of risk potential; CL: Knowledge

37. 1. Clients infected with HIV are unable to breast-feed because HIV has been isolated in breast milk and could be transmitted to the infant. Taking zidovudine doesn't prevent transmission. The risk of breast-feeding is associated not with direct contact with the breast but with the possibility of HIV contained in the breast milk.
NP: Implementation; CN: Health promotion and maintenance; CNS: Prevention and early detection of disease; CL: Application

38. Which finding aids in diagnosing human immunodeficiency virus infection in children?
1. Excessive weight gain
2. Frequent ear infections
3. Intermittent diarrhea
4. Tolerance of feedings

38. 3. A differential diagnosis may be based on the presence of an underlying cellular immunodeficiency-related disease; symptoms include intermittent episodes of diarrhea, repeated respiratory infections, and the inability to tolerate feedings. Poor weight gain and failure to thrive are objective assessment findings that result from intolerance of feedings and frequent infections.
NP: Data collection; CN: Physiological integrity; CNS: Physiological adaptation; CL: Knowledge

39. Which of the following factors makes it difficult to diagnose human immunodeficiency virus infection during the first 15 months of life?
1. Presence of maternal antibody
2. Unavailability of proper testing
3. Onset of symptoms at age 2
4. Transmission occurring after birth

Antibody can see this is an important question. (Oops! I gave you a hint, didn't I?)

39. 1. A confirmed diagnosis is difficult during the first 15 months because of the presence of maternal antibody. Onset of symptoms can occur before age 1, and the prognosis is poorer than if the onset of symptoms occurs between ages 2 and 3. Transmission of the virus occurs across the placenta in more than 50% of pregnancies in which the mother is infected.
NP: Data collection; CN: Health promotion and maintenance; CNS: Prevention and early detection of disease; CL: Knowledge

40. Which finding suggests Kawasaki disease?
1. Fever response to antipyretics
2. Vesicular rash on arms and legs
3. Edema and erythema of the hands and feet
4. Inflamed conjunctivitis with watery drainage

40. 3. The clinical manifestations of Kawasaki disease include changes in the extremities, such as peripheral edema, peripheral erythema, and desquamation of the palms and the soles of the feet. Kawasaki disease is an acute febrile illness of unknown etiology that doesn't respond well to antipyretics. Rash associated with Kawasaki disease is usually located primarily on the trunk and is nonvesicular. Inflamed, watery conjunctivitis is suggestive of viral conjunctivitis.
NP: Data collection; CN: Physiological integrity; CNS: Physiological adaptation; CL: Knowledge

Groovy! You're surfing through this test. Keep it up!

41. Sickle cell anemia occurs primarily in which of the following ethnic groups?
1. African-American
2. Asian
3. Caucasian
4. Hispanic

41. 1. Sickle cell anemia is primarily found in African Americans. Infrequently, it affects Whites, especially those of Mediterranean descent. Among African Americans, the incidence is reported to be as high as 45%. It's estimated that 1 in 12 African Americans carries the trait. Sickle cell anemia is rare among Asians and Hispanics.
NP: Data collection; CN: Physiological integrity; CNS: Physiological adaptation; CL: Knowledge

42. A child comes to the emergency department suspected of being in vaso-occlusive crisis. Which of the following assessment findings would indicate that the client is having a vaso-occlusive crisis?
1. Precipitous drop in blood volume
2. Decreased red blood cell (RBC) production
3. Anemia, jaundice, and reticulocytosis
4. Acute abdominal pain and hand-foot syndrome

Question 42 is testing your knowledge of the four types of episodic crisis.

42. 4. There are four types of episodic crises in sickle cell anemia: vaso-occlusive, splenic sequestration, aplastic, and hyperhemolytic. Vaso-occlusive crises are the most common and the only painful ones. They're the result of sickled cells obstructing the blood vessels. The major symptoms are fever, acute abdominal pain from visceral hypoxia, hand-foot syndrome, and arthralgia, without an exacerbation of anemia. A precipitous drop in blood volume is indicative of a splenic sequestration crisis. Aplastic crisis exhibits diminished RBC production. Hyperhemolytic crisis is characterized by anemia, jaundice, and reticulocytosis.
NP: Data collection; CN: Physiological integrity; CNS: Physiological adaptation; CL: Knowledge

43. Which of the following responses would be appropriate for the parent inquiring about a child who tests positive for sickle cell trait?
1. Your child has sickle cell anemia.
2. Your child is a carrier of the disorder but doesn't have sickle cell anemia.
3. Your child is a carrier of the disease and will pass the disease to any offspring.
4. Your child doesn't have the disease at present but may show evidence of the disease as he gets older.

Know the difference between a positive test for a sickle cell trait and a positive test for sickle cell anemia.

43. 2. A child with sickle cell trait is only a carrier and may never show any symptoms, except under special hypoxic conditions. A child with sickle cell trait doesn't have the disease and will never test positive for sickle cell anemia. Sickle cell anemia would be transmitted to offspring only as the result of a union between two individuals who are positive for the trait.
NP: Implementation; CN: Health promotion and maintenance; CNS: Prevention and early detection of disease; CL: Application

44. What is the primary nursing objective in caring for a client with sickle cell anemia in vaso-occlusive crisis?
1. Managing pain
2. Promoting activity
3. Promoting sickling
4. Preventing tissue oxygenation

44. 1. Pain management is an important aspect in the care of a client with sickle cell anemia in vaso-occlusive crisis. The goal is to prevent sickling. This can be accomplished by promoting tissue oxygenation, hydration, and rest, which minimize energy expenditure and oxygen utilization.
NP: Evaluation; CN: Physiological integrity; CNS: Basic care and comfort; CL: Analysis

45. Which of the following complications is the reason for not palpating the abdomen of a child in vaso-occlusive crisis?
1. Risk of splenic rupture
2. Risk of inducing vomiting
3. Presence of abdominal pain
4. Risk of blood cell destruction

Be careful! The word not is important to the understanding of this question.

45. 1. Palpating a child's abdomen in vaso-occlusive crisis should be avoided because sequestered red blood cells may precipitate splenic rupture. Abdominal pain alone wouldn't be a reason to avoid palpation. Vomiting or blood cell destruction wouldn't occur from palpation of the abdomen.
NP: Implementation; CN: Physiological integrity; CNS: Reduction of risk potential; CL: Knowledge

46. Which nursing measure would be included in the care of a client in vaso-occlusive crisis to maximize tissue perfusion?
1. Administering analgesics
2. Monitoring fluid restrictions
3. Encouraging activity as tolerated
4. Administering oxygen as prescribed

Think: Maximize.

46. 4. Administering oxygen on a short-term basis helps to prevent hypoxia, which leads to metabolic acidosis, causing sickling. Long-term oxygen therapy will depress erythropoiesis. Analgesics are used to control pain. Hydration is essential to promote hemodilution and maintain electrolyte balance. Bed rest should be promoted to reduce oxygen utilization.

NP: Implementation; CN: Physiological integrity; CNS: Reduction of risk potential; CL: Application

47. Which of the following nursing measures is most important to decrease the surgical risks of a client with sickle cell anemia?
1. Increasing fluids
2. Preparing the child psychologically
3. Discouraging coughing
4. Limiting the use of analgesics

47. 1. The main surgical risk from anesthesia is hypoxia; however, emotional stress, demands of wound healing, and the potential for infection can each increase the sickling phenomenon. Increased fluids are encouraged because keeping the child well-hydrated is important for hemodilution to prevent sickling. Preparing the child psychologically to decrease fear will minimize undue emotional stress. Deep coughing is encouraged to promote pulmonary hygiene and prevent respiratory tract infection. Analgesics are used to control wound pain and to prevent abdominal splinting and decreased ventilation.

NP: Implementation; CN: Health promotion and maintenance; CNS: Prevention and early detection of disease; CL: Application

Don't let question 48 trip you up! It's asking you to prioritize.

48. Which of the following factors should be included as a priority in teaching parents about prevention of infection in children with sickle cell anemia?
1. Providing adequate nutrition
2. Avoiding emotional stress
3. Visiting the physician when sick
4. Avoiding strenuous physical exertion

48. 1. The nurse must stress adequate nutrition. Frequent medical supervision is imperative to prevention because infection is often a predisposing factor toward development of a crisis. Avoiding strenuous physical exertion and emotional stress is an important aspect to prevent sickling, but adequate nutrition remains a priority.

NP: Implementation; CN: Health promotion and maintenance; CNS: Prevention and early detection of disease; CL: Application

49. Thalassemia is seen predominantly in which of the following ethnic groups?
1. Blacks
2. Asians
3. Greeks
4. Hispanics

49. 3. Thalassemia is highly prevalent in people living near the Mediterranean Sea, namely Greeks, Italians, and Syrians. Evidence suggests that the high prevalence among these groups is a result of selective damage of the trait as a result of malaria.

NP: Data collection; CN: Health promotion and maintenance; CNS: Prevention and early detection of disease; CL: Knowledge

50. Which finding would indicate vaso-occlusive crisis?

1. Pain upon urination
2. Pain with ambulation
3. Throat pain
4. Fever with associated rash

51. The nurse is working overnight in the emergency department when a client is admitted in sickle cell crisis. Which intervention should the nurse expect to perform?

1. Giving blood transfusions
2. Giving antibiotics
3. Increasing fluid intake and giving analgesics
4. Preparing the client for a splenectomy

52. What is the nurse's role with the parents of a client who has been diagnosed with sickle cell anemia?

1. Encouraging selective birth methods or abortion
2. Referring only sickle cell positive parents for counseling
3. Rendering support to parents of newly diagnosed clients
4. Reinforcing that transmission is unlikely in subsequent pregnancies

53. A client is brought to the emergency department in hypovolemic shock secondary to vaso-occlusive crisis. Which of the following types of crises can lead to shock?

1. Aplastic
2. Hyperhemolytic
3. Splenic sequestration
4. Vaso-occlusive

Careful! The context of this question is telling you to prioritize!

Excellent! You're over halfway finished! Keep on truckin'!

50. 2. Bone pain is one of the major symptoms of vaso-occlusive crisis. Hand-foot syndrome, characterized by edematous painful extremities, is usually exhibited in the refusal of the child to bear weight and ambulate. Painful urination doesn't occur, but sickle cell anemia can cause kidney abnormalities. Throat pain isn't a symptom of vaso-occlusive crisis. Fever is one of the major symptoms of vaso-occlusive crisis but isn't associated with rash.

NP: Data collection; CN: Physiological integrity; CNS: Physiological adaptation; CL: Knowledge

51. 3. The primary therapy for sickle cell crisis is to increase fluid intake according to age and to give analgesics. Blood transfusions are only given conservatively to avoid iron overload. Antibiotics are given to clients with fever. Routine splenectomy isn't recommended. Splenectomy in clients with sickle cell anemia is controversial.

NP: Implementation; CN: Physiological integrity; CNS: Physiological adaptation; CL: Application

52. 3. The nurse can be instrumental in providing genetic counseling. She can give parents correct information about the disease and render support to parents of newly diagnosed clients. Alternative birth methods are discussed, but parents make their own decisions. All heterozygous, or trait-positive, parents should be referred for genetic counseling. The risk of transmission in subsequent pregnancies remains the same.

NP: Implementation; CN: Health promotion and maintenance; CNS: Prevention and early detection of disease; CL: Application

53. 3. Splenic sequestration crises are caused by large quantities of blood pooling in the spleen, causing a precipitous drop in blood volume and ultimately shock. Aplastic crises occur when red blood cell production is diminished. Hyperhemolytic crises are characterized by anemia, jaundice, and reticulocytosis. Vaso-occlusive crises occur without pooling of blood and without an exacerbation of anemia.

NP: Evaluation; CN: Physiological integrity; CNS: Physiological adaptation; CL: Knowledge

54. The nurse is administering a blood transfusion to a client with sickle cell anemia. Which of the following assessment findings would indicate that the client is having a transfusion reaction?
1. Diaphoresis and hot flashes
2. Urticaria, flushing, and wheezing
3. Fever, urticaria, and red, raised rash
4. Fever, disorientation, and abdominal pain

55. How often should a child receive the influenza virus vaccine?
1. Annually
2. Twice a year
3. Contraindicated in children
4. Only with the outbreak of illness

56. A 3-year-old sister of a neonate is diagnosed with pertussis. The mother gives a history of having been immunized as a child. Which of the following information should be included in teaching the mother about possible infection of her neonate?
1. The neonate will inevitably contract pertussis.
2. Immune globulin is effective in protecting the neonate.
3. The risk to the neonate depends on the mother's immune status.
4. Erythromycin should be administered prophylactically to the neonate.

57. An increased concentration of hemoglobin A_2 is found in children with which of the following disorders?
1. Beta-thalassemia trait
2. Iron deficiency
3. Lead poisoning
4. Sickle cell anemia

One wrong part of an option makes the entire option wrong.

Prevention is often the best medicine!

You beta get this one right!

54. 2. Allergic reactions may occur when the recipient reacts to allergens in the donor's blood; this reaction causes urticaria, flushing, and wheezing. A febrile reaction can occur, causing fever and urticaria but it isn't accompanied by rash. Diaphoresis, hot flashes, disorientation, and abdominal pain aren't symptoms of a transfusion reaction.
NP: Data collection; CN: Physiological integrity; CNS: Reduction of risk potential; CL: Knowledge

55. 1. The influenza virus vaccine is usually administered annually. The vaccine isn't contraindicated in children but is targeted at clients with chronic cardiac, pulmonary, hematologic, and neurologic problems. The vaccine is given to prevent the onset of illness before an outbreak occurs.
NP: Planning; CN: Health promotion and maintenance; CNS: Prevention and early detection of disease; CL: Knowledge

56. 4. In exposed, high-risk clients, such as neonates, erythromycin may be effective in preventing or lessening the severity of the disease if administered during the preparoxysmal stage. Immune globulin isn't indicated, it's used as an immunization against hepatitis A. Neonates exposed to pertussis are at considerable risk for infections, regardless of the mother's immune status; however, infection isn't inevitable.
NP: Implementation; CN: Health promotion and maintenance; CNS: Prevention and early detection of disease; CL: Application

57. 1. The concentration of hemoglobin A_2 is increased with beta-thalassemia trait. In severe iron deficiency, hemoglobin A_2 may be decreased. The hemoglobin A_2 level is normal in sickle cell anemia and in lead poisoning.
NP: Data collection; CN: Physiological integrity; CNS: Physiological adaptation; CL: Knowledge

58. A 4-year-old client has a petechial rash but is otherwise well. The platelet count is 20,000/µl and the hemoglobin level and white blood cell (WBC) count are normal. Which of the following diagnoses is most likely?

1. Acute lymphoblastic leukemia (ALL)
2. Disseminated intravascular coagulation (DIC)
3. Idiopathic thrombocytopenic purpura (ITP)
4. Systemic lupus erythematosus (SLE)

59. Which of the following instructions would be included in the nurse's discharge teaching for the parents of a neonate diagnosed with sickle cell anemia?

1. Stressing the importance of iron supplementation
2. Stressing the importance of monthly vitamin B$_{12}$ injections
3. Demonstrating how to take a temperature
4. Explaining that polyvalent pneumococcal vaccine is contraindicated

60. Which of the following findings yields a poor prognosis for a client with leukemia?

1. Presence of a mediastinal mass
2. Late central nervous system leukemia
3. Normal white blood cell (WBC) count at diagnosis
4. Disease presents between ages 2 and 10

Question 60 asks about a *prognosis,* not a diagnosis.

61. A 1-year-old male client in the pediatrician's office for an examination is noted to be pale. He's in the 75th percentile for weight and the 25th percentile for length. His physical examination is normal, but his hematocrit is 24%. Which of the following questions would help to establish a diagnosis of anemia?

1. Is the client on any medications?
2. What is the client's usual daily diet?
3. Did the client receive phototherapy for jaundice?
4. What is the pattern and appearance of bowel movements?

58. 3. The onset of ITP typically occurs between ages 1 and 6. Clients look well, except for petechial rash. ALL is associated with a low platelet count but an *abnormal* hemoglobin level and WBC count. DIC is secondary to a severe underlying disease. SLE is rare in a 4-year-old child.
NP: Data collection; CN: Physiological integrity; CNS: Physiological adaptation; CL: Knowledge

59. 3. A temperature of 101.3° to 102.2° F (38.5° to 39° C) calls for emergency evaluation, even if the neonate appears well. Folic acid requirement is increased; therefore, supplementation is prudent. Vitamin B$_{12}$ supplementation and iron supplementation aren't necessary. Pneumococcal vaccine is used because children with sickle cell anemia are prone to infection with *Streptococcus pneumoniae.*
NP: Implementation; CN: Health promotion and maintenance; CNS: Prevention and early detection of disease; CL: Knowledge

60. 1. The presence of a mediastinal mass indicates a poor prognosis for clients with leukemia. The prognosis is poorer if age at onset is less than 2 years or greater than 10 years. A WBC count of 100,000/µl or higher and early central nervous system leukemia also indicate a poor prognosis for a client with leukemia.
NP: Data collection; CN: Physiological integrity; CNS: Physiological adaptation; CL: Knowledge

61. 2. Iron deficiency anemia is the most common nutritional deficiency in clients between ages 9 months and 15 months. Anemia in a 1-year-old client is mostly nutritional in origin, and its cause will be suggested by a detailed nutritional history. None of the other selections would be helpful in diagnosing anemia.
NP: Data collection; CN: Health promotion and maintenance; CNS: Prevention and early detection of disease; CL: Knowledge

62. Which of the following statements regarding Hodgkin's disease is true?
1. Staging laparotomy is mandatory for every client.
2. Diaphoresis and chills can be symptoms.
3. Hodgkin's disease is rare before age 5.
4. Incidence of Hodgkin's disease peaks between ages 11 and 15.

63. Which of the following percentages represents the incidence of perinatal transmission of human immunodeficiency virus (HIV) infection from an infected client to her neonate?
1. 10% to 25%
2. 25% to 50%
3. 50% to 75%
4. 100%

64. A 14-year-old female client is seen in the pediatrician's office with a history of mild sore throat, low-grade fever, and a diffuse maculopapular rash. She now complains of swelling of her wrists and redness in her eyes. The client is exhibiting symptoms of which condition?
1. Rubella
2. Rubeola
3. Roseola
4. Erythema infectiosum

65. Which treatment would be most commonly recommended for a client with iron deficiency anemia?
1. Blood transfusion
2. Oral ferrous sulfate
3. An iron-fortified cereal
4. Intramuscular iron dextran

Read this question carefully. It uses such terms as incidence and perinatal and could be confusing.

Don't give up. You know the answer.

62. 3. Hodgkin's disease is rare before age 5. Systemic symptoms of Hodgkin's disease include fever, night sweats, malaise, weight loss, and pruritus. The peak incidence of Hodgkin's disease occurs in late adolescence and young adulthood (ages 15 to 34). Staging laparotomy isn't recommended for clients who have obvious intra-abdominal disease easily diagnosed by noninvasive studies.
NP: Evaluation; CN: Physiological integrity; CNS: Physiological adaptation; CL: Knowledge

63. 2. The risk of HIV infection in the neonate of a client who is HIV seropositive is in the range of 20% to 40%, if the client hasn't previously delivered an HIV-infected neonate. If the client has had a previous neonate with HIV infection, the risk to a subsequent neonate may be higher.
NP: Evaluation; CN: Health promotion and maintenance; CNS: Prevention and early detection of disease; CL: Knowledge

64. 1. Symptoms of rubella include a diffuse maculopapular rash lasting 3 days, mild sore throat, low-grade fever and, occasionally, conjunctivitis, arthralgia, or arthritis. Rubeola is associated with high fever, which reaches its peak at the height of a generalized macular rash and typically lasts for 5 days. Roseola involves high fever and is abruptly followed by a rash. Erythema infectiosum (fifth disease) begins with bright erythema on the cheeks ("slapped cheek" sign), followed by a red maculopapular rash on the trunk and extremities that fades centrally at first.
NP: Data collection; CN: Physiological integrity; CNS: Physiological adaptation; CL: Knowledge

65. 2. A prompt rise in hemoglobin level and hematocrit follows the administration of oral ferrous sulfate. Blood transfusion is rarely indicated unless a child becomes symptomatic or is further compromised by a superimposed infection. Dietary modifications are appropriate long-term measures, but they won't make enough iron available to replenish iron stores. Intramuscular iron dextran is reserved for situations in which compliance can't be achieved because it's expensive, painful, and no more effective than oral iron.
NP: Evaluation; CN: Physiological integrity; CNS: Pharmacological therapies; CL: Knowledge

66. Which statement by a caregiver indicates that a 10-month-old client is at high risk for iron deficiency anemia?
1. "The baby is sleeping through the night without a bottle."
2. "The baby drinks about five 8-oz bottles of milk per day."
3. "The baby likes egg yolk in his cereal."
4. "The baby dislikes some vegetables, especially carrots."

66. 2. The recommended intake of milk, which doesn't contain iron, is 24 oz per day; 40 oz per day exceeds the recommended allotment and may reduce iron intake from solid food sources, risking iron deficiency anemia. Sleeping through the night without a bottle is an anticipated behavior at this age. Egg yolk is a good source of iron and would minimize any risk factor related to nutritional anemia. Because only dark green vegetables are good sources of iron, a dislike of carrots wouldn't be significant.
NP: Data collection; CN: Physiological integrity; CNS: Basic care and comfort; CL: Analysis

67. A 6-year-old client has been diagnosed with Rocky Mountain spotted fever. In teaching the parents about the cause of the illness, the nurse would be correct in telling them that a bite by which of the following animals or insects caused the illness?
1. Cat
2. Mosquito
3. Spider
4. Tick

67. 4. Rocky Mountain spotted fever is caused by *Rickettsia rickettsii*, which is transmitted by the bite of a tick. Spider, mosquito, and cat bites haven't been known to transmit *R. rickettsii*.
NP: Implementation; CN: Safe, effective care environment; CNS: Safety and infection control; CL: Knowledge

68. An iron dextran (Imferon) injection has been ordered for an 8-month-old client with iron deficiency anemia whose parents haven't been compliant with oral supplements. What is the correct method of injection for Imferon?
1. Intradermally
2. Subcutaneously
3. Intramuscularly
4. Intramuscularly, using the Z-track method

68. 4. If iron dextran is ordered, it must be injected deeply into a large muscle mass, using the Z-track method to minimize skin staining and irritation. Neither a subcutaneous nor an intradermal injection would inject the dextran into the muscle. The Z-track method is preferred over a normal intramuscular injection.
NP: Implementation; CN: Physiological integrity; CNS: Pharmacological therapies; CL: Application

69. Which of the following foods are appropriate sources of dietary iron for the prevention of nutritional anemia?
1. Citrus fruits
2. Fish
3. Green vegetables
4. Milk products

69. 3. Green vegetables are good sources of iron. Citrus foods aren't sources of iron but help with the absorption of iron. Fish isn't a good source of dietary iron. Milk is deficient in iron and should be limited.
NP: Implementation; CN: Health promotion and maintenance; CNS: Prevention and early detection of disease; CL: Knowledge

70. Which of the following instructions should the nurse provide when teaching parents the proper administration of oral iron supplements?
1. Give the supplements with food.
2. Stop medication if vomiting occurs.
3. Decrease dose if constipation occurs.
4. Give the medicine via a dropper or through a straw.

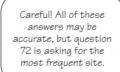

You're already at question 70?! Super! How about a snack?

70. 4. Liquid iron preparations may temporarily stain the teeth; therefore, the drug should be given by dropper or through a straw. Supplements should be given between meals, when the presence of free hydrochloric acid is greatest. Constipation can be decreased by increasing intake of fruits and vegetables. If vomiting occurs, supplementation shouldn't be stopped; instead, it should be administered with food.
NP: Implementation; CN: Physiological integrity; CNS: Pharmacological therapies; CL: Knowledge

71. Which of the following symptoms is the primary clinical manifestation of hemophilia?
1. Petechiae
2. Prolonged bleeding
3. Decreased clotting time
4. Decreased white blood cell (WBC) count

71. 2. The effect of hemophilia is prolonged bleeding, anywhere from or within the body. With severe deficiencies, hemorrhage can occur as a result of minor trauma. Petechiae are uncommon in persons with hemophilia because repair of small hemorrhages depends on platelet function, not on blood clotting mechanisms. Clotting time is increased in a client with hemophilia. A decrease in WBCs is *not* indicative of hemophilia.
NP: Data collection; CN: Physiological integrity; CNS: Physiological adaptation; CL: Knowledge

72. Which of the following areas is the most frequent site of internal bleeding associated with hemophilia?
1. Brain tissue
2. GI tract
3. Joint cavities
4. Spinal cord

Careful! All of these answers may be accurate, but question 72 is asking for the most frequent site.

72. 3. The joint cavities, especially the knees, ankles, and elbows, are the most frequent site of internal bleeding. This bleeding often results in bone changes and crippling, disabling deformities. Intracranial hemorrhage occurs less frequently than expected because the brain tissue has a high concentration of thromboplastin. Hemorrhage along the GI tract and spinal cord can occur but is less common.
NP: Data collection; CN: Physiological integrity; CNS: Physiological adaptation; CL: Knowledge

73. Which of the following measures should parents of a hemophilic client be taught to prepare them to initiate immediate treatment before blood loss is excessive?
1. Apply heat to the area.
2. Don't give factor replacement.
3. Apply pressure for at least 5 minutes.
4. Immobilize and elevate the affected area.

73. 4. Elevating the area above the level of the heart will decrease blood flow. Pressure should be applied to the area for at least 10 to 15 minutes to allow clot formation. Cold, not heat, should be applied to promote vasoconstriction. Factor replacement should *not* be delayed.
NP: Implementation; CN: Physiological integrity; CNS: Physiological adaptation; CL: Application

74. A nurse is caring for a client with leukemia who has an absolute granulocyte count of 400 µl. Which of the following interventions should the nurse implement?
  1. Place the client in strict isolation.
  2. Notify the physician immediately.
  3. Restrict visitors with active infections.
  4. Begin antibiotics per protocol.

Read carefully. Question 75 is looking for a *preventive* measure.

75. Which of the following nursing measures is an important aid in prevention of the crippling effects of joint degeneration caused by hemophilia?
  1. Avoiding the use of analgesics
  2. Using aspirin for pain relief
  3. Administering replacement factor
  4. Using active range-of-motion (ROM) exercises

76. Increased tendency to bleed in which of the following areas differentiates von Willebrand's disease from hemophilia?
  1. Brain tissue
  2. GI tract
  3. Mucous membranes
  4. Spinal cord

Epistaxis is a term from early in your schooling. Remember?

77. Which of the following nursing measures should be implemented for a client with von Willebrand's disease who is having epistaxis?
  1. Lying the client supine
  2. Avoiding packing of nostrils
  3. Avoiding pressure to the nose
  4. Applying ice to the bridge of the nose

74. 3. When the absolute granulocyte count is low, a client has difficulty fighting an infection. Visitors with active infections should be restricted to prevent the client from developing an infection. The client should be placed in protective, not strict, isolation. Antibiotics shouldn't be started without a septic workup first. The physician must also be notified of the client's condition so that appropriate medical management is initiated.
NP: Planning; CN: Safe, effective care environment; CNS: Safety and infection control; CL: Application

75. 3. Prevention of bleeding is the ideal goal and is achieved by factor replacement therapy. Passive ROM exercises are usually instituted *after* the acute phase. Acetaminophen should be used for pain relief because aspirin has anticoagulant effects. Analgesics should be administered before physical therapy to control pain and provide the maximum benefit.
NP: Implementation; CN: Physiological integrity; CNS: Physiological adaptation; CL: Application

76. 3. The most characteristic clinical feature of von Willebrand's disease is an increased tendency to bleed from mucous membranes, which may be seen as frequent nosebleeds or menorrhagia. In hemophilia, the joint cavities are the most common site of internal bleeding. Bleeding into the GI tract, spinal cord, and brain tissue can occur, but these are *not* the most common sites for bleeding.
NP: Evaluation; CN: Physiological integrity; CNS: Physiological adaptation; CL: Knowledge

77. 4. Applying ice to the bridge of the nose causes vasoconstriction and may stop bleeding. The client should be instructed to sit up and lean forward to avoid aspiration of blood. Pressure should be maintained for at least 10 minutes to allow clotting to occur. Packing with tissue or cotton may be used to help stop bleeding, although care must be taken in removing packing to avoid dislodging the clot.
NP: Implementation; CN: Physiological integrity; CNS: Physiological adaptation; CL: Application

78. What are the three most important prognostic factors in determining long-term survival for children with acute leukemia?
1. Histologic type of disease, initial platelet count, and type of treatment
2. Type of treatment, stage at time of diagnosis, and client's age at diagnosis
3. Histologic type of disease, initial white blood cell (WBC) count, and client's age at diagnosis
4. Progression of illness, WBC at time of diagnosis, and client's age at time of diagnosis

Some of the same factors appear in more than one answer. Select your final answer carefully.

79. Which of the following complications are three main consequences of leukemia?
1. Bone deformities, anemia, and infection
2. Anemia, infection, and bleeding tendencies
3. Pneumonia, thrombocytopenia, and alopecia
4. Polycythemia, decreased clotting time, and infection

80. A client is seen in the pediatrician's office for complaints of bone and joint pain. Which other finding may suggest leukemia?
1. Abdominal pain
2. Increased activity level
3. Increased appetite
4. Petechiae

You, my friend, are doing extraordinarily well. Keep up the good work!

81. Which of the following assessment findings in a client with leukemia would indicate that the cancer has invaded the brain?
1. Headache and vomiting
2. Restlessness and tachycardia
3. Normal level of consciousness
4. Increased heart rate and decreased blood pressure

78. 3. The factor whose prognostic value is considered to be of greatest significance in determining long-range outcome is the histologic type of leukemia. Clients with a normal or low WBC count appear to have a much better prognosis than those with a high WBC count. Clients diagnosed between ages 2 and 10 have consistently demonstrated a better prognosis than those diagnosed before age 2 or after age 10.
NP: Evaluation; CN: Physiological integrity; CNS: Physiological adaptation; CL: Knowledge

79. 2. The three main consequences of leukemia are anemia, caused by decreased erythrocyte production, secondary to neutropenia; and infection and bleeding tendencies, from decreased platelet production. Pneumonia isn't a main consequence of leukemia. Alopecia occurs as an adverse effect of therapy. Bone deformities don't occur with leukemia. Anemia, not polycythemia, occurs. Clotting times would be prolonged.
NP: Data collection; CN: Physiological integrity; CNS: Physiological adaptation; CL: Knowledge

80. 4. The most frequent signs and symptoms of leukemia are a result of infiltration of the bone marrow. These include fever, pallor, fatigue, anorexia, and petechiae, along with bone and joint pain. Increased appetite can occur, but it usually isn't a presenting symptom. Abdominal pain is caused by areas of inflammation from normal flora within the GI tract.
NP: Data collection; CN: Physiological integrity; CNS: Physiological adaptation; CL: Knowledge

81. 1. The usual effect of leukemic infiltration of the brain is increased intracranial pressure. The proliferation of cells interferes with the flow of cerebrospinal fluid in the subarachnoid space and at the base of the brain. The increased fluid pressure causes dilation of the ventricles, which creates symptoms of severe headache, vomiting, irritability, lethargy, increased blood pressure, decreased heart rate and, eventually, coma.
NP: Data collection; CN: Physiological integrity; CNS: Physiological adaptation; CL: Knowledge

82. Which of the following types of leukemia carries the best prognosis?
1. Acute lymphoblastic leukemia
2. Acute myelogenous leukemia
3. Basophilic leukemia
4. Eosinophilic leukemia

82. 1. Acute lymphoblastic leukemia, which accounts for more than 80% of all childhood cases, carries the best prognosis. Acute myelogenous leukemia, with several subtypes, accounts for most of the other leukemias affecting children. Basophilic and eosinophilic leukemia are named for the specific cells involved. These are rarer and carry a poorer prognosis.
NP: Evaluation; CN: Physiological integrity; CNS: Physiological adaptation; CL: Knowledge

83. Which of the following is the reason to perform a spinal tap on a client newly diagnosed with leukemia?
1. To rule out meningitis
2. To decrease intracranial pressure
3. To aid in classification of the leukemia
4. To assess for central nervous system infiltration

83. 4. A spinal tap is performed to assess for central nervous system infiltration. It wouldn't be done to decrease intracranial pressure nor does it aid in the classification of the leukemia. A spinal tap can be done to rule out meningitis, but this isn't the indication for the test on a client with leukemia.
NP: Data collection; CN: Physiological integrity; CNS: Physiological adaptation; CL: Knowledge

Think back to basic physiology. Which organs metabolize drugs?

84. Which of the following tests is performed on a client with leukemia before initiation of therapy to evaluate his ability to metabolize chemotherapeutic agents?
1. Lumbar puncture
2. Liver function studies
3. Complete blood count (CBC)
4. Peripheral blood smear

84. 2. Liver and kidney function studies are done before initiation of chemotherapy to evaluate the child's ability to metabolize the chemotherapeutic agents. A CBC is performed to assess for anemia. A peripheral blood smear is done to assess the level of immature white blood cells (blastocytes). A lumbar puncture is performed to assess for central nervous system infiltration.
NP: Data collection; CN: Physiological integrity; CNS: Pharmacological therapies; CL: Knowledge

85. Which type of leukemia accounts for most cases of childhood leukemia?
1. Acute lymphocytic leukemia (ALL)
2. Acute myelogenous leukemia (AML)
3. Chronic myelogenous leukemia (CML)
4. Chronic lymphocytic leukemia (CLL)

One of these is much more common in children than the others. Do you know which one?

85. 1. The most common subtype, ALL, accounts for 75% to 80% of all childhood cases, with AML (myelocytic, myelogenous, or non-lymphoblastic) comprising approximately 20%, and CML approximately 2%. CLL occurs in older clients; 90% of cases are persons older than age 50.
NP: Data collection; CN: Physiological integrity; CNS: Physiological adaptation; CL: Knowledge

86. Which of the following medications usually is given to a client with leukemia as prophylaxis against *Pneumocystis carinii* pneumonia?
1. Co-trimoxazole (Bactrim)
2. Oral nystatin suspension
3. Prednisone (Deltasone)
4. Vincristine (Oncovin)

86. 1. The most frequent cause of death from leukemia is overwhelming infection. *P. carinii* infection is lethal to a child with leukemia. As prophylaxis against *P. carinii* pneumonia, continuous low dosages of co-trimoxazole are frequently prescribed. Oral nystatin suspension would be indicated for the treatment of thrush. Prednisone isn't an antibiotic and increases susceptibility to infection. Vincristine is an antineoplastic agent.
NP: Implementation; CN: Physiological integrity; CNS: Pharmacological therapies; CL: Knowledge

87. In which of the following diseases would bone marrow transplantation be least indicated in a newly diagnosed client?
1. Acute lymphocytic leukemia
2. Chronic myeloid leukemia
3. Severe aplastic anemia
4. Severe combined immune deficiency

Caution: The word *least* makes this almost a negative question.

87. 1. For the first episode of acute lymphocytic leukemia, conventional therapy is superior to bone marrow transplantation. In severe combined immune deficiency and in severe aplastic anemia, bone marrow transplantation has been employed to replace abnormal stem cells with healthy cells from the donor's marrow. In myeloid leukemia, bone marrow transplantation is done after chemotherapy to infuse healthy marrow and to replace marrow stem cells ablated during chemotherapy.
NP: Implementation; CN: Physiological integrity; CNS: Pharmacological therapies; CL: Knowledge

88. A 4-year-old male client is diagnosed as having acute lymphocytic leukemia. His white blood cell (WBC) count, especially the neutrophil count, is low. Which nursing action would be best for the client?
1. Protect the client from falls because of his increased risk of bleeding.
2. Protect the client from infections because his resistance to infection is decreased.
3. Provide rest periods because the oxygen-carrying capacity of the client's blood is diminished.
4. Treat constipation, which frequently accompanies a decrease in WBCs.

88. 2. One of the complications of both acute lymphocytic leukemia and its treatment is a decreased WBC count, specifically a decreased absolute neutrophil count. Because neutrophils are the body's first line of defense against infection, the client must be protected from infection. Bleeding is a risk factor if platelets or other coagulation factors are decreased. A decreased hemoglobin level, hematocrit, or both would reduce the oxygen-carrying capacity of the client's blood. Constipation isn't related to the WBC count.
NP: Implementation; CN: Safe, effective care environment; CNS: Safety and infection control; CL: Application

89. Nausea and vomiting are common adverse effects of radiation and chemotherapy. When should a nurse administer antiemetics?
1. 30 minutes before initiation of therapy
2. With the administration of therapy
3. Immediately after nausea begins
4. When therapy is completed

89. 1. Antiemetics are most beneficial if given before the onset of nausea and vomiting. To calculate the optimum time for administration, the first dose is given 30 minutes to 1 hour before nausea is expected, and then every 2, 4, or 6 hours for approximately 24 hours after chemotherapy. If the antiemetic were given with the medication or after the medication, it could lose its maximum effectiveness when needed.
NP: Implementation; CN: Physiological integrity; CNS: Pharmacological therapies; CL: Application

90. Which of the following factors would <u>contraindicate</u> the use of the suppository form of an antiemetic in a client with leukemia?
1. Constipation
2. Diarrhea
3. Rectal ulcers
4. Wouldn't be contraindicated

This question is looking for a contraindication, not an indication.

90. 3. The oral and suppository forms of antiemetics are recommended for children. Rectal ulcers would prohibit use of the rectal form because of the risk of increased tissue trauma and bleeding. Diarrhea and constipation aren't contraindications to receiving the rectal form of an antiemetic, but many facilities have protocols that prohibit rectal medications or temperatures for all pediatric oncology clients.
NP: Implementation; CN: Physiological integrity; CNS: Pharmacological therapies; CL: Application

91. Which of the following nursing measures are helpful when mouth ulcers develop as an adverse effect of chemotherapy?
1. Using lemon glycerin swabs
2. Administering milk of magnesia
3. Providing a bland, moist, soft diet
4. Frequently washing the mouth with hydrogen peroxide

91. 3. Oral ulcers are red, eroded, and painful. Providing a bland, moist, soft diet will make chewing and swallowing less painful. The use of lemon glycerin swabs and milk of magnesia should be avoided. Glycerin, a trihydric alcohol, absorbs water and dries the membranes. Milk of magnesia also has a drying effect because unabsorbed magnesium salts exert an osmotic pressure on tissue fluids. Many children also find the taste unpleasant. Frequent mouthwashes are indicated, but peroxide should be diluted to a 1:4 ratio with normal saline solution.
NP: Implementation; CN: Physiological integrity; CNS: Basic care and comfort; CL: Application

Don't stop now! You're almost done!

92. Which instruction should the nurse include when discharging a client receiving chemotherapy?
1. "Decrease spicy food intake to prevent stomach upset."
2. "Decrease fluid intake to prevent overworking the kidneys."
3. "Be sure to include such foods as liver to help prevent anemia."
4. "Eat foods low in purines to prevent increased urate deposits."

92. 4. The client is at risk for hyperuricemia from the rapid lysis of neoplastic cells. The breakdown of purines produces uric acid, which can add to the already high levels of uric acid and urate deposits in the kidneys, leading to renal failure. There is no chemotherapeutic or physiologic reason to address intake of spicy foods. Increased fluid intake is essential to prevent urate deposits and calculi formation. Liver contains purines and should be avoided.
NP: Implementation; CN: Health promotion and maintenance; CNS: Prevention and early detection of disease; CL: Comprehension

93. Which of the following interventions can prevent hemorrhagic cystitis caused by bladder irritation from chemotherapeutic medications?
1. Giving antacids
2. Giving antibiotics
3. Restricting fluid intake
4. Increasing fluid intake

93. 4. Sterile hemorrhagic cystitis is an adverse effect of chemical irritation of the bladder from cyclophosphamide. It can be prevented by liberal fluid intake (at least 1½ times the recommended daily fluid requirement). Antibiotics don't aid in the prevention of sterile hemorrhagic cystitis. Restricting fluids would only increase the risk of developing cystitis. Antacids wouldn't be indicated for treatment.
NP: Implementation; CN: Physiological integrity; CNS: Reduction of risk potential; CL: Application

94. Which of the following characteristics explains why aspirin is <u>contraindicated</u> for pain relief for a child with leukemia?
1. Decreases platelet production
2. Promotes bleeding tendencies
3. Not a strong enough analgesic
4. Decreases the effects of methotrexate

> What other conditions is aspirin used for?

94. 2. Aspirin would be contraindicated because it promotes bleeding. Aspirin use has also been associated with Reye's syndrome in children. For home use, acetaminophen is recommended for mild to moderate pain. Aspirin enhances the effects of methotrexate and has no effect on platelet production. Nonnarcotic analgesia has been effective for mild to moderate pain in clients with leukemia.
NP: Implementation; CN: Physiological integrity; CNS: Pharmacological therapies; CL: Application

95. Which of the following nursing measures helps prepare the parent and child for alopecia, a common adverse effect of several chemotherapeutic agents?
1. Introducing the idea of a wig after hair loss occurs
2. Explaining that hair begins to regrow in 6 to 9 months
3. Stressing that hair loss during a second treatment with the same medication will be more severe
4. Explaining that, as hair thins, keeping it clean, short, and fluffy may camouflage partial baldness

95. 4. The nurse must prepare parents and children for possible hair loss. Cutting the hair short lessens the impact of seeing large quantities of hair on bed linens and clothing. Sometimes, keeping the hair short and fuller can make a wig unnecessary. Hair regrows in 3 to 6 months. A child should be encouraged to pick out a wig similar to his own hair style and color before the hair falls out, to foster adjustment to hair loss. Hair loss during a second treatment with the same medication is usually less severe.
NP: Implementation; CN: Physiological integrity; CNS: Coping and adaptation; CL: Application

> You're doing swell! Keep going!

96. Which of the following areas is the most common site of cancer in children?
1. Lungs
2. Genitalia
3. Bone marrow
4. GI tract

96. 3. Childhood cancers occur most frequently in rapidly growing tissue, especially in the bone marrow. Tumors of the lungs, genitalia, and GI tract are rarely seen in children.
NP: Data collection; CN: Physiological integrity; CNS: Physiological adaptation; CL: Knowledge

97. Which of the following symptoms are early warning signs of childhood cancer?
1. Difficulty in swallowing
2. Nagging cough or hoarseness
3. Slight change in bowel and bladder habits
4. Swellings, lumps, or masses anywhere on the body

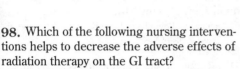

A nurse's keen observation is important.

97. 4. By being aware of early signs of childhood cancer, nurses can refer children for further evaluation. Swellings, lumps, or masses anywhere on the body are early warning signals of childhood cancer. Difficulty swallowing, cough, and hoarseness are early signs of cancer in adults. Usually, there's also a marked change in bowel or bladder habits.
NP: Data collection; CN: Health promotion and maintenance; CNS: Prevention and early detection of disease; CL: Knowledge

98. Which of the following nursing interventions helps to decrease the adverse effects of radiation therapy on the GI tract?
1. Avoiding the use of antispasmodics
2. Encouraging fluids and a soft diet
3. Giving antiemetics when nausea or vomiting occurs
4. Avoiding mouthwashes to prevent irritation of mouth ulcers

98. 2. Radiation therapy can cause such adverse effects as nausea and vomiting, anorexia, mucosal ulceration, and diarrhea. Encouraging fluids and a soft diet will help with anorexia. Antiemetics should be given before the onset of nausea. Antispasmodics are used to help reduce diarrhea. Frequent mouthwashes are indicated to prevent mycosis.
NP: Implementation; CN: Physiological integrity; CNS: Pharmacological therapies; CL: Application

99. Which of the following types of cancer is most common in children?
1. Hodgkin's disease
2. Leukemia
3. Osteogenic sarcoma
4. Wilms' tumor

99. 2. Leukemia, cancer of the blood-forming tissues, is the most common form of childhood cancer. The annual incidence in children under age 15 is approximately 4 in every 100,000. Hodgkin's disease occurs almost as frequently as leukemia. Osteogenic sarcoma is the most frequently encountered malignant bone cancer in children, but the incidence isn't as high as leukemia. Wilms' tumor is the most frequent intra-abdominal tumor of childhood, but leukemia is most prevalent when comparing all types of childhood cancers.
NP: Evaluation; CN: Health promotion and maintenance; CNS: Prevention and early detection of disease; CL: Knowledge

100. Short-term steroid therapy is used in clients with leukemia to promote which of the following reactions?
1. Increased appetite
2. Altered body image
3. Increased platelet production
4. Decreased susceptibility to infection

One hundred questions! You must be proud! Only a few more to go!

100. 1. Short-term steroid therapy produces no acute toxicities and results in two beneficial reactions: increased appetite and a sense of well-being. Physical changes, such as "moon-face," a result of steroid use, can cause alterations in body image and can be extremely distressing to children. Prednisone (steroid therapy) has no effect on platelet production or susceptibility to infection.
NP: Planning; CN: Physiological integrity; CNS: Pharmacological therapies; CL: Knowledge

101. Which of the following interventions is a priority for a hemophiliac client who has fallen and badly bruised his leg?
1. Appropriate dose of aspirin and rest
2. Immobilization of the leg and a dose of ibuprofen
3. Heating pad and administration of factor VIII concentrate
4. Pressure on the site and administration of factor VIII concentrate

This question is asking you to prioritize.

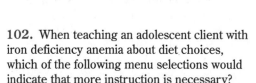

101. 4. With any bleeding injury in a hemophilic client, the first line of treatment is always to replace the clotting factor. Pressure is applied along with cool compresses, and the extremity is immobilized. Heat isn't used because it increases bleeding. Aspirin isn't used because of its anticoagulant properties and the risk of Reye's syndrome in children. Immobilizing the leg and giving ibuprofen would be done after applying pressure and administering factor VIII.
NP: Planning; CN: Physiological integrity; CNS: Physiological adaptation; CL: Application

102. When teaching an adolescent client with iron deficiency anemia about diet choices, which of the following menu selections would indicate that more instruction is necessary?
1. Caesar salad and pretzels
2. Cheeseburger and a milkshake
3. Red beans and rice with sausage
4. Egg sandwich and snack peanuts

Yummmmmm!!!

102. 1. Caesar salad and pretzels aren't foods high in iron and protein. Meats (especially organ meats), eggs, and nuts have high protein and iron.
NP: Evaluation; CN: Health promotion and maintenance; CNS: Prevention and early detection of disease; CL: Analysis

103. The nurse is speaking to the mother of a child with leukemia who wants to know why her child is so susceptible to infection if he has too many white blood cells (WBCs). Which of the following responses would be most accurate?
1. This is an adverse effect of the medication he has to take.
2. He hasn't been able to eat a proper diet since he's been sick.
3. Leukemia is a problem of tumors in the internal organs that prevent the ability to fight infection.
4. Leukemia causes production of too many immature WBCs, which can't fight infection very well.

103. 4. Leukemia is an unrestricted proliferation of immature WBCs, which don't function properly and are a poor defense against infection. Diet contributes to overall health but doesn't cause the overproduction of WBCs. There are no solid tumors in the internal organs in leukemia. Medications, such as chemotherapy, can diminish the immune system's effectiveness; however, they don't cause the overproduction of immature WBCs and the poor resistance to infection that the mother asked about.
NP: Implementation; CN: Physiological integrity; CNS: Physiological adaptation; CL: Application

Congratulations! You finished! Great job!

From the simple otitis media to the uncommon and dangerous epiglottiditis, this chapter covers a wide variety of respiratory disorders in children. So, take a deep breath and go for it!

Chapter 29
Respiratory disorders

1. Which of the following definitions best describes the etiology of sudden infant death syndrome (SIDS)?
  1. Cardiac arrhythmias
  2. Apnea of prematurity
  3. Unexplained death of an infant
  4. Apparent life-threatening event

2. Which of the following children has an increased risk of sudden infant death syndrome (SIDS)?
  1. Premature infant with low birth weight
  2. A healthy 2-year-old
  3. Infant hospitalized for fever
  4. Firstborn child

3. A 6-week-old infant who isn't breathing is brought to the emergency department; a preliminary finding of sudden infant death syndrome (SIDS) is made to the parents. Which of the following interventions should the nurse take initially?
  1. Call their spiritual advisor.
  2. Explain the etiology of SIDS.
  3. Allow them to see their infant.
  4. Collect the infant's belongings and give them to the parents.

Careful. This question is asking you to prioritize.

1. 3. SIDS can best be defined as the sudden death of an infant under age 1 that remains unexplained after autopsy. Apnea of prematurity occurs in infants less than 32 weeks' gestation who have periodic breathing lapses for 20 seconds or more. Apparent life-threatening events usually have some combination of apnea, color change, marked change in muscle tone, choking, or gagging.
NP: Data collection; CN: Physiological integrity; CNS: Physiological adaptation; CL: Knowledge

2. 1. Premature infants, especially those with low birth weight, have an increased risk for SIDS. Hospitalization for fever is insignificant. Infants with apnea, central nervous system disorders, or respiratory disorders have a higher risk of SIDS. Peak age for SIDS is 2 to 4 months. There's an increased risk of SIDS in subsequent siblings of two or more SIDS victims.
NP: Data collection; CN: Physiological integrity; CNS: Reduction of risk potential; CL: Knowledge

3. 3. The parents need time with their infant to assist with the grieving process. Calling their pastor and collecting the infant's belongings are also important steps in the care plan but aren't priorities. The parents will be too upset to understand an explanation of SIDS at this time.
NP: Implementation; CN: Physiological integrity; CNS: Physiological adaptation; CL: Application

NP: Nursing process   CN: Client needs category   CNS: Client needs subcategory   CL: Cognitive level

4. Which of the following risk factors is related to sudden infant death syndrome (SIDS)?
1. Feeding habits
2. Gestational age of 42 weeks
3. Immunizations
4. Multiple birth

It's important to be sensitive to a family's feelings when they've lost a child to SIDS.

4. 4. Multiple births, prematurity, and low birth weight are important risk factors associated with SIDS. Immunizations have been disproved to be associated with the disorder. Feeding habits and a gestational age of 42 weeks aren't significant.
NP: Data collection; CN: Physiological integrity; CNS: Reduction of risk potential; CL: Knowledge

5. An infant is brought to the emergency department and pronounced dead with the preliminary finding of sudden infant death syndrome (SIDS). Which of the following questions to the parents is appropriate?
1. Did you hear the infant cry out?
2. Was the infant's head buried in a blanket?
3. Were any of the siblings jealous of the new baby?
4. How did the infant look when you found him?

5. 4. Only factual questions should be asked during the initial history in the emergency department. The other questions imply blame, guilt, or neglect.
NP: Implementation; CN: Physiological integrity; CNS: Physiological adaptation; CL: Application

6. Which of the following interventions is not recommended for children with an increased risk of sudden infant death syndrome (SIDS)?
1. Pneumogram
2. Home apnea monitor
3. Respiratory stimulant drugs
4. Chest X-ray at 1 month of age

Careful. This is a negative question.

6. 4. No diagnostic test such as a chest X-ray can predict which infants will survive or die from SIDS. A home apnea monitor or the use of respiratory stimulant drugs is recommended. A pneumogram may be ordered to check for periods of apnea or reflux during sleep.
NP: Data collection; CN: Physiological integrity; CNS: Reduction of risk potential; CL: Knowledge

7. Which of the following reactions are usually exhibited by the family of an infant who has died from sudden infant death syndrome (SIDS)?
1. Feelings of blame or guilt
2. Acceptance of the diagnosis
3. Requests for the infant's belongings
4. Questions regarding the etiology of the diagnosis

7. 1. During the first few moments, the parents often are in shock and have overwhelming feelings of blame or guilt. Acceptance of the diagnosis and questions regarding the etiology may not occur until the parents have had time to see the child. The infant's belongings are usually packaged for the family to take home but some parents may see this as a painful reminder.
NP: Data collection; CN: Psychosocial integrity; CNS: Coping and adaptation; CL: Application

8. Sudden infant death syndrome (SIDS) is confirmed by which of the following procedures?
1. Autopsy
2. Chest X-ray
3. Skeletal survey
4. Laboratory analysis

8. 1. Autopsies reveal consistent pathological findings, such as pulmonary edema and intrathoracic hemorrhages, that confirm the diagnosis of SIDS. Chest X-rays are used to diagnose respiratory complications. Skeletal surveys are used with cases of suspected child abuse. Laboratory analysis will show no characteristics to confirm the diagnosis of SIDS.
NP: Evaluation; CN: Physiological integrity; CNS: Physiological adaptation; CL: Knowledge

NP: Nursing process   CN: Client needs category   CNS: Client needs subcategory   CL: Cognitive level

9. Which of the following plans is most appropriate for a discharge home visit to parents who lost an infant to sudden infant death syndrome (SIDS)?

1. One visit in 2 weeks
2. No visit is necessary
3. As soon after death as possible
4. One visit with parents only, no siblings

Make sure you understand the grieving process necessary for SIDS parents.

9. 3. When parents return home, a visit is necessary as soon after the death as possible. The nurse should assess what the parents have been told, what they think happened, and how they've explained this to the other siblings. Not all of these issues will be resolved in one visit. The number of visits and plan for intervention must be flexible. The needs of the siblings must always be considered.
NP: Planning; CN: Psychosocial integrity; CNS: Coping and adaptation; CL: Application

10. A few days after the death of an infant from sudden infant death syndrome (SIDS), which of the following behaviors would the nurse expect to observe in a parent?

1. Disorganized thinking
2. Feelings of guilt
3. Repressed thoughts
4. Structured thinking

10. 1. Within a day or two of the infant's death, the parents enter the impact phase of crisis, which consists of disorganized thoughts in which they can't deal with the crisis in concrete terms. Almost immediately, at the time of death, parents may have repressed thoughts and feelings of guilt or blame. One week after the death, the parent of the infant would most likely be in the turmoil phase. In the turmoil phase, structured thinking is common.
NP: Data collection; CN: Psychosocial integrity; CNS: Coping and adaptation; CL: Comprehension

11. Which of the following positions is recommended for placing an infant to sleep?

1. Prone position
2. Supine position
3. Side-lying position
4. With head of bed elevated 30 degrees

11. 2. The American Academy of Pediatrics endorses placing infants face-up in their cribs as a way to reduce the risk of sudden infant death syndrome (SIDS). Raising the head of the bed 30 degrees is recommended for infants with gastroesophageal reflux. The side-lying position promotes gastric emptying. Placing infants on their stomachs is thought to make an attack of apnea harder to fight off but how exactly the sleeping position predisposes a child to SIDS is still unclear.
NP: Implementation; CN: Health promotion and maintenance; CNS: Growth and development through the life span; CL: Application

Note that this question is looking for a long-term response.

12. Which of the following activities should be recommended for long-term support of parents who have lost an infant due to sudden infant death syndrome (SIDS)?

1. Attending support groups
2. Attending church regularly
3. Attending counseling sessions
4. Discussing feelings with family and friends

12. 1. The best support will come from parents who have had the same experience. Attending church and discussing feelings with family and friends can offer support, but they may not understand the experience. Counseling sessions are usually a short-term support.
NP: Evaluation; CN: Psychosocial integrity; CNS: Coping and adaptation; CL: Application

13. Which of the following interventions is best to help a 2-year-old child adapt to a mist tent?
1. Tape the tent closed.
2. Place the child's favorite toys in the tent.
3. Allow the child to play with the tent first.
4. Ask one or both parents to stay with the child.

13. 4. The most important factor in helping a child cope with new and strange surroundings is to encourage feelings of security by having the parents present. This is the hallmark of family-centered care. Placing the child's favorite toys in the tent provides distraction and allows the child to have something of his own with him but may *not* alleviate fears. Allowing the child to play with the tent may pose a safety hazard and isn't appropriate for a child in respiratory distress. Taping the tent closed is inappropriate.
NP: Implementation; CN: Psychosocial integrity; CNS: Coping and adaptation; CL: Application

14. A 2-year-old child comes to the emergency department with inspiratory stridor and a barking cough. A preliminary diagnosis of croup has been made. Which of the following actions should be an <u>initial</u> intervention?
1. Administer I.V. antibiotics.
2. Provide oxygen by facemask.
3. Establish and maintain the airway.
4. Ask the mother to go to the waiting room.

Congratulations! You've finished the first 15 questions! Good job!

14. 3. The initial priority is to establish and maintain the airway. Edema and accumulation of secretions may contribute to airway obstruction. Oxygen should be administered by tent as soon as possible to decrease the child's distress. Allowing the child to stay with the mother reduces anxiety and distress. Antibiotics aren't indicated for viral illnesses.
NP: Data collection; CN: Physiological integrity; CNS: Physiological adaptation; CL: Application

15. Which of the following is the best intervention for parents to take if their child is experiencing an episode of "midnight croup," or acute spasmodic laryngitis?
1. Give warm liquids.
2. Raise the heat on the thermostat.
3. Provide humidified air with cool mist.
4. Take the child into the bathroom with a warm, running shower.

15. 3. High humidity with cool mist provides the most relief. Cool liquids would be best for the child. If unable to take liquid, the child needs to be in the emergency unit. Raising the heat on the thermostat will result in dry, warm air, which may cause secretions to adhere to the airway wall. A warm, running shower provides a mist that may be helpful to moisten and decrease the viscosity of airway secretions and may also decrease laryngeal spasm.
NP: Evaluation; CN: Physiological integrity; CNS: Physiological adaptation; CL: Application

16. Which of the following signs is <u>most</u> <u>characteristic</u> of a child with croup?
1. "Barking" cough
2. High fever
3. Low heart rate
4. Severe respiratory distress

This question is asking for the symptom most commonly associated with croup.

16. 1. A resonant cough described as "barking" is the most characteristic sign of croup. The child may have varying degrees of respiratory distress related to swelling or obstruction. The child may present with a low-grade or high fever depending on whether the etiologic agent is viral or bacterial. Usually the heart rate is rapid.
NP: Data collection; CN: Physiological integrity; CNS: Physiological adaptation; CL: Knowledge

17. Which of the following characteristics best describes croup?
1. Inflammation of the palatine tonsils
2. Acute, highly contagious respiratory infection
3. Infection of supraglottic area with involvement of the epiglottis
4. Clinical syndrome of laryngitis and laryngotracheobronchitis

18. Which of the following interventions is the most important goal for a child with ineffective airway clearance?
1. Reducing the child's anxiety
2. Maintaining a patent airway
3. Providing adequate oral fluids
4. Administering medications as ordered

19. A 2-year-old child with croup suffering respiratory distress may show which of the following signs on initial assessment?
1. Capillary refill time less than 3 seconds
2. Low temperature
3. Grunting or head bobbing
4. Respiratory rate of 28 breaths/minute

20. Which of the following precautions is recommended when caring for children with respiratory infections such as croup?
1. Enforce hand washing.
2. Place the child in isolation.
3. Teach children to use tissues.
4. Keep siblings in the same room.

21. What's the best time a nebulizer treatment should be administered to a child with croup?
1. During naptime
2. During playtime
3. After the child eats
4. After the parents leave

Prioritize!

17. 4. Croup is a general term referring to acute infections affecting varying degrees of the larynx, trachea, and bronchi. Epiglottiditis is a form of croup affecting the supraglottic area. Inflammation of the palatine tonsils is consistent with tonsillitis. Pertussis, or whooping cough, is an acute, highly contagious respiratory infection.
NP: Data collection; CN: Physiological integrity; CNS: Physiological adaptation; CL: Knowledge

18. 2. The most important goal is to maintain a patent airway. Reducing anxiety and administering medications will follow after the airway is secure. The child shouldn't be allowed to eat or drink anything to avoid the risk of aspiration.
NP: Data collection; CN: Physiological integrity; CNS: Physiological adaptation; CL: Application

19. 3. Grunting or head bobbing is seen in a child in respiratory distress. A respiratory rate of 28 breaths/minute and capillary refill time of less than 3 seconds are normal findings. The child's temperature may be elevated if infection is present.
NP: Data collection; CN: Physiological integrity; CNS: Physiological adaptation; CL: Knowledge

20. 1. Hand washing helps prevent the spread of infections. Ill children should be placed in separate bedrooms if possible but don't need to be isolated. Teaching children to use tissues properly is important, but the key is disposal and hand washing after use.
NP: Implementation; CN: Health promotion and maintenance; CNS: Prevention and early detection of disease; CL: Application

21. 1. The nurse should administer nebulizer treatments at prescribed intervals. During naptime allows for as little disruption as possible. A child should be given a treatment before eating so the airway will be open and the work of eating will be decreased. Administering treatment during playtime will disrupt the child's daily pattern. Parents are usually helpful when administering treatments. The child can sit on the parents' lap to help decrease anxiety or fear.
NP: Planning; CN: Physiological integrity; CNS: Pharmacological therapies; CL: Application

22. During the recovery stages of croup, the nurse should explain which of the following interventions to parents?
1. Limiting oral fluid intake
2. Recognizing signs of respiratory distress
3. Providing three nutritious meals per day
4. Allowing the child to go to the playground

Adequate parent teaching is essential for managing a child with croup.

22. 2. Although most children recover without complications, the parents should be able to recognize signs and symptoms of respiratory distress and know how to access emergency services. Oral fluids should be encouraged because fluids help to thin secretions. Although nutrition is important, frequent small nutritious snacks are usually more appealing than an entire meal. Children should have optimal rest and engage in quiet play. A comfortable environment free of noxious stimuli lessens respiratory distress.

NP: Evaluation; CN: Physiological integrity; CNS: Physiological adaptation; CL: Application

23. Which of the following agents is related to acute epiglottiditis?
1. Allergen
2. Bacteria
3. Virus
4. Yeast

23. 2. *Haemophilus influenzae* type B is usually the bacterial agent responsible for acute epiglottiditis. Viral and allergen agents can be seen with acute laryngotracheobronchitis and acute spasmodic croup. Yeast infections in the oral cavity cause thrush.

NP: Data collection; CN: Physiological integrity; CNS: Physiological adaptation; CL: Knowledge

24. Which of the following clinical observations indicates epiglottiditis?
1. Decreased secretions
2. Drooling
3. Low-grade fever
4. Spontaneous cough

24. 2. Drooling is common due to the pain of swallowing, excessive secretions, and sore throat. The child usually has a high fever and the absence of a spontaneous cough. The classic picture is the child in a tripod position with mouth open and tongue protruding.

NP: Data collection; CN: Physiological integrity; CNS: Physiological adaptation; CL: Knowledge

25. Which of the following strategies is the best care plan for a child with acute epiglottiditis?
1. Encourage oral fluids for hydration.
2. Maintain the client in semi-Fowler's position.
3. Administer I.V. antibiotic therapy.
4. Maintain respiratory isolation for 48 hours.

Another 10 down! Way to go! Keep at it!

25. 3. The etiologic agent for epiglottiditis is usually bacterial; therefore, the treatment consists of I.V. antibiotic therapy. The client shouldn't be allowed anything by mouth during the initial phases of the infection to prevent aspiration. The client should be placed in Fowler's position or any position that provides the most comfort and security. Respiratory isolation isn't required.

NP: Planning; CN: Physiological integrity; CNS: Physiological adaptation; CL: Application

NP: Nursing process   CN: Client needs category   CNS: Client needs subcategory   CL: Cognitive level

26. A 2-year-old child is found on the floor next to his toy chest. After first determining unresponsiveness and calling for help, which of the following steps should be taken next?
1. Start mouth-to-mouth resuscitation.
2. Begin chest compressions.
3. Check for a pulse.
4. Open the airway.

27. A 10-month-old infant is found in respiratory arrest and cardiopulmonary resuscitation is started. Which of the following sites is best to check for a pulse?
1. Brachial
2. Carotid
3. Femoral
4. Radial

28. When giving rescue breathing to an infant under age 1, what's the ratio of breaths per second?
1. 1 breath every 2 seconds
2. 1 breath every 3 seconds
3. 1 breath every 4 seconds
4. 1 breath every 5 seconds

29. When performing chest compressions on a 2-year-old child, which of the following depths is correct?
1. ½" to 1" (1.3 to 2.5 cm)
2. 1" to 1½" (2.5 to 3.8 cm)
3. 1½" to 2" (3.8 to 5 cm)
4. 2" to 2½" (5 to 6.4 cm)

30. Chest compressions must be coordinated with ventilations. Which of the following ratios is accurate for a 3-year-old child?
1. 5 compressions to 1 ventilation
2. 5 compressions to 2 ventilations
3. 15 compressions to 1 ventilation
4. 15 compressions to 2 ventilations

It's important to prioritize in an emergency situation.

The procedure for rescue breathing for infants is different from that for adults.

26. 4. The airway should be opened by using the head-tilt, chin-lift maneuver and breathlessness should be determined at the start of cardiopulmonary resuscitation. The sequence of airway, breathing, and circulation must be followed.
NP: Implementation; CN: Physiological integrity; CNS: Physiological adaptation; CL: Application

27. 1. Palpation of the brachial artery is recommended. The short, chubby neck of infants makes rapid location of the carotid artery difficult. After age 1, the carotid would be used. The femoral pulse, often palpated in a hospital setting, may be difficult to assess because of the infant's position, fat folds, and clothing. The radial pulse isn't a good indicator of central artery perfusion.
NP: Implementation; CN: Physiological integrity; CNS: Physiological adaptation; CL: Application

28. 2. Rescue breathing should be performed once every 3 seconds until spontaneous breathing resumes. This provides 20 breaths/minute. One breath every 5 seconds is recommended for adults. One breath every 2 seconds may cause gastric distention.
NP: Implementation; CN: Physiological integrity; CNS: Physiological adaptation; CL: Application

29. 2. The chest compressions should equal approximately one-third to one-half the total depth of the chest. This corresponds to about 1½" in a child age 1 to 8, ½" to 1" for an infant younger than age 1, and 1½" to 2" for an adult.
NP: Implementation; CN: Physiological integrity; CNS: Physiological adaptation; CL: Application

30. 1. A ratio of 5:1 is recommended for children under age 8. A ratio of 15:2 is recommended for adults. Ratios of 5:2 and 15:1 won't provide optimal compression and ventilation.
NP: Implementation; CN: Physiological integrity; CNS: Physiological adaptation; CL: Application

31. A 10-month-old child is found choking and soon becomes unconscious. Which of the following interventions should the nurse attempt first after opening the airway?
1. Look inside the infant's mouth for a foreign object.
2. Give five back blows and five chest thrusts.
3. Attempt a blind finger sweep.
4. Attempt rescue breathing.

Prioritize!

31. 1. After the airway is open, the nurse should check for the foreign object and remove it with a finger sweep if it can be visualized. After this step, rescue breathing should be attempted. If ventilation is unsuccessful, the nurse should then give five back blows and five chest thrusts in an attempt to dislodge the object. Blind finger sweeps should never be performed because this may push the object further back into the airway.

NP: Implementation; CN: Physiological integrity; CNS: Physiological adaptation; CL: Application

32. Using which part of the hands is appropriate when performing chest compressions on a child between ages 1 and 8?
1. Heels of both hands
2. Heel of one hand
3. Index and middle fingers
4. Thumbs of both hands

32. 2. The heel of one hand is recommended for performing chest compressions on children between ages 1 and 8. Two hands are used for adult cardiopulmonary resuscitation. Chest thrusts administered with the middle and third fingers, and in some cases the thumbs of each hand, are used on infants under age 1.

NP: Implementation; CN: Physiological integrity; CNS: Physiological adaptation; CL: Application

33. A 3-year-old child who isn't breathing and is cyanotic and lethargic is brought to the emergency department. The mother states that she thinks he swallowed a penny. Which of the following interventions should the nurse take first?
1. Give 100% oxygen.
2. Administer five back blows.
3. Attempt a blind finger sweep.
4. Administer abdominal thrusts.

Again, you're being asked to prioritize.

33. 4. A child between ages 1 and 8 should receive abdominal thrusts to help dislodge the object. Blind finger sweeps should never be performed because this could push the object further back into the airway. Administering 100% oxygen won't help if the airway is occluded. Infants younger than age 1 should receive back blows before chest thrusts.

NP: Implementation; CN: Physiological integrity; CNS: Physiological adaptation; CL: Application

34. A 1-year-old child is brought to the emergency department with a mild respiratory infection and a temperature of 101.3° F (38.5° C). Otitis media is diagnosed. Which of the following signs is characteristic of otitis media?
1. Excessive drooling
2. Tugging on the ears
3. High-pitched, barking cough
4. Pearl-gray tympanic membrane

34. 2. Tugging on the ears is a common sign for a child with ear pain. Pearl-gray tympanic membranes are a normal finding. A child with otitis media usually exhibits a discolored membrane (bright red, yellow, or dull gray). A high-pitched, barking cough and excessive drooling indicate croup.

NP: Data collection; CN: Physiological integrity; CNS: Physiological adaptation; CL: Knowledge

35. A 7-month-old child is diagnosed with otitis media; the physician orders amoxicillin 40 mg/kg/day to be administered three times per day. The child weighs 9 kg. How much amoxicillin should the child receive per dose?
  1. 120 mg
  2. 180 mg
  3. 200 mg
  4. 360 mg

Make sure you understand this formula. You'll use it again later in this test.

35. 1. The child should receive 120 mg per dose. Here are the calculations: 40 mg × 9 kg = 360 mg/day; 360 mg/3 doses = 120 mg/dose.

NP: Implementation; CN: Physiological integrity; CNS: Pharmacological therapies; CL: Knowledge

36. Children with chronic otitis media commonly require surgery for a myringotomy and ear tube placement. Which of the following management strategies explains the purpose of the ear tubes?
  1. To administer antibiotics
  2. To flush the middle ear
  3. To increase pressure
  4. To drain fluid

36. 4. Ear tubes allow normal fluid to drain (not flush) from the middle ear. They also allow ventilation. The purpose isn't to administer medication. The tubes also allow pressure to equalize in the middle ear.

NP: Implementation; CN: Physiological integrity; CNS: Physiological adaptation; CL: Application

37. The nurse is discharging a 10-month-old client with eardrops. Which of the following information should she give the parent about how to administer the drops?
  1. Pull the earlobe upward.
  2. Pull the earlobe up and back.
  3. Pull the earlobe down and back.
  4. Pull the earlobe down and forward.

You're doing great! Keep it up!

37. 3. For infants, the parent should be told to gently pull the earlobe down and back to visualize the external auditory canal. For children over age 3 and for adults, the earlobe is gently pulled slightly up and back.

NP: Evaluation; CN: Physiological integrity; CNS: Pharmacological therapies; CL: Knowledge

38. A child is diagnosed as having right chronic otitis media. After the child returns from surgery for myringotomy and placement of ear tubes, which of the following interventions is appropriate?
  1. Apply gauze dressings.
  2. Position the child on the left side.
  3. Position the child on the right side.
  4. Apply warm compresses to both ears.

Question 39 is testing your knowledge of the different types of otitis media.

38. 3. The child should be positioned on the right side to facilitate drainage. The left side isn't an area of concern for drainage. Warm compresses may help to facilitate drainage only when used on the affected ear. Gauze dressings aren't necessary after surgery. Some physicians may prefer a loose cotton wick.

NP: Implementation; CN: Physiological integrity; CNS: Physiological adaptation; CL: Application

39. Which of the following durations would cause otitis media to be classified as acute?
  1. Approximately 2 weeks
  2. Approximately 4 weeks
  3. 3 weeks to 3 months
  4. Longer than 3 months

39. 1. Acute otitis media has a rapid onset and lasts approximately 2 to 3 weeks. At approximately 4 weeks' duration the otitis media would be considered subacute rather than acute. Otitis media is considered chronic when it lasts longer than 3 months. Subacute otitis media lasts 3 weeks to 3 months.

NP: Evaluation; CN: Physiological integrity; CNS: Physiological adaptation; CL: Knowledge

40. Which of the following descriptions best describes expected findings of otoscopy in acute otitis media?
1. Pearl-gray tympanic membrane
2. Bright red, bulging tympanic membrane
3. Dull gray membrane with fluid behind the eardrum
4. Bright red or yellow bulging or retracted tympanic membrane

40. 4. With acute otitis media, the tympanic membrane may present as bright red or yellow, bulging or retracted. Dull gray membrane with fluid is consistent with subacute or chronic otitis media. A pearl-gray tympanic membrane is a normal finding.
NP: Data collection; CN: Physiological integrity; CNS: Physiological adaptation; CL: Knowledge

41. Which of the following conditions is a predisposing factor for development of otitis media in children?
1. The cartilage lining is overdeveloped.
2. An infant sitting up contributes to the pooling of fluid.
3. Humoral defense mechanisms decrease the risk of infection.
4. Eustachian tubes are short, wide, and straight and lie in a horizontal plane.

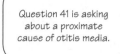

Question 41 is asking about a proximate cause of otitis media.

41. 4. In an infant or child, the eustachian tubes are short, wide, and straight and lie in a horizontal plane, allowing them to be more easily blocked by conditions such as large adenoids and infections. Until the eustachian tubes change in size and angle, children are more susceptible to otitis media. Cartilage lining is underdeveloped, making the tubes more distensible and more likely to open inappropriately. Immature humoral defense mechanisms increase the risk of infection. The usual lying-down position of infants contributes to the pooling of fluid such as formula in the pharyngeal cavity.
NP: Data collection; CN: Physiological integrity; CNS: Physiological adaptation; CL: Knowledge

42. Which of the following complications is most commonly related to acute otitis media?
1. Eardrum perforation
2. Hearing loss
3. Meningitis
4. Tympanosclerosis

42. 1. Eardrum perforation is the most common complication of acute otitis media as the exudate accumulates and pressure increases. Hearing loss in most cases is conductive in nature and mild in severity but is less common than eardrum perforation. In addition, hearing tests aren't usually performed during episodes of otitis media. Tympanosclerosis and meningitis are possible but uncommon when adequate antibiotic therapy is implemented.
NP: Data collection; CN: Physiological integrity; CNS: Physiological adaptation; CL: Application

43. Which of the following interventions may be recommended for children with chronic otitis media with effusion?
1. Antihistamines
2. Corticosteroids
3. Decongestants
4. Surgical intervention

You're flying high now. Keep going!

43. 4. The most common treatment for chronic otitis media is the surgical placement of tympanostomy tubes. They allow drainage of fluid and ventilation of the middle ear. Decongestants and antihistamines are used to shrink mucous membranes but are of unproven benefit. Corticosteroids have also proved to be of limited value.
NP: Implementation; CN: Physiological integrity; CNS: Pharmacological therapies; CL: Knowledge

NP: Nursing process   CN: Client needs category   CNS: Client needs subcategory   CL: Cognitive level

44. A 2-year-old child is diagnosed with epiglottiditis. Ampicillin is ordered 300 mg/kg/day in 6 divided doses. The client weighs 12 kg. How much ampicillin is given per dose?

    1. 300 mg
    2. 600 mg
    3. 900 mg
    4. 1,200 mg

45. A 3-year-old child is receiving ampicillin for acute epiglottiditis. Which of the following adverse effects is common with ampicillin?

    1. Constipation
    2. Generalized rash
    3. Increased appetite
    4. Low-grade temperature

46. A 3-year-old child is given a preliminary diagnosis of acute epiglottiditis. Which of the following nursing interventions is appropriate?

    1. Obtain a throat culture immediately.
    2. Place the child in a side-lying position.
    3. Don't attempt to visualize the epiglottis.
    4. Use a tongue depressor to look inside the throat.

47. Administration of which childhood vaccination assists in decreasing a child's incidence of developing of epiglottiditis?

    1. Diphtheria vaccine
    2. *Haemophilus influenzae* type B (Hib) vaccine
    3. Measles vaccine
    4. Oral poliovirus vaccine (OPV)

What did I tell you? Here's that formula again!

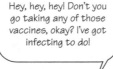

Hey, hey, hey! Don't you go taking any of those vaccines, okay? I've got infecting to do!

44. 2. The child should receive 600 mg per dose. Here are the calculations: 300 mg × 12 kg = 3,600 mg/day; 3,600 mg/6 doses = 600 mg/dose

NP: Implementation; CN: Physiological integrity; CNS: Pharmacological therapies; CL: Knowledge

45. 2. Some clients may develop an erythematous or maculopapular rash after 3 to 14 days of therapy; however, this complication doesn't necessitate discontinuing the drug. Nausea, vomiting, epigastric pain, and diarrhea are adverse effects that may necessitate discontinuation of the drug.

NP: Implementation; CN: Physiological integrity; CNS: Pharmacological therapies; CL: Knowledge

46. 3. The nurse shouldn't attempt to visualize the epiglottis. The use of tongue blades or throat culture swabs may cause the epiglottis to spasm and totally occlude the airway. Throat inspection should be attempted only when immediate intubation or tracheostomy can be performed in the event of further or complete obstruction. The child should always remain in the position that provides the most comfort and security and ease of breathing.

NP: Data collection; CN: Physiological integrity; CNS: Physiological adaptation; CL: Application

47. 2. Epiglottiditis is caused by the bacterial agent *H. influenzae*. The American Academy of Pediatrics recommends that, beginning at age 2 months, children receive the Hib conjugate vaccine. A decline in the incidence of epiglottiditis has been seen as a result of this vaccination regimen. The OPV, measles vaccine, and diphtheria vaccine are preventive for those diseases, not epiglottiditis.

NP: Planning; CN: Health promotion and maintenance; CNS: Prevention and early detection of disease; CL: Knowledge

48. In addition to an increasing respiratory rate, which of the following signs in a 3-year-old child with acute epiglottiditis indicates that the client's respiratory distress is increasing?
1. Progressive, barking cough
2. Increasing irritability
3. Increasing heart rate
4. Productive cough

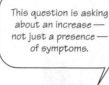

This question is asking about an increase — not just a presence — of symptoms.

48. 3. Increasing heart rate is an early sign of hypoxia. A progressive, barking cough is characteristic of spasmodic croup. A child in respiratory distress will be irritable and restless. As distress increases, the child will become lethargic related to the work of breathing and impending respiratory failure. A productive cough shows that secretions are moving and the child is able to effectively clear them.

NP: Data collection; CN: Physiological integrity; CNS: Physiological adaptation; CL: Application

49. While examining a child with acute epiglottiditis, the nurse should have which of the following items available?
1. Cool mist tent
2. Intubation equipment
3. Tongue depressors
4. Viral culture medium

49. 2. Emergency intubation equipment should be at the bedside to secure the airway if examination precipitates further or complete obstruction. Tongue depressors are contraindicated and may cause the epiglottis to spasm. Viral culture medium and cool mist tents are recommended for the diagnosis and treatment of croup.

NP: Data collection; CN: Physiological integrity; CNS: Physiological adaptation; CL: Application

50. A 2-year-old child is brought to the emergency department in respiratory distress. The child refuses to lie down and prefers to sit upright and lean forward with chin thrust out, mouth open, and tongue protruding (tripod position). This is a classic sign of which of the following illnesses?
1. Acute epiglottiditis
2. Croup
3. Pertussis
4. Pharyngitis

50. 1. The onset of acute epiglottiditis is abrupt and rapidly progresses to severe respiratory distress. The child prefers the tripod position due to the difficulty swallowing, excessive drooling, and sore throat. A barking cough and stridor represent croup. Pertussis is characterized by its whooping cough. Pharyngitis presents with a fever and sore throat.

NP: Data collection; CN: Physiological integrity; CNS: Physiological adaptation; CL: Knowledge

51. When is the best time to obtain a sputum specimen for culture and sensitivity testing from a 12-year-old child?
1. 7 a.m.
2. 11 a.m.
3. 3 p.m.
4. 8 p.m.

Congratulations! You've finished more than 50 questions! Keep up the good work!

51. 1. A sputum specimen is best obtained early in the morning because a higher volume of secretions is likely to have accumulated throughout the night. The other times aren't optimal for obtaining the specimen.

NP: Implementation; CN: Physiological integrity; CNS: Reduction of risk potential; CL: Comprehension

---

NP: Nursing process   CN: Client needs category   CNS: Client needs subcategory   CL: Cognitive level

52. An 18-month-old boy is being evaluated at the clinic. In reviewing his chart, the nurse notices that he has missed many appointments and, consequently, his immunizations aren't up-to-date. While discussing communicable disease and the importance of immunizations with his mother, the nurse explains that meningitis can be prevented with:

1. corticosteroid therapy.
2. tetanus toxoid vaccination.
3. tine testing.
4. *Haemophilus influenzae* vaccination.

I'm positive you'll do well on this test!

52. 4. *H. influenzae* vaccination is recommended for all children as a preventive measure for meningitis. Corticosteroids such as prednisone weaken the immune system and increase susceptibility to infectious disease. Tetanus toxoid vaccination provides immunity against tetanus or "lockjaw." A tine test is a screening measure for tuberculosis.

NP: Implementation; CN: Health promotion and maintenance; CNS: Prevention and early detection of disease; CL: Application

53. Which nursing action would relieve respiratory distress and dyspnea in a 2-year-old boy with laryngotracheobronchitis?

1. Stimulating the child to keep him awake
2. Providing an atmosphere of high humidity
3. Offering frequent oral feedings
4. Administering frequent sedatives

53. 2. High humidity reduces mucosal edema and prevents drying of secretions, thus helping to maintain an open airway. Keeping the child calm, not stimulated, helps to reduce oxygen need. Oral feedings may need to be withheld in a child experiencing respiratory distress because eating may interfere with his ability to breathe. Sedation is generally contraindicated because it may cause respiratory depression and mask anxiety, a sign of respiratory distress.

NP: Implementation; CN: Physiological integrity; CNS: Physiological adaptation; CL: Application

54. When assessing a child for increased laryngotracheal edema and early signs of impending airway obstruction, the nurse should observe for which of the following warning signs?

1. Decreased heart and respiratory rates and a high peak flow rate.
2. Increased heart and respiratory rates, retractions, and restlessness
3. Decreased blood pressure
4. Increased temperature

54. 2. Increased heart and respiratory rates, retractions, and restlessness are classic indicators of hypoxia. The heart and respiratory rates increase to enable increased oxygenation. Accessory breathing muscles are used, causing retractions in substernal, suprasternal, and intercostal areas. Reduced oxygen to the brain causes restlessness initially and altered level of consciousness later. A decrease in heart and respiratory rates would be a late ominous sign of decompensation. Peak flow rate is a test used in asthma. A decrease, not increase, in peak flow is diagnostic of disease. A drop in blood pressure would be a late sign of hypoxia. An increase in temperature is more indicative of infection or inflammation than respiratory distress.

NP: Data collection; CN: Physiological integrity; CNS: Physiological adaptation; CL: Application

55. Which of the following strategies is recommended when caring for an infant with bronchopulmonary dysplasia?
1. Provide frequent playful stimuli.
2. Decrease oxygen during feedings.
3. Place the infant on a set schedule.
4. Place the infant in an open crib.

55. 3. Timing care activities with rest periods to avoid fatigue and to decrease respiratory effort is essential. Early stimulation activities are recommended but the infant will have limited tolerance for them because of the illness. Oxygen is usually increased during feedings to help decrease respiratory and energy requirements. Thermoregulation is important because both hypothermia and hyperthermia will increase oxygen consumption and may increase oxygen requirements. These infants are usually maintained on warmer beds or in Isolettes.
NP: Implementation; CN: Safe, effective care environment; CNS: Coordinated care; CL: Application

Be sure not to confuse the prefixes hyper and hypo!

56. Which of the following symptoms is seen in a child with bronchopulmonary dysplasia?
1. Minimal work of breathing
2. Tachypnea and dyspnea
3. Easily consolable
4. Hypotension

56. 2. Tachypnea, dyspnea, and wheezing are intermittently or chronically present secondary to airway obstruction and increased airway resistance. These infants usually show increased work of breathing and increased use of accessory muscles. They're frequently described as irritable and difficult to comfort. Pulmonary hypertension is a common finding resulting from fibrosis and chronic hypoxia.
NP: Data collection; CN: Physiological integrity; CNS: Physiological adaptation; CL: Knowledge

57. Which of the following interventions is most appropriate for helping parents to cope with a child newly diagnosed with bronchopulmonary dysplasia?
1. Teach cardiopulmonary resuscitation.
2. Refer them to support groups.
3. Help parents identify necessary lifestyle changes.
4. Evaluate and assess parents' stress and anxiety levels.

Prioritize. Prioritize. Prioritize.

57. 4. The emotional impact of bronchopulmonary dysplasia is clearly a crisis situation. The parents are experiencing grief and sorrow over the loss of a "healthy" child. The other strategies are more appropriate for long-term intervention.
NP: Evaluation; CN: Psychosocial integrity; CNS: Coping and adaptation; CL: Application

58. Infants with bronchopulmonary dysplasia are commonly treated with bronchodilators such as theophylline. Which of the following adverse reactions should the nurse watch for in the client taking this drug?
1. Lethargy
2. Decreased calcium level
3. Increased heart rate
4. Decreased serum potassium level

58. 3. Theophylline dilates the bronchi and relaxes bronchial smooth muscle cells. It also increases both cardiac contractile force and heart rate. Decreased potassium and calcium levels can be seen with the use of diuretics. Theophylline is a central nervous system stimulant and produces hyperactivity or irritability.
NP: Evaluation; CN: Physiological integrity; CNS: Pharmacological therapies; CL: Knowledge

---

NP: Nursing process   CN: Client needs category   CNS: Client needs subcategory   CL: Cognitive level

59. Theophylline is ordered for a 1-year-old client with bronchopulmonary dysplasia. The recommended dosage is 24 mg/kg/day. The client weighs 10 kg. How much is given per dose when administered 4 times per day?
1. 60 mg/dose
2. 80 mg/dose
3. 120 mg/dose
4. 240 mg/dose

60. Infants with bronchopulmonary dysplasia require frequent, prolonged rest periods. Which of the following signs indicates overstimulation?
1. Increased alertness
2. Good eye contact
3. Cyanosis
4. Lethargy

61. Which of the following parental care outcomes should be anticipated for a child with bronchopulmonary dysplasia?
1. Reports increased levels of stress
2. Only makes safe decisions with professional assistance
3. Participates in routine, but not complex, caretaking activities
4. Verbalizes the causes, risks, therapy options, and nursing care

62. Bronchopulmonary dysplasia can be classified into four categories. Which of the following characteristics is commonly noted during the early or first stage of the disease?
1. Interstitial fibrosis
2. Signs of emphysema
3. Hyperexpansion on chest X-ray
4. Resemblance to respiratory distress syndrome

63. Which of the following adverse reactions to furosemide (Lasix) would not be commonly associated with administration of this drug to a client with bronchopulmonary dysplasia?
1. Hypercalcemia
2. Hypokalemia
3. Hyponatremia
4. Hypocalcemia

Here's that formula yet again!

Hooray! You're more than halfway finished! Keep going!

59. 1. The child should receive 60 mg/dose. Here are the calculations: 24 mg/kg × 10 kg = 240 mg/day; 240 mg/4 doses = 60 mg/dose.
NP: Evaluation; CN: Physiological integrity; CNS: Pharmacological therapies; CL: Application

60. 3. Signs of overstimulation in an immature child include cyanosis, avoidance of eye contact, vomiting, diaphoresis, or falling asleep. The child may also become irritable and show signs of respiratory distress.
NP: Data collection; CN: Psychosocial integrity; CNS: Coping and adaptation; CL: Application

61. 4. The parents should understand the causes, risks, therapy options, and care of their infant by the time of discharge. Having the parents verbalize this information is the only way to assess their understanding. The parents should report decreased levels of stress, be capable of making decisions independently, and participate in routine and complex care.
NP: Evaluation; CN: Psychosocial integrity; CNS: Psychosocial adaptation; CL: Application

62. 4. Stage I can be characterized by early interstitial changes and resembles respiratory distress syndrome. Stage III shows signs of the beginning of chronic disease with interstitial edema, signs of emphysema, and pulmonary hypertension. Stage IV shows interstitial fibrosis and hyperexpansion on chest X-ray.
NP: Data collection; CN: Physiological integrity; CNS: Physiological adaptation; CL: Knowledge

63. 1. Hypercalcemia isn't associated with furosemide. Furosemide can cause hypokalemia, dilution hyponatremia, hypocalcemia, and hypomagnesemia.
NP: Evaluation; CN: Physiological integrity; CNS: Pharmacological therapies; CL: Knowledge

64. A pediatric client is to receive furosemide (Lasix) 4 mg/kg/day in one daily dose. The client weighs 20 kg. How many milligrams should be administered in each dose?
   1. 20 mg
   2. 40 mg
   3. 80 mg
   4. 160 mg

65. A 2-year-old child with bronchopulmonary dysplasia is placed on furosemide (Lasix) once per day. The nurse is educating the parents on foods that are rich in potassium. Which of the following foods should the nurse recommend?
   1. Apples
   2. Oranges
   3. Peaches
   4. Raisins

66. Which of the following scenarios necessitates tracheostomy tube placement in long-term care of an infant with bronchopulmonary dysplasia?
   1. Increased risk of tracheomalacia
   2. Prolonged dependence on the ventilator
   3. Need to allow for gastrostomy tube feedings
   4. Increased signs of respiratory distress

Might I recommend something from our fruit tray?

Let's see, what would I want most if the worst occurred?

67. A 1-year-old infant with bronchopulmonary dysplasia has just received a tracheostomy. Which of the following interventions is appropriate?
   1. Keep extra tracheostomy tubes at the bedside.
   2. Secure ties at the side of the neck for easy access.
   3. Change the tracheostomy tube 2 weeks after surgery.
   4. Secure the tracheostomy ties tightly to prevent dislodgment of the tube.

64. 3. The child should receive 80 mg per dose. Here is the calculation: 4 mg/kg × 20 kg = 80 mg.
NP: Evaluation; CN: Physiological integrity; CNS: Pharmacological therapies; CL: Application

65. 4. Raisins, dates, figs, and prunes are among the highest potassium-rich foods. They average 17 to 20 mEq of potassium. Apples, oranges, and peaches have very low amounts of potassium. They average 3 to 4 mEq.
NP: Evaluation; CN: Physiological integrity; CNS: Pharmacological therapies; CL: Knowledge

66. 2. Tracheostomy may be required after a child has been ventilator dependent for 6 to 8 weeks and is unable to wean from the ventilator. Tracheomalacia can be a complication of prolonged tracheal intubation, not a condition that necessitates a tracheostomy. The need for gastrostomy feedings wouldn't necessitate putting in a tracheostomy tube. Increased signs of respiratory distress may indicate the need for endotracheal intubation and mechanical ventilation, not tracheostomy.
NP: Planning; CN: Physiological integrity; CNS: Physiological adaptation; CL: Application

67. 1. Extra tracheostomy tubes should be kept at the bedside in case of an emergency, including one size smaller in case the appropriate size doesn't fit due to edema. The ties should be placed securely but allow the width of a little finger for room to prevent excessive pressure or skin breakdown. The first tracheostomy tube change is usually performed by the physician after 7 days. Ties are usually placed at the back of the neck.
NP: Implementation; CN: Physiological integrity; CNS: Physiological adaptation; CL: Application

68. An 11-month-old infant with bronchopulmonary dysplasia and a tracheostomy experiences a decline in oxygen saturation from 97% to 88%. He appears anxious and his heart rate is 180 beats/minute. Which of the following interventions is most appropriate?
1. Change the tracheostomy tube.
2. Suction the tracheostomy tube.
3. Obtain an arterial blood gas (ABG) level.
4. Increase the oxygen flow rate.

69. Which of the following interventions is most appropriate when suctioning a tracheostomy tube?
1. Hypoventilate the child before suctioning.
2. Repeat the suctioning process for two intervals.
3. Insert the catheter 1 to 2 cm below the tracheostomy tube.
4. Inject a small amount of normal saline solution into the tube before suctioning.

70. Which of the following characteristics distinguishes allergies from colds?
1. Skin tests can diagnose a cold.
2. Allergies are accompanied by fever.
3. Colds cause itching of the eyes and nose.
4. Allergies trigger constant and consistent bouts of sneezing.

71. Which of the following definitions best describes asthma?
1. Inflammation of the pulmonary parenchyma
2. Chronic lung disease caused by damaged alveoli
3. Infection of the lower airway, most often caused by a viral agent
4. Disease of the airways characterized by hyperactivity of the bronchi

Pay attention! This question is asking you to distinguish between two disorders.

68. 2. Tracheostomy tubes, particularly in small children, require frequent suctioning to remove mucus plugs and excessive secretions. The tracheostomy tube can be changed if suctioning is unsuccessful. Obtaining an ABG level may be beneficial if oxygen saturation remains low and the child appears to be in respiratory distress. Increasing the oxygen flow rate will only help if the airway is patent.
NP: Implementation; CN: Physiological integrity; CNS: Physiological adaptation; CL: Application

69. 4. Injecting a small amount of normal saline solution helps to loosen secretions for easier aspiration. Preservative-free normal saline solution should be used. The child should be hyperventilated before and after suctioning to prevent hypoxia. The suctioning process should be repeated until the trachea is clear. The catheter should be inserted 0.5 cm beyond the tracheostomy tube. If the catheter is inserted too far, it will irritate the carina and may cause blood-tinged secretions.
NP: Implementation; CN: Physiological integrity; CNS: Physiological adaptation; CL: Application

70. 4. Allergies elicit consistent bouts of sneezing, are seldom accompanied by fever, and tend to cause itching of the eyes and nose. Colds are accompanied by fever and are characterized by sporadic sneezing. Skin testing is performed to determine the client's sensitivity to specific allergens.
NP: Data collection; CN: Health promotion and maintenance; CNS: Prevention and early detection of disease; CL: Application

71. 4. Asthma is an obstructive disease that shows hyperactivity of the bronchi and edema of the mucous membranes. Bronchopulmonary dysplasia is a chronic lung disease with damaged alveoli. Pneumonia is characterized by an inflammation of the pulmonary parenchyma. Lower airway infection caused by a viral agent describes bronchitis.
NP: Evaluation; CN: Health promotion and maintenance; CNS: Prevention and early detection of disease; CL: Knowledge

72. A 2-year-old child has been diagnosed with asthma. Which of the following allergens can be considered one of the most common asthma triggers?

1. Weather
2. Peanut butter
3. The cat next door
4. One parent with asthma

Some things trigger asthma in me more than others.

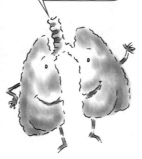

73. Which of the following symptoms is common in asthma?

1. Barking cough
2. Bradycardia
3. Dry, productive cough
4. Wheezing

Pay special attention to questions asking about such a serious disorder. Read them extra carefully.

74. Presence of which of the following factors would place a child at increased risk for an asthma-related death?

1. Use of an inhaler at home
2. One admission for asthma last year
3. Prior admission to the general pediatric floor
4. Prior admission to an intensive care unit for asthma

75. Which of the following characteristics may be present in a client with status asthmaticus?

1. Several attacks per month
2. Less than six attacks per year
3. Little or no response to bronchodilators
4. Constant attacks unrelieved by bronchodilators

This question is asking you to distinguish between varying degrees of asthma.

72. 1. Excessively cold air, wet or humid changes in weather and seasons, and air pollution are some of the most common asthma triggers. Household pets are also a trigger. Evidence suggests that asthma is partly hereditary in nature but heredity isn't an allergen. Food allergens are rarely responsible for airway reactions in children.
NP: Data collection; CN: Physiological integrity; CNS: Physiological adaptation; CL: Knowledge

73. 4. Asthma commonly occurs with wheezing and coughing. Airway inflammation and edema increase mucus production. Other signs include dyspnea, tachycardia (not bradycardia), and tachypnea. A barking cough is associated with croup. A dry, nonproductive cough is an asthma variant often seen in older children and teenagers.
NP: Data collection; CN: Physiological integrity; CNS: Physiological adaptation; CL: Knowledge

74. 4. Asthma results in varying degrees of respiratory distress. A prior admission to an intensive care unit marks an increased severity and need of immediate therapy. Two or more hospitalizations for asthma, a recent hospitalization or emergency department visit in the past month, or three or more emergency department visits in the past year puts a child at high risk for asthma-related death. Although current use of systemic steroids would also be a risk factor, not all inhalers contain steroids.
NP: Data collection; CN: Physiological integrity; CNS: Reduction of risk potential; CL: Application

75. 4. Status asthmaticus can best be described as constant attacks unrelieved by bronchodilators. Mild asthma is less than 6 attacks per year. Moderate asthma is characterized by several attacks per month. Little or no response to bronchodilators would describe severe asthma.
NP: Data collection; CN: Physiological integrity; CNS: Physiological adaptation; CL: Knowledge

76. A 2-year-old child with status asthmaticus is admitted to the pediatric unit and begins to receive continuous treatment with albuterol, given by nebulizer. Which of the following adverse effects is common with albuterol?
1. Bradycardia
2. Lethargy
3. Tachycardia
4. Tachypnea

77. A 10-year-old child is admitted with asthma. The physician orders an aminophylline infusion. A loading dose of 6 mg/kg is ordered. The client weighs 30 kg. How much aminophylline is contained in the loading dose?
1. 60 mg
2. 90 mg
3. 120 mg
4. 180 mg

78. Which complication is possible for a pediatric client receiving mechanical ventilation?
1. Pneumothorax
2. High cardiac output
3. Polycythemia
4. Hypovolemia

79. Which of the following X-ray findings would you expect for a child with asthma?
1. Atelectasis
2. Hemothorax
3. Infiltrates
4. Pneumothoraces

80. Which of the following interventions is most appropriate for a client with atelectasis?
1. Perform chest physiotherapy.
2. Give increased I.V. fluids.
3. Administer oxygen.
4. Obtain arterial blood gas (ABG) levels.

Wheee! Don'tcha just love math?

You're doing great! 80 down, not many more to go!

76. 3. Albuterol is a rapid-acting bronchodilator. Common adverse effects include tachycardia, nervousness, tremors, insomnia, irritability, and headache.
NP: Data collection; CN: Physiological integrity; CNS: Pharmacological therapies; CL: Knowledge

77. 4. The child should receive 180 mg per dose. Here is the calculation: 6 mg/kg × 30 kg = 180 mg.
NP: Evaluation; CN: Physiological integrity; CNS: Pharmacological therapies; CL: Application

78. 1. Mechanical ventilation can cause barotrauma, as occurs with pneumothorax; clients receiving mechanical ventilation must be carefully monitored. Mechanical ventilation decreases, not increases, cardiac output. Polycythemia is the result of chronic hypoxia, not mechanical ventilation. Mechanical ventilation can cause fluid overload, not dehydration.
NP: Evaluation; CN: Physiological integrity; CNS: Physiological adaptation; CL: Application

79. 1. Hyperexpansion, atelectasis, and a flattened diaphragm are typical X-ray findings for a child with asthma. Air becomes trapped behind the narrowed airways and the residual capacity rises, leading to hyperinflation. Hypoxemia results from areas of the lung not being well perfused. Infiltrates and pneumothoraces are uncommon. A hemothorax isn't a finding related to asthma.
NP: Data collection; CN: Physiological integrity; CNS: Physiological adaptation; CL: Application

80. 1. Chest physiotherapy and incentive spirometry help to enhance the clearance of mucus and open the alveoli. Administration of oxygen won't provide enough pressure to open the alveoli. I.V. and oral fluids are recommended to help liquefy and thin secretions. Obtaining ABG levels isn't necessary.
NP: Implementation; CN: Physiological integrity; CNS: Physiological adaptation; CL: Application

81. The parents of a 10-year-old child recently diagnosed with asthma ask if the child can continue to play sports. Which of the following responses is most appropriate?
1. Sports can cause asthma attacks.
2. You should limit activities to quiet play.
3. It's okay to play some sports but swimming isn't recommended.
4. Physical activity and sports are encouraged as long as the asthma is under control.

82. A client with thoracic water-seal drainage is on the elevator. The transport aide has placed the drainage system on the stretcher. What action should the nurse on the elevator take?
1. Assist the aide in placing the drainage system lower than the client's chest.
2. Report the incident to the registered nurse when she returns to the unit.
3. Clamp the drainage tubing with a hemostat.
4. Immediately take the client's respiratory and pulse rates.

83. Which of the following nursing interventions is appropriate to correct dehydration for a 2-year-old client with asthma?
1. Give warm liquids.
2. Give cold juice or ice pops.
3. Provide three meals and three snacks.
4. Provide small, frequent meals and snacks.

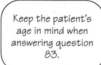

Keep the patient's age in mind when answering question 83.

84. Which of the following interventions by the parents can help reduce allergens in the home of a child with asthma?
1. Cover floors with carpeting.
2. Designate the basement as the play area.
3. Dust and clean the house thoroughly twice a month.
4. Use foam rubber pillows and synthetic blankets.

"Allergy proofing" a home can be just as important as "kid proofing" it.

81. 4. Participation in sports is encouraged but should be evaluated on an individual basis as long as the asthma is under control. Exercise-induced asthma is an example of the airway hyperactivity common to asthmatics. Exclusion from sports or activities may hamper peer interaction. Swimming is well tolerated because of the type of breathing involved and the moisture in the air.
NP: Evaluation; CN: Physiological integrity; CNS: Physiological adaptation; CL: Application

82. 1. The drainage device must be kept below the level of the chest to maintain straight gravity drainage. Placing it on the stretcher may cause a backflow of drainage into the thoracic cavity, which could collapse the partially expanded lung. Reporting the incident is indicated, but the immediate safety of the client takes priority. Clamping the tubing would place the client at risk for a tension pneumothorax. After the drainage system has been properly repositioned, the client's respiratory and pulse rates may be taken.
NP: Implementation; CN: Physiological integrity; CNS: Basic care and comfort; CL: Application

83. 1. Liquids are best tolerated if they're warm. Cold liquids may cause bronchospasm and should be avoided. Dehydration should be corrected slowly. Overhydration may increase interstitial pulmonary fluid and exacerbate small airway obstruction. Small, frequent meals should be provided to avoid abdominal distention that may interfere with diaphragm excursion.
NP: Implementation; CN: Physiological integrity; CNS: Physiological adaptation; CL: Application

84. 4. Bedding should be free from allergens with nonallergenic covers. Unnecessary rugs should be removed, and floors should be bare and mopped a few times a week to reduce dust. Dusting and cleaning should occur daily or at least weekly. Basements or cellars should be avoided to lessen the child's exposure to molds and mildew.
NP: Planning; CN: Physiological integrity; CNS: Physiological adaptation; CL: Knowledge

85. Which of the following nursing diagnoses is appropriate for a client with acute asthma?
1. Imbalanced nutrition: More than body requirements
2. Excess fluid volume
3. Activity intolerance
4. Constipation

86. Which of the following definitions best describes bronchiolitis?
1. Acute inflammation and obstruction of the bronchioles
2. Airway obstruction from aspiration of a solid object
3. Inflammation of the pulmonary parenchyma
4. Acute highly contagious crouplike syndrome

87. A 2-month-old infant is given a preliminary diagnosis of bronchiolitis. Which of the following symptoms would the nurse expect to find?
1. Bradycardia
2. Increased appetite
3. Wheezing on auscultation
4. No signs of an upper respiratory infection

88. In most cases, bronchiolitis is caused by a viral agent, most commonly respiratory syncytial virus (RSV). Which of the following characteristics is associated with RSV?
1. Prevalence in the summer and fall months
2. High affinity for respiratory tract mucosa
3. Most common in children over age 5
4. Infected children aren't contagious

This is a cute question. Get it? A cute?

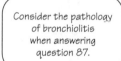

Consider the pathology of bronchiolitis when answering question 87.

85. 3. Ineffective oxygen supply and demand may lead to activity intolerance. The nurse should promote rest and encourage developmentally appropriate activities. Nutrition may be decreased, not increased, due to respiratory distress and GI upset. Dehydration, not overhydration, is common due to diaphoresis, insensible water loss, and hyperventilation. Medications given to treat asthma may cause nausea, vomiting, and diarrhea, not constipation.
NP: Data collection; CN: Physiological integrity; CNS: Physiological adaptation; CL: Application

86. 1. Bronchiolitis is an infection of the bronchioles, causing the mucosa to become edematous, inflamed, and full of mucus. Pneumonia is characterized by inflammation of the pulmonary parenchyma. Lower airway obstruction from a solid object is a form of foreign body aspiration. Croup syndromes are generally upper airway infections or obstructions.
NP: Data collection; CN: Physiological integrity; CNS: Physiological adaptation; CL: Knowledge

87. 3. In bronchiolitis, the bronchioles become narrowed and edematous, which can cause wheezing. These infants typically have a 2- to 3-day history of an upper respiratory infection and feeding difficulties with loss of appetite due to nasal congestion and increased work of breathing. This combination leads to respiratory distress with tachypnea and tachycardia.
NP: Data collection; CN: Physiological integrity; CNS: Physiological adaptation; CL: Application

88. 2. RSV attacks the respiratory tract mucosa. The virus is most prevalent in the winter and early spring months. Most children develop the infection between ages 2 and 6 months, and RSV generally occurs during the first 3 years of life. RSV is a highly contagious respiratory virus.
NP: Data collection; CN: Physiological integrity; CNS: Physiological adaptation; CL: Knowledge

89. Which of the following precautions should a nurse caring for a 2-month-old infant with respiratory syncytial virus (RSV) take to prevent the spread of infection?
1. Gloves only
2. Gown, gloves, and mask
3. No precautions are required; the virus isn't contagious
4. Proper hand washing between clients

90. Which of the following tests positively diagnoses respiratory syncytial virus (RSV)?
1. Blood test
2. Nasopharyngeal washings
3. Sputum culture
4. Throat culture

91. Which of the following children would be at increased risk for a respiratory syncytial virus (RSV) infection?
1. 2-month-old child managed at home
2. 2-month-old child with bronchopulmonary dysplasia
3. 3-month-old child requiring low-flow oxygen
4. 2-year-old child

92. Which nursing diagnosis would the nurse expect to find on the care plan of a 10-month-old infant to promote coping during hospitalization?
1. Toileting self-care deficit related to the child's age
2. Powerlessness related to hospital environment
3. Deficient diversional activity related to hospital environment
4. Anxiety related to separation from parents

What precautions should you take when treating a highly contagious agent such as my none-too-lovable self?

This question is asking you to prioritize the risk associated with RSV infection.

89. 2. RSV is highly contagious and is spread through direct contact with infectious secretions via hands, droplets, and fomites. Gowns, gloves, and masks should be worn for client care to prevent the spread of infection.
NP: Implementation; CN: Safe, effective care environment; CNS: Safety and infection control; CL: Application

90. 2. RSV can only be diagnosed with direct aspiration of nasal secretions or nasopharyngeal washings. Positive identification is accomplished using the enzyme-linked immunosorbent assay. Blood, throat, and sputum cultures can't definitively diagnose RSV.
NP: Data collection; CN: Physiological integrity; CNS: Physiological adaptation; CL: Knowledge

91. 2. Infants with cardiac or pulmonary conditions are at highest risk for RSV. Because of their underlying conditions, they're more likely to require mechanical ventilation. Many infants can be managed at home; few require hospitalization. A 2-year-old child has built up the immune system and can tolerate the infection without major problems. A 3-month-old on low-flow oxygen has some risks of progression but isn't at high risk.
NP: Data collection; CN: Physiological integrity; CNS: Reduction of risk potential; CL: Application

92. 4. Attachment is critical in infancy, and prolonged separation has been well documented as a risk factor that compromises normal infant development. Toilet training wouldn't be an issue for a 10-month-old. Powerlessness is a concern after the toddler stage, when a child develops autonomy and independence. Diversion won't be an issue until the acute phase of the illness has passed. Providing diversion for infants is easily accomplished by the use of age-appropriate toys and play activities.
NP: Planning; CN: Psychosocial integrity; CNS: Coping and adaptation; CL: Application

NP: Nursing process CN: Client needs category CNS: Client needs subcategory CL: Cognitive level

93. Which of the following medications is an antiviral agent used to treat bronchiolitis caused by respiratory syncytial virus (RSV)?

1. Albuterol
2. Aminophylline
3. Cromolyn sodium
4. Ribavirin (Virazole)

94. Which of the following interventions is most important when monitoring dehydration in an infant with bronchiolitis?

1. Measurement of intake and output
2. Blood levels every 4 hours
3. Urinalysis every 8 hours
4. Weighing each diaper

This one is tricky. Pay attention to the frequency of the measurements.

95. Which of the following nursing diagnoses is appropriate for an infant with bronchiolitis?

1. Imbalanced nutrition: More than body requirements
2. Deficient diversional activity
3. Impaired gas exchange
4. Social isolation

96. Which of the following items is an essential skill for parents managing bronchiolitis at home?

1. Place the child in a prone position for comfort.
2. Use warm mist to replace insensible fluid loss.
3. Recognize signs of increasing respiratory distress.
4. Engage the child in many activities to prevent developmental delay.

93. 4. Ribavirin is an antiviral agent sometimes used to reduce the severity of bronchiolitis caused by RSV. Aminophylline and albuterol are bronchodilators and haven't been proven effective in treating viral bronchiolitis. Cromolyn sodium is an inhaled anti-inflammatory agent.

NP: Evaluation; CN: Physiological integrity; CNS: Pharmacological therapies; CL: Application

94. 1. Accurate measurement of intake and output is essential to assess for dehydration. Blood levels may be obtained daily or every other day. A urinalysis every 8 hours isn't necessary. Urine specific gravities are recommended but can be obtained with diaper changes. Weighing diapers is a way of measuring output only.

NP: Data collection; CN: Physiological integrity; CNS: Physiological adaptation; CL: Application

95. 3. Infants with bronchiolitis will have impaired gas exchange related to bronchiolar obstruction, atelectasis, and hyperinflation. Nutrition may be seen as less than body requirements. If respiratory distress is present, these infants should have nothing by mouth and fluids given I.V. only. Deficient diversional activity deficit and social isolation usually aren't priorities. These infants are too uncomfortable to respond to social stimuli and need quiet, soothing activities that minimize energy.

NP: Planning; CN: Physiological integrity; CNS: Physiological adaptation; CL: Application

96. 3. It's essential for parents to be able to recognize signs of increasing respiratory distress and know how to count the respiratory rate. The child should be positioned with the head of the bed elevated for comfort and to facilitate removal of secretions. Use of cool mist may help to replace insensible fluid loss. Quiet play activities are required only as the child's energy level permits. These infants show clinical improvement in 3 to 4 days; therefore, developmental delay isn't an issue.

NP: Planning; CN: Physiological integrity; CNS: Physiological adaptation; CL: Application

97. Which of the following definitions best describes pneumonia?
1. Inflammation of the large airways
2. Severe infection of the bronchioles
3. Inflammation of the pulmonary parenchyma
4. Acute viral infection with maximum effect at the bronchiolar level

98. Pneumonias can be classified by four etiologic processes. Which of the following causative agents is responsible for bacterial pneumonia?
1. *Mycoplasma*
2. Parainfluenza virus
3. Pneumococci
4. Respiratory syncytial virus (RSV)

99. Which of the following types of pneumonia is most common in children ages 5 to 12?
1. Enteric bacilli
2. *Mycoplasma* pneumonia
3. Staphylococcal pneumonia
4. Streptococcal pneumonia

100. Which of the following characteristics is commonly seen in a child with pertussis?
1. Barking cough
2. Whooping cough
3. Abrupt high fever
4. Inspiratory stridor

101. Which of the following tests is the definitive means of diagnosing tuberculosis (TB) in children?
1. Chest X-ray
2. Sputum sample
3. Tuberculin test
4. Urine culture

Whoopee! You reached 100! Can you believe it?

Psssst! Remember that *definitive* means best!

97. 3. Pneumonia is an inflammation of the pulmonary parenchyma. Bronchitis is inflammation of the large airways. Bronchiolitis is a severe infection of the bronchioles. Bronchiolitis and respiratory syncytial virus are types of acute viral infection with maximum effect at the bronchiolar level.
NP: Data collection; CN: Physiological integrity; CNS: Physiological adaptation; CL: Knowledge

98. 3. Pneumococcal pneumonia is the most common causative agent accounting for about 90% of bacterial pneumonia. *Mycoplasma* is a causative agent for primary atypical pneumonia. Parainfluenza virus and RSV account for viral pneumonia.
NP: Data collection; CN: Physiological integrity; CNS: Physiological adaptation; CL: Knowledge

99. 2. *Mycoplasma* pneumonia is a primary atypical pneumonia seen in children between ages 5 and 12. Streptococcal pneumonia, enteric bacilli, and staphylococcal pneumonia are mostly seen in children in the 3-month to 5-year age-group.
NP: Data collection; CN: Physiological integrity; CNS: Physiological adaptation; CL: Knowledge

100. 2. Pertussis is characterized by consistent short, rapid coughs followed by a sudden inspiration with a high-pitched whooping sound. A barking cough and inspiratory stridor are noted with croup. Pertussis usually is accompanied by a low-grade fever.
NP: Data collection; CN: Physiological integrity; CNS: Physiological adaptation; CL: Knowledge

101. 2. A sputum culture is the definitive test for TB in children. The tuberculin test is the most accurate but not necessarily the most reliable test for TB in children. X-rays usually appear normal in children with TB. Stool culture, not urine culture, and gastric washings will show positive results on acid-fast smears but aren't specific for *Mycobacterium tuberculosis*. Sputum samples are difficult to obtain from children, so gastric washings often replace them.
NP: Data collection; CN: Physiological integrity; CNS: Physiological adaptation; CL: Application

NP: Nursing process CN: Client needs category CNS: Client needs subcategory CL: Cognitive level

102. Which of the following symptoms is a characteristic sign of tuberculosis (TB)?
1. Chills
2. Hyperactivity
3. Lymphadenitis
4. Weight gain

103. Which of the following adverse effects can be expected by the parents of a 2-year-old child who has been started on rifampin (Rifadin) after testing positive for tuberculosis?
1. Hyperactivity
2. Orange body secretions
3. Decreased bilirubin levels
4. Decreased levels of liver enzymes

Orange you glad you're almost finished! (Hahahahaha! I'm too much!)

104. Children under age 3 are prone to aspirating foreign bodies. Which of the following actions is recommended to prevent aspiration?
1. Cut hot dogs in half.
2. Limit popcorn and peanuts.
3. Cut grapes into small pieces.
4. Limit hard candy to special occasions.

105. Which of the following groups of findings suggests a foreign body located in the trachea?
1. Cough, dyspnea, and drooling
2. Cough, stridor, and changes in phonation
3. Expiratory wheeze and inspiratory stridor
4. Cough, asymmetric breath sounds, and wheeze

106. Which of the following activities is recommended to prevent foreign body aspiration during meals?
1. Insist that children are seated.
2. Give children toys to play with.
3. Allow children to watch television.
4. Allow children to eat in a separate room.

Keep pumping out those answers!

102. 3. Children with TB are usually asymptomatic and often don't manifest the usual pulmonary symptoms, but lymphadenitis is more likely in infants and children than in adults. Weight loss, anorexia, night sweats, fatigue, and malaise are general responses to the disease.
NP: Data collection; CN: Physiological integrity; CNS: Physiological adaptation; CL: Knowledge

103. 2. Rifampin and its metabolites will turn urine, feces, sputum, tears, and sweat an orange color. This isn't a serious adverse effect. Rifampin may also cause GI upset, headache, drowsiness, dizziness, visual disturbances, and fever. Liver enzyme and bilirubin levels increase because of hepatic metabolism of the drug. Parents should be taught the signs and symptoms of hepatitis and hyperbilirubinemia such as jaundice of the sclera or skin.
NP: Evaluation; CN: Physiological integrity; CNS: Pharmacological therapies; CL: Application

104. 3. Grapes, hot dogs, and sausage should be cut into many small pieces. Hard candy, raisins, popcorn, and peanuts should be avoided for children age 4 and younger.
NP: Evaluation; CN: Physiological integrity; CNS: Reduction of risk potential; CL: Application

105. 3. Expiratory and inspiratory noise indicates that the foreign body is in the trachea. Cough, dyspnea, drooling, and gagging indicate supraglottic obstruction. A cough with stridor and changes in phonation would occur if the foreign body were in the larynx. Asymmetric breath sounds indicate that the object may be located in the bronchi.
NP: Data collection; CN: Physiological integrity; CNS: Physiological adaptation; CL: Knowledge

106. 1. Children should remain seated while eating. The risk of aspiration increases if the child is running, jumping, or talking with food in their mouth. Television and toys are a dangerous distraction to toddlers and young children and should be avoided. Children need constant supervision and should be monitored while eating snacks and meals.
NP: Planning; CN: Safe, effective care environment; CNS: Safety and infection control; CL: Application

107. Which of the following diagnostic tools is best for general diagnosis of foreign body aspiration?
1. Bronchoscopy
2. Chest X-ray
3. Fluoroscopy
4. Lateral neck X-ray

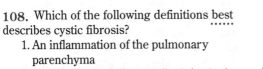

107. 1. Bronchoscopy can give a definitive diagnosis of the presence of foreign bodies and is also the best choice for removal of the object with direct visualization. Chest X-ray and lateral neck X-ray may also be used but findings vary. Some films may appear normal or show changes such as inflammation related to the presence of the foreign body. Fluoroscopy is valuable in detecting and localizing foreign bodies in the bronchi.
NP: Data collection; CN: Physiological integrity; CNS: Physiological adaptation; CL: Knowledge

Here are two questions that use the word best in different ways.

108. Which of the following definitions best describes cystic fibrosis?
1. An inflammation of the pulmonary parenchyma
2. A chromosomal abnormality inherited as an autosomal-dominant trait
3. A multisymptom disorder affecting the exocrine or mucus-producing glands
4. A chronic lung disease related to high concentrations of oxygen and ventilation

108. 3. Cystic fibrosis affects many organs as well as the exocrine or mucus-producing glands. In cystic fibrosis, an autosomal-recessive chromosomal abnormality, the child inherits defective genes from both parents. Inflammation of the pulmonary parenchyma describes pneumonia. Bronchopulmonary dysplasia is related to high concentrations of oxygen and ventilation.
NP: Data collection; CN: Physiological integrity; CNS: Physiological adaptation; CL: Knowledge

109. Which of the following clinical features would be found in a child with cystic fibrosis?
1. Increase in pancreatic secretions of bicarbonate and chloride
2. Decrease in sodium and chloride in both saliva and sweat
3. Increased viscosity of mucous gland secretions
4. Decreased viscosity of mucous gland secretions

109. 3. A primary feature of children with cystic fibrosis includes the increased viscosity of mucous gland secretions. Instead of thin, free-flowing secretions, the mucous glands produce a thick mucoprotein that accumulates and dilates them. Pancreatic enzymes are decreased because mucus secretions block the pancreatic ducts. The electrolytes sodium and chloride are increased in sweat; this forms the basis of the diagnosis of the disorder.
NP: Data collection; CN: Physiological integrity; CNS: Physiological adaptation; CL: Knowledge

Which of these options would be most likely?

110. Which of the following signs would the nurse most likely see on assessment of an infant with cystic fibrosis?
1. Constipation
2. Decreased appetite
3. Hyperalbuminemia
4. Meconium ileus

110. 4. Meconium ileus is a common early sign of cystic fibrosis. Thick, mucilaginous meconium blocks the lumen of the small intestine, causing intestinal obstruction, abdominal distention, and vomiting. These infants may have an increased appetite related to poor absorption from the intestine. Large-volume, loose, frequent, foul-smelling stools are common. The undigested food is excreted, increasing the bulk of feces. Hypoalbuminemia is a common result from the decreased absorption of protein.
NP: Data collection; CN: Physiological integrity; CNS: Physiological adaptation; CL: Knowledge

NP: Nursing process CN: Client needs category CNS: Client needs subcategory CL: Cognitive level

111. Which of the following tools are most commonly used to diagnose cystic fibrosis?
1. Chest X-ray
2. Pulmonary function test
3. Stool culture
4. Sweat test

Keep at it! You're almost there!

112. A client is receiving oxygen via a nasal cannula at 2 L/minute. What percentage of oxygen concentration is coming through the cannula?
1. 23% to 30%
2. 30% to 40%
3. 40% to 60%
4. 50% to 75%

113. Which of the following diets is recommended for a child with cystic fibrosis?
1. Fat-restricted diet
2. High-calorie diet
3. Low-protein diet
4. Sodium-restricted diet

In light of the pathology of cystic fibrosis, which is the only diet that makes sense?

114. Which of the following statements concerning pancreatic enzymes for a cystic fibrosis client is correct?
1. Capsules may not be opened.
2. Microcapsules can be crushed.
3. Encourage eating throughout the day.
4. Administer enzymes at each meal and with snacks.

111. 4. A sweat test is the most reliable diagnostic procedure for cystic fibrosis. It involves stimulating the production of sweat, collecting the sweat, and measuring electrolytes in the sweat. Two separate samples are collected to assure reliability of the test. Chest X-rays can show characteristic atelectasis and obstructive emphysema. Pulmonary function tests can show lung function and abnormal small airway function in cystic fibrosis. Stool analysis requires a 72-hour sample with an accurate food intake record.
NP: Data collection; CN: Physiological integrity; CNS: Physiological adaptation; CL: Knowledge

112. 1. The percentage of oxygen concentration as it passes out of the nasal cannula at 2 L/minute is 23% to 30%. The oxygen concentration via cannula at 3 to 5 L/minute is 30% to 40%. A simple mask at 6 to 8 L/minute delivers 40% to 60% oxygen. A partial rebreather mask at 8 to 11 L/minute delivers 50% to 75% oxygen.
NP: Data collection; CN: Physiological integrity; CNS: Reduction of risk potential; CL: Comprehension

113. 2. A well-balanced, high-calorie, high-protein diet is recommended for a child with cystic fibrosis due to impaired intestinal absorption. Fat restriction isn't required because digestion and absorption of fat in the intestine are impaired. The child usually increases enzyme intake when high-fat foods are eaten. Low-sodium foods can lead to hyponatremia; therefore, high-salt foods are recommended, especially during hot weather or when the child has a fever.
NP: Planning; CN: Safe, effective care environment; CNS: Coordinated care; CL: Application

114. 4. Enzymes are administered with each feeding, meal, and snack to optimize absorption of the nutrients consumed. Microcapsules shouldn't be crushed due to the enteric coating. Regular capsules may be opened and the contents mixed with a small amount of applesauce or other nonalkaline food. Eating throughout the day should be discouraged. Three meals and two or three snacks per day are recommended.
NP: Planning; CN: Physiological integrity; CNS: Physiological adaptation; CL: Application

115. Ranitidine (Zantac) is ordered for a 2-year-old child with cystic fibrosis. Which of the following classifications accurately describes this drug?
1. Aluminum salt
2. Antacid
3. Anticholinergic
4. Histamine-receptor antagonist

116. The nurse is caring for a client with cystic fibrosis. Ranitidine (Zantac) 4 mg/kg/day every 12 hours is ordered. The child weighs 20 kg. How many milligrams are given per dose?
1. 16 mg
2. 20 mg
3. 40 mg
4. 80 mg

117. Which of the following interventions is appropriate for care of the child with cystic fibrosis?
1. Decrease exercise and limit physical activity.
2. Administer cough suppressants and antihistamines.
3. Administer chest physiotherapy two to four times per day.
4. Administer bronchodilator or nebulizer treatments after chest physiotherapy.

118. Which of the following statements is appropriate for the nurse to make to the parents of a child with cystic fibrosis who are planning to have a second child?
1. Genetic counseling is recommended.
2. There's a 50% chance that the child will be normal.
3. There's a 50% chance that the child will be affected.
4. There's a 25% chance that the child will only be a carrier.

Don't antagonize me, now, I'm just not in the mood! (Hey, that's a hint!)

115. 4. Ranitidine acts as a histamine-2 receptor antagonist. It helps to decrease duodenal acidity and enhance pancreatic enzyme activity. Aluminum salts are used as antacids, which are used to neutralize gastric acid. Anticholinergics may be used to decrease gastric emptying and to decrease gastric acid secretions.
NP: Evaluation; CN: Physiological integrity; CNS: Pharmacological therapies; CL: Knowledge

116. 3. The child should receive 40 mg per dose. Here are the calculations: 20 kg × 4 mg/kg = 80 mg; 24 hours/12 hours = 2 doses; 80 mg/2 doses = 40 mg.
NP: Evaluation; CN: Physiological integrity; CNS: Pharmacological therapies; CL: Application

117. 3. Chest physiotherapy is recommended two to four times per day to help loosen and move secretions to facilitate expectoration. Exercise and physical activity is recommended to stimulate mucus secretion and to establish a good habitual breathing pattern. Cough suppressants and antihistamines are contraindicated. The goal is for the child to be able to cough and expectorate mucus secretions. Bronchodilator or nebulizer treatments are given before chest physiotherapy to help open the bronchi for easier expectoration.
NP: Implementation; CN: Safe, effective care environment; CNS: Coordinated care; CL: Application

118. 1. Genetic counseling should be recommended. Cystic fibrosis is an autosomal-recessive disease. Therefore, there's a 25% chance of the child having the disease, a 25% chance of the child being normal, and a 50% chance of the child being a carrier.
NP: Planning; CN: Health promotion and maintenance; CNS: Prevention and early detection of disease; CL: Application

119. Which of the following statements best describes an autosomal-recessive disorder such as cystic fibrosis?

    1. The genetic disorder is carried on the X chromosome.

    2. Both parents must pass the defective gene or set of genes.

    3. Only one defective gene or set of genes is passed by one parent.

    4. The child has an extra chromosome, resulting in an XXY karyotype.

120. The physician orders chest physiotherapy for a pediatric client. The nurse shouldn't perform chest physiotherapy when the client is experiencing:

    1. a productive cough.

    2. retained secretions.

    3. acute bronchoconstriction.

    4. hypoxia.

121. Ceftazidime (Fortaz) has been ordered for a client with cystic fibrosis. The order states to give 40 mg/kg every 8 hours. The child is 2 years old and weighs 38½ lb. How many milligrams of the ceftazidime is given in one dose?

    1. 116 mg

    2. 233 mg

    3. 260 mg

    4. 466 mg

122. The nurse is caring for a 5-year-old male child admitted to the pediatric unit with cystic fibrosis. Stools of a child with cystic fibrosis are characteristically:

    1. black and tarry.

    2. frothy, foul-smelling, and steatorrheaic.

    3. clay-colored.

    4. orange or green.

Here's a hint: X doesn't mark the spot on this one!

Getting closer! Keep it up!

Pay attention to the characteristics when answering this one!

119. 2. In recessive disorders such as cystic fibrosis, both parents must pass the defective gene or set of genes to the child. Dominant disorders are characterized by only one defective gene or set of genes passed by one parent. Sex-linked genetic disorders are carried on the X chromosome. A child with an XXY karyotype would have Klinefelter's syndrome.
NP: Data collection; CN: Health promotion and maintenance; CNS: Prevention and early detection of disease; CL: Knowledge

120. 3. The nurse shouldn't administer chest physiotherapy during episodes of acute bronchoconstriction or airway edema because loosening mucus plugs could cause airway obstruction. Chest physiotherapy aids in the elimination of secretions and reexpansion of lung tissue. Successful treatment with chest physiotherapy produces improved oxygenation and increased sputum production and airflow. Therefore, it should be performed when a productive cough, retained secretions, or hypoxia is present.
NP: Implementation; CN: Physiological integrity; CNS: Reduction of risk potential; CL: Application

121. 2. The child should receive 233 mg per dose. Here are the calculations: 38.5 lb/2.2 kg = 17.5 kg (1 lb equals 2.2 kg); 40 mg/kg × 17.5 kg = 700 mg; 24 hours/8 hours = 3 doses; 700 mg/3 doses = 233 mg.
NP: Evaluation; CN: Physiological integrity; CNS: Pharmacological therapies; CL: Application

122. 2. Clients with cystic fibrosis have an abnormal electrolyte transport system in the cells that eventually blocks the pancreas, preventing the secretion of enzymes that digest certain foods such as protein and fats. This results in foul-smelling, fatty (steatorrheaic) stools. Black, tarry stools are observed in clients who have upper GI bleeding, are on iron medications, or who consume diets high in red meat and dark-green vegetables. Clay-colored stools indicate possible bile obstruction. Orange or green stools may indicate intestinal infection.
NP: Data collection; CN: Physiological integrity; CNS: Physiological adaptation; CL: Comprehension

123. Which of the following complications of cystic fibrosis may eventually lead to death?
 1. Rectal prolapse
 2. Pulmonary obstruction
 3. Gastroesophageal reflux
 4. Reproductive system obstruction

123. 2. Pulmonary obstruction related to thickened mucus secretions can lead to a progressive pulmonary disturbance and secondary infections that can lead to death. Rectal prolapse is managed with enzyme replacement therapy and manipulation of the rectum back into place. Gastroesophageal reflux can be managed with medications and proper reflux precautions. Obstruction of the reproductive system can lead to infertility due to increased mucus blocking sperm entry in the female or blockage of the vas deferens in the male.
NP: Evaluation; CN: Physiological integrity; CNS: Physiological adaptation; CL: Application

124. Which of the following methods is best for evaluation of a 6-year-old child with cystic fibrosis who has been placed on an aerosol inhaler?
 1. Ask if the parents have any questions.
 2. Ask if the child can explain the procedure.
 3. Ask the parents if they understand the usage.
 4. Ask the client to perform a return demonstration.

124. 4. A return demonstration is the best evaluation. It will show if the client can repeat the steps shown and appropriately use the inhaler. The child may have difficulty explaining the procedure at age 6. The parents should understand how the inhaler should be used and ask questions, but the child must be able to correctly demonstrate usage first.
NP: Evaluation; CN: Physiological integrity; CNS: Pharmacological therapies; CL: Application

125. Which of the following interventions is appropriate for a 2-year-old client with chest trauma who has a left lower chest tube in place?
 1. Stripping or milking the tubing
 2. Requiring routine dressing changes
 3. Clamping the chest tube during transport
 4. Inspecting tubing for kinks or obstructions

125. 4. Tubing should be inspected for kinks or obstructions so that drainage can flow freely. Manipulation of the tubing should be avoided. The pressure created from stripping can damage the pleural space or mediastinum. There's no need for routine dressing changes if the dressing isn't soiled and there's no evidence of infection. Inspect and palpate around the dressing routinely. The chest tube should never be clamped because it may lead to tension pneumothorax. Waterseal will protect the client during transport.
NP: Implementation; CN: Physiological integrity; CNS: Physiological adaptation; CL: Application

You did it! Congratulations! Oh, you can breathe now. Whew!

Here are three internet sites you can check for more information about neurosensory disorders in children: **www.add.org** (Attention Deficit Disorder Association), **www.ndss.org** (National Down Syndrome Society), and **www.sbaa.org** (Spina Bifida Association of America).

Chapter 30
Neurosensory disorders

1. The mother of a 3-year-old with a myelomeningocele is thinking about having another baby. The nurse should inform the woman that she should increase her intake of which of the following acids?
1. Folic acid to 0.4 mg/day
2. Folic acid to 4.0 mg/day
3. Ascorbic acid to 0.4 mg/day
4. Ascorbic acid to 4.0 mg/day

2. Which of the following nursing diagnoses is most relevant in the first 12 hours of life for a neonate born with a myelomeningocele?
1. Risk for infection
2. Constipation
3. Impaired physical mobility
4. Delayed growth and development

3. Which of the following conditions would the nurse expect when assessing a neonate for hydrocephalus?
1. Bulging fontanel, low-pitched cry
2. Depressed fontanel, low-pitched cry
3. Bulging fontanel, eyes rotated downward
4. Depressed fontanel, eyes rotated downward

Most relevant — that's the key phrase for question 2.

1. 2. The American Academy of Pediatrics recommends that a woman who has had a child with a neural tube defect increase her intake of folic acid to 4.0 mg per day 1 month before becoming pregnant and continue this regimen through the first trimester. A woman who has no family history of neural tube defects should take 0.4 mg/day. All women of childbearing age should be encouraged to take a folic acid supplement because the majority of pregnancies in the United States are unplanned. Ascorbic acid hasn't been shown to have any effect on preventing neural tube defects.
NP: Implementation; CN: Health promotion and maintenance; CNS: Prevention and early detection of disease; CL: Application

2. 1. All of these diagnoses are important for a child with a myelomeningocele. However, during the first 12 hours of life, the most life-threatening event would be an infection. The other diagnoses will be addressed as the child develops.
NP: Planning; CN: Physiological integrity; CNS: Reduction of risk potential; CL: Application

3. 3. Hydrocephalus is caused by an alteration in circulation of the cerebrospinal fluid (CSF). The amount of CSF increases, causing the fontanel to bulge. This also causes an increase in intracranial pressure. This increase in pressure causes the neonate's eyes to deviate downward (the "setting sun sign"), and the neonate's cry becomes high-pitched.
NP: Data collection; CN: Health promotion and maintenance; CNS: Prevention and early detection of disease; CL: Knowledge

NP: Nursing process   CN: Client needs category   CNS: Client needs subcategory   CL: Cognitive level

4. A 2-year-old child is admitted to the hospital for revision of a ventriculoperitoneal shunt. Which of the following complications is the most common reason for a revision?
1. Shunt infection
2. A broken shunt
3. Growth of the child
4. Heart failure

5. Which of the following nursing actions is appropriate when a child has a seizure?
1. Inserting a nasogastric tube to prevent emesis
2. Restraining the extremities with a pillow or blanket
3. Inserting a tongue blade to prevent injury to the tongue
4. Padding the side rails of the bed to protect the child from injury

6. A mother brings her infant to the emergency department and says he had a seizure. While the nurse is obtaining a history, the mother says she was running out of formula so she stretched the formula by adding three times the normal amount of water. Electrolytes and blood glucose levels are drawn on the infant. The nurse would expect which of the following laboratory values?
1. Blood glucose: 120 mg/dl
2. Chloride: 104 mmol/L
3. Potassium: 4 mmol/L
4. Sodium: 125 mmol/L

7. For which of the following symptoms will the nurse assess a neonate diagnosed with bacterial meningitis?
1. Hypothermia, irritability, and poor feeding
2. Positive Babinski's reflex, mottling, and pallor
3. Headache, nuchal rigidity, and developmental delays
4. Positive Moro's embrace reflex, hyperthermia, and sunken fontanel

Remember, in emergency situations, your first priority is to ensure the client's safety.

For the NCLEX, you need to be familiar with normal lab values.

4. 3. This type of shunt has to be replaced periodically as the child grows and the tubing becomes too short. Shunt infection usually occurs within 2 to 3 months of insertion, not at 2 years. The shunts rarely break, unless there's head trauma. The fluid from a ventriculoperitoneal shunt goes into the peritoneal cavity and not into the heart; therefore, heart failure shouldn't occur.
NP: Data collection; CN: Health promotion and maintenance; CNS: Growth and development through the life span; CL: Knowledge

5. 4. A child having a seizure could fall out of bed or injure himself on anything, including the side rails of the bed. Attempts to insert anything into the child's mouth may injure the child. Attempting to restrain the child can't stop seizures. In fact, tactile stimulation may increase the seizure activity; therefore, it must be limited as much as possible.
NP: Implementation; CN: Safe, effective care environment; CNS: Safety and infection control; CL: Application

6. 4. Diluting formula in a different manner than is recommended alters the infant's electrolyte levels. Normal serum sodium for an infant is 135 to 145 mmol/L. When formula is diluted, the sodium in it is also diluted and will decrease the infant's sodium level. Hyponatremia is one of the causes of seizures in infants. The other values are all within normal limits.
NP: Evaluation; CN: Physiological integrity; CNS: Physiological adaptation; CL: Application

7. 1. The clinical appearance of a neonate with meningitis is different from that of a child or an adult. Neonates may be either hypothermic or hyperthermic. The irritation to the meninges causes the neonates to be irritable and to have a decreased appetite. They may be pale and mottled with a bulging, full fontanel. Normal neonates have positive Moro's embrace and Babinski's reflexes. Older children and adults with meningitis have headaches, nuchal rigidity, and hyperthermia as clinical manifestations.
NP: Data collection; CN: Physiological integrity; CNS: Physiological adaptation; CL: Comprehension

8. Which of the following types of behavior demonstrated by a 6-year-old would help the nurse recognize a learning disability as opposed to attention deficit hyperactivity disorder (ADHD)?
1. The child reverses letters and words while reading.
2. The child is easily distracted and reacts impulsively.
3. The child is always getting into fights during recess.
4. The child has a difficult time reading a chapter book.

9. Which of the following characteristics is true of cerebral palsy?
1. It's reversible.
2. It's progressive.
3. It results in mental retardation.
4. It appears at birth or during the first 2 years of life.

10. While assessing a full-term neonate, which of the following symptoms would cause the nurse to suspect a neurologic impairment?
1. A weak sucking reflex
2. A positive rooting reflex
3. A positive Babinski's reflex
4. Startle reflex in response to a loud noise

You're through the first 10! Keep going!

11. A mother reports that her school-age child has been reprimanded for daydreaming during class. This is a new behavior, and the child's grades are dropping. The nurse should suspect which of the following problems?
1. The child may have a hearing problem and needs to have his ears checked.
2. The child may have a learning disability and needs referral to the special education department.
3. The child may have attention deficit hyperactivity disorder (ADHD) and needs medication.
4. The child may be having absence seizures and needs to see his primary health care provider for evaluation.

8. 1. Children who reverse letters and words while reading have dyslexia. Two of the most common characteristics of children with ADHD include inattention and impulsiveness. Although aggressiveness may be common in children with ADHD, it isn't a characteristic that will aid in the diagnosis of this disorder. Six-year-old children aren't usually cognitively ready to read a chapter book.
NP: Data collection; CN: Health promotion and maintenance; CNS: Prevention and detection of disease; CL: Application

9. 4. Cerebral palsy is an irreversible, nonprogressive disorder that results from damage to the developing brain during the prenatal, perinatal, or postnatal period. Although some children with cerebral palsy are mentally retarded, many have normal intelligence.
NP: Implementation; CN: Health promotion and maintenance; CNS: Prevention and early detection of disease; CL: Knowledge

10. 1. Normal neonates have a strong, vigorous sucking reflex. A positive Babinski's reflex is present at birth and disappears by the time the infant is age 2. The rooting reflex is present at birth and disappears when the infant is between ages 3 and 4 months. The startle reflex is present at birth and disappears when the infant is approximately age 4 months.
NP: Data collection; CN: Health promotion and maintenance; CNS: Growth and development through the life span; CL: Knowledge

11. 4. Absence seizures are often misinterpreted as daydreaming. The child loses awareness but no alteration in motor activity is exhibited. ADHD isn't characterized by episodes of quietness. There isn't enough information to indicate a learning disability. A mild hearing problem usually is exhibited as leaning forward, talking louder, listening to louder TV and music than usual, and a repetitive "what?" from the child.
NP: Data collection; CN: Physiological integrity; CNS: Physiological adaptation; CL: Application

12. A 2-month-old infant is brought to the well-baby clinic for a first check-up. On initial assessment, the nurse notes the infant's head circumference is at the 95th percentile. Which of the following actions would the nurse take initially?

1. Assess vital signs.
2. Measure the head again.
3. Assess neurologic signs.
4. Notify the primary health care provider.

13. An 18-month-old infant has been diagnosed with bilateral otitis media; amoxicillin suspension has been prescribed. Which of the following symptoms is a common adverse effect the nurse should teach the family about?

1. Diarrhea
2. Nausea
3. Petechiae
4. Rash

14. A 2-year-old child is admitted to the pediatric unit with the diagnosis of bacterial meningitis. Which of the following diagnostic measures would be appropriate for the nurse to perform first?

1. Obtain a urine specimen.
2. Draw ordered laboratory tests.
3. Place the toddler in respiratory isolation.
4. Explain the treatment plan to the parents.

15. A 10-month-old boy with bacterial meningitis was just started on antibiotic therapy. Which nursing action is especially important in this situation?

1. Wearing a mask while providing care
2. Flexing the child's neck every 4 hours to maintain range of motion
3. Administering oral gentamicin (Garamycin)
4. Encouraging the child to drink 3,000 ml of fluid per day

First: prioritize. Second: prioritize. Think prioritize.

Key word here: diagnostic. Y'all keep that in mind when reviewing the options, y'hear?

12. 2. Whenever there's a question about vital signs or assessment data, the first logical step is to reassess to determine if an error has been made initially. Notifying the primary health care provider and assessing neurologic and vital signs are important and would follow the reassessment, if warranted.

NP: Implementation; CN: Health promotion and maintenance; CNS: Prevention and early detection of disease; CL: Comprehension

13. 1. Diarrhea is a common adverse effect of amoxicillin suspension. Red rash and petechiae occur less commonly. Nausea also occurs, but it's much harder to determine in an 18-month-old infant.

NP: Implementation; CN: Physiological integrity; CNS: Pharmacological therapies; CL: Comprehension

14. 3. Nurses should take necessary precautions to protect themselves and others from possible infection from the bacterial organism causing meningitis. The affected child should immediately be placed in respiratory isolation; then the parents can be informed about the treatment plan. This should be done before laboratory tests are performed.

NP: Implementation; CN: Safe, effective care environment; CNS: Safety and infection control; CL: Application

15. 1. With bacterial meningitis, respiratory isolation must be maintained for at least 24 hours after beginning antibiotic therapy. Wearing a mask is an important part of respiratory isolation. Moving the child's head would cause pain because his meninges are inflamed. Gentamicin is never administered orally. This amount of fluid would cause overhydration in a 10-month-old infant and place him at risk for increased intracranial pressure.

NP: Implementation; CN: Safe, effective care environment; CNS: Safety and infection control; CL: Application

16. A preschool-age child has just been admitted to the pediatric unit with a diagnosis of bacterial meningitis. The nurse would include which of the following recommendations in the nursing plan?
1. Take vital signs every 4 hours.
2. Monitor temperature every 4 hours.
3. Decrease environmental stimulation.
4. Encourage the parents to hold the child.

16. 3. A child with the diagnosis of meningitis is much more comfortable with decreased environmental stimuli. Noise and bright lights stimulate the child and can be irritating, causing the child to cry, in turn increasing intracranial pressure. Vital signs would be taken initially every hour and temperature monitored every two hours. Children with bacterial meningitis are usually much more comfortable if allowed to lie flat because this position doesn't cause increased meningeal irritation.
NP: Planning; CN: Physiological integrity; CNS: Physiological adaptation; CL: Application

17. A child has just returned to the pediatric unit following ventriculoperitoneal shunt placement for hydrocephalus. Which of the following interventions would the nurse perform first?
1. Monitor intake and output.
2. Place the child on the side opposite the shunt.
3. Offer fluids because the child has a dry mouth.
4. Administer pain medication by mouth as ordered.

This question requires prioritizing.

17. 2. Following shunt placement surgery, the child should be placed on the side opposite the surgical site to prevent pressure on the shunt valve. Intake and output will be monitored, but that isn't the priority nursing intervention. Pain medication should initially be administered by an I.V. route postoperatively. Many children are nauseated after a general anesthetic, and ice chips or clear liquids would be introduced after the nurse had determined if the child was nauseated.
NP: Implementation; CN: Physiological integrity; CNS: Basic care and comfort; CL: Application

18. An otherwise healthy 18-month-old child with a history of febrile seizures is in the well-child clinic. Which of the following statements by the father would indicate to the nurse that additional teaching should be done?
1. "I have ibuprofen available in case it's needed."
2. "My child will outgrow these seizures by age 5."
3. "I always keep phenobarbital with me in case of a fever."
4. "The most likely time for a seizure is when the fever is rising."

Note that, for question 18, you're looking for an incorrect response from the father.

18. 3. Antiepileptics, such as phenobarbital, are administered to children with prolonged seizures or neurologic abnormalities. Ibuprofen, not phenobarbital, is given for fever. Febrile seizures usually occur after age 6 months and are unusual after age 5. Treatment is to decrease the temperature because seizures occur as the temperature rises.
NP: Evaluation; CN: Health promotion and maintenance; CNS: Prevention and early detection of disease; CL: Application

19. When checking a 5-month-old infant, which of the following symptoms would alert the nurse that the infant needs further follow-up?
1. Absent grasp reflex
2. Rolls from back to side
3. Balances head when sitting
4. Presence of Moro's embrace reflex

19. 4. Moro's embrace reflex should be absent at 4 months. Grasp reflex begins to fade at 2 months and should be absent at 3 months. A 4-month-old infant should be able to roll from back to side and balance his head when sitting.
NP: Data collection; CN: Health promotion and maintenance; CNS: Prevention and early detection of disease; CL: Application

20. An adolescent is started on valproic acid to treat seizures. Which of the following statements should be included when educating the adolescent?
1. This medication has no adverse effects.
2. A common adverse effect is weight gain.
3. Drowsiness and irritability commonly occur.
4. Early morning dosing is recommended to decrease insomnia.

20. 2. Weight gain is a common adverse effect of valproic acid. Drowsiness and irritability are adverse effects more commonly associated with phenobarbital. Felbamate (Felbatol) more commonly causes insomnia.

NP: Planning; CN: Physiological integrity; CNS: Pharmacological therapies; CL: Application

21. Which of the following statements about cerebral palsy would be accurate?
1. Cerebral palsy is a condition that runs in families.
2. Cerebral palsy means there will be many disabilities.
3. Cerebral palsy is a condition that doesn't get worse.
4. Cerebral palsy occurs because of too much oxygen to the brain.

21. 3. By definition, cerebral palsy is a nonprogressive neuromuscular disorder. It can be mild or quite severe and is believed to be the result of a hypoxic event during pregnancy or the birth process.

NP: Evaluation; CN: Physiological integrity; CNS: Basic care and comfort; CL: Application

For question 22, keep in mind the difference between the nurse's and physician's notes.

22. An older child has a craniotomy for removal of a brain tumor. Which of the following statements would be appropriate for the nurse to say to the parents?
1. "Your child really had a close call."
2. "I'm sure your child will be back to normal soon."
3. "I'm so glad to hear your child doesn't have cancer."
4. "It may take some time for your child to return to normal."

22. 4. After a craniotomy, it usually takes several weeks or longer before the child is back to normal. When comforting parents, it's best to first ascertain what the primary health care provider has told them about the tumor. Final pathology results won't be available for several days, so refrain from making premature statements about whether the tumor is malignant or not.

NP: Implementation; CN: Psychosocial integrity; CNS: Coping and adaptation; CL: Application

23. A 6-month-old infant is admitted with a diagnosis of bacterial meningitis. The nurse would place the infant in which of the following rooms?
1. A room with a 12-month-old infant with a urinary tract infection
2. A room with an 8-month-old infant with failure to thrive
3. An isolation room near the nurses' station
4. A two-bed room in the middle of the hall

23. 3. A child who has the diagnosis of bacterial meningitis will need to be placed in isolation near the nurses' station until that child has received I.V. antibiotics for 24 hours because the child is considered contagious. Additionally, bacterial meningitis can be quite serious; therefore, the child should be placed near the nurses' station for close monitoring and easier access in case of a crisis.

NP: Implementation; CN: Safe, effective care environment; CNS: Coordinated care; CL: Application

24. In caring for a child immediately after a head injury, the nurse notes a blood pressure of 110/60, a heart rate of 78, dilated and nonreactive pupils, minimal response to pain, and slow response to name. Which of the following symptoms would cause the nurse the most concern?
1. Vital signs
2. Nonreactive pupils
3. Slow response to name
4. Minimal response to pain

25. Which nursing action should be included in the care plan to promote comfort in a 4-year-old child hospitalized with meningitis?
1. Avoid making noise when in the child's room.
2. Rock the child frequently.
3. Have the child's 2-year-old brother stay in the room.
4. Keep the lights on brightly so that he can see his mother.

26. Which developmental milestone would the nurse expect an 11-month-old infant to have achieved?
1. Sitting independently
2. Walking independently
3. Building a tower of four cubes
4. Turning a doorknob

27. What's the nurse's priority when caring for a 10-month-old infant with meningitis?
1. Maintaining an adequate airway
2. Maintaining fluid and electrolyte balance
3. Controlling seizures
4. Controlling hyperthermia

Remember to watch those signs!

I can see you're up to snuff on your casework. Great job!

24. 2. Dilated and nonreactive pupils indicate that anoxia or ischemia of the brain has occurred. If the pupils are also fixed (don't move), then herniation of the brain through the tentorium has occurred. The vital signs are normal. Slow response to name can be normal after a head injury. Minimal response to pain is an indication of the child's level of consciousness.
NP: Planning; CN: Physiological integrity; CNS: Basic care and comfort; CL: Application

25. 1. Meningeal irritation may cause seizures and heightens a child's sensitivity to all stimuli, including noise, lights, movement, and touch. Frequent rocking, presence of a younger sibling, and bright lights would increase stimulation.
NP: Planning; CN: Physiological integrity; CNS: Basic care and comfort; CL: Application

26. 1. Infants typically sit independently, without support, by age 8 months. Walking independently may be accomplished as late as age 15 months and still be within the normal range. Few infants walk independently by age 11 months. Building a tower of three or four blocks is a milestone of an 18-month-old. Turning a doorknob is a milestone of a 24-month-old.
NP: Evaluation; CN: Health promotion and maintenance; CNS: Growth and development through the life span; CL: Application

27. 1. Maintaining an adequate airway is always a top priority. Maintaining fluid and electrolyte balance and controlling seizures and hyperthermia are all important but not as important as an adequate airway.
NP: Implementation; CN: Physiological integrity; CNS: Reduction of risk potential; CL: Application

28. Which of the following interventions prevents a 17-month-old child with spastic cerebral palsy from going into a scissoring position?
1. Keep the child in leg braces 23 hours per day.
2. Let the child lay down as much as possible.
3. Try to keep the child as quiet as possible.
4. Place the child on your hip.

28. 4. To interrupt the scissoring position, flex the knees and hips. Placing the child on your hip is an easy way to stop this common spastic positioning. Trying to keep the child quiet and flat are inappropriate measures. This child needs stimulation and movement to reach the goal of development to the fullest potential. Wearing leg braces 23 hours per day is inappropriate and doesn't allow the child to move freely.

NP: Implementation; CN: Physiological integrity; CNS: Basic care and comfort; CL: Application

29. The mother of a child with a ventriculoperitoneal shunt says her child has a temperature of 101.2° F (38.4° C), a blood pressure of 108/68 mm Hg, and a pulse of 100. The child is lethargic and vomited the night before. Other children in the family have had similar symptoms. Which of the following nursing interventions is most appropriate?
1. Provide symptomatic treatment.
2. Advise the mother that this is a viral infection.
3. Consult the primary health care provider.
4. Have the mother bring the child to the primary health care provider's office.

29. 4. One of the complications of a ventriculoperitoneal shunt is a shunt infection. Shunt infections can have similar symptoms as a viral infection so it's best to have the child examined. These symptoms may be due to the same viral infection that the siblings have, but it's better to have the child examined to rule out a shunt infection because infection can progress quickly to a very serious illness.

NP: Implementation; CN: Physiological integrity; CNS: Basic care and comfort; CL: Application

Make sure both parents understand the rationale for treatment of a sick child.

30. The mother of a 10-year-old child with attention deficit hyperactivity disorder says her husband won't allow their child to take more than 5 mg of methylphenidate every morning. The child isn't doing better in school. Which of the following recommendations would the nurse make to the mother?
1. Sneak the medication to the child anyway.
2. Put the child in charge of administering the medication.
3. Bring the child's father to the clinic to discuss the medication.
4. Have the school nurse give the child the rest of the medication.

30. 3. Bringing the father to the clinic for a teaching session about the medication should assist him in understanding why it's necessary for the child to receive the full dose. A nurse shouldn't advise dishonesty to a client or family. The father should be included in the treatment as much as possible.

NP: Planning; CN: Physiological integrity; CNS: Pharmacological therapies; CL: Application

31. A hospitalized child is to receive 75 mg of acetaminophen for fever control. How much will the nurse administer if the acetaminophen is 40 mg per 0.4 ml?
1. 0.37 ml
2. 0.75 ml
3. 1.12 ml
4. 1.5 ml

31. 2. The nurse will administer 0.75 ml because if 10 mg equals 0.1 ml, then 75 mg equals 0.75 ml.

NP: Planning; CN: Physiological integrity; CNS: Pharmacological therapies; CL: Application

NP: Nursing process   CN: Client needs category   CNS: Client needs subcategory   CL: Cognitive level

32. The nurse is observing a child who may have acute bacterial meningitis. Which of the following findings might the nurse look for?
1. Flat fontanel
2. Irritability, fever, and vomiting
3. Jaundice, drowsiness, and refusal to eat
4. Negative Kernig's sign

You're looking for the correct findings in question 32.

32. 2. Findings associated with acute bacterial meningitis may include irritability, fever, and vomiting along with seizure activity. Fontanels would be bulging as intracranial pressure rises, and Kernig's sign would be present due to meningeal irritation. Jaundice, drowsiness, and refusal to eat indicate a GI disturbance rather than meningitis.
NP: Data collection; CN: Physiological integrity; CNS: Physiological adaptation; CL: Knowledge

33. When caring for a school-aged child who has had a brain tumor removed, the nurse makes the following assessment: pupils equal and reactive to light; motor strength equal; knows name, date, but not location; and complains of a headache. Which of the following nursing interventions would be most appropriate?
1. Provide medication for the headache.
2. Immediately notify the primary health care provider.
3. Check what the child's level of consciousness has been.
4. Call the child's parents to come and sit at the child's bedside.

33. 3. When there's an abnormality in current assessment data, it's vital to determine what the client's previous status was. Determine whether the status has changed or remained the same. Providing medication for the headache would be done after ascertaining the previous level of consciousness. Contacting the primary health care provider and the child's parents isn't necessary before a final assessment has been made.
NP: Planning; CN: Physiological integrity; CNS: Physiological adaptation; CL: Application

34. Following a craniotomy on a child, I.V. fluids are ordered to run at 27 ml/hour. The tubing delivers 60 ml/hour. How many drops per minute should the nurse set the pump for?
1. 14
2. 27
3. 54
4. 60

34. 2. Tubing that delivers 60 ml/60 minutes would deliver 1 ml/minute. To deliver 27 ml/hour, the nurse would set the pump at 27.
NP: Planning; CN: Physiological integrity; CNS: Pharmacological therapies; CL: Application

Hint! You need to prioritize here.

35. A nurse is caring for a child with cerebral palsy. The grandmother, the primary caregiver, says she chews the food first and then puts it in the child's mouth because the child is difficult to feed. Which of the following statements would be most appropriate to make first?
1. "You need to make an appointment to talk to the hospital dietitian."
2. "The child needs to eat different foods than what you're chewing."
3. "Many of the food's nutrients are gone because of the prechewing."
4. "You should offer the child a normal diet instead of what you're doing."

35. 3. Many of the nutrients are lost to the infant by the prechewing method. The other responses are appropriate but not what the nurse should say first.
NP: Implementation; CN: Health promotion and maintenance; CNS: Prevention and early detection of disease; CL: Application

36. A 10-year-old child with a concussion is admitted to the pediatric unit. The nurse would place this child in a room with which of the following roommates?

1. A 6-year-old child with osteomyelitis
2. An 8-year-old child with gastroenteritis
3. A 10-year-old child with rheumatic fever
4. A 12-year-old child with a fractured femur

37. The nurse is caring for a child with spina bifida. Which of the following factors determines the extent of sensory and motor function loss in the lower limbs of the child?

1. Maternal age at conception
2. Degree of spinal cord abnormality
3. Uterine environmental factors such as cloudy amniotic fluid
4. Timeframe from diagnosis to birth of the infant

38. An 11-year-old boy with a head injury has been in the hospital for 16 days. He's receiving physical, occupational, and speech therapies. He can swallow and has adequate oral intake; however, his speech is slow and he sometimes makes inappropriate statements. He's usually cooperative but occasionally has combative and violent outbursts. His parents are upset and want to know what's wrong with him. What's the best response by the nurse?

1. "He probably didn't receive enough discipline growing up and is throwing tantrums."
2. "He needs to be restrained during these episodes."
3. "This is a stage of healing for him."
4. "He'll need to be on life-long medication to control his temper."

You're more than halfway finished! Looking good!

36. 4. A child with a concussion should be placed with a roommate who's free from infection and close to the child's age. Osteomyelitis, gastroenteritis, and rheumatic fever involve infection.

NP: Planning; CN: Safe, effective care environment; CNS: Coordinated care; CL: Application

37. 2. The extent of motor and sensory loss primarily depends on the degree of spinal cord abnormality. Secondarily, it depends on traction or stretch resulting from an abnormally tethered cord, trauma to exposed neural tissue during delivery, and postnatal damage resulting from drying or infection of the neural plate. Maternal age and uterine environment haven't been identified as factors. The time from diagnosis to birth isn't related.

NP: Data collection; CN: Physiological integrity; CNS: Physiological adaptation; CL: Knowledge

38. 3. Clients with head injuries may pass through eight stages during their recovery. Stage 1, marked by unresponsiveness, is the worst stage. Stage 8, characterized by purposeful, appropriate behavior, is the final stage of healing. This client is somewhere between stage 4 (confused, agitated behavior) and stage 6 (confused, appropriate behavior) because sometimes he can answer appropriately but at other times becomes confused and angry, resorting to violent behavior. The other statements are inappropriate.

NP: Implementation; CN: Health promotion and maintenance; CNS: Prevention and early detection of disease; CL: Application

---

NP: Nursing process   CN: Client needs category   CNS: Client needs subcategory   CL: Cognitive level

39. Which of the following mechanisms causes the severe headache that accompanies an increase in intracranial pressure (ICP)?
1. Cervical hyperextension
2. Stretching of the meninges
3. Cerebral ischemia related to altered circulation
4. Reflex spasm of the neck extensors to splint the neck against cervical flexion

39. 2. The mechanism producing the headache that accompanies increased ICP may be the stretching of the meninges and pain fibers associated with blood vessels. Cerebral ischemia occurs because of vascular obstruction and decreased perfusion of the brain tissue. With nuchal rigidity, cervical flexion is painful due to the stretching of the inflamed meninges and the pain triggers a reflex spasm of the neck extensors to splint the area against further cervical flexion. It occurs in response to the pain; it doesn't cause it.
NP: Data collection; CN: Physiological integrity; CNS: Physiological adaptation; CL: Application

40. The nurse at a family health clinic is teaching a group of parents about normal infant development. What patterns of communication should the nurse tell parents to expect from an infant at age 1?
1. Squeals and makes pleasure sound
2. Understands "no" and other simple commands
3. Uses speechlike rhythm when talking with an adult
4. Uses multisyllabic babbling

40. 2. At age 1, most babies understand the word "no" and other simple commands. Children at this age also learn one or two other words. Babies squeal, make pleasure sounds, and use multisyllabic babbling at age 3 to 6 months. Using speechlike rhythm when talking with an adult usually occurs between ages 6 to 9 months.
NP: Data collection; CN: Health promotion and maintenance; CNS: Growth and development through the life span; CL: Knowledge

So many signs, so little time!

41. Which of the following assessment data indicates nuchal rigidity?
1. Positive Kernig's sign
2. Negative Brudzinski's sign
3. Positive Homans' sign
4. Negative Kernig's sign

41. 1. A positive Kernig's sign indicates nuchal rigidity, caused by an irritative lesion of the subarachnoid space. Brudzinski's sign is also indicative of the condition. Homans' sign indicates venous inflammation of the lower leg, not nuchal rigidity.
NP: Data collection; CN: Physiological integrity; CNS: Physiological adaptation; CL: Knowledge

42. The nurse is collecting data from a child who may have a seizure disorder. Which of the following is a description of an absence seizure?
1. Sudden, momentary loss of muscle tone, with a brief loss of consciousness
2. Muscle tone maintained and child frozen in position
3. Brief, sudden contracture of a muscle or muscle group
4. Minimal or no alteration in muscle tone, with a brief loss of consciousness

42. 4. Absence seizures are characterized by a brief loss of responsiveness with minimal or no alteration in muscle tone. They may go unrecognized because the child's behavior changes very little. A sudden loss of muscle tone describes atonic seizures. "Frozen positions" describe akinetic seizures. A brief, sudden contraction of muscles describes a myoclonic seizure.
NP: Data collection; CN: Physiological integrity; CNS: Physiological adaptation; CL: Knowledge

43. A child with a diagnosis of meningococcal meningitis develops signs of sepsis and a purpuric rash over both lower extremities. The primary health care provider should be notified immediately because these signs could be indicative of which of the following complications?
1. A severe allergic reaction to the antibiotic regimen with impending anaphylaxis
2. Onset of the syndrome of inappropriate antidiuretic hormone (SIADH)
3. Fulminant meningococcemia (Waterhouse-Friderichsen syndrome)
4. Adhesive arachnoiditis

Check these signs carefully. They're critical for answering question 43.

43. 3. Meningococcemia is a serious complication usually associated with meningococcal infection. When onset is severe, sudden, and rapid (fulminant), it's known as Waterhouse-Friderichsen syndrome. Anaphylactic shock would need to be differentiated from septic shock. SIADH can be an acute complication but wouldn't be accompanied by the purpuric rash. Adhesive arachnoiditis occurs in the chronic phase of the disease and leads to obstruction of the flow of cerebrospinal fluid.
NP: Implementation; CN: Physiological integrity; CNS: Reduction of risk potential; CL: Application

44. Which of the following pathologic processes is often associated with aseptic meningitis?
1. Ischemic infarction of cerebral tissue
2. Childhood virus, such as mumps
3. Brain abscess caused by a variety of pyogenic organisms
4. Cerebral ventricular irritation from a traumatic brain injury

I see you've finished 44 questions. Super! Are we having fun yet? I know I am.

44. 2. Aseptic meningitis is caused principally by viruses and is often associated with other diseases, such as measles, mumps, herpes, and leukemia. Incidence of brain abscess is high in bacterial meningitis, and ischemic infarction of cerebral tissue can occur with tubercular meningitis. Traumatic brain injury could lead to bacterial (not viral) meningitis.
NP: Data collection; CN: Physiological integrity; CNS: Physiological adaptation; CL: Knowledge

45. To alleviate the child's pain and fear of lumbar puncture, which of the following interventions should the nurse perform?
1. Sedate the child with fentanyl (Sublimaze).
2. Apply a topical anesthetic to the skin 5 to 10 minutes before the puncture.
3. Have a parent hold the child in their lap during the procedure.
4. Have the child inhale small amounts of nitrous oxide gas before the puncture.

45. 1. Sedation with fentanyl or other drugs can alleviate the pain and fear associated with a lumbar puncture. A topical anesthetic can be applied, but it should be done 1 hour before the procedure to be fully effective. Having a parent holding a child in their lap increases the risk of neurologic injury due to the inability to assume and maintain the proper anatomic position required for a safe lumbar puncture. Use of nitrous oxide gas isn't recommended.
NP: Implementation; CN: Physiological integrity; CNS: Basic care and comfort; CL: Application

46. Antibiotic therapy to treat meningitis should be instituted immediately after which of the following events?
1. Admission to the nursing unit
2. Initiation of I.V. therapy
3. Identification of the causative organism
4. Collection of cerebrospinal fluid (CSF) and blood for culture

Hear ye, hear ye, hear ye! Prioritize, my fellow countrypeoples, prioritize!

46. 4. Antibiotic therapy should always begin immediately after the collection of CSF and blood cultures. After the specific organism is identified, bacteria-specific antibiotics can be administered if the initial choice of antibiotic therapy isn't appropriate. Admission and initiation of I.V. therapy aren't, by themselves, appropriate times to begin antibiotic therapy.
NP: Planning; CN: Physiological integrity; CNS: Pharmacological therapies; CL: Application

47. An infant who appears to be younger than age 1 is admitted to the emergency department. The nurse observes that the baby is lethargic and drowsy and is having respiratory difficulty. The parent describes seizure activity, which occurred at home just before their arrival. There are no external signs of injury. Which condition should the nurse suspect?
1. Epilepsy
2. Shaken baby syndrome
3. Failure to thrive
4. Infantile spasms

47. 2. Signs and symptoms of shaken baby syndrome include apnea, seizures, lethargy, drowsiness, bradycardia, respiratory difficulty, and coma. Subdural and retinal hemorrhages in the absence of external signs of injury are strong indicators of the syndrome. Epilepsy is a recurrent and chronic seizure disorder with onset usually between ages 4 and 8. Infants with failure to thrive show delayed development without physical cause. Infantile spasms are sudden brief, symmetric muscle contractions accompanied by rolling of the eyes.
NP: Data collection; CN: Physiological integrity; CNS: Physiological adaptation; CL: Comprehension

48. A 1-month-old male infant is admitted to the pediatric unit and diagnosed with bacterial meningitis. Which signs or symptoms does the nurse observe in this client?
1. Hemorrhagic rash, first appearing as petechiae
2. Photophobia
3. Fever, change in feeding pattern, vomiting, or diarrhea
4. Fever, lethargy, and purpura or large necrotic patches

48. 3. Fever, change in feeding patterns, vomiting, and diarrhea are often observed in children with bacterial meningitis. Hemorrhagic rashes, petechiae, photophobia, fever, lethargy, and purpura are observed in older children with meningitis.
NP: Data collection; CN: Physiological integrity; CNS: Physiological adaptation; CL: Comprehension

I can see the finish line from here. Keep going!

49. Which of the following goals of nursing care is the most difficult to accomplish in caring for a child with meningitis?
1. Protecting self and others from possible infection
2. Avoiding actions that increase discomfort, such as lifting the head
3. Keeping environmental stimuli to a minimum, such as reduced light and noise
4. Maintaining I.V. infusion to administer adequate antimicrobial therapy

49. 4. One of the most difficult problems in the nursing care of children with meningitis is maintaining the I.V. infusion for the length of time needed to provide adequate therapy. All of the other options are important aspects in the provision of care for the child with meningitis, but they're secondary to antimicrobial therapy.
NP: Implementation; CN: Physiological integrity; CNS: Basic care and comfort; CL: Application

All of the choices in question 50 may be correct. So you need to prioritize.

50. Which of the following nursing assessment data should be given the highest priority for a child with clinical findings related to tubercular meningitis?
1. Onset and character of fever
2. Degree and extent of nuchal rigidity
3. Signs of increased intracranial pressure (ICP)
4. Occurrence of urine and fecal contamination

50. 3. Assessment of fever and evaluation of nuchal rigidity are important aspects of care, but assessment for signs of increasing ICP should be the highest priority due to the life-threatening implications. Urinary and fecal incontinence can occur in a child who's ill from nearly any cause but don't pose a great danger to life.
NP: Implementation; CN: Physiological integrity; CNS: Reduction of risk potential; CL: Application

51. The clinical manifestations of acute bacterial meningitis are dependent on which of the following factors?

1. Age of the child
2. Length of the prodromal period
3. Time span from bacterial invasion to onset of symptoms
4. Degree of elevation of cerebrospinal fluid (CSF) glucose compared to serum glucose level

52. Which of the following responses is most appropriate for a mother who asks why her son, who has a history of closed head injury, would begin having seizures without warning?

1. Clonic seizure activity is usually interpreted as falling.
2. A delay between head injury and the development of seizures isn't unusual.
3. Focal discharge in the brain may lead to absence seizures that go unnoticed.
4. The epileptogenic focus in the brain needs multiple stimuli before it will discharge to cause a seizure.

53. The nurse is assessing a 1-month-old male infant during a routine examination at a family health center. Which method does the nurse use to test for Babinski's sign?

1. Raise the child's leg with the knee flexed and then extend the child's leg at the knee to determine if resistance is noted.
2. With the knee flexed, dorsiflex the foot to determine if there's pain in the calf of the leg.
3. Flex the child's head while he's in a supine position to determine if the knees or hips flex involuntarily.
4. Stroke the bottom of the foot to determine if there's fanning and dorsiflexion of the big toe.

You've got to know not only what drug interactions occur but also why.

51. 1. Clinical manifestations of acute bacterial meningitis depend largely on the age of the child. Clinical manifestations aren't dependent on the prodromal or initial period of the disease nor the time from invasion of the host to the onset of symptoms. The glucose level of the CSF is reduced, not elevated.
NP: Data collection; CN: Physiological integrity; CNS: Physiological adaptation; CL: Application

52. 2. Stimuli from an earlier injury may eventually elicit seizure activity, a process known as kindling. Focal seizures are partial seizures; absence seizures are generalized seizures. Focal seizures don't lead to absence seizures. Atonic seizures, not clonic, are frequently accompanied by falling. The epileptogenic focus consists of a group of hyperexcitable neurons responsible for initiating synchronous, high-frequency discharges that lead to a seizure and don't need multiple stimuli.
NP: Data collection; CN: Physiological integrity; CNS: Physiological adaptation; CL: Comprehension

53. 4. To test for Babinski's sign, stroke the bottom of the foot to determine if there's fanning and dorsiflexion of the big toe. Raising the child's leg with the knee flexed and then extending the leg at the knee to determine if resistance is noted tests for Kernig's sign. Dorsiflexion of the foot with the knee flexed to determine if there's pain in the calf of the leg tests for Homans' sign. Flexing the child's head while he's in a supine position to determine if the knees or hips flex involuntarily tests for Brudzinski's sign.
NP: Implementation; CN: Physiological integrity; CNS: Physiological adaptation; CL: Knowledge

54. During the trial period to determine the efficacy of an anticonvulsant drug, which of the following cautions should be explained to the parents?

1. Plasma levels of the drug will be monitored on a daily basis.
2. Drug dosage will be adjusted depending on the frequency of seizure activity.
3. The drug must be discontinued immediately if even the slightest problem occurs.
4. The child shouldn't participate in activities that could be hazardous if a seizure occurs.

Let the trial period begin!

55. What behavioral responses to pain would a nurse observe from an infant under age 1?

1. Localized withdrawal and resistance of the entire body
2. Passive resistance, clenching fists, and holding body rigid
3. Reflex withdrawal to stimulus and facial grimacing
4. Low frustration level and striking out physically

56. Maintaining plasma phenytoin levels within the therapeutic range is difficult for which of the following reasons?

1. A drop in the plasma drug level will lead to a toxic state.
2. The capacity to metabolize the drug becomes overwhelmed over time.
3. Small increments in dosage lead to sharp increases in plasma drug levels.
4. Large increments in dosage lead to a more rapid stabilizing therapeutic effect.

57. Client teaching should stress which of the following rules in relation to the differences in bioavailability of different forms of phenytoin?

1. Use the cheapest formulation the pharmacy has on hand at the time of refill.
2. Shop around to get the least expensive formulation.
3. There's no difference between one formulation and another, regardless of price.
4. Avoid switching formulations without the primary health care provider's approval.

54. 4. Until seizure control is certain, clients shouldn't participate in activities (such as riding a bicycle) that could be hazardous if a seizure were to occur. Plasma levels need to be monitored periodically over the course of drug therapy; daily monitoring isn't necessary. Dosage changes are usually based on plasma drug levels as well as seizure control. Anticonvulsant drugs should be withdrawn over a period of 6 weeks to several months, never immediately, as doing so could precipitate status epilepticus.

NP: Implementation; CN: Physiological integrity; CNS: Pharmacological therapies; CL: Application

55. 3. Infants under age 1 become irritable and exhibit reflex withdrawal to the painful stimulus. Facial grimacing also occurs. Localized withdrawal is experienced by toddlers ages 1 to 3 in response to pain. The nurse would observe passive resistance in school-age children. Preschoolers show a low frustration level and strike out physically.

NP: Data collection; CN: Physiological integrity; CNS: Physiological adaptation; CL: Comprehension

56. 3. Within the therapeutic range for phenytoin, small increments in dosage produce sharp increases in plasma drug levels. The capacity of the liver to metabolize phenytoin is affected by slight changes in the dosage of the drug, not necessarily the length of time the client has been taking the drug. Large increments in dosage will greatly increase plasma levels leading to drug toxicity.

NP: Implementation; CN: Physiological integrity; CNS: Pharmacological therapies; CL: Comprehension

57. 4. Differences in bioavailability exist among different formulations (tablets and capsules) and among the same formulations produced by different manufacturers. Clients shouldn't switch from one formulation to another or from one brand to another without primary health care provider approval and supervision.

NP: Implementation; CN: Physiological integrity; CNS: Pharmacological therapies; CL: Application

58. In contrast to other barbiturates, the anticonvulsant barbiturate phenobarbital ultimately has the action of seizure suppression without which of the following adverse effects?
1. Causing sedation
2. Development of ataxia and nystagmus
3. Development of significant dependence
4. Risk of developing acute intermittent porphyria

Be careful! Question 58 is negative.

58. 1. Phenobarbital is able to suppress seizure activity at doses that produce minimal disruption to central nervous system function. During the initial phase, some drowsiness may occur, but with continued use, tolerance to sedation occurs. Ataxia and nystagmus can be signs of overdose or toxicity. As with all other barbiturates, physical dependence on phenobarbital can develop but is dosage-related. Exacerbation of intermittent porphyria is an adverse effect.
NP: Planning; CN: Physiological integrity; CNS: Pharmacological therapies; CL: Comprehension

59. Which of the following instructions should be included in client teaching specifically related to anticonvulsant drug efficacy?
1. Wear a medical identification bracelet.
2. Maintain a seizure frequency chart.
3. Avoid potentially hazardous activities.
4. Discontinue the drug immediately if adverse effects are suspected.

59. 2. Ongoing evaluation of therapeutic effects can be accomplished by maintaining a frequency chart that indicates the date, time, and nature of all seizure activity. These data may be helpful in making dosage alterations and specific drug selection. Wearing a medical identification bracelet and avoiding hazardous activities are ways to minimize danger related to seizure activity, but these factors don't affect drug efficacy. Anticonvulsant drugs should never be discontinued abruptly due to the potential for development of status epilepticus.
NP: Planning; CN: Physiological integrity; CNS: Pharmacological therapies; CL: Comprehension

60. A 13-year-old with structural scoliosis has Harrington rods inserted. Which of the following positions would be best during the postoperative period?
1. Supine in bed
2. Side-lying
3. Semi-Fowler's
4. High Fowler's

60. 1. After placement of Harrington rods, the client must remain flat in bed. The gatch on a manual bed should be taped and electric beds should be unplugged to prevent the client from raising the head or foot of the bed. Other positions, such as the side-lying, semi-Fowler, and high Fowler positions, could prove damaging because the rods couldn't maintain the spine in a straight position.
NP: Implementation; CN: Physiological integrity; CNS: Reduction of risk potential; CL: Application

Hint, hint, hint!

61. At which of the following times is seizure activity most likely to occur?
1. During the rapid eye movement (REM) stage of sleep
2. During long periods of excitement
3. While falling asleep and on awakening
4. While eating, particularly if the client is hurried

61. 3. Falling asleep or awakening from sleep are periods of functional instability of the brain; seizure activity is more likely to occur during these times. Eating quickly, excitement without undo fatigue, and REM sleep haven't been identified as contributing factors.
NP: Planning; CN: Physiological integrity; CNS: Physiological adaptation; CL: Knowledge

NP: Nursing process   CN: Client needs category   CNS: Client needs subcategory   CL: Cognitive level

62. When educating the family of a child with seizures, it's appropriate to tell them to call emergency medical services in the event of a seizure if which of the following complications occurs?

1. Continuous vomiting for 30 minutes following the seizure
2. Stereotypic or automatous body movements during onset
3. Lack of expression, pallor, or flushing of the face during the seizure
4. Unilateral or bilateral posturing of one or more extremities during onset

63. Identifying factors that trigger seizure activity could lead to which of the following alterations in the child's environment or activities of daily living?

1. Avoiding striped wallpaper and ceiling fans
2. Having the child sleep alone to prevent sleep interruption
3. Including extended periods of intense physical activity daily
4. Allowing the child to drink soda only between noon and 5 p.m.

64. Which of the following nursing interventions would be included to support the goal of avoiding injury, respiratory distress, or aspiration during a seizure?

1. Positioning the child with the head hyperextended
2. Placing a hand under the child's head for support
3. Using pillows to prop the child into the sitting position
4. Working a padded tongue blade or small plastic airway between the teeth

65. Which of the following factors can aid in neural tube defect detection?

1. Flat plate of the lower abdomen after the 23rd week of gestation
2. Significant level of alpha-fetoprotein present in amniotic fluid
3. Amniocentesis for lecithin-sphingomyelin (L/S) ratio
4. Presence of high maternal levels of albumin after 12th week of gestation

It's important to know which activities and environmental factors can trigger seizures.

Hang on! You're almost finished!

62. 1. Continuous vomiting after a seizure has ended can be a sign of an acute problem and indicates that the child requires an immediate medical evaluation. All of the other manifestations are normally present in various types of seizure activity and don't indicate a need for immediate medical evaluation.

NP: Planning; CN: Physiological integrity; CNS: Reduction of risk potential; CL: Application

63. 1. Striped wallpaper, ceiling fans, and blinking lights on a Christmas tree can all be triggers to seizure activity if the child is photosensitive. Sleep interruption hasn't been identified as a triggering factor. Avoidance of fatigue can reduce seizure activity; therefore, intense physical activity for extended periods of time should be avoided. Restricting caffeine intake by using caffeine-free soda is a dietary modification that may prevent seizures.

NP: Implementation; CN: Physiological integrity; CNS: Physiological adaptation; CL: Application

64. 2. Placing a hand or a small cushion or blanket under the child's head will help prevent injury. Position the child with the head in midline, not hyperextended, to promote a good airway and adequate ventilation. Don't attempt to prop the child up into a sitting position, but ease him to the floor to prevent falling and unnecessary injury. Don't put anything in the child's mouth because it could cause infection or obstruct the airway.

NP: Planning; CN: Physiological integrity; CNS: Reduction of risk potential; CL: Application

65. 2. Screening for significant levels of alpha-fetoprotein is 90% effective in detecting neural tube defects. Prenatal screening includes a combination of maternal serum and amniotic fluid levels, amniocentesis, amniography, and ultrasonography and has been relatively successful in diagnosing the defect. Flat plate X-rays of the abdomen, L/S ratio, and maternal serum albumin levels aren't diagnostic for the defect.

NP: Data collection; CN: Health promotion and maintenance; CNS: Prevention and early detection of disease; CL: Knowledge

66. Spina bifida, a congenital spinal cord injury, may be characterized by which of the following descriptions?
1. It has little influence on the intellectual and perceptual abilities of the child.
2. It's a simple neurologic defect that's completely corrected surgically within 1 to 2 days after birth.
3. Its presence indicates that many areas of the central nervous system (CNS) may not develop or function adequately.
4. It's a complex neurologic disability that involves a collaborative health team effort for the entire 1st year of life.

66. 3. When a spinal cord lesion exists at birth, it often leads to altered development or function of other areas of the CNS. Spina bifida is a complex neurologic defect that heavily impacts the physical, cognitive, and psychosocial development of the child and involves collaborative, life-long management due to the chronicity and multiplicity of the problems involved.

NP: Data collection; CN: Physiological integrity; CNS: Physiological adaptation; CL: Knowledge

67. Common deformities occurring in the child with spina bifida are related to the muscles of the lower extremities that are active or inactive. These may include which of the following complications?
1. Club feet
2. Hip extension
3. Ankylosis of the knee
4. Abduction and external rotation of the hip

67. 1. The type and extent of deformity in the lower extremities depends on the muscles that are active or inactive. Passive positioning in utero may result in deformities of the feet such as equinovarus (club foot), knee flexion and extension contractures, and hip flexion with adduction and internal rotation leading to subluxation or dislocation of the hip.

NP: Data collection; CN: Physiological integrity; CNS: Physiological adaptation; CL: Application

68. A mother asks the nurse what role she had in causing her child's spina bifida. Which of the following explanations is most appropriate?
1. Exact causative factors of this defect remain in question.
2. This defect is a hereditary problem transmitted through the autosomal recessive process.
3. Incidence of spina bifida is associated with a consistently high intake of alcohol during pregnancy.
4. Development of this defect is related to some type of maternal traumatic injury during pregnancy.

68. 1. The exact cause of spina bifida remains unknown although environmental factors such as hyperthermia in the first weeks of pregnancy or dietary factors, such as canned meats, potatoes, and peas have been implicated but not substantiated. Deficiencies of such nutrients as folic acid and vitamin A have also been implicated. There's no evidence to support a connection with intake of alcohol or maternal trauma. A genetic predisposition, along with certain environmental factors, may trigger the development; currently, there's no definitive evidence to support this theory.

NP: Implementation; CN: Health promotion and maintenance; CNS: Prevention and early detection of disease; CL: Application

---

NP: Nursing process   CN: Client needs category   CNS: Client needs subcategory   CL: Cognitive level

69. Infants with myelomeningocele must be closely observed for manifestations of Chiari II malformation, including which of the following symptoms?

1. Rapidly progressing scoliosis
2. Changes in urologic functioning
3. Back pain below the site of the sac closure
4. Respiratory stridor, apneic periods, and difficulty swallowing

70. A child with myelomeningocele and hydrocephalus may demonstrate problems related to damage of the white matter caused by ventricular enlargement. This damage may manifest itself in which of the following conditions?

1. Inability to speak
2. Early hand dominance
3. Impaired intellectual functions
4. Flaccid paralysis of the lower extremities

71. Myelomeningocele requires nearly immediate surgical repair for which of the following purposes?

1. Rapid restoration of the neural pathways to the legs
2. Decreased possibility of infection and further cord damage
3. Exposure of the spinal cord defect to individualize the therapeutic strategy
4. Removal of excess nerve tissue from the vertebral canal to decrease pressure on the cord

For question 70, it helps to know that my white matter is also known as the "association area."

Oh, my. Only four questions left. One more turn of the page and you're home!

69. 4. Children with a myelomeningocele have a 90% chance of having a Chiari II malformation. This may lead to a possibility of respiratory function problems, such as respiratory stridor associated with paralysis of the vocal cords, apneic episodes of unknown cause, difficulty swallowing, and an abnormal gag reflex. Scoliosis and urologic function changes occur with myelomeningocele, but these complications aren't specifically related to Chiari II malformation. Lower back pain doesn't occur due to the loss of sensory function related to the cord defect.

NP: Data collection; CN: Physiological integrity; CNS: Reduction of risk potential; CL: Knowledge

70. 3. Damage to the white matter (association area) caused by ventricular enlargement has been linked to impairment of intellectual and perceptual abilities often seen in children with spina bifida. It hasn't been related to hand dominance development, flaccid paralysis of the lower extremities, or the ability to speak, though it may affect the semantics of speech dependent upon the association areas.

NP: Data collection; CN: Physiological integrity; CNS: Physiological adaptation; CL: Application

71. 2. The myelomeningocele sac presents a dynamic disability and is treated as a life-threatening situation with surgical sac closure taking place within the first 24 to 48 hours after birth. This early management decreases the possibility of infection and further injury to the exposed neural cord. There's complete loss of nervous function below the level of the spinal cord lesion. The aim of surgery is to replace the nerve tissue into the vertebral canal, cover the spinal defect, and achieve a watertight sac closure.

NP: Implementation; CN: Physiological integrity; CNS: Reduction of risk potential; CL: Comprehension

72. One of the most important aspects of pre-operative care with myelomeningocele is the positioning of the infant. Which of the following positions is the most appropriate?
1. Prone position with head turned to the side for feeding
2. Side-lying position with the head at a 30-degree angle to the feet
3. Prone position with a nasogastric (NG) tube inserted for feedings
4. Supported by diaper rolls, both anterior and posterior, in the side-lying position

Key words here: *most appropriate*.

72. 1. Prone position is used preoperatively because it minimizes tension on the sac and the risk of trauma. The head is turned to one side for feeding. Side-lying or partial side-lying positions are better used after the repair has been made unless it permits undesirable hip flexion. Although feeding can be a problem in the prone position, it can be accomplished without the need for an NG tube. There's no advantage to positioning the body with a 30-degree head elevation.

NP: Implementation; CN: Physiological integrity; CNS: Basic care and comfort; CL: Application

73. A parent brings a toddler, age 19 months, to the clinic for a regular check-up. When palpating the toddler's fontanels, what should the nurse expect to find?
1. Closed anterior fontanel and open posterior fontanel
2. Open anterior fontanel and closed posterior fontanel
3. Closed anterior and posterior fontanels
4. Open anterior and posterior fontanels

73. 3. By age 18 months, the anterior and posterior fontanels should be closed. The diamond-shaped anterior fontanel normally closes between ages 9 and 18 months. The triangular posterior fontanel normally closes between ages 2 and 3 months.

NP: Data collection; CN: Health promotion and maintenance; CNS: Growth and development through the life span; CL: Knowledge

74. The nurse is preparing a toddler for a lumbar puncture. For this procedure, the nurse should place the child in which position?
1. Lying prone, with the neck flexed
2. Sitting up, with the back straight
3. Lying on one side, with the back curved
4. Lying prone, with the feet higher than the head

Congratulations! You finished! Great job!

74. 3. A lumbar puncture involves placing a needle between the lumbar vertebrae into the subarachnoid space. For this procedure the nurse should position the client on one side with the back curved because curving the back maximizes the space between the lumbar vertebrae, facilitating needle insertion. Prone and seated positions don't achieve maximum separation of the vertebrae.

NP: Implementation; CN: Physiological integrity; CNS: Reduction of risk potential; CL: Application

75. Which of the following explanations of how to decrease the chances of having a second child with spina bifida is most accurate?
1. There's no known way to avoid it; adoption is recommended.
2. A previous pregnancy affected by a neural tube defect isn't a factor.
3. Prepregnancy intake of 4 mg of folic acid daily reduces the recurrence rate.
4. Aerobic exercise in the first trimester to decrease the chances of a positive alpha-fetoprotein (AFP).

75. 3. Studies have shown that women at high risk for having an infant with a neural tube defect (demonstrated by a previously delivered infant or fetus with spina bifida) who took supplements of folic acid before conception significantly reduced the recurrence rate. The chances of having a second affected child are low (between 1% and 2%) but still greater than the chances of the general population. Aerobic exercise won't decrease the chances of a positive AFP.

NP: Implementation; CN: Health promotion and maintenance; CNS: Prevention and early detection of disease; CL: Application

NP: Nursing process   CN: Client needs category   CNS: Client needs subcategory   CL: Cognitive level

Challenge yourself with these questions on musculoskeletal system disorders in children. I'm betting you'll have a blast!

1. Which of the following definitions best describes the form of clubfoot called talipes varus?

1. Inversion of the foot
2. Eversion of the foot
3. Plantar flexion
4. Dorsiflexion

2. Which of the following types of clubfoot is the one most commonly treated by primary health care providers?

1. Talipes calcaneus
2. Talipes equinovarus
3. Talipes valgus
4. Talipes varus

3. The mother of a neonate with clubfoot feels guilty because she believes she did something to cause the condition. The nurse should explain that the cause of clubfoot in neonates is due to which of the following factors?

1. Unknown
2. Hereditary
3. Restricted movement in utero
4. Anomalous embryonic development

4. Which of the following statements about clubfoot is true?

1. It's hereditary.
2. More girls are affected.
3. More boys are affected.
4. It occurs once in every 500 births.

Knowing how to treat a disorder is only part of the NCLEX. Question 3 asks you for the cause of the disorder as well.

1. 1. Talipes varus is an inversion of the foot. Talipes valgus is an eversion of the foot. Talipes equinus is plantar flexion of the foot and talipes calcaneus is dorsiflexion of the foot.

NP: Data collection; CN: Physiological integrity; CNS: Physiological adaptation; CL: Knowledge

2. 2. Ninety-five percent of treated clubfoot cases are for feet that point downward and inward (talipes equinovarus) in varying degrees. Talipes valgus, talipes calcaneus, and talipes varus aren't as common.

NP: Data collection; CN: Physiological integrity; CNS: Physiological adaptation; CL: Knowledge

3. 1. The definitive cause of clubfoot is unknown. In some families, there's an increased incidence. Some postulate that anomalous embryonic development or restricted fetal movement are the reasons. Currently, there's no way to predict the occurrence of clubfoot.

NP: Data collection; CN: Psychosocial integrity; CNS: Coping and adaptation; CL: Knowledge

4. 3. Boys are affected twice as often as girls. It isn't known if the condition is hereditary. It occurs once in every 700 to 1,000 births.

NP: Data collection; CN: Physiological integrity; CNS: Physiological adaptation; CL: Knowledge

5. Congenital torticollis usually involves which of the following muscles?
   1. Platysma
   2. Lower trapezius
   3. Middle trapezius
   4. Sternocleidomastoid

There sure is a lot of muscle in this question!

6. A 3-month-old infant who has severe torticollis presents with the head rotated to the left and the side bent to the right. Which of the following muscles is shortened?
   1. Left upper trapezius
   2. Right middle trapezius
   3. Left sternocleidomastoid
   4. Right sternocleidomastoid

Muscle lengthening? Maybe I'd better lengthen my study sessions!

7. A 9-month-old infant has torticollis with rotation of the head to the left and side bending to the right. Placing the infant in which of the following positions would be most effective for developing muscle lengthening?
   1. Prone
   2. Supine
   3. Left side-lying
   4. Right side-lying

8. Scoliosis can be defined as:
   1. an increase in the lumbar lordosis.
   2. a decrease in the thoracic kyphosis.
   3. lateral curves in the spinal column described as right or left concavities.
   4. lateral curves in the spinal column described as right or left convexities.

9. Which of the following surgical interventions could be used in the treatment of scoliosis?
   1. Segmental diskectomy
   2. Austin Moore procedure
   3. Segmented spine instrumentation
   4. Thoracic and lumbar laminectomy

5. 4. Congenital torticollis usually involves a shortening of the sternocleidomastoid muscle. Rotation is away from the side of shortening and side bending is toward the side of the contraction. The platysma muscle is involved in flexion and is shortened if the head is held in flexion, not contraction. The middle and lower trapezius aren't associated with torticollis.
NP: Data collection; CN: Physiological integrity; CNS: Physiological adaptation; CL: Knowledge

6. 4. The right sternocleidomastoid is shortened with the head in this position. The left upper trapezius isn't shortened, the right one is. The middle trapezius isn't affected and the left sternocleidomastoid is in a lengthened position.
NP: Data collection; CN: Physiological integrity; CNS: Physiological adaptation; CL: Knowledge

7. 3. The left side-lying position will assist with lengthening of the muscles because this position will make it easier to stretch the sternocleidomastoid and upper trapezius. No other positions will assist in increasing muscle length.
NP: Implementation; CN: Physiological integrity; CNS: Reduction of risk potential; CL: Application

8. 4. Scoliosis is defined as curves along the longitudinal axis of the body and curves are described according to the convexity, not the concavity. Lordosis and kyphosis describe the back in the frontal plane.
NP: Data collection; CN: Physiological integrity; CNS: Physiological adaptation; CL: Knowledge

9. 3. In segmented spine instrumentation, flexible rods are placed on the transverse processes of the spine and wires are threaded through the laminae so the spinal column is stabilized. Austin Moore procedure is used in fractured hips. Thoracic and lumbar laminectomies are performed for herniated disks. Segmental diskectomy won't cure or help scoliosis.
NP: Implementation; CN: Physiological integrity; CNS: Basic care and comfort; CL: Knowledge

NP: Nursing process   CN: Client needs category   CNS: Client needs subcategory   CL: Cognitive level

10. Upon physical examination, which of the following conditions may be a warning sign of scoliosis?
1. Scapula winging
2. Forward head posture
3. Raised right iliac crest
4. Forward flexion of the cervical spine

11. Which of the following braces is used in the treatment of scoliosis?
1. Don Joy braces
2. Long leg braces
3. Milwaukee brace
4. Knee-ankle-foot orthosis

12. The Milwaukee brace often is used in the treatment of scoliosis. Which of the following positions best describes the placement of the pressure rods?
1. Laterally on the convex portion of the curve
2. Laterally on the concave portion of the curve
3. Posteriorly on the convex portion of the curve
4. Posteriorly along the spinal column at the exact level of the curve

13. Strengthening which of the following muscle groups is important in a client diagnosed with talipes equinovarus?
1. Evertors
2. Invertors
3. Plantar flexors
4. Plantar fascia musculature

14. Torticollis describes scoliosis in which of the following areas of the spine?
1. Cervical
2. Lumbar
3. Sacral
4. Thoracic

Congratulations! You've finished the first 10 questions!

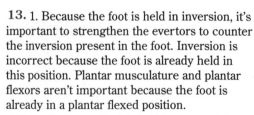

All this studying is strengthening my muscle groups. Oh, goody.

10. 3. A raised iliac crest may be a warning sign of some curvature secondary to the attachment of the pelvis to the spine. Scapula winging could be caused by muscle paralysis. Forward head posture isn't a result of scoliosis. Forward flexion doesn't describe lateral curves in the spine.
NP: Data collection; CN: Physiological integrity; CNS: Physiological adaptation; CL: Knowledge

11. 3. The Milwaukee brace is used to place lateral pressure on the back in hopes of halting or slowing the progression of the curves in the back. It isn't curative. Long leg braces are used for lower extremity weakness or partial paralysis. Don Joy braces are used following knee reconstruction. Knee-ankle-foot orthosis is used for lower extremity and foot weakness.
NP: Data collection; CN: Physiological integrity; CNS: Reduction of risk potential; CL: Knowledge

12. 1. Lateral pressure applied to the convex portion of the curve will help best in reducing the curvature. Pressure pads applied posteriorly will help maintain erect posture. Pressure applied to the concave portion of the curve will increase the lordosis.
NP: Data collection; CN: Physiological integrity; CNS: Reduction of risk potential; CL: Application

13. 1. Because the foot is held in inversion, it's important to strengthen the evertors to counter the inversion present in the foot. Inversion is incorrect because the foot is already held in this position. Plantar musculature and plantar flexors aren't important because the foot is already in a plantar flexed position.
NP: Data collection; CN: Physiological integrity; CNS: Reduction of risk potential; CL: Analysis

14. 1. Torticollis describes only the cervical spine. No other areas apply.
NP: Data collection; CN: Physiological integrity; CNS: Physiological adaptation; CL: Knowledge

15. Which of the following conditions describes a structural scoliosis?
1. Muscular dystrophy with muscle weakness
2. Cerebral palsy with muscle weakness
3. Leg length discrepancy
4. Spina bifida

16. Nonstructural scoliosis can be caused by which of the following complications?
1. Wedge vertebrae
2. Hemivertebrae
3. Poor posture
4. Tumor

17. Idiopathic scoliosis accounts for which of the following percentages of cases of scoliosis?
1. 50% to 60%
2. 65% to 75%
3. 75% to 85%
4. 85% to 95%

18. Which of the following techniques may assist a 3-month-old client diagnosed with torticollis?
1. Lying supine
2. Gentle massage
3. Range-of-motion exercises
4. Lying on the side

19. A physical therapist has instructed the nursing staff in range-of-motion exercises for an infant with torticollis. Which of the following interventions should the nurse perform if she feels uncomfortable performing the stretches that result in crying and grimacing of the client?
1. Check the primary health care provider's orders.
2. Call the primary health care provider.
3. Call the physical therapist.
4. Discontinue the exercises.

Nonstructural is the key word in question 16.

15. 3. Leg length discrepancy is a structural deformity that can be associated with scoliosis. Cerebral palsy, muscular dystrophy, and spina bifida are disease processes that may cause scoliosis.
NP: Data collection; CN: Physiological integrity; CNS: Physiological adaptation; CL: Knowledge

16. 3. Poor posture is a nonstructural problem that may be a cause of scoliosis. Hemivertebrae, wedge vertebrae, and tumor are all structural causes of scoliosis.
NP: Data collection; CN: Physiological integrity; CNS: Physiological adaptation; CL: Knowledge

17. 3. Studies have shown that idiopathic scoliosis accounts for 75% to 85% of all cases.
NP: Data collection; CN: Physiological integrity; CNS: Physiological adaptation; CL: Knowledge

18. 4. Side-lying opposite the affected side may help elongate shortened muscles. Gentle massage won't assist with elongation of muscles. Range-of-motion exercises won't assist with shortened muscles unless in specific patterns and with stretching. Lying supine won't assist with elongation of muscles.
NP: Implementation; CN: Health promotion and maintenance; CNS: Growth and development through the life span; CL: Application

19. 3. The only cure for the torticollis is exercise or surgery. The therapist is the expert in exercise and should be called for assistance in this situation. The primary health care provider would be called only if there was concern over the orders written or an abnormal development in the child.
NP: Implementation; CN: Physiological integrity; CNS: Physiological adaptation; CL: Application

NP: Nursing process   CN: Client needs category   CNS: Client needs subcategory   CL: Cognitive level

20. A client has developed a right torticollis with side-bending to the right and rotation to the left. Which of the following exercises may assist in reduction of the torticollis?
1. Rotation exercises to the right
2. Rotation exercises to the left
3. Cervical extension exercises
4. Cervical flexion exercises

21. Which of the following complications may occur due to severe scoliosis?
1. Increased vital capacity
2. Increased oxygen uptake
3. Diminished vital capacity
4. Decreased residual volume

22. A 4-year-old child is diagnosed with cerebral palsy and resultant thoracic scoliosis. Which of the following conditions may be the cause of the scoliosis?
1. Hypotonia
2. Mental retardation
3. Autonomic dysreflexia
4. Increased thoracic kyphosis

23. Which of the following disease processes is myopathic in nature?
1. Cerebral palsy
2. Marfan syndrome
3. Muscular dystrophy
4. Spinal muscular atrophy

24. During a scoliosis screening, the school nurse notices a raised iliac crest, indicating which of the following conditions?
1. Forward head posture
2. Leg length discrepancy
3. Increased lumbar lordosis
4. Increased thoracic kyphosis

You've done 20 questions already. Great job!

Do you know what myopathic means? Knew you would!

20. 1. Performing rotation exercises to the right will help increase the length of the shortened right sternocleidomastoid. Cervical extension exercises won't lengthen tightened muscles. Rotation to the left will just add to the torticollis as the head is already rotated in that direction. Cervical flexion will add to shortening of the muscles.
NP: Implementation; CN: Physiological integrity; CNS: Reduction of risk potential; CL: Application

21. 3. Scoliosis of greater then 60 degrees can cause shifting of organs and decreased ability for the ribs to expand, thus decreasing vital capacity. An increase in oxygen uptake won't occur secondary to a decrease in chest expansion. An increase in vital capacity also won't occur secondary to a decrease in chest expansion. Residual volume will increase secondary to decreased ability of the lungs to expel air.
NP: Data collection; CN: Physiological integrity; CNS: Physiological adaptation; CL: Knowledge

22. 1. Cerebral palsy is usually associated with some degree of hypotonia or hypertonia. Poor muscle tone may result in scoliosis. Increased thoracic kyphosis won't result in scoliosis. Autonomic dysreflexia is described in spinal cord injury and involves abnormal muscle spasms secondary to abnormal inhibitory neurons present during stretch reflexes. Mental retardation isn't a cause of scoliosis.
NP: Data collection; CN: Physiological integrity; CNS: Physiological adaptation; CL: Knowledge

23. 3. Muscular dystrophy is myopathic in nature. Cerebral palsy occurs from prenatal, perinatal, or postnatal central nervous system damage. An upper motor neuron lesion may be the cause. Marfan syndrome is mesenchymal in nature. Spinal muscular atrophy is related to lower motor neurons.
NP: Data collection; CN: Physiological integrity; CNS: Physiological adaptation; CL: Knowledge

24. 2. A raised iliac crest may be indicative of a leg length discrepancy or a curvature in the lumbar spine. It isn't indicative of forward head posture, lumbar lordosis, or thoracic kyphosis.
NP: Data collection; CN: Physiological integrity; CNS: Physiological adaptation; CL: Knowledge

25. Scoliosis is most common in which of the following groups?

1. Preteenage girls
2. Preteenage boys
3. Adolescent girls
4. Adolescent boys

26. In caring for a child with a Harrington rod placement, which of the following symptoms would be of greatest concern two days postoperatively?

1. Fever of 99.5° F (37.5° C)
2. Pain along the incision
3. Decreased urinary output
4. Hypoactive bowel sounds

27. Observing which of the following structures will serve the nurse best when screening a child for scoliosis?

1. Iliac crests
2. Spinous processes
3. Acromion processes
4. Posterior superior iliac spines

28. Which of the following interventions may be a possible treatment choice for talipes equinovarus?

1. Traction
2. Serial casting
3. Short leg braces
4. Inversion range-of-motion exercises

29. When performing stretches with a child who has scoliosis, which of the following techniques should be used?

1. Slow and sustained
2. Until a change in muscle length is seen
3. Quick movements to the end range of pain
4. Slow movements for brief, 3- to 4-second periods

In question 26, you've got to determine the symptom that causes the greatest concern.

You might need a stretch right now. Why not take a brief one and then get right back to work?

25. 3. Scoliosis is eight times more prominent in adolescent girls than boys.
NP: Data collection; CN: Physiological integrity; CNS: Physiological adaptation; CL: Knowledge

26. 3. Due to extensive blood loss during surgery and possible renal hypoperfusion, decreased urinary output could indicate decreased renal function. A fever of 99.5° F is of concern but may be due to decreased chest expansion secondary to anesthesia, surgery, and pain. A paralytic ileus is common after this surgery and the client may have a nasogastric tube for the first 48 hours.
NP: Data collection; CN: Physiological integrity; CNS: Reduction of risk potential; CL: Application

27. 2. Spinous processes are the best bony landmark to identify when attempting to screen for scoliosis because this will show lateral deviation of the column. Abnormalities in the acromion process, iliac crests, and posterior superior iliac spines may not be indicative of scoliosis.
NP: Data collection; CN: Physiological integrity; CNS: Physiological adaptation; CL: Knowledge

28. 2. Serial casting is a treatment choice in attempts to change the length of soft tissue. Traction isn't an option. Corrective shoes are used instead of short leg braces. Inversion exercises won't help; eversion exercise will.
NP: Data collection; CN: Physiological integrity; CNS: Reduction of risk potential; CL: Knowledge

29. 1. Stretches should be slow and sustained. It's difficult to see changes in muscle length. Stretches shouldn't be performed with quick movements and should be performed for longer than a few seconds.
NP: Implementation; CN: Safe, effective care environment; CNS: Coordinated care; CL: Application

30. Congenital hip dislocation is most commonly found in which of the following groups?
1. Males
2. Females
3. First-born males
4. First-born females

31. Diagnosis of congenital hip dislocation can best be <u>confirmed</u> by which of the following diagnostic techniques?
1. X-ray
2. Positive Ortolani's sign
3. Positive Trendelenburg gait
4. Audible clicking with adduction

32. Which of the following hip positions should be avoided in an 8-month-old infant who has been diagnosed with congenital hip dysplasia?
1. Extension
2. Abduction
3. Internal rotation
4. External rotation

33. Based on knowledge of the progression of muscular dystrophy, which of the following activities would a nurse anticipate the client having difficulty with <u>first?</u>
1. Breathing
2. Sitting
3. Standing
4. Swallowing

34. Barlow's test is used to diagnose hip dislocation in which of the following age groups?
1. 0 to 3 months
2. 3 to 18 months
3. 18 to 36 months
4. 0 to 4 years

In question 31, you're looking for a test that will confirm the diagnosis, not just indicate it.

All of the choices in question 33 may be correct. You must prioritize for this one.

30. 4. Studies have shown that first-born females are six times more likely to have a congenital hip dislocation than males.
NP: Data collection; CN: Physiological integrity; CNS: Physiological adaptation; CL: Knowledge

31. 1. X-ray will confirm the diagnosis of congenital hip dislocation. All of the options are positive signs of dislocation, but only the X-ray will confirm the diagnosis.
NP: Data collection; CN: Physiological integrity; CNS: Physiological adaptation; CL: Knowledge

32. 3. Internal rotation of the hip is an unstable position and should be avoided in infants with hip instability. Hip extension is a relatively stable position. External rotation isn't necessarily an unstable position, as long as it isn't externally rotated too far. Typically, the child is placed in slight abduction while in a hip-spica cast.
NP: Implementation; CN: Physiological integrity; CNS: Reduction of risk potential; CL: Knowledge

33. 3. Muscular dystrophy usually affects postural muscles of the hip and shoulder first. Swallowing and breathing are usually affected last. Sitting may be affected, but a client would have difficulty standing before having difficulty sitting.
NP: Data collection; CN: Physiological integrity; CNS: Physiological adaptation; CL: Knowledge

34. 1. Barlow's test is most effective up to 3 months of age. After 3 months, development of adduction contractures causes the hip "click" to disappear.
NP: Data collection; CN: Physiological integrity; CNS: Physiological adaptation; CL: Knowledge

35. The nurse is trying to weigh a 3-year-old child who's irritable and refuses to stand on the scale. What's the best way to obtain an accurate weight?

1. Ask the mother to approximate the weight.
2. Ask the mother to hold the child and record the combined weight.
3. Weigh the mother and child and subtract the mother's weight from the combined weight.
4. Obtain the admission weight and add 2 oz per day.

36. A young child sustains a dislocated hip as well as a subcapital fracture. Which of the following complications is of greatest concern?

1. Avascular necrosis
2. Postsurgical infection
3. Hemorrhage during surgery
4. Poor postsurgical ambulation

Question 36? Another chance to prioritize.

37. Which of the following positions of the femur is accurate in relation to the acetabulum in a child with congenital hip dislocation?

1. Anterior
2. Inferior
3. Posterior
4. Superior

38. Which of the following definitions best describes muscular dystrophy?

1. A demyelinating disease
2. Lesions of the brain cortex
3. Upper motor neuron lesions
4. Degeneration of muscle fibers

39. Which of the following muscles are affected first with muscular dystrophy?

1. Muscles of the hip
2. Muscles of the foot
3. Muscles of the hand
4. Muscles of respiration

35. 3. Subtracting the mother's weight from the combined weight will yield the child's weight. Weight is an important parameter used in calculating drug dosages based on kilograms per body weight. It also provides the most accurate information about a client's fluid balance. For these reasons, it should never be approximated. Combining the weights would be inaccurate.

NP: Data collection; CN: Health promotion and maintenance; CNS: Growth and development through the life span; CL: Application

36. 1. Avascular necrosis is common with fractures to the subcapital region secondary to possible compromise of blood supply to the femoral head. Postsurgical infection is always a concern but not a priority at first. Hemorrhage shouldn't occur. Poor postsurgical ambulation is of concern but not as much as the possibility of avascular necrosis.

NP: Data collection; CN: Physiological integrity; CNS: Reduction of risk potential; CL: Knowledge

37. 1. The head of the femur is anterior to the acetabulum in a congenital hip dislocation. All other positions are inaccurate.

NP: Data collection; CN: Physiological integrity; CNS: Physiological adaptation; CL: Knowledge

38. 4. Degeneration of muscle fibers with progressive weakness and wasting best describes muscular dystrophy. Demyelination of myelin sheaths is a description of multiple sclerosis. Lesions within the cortex and in upper motor neurons suggest a neurologic, not a muscular, disease.

NP: Data collection; CN: Physiological integrity; CNS: Physiological adaptation; CL: Knowledge

39. 1. Positional muscles of the hip and shoulder are affected first. Progression later advances to muscles of the foot and hand. Involuntary muscles, such as the muscles of respiration, are affected last.

NP: Data collection; CN: Physiological integrity; CNS: Physiological adaptation; CL: Knowledge

NP: Nursing process   CN: Client needs category   CNS: Client needs subcategory   CL: Cognitive level

40. Which of the following tests is most commonly used to help diagnose muscular dystrophy?

1. X-ray
2. Muscle biopsy
3. Electroencephalogram
4. Assessment of ambulation

41. Which of the following laboratory values would be most <u>abnormal</u> in a client diagnosed with muscular dystrophy?

1. Bilirubin
2. Creatinine
3. Serum potassium
4. Sodium

42. Which of the following forms of muscular dystrophy is the most common?

1. Duchenne's
2. Becker's
3. Limb girdle
4. Myotonic

43. Which of the following conditions may be indicative of a child suffering from muscular dystrophy?

1. Hypertonia of extremities
2. Increased lumbar lordosis
3. Upper extremity spasticity
4. Hyperactive lower extremity reflexes

44. Through which of the following mechanisms is Duchenne's muscular dystrophy acquired?

1. Virus
2. Hereditary
3. Autoimmune factors
4. Environmental toxins

Careful. Question 41 asks for an abnormal value.

You just keep going and going! Good work!

40. 2. A muscle biopsy shows the degeneration of muscle fibers and infiltration of fatty tissue. It's used for diagnostic confirmation. X-ray is best for osseous deformity. Ambulation assessment alone wouldn't confirm diagnosis. Electroencephalogram wouldn't be appropriate in this case.

NP: Data collection; CN: Health promotion and maintenance; CNS: Prevention and early detection of disease; CL: Knowledge

41. 2. Creatinine is a by-product of muscle metabolism as the muscle hypertrophies. Bilirubin is a by-product of liver function. Potassium and sodium levels can change due to a variety of factors and aren't indicators of muscular dystrophy.

NP: Data collection; CN: Health promotion and maintenance; CNS: Prevention and early detection of disease; CL: Knowledge

42. 1. Duchenne's accounts for 50% of all cases of muscular dystrophy.

NP: Data collection; CN: Physiological integrity; CNS: Physiological adaptation; CL: Knowledge

43. 2. An increased lumbar lordosis occurs secondarily to paralysis of lower lumbar postural muscles; it also occurs to increase lower extremity support. Hypertonia isn't seen in this disease. Upper extremity spasticity isn't seen because this disease isn't caused by upper motor neuron lesions. Hyperactive reflexes aren't indications of muscular dystrophy.

NP: Data collection; CN: Physiological integrity; CNS: Physiological adaptation; CL: Knowledge

44. 2. Muscular dystrophy is hereditary and acquired through a recessive sex-linked trait. Therefore, it isn't viral, autoimmune, or caused by toxins.

NP: Data collection; CN: Physiological integrity; CNS: Physiological adaptation; CL: Knowledge

45. A client with muscular dystrophy has lost complete control of his lower extremities. He has some strength bilaterally in the upper extremities, but poor trunk control. Which of the following mechanisms would be the most important to have on the wheelchair?
1. Antitip device
2. Extended breaks
3. Headrest support
4. Wheelchair belt

46. A 2-year-old child has muscular dystrophy. His legs are held together with the knees touching. Which of the following muscles are contracted?
1. Hip abductors
2. Hip adductors
3. Hip extensors
4. Hip flexors

47. A 12-year-old child diagnosed with muscular dystrophy is hospitalized secondary to a fall. Surgery is necessary as well as skeletal traction. Which of the following complications would be of greatest concern to the nursing staff?
1. Skin integrity
2. Infection of pin sites
3. Respiratory infection
4. Nonunion healing of the fracture

48. Which problem is most commonly encountered by adolescent females with scoliosis?
1. Respiratory distress
2. Poor self-esteem
3. Poor appetite
4. Renal difficulty

Question 45 tests your ability to ensure the client's safety.

Prioritize? What a surprise!

45. 4. This client has poor trunk control; a belt will prevent him from falling out of the wheelchair. Antitip devices, head rest supports, and extended breaks are all important options but aren't the best choice in this situation.
NP: Planning; CN: Safe, effective care environment; CNS: Safety and infection control; CL: Application

46. 2. The hip adductors are in a shortened position. The abductors are in a lengthened position. This position isn't indicative of hip flexor or hip extensor shortening.
NP: Data collection; CN: Physiological integrity; CNS: Physiological adaptation; CL: Knowledge

47. 3. Respiratory infection can be fatal for clients with muscular dystrophy due to poor chest expansion and decreased ability to mobilize secretions. Skin integrity, infection of pin sites, and nonunion healing are important but not as important as prevention of respiratory infection.
NP: Data collection; CN: Physiological integrity; CNS: Reduction of risk potential; CL: Application

48. 2. Poor self-esteem is a major issue with many adolescents. The use of orthopedic appliances such as those used to treat scoliosis make this issue much more significant for adolescents with scoliosis. Although respiratory distress and poor appetite may surface, they aren't as common as self-esteem problems. Renal problems aren't usually an issue in adolescents with scoliosis.
NP: Planning; CN: Health promotion and maintenance; CNS: Growth and development through the life span; CL: Application

49. A 10-year-old girl has femur fractures in both legs and is in bilateral skeletal traction. Five days after being admitted, she begins ringing for the nurse every 30 minutes and requesting the bedpan. She voids only 20 to 40 ml each time, but there's no evidence of a urinary tract infection. The nurse teaches the child to perform Kegel exercises. The purpose of these exercises is to help:
1. strengthen the child's arms so that she can better use the trapeze to lift up for bedpan placement and removal.
2. strengthen the child's calf muscles so that she's less likely to get leg cramps.
3. distract the child.
4. maintain good perineal muscle tone by tightening the pubococcygeus muscle.

50. A child has developed difficulty ambulating and tends to walk on his toes. Which of the following surgical techniques may benefit the client?
1. Adductor release
2. Hamstring release
3. Plantar fascia release
4. Achilles tendon release

51. Muscular dystrophy is a result of which of the following causes?
1. Gene mutation
2. Chromosomal aberration
3. Unknown nongenetic origin
4. Genetic and environmental factors

52. Evidence of muscle weakness associated with muscular dystrophy usually appears at which of the following ages?
1. Age 1 year
2. Age 2 years
3. Age 3 years
4. Age 4 years

53. Which of the following muscle groups is most important in maintaining maximum function through lower extremity strength in a client diagnosed with muscular dystrophy?
1. Gastrocnemius
2. Gluteus maximus
3. Hamstrings
4. Quadriceps

Looking good! More than halfway finished!

49. 4. Kegel exercises involve tightening the perineal muscles to help strengthen the pubococcygeus muscle and increase its elasticity. This helps to keep the child from becoming incontinent. None of the other options are related to Kegel exercises.
NP: Implementation; CN: Physiological integrity; CNS: Basic care and comfort; CL: Application

50. 4. A shortened Achilles tendon may cause a child to walk on his toes. A release of the tendon may assist the child in walking. An adductor release is often performed if the legs are held together. A plantar fascia release won't help and a hamstring release is done only when there's a knee flexion contracture.
NP: Evaluation; CN: Physiological integrity; CNS: Reduction of risk potential; CL: Application

51. 1. Muscular dystrophy is a result of a gene mutation. It isn't from a chromosome aberration or environmental factors. It's genetic and there's a known origin of the disease.
NP: Data collection; CN: Physiological integrity; CNS: Physiological adaptation; CL: Knowledge

52. 3. Studies have shown that children diagnosed with muscular dystrophy usually show some form of weakness around 3 years of age.
NP: Data collection; CN: Health promotion and maintenance; CNS: Prevention and early detection of disease; CL: Knowledge

53. 2. Gluteus maximus is the strongest muscle in the body and is important for standing as well as for transfers. All of the named muscles are important, but the maintenance of the gluteus maximus will enable maximum function.
NP: Data collection; CN: Health promotion and maintenance; CNS: Growth and development through the life span; CL: Knowledge

54. Which of the following strategies would be the last choice in attempting to maximize function in a child with muscular dystrophy?
1. Long leg braces
2. Straight cane
3. Wheelchair
4. Walker

Question 54 is another priority question — but this time you're looking for the least appropriate choice.

54. 3. A wheelchair provides decreased independence and decreased use of upper extremity and lower extremity function. It should be used only when absolutely necessary. A straight cane, long leg braces, and a walker are still functional assistive devices, much more so than a wheelchair.
NP: Evaluation; CN: Physiological integrity; CNS: Basic care and comfort; CL: Application

55. A child is having increased difficulty getting out of his chair at school. Which of the following recommendations may the nurse make to assist the child?
1. A seat cushion
2. Long leg braces
3. Powered wheelchair
4. Removable arm rests on wheelchair

55. 1. A seat cushion will put the hip extensors at an advantage and make it somewhat easier to get up. Long leg braces wouldn't be the first choice. A powered wheelchair wouldn't be important in assisting with the transfer. Removable armrests have no bearing on assisting the client.
NP: Implementation; CN: Physiological integrity; CNS: Basic care and comfort; CL: Application

56. Muscles in a child with muscular dystrophy will have which of the following characteristics?
1. Soft on palpation
2. Firm or woody on palpation
3. Extremely hard on palpation
4. No muscle consistency on palpation

56. 2. Muscles will often be firm on palpation secondary to the infiltration of fatty tissue and connective tissue into the muscle. The muscles won't be soft secondary to the infiltration and won't be hard upon palpation. There's some consistency to the muscle although, in advanced stages, atrophy is present.
NP: Data collection; CN: Physiological integrity; CNS: Physiological adaptation; CL: Knowledge

57. Nurses should instruct wheelchair-bound clients with muscular dystrophy in which of the following exercises to best prevent skin breakdown?
1. Wheelchair push-ups
2. Leaning side-to-side
3. Leaning forward
4. Gluteal sets

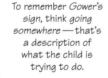

To remember Gower's sign, think going somewhere — that's a description of what the child is trying to do.

57. 1. A wheelchair push-up will alleviate the most pressure off the buttocks. Leaning side-to-side will help but not as much as wheelchair push-ups. Gluteal sets won't help with pressure relief.
NP: Planning; CN: Health promotion and maintenance; CNS: Prevention and early detection of disease; CL: Application

58. Which of the following definitions best describes Gower's sign?
1. A transfer technique
2. A waddling-type gait
3. The pelvis position during gait
4. Muscle twitching present during a quick stretch

58. 1. Gower's sign is a description of a transfer technique present during some phases of muscular dystrophy. The child turns on the side or abdomen, extends the knees, and pushes on the torso to an upright position by walking his hands up the legs. The position of the pelvis during gait isn't involved in Gower's sign. Waddling-type gait doesn't describe Gower's sign. Muscle twitching present after a quick stretch is described as clonus.
NP: Data collection; CN: Physiological integrity; CNS: Physiological adaptation; CL: Knowledge

59. Which of the following definitions best describes pseudohypertrophy?
1. Increased muscle hypertrophy secondary to increased muscle mass
2. Increased muscle hypertrophy secondary to fat infiltration
3. Decreased muscle secondary to muscle degeneration
4. Decreased muscle mass secondary to disease

Focus in on the right definition — and on the prefix *pseudo.*

59. 2. Pseudohypertrophy is present secondary to fat infiltration. Increased muscle mass is called hypertrophy. Pseudohypertrophy isn't due to degeneration or decreased muscle mass.
NP: Data collection; CN: Physiological integrity; CNS: Physiological adaptation; CL: Knowledge

60. A client with bilateral fractured femurs is scheduled for a double-hip spica cast. She says to the nurse, "Only 3 more months, and I can go home." Further investigation reveals that the client and her family believe she'll be hospitalized until the cast comes off. The nurse should explain to the client and her family that she:
1. may be hospitalized 2 to 4 months.
2. will go home 2 to 4 days after casting.
3. will go home a week after casting.
4. will go home as soon as she can move around.

60. 2. The cast will dry fairly rapidly with the use of fiberglass casting material. The time spent in the hospital after casting, typically 2 to 4 days, will be for teaching the client and family how to care for her at home and for evaluating the client's skin integrity and neurovascular status before discharge. The time frames in the other options given are inaccurate for a double-hip spica cast.
NP: Implementation; CN: Health promotion and maintenance; CNS: Prevention and early detection of disease; CL: Application

61. A toddler is hospitalized for treatment of multiple injuries. The parents state that the injuries occurred when their child fell down the stairs. However, inconsistencies between the history and physical findings suggest child abuse. What should the nurse do next?
1. Refer the parents to Parents Anonymous.
2. Prepare the child for foster care placement.
3. Prevent the parents from seeing their child.
4. Report the incident to the proper authorities.

61. 4. The law requires the nurse to report all cases of suspected child abuse. Therefore, the nurse's first action should be to report this incident. Once the authorities have been notified, steps can be taken toward protective custody, if appropriate, when the child is medically stable. Later, the nurse can refer the parents to Parents Anonymous, if needed. The nurse should give the parents opportunities to visit and help care for their child. During these visits, the nurse can reinforce positive parenting behaviors.
NP: Implementation; CN: Safe, effective care environment; CNS: Coordinated care; CL: Application

62. Which of the following forms of muscular dystrophy is the most severe?
1. Duchenne's
2. Facioscapulohumeral
3. Limb girdle
4. Myotonia

62. 1. Studies have shown that Duchenne's is the most severe form of muscular dystrophy. Myotonia isn't a form of the disease; it's a symptom.
NP: Data collection; CN: Physiological integrity; CNS: Physiological adaptation; CL: Knowledge

63. Which of the following definitions best describes acetabular dysplasia?
1. Partial dislocation of the head of the femur
2. Delay in acetabular development
3. Ligamentous laxity of the joint
4. Audible clicking of the femur

63. 2. Acetabular dysplasia is characterized by an underdevelopment of the acetabular ridge. Partial dislocation is described as a subluxation. Ligamentous laxity can be described in a number of ways but isn't appropriate here. Audible clicking is indicative of some degree of subluxation or dislocation.
NP: Data collection; CN: Physiological integrity; CNS: Physiological adaptation; CL: Knowledge

64. Which of the following complications accounts for the greatest number of cases of congenital hip dysplasia?
1. Dislocation
2. Subluxation
3. Acetabular dysplasia
4. Dislocation with fracture

Key here: greatest number. Use it, or lose it.

64. 2. Studies show that subluxation accounts for the greatest number of cases of congenital hip dysplasia.
NP: Data collection; CN: Physiological integrity; CNS: Physiological adaptation; CL: Knowledge

65. Which of the following positions of the hip best describes the condition of a client diagnosed with congenital hip dysplasia?
1. Ligamentum teres is shortened.
2. Femoral head loses contact with acetabulum and is displaced inferiorly.
3. Femoral head loses contact with acetabulum and is displaced posteriorly.
4. Femoral head maintains contact with acetabulum but there's noted capsular rupture.

I hope you aren't dysplased with your answer! Yuk, yuk, yuk.

65. 3. In congenital hip dysplasia, the femoral head loses contact with the acetabulum and is displaced posteriorly, not inferiorly, and the ligamentum teres is lengthened.
NP: Data collection; CN: Physiological integrity; CNS: Physiological adaptation; CL: Knowledge

66. Which of the following percentages represents cases of congenital hip dysplasia involving both hips?
1. 10%
2. 25%
3. 50%
4. 75%

66. 2. Studies show that 25% of all cases of congenital hip dysplasia involve both hips.
NP: Data collection; CN: Physiological integrity; CNS: Physiological adaptation; CL: Knowledge

67. Which of the following choices is best for handling a client's hip-spica cast that has been soiled?
1. Clean with damp cloth and dry cleanser.
2. Clean with soap and water.
3. Don't do anything.
4. Change the cast.

67. 1. A damp cloth is best to use rather than water. Water will break the cast down. Changing the cast isn't an option. If nothing is done, the cast will give off an odor.
NP: Data collection; CN: Physiological integrity; CNS: Basic care and comfort; CL: Application

NP: Nursing process   CN: Client needs category   CNS: Client needs subcategory   CL: Cognitive level

68. Which of the following positions is best for a child in a hip-spica cast who needs to be toileted?

1. Supine
2. Sitting in a toilet chair
3. Shoulder lower than buttocks
4. Buttocks lower than shoulder

69. Congenital hip dysplasia is most common in which of the following ethnic groups?

1. Blacks
2. Whites
3. Native Americans
4. Chinese

70. A toddler is immobilized with traction to the legs. Which play activity would be appropriate for this child?

1. Pounding board
2. Tinker toys
3. Pull toy
4. Games

71. Which of the following complications involving leg length is seen with a congenital hip dislocation?

1. Increased hip abduction
2. Increased leg length on the affected side
3. Decreased leg length on the affected side
4. No change in muscle length or leg length

72. Which of the following positions should be avoided in a child suspected of congenital hip dysplasia?

1. Hip abduction
2. Knee extension
3. External rotation
4. Tightly wrapped blankets

Knowing the shape of a spica cast can help answer this question.

You've completed 70 questions. You're awesome!

The word avoided makes this a negative question.

68. 4. The buttocks need to be lowered to toilet the child. This will keep the cast from being soiled. Supine will cause soiling of the cast. The child isn't able to use a toilet chair.

NP: Planning; CN: Physiological integrity; CNS: Basic care and comfort; CL: Knowledge

69. 3. Native Americans traditionally wrap their neonates in tight blankets that increase internal rotation in the hip. This condition isn't as common in other ethnic groups.

NP: Data collection; CN: Physiological integrity; CNS: Physiological adaptation; CL: Knowledge

70. 1. A pounding board is appropriate for an immobilized toddler because it promotes physical development and provides an acceptable energy outlet. Toys with small parts, such as tinker toys, aren't suitable because a toddler may swallow the parts. A pull toy is suitable for most toddlers but not for one who's immobilized. Games are too advanced for the developmental skills of a toddler.

NP: Planning; CN: Health promotion and maintenance; CNS: Growth and development through the life span; CL: Application

71. 3. Internal rotation with subsequent dislocation will cause the leg to be shorter, not longer. There's often decreased, not increased, abduction as well as muscle and leg length changes.

NP: Data collection; CN: Physiological integrity; CNS: Physiological adaptation; CL: Knowledge

72. 4. Tightly wrapped blankets force the hip into internal rotation. Abduction, external rotation, and knee extension won't increase the risk of dislocation.

NP: Implementation; CN: Health promotion and maintenance; CNS: Prevention and early detection of disease; CL: Application

73. Immediately after a spinal fusion, which of the following restrictions is usually put on the child's activity?
1. Bed rest
2. Non-weight bearing
3. No restriction
4. Limited weight bearing

74. Which of the following interventions would a nurse expect to use to prevent venous stasis after skeletal traction application?
1. Bed rest only
2. Eggcrate mattress
3. Vigorous pulmonary care
4. Antiembolism stockings or an intermittent compression device

75. A 13-year-old girl is suspected of having structural scoliosis by her school nurse. What should the nurse ask the girl to do to help confirm her suspicion?
1. Bend over and touch her toes while the nurse observes from the back.
2. Stand sideways while the nurse observes her profile.
3. Assume a knee-chest position on the examination table.
4. Arch her back while the nurse observes her from the back.

76. At the scene of a trauma, which of the following nursing interventions is appropriate for a child with a suspected fracture?
1. Never move the child.
2. Sit the child up to facilitate breathing.
3. Move the child to a safe place immediately.
4. Immobilize the extremity and then move child to a safe place.

77. Which of the following statements is true about fracture reduction?
1. All fractures can be reduced.
2. Fracture reduction restores alignment.
3. Undisplaced fractures may be reduced.
4. Fracture reduction is usually performed with minimal discomfort.

At the rate you're going, the sky's the limit!

For questions involving a trauma scene, think safety first!

73. 1. After a spinal fusion, the child is usually placed on bed rest and ordered to lie flat. In 2 to 4 days, the child is allowed to sit up in and get out of bed. Other activities are gradually reintroduced.
NP: Implementation; CN: Physiological integrity; CNS: Basic care and comfort; CL: Knowledge

74. 4. To prevent venous stasis after skeletal traction application, antiembolism stockings or an intermittent compression device is used on the unaffected leg. Eggcrate mattresses and pulmonary care don't prevent venous stasis. Bed rest can *cause* venous stasis.
NP: Implementation; CN: Health promotion and maintenance; CNS: Prevention and early detection of disease; CL: Application

75. 1. As the child bends over, the curvature of the spine is more apparent. The scapula on one side becomes more prominent, and the opposite side hollows. Scoliosis can't be properly assessed from the side or the front. The knee-chest position is used for lumbar puncture.
NP: Data collection; CN: Health promotion and maintenance; CNS: Prevention and early detection of disease; CL: Comprehension

76. 4. At the scene of a trauma, the nurse should immobilize the extremity of a child with a suspected fracture and then move him to safety. If the child is already in a safe place, don't attempt to move him. Never try to sit the child up; this could make the fracture worse.
NP: Implementation; CN: Safe, effective care environment; CNS: Safety and infection control; CL: Application

77. 2. Fracture reduction restores alignment. Some fractures, such as undisplaced fractures, can't be reduced. Fracture reduction is usually painful.
NP: Implementation; CN: Physiological integrity; CNS: Reduction of risk potential; CL: Knowledge

78. A child in skeletal traction for a fracture of the right femur exhibits a positive Homans' sign, complains of left-sided leg pain, and has edema in the left leg. Which condition should be expected?

1. A fat emboli
2. An infection
3. A pulmonary embolism
4. Deep vein thrombosis (DVT)

This question is deep, man. (Tee-hee!)

79. Nursing care for a child in traction may include which of the following interventions?

1. Assessing pin sites every shift and as needed
2. Ensuring that the rope knots catch on the pulley
3. Adding and removing weights per the client's request
4. Placing all joints through range of motion every shift

80. After assisting the primary health care provider in applying a cast, the nurse should include which of the following interventions in her immediate cast care?

1. Rest the cast on the bedside table.
2. Dispose of the plaster water in the sink.
3. Support the cast with the palms of her hand.
4. Wait until the cast dries before cleansing surrounding skin.

81. Which of the following time periods best represents approximately how long it takes a synthetic cast to set?

1. Immediately
2. 20 minutes
3. 45 minutes
4. 2 hours

Only 12 more to go! Can you spot the finish line ahead?

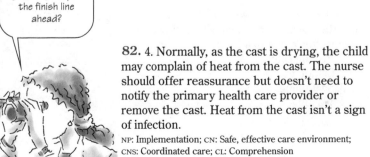

82. Which of the following nursing interventions should be taken if, as a cast is drying, the child complains of heat from the cast?

1. Remove the cast immediately.
2. Notify the primary health care provider.
3. Assess the child for other signs of infection.
4. Explain to the child that this is a normal sensation.

78. 4. Unilateral leg pain and edema with a positive Homans' sign (not always present) should lead you to suspect DVT. Symptoms of fat emboli include restlessness, tachypnea, and tachycardia and are more common in long bone injuries. It's unlikely that an infection would occur on the opposite side of the fracture without cause. Tachycardia, chest pain, and shortness of breath may be symptoms of a pulmonary embolism.
NP: Implementation; CN: Physiological integrity; CNS: Reduction of risk potential; CL: Knowledge

79. 1. Nursing care for a child in traction may include assessing pin sites every shift and as needed and ensuring that the knots in the rope don't catch on the pulley. Weights should be added and removed per the primary health care provider's order, and all joints, except those immediately proximal and distal to the fracture, should be placed through range of motion every shift.
NP: Implementation; CN: Physiological integrity; CNS: Basic care and comfort; CL: Application

80. 3. After a cast has been applied, it should be immediately supported with the palms of the nurse's hands. Later, the nurse should dispose of the plaster water in a sink with a plaster trap or in a garbage bag, cleanse the surrounding skin before the cast dries, and make sure that the cast isn't resting on a hard or sharp surface.
NP: Implementation; CN: Safe, effective care environment; CNS: Coordinated care; CL: Comprehension

81. 2. Synthetic casts take about 20 minutes to set.
NP: Implementation; CN: Safe, effective care environment; CNS: Coordinated care; CL: Knowledge

82. 4. Normally, as the cast is drying, the child may complain of heat from the cast. The nurse should offer reassurance but doesn't need to notify the primary health care provider or remove the cast. Heat from the cast isn't a sign of infection.
NP: Implementation; CN: Safe, effective care environment; CNS: Coordinated care; CL: Comprehension

83. Which of the following nursing interventions can be implemented to <u>prevent</u> foot drop in a casted leg?
1. Encourage bed rest.
2. Support the foot with 45 degrees of flexion.
3. Support the foot with 90 degrees of flexion.
4. Place a stocking on the foot to provide warmth.

In question 83, you're trying to prevent a problem, not react to it.

83. 3. To prevent foot drop in a casted leg, the foot should be supported with 90 degrees of flexion. Bed rest can cause foot drop. Keeping the extremity warm won't prevent foot drop.
NP: Implementation; CN: Health promotion and maintenance; CNS: Prevention and early detection of disease; CL: Comprehension

84. A child with a hip-spica cast should avoid gas forming foods for which of the following reasons?
1. To prevent flatus
2. To prevent diarrhea
3. To prevent constipation
4. To prevent abdominal distention

84. 4. A child with a hip-spica cast should avoid gas forming foods to prevent abdominal distention. Gas forming foods may cause flatus, but that isn't a reason to avoid them. Gas forming foods don't generally cause diarrhea or constipation.
NP: Implementation; CN: Physiological integrity; CNS: Reduction of risk potential; CL: Comprehension

85. Which of the following actions does touchdown weight bearing allow the child to perform?
1. Allows full weight bearing on the affected extremity
2. Allows 30% to 50% weight bearing on the affected extremity
3. Allows no weight on the extremity, which must be kept elevated at all times
4. Allows no weight on the extremity, but the client may touch the floor with the affected extremity

Warming up, pacing activity, moderating intensity — those sound like study instructions!

85. 4. Touch-down weight bearing allows the child to put no weight on the extremity, but the child may touch the floor with the affected extremity. Full weight bearing allows for full weight bearing on the affected extremity. Partial weight bearing allows for only 30% to 50% weight bearing on the affected extremity. Non-weight bearing is no weight on the extremity, and it must remain elevated.
NP: Implementation; CN: Safe, effective care environment; CNS: Coordinated care; CL: Knowledge

86. Which of the following strategies is best for preventing sports-related injuries?
1. Warming up
2. Pacing activity
3. Building strength
4. Moderating intensity

86. 1. To prevent sports-related injuries, instruct your client that the best prevention is warming up. Pacing activity, building strength, and using moderate intensity are also prevention measures.
NP: Planning; CN: Safe, effective care environment; CNS: Coordinated care; CL: Knowledge

87. Which of the following activities may be most therapeutic for a child who's allowed full activity after repair of a clubfoot?
1. Playing catch
2. Standing
3. Swimming
4. Walking

87. 4. Walking will stimulate all of the involved muscles and help with strengthening. All of the options are good exercises, but walking is the best choice.
NP: Planning; CN: Physiological integrity; CNS: Physiological adaptation; CL: Application

---

NP: Nursing process   CN: Client needs category   CNS: Client needs subcategory   CL: Cognitive level

88. Which of the following joints is most involved with clubfoot?
1. Midtarsal
2. Subtalar
3. Talocrural
4. Transmetatarsal

89. Which of the following complications may occur secondary to a mild talipes equinovarus deformity?
1. Medial ankle sprain
2. Irritation of the fifth metatarsal
3. Lengthening of the Achilles tendon
4. Lengthening of medial joint structures

90. Which of the following diseases may be associated with a congenital clubfoot?
1. Spina bifida
2. Muscular dystrophy
3. Osteogenesis imperfecta
4. Juvenile rheumatoid arthritis

91. Which of the following history findings is most significant related to congenital hip dislocation?
1. Mother's activity during the third trimester
2. Breech presentation at birth
3. Infant's serum calcium level at birth
4. Apgar score of 4 at 1 minute and 6 at 5 minutes

92. The dietitian has developed a dietary teaching plan for a child with Duchenne's muscular dystrophy. Which elements are most important to include when explaining the child's diet to the parents?
1. Low calorie, high protein, and high fiber
2. Low calorie, high protein, and low fiber
3. High calorie, high protein, and restricted fluids
4. High calorie, high protein, and high fiber

93. A child with structural scoliosis has been fitted for a Milwaukee brace. How many hours a day should the nurse tell the child that the brace must be worn?
1. 8 hours
2. 12 hours
3. 23 hours
4. 24 hours

Watch the words most significant. They're hints!

Let's see, if it's 10 o'clock in Milwaukee, how long does the child need to wear the brace?

88. 2. The subtalar joint is most involved with clubfoot. It's primarily responsible for adduction and inversion of the foot, the greatest defect. The midtarsal and transmetatarsal joints aren't affected.
NP: Data collection; CN: Physiological integrity; CNS: Physiological adaptation; CL: Knowledge

89. 2. Irritation of the fifth metatarsal is common because of increased weight bearing on the lateral side of the foot. Lateral, not medial, ankle sprains are common. The Achilles tendon is shortened as are the medial structures.
NP: Data collection; CN: Physiological integrity; CNS: Physiological adaptation; CL: Knowledge

90. 1. Studies have shown that there's an increased incidence of clubfoot with spina bifida. Clubfoot doesn't occur at an increased rate with these other conditions.
NP: Data collection; CN: Physiological integrity; CNS: Physiological adaptation; CL: Knowledge

91. 2. Breech presentation is a factor frequently associated with congenital hip dislocation. The mother's activity during the third trimester, the infant's serum calcium level at birth, and Apgar scores have no bearing on hip dislocation.
NP: Data collection; CN: Health promotion and maintenance; CNS: Growth and development through the life span; CL: Application

92. 1. A child with Duchenne's muscular dystrophy is prone to constipation and obesity, so dietary intake should include a diet low in calories, high in protein, and high in fiber. Adequate fluid intake should also be encouraged.
NP: Planning; CN: Physiological integrity; CNS: Basic care and comfort; CL: Knowledge

93. 3. The brace can be removed only 1 hour per day for bathing and hygiene; otherwise, it must remain in place. Wearing the brace for 8 or 12 hours per day isn't enough time to provide the necessary correction. Wearing the brace 24 hours per day doesn't allow for bathing or skin integrity checks.
NP: Implementation; CN: Physiological integrity; CNS: Reduction of risk potential; CL: Knowledge

94. A 6-month-old male with hip dislocation has been treated for the past 6 weeks with a Frejka splint, which maintains abduction through padding of the diaper area. At his follow-up visit, the client's mother reports that she removes the splint when he gets too fussy and that he settles down and sleeps well for several hours after the padding is removed. Which of the following responses by the nurse would be most appropriate?

1. "I can tell you're concerned about his comfort, but he must wear the padded splint except during the three times a day when you perform range-of-motion exercises on his legs."
2. "I'm pleased that you recognize that the padding is too thick and have adjusted it so he can sleep comfortably."
3. "I realize that seeing him uncomfortable is difficult for you, but he needs to keep his splint on except when you bathe him or change his diaper."
4. "If he seems uncomfortable while wearing the splint, it's important that you call us immediately."

94. 3. Soft abduction devices, such as the Frejka splint, must be worn continually except for diaper changes and skin care. The abduction position must be maintained to establish a deep hip socket. Discomfort is anticipated; appropriate responses including changing position, holding, cuddling, and providing diversion.

NP: Evaluation; CN: Physiological integrity; CNS: Physiological adaptation; CL: Application

Congratulations! You finished! Great job!

I'll bet that when you started nursing school, you had no idea kids could be subject to so many GI disorders. I know I didn't. This chapter tests you on the most common ones. Good luck!

Chapter 32
Gastrointestinal disorders

1. The nurse is caring for a 17-year-old girl who's receiving total parenteral nutrition (TPN). The nurse understands that this solution is given:

    1. directly into a superficial vein.

    2. directly into the superior vena cava.

    3. through a gastrostomy tube.

    4. orally as part of the prescribed diet.

2. Which of the following goals is most important when teaching the parents of a child diagnosed with celiac disease?

    1. Promote a normal life for the child.

    2. Stress the importance of good health in preventing infection.

    3. Introduce the parents and child to another peer with celiac disease.

    4. Help the parents and child follow the prescribed dietary restrictions.

3. Which of the following characteristics or conditions would the nurse expect to find in a child diagnosed with celiac disease?

    1. Constipation

    2. Pleasant disposition

    3. Proper weight gain

    4. Steatorrhea

4. A client with celiac disease is being discharged from the hospital. Which of the following food items would be included in his diet?

    1. Cereal

    2. Luncheon meat

    3. Pizza

    4. Rice

Nutrition is served!

In question 2, the words most important guide you to the right answer.

Gluten-free?

1. 2. TPN is given through a catheter passed directly into the superior vena cava by way of the jugular or subclavian vein. A cellulose membrane filter is used to filter out bacteria. A superficial vein, gastrostomy tube, and oral route are never used for TPN administration.
NP: Implementation; CN: Physiological integrity; CNS: Pharmacological therapies; CL: Comprehension

2. 4. It takes a long time to describe the disease process, the specific role of gluten, and the foods that must be restricted. Gluten is added to many foods but is obscurely listed on labels. To avoid hidden sources of gluten, parents need to read labels carefully. Promoting a normal life for the child, stressing good health in preventing infection, and meeting a peer with celiac disease are also important nursing considerations but would come after education about the dietary means of dealing with this chronic disease.
NP: Planning; CN: Psychosocial integrity; CNS: Coping and adaptation; CL: Knowledge

3. 4. Steatorrhea (fatty, foul-smelling, frothy, bulky stools) is common due to the inability to absorb fat. Profuse watery diarrhea, not constipation, is usually a sign of celiac crisis. Behavior changes, such as irritability, uncooperativeness, and apathy, are common and they usually aren't pleasant. Poor weight gain would be a symptom of celiac disease because impaired absorption leads to malnutrition.
NP: Data collection; CN: Physiological integrity; CNS: Physiological adaptation; CL: Knowledge

4. 4. Sources of gluten found in wheat, rye, barley, and oats should be avoided. Rice and corn are suitable substitutes because they don't contain gluten. Pizza, luncheon meat, and cereal contain gluten and, when broken down, can't be digested by people with celiac disease.
NP: Planning; CN: Psychosocial integrity; CNS: Coping and adaptation; CL: Knowledge

NP: Nursing process   CN: Client needs category   CNS: Client needs subcategory   CL: Cognitive level

5. To help promote a normal life for children with celiac disease, which of the following interventions should the parents use?
1. Treat the child differently than other siblings.
2. Focus on restrictions that make him feel different.
3. Introduce the child to another peer with celiac disease.
4. Don't allow the child to express doubt in keeping with dietary restrictions.

6. Which of the following conditions would be considered a malabsorption disease of the GI system?
1. Addison's disease
2. Celiac disease
3. Crohn's disease
4. Hirschsprung's disease

You'll be finished before you know it!

7. Within a day or two after starting their prescribed diet, most children with celiac disease show which of the following characteristics?
1. Diarrhea
2. Foul-smelling stools
3. Improved appetite
4. Weight loss

8. In caring for an neonate with cleft lip and palate, which of the following issues is first encountered by the nurse?
1. Feeding difficulties
2. Operative care
3. Pain management
4. Parental reaction

Here's a key word to consider.

9. To prevent trauma to the suture line of an infant who underwent cleft lip repair, the nurse would perform which of the following interventions?
1. Place mittens on the infant's hands.
2. Maintain arm restraints.
3. Don't allow the parents to touch the infant.
4. Remove the lip device from the infant after surgery.

5. 3. Introducing the child to another child with celiac disease will let him know he isn't alone. It will show him how other people live a normal life with similar restrictions. Instead of focusing on restrictions that make him feel different, the parents should focus on ways he can be normal. They should treat the child no differently than other siblings but stress appropriate limit setting. Allow the child with celiac disease to express his feelings about dietary restrictions.
NP: Implementation; CN: Psychosocial integrity; CNS: Coping and adaptation; CL: Application

6. 2. In celiac disease, the absorptive surface of the small intestine is impaired. Crohn's disease is an inflammatory disease of the bowel. Addison's disease involves dysfunction of the adrenal cortex. Hirschsprung's disease is an obstructive defect in part of the intestine.
NP: Data collection; CN: Physiological integrity; CNS: Physiological adaptation; CL: Knowledge

7. 3. Within a day or two of starting their diet, most children show improved appetite, weight gain, and disappearance of diarrhea. Steatorrhea (fatty, oily, foul-smelling stools) usually doesn't occur for several days or weeks.
NP: Evaluation; CN: Physiological integrity; CNS: Physiological adaptation; CL: Application

8. 4. Parents often show strong negative responses to this deformity. They may mourn the loss of the perfect child. Helping the parents cope with their child's condition is the first step. Feeding issues are important, but parents must first cope with the reality of their neonate's condition. Surgical repair is usually delayed until 6 to 12 weeks of age. This deformity isn't painful.
NP: Data collection; CN: Psychosocial integrity; CNS: Coping and adaptation; CL: Application

9. 2. Arm restraints are used to prevent the infant from rubbing the sutures. Placing mittens alone won't prevent the infant from rubbing the suture line. Parental contact will increase the infant's comfort. The lip device shouldn't be removed.
NP: Implementation; CN: Physiological integrity; CNS: Reduction of risk potential; CL: Analysis

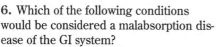

NP: Nursing process CN: Client needs category CNS: Client needs subcategory CL: Cognitive level

10. To prevent tissue infection and breakdown after cleft palate or lip repair, the nurse would use which of the following interventions?
   1. Keep the suture line moist at all times.
   2. Allow the infant to suck on his pacifier.
   3. Rinse the infant's mouth with water after each feeding.
   4. Follow orders from the physician to not feed the infant by mouth.

11. Which of the following structural defects involves the postoperative use of the Logan bow?
   1. Cleft lip or palate
   2. Esophageal atresia
   3. Hiatal hernia
   4. Tracheoesophageal fistula

12. When bottle-feeding an infant with a cleft palate or lip, gentle steady pressure should be applied to the base of the bottle for which of the following reasons?
   1. Reduce the risk for choking or coughing.
   2. Prevent further damage to the affected area.
   3. Decrease the amount of formula lost while eating.
   4. Decrease the amount of noise the infant makes when eating.

13. Which of the following nursing interventions should be used when feeding an infant with cleft lip and palate?
   1. Burp the infant often.
   2. Limit the amount the infant eats.
   3. Feed the infant at scheduled times.
   4. Remove the nipple if the infant is making loud noises.

Keep going to the finish line! Don't give up!

What's the best way to feed an infant with a cleft palate?

10. 3. To prevent formula buildup around the suture line, the mouth is usually rinsed. The sutures should be kept dry at all times. Objects placed in the mouth are generally avoided after surgery. Infants are fed by mouth using the Asepto-syringe technique.
NP: Implementation; CN: Physiological integrity; CNS: Physiological adaptation; CL: Application

11. 1. Immediately after surgery for cleft lip or palate, the Logan bow, a thin arched metal device, is used to protect the suture line from tension. Esophageal atresia, hiatal hernia, or tracheoesophageal fistula repairs don't need a device to protect sutures after surgery.
NP: Implementation; CN: Physiological integrity; CNS: Reduction of risk potential; CL: Knowledge

12. 1. Children with cleft palate or lip have a greater risk for choking while eating, so all measures are used to reduce this risk. Steady pressure creates a seal when the nipple is against the cleft palate or lip, reducing the risk of aspiration. The nurse can't cause more damage to an infant's cleft lip or palate unless proper precautions aren't followed postoperatively. If the nipple is cut correctly and proper procedures are followed, the infant won't lose a lot of formula during a feeding. Infants with cleft palate or lip often make more noise while eating.
NP: Implementation; CN: Physiological integrity; CNS: Reduction of risk potential; CL: Knowledge

13. 1. Infants with cleft lip and palate have a tendency to swallow an excessive amount of air. The amount of formula they eat at each feeding is the same as an infant without cleft lip or palate, and scheduled feedings aren't necessary. Loud noises are common when these infants eat.
NP: Implementation; CN: Physiological integrity; CNS: Physiological adaptation; CL: Knowledge

14. Nursing care for an infant with cleft lip or palate would include which of the following interventions?
1. Discourage breast-feeding.
2. Hold the infant flat while feeding.
3. Involve the parents as soon as possible.
4. Use a normal nursery nipple for feedings.

14. 3. The sooner the parents become involved, the quicker they're able to determine the method of feeding best suited for them and the infant. Breast-feeding, like bottle-feeding, may be difficult but can be facilitated if the mother intends to breast-feed. Sometimes, especially if the cleft isn't severe, breast-feeding may be easier because the human nipple conforms to the shape of the infant's mouth. Feedings are usually given in the upright position to prevent formula from coming through the nose. There are a variety of special nipples devised for infants with cleft lip or palate; a normal nursery nipple isn't effective.
NP: Implementation; CN: Physiological integrity; CNS: Physiological adaptation; CL: Knowledge

> Here's where to show your compassion.

15. The parents of an infant born with cleft lip and palate are seeing the infant for the first time. The nurse caring for the infant should focus on which of the following areas?
1. The infant's positive features
2. Irritation with how the infant eats
3. Ambivalence in caring for an infant with this defect
4. Dissatisfaction with the infant's physical appearance

15. 1. To relieve the parents' anxiety, positive aspects of the infant's physical appearance need to be emphasized. Showing optimism toward surgical correction and showing a photograph of possible cosmetic improvements may be helpful. The other responses are inappropriate.
NP: Implementation; CN: Psychosocial integrity; CNS: Coping and adaptation; CL: Application

16. Which of the following long-term physical problems might occur after cleft palate surgery?
1. Deviated septum
2. Recurring tonsillitis
3. Tooth decay
4. Varying degrees of hearing loss

16. 4. Improper draining of the middle ear causes recurrent otitis media and scarring of the tympanic membrane, which lead to varying degrees of hearing loss. Cleft palate doesn't cause problems with the tonsils. Improper tooth alignment is common, not tooth decay. The septum remains intact with cleft palate repair.
NP: Evaluation; CN: Physiological adaptation; CNS: Reduction of risk potential; CL: Knowledge

> Prioritize!

17. The mother of a neonate born with a cleft lip and palate is preparing to bottle-feed her neonate for the first time. When giving her instructions, which of the following interventions should be taught first?
1. Burp the neonate.
2. Clean the mouth.
3. Hold the neonate in an upright position.
4. Prepare the bottle using a normal nursery nipple.

17. 3. When neonates are held in the upright position, the formula is less likely to leak out the nose or mouth. Neonates need to be burped frequently but not before a feeding. There's no need to clean the mouth before eating. After surgical repair, the mouth is cleaned at the suture site to prevent infection. The bottle should be prepared using a special nipple or feeding device.
NP: Planning; CN: Physiological integrity; CNS: Physiological adaptation; CL: Comprehension

NP: Nursing process   CN: Client needs category   CNS: Client needs subcategory   CL: Cognitive level

18. An infant returns from surgery after repair of a cleft palate. Which of the following nursing interventions should be done first?

1. Offer a pacifier for comfort.
2. Position the infant on his side.
3. Suction all secretions from the mouth and nose.
4. Remove the arm restraints placed on the infant after surgery.

19. A small child has just had surgical repair of a cleft palate. Which of the following instructions should be included in the discharge teaching to the parents?

1. Continue a normal diet.
2. Continue using arm restraints at home.
3. Don't allow the child to drink from a cup.
4. Establish good mouth care and proper brushing.

20. Most cleft palates are repaired at what age?

1. Immediately after birth
2. 1 to 2 months
3. 3 to 4 months
4. 1 to 2 years

21. After an infant with a cleft lip has surgical repair and heals, the parents can expect to see which of the following results?

1. A large scar
2. An abnormally large upper lip
3. Distorted jaw
4. Some scarring

22. Which of the following specialists are involved in the management of a neonate born with cleft lip or palate?

1. Cardiologist
2. Neurologist
3. Nutritionist
4. Otolaryngologist

Client teaching includes the family.

Way to go!

18. 2. The infant should be positioned on his side to allow oral secretions to drain from the mouth so suctioning is avoided. Pacifiers shouldn't be used since they can damage the suture line. Arm restraints should be kept on to protect the suture line. The restraints should be removed periodically to allow for full range of motion during this time. Only one restraint should be removed at a time, and the infant should be closely supervised.

NP: Implementation; CN: Physiological integrity; CNS: Reduction of risk potential; CL: Application

19. 2. Arm restraints are also used at home to keep the child's hands away from the mouth until the palate is healed. A soft diet is recommended. No food harder than mashed potatoes can be eaten. Fluids are best taken from a cup. Proper mouth care is encouraged after the palate is healed.

NP: Planning; CN: Physiological integrity; CNS: Physiological adaptation; CL: Application

20. 4. Most surgeons will correct the cleft at age 1 to 2, before faulty speech patterns develop. To take advantage of palatal changes, surgical repair is usually postponed until this time.

NP: Data collection; CN: Physiological integrity; CNS: Physiological adaptation; CL: Knowledge

21. 4. If there's no trauma or infection to the site, healing occurs with little scar formation. There may be some inflammation right after surgery, but after healing, the lip is a normal size. No jaw malformation occurs with cleft lip repair.

NP: Evaluation; CN: Psychosocial integrity; CNS: Coping and adaptation; CL: Comprehension

22. 4. An otolaryngologist is used because ear infections are common, along with hearing loss. Brain and cardiac function are usually normal. A nutritionist isn't needed unless the neonate becomes malnourished.

NP: Implementation; CN: Safe, effective care environment; CNS: Coordinated care; CL: Knowledge

23. Which of the following conditions or factors is the major cause of esophageal atresia and tracheoesophageal fistula?
1. Genetic
2. Prematurity
3. Poor nutrition during pregnancy
4. Unknown

The word *major* clarifies the answer.

23. 4. The cause of these malformations is unknown. Genetics isn't an issue with these structural defects. A premature neonate isn't necessarily born with either malformation. These defects occur much earlier in the process of development than viability (23 to 24 weeks' gestation). Poor nutrition could contribute to other factors in a neonate such as iron deficiency anemia.
NP: Data collection; CN: Physiological integrity; CNS: Physiological adaptation; CL: Knowledge

24. Which of the following types of tracheoesophageal fistula and esophageal atresia is the most commonly encountered?
1. A cleft from the trachea to the upper esophagus
2. A normal trachea and esophagus connected by a common fistula
3. A blind pouch at each end, widely separated, with no involvement of the trachea
4. Proximal esophageal segment terminated in a blind pouch, distal segment connected to the trachea

24. 4. In 80% to 90% of all cases, these malformations have the proximal esophageal segment terminated in a blind pouch and the distal segment connected to the trachea. A cleft doesn't occur in the trachea. A normal trachea and esophagus connected by a common fistula represents the rarest form. A blind pouch at each end, widely separated, with no involvement of the trachea is the second most common type of tracheoesophageal fistula, making up 5% to 8% of all cases.
NP: Data collection; CN: Physiological integrity; CNS: Physiological adaptation; CL: Knowledge

Another hint.

25. Which of the following findings is common in neonates born with esophageal atresia?
1. Cyanosis
2. Decreased production of saliva
3. Inability to cough
4. Inadequate swallow

25. 1. Cyanosis occurs when fluid from the blind pouch is aspirated into the trachea. Increased drooling is common, along with choking, coughing, and sneezing. The ability to swallow isn't affected by this disorder.
NP: Data collection; CN: Physiological integrity; CNS: Physiological adaptation; CL: Knowledge

26. For a neonate suspected of having esophageal atresia, a definitive diagnostic evaluation would include which of the following factors?
1. Decreased breath sounds
2. Absence of bowel sounds
3. How a neonate tolerates eating
4. Ability to pass a catheter down the esophagus

26. 4. A moderately stiff catheter will meet resistance if the esophagus is blocked and will pass unobstructed if the esophagus is patent. Breath sounds are normal unless aspiration occurs. The intestinal tract isn't affected with this anomaly, so bowel sounds are present. If a neonate doesn't tolerate eating, it doesn't mean he has an esophageal atresia.
NP: Data collection; CN: Physiological integrity; CNS: Physiological adaptation; CL: Knowledge

NP: Nursing process   CN: Client needs category   CNS: Client needs subcategory   CL: Cognitive level

27. For a neonate diagnosed with a tracheoesophageal fistula, which of the following interventions would be needed?

1. Initiate antibiotic therapy.
2. Keep the neonate lying flat.
3. Continue feedings.
4. Remove the diagnostic catheter from the esophagus.

Question 28 asks you to prioritize your care.

28. When tracheoesophageal fistula or esophageal atresia is suspected, which of the following nursing interventions would be done first?

1. Give oxygen.
2. Tell the parents.
3. Put the neonate in an Isolette or on a radiant warmer.
4. Report the suspicion to the physician.

29. Which of the following complications may follow the surgical repair of a tracheoesophageal fistula?

1. Atelectasis
2. Choking during feeding attempts
3. Damaged vocal cords
4. Infection

Here's another hint.

30. Which of the following signs of esophageal atresia with distal tracheoesophageal fistula would appear the earliest?

1. Abdominal distention
2. Decreased oral secretions
3. Normal respiratory effort
4. Scaphoid abdomen

27. 1. Antibiotic therapy is started because aspiration pneumonia is inevitable and appears early. The neonate's head is usually kept in an upright position to prevent aspiration. I.V. fluids are started, and the neonate isn't allowed oral intake. The catheter is left in the upper esophageal pouch to easily remove fluid that collects there.
NP: Implementation; CN: Physiological integrity; CNS: Pharmacological therapies; CL: Application

28. 4. The physician needs to be told so immediate diagnostic tests can be done for a definitive diagnosis and surgical correction. Oxygen should be given only after notifying the physician except in the case of an emergency. It isn't the nurse's responsibility to inform the parents of the suspected finding. By the time tracheoesophageal fistula or esophageal atresia is suspected, the neonate would have already been placed in an Isolette or a radiant warmer.
NP: Implementation; CN: Physiological integrity; CNS: Physiological adaptation; CL: Application

29. 1. Respiratory complications (atelectasis) are a threat to the neonate's life preoperatively and postoperatively due to the continual risk for aspiration. Choking is more likely to occur preoperatively, although careful attention is paid postoperatively when neonates begin to eat to make sure they can swallow without choking. Vocal cord damage isn't common after this repair. The neonate is generally given antibiotics preoperatively to prevent infection.
NP: Data collection; CN: Physiological integrity; CNS: Physiological adaptation; CL: Application

30. 1. Crying may force air into the stomach, causing distention. Secretions in a client with this condition may be more visible, though normal in quantity, due to the client's inability to swallow effectively. Respiratory effort is usually more difficult. When no distal fistula is present, the abdomen will appear scaphoid.
NP: Data collection; CN: Physiological integrity; CNS: Physiological adaptation; CL: Knowledge

31. Dietary management in a child diagnosed with ulcerative colitis would include which of the following diets?
1. High-calorie diet
2. High-residue diet
3. Low-protein diet
4. Low-salt diet

32. A neonate returns from the operating room after surgical repair of a tracheoesophageal fistula and esophageal atresia. Which of the following interventions is done immediately?
1. Maintain a patent airway.
2. Start feedings right away.
3. Let the parents hold the neonate right away.
4. Suction the endotracheal tube stopping when resistance is met.

33. Before discharging a neonate with a repaired tracheoesophageal fistula and esophageal atresia, the nurse would give the parents or caregivers instructions in which of the following areas?
1. Giving antibiotics
2. Preventing infection
3. Positioning techniques
4. Giving solid food as soon as possible

34. Which of the following nursing interventions would be done postoperatively for a neonate after repair of tracheoesophageal fistula and esophageal atresia?
1. Withhold mouth care
2. Offer a pacifier frequently
3. Decrease tactile stimulation
4. Use restraints to prevent injury to the repair

35. Which of the following conditions would be a long-term postoperative complication of a tracheoesophageal fistula and esophageal atresia repair?
1. Oral aversion
2. Gastroesophageal reflux
3. Inability to tolerate feedings
4. Strictures

Immediately means, you know, right away!

Yet another classic clue. Don'tcha just love 'em?

31. 1. A high-calorie diet is given to combat weight loss and restore nitrogen balance. A low-residue or residue-free diet is encouraged to decrease bowel irritation. A high-protein diet is encouraged. Salt intake isn't a factor in this disease.
NP: Implementation; CN: Physiological integrity; CNS: Physiological adaptation; CL: Analysis

32. 1. Maintaining a patent airway is essential until sedation from surgery wears off. Feedings usually aren't started for at least 48 hours after surgery. The catheter should be measured before suctioning so the tube doesn't meet resistance, which could cause damage. Parents are encouraged to participate in the neonate's care but not immediately after surgery.
NP: Implementation; CN: Physiological integrity; CNS: Physiological adaptation; CL: Knowledge

33. 3. Positioning instructions should be given during hospitalization and before the neonate returns home. Antibiotics are usually discontinued before discharge. Preventing infection, especially at the operative sites, is a responsibility of the nurse postoperatively. For optimum effective respiration and because gastroesophageal reflux is a common complication, solid food usually isn't started until liquid feedings are tolerated.
NP: Planning; CN: Psychosocial integrity; CNS: Coping and adaptation; CL: Comprehension

34. 2. Meeting the neonate's oral needs is important because he's unable to drink from a bottle. Restraints should be avoided if possible. The nurse should give mouth care to these neonates. The nurse should provide tactile stimulation.
NP: Implementation; CN: Physiological integrity; CNS: Physiological adaptation; CL: Application

35. 4. Strictures of the anastomosis occur in 40% to 50% of cases. Oral aversion can be a problem, but it occurs quickly after surgery. Reflux is a common complication but appears when feedings are started. If the neonate is having problems tolerating feedings, it's quickly noted.
NP: Data collection; CN: Physiological integrity; CNS: Physiological adaptation; CL: Knowledge

NP: Nursing process   CN: Client needs category   CNS: Client needs subcategory   CL: Cognitive level

36. Which of the following structural defects is suspected when a neonate has excessive salivation and drooling, accompanied by coughing, choking, and sneezing?
1. Cleft lip
2. Cleft palate
3. Gastroschisis
4. Tracheoesophageal fistula and esophageal atresia

37. Esophageal atresia and tracheoesophageal fistula are usually associated with which of the following factors?
1. Female sex
2. Higher then average birth weight
3. Male sex
4. Prematurity

38. Which of the following structural defects involves a portion of an organ protruding through an abnormal opening?
1. Cleft lip
2. Cleft palate
3. Gastroschisis
4. Tracheoesophageal fistula

39. When an infant is diagnosed with a diaphragmatic hernia on the left side, which of the following abdominal organs may be found in the thorax?
1. Appendix
2. Descending colon
3. Right kidney
4. Spleen

40. In which direction does the mediastinum shift in an infant diagnosed with a diaphragmatic hernia?
1. No shift
2. Shifts to the affected side
3. Shifts to the unaffected side
4. Partially shifts to the affected or unaffected sides

Looking for the end of these questions? Hang in there!

There's nothing tricky about this hint.

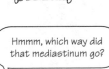

Hmmm, which way did that mediastinum go?

36. 4. Because of an ineffective swallow, saliva and secretions appear in the mouth and around the lips. Coughing, choking, and sneezing occur for the same reason and usually after an attempt at eating. Cleft lip and palate don't produce excessive salivation. None of these symptoms occur with gastroschisis.
NP: Data collection; CN: Physiological integrity; CNS: Physiological adaptation; CL: Knowledge

37. 4. The cause of these defects is unknown, but an unusually high percentage of cases occurs in premature infants. The birth weight is significantly less than average. There are no sex differences noted.
NP: Data collection; CN: Physiological integrity; CNS: Physiological adaptation; CL: Knowledge

38. 3. Gastroschisis is a herniation of the bowel through an abnormal opening in the abdominal wall. Tracheoesophageal fistula is a malformation of the trachea and esophagus. Cleft lip and palate are facial malformations, not herniations.
NP: Data collection; CN: Physiological integrity; CNS: Physiological adaptation; CL: Knowledge

39. 4. The spleen has often been seen in the thorax of infants with this defect. The right kidney wouldn't be seen with a left-sided defect. The appendix and descending colon usually don't protrude into the thorax due to limited space from the other organs present.
NP: Data collection; CN: Physiological integrity; CNS: Physiological adaptation; CL: Knowledge

40. 3. The increased volume in the chest cavity from the abdominal organs causes the mediastinum to shift to the unaffected side, which causes a partial collapse of that lung. Due to the increased volume on the affected side, the mediastinum can't shift that way.
NP: Data collection; CN: Physiological integrity; CNS: Physiological adaptation; CL: Application

41. Which action confirms that a nasogastric (NG) tube is properly positioned in a client's stomach?
1. Inverting the tube into a glass of water and observing for bubbling
2. Instilling 10 ml of air into the tube and listening with a stethoscope for air entering the stomach
3. Clamping the tube for 10 minutes and listening with a stethoscope for increased peristalsis
4. Instilling 30 ml of normal saline solution and observing the client's response

Which position would help us expand?

42. Before surgery, which of the following interventions would be used for an infant with a diaphragmatic hernia?
1. Feed the infant.
2. Provide tactile stimulation.
3. Prevent the infant from crying.
4. Place the infant on the unaffected side.

43. The nursing care of a neonate born with an omphalocele would include which of the following interventions?
1. Keep the malformation dry.
2. Don't let the parents see it.
3. Carefully position and handle the neonate.
4. Touch it often to assess any changes.

You're doing a great job!

44. What's the treatment for a diagnosed omphalocele?
1. Immediate surgical repair
2. Surgical repair after the sac ruptures
3. Using sterile technique and manual manipulation
4. No treatment as it goes away by itself

41. 2. The "whooshing" sound of air entering the stomach confirms that the NG tube is in the stomach. To further verify positioning, the pH of the gastric aspirate is approximately 3.0 (acidic), whereas the pH of respiratory aspirate is 7.0 or greater (alkaline). Inverting the tube into a glass of water and observing for bubbling would be done to verify that an NG tube is in the respiratory tract. Clamping the tube and listening for increased peristalsis provides no information on the location of the tube. Instilling normal saline is dangerous. If the tube is in the respiratory tract, this action will cause coughing, choking and, possibly, cyanosis.

NP: Data collection; CN: Physiological integrity; CNS: Reduction of risk potential; CL: Comprehension

42. 3. The stomach and intestine in the chest cavity become distended with swallowed air from crying. Negative pressure from crying pulls the intestines into the chest cavity, increasing the amount of distention. The infant usually isn't fed until after surgery. Tactile stimulation is limited because it may disturb the infant's fragile condition. The infant is always placed on the affected side.

NP: Data collection; CN: Physiological integrity; CNS: Physiological adaptation; CL: Knowledge

43. 3. Careful positioning and handling of the neonate prevents infection and rupture of the sac. The omphalocele is kept moist until the neonate is taken to the operating room. The parents should be allowed to see the defect if they choose. Touching it often increases the risk for infection.

NP: Planning; CN: Physiological integrity; CNS: Physiological adaptation; CL: Knowledge

44. 1. Surgical repair is done immediately to prevent infection and possible tissue damage. The omphalocele is covered and kept moist until the infant goes to the operating room. Careful positioning and handling techniques are used to prevent rupture of the sac and damage to the abdominal contents.

NP: Planning; CN: Physiological integrity; CNS: Physiological adaptation; CL: Knowledge

NP: Nursing process   CN: Client needs category   CNS: Client needs subcategory   CL: Cognitive level

45. Which of the following abdominal defects <u>isn't</u> covered by a protective sac and <u>doesn't</u> cause damage to the umbilical cord?
 1. Diaphragmatic hernia
 2. Gastroschisis
 3. Omphalocele
 4. Umbilical hernia

Is this a double negative? Look out!

45. 2. Gastroschisis is always located to the right of an intact umbilical cord and isn't enclosed in a protective sac. Diaphragmatic hernia is a protrusion of abdominal organs into the thoracic cavity. The omphalocele is covered by only a translucent sac of amnion. Umbilical hernia is a protrusion of the intestine into the umbilicus, which is covered by skin.
NP: Data collection; CN: Physiological integrity; CNS: Physiological adaptation; CL: Knowledge

46. Which of the following statements is true about pyloric stenosis?
 1. It's more common in girls.
 2. It's diagnosed by severe diarrhea.
 3. It's more frequent in Blacks and Asians.
 4. It's more common in full-term neonates.

46. 4. Pyloric stenosis is more likely to affect a full-term neonate than a preterm one. It's more common in boys and is usually diagnosed by vomiting. It occurs more frequently in white neonates.
NP: Data collection; CN: Physiological integrity; CNS: Physiological adaptation; CL: Knowledge

47. Which of the following factors is the cause of pyloric stenosis?
 1. Unknown
 2. Hereditary
 3. Poor nutrition in pregnancy
 4. Usually directly related to poor muscle development in the stomach

It's OK to say you don't know.

47. 1. The cause of the narrowing of the pyloric musculature is unknown. A hereditary factor hasn't been established. Poor nutrition in pregnancy and poor muscle development in the stomach may relate to this defect, but at present, haven't been established as a definitive cause.
NP: Data collection; CN: Physiological integrity; CNS: Physiological adaptation; CL: Knowledge

48. In which of the following areas does the pyloric canal narrow in clients with pyloric stenosis?
 1. Stomach and esophagus
 2. Stomach and duodenum
 3. Both the stomach and esophagus and the stomach and duodenum
 4. Neither the stomach and esophagus nor the stomach and duodenum

48. 2. The narrowing of the pyloric canal occurs between the stomach and duodenum, where the pyloric sphincter is located. Hyperplasia and hypertrophy cause narrowing and possibly obstruction of the circular muscle of the pylorus.
NP: Data collection; CN: Physiological integrity; CNS: Physiological adaptation; CL: Knowledge

49. Which of the following signs and symptoms is classic for an infant with pyloric stenosis?
 1. Loss of appetite
 2. Chronic diarrhea
 3. Projectile vomiting
 4. Occasional nonprojectile vomiting

49. 3. The obstruction doesn't allow food to pass through to the duodenum. Once the stomach becomes full, the infant vomits for relief. Occasional nonprojectile vomiting may occur initially if the obstruction is only partial. Chronic hunger is commonly seen. There's no diarrhea because food doesn't pass the stomach.
NP: Data collection; CN: Physiological integrity; CNS: Physiological adaptation; CL: Knowledge

50. When assessing a neonate, the nurse notes visible peristaltic waves across the epigastrium. This characteristic is indicative of which of the following disorders?
1. Hypertrophic pyloric stenosis
2. Imperforate anus
3. Intussusception
4. Short-gut syndrome

51. After surgical repair of pyloric stenosis, at which of the following times will the infant return to a normal feeding regimen?
1. 4 to 6 hours after surgery
2. 24 hours after surgery
3. 48 hours after surgery
4. 1 week after surgery

52. A nurse admits an infant diagnosed with pyloric stenosis. Which of the following nursing interventions would most likely be done first?
1. Weigh the infant.
2. Check urine specific gravity.
3. Place an I.V. catheter.
4. Change the infant and weigh the diaper.

53. A nurse is caring for an infant with pyloric stenosis. After feeding the infant, he should be placed in which of the following positions?
1. Prone in Fowler's position
2. On his back without elevation
3. On the left side in Fowler's position
4. Slightly on the right side in high semi-Fowler's position

Not this kind of wave!

Put your priorities in order.

Here's that important positioning again!

50. 1. The diagnosis of pyloric stenosis can be established from a finding of hypertrophic pyloric stenosis. Imperforate anus, intussusception, and short-gut syndrome are diagnosed by other characteristics.
NP: Data collection; CN: Physiological integrity; CNS: Physiological adaptation; CL: Knowledge

51. 3. Small frequent feedings of clear fluids are usually started 4 to 6 hours after surgery. If clear fluids are tolerated, formula feedings are started 24 hours after surgery, in gradually increasing amounts. It usually takes 48 hours to reach a normal full feeding regimen in this manner. The infant usually goes home on the fourth postoperative day.
NP: Planning; CN: Physiological integrity; CNS: Physiological adaptation; CL: Knowledge

52. 1. Weighing the infant would be done first so a baseline weight can be established and weight changes can be assessed. After a baseline weight is obtained, an I.V. catheter can be placed because oral feedings generally aren't given. These infants are usually dehydrated, so while checking the diaper and specific gravity are important tools to help assess their status, they aren't the first priority.
NP: Implementation; CN: Physiological integrity; CNS: Physiological adaptation; CL: Application

53. 4. Positioning the infant slightly on the right side in high semi-Fowler's position will help facilitate gastric emptying. The other positions won't facilitate gastric emptying and may cause the infant to vomit.
NP: Implementation; CN: Physiological integrity; CNS: Physiological adaptation; CL: Knowledge

54. When preparing to feed an infant with pyloric stenosis before surgical repair, which of the following interventions is important?
1. Give feedings quickly.
2. Burp the infant frequently.
3. Discourage parental participation.
4. Don't give more feedings if the infant vomits.

55. Which of the following symptoms is common up to 48 hours after the surgical repair of pyloric stenosis?
1. Dysuria
2. Oral aversion
3. Scaphoid abdomen
4. Vomiting

56. Which of the following interventions will help prevent vomiting in an infant diagnosed with pyloric stenosis?
1. Hold the infant for 1 hour after feeding.
2. Handle the infant minimally after feedings.
3. Space feedings out and give large amounts.
4. Lay the infant prone with the head of the bed elevated.

57. It's an important nursing function to give support to the parents of an infant diagnosed with pyloric stenosis. Which of the following nursing interventions best serves that purpose?
1. Keep the parents informed of the infant's progress.
2. Provide all care for the infant, even when the parents visit.
3. Tell the parents to minimize handling of the infant at all times.
4. Tell the physician to keep the parents informed of the infant's progress.

Don't forget us — the family — when caring for an infant.

54. 2. These infants usually swallow a lot of air from sucking on their hands and fingers because of their intensive hunger (feedings aren't easily tolerated). Burping frequently will lessen gastric distention and increase the likelihood the infant will retain the feeding. Feedings are given slowly with the infant lying in a semi-upright position. Parental participation should be encouraged and allowed to the extent possible. Record the type, amount, and character of the vomit as well as it's relation to the feeding. The amount of feeding volume lost is usually refed.
NP: Implementation; CN: Physiological integrity; CNS: Physiological adaptation; CL: Application

55. 4. Even with successful surgery, most infants have some vomiting during the first 24 to 48 hours after surgery. Dysuria isn't a complication with this surgical procedure. Oral aversion doesn't occur because these infants may be fed until surgery. Scaphoid abdomen isn't characteristic of this condition. The abdomen may appear distended, not scaphoid.
NP: Data collection; CN: Physiological integrity; CNS: Physiological adaptation; CL: Knowledge

56. 2. Minimal handling, especially after a feeding, will help prevent vomiting. Holding the infant would provide too much stimulation, which might increase the risk for vomiting. Feedings are given frequently and slowly in small amounts. An infant should be positioned in a semi-Fowler's position and slightly on the right side after a feeding.
NP: Implementation; CN: Physiological integrity; CNS: Physiological adaptation; CL: Knowledge

57. 1. Keeping the parents informed will decrease their anxiety. The nurse should encourage the parents to be involved with the infant's care. Telling the parents to minimize handling of the infant isn't appropriate because parent-child contact is important. The physician is responsible for updating the parents on the infant's medical condition, and the nurse is responsible for updating the parents on the day-to-day activities of the infant and his improvement with the day's activities.
NP: Planning; CN: Psychosocial integrity; CNS: Coping and adaptation; CL: Application

58. Which of the following symptoms would be likely in an infant diagnosed with pyloric stenosis?
1. Apathy
2. Arrhythmia
3. Dry lips and skin
4. Hypothermia

59. When assessing an infant diagnosed with pyloric stenosis, which of the following findings would be considered normal?
1. Decreased or diminished bowel sounds
2. Heart murmur
3. Normal respiratory effort
4. Positive bowel sounds

60. Which of the following abdominal organs is directly affected when pyloric stenosis is diagnosed?
1. Colon and rectum
2. Stomach and duodenum
3. Stomach and esophagus
4. Liver and bile ducts

61. Which of the following nursing interventions is the most important in dealing with a child who has been poisoned?
1. Stabilize the child.
2. Notify the parents.
3. Identify the poison.
4. Determine when the poisoning took place.

Don't be fooled. Question 59 asks what's normally found with a disease, not what's normal in a healthy infant.

Practice setting priorities because it's part of nursing practice.

58. 3. Dry lips and skin are signs of dehydration, which is common in infants with pyloric stenosis. These infants are constantly hungry due to their inability to retain feedings. Apathy, arrhythmias, and hypothermia aren't clinical findings with pyloric stenosis.
NP: Data collection; CN: Physiological integrity; CNS: Physiological adaptation; CL: Application

59. 1. Bowel sounds decrease because food can't pass into the intestines. Normal respiratory effort is affected due to the abdominal distension that pushes the diaphragm up into the pleural cavity. Heart murmurs may be present but aren't directly associated with pyloric stenosis.
NP: Data collection; CN: Physiological integrity; CNS: Physiological adaptation; CL: Knowledge

60. 2. This defect occurs at the pyloric sphincter, which is located between the stomach and duodenum. The stomach and esophagus, liver and bile ducts, and colon and rectum aren't affected by this obstructive disorder.
NP: Data collection; CN: Physiological integrity; CNS: Physiological adaptation; CL: Knowledge

61. 1. Stabilization and the initial emergency treatment of the child (such as respiratory assistance, circulatory support, or control of seizures) will prevent further damage to the body from the poison. If the parents didn't bring the child in, they can be notified as soon as the child is stabilized or treated. Although identification of the poison is crucial and should begin at the same time as the stabilization of the child, the initial ABCs should be assessed first. Determining when the poisoning took place is an important consideration, but emergency stabilization and treatment are priorities.
NP: Data collection; CN: Physiological integrity; CNS: Physiological adaptation; CL: Application

62. For a child who has ingested a poisonous substance, the initial step in emergency treatment is to stop the exposure to the substance. Which of the following methods would best achieve this?
1. Make the child vomit.
2. Call 911 as soon as possible.
3. Give large amounts of water to flush the system.
4. Empty the mouth of pills, plant parts, or other material.

62. 4. Emptying the mouth of pills, plant parts, or other material will stop exposure to the poison. Making the child vomit is important, but won't remove exposure to the substance; it's also contraindicated with some poisons. Calling 911 is important, but removing any further sources of the poison would come first. Only small amounts of water are recommended so the poison is confined to the smallest volume. Large amounts of water will let the poison pass the pylorus. The small intestines absorb fluid rapidly, increasing the potential toxicity.
NP: Implementation; CN: Physiological integrity; CNS: Physiological adaptation; CL: Application

63. An ingested poison can be removed by inducing vomiting. Which of the following methods is preferred?
1. Give syrup of ipecac.
2. Have the child smell something offensive to cause vomiting.
3. Put a finger down the throat of the child who ingested the poison.
4. Do nothing because ingesting the poison may cause vomiting in some children.

Yo! Take a hint, will ya?

63. 1. If vomiting is indicated, syrup of ipecac should be given instead of waiting for the child to vomit by himself. This syrup causes stimulation of the vomiting center and an irritant effect on gastric mucosa. Vomiting should occur within 20 minutes. Smelling something offensive may not work because it's difficult to get the child to comply. Putting a finger down his throat could cause damage to the oropharynx. Waiting for the child to vomit may allow the poison to enter the systemic circulation.
NP: Implementation; CN: Physiological integrity; CNS: Pharmacological therapies; CL: Application

64. Under which of the following circumstances is emetic treatment appropriate for a child who ingested a poison?
1. The poison is a salicylate (aspirin).
2. The child is in severe shock.
3. The child is experiencing seizures.
4. The poison is a low-viscosity hydrocarbon.

You're moving right along!

64. 1. If the child has digested a salicylate (aspirin), inducing vomiting with an emetic is indicated. Severe shock increases the risk for aspiration, so emetic treatment is contraindicated when a child is in severe shock. An emetic shouldn't be administered to a child experiencing seizures because the child could aspirate the poison if he vomits while having a seizure. If the poison is a low-viscosity hydrocarbon, it can cause severe chemical pneumonitis if aspirated; therefore, inducing vomiting is contraindicated.
NP: Planning; CN: Physiological integrity; CNS: Physiological adaptation; CL: Knowledge

65. Which of the following inflammatory diseases most commonly affects the terminal ileum?
1. Acute appendicitis
2. Crohn's disease
3. Meckel's diverticulum
4. Ulcerative colitis

65. 2. Crohn's disease affects the terminal ileum. Ulcerative colitis affects the entire bowel. Acute appendicitis affects the blind sac at the end of the cecum. Meckel's diverticulum is a sac that becomes inflamed.
NP: Data collection; CN: Physiological integrity; CNS: Physiological adaptation; CL: Knowledge

66. If a child ingests poisonous hydrocarbons, an important nursing intervention would include which of the following actions?
1. Induce vomiting.
2. Keep the child calm and relaxed.
3. Scold the child for the wrongdoing.
4. Keep the parents away from the child.

The right position helps.

66. 2. Keeping the child calm and relaxed will help prevent vomiting. If vomiting occurs, there's a great chance the esophagus will be damaged from regurgitation of the gastric poison. Additionally, the risk for chemical pneumonitis exists if vomiting occurs. Scolding the child may upset him. The parents should remain with the child to help keep him calm.
NP: Implementation; CN: Physiological integrity; CNS: Physiological adaptation; CL: Application

67. Shock is a complication of several types of poisoning. Which of the following measures would help reduce the risk for shock?
1. Keep the child on his right side.
2. Let the child maintain normal activity as possible.
3. Elevate the head and legs to the level of the heart.
4. Keep the head flat and raise the legs to the level of the heart.

67. 3. Elevating the head and legs to the level of the heart will promote venous drainage and decrease the chance of the child going into shock. The child may safely lie on the side he prefers. The child should be encouraged to get plenty of rest.
NP: Implementation; CN: Physiological integrity; CNS: Physiological adaptation; CL: Knowledge

68. A 7-year-old child ingested several leaves of a poinsettia plant. After arrival in the emergency department, which of the following interventions would be the main nursing function for this client?
1. Begin teaching accident prevention.
2. Provide emotional support to the child.
3. Be prepared for immediate intervention.
4. Provide emotional support to the parents.

A main hint.

68. 3. Time and speed are critical factors in recovery from poisonings. The remaining three answers are important nursing functions but don't require the immediate attention that first stabilizing the child does.
NP: Planning; CN: Physiological integrity; CNS: Reduction of risk potential; CL: Application

69. Which of the following drugs is most commonly ingested by children?
1. Acetaminophen (Tylenol)
2. Aspirin
3. Ibuprofen (Motrin)
4. Acetaminophen and oxycodone (Percocet)

Is this how to tell the most common?

69. 2. Aspirin is the most readily available drug in many homes, thus increasing the likelihood of poisoning by this drug. Acetaminophen would be the second most readily available drug. Percocet isn't a common household drug. Motrin may be used in the home.
NP: Data collection; CN: Physiological integrity; CNS: Physiological adaptation; CL: Knowledge

70. Which of the following symptoms helps in the initial diagnosis of ulcerative colitis?
1. Constipation
2. Diarrhea
3. Vomiting
4. Weight loss

70. 2. Recurrent or persistent diarrhea is a common feature of ulcerative colitis. Constipation doesn't occur because the bowel becomes smooth and inflexible. Vomiting isn't common in this disease. Weight loss will occur after or during the episode but not initially.
NP: Data collection; CN: Physiological integrity; CNS: Physiological adaptation; CL: Knowledge

71. If a child ingests poisonous amounts of salicylates, how soon after ingestion would signs of toxicity become obvious?

1. Immediately
2. 2 to 4 hours after ingestion
3. 6 hours after ingestion
4. 18 hours after ingestion

72. In extreme cases of salicylate poisoning, which of the following treatments is used?

1. Forced emesis
2. Hypothermia blankets
3. Peritoneal dialysis
4. Vitamin K injection

73. When a child has been poisoned, identifying the ingested poison is an important treatment goal. Which of the following actions would help determine which poison was ingested?

1. Call the local poison control center.
2. Ask the child.
3. Ask the parents.
4. Save all evidence of poison.

74. One of the most important nursing responsibilities to help prevent salicylate poisoning would include which of the following actions?

1. Identify salicylate overdose.
2. Teach children the hazards of ingesting nonfood items.
3. Decrease curiosity; teach parents to keep aspirin and drugs in clear view.
4. Teach parents to keep large amounts of drugs on hand but out of reach of children.

Timing is important.

Wait, there's more on the next page. Hooray!

Know your nursing responsibilities.

71. 3. There's usually a delay of 6 hours before evidence of toxicity is noted. Toxic evidence is rarely immediate. Aspirin will exert its peak effect in 2 to 4 hours. The effect of aspirin may last as long as 18 hours.

NP: Data collection; CN: Health promotion and maintenance; CNS: Prevention and early detection of disease; CL: Knowledge

72. 3. Peritoneal dialysis is usually reserved for cases of life-threatening salicylism. Forced emesis is the immediate treatment for salicylate poisoning because the stomach contents and salicylates will move from the stomach to the remainder of the GI tract, where vomiting will no longer result in the removal of the poison. Hypothermia blankets may be used to reduce the possibility of seizures. Vitamin K may be used to decrease bleeding tendencies but only if evidence of this exists.

NP: Planning; CN: Physiological integrity; CNS: Physiological adaptation; CL: Knowledge

73. 4. Saving all evidence of poison (container, vomitus, urine) will help determine which drug was ingested and how much. Calling the local poison control center may help get information on specific poisons or if a certain household placed a call, although rarely can they help determine which poison has been ingested. Asking the child may help, but the child may fear punishment and may not be honest about the incident. The parent may be helpful in some instances, although the parent may not have been home or with the child when the ingestion occurred.

NP: Data collection; CN: Physiological integrity; CNS: Reduction of risk potential; CL: Application

74. 2. Teaching children the hazards of ingesting nonfood items will help prevent ingestion of poisonous substances. Identifying the overdose won't prevent it from occurring. Aspirin and drugs should be kept out of the sight of children. Parents should be warned about keeping large amounts of drugs on hand.

NP: Implementation; CN: Health promotion and maintenance; CNS: Prevention and early detection of disease; CL: Comprehension

75. Which of the following conditions can result from an acute overdose of acetaminophen?
1. Brain damage
2. Heart failure
3. Hepatic damage
4. Kidney damage

76. Which of the following symptoms is a sign of acetaminophen poisoning?
1. Hyperthermia
2. Increased urine output
3. Profuse sweating
4. Rapid pulse

77. In a client diagnosed with acetaminophen poisoning, the initial therapeutic management would include which of the following actions?
1. Induce vomiting.
2. Obtain blood work.
3. Give I.V. fluid.
4. Use activated charcoal.

78. Which of the following ethnic groups has the highest prevalence of lead poisoning in children?
1. Black
2. Hispanic
3. Asian
4. White

79. Which of the following factors influence the ingestion of lead-containing substances?
1. Child's age
2. Child's sex
3. Child's nationality
4. A parent with the same habit

You're right on target!

Priority work!

Who's at highest risk?

75. 3. The damage to the hepatic system isn't from the drug but from one of its metabolites. This metabolite binds to liver cells in large quantities. Brain damage, heart failure, and kidney damage may develop but not initially.
NP: Data collection; CN: Physiological integrity; CNS: Physiological adaptation; CL: Knowledge

76. 3. During the first 12 to 24 hours, profuse sweating is a significant sign of acetaminophen poisoning. Weak pulse, hypothermia, and decreased urine output are also common findings.
NP: Data collection; CN: Physiological integrity; CNS: Physiological adaptation; CL: Knowledge

77. 1. Inducing vomiting using syrup of ipecac would be the initial treatment after acetaminophen poisoning has been confirmed. Blood work would be obtained but wouldn't be the first priority. Caution is taken when fluids are administered. Activated charcoal isn't used because it interferes with the antidote N-acetylcysteine, which helps protect the liver.
NP: Planning; CN: Physiological integrity; CNS: Physiological adaptation; CL: Knowledge

78. 1. Black children have a six times greater risk for lead poisoning compared with white children. Lead poisoning isn't common in Asian or Hispanic populations.
NP: Data collection; CN: Health promotion and maintenance; CNS: Prevention and early detection of disease; CL: Knowledge

79. 1. The highest risk for lead poisoning occurs in young children who have a tendency to put things in their mouths. In older homes that contain lead-based paint, paint chips may be eaten directly by the child or they may cling to toys or hands that are then put into the child's mouth. Poisoning isn't sex-related. Blacks have a higher incidence of lead poisoning, but it can happen in any race. Children of low socioeconomic status are more likely to eat lead-based paint chips. Most parents don't eat lead-based paint on purpose.
NP: Data collection; CN: Health promotion and maintenance; CNS: Prevention and early detection of disease; CL: Knowledge

80. Lead retained in the body is largely stored in which of the following organs?
1. Bone
2. Brain
3. Kidney
4. Liver

81. Which of the following conditions is one of the initial signs of lead poisoning?
1. Anemia
2. Diarrhea
3. Overeating
4. Paralysis

Late, schmate. Question 81 asks for the first signs.

82. The most serious and irreversible adverse effects of lead poisoning affect which of the following systems?
1. Central nervous system
2. Hematologic system
3. Renal system
4. Respiratory system

83. Black lines along the gums indicate which if the following types of poisoning?
1. Acetaminophen
2. Lead
3. Plants
4. Salicylates

Another milestone! You're more than three-quarters finished.

84. Which of the following procedures is the main treatment for lead poisoning?
1. Blood transfusion
2. Bone marrow transplant
3. Chelation therapy
4. Dialysis

Watch for these words.

85. Which of the following nursing objectives would be the most important for a child with lead poisoning who must undergo chelation therapy?
1. Prepare the child for complete bed rest.
2. Prepare the child for I.V. fluid therapy.
3. Prepare the child for an extended hospital stay.
4. Prepare the child for a large number of injections.

80. 1. Ingested lead is initially absorbed by bone. If chronic ingestion occurs, then the hematologic, renal, and central nervous systems are affected.
NP: Data collection; CN: Physiological integrity; CNS: Physiological adaptation; CL: Knowledge

81. 1. Lead is dangerously toxic to the biosynthesis of heme. The reduced heme molecule in red blood cells causes anemia. Constipation, not diarrhea, and a poor appetite and vomiting, not overeating, are signs of lead poisoning. Paralysis may occur as toxic damage to the brain progresses but isn't an initial sign.
NP: Data collection; CN: Physiological integrity; CNS: Physiological adaptation; CL: Knowledge

82. 1. Damage that occurs to the central nervous system is difficult to repair. Damage to the renal and hematologic systems can be reversed if treated early. The respiratory system isn't affected until coma and death occur.
NP: Data collection; CN: Physiological integrity; CNS: Physiological adaptation; CL: Application

83. 2. One diagnostic characteristic of lead poisoning is black lines along the gums. Black lines don't occur along the gums with acetaminophen, plant, or salicylate poisoning.
NP: Data collection; CN: Physiological integrity; CNS: Physiological adaptation; CL: Knowledge

84. 3. Chelation therapy involves the removal of metal by combining it with another substance. Sometimes exchange transfusions are used to rid the blood of lead quickly. Bone marrow transplants usually aren't needed. Dialysis usually isn't part of treatment.
NP: Implementation; CN: Physiological integrity; CNS: Physiological adaptation; CL: Knowledge

85. 4. Chelation therapy involves receiving a large number of injections in a relatively short period of time. It's traumatic to the majority of children, and they need some preparation for the treatment. The other components of the treatment plan are important but not as likely to cause the same anxiety as multiple injections. Receiving I.V. fluid isn't as traumatizing as 60 injections. Physical activity is usually limited. Allowing adequate rest to not aggravate the painful injection sites is important.
NP: Implementation; CN: Physiological integrity; CNS: Physiological adaptation; CL: Application

86. The nurse is assessing a 5-year-old male child who exhibits signs of possible lead poisoning. Signs and symptoms of lead poisoning include:
1. nausea, vomiting, seizures, and coma.
2. jaundice, confusion, and coagulation abnormalities.
3. insomnia, weight loss, diarrhea, and gingivitis.
4. general fatigue, difficulty concentrating, tremors, and headache.

87. Which of the following interventions is the best way to prevent lead poisoning in children?
1. Educate the child.
2. Educate the public.
3. Identify high-risk groups.
4. Provide home chelation kits.

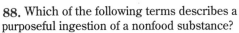

88. Which of the following terms describes a purposeful ingestion of a nonfood substance?
1. Patent ductus arteriosus (PDA)
2. RDSI
3. Pica
4. Plumbism

89. Certain forms of pica are caused by a deficiency of which of the following nutrients?
1. Minerals
2. Vitamin B complex
3. Vitamin C
4. Vitamin D

90. Which of the following helminthic infections caused by nematodes is the most common?
1. Flukes
2. Hookworms
3. Roundworms
4. Tapeworms

86. 4. Signs and symptoms of lead poisoning depend on the degree of toxicity. General fatigue, difficulty concentrating, tremors, and headache indicate moderate toxicity. Nausea, vomiting, seizures, and coma are observed in salicylate and iron poisoning. Jaundice, confusion, and coagulation abnormalities are observed in acetaminophen poisoning. Insomnia, weight loss, diarrhea, and gingivitis are observed in mercury poisoning.
NP: Data collection; CN: Physiological integrity; CNS: Physiological adaptation; CL: Comprehension

87. 2. By educating others about lead poisoning, including danger signs, symptoms, and treatment, identification can be determined quickly. Young children may not understand the dangers of lead poisoning. Identifying high-risk groups will help but won't prevent the poisoning. Home chelation kits currently aren't available.
NP: Implementation; CN: Health promotion and maintenance; CNS: Prevention and early detection of disease; CL: Application

88. 3. Pica is a Latin word for magpie, a bird with a voracious appetite. Today, pica refers to the purposeful ingestion of a nonfood substance. PDA is a term for a heart murmur. RDSI stands for revised developmental screening inventory, a developmental screening test. Plumbism is another term for lead poisoning.
NP: Data collection; CN: Health promotion and maintenance; CNS: Prevention and early detection of disease; CL: Knowledge

89. 1. Eating clay is related to zinc deficiency, and eating chalk to calcium deficiency. Vitamin deficiencies aren't related to pica.
NP: Data collection; CN: Health promotion and maintenance; CNS: Prevention and early detection of disease; CL: Knowledge

90. 3. Roundworms are caused by nematodes. Hookworm is caused by *Nectator americanus*. Tapeworms are caused by cestodes, and flukes by trematodes; both live throughout North America.
NP: Data collection; CN: Health promotion and maintenance; CNS: Prevention and early detection of disease; CL: Knowledge

NP: Nursing process   CN: Client needs category   CNS: Client needs subcategory   CL: Cognitive level

91. Which of the following inflammatory conditions requiring surgery during childhood would be the most common?
1. Acute appendicitis
2. Necrotizing enterocolitis
3. Tonsillitis
4. Ulcerative colitis

92. Which of the following symptoms is the most common for acute appendicitis?
1. Bradycardia
2. Fever
3. Pain descending to the lower left quadrant
4. Pain radiating down the legs

93. Which of the following nursing interventions would be important to do preoperatively in a child with appendicitis?
1. Give clear fluids.
2. Apply heat to the abdomen.
3. Maintain complete bed rest.
4. Administer an enema, if ordered.

94. Postoperative care of a child with a ruptured appendix would include which of the following treatments or interventions?
1. Liquid diet
2. Oral antibiotics for 7 to 10 days
3. Positioning the child on the left side
4. Parenteral antibiotics for 7 to 10 days

95. After surgical repair of a ruptured appendix, which of the following positions would be the most appropriate?
1. High-Fowler's position
2. Left side-lying
3. Semi-Fowler's position
4. Supine

Look out for these words.

Keep up the good work!

Believe me, position counts.

91. 1. Acute appendicitis is the most common inflammatory condition requiring surgery in childhood. Ulcerative colitis occurs in prepubescent and adolescent children and generally doesn't require surgery. Necrotizing enterocolitis is seen in ill neonates. Tonsillitis can occur at any age but doesn't always require surgery.
NP: Data collection; CN: Physiological integrity; CNS: Physiological adaptation; CL: Knowledge

92. 2. Fever, abdominal pain, and tenderness are the first signs of appendicitis. Tachycardia, not bradycardia, is seen. Pain can be generalized or periumbilical. It usually descends to the lower right quadrant, not the left.
NP: Data collection; CN: Physiological integrity; CNS: Physiological adaptation; CL: Application

93. 3. Bed rest will prevent aggravating the condition. Clients with appendicitis aren't allowed anything by mouth. Cold applications are placed on the abdomen as heat would increase blood flow to the area and possibly spread infectious disease. Enemas may aggravate the condition.
NP: Implementation; CN: Physiological integrity; CNS: Physiological adaptation; CL: Application

94. 4. Parenteral antibiotics are used for 7 to 10 days postoperatively to help prevent the spread of infection. The child is kept on I.V. fluids and isn't allowed anything by mouth. Oral antibiotics may continue after the parenteral antibiotics are discontinued. The child is positioned on the right side after surgery.
NP: Implementation; CN: Physiological integrity; CNS: Physiological adaptation; CL: Application

95. 3. Using the semi-Fowler's or right side-lying positions will facilitate drainage from the peritoneal cavity and prevent the formation of a subdiaphragmatic abscess. High-Fowler's, left side-lying, and supine positions won't facilitate drainage from the peritoneal cavity.
NP: Implementation; CN: Physiological integrity; CNS: Physiological adaptation; CL: Application

96. In which of the following inflammatory diseases do the majority of children show signs of painless bright or dark red rectal bleeding?
1. Crohn's disease
2. Meckel's diverticulum
3. Ruptured appendix
4. Ulcerative colitis

97. The nurse is assessing the laboratory results for a 9-year-old child hospitalized with severe vomiting and diarrhea. What's the normal serum potassium range in a child?
1. 4.5 to 7.2 mmol/L
2. 3.7 to 5.2 mmol/L
3. 3.5 to 5.8 mmol/L
4. 3.5 to 5.5 mmol/L

98. A neonate has been diagnosed with a unilateral complete cleft lip and cleft palate. The nurse formulating the care plan for this neonate will have which of the following nursing diagnoses as a priority?
1. Risk for infection
2. Impaired skin integrity
3. Risk for aspiration
4. Delayed growth and development

99. A neonate is suspected of having a tracheoesophageal fistula (type III/C). Which of the following symptoms would be seen on the initial assessment?
1. Excessive drooling
2. Excessive vomiting
3. Mottling
4. Polyhydramnios

Only 11 more questions!

Here's that word *initial* again. It's still a hint!

96. 2. Acute and sometimes massive hemorrhage may occur in Meckel's diverticulum. There's no rectal bleeding with a ruptured appendix or Crohn's disease. Children with ulcerative colitis present with recurring diarrhea.
NP: Data collection; CN: Physiological integrity; CNS: Physiological adaptation; CL: Knowledge

97. 3. The normal potassium level in children ranges from 3.5 to 5.8 mmol/L. Levels of 4.5 to 7.2 mmol/L are observed in premature infants. Potassium levels of 3.7 to 5.2 mmol/L are observed in full-term infants. Potassium levels of 3.5 to 5.5 mmol/L are usually seen in adults.
NP: Data collection; CN: Physiological integrity; CNS: Reduction of risk potential; CL: Knowledge

98. 3. Although all these diagnoses are important for the neonate with a cleft lip and cleft palate, the most important diagnosis relates to the airway. Neonates with a cleft lip and a cleft palate may have an excessive amount of saliva and usually have a difficult time with feedings. Special feeding techniques, such as using a flanged nipple, may be necessary to prevent aspiration.
NP: Planning; CN: Physiological integrity; CNS: Reduction of risk potential; CL: Application

99. 1. In type III/C tracheoesophageal fistula, the proximal end of the esophagus ends in a blind pouch and a fistula connects the distal end of the esophagus to the trachea. Saliva will pool in this pouch and cause the child to drool. Because the distal end of the esophagus is connected to the trachea, the neonate can't vomit, but he can aspirate, and stomach acid may go into the lungs through this fistula, causing pneumonitis. Mottling is a netlike reddish blue discoloration of the skin usually due to vascular contraction in response to hypothermia. The mother of a neonate with tracheoesophageal fistula may have had polyhydramnios.
NP: Data collection; CN: Physiological integrity; CNS: Physiological adaptation; CL: Application

100. When assessing a client suspected of having pyloric stenosis, which of the following findings would be seen?

1. An "olive" mass in the right upper quadrant
2. An "olive" mass in the left upper quadrant
3. A "sausage" mass in the right upper quadrant
4. A "sausage" mass in the left upper quadrant

101. The nurse caring for an infant with pyloric stenosis would expect which of the following laboratory values?

1. pH, 7.30; chloride, 120 mEq/L
2. pH, 7.38; chloride, 110 mEq/L
3. pH, 7.43; chloride, 100 mEq/L
4. pH, 7.49; chloride, 90 mEq/L

Lab results will tell you waaaay more than these cards.

102. A 1-day-old neonate hasn't passed meconium. The nurse would assess this neonate for which of the following conditions?

1. Biliary atresia
2. Gastroesophageal reflux
3. Gastroschisis
4. Hirschsprung's disease

The nursing diagnosis is so important.

103. A nurse caring for a client with necrotizing enterocolitis would expect which of the following findings?

1. Abdominal distention and gastric retention
2. Gastric retention and guaiac-negative stools
3. Metabolic alkalosis and abdominal distention
4. Guaiac-negative stools and metabolic alkalosis

100. 1. Pyloric stenosis involves hypertrophy of the circular muscle fibers of the pylorus. This hypertrophy is palpable in the right upper quadrant of the abdomen. A "sausage" mass is palpable in the right upper quadrant in children with intussusception. A "sausage" mass in the left upper quadrant wouldn't indicate pyloric stenosis.

NP: Data collection; CN: Physiological integrity; CNS: Physiological adaptation; CL: Comprehension

101. 4. Infants with pyloric stenosis vomit hydrochloric acid. This causes them to become alkalotic and hypochloremic. Normal serum pH is 7.35 to 7.45; levels above 7.45 represent alkalosis. The normal serum chloride level is 99 to 111 mEq/L; levels below 99 mEq/L represent hypochloremia.

NP: Data collection; CN: Physiological integrity; CNS: Physiological adaptation; CL: Application

102. 4. Hirschsprung's disease is an absence of the autonomic parasympathetic ganglion cells of a section of colon. The lack of neural intervention prevents peristalsis, so the neonate can't pass meconium. Biliary atresia is a malformation of the biliary drainage system, causing the baby to eventually develop liver failure; the neonate appears normal at birth. Gastroesophageal reflux is caused by stomach contents entering the esophagus secondary to a weak lower esophageal sphincter. Gastroschisis is an abdominal wall defect through which the intestines herniate. This defect is obvious at birth.

NP: Data collection; CN: Physiological integrity; CNS: Physiological adaptation; CL: Comprehension

103. 1. Necrotizing enterocolitis is an ischemic disorder of the gut. The cause is unknown, but it's more common in premature neonates who had a hypoxic episode. The neonate's intestines become dilated and necrotic and the abdomen becomes very distended. Paralytic ileus develops, causing the neonate to have gastric retention. These retained gastric contents, along with any passed stool, will be guaiac-positive. The neonate also develops metabolic acidosis.

NP: Data collection; CN: Physiological integrity; CNS: Physiological adaptation; CL: Comprehension

104. An infant has been admitted to the hospital with gastroenteritis. The nursing care plan for this infant will consider which of the following nursing diagnoses first?
1. Acute pain
2. Diarrhea
3. Deficient fluid volume
4. Imbalanced nutrition: Less than body requirements

105. Which of the following symptoms are consistent with the diagnosis of pyloric stenosis?
1. Projectile vomiting, metabolic alkalosis, hunger
2. Projectile vomiting, metabolic acidosis, dehydration, hunger
3. Frequent vomiting of formula, respiratory alkalosis, weight gain
4. Frequent vomiting of bilious emesis, metabolic acidosis, poor appetite

106. A 3-week-old infant is diagnosed with pyloric stenosis and is admitted to the hospital during a vomiting episode. Which action by the nurse is most appropriate?
1. Place the infant on his back to sleep.
2. Weigh the infant every 12 hours.
3. Position the infant on his right side.
4. Take the infant's vital signs every 8 hours.

107. Which of the following facts should be emphasized in the teaching plan for the parents of a child with celiac disease?
1. The gluten-free diet alterations must be continued for a lifetime.
2. The diet needs to be free of lactose because the child is intolerant.
3. Diet alterations are necessary when the child reports cramping and bloating.
4. The diet needs to be low in fat because of the malabsorption problem in the intestines.

This one's tricky. Take your time.

You're done with all those questions?! Knew you could do it!

104. 3. Young children with gastroenteritis are at high risk for developing deficient fluid volume. Their intestinal mucosa allows for more fluid and electrolytes to be lost when they have gastroenteritis. The main goal of the health care team should be to rehydrate the infant. The other nursing diagnoses are important, but deficient fluid volume is more life-threatening.
NP: Planning; CN: Physiological integrity; CNS: Physiological adaptation; CL: Application

105. 1. Classic signs of pyloric stenosis include projectile vomiting of nonbilious emesis, metabolic alkalosis (because acid is lost in vomit), hunger, dehydration, and failure to thrive. The problem is a too-tight pylorus, which causes projectile nonbilious vomiting. The infants are hungry because they continue to lose their feedings; it isn't an absorption or digestion problem. The problem results in metabolic imbalances, not problems with the respiratory system, and weight loss instead of weight gain.
NP: Data collection; CN: Physiological integrity; CNS: Physiological adaptation; CL: Comprehension

106. 3. The nurse should position the infant on his right side to prevent aspiration. The infant should be weighed daily, not every 12 hours. Vital signs should be monitored every 4 hours, not every 8 hours.
NP: Implementation; CN: Physiological integrity; CNS: Reduction of risk potential; CL: Application

107. 1. Celiac disease is the inability to digest gluten. The treatment is a gluten-free diet for life. It's important that the diet is continued to avoid symptoms and the associated risk for colon cancer. The disease isn't caused by lactose intolerance or a problem digesting fats.
NP: Planning; CN: Health promotion and maintenance; CNS: Prevention and early detection of disease; CL: Knowledge

Caring for a child with an endocrine system disorder can be overwhelming. To get started on the right track, check out the Internet site of the Juvenile Diabetes Research Foundation at **www.jdrf.org/**. Go for it!

Chapter 33
Endocrine disorders

1. When explaining the causes of hypothyroidism to the parents of a newly diagnosed infant, the nurse should recognize that further education is needed when the parents ask which of the following questions?
   1. "So, hypothyroidism can be only temporary, right?"
   2. "Are you saying that hypothyroidism is caused by a problem in the way the thyroid gland develops?"
   3. "Do you mean that hypothyroidism may be caused by a problem in the way the body makes thyroxine?"
   4. "So, hypothyroidism can be treated by exposing our baby to a special light, right?"

2. The nurse understands that transient hypothyroidism can result from which of the following factors?
   1. Intrauterine transfer of insulin
   2. Placental transfer of antibodies
   3. Placental transfer of teratogens
   4. Intrauterine transfer of expectorants

3. When a nurse is teaching parents of a neonate newly diagnosed with hypothyroidism, which of the following statements should be included?
   1. A large goiter in a neonate doesn't present a problem.
   2. Preterm neonates usually aren't affected by hypothyroidism.
   3. Usually, the neonate exhibits obvious signs of hypothyroidism.
   4. The severity of the disorder depends on the amount of thyroid tissue present.

1. 4. Congenital hypothyroidism can be permanent or transient and may result from a defective thyroid gland or an enzymatic defect in thyroxine synthesis. Only the last question, which refers to phototherapy for physiologic jaundice, indicates that the parents need more information.

NP: Implementation; CN: Psychosocial integrity; CNS: Coping and adaptation; CL: Knowledge

Don't let your studying be transient. Concentrate and know you can succeed.

2. 4. Intrauterine transfer of antithyroid drugs and expectorants given for asthma are factors associated with transient hypothyroidism. The other choices don't affect hypothyroidism.

NP: Data collection; CN: Health promotion and maintenance; CNS: Prevention and early detection of disease; CL: Knowledge

3. 4. The severity of the disorder depends on the amount of thyroid tissue present. The more thyroid tissue present, the less severe the disorder. Usually, the neonate doesn't exhibit obvious signs of the disorder due to maternal circulation. A large goiter in a neonate could possibly occlude the airway and lead to obstruction. Preterm neonates are usually affected by hypothyroidism due to hypothalamic and pituitary immaturity.

NP: Implementation; CN: Health promotion and maintenance; CNS: Prevention and early detection of disease; CL: Application

NP: Nursing process   CN: Client needs category   CNS: Client needs subcategory   CL: Cognitive level

4. Which of the following conditions is a subtle sign of hypothyroidism?
  1. Diarrhea
  2. Lethargy
  3. Severe jaundice
  4. Tachycardia

5. When observing a neonate with congenital hypothyroidism, the nurse understands that which of the following complications is the most serious consequence of this condition?
  1. Anemia
  2. Cyanosis
  3. Retarded bone age
  4. Delayed central nervous system (CNS) development

6. When counseling parents of a neonate with congenital hypothyroidism, the nurse understands that the severity of the intellectual deficit is related to which of the following parameters?
  1. Duration of condition before treatment
  2. Degree of hypothermia
  3. Cranial malformations
  4. T_4 level at diagnosis

7. Which of the following statements should be included in an explanation of the diagnostic evaluation of neonates for congenital hypothyroidism?
  1. Tests are mandatory in all states.
  2. An arterial blood test is preferred.
  3. Tests shouldn't be performed until after discharge.
  4. Blood tests should be done after the first month of life.

In question 5, the phrase *most serious* means you must prioritize.

4. 2. Subtle signs of this disorder that may be seen shortly after birth include lethargy, poor feeding, prolonged jaundice, respiratory difficulty, cyanosis, constipation, and bradycardia. Diarrhea in the neonate isn't normal and isn't associated with this disorder. Severe jaundice needs immediate attention by the primary health care provider and isn't a subtle sign. Tachycardia typically occurs in hyperthyroidism, not hypothyroidism.

NP: Implementation; CN: Health promotion and maintenance; CNS: Prevention and early detection of disease; CL: Knowledge

5. 4. The most serious consequence of congenital hypothyroidism is delayed development of the CNS, which leads to severe mental retardation. The other choices occur, but aren't the most serious consequences.

NP: Data collection; CN: Safe, effective care environment; CNS: Coordinated care; CL: Knowledge

6. 1. The severity of the intellectual deficit is related to the degree of hypothyroidism and the duration of the condition before treatment. Cranial malformations don't affect the severity of the intellectual deficit nor does the degree of hypothermia as it relates to hypothyroidism. It isn't the specific T_4 level at diagnosis that affects the intellect but how long the client has been hospitalized.

NP: Implementation; CN: Health promotion and maintenance; CNS: Prevention and early detection of disease; CL: Application

7. 1. Heel stick blood tests are mandatory in all states and are usually done on neonates between ages 2 and 6 days. Typically, specimens are taken before the neonate is discharged from the hospital; the test is included with other tests that screen the neonate for errors of metabolism.

NP: Implementation; CN: Health promotion and maintenance; CNS: Prevention and early detection of disease; CL: Application

NP: Nursing process   CN: Client needs category   CNS: Client needs subcategory   CL: Cognitive level

8. Which of the following results would indicate to the nurse the possibility that a neonate has congenital hypothyroidism?
1. High T_4 level and low TSH level
2. Low T_4 level and high TSH level
3. Normal TSH level and high T_4 level
4. Normal T_4 level and low TSH level

9. The nurse is teaching parents about therapeutic management of their neonate diagnosed with congenital hypothyroidism. Which of the following responses by a parent would indicate the need for further teaching?
1. "My baby will need regular measurements of his thyroxine levels."
2. "Treatment involves lifelong thyroid hormone replacement therapy."
3. "Treatment should begin as soon as possible after diagnosis is made."
4. "As my baby grows, his thyroid gland will mature and he won't need medications."

In question 9, the phrase further teaching indicates that you're looking for an incorrect statement.

10. Which of the following comments made by the mother of a neonate at her 2-week office visit should alert the nurse to suspect congenital hypothyroidism?
1. "My baby is unusually quiet and good."
2. "My baby seems to be a yellowish color."
3. "After feedings, my baby pulls her legs up and cries."
4. "My baby seems to study my face during feeding time."

You've finished 10 questions already! Congratulations!

11. Which of the following statements should be included when educating a mother about giving levothyroxine (Synthroid) to her neonate after a diagnosis of hypothyroidism is made?
1. The drug has a bitter taste.
2. The pill shouldn't be crushed.
3. Never put the medication in formula or juice.
4. If a dose is missed, double the dose the next day.

8. 2. Screening results that show a low T_4 level and a high TSH level indicate congenital hypothyroidism and the need for further tests to determine the cause of the disease.
NP: Implementation; CN: Physiological integrity; CNS: Reduction of risk potential; CL: Application

9. 4. Treatment involves lifelong thyroid hormone replacement therapy that begins as soon as possible after diagnosis to abolish all signs of hypothyroidism and to reestablish normal physical and mental development. The drug of choice is synthetic levothyroxine (Synthroid or Levothroid). Regular measurements of thyroxine levels are important in ensuring optimum treatment.
NP: Evaluation; CN: Health promotion and maintenance; CNS: Prevention and early detection of disease; CL: Application

10. 1. Parental remarks about an unusually "quiet and good" neonate together with any of the early physical manifestations should lead to a suspicion of hypothyroidism, which requires a referral for specific tests. The neonate likes looking at the human face and should show this interest at age 2 weeks. If the neonate is pulling her legs up and crying after feedings, she might be showing signs of colic. If a neonate begins to look yellow in color, hyperbilirubinemia may be the cause.
NP: Data collection; CN: Health promotion and maintenance; CNS: Prevention and early detection of disease; CL: Application

11. 4. If a dose is missed, twice the dose should be given the next day. The importance of compliance with the drug regimen must be emphasized in order for the neonate to achieve normal growth and development. Because the drug is flavorless, it can be crushed and added to formula, water, or food.
NP: Implementation; CN: Physiological integrity; CNS: Pharmacological therapies; CL: Application

12. When teaching the parents about signs that indicate levothyroxine (Synthroid) overdose, which of the following comments by a parent indicates the need for <u>further teaching</u>?
1. "Irritability is a sign of overdose."
2. "If my baby's heartbeat is fast, I should count it."
3. "If my baby loses weight, I should be concerned."
4. "I shouldn't worry if my baby doesn't sleep very much."

Watch out! Question 12 is another further teaching question.

12. 4. Parents need to be aware of signs indicating overdose, such as rapid pulse, dyspnea, irritability, insomnia, fever, sweating, and weight loss. The parents would be given acceptable parameters for the heart rate and weight loss or gain. If the baby is experiencing a heart rate or weight loss outside of the acceptable parameters, the primary health care provider should be called.
NP: Evaluation; CN: Physiological integrity; CNS: Pharmacological therapies; CL: Analysis

13. A nurse should recognize that exophthalmos (protruding eyeballs) may occur in children with which of the following conditions?
1. Hypothyroidism
2. Hyperthyroidism
3. Hypoparathyroidism
4. Hyperparathyroidism

13. 2. Exophthalmos occurs when there's an overproduction of thyroid hormone. This sign should alert the primary health care provider to follow up with further testing.
NP: Data collection; CN: Health promotion and maintenance; CNS: Prevention and early detection of disease; CL: Application

14. A nurse should understand that which of the following symptoms is a common clinical manifestation of juvenile hypothyroidism?
1. Accelerated growth
2. Diarrhea
3. Dry skin
4. Insomnia

14. 3. Children with hypothyroidism have dry skin. The other choices aren't evident in children with juvenile hypothyroidism.
NP: Data collection; CN: Health promotion and maintenance; CNS: Prevention and early detection of disease; CL: Knowledge

15. A nurse should understand that manifestations of thyroid hormone deficiency in infancy include which of the following symptoms?
1. Tachycardia, profuse perspiration, and diarrhea
2. Lethargy, feeding difficulties, and constipation
3. Hypertonia, small fontanels, and moist skin
4. Dermatitis, dry skin, and round face

Question 16 is asking for the *most appropriate* behavior that should be encouraged. In other words, *prioritize!*

15. 2. Hypothyroidism results from inadequate thyroid production to meet an infant's needs. Clinical signs include feeding difficulties, prolonged physiologic jaundice, lethargy, and constipation.
NP: Data collection; CN: Health promotion and maintenance; CNS: Prevention and early detection of disease; CL: Comprehension

16. When counseling parents of a neonate with congenital hypothyroidism, the nurse should encourage which of the following behaviors?
1. Seeking professional genetic counseling
2. Retracing the family tree for others born with this condition
3. Talking to relatives who have gone through a similar experience
4. Waiting until the neonate is age 1 year before obtaining counseling

16. 1. Seeking professional genetic counseling is the best option for parents who have a neonate with a genetic disorder. Retracing the family tree and talking to relatives won't help the parents to become better educated about the disorder. Education about the disorder should occur as soon as the parents are ready so they'll understand the genetic implications for future children.
NP: Implementation; CN: Safe, effective care environment; CNS: Coordinated care; CL: Application

NP: Nursing process   CN: Client needs category   CNS: Client needs subcategory   CL: Cognitive level

17. While teaching an adolescent about giving insulin injections, the adolescent questions the nurse about the reuse of disposable needles and syringes. Which of the following responses from the nurse is most appropriate?
　　1. This is an unsafe practice.
　　2. This is acceptable for up to 7 days.
　　3. This is acceptable for only 48 hours.
　　4. This is acceptable only if the family has very limited resources.

17. 2. It has become acceptable practice for clients to reuse their own disposable needles and syringes for up to 7 days. Bacteria counts are unaffected, and cost savings are considerable. If this method is approved, it's imperative to stress the importance of vigorous handwashing before handling equipment as well as capping the syringe immediately after use and storing it in the refrigerator to decrease the growth of organisms.

NP: Implementation; CN: Safe, effective care environment; CNS: Safety and infection control; CL: Application

18. When caring for a child with diabetes, the nurse should expect which changes when the child is more physically active?
　　1. Increased food intake
　　2. Decreased food intake
　　3. Decreased risk of insulin shock
　　4. Increased risk of hyperglycemia

18. 1. If a child is more active at one time of the day than another, food or insulin should be altered to meet the child's activity pattern. Food intake should be increased when a child with diabetes is more physically active. There would be an increased risk of insulin shock if the child didn't take in more food and the child would become hypoglycemic, not hyperglycemic.

NP: Planning; CN: Physiological integrity; CNS: Reduction of risk potential; CL: Knowledge

19. When the nurse is helping the adolescent deal with diabetes, which of the following characteristics of adolescence should be considered?
　　1. Wanting to be an individual
　　2. Needing to be like their peers
　　3. Being preoccupied with future plans
　　4. Teaching peers that this is a serious disease

19. 2. Adolescents appear to have the most difficulty adjusting to diabetes. Adolescence is a time when being "perfect" and being like one's peers are emphasized and, to adolescents, having diabetes means they're different.

NP: Implementation; CN: Health promotion and maintenance; CNS: Growth and development through the life span; CL: Knowledge

20 questions completed? That's cause for celebration!

20. An adolescent with diabetes tells the community nurse that he has recently started drinking alcohol on the weekends. Which of the following actions would be most appropriate for the nurse to take?
　　1. Recommend referral to counseling.
　　2. Make the adolescent promise to stop drinking.
　　3. Discuss with the adolescent why he has started drinking.
　　4. Teach the adolescent about the effects of alcohol on diabetes.

20. 4. Confusion about the effects of alcohol on blood glucose is common. Teenagers may believe that alcohol will increase blood glucose levels, when in fact the opposite occurs. Ingestion of alcohol inhibits the release of glycogen from the liver, resulting in hypoglycemia. Teens who drink alcohol may become hypoglycemic, but are then treated as if they were intoxicated. Behaviors may be similar, such as shakiness, combativeness, slurred speech, and loss of consciousness.

NP: Implementation; CN: Health promotion and maintenance; CNS: Growth and development through the life span; CL: Application

21. A child has experienced symptoms of hypoglycemia and has eaten sugar cubes. The nurse should follow this rapid-releasing sugar with which of the following foods?
1. Fruit juices
2. Six glasses of water
3. Foods that are high in protein
4. Complex carbohydrates and protein

If you know the definitions of hypo and hyper, it can help you with a bunch of questions.

22. Which of the following symptoms is indicative of hypoglycemia?
1. Irritability
2. Drowsiness
3. Abdominal pain
4. Nausea and vomiting

23. Which of the following observations is accurate for a child with diabetes who develops ketoacidosis?
1. Normal outcome
2. Life-threatening situation
3. Situation that can easily be treated at home
4. Situation best treated in the pediatrician's office

24. Which of the following guidelines is appropriate when teaching an 11-year-old child recently diagnosed with diabetes about insulin injections?
1. The parents don't need to be involved in learning this procedure.
2. Self-injection techniques aren't usually taught until the child reaches age 16.
3. At age 11, the child should be old enough to give most of his own injections.
4. Self-injection techniques should only be taught when the child can reach all injection sites.

21. 4. When a child exhibits signs of hypoglycemia, most cases can be treated with a simple concentrated sugar, such as honey or sugar cubes, that can be held in the mouth for a short time. This will elevate the blood glucose level and alleviate the symptoms. The simpler the carbohydrate, the more rapidly it will be absorbed. A complex carbohydrate and protein, such as a slice of bread or a cracker spread with peanut butter should follow the rapid-releasing sugar or the client may become hypoglycemic again.
NP: Evaluation; CN: Health promotion and maintenance; CNS: Prevention and early detection of disease; CL: Application

22. 1. Signs of hypoglycemia include irritability, shaky feeling, hunger, headache, and dizziness. Drowsiness, abdominal pain, nausea, and vomiting are signs of *hyper*glycemia.
NP: Data collection; CN: Physiological integrity; CNS: Physiological adaptation; CL: Knowledge

23. 2. Diabetic ketoacidosis, the most complete state of insulin deficiency, is a life-threatening situation. The child should be admitted to an intensive care facility for management, which consists of rapid assessment, adequate insulin to reduce the elevated blood glucose level, fluids to overcome dehydration, and electrolyte replacement (especially potassium).
NP: Evaluation; CN: Safe, effective care environment; CNS: Coordinated care; CL: Knowledge

24. 3. The parents must supervise and manage the child's therapeutic program, but the child should assume responsibility for self-management as soon as he can. Children can learn to collect their own blood for glucose testing at a relatively young age (4 to 5 years), and most can check their blood glucose level and administer insulin at about age 9. Some children can do it earlier.
NP: Implementation; CN: Health promotion and maintenance; CNS: Growth and development through the life span; CL: Application

NP: Nursing process   CN: Client needs category   CNS: Client needs subcategory   CL: Cognitive level

25. A nurse should understand that hyperglycemia associated with diabetic ketoacidosis is defined as a blood glucose measurement equal to or greater than which of the following values?
  1. 150 mg/dl
  2. 300 mg/dl
  3. 450 mg/dl
  4. 600 mg/dl

26. A nurse should recognize which of the following symptoms as a cardinal sign of diabetes mellitus?
  1. Nausea
  2. Seizure
  3. Hyperactivity
  4. Frequent urination

Knowing blood glucose values is a must for taking the NCLEX.

27. The parent of a child with diabetes asks a nurse why blood glucose monitoring is needed. The nurse should base her reply on which of the following premises?
  1. This is an easier method of testing.
  2. This is a less expensive method of testing.
  3. This allows children the ability to better manage their diabetes.
  4. This gives children a greater sense of control over their diabetes.

28. To increase the adolescent's compliance with treatment for diabetes mellitus, the nurse should attempt which of the following strategies?
  1. Provide for a special diet in the high school cafeteria.
  2. Clarify the adolescent's values to promote involvement in care.
  3. Identify energy requirements for participation in sports activities.
  4. Educate the adolescent about long-term consequences of poor metabolic control.

Teaching clients how to help manage their own conditions is a common subject on the NCLEX.

25. 2. Diabetic ketoacidosis is determined by the presence of hyperglycemia (blood glucose measurement of 300 mg/dl or higher), accompanied by acetone breath, dehydration, weak and rapid pulse, and decreased level of consciousness.
NP: Data collection; CN: Physiological integrity; CNS: Physiological adaptation; CL: Knowledge

26. 4. Polyphagia, polyuria, polydipsia, and weight loss are cardinal signs of diabetes mellitus. Other signs include irritability, shortened attention span, lowered frustration tolerance, fatigue, dry skin, blurred vision, sores that are slow to heal, and flushed skin.
NP: Data collection; CN: Health promotion and maintenance; CNS: Prevention and early detection of disease; CL: Knowledge

27. 3. Blood glucose monitoring improves diabetes management and is used successfully by children from the onset of their diabetes. By testing their own blood, children can change their insulin regimen to maintain their glucose level in the normoglycemic range of 80 to 120 mg/dl. This allows them to better manage their diabetes.
NP: Implementation; CN: Health promotion and maintenance; CNS: Prevention and early detection of disease; CL: Application

28. 2. Adolescent compliance with diabetes management may be hampered by dependence versus independence conflicts and ego development. Attempts to have the adolescent clarify personal values fosters compliance. Providing for a special meal in the school cafeteria isn't feasible and doesn't guarantee that the adolescent would eat it. The question doesn't provide sufficient information about the adolescent's sports activity. An adolescent is only concerned with the present and not the future.
NP: Planning; CN: Health promotion and maintenance; CNS: Growth and development through the life span; CL: Application

29. A child with type 1 diabetes tells the nurse she feels shaky. The nurse assesses the child's skin to be pale and sweaty. Which of the following actions should the nurse initiate immediately?

1. Give supplemental insulin.
2. Have the child eat a glucose tablet.
3. Administer glucagon subcutaneously (S.C.)
4. Offer the child a complex carbohydrate snack.

The word *immediately* signals a need for you to prioritize.

29. 2. These are symptoms of hypoglycemia. Rapid treatment involves giving the alert child a glucose tablet (4 mg of dextrose) or, if unavailable, a glass of glucose-containing liquid. It would be followed by a complex carbohydrate snack and protein. Giving supplemental insulin would be contraindicated because that would lower the blood glucose even more. Glucagon would be given only if there was a risk of aspiration with oral glucose, such as if the child was semiconscious.
NP: Implementation; CN: Safe, effective care environment; CNS: Coordinated care; CL: Application

30. The parents of a child diagnosed with diabetes ask the nurse about maintaining metabolic control during a minor illness with loss of appetite. Which of the following nursing responses is appropriate?

1. "Decrease the child's insulin by one-half of the usual dose during the course of the illness."
2. "Call your primary health care provider to arrange hospitalization."
3. "Give increased amounts of clear liquids to prevent dehydration."
4. "Substitute calorie-containing liquids for uneaten solid food."

Another 10 down! You're cruisin' now!

30. 4. Calorie-containing liquids can help to maintain more normal blood glucose levels as well as decrease the danger of dehydration. The child with diabetes should always take *at least* the usual dose of insulin during an illness based on more frequent blood glucose checks. During an illness where there's vomiting or loss of appetite, NPH insulin is cut in half or stopped altogether and regular insulin is given according to home glucose monitoring results.
NP: Implementation; CN: Safe, effective care environment; CNS: Coordinated care; CL: Application

31. Which of the following criteria would the nurse use to measure good metabolic control in a child with diabetes mellitus?

1. Fewer than eight episodes of severe hyperglycemia in a month
2. Infrequent occurrences of mild hypoglycemia reactions
3. Hemoglobin A values less than 12%
4. Growth below the 15th percentile

31. 2. Criteria for good metabolic control generally includes few episodes of hypoglycemia or hyperglycemia, hemoglobin A values less than 8%, and normal growth and development.
NP: Data collection; CN: Health promotion and maintenance; CNS: Prevention and early detection of disease; CL: Application

32. Which of the following congenital anomalies is most commonly associated with poorly controlled maternal diabetes?

1. Cataracts
2. Low-set ears
3. Cardiac malformations
4. Cleft lip and palate deformities

32. 3. Cardiac and central nervous system anomalies, along with neural tube defects, skeletal abnormalities, and GI anomalies, are most likely to occur in uncontrolled maternal diabetes.
NP: Data collection; CN: Health promotion and maintenance; CNS: Prevention and early detection of disease; CL: Knowledge

33. Which of the following conditions could possibly cause hypoglycemia?
1. Too little insulin
2. Mild illness with fever
3. Excessive exercise without a carbohydrate snack
4. Eating ice cream and cake to celebrate a birthday

34. A nurse should recognize which of the following factors as one of the best indicators of a client's control of his diabetes during the preceding 2 to 3 months?
1. Fasting glucose level
2. Oral glucose tolerance test
3. Glycosylated hemoglobin level
4. Client's verbal report of his symptoms

35. A client has received diet instruction as part of his treatment plan for type 1 diabetes. Which of the following statements by the client indicates to the nurse that he needs additional instructions?
1. "I will need a bedtime snack because I take an evening dose of NPH insulin."
2. "I can eat whatever I want as long as I cover the calories with sufficient insulin."
3. "I can have an occasional low-calorie drink as long as I include it in my meal plan."
4. "I should eat meals as scheduled, even if I'm not hungry, to prevent hypoglycemia."

36. Which of the following symptoms is a sign of hyperglycemia?
1. Rapid heart rate
2. Headache
3. Hunger
4. Thirst

Comparing signs of hypoglycemia and hyperglycemia is a common NCLEX subject.

Additional instructions is another way of saying further teaching. Both types of questions ask you to find an incorrect statement.

What did I tell you? Another hyperglycemia versus hypoglycemia question.

33. 3. Excessive exercise without a carbohydrate snack could cause hypoglycemia. The other options describe situations that cause *hyper*glycemia.
NP: Evaluation; CN: Health promotion and maintenance; CNS: Prevention and early detection of disease; CL: Application

34. 3. A glycosylated hemoglobin level provides an overview of a person's blood glucose level over the previous 2 to 3 months. Glycosylated hemoglobin values are reported as a percentage of the total hemoglobin within an erythrocyte. The time frame is based on the fact that the usual life span of an erythrocyte is 2 to 3 months; a random blood sample, therefore, will theoretically give samples of erythrocytes for this same period. The other options won't indicate a true picture of the person's blood glucose level over the previous 2 to 3 months.
NP: Evaluation; CN: Health promotion and maintenance; CNS: Prevention and early detection of disease; CL: Knowledge

35. 2. The goal of diet therapy in diabetes mellitus is to attain and maintain ideal body weight. Each client will be prescribed a specific caloric intake and insulin regimen to help accomplish this goal.
NP: Evaluation; CN: Physiological integrity; CNS: Basic care and comfort; CL: Analysis

36. 4. Thirst (polydipsia) is one of the symptoms of hyperglycemia. Rapid heart rate, headache, and hunger are signs and symptoms of *hypo*glycemia.
NP: Evaluation; CN: Physiological integrity; CNS: Physiological adaptation; CL: Knowledge

37. A client is learning to mix regular insulin and NPH insulin in the same syringe. Which of the following actions, if performed by the client, would indicate the need for further teaching?
 1. Withdraws the NPH insulin first
 2. Injects air into NPH insulin bottle first
 3. After drawing up first insulin, removes air bubbles
 4. Injects an amount of air equal to the desired dose of insulin

In question 37, you're looking for an incorrect action, rather than an incorrect statement.

37. 1. Regular insulin is *always* withdrawn first so it won't become contaminated with NPH insulin. The client is instructed to inject air into the NPH insulin bottle equal to the amount of insulin to be withdrawn because there will be regular insulin in the syringe and he won't be able to inject air when he needs to withdraw the NPH. It's necessary to remove the air bubbles to be assured of a correct dosage before drawing up the second insulin.

NP: Data collection; CN: Physiological integrity; CNS: Pharmacological therapies; CL: Application

38. A client is diagnosed with type 1 diabetes. The primary health care provider prescribes an insulin regimen of regular insulin and NPH insulin administered S.C. each morning. How soon after administration will the onset of regular insulin begin?
 1. Within 5 minutes
 2. ½ to 1 hour
 3. 1 to 1½ hours
 4. 4 to 8 hours

It's important to know how and when different types of insulin react.

38. 2. Regular insulin's onset is ½ to 1 hour, peak is 2 to 4 hours, and duration is 8 to 12 hours. Lispro insulin has an onset within 5 minutes. NPH insulin has an onset within 1 to 1½ hours and extended insulin zinc suspension (Ultralente) is the longest acting with an onset of 4 to 8 hours.

NP: Evaluation; CN: Physiological integrity; CNS: Pharmacological therapies; CL: Application

39. When collecting data on a neonate for signs of diabetes insipidus, a nurse should recognize which of the following symptoms as signs of this disorder?
 1. Hyponatremia
 2. Jaundice
 3. Polyuria and polydipsia
 4. Hypochloremia

39. 3. The cardinal signs of diabetes insipidus are polyuria and polydipsia. Hypernatremia occurs with diabetes insipidus not hyponatremia. Jaundice occurs because of abnormal bilirubin metabolism not diabetes insipidus. Hyperchloremia not hypochloremia occurs with diabetes insipidus.

NP: Data collection; CN: Physiological integrity; CNS: Physiological adaptation; CL: Application

40. Which of the following complications isn't an initial symptom of diabetes insipidus in the neonate?
 1. Dehydration
 2. Soaking wet diaper
 3. Diarrhea
 4. Fever

Looking good, champ! Hang in there!

40. 3. Constipation and hard pebblelike stools as well as signs of dehydration, fever, and soaking wet diapers are initial signs of diabetes insipidus in the neonate.

NP: Data collection; CN: Health promotion and maintenance; CNS: Prevention and early detection of disease; CL: Knowledge

NP: Nursing process   CN: Client needs category   CNS: Client needs subcategory   CL: Cognitive level

41. A nurse is helping parents understand when treatments of growth hormone replacement will end. Which of the following statements should be included?
1. The dosage of growth hormone will decrease as the child's age increases.
2. The dosage of growth hormone will increase as time of epiphyseal closure nears.
3. After giving growth hormone replacement for 1 year, the dose will be tapered down.
4. Growth hormone replacement can't be abruptly stopped. Decreasing growth hormone replacement must be spread out over several months.

41. 2. Dosage of growth hormone is increased as the time of epiphyseal closure nears to gain the best advantage of the growth hormone. The medication is then stopped. There's no tapering off of the dose.
NP: Evaluation; CN: Physiological integrity; CNS: Pharmacological therapies; CL: Application

Question 42 asks you to identify a knowledgeable response.

42. The nurse is explaining diabetes insipidus to an infant's parents.. When explaining the diagnostic test that's used, which of the following comments by the parents would indicate an understanding of the diagnostic test?
1. "Fluids will be offered every 2 hours."
2. "My infant's fluid intake will be restricted."
3. "I won't change anything about my infant's intake."
4. "Formula will be restricted, but glucose water is okay."

42. 2. The simplest test used to diagnose diabetes insipidus is restriction of oral fluids and observation of consequent changes in urine volume and concentration. A weight loss of 3% to 5% indicates severe dehydration and the test should be terminated at this point. This is done in the hospital and the infant is watched closely.
NP: Implementation; CN: Health promotion and maintenance; CNS: Prevention and early detection of disease; CL: Application

43. A nurse should anticipate which of the following physiologic responses in an infant being tested for diabetes insipidus?
1. Increase in urine output
2. Decrease in urine output
3. No effect on urine output
4. Increase in urine specific gravity

43. 3. In diabetes insipidus, fluid restriction for diagnostic testing has little or no effect on urine formation but causes weight loss from dehydration.
NP: Data collection; CN: Physiological integrity; CNS: Physiological adaptation; CL: Application

We're nearing the halfway point. Let's keep going!

44. If an infant has a positive test result for diabetes insipidus, the nurse should expect the primary health care provider to order a test dose of which of the following medications?
1. Human placental lactogen (HPL)
2. Glucose loading test
3. Corticotropin
4. Aqueous vasopressin (Pitressin)

44. 4. If the fluid restriction test is positive, the child should be given a test dose of injected aqueous vasopressin, which should alleviate the polyuria and polydipsia. Unresponsiveness to exogenous vasopressin usually indicates nephrogenic diabetes insipidus. The other choices are used to determine other types of endocrine disorders. HPL is a hormone secreted by the placenta and its level in the blood assesses placental function. Glucose loading test or growth hormone suppression test evaluates baseline levels of growth hormone. Administration of corticotropin measures pituitary gland function.
NP: Evaluation; CN: Physiological integrity; CNS: Pharmacological therapies; CL: Knowledge

45. In teaching the parents of an infant diagnosed with diabetes insipidus, the nurse should include which of the following treatments?
1. Antihypertensive medications
2. The need for blood products
3. Hormone replacement
4. Fluid restrictions

45. 3. The usual treatment for diabetes insipidus is hormone replacement with vasopressin or desmopressin acetate (DDAVP). Blood products shouldn't be needed. No problem with hypertension is associated with this condition, and fluids shouldn't be restricted.
NP: Implementation; CN: Health promotion and maintenance; CNS: Prevention and early detection of disease; CL: Knowledge

Although the word *deter* sounds like a negative, question 46 actually asks you to identify a true characteristic of nasal sprays.

46. When providing information about treatment of diabetes insipidus to parents, a nurse explains the use of nasal spray and injections. Which of the following indications might deter a parent from choosing nasal spray treatment?
1. Applications must be repeated every 8 to 12 hours.
2. Applications must be repeated every 2 to 4 hours.
3. Nasal sprays can't be used in infants.
4. Measurements are too difficult.

46. 1. Applications of nasal spray used to treat diabetes insipidus must be repeated every 8 to 12 hours; injections, although quite painful, last for 48 to 72 hours. The nasal spray must be timed for adequate night sleep. Nasal sprays have been used in infants with diabetes insipidus and are dispensed in premeasured intranasal inhalers, eliminating the need for measuring doses.
NP: Implementation; CN: Physiological integrity; CNS: Pharmacological therapies; CL: Analysis

---

NP: Nursing process   CN: Client needs category   CNS: Client needs subcategory   CL: Cognitive level

47. A nurse is teaching the parents of an infant with diabetes insipidus about an injectable drug used to treat the disorder. Which of the following statements made by a parent would indicate the need for further teaching?

1. "I must hold the medication under warm running water for 10 to 15 minutes before administering it."
2. "The medication must be shaken vigorously before being drawn up into the syringe."
3. "Small brown particles must be seen in the suspension."
4. "I will store this medication in the refrigerator."

Caution! Question 47 is looking for an inaccurate statement...

47. 4. The medication should be stored at room temperature. When giving injectable vasopressin, it must be thoroughly resuspended in the oil by being held under warm running water for 10 to 15 minutes and shaken vigorously before being drawn into the syringe. If this isn't done, the oil may be injected minus the drug. Small brown particles, which indicate drug dispersion, must be seen in the suspension.

NP: Implementation; CN: Physiological integrity; CNS: Pharmacological therapies; CL: Application

48. When teaching parents of an infant newly diagnosed with diabetes insipidus, which of the following statements by a parent indicates a good understanding of this condition?

1. "When my infant stabilizes, I won't have to worry about giving hormone medication."
2. "I don't have to measure the amount of fluid intake that I give my infant."
3. "I realize that treatment for diabetes insipidus is lifelong."
4. "My infant will outgrow this condition."

...While question 48 asks you to identify a true statement.

48. 3. Diabetes insipidus requires lifelong treatment. The amount of fluid intake is important and must be measured with the infant's output to monitor the medication regimen. The infant won't outgrow this condition.

NP: Implementation; CN: Safe, effective care environment; CNS: Coordinated care; CL: Application

49. The nurse understands that diabetes insipidus involves which of the following glandular dysfunctions?

1. Thyroid hyperfunction
2. Pituitary hypofunction
3. Pituitary hyperfunction
4. Parathyroid hypofunction

49. 2. Diabetes insipidus is the principal disorder caused by posterior pituitary hypofunction, not hyperfunction. The disorder results from hyposecretion of antidiuretic hormone, producing a state of uncontrolled diuresis. Diabetes insipidus doesn't involve the thyroid or parathyroid glands.

NP: Data collection; CN: Health promotion and maintenance; CNS: Prevention and early detection of disease; CL: Knowledge

50. After the nurse has explained the causes of diabetes insipidus to the parents, which of the following statements made by a parent indicates the need for further teaching?

1. "This condition could be familial or congenital."
2. "Drinking alcohol during my pregnancy caused this condition."
3. "My child might have a tumor that's causing these symptoms."
4. "An infection such as meningitis may be the reason my child has diabetes insipidus."

Careful! Here's one of those further teaching questions.

50. 2. Drinking alcohol during pregnancy can lead to a neonate born with fetal alcohol syndrome, but has no known effect on diabetes insipidus. The other options are possible causes of diabetes insipidus.

NP: Implementation; CN: Health promotion and maintenance; CNS: Prevention and early detection of disease; CL: Application

51. Which of the following assessment findings would alert the nurse to change the intranasal route for vasopressin administration?

1. Mucous membrane irritation
2. Severe coughing
3. Nosebleeds
4. Pneumonia

Question 51 asks for a common adverse effect of vasopressin administration.

51. 1. Mucous membrane irritation caused by a cold or allergy renders the intranasal route unreliable. Severe coughing, pneumonia, or nosebleeds shouldn't interfere with the intranasal route.

NP: Data collection; CN: Physiological integrity; CNS: Pharmacological therapies; CL: Application

52. The nurse should include which of the following in-home management instructions for a child who's receiving desmopressin acetate (DDAVP) for symptomatic control of diabetes insipidus?

1. Give DDAVP only when urine output begins to decrease.
2. Clean skin with alcohol before application of the DDAVP dermal patch.
3. Increase the DDAVP dose if polyuria occurs just before the next scheduled dose.
4. Call the primary health care provider for an alternate route of DDAVP when the child has an upper respiratory infection (URI) or allergic rhinitis.

52. 4. Excessive nasal mucus associated with URI or allergic rhinitis may interfere with DDAVP absorption because it's given intranasally. Parents should be instructed to contact the primary health care provider for advice in altering the hormone dose during times when nasal mucus may be increased. To avoid overmedicating the child, the DDAVP dose should remain unchanged, even if there's polyuria just before the next dose.

NP: Implementation; CN: Safe, effective care environment; CNS: Coordinated care; CL: Application

53. The nurse is collecting data on a client with suspected hypopituitarism. Which of the following conditions is the chief complaint associated with this condition?

1. Sleep disturbance
2. Polyuria
3. Polydipsia
4. Short stature

53. 4. The chief complaint in most instances of hypopituitarism is short stature. Sleep disturbance may indicate thyrotoxicosis. Polydipsia and polyuria may be signs of diabetes mellitus or diabetes insipidus.

NP: Data collection; CN: Health promotion and maintenance; CNS: Prevention and early detection of disease; CL: Knowledge

54. Which of the following statements made to the nurse by the parents of a child with hypopituitarism would indicate the need for *further teaching*?
1. "This disorder may be familial."
2. "There's no genetic basis for this disorder."
3. "This disorder might be secondary to hypothalamic deficiency."
4. "There may be other disorders related to pituitary hormone deficiencies."

55. A nurse is teaching health to a class of fifth graders. Which of the following statements related to growth should be included?
1. "There's nothing that you can do to influence your growth."
2. "Intensive physical activity that begins before puberty might stunt growth."
3. "Children who are short in stature also have parents who are short in stature."
4. "Because this is a time of tremendous growth, being concerned about calorie intake isn't important."

56. A nurse should understand that familial short stature refers to which of the following children?
1. Children who are members of a very large family with limited resources
2. Children who have no siblings who moved a great deal during their early childhood
3. Children with delayed linear growth and skeletal and sexual maturation that's behind that of age-mates
4. Children who have ancestors with adult height in the lower percentiles and whose height during childhood is appropriate

54. 2. The cause of idiopathic growth hormone deficiency is unknown. The condition is typically associated with other pituitary hormone deficiencies, such as deficiencies of thyroid stimulating hormone and corticotropin, and so may be secondary to hypothalamic deficiency. There's also a higher-than-average occurrence of the disorder in some families, which indicates a possible genetic cause.
NP: Implementation; CN: Psychosocial integrity; CNS: Coping and adaptation; CL: Application

55. 2. Intensive physical activity (greater than 18 hours per week) that begins before puberty may stunt growth so that the child doesn't reach full adult height. During the school-age years, growth slows and doesn't accelerate again until adolescence. Nutrition and environment influence a child's growth. Children who are short in stature don't necessarily have parents who are short in stature.
NP: Implementation; CN: Health promotion and maintenance; CNS: Growth and development through the life span; CL: Application

56. 4. Familial short stature refers to otherwise healthy children who have ancestors with adult height in the lower percentiles and whose height during childhood is appropriate for genetic background. Children with delayed linear growth and skeletal and sexual maturation behind that of age-mates are said to have constitutional growth delay. Children who are members of very large families with limited resources or who are only children don't fit the description of familial short stature.
NP: Data collection; CN: Health promotion and maintenance; CNS: Prevention and early detection of disease; CL: Knowledge

I'm tired, but I can't stop now — and neither should you! Let's keep going!

I know practicing for the NCLEX will make me feel good in the long run. I just have to keep my eyes on the prize!

NCLEX-PN Made Incredibly E-Z

57. When evaluating a 2-year-old child, which of the following findings would indicate to the nurse the possibility of growth hormone deficiency?

1. The child had normal growth during the first year of life but showed a slowed growth curve below the 3rd percentile for the 2nd year of life.
2. The child fell below the 5th percentile for growth during the first year of life but, at this checkup, only falls below the 50th percentile.
3. There has been a steady decline in growth over the 2 years of this infant's life that has accelerated during the past 6 months.
4. There was delayed growth below the 5th percentile for the first and second years of life.

57. 1. Children generally grow normally during the first year and then follow a slowed growth curve that's below the 3rd percentile. Growth consistently below the 5th percentile, may be an indication of failure to thrive.

NP: Data collection; CN: Health promotion and maintenance; CNS: Prevention and early detection of disease; CL: Application

58. When collecting data on a child with growth hormone deficiency, the nurse would expect to observe which of the following characteristics?

1. Decreased weight with no change in height
2. Decreased weight with increased height
3. Increased weight with decreased height
4. Increased weight with increased height

58. 3. Height may be retarded more than weight because, with good nutrition, children with growth hormone deficiency can become overweight or even obese. Their well-nourished appearance is an important diagnostic clue to differentiation from other disorders such as failure to thrive.

NP: Data collection; CN: Health promotion and maintenance; CNS: Prevention and early detection of disease; CL: Knowledge

59. The nurse should find which of the following characteristics in her observations of a child with growth hormone deficiency?

1. Normal skeletal proportions
2. Abnormal skeletal proportions
3. Child appearing older than his age
4. Longer than normal upper extremities

59. 1. Skeletal proportions are normal for the age, but these children appear younger than their chronological age. However, later in life, premature aging is evident.

NP: Data collection; CN: Physiological integrity; CNS: Physiological adaptation; CL: Application

Which of these sports would be best for a child with growth hormone deficiency?

60. When counseling the parents of a child with growth hormone deficiency, the nurse should encourage which of the following sports?

1. Basketball
2. Field hockey
3. Football
4. Gymnastics

60. 4. Children with growth hormone deficiency can be no less active than other children if directed to size-appropriate sports, such as gymnastics, swimming, wrestling, or soccer.

NP: Planning; CN: Psychosocial integrity; CNS: Coping and adaptation; CL: Application

---

NP: Nursing process   CN: Client needs category   CNS: Client needs subcategory   CL: Cognitive level

61. In explaining to parents the social behavior of children with hypopituitarism, a nurse should recognize which of the following statements as a need for further teaching?

    1. "I realize that my child might have school anxiety and low self-esteem."

    2. "Because my child is short in stature, people expect less of him than his peers."

    3. "Because of my child's short stature, he may not be pushed to perform at his chronological age by others."

    4. "My child's vocabulary is very well developed, so even though he's short in stature, no one will treat him differently."

It's important that parents have realistic expectations about their child's disorder.

62. The mother of a child diagnosed with hypopituitarism states to the nurse that she feels guilty because she feels that she should have recognized this disorder. Which of the following statements by the nurse about children with hypopituitarism would be the most helpful?"

    1. "They're usually large for gestational age at birth."

    2. "They're usually small for gestational age at birth."

    3. "They usually exhibit signs of this disorder soon after birth."

    4. "They're usually of normal size for gestational age at birth."

You're more than two-thirds finished! Outstanding!

63. When plotting height and weight on a growth chart, which of the following observations would indicate that a child who's 4 years old has a growth hormone deficiency?

    1. Upward shift of 1 percentile or more

    2. Upward shift of 5 percentiles or more

    3. Downward shift of 2 percentiles or more

    4. Downward shift of 5 percentiles or more

64. When reviewing the results of radiographic examinations of a child with hypopituitarism, which of the following characteristics should the nurse expect to observe?

    1. Bone age near normal

    2. Epiphyseal maturation normal

    3. Epiphyseal maturation retarded

    4. Bone maturation greatly retarded

61. 4. Height discrepancy has been significantly correlated with emotional adjustment problems and may be a valuable predictor of the extent to which growth hormone-delayed children will have trouble with anxiety, social skills, and positive self-esteem. Also, academic problems aren't uncommon. These children aren't usually pushed to perform at their chronological age, but are typically subjected to juvenilization (related to an infantile or childish manner).

NP: Evaluation; CN: Psychosocial integrity; CNS: Psychosocial adaptation; CL: Application

62. 4. Children with hypopituitarism are usually of normal size for gestational age at birth. Clinical features develop slowly and vary with the severity of the disorder and the number of deficient hormones.

NP: Implementation; CN: Psychosocial integrity; CNS: Coping and adaptation; CL: Knowledge

63. 3. When the primary health care provider evaluates the results of plotting height and weight, upward or downward shifts of 2 percentiles or more in children older than age 3 may indicate a growth abnormality.

NP: Data collection; CN: Health promotion and maintenance; CNS: Prevention and early detection of disease; CL: Analysis

64. 3. Epiphyseal maturation is retarded in hypopituitarism consistent with retardation in height. This is in contrast to hypothyroidism, in which bone maturation is greatly retarded, or Turner's syndrome, in which bone age is near normal.

NP: Data collection; CN: Health promotion and maintenance; CNS: Prevention and early detection of disease; CL: Knowledge

65. A nurse should understand that which of the following tests is used for a definitive diagnosis of hypopituitarism?
1. Hypersecretion of thyroid hormone
2. Increased reserves of growth hormone
3. Hyposecretion of antidiuretic hormone
4. Decreased reserves of growth hormone

66. The parents of a child who's going through testing for hypopituitarism ask the nurse what test results they should expect. The nurse's response should be based on which of the following factors?
1. Measurement of growth hormone will occur only one time.
2. Growth hormone levels are decreased after strenuous exercise.
3. There will be increased overnight urine growth hormone concentration.
4. Growth hormone levels are elevated 45 to 90 minutes following the onset of sleep.

67. Which of the following methods is considered the definitive treatment for hypopituitarism due to growth hormone deficiency?
1. Treatment with desmopressin acetate (DDAVP)
2. Replacement of antidiuretic hormone
3. Treatment with testosterone or estrogen
4. Replacement with biosynthetic growth hormone

All of these treatments may be used but only one is *definitive*.

68. When obtaining information about a child, which of the following comments made by a parent to the nurse would indicate the possibility of hypopituitarism in a child?
1. "I can pass down my child's clothes to his younger brother."
2. "Usually my child wears out his clothes before his size changes."
3. "I have to buy bigger-sized clothes for my child about every 2 months."
4. "I have to buy larger shirts more frequently than larger pants for my child."

65. 4. Definitive diagnosis is based on absent or subnormal levels of pituitary growth hormone. Antidiuretic hormone and thyroid hormone levels aren't affected.
NP: Data collection; CN: Health promotion and maintenance; CNS: Prevention and early detection of disease; CL: Knowledge

66. 4. Growth hormone levels are elevated 45 to 90 minutes following the onset of sleep. Low growth hormone levels following the onset of sleep would indicate the need for further evaluation. Exercise is a natural and benign stimulus for growth hormone release, and elevated levels can be detected after 20 minutes of strenuous exercise in normal children. Also, growth hormone levels will need to be checked frequently related to the type of therapy instituted.
NP: Evaluation; CN: Physiological integrity; CNS: Physiological adaptation; CL: Application

67. 4. The definitive treatment of growth hormone deficiency is replacement of growth hormone and is successful in 80% of affected children. DDAVP is used to treat diabetes insipidus. Antidiuretic hormone deficiency causes diabetes insipidus and isn't related to hypopituitarism. Testosterone or estrogen may be given during adolescence for normal sexual maturation but aren't the definitive treatment for hypopituitarism.
NP: Evaluation; CN: Physiological integrity; CNS: Pharmacological therapies; CL: Application

68. 2. Parents of children with hypopituitarism usually comment that the child wears out clothes before growing out of them or that, if the clothing fits the body, it's typically too long in the sleeves or legs.
NP: Data collection; CN: Health promotion and maintenance; CNS: Growth and development through the life span; CL: Application

NP: Nursing process   CN: Client needs category   CNS: Client needs subcategory   CL: Cognitive level

69. In helping parents who are planning to give growth hormone at home, the nurse should explain that optimum dosing is achieved when growth hormone is administered at which of the following times?
1. At bedtime
2. After dinner
3. In the middle of the day
4. First thing in the morning

70. In educating parents of a child with hypopituitarism about realistic expectations of height for their child who's successfully responding to growth hormone replacement, a nurse should include which of the following statements?
1. "Your child will never reach a normal adult height."
2. "Your child will attain his eventual adult height at a faster rate."
3. "Your child will attain his eventual adult height at a slower rate."
4. "The rate of your child's growth will be the same as children without this disorder."

71. Which of the following statements made by a parent of a child with short stature would indicate to the nurse the need for further teaching?
1. "Obtaining blood studies won't aid in proper diagnosis."
2. "A history of my child's growth patterns should be discussed."
3. "X-rays should be included in my child's diagnostic procedures."
4. "A family history is important information for me to share with my primary health care provider."

72. If hypersecretion of growth hormone occurs after epiphyseal closure, which of the following conditions might be observed by the nurse?
1. Acromegaly
2. Cretinism
3. Dwarfism
4. Gigantism

The point at which a drug is most effective is information to know for the NCLEX.

Careful — here's another further teaching question.

69. 1. Optimum dosing is usually achieved when growth hormone is administered at bedtime. Pituitary release of growth hormone occurs during the first 45 to 90 minutes after the onset of sleep so normal physiological release is mimicked with bedtime dosing.
NP: Implementation; CN: Physiological integrity; CNS: Pharmacological therapies; CL: Knowledge

70. 3. Even when hormone replacement is successful, these children attain their eventual adult height at a slower rate than their peers do; therefore, they need assistance in setting realistic expectations regarding improvement.
NP: Implementation; CN: Psychosocial adaptation; CNS: Coping and adaptation; CL: Application

71. 1. A complete diagnostic evaluation should include a family history, a history of the child's growth patterns and previous health status, physical examination, physical evaluation, radiographic survey, and endocrine studies that may involve blood samples.
NP: Evaluation; CN: Health promotion and maintenance; CNS: Prevention and early detection of disease; CL: Application

72. 1. If excessive growth hormone is evident after epiphyseal closure, growth is in the transverse direction, producing a condition known as acromegaly. The other options aren't seen as a result of excess growth hormone after epiphyseal closure.
NP: Data collection; CN: Physiological integrity; CNS: Physiological adaptation; CL: Knowledge

73. Which of the following metabolic alteration characteristics might be associated with growth hormone deficiency?

1. Galactosemia
2. Homocystinuria
3. Hyperglycemia
4. Hypoglycemia

73. 4. The development of hypoglycemia is a characteristic finding related to growth hormone deficiency. Galactosemia is a rare autosomal recessive disorder with an inborn error of carbohydrate metabolism. Homocystinuria is an indication of amino acid transport or metabolism problems. Hyperglycemia isn't a problem in hypopituitarism.

NP: Evaluation; CN: Health promotion and maintenance; CNS: Prevention and early detection of disease; CL: Application

I can't weight to see how well you do on this one!

74. A child is admitted to the medical-surgical unit with complaints of weight loss and lack of energy. The child's ears and cheeks are flushed and the nurse observes an acetone odor to the client's breath. The blood glucose level is 325. The blood pressure is 104/60 mm Hg, pulse is 88, and respirations are 16 breaths/minute. Which of the following does the nurse expect the physician to order first?

1. Subcutaneous (S.C.) administration of glucagon
2. Administration of regular insulin by continuous infusion pump
3. S.C. administration of regular insulin every 4 hours as needed by sliding scale insulin
4. Administration of I.V. fluids in boluses of 20 ml/kg

74. 2. Weight loss, lack of energy, acetone odor to the breath, and a blood glucose level of 325 indicate diabetic ketoacidosis. Insulin is given by continuous infusion pump at a rate not to exceed 100 mg/dl/hour. Faster reduction of hypoglycemia could be related to the development of cerebral edema. Glucagon is administered for mild hypoglycemia. Sliding scale insulin isn't as effective as the administration of insulin by continuous infusion pump in the treatment of diabetic ketoacidosis. Administration of I.V. fluids in boluses of 20 mg/kg is recommended for the treatment of shock.

NP: Implementation; CN: Physiological integrity; CNS: Physiological adaptation; CL: Application

75. When assessing a neonate diagnosed with diabetes insipidus, which of the following findings would indicate the need for intervention?

1. Edema
2. Increased head circumference
3. Weight gain
4. Weight loss

75. 4. Diabetes insipidus usually appears gradually. Weight loss from a large loss of fluid occurs. A normal neonate should gain weight as he grows. There should be an increase in his head circumference with treatment. Edema isn't evident in the neonate with diabetes insipidus.

NP: Data collection; CN: Physiological integrity; CNS: Reduction of risk potential; CL: Application

76. In a client with diabetes insipidus, a nurse could expect which of the following characteristics of the urine?

1. Pale; specific gravity less than 1.006
2. Concentrated; specific gravity less than 1.006
3. Concentrated; specific gravity less than 1.030
4. Pale; specific gravity more than 1.030

76. 1. With diabetes insipidus, the client has difficulty with excessive urine output; therefore, the urine will be pale and the specific gravity will fall below the low normal of 1.010.

NP: Data collection; CN: Physiological integrity; CNS: Physiological adaptation; CL: Analysis

NP: Nursing process   CN: Client needs category   CNS: Client needs subcategory   CL: Cognitive level

77. In a child with diabetes insipidus, which of the following characteristics would most likely be present in the child's health history?

1. Delayed closure of the fontanels, coarse hair, and hypoglycemia in the morning
2. Gradual onset of personality changes, lethargy, and blurred vision
3. Vomiting early in the morning, headache, and decreased thirst
4. Abrupt onset of polyuria, nocturia, and polydipsia

78. A client is on fluid restriction before diagnostic testing for diabetes insipidus. Which of the following conditions would indicate to the nurse the need to discontinue fluid restriction?

1. Weight gain of 3% to 5%
2. Weight loss of 3% to 5%
3. Increase in urine output
4. Generalized edema

79. When a child with diabetes insipidus has a viral illness that includes congestion, nausea, and vomiting, the nurse should instruct the parents to take which of the following actions?

1. Make no changes in the medication regimen.
2. Give medications only once per day.
3. Obtain an alternate route for desmopressin acetate (DDAVP) administration.
4. Give medication 1 hour after vomiting has occurred.

80. A nurse is preparing for discharge a child with diabetes insipidus who'll be taking injectable vasopressin. Which of the following teaching strategies is best for the nurse regarding injection techniques?

1. Teach injection techniques to the primary caregiver.
2. Teach injection techniques to anyone who'll provide care for the child.
3. Teach injection techniques to anyone who'll provide care for the child as well as to the child if he's old enough to understand.
4. Provide information about the nearest home health agency so the parents can arrange for the home health nurse to come and give the injection.

Question 78 is asking for an adverse reaction.

77. 4. Diabetes insipidus is characterized by deficient secretion of antidiuretic hormone leading to diuresis. Most children with this disorder experience an abrupt onset of symptoms, including polyuria, nocturia, and polydipsia. The other choices reflect symptoms of pituitary hyperfunction.
NP: Evaluation; CN: Health promotion and maintenance; CNS: Prevention and early detection of disease; CL: Application

78. 2. A weight loss of 3% to 5% indicates significant dehydration and requires termination of fluid restriction. Weight gain would be a good sign. Generalized edema wouldn't occur with fluid restriction, nor would increased urine output.
NP: Data collection; CN: Physiological integrity; CNS: Physiological adaptation; CL: Analysis

79. 3. An alternate route for administration of DDAVP would be needed for absorption due to nasal congestion. The other options reflect actions that need to be covered by a medical order.
NP: Implementation; CN: Health promotion and maintenance; CNS: Prevention and early detection of disease; CL: Application

80. 3. The best strategy is to teach all those who'll provide care for the child. The child should be included if age-appropriate. It's unrealistic to arrange home health nurses to give injections that are required throughout the life span.
NP: Implementation; CN: Health promotion and maintenance; CNS: Prevention and early detection of disease; CL: Application

Good work! You're flying through this test!

81. When providing care for a school-age client with diabetes insipidus, the nurse understands that which of the following behaviors might be difficult related to this child's growth and development?

1. Taking his medication at school
2. Taking his medication before bedtime
3. Letting his mother administer his medication
4. Giving himself a vasopressin injection before school starts

82. Which of the following monitoring methods would be best for a client newly diagnosed with diabetes insipidus?

1. Measuring abdominal girth every day
2. Measuring intake, output, and urine specific gravity
3. Checking daily weight and measuring intake
4. Checking for pitting edema in the lower extremities

Question 82 asks you to rank monitoring methods for appropriateness. Can you do it? I knew you could!

83. When observing a neonate for signs of congenital hypothyroidism, which of the following characteristics would the nurse observe?

1. Hyperreflexia
2. Long forehead
3. Puffy eyelids
4. Small tongue

Question 84 is almost a negative question. It asks about adverse effects.

84. Which of the following factors is most significant in adversely affecting eventual intelligence of a neonate with hypothyroidism?

1. Overtreatment
2. Inadequate treatment
3. Educational level of the parents
4. Socioeconomic level of the family

81. 1. Anything that singles a child out and makes him feel different from his peers will result in possible noncompliance with the medical regimen. It's important for the nurse to help the client schedule the need for medications around the times he'll be in school.
NP: Implementation; CN: Health promotion and maintenance; CNS: Growth and development through the life span; CL: Application

82. 2. Measuring intake and output with related specific gravity results will enable the nurse to closely monitor the client's condition along with daily weight. All other options aren't as accurate for a child with diabetes insipidus.
NP: Data collection; CN: Health promotion and maintenance; CNS: Prevention and early detection of disease; CL: Application

83. 3. Assessment findings would include depressed nasal bridge, short forehead, puffy eyelids, and large tongue; thick, dry, mottled skin that feels cold to the touch; coarse, dry, lusterless hair; abdominal distention; umbilical hernia; hyporeflexia; bradycardia; hypothermia; hypotension; anemia; and wide cranial sutures.
NP: Data collection; CN: Health promotion and maintenance; CNS: Prevention and early detection of disease; CL: Knowledge

84. 2. The most significant factor adversely affecting eventual intelligence is inadequate treatment, which may be related to noncompliance. Parental factors, such as educational and socioeconomic level, affect only the environmental stimulation, not the child's basic intellect. Although overtreatment could cause physical problems and, possibly, death, these adverse reactions would occur before the intellect was affected.
NP: Implementation; CN: Health promotion and maintenance; CNS: Prevention and early detection of disease; CL: Application

NP: Nursing process   CN: Client needs category   CNS: Client needs subcategory   CL: Cognitive level

85. Which of the following nursing objectives is most important when working with neonates who are suspected of having congenital hypothyroidism?
1. Early identification
2. Promoting bonding
3. Allowing rooming in
4. Encouraging fluid intake

Question 85 is asking you to prioritize!

85. 1. The most important nursing objective is early identification of the disorder. Nurses caring for neonates must be certain that screening is performed, especially in neonates who are preterm, discharged early, or born at home. Promoting bonding, allowing rooming in, and encouraging fluid intake are all important but are less important than early identification.
NP: Data collection; CN: Physiological integrity; CNS: Basic care and comfort; CL: Knowledge

86. When the parents of an infant diagnosed with hypothyroidism have been taught to count the pulse, which of the following interventions should the nurse teach them in case they obtain a high pulse rate?
1. Allow the infant to take a nap and then give the medication.
2. Withhold the medication and give a double dose the next day.
3. Withhold the medication and call the primary health care provider.
4. Give the medication and then consult the primary health care provider.

86. 3. If parents have been taught to count the infant's pulse, they should be instructed to withhold the dose and consult their primary health care provider if the pulse rate is above a certain value.
NP: Evaluation; CN: Health promotion and maintenance; CNS: Prevention and early detection of disease; CL: Application

87. In an infant receiving inadequate treatment for congenital hypothyroidism, the nurse would expect to observe which of the following symptoms?
1. Irritability and jitteriness
2. Fatigue and sleepiness
3. Increased appetite
4. Diarrhea

87. 2. Signs of inadequate treatment are fatigue, sleepiness, decreased appetite, and constipation.
NP: Data collection; CN: Health promotion and maintenance; CNS: Prevention and early detection of disease; CL: Application

88. Which of the following characteristics best describes congenital hypothyroidism?
1. It's sex-linked.
2. It has no genetic basis.
3. It's an autosomal dominant gene.
4. It's caused by an inborn error of metabolism.

Just a few more questions to go. Have I told you lately how proud of you I am? Well, I am!

88. 4. The disorder is caused by an inborn error of thyroid hormone synthesis, which is autosomal recessive. Therefore, genetic counseling is important. There's no evidence that this disorder is sex-linked.
NP: Data collection; CN: Health promotion and maintenance; CNS: Prevention and early detection of disease; CL: Knowledge

89. In an adolescent with type 1 diabetes, which of the following recommendations for preventing hypoglycemia should the nurse make?
 1. Limit participation in planned exercise activities that involve competition.
 2. Carry crackers or fruit to eat before or during periods of increased activity.
 3. Increase the insulin dosage before planned or unplanned strenuous exercise.
 4. Check blood glucose before exercising and eat a protein snack if the level is elevated.

90. Which of the following statements accurately describes the incidence of type 1 diabetes mellitus?
 1. Diabetes mellitus is the most common endocrine disease of childhood.
 2. Diabetes mellitus is an inherited disease caused by a recessive gene.
 3. Diabetes mellitus is more commonly seen in children who are obese.
 4. The prevalence of diabetes mellitus is decreasing due to early detection.

91. An adolescent girl is admitted to the hospital with type 1 diabetes and unstable blood glucose levels. Which of the following questions is most important to include in the history?
 1. Does she play any team sports?
 2. Does she refrigerate her insulin?
 3. Is she satisfied with her weight?
 4. Does she use recreational drugs?

Let's discuss some steps for preventing hypoglycemia.

89. 2. Hypoglycemia can usually be prevented if a child with diabetes eats more food before or during exercise. Because exercise with adolescents is usually unplanned, carrying additional carbohydrate foods is a good preventive measure.
NP: Implementation; CN: Health promotion and maintenance; CNS: Prevention and early detection of disease; CL: Application

90. 1. Diabetes mellitus is the most common endocrine disease in childhood. There's a higher incidence of type 2 diabetes when there's a strong family history of diabetes, but the genetic predisposition for type 1 diabetes isn't known. Obesity is also a predisposing factor for type 2 diabetes; there's no link to obesity in type 1 diabetes. Early detection of a disease doesn't affect its prevalence.
NP: Data collection; CN: Health promotion and maintenance; CNS: Prevention and early detection of disease; CL: Knowledge

91. 3. It's important to ascertain the adolescent's feelings about her body — in particular, her weight. Some adolescents skip their insulin because they know it will result in weight loss. The other issues of sports, drug use, and insulin administration technique are relevant but not as important as knowing what she's thinking about her own body.
NP: Implementation; CN: Psychosocial integrity; CNS: Coping and adaptation; CL: Application

Hooray! Another test finished! Good job!

This chapter covers altered patterns of urinary elimination in children and includes glomerulonephritis, hypospadias, and — oh, a whole lot of other conditions. Ready? Let's go!

Chapter 34
Genitourinary disorders

1. A child with acute glomerulonephritis has a nursing diagnosis of impaired urinary elimination related to fluid retention and impaired glomerular filtration. The child should have which of the following expected outcomes?
  1. Exhibits no evidence of infection
  2. Engages in activities appropriate to capabilities
  3. Demonstrates no periorbital, facial, or body edema
  4. Maintains a fluid intake of more than 2,000 ml in 24 hours

1. 3. The goal of this diagnosis involves interventions, such as decreased fluid and salt intake, designed to minimize or prevent fluid retention and edema. These interventions may be evaluated through observations for edema. The other options are appropriate outcomes for other nursing diagnoses, not the diagnosis in question.
NP: Evaluation; CN: Health promotion and maintenance; CNS: Prevention and early detection of disease; CL: Analysis

2. An important nursing intervention to support the therapeutic management of the child with acute glomerulonephritis would include which of the following?
  1. Measuring daily weight
  2. Increasing oral fluid intake
  3. Providing sodium supplements
  4. Monitoring the client for signs of hypokalemia

If you're having trouble deciding on an answer, begin by eliminating the ones you know are incorrect.

2. 1. The child with acute glomerulonephritis should be monitored for fluid imbalance, which is done through daily weights. Increasing oral intake, providing sodium supplements, and monitoring for hypokalemia aren't part of the therapeutic management of acute glomerulonephritis.
NP: Implementation; CN: Physiological integrity; CNS: Basic care and comfort; CL: Application

3. A child has been diagnosed with acute glomerulonephritis. Which of the following components should the nurse expect the child's urine to contain?
  1. Red blood cell (RBC) casts
  2. Calcium casts
  3. Cystine crystals
  4. Glucose

3. 1. Urinalysis findings consistent with glomerulonephritis should include specific gravity less than 1.030, proteinuria, hematuria, cylindruria, and RBC casts. Calcium casts and glucose aren't usually found in the urine of a client with acute glomerulonephritis. The presence of cystine crystals typically indicates a congenital metabolic problem.
NP: Data collection; CN: Physiological integrity; CNS: Physiological adaptation; CL: Knowledge

NP: Nursing process CN: Client needs category CNS: Client needs subcategory CL: Cognitive level

4. The nurse is taking frequent blood pressure readings on a child diagnosed with acute glomerulonephritis. The parents ask the nurse why this is necessary. The nurse's reply should be based on which of the following factors?
1. Blood pressure fluctuations are a sign that the condition has become chronic.
2. Blood pressure fluctuations are a common adverse effect of antibiotic therapy.
3. Hypotension leading to sudden shock can develop at any time.
4. Acute hypertension must be anticipated and identified.

5. When evaluating the urinalysis report of a child with acute glomerulonephritis, the nurse would expect which of the following results?
1. Proteinuria and decreased specific gravity
2. Bacteriuria and increased specific gravity
3. Hematuria and proteinuria
4. Bacteriuria and hematuria

6. Which of the following statements by the nurse would be the best response to a mother who wants to know the first indication that acute glomerulonephritis is improving?
1. Urine output will increase.
2. Urine will be protein-free.
3. Blood pressure will stabilize.
4. The child will have more energy.

7. Which of the following statements best describes acute glomerulonephritis?
1. This disease occurs after a urinary tract infection.
2. This disease is associated with renal vascular disorders.
3. This disease occurs after an antecedent streptococcal infection.
4. This disease is associated with structural anomalies of the genitourinary tract.

More than one answer may seem correct. It's your job to choose the best answer.

4. 4. Regular measurement of vital signs, body weight, and intake and output is essential to monitor the progress of the disease and to detect complications that may appear at any time during the course of the disease. Hypertension is more likely to occur with glomerulonephritis than hypotension. Blood pressure fluctuations don't indicate that the condition has become chronic and aren't common adverse reactions to antibiotic therapy.
NP: Data collection; CN: Safe, effective care environment; CNS: Coordinated care; CL: Comprehension

5. 3. Urinalysis during the acute phase of this disease characteristically shows hematuria, proteinuria, and increased specific gravity.
NP: Evaluation; CN: Physiological integrity; CNS: Physiological adaptation; CL: Analysis

6. 1. One of the first signs of improvement during the acute phase of glomerulonephritis is an increase in urine output. It will take time for the urine to be protein-free. Antihypertensive drugs may be needed to stabilize blood pressure. Children generally don't have much energy during the acute phase of this disease.
NP: Data collection; CN: Health promotion and maintenance; CNS: Prevention and early detection of disease; CL: Analysis

7. 3. Acute glomerulonephritis is an immune-complex disease that occurs as a by-product of an antecedent streptococcal infection. Certain strains of the infection are usually beta-hemolytic streptococci.
NP: Data collection; CN: Health promotion and maintenance; CNS: Prevention and early detection of disease; CL: Knowledge

8. When obtaining a client's daily weight, the nurse notes that the child has lost 6 lb (2.7 kg) after 3 days of hospitalization for acute glomerulonephritis. This is most likely the result of which of the following factors?
1. Poor appetite
2. Reduction of edema
3. Decreased salt intake
4. Restriction to bed rest

9. A nurse should make which of the following dietary recommendations to a client who has been newly diagnosed with acute glomerulonephritis?
1. Reduce calories.
2. Increase potassium.
3. Severely restrict sodium.
4. Moderately restrict sodium.

10. The nurse is evaluating a group of children for acute glomerulonephritis. Which of the following clients would be most likely to develop the disease?
1. A client who had pneumonia a month ago
2. A client who was bitten by a brown spider
3. A client who shows no signs of periorbital edema
4. A client who had a streptococcal infection 2 weeks ago

11. At which of the following ages would the nurse observe a higher incidence of acute glomerulonephritis?
1. 1 to 2 years
2. 6 to 7 years
3. 12 to 13 years
4. 18 to 20 years

12. In understanding the recurrence of glomerulonephritis, the nurse should understand which of the following characteristics?
1. Second attacks are quite common.
2. A recessive gene transfers this disease.
3. Multiple cases tend to occur in families.
4. Overcrowding in the schoolroom leads to higher incidence.

10 questions down! That's a good start!

8. 2. When edema is reduced, the client will lose weight. This should normally occur after treatment for acute glomerulonephritis has been followed for several days. It will take longer for the child's appetite to improve, but this shouldn't lead to such a dramatic weight loss in a child this age.
NP: Evaluation; CN: Physiological integrity; CNS: Basic care and comfort; CL: Application

9. 4. Moderate sodium restriction with a diet that has no added salt after cooking is usually effective. Reduced calorie consumption and increased potassium consumption aren't necessary due to the decrease in urine output. *Severe* sodium restriction isn't needed and will make it more difficult to ensure adequate nutrition.
NP: Implementation; CN: Physiological integrity; CNS: Basic care and comfort; CL: Application

10. 4. A latent period of 10 to 14 days occurs between the streptococcal infection of the throat or skin and the onset of clinical manifestations. The peak incidence of disease corresponds to the incidence of streptococcal infections. Pneumonia isn't a precursor to glomerulonephritis, nor is a bite from a brown spider. A sign of periorbital edema would lead the nurse to investigate the possibility of glomerulonephritis, especially if reported to be worse in the morning.
NP: Evaluation; CN: Safe, effective care environment; CNS: Coordinated care; CL: Analysis

11. 2. Acute glomerulonephritis can occur at any age, but it primarily affects early school-age children with a peak age of onset of 6 to 7 years. It's uncommon in children younger than 2 years old.
NP: Data collection; CN: Health promotion and maintenance; CNS: Growth and development through the life span; CL: Knowledge

12. 3. Multiple cases of glomerulonephritis tend to occur in families. Second attacks are rare. Acute glomerulonephritis isn't transmitted through a recessive gene and overcrowding in the schoolroom should have no influence on this disease.
NP: Data collection; CN: Health promotion and maintenance; CNS: Prevention and early detection of disease; CL: Application

13. When teaching families of children with acute glomerulonephritis about complications, which of the following comments made by a parent would indicate to the nurse the need for further education?

1. "Dizziness is expected and I should have my child lie down."
2. "I should let the nurse know every time my child urinates."
3. "I need to ask my child if he has a headache."
4. "I shouldn't force my child to eat."

14. Which of the following tests is the most familiar and most readily available test for streptococcal antibodies?

1. Antistreptolysin-O test (ASOT)
2. Blood culture
3. Blood urea nitrogen (BUN)
4. Mono spot

15. When teaching an 8-year-old child to obtain a clean-catch urine specimen, which of the following techniques should be included by the nurse?

1. Collect the specimen right after a nap.
2. Never use the first voided specimen of the day.
3. Collect the specimen at the beginning of urination.
4. You don't need to wash your perineal area before collecting the specimen.

The phrase *further education* indicates that question 13 is looking for an inaccurate statement.

Don't a-*void* this question. You're doing great!

13. 1. Dizziness is a sign of encephalopathy and must be reported to the nurse. Hypertensive encephalopathy, acute cardiac decompensation, and acute renal failure are the major complications that tend to develop during the acute phase of glomerulonephritis.
NP: Implementation; CN: Physiological integrity; CNS: Reduction of risk potential; CL: Application

14. 1. The ASOT is the most familiar and most readily available test for streptococcal antibodies. ASO appears in the serum approximately 10 days after the initial infection; however, there's no correlation between the degree of elevation and the severity or prognosis of the glomerulonephritis. The mono spot test is done to detect mononucleosis. Blood cultures are drawn to determine sepsis and a BUN will indicate renal function.
NP: Planning; CN: Physiological integrity; CNS: Physiological adaptation; CL: Comprehension

15. 2. When collecting a clean-catch urine specimen, the first voided specimen of the day should never be used because of urinary stasis; this also applies after a nap. The specimen should be collected mid-stream and shouldn't include the first voided specimen. Washing the perineal area before collecting a specimen is important to make sure there are no contaminants from the skin in the specimen.
NP: Implementation; CN: Health promotion and maintenance; CNS: Growth and development through the life span; CL: Application

16. In explaining treatment for glomerulo-nephritis, the nurse should include which of the following statements?

1. All children who have signs of glomerulonephritis are hospitalized for approximately 1 week.
2. Parents should expect children to have a normal energy level during the acute phase.
3. Children who have normal blood pressure and a satisfactory urine output can generally be treated at home.
4. Children with gross hematuria and significant oliguria should be brought to the primary health care provider's office about every 2 days for monitoring.

16. 3. Children who have normal blood pressure and a satisfactory urine output can generally be treated at home. Those with gross hematuria and significant oliguria will probably be hospitalized for monitoring. Parents should expect children to have a decrease in energy levels during the acute phase of the disease.
NP: Implementation; CN: Health promotion and maintenance; CNS: Prevention and early detection of disease; CL: Comprehension

Careful. Question 17 is a negative question.

17. Which of the following reasons accounts for why bed rest <u>isn't recommended</u> during the acute phase of glomerulonephritis?

1. It's too difficult to keep a child on bed rest.
2. Children on bed rest lose too much muscle tone due to lack of movement.
3. Parents find enforcing bed rest causes them to feel guilty about the disease.
4. Ambulation doesn't seem to have an adverse effect on the course of the disease.

17. 4. Ambulation doesn't seem to have an adverse effect on the course of the disease after the gross hematuria, edema, hypertension, and azotemia have abated. Because they're generally listless and experience fatigue and malaise, most children voluntarily restrict their activities during the most active phase of the disease. Children on short-term bed rest don't lose muscle tone because they usually move around in the bed. Parents don't feel guilty enforcing bed rest; but they may find it challenging.
NP: Data collection; CN: Physiological integrity; CNS: Reduction of risk potential; CL: Application

Question 18 asks what foods the client with acute glomerulonephritis should avoid.

18. Which of the following foods should the nurse <u>eliminate</u> from the diet of a child who's diagnosed with acute glomerulonephritis?

1. Turkey sandwich with mayonnaise
2. Hot dog with ketchup and mustard
3. Chocolate cake with white icing
4. Apple with peanut butter

18. 2. Foods that are high in sodium should be eliminated from the child's diet. Such snacks as pretzels and potato chips should be discouraged. Any other foods that the child likes should be encouraged. Because hot dogs contain a great deal of sodium, they should be eliminated from the child's diet.
NP: Evaluation; CN: Physiological integrity; CNS: Basic care and comfort; CL: Application

19. Which of the following therapies should the nurse expect to incorporate into the care of the child with acute glomerulonephritis?

1. Antibiotic therapy
2. Dialysis therapy
3. Diuretic therapy
4. Play therapy

19. 4. Play therapy is a very important aspect of care to help the child understand what's happening to him. Unless the child can express concerns and fears, he may have night terrors and regress in his stage of growth and development.
NP: Implementation; CN: Health promotion and maintenance; CNS: Prevention and early detection of disease; CL: Application

20. After the acute phase of glomerulonephritis is over, which of the following discharge instructions should the nurse include?
1. Every 6 months, a cystogram will be needed for evaluation of progress.
2. Weekly visits to the primary health care provider may be needed for evaluation.
3. It will be acceptable to keep the regular yearly checkup appointment for the next evaluation.
4. There's no need to worry about further evaluations by the primary health care provider related to this disease.

Twenty questions done! Keep plugging along!

20. 2. Weekly or monthly visits to the primary health care provider will be needed for evaluation of improvement and will usually involve the collection of a urine specimen for urinalysis. A cystogram isn't helpful in determining the progression of this disease; it's used to review the anatomic structures of the urinary tract.
NP: Implementation; CN: Physiological integrity; CNS: Reduction of risk potential; CL: Application

21. The nurse understands that hypospadias refers to which of the following conditions?
1. Absence of a urethral opening
2. Penis shorter than usual for age
3. Urethral opening along the dorsal surface of the penis
4. Urethral opening along the ventral surface of the penis

21. 4. Hypospadias refers to a condition in which the urethral opening is located below the glans penis or anywhere along the ventral surface of the penile shaft. Hypospadias refers to a malposition of the opening, not absence of the opening, and has nothing to do with the size of the penis.
NP: Data collection; CN: Health promotion and maintenance; CNS: Prevention and early detection of disease; CL: Knowledge

22. The nurse understands that chordee refers to which of the following conditions?
1. Ventral curvature of the penis
2. Dorsal curvature of the penis
3. No curvature of the penis
4. Misshapen penis

Knowing common accompanying conditions is important for taking the NCLEX.

22. 1. Chordee, or ventral curvature of the penis, results from the replacement of normal skin with a fibrous band of tissue and usually accompanies more severe forms of hypospadias.
NP: Data collection; CN: Health promotion and maintenance; CNS: Prevention and early detection of disease; CL: Knowledge

23. Which of the following anomalies commonly accompanies hypospadias?
1. Undescended testes
2. Ambiguous genitalia
3. Umbilical hernias
4. Inguinal hernias

23. 1. Because undescended testes may also be present, the small penis may appear to be an enlarged clitoris. This shouldn't be mistaken for ambiguous genitalia. If there's any doubt, more tests should be performed. Hernias don't generally accompany hypospadias.
NP: Data collection; CN: Health promotion and maintenance; CNS: Prevention and early detection of disease; CL: Knowledge

24. Which of the following reasons explains why surgical repair of hypospadias is done as early as possible?
1. Prevent separation anxiety.
2. Prevent urinary complications.
3. Promote acceptance of hospitalization.
4. Promote development of normal body image.

25. A child is undergoing hypospadias repair. Which of the following statements made by the child's parents about the principal objective of surgical correction implies a need for further teaching?
1. "The purpose is to improve the physical appearance of the genitalia for psychological reasons."
2. "The purpose is to enhance the child's ability to void in the standing position."
3. "The purpose is to decrease the chances of urinary tract infections."
4. "The purpose is to preserve a sexually adequate organ."

Question 25 is looking for the one inaccurate statement.

26. The nurse should counsel parents to postpone which of the following actions until after their son's hypospadias has been repaired?
1. Circumcision
2. Infant baptism
3. Getting hepatitis B vaccine
4. Checking blood for inborn errors of metabolism

27. Which of the following nursing interventions should be included in the care plan for a male infant following surgical repair of hypospadias?
1. Sterile dressing changes every 4 hours
2. Frequent inspection of the tip of the penis
3. Removal of the suprapubic catheter on the 2nd postoperative day
4. Urethral catheterization if voiding doesn't occur over an 8-hour period

24. 4. Whenever there are defects of the genitourinary tract, surgery should be performed early to promote development of a normal body image. Hypospadias doesn't put the child at a greater risk for urinary complications. A child with normal emotional development shows separation anxiety at 7 to 9 months. Within a few months, he understands the mother's permanence, and separation anxiety diminishes.
NP: Implementation; CN: Health promotion and maintenance; CNS: Prevention and early detection of disease; CL: Application

25. 3. A child with hypospadias isn't at greater risk for urinary tract infections. The principal objectives of surgical corrections are to enhance the child's ability to void in the standing position with a straight stream, to improve the physical appearance of the genitalia for psychological reasons, and to preserve a sexually adequate organ.
NP: Implementation; CN: Health promotion and maintenance; CNS: Prevention and early detection of disease; CL: Application

26. 1. Circumcision shouldn't be performed until after the hypospadias has been repaired. The foreskin might be needed to help in the repair of hypospadias. None of the other choices have any bearing on the repair of hypospadias.
NP: Evaluation; CN: Health promotion and maintenance; CNS: Prevention and early detection of disease; CL: Application

27. 2. Following hypospadias repair, a pressure dressing is applied to the penis to reduce bleeding and tissue swelling. The penile tip should then be assessed frequently for signs of circulatory impairment. The dressing around the penis shouldn't be changed as frequently as every 4 hours. The primary health care provider will determine when the suprapubic catheter will be removed. Urethral catheterization should be avoided after repair of hypospadias to prevent injury to the urethra.
NP: Data collection; CN: Physiological integrity; CNS: Basic care and comfort; CL: Application

28. When explaining to the parents the optimal time for repair of hypospadias, the nurse should indicate which of the following as the age of choice?

1. 1 week
2. 6 to 18 months
3. 2 years
4. 4 years

28. 2. The preferred time for surgical repair is ages 6 to 18 months, before the child has developed body image and castration anxiety. Surgical repair of hypospadias as early as age 3 months has been successful but with a high incidence of complications.

NP: Implementation; CN: Health promotion and maintenance; CNS: Prevention and early detection of disease; CL: Comprehension

29. When providing discharge information to parents of a child with hypospadias repair, which of the following instructions would be inappropriate?

1. Irrigation techniques if indicated
2. Techniques for providing tub baths
3. Care for the indwelling catheter or stent
4. Techniques for avoiding kinking, twisting, or blockage of the catheter or stent

Think question 29 is a negative question? You're right!

29. 2. A tub bath should be avoided to prevent infection until the stent has been removed. Parents are taught to care for the indwelling catheter or stent and irrigation techniques if indicated. They need to know how to empty the urine bag and how to avoid kinking, twisting, or blockage of the catheter or stent.

NP: Evaluation; CN: Safe, effective care environment; CNS: Coordinated care; CL: Analysis

30. When providing discharge instructions to parents of an older child who has had hypospadias repair, which of the following activities should be avoided?

1. Finger painting
2. Playing in sandboxes
3. Increased fluid intake
4. Playing with the family pet

30. 2. Sandboxes, straddle toys, swimming, and rough activities are avoided until allowed by the surgeon. The family is advised to encourage the child to increase fluid intake. Quiet, nonstrenuous activities are encouraged.

NP: Implementation; CN: Physiological integrity; CNS: Physiological adaptation; CL: Application

31. The mother of a neonate born with hypospadias is sharing her feelings of guilt about this anomaly with a nurse. The nurse should explain which of the following facts about the defect?

1. It occurs around the 3rd month of fetal development.
2. It occurs around the 6th month of fetal development.
3. It's carried by an autosomal recessive gene.
4. It's hereditary.

31. 1. The defect of hypospadias occurs around the end of the 3rd month of fetal development. Many women don't even know that they're pregnant at this time. This defect isn't hereditary, nor is it carried by an autosomal recessive gene.

NP: Implementation; CN: Health promotion and maintenance; CNS: Prevention and early detection of disease; CL: Application

32. The discovery of hypospadias is usually made by which of the following people?
1. By the primary health care provider when doing a neonatal assessment
2. By the primary health care provider, just before circumcision
3. By the mother when she sees her neonate for the first time
4. By the nurse doing the neonatal assessment

33. A 1-year-old underwent hypospadias repair yesterday; he has a suprapubic catheter in place and an I.V. Which of the following rationales is appropriate for administering propantheline (Pro-Banthine) on an as-needed basis?
1. To decrease the chance of infection at the suture line
2. To decrease the number of organisms in the urine
3. To prevent bladder spasms while the catheter is present
4. To increase urine flow from the kidney to the ureters

34. Which of the following interventions by the nurse would be most helpful when discussing hypospadias with the parents of an infant with this defect?
1. Refer the parents to a counselor.
2. Be there to listen to the parents' concerns.
3. Notify the primary health care provider, and have him talk to the parents.
4. Suggest a support group of other parents who have gone through this experience.

35. The nurse should understand that hypospadias defects take the greatest emotional toll on which of the following persons?
1. The father
2. The mother
3. The grandfather
4. The grandmother

Question 32 asks you to think about the typical sequence of events for a neonate.

Question 34 asks you to prioritize.

32. 4. After delivery, neonates bond with their mothers for a period of time and then are taken to the neonate nursery. The nurse who admits the neonate does a thorough assessment and should recognize hypospadias and alert the primary health care provider. The pediatrician who examines the neonate may not come to the nursery for hours after the delivery.
NP: Data collection; CN: Health promotion and maintenance; CNS: Prevention and early detection of disease; CL: Knowledge

33. 3. Propantheline is an antispasmodic that works effectively on children. It isn't an antibiotic and therefore won't decrease the chance of infection or the number of organisms in the urine. The drug has no diuretic effect and won't increase urine flow.
NP: Planning; CN: Physiological integrity; CNS: Pharmacological therapies; CL: Application

34. 2. The nurse must recognize that parents are going to grieve the loss of the normal child when they have a neonate born with a birth defect. Initially, the parents need to have a nurse who will listen to their concerns for their neonate's health. Suggesting a support group or referring the parents to a counselor might be good actions, but not initially. The primary health care provider will need to spend time with the parents but, again, the nurse is in the best position to allow the parents to vent their grief and anger.
NP: Evaluation; CN: Psychosocial integrity; CNS: Coping and adaptation; CL: Application

35. 1. Because the penis is involved, studies have shown that fathers have a great deal of difficulty dealing with a birth defect like hypospadias.
NP: Data collection; CN: Psychosocial integrity; CNS: Coping and adaptation; CL: Knowledge

36. The difference between hypospadias and epispadias is defined by which of the following characteristics?
  1. Epispadias defects can only occur in males.
  2. The difference between the defects is the length of the urethra.
  3. Hypospadias is an abnormal opening on the ventral side of the penis; epispadias is an abnormal opening on the dorsal side.
  4. Hypospadias is an abnormal opening on the dorsal side of the penis; epispadias is an abnormal opening on the ventral side.

37. Which of the following nursing diagnoses would be most appropriate for a client with hypospadias?
  1. Deficient fluid volume
  2. Impaired urinary elimination
  3. Delayed growth and development
  4. Risk for infection

38. When a nurse is teaching the parent how to care for the penis after hypospadias repair with a skin graft, which of the following statements made by the parent would indicate the need for further teaching?
  1. "My infant will be able to take baths."
  2. "I'll change the dressing around the penis daily."
  3. "I'll make sure I change my infant's diaper often."
  4. "If there's a color change in the penis, I'll notify my primary care provider."

39. The nurse is preparing the parents of an infant with hypospadias for surgery. Which of the following statements made by the parents would indicate the need for further teaching?
  1. "Skin grafting might be involved in my infant's repair."
  2. "After surgery, my infant's penis will look perfectly normal."
  3. "Surgical repair may need to be performed in several stages."
  4. "My infant will probably be in some pain after the surgery and might need to take some medication for relief."

Don't alter your course. You're on the road to success!

Lots of negative questions here. Remember to look for incorrect statements.

36. 3. Hypospadias results from the incomplete closure of the urethral folds along the ventral surface of the developing penis. Epispadias results when the urinary meatus is on the dorsal surface of the penis. Epispadias defects can occur in males and females. The difference is where the opening of the urinary meatus is located, not the urethra's length.
NP: Data collection; CN: Health promotion and maintenance; CNS: Prevention and early detection of disease; CL: Comprehension

37. 2. The most appropriate diagnosis for a client with hypospadias is *Impaired urinary elimination.* A client with hypospadias should have no problems with the ingestion of fluids. The child's growth and development isn't affected with this defect, and he doesn't have any problem with infection until possibly after hypospadias repair is performed.
NP: Planning; CN: Health promotion and maintenance; CNS: Prevention and early detection of disease; CL: Analysis

38. 2. Dressing changes after a hypospadias repair with a skin graft are generally performed by the primary health care provider and aren't performed every day because the skin graft needs time to heal and adhere to the penis. Changing the infant's diapers usually helps keep the penis dry. Baths aren't given until postoperative healing has taken place. If the penis color changes, it might be evidence of circulation problems and should be reported.
NP: Evaluation; CN: Psychosocial integrity; CNS: Psychosocial adaptation; CL: Analysis

39. 2. It's important to stress to the parents that, even after a repair of hypospadias, the outcome isn't a completely "normal-looking" penis. The goals of surgery are to allow the child to void from the tip of his penis, void with a straight stream, and stand up while voiding.
NP: Evaluation; CN: Psychosocial integrity; CNS: Coping and adaptation; CL: Application

NP: Nursing process   CN: Client needs category   CNS: Client needs subcategory   CL: Cognitive level

40. Which of the following data collected by the nurse would indicate to the primary care provider the need for a staged repair of a hypospadias rather than a single repair?
1. Chordee is present with the hypospadias.
2. The urinary meatus opens close to the scrotum.
3. The urinary meatus is just below the tip of the penis.
4. The infant has been circumcised before the defect was discovered.

40. 2. Increased surgical experience and improvements in technique have reduced the number of staged procedures applied to hypospadias defects; however, a staged procedure is indicated in particularly severe defects with marked deficits of available skin for mobilization of flaps. If an infant has a relatively minor hypospadias or has been circumcised, the repair can still occur in one stage. Having chordee present doesn't require a staged hypospadias repair.

NP: Data collection; CN: Physiological integrity; CNS: Physiological adaptation; CL: Analysis

41. A nurse should understand that prevention of pelvic inflammatory disease (PID) in adolescents is important for which of the following reasons?
1. PID is easily prevented by proper personal hygiene.
2. PID is easily prevented by compliance with any form of contraception.
3. PID can have devastating effects on the reproductive tract of affected adolescents.
4. PID can cause life-threatening defects in infants born to affected adolescents.

41. 3. Long-term complications of PID include abscess formation in the fallopian tubes and adhesion formation leading to increased risk for ectopic pregnancy or infertility. It isn't prevented by proper personal hygiene or any form of contraception, even though some forms of contraception such as the male or female condom do help to decrease the incidence.

NP: Evaluation; CN: Health promotion and maintenance; CNS: Prevention and early detection of disease; CL: Comprehension

42. After the nurse has completed discharge teaching, which of the following statements made by the client treated for a sexually transmitted disease would indicate that discharge instructions were understood?
1. "I don't need those condoms because I'm not allergic to penicillin and I'll come for a shot at the first sign of infection."
2. "I will notify my sex partners and not have unprotected sex from now on."
3. "I will be careful not to have intercourse with someone who isn't clean."
4. "If you're going to get it, you're going to get it."

Hmmm. Question 42 is looking for a correct statement.

42. 2. Goal achievement is indicated by the client's ability to describe preventive behaviors and health practices. The other options indicate that the client doesn't understand the need to take preventive measures.

NP: Evaluation; CN: Health promotion and maintenance; CNS: Prevention and early detection of disease; CL: Analysis

43. Besides the fact that it's highly infectious, the nurse should teach a client with gonorrhea about which of the following characteristics?
1. It occurs rarely.
2. It can produce sterility.
3. It's always easily cured.
4. It's limited to the external genitalia.

43. 2. Gonorrhea may destroy the epididymis in a male or tubal mucosa in a female, which can cause sterility. Gonorrhea, a common sexually transmitted disease, occasionally affects the rectum, pharynx, and eyes. Treatment is effective when the client complies.

NP: Implementation; CN: Health promotion and maintenance; CNS: Prevention and early detection of disease; CL: Knowledge

44. Before a client with syphilis can be treated, the nurse must determine which of the following factors?

1. Portal of entry
2. Size of the chancre
3. Names of sexual contacts
4. Existence of medication allergies

45. Which of the following statements regarding chlamydial infections is correct?

1. The treatment of choice is oral penicillin.
2. The treatment of choice is nystatin or miconazole.
3. Clinical manifestations include dysuria and urethral itching in males.
4. Clinical manifestations include small, painful vesicles on genital areas.

46. Which of the following techniques should the nurse consider when she's discussing sex and sexual activities with adolescents?

1. Break down all the information into scientific terminology.
2. Refer adolescents to their parents for sexual information.
3. Only answer questions that are asked; don't present any other content.
4. Present sexual information using the proper terminology and in a straightforward manner.

The NCLEX tests your ability to teach clients at different life stages.

47. Without proper treatment, anogenital warts caused by the human papillomavirus (HPV) increase the risk of which of the following illnesses in adolescent females?

1. Gonorrhea
2. Cervical cancer
3. Chlamydial infections
4. Urinary tract infections

44. 4. The treatment of choice for syphilis is penicillin; clients allergic to penicillin must be given another antibiotic. The other choices aren't necessary before treatment can begin.
NP: Data collection; CN: Health promotion and maintenance; CNS: Prevention and early detection of disease; CL: Comprehension

45. 3. Clinical manifestations of chlamydia include meatal erythema, tenderness, itching, dysuria, and urethral discharge in the male and mucopurulent cervical exudate with erythema, edema, and congestion in the female. The treatment of choice is doxycycline or azithromycin. Small, painful vesicles in the genital area refer to herpes.
NP: Implementation; CN: Health promotion and maintenance; CNS: Prevention and early detection of disease; CL: Application

46. 4. Although many adolescents have received sex education from parents and school throughout childhood, they aren't always adequately prepared for the impact of puberty. A large portion of their knowledge is acquired from peers, television, movies, and magazines. Consequently, much of the sex information they have is incomplete, inaccurate, riddled with cultural and moral values, and not very helpful. The public perceives nurses as having authoritative information and being willing to take time with adolescents and their parents. To be effective teachers, nurses need to be honest and open with sexual information.
NP: Implementation; CN: Health promotion and maintenance; CNS: Growth and development through the life span; CL: Application

47. 2. All external lesions are treated because of concern regarding the relationship of HPV to cancer. HPV doesn't increase the risk for gonorrhea, chlamydia, or urinary tract infections.
NP: Implementation; CN: Health promotion and maintenance; CNS: Prevention and early detection of disease; CL: Application

---

NP: Nursing process   CN: Client needs category   CNS: Client needs subcategory   CL: Cognitive level

48. Which of the following statements should the nurse include when teaching an adolescent about gonorrhea?

1. It's caused by *Treponema pallidum.*
2. Treatment of sexual partners is an essential part of treatment.
3. It's usually treated by multidose administration of penicillin.
4. It may be contracted through contact with a contaminated toilet bowl.

49. When planning sex education and contraceptive teaching for adolescents, which of the following factors should the nurse consider?

1. Neither sexual activity nor contraception requires planning.
2. Most teenagers today are knowledgeable about reproduction.
3. Most teenagers use pregnancy as a way to rebel against their parents.
4. Most teenagers are open about contraception but inconsistently use birth control.

50. A sexually active teenager seeks counseling from the school nurse about prevention of sexually transmitted diseases (STDs). Which of the following contraceptive measures should the nurse recommend?

1. Rhythm method
2. Withdrawal method
3. Prophylactic antibiotic use
4. Condom and spermicide use

51. The nurse understands that which of the following developmental rationales explains risk-taking behavior in adolescents?

1. Adolescents are concrete thinkers and concentrate only on what's happening at the time.
2. Belief in their own invulnerability persuades adolescents that they can take risks safely.
3. Risk of parents' anger and disappointment usually deters adolescents from risky behavior.
4. Peer pressure usually doesn't play an important part in a adolescent's decision to become sexually active.

Nice work! You've finished nearly 50 questions already!

Adolescents may engage in sex as part of risky behavior.

48. 2. Adolescents should be taught that treatment is needed for all sexual partners. The medication of choice is a single dose of I.M. ceftriaxone (Rocephin) in males and a single oral dose of cefixime (Suprax) in females. Gonorrhea can't be contracted from a contaminated toilet bowl.

NP: Implementation; CN: Health promotion and maintenance; CNS: Prevention and early detection of disease; CL: Application

49. 4. Most teenagers today are very open about discussing contraception and sexuality, but may get caught up in the moment of sexuality and forget about birth control measures. Adolescents receive most of their information on reproduction and sexuality from their peers, who generally don't have correct information. Teenagers generally become pregnant because they fail to use birth control for other reasons than rebelling against their parents. Contraception should always be part of sex education and requires planning.

NP: Planning; CN: Physiological integrity; CNS: Reduction of risk potential; CL: Analysis

50. 4. Prevention of STDs is the primary concern of health care professionals. Barrier contraceptive methods, such as condoms with the addition of spermicide, seem to offer the best protection for preventing STDs and their serious complications. The other contraceptive choices don't prevent the transmission of an STD. Antibiotics can't be taken throughout the life span.

NP: Implementation; CN: Health promotion and maintenance; CNS: Prevention and early detection of disease; CL: Application

51. 2. Understanding the growth and development of an adolescent helps the nurse see that they feel they're invulnerable. Peer pressure plays an important role in risk-taking behaviors, more so than fear of parents' anger or disappointment. Adolescents can and do think about the future but are willing to take risks that more mature adults might not take.

NP: Evaluation; CN: Health promotion and maintenance; CNS: Growth and development through the life span; CL: Analysis

52. Statistics about sexually transmitted diseases (STDs) may not be reliable for which of the following reasons?

1. Most adolescents seek out treatment for their STD.
2. Adolescents are usually honest with their parents about their sexual behavior.
3. All STDs must be reported to the Centers for Disease Control and Prevention (CDC).
4. Chlamydial infections and human papillomavirus (HPV) infections aren't required to be reported to the CDC.

53. Which of the following statements by an adolescent would alert the nurse that more education about sexually transmitted diseases (STDs) is needed?

1. "You always know when you've got gonorrhea."
2. "The most common STD in kids my age is chlamydia infection."
3. "Most of the girls who have Chlamydia don't even know it."
4. "If you have symptoms of gonorrhea, they can show up a day or a couple of weeks after you got the infection to begin with."

54. Which of the following approaches describes the method of preventing sexually transmitted diseases (STDs) by avoiding exposure?

1. The least accepted and most difficult approach
2. The least expensive and most effective approach
3. The most expensive and least effective approach
4. The most difficult and most time-consuming approach

55. It's very important for the nurse to include which of the following statements in discharge education for the client who's taking metronidazole (Flagyl) to treat trichomoniasis?

1. Sexual intercourse should stop.
2. Alcohol shouldn't be consumed.
3. Milk products should be avoided.
4. Exposure to sunlight should be limited.

The phrase *more education is needed* indicates that question 53 asks for an incorrect statement.

Keep at it! You're more than halfway finished!

52. 4. Chlamydial infections and HPV infections aren't required to be reported to the CDC. Most teenagers are afraid to seek out health care for sexual diseases or are unaware of the signs and symptoms of STDs. Teenagers find this a very difficult topic to discuss with their parents and will usually seek out a peer or another adult to obtain information.
NP: Evaluation; CN: Safe, effective care environment; CNS: Safety and infection control; CL: Application

53. 1. Gonorrhea can occur with or without symptoms. There are four main forms of the disease: asymptomatic, uncomplicated symptomatic, complicated symptomatic, and disseminated disease. All of the other statements are accurate.
NP: Data collection; CN: Health promotion and maintenance; CNS: Prevention and early detection of disease; CL: Application

54. 2. Primary prevention of STDs by avoiding exposure is the least expensive and most effective approach. The nurse can play a role in offering this education to young people before they initiate sexual intercourse.
NP: Planning; CN: Safe, effective care environment; CNS: Safety and infection control; CL: Application

55. 2. While taking metronidazole to treat trichomoniasis, clients shouldn't consume alcohol for at least 48 hours following the last dose. The other choices have no effect on the client while taking this medication.
NP: Implementation; CN: Physiological integrity; CNS: Pharmacological therapies; CL: Application

56. The nurse should include which of the following facts when teaching an adolescent group about the human immunodeficiency virus (HIV)?
1. The incidence of HIV in the adolescent population has declined since 1995.
2. The virus can be spread through many routes, including sexual contact.
3. Knowledge about HIV spread and transmission has led to a decrease in the spread of the virus among adolescents.
4. About 50% of all new HIV infections in the United States occurs in people under age 22.

56. 2. HIV can be spread through many routes, including sexual contact and contact with infected blood or other body fluids. The incidence of HIV in the adolescent population has *increased* since 1995, even though more information about the virus is targeted to reach the adolescent population. Only about 25% of all new HIV infections in the United States occurs in people under age 22.
NP: Implementation; CN: Health promotion and maintenance; CNS: Prevention and early detection of disease; CL: Application

57. When planning a program to teach adolescents about acquired immunodeficiency syndrome (AIDS), which of the following actions might lead to better success of the program?
1. Surveying the community to evaluate the level of education
2. Obtaining peer educators to provide information about AIDS
3. Setting up clinics in community centers and having condoms readily available
4. Having primary health care providers host workshops in community centers

57. 2. Peer education programs have noted that teens are more likely to ask questions of peer educators than of adults and that peer education can change personal attitudes and the perception of the risk of HIV infection. The other approaches would be helpful, but not necessarily make the outreach program more successful.
NP: Evaluation; CN: Safe, effective care environment; CNS: Coordinated care; CL: Analysis

58. Which of the following clients would the nurse consider at greater risk for developing acquired immunodeficiency syndrome (AIDS)?
1. Clients who live in crowded housing with poor ventilation
2. A young sexually active client with multiple partners
3. Homeless adolescents who live in shelters
4. A young sexually active client with one partner

58. 2. The younger the client when sexual activity begins, the higher the incidence of HIV and AIDS. Also, the more sexual partners, the higher the incidence. Neither crowded living environments nor homeless environments by themselves lead to an increase in the incidence of AIDS.
NP: Data collection; CN: Health promotion and maintenance; CNS: Prevention and early detection of disease; CL: Knowledge

Questions about data collection skills are common on the NCLEX.

59. When collecting data from an adolescent with pelvic inflammatory disease (PID), which of the following signs and symptoms should the nurse expect to see?
1. A hard, painless, red defined lesion
2. Small vesicles on the genital area with itching
3. Cervical discharge with redness and edema
4. Lower abdominal pain and urinary tract symptoms

59. 4. PID is an infection of the upper female genital tract most commonly caused by sexually transmitted diseases. Initial symptoms in the adolescent may be generalized, with fever, abdominal pain, urinary tract symptoms, and vague influenza-like symptoms. Small vesicles on the genital area with itching indicates herpes genitalis. Cervical discharge with redness and edema indicates chlamydia. A hard, painless, red defined lesion indicates syphilis.
NP: Data collection; CN: Health promotion and maintenance; CNS: Prevention and early detection of disease; CL: Application

60. Which of the following statements should the nurse include when teaching adolescents about syphilis?
1. Syphilis is rarely transmitted sexually.
2. There's no known cure or treatment for syphilis.
3. The viability of the organism outside the body is long.
4. Affected persons are most infectious during the first year.

60. 4. Affected persons are most infectious during the first year of the disease. Syphilis is treated with penicillin or doxycycline. About 95% of the cases of syphilis are sexually transmitted. Syphilis can be transmitted to a fetus. The viability of the organism outside the body is short.
NP: Implementation; CN: Health promotion and maintenance; CNS: Prevention and early detection of disease; CL: Knowledge

61. In teaching a group of parents about monitoring for urinary tract infection (UTI) in preschoolers, which of the following symptoms would indicate that a child should be evaluated?
1. Voids only twice in any 6-hour period
2. Exhibits incontinence after being toilet trained
3. Has difficulty sitting still for more than a 30-minute period of time
4. Urine smells strongly of ammonia after standing for more than 2 hours

61. 2. A child who exhibits incontinence after being toilet trained should be evaluated for UTI. Most urine smells strongly of ammonia after standing for more than 2 hours, so this doesn't necessarily indicate UTI. The other options aren't reasons for parents to suspect problems with their child's urinary system.
NP: Evaluation; CN: Safe, effective care environment; CNS: Coordinated care; CL: Application

62. Which of the following instructions should the nurse include in the teaching plan for a client receiving co-trimoxazole (Septra) for a repeated urinary tract infection with *Escherichia coli?*
1. "For the drug to be effective, keep your urine acidic by drinking at least a quart of cranberry juice per day."
2. "Take the medication for 10 days even if your symptoms improve in a few days."
3. "Return to the clinic in 3 days for another urine culture."
4. "Take two pills each day now, but keep the rest of the pills to take if the symptoms reappear within 2 weeks."

Teaching about medications — that's another common NCLEX subject.

62. 2. Discharge instructions for clients receiving an anti-infective medication should include taking all of the prescribed medication for the prescribed time. The child won't need to have a culture repeated until the medication is completed. Drinking highly acidic juices such as cranberry juice may help maintain urinary health, but won't get rid of an already present infection.
NP: Planning; CN: Physiological integrity; CNS: Pharmacological therapies; CL: Application

63. The nurse should include which of the following facts when teaching parents about handling a child with recurrent urinary tract infection (UTI)?
1. Antibiotics should be discontinued 48 hours after symptoms subside.
2. Recurrent symptoms should be treated by renewing the antibiotic prescription.
3. Complicated UTIs are related to poor perineal hygiene practice.
4. Follow-up urine cultures are necessary to detect recurrent infections and antibiotic effectiveness.

63. 4. A routine follow-up urine specimen is usually obtained 2 or 3 days after the completion of the antibiotic treatment. All of the antibiotic should be taken as ordered and not stopped when symptoms disappear. If recurrent symptoms appear, a urine culture should be obtained to see if the infection is resistant to antibiotics. Simple, not complicated, UTIs are generally caused by poor perineal hygiene.
NP: Implementation; CN: Health promotion and maintenance; CNS: Prevention and early detection of disease; CL: Application

NP: Nursing process CN: Client needs category CNS: Client needs subcategory CL: Cognitive level

64. When monitoring a child with vesicoureteral reflux, the nurse should understand that this client is at risk for developing which of the following complications?
1. Glomerulonephritis
2. Hemolytic uremia syndrome
3. Nephrotic syndrome
4. Renal damage

64. 4. Reflux of urine into the ureters and then back into the bladder after voiding sets up the client for a urinary tract infection, which can lead to renal damage due to scarring of the parenchyma. Glomerulonephritis is an autoimmune reaction to a beta-hemolytic streptococcal infection. Eighty percent of nephrotic syndrome cases are idiopathic. Hemolytic uremia syndrome may be the result of genetic factors.
NP: Data collection; CN: Health promotion and maintenance; CNS: Prevention and early detection of disease; CL: Knowledge

65. When reviewing the results of a clean-voided urine specimen, which of the following results would indicate to the nurse that the child may have a urinary tract infection (UTI)?
1. A specific gravity of 1.020
2. Cloudy color without odor
3. A large amount of casts present
4. 100,000 bacterial colonies per milliliter

65. 4. The diagnosis of UTI is determined by the detection of bacteria in the urine. Infected urine usually contains over 100,000 colonies/ml, usually of a single organism. The urine is usually cloudy, hazy, and may have strands of mucus. It also has a foul, fishy odor even when fresh. Casts and increased specific gravity aren't specific to UTI.
NP: Evaluation; CN: Health promotion and maintenance; CNS: Prevention and early detection of disease; CL: Knowledge

Keep pushing! You'll be finished soon!

66. Nurses should understand that which of the following factors contributes to the increased incidence of urinary tract infections (UTI) in girls?
1. Vaginal secretions are too acidic.
2. Girls can't be protected by circumcision like boys can.
3. The urethra is in close proximity to the anus.
4. Girls touch their genitalia more often than boys do.

66. 3. Girls are especially at risk for bacterial invasion of the urinary tract because of basic anatomical differences; the urethra is shorter and closer to the anus. Vaginal secretions are normally acidic, which decreases the risk of infection, circumcision doesn't protect boys from UTI, and there's no documented research that supports that girls touch their genitalia more often than boys do..
NP: Evaluation; CN: Health promotion and maintenance; CNS: Growth and development through the life span; CL: Knowledge

67. Which of the following factors should the nurse recognize as predisposing the urinary tract to infection?
1. Increased fluid intake
2. Short urethra
3. Ingestion of highly acidic juices
4. Frequent emptying of the bladder

67. 2. A short urethra contributes to infection because bacteria have to travel a shorter distance to the urinary tract. The risk of infection is higher in women because women have shorter urethras than men (¾″ [2 cm] in young women, 1½″ [4 cm] in mature women, 7¾″ [20 cm] in adult men). Increased fluid intake would help flush the urinary tract system and frequent emptying of the bladder would decrease the risk of urinary tract infection. Drinking highly acidic juices such as cranberry juice may help maintain urinary health.
NP: Data collection; CN: Health promotion and maintenance; CNS: Prevention and early detection of disease; CL: Comprehension

68. A child has been sent to the school nurse for wetting her pants three times in the past 2 days. The nurse should recommend that this child be evaluated for which of the following complications?

1. School phobia
2. Emotional trauma
3. Urinary tract infection
4. Structural defect of the urinary tract

69. Which of the following interventions should a nurse recommend to parents of young girls to help prevent urinary tract infections (UTIs)?

1. Limit bathing as much as possible.
2. Increase fluids and decrease salt intake.
3. Have the child wear cotton underpants.
4. Have the child clean her perineum from back to front.

70. The nurse understands that which of the following characteristics is the single most important factor influencing the occurrence of urinary tract infections (UTI)?

1. Urinary stasis
2. Frequency of baths
3. Uncircumcised males
4. Amount of fluid intake

71. When the nurse is teaching parents of children about recurrent urinary tract infections, which of the following goals should be included as the most important?

1. Detection
2. Education
3. Prevention
4. Treatment

Again, you're being asked to prioritize.

Prioritize. Prioritize. Prioritize.

68. 3. Frequent urinary incontinence should be evaluated by the primary health care provider, with the first action being checking the urine for infection. Children exhibit signs of school phobia by complaining of an ailment before school starts and getting better after they're allowed to miss school. After infection, structural defect, and diabetes mellitus have been ruled out, emotional trauma should be investigated.

NP: Implementation; CN: Health promotion and maintenance; CNS: Growth and development through the life span; CL: Application

69. 3. Cotton is a more breathable fabric and allows for dampness to be absorbed from the perineum. Increasing fluids would be helpful, but decreasing salt isn't necessary. Bathing shouldn't be limited; however, the use of bubble bath or whirlpool baths should. However, if the child has frequent UTIs, taking a bath should be discouraged and taking a shower encouraged. The perineum should always be cleaned from front to back.

NP: Implementation; CN: Health promotion and maintenance; CNS: Prevention and early detection of disease; CL: Knowledge

70. 1. Ordinarily, urine is sterile. However, at 98.6° F (37° C), it provides an excellent culture medium. Under normal conditions, the act of completely and repeatedly emptying the bladder flushes away organisms before they have an opportunity to multiply and invade surrounding tissue. There's an increased incidence of UTI in uncircumcised infants under age 1 but not after that age. Baths and fluid intake are factors in the development of UTI, but aren't the most important.

NP: Data collection; CN: Physiological integrity; CNS: Reduction of risk potential; CL: Application

71. 3. Prevention is the most important goal in primary and recurrent infection; most preventive measures are simple, ordinary hygienic habits that should be a routine part of daily care. Treatment, detection, and education are all important, but not the most important goal.

NP: Implementation; CN: Physiological integrity; CNS: Reduction of risk potential; CL: Analysis

72. When evaluating infants and young toddlers for signs of urinary tract infections (UTIs), the nurse should demonstrate knowledge that which of the following symptoms would be most common?
1. Abdominal pain
2. Feeding problems
3. Frequency
4. Urgency

72. 2. In infants and children under age 2, the signs are characteristically nonspecific, and feeding problems are usually the first indication. Symptoms more nearly resemble GI tract disorders. Abdominal pain, urgency, and frequency are signs that would be observed in the older child with a UTI.
NP: Evaluation; CN: Physiological integrity; CNS: Physiological adaptation; CL: Knowledge

73. When obtaining a urine specimen for culture and sensitivity, the nurse should understand that which of the following methods of collection is best?
1. Bagged urine specimen
2. Clean-catch urine specimen
3. First-voided urine specimen
4. Catheterized urine specimen

73. 4. The most accurate tests of bacterial content are suprapubic aspiration (for children under age 2) and properly performed bladder catheterization. The other methods of obtaining a specimen have a high incidence of contamination not related to infection.
NP: Implementation; CN: Physiological integrity; CNS: Basic care and comfort; CL: Application

More than one answer for question 74 may seem correct. Choose the best answer.

74. After collecting a urine specimen, which of the following actions by the nurse is the most appropriate?
1. Take the specimen to the laboratory immediately.
2. Send the specimen to the laboratory on the scheduled run.
3. Take the specimen to the laboratory on the nurse's next break.
4. Keep the specimen in the refrigerator until it can be taken to the laboratory.

74. 1. Care of urine specimens obtained for culture is an important nursing aspect related to diagnosis. Specimens should be taken to the laboratory for culture immediately. If the culture is delayed, the specimen can be placed in the refrigerator, but storage can result in a loss of formed elements, such as blood cells and casts.
NP: Implementation; CN: Physiological integrity; CNS: Basic care and comfort; CL: Application

75. When teaching parents of a child with a urinary tract infection (UTI) about fluid intake, which of the following statements by a parent would indicate the need for further teaching?
1. "I should encourage my child to drink about 50 ml per pound of body weight daily."
2. "Clear liquids should be the primary liquids that my child drinks."
3. "I should offer my child carbonated beverages about every 2 hours."
4. "My child should avoid drinking caffeinated beverages."

Wow! You're whippin' through these questions! Whoopee!

75. 3. Carbonated or caffeinated beverages are avoided because of their potentially irritating effect on the bladder mucosa. Adequate fluid intake is always indicated during an acute UTI. It's recommended that a person drink approximately 50 ml/lb of body weight daily. The client should primarily drink clear liquids.
NP: Evaluation; CN: Physiological integrity; CNS: Basic care and comfort; CL: Analysis

76. Which of the following treatments should the nurse anticipate in a child who has a history of recurrent urinary tract infections (UTIs)?
1. Frequent catheterizations
2. Prophylactic antibiotics
3. Limited activities
4. Surgical intervention

77. When teaching parents about giving medications to children for recurrent urinary tract infections, which of the following instructions should be included?
1. The medication should be given first thing in the morning.
2. The medication should be given right before bedtime.
3. The medication is generally given four times per day.
4. It doesn't matter when the medication is given.

78. The nurse understands that the potential for progressive renal injury is greatest when urinary tract infections (UTIs) occur in which of the following age-groups or situations?
1. A school-age child who must get permission to go to the bathroom
2. An adolescent female who has started menstruation
3. Children who participate in competitive sports
4. Young infants and toddlers

79. Which of the following statements should the nurse make to help parents understand the recovery period after a child has had surgery to remove a Wilms' tumor?
1. "Children will easily lie in bed and restrict their activities."
2. "Recovery is usually fast in spite of the abdominal incision."
3. "Recovery usually takes a great deal of time due to the large incision."
4. "Parents need to perform activities of daily living for about 2 weeks after surgery."

Keep climbing! You're almost at the top!

76. 2. Children who experience recurrent UTI may require antibiotic therapy for months or years. Recurrent UTI would be investigated for anatomic abnormalities and surgical intervention may be indicated, but the client would also be placed on antibiotics before the tests. The child's activities aren't limited, and frequent catheterization predisposes a child to infection.
NP: Planning; CN: Physiological integrity; CNS: Basic care and comfort; CL: Application

77. 2. Medication is commonly administered once per day, and the client and parents are advised to give the antibiotic before sleep because this represents the longest period without voiding.
NP: Implementation; CN: Physiological integrity; CNS: Pharmacological therapies; CL: Application

78. 4. The hazard of progressive renal injury is greatest when infection occurs in young children, especially those under age 2. The first two options might lead to a simple UTI that would need to be treated. Competitive sports have no bearing on a UTI.
NP: Evaluation; CN: Safe, effective care environment; CNS: Safety and infection control; CL: Comprehension

79. 2. Children generally recover very quickly from surgery to remove a Wilms' tumor, even though they may have a large abdominal incision. Children like to get back into the normalcy of being a child, which is through play. Parents need to encourage their children to do as much for themselves as possible, although some regression is expected.
NP: Implementation; CN: Psychosocial integrity; CNS: Coping and adaptation; CL: Analysis

NP: Nursing process   CN: Client needs category   CNS: Client needs subcategory   CL: Cognitive level

80. When teaching parents about administering co-trimoxazole (Septra) to a child for treatment of a urinary tract infection, the nurse should include which of the following instructions?
1. Give the medication with food.
2. Give the medication with water.
3. Give the medication with a cola beverage.
4. Give the medication 2 hours after a meal.

81. The nurse understands that which of the following characteristics is true of the incidence of Wilms' tumor?
1. Peak incidence occurs at age 10.
2. It's the least common type of renal cancer.
3. It's the most common type of renal cancer.
4. It has a decreased incidence among siblings.

82. Which of the following initial signs is most common with Wilms' tumor?
1. Pain in the abdomen
2. Fever greater than 104° F (40° C)
3. Decreased blood pressure
4. Swelling within the abdomen

83. When the nurse is explaining the diagnosis of Wilms' tumor to parents, which of the following statements by a parent would indicate the need for further teaching?
1. "Wilms' tumor usually involves both kidneys."
2. "Wilms' tumor is slightly more common in the left kidney."
3. "Wilms' tumor is staged during surgery for treatment planning."
4. "Wilms' tumor stays encapsulated for an extended time."

Question 81 is looking for accurate information about a condition.

Keep it up! You're doing swell!

80. 2. When giving Septra, the medication should be administered with a full glass of water on an empty stomach. If nausea and vomiting occur, giving the drug with food may decrease gastric distress. Carbonated beverages should be avoided because they irritate the bladder.
NP: Implementation; CN: Physiological integrity; CNS: Pharmacological therapies; CL: Application

81. 3. Wilms' tumor is the most common intraabdominal tumor of childhood and the most common type of renal cancer. The peak incidence is age 3, and there's an increased incidence among siblings and identical twins.
NP: Data collection; CN: Health promotion and maintenance; CNS: Prevention and early detection of disease; CL: Knowledge

82. 4. The most common initial sign is a swelling or mass within the abdomen. The mass is characteristically firm, nontender, confined to one side, and deep within the flank. A high fever isn't an initial sign of Wilms' tumor. Blood pressure is characteristically increased, not decreased.
NP: Data collection; CN: Health promotion and maintenance; CNS: Prevention and early detection of disease; CL: Application

83. 1. Wilms' tumor usually involves only one kidney and is usually staged during surgery so an effective course of treatment can be established. Wilms' tumor has a slightly higher occurrence in the left kidney and it stays encapsulated for an extended time.
NP: Implementation; CN: Health promotion and maintenance; CNS: Prevention and early detection of disease; CL: Knowledge

84. A parent asks the nurse about the prognosis of her child diagnosed with Wilms' tumor. The nurse should base her response on which of the following factors?
1. Usually children with Wilms' tumor only need surgical intervention.
2. Survival rates for Wilms' tumor are the lowest among childhood cancers.
3. Survival rates for Wilms' tumor are the highest among childhood cancers.
4. Children with localized tumor have only a 30% chance of cure with multimodal therapy.

84. 3. Survival rates for Wilms' tumor are the highest among childhood cancers. Usually children with Wilms' tumor who have stage I or II localized tumor have a 90% chance of cure with multimodal therapy.
NP: Implementation; CN: Health promotion and maintenance; CNS: Prevention and early detection of disease; CL: Application

Hmmm, might the treatment for both kidneys differ from that for one?

85. If both kidneys are involved with Wilms' tumor, the nurse should understand that treatment prior to surgery might include which of the following methods?
1. Peritoneal dialysis
2. Abdominal gavage
3. Radiation and chemotherapy
4. Antibiotics and I.V. fluid therapy

85. 3. If both kidneys are involved, the child may be treated with radiation therapy or chemotherapy preoperatively to shrink the tumor, allowing more conservative therapy. Antibiotics aren't needed because Wilms' tumor isn't an infection. Peritoneal dialysis would only be needed if the kidneys aren't functioning. Abdominal gavage wouldn't be indicated.
NP: Implementation; CN: Safe, effective care environment; CNS: Coordinated care; CL: Application

86. When caring for the child with Wilms' tumor preoperatively, which of the following nursing interventions would be most important?
1. Avoid abdominal palpation.
2. Closely monitor arterial blood gases (ABGs).
3. Prepare child and family for long-term dialysis.
4. Prepare child and family for renal transplantation.

Note the word preoperatively in question 86.

86. 1. After the diagnosis of Wilms' tumor is made, the abdomen shouldn't be palpated. Palpation of the tumor might lead to rupture, which will cause the cancerous cells to spread throughout the abdomen. ABGs shouldn't be affected. If surgery is successful, there won't be a need for long-term dialysis or renal transplantation.
NP: Implementation; CN: Physiological integrity; CNS: Reduction of risk potential; CL: Application

87. A child is scheduled for surgery to remove a Wilms' tumor from one kidney. The parents ask the nurse what treatment, if any, they should expect after their child recovers from surgery. Which of the following responses would be most accurate?
1. "Chemotherapy may be necessary."
2. "Kidney transplant is indicated eventually."
3. "No additional treatments are usually necessary."
4. "Chemotherapy with or without radiation therapy is indicated."

87. 4. Because radiation therapy and chemotherapy are usually begun immediately after surgery, parents need an explanation of what to expect, such as major benefits and adverse effects. Kidney transplant isn't usually necessary.
NP: Implementation; CN: Safe, effective care environment; CNS: Coordinated care; CL: Application

NP: Nursing process   CN: Client needs category   CNS: Client needs subcategory   CL: Cognitive level

88. A toddler is admitted to the hospital with nephrotic syndrome. The nurse carefully monitors the toddler's fluid intake and output and checks urine specimens regularly with a reagent strip (Labstix). Which finding is the nurse most likely to report?

    1. Proteinuria
    2. Glucosuria
    3. Ketonuria
    4. Polyuria

89. The parent of a child with Wilms' tumor asks the nurse about surgery. Which of the following statements best explains the need for surgery for Wilms' tumor?

    1. Surgery isn't indicated in children with Wilms' tumor.
    2. Surgery is usually performed within 24 to 48 hours of admission.
    3. Surgery is the least favorable therapy for the treatment of Wilms' tumor.
    4. Surgery will be delayed until the client's overall health status improves.

90. A 3-year-old client has had surgery to remove a Wilms' tumor. Which of the following actions should the nurse take first when the mother asks for pain medication for the child?

    1. Get the pain medication ready for administration.
    2. Assess the client's pain using a pain scale of 1 to 10.
    3. Assess the client's pain using a smiley face pain scale.
    4. Check for the last time pain medication was administered.

91. A child has been diagnosed with Wilms' tumor. Because of the parents' religious beliefs, they choose not to treat the child. Which of the following statements by the nurse indicates the need for further discussion?

    1. "I know this is a lot of information in a short period of time."
    2. "I don't think parents have the legal right to make these kinds of decisions."
    3. "These parents just don't understand how easily treated a Wilms' tumor is."
    4. "I think the parents are in shock."

You're getting ready to cross the finish line!

Question 91 asks for an inaccurate assessment by the nurse.

88. 1. In nephrotic syndrome, the glomerular membrane of the kidneys becomes permeable to proteins. This results in massive proteinuria, which the nurse can detect with a reagent strip. Nephrotic syndrome typically doesn't cause glucosuria or ketonuria. Because the syndrome causes fluids to shift from plasma to interstitial spaces, it's more likely to decrease urine output than to cause polyuria (excessive urine output).
NP: Data collection; CNS: Physiological integrity; CN: Reduction of risk potential; CL: Application

89. 2. Surgery is the preferred treatment and is scheduled as soon as possible after confirmation of a renal mass, usually within 24 to 48 hours of admission, to make sure the encapsulated tumor remains intact.
NP: Implementation; CN: Safe, effective care environment; CNS: Coordinated care; CL: Comprehension

90. 3. The first action of the nurse should be to assess the client for pain. A 3-year-old child is too young to use a pain scale from 1 to 10 but can easily use the smiley face pain scale. After assessing the pain, the nurse should then investigate the time the pain medication was last given and administer the medication accordingly.
NP: Data collection; CN: Physiological integrity; CNS: Pharmacological therapies; CL: Application

91. 2. Parents *do* have the legal right to make decisions regarding the health issues for their child. Religion plays an important role in many people's lives and decisions about surgery and treatment for cancer are sometimes made that scientifically don't make sense to the health care provider. The parents are probably in a state of shock because a lot of information has been given and this is a cancer that requires decisions to be made quickly, especially surgical intervention.
NP: Evaluation; CN: Psychosocial integrity; CNS: Coping and adaptation; CL: Analysis

92. A child with a Wilms' tumor has had surgery to remove a kidney and has received chemotherapy. The nurse should include which of the following instructions at discharge?
　　1. Avoid contact sports.
　　2. Decrease fluid intake.
　　3. Decrease sodium intake.
　　4. Avoid contact with other children.

You're about to make contact with the end of the test.

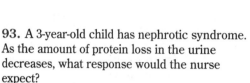

92. 1. Because the child is left with only one kidney, certain precautions such as avoiding contact sports are recommended to prevent injury to the remaining kidney. Decreasing fluid intake wouldn't be indicated; fluid intake is essential for renal function. Avoiding other children is unnecessary and will make the child feel self-conscious and may lead to regressive behavior. The child's sodium intake shouldn't be reduced.

NP: Implementation; CN: Health promotion and maintenance; CNS: Prevention and early detection of disease; CL: Application

93. A 3-year-old child has nephrotic syndrome. As the amount of protein loss in the urine decreases, what response would the nurse expect?
　　1. Weight gain
　　2. Increased hyperlipidemia
　　3. Decreased edema
　　4. Decreased appetite

I must say, you're doing an excellent job of answering these questions.

93. 3. The loss of protein in the urine results in decreased osmotic pressure in the vascular system, causing a fluid shift to the extravascular compartments and producing edema. As the proteinuria decreases, the edema decreases. Weight gain is consistent with proteinuria and a fluid shift from the vascular to the interstitial spaces. Hyperlipidemia and anorexia are associated with nephrotic syndrome but decrease as the disease resolves.

NP: Evaluation; CN: Physiological integrity; CNS: Reduction of risk potential; CL: Comprehension

94. When caring for a child after removal of a Wilms' tumor, which of the following findings would indicate the need to notify the primary care provider?
　　1. Fever of 101° F (38.3° C)
　　2. Absence of bowel sounds
　　3. Slight congestion in the lungs
　　4. Complaints of pain when moving

94. 2. After tumor removal, the child is at risk for intestinal obstruction. GI abnormalities require notification of the primary health care provider. A slight fever following surgery isn't uncommon, nor are slight congestion in the lungs and complaints of pain.

NP: Data collection; CN: Physiological integrity; CNS: Reduction of risk potential; CL: Application

95. A child with nephrotic syndrome develops generalized edema as a result of his nephrosis. Which goal would be included in the plan of care to prevent complications of edema?
　　1. Continually support the scrotum.
　　2. Change the child's position every 2 hours.
　　3. The child's skin will remain intact during hospitalization.
　　4. Maintain continuous bed rest.

95. 3. This is the only option written as a goal. The other options are interventions recommended to maintain skin integrity, not goals.

NP: Planning; CN: Physiological integrity; CNS: Basic care and comfort; CL: Application

96. In providing psychosocial care to a 6-year-old client who has had abdominal surgery for Wilms' tumor, which of the following activities initiated by the nurse would be the most appropriate?

1. Allowing the child to watch a 2-hour movie without interruptions
2. Giving the child a puzzle with 5 pieces to encourage him to move while in bed
3. Telling the child that you can give him enough medication so that he feels no pain
4. Providing the child with puppets and supplies and asking him to draw how he feels

97. A nurse should understand that staging of a Wilms' tumor helps to determine which of the following parameters?

1. Size of tumor
2. Level of treatment
3. Length of incision
4. Amount of anesthesia

98. The nurse is educating parents about Wilms' tumor. Which of the following statements made by a parent would indicate the need for further teaching?

1. "My child could have inherited this disease."
2. "Wilms' tumor can be associated with other congenital anomalies."
3. "This disease could have been a result of trauma to the baby in utero."
4. "There's no method to identify gene carriers of Wilms' tumor."

99. A 6-year-old boy's indwelling urinary catheter was removed at 6 a.m. At noon, the child still hasn't voided. He appears uncomfortable, and the nurse palpates slight bladder distention. Which action should the nurse take first?

1. Insert a straight catheter, as ordered, for urine retention.
2. Consult the physician about replacing the indwelling urinary catheter.
3. Wait awhile longer to see if the client can void on his own.
4. Turn on the water faucet and provide privacy.

This item is looking for the most appropriate answer. In other words, prioritize.

You didn't think you'd finish this test without another further teaching question, did you?

96. 4. A movie is a good diversion, but giving puppets and encouraging the child to draw his feelings is a better outlet. Many procedures have been performed on this client since admission. You probably can't give enough pain medication so that a person who has had surgery will feel no pain. A puzzle with only 5 pieces is too basic for a 6-year-old child and wouldn't hold his interest.
NP: Evaluation; CN: Psychosocial integrity; CNS: Coping and adaptation; CL: Application

97. 2. Staging the tumor helps to determine the level of treatment because it provides information about the level of involvement. The other choices aren't influenced by staging the tumor.
NP: Planning; CN: Health promotion and maintenance; CNS: Prevention and early detection of disease; CL: Knowledge

98. 3. Wilms' tumor isn't a result of trauma to the fetus in utero. Wilms' tumor can be genetically inherited and is associated with other congenital anomalies. There is, however, no method to identify gene carriers of Wilms' tumor at this time.
NP: Implementation; CN: Psychosocial integrity; CNS: Psychosocial adaptation; CL: Application

99. 4. Urine retention can result from many factors, including stress and use of opiates. Initially, the nurse should use independent nursing actions, such as providing the client with privacy, placing him in a sitting or standing position to enlist the aid of gravity and increase intra-abdominal pressure, and turning on the water faucet. If these measures are unsuccessful and the physician has left standing orders for straight catheterization, the nurse can proceed with the catheterization. Consulting the physician would involve the use of a dependent nursing action; independent actions should be attempted first. Waiting longer will only increase the child's distention and pain.
NP: Implementation; CN: Physiological integrity; CNS: Basic care and comfort; CL: Application

100. Which of the following actions by the nurse would be appropriate to take in a child diagnosed with a Wilms' tumor?
1. Take blood pressure in the right arm only.
2. Only offer clear liquids at room temperature.
3. Post a sign over the bed that reads, "Don't palpate abdomen."
4. Allow the child to participate in group activities in the playroom.

101. A 6-year-old girl has a history of repeated urinary tract infections (UTIs). She has been diagnosed with vesicoureteral reflux. Which of the following nursing responses would be most accurate to the mother who asks what the major complications are?
1. Damage to the ovaries that could lead to fertility problems
2. Minimal change nephrosis, which results in kidney damage and hypertension
3. Development of pyelonephritis and possible renal damage from the reflux of urine
4. Hemolytic syndrome that results in damage to the kidneys and abnormal spilling of protein

102. Which of the following treatments is administered for the correction of minimal change nephrosis?
1. Diuretics are given until the urine is protein-free.
2. Steroids are administered until the urine is protein-free.
3. Antihypertensives are given until the hypertension is resolved.
4. Ibuprofen is administered for the anti-inflammatory effect until the hematuria resolves.

103. Which of the following play activities would be appropriate for a 2-year-old child waiting for surgery for the removal of a Wilms' tumor?
1. Playing a group game in the playroom
2. Riding the toy scooter in the playroom
3. Playing in the playroom on the jungle gym
4. Playing with clay and cookie cutters in his room

100. 3. To reinforce the need for caution, it may be necessary to post a sign over the bed that reads, "Don't palpate abdomen." Careful bathing and handling are also important in preventing trauma to the tumor site; thus, group activities should be discouraged. The blood pressure could be taken in any extremity and, prior to surgery, there are usually no dietary restrictions.
NP: Data collection; CN: Health promotion and maintenance; CNS: Prevention and early detection of disease; CL: Application

101. 3. Vesicoureteral reflux is the abnormal reflux of urine from the bladder back up the ureters and possibly into the kidneys. Residual urine that isn't voided promotes growth of bacteria and results in a high incidence of UTI. This infected urine could reflux into the kidneys, resulting in pyelonephritis. Vesicoureteral reflux isn't associated with hemolytic syndrome, minimal change nephrosis, or damage to the ovaries.
NP: Implementation; CN: Physiological integrity; CNS: Physiological adaptation; CL: Knowledge

102. 2. Nephrotic syndrome is characteristic of glomerular damage that results in increased permeability to protein. Steroids are given until the urine is protein-free for several weeks. Ibuprofen doesn't treat hematuria. Diuretics don't treat glomerular permeability to protein. Antihypertensives are given as needed, but the evaluation criterion is protein in the urine.
NP: Planning; CN: Physiological integrity; CNS: Physiological adaptation; CL: Knowledge

Congratulations! You finished all 103 questions! Fantastic!

103. 4. A child with a diagnosis of a Wilms' tumor must be treated with extreme caution to prevent injury to the encapsulated tumor. Even palpation of the abdomen could result in spreading of the tumor. Therefore, the least active choice is playing with clay. The other choices are more physical activities that would carry a high risk of injury or falling.
NP: Planning; CN: Physiological integrity; CNS: Reduction of risk potential; CL: Application

NP: Nursing process   CN: Client needs category   CNS: Client needs subcategory   CL: Cognitive level

Chapter 35
Integumentary disorders

1. A 3-year-old child gets a burn at the angle of the mouth from chewing on an electrical cord. Which of the following findings should the nurse expect 10 days after the injury?
    1. Normal granular tissue
    2. Contracture of the injury site
    3. Ulceration with serous drainage
    4. Profuse bleeding from the injury site

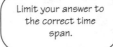

Limit your answer to the correct time span.

2. Providing adequate nutrition is essential for a burn client. Which of the following statements best describes a burn client's nutritional needs?
    1. A child needs 100 cal/kg during hospitalization.
    2. The hypermetabolic state after a burn injury leads to poor healing.
    3. Caloric needs can be lowered by controlling environmental temperature.
    4. Maintaining a hypermetabolic rate will lower the child's risk of infection.

Measuring burns in children is different from measuring them in adults.

3. A 1-year-old child is treated in the clinic for a burn to the left leg. Which of the following ways to measure burn size would be accurate for this child?
    1. The Rule of Nines
    2. Percentage based on the child's weight
    3. The child's palm equals 1% of the child's body surface area
    4. Percentage can't be determined without knowing the type of burn

1. 4. Ten days after oral burns from electrical cords, the eschar falls off, exposing arteries and veins. Burns to the oral cavity heal rapidly, but with contractures and scarring. Although contractures are likely, they aren't seen 10 days postinjury.
NP: Data collection; CN: Physiological integrity; CNS: Physiological adaptation; CL: Knowledge

2. 2. A burn injury causes a hypermetabolic state leading to protein and lipid catabolism, which affects wound healing. Caloric intake should be 1½ to 2 times the basal metabolic rate, with a minimum of 1.5 to 2 g/kg of body weight of protein daily. High metabolic rates increase the risk of infection. Keeping the temperature within a normal range lets the body function efficiently and use calories for healing and normal physiological processes. If the temperature is too warm or too cold, energy must be used for warming or cooling, taking energy away from tissue repair.
NP: Planning; CN: Physiological integrity; CNS: Basic care and comfort; CL: Knowledge

3. 3. The palm of a child is equal to 1% of that child's body surface. The Rule of Nines is used for children age 14 and older. The child's weight is important to calculate fluid replacement for extensive burns, not to estimate total body surface area. Burn type doesn't determine the percentage of body surface involved.
NP: Data collection; CN: Physiological integrity; CNS: Physiological adaptation; CL: Knowledge

NP: Nursing process   CN: Client needs category   CNS: Client needs subcategory   CL: Cognitive level

4. An 18-month-old child is admitted to the hospital for full-thickness burns to the anterior chest. The mother asks how the burn will heal. Which of the following statements is accurate about healing for full-thickness burns?
1. Surgical closure and grafting are usually needed.
2. Healing takes 10 to 12 days, with little or no scarring.
3. Pigment in a black client will return to the injured area.
4. Healing can take up to 6 weeks, with a high incidence of scarring.

You don't need a crystal ball to answer question 4.

4. 1. Full-thickness burns usually need surgical closure and grafting for complete healing. Deep partial-thickness burns heal in 6 weeks, with scarring. Healing in 10 to 12 weeks with little or no scarring is associated with superficial partial-thickness burns. With superficial partial-thickness burns, pigment is expected to return to the injured area after healing.
NP: Evaluation; CN: Physiological integrity; CNS: Physiological adaptation; CL: Knowledge

5. A 9-year-old child is admitted to the hospital with deep partial-thickness burns to 25% of his body. Which of the following findings is consistent with a deep partial-thickness burn?
1. Erythema and pain
2. Minimal damage to the epidermis
3. Necrosis through all layers of skin
4. Tissue necrosis through most of the dermis

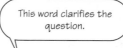

This word clarifies the question.

5. 4. A client with a deep partial-thickness burn will have tissue necrosis to the epidermis and dermis layers. Necrosis through all skin layers is seen with full-thickness injuries. Erythema and pain are characteristic of superficial injury. With deep burns, the nerve fibers are destroyed and the client won't feel pain in the affected area. Superficial burns are characteristic of slight epidermal damage.
NP: Data collection; CN: Physiological integrity; CNS: Physiological adaptation; CL: Knowledge

6. A 4-year-old child is admitted to the burn unit with a circumferential burn to the left forearm. Which of the following findings should be reported to the physician?
1. Numbness of fingers
2. +2 radial and ulnar pulses
3. Full range of motion and no pain
4. Bilateral capillary refill less than 2 seconds

6. 1. Circumferential burns can compromise blood flow to an extremity, causing numbness. Capillary refill less than 2 seconds indicates a normal vascular blood flow. Absence of pain and full range of motion imply good tissue oxygenation from intact circulation. +2 pulses indicate normal circulation.
NP: Data collection; CN: Physiological integrity; CNS: Physiological adaptation; CL: Knowledge

7. Which of the following facts should be given to the parents of a child with erythema infectiosum (fifth disease)?
1. There's a possible reappearance of the rash for up to 1 week.
2. Isolation of high-risk contacts should be avoided for 4 to 10 days.
3. Pregnant clients are at risk for fetal death if infected with fifth disease.
4. Children with fifth disease are contagious only while the rash is present.

7. 3. There's a 3% to 5% risk of fetal death from hydrops fetalis if a pregnant client is exposed during the first trimester. The cutaneous eruption of fifth disease can reappear for up to 4 months. A child with fifth disease is contagious during the first stage, when symptoms of headache, body aches, fever, and chills are present, not after the rash. The child should be isolated from pregnant women, immunocompromised clients, and clients with chronic anemia for up to 2 weeks.
NP: Implementation; CN: Safe, effective care environment; CNS: Safety and infection control; CL: Knowledge

NP: Nursing process CN: Client needs category CNS: Client needs subcategory CL: Cognitive level

8. A mother is concerned that her 3-year-old daughter has been exposed to erythema infectiosum (fifth disease). Which of the following symptoms are characteristic of this viral infection?

    1. A fine, erythematous rash with a sandpaper-like texture

    2. Intense redness of both cheeks that may spread to the extremities

    3. Low-grade fever, followed by vesicular lesions of the trunk, face, and scalp

    4. Three- to five-day history of sustained fever, followed by a diffuse erythematous maculopapular rash

Different symptoms indicate different diagnoses.

8. 2. The classic symptoms of erythema infectiosum begin with intense redness of both cheeks. An erythematous rash after a fever is characteristic of roseola. Children with varicella typically have vesicular lesions of the trunk, face, and scalp after a low-grade fever. An erythematous rash with a sandpaper-like texture is associated with scarlet fever, which is a bacterial infection.

NP: Data collection; CN: Physiological integrity; CNS: Physiological adaptation; CL: Knowledge

9. A family that recently went camping brings their child to the clinic with a complaint of a rash after a tick bite. Lyme disease is suspected. Which of the following findings would be seen with Lyme disease?

    1. Erythematous rash surrounding a necrotic lesion

    2. Bright rash with red outer border circling the bite site

    3. Onset of a diffuse rash over the entire body 2 months after exposure

    4. A linear rash of papules and vesicles that occur 1 to 3 days after exposure

9. 2. A bull's eye rash is a classic symptom of Lyme disease. Necrotic, painful rashes are associated with the bite of a brown recluse spider. In Lyme disease, the rash is located primarily at the site of the bite. A linear, papular, vesicular rash indicates exposure to the leaves of poison ivy.

NP: Data collection; CN: Physiological integrity; CNS: Physiological adaptation; CL: Knowledge

Read the results at the right time!

10. A Mantoux test is ordered for a 6-year-old child. Which of the following directions about this test is accurate?

    1. Read results within 24 hours.

    2. Read results 48 to 72 hours later.

    3. Use the large muscle of the upper leg.

    4. Massage the site to increase absorption.

10. 2. The test should be read 48 to 72 hours after placement by measuring the diameter of the induration that develops at the site. The purified protein derivative is injected intradermally on the volar surface of the forearm. Massaging the site could cause leakage from the injection site.

NP: Data collection; CN: Physiological integrity; CNS: Reduction of risk potential; CL: Knowledge

11. Which of the following statements about Kawasaki disease is true?

    1. It's highly contagious.

    2. It's an afebrile condition with cardiac involvement.

    3. It usually occurs in children older than 5 years.

    4. Prolonged fever, with peeling of the fingers and toes, are the initial symptoms.

11. 4. To be diagnosed with Kawasaki disease, the child must have a fever for 5 days or more, plus four of the following five symptoms: bilateral conjunctivitis, changes in the oral mucosa and peripheral extremities, rash, and lymphadenopathy. Kawasaki disease is more likely to occur in children younger than age 5. It isn't contagious.

NP: Data collection; CN: Physiological integrity; CNS: Physiological adaptation; CL: Knowledge

12. A 22-lb child is diagnosed with Kawasaki disease and started on gamma globulin therapy. The physician orders an I.V. infusion of gamma globulin, 2 g/kg, to run over 12 hours. Which of the following doses is correct?
1. 11 g
2. 20 g
3. 22 g
4. 44 g

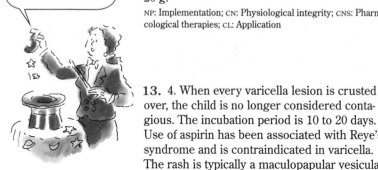

You need more than magic to find the correct dosage. You need math.

12. 2. One kilogram equals 2.2 pounds, so a 22-pound child weighs 10 kg. 2 g × 10 kg = 20 g. Convert the weight to kilograms, 22 (lb) ÷ 2.2 = 10 (kg). Calculate the dose: 2 g × 10 = 20 g.
NP: Implementation; CN: Physiological integrity; CNS: Pharmacological therapies; CL: Application

13. A mother is concerned because her child was exposed to varicella in day care. Which of the following statements is true about varicella?
1. The rash is nonvesicular.
2. The treatment of choice is aspirin.
3. Varicella has an incubation period of 5 to 10 days.
4. A child is no longer contagious after the rash has crusted over.

13. 4. When every varicella lesion is crusted over, the child is no longer considered contagious. The incubation period is 10 to 20 days. Use of aspirin has been associated with Reye's syndrome and is contraindicated in varicella. The rash is typically a maculopapular vesicular rash.
NP: Data collection; CN: Physiological integrity; CNS: Physiological adaptation; CL: Knowledge

14. Which of the following statements is correct about the rash associated with varicella?
1. It's diagnostic in the presence of Koplik's spots in the oral mucosa.
2. It's a maculopapular rash starting on the scalp and hairline and spreading downward.
3. It's a vesicular maculopapular rash that appears abruptly on the trunk, face, and scalp.
4. It appears as yellow ulcers surrounded by red halos on the surface of the hands and feet.

You're doing great! Keep up the outstanding work!

14. 3. In varicella, teardrop vesicles on an erythematous base generally begin on the trunk, face, and scalp, with minimal involvement of the extremities. A descending maculopapular rash is characteristic of rubeola and Koplik's spots are diagnostic of the disease. Yellow ulcers of the hands and feet are associated with Coxsackie virus, or hand-foot-and-mouth disease.
NP: Data collection; CN: Physiological integrity; CNS: Physiological adaptation; CL: Knowledge

15. A child is brought to the emergency department after an extended period of sledding. Frostbite of the hands is suspected. Which of the following statements is true about frostbite?
1. The skin is white.
2. The skin looks deeply flushed and red.
3. Frostbite is helped by rubbing to increase circulation.
4. Slow gradual rewarming of the extremities with hot water is needed.

15. 1. Signs and symptoms of frostbite include tingling, numbness, burning sensation, and white skin. Gradual rewarming by exposure to hot water can lead to more tissue damage. Treatment includes very gentle handling of the affected area. Rubbing is contraindicated as it can damage fragile tissue.
NP: Data collection; CN: Physiological integrity; CNS: Physiological adaptation; CL: Knowledge

---

NP: Nursing process CN: Client needs category CNS: Client needs subcategory CL: Cognitive level

16. A mother brings her child to the physician's office because the child complains of pain, redness, and tenderness of the left index finger. The child is diagnosed with paronychia. Which of the following organisms is the most likely cause of this superficial abscess of the cuticle?

    1. *Borrelia burgdorferi*
    2. *Escherichia coli*
    3. *Pseudomonas* species
    4. *Staphylococcus* species

17. Which of the following treatments for paronychia would be the most appropriate?

    1. Drain the abscess and give warm soaks.
    2. Splint and put ice on the affected finger.
    3. Allow the infection to resolve without treatment.
    4. Admit the child to the hospital for I.V. antibiotic therapy.

18. A mother is concerned that her 9-month-old infant has scabies. Which of the following assessment findings is associated with this infestation?

    1. Diffuse pruritic wheals
    2. Oval white dots stuck to the hair shafts
    3. Pain, erythema, and edema with an embedded stinger
    4. Pruritic papules, pustules, and linear burrows of the finger and toe webs

19. After treatment with permethrin (Elimite) for scabies, the mother of a 16-month-old child is concerned the cream didn't work because the child is still scratching. Which of the following explanations or instructions would be correct?

    1. Continue the application daily until the rash disappears.
    2. Pruritus caused by secondary reactions to the mites can be present for weeks.
    3. Stop treatment because the cream is unsafe for children younger than age 2.
    4. Pruritus caused by permethrin is usually present in children younger than age 5.

Which germ are you?

Assess the symptoms.

16. 4. Paronychia is a localized infection of the nail bed caused by either staphylococci or streptococci. *Pseudomonas* species are associated with ecthyma. *Escherichia coli* is associated with urinary tract infections. *Borrelia burgdorferi* is responsible for Lyme disease.
NP: Data collection; CN: Physiological integrity; CNS: Physiological adaptation; CL: Knowledge

17. 1. Draining the abscess and giving warm soaks is the treatment of choice for paronychia. Splinting and icing aren't indicated. Untreated, the local abscess can spread beneath the nail bed, called secondary lymphangitis. I.V. antibiotic therapy isn't needed if the abscess is kept from spreading.
NP: Planning; CN: Physiological integrity; CNS: Physiological adaptation; CL: Knowledge

18. 4. Pruritic papules, vesicles, and linear burrows are diagnostic for scabies. Urticaria is associated with an allergic reaction. Nits, seen as white oval dots, are characteristic of head lice. Bites from honeybees are associated with a stinger, pain, and erythema.
NP: Data collection; CN: Physiological integrity; CNS: Physiological adaptation; CL: Knowledge

19. 2. Sensitization of the host is the cause of the intense itching and can last for weeks. Permethrin is the recommended treatment for scabies in infants as young as 2 months. It can safely be repeated after 2 weeks.
NP: Evaluation; CN: Physiological integrity; CNS: Pharmacological therapies; CL: Knowledge

20. The mother of a 5-month-old infant is planning a trip to the beach and asks for advice about sunscreen. Which of the following instructions should the nurse give the mother?
　1. The sun protection factor (SPF) of the sunscreen should be at least 10.
　2. Apply sunscreen to the exposed areas of the skin.
　3. Sunscreen shouldn't be applied to infants younger than age 6 months.
　4. Sunscreen needs to be applied heavily only once, 30 minutes before going out in the sun.

21. An infant is being treated with antibiotic therapy for otitis media and develops an erythematous, fine, raised rash in the groin and suprapubic area. Which of the following instructions or explanations will most likely be given to the mother?
　1. The infant has candidiasis.
　2. Change the brand of diapers.
　3. Use an over-the-counter diaper remedy.
　4. Stop the antibiotic therapy immediately.

22. The skin in the diaper area of a 6-month-old infant is excoriated and red. Which of the following instructions would the nurse give to the mother?
　1. Change the diaper more often.
　2. Apply talcum powder with diaper changes.
　3. Wash the area vigorously with each diaper change.
　4. Decrease the infant's fluid intake to decrease saturating diapers.

23. A 9-year-old child is being discharged from the hospital after severe urticaria caused by allergy to nuts. Which of the following instructions would be included in discharge teaching for the child's parents?
　1. Use emollient lotions and baths.
　2. Apply topical steroids to the lesions as needed.
　3. Apply over-the-counter products such as diphenhydramine (Benadryl).
　4. Use an epinephrine administration kit and follow up with an allergist.

Teaching parents keeps the children healthy.

Whole lotta client teaching going on!

20. 3. Sunscreen isn't recommended for use in infants younger than age 6 months. These children should be dressed in cool light clothes and kept in the shade. Sunscreen should be applied evenly throughout the day and each time the child is in the water. The SPF for children should be 15 or greater. Sunscreen should be applied to all areas of the skin.
NP: Implementation; CN: Health promotion and maintenance; CNS: Prevention and early detection of disease; CL: Knowledge

21. 1. Candidiasis, caused by yeastlike fungi, can occur with the use of antibiotics. Changing the brand of diapers or suggesting that the parent use an over-the-counter remedy would be appropriate for treating diaper rash, not candidiasis. Antibiotic therapy shouldn't be stopped. The treatment for candidiasis is topical nystatin ointment.
NP: Data collection; CN: Physiological integrity; CNS: Physiological adaptation; CL: Knowledge

22. 1. Simply decreasing the amount of time the skin comes in contact with wet, soiled diapers will help heal the irritation. Talc is contraindicated in children because of the risks associated with inhalation of the fine powder. Gentle cleaning of the irritated skin should be encouraged. Infants shouldn't have fluid intake restrictions.
NP: Data collection; CN: Safe, effective care environment; CNS: Safety and infection control; CL: Application

23. 4. Children who have urticaria in response to nuts, seafood, or bee stings should be warned about the possibility of anaphylactic reactions to future exposure. The use of epinephrine pens should be taught to the parents and older children. Other treatment choices, such as Benadryl, topical steroids, and emollients, are for the treatment of mild urticaria.
NP: Planning; CN: Health promotion and maintenance; CNS: Prevention and early detection of disease; CL: Application

24. When examining a nursery school-age child, the nurse finds multiple contusions over the body. Child abuse is suspected. Which of the following statements indicates the findings that should be documented?
　　1. Contusions confined to one body area are typically suspicious.
　　2. All lesions, including location, shape, and color, should be documented.
　　3. Natural injuries usually have straight linear lines, while injuries from abuse have multiple curved lines.
　　4. The depth, location, and amount of bleeding that initially occurs is constant, but the sequence of color change is variable.

25. A 7-year-old child is diagnosed with head lice. The mother asks how to get the nits out of the hair. The nits represent which of the following parts of the louse's life cycle?
　　1. Adult
　　2. Empty eggshells
　　3. Newly laid eggs
　　4. Nymph

26. For which of the following reasons is lindane shampoo used only as a second-line treatment for lice?
　　1. Lindane causes alopecia.
　　2. Lindane causes hypertension.
　　3. Lindane is associated with seizures.
　　4. Lindane increases liver function test (LFT) results.

27. Which of the following instructions should be given to the parents about the treatment of hair lice?
　　1. The treatment should be repeated in 7 to 12 days.
　　2. Treatment should be repeated every day for 1 week.
　　3. If treated with a shampoo, combing to remove eggs isn't necessary.
　　4. All contacts with the infested child should be treated even without evidence of infestation.

Be aware of the signs of child abuse. It's the nurse's responsibility to report it.

What do we need to know about lice treatment?

24. 2. An accurate precise examination must be properly documented as a legal document. Contusions that result from falls are typically confined to a single body area and are considered a reasonable finding of a child still learning to walk. Injuries from normal falls are usually not linear in nature. Bleeding can cause variations, but color change is consistent.
NP: Data collection; CN: Health promotion and maintenance; CNS: Prevention and early detection of disease; CL: Application

25. 2. The mother is finding empty eggshells in the child's hair. Adults are the last stage of development, living about 30 days. Nymphs are the newly hatched lice and become adults in 8 to 9 days. Newly laid eggs are small, translucent, and difficult to see.
NP: Data collection; CN: Physiological integrity; CNS: Physiological adaptation; CL: Knowledge

26. 3. Lindane is associated with seizures after absorption with topical use. Alopecia, increased LFT results, and hypertension aren't associated with the use of lindane.
NP: Implementation; CN: Physiological integrity; CNS: Pharmacological therapies; CL: Knowledge

27. 1. Treatment should be repeated in 7 to 12 days to ensure that all eggs are killed. Combing the hair thoroughly is necessary to remove the lice eggs. People exposed to head lice should be examined to assess the presence of infestation before treatment.
NP: Planning; CN: Physiological integrity; CNS: Physiological adaptation; CL: Knowledge

28. A mother reports that her 4-year-old child has been scratching at his rectum recently. Which of the following infestations or conditions would the nurse suspect?
1. Anal fissure
2. Lice
3. Pinworms
4. Scabies

29. Diagnosing pinworms by the clear cellophane tape test is preferred. How many tests are necessary to detect infestations at virtually 100% accuracy?
1. One
2. Three
3. Five
4. Ten

30. Each member of the family of a child diagnosed with pinworms is prescribed a single dose of pyrantel (Antiminth). Which of the following statements is correct about pyrantel?
1. The drug may stain the feces red.
2. The dose may be repeated in 2 weeks.
3. Fever and rash are common adverse effects.
4. The medicine will kill the eggs in about 48 hours.

31. A large dog bit the hand of a child. Which of the following types of injury is associated with bites from large dogs?
1. Abrasion
2. Crush injury
3. Fracture
4. Puncture wound

32. The nurse understands that bites from dogs heighten the risk of infection. Which of the following interventions should be done to help prevent infection?
1. Give the rabies vaccine.
2. Give antibiotics immediately.
3. Clean and irrigate the wounds.
4. Nothing; bites from dogs have a low incidence of infection.

You've finished more than one-quarter of the questions.

If I come from a dog, grrr, watch out!

28. 3. The clinical sign of pinworms is perianal itching that increases at night. Anal fissures are associated with rectal bleeding and pain with bowel movements. Lice are infestations of the hair. Scabies are associated with a pruritic rash characterized as linear burrows of the webs of the fingers and toes.
NP: Data collection; CN: Physiological integrity; CNS: Physiological adaptation; CL: Knowledge

29. 3. Detection is virtually 100% accurate with five tests. Three tests should detect infestations at about 90% accuracy. One test is only 50% accurate.
NP: Data collection; CN: Health promotion and maintenance; CNS: Prevention and early detection of disease; CL: Application

30. 2. Pyrantel is effective against the adult worms only, so treatment can be repeated to eradicate any emerging parasites in 2 weeks. Staining the feces isn't associated with pyrantel. Common adverse effects are headaches and abdominal complaints.
NP: Data collection; CN: Physiological integrity; CNS: Pharmacological therapies; CL: Knowledge

31. 2. Although the bite of a large dog can exert pressure of 150 to 400 lb per square inch, the bite causes crush injuries, not fractures. Abrasions are associated with friction injuries. Puncture wounds are associated with smaller animals such as cats.
NP: Data collection; CN: Physiological integrity; CNS: Physiological adaptation; CL: Knowledge

32. 3. Not every dog bite requires antibiotic therapy, but cleaning the wound is necessary for all injuries involving a break in the skin. Rabies vaccine is used if the dog is suspected of having rabies. The infection rate for dog bites has been reported to be as high as 50%.
NP: Implementation; CN: Physiological integrity; CNS: Reduction of risk potential; CL: Application

33. When collecting data from a 6-year-old burn client who has a 20% deep partial-thickness (second-degree) burn of the arms and trunk, the nurse understands that the client has damage to what layers of skin?
1. Epidermis
2. Epidermis and part of the dermis
3. Epidermis and all of the dermis
4. Dermis and subcutaneous tissue

34. A child is brought to the office for multiple scratches and bites from a kitten and is being evaluated for cat-scratch disease. While collecting data, which of the following symptoms would the nurse expect to find?
1. Abdominal pain
2. Adenitis
3. Fever
4. Pruritus

35. In which of the following populations is giardiasis the most common parasitic intestinal infection in the United States?
1. Children riding a school bus
2. Children playing on a playground
3. Children attending a sporting event
4. Children attending group day care or nursery school

36. Which of the following characteristics would apply to such lesions as papules, nodules, and tumors?
1. Palpable elevated masses
2. Loss of the epidermal layer
3. Fluid-filled elevations of the skin
4. Nonpalpable flat changes in skin color

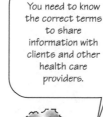

You need to know the correct terms to share information with clients and other health care providers.

37. A child is diagnosed with impetigo. Pustules are the primary lesions found on this child. Which of the following descriptions is correct for pustules?
1. Lesion filled with pus
2. Superficial area of localized edema
3. Serous-filled lesion less than 0.5 cm
4. Serous-filled lesion greater than 0.5 cm

33. 2. A deep partial-thickness burn affects the epidermis and part of the dermis. A superficial partial-thickness (first-degree) burn affects the epidermis and all the dermis; it may also affect the subcutaneous tissue.
NP: Data collection; CN: Physiological integrity; CNS: Physiological adaptation; CL: Application

34. 2. Adenitis is the primary feature of cat-scratch disease. Although low-grade fever has been associated with cat-scratch disease, it's only present 25% of the time. Pruritus and abdominal pain aren't symptoms of cat-scratch disease.
NP: Data collection; CN: Physiological integrity; CNS: Physiological adaptation; CL: Knowledge

35. 4. The most common intestinal parasitic infection in the United States is giardiasis, prevalent among children attending group day care or nursery school. Playgrounds, sporting events, and school buses don't present unusual risk of giardiasis.
NP: Data collection; CN: Safe, effective care environment; CNS: Safety and infection control; CL: Knowledge

36. 1. Papules are elevated up to 0.5 cm. Nodules and tumors are elevated more than 0.5 cm. Erosions are characterized as loss of the epidermal layer. Fluid-filled lesions are vesicles and pustules. Macules and patches are described as nonpalpable flat changes in skin color.
NP: Data collection; CN: Health promotion and maintenance; CNS: Prevention and early detection of disease; CL: Knowledge

37. 1. Pustules are pus-filled lesions, such as acne and impetigo. Bullae are serous-filled lesions greater than 0.5 cm in diameter. A wheal is a superficial area of localized edema. Vesicles are serous-filled lesions up to 0.5 cm in diameter.
NP: Data collection; CN: Physiological integrity; CNS: Physiological adaptation; CL: Knowledge

38. A 5-year-old male sustained third-degree burns to the right upper extremity after tipping over a frying pan. The nurse explains to the mother that a third-degree burn involves the:
   1. epidermis only.
   2. epidermis and dermis.
   3. skin layers and nerve endings.
   4. skin layers, nerve endings, muscles, tendons, and bone.

39. A child is brought to the physician's office for treatment of a rash. Many petechiae are seen over his entire body. Multiple petechiae would be consistent with which of the following conditions or symptoms?
   1. Bleeding disorder
   2. Scabies
   3. Varicella
   4. Vomiting

40. A child fell at camp and sustained a bruise to his thigh. Which of the following descriptions would accurately describe the bruise after 1 week?
   1. Resolved
   2. Reddish blue
   3. Greenish yellow
   4. Dark blue to bluish brown

41. Which of the following factors about contusions suggest child abuse?
   1. Multiple contusions of the shins
   2. Contusions of the back and buttocks
   3. Contusions at the same stages of healing
   4. Large contusion and hematoma of the forehead

42. Which of the following statements about salmon patches (stork bites) is correct?
   1. They're benign and usually fade in adult life.
   2. They're usually associated with syndromes of the neonate.
   3. They can cause mild hypertrophy of the muscle associated with the lesion.
   4. They're treatable with laser pulse surgery in late adolescence and adulthood.

You need to look into these questions to get the correct answers.

Don't fall down now! Keep at it!

Know what to tell a family when a neonate is born with salmon patches.

38. 3. A third-degree burn involves all of the skin layers and the nerve endings. First-degree burns involve only the epidermis. Second-degree burns affect the epidermis and dermis. Fourth-degree burns involve all skin layers, nerve endings, muscles, tendons, and bone.
NP: Evaluation; CN: Physiological integrity; CNS: Physiological adaptation; CL: Knowledge

39. 1. Petechiae are caused by blood outside a vessel, associated with low platelet counts and bleeding disorders. Petechiae aren't found with varicella disease or scabies. Petechiae can be associated with vomiting, but they would be present on the face, not the entire body.
NP: Data collection; CN: Physiological integrity; CNS: Physiological adaptation; CL: Knowledge

40. 3. After 7 to 10 days, the bruise becomes greenish yellow. Resolution can take up to 2 weeks. Initially after the fall, there's a reddish blue discoloration, followed by dark blue to bluish brown at days 1 to 3.
NP: Evaluation; CN: Physiological integrity; CNS: Physiological adaptation; CL: Application

41. 2. Contusions of the back and buttocks are highly suggestive of abuse related to punishment. Contusions at various stages of healing are red flags to potential abuse. Contusions of the shins and forehead are usually related to an active toddler falling and bumping into objects.
NP: Data collection; CN: Health promotion and maintenance; CNS: Prevention and early detection of disease; CL: Knowledge

42. 1. Salmon patches occur over the back of the neck in 40% of neonates and are harmless, needing no intervention. Port-wine stains are associated with syndromes of neonates, such as Sturge-Weber syndrome. Port-wine stains found on the face or extremities may be associated with soft tissue and bone hypertrophy. Laser pulse surgery isn't recommended for salmon patches because they typically fade on their own in adulthood.
NP: Data collection; CN: Health promotion and maintenance; CNS: Growth and development through the life span; CL: Knowledge

NP: Nursing process   CN: Client needs category   CNS: Client needs subcategory   CL: Cognitive level

43. A neonate is born with a blue-black macular lesion over the lower lumbar sacral region. Which of the following terms applies to this lesion?
1. Café-au-lait spots
2. Mongolian spots
3. Nevis of Ota
4. Stork bites

The word *severe* is a clue to the correct answer.

43. 2. Mongolian spots are large blue-black macular lesions generally located over the lumbosacral areas, buttocks, and limbs. Café-au-lait spots occur between ages 2 and 16, not in infancy. Nevis of Ota is found surrounding the eyes. Stork bites, or salmon patches, occur at the neck and hairline area.

NP: Data collection; CN: Health promotion and maintenance; CNS: Growth and development through the life span; CL: Knowledge

44. Which of the following findings indicate severe dehydration in a child?
1. Gray skin and decreased tears
2. Capillary refill less than 2 seconds
3. Mottled skin and tenting of the skin
4. Pale skin with dry mucous membranes

44. 3. Severe dehydration is associated with mottling and tenting of the skin. Pale skin with dry mucous membranes is a sign of mild dehydration. Malnutrition is characterized by gray skin and tenting of the skin. Capillary refill less than 2 seconds is normal.

NP: Data collection; CN: Health promotion and maintenance; CNS: Prevention and early detection of disease; CL: Knowledge

45. Isotretinoin (Accutane) is associated with which of the following adverse effects?
1. GI upset
2. Gram-negative folliculitis
3. Teratogenesis
4. Vaginal candidiasis

45. 3. The use of even small amounts of isotretinoin has been associated with severe birth defects. Most female clients are prescribed oral contraceptives. Clindamycin (Cleocin T), another medicine used in the treatment of acne, is associated with both diarrhea and gram-negative folliculitis. Tetracycline (Sumycin) is associated with yeast infections.

NP: Data collection; CN: Health promotion and maintenance; CNS: Growth and development through the life span; CL: Knowledge

46. Which of the following characteristics is commonly responsible for the failure of treatment of acne in teenagers?
1. Topical treatment
2. Systemic treatment
3. A dominant parent who wants treatment and a passive teenager who doesn't
4. A dominant teenager who wants treatment and a passive, uninterested parent

46. 3. The active participation of a teenager is needed for the successful treatment of acne. Systemic and topical therapy are needed in most acne treatment.

NP: Data collection; CN: Health promotion and maintenance; CNS: Growth and development through the life span; CL: Knowledge

47. Which of the following information should be given to a teenager about acne?
1. Acne is caused by diet.
2. Acne is related to gender.
3. Acne is caused by poor hygiene.
4. Acne is caused by hormonal changes.

47. 4. Acne is caused by hormonal changes in sebaceous gland anatomy and the biochemistry of the glands. These changes lead to a blockage in the follicular canal and cause an inflammatory response. Diet, hygiene, and the client's gender don't cause acne.

NP: Planning; CN: Health promotion and maintenance; CNS: Growth and development through the life span; CL: Knowledge

48. When teaching a client about tetracycline for severe inflammatory acne, which of the following instructions must be given?
1. Take the drug with or without meals.
2. Take the drug with milk and milk products.
3. Take the drug on an empty stomach with small amounts of water.
4. Take the drug 1 hour before or 2 hours after meals with large amounts of water.

49. When advising parents about the prevention of burns from tap water, which of the following instructions should be given?
1. Set the water-heater temperature at 130° F (54.4° C) or less.
2. Run the hot water first, then adjust the temperature with cold water.
3. Before you put your infant in the tub, first test the water with your hand.
4. Supervise an infant in the bathroom, only leaving him for a few seconds if needed.

50. A 14-year-old male client is brought to the hospital with smoke inhalation due to a house fire. The nurse's major intervention for this client is to:
1. check the oral mucous membranes.
2. check for any burned areas.
3. obtain a medical history.
4. ensure a patent airway.

51. A 15-month-old child is diagnosed with pediculosis of the eyebrows. Which of the following interventions are included in the treatment?
1. Use lindane.
2. Use petroleum jelly.
3. Shave the eyebrows.
4. No treatment is needed.

52. A 13-year-old client has received third-degree burns over 20% of his body. When observing this client 72 hours after the burn, which of the following findings should the nurse expect?
1. Increased urine output
2. Severe peripheral edema
3. Respiratory distress
4. Absent bowel sounds

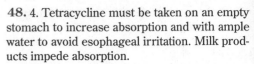

Parent teaching prevents future mishaps.

48. 4. Tetracycline must be taken on an empty stomach to increase absorption and with ample water to avoid esophageal irritation. Milk products impede absorption.
NP: Implementation; CN: Physiological integrity; CNS: Pharmacological therapies; CL: Application

49. 3. The cold water should be run first and then adjusted with hot water. Instruct the parents to fill the tub with water first and then test all of the water in the tub with their hand for hot spots. Water heaters should be set at 120° F (48.9° C). Never leave a infant alone in the bathroom, even for a second.
NP: Planning; CN: Health promotion and maintenance; CNS: Prevention and early detection of disease; CL: Knowledge

Hip! Hip! Hurray for you! You're about halfway there.

50. 4. The nurse's top priority is to make sure the airway is open and the client is breathing. Checking the mucous membranes and burned areas is important but not as vital as maintaining a patent airway. Obtaining a medical history can be pursued after ensuring a patent airway.
NP: Evaluation; CN: Physiological integrity; CNS: Physiological adaptation; CL: Application

51. 2. Petroleum jelly should be applied twice daily for 8 days, followed by manual removal of nits. Lindane is contraindicated because of the risk of seizures. The eyebrow should never be shaved because of the uncertainty of hair return.
NP: Implementation; CN: Physiological integrity; CNS: Physiological adaptation; CL: Knowledge

52. 1. During the resuscitative-emergent phase of a burn, fluids shift back into the interstitial space resulting in the onset of diuresis. Edema resolves during the emergent phase, when fluid shifts back to the intravascular space. Respiratory rate increases during the first few hours as a result of edema. When edema resolves, respirations return to normal. Absent bowel sounds occur in the initial stage.
NP: Data collection; CN: Physiological integrity; CNS: Physiological adaptation; CL: Knowledge

NP: Nursing process CN: Client needs category CNS: Client needs subcategory CL: Cognitive level

53. Which of the following changes in the mouth is consistent with Kawasaki disease?
1. Koplik's spots
2. Tonsillar exudate
3. Vesicular lesions
4. Dry cracked lips, strawberry tongue

Which symptom goes with which diagnosis?

53. 4. Oral changes associated with Kawasaki disease include an injected pharynx, injected lips, dry fissured lips, and strawberry tongue. Koplik's spots are consistent with measles. Vesicular lesions are associated with Coxsackie virus. Tonsillar exudate is consistent with pharyngitis caused by group A beta-hemolytic streptococci.

NP: Data collection; CN: Physiological integrity; CNS: Physiological adaptation; CL: Knowledge

54. The nurse is assessing a 3-month-old male infant during a routine examination in a family health center. The nurse notes the presence of bluish discolorations of the skin. Such markings, which are common in babies of Black, Native American, and Mediterranean descent, are called:
1. milia.
2. Mongolian spots.
3. lanugo.
4. vernix caseosa.

54. 2. Bluish discolorations of the skin, which are common in babies of Black, Native American, and Mediterranean races, are called Mongolian spots. Pinpoint pimples caused by obstruction of sebaceous glands are called milia. The fine hair covering the body of a neonate is called lanugo. Vernix caseosa is a cheeselike substance that covers the skin of a neonate.

NP: Data collection; CN: Health promotion and maintenance; CNS: Growth and development through the life span; CL: Knowledge

55. Topical treatment with 2.5% hydrocortisone is prescribed for a 6-month-old infant with eczema. The mother is instructed to use the cream for no more than 1 week. Why is this time limit appropriate?
1. The drug loses its efficacy after prolonged use.
2. Excessive use can have adverse effects, such as skin atrophy and fragility.
3. If no improvement is seen, a stronger concentration will be prescribed.
4. If no improvement is seen after 1 week, an antibiotic will be prescribed.

Teach the parents why the time limit is important.

55. 2. Hydrocortisone cream should be used for brief periods to decrease adverse effects, such as atrophy of the skin. The drug doesn't lose efficacy after prolonged use, a stronger concentration may not be prescribed if no improvement is seen, and an antibiotic would be inappropriate in this instance.

NP: Implementation; CN: Physiological integrity; CNS: Pharmacological therapies; CL: Knowledge

56. A 1-year-old infant is hospitalized with a diagnosis of eczema. Which signs and symptoms does the nurse expect to observe?
1. Exudative, crusty, papulovesicular, erythematous lesions on the cheeks, scalp, forehead, and arms
2. Erythematous, dry, scaly, well-circumscribed, papular, thickened, and lichenified pruritic lesions on wrists, hands, and neck
3. Large, thickened, lichenified plaques on the face, neck, and back
4. Erythematous papules with oozing, crusting, and edema

56. 1. Exudative, crusty, papulovesicular, erythematous lesions on the cheeks, scalp, forehead, and arms are observed in children ages 2 months to 2 years with a diagnosis of eczema. Erythematous, dry, scaly, well-circumscribed, papular, thickened, lichenified, pruritic lesions on the wrists, hands, and neck are observed in children ages 2 years to puberty. In adolescents, lesions on the face, neck, and back consist of large plaques that are thickened and lichenified. Erythematous papules with oozing, crusting, and edema are characteristic of contact dermatitis.

NP: Data collection; CN: Physiological integrity; CNS: Physiological adaptation; CL: Comprehension

57. A 4-year-old child had a subungual hemorrhage of the toe after a jar fell on his foot. Electrocautery is performed. For which of the following reasons is electrocautery done?
1. Prevent loss of nail growth
2. Prevent loss of the nail
3. Relieve pain and reduce the risk of infection
4. Prevent permanent discoloration of the nail bed

58. The nurse is caring for a 12-year-old child with a diagnosis of eczema. Which nursing interventions are appropriate for a child with eczema?
1. Administer antibiotics as prescribed.
2. Administer antifungals as ordered.
3. Administer tepid baths and pat dry or air-dry the affected areas.
4. Administer hot baths and use moisturizers immediately after the bath.

59. A 9-year-old child is brought to the emergency department with extensive burns received in a restaurant fire. What's the most important aspect of caring for the burned child?
1. Administering antibiotics to prevent superimposed infections
2. Conducting wound management
3. Administering liquids orally to replace fluid
4. Administering frequent small meals to support nutritional requirements

Number 59 is asking you to prioritize!

60. A mother of a 4-month-old infant asks about the strawberry hemangioma on his cheek. Which of the following statements is correct about strawberry hemangiomas?
1. The lesion will continue to grow for 3 years, then need surgical removal.
2. If the lesion continues to enlarge, referral to a pediatric oncologist is warranted.
3. Surgery is indicated before age 12 months if the diameter of the lesion is greater then 3 cm.
4. The lesion will continue to grow until age 12 months, then begin to resolve by age 2 to 3 years.

57. 3. The hematoma is treated with electrocautery to relieve pain and reduce the risk of infection. The discoloration seen with subungual hemorrhage is from the collection of blood under the nail bed. It isn't permanent and doesn't affect nail growth. Electrocautery doesn't prevent the loss of the nail.
NP: Data collection; CN: Physiological integrity; CNS: Physiological adaptation; CL: Knowledge

58. 3. Tepid baths and moisturizers are indicated to keep the infected areas clean and minimize itching. Antibiotics are given only when superimposed infection is present. Antifungals aren't usually administered in the treatment of eczema. Hot baths can exacerbate the condition and increase itching.
NP: Implementation; CN: Physiological integrity; CNS: Physiological adaptation; CL: Application

59. 2. The most important aspect of caring for a burned child is wound management. The goals of wound care are to speed debridement, protect granulation tissue and new grafts, and conserve body heat and fluids. Antibiotics aren't always administered prophylactically. Fluids are administered I.V. according to the child's body weight to replace volume. Enteral feedings, rather than meals, are initiated within the first 24 hours after the burn to support the child's increased nutritional requirements.
NP: Implementation; CN: Physiological integrity; CNS: Physiological adaptation; CL: Application

60. 4. These rapidly growing vascular lesions reach maximum growth by age 1 year. The growth period is then followed by an involution period of 6 to 12 months. Lesions show complete involution by age 2 or 3 years. These benign lesions don't need surgical or oncologic referrals.
NP: Data collection; CN: Health promotion and maintenance; CNS: Growth and development through the life span; CL: Knowledge

61. A 3-year-old child is being discharged from the emergency department after receiving three sutures for a scalp laceration. In how many days would the nurse tell the family to return for suture removal?
1. 1 to 3 days
2. 5 to 7 days
3. 8 to 10 days
4. 10 to 14 days

62. Which of the following symptoms are early signs of infection of a laceration?
1. Fever
2. Copious drainage
3. Excessive discomfort
4. Local nodal enlargement

These words point to the correct answer.

63. The nurse is teaching a 17-year-old client who'll be discharged soon how to change a sterile dressing on the right leg. During the teaching session, the nurse notices redness, swelling, and induration at the wound site. What do these signs suggest?
1. Infection
2. Dehiscence
3. Hemorrhage
4. Evisceration

64. When being examined, a 6-year-old child was noted to have a papulovesicular eruption on the left anterior lateral chest, with complaints of pain and tenderness of the lesion. Which of the following disorders would these findings indicate?
1. Contusion
2. Herpes zoster
3. Scabies
4. Varicella

61. 2. The recommended healing time for this type of laceration is 5 to 7 days. Sutures need longer than 1 to 3 days to form an effective bond. Sutures of the fingertips and feet need 8 to 10 days, and 10 to 14 days is the recommended time for extensor surfaces of the knees and elbows.
NP: Planning; CN: Physiological integrity; CNS: Physiological adaptation; CL: Knowledge

62. 3. The first sign of infection is usually excessive discomfort. Fever, copious drainage, and nodal enlargement are advanced signs of infection.
NP: Data collection; CN: Health promotion and maintenance; CNS: Prevention and early detection of disease; CL: Knowledge

63. 1. Infection produces such signs as redness, swelling, induration, warmth and, possibly, drainage. Dehiscence may cause unexplained fever and tachycardia, unusual wound pain, prolonged paralytic ileus, and separation of the surgical incision. Hemorrhage can result in increased pulse and respiratory rate, decreased blood pressure, restlessness, thirst, and cold, clammy skin. Evisceration produces visible protrusion of organs, usually through an incision.
NP: Data collection; CN: Physiological integrity; CNS: Physiological adaptation; CL: Knowledge

64. 2. Herpes zoster is caused by the varicella zoster virus. It has papulovesicular lesions that erupt along a dermatome, usually with hyperesthesia, pain, and tenderness. Contusions aren't found with papulovesicular lesions. Scabies appear as linear burrows of the fingers and toes caused by a mite. The papulovesicular lesions of varicella are distributed over the entire trunk, face, and scalp and don't follow a dermatome.
NP: Data collection; CN: Physiological integrity; CNS: Physiological adaptation; CL: Knowledge

65. During an examination of a 5-month-old infant, a flat, dull pink, macular lesion is noted on the infant's forehead. Which of the following conditions does this lesion indicate?
1. Cavernous hemangioma
2. Nevus flammeus
3. Salmon patch
4. Strawberry hemangioma

Fish? Fruit? Caverns? What's going on here?

65. 3. Salmon patches are common vascular lesions in infants. They appear as flat, dull pink, macular lesions in various regions of the face and head. When they appear on the nape of the neck, they're commonly called "stork bites." These lesions fade by the first year of life. Nevus flammeus, or port-wine stains, are reddish purple lesions that don't fade. Strawberry and cavernous hemangiomas are raised lesions.
NP: Data collection; CN: Physiological integrity; CNS: Physiological adaptation; CL: Knowledge

66. A child's parents ask for advice on the use of an insect repellent that contains DEET. Which of the following statements would be correct?
1. Spray the child's clothing instead of the skin.
2. The repellent works better as the temperature increases.
3. The repellent isn't effective against the ticks responsible for Lyme disease.
4. Apply insect repellent as you would sunscreen, with frequent applications during the day.

You're almost at question 70. That was quick!

66. 1. DEET spray has been approved for use on children. It should be used sparingly on all skin surfaces. By concentrating the spray on clothing and camping equipment, the adverse effects and potential toxic buildup is significantly reduced. Repellent is lost to evaporation, wind, heat, and perspiration. With each 10° F increase in temperature, it leads to as much as a 50% reduction in protection time. DEET is very effective as a tick repellent.
NP: Planning; CN: Physiological integrity; CNS: Reduction of risk potential; CL: Knowledge

67. DEET-containing products such as insect repellent should be used in which of the following concentrations on a child's skin for optimal results?
1. 10%
2. 15%
3. 20%
4. 30%

67. 1. The highest concentration approved by the Food and Drug Administration for children is 10%. Because of thinner skin and a greater surface-area-to-mass ratio in children, parents should use DEET products sparingly.
NP: Data collection; CN: Physiological integrity; CNS: Reduction of risk potential; CL: Knowledge

68. Which of the following statements about warts is correct?
1. Cutting the wart is the preferred treatment for children.
2. No treatment exists that specifically kills the wart virus.
3. Warts are caused by a virus affecting the inner layer of skin.
4. Warts are harmless and usually last 2 to 4 years if untreated.

68. 2. The goal of treatment is to kill the skin that contains the wart virus. Cutting the wart is likely to spread the virus. The virus that causes warts affects the outer layer of the skin. Warts are harmless and last 1 to 2 years if untreated.
NP: Data collection; CN: Physiological integrity; CNS: Physiological adaptation; CL: Knowledge

69. Which of the following factors causes tooth decay and gum disease when allowed to remain on the teeth for prolonged periods?

1. Breast milk
2. Pacifiers
3. Thumb or other fingers
4. Formula

70. What are the classic signs of cellulitis in a child?

1. Pale, irritated, and cold to touch
2. Vesicular blisters at the site of the injury
3. Fever, edema, tenderness, and warmth at the site
4. Swelling and redness with well-defined borders

71. A 2-year-old child has cellulitis of the finger. Which of the following organisms or conditions is the most likely cause of the infection?

1. Parainfluenza virus
2. Respiratory syncytial virus
3. *Escherichia coli*
4. *Streptococcus*

72. A child has a desquamative rash of the hands and feet. Which of the following symptoms best describes desquamative rash?

1. Peeling skin
2. Thin, reddened layers of epidermis
3. Thick skin with deep, visible burrows
4. Thinning skin that may appear translucent

73. Which of the following instructions about the administration of nystatin oral solution is correct?

1. Give the solution immediately after feedings.
2. Give the solution immediately before feedings.
3. Mix the solution with small amounts of the feeding.
4. Give half the solution before and half the solution after the feeding.

Here's another hint.

Note that this item asks about most likely causes, not the only ones.

69. 4. Tooth decay and gum disease result when the carbohydrates in commercial formula, cow's milk, and fruit juices are allowed to remain on the teeth for a prolonged period. Studies have shown that breast milk only contributes to dental caries when sugar is already present on the teeth. Breast milk alone actually promotes enamel growth. Pacifiers and fingers don't cause tooth decay and gum disease, although they may contribute to malocclusion.
NP: Data collection; CN: Health promotion and maintenance; CNS: Prevention and early detection of disease; CL: Knowledge

70. 3. Cellulitis is a deep, locally diffuse infection of the skin. It's associated with redness, fever, edema, tenderness, and warmth at the site of the injury. Vesicular blisters suggest impetigo. Cellulitis has no well-defined borders.
NP: Data collection; CN: Physiological integrity; CNS: Physiological adaptation; CL: Knowledge

71. 4. *Streptococcus* cause most cases of cellulitis. Parainfluenza and respiratory syncytial virus cause infections of the respiratory tract. *E. coli* is a cause of bladder infections.
NP: Data collection; CN: Physiological integrity; CNS: Physiological adaptation; CL: Knowledge

72. 1. Desquamation is characteristic in diseases such as Stevens-Johnson syndrome. Scaling is thin, reddened layers of epidermis. Thickening of the skin with burrows is defined as lichenification. Thinning skin is best described as atrophy of the skin.
NP: Data collection; CN: Physiological integrity; CNS: Physiological adaptation; CL: Knowledge

73. 1. Nystatin oral solution should be swabbed onto the mouth after feedings to allow for optimal contact with mucous membranes. Before meals and with meals doesn't give the best contact with the mucous membranes.
NP: Implementation; CN: Physiological integrity; CNS: Pharmacological therapies; CL: Application

74. An infant is examined and found to have a petechial rash. Which of the following descriptions is correct for petechiae?
1. Purple macular lesions larger than 1 cm in diameter
2. Purple to brown bruises, macular or papular, of various sizes
3. Collection of blood from ruptured blood vessels and larger than 1 cm in diameter
4. Pinpoint, pink to purple, nonblanching, macular lesions that are 1 to 3 mm in diameter

Petechiae. Even the word sounds small, doesn't it? Oops, I gave you a hint. Oh, well.

74. 4. Petechiae are small 1- to 3-mm macular lesions. Purple macular lesions greater than 1 cm are defined as purpura. A bruise is defined as ecchymosis. A hematoma is a collection of blood.
NP: Data collection; CN: Physiological integrity; CNS: Physiological adaptation; CL: Knowledge

75. A child has a red rash in a circular shape on his legs. The lesions aren't connected. Which of the following classifications is the most appropriate for this rash?
1. A linear rash
2. A diffuse rash
3. An annular rash
4. A confluent rash

75. 3. An annular rash is ring-shaped. Linear rashes are lesions arranged in a line. A diffuse rash usually has scattered, widely distributed lesions. Confluent rash has lesions that are touching or adjacent to each other.
NP: Data collection; CN: Physiological integrity; CNS: Physiological adaptation; CL: Knowledge

76. A mother noticed her teenager is losing hair in small round areas on the scalp. This symptom is characteristic of which of the following conditions?
1. Alopecia
2. Amblyopia
3. Exotropia
4. Seborrhea dermatitis

Watch out for question 77. It's a negative question.

76. 1. Alopecia is the correct term for thinning hair loss. Amblyopia and exotropia are eye disorders. Seborrhea dermatitis is cradle cap and occurs in infants.
NP: Data collection; CN: Physiological integrity; CNS: Physiological adaptation; CL: Knowledge

77. Which of the following rashes doesn't show dermatologic changes of the palms?
1. Coxsackie virus
2. Measles
3. Rocky Mountain spotted fever
4. Syphilis

77. 2. Rocky Mountain spotted fever, syphilis, and Coxsackie virus show changes on the palms and soles of the feet. The rash in measles occurs on the face, trunk, and extremities.
NP: Data collection; CN: Physiological integrity; CNS: Physiological adaptation; CL: Knowledge

78. Atopic dermatitis can be characterized as which of the following types of disorder?
1. Fungal infection
2. Hereditary disorder
3. Sex-linked disorder
4. Viral infection

78. 2. Atopic dermatitis is a hereditary disorder associated with a family history of asthma, allergic rhinitis, or atopic dermatitis. Viral and fungal infections don't cause atopic dermatitis.
NP: Data collection; CN: Physiological integrity; CNS: Physiological adaptation; CL: Knowledge

79. The mother of a 6-month-old infant with atopic dermatitis asks for advice on bathing the child. Which of the following instructions or information should be given to her?
1. Bathe the infant twice daily.
2. Bathe the infant every other day.
3. Use bubble baths to decrease itching.
4. The frequency of the infant's baths isn't important in atopic dermatitis.

80. Discharge instructions for a child with atopic dermatitis include keeping the fingernails cut short. Which of the following statements would explain the reason for this intervention?
1. Prevent infection of the nail bed
2. Prevent the spread of the disorder
3. Prevent the child from causing a corneal abrasion
4. Reduce breaks in skin from scratching that may lead to secondary bacterial infections

81. A 10-year-old child is being treated for common warts. Which of the following viruses causes this condition?
1. Coxsackie virus
2. Human herpesvirus (HHV)
3. Human immunodeficiency virus (HIV)
4. Human papillomavirus (HPV)

82. The nurse is caring for an 11-year-old client with cerebral palsy with a pressure ulcer on the sacrum. When teaching the mother about dietary intake, which foods should the nurse plan to emphasize?
1. Legumes and cheese
2. Whole grain products
3. Fruits and vegetables
4. Lean meats and low-fat milk

Here's another question pointing out the importance of client teaching.

79. 2. Bathing removes lipoprotein complexes that hold water in the stratum corneum and increase water loss. Decreasing bathing to every other day can help prevent the removal of lipoprotein complexes. Soap and bubble bath should be used sparingly while bathing the child.
NP: Planning; CN: Physiological integrity; CNS: Basic care and comfort; CL: Application

80. 4. Keeping fingernails cut short will prevent breaks in the skin when a child scratches. Cutting fingernails too short or cutting the skin around the nail can increase the risk for infection. Atopic dermatitis can be found in various areas of the skin, but isn't spread from one area to another. Keeping fingernails short is a good way to reduce corneal abrasions, but doesn't apply to atopic dermatitis
NP: Implementation; CN: Physiological integrity; CNS: Physiological adaptation; CL: Knowledge

81. 4. HPV is responsible for various forms of warts. Coxsackie virus is associated with hand-foot-mouth disease. HHV is associated with varicella and herpes zoster. HIV infections aren't associated with epithelial tumors known as warts.
NP: Data collection; CN: Physiological integrity; CNS: Physiological adaptation; CL: Knowledge

82. 4. Although the client should eat a balanced diet with foods from all food groups, the diet should emphasize foods that supply complete protein, such as lean meats and low-fat milk. Protein helps build and repair body tissue, which promotes healing. Legumes provide incomplete protein. Cheese contains complete protein but also fat, which should be limited to 30% or less of caloric intake. Whole grain products supply incomplete proteins and carbohydrates. Fruits and vegetables provide mainly carbohydrates.
NP: Planning; CN: Physiological integrity; CNS: Basic care and comfort; CL: Application

83. An adolescent says his feet itch, sweat a lot, and have a foul odor. Which of the following terms applies to this condition?
  1. Candidiasis
  2. Tinea corporis
  3. Tinea pedis
  4. Molluscum contagiosum

Here's another hint. Don't you just love 'em?

84. Which of the following treatments would best control hypertrophic scarring?
  1. Compression garments
  2. Moisturizing creams
  3. Physiotherapy
  4. Splints

85. During a physical examination, a child is noted to have nails with "ice-pick" pits and ridges. The nails are thick and discolored and have splintered hemorrhages easily separated from the nail bed. Which of the following conditions would cause this condition?
  1. Paronychia
  2. Psoriasis
  3. Scabies
  4. Seborrhea

86. A neonate is examined and noted to have bruising on the scalp, along with diffuse swelling of the soft tissue that crosses over the suture line. Which of the following assessments is most accurate?
  1. Caput succedaneum
  2. Cephalohematoma
  3. Craniotabes
  4. Hydrocephalus

The words most accurate give you a bit of a clue to this question.

83. 3. Tinea pedis is a superficial fungal infection on the feet, commonly called athlete's foot. Candidiasis is a fungal infection of the skin or mucous membranes commonly found in the oral, vaginal, and intestinal mucosal tissue. Tinea corporis, or ringworm, is a flat, scaling, papular lesion with raised borders. Molluscum contagiosum is a viral skin infection with lesions that are small, red papules.
NP: Data collection; CN: Physiological integrity; CNS: Physiological adaptation; CL: Knowledge

84. 1. Compression garments are worn for up to 1 year to control hypertrophic scarring. Moisturizing creams help decrease hyperpigmentation. Physiotherapy and splints help keep joints and limbs supple.
NP: Data collection; CN: Physiological integrity; CNS: Physiological adaptation; CL: Knowledge

85. 2. Psoriasis is a chronic skin disorder with an unknown cause that shows these characteristic skin changes. A paronychia is a bacterial infection of the nail bed. Scabies are mites that burrow under the skin, usually between the webbing of the fingers and toes. Seborrhea is a chronic inflammatory dermatitis or cradle cap.
NP: Data collection; CN: Physiological integrity; CNS: Physiological adaptation; CL: Knowledge

86. 1. Caput succedaneum originates from trauma to the neonate while descending through the birth canal. It's usually a benign injury that spontaneously resolves over time. Cephalohematoma is a collection of blood in the periosteum of the scalp that doesn't cross over the suture line. Craniotabes is the thinning of the bone of the scalp. Hydrocephalus is an increased volume of cerebrospinal fluid (CSF) or the obstruction of the flow of the CSF and isn't related to soft-tissue swelling.
NP: Data collection; CN: Physiological integrity; CNS: Physiological adaptation; CL: Knowledge

NP: Nursing process CN: Client needs category CNS: Client needs subcategory CL: Cognitive level

87. A child has a healed wound from a traumatic injury. A lesion formed over the wound is pink, thickened, smooth, and rubbery in nature. Which of the following conditions best describes this wound?
1. Erosion
2. Fissure
3. Keloids
4. Striae

88. An infant's mother gives a history of poor feeding for a few days. A complete physical examination shows white plaques in the infant's mouth with an erythematous base. The plaques stick to the mucous membranes tightly and bleed when scraped. Which of the following conditions best describes the plaques?
1. Chickenpox
2. Herpes lesions
3. Measles
4. Oral candidiasis

89. A child was found unconscious at home and brought to the emergency department by the fire and rescue unit. While collecting data, the nurse notes cherry-red mucous membranes, nail beds, and skin. Which of the following causes is the most likely explanation for the child's condition?
1. Aspirin ingestion
2. Carbon monoxide poisoning
3. Hydrocarbon ingestion
4. Spider bite

90. A teenager asks advice about getting a tattoo. Which of the following statements about tattoos is a common misperception?
1. Human immunodeficiency virus (HIV) is a possible risk factor.
2. Hepatitis B is a possible risk factor.
3. Tattoos are easily removed with laser surgery.
4. Allergic response to pigments is a possible risk factor.

Hmmm, white plaques with an erythematous base. What could that mean?

Be careful! This question asks for what isn't correct.

87. 3. Keloids are an exaggerated connective tissue response to skin injury. Striae are linear depressions of the skin. An erosion is a depressed vesicular lesion. A fissure is a cleavage in the surface of skin.
NP: Data collection; CN: Physiological integrity; CNS: Physiological adaptation; CL: Knowledge

88. 4. Oral candidiasis, or thrush, is a painful inflammation that can affect the tongue, soft and hard palates, and buccal mucosa. Chickenpox, or varicella, causes open ulcerations of the mucous membranes. Herpes lesions are usually vesicular ulcerations of the oral mucosa around the lips. Measles that form Koplik's spots can be identified as pinpoint, white, elevated lesions.
NP: Data collection; CN: Physiological integrity; CNS: Physiological adaptation; CL: Knowledge

89. 2. Cherry-red skin changes are seen when a child has been exposed to high levels of carbon monoxide. Nausea and vomiting and pale skin are symptoms of aspirin ingestion. A hydrocarbon or petroleum ingestion usually results in respiratory symptoms and tachycardia. Spider-bite reactions are usually localized to the area of the bite.
NP: Data collection; CN: Physiological integrity; CNS: Physiological adaptation; CL: Knowledge

90. 3. The removal of tattoos isn't easily done, and most people are left with a significant scar. The cost is expensive and not covered by insurance. Because of the moderate amount of bleeding with a tattoo, both hepatitis B and HIV are potential risks if proper techniques aren't followed. Allergic reactions have been seen when establishments don't use Food and Drug Administration-approved pigments for tattoo coloring.
NP: Data collection; CN: Health promotion and maintenance; CNS: Prevention and early detection of disease; CL: Knowledge

91. Which of the following terms describes a fungal infection found on the upper arm?
1. Tinea capitis
2. Tinea corporis
3. Tinea cruris
4. Tinea pedis

92. A 15-kg toddler is started on amoxicillin and clavulanate (Augmentin) therapy, 200 mg/5 ml, for cellulitis. The dose is 40 mg/kg over 24 hours given three times daily. How many milliliters would be given for each dose?
1. 2.5 ml
2. 5 ml
3. 15 ml
4. 20 ml

93. At what stage of acquired syphilis are chancres found?
1. Tertiary syphilis
2. Primary acquired syphilis
3. Secondary acquired syphilis
4. Chancres aren't found in syphilis

94. A 4-year-old child has a tick embedded in the scalp. Which of the following methods is the preferred way of removing the tick?
1. Burning the tick at the skin surface
2. Surgically removing the tick
3. Grasping the tick with tweezers and applying slow, outward pressure
4. Grasping the tick with tweezers and quickly pulling the tick out

95. An infant with hives is prescribed diphenhydramine (Benadryl) 5 mg/kg over 24 hours in divided doses every 6 hours. The child weighs 8 kg. How many milligrams should be given with each dose?
1. 4.5 mg
2. 10 mg
3. 22 mg
4. 40 mg

Practice with numbers makes perfect.

91. 2. Tinea corporis describes fungal infections of the body. Tinea capitis describes fungal infections of the scalp. Tinea cruris is used to describe fungal infections of the inner thigh and inguinal creases. Tinea pedis is the term for fungal infections of the foot.
NP: Data collection; CN: Physiological integrity; CNS: Physiological adaptation; CL: Application

92. 2. The dose is first calculated by multiplying the weight and the milligrams and then dividing into three even doses. The milligrams are then used to determine the milliliters based on the concentration of the medicine. 40 mg × 15 kg = 600/3 doses = 200 mg/dose. The concentration is 200 mg in every 5 ml.
NP: Implementation; CN: Physiological integrity; CNS: Pharmacological therapies; CL: Knowledge

93. 2. A chancre is a painless, shallow ulcer that develops in primary syphilis and appears 3 weeks after exposure. In secondary syphilis, nodular, pustular, and annular lesions are seen. Tertiary stages are rare in children.
NP: Data collection; CN: Physiological integrity; CNS: Physiological adaptation; CL: Knowledge

94. 3. Applying gentle outward pressure prevents injury to the skin and leaving parts of the tick in the skin. Surgical removal is indicated if portions of the tick remain in the skin. Burning the tick and quickly pulling the tick out may cause injury to the skin and should be avoided.
NP: Data collection; CN: Physiological integrity; CNS: Physiological adaptation; CL: Knowledge

95. 2. Multiplying 5 mg by the weight (8 kg) gives the amount of milligrams for 24 hours (40 mg). Divide this by the number of doses per day (4), giving 10 mg/dose. 5 mg × 8 kg = 40 mg/4 doses = 10 mg/dose.
NP: Implementation; CN: Physiological integrity; CNS: Pharmacological therapies; CL: Application

96. An 8-year-old child arrives at the emergency department with chemical burns to both legs. Which of the following treatments has priority for this child?
1. Dilute the burns.
2. Apply sterile dressings.
3. Apply topical antibiotics.
4. Debride and graft the burns.

97. A 12-year-old child with full-thickness, circumferential burns to the chest has difficulty breathing. Which of the following procedures will most likely be performed?
1. Chest tube insertion
2. Escharotomy
3. Intubation
4. Needle thoracocentesis

98. A 6-year-old child is evaluated after sustaining burns to his left shoulder. The parents are instructed to use moisturizing cream and protect the burn from sunlight. What's the purpose of this treatment?
1. To avoid keloids
2. To avoid scarring
3. To avoid hypopigmentation
4. To avoid hyperpigmentation

99. A child arrives in the emergency department 20 minutes after sustaining a major burn injury to 40% of his body. After initiating an I.V. line, which of the following interventions would the nurse in the emergency department perform next for this child?
1. Insert an indwelling catheter.
2. Apply silver sulfadiazine (Silvadene) cream to the burn.
3. Shave the hair around the burn wound.
4. Obtain cultures from the deepest area burned.

Priority keeps popping up!

Knowing why something happens allows you to teach more effectively.

There's that priority again!

96. 1. Diluting the chemical is the first treatment. It will help remove the chemical and stop the burning process. The remaining treatments are initiated after dilution.
NP: Data collection; CN: Physiological integrity; CNS: Physiological adaptation; CL: Knowledge

97. 2. Escharotomy is a surgical incision used to relieve pressure from edema. It's needed with circumferential burns that prevent chest expansion or cause circulatory compromise. Intubation is performed to maintain a patent airway. Insertion of a chest tube and needle thoracocentesis are performed to relieve a pneumothorax.
NP: Data collection; CN: Physiological integrity; CNS: Physiological adaptation; CL: Knowledge

98. 4. Healed or grafted burns would require creams and protection from the sun to decrease hyperpigmentation. Scarring, hypopigmentation, and keloids aren't treated with moisturizing creams and avoidance of sunlight.
NP: Implementation; CN: Physiological integrity; CNS: Physiological adaptation; CL: Knowledge

99. 1. I.V. fluids must be started immediately on children who sustain a major burn injury to prevent them from going into hypovolemic shock. The fluids are titrated based on urine output. To monitor this output exactly, an indwelling urinary catheter must be inserted. The other interventions will be performed, but not immediately.
NP: Implementation; CN: Physiological integrity; CNS: Reduction of risk potential; CL: Application

100. A 12-year-old child sustains a moderate burn injury. The mother reports that the child last received a tetanus injection when he was 5 years old. An appropriate nursing intervention would be to administer which of the following immunizations?

1. 0.5 ml of tetanus toxoid I.M.
2. 0.5 ml of tetanus toxoid I.V.
3. 250 U of tetanus immunoglobulin (Hyper-Tet) I.M.
4. 250 U of tetanus immunoglobulin (Hyper-Tet) I.V.

100. 1. Tetanus prophylaxis is given to clients with moderate to severe burn injuries if more than 5 years have elapsed since the last immunization or if there's no history of immunization. The correct dosage is 0.5 ml I.M. one time if the child was immunized within 10 years. If it's been more than 10 years or the child hasn't received tetanus immunization, the dosage is 250 U of Hyper-Tet one time. No I.V. form of tetanus immunization is available.

NP: Implementation; CN: Physiological integrity; CNS: Pharmacological therapies; CL: Analysis

101. In a 10-year-old client who has been burned, which medication should the nurse expect to use to prevent infection?

1. lindane
2. diazepam (Valium)
3. mafenide (Sulfamylon)
4. meperidine (Demerol)

101. 3. The topical antibiotic mafenide is prescribed to prevent infection in clients with second- and third-degree burns. Lindane is a pediculicide used to treat lice infestation. Diazepam is an antianxiety agent that may be administered to clients with burns, but not to prevent infection. The narcotic analgesic meperidine is used to help control pain in clients with burns.

NP: Planning; CN: Physiological integrity; CNS: Pharmacological therapies; CL: Comprehension

102. In a 7-year-old client with burns on the legs, which nursing intervention helps prevent contractures?

1. Applying knee splints
2. Elevating the foot of the bed
3. Hyperextending the client's palms
4. Performing shoulder range-of-motion (ROM) exercises

102. 1. Applying knee splints prevents leg contractures by holding the joints in a position of function. Elevating the foot of the bed can't prevent contractures because this action doesn't hold the joints in a position of function. Hyperextending a body part for an extended time is inappropriate because it can cause contractures. Performing shoulder ROM exercises can prevent contractures in the shoulders, but not in the legs.

NP: Implementation; CN: Physiological integrity; CNS: Reduction of risk potential; CL: Application

Success! And you thought you couldn't do it. Super job!

Part VI Coordinated care

36 Concepts of management
    & supervision   679

37 Ethical & legal issues   683

Knowing key concepts of management and supervision are just as important as your clinical knowledge for a successful nursing career. Test your knowledge of these concepts with the following questions. Have fun!

Chapter 36
Concepts of management & supervision

1. The nurse and a new licensed practical nurse she's supervising are reviewing proper procedures for administering medications. What are the 5 "rights" of medication administration?

1. Right client, right medication, right dose, right time, and right route
2. Right client, right medication, right dose, right time, and right reason
3. Right client, right medication, right dose, right time, and right physician
4. Right client, right medication, right dose, right time, and right quantity

2. The nurse is caring for a 52-year-old female client diagnosed with right-sided stroke who has expressive aphasia and left-sided weakness. When planning care for this client, which intervention should the nurse delegate to a nursing assistant?

1. Accompany the client to speech therapy.
2. Perform active range-of-motion (ROM) exercises for the client's upper extremities.
3. Turn and position the client every 2 hours.
4. Begin teaching the client simple sign language phrases.

3. The nurse finds a suicidal client trying to hang himself in his room. In order to preserve self-esteem and safety, the nurse should:

1. place the client in seclusion with checks every 15 minutes.
2. assign a nursing staff member to remain with the client at all times.
3. make the client stay with the group at all times.
4. refuse to let the client in his room.

It's important to know the limits of a nursing assistant's skills and responsibilities.

1. 1. Before administering medication, the nurse should make sure she has the right client by checking the identification band, check the physician's order for dosage and frequency, check that the medication is ordered for the right route, and make sure she administers the medication at the right time. The right reason, right physician, and right quantity aren't part of this process.
NP: Implementation; CN: Physiological integrity; CNS: Pharmacological therapies; CL: Application

2. 3. Nursing assistants are taught proper positioning skills, although this activity should still be supervised. It isn't necessary to accompany the client to speech therapy and would take the assistant off the unit, reducing available help. Not all nursing assistants are taught to perform active ROM exercises. It wouldn't be necessary to teach the client sign language.
NP: Planning; CN: Safe, effective care environment; CNS: Coordinated care; CL: Comprehension

3. 2. Implementing a one-on-one staff-to-client ratio is the nurse's highest priority. This allows the client to maintain his self-esteem and keeps him safe. Seclusion would damage the client's self-esteem. Forcing the client to stay with the group and refusing to let him in his room don't guarantee his safety.
NP: Implementation; CN: Psychosocial integrity; CNS: Psychosocial adaptation; CL: Application

---

NP: Nursing process CN: Client needs category CNS: Client needs subcategory CL: Cognitive level

4. The staff of an outpatient clinic has formed a task force to develop new procedures for swift, safe evacuation of the unit. These new procedures haven't yet been reviewed, approved, or shared with all personnel. The nurse-manager has been informed of a bomb threat and her task force members are pushing her to evacuate the unit using the new procedures. The nurse-manager should:

1. determine that the procedures currently in place must be followed and direct staff to follow them without question.
2. tell staff members to use whatever procedures they feel are best.
3. ask staff members to quickly meet among themselves and decide what procedures to follow.
4. tell staff members to assemble in the staff lounge to quickly offer their opinions about evacuation procedures before deciding what to do.

Keep going, you're doing great!

4. 1. In an emergency situation the nurse manager must determine the best course of action for the safety and welfare of clients and staff. In this particular situation there's no time for hesitation. Allowing staff members to do whatever they think best will cause confusion and inefficient client evacuation because following different procedures won't allow them to function effectively as a team during this crisis. Taking the time to have staff meet with each other and the nurse-manager is wasting valuable time during a life-or-death crisis.

NP: Implementation; CN: Safe, effective care environment; CNS: Coordinated care; CL: Analysis

5. The nursing assistant reports to the nurse that a client became short of breath while being bathed but is breathing better now. Which approach should the nurse take initially?

1. Instruct the nursing assistant to observe the client for any further shortness of breath.
2. Check the client and gather subjective and objective data related to shortness of breath.
3. Tell the physician about the client's episode of shortness of breath.
4. Instruct the nursing assistant to complete the bath after allowing the client to rest.

Don't gamble with client safety when you suspect your coworker of substance abuse.

5. 2. The nurse must assess the client herself to determine what might have precipitated the episode and obtain a pulse oximetry reading if indicated. Instructing the nursing assistant to observe the client for further shortness of breath would be appropriate after the nurse has checked the client. It wouldn't be necessary to inform the physician about the client's episode of shortness of breath. After checking the client, the nurse may ask the nursing assistant to complete the bath after allowing the client to rest.

NP: Implementation; CN: Safe, effective care environment; CNS: Coordinated care; CL: Application

6. While discussing a client's care with a nursing assistant, the nurse detects an odor of alcohol on the assistant's breath. The nurse should plan to:

1. report her observations to the charge nurse.
2. monitor the assistant closely to determine if performance is impaired.
3. tell the assistant she'll have to leave the unit immediately.
4. warn the assistant that she could lose her certification.

6. 1. The nurse is obligated to report suspected substance abuse. Allowing the assistant to continue to work could jeopardize client care. It isn't the practical nurse's role to decide that the assistant must leave the unit immediately. Warning the assistant that she could lose her certification doesn't address the issue sufficiently.

NP: Planning; CN: Safe, effective, care environment; CNS: Coordinated care; CL: Comprehension

NP: Nursing process   CN: Client needs category   CNS: Client needs subcategory   CL: Cognitive level

7. The nurse notices that a 33-year-old male client with borderline personality disorder is very manipulative and plays one staff member against another. The best way to deal with this behavior is to:
1. consistently enforce limits when the client attempts to manipulate.
2. seek the client's approval for any change in routine on the unit.
3. allow the client to bend the unit rules rather than entering into power struggles.
4. assign the client to the same staff member to maintain consistency.

8. The nurse is concerned about another nurse's relationship with the members of a family and their ill preschooler. Which of the following behaviors would be most worrisome and should be brought to the attention of the nurse-manager?
1. The nurse keeps communication channels open among herself, the family, physicians, and other health care providers.
2. The nurse attempts to influence the family's decisions by presenting her own thoughts and opinions.
3. The nurse works with the family members to find ways to decrease their dependence on health care providers.
4. The nurse has developed teaching skills to instruct the family members so they can accomplish tasks independently.

9. The manager of an outpatient clinic is explaining the various health care delivery systems to a client. The client is interested in joining a system that has a reasonable fixed capitation rate. The manager realizes that the client is primarily interested in joining:
1. a preferred provider organization (PPO).
2. a managed care organization.
3. a health maintenance organization (HMO).
4. a privately funded insurance company.

Eliminating the wrong answers can be as important as selecting the right one.

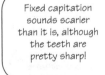

Fixed capitation sounds scarier than it is, although the teeth are pretty sharp!

7. 1. The staff must be consistent in setting limits on negative behaviors. Situations can escalate to crisis level when limit setting is inconsistent. The nurse should inform the client about changes in unit routines but shouldn't seek his approval. Allowing the client to bend the unit rules is a form of manipulation and shouldn't be allowed. Staff members should be rotated so that the client can learn to relate to more than one person.
NP: Implementation; CN: Psychosocial integrity; CNS: Psychosocial adaptation; CL: Application

8. 2. When a nurse attempts to influence a family's decision with her own opinions and values, the situation becomes one of over-involvement on the nurse's part and a nontherapeutic relationship. When a nurse keeps communication channels open, works with family members to decrease their dependence on health care providers, and instructs family members so they can accomplish tasks independently, she has developed an appropriate therapeutic relationship.
NP: Evaluation; CN: Safe, effective care environment; CNS: Coordinated care; CL: Analysis

9. 3. An HMO provides comprehensive health services for a fixed rate of payment or capitation. A PPO pays healthcare expenses for members if they use a provider who's under contract to that PPO. Managed care provides beneficiaries with various services for an established, agreed-upon payment. A privately funded insurance company won't offer services for a fixed rate.
NP: Planning; CN: Safe, effective care environment; CNS: Coordinated care; CL: Application

10. The nurse receives an assignment to provide care to 15 clients. Two of them have had kidney transplantation surgery within the last 36 hours. The nurse feels overwhelmed with the number of clients. In addition, the nurse has never cared for a client who has undergone recent transplantation surgery. What's the most appropriate action for the nurse to take?

1. Speak to the manager and document in writing all concerns related to the assignment.
2. Refuse the assignment.
3. Ignore the assignment and leave the unit.
4. Trade assignments with another nurse.

10. 1. When a nurse feels incapable of performing an assignment safely, the appropriate action is to speak to the manager or nurse in charge. The nurse should also document the concerns in writing and ask that the assignment be changed. In the event that the manager chooses to leave the assignment as given, the nurse should accept the assignment. The nurse should never abandon the assigned clients by leaving the workplace or asking another nurse to take care of them. The nurse may, however, refuse to perform a task outside the scope of practice.

NP: Implementation; CN: Safe, effective care environment; CNS: Coordinated care; CL: Analysis

Congratulations! You're done with the chapter and your review is complete!

NP: Nursing process   CN: Client needs category   CNS: Client needs subcategory   CL: Cognitive level

Chapter 37
Ethical & legal issues

1. An elderly client has been admitted to the medical-surgical unit from the postanesthesia care unit. While the nurse is off the floor, the client falls out of bed resulting in a fracture of his right leg and right wrist. The nurse finding him states that the "side rails were left down and the bed was in the high position." Legal charges are filed against the nurse and the hospital. What's the most likely charge for her actions?
　　1. Collective liability
　　2. Comparative negligence
　　3. Battery
　　4. Negligence

Here's a hint: Only three of these four choices refer to "time frames." Even eliminating one option makes choosing the correct answer easier.

2. The time frame in which a client's attorney has to file a lawsuit is known as:
　　1. discovery rule.
　　2. statute of limitation.
　　3. grace period.
　　4. alternative dispute resolution.

3. What are the elements that must be proven by a client's attorney in the case of a professional negligence action?
　　1. Duty, breach of duty, damages, and causation
　　2. Duty, damages, and causation
　　3. Duty, breach of duty, and damages
　　4. Breach of duty, damages, and causation

1. 4. Negligence is a general term that denotes conduct lacking in due care and is commonly interpreted as a deviation from the standard of care that a reasonable person would use in a particular set of circumstances. Collective liability stems from cooperation by several manufacturers in a wrongful activity that by its nature requires group participation. Comparative negligence is a defense that holds injured parties accountable for their fault in the injury. Battery involves a harmful or unwarranted contact with the client.
NP: Evaluation; CN: Safe, effective care environment; CNS: Safety and infection control; CL: Analysis

2. 2. Statute of limitation is the time interval during which a case must be filed or the injured party is barred from bringing the lawsuit. The statute of limitations typically gives clients 2 years from the time of discovery to file a lawsuit, however this may vary from state to state. Statutes of limitations are set by state legislature. Discovery rule is the term for when the client has discovered the injury. A grace period refers to any period specified in a contract during which payment is permitted without penalty, beyond the due date of the debt. Alternative dispute resolution refers to any means of settling disputes outside the courtroom setting.
NP: Data collection; CN: Safe, effective care environment; CNS: Safety and infection control; CL: Comprehension

3. 1. Any professional negligence action must meet these demands in order to be considered negligence and result in legal action. They're commonly known as the four D's: duty of the health care professional to provide care to the person making the claim, a dereliction (breach) of that duty, damages resulting from that breach of duty, and evidence that damages were directly due to negligence (causation).
NP: Evaluation; CN: Safe, effective care environment; CNS: Coordinated care; CL: Application

NP: Nursing process CN: Client needs category CNS: Client needs subcategory CL: Cognitive level

4. The scope of nursing professional practice defines what a nurse can and can't do. What defines and regulates the scope of nursing professional practice?
  1. Nursing process
  2. Facilities' policies and procedures
  3. Standards of care
  4. Nurse Practice Act

4. 4. The Nurse Practice Act is a series of statutes enacted by each state to outline the legal scope of nursing practice within that state. State boards of nursing oversee this statutory law. Nurse practice acts set educational requirements for the nurse, distinguish between nursing practice and medical practice, and define the scope of nursing practice. Nursing process is an organizational framework for nursing practice, encompassing all major steps a nurse takes when caring for a patient. Facility policies govern the practice in that particular facility. Standards of care are criteria that serve as a basis for comparison when evaluating the quality of nursing practice. Standards of care are established by federal, state, professional, and accreditation organizations.
NP: Evaluation; CN: Safe, effective care environment; CNS: Coordinated care; CL: Application

5. A Jehovah's Witness client refuses a blood transfusion based on his religious beliefs and practices. His decision must be followed based on what ethical principle?
  1. The right to die
  2. Advance directive
  3. The right to refuse treatment
  4. Substituted judgment

Peace, love, and understanding. That's what question 5 is all about, man.

5. 3. The right to refuse treatment is grounded in the ethical principle of respect for autonomy of the individual. The client has the right to refuse treatment as long as he's competent and is made aware of the risks and complications associated with refusal of treatment. Substituted judgment is an ethical principle used when the decision is made for an incapacitated client based on what's best for the client. An advance directive is a document used as a guideline for life-sustaining medical care of a client with a terminal disease or disability who can no longer indicate his own wishes.
NP: Implementation; CN: Safe, effective care environment; CNS: Coordinated care; CL: Application

6. A nurse gives a client the wrong medication. After assessment of the client, the nurse completes an incident report. Which of the following scenarios describes what should occur next?
  1. The incident would be reported to the state board of nursing for disciplinary action.
  2. The incident would be documented in the nurse's personnel file.
  3. The medication error would result in the nurse being suspended and possibly terminated from employment at the facility.
  4. The incident report would be used to promote quality care and risk management.

6. 4. Unusual occurrences and deviations from care are documented on incident reports. Incident reports are internal to the facility and are used to evaluate the care, determine potential risks, and identify possible system problems that could have attributed to the error. This type of error wouldn't result in suspension of the nurse or reporting to the state board of nursing. Some facilities do trend and track the number of errors that take place on particular units or by individual nurses for educational purposes and as a way to improve the process.
NP: Evaluation; CN: Safe, effective care environment; CNS: Coordinated care; CL: Analysis

NP: Nursing process   CN: Client needs category   CNS: Client needs subcategory   CL: Cognitive level

7. A client has suffered an extensive brain injury and can't make his own treatment choices. Which of the following documents doesn't provide directions concerning the provision of care in this situation?
1. Advance directive
2. Living will
3. Durable power of attorney
4. Occurrence policy

7. 4. An occurrence policy is a professional liability policy that protects against an error of omission that occurs during a policy period, which doesn't involve provision of care. A living will is a document prepared by a competent adult that provides direction regarding medical care in the event the person can't make his own decisions. A durable power of attorney is an authorization that enables any competent individual to name someone to exercise decision-making authority on the individual's behalf under specific circumstances. A living will and a durable power of attorney are examples of advance directives.

NP: Data collection; CN: Safe, effective care environment; CNS: Coordinated care; CL: Knowledge

8. In what way do nurses play a key role in error prevention?
1. Identifying incorrect dosages or potential interactions of prescribed medications
2. Never questioning the order of a physician because he's ultimately responsible for the client outcome
3. Notifying the Occupational Safety and Health Association (OSHA) of violations in the workplace
4. Informing the client of his rights as a client

Nurses have lots of roles and responsibilities but question 8 is asking specifically about error prevention. Which answer fits best?

8. 1. Nurses must be knowledgeable about drug dosages and possible interactions when administering medications and be able to follow the appropriate policy to correct the situation. The nurse is responsible for questioning an unclear or ambiguous order from the physician and should never carry out an order she's uncomfortable with. Notifying OSHA doesn't solve medication errors. OSHA establishes comprehensive safety and health standards, inspects workplaces, and orders employers to eliminate safety hazards. The client should be aware of his rights as a client but that doesn't play a key role in error prevention.

NP: Evaluation; CN: Safe, effective care environment; CNS: Safety and infection control; CL: Application

9. Which of the following statements is true concerning informed consent?
1. Minors are permitted to give informed consent.
2. The professional nurse and physician may both obtain informed consent.
3. The client must be fully informed regarding treatment, tests, surgery, risks, and benefits before obtaining informed consent.
4. Mentally competent and incompetent clients can legally give informed consent.

9. 3. When the professional nurse is involved in the informed consent process, the nurse is only witnessing the consent process and doesn't actually obtain the consent. Only a minor who's married or emancipated can give informed consent. Obtaining the consent is the physician's responsibility. Legally, the client must be mentally competent to give consent for procedures.

NP: Planning; CN: Safe, effective care environment; CNS: Coordinated care; CL: Comprehension

10. The Omnibus Reconciliation Act of 1986 states that:

1. all families of clients who are nearing death, or have died, must be approached with the option of organ and tissue donation.
2. the medical examiner should be notified of all potential organ donors.
3. a request must be made to the family regarding release or remains of donors.
4. hospitals aren't responsible for establishing designated requesters for donation.

Knowing the facts about state and federal mandates is essential for your clinical practice and the NCLEX!

11. When approaching a family for organ or tissue donation, it's important that:

1. it's done with a physician's approval and written order.
2. the requester doesn't have to believe in the benefits of organ donation or support the process with a positive attitude.
3. the requester is knowledgeable about the basics of organ and tissue donation and can educate the family members about brain death early in the organ donation process.
4. the family is offered an opportunity to speak with an organ procurement coordinator.

Question 12 asks about stages — in other words, a progression. Only one of these answer choices offers a logical progression. Can you figure it out?

12. The stages of grief that the grieving client or family member will go through are:

1. acceptance, depression, anger, bargaining, and denial
2. denial, anger, decreased interaction, depression, and mourning
3. acceptance, anger, denial, and bargaining
4. denial, anger, bargaining, depression, and acceptance

10. 1. The federal Omnibus Reconciliation Act of 1986 mandates that hospitals establish written protocols for the identification of potential organ and tissue donors. This law was enacted to attempt to correct the disparity between the organs that could be transplanted and the number of people waiting for them. The medical examiner should be notified if the client is a potential organ or tissue donor only in the event that the medical examiner is involved in the case. Requesters for donation are health care professionals who have received special training on how to properly approach family members regarding organ or tissue donation.
NP: Implementation; CN: Safe, effective care environment; CNS: Coordinated care; CL: Comprehension

11. 4. The family should be offered an opportunity to speak with an organ procurement coordinator. An organ procurement coordinator is very knowledgeable about the organ donation process and should have exceptional interpersonal skills for dealing with grieving family members. Physician support in the process is desirable, but consent or written orders aren't necessary for a referral to the organ procurement organization. The requester has to believe in the benefits of organ donation and support the process with a positive attitude. Approaching the family should only occur when the family members are made aware of the client's condition and prognosis. Approaching a family member when he believes that there's still hope for recovery will only result in a negative outcome.
NP: Evaluation; CN: Safe, effective care environment; CNS: Coordinated care; CL: Analysis

12. 4. Denial is the avoidance of death's inevitability and is the first step of the grieving process. Anger, the most intense grief reaction, arises when people realize that death and loss will actually occur or has occurred for a family member. Bargaining happens when family members attempt to stall or to manipulate the outcome or death. Depression is a response to loss that's expressed as profound sadness or deep suffering. Acceptance, the final stage, is the ability to overcome the grief and accept what has happened.
NP: Data collection; CN: Psychosocial integrity; CNS: Coping and adaptation; CL: Comprehension

13. The Licensed Practical Nurse (LPN) has been assigned to a client who's scheduled to receive several medications and treatments. The LPN must request assistance from the Registered Nurse in completing which of the following?

1. Checking blood glucose level with a fingerstick
2. Administering furosemide (Lasix) 40 mg I.V. push
3. Monitoring an I.V. infusion of lactated Ringer's solution
4. Changing a wet-to-dry dressing involving deep wound packing

14. Although living will laws vary from state to state, they generally include which of the following provisions?

1. Instructions on when and how the living will should be executed
2. Who may uphold a living will declaration
3. How long the living will is in effect
4. What will happen to the client's valuables after death

15. The role of the nurse in a domestic abuse situation is to:

1. document the situation and provide support for the victim.
2. protect the client' privacy by not documenting the abuse.
3. provide counseling to the person committing the abuse.
4. provide counseling for the victim.

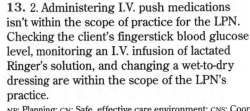

13. 2. Administering I.V. push medications isn't within the scope of practice for the LPN. Checking the client's fingerstick blood glucose level, monitoring an I.V. infusion of lactated Ringer's solution, and changing a wet-to-dry dressing are within the scope of the LPN's practice.
NP: Planning; CN: Safe, effective care environment; CNS: Coordinated care; CL: Knowledge

14. 1. Living wills include instructions on when and how a living will should be executed, the witness and testator requirements, immunity from liability for following a living will directive, documentation requirements, and under what circumstances the living will takes effect. A living will doesn't state how long it's in effect; it's assumed that the living will is in effect indefinitely or until revised. A living will doesn't dictate what will happen to the client's valuables; this is the scope of a regular will.
NP: Data collection; CN: Safe, effective care environment; CNS: Coordinated care; CL: Knowledge

15. 1. The nurse must carefully and adequately document the data she collects from the abused victim. It must include statements from the victim, physical and psychological assessment findings, and observations relative to the abuse situation. The victim should be provided with local community resources, social agencies, and legal services as necessary to prevent recurrence of physical abuse. The professional nurse isn't qualified to counsel the abuser or the victim. The abuser as well as the victim should be referred to a professional counselor who's trained in dealing with domestic violence therapy.
NP: Data collection; CN: Psychosocial integrity; CNS: Psychosocial adaptation; CL: Analysis

Appendices and index

Comprehensive test 1   **691**

Comprehensive test 2   **712**

Comprehensive test 3   **733**

Comprehensive test 4   **755**

Index   **776**

This comprehensive test, the first of four, is just like a real NCLEX exam. It's a great way to practice!

COMPREHENSIVE
Test 1

1. A 43-year-old client with blunt chest trauma from a motor vehicle accident has sinus tachycardia, is hypotensive, and has developed muffled heart sounds. There are no obvious signs of bleeding. Which of the following conditions is suspected?
1. Heart failure
2. Pneumothorax
3. Cardiac tamponade
4. Myocardial infarction

2. Which of the following types of shock is associated with tamponade?
1. Anaphylactic
2. Cardiogenic
3. Hypovolemic
4. Septic

3. Which of the following diagnostic tests is used to detect cardiac tamponade?
1. Chest X-ray
2. Echocardiography
3. Electrocardiogram (ECG)
4. Pulmonary artery pressure monitoring

1. 3. Cardiac tamponade results in signs of obvious shock and muffled heart sounds. In a myocardial infarction, the client may complain of chest pain. Heart failure would result in inspiratory rales, pulmonary edema, and jugular venous distention. Pneumothorax would result in diminished breath sounds in the affected lung, respiratory distress, and tracheal displacement. An electrocardiogram could confirm changes consistent with a myocardial infarction.
NP: Data collection; CN: Physiological integrity; CNS: Physiological adaptation; CL: Analysis

2. 2. Fluid accumulates in the pericardial sac, hindering motion of the heart muscle and causing it to pump inefficiently, resulting in signs of cardiogenic shock. Anaphylactic and septic shock are types of distributive shock in which fluid is displaced from the capillaries and leaks into surrounding tissues. Hypovolemic shock involves the actual loss of fluid.
NP: Data collection; CN: Physiological integrity; CNS: Physiological adaptation; CL: Knowledge

3. 1. Chest X-rays show a slightly widened mediastinum and enlarged cardiac silhouette. Echocardiography records pericardial effusion with signs of right ventricular and atrial compression. An ECG can rule out other cardiac disorders. Pulmonary artery pressure monitoring shows increased right atrial or central venous pressure and right ventricular diastolic pressure.
NP: Data collection; CN: Physiological integrity; CNS: Reduction of risk potential; CL: Knowledge

NP: Nursing process   CN: Client needs category   CNS: Client needs subcategory   CL: Cognitive level

4. Which of the following interventions or medications is the emergency treatment for cardiac tamponade?
1. Surgery
2. Dopamine
3. Blood transfusion
4. Pericardiocentesis

4. 4. Pericardiocentesis, or needle aspiration of the pericardial cavity, is done to relieve tamponade. An opening is created surgically if the client continues to have recurrent episodes of tamponade. Dopamine is used to restore blood pressure in normovolemic individuals. Blood transfusions may be given if the client is hypovolemic from blood loss.
NP: Implementation; CN: Physiological integrity; CNS: Physiological adaptation; CL: Knowledge

5. A nurse is teaching a 50-year-old client how to decrease risk factors for coronary artery disease. He's an executive who smokes, has a type A personality, and is hypertensive. Which of the following risk factors can't be changed?
1. Age
2. Hypertension
3. Personality
4. Smoking

5. 1. Age is a risk factor that can't be changed. Type A personality, hypertension, and smoking factors can be controlled.
NP: Data collection; CN: Health promotion and maintenance; CNS: Prevention and early detection of disease; CL: Comprehension

6. A client says he's stressed by his job but enjoys the challenge. Which of the following suggestions is best to help the client?
1. Switch job positions.
2. Take stress management classes.
3. Spend more time with his family.
4. Don't take his work home with him.

6. 2. Stress management classes will teach the client how to better manage the stress in his life, after identifying the factors that contribute to it. Alternatives may be found to leaving his job, which he enjoys. Not spending enough time with his family and not taking his job home with him haven't yet been identified as contributing factors.
NP: Implementation; CN: Physiological integrity; CNS: Reduction of risk potential; CL: Application

7. Which of the following nursing diagnoses is correctly worded?
1. Anger related to terminal illness
2. Pain related to alteration in comfort
3. Red sacrum related to improper positioning
4. Social isolation related to inability to speak because of laryngectomy

7. 4. This is a correctly worded nursing diagnosis. The first option identifies anger as an unhealthy response when it may be an appropriate and socially acceptable response. The second option is incorrect because both parts of this diagnosis relate to pain and say the same thing. The third option is improperly written; it's a legally inadvisable statement.
NP: Planning; CN: Safe, effective care environment; CNS: Coordinated care; CL: Application

---

NP: Nursing process   CN: Client needs category   CNS: Client needs subcategory   CL: Cognitive level

8. A 28-year-old client with human immunodeficiency virus (HIV) is admitted to the hospital with flulike symptoms. He has dyspnea and a cough. He's placed on a 100% nonrebreather mask and arterial blood gases are drawn. Which of the following results indicates the client needs intubation?
  1. Pao_2, 90 mm Hg; $Paco_2$, 40 mm Hg
  2. Pao_2, 85 mm Hg; $Paco_2$, 45 mm Hg
  3. Pao_2, 80 mm Hg; $Paco_2$, 45 mm Hg
  4. Pao_2, 70 mm Hg; $Paco_2$, 55 mm Hg

9. Which of the following substances transmits the human immunodeficiency virus (HIV)?
  1. Blood
  2. Feces
  3. Saliva
  4. Urine

10. Which of the following opportunistic diseases is caused by protozoa in clients with acquired immunodeficiency syndrome?
  1. Tuberculosis (TB)
  2. Histoplasmosis
  3. Kaposi's sarcoma
  4. *Pneumocystis carinii* infection

11. A client with acquired immunodeficiency syndrome is intubated, leaving him prone to skin breakdown from the endotracheal tube. Which of the following interventions is best to prevent this?
  1. Use lubricant on the lips.
  2. Provide oral care every 2 hours.
  3. Suction the oral cavity every 2 hours.
  4. Reposition the endotracheal tube every 24 hours.

Ten questions finished! Keep it up!

8. 4. A decreasing Pao_2 and an increasing $Paco_2$ indicate poor oxygen perfusion. Normal Pao_2 levels are 80 to 100 mm Hg and normal $Paco_2$ levels are 35 to 45 mm Hg.
NP: Data collection; CN: Physiological integrity; CNS: Physiological adaptation; CL: Analysis

9. 1. HIV is transmitted by contact with infected blood. It exists in all body fluids but transmission through saliva, urine, and feces hasn't been shown to occur.
NP: Data collection; CN: Safe, effective care environment; CNS: Safety and infection control; CL: Knowledge

10. 4. *P. carinii* infection is caused by protozoa. TB is caused by bacteria. Histoplasmosis is a fungal infection. Kaposi's sarcoma is a neoplasm.
NP: Data collection; CN: Physiological integrity; CNS: Physiological adaptation; CL: Knowledge

11. 4. Pressure causes skin breakdown. However, repositioning the endotracheal tube from one side of the mouth to the other or to the center of the mouth can relieve pressure in one area for a time. Extreme care must be taken to move the tube only laterally; it must not be pushed in or pulled out. The tape securing the tube must be changed daily. Two nurses should perform this procedure. Oral care, suctioning, and lubricant help keep skin clean and intact and reduce the risk of further infection.
NP: Implementation; CN: Physiological integrity; CNS: Basic care and comfort; CL: Application

12. A client requiring the highest possible concentration of oxygen will need which of the following delivery systems?
1. Face tent
2. Venturi mask
3. Nasal cannula
4. Mask with reservoir bag

12. 4. A mask with a reservoir bag administers 70% to 100% oxygen at flow rates of 8 to 10 L/minute. The nasal cannula maximum rate is 44% at 6 L/minute, the Venturi mask maximum rate is 24% to 55%, and a face tent maximum delivery is 22% to 34%.
NP: Implementation; CN: Physiological integrity; CNS: Pharmacological therapies; CL: Comprehension

13. A client with difficulty breathing has a respiratory rate of 34 breaths/minute and seems anxious. She's refusing all her medications, claiming they're making her worse. Which of the following nursing actions is best?
1. Notify the physician of the status of this client.
2. Hold the medication until the next scheduled dose.
3. Encourage the client to take some of her medications.
4. Put the medicine in applesauce to give it without the client's knowledge.

13. 1. Notifying the physician of the client's condition and her refusal to take her medications allows the physician to decide what alternatives should be instituted. Holding a medication requires the physician to be notified. Giving medications in applesauce destroys trust between the nurse and client. Even if the client takes some of the medications, the physician will still need to be notified. It needs to be explored why the client believes the medications are making her worse.
NP: Implementation; CN: Physiological integrity; CNS: Pharmacological therapies; CL: Application

14. A nurse caring for a client with acquired immunodeficiency syndrome is working with a nursing student. She notes the student doesn't attempt to suction or assist with care of the client. Which of the following actions is appropriate?
1. Talk to the student.
2. Talk to the charge nurse.
3. Address a coworker with the concerns.
4. Seek advice from the student's instructor.

14. 1. The nurse should approach the student to determine her feelings and experience in caring for this client. The charge nurse and coworkers aren't familiar with the student's abilities, but the instructor may be approached if the nurse can't communicate with the student.
NP: Implementation; CN: Safe, effective care environment; CNS: Coordinated care; CL: Analysis

15. A client's significant other is tearful over the client's condition and lack of improvement. He says he feels very powerless and unable to help his friend. Which of the following responses by the nurse is the best?
1. Agree with the person.
2. Tell him there's nothing he can do.
3. State she understands how he must feel.
4. Ask if he would like to help with some comfort measures.

15. 4. The significant other expresses a need to help and the nurse can encourage him to do whatever he feels comfortable with, such as putting lubricant on lips, moist cloth on forehead, or lotion on skin. The nurse may not understand his situation, and agreeing with a person doesn't diminish powerlessness. There are many ways the significant other can help if he wants to.
NP: Implementation; CN: Psychosocial integrity; CNS: Coping and adaptation; CL: Analysis

NP: Nursing process   CN: Client needs category   CNS: Client needs subcategory   CL: Cognitive level

16. A 31-year-old client is admitted to the hospital with respiratory failure. He's intubated in the emergency department, placed on 100% F_{IO_2}, and is coughing up copious secretions. Which of the following interventions has priority?

    1. Get an X-ray.
    2. Suction the client.
    3. Restrain the client.
    4. Obtain an arterial blood gas (ABG) analysis.

16. 2. Secretions can cut off the oxygen supply to the client and result in hypoxia. After the client has acclimated to his ventilator settings, ABG levels can be drawn. X-rays are a priority to check placement of the endotracheal tube. Restraints are warranted only if the client is a threat to his safety.

NP: Implementation; CN: Physiological integrity; CNS: Reduction of risk potential; CL: Analysis

17. A client with an endotracheal tube has copious, brown-tinged secretions. Which of the following interventions is a priority?

    1. Use a trap to obtain a specimen.
    2. Instill saline to break up secretions.
    3. Culture the specimen with a culturette swab.
    4. Obtain an order for a liquefying agent for the sputum.

17. 1. Suspicious secretions should be sent for culture and sensitivity testing using a sterile technique such as a trap. Swab culturettes are useful for wound cultures — not endotracheal cultures. Saline would dilute the specimen. Various agents are available to help break up secretions, and respiratory therapists can usually help recommend the right agent, but this isn't a priority.

NP: Implementation; CN: Safe, effective care environment; CNS: Safety and infection control; CL: Analysis

18. An X-ray shows that an endotracheal (ET) tube is ¾″ (2 cm) above the carina, and there are nodular lesions and patchy infiltrates in the upper lobe. Based on this report, the nurse can expect which of the following conclusions?

    1. The X-ray is inconclusive.
    2. The client has a disease process going on.
    3. The ET tube needs to be advanced.
    4. The ET tube needs to be pulled back.

18. 2. The X-ray suggests tuberculosis. At ¾″, the ET tube is at an adequate level in the trachea and doesn't have to be advanced or pulled back.

NP: Evaluation; CN: Health promotion and maintenance; CNS: Prevention and early detection of disease; CL: Analysis

19. A client has copious secretions. X-ray results indicate tuberculosis (TB). Which of the following interventions should be performed first?

    1. Repeat X-ray
    2. Tracheostomy
    3. Bronchoscopy
    4. Arterial blood gas (ABG) analysis

19. 3. Bronchoscopy can help diagnose TB and obtain specimens while clearing the bronchial tree of secretions. Tracheostomy may be done if the client remains on the ventilator for a prolonged period. A change in condition or treatment may require an ABG analysis. X-rays may be repeated periodically to determine lung and endotracheal tube status.

NP: Planning; CN: Physiological integrity; CNS: Reduction of risk potential; CL: Comprehension

20. A nurse is aware that family members of a client diagnosed with tuberculosis, as well as staff members, may have been exposed to the disease. A tuberculin skin test may show which of the following conditions?

    1. Active disease
    2. Recent infection
    3. Extent of the infection
    4. Infection at some point

Looking good! Keep at it!

20. 4. A tuberculin skin test shows the presence of infection at some point; a positive skin test doesn't guarantee that an infection is *currently* present, however. Some people have false-positive results. Active disease may be viewed on a chest X-ray. Computed tomography scan or magnetic resonance imaging can evaluate the extent of lung damage.
NP: Data collection; CN: Safe, effective care environment; CNS: Safety and infection control; CL: Knowledge

21. A client is diagnosed with tuberculosis (TB). In addition to recommending skin testing of the family members, TB should be reported to which of the following individuals or agencies?

    1. Centers for Disease Control and Prevention (CDC)
    2. Local health department
    3. Infection-control nurse
    4. Client's physician

21. 2. The local health department must be informed of an outbreak of TB because it's a reportable disease. They, in turn, inform the CDC. The infection-control nurse or local health department may request that staff be tested if exposed. Generally, the client's family can inform his physician.
NP: Planning; CN: Safe, effective care environment; CNS: Safety and infection control; CL: Application

22. The usual treatment for tuberculosis (TB) includes the use of isoniazid (INH) and which of the following therapies?

    1. Theophylline inhaler
    2. I.M. penicillin
    3. Three other antibacterial agents
    4. Aerosol treatments with pentamidine

22. 3. Because TB has become resistant to many antibacterial agents, the initial treatment includes the use of multiple antitubercular or antibacterial drugs. These may include rifampin, ethambutol, pyrazinamide, cycloserine, clofazimine, and streptomycin. Pentamidine is used in the treatment of *Pneumocystis carinii* pneumonia. Theophylline is a bronchodilator used to treat asthma and chronic obstructive pulmonary disease. Penicillins are used to treat *Staphylococcus aureus* infection — not TB.
NP: Implementation; CN: Physiological integrity; CNS: Pharmacological therapies; CL: Knowledge

23. How long do most clients receive treatment for tuberculosis (TB)?

    1. 2 to 4 months
    2. 9 to 12 months
    3. 18 to 24 months
    4. More than 2 years

23. 2. Treatment for TB is usually continued for 9 to 12 months.
NP: Implementation; CN: Physiological integrity; CNS: Pharmacological therapies; CL: Knowledge

NP: Nursing process   CN: Client needs category   CNS: Client needs subcategory   CL: Cognitive level

24. For what length of time is a client considered infectious after treatment for tuberculosis is started?
  1. 72 hours
  2. 1 week
  3. 2 weeks
  4. 4 weeks

24. 4. After 4 weeks, the disease is no longer infectious but the client must continue to take the medication.
NP: Implementation; CN: Safe, effective care environment; CNS: Safety and infection control; CL: Knowledge

25. A client tells his nurse that his tuberculosis medications are so expensive that he can't afford to take them. Which of the following interventions by the nurse is best?
  1. Refer the client to social services.
  2. Tell the client to apply for Medicaid.
  3. Refer the client to the local or county health department.
  4. Tell the client to follow his insurance rules and regulations.

25. 3. The local and county health departments provide treatment and follow-up free of charge for all residents to ensure proper care. Insurance can be an alternative source to help pay for treatment, but the client may not be insured or the policy may not cover prescriptions. Medicaid or medical assistance is another avenue for the client, if he qualifies. Social services can help seek alternative methods of payment and reimbursement but would probably first refer the client to the local and county health departments.
NP: Implementation; CN: Safe, effective care environment; CNS: Coordinated care; CL: Knowledge

26. A 62-year-old client is admitted to the hospital with pneumonia. He has a history of Parkinson's disease, which his family says is progressively worsening. Which of the following assessments is expected?
  1. Impaired speech
  2. Muscle flaccidity
  3. Pleasant and smiling demeanor
  4. Tremors in the fingers that increase with purposeful movement

26. 1. In Parkinson's disease, dysarthria (impaired speech) is due to a disturbance in muscle control. Muscle rigidity results in resistance to passive muscle stretching. The client may have a masklike appearance. Tremors should decrease with purposeful movement and sleep.
NP: Data collection; CN: Physiological integrity; CNS: Physiological adaptation; CL: Application

27. Which of the following terms describes a clinical judgment that an individual, family, or community is more vulnerable to develop a certain problem than others in the same or similar situation?
  1. Risk nursing diagnosis
  2. Actual nursing diagnosis
  3. Possible nursing diagnosis
  4. Wellness nursing diagnosis

27. 1. Risk nursing diagnosis refers to the vulnerability of a client, family, or community to health problems. An actual nursing diagnosis describes a human response to a health problem being manifested. Possible nursing diagnoses are made when there isn't enough evidence to support the presence of the problem, but the nurse believes the problem is highly probable and wants to collect more data. A wellness nursing diagnosis is a diagnostic statement describing the human response to levels of wellness in an individual, family, or community that have a potential for enhancement to a higher state.
NP: Planning; CN: Safe, effective care environment; CNS: Coordinated care; CL: Knowledge

28. Which of the following interventions is best to decrease a client's risk of skin breakdown?

1. Use a specialty mattress.
2. Position the client in alignment.
3. Reposition the client every 4 hours.
4. Massage bony prominences every shift.

28. 1. Specialty beds, such as fluid, air, and egg crate mattresses, can protect pressure areas on the client. Pressure areas on the client should be padded to prevent skin breakdown. The client should be turned every 2 hours. Massaging bony prominences causes friction and may irritate tissues.

NP: Implementation; CN: Physiological integrity; CNS: Reduction of risk potential; CL: Application

29. A client with Parkinson's disease is frequently incontinent of urine. Which of the following interventions is appropriate?

1. Diaper the client.
2. Apply a condom catheter.
3. Insert an indwelling urinary catheter.
4. Provide skin care every 4 hours.

29. 2. A condom catheter uses a condom-type device to drain urine away from the client. Diapering the client may keep urine away from the body but may also be demeaning if the client is alert or the family objects. Because the client with Parkinson's disease is already prone to urinary tract infections, an indwelling urinary catheter should be avoided because it may promote this. Skin care must be provided as soon as the client is incontinent to prevent skin maceration and breakdown.

NP: Implementation; CN: Physiological integrity; CNS: Basic care and comfort; CL: Analysis

30. Family members report exhaustion and difficulty taking care of a dependent family member. Which of the following approaches is in the best interest of the client?

1. Ask the client what he wishes.
2. Have the family members discuss it among themselves.
3. Tell the family the client should go to a nursing care facility.
4. Call a family conference and ask social services for assistance.

30. 4. A family conference with social services can enlighten the family to all prospects of care available to them. The family may not be aware of alternative care measures for the client so a discussion among themselves may not be helpful. The client should supply input if he's able but this may not help solve the problems of exhaustion and care difficulties. The client may not qualify for a nursing care facility because of stringent criteria.

NP: Planning; CN: Safe, effective care environment; CNS: Coordinated care; CL: Analysis

31. A 30-year-old primagravida in her second trimester tells a nurse her fingers feel tight and sometimes she feels as though her heart skips a beat. She has a history of rheumatic fever. Which of the following indicates the client may be experiencing cardiovascular disease?

1. Clear lungs
2. Sinus tachycardia
3. Increased dyspnea on exertion
4. Runs of paroxysmal atrial tachycardia

31. 3. Increasing dyspnea on exertion should alert the nurse to cardiovascular compromise. Cardiac arrhythmias (other than sinus tachycardia or paroxysmal atrial tachycardia) and persistent crackles at the bases are also symptoms of cardiovascular disease.

NP: Data collection; CN: Health promotion and maintenance; CNS: Growth and development through the life span; CL: Analysis

NP: Nursing process   CN: Client needs category   CNS: Client needs subcategory   CL: Cognitive level

32. Which of the following diagnostic tests determines the extent of cardiovascular disease during pregnancy?
1. Stress test
2. Chest X-ray
3. Echocardiography
4. Cardiac catheterization

You're cruisin'! Keep up the good work!

33. Which of the following factors is assessed to determine the effect of a client's heart condition on the fetus?
1. Urinalysis
2. Fetal heart tones
3. Laboratory test results of the mother
4. Other signs and symptoms of the client

34. Which of the following factors is an example of subjective data in a nursing assessment?
1. Laboratory study results
2. Physical assessment data
3. Report of a diagnostic procedure
4. Client's feelings and statements about health problems

35. Which of the following classifications of medication may be used safely for a pregnant client with cardiovascular disease?
1. Antibiotics
2. Warfarin (Coumadin)
3. Cardiac glycosides
4. Diuretics

36. A client arrives at the emergency department in her third trimester with painless vaginal bleeding. Which of the following conditions is suspected?
1. Placenta previa
2. Premature labor
3. Abruptio placentae
4. A sexually transmitted disease

32. 3. Echocardiography is less invasive than X-rays and other methods and provides the information needed to determine cardiovascular disease, especially valvular disorders. Cardiac catheterization and stress tests may be postponed until after delivery.
NP: Planning; CN: Physiological integrity; CNS: Physiological adaptation; CL: Knowledge

33. 2. Fetal heart tones show how the fetus is responding to the environment. Assessing other signs and symptoms of the mother, including laboratory test results and urinalysis, can only determine the effect on the mother.
NP: Data collection; CN: Health promotion and maintenance; CNS: Growth and development through the life span; CL: Comprehension

34. 4. Subjective data, also known as symptoms or covert cues, include the client's own verbatim statements about health problems. Physical assessment data, laboratory study results, and diagnostic procedure reports are observable, perceptible, and measurable and can be verified and validated by others.
NP: Data collection; CN: Safe, effective care environment; CNS: Coordinated care; CL: Comprehension

35. 3. Cardiac glycosides and common antiarrhythmics such as procainamide hydrochloride and quinidine may be used. If anticoagulants are needed, heparin is the drug of choice — not warfarin sodium. Prophylactic antibiotics are reserved for clients susceptible to endocarditis. Diuretics should be used with extreme caution, if at all, because of the potential for causing uterine contractions.
NP: Planning; CN: Physiological integrity; CNS: Pharmacological therapies; CL: Comprehension

36. 1. Placenta previa presents with painless vaginal bleeding. Abruptio placentae usually includes vague abdominal discomfort and tenderness. Sexually transmitted diseases and premature labor usually don't cause bleeding.
NP: Data collection; CN: Health promotion and maintenance; CNS: Prevention and early detection of disease; CL: Knowledge

37. After assessing vital signs and applying an external monitor, which of the following interventions is a priority for a client with suspected placenta previa?

1. Insert an indwelling urinary catheter.
2. Plan for an immediate cesarean delivery.
3. Place the client in Trendelenburg's position.
4. Start I.V. catheters and obtain blood work.

37. 4. Blood for hemoglobin, hematocrit, type, and crossmatch should be collected and I.V. catheters inserted. The nurse shouldn't attempt Trendelenburg's positioning or urinary catheterization. The client may be placed on her left side. Depending on the degree of bleeding and fetal maturity, a cesarean delivery may be required.

NP: Implementation; CN: Physiological integrity; CNS: Reduction of risk potential; CL: Application

38. A pregnant client with vaginal bleeding asks a nurse how the fetus is doing. Which of the following responses is best?

1. "I don't know for sure."
2. "I can't answer that question."
3. "It's too early to tell anything."
4. Tell the client what the monitors show.

38. 4. The client deserves a truthful answer and the nurse should be objective without giving opinions. Vague answers may be misleading and aren't therapeutic.

NP: Implementation; CN: Psychosocial integrity; CNS: Coping and adaptation; CL: Analysis

39. A client with placenta previa is hospitalized, and a cesarean delivery is planned. In addition to the routine neonatal assessment, the neonate should be assessed for which of the following conditions?

1. Prematurity
2. Congenital anomalies
3. Respiratory distress
4. Aspiration pneumonia

39. 3. Hypoxia, resulting in respiratory distress, is a potential risk due to decreased blood volume and prematurity. Congenital anomalies aren't necessarily associated with placenta previa. Aspiration pneumonia isn't considered a threat unless the amniotic fluid is meconium-stained. The age of maturity can be determined through established maternal dates.

NP: Data collection; CN: Physiological integrity; CNS: Reduction of risk potential; CL: Application

40. A neonate requires blood transfusions after birth. Which of the following cannulation sites is most preferred?

1. Scalp veins
2. Intraosseous
3. Umbilical cord
4. Subclavian cutdown

40. 3. The umbilical cord may be easily cannulated and is the preferred site. Scalp veins may also be used. Intraosseous cannulation is attempted if two attempts at other sites prove inaccessible. A subclavian cutdown takes a prolonged time and is the least desired.

NP: Implementation; CN: Health promotion and maintenance; CNS: Growth and development through the life span; CL: Knowledge

41. A nurse working in the triage area of an emergency department sees that several pediatric clients arrive simultaneously. Which of the following clients is treated first?
  1. A crying 4-year-old child with a laceration on his scalp
  2. A 3-year-old child with a barking cough and flushed appearance
  3. A 3-year-old child with Down syndrome who's pale and asleep in his mother's arms
  4. A 2-month-old infant with stridorous breath sounds, sitting up in his mother's arms and drooling

41. 4. The 2-month-old infant with the airway emergency should be treated first because of the risk of epiglottiditis. The 3-year-old with the barking cough and fever should be suspected of having croup and should be seen promptly, as should the child with the laceration. The nurse would need to gather information about the child with Down syndrome to determine the priority of care.

NP: Data collection; CN: Safe, effective care environment; CNS: Coordinated care; CL: Analysis

42. A 2-year-old child is being examined in the emergency department for epiglottiditis. Which of the following assessment findings supports this diagnosis?
  1. Mild fever
  2. Clear speech
  3. Tripod position
  4. Gradual onset of symptoms

42. 3. The tripod position (sitting up and leaning forward) facilitates breathing. Epiglottiditis presents with a sudden onset of symptoms, high fever, and muffled speech. Additional symptoms are inspiratory stridor and drooling.

NP: Data collection; CN: Physiological integrity; CNS: Physiological adaptation; CL: Application

Way to go! You're more than halfway finished!

43. Which of the following methods is best when approaching a 2-year-old child to listen to breath sounds?
  1. Tell the child it's time to listen to his lungs now.
  2. Tell the child to lie down while the nurse listens to his lungs.
  3. Ask the caregiver to wait outside while the nurse listens to his lungs.
  4. Ask if he would like the nurse to listen to the front or the back of his chest first.

43. 4. The 2-year-old child needs to feel in control, and this approach best supports the child's independence. Giving the child no choice may make him uncooperative. The caregiver should be allowed to remain with the child because fear of separation is common in 2-year-olds. The child should be allowed to remain in the tripod position to facilitate breathing.

NP: Implementation; CN: Health promotion and maintenance; CNS: Growth and development through the life span; CL: Application

44. A mother says a 2-year-old child is up to date with his vaccines. Which of the following immunizations should be included?
  1. Diphtheria-pertussis-tetanus (DTP), oral polio (OPV), measles-mumps-rubella (MMR)
  2. DTP, OPV, MMR, hepatitis B
  3. DTP, hepatitis B, OPV
  4. MMR, OPV, hepatitis B

44. 2. By the age of 2, the DTP, OPV, MMR, and hepatitis B vaccines should have been received. The nurse should clarify this with the mother or caregiver.

NP: Data collection; CN: Safe, effective care environment; CNS: Safety and infection control; CL: Knowledge

45. A child with epiglottiditis is at risk for which of the following conditions?
1. Airway obstruction
2. Dehydration
3. Malnutrition
4. Seizures

45. 1. The biggest threat to the child is airway obstruction because of the inflammation and swelling of the epiglottis and surrounding tissue. Dehydration can be prevented with I.V. therapy and seizures averted by decreasing the fever. Malnutrition is least likely to occur because epiglottiditis is a short-lived situation.
NP: Planning; CN: Physiological integrity; CNS: Reduction of risk potential; CL: Comprehension

46. Which of the following methods is used to diagnose epiglottiditis?
1. Lateral neck X-ray
2. Direct visualization
3. History of sudden onset
4. Presenting signs and symptoms

46. 4. The presenting symptoms are diagnostic of epiglottiditis. Lateral neck X-rays aren't necessary. Only an anesthesiologist or physician skilled in intubation should do direct visualization. History of sudden onset helps support the assessment, but a history alone wouldn't be sufficient to make a diagnosis.
NP: Data collection; CN: Health promotion and maintenance; CNS: Prevention and early detection of disease; CL: Knowledge

47. A student nurse working with a registered nurse is assessing a child with epiglottiditis. The student tells the client she needs to look at his throat. Which of the following interventions by the registered nurse is best?
1. Hand her a flashlight and tongue blade.
2. Give her a sterile tongue blade and culturette swab.
3. Tell the student that the registered nurse will visualize the child's throat.
4. Tell the student visualization will be done by the anesthesiologist.

47. 4. Direct visualization of the epiglottis can trigger a complete airway obstruction and should only be done in a controlled environment by an anesthesiologist or a physician skilled in pediatric intubation.
NP: Implementation; CN: Safe, effective care environment; CNS: Coordinated care; CL: Application

48. The mother of a 2-year-old child with epiglottiditis says she needs to pick up her older child from school. The 2-year-old child begins to cry and appears more stridorous. Which of the following interventions by the nurse is best?
1. Ask the mother how long she may be gone.
2. Tell the 2-year-old everything will be all right.
3. Tell the 2-year-old the nurse will stay with him.
4. Ask the mother if there is anyone else who can meet the older child.

48. 4. Increased anxiety and agitation should be avoided in the child to prevent airway obstruction. A 2-year-old child fears separation from parents, so the mother should be encouraged to stay. Other means of picking up the older child need to be found. The mother is the primary caregiver and important to the child for emotional and security reasons.
NP: Implementation; CN: Health promotion and maintenance; CNS: Growth and development through the life span; CL: Analysis

NP: Nursing process   CN: Client needs category   CNS: Client needs subcategory   CL: Cognitive level

49. A father arrives in a busy emergency department and is upset with his wife for bringing their 2-year-old child with epiglottiditis in for treatment. Which of the following interventions by the nurse is best?
  1. Leave the room.
  2. Call for security.
  3. Recognize the father's behavior as his attempt to cope with the situation.
  4. Tell both parents to leave because they're upsetting the child.

50. A 40-year-old client is being treated for GI bleeding. On his 5th day of hospitalization, he begins to have tremors, is agitated, and is experiencing hallucinations. These signs suggest which of the following conditions?
  1. Alcohol withdrawal
  2. Allergic response
  3. Alzheimer's disease
  4. Hypoxia

51. If a nurse suspects a client is experiencing alcohol withdrawal syndrome, which of the following actions is appropriate?
  1. Verify it with family.
  2. Inform social services.
  3. Ask the client about his drinking.
  4. Tell the client everything will be all right.

52. A client experiencing alcohol withdrawal syndrome says he sees cockroaches on the ceiling. Which of the following responses is appropriate?
  1. Ask the client where he sees them.
  2. Ask the client if the cockroaches are still there.
  3. Tell the client there are no cockroaches on the ceiling.
  4. Tell the client it's dim in the room and turn on the overhead lights.

49. 3. Lack of control over his son's situation results in irrational behavior. The nurse should try to calm both parents and let them know they did the right thing due to the seriousness of their child's situation. Calling for security, sending the parents out, or leaving the room won't help the child, nor will it reduce frustration or inappropriate behavior.
NP: Implementation; CN: Psychosocial integrity; CNS: Coping and adaptation; CL: Analysis

50. 1. These are signs of alcohol withdrawal syndrome, which usually begins 5 to 7 days after the last drink. An allergic reaction would cause difficulty breathing, skin rash, or edema as primary symptoms. Alzheimer's disease occurs in older individuals and has other psychosocial signs, such as a masklike face and altered mentation. Hypoxia would cause symptoms of respiratory distress.
NP: Data collection; CN: Psychosocial integrity; CNS: Psychosocial adaptation; CL: Analysis

51. 3. Confirming suspicions with the client is the most beneficial way to help in diagnosis and treatment. If the client isn't cooperative, verification can be sought with the family. Social services aren't required at this time but may be helpful in discharge planning. Giving false reassurance isn't therapeutic for the client.
NP: Data collection; CN: Psychosocial integrity; CNS: Psychosocial adaptation; CL: Application

52. 4. Try to reorient the client to reality and minimize distortions. Don't support the client's hallucinations or place the client on the defensive but try to present reality gently without agitating the client.
NP: Implementation; CN: Psychosocial integrity; CNS: Psychosocial adaptation; CL: Application

53. A client experiencing alcohol withdrawal syndrome says he's itching everywhere from the bugs on his bed. Which of the following responses is appropriate?
1. Examine the client's skin.
2. Ask what kind of bugs he thinks they are.
3. Tell the client there are no bugs on his bed.
4. Tell the client he's having tactile hallucinations.

54. A client with alcohol withdrawal syndrome is pulling at his central venous catheter saying he's swatting the spiders crawling over him. Which of the following interventions is appropriate?
1. Encourage the client to rest.
2. Protect the client from harm.
3. Tell the client there are no spiders.
4. Tell the client he's pulling the I.V. tubing.

55. A client who experienced alcohol withdrawal syndrome is no longer having hallucinations or tremors and says he would like to enter a rehabilitation facility to stop drinking. Which of the following interventions is appropriate?
1. Ask about his insurance.
2. Tell him he should talk with his family.
3. Refer him to Alcoholics Anonymous (AA).
4. Promote participation in a treatment program.

56. A 72-year-old man with cirrhosis is admitted to the hospital in a hepatic coma. Which of the following nursing interventions has priority?
1. Perform a neurologic check.
2. Complete the client admission.
3. Orient the client to his environment.
4. Check airway, breathing, and circulation.

Only 30 more to go. You can do it! I know you can!

53. 1. Make sure the client doesn't have a rash, skin allergy, or something on his skin (such as crumbs) causing his discomfort. Reality should then be presented to the client gently without being derogatory. The nurse shouldn't support the client's hallucinations.
NP: Implementation; CN: Psychosocial integrity; CNS: Coping and adaptation; CL: Application

54. 2. If the client dislodges the central venous catheter, he may incur an air embolus, which can be life-threatening. The client may need to be restrained if continued observation during this time isn't available. Reality should be presented to the client. During periods of alcohol withdrawal syndrome, the client needs to be protected from harm; telling him to rest may not be heeded.
NP: Implementation; CN: Psychosocial integrity; CNS: Psychosocial adaptation; CL: Analysis

55. 4. The client should be encouraged to enter a facility if that's in his best interest. Arrangements can be made and discussed with the social service coordinator and his physician. The client can inform his family, and support should be encouraged. Referral to AA should be considered after rehabilitation takes place.
NP: Planning; CN: Psychosocial integrity; CNS: Psychosocial adaptation; CL: Application

56. 4. Priorities include airway, breathing, and circulation. Once these are ensured, a neurologic check is needed to determine status. Depending on the client's alertness, orientation to the environment may need to be kept simple (where he is, date, person, time). General orientation and completing the admission may require the help and affirmation of family members.
NP: Data collection; CN: Physiological integrity; CNS: Reduction of risk potential; CL: Comprehension

NP: Nursing process   CN: Client needs category   CNS: Client needs subcategory   CL: Cognitive level

57. A client with cirrhosis is restless and at times tries to climb out of bed. Which of the following interventions is best to promote safety?
1. Use leather restraints.
2. Use soft wrist restraints.
3. Use a vest restraint device.
4. Use a sheet tied across the client's chest.

57. 3. The client may require gentle reminders not to get out of bed to prevent a fall. The vest restraint would help in this endeavor. Soft wrist restraints may not stop the client from sitting up or trying to swing his legs over the bed rails. Leather restraints are only warranted for extremely combative and unsafe clients. A sheet tied across the client's chest can hamper breathing or may asphyxiate the client if he slides down in the bed.
NP: Implementation; CN: Safe, effective care environment; CNS: Coordinated care; CL: Analysis

58. Which of the following assessment findings in a client with cirrhosis indicates late-stage symptoms?
1. Constipation
2. Diarrhea
3. Hypoxia
4. Vomiting

58. 3. Fluid in the lungs and weak chest expansion can lead to hypoxia. Diarrhea, vomiting, and constipation are early signs and symptoms of cirrhosis.
NP: Data collection; CN: Physiological integrity; CNS: Physiological adaptation; CL: Knowledge

59. A client with cirrhosis is jaundiced and edematous. He's experiencing severe itching and dryness. Which of the following interventions is best to help the client?
1. Put mitts on his hands.
2. Use alcohol-free body lotion.
3. Lubricate the skin with baby oil.
4. Wash the skin with soap and water.

59. 2. Alcohol-free body lotion applied to the skin can help relieve dryness and is absorbed without oiliness. Mitts may help keep the client from scratching his skin open. Soap dries out the skin. Baby oil doesn't allow excretions through the skin and may block pores.
NP: Implementation; CN: Physiological integrity; CNS: Basic care and comfort; CL: Knowledge

60. A 20-year-old client with a spinal cord injury sustained in a previous motorcycle accident is hospitalized for renal calculi. To reduce the client's risk for developing recurrent kidney stones, which of the following instructions is correct?
1. Eat yogurt daily.
2. Drink cranberry juice.
3. Eat more fresh fruits and vegetables.
4. Increase the intake of dairy products.

60. 2. Acid urine decreases the potential for renal calculi. The majority of renal calculi form in alkaline urine. Cranberries, prunes, and plums promote acidic urine. Yogurt helps restore pH balance to secretions in yeast infections. Fruits and vegetables increase fiber in the diet and promote alkaline urine. Dairy products may contribute to the formation of renal calculi.
NP: Implementation; CN: Physiological integrity; CNS: Reduction of risk potential; CL: Application

61. A client with a spinal cord injury says he has difficulty recognizing the symptoms of urinary tract infection (UTI) before it's too late. Which of the following symptoms is an early sign of UTI?
1. Lower back pain
2. Burning on urination
3. Frequency of urination
4. Fever and change in the clarity of urine

61. 4. The client with a spinal cord injury should recognize fever and change in the clarity of urine as early signs of UTI. Low back pain is a late sign. The client with a spinal cord injury may not have burning or frequency of urination.
NP: Implementation; CN: Physiological integrity; CNS: Reduction of risk potential; CL: Application

62. A client tells a nurse he boils his urinary catheters to keep them sterile. Which of the following questions should the nurse ask?

1. "What technique is used for sterilization?"
2. "What temperature are the catheters boiled at?"
3. "Why aren't prepackaged sterile catheters used?"
4. "Are the catheters dried and stored in a clean, dry place?"

You're doing great! Keep pluggin' away!

62. 1. The client should describe his procedure to make sure sterile technique is used. Water boils at 212° F (100° C), but the nurse should make sure the client is boiling the catheters for an appropriate amount of time. Catheters should be boiled just before use and allowed to cool before using. Prepackaged sterile catheters aren't necessary if the proper sterilization techniques are used.

NP: Evaluation; CN: Physiological integrity; CNS: Reduction of risk potential; CL: Analysis

63. A 60-year-old client had a colostomy 4 days ago due to rectal cancer and is having trouble adjusting to it. Which of the following conditions is most common?

1. Anxiety
2. Low self-esteem
3. Alteration in comfort
4. Alteration in body image

63. 4. Alteration in body image is most common with a new colostomy and dealing with its care. Low self-esteem may also be a concern for the client. The client should be having less discomfort postoperatively. The client shouldn't have signs of anxiety, but he may not be comfortable caring for the colostomy.

NP: Data collection; CN: Psychosocial integrity; CNS: Coping and adaptation; CL: Comprehension

64. A nurse approaches a client who recently had a colostomy and finds him tearful. Which of the following actions is appropriate?

1. State she'll come back another time.
2. Ask the client if he's having pain or discomfort.
3. Tell the client she needs to take her vital signs.
4. Sit down with the client and ask if he'd like to talk about anything.

64. 4. Asking open-ended questions and appearing interested in what the client has to say will encourage verbalization of feelings. Leaving the client may make him feel unaccepted. Asking closed-ended questions won't encourage verbalization of feelings. Ignoring the client's present state isn't therapeutic for the client.

NP: Implementation; CN: Psychosocial integrity; CNS: Coping and adaptation; CL: Application

65. After a review of colostomy care, a client says she doesn't know if she'll be able to care for herself at home without help. Which of the following nursing interventions is most appropriate?

1. Review care with the client again.
2. Provide written instructions for the client.
3. Ask the client if there's anyone who can help.
4. Arrange for home health care to visit the client.

65. 4. Home health care should be contacted to ensure continuity of appropriate care. Because of the complexity of care, family members can help if possible and should be present for teaching. Follow a detailed teaching plan to educate the client before discharge.

NP: Data collection; CN: Safe, effective care environment; CNS: Coordinated care; CL: Application

NP: Nursing process   CN: Client needs category   CNS: Client needs subcategory   CL: Cognitive level

66. A client is experiencing mild diarrhea through his colostomy. Which of the following instructions is correct?
1. Eat prunes.
2. Drink apple juice.
3. Increase lettuce intake.
4. Increase intake of bananas.

67. A client reports a lot of gas in her colostomy bag. Which of the following instructions is best?
1. Burp the bag.
2. Eat fewer beans.
3. Replace the bag.
4. Put a tiny hole in the top of the bag.

68. A client at a routine blood glucose screening for diabetes mellitus tells a nurse she has excessive urination and excessive thirst. The nurse should ask about which of the following symptoms first?
1. Weakness
2. Weight loss
3. Vision changes
4. Excessive hunger

69. Which of the following factors places a client at greatest risk for developing diabetes mellitus?
1. Obesity
2. Japanese descent
3. A great-grandparent with diabetes mellitus
4. Delivery of a neonate weighing more than 10 lb

70. Which of the following blood glucose levels is considered within normal limits?
1. 70 to 125 mg/dl
2. 130 to 135 mg/dl
3. 135 to 140 mg/dl
4. 140 to 145 mg/dl

66. 4. Bananas help make formed stool and aren't irritating to the bowel. Apple juice and prunes can increase the frequency of diarrhea. Lettuce acts as a fiber and can increase the looseness of stools.
NP: Implementation; CN: Physiological integrity; CNS: Basic care and comfort; CL: Application

67. 1. Letting air out of the bag by opening it and burping it is the best solution. Replacing the bag is costly. Putting a hole in the bag will also cause fluids to leak out. The client can be encouraged to note which foods are causing gas and to eat less gas-forming foods.
NP: Implementation; CN: Physiological integrity; CNS: Basic care and comfort; CL: Knowledge

68. 4. Polyuria, polydipsia, and polyphagia are the three hallmark signs of diabetes mellitus. Weight loss, weakness, and vision changes also occur with diabetes mellitus.
NP: Data collection; CN: Health promotion and maintenance; CNS: Prevention and early detection of disease; CL: Knowledge

69. 1. Obesity is a risk factor associated with diabetes mellitus. Delivery of a neonate weighing more than 9 lb, a family history of diabetes mellitus (mother, father, or sibling), and those of Native American, Black, Asian, or Hispanic descent are at high risk for developing diabetes mellitus, but obesity puts the client at greatest risk.
NP: Data collection; CN: Health promotion and maintenance; CNS: Prevention and early detection of disease; CL: Knowledge

70. 1. A blood glucose level more than 126 mg/dl should prompt the nurse to refer the client to the physician for follow-up due to the risk of diabetes mellitus.
NP: Data collection; CN: Health promotion and maintenance; CNS: Prevention and early detection of disease; CL: Knowledge

71. A client asks what diabetes mellitus does to the body over time. Which of the following conditions is a common chronic complication of diabetes mellitus?
1. Vaginal infections
2. Diabetic ketoacidosis
3. Cardiovascular disease
4. Hyperosmolar hyperglycemic nonketotic syndrome (HHNS)

72. A client asks how she might decrease the risk of developing diabetes mellitus, which runs in her family. Which of the following responses is appropriate?
1. "Eat only poultry and fish."
2. "Omit carbohydrates from your diet."
3. "Start a moderate exercise program."
4. "Check blood glucose levels every month."

73. An 83-year-old client fractured a hip after a fall in her home. Because of her extensive cardiac history and chronic obstructive pulmonary disease, surgery isn't an option. The client tells a nurse she doesn't know how she's going to get better. Which of the following responses is best?
1. "You're doing fine."
2. "What's your biggest concern right now?"
3. "Just give it some time and you'll be okay."
4. "You don't believe you're doing well?"

74. A client says she slipped on a throw rug while going to the bathroom at night. Which of the following factors needs assessment?
1. If the home is safe
2. If the client is confused
3. If the client hit her head
4. If the client has a urinary tract infection (UTI)

71. 3. Cardiovascular disease is a common chronic complication of diabetes mellitus. Diabetic ketoacidosis and HHNS are acute complications that can occur. Vaginal discomfort is an early sign of diabetes mellitus.
NP: Data collection; CN: Health promotion and maintenance; CNS: Prevention and early detection of disease; CL: Knowledge

72. 3. Exercise and weight control are the goals in preventing and treating diabetes mellitus. Red meat can be eaten but should be limited because it contributes to cardiovascular disease. Complex carbohydrates account for a large portion of the diabetic diet and shouldn't be omitted. Checking blood glucose levels will help monitor the development of diabetes mellitus but won't prevent or decrease the chance of it occurring.
NP: Implementation; CN: Physiological integrity; CNS: Reduction of risk potential; CL: Knowledge

73. 2. Open-ended questions allow the client to have control over what she wants to discuss and help the nurse determine care needs. Telling the client she's doing fine or that she just needs more time doesn't encourage her to verbalize concerns. A reiteration of the client's concerns may not be helpful in encouraging the client to verbalize feelings.
NP: Implementation; CN: Health promotion and maintenance; CNS: Growth and development through the life span; CL: Comprehension

74. 1. A safety assessment of the home can determine if changes need to be made to ensure the client doesn't fall again. Going to the bathroom at night isn't necessarily a sign of a UTI. The nurse may determine if the client has experienced a head injury or confusion by asking how the accident occurred.
NP: Data collection; CN: Safe, effective care environment; CNS: Safety and infection control; CL: Application

NP: Nursing process   CN: Client needs category   CNS: Client needs subcategory   CL: Cognitive level

75. A client in her third trimester of pregnancy is having contractions 5 minutes apart that began suddenly. The nurse identifies that it's the client's seventh month. She's admitted directly to the obstetric department. Which of the following interventions has priority?
    1. Call the obstetrician.
    2. Time the contractions.
    3. Check fetal heart tones.
    4. Call the client's husband.

76. Which information about vital signs should be reported to the physician?
    1. Blood pressure of 120/72 mm Hg in a healthy man
    2. Pulse of 110 beats/minute on awakening in the morning
    3. Blood pressure of 110/68 mm Hg in a healthy woman
    4. Pulse of 120 beats/minute after 30 minutes of aerobic exercise

77. A 40-year-old client is scheduled to have elective facial surgery later in the morning. The nurse notes the pulse rate is 130 beats/ minute. Which of the following reasons best explains the tachycardia?
    1. Age
    2. Anxiety
    3. Exercise
    4. Pain

78. A client on a psychiatric unit asks a nurse about the medications another client takes. Which of the following responses is best?
    1. "How close are the two of you?"
    2. "I can't give you that information, I must protect her privacy."
    3. "Let me ask her if it's OK for me to tell you about her condition and medications."
    4. "The client is taking insulin for her diabetes and digoxin for her heart condition."

75. 3. The nurse should check fetal heart tones and assess the client's vital signs. The client should be placed on a monitor to check contractions and for continuous fetal monitoring. The obstetrician and husband should be notified as soon as possible.
NP: Implementation; CN: Health promotion and maintenance; CNS: Growth and development through the life span; CL: Application

76. 2. The normal range for a pulse is 60 to 100 beats/minute and, in the morning, the rate is at its lowest. Blood pressures of 120/72 mm Hg for a healthy man and 110/68 mm Hg for a healthy woman are normal. Aerobic exercise increases the heart rate over the normal range of 60 to 100 beats/minute. The formula for maximum aerobic heart rate is: 210 − age × 80%. A person shouldn't go over the maximum heart rate during aerobic exercise.
NP: Data collection; CN: Physiological integrity; CNS: Physiological adaptation; CL: Comprehension

77. 2. Anxiety tends to increase heart rate, temperature, and respirations. Exercise will increase the heart rate but most likely won't occur preoperatively. The normal heart rate for a client this age is 60 to 100 beats/minute. The client shouldn't be in any pain preoperatively.
NP: Data collection; CN: Physiological integrity; CNS: Physiological adaptation; CL: Comprehension

78. 2. Revealing one client's medication to another client is violating procedures of client confidentiality. Seeking the client's permission to release confidential information is an inappropriate action. Asking the client the nature of his relationship to the other client won't help the client understand the purpose of protecting confidentiality. Assuring the client that the facility has an obligation to protect not only his confidentiality but that of others will provide the client with a sense of comfort.
NP: Implementation; CN: Psychosocial integrity; CNS: Psychosocial adaptation; CL: Application

79. The vital signs of a 56-year-old client are: temperature, 98.6° F (37° C) orally; pulse, 80 beats/minute; and respirations, 30 breaths/minute. Which of the following interpretations of these values is correct?
1. Pulse is above normal range.
2. Temperature is above normal range.
3. Respirations are above normal range.
4. Respirations and pulse are above normal range.

80. A nurse notes a client's pulse is regular and readily palpable at a rate of 84 beats/minute. Which of the following terms best describes this pulse?
1. Arrhythmia
2. Bradycardia
3. Regular
4. Tachycardia

81. Which of the following actions is correct for performing tracheal suctioning?
1. Apply suction during insertion of the catheter.
2. Limit suctioning to 10 to 15 seconds' duration.
3. Resterilize the suction catheter in alcohol after use.
4. Repeat suctioning intervals every 15 minutes until clear.

82. Which of the following positions is optimal for a client having a nasogastric tube inserted?
1. Fowler's
2. Prone
3. Side-lying
4. Supine

Five more to go! Ooooh, I'm getting excited!

79. 3. Normal vital signs for an adult client are: temperature, 96.6° to 99° F (35.9° to 37.2° C); pulse, 60 to 100 beats/minute; respirations, 16 to 20 breaths/minute.
NP: Evaluation; CN: Health promotion and maintenance; CNS: Growth and development through the life span; CL: Knowledge

80. 3. The pulse is regular when it's rhythmic, easily palpable, and between the rate of 60 to 100 beats/minute. Tachycardia is a heart rate faster than 100 beats/minute. Arrhythmia is a heart rate with either irregular rate or rhythm. Bradycardia is a heart rate slower than 60 beats/minute.
NP: Evaluation; CN: Health promotion and maintenance; CNS: Growth and development through the life span; CL: Knowledge

81. 2. The length of time a client should be able to tolerate the suction procedure is 10 to 15 seconds. Any longer may cause hypoxia. Suctioning during insertion can cause trauma to the mucosa and removes oxygen from the respiratory tract. Suctioning intervals with supplemental oxygen between suctions is performed after at least 1-minute intervals to allow the client to rest. Suction catheters are disposed of after each use and are cleaned in normal saline solution after each pass.
NP: Implementation; CN: Physiological integrity; CNS: Physiological adaptation; CL: Application

82. 1. The upright position is more natural for swallowing and protects against aspiration. Positioning on the client's back, stomach, or side places the client at risk for aspiration if he should gag. It's also difficult to swallow in these positions.
NP: Implementation; CN: Physiological integrity; CNS: Basic care and comfort; CL: Comprehension

NP: Nursing process   CN: Client needs category   CNS: Client needs subcategory   CL: Cognitive level

83. Which of the following steps should the nurse take first when preparing to insert a nasogastric (NG) tube?
1. Wash hands.
2. Apply sterile gloves.
3. Apply a mask and gown.
4. Open all necessary kits and tubing.

83. 1. The first intervention before a procedure is hand washing. Clean gloves are used because the mouth and nasopharynx aren't considered sterile. A mask and gown aren't required. Opening all the equipment is the next step before inserting the NG tube.

NP: Implementation CN: Safe, effective care environment; CNS: Safety and infection control; CL: Comprehension

84. As a nurse is inserting a nasogastric tube, the client begins to gag. Which action should the nurse take?
1. Remove the inserted tube and notify the physician of the client's status.
2. Stop the insertion, allow the client to rest, then continue inserting the tube.
3. Encourage the client to take deep breaths through the mouth while the tube is being inserted.
4. Pause until the gagging stops, tell the client to take a few sips of water and swallow as the tube is inserted.

84. 4. Swallowing helps advance the tube by causing the epiglottis to cover the opening of the trachea, thus helping to eliminate gagging and coughing. Removing the tube is unnecessary as gagging is an expected response to this procedure. Deep breathing opens the trachea, allowing the tube to possibly advance into the lungs.

NP: Evaluation; CN: Safe, effective care environment; CNS: Safety and infection control; CL: Application

85. Which of the following steps, if taken by the nurse after insertion of a nasogastric (NG) tube, could harm the client?
1. Affix the NG tube to the nose with tape.
2. Check tube placement by aspirating stomach contents using a piston syringe.
3. Check tube placement by instilling 100 ml of water into the tube to check for stomach filling.
4. Document in the chart the insertion, method used to check tube placement, and client's response to the procedure.

85. 3. Should the tube be located in the lungs, instilling water would flood the lungs, precipitating choking, coughing, hypoxemia and, possibly, pneumonia. Documentation is required for any procedure. Withdrawing stomach contents from the NG tube double checks the correct placement. Anchoring the tube after placement to the nose with tape or a manufactured device prevents the tube from becoming dislodged.

NP: Evaluation; CN: Safe, effective care environment; CNS: Coordinated care; CL: Comprehension

I knew you could do it! For your first comprehensive test, that was outstanding! Congratulations!

Here's another comprehensive test to help you get ready to take the NCLEX exam. Good luck!

COMPREHENSIVE
Test 2

1. A young client with anorexia nervosa is being treated using family therapy. Which of the following descriptions applies to anorexia nervosa, according to the social theory of causation?
1. A repressed issue
2. A learned maladaptive behavior
3. A reaction related to poor school achievements
4. A sign of pain within the whole family system

2. Which of the following diagnostic tests is performed first to detect transposition of the great vessels (TGV)?
1. Blood cultures
2. Cardiac catheterization
3. Chest X-ray
4. Echocardiogram

3. Four 6-month-old children arrive at the clinic for diphtheria-pertussis-tetanus (DPT) immunization. Which of the following children can safely receive the immunization at this time?
1. The child with a temperature of 103° F (39.4° C)
2. The child with a runny nose and cough
3. The child taking prednisone for the treatment of leukemia
4. The child with difficulty breathing after the last immunization

4. A nurse is giving discharge instructions to parents of a child who had a tonsillectomy. Which of the following instructions is the most important?
1. The child should drink extra milk.
2. The child shouldn't drink from straws.
3. Orange juice should be given to provide pain control.
4. Rinse the mouth with salt water to provide pain relief.

1. 4. The social theory of causation recognizes that difficulties within families are generally expressed by one member showing symptoms. Behavioral theory considers anorexia nervosa a learned maladaptive behavior, and psychoanalytic theory considers it a sign of repression. Anorexia nervosa is associated with a high level of achievement in school.
NP: Implementation; CN: Psychosocial integrity; CNS: Psychosocial adaptation; CL: Comprehension

2. 3. Chest X-ray would be done first to visualize congenital heart diseases such as TGV. Cardiac catheterization and an echocardiogram would be done after TGV is seen on the chest X-ray. Blood cultures won't diagnose TGV.
NP: Planning; CN: Health promotion and maintenance; CNS: Prevention and early detection of disease; CL: Application

3. 2. Children with cold symptoms can safely receive DPT immunization. Children with a temperature more than 102° F (38.9° C), serious reactions to previous immunizations, or immunosuppressive therapy shouldn't receive DPT immunization.
NP: Data collection; CN: Health promotion and maintenance; CNS: Prevention and early detection of disease; CL: Analysis

4. 2. Straws and other sharp objects inserted into the mouth could disrupt the clot at the operative site. Extra milk wouldn't promote healing and may encourage mucus production. Although drinking orange juice and rinsing with salt water will irritate the tissue at the operative site, irritation doesn't pose the same level of danger as clot disruption.
NP: Implementation; CN: Physiological adaptation; CNS: Basic care and comfort; CL: Application

NP: Nursing process   CN: Client needs category   CNS: Client needs subcategory   CL: Cognitive level

5. A 2-year-old child is diagnosed with bronchiolitis caused by respiratory syncytial virus (RSV). The client has an 8-year-old sibling. Which of the following statements is correct?
  1. RSV isn't highly communicable in infants.
  2. RSV isn't communicable to older children and adults.
  3. The 2-year-old client must be admitted to the hospital for isolation.
  4. The children should be separated to prevent the spread of the infection.

5. 4. RSV is communicable among children and adults. Older children and adults may have mild symptoms of the disorder. Hospitalization is indicated only for children who need oxygen and I.V. therapy. Toddlers easily transmit and contract RSV.
NP: Planning; CN: Safe, effective care environment; CNS: Safety and infection control; CL: Analysis

6. A child with asthma uses a peak expiratory flowmeter in school. The results indicate his peak flow is in the yellow zone. Which of the following interventions by the school nurse is appropriate?
  1. Follow the child's routine asthma treatment plan.
  2. Monitor the child for signs and symptoms of an acute attack.
  3. Call 911 and prepare for transport to the nearest emergency department.
  4. Call the child's mother to take the child to the family physician immediately.

6. 2. The routine treatment plan may be insufficient when the peak flow is in the yellow zone (50% to 80% of personal best). The child should be monitored to determine if an asthma attack is imminent. There's no immediate need to see the physician if the child is asymptomatic. This isn't an emergency situation.
NP: Implementation; CN: Physiological integrity; CNS: Reduction of risk potential; CL: Application

7. Parents of a child with asthma are trying to identify possible allergens in their household. Which of the following inhaled allergens is the most common?
  1. Perfume
  2. Dust mites
  3. Passive smoke
  4. Dog or cat dander

7. 2. The household dust mite is the most commonly inhaled allergen that can cause an asthma attack. Animal dander, passive smoke, and perfume are sometimes allergens causing asthma attacks but aren't as common as dust mites.
NP: Implementation; CN: Physiological integrity; CNS: Reduction of risk potential; CL: Application

8. A nurse is verifying orders from a physician. Which of the following diets is correct for a child newly diagnosed with celiac disease?
  1. Low-fat diet
  2. No-gluten diet
  3. High-protein diet
  4. No-phenylalanine diet

8. 2. The intestinal cells of individuals with celiac disease become inflamed when the child eats products containing gluten, such as wheat, rye, barley, or oats. The child with celiac disease needs normal amounts of fat and protein in the diet for growth and development. Omitting phenylalanine products would be appropriate for the client with phenylketonuria.
NP: Planning; CN: Physiological integrity; CNS: Reduction of risk potential; CL: Application

9. A client is undergoing a thoracentesis at the bedside. A nurse assists the client to an upright position with a table and pillow in front of him to support his arms. Which of the following rationales for this intervention is correct?
1. Fluid will accumulate at the base.
2. There's less chance to injure lung tissue.
3. It allows for better expansion of the lung.
4. It's less painful for the client in this position.

You're doing great! Keep at it!

9. 1. Fluids will drain and collect in the dependent positions. There's a risk of pneumothorax regardless of the client's position. This procedure is done using local anesthesia, so it isn't painful.
NP: Planning; CN: Physiological integrity; CNS: Physiological adaptation; CL: Comprehension

10. Which of the following leisure activities is recommended for a school-age child with hemophilia?
1. Baseball
2. Cross-country running
3. Football
4. Swimming

10. 4. Swimming is a noncontact sport with low risk of traumatic injury. Baseball, cross-country running, and football all involve a risk of trauma from falling, sliding, or contact.
NP: Implementation; CN: Physiological integrity; CNS: Physiological adaptation; CL: Application

11. Which of the following findings is important for an infant in sickle cell crisis?
1. The infant has no bruises.
2. The infant has normal skin turgor.
3. The infant participates in exercise.
4. The infant maintains bladder control.

11. 2. Normal skin turgor indicates the infant isn't severely dehydrated. Dehydration may cause sickle cell crisis or worsen a crisis. Bruising isn't associated with sickle cell crisis. Bed rest is preferable during a sickle cell crisis. Bladder control may be lost when oral or I.V. fluid intake is increased during a sickle cell crisis.
NP: Planning; CN: Physiological integrity; CNS: Physiological adaptation; CL: Analysis

12. A client has arterial blood gases drawn. The results are as follows: pH, 7.52; Pao_2, 50 mm Hg; $Paco_2$, 28 mm Hg; HCO_3^-, 24 mEq/L. Which of the following conditions is indicated?
1. Metabolic acidosis
2. Metabolic alkalosis
3. Respiratory acidosis
4. Respiratory alkalosis

12. 4. A pH greater than 7.45 and a $Paco_2$ less than 35 mm Hg indicate respiratory alkalosis. A pH less than 7.35 and an HCO_3^- less than 22 mEq/L indicate metabolic acidosis. A pH greater than 7.45 and an HCO_3^- greater than 24 mEq/L indicate metabolic alkalosis. A pH less than 7.35 and a $Paco_2$ greater than 45 mm Hg indicate respiratory acidosis.
NP: Data collection; CN: Physiological integrity; CNS: Physiological adaptation; CL: Knowledge

13. A client with chronic alcohol abuse is admitted to the hospital for detoxification. Later that day, his blood pressure increases and he is given an as-needed medication to prevent which of the following complications?
1. Stroke
2. Seizure
3. Fainting
4. Anxiety reaction

13. 2. During detoxification from alcohol, changes in the client's physiologic status, especially an increase in blood pressure, may indicate a possible seizure. Clients are treated with benzodiazepines such as chlordiazepoxide (Librium) to prevent this occurrence. Stroke, fainting, and anxiety aren't the primary concerns when withdrawing from alcohol.
NP: Implementation; CN: Physiological integrity; CNS: Physiological adaptation; CL: Application

NP: Nursing process   CN: Client needs category   CNS: Client needs subcategory   CL: Cognitive level

14. An adolescent client ingests a large number of acetaminophen tablets in an attempt to commit suicide. Which of the following laboratory results is associated with acetaminophen overdose?
　　1. Metabolic acidosis
　　2. Elevated liver enzyme levels
　　3. Increased serum creatinine level
　　4. Increased white blood cell (WBC) count

15. A nurse is caring for a client recently diagnosed with acute pancreatitis. Which of the following statements indicates that a short-term goal of nursing care has been met?
　　1. The client denies abdominal pain.
　　2. The client doesn't complain of thirst.
　　3. The client denies pain at McBurney's point.
　　4. The client swallows liquids without coughing.

16. A physician prescribes acetaminophen gr X (10 grains) as necessary every 4 hours for pain for a client in a long-term care facility. How many milligrams of acetaminophen should the nurse give?
　　1. 10 mg
　　2. 325 mg
　　3. 650 mg
　　4. 1,000 mg

17. A man stepped on a piece of sharp glass while walking barefoot. He comes to the emergency department with a deep laceration on the bottom of his foot. Which of the following questions is the most important for the nurse to ask?
　　1. "Was the glass dirty?"
　　2. "Are you immune to tetanus?"
　　3. "When did you have your last tetanus shot?"
　　4. "How many diphtheria-pertussis-tetanus (DPT) shots did you receive as a child?"

14. 2. Elevated liver enzyme levels, which could indicate liver damage, are associated with acetaminophen overdose. Metabolic acidosis isn't associated with acetaminophen overdose. An increased serum creatinine level may indicate renal damage. An increased WBC count indicates infection.
NP: Data collection; CN: Physiological integrity; CNS: Pharmacological therapies; CL: Comprehension

15. 1. Pancreatitis is accompanied by acute pain from autodigestion by pancreatic enzymes. When the client denies abdominal pain, the short-term goal of pain control is met. Clients with acute pancreatitis receive I.V. fluids and may not have a sensation of thirst. Pain at McBurney's point accompanies appendicitis. Clients with acute pancreatitis receive nothing by mouth during initial therapy.
NP: Evaluation; CN: Physiological integrity; CNS: Physiological adaptation; CL: Application

16. 3. One grain = 65 mg. Ten times 65 = 650 mg.
NP: Planning; CN: Physiological integrity; CNS: Pharmacological therapies; CL: Application

17. 3. Questioning the client about the date of his last tetanus immunization is important because the booster immunization should be received every 10 years in adulthood or at the time of the injury if the last booster immunization was given more than 5 years before the injury. A client wouldn't know his tetanus immune status. DPT immunizations in childhood don't give lifelong immunization to tetanus. Whether the client noticed dirt on the glass is immaterial because all deep lacerations require a tetanus immunization or booster.
NP: Data collection; CN: Safe, effective care environment; CNS: Safety and infection control; CL: Application

18. A postmenopausal client asks a nurse how to prevent osteoporosis. Which of the following responses is best?
1. "Take a multivitamin daily."
2. "After menopause, there's no way to prevent osteoporosis."
3. "Drink two glasses of milk each day and swim three times per week."
4. "Do weight-bearing exercises regularly."

18. 4. Weight-bearing exercises are recommended for the prevention of osteoporosis. Two glasses of milk per day don't provide the daily requirements for adult women, and swimming isn't a weight-bearing exercise. A multivitamin doesn't provide adequate calcium for a postmenopausal woman, and calcium alone won't prevent osteoporosis.
NP: Implementation; CN: Health promotion and maintenance; CNS: Prevention and early detection of disease; CL: Application

19. A client diagnosed with cardiomyopathy saw a posting on the Internet describing research about a new herbal treatment for the disorder. When the client asks about this research, which of the following responses is most appropriate?
1. "Herbs are often used to treat cardiomyopathy."
2. "Cardiomyopathy can be treated only by heart surgery."
3. "The Internet is a reliable source of research, so try this treatment."
4. "Research found on the Internet should be verified with a physician."

19. 4. Although the Internet contains some valid medical research, there's no control over the validity of information posted on it. The research should be discussed with a physician, who has access to medical research and can verify the accuracy of the information. Cardiomyopathy is treatable with drugs or surgery. Herbs aren't standard treatment for cardiomyopathy.
NP: Implementation; CN: Safe, effective care environment; CNS: Coordinated care; CL: Application

20. A young adult client received her first chemotherapy treatment for breast cancer. Which of the following statements by the client requires further exploration by the nurse?
1. "I'm thinking about joining a dance club."
2. "I don't think I'm going to work tomorrow."
3. "I don't care about the side effects of drugs."
4. "I want to return to school for a college degree."

You've finished 20 questions already? Wow, super!

20. 3. Adverse effects of chemotherapy may occur after treatment and should be discussed with the client because some can be treated, controlled, or prevented. The nurse needs to explore what the client means by this statement. The client may feel poorly after chemotherapy and may want to take time off from work until feeling better. Joining social clubs is typical behavior for a young adult. Returning to school is also typical of a young adult.
NP: Data collection; CN: Health promotion and maintenance; CNS: Growth and development through the life span; CL: Analysis

21. A client with long-standing rheumatoid arthritis has frequent complaints of joint pain. The nurse's care plan is based on the understanding that chronic pain is most effectively relieved when analgesics are administered in which of the following ways?
1. Conservatively
2. I.M.
3. On an as-needed basis
4. At regularly scheduled intervals

21. 4. To control chronic pain and prevent cycled pain, regularly scheduled intervals are most effective. As-needed and conservative methods aren't effective means to manage chronic pain because the pain isn't relieved regularly. I.M. administration isn't practical on a long-term basis.
NP: Data collection; CN: Physiological integrity; CNS: Pharmacological therapies; CL: Application

NP: Nursing process   CN: Client needs category   CNS: Client needs subcategory   CL: Cognitive level

22. Which of the following nursing interventions is appropriate for an adult client with chronic renal failure?
1. Weigh the client daily before breakfast.
2. Offer foods high in calcium and phosphorus.
3. Serve the client large meals and a bedtime snack.
4. Encourage the client to drink large amounts of fluids.

23. During an admission interview, the nurse gathers data from a male client. Which of the following findings is a risk factor for testicular cancer?
1. Obesity
2. Cryptorchidism
3. History of alcoholism
4. History of cigarette smoking

24. Which of the following findings indicates an increased risk of skin cancer?
1. A deep sunburn
2. A dark mole on the client's back
3. An irregular scar on the client's abdomen
4. White irregular patches on the client's arm

25. Which of the following behaviors is consistent with the diagnosis of conduct disorder in a child?
1. Enuresis
2. Suicidal ideation
3. Cruelty to animals
4. Fear of going to school

26. Which of the following symptoms is associated with a genital chlamydia infection?
1. Genital warts
2. No symptoms
3. Purulent discharge
4. Fluid-filled blisters

22. 1. Daily weights are obtained to monitor fluid retention. Fluids should be restricted for the client with chronic renal failure. To improve food intake, meals and snacks should be given in small portions. Calcium intake is encouraged, but clients with chronic renal failure have difficulty excreting phosphorus. Therefore, phosphorus must be restricted.
NP: Implementation; CN: Physiological integrity; CNS: Physiological adaptation; CL: Application

23. 2. Cryptorchidism, or an undescended testicle, is a risk factor for testicular cancer. Cigarette smoking, obesity, and alcoholism aren't risk factors for developing testicular cancer.
NP: Data collection; CN: Health promotion and maintenance; CNS: Prevention and early detection of disease; CL: Comprehension

24. 1. A deep sunburn is a risk factor for skin cancer. A dark mole or an irregular scar are benign findings. White irregular patches are abnormal but aren't a risk factor for skin cancer.
NP: Data collection; CN: Health promotion and maintenance; CNS: Prevention and early detection of disease; CL: Application

25. 3. Cruelty to animals is a symptom of conduct disorder. Fear of going to school is school phobia. Enuresis and suicidal ideation aren't usually associated with conduct disorder.
NP: Data collection; CN: Psychosocial integrity; CNS: Psychosocial adaptation; CL: Application

26. 3. Purulent discharge from the cervix, urethra, or Bartholin's gland is associated with several sexually transmitted diseases, including chlamydia. Although some women with genital chlamydia infection are asymptomatic, this isn't the usual course of this condition. Genital warts are a sign of human papillomavirus. Fluid-filled blisters are a sign of herpes infection.
NP: Data collection; CN: Health promotion and maintenance; CNS: Prevention and early detection of disease; CL: Comprehension

27. Which of the following outcomes is appropriate for a client with a diagnosis of depression and attempted suicide?
1. The client will never feel suicidal again.
2. The client will find a group home to live in.
3. The client will remain hospitalized for at least 6 months.
4. The client will verbalize an absence of suicidal ideation, plan, and intent.

28. Which of the following actions is correct when collecting a urine specimen from a client's indwelling urinary catheter?
1. Collect urine from the drainage collection bag.
2. Disconnect the catheter from the drainage tubing to collect urine.
3. Remove the indwelling catheter and insert a sterile straight catheter to collect urine.
4. Insert a sterile needle with syringe through a tubing drainage port cleaned with alcohol to collect the specimen.

29. A registered nurse is supervising a licensed practical nurse (LPN). The LPN is caring for a client diagnosed with a terminal illness. Which of the following statements by the LPN should be corrected by the registered nurse?
1. "Some clients write a living will indicating their end-of-life preferences."
2. "The law says you have to write a new living will each time you go to the hospital."
3. "You could designate another person to make end-of-life decisions when you can't make them yourself."
4. "Some people choose to tell their physician they don't want to have cardiopulmonary resuscitation."

27. 4. An appropriate outcome is that the client will verbalize that he no longer feels suicidal. It's unrealistic to ask that he'll never feel suicidal. There's no reason for a group home or 6 months of hospitalization.
NP: Planning; CN: Psychosocial integrity; CNS: Coping and adaptation; CL: Application

28. 4. Wearing clean gloves, cleaning the port with alcohol, and then obtaining the specimen with a sterile needle ensures the specimen and the closed urinary drainage system won't be contaminated. A urine sample must be new urine, and the urine in the bag could be several hours old and growing bacteria. The urinary drainage system must be kept closed to prevent microorganisms from entering. A straight catheter is used to relieve urinary retention, obtain sterile urine specimens, measure the amount of postvoid residual urine, and empty the bladder for certain procedures. It isn't necessary to remove an indwelling catheter to obtain a sterile urine specimen unless the physician requests that the whole system be changed.
NP: Implementation; CN: Safe, effective care environment; CNS: Safety and infection control; CL: Comprehension

29. 2. One living will is sufficient for all hospitalizations unless the client wishes to make changes. The "No Code" or "Do Not Resuscitate" status is discussed with the physician, who then enters this in the client's chart. A living will explains a person's end-of-life preferences. A durable power of attorney for health care can be written to designate who will make health care decisions for the client in the event the client can't make decisions for himself.
NP: Evaluation; CN: Safe, effective care environment; CNS: Coordinated care; CL: Analysis

30. An elderly client's husband tells the nurse he's concerned because his wife insists on talking about events that happened to her years in the past. The nurse finds the client alert, oriented, and answering questions appropriately. Which of the following statements made to the husband is correct?
1. "Your wife is reviewing her life."
2. "A spiritual advisor should be notified."
3. "Your wife should be discouraged from talking about the past."
4. "Your wife is regressing to a more comfortable time in the past."

30. 1. Life review or reminiscing is characteristic of elderly people and the dying. A spiritual advisor might comfort the client but isn't necessary for a life review. Discouraging the client from talking would block communication. Regression occurs when a client returns to behaviors typical of another developmental stage.
NP: Implementation; CN: Health promotion and maintenance; CNS: Growth and development through the life span; CL: Application

31. A client with a new colostomy asks the nurse how to avoid leakage from the ostomy bag. Which of the following instructions is correct?
1. Limit fluid intake.
2. Eat more fruits and vegetables.
3. Empty the bag when it's about half full.
4. Tape the end of the bag to the surrounding skin.

31. 3. Emptying the bag when partially full will prevent the bag from becoming heavy and detaching from the skin or skin barrier. Limiting fluids may cause constipation but won't prevent leakage. Increasing fruits and vegetables in the diet will help prevent constipation, not leakage. Taping the bag to the skin will secure the bag to the skin but won't prevent detachment.
NP: Implementation; CN: Physiological integrity; CNS: Basic care and comfort; CL: Application

32. A nurse must obtain the blood pressure of a client on airborne precautions. Which of the following methods is best to prevent transmission of infection to other clients by the equipment?
1. Dispose of the equipment after each use.
2. Wear gloves while handling the equipment.
3. Use the equipment only with other clients in airborne isolation.
4. Leave the equipment in the room for use only with that client.

Going great so far. Keep it up!

32. 4. Leaving equipment in the room is appropriate to avoid organism transmission by inanimate objects. Disposing of equipment after each use prevents the transmission of organisms but isn't cost-effective. Wearing gloves protects the nurse, not other clients. Using equipment for other clients spreads infectious organisms among clients.
NP: Implementation; CN: Safe, effective care environment; CNS: Safety and infection control; CL: Application

33. To prevent circulatory impairment in an arm when applying an elastic bandage, which of the following methods is best?
1. Wrap the bandage around the arm loosely.
2. Apply the bandage while stretching it slightly.
3. Apply heavy pressure with each turn of the bandage.
4. Start applying the bandage at the upper arm and work toward the lower arm.

33. 2. Stretching the bandage slightly maintains uniform tension on the bandage. Wrapping the bandage loosely wouldn't secure the bandage on the arm. Using heavy pressure would cause circulatory impairment. Beginning the wrapping at the upper arm would cause uneven application of the bandage. For example, elastic stockings are applied distal to proximal to promote venous return.
NP: Implementation; CN: Physiological integrity; CNS: Reduction of risk potential; CL: Application

34. A client needs to use an incentive spirometer after abdominal surgery. Which of the following statements about incentive spirometry is correct?
  1. It's a substitute for early postoperative ambulation.
  2. It's better than deep breathing to prevent atelectasis.
  3. It causes less discomfort for the client than deep breathing.
  4. It helps the client visualize deep breathing to prevent atelectasis.

34. 4. Incentive spirometry helps the client see inspiratory effort using floating balls, lights, or bellows. Early ambulation is still indicated for this postoperative client. Incentive spirometry is no more effective than deep breathing without equipment. Deep breathing and incentive spirometry cause equal discomfort during inspiration.
NP: Implementation; CN: Physiological integrity; CNS: Reduction of risk potential; CL: Comprehension

35. A client complains of an inability to sleep while on the medical unit. Which intervention to promote sleep has priority?
  1. Offer a sedative routinely at bedtime.
  2. Give the client a backrub before bedtime.
  3. Question the client about sleeping habits.
  4. Move the client to a bed farthest from the nurses' station.

35. 3. Interviewing the client about sleeping habits may give more information about the causes of the inability to sleep. Sedatives should be given as a last option. A backrub may promote sleep but may not address this client's problem. Moving the client may not address the client's specific problem.
NP: Implementation; CN: Physiological integrity; CNS: Basic care and comfort; CL: Application

36. To test the function of the optic nerve, which of the following tools is used?
  1. Finger, to test the cardinal fields
  2. Flashlight, to test corneal reflexes
  3. Snellen chart, to test visual acuity
  4. Piece of cotton, to test corneal reflexes

36. 3. The Snellen chart is used to test the function of the optic nerve. Testing the cardinal fields assesses the oculomotor, trochlear, and abducens nerves. Corneal light reflex indicates the function of the oculomotor nerve. Corneal sensitivity is controlled by the trigeminal and facial nerves.
NP: Data collection; CN: Physiological integrity; CNS: Basic care and comfort; CL: Comprehension

37. Which of the following descriptions is correct for the prone position?
  1. Sitting upright
  2. Lying on the side
  3. Lying on the back
  4. Lying on the abdomen

37. 4. Prone position is lying on the abdomen. Sitting upright is known as Fowler's or high-Fowler's position. Side lying is also known as the lateral position. Lying on the back is the supine position.
NP: Implementation; CN: Safe, effective care environment; CNS: Coordinated care; CL: Comprehension

NP: Nursing process   CN: Client needs category   CNS: Client needs subcategory   CL: Cognitive level

38. Which of the following interventions is best to prevent bladder infections for a client with an indwelling urinary catheter?
1. Limit fluid intake.
2. Encourage showers rather than tub baths.
3. Open the drainage system to obtain a urine specimen.
4. Irrigate the catheter twice daily with sterile saline solution.

39. A nurse wants to use a waist restraint for a client who wanders at night. Which of the following factors or interventions should be considered before applying the restraint?
1. The nurse's convenience
2. The client's reason for getting out of bed
3. A sleeping medication ordered as needed at bedtime
4. The lack of nursing assistants on the night shift

40. Six months after the death of her infant son, a client is diagnosed with dysfunctional grieving. Which of the following behaviors would the nurse expect to find?
1. She goes to the infant's grave weekly.
2. She cries when talking about the loss.
3. She's overactive without a sense of loss.
4. She states the infant will always be part of the family.

41. A nurse notices a client has been crying. Which of the following responses is the most therapeutic?
1. None; this is a private matter.
2. "You seem sad, would you like to talk?"
3. "Why are you crying and upsetting yourself?"
4. "It's hard being in the hospital, but you must keep your chin up."

38. 2. A shower would prevent bacteria in the bath water from sustaining contact with the urinary meatus and the catheter, while a tub bath may allow easier transit of bacteria into the urinary tract. Increased — not limited — fluid intake is recommended for a client with an indwelling urinary catheter. Opening the drainage system would provide a pathway for the entry of bacteria. Catheter irrigation is performed only with an order from the physician to keep the catheter patent.
NP: Implementation; CN: Physiological integrity; CNS: Reduction of risk potential; CL: Comprehension

39. 2. The nurse should question the client's reason for getting out of bed because the client may be looking for a bathroom. Lack of adequate staffing and convenience aren't reasons for applying restraints. Sleeping medications are chemical restraints that should be used only if the client is unable to go to sleep and stay asleep.
NP: Evaluation; CN: Safe, effective care environment; CNS: Safety and infection control; CL: Application

40. 3. One of the signs of dysfunctional grieving is overactivity without a sense of loss. Going to the grave, tears, and including the infant as a part of the family are all normal responses.
NP: Data collection; CN: Psychosocial integrity; CNS: Coping and adaptation; CL: Comprehension

41. 2. Therapeutic communication is a primary tool of nursing. The nurse must recognize the client's nonverbal behaviors indicate a need to talk. Asking "why" is often interpreted as an accusation. Ignoring the client's nonverbal cues or giving opinions and advice are barriers to communication.
NP: Implementation; CN: Psychosocial integrity; CNS: Psychosocial adaptation; CL: Application

42. A nurse gives the wrong medication to a client. Another nurse employed by the hospital as a risk manager will expect to receive which of the following communications?
　　1. Incident report
　　2. Oral report from the nurse
　　3. Copy of the medication Kardex
　　4. Order change signed by the physician

42. 1. Incident reports are tools used by risk managers when a client might be harmed. They're used to determine how future problems can be avoided. An oral report won't serve as legal documentation. A copy of the medication Kardex wouldn't be sent with the incident report to the risk manager. A physician won't change an order to cover the nurse's mistake.
NP: Implementation; CN: Safe, effective care environment; CNS: Coordinated care; CL: Application

43. Performing a procedure on a client in the absence of informed consent can lead to which of the following charges?
　　1. Fraud
　　2. Harassment
　　3. Assault and battery
　　4. Breach of confidentiality

43. 3. Performing a procedure on a client without informed consent can be grounds for charges of assault and battery. Breach of confidentiality refers to conveying information about the client, harassment means to annoy or disturb, and fraud is to cheat.
NP: Implementation; CN: Safe, effective care environment; CNS: Coordinated care; CL: Comprehension

44. A surgical client newly diagnosed with cancer tells a nurse she knows the laboratory made a mistake about her diagnosis. Which of the following terms describes this reaction?
　　1. Denial
　　2. Intellectualization
　　3. Regression
　　4. Repression

44. 1. Cancer clients often deny this diagnosis when first made. Such a response may benefit the client in that it allows energy for surgical healing. Repression describes not remembering being diagnosed, regression describes childlike behavior, and intellectualization describes speaking of the disease as if reading a textbook.
NP: Data collection; CN: Psychosocial integrity; CNS: Coping and adaptation; CL: Comprehension

45. A client who's single delivers a premature neonate. Which of the following interventions would be included in her treatment plan?
　　1. An early postpartum physician visit
　　2. Referral to the health department
　　3. Request for a social service visit in the hospital
　　4. Request for a home health visit the day after discharge

45. 3. Due to the client's marital status and premature condition of the neonate, a social service visit is appropriate. The social service visit will determine if there's a need for a referral to the health department. The mother has no physical indications for an early postpartum visit or need for an early home visit.
NP: Planning; CN: Safe, effective care environment; CNS: Coordinated care; CL: Analysis

NP: Nursing process　CN: Client needs category　CNS: Client needs subcategory　CL: Cognitive level

46. A nurse is helping another nurse transfer a client from the bed to a stretcher. Which of the following principles of body mechanics is correct?

1. Bend from the waist.
2. Pull rather than push.
3. Stretch to reach an object.
4. Use large muscles in the legs for leverage.

46. 4. Keeping one's back straight and using the large muscles in the legs will help avoid back injury, as the muscles in one's back are relatively small compared with the larger muscles of the thighs. Bending from the waist can cause stress on the back muscles, causing a potential injury. Pulling isn't the best option and may cause straining. When feasible, one should push an object rather than pull it. Stretching to reach an object increases the risk of injury.

NP: Planning; CN: Safe, effective care environment; CNS: Safety and infection control; CL: Comprehension

47. A 6-year-old child needs diabetic teaching. Which of the following factors is considered when the nurse plans the teaching?

1. Another child with diabetes can teach the client.
2. The child can teach his parents after the nurse teaches him.
3. The child and parents should be recipients of teaching.
4. Teaching should be directed to the parents, who then can teach the child.

47. 3. The parents and child should participate in the nurse's teaching to ensure accuracy of teaching and that the child has adult caregivers who are educated. Another school-aged child couldn't be entrusted to teach this child, although their input would be valuable. Parents should be included in the teaching plan but shouldn't be responsible for the teaching. The school-aged child shouldn't be the sole provider of teaching to the parents.

NP: Planning; CN: Health promotion and maintenance; CNS: Growth and development through the life span; CL: Application

48. A client who just gave birth is concerned about her neonate's Apgar scores of 7 and 8. She says she's been told scores lower than 9 are associated with learning difficulties in later life. Which of the following responses is best?

1. "You shouldn't worry so much, your infant is perfectly fine."
2. "You should ask about placing the infant in a follow-up diagnostic program."
3. "You're right in being concerned, but there are good special education programs available."
4. "Apgar scores are used to indicate a need for resuscitation at birth. Scores of 7 and above indicate no problem."

48. 4. Apgar scores don't indicate future learning difficulties; they're for rapid assessment of the need for resuscitation. An Apgar score of 7 and 8 is normal and doesn't indicate a need for intervention. It's inappropriate to just tell a client not to worry.

NP: Implementation; CN: Health promotion and maintenance; CNS: Growth and development through the life span; CL: Application

49. After delivering a neonate with a cleft palate and cleft lip, a client has minimal contact with her neonate. She asks the nurse to do most of the neonate's care. Which of the following nursing diagnoses is appropriate?
1. Anxiety related to fear of harming the neonate
2. Deficient knowledge related to neonate's health status
3. Risk for impaired parenting related to birth defect
4. Ineffective coping related to birth defect

50. A breast-feeding client asks how she can do breast self-examination (BSE) while nursing. Which of the following responses is best?
1. "You should do BSE after the infant has emptied the breast."
2. "You don't have to do BSE until after you stop breast-feeding."
3. "You should continue to do BSE the way you did before becoming pregnant."
4. "Your physician will examine your breasts until after you stop breast-feeding."

51. A prenatal client says she can't believe she has such mixed feelings about being pregnant. She tried for 10 years to become pregnant and now she feels guilty for her conflicting reactions. Which of the following responses is best?
1. "You need to talk to your midwife about these feelings."
2. "You're experiencing the normal ambivalence pregnant mothers feel."
3. "These feelings are expected only in women who have had difficulty becoming pregnant."
4. "Let's make an appointment with a counselor."

You've now finished 50 questions. Groovy!

49. 3. Neonates born with birth defects are at risk for impaired parenting. The parents must work through issues of not producing the perfect dream child and guilt associated with this. There's nothing in the question that indicates the client felt anxious about caring for the neonate or had ineffective coping problems or knowledge deficit.
NP: Planning; CN: Health promotion and maintenance; CNS: Growth and development through the life span; CL: Application

50. 1. During breast-feeding, the client should examine each breast after the neonate has emptied the breast. Women must continue to examine their breasts, even if they're lactating. Contrary to how it's performed before pregnancy, BSE should be done on the same day of the month until the menstrual cycle returns. Breast examination shouldn't be done solely by the physician.
NP: Implementation; CN: Health promotion and maintenance; CNS: Prevention and early detection of disease; CL: Comprehension

51. 2. Conflicting, ambivalent feelings regarding pregnancy are normal for all pregnant women. These feelings don't call for counseling or other professional interventions. Ambivalence is felt by most pregnant women, not only mothers who had difficulty becoming pregnant.
NP: Implementation; CN: Psychosocial integrity; CNS: Coping and adaptation; CL: Application

---

NP: Nursing process   CN: Client needs category   CNS: Client needs subcategory   CL: Cognitive level

52. A client with terminal cancer tells a nurse, "I've given up. I have no hope left. I'm ready to die." Which of the following responses is most therapeutic?
   1. "You've given up hope?"
   2. "We should talk about dying to a social worker."
   3. "You should talk to your physician about your fears of dying so soon."
   4. "Now, you shouldn't give up hope. There are cures for cancer found every day."

52. 1. The use of reflection invites the client to talk more about his concerns. Deferring the conversation to a social worker or physician closes the conversation. Telling the client the cure for cancer is right around the corner gives false hope.

NP: Implementation; CN: Psychosocial integrity; CNS: Coping and adaptation; CL: Comprehension

53. Three days after discharge, a client bottle-feeding her neonate calls the postpartum floor, asking what she can do for breast engorgement. Which of the following instructions is correct?
   1. Wear a supportive bra.
   2. Get under a warm shower and let the water flow on her breasts.
   3. Stop drinking milk because it contributes to breast engorgement.
   4. Contact her physician; she shouldn't be engorged at this late date.

53. 1. A supportive bra is recommended for the client bottle-feeding her neonate to reduce engorgement. A warm shower will stimulate milk production. It's normal to become engorged during the first few days after delivery; drinking milk isn't the cause.

NP: Implementation; CN: Physiological integrity; CNS: Basic care and comfort; CL: Application

54. A client complains of excessive flatulence. Which food, reported by the client as consumed regularly, may be responsible for this?
   1. Cauliflower
   2. Ice cream
   3. Meat
   4. Potatoes

54. 1. Cauliflower is the only food listed that often results in flatulence.

NP: Data collection; CN: Physiological integrity; CNS: Basic care and comfort; CL: Comprehension

55. A client is being treated for premature labor with ritodrine (Yutopar). After receiving this medication for 12 hours, her blood pressure is slightly elevated, her chest is clear, and her pulse is 120 beats/minute. She complains of a little nausea, and the fetal heart rate is 145 beats/minute. Which of the following interventions is correct?
   1. Continue routine monitoring.
   2. Contact the physician immediately.
   3. Turn the client on her left side and give oxygen.
   4. Increase the flow rate of the I.V. and give oxygen.

55. 1. These findings are normal adverse effects to the medication and don't call for interventions at this time except to continue close monitoring. Contacting the physician, placing the client on her left side, changing the I.V. flow rate, and giving oxygen are all interventions for abnormal assessment findings.

NP: Implementation; CN: Physiological integrity; CNS: Pharmacological therapies; CL: Analysis

56. At 6 cm of dilation, the client in labor receives a lumbar epidural for pain control. Which of the following nursing diagnoses is possible?
　1. Risk for injury related to rapid delivery
　2. Acute pain related to wearing off of anesthesia
　3. Hyperthermia related to effects of anesthesia
　4. Ineffective tissue perfusion related to effects of anesthesia

56. 4. A disadvantage of a lumbar epidural is the risk for hypotension. Epidurals are associated with a longer labor and hypothermia. There's no pain involved with the anesthesia wearing off.
NP: Planning; CN: Physiological integrity; CNS: Pharmacological therapies; CL: Application

57. When assessing a client who just gave birth, a nurse finds the following: blood pressure, 110/70 mm Hg; pulse, 60 beats/minute; respirations, 16 breaths/minute; lochia, moderate rubra; fundus, above the umbilicus to the right; and negative Homan's sign. Which of the following interventions is correct?
　1. Do nothing; all findings are normal.
　2. Have the client void and recheck the fundus.
　3. Turn the client on her left side to decrease the blood pressure.
　4. Rub the fundus to decrease lochia flow and prevent hemorrhage.

57. 2. A fundus up and to the right indicates a full bladder. The client should empty her bladder and be rechecked. Lochia flow and blood pressure are normal. Placement of the uterus at the umbilicus and to the right isn't a normal finding.
NP: Implementation; CN: Physiological integrity; CNS: Reduction of risk potential; CL: Analysis

58. A client with diabetes delivers a 9-lb, 6-oz neonate. The nurse should be alert for which of the following?
　1. Hyperglycemia
　2. Hypoglycemia
　3. Hyperthermia
　4. Hypothermia

58. 2. Neonates of mothers with diabetes and large neonates are at risk for hypoglycemia related to increased production of insulin by the neonate in utero. Hypothermia, hyperthermia, and hyperglycemia aren't primary concerns.
NP: Data collection; CN: Physiological integrity; CNS: Reduction of risk potential; CL: Application

59. A prenatal client, age 13, asks about getting fat while she's pregnant. A nurse tells her she needs to gain enough weight to be in the upper portions of her recommended weight due to her age to prevent which of the following conditions?
　1. A premature neonate
　2. A difficult delivery
　3. A low birth-weight neonate
　4. Pregnancy-induced hypertension (PIH)

59. 3. Adolescent girls, especially those younger than age 15, are at higher risk for delivering low birth-weight neonates unless they gain adequate weight during pregnancy. Gaining weight isn't associated with preventing a difficult delivery, risk of PIH, or a premature neonate.
NP: Implementation; CN: Physiological integrity; CNS: Reduction of risk potential; CL: Application

60. A nurse is preparing to bathe a client hospitalized for emphysema. Which of the following nursing interventions is correct?
1. Remove the oxygen and proceed with the bath.
2. Increase the flow of oxygen to 6 L/minute by nasal cannula.
3. Keep the head of the bed slightly elevated during the procedure.
4. Lower the head of the bed and roll the client to his left side to increase oxygenation.

61. A client who is 36 weeks pregnant chokes on her food while eating at a restaurant. Which of the following statements is correct about performing the Heimlich maneuver on a pregnant client?
1. Chest thrusts are used when the client is pregnant.
2. Only back thrusts are used when the client is pregnant.
3. The Heimlich is performed the same as when not pregnant.
4. The Heimlich maneuver can't be performed on a pregnant client.

62. A nurse works in a mental health facility that uses a therapeutic community (milieu) approach to client care. Which of the following statements describes the nurse's role in this facility?
1. Primary caregiver
2. Member of the milieu
3. Supervision more than counseling
4. Distinctly separate from the psychiatrist

63. A client with a substance abuse problem is being discharged from the state mental hospital. His discharge plans should include which of the following interventions?
1. Referral to Al-Anon
2. Weekly urine testing for drug use
3. Day hospital treatment for 6 months
4. Participation in a support group like Alcoholics Anonymous (AA)

60. 3. The elasticity of the lungs is lost for clients with emphysema. Therefore, these clients can't tolerate lying flat because the abdominal organs compress the lungs. The best position is one with the head slightly elevated. Discontinuing oxygen or altering the oxygen delivery rate should never be done without an order from the physician. Increasing oxygen flow on a client with emphysema may also suppress the hypoxic drive to breathe. Positioning the client on his left side with the head of the bed flat would decrease oxygenation.
NP: Data collection; CN: Physiological integrity; CNS: Physiological adaptation; CL: Application

61. 1. During pregnancy, chest thrusts are used instead of abdominal thrusts. Abdominal thrusts compress the abdomen, which would harm the fetus. Because of this, the Heimlich is adjusted for the pregnant woman. A fist is made with one hand, placing thumb side against the center of the breastbone. The fist is grabbed with the other hand and thrust inward. Avoid the lower tip of the breastbone. Back thrusts aren't done as they may result in dislodgment of the obstruction, further obstructing the airway.
NP: Implementation; CN: Physiological integrity; CNS: Reduction of risk potential; CL: Application

62. 2. In a therapeutic community, everything focuses on the client's treatment. Staff and clients work together as a team or member of the milieu. The nurse wouldn't be a primary caregiver, but would work with the psychiatrist. The nurse's role could be that of supervision as well as counseling.
NP: Implementation; CN: Safe, effective care environment; CNS: Coordinated care; CL: Comprehension

63. 4. AA is a major support group for alcoholics after treatment. Membership in AA is associated with relapse prevention. Al-Anon is a support group for the family of the abuser of alcohol. Weekly urine testing or day hospital treatment isn't usual.
NP: Planning; CN: Safe, effective care environment; CNS: Coordinated care; CL: Application

64. A nurse is removing an indwelling urinary catheter. Which of the following actions is appropriate?
  1. Wear sterile gloves.
  2. Cut the lumen of the balloon.
  3. Document the time of removal.
  4. Position the client on the left side.

64. 3. The client should void within 8 hours of the removal of an indwelling urinary catheter. Documenting the time of removal allows the nurse and physician to verify the duration of elapsed time since removal, thus contributing to continuity of care. Clean, disposable gloves are required because it isn't a sterile procedure. The client should be positioned comfortably on his back, and privacy should be provided. The catheter may retrograde into the bladder, requiring surgical removal, if the balloon is cut from the lumen and the catheter isn't secured.

NP: Implementation; CN: Safe, effective care environment; CNS: Safety and infection control; CL: Comprehension

Only 20 more to go! Oh, my, I just can't wait!

65. A client is scheduled to retire in the next month. He phones his nurse therapist and says he can't cope; his whole world is falling apart. The therapist recognizes this reaction as which of the following conditions?
  1. Panic reaction
  2. Situational crisis
  3. Normal separation anxiety
  4. Maturational crisis

65. 4. A maturational (developmental) crisis is one that occurs at a predictable milestone during a life span; birth, marriage, and retirement are examples. Separation anxiety is a childhood disorder; a situational crisis is caused by events such as an earthquake. A panic reaction would also involve physical symptoms.

NP: Evaluation; CN: Health promotion and maintenance; CNS: Growth and development through the life span; CL: Analysis

66. A client with a phobic condition is being treated with behavior modification therapy. Which of the following treatments is expected?
  1. Dream analysis
  2. Free association
  3. Systematic desensitization
  4. Electroconvulsive therapy (ECT)

66. 3. Systematic desensitization is a behavior therapy used in the treatment of phobias. ECT is used with depression. Dream analysis and free association are techniques used in psychoanalytic therapy.

NP: Planning; CN: Psychosocial integrity; CNS: Psychosocial adaptation; CL: Comprehension

67. A severely depressed client rarely leaves her chair. To prevent physiologic complications associated with psychomotor retardation, which of the following steps is appropriate?
  1. Restricting coffee intake
  2. Increasing calcium intake
  3. Resting in bed three times per day
  4. Emptying the bladder on a schedule

67. 4. To prevent bladder infections associated with stasis of urine, the client should be encouraged to routinely empty her bladder. Resting in bed is another form of psychomotor retardation. Neither calcium nor coffee intake are directly related to the psychological effects associated with this condition.

NP: Planning; CN: Health promotion and maintenance; CNS: Prevention and early detection of disease; CL: Application

---

NP: Nursing process CN: Client needs category CNS: Client needs subcategory CL: Cognitive level

68. During the termination phase of a therapeutic nurse-client relationship, which of the following interventions is avoided?
1. Referring the client to support groups.
2. Addressing new issues with the client.
3. Reviewing what has been accomplished during this relationship.
4. Having the client express sadness that the relationship is ending.

69. The behavior of a client with borderline personality disorder causes a nurse to feel angry toward the client. Which of the following responses by the nurse is the most therapeutic?
1. Ignore the client's irritating behavior.
2. Restrict the client to her room until supper.
3. Report her feelings to the client's physician.
4. Tell the client how her behavior makes the nurse feel.

70. During a manic state, a client paced around the dayroom for 3 days. He talked to the furniture, proclaimed he was a king, and refused to partake in unit activities. Which of the following nursing diagnoses has priority?
1. Hypertension related to hyperactivity
2. Risk for other-directed violence related to manic state
3. Imbalanced nutrition: Less than body requirements, related to hyperactivity
4. Ineffective coping related to manic state

71. A client with a panic disorder is having difficulty falling asleep. Which of the following nursing interventions should be performed first?
1. Call the client's psychotherapist.
2. Teach the client progressive relaxation.
3. Allow the client to stay up and watch television.
4. Obtain an order for a sleeping medication as needed.

68. 2. During the termination phase, new issues shouldn't be explored. It's appropriate to refer the client to support groups. To review what has been accomplished is a goal of this phase. Sadness is a normal response.
NP: Implementation; CN: Psychosocial integrity; CNS: Coping and adaptation; CL: Application

69. 4. A nursing intervention used with personality disorders is to help the client recognize how his behavior affects others. Restricting the client, ignoring the client, and reporting feelings to the physician aren't appropriate interventions at this time.
NP: Implementation; CN: Psychosocial integrity; CNS: Psychosocial adaptation; CL: Application

70. 3. During a manic state, clients are at risk for malnutrition due to not taking in enough calories for the energy they're expending. This client isn't showing violent behavior. Coping issues are the primary concern at this time. Hypertension isn't an approved nursing diagnosis.
NP: Planning; CN: Physiological integrity; CNS: Basic care and comfort; CL: Application

71. 2. Relaxation techniques work very well with a client showing anxiety. If this doesn't work, then pharmacologic interventions, diversion activities, and contacting the psychotherapist would be in order.
NP: Implementation; CN: Psychosocial integrity; CNS: Coping and adaptation; CL: Application

72. A 65-year-old client with major depression hasn't responded to antidepressants. Which of the following interventions used to treat major depression might be added to the treatment plan?
    1. Electroconvulsive therapy (ECT)
    2. Electroencephalography (EEG)
    3. Electromyography (EMG)
    4. Tranquilizers

72. 1. ECT is commonly used for treatment of major depression for clients who haven't responded to antidepressants or who have medical problems that contraindicate the use of antidepressants. EEG is a technique used to treat clients with anxiety, general tension, stuttering, insomnia, and chronic pain. EMG is used in biofeedback. Major tranquilizers are used to treat schizophrenia or anxiety disorders.
NP: Implementation; CN: Physiological integrity; CNS: Physiological adaptation; CL: Comprehension

73. A client diagnosed with bipolar disease is receiving a maintenance dosage of lithium carbonate. His wife calls the community mental health nurse to report that her husband is hyperactive and hyperverbal. Which of the following interventions is appropriate?
    1. Mental status examination
    2. Measurement of lithium blood levels
    3. Evaluation at the local emergency department
    4. Admission to the hospital for observation

73. 2. Increased activity might indicate a need for an increased dose of lithium or that the client isn't taking his medications; blood levels will determine this. He doesn't need to go to the emergency department, have a mental status examination, or be admitted to the hospital at this time.
NP: Implementation; CN: Physiological integrity; CNS: Pharmacological therapies; CL: Analysis

74. After electroconvulsive therapy (ECT), which of the following nursing interventions is correct?
    1. Assess the client's vital signs.
    2. Let the client sleep undisturbed.
    3. Allow the family to visit immediately.
    4. Restrain the client until completely awake.

74. 1. Vital signs are monitored carefully for approximately 1 hour after ECT or until stable. The client shouldn't be restrained or left alone. Visitors should be allowed when the client is awake and ready.
NP: Implementation; CN: Physiological integrity; CNS: Reduction of risk potential; CL: Application

75. Which of the following statements by a client who had nasal surgery indicates that the client needs further teaching about postoperative care?
    1. "I'll do frequent mouth care."
    2. "I'll eat two oranges per day."
    3. "I'll eat two bananas per day."
    4. "I'll drink at least 8 glasses of fluid per day."

75. 3. After nasal surgery, the client shouldn't strain or bear down as this will increase the risk for bleeding. Bananas can cause severe constipation, which could lead to straining. The other interventions would be appropriate postoperative care for this client.
NP: Evaluation; CN: Physiological integrity; CNS: Reduction of risk potential; CL: Comprehension

76. A client recently placed on a cardiac monitor has a heart rate of 170 beats/minute, with frequent premature contractions. Which of the following nursing actions is best?
1. Call the client's physician immediately.
2. See the client and make a full assessment.
3. Delegate one of the nursing assistants to take the client's vital signs.
4. Notify the supervisor about the change in the client's condition.

77. A client visits a physician's office seeking treatment for depression, feelings of hopelessness, poor appetite, insomnia, low self-esteem, and difficulty making decisions. The client says that these symptoms began at least 2 years ago. Which of the following disorders is suspected?
1. Major depression
2. Dysthymic disorder
3. Cyclothymic disorder
4. Atypical affective disorder

78. Which of the following techniques is correct for postoperative coughing and deep-breathing exercises?
1. Splint the incision and cough.
2. Splint the incision, take a deep breath, and then cough.
3. Lie prone, splint the incision, take a deep breath, and then cough.
4. Lie supine, splint the incision, take a deep breath, and then cough.

79. Which of the following actions is an example of developmentally based care?
1. Provide books to a 9-year-old client.
2. Walk a 10-year-old client according to written orders.
3. Provide a pureed diet to a postoperative 13-year-old client.
4. Change a surgical dressing on a 15-year-old client every 4 hours as ordered.

76. 2. Because a change has occurred in the client's status, the nurse must assess the client first. This shouldn't be delegated to unlicensed personnel. Before the physician or supervisor is notified, a full assessment must be made.
NP: Implementation; CN: Health promotion and maintenance; CNS: Reduction of risk potential; CL: Application

77. 2. Dysthymic disorder is marked by feelings of depression lasting at least 2 years, accompanied by at least two of the following symptoms: sleep disturbance, appetite disturbance, low energy or fatigue, low self-esteem, poor concentration, difficulty making decisions, and hopelessness. Cyclothymic disorder is a chronic mood disturbance of at least 2 years' duration marked by numerous periods of depression and hypomania. Manic signs and symptoms characterize atypical affective disorder. Major depression is a recurring, persistent sadness or loss of interest or pleasure in almost all activities, with signs and symptoms recurring for at least 2 weeks.
NP: Data collection; CN: Psychosocial integrity; CNS: Psychosocial adaptation; CL: Knowledge

78. 2. Splinting the incision with a pillow will protect the incision while the client coughs. Taking a deep breath will help open the alveoli, which promotes oxygen exchange and prevents atelectasis. Coughing and deep-breathing exercises are best accomplished in a sitting or semisitting position. Expectoration of secretions will be facilitated in a sitting position, as will splinting and taking deep breaths.
NP: Planning; CN: Physiological integrity; CNS: Reduction of risk potential; CL: Application

79. 1. Providing books to a 9-year-old client facilitates his reading skills and helps him grow developmentally. Changing a surgical dressing, walking a client, and providing a pureed diet are routine care tasks, which don't necessarily promote further development of the individual.
NP: Data collection; CN: Health promotion and maintenance; CNS: Growth and development through the life span; CL: Comprehension

80. A client on complete bed rest complains of excessive flatulence. Which of the following positions would be helpful to the client?
1. Fowler's
2. Knee-chest
3. Semi-Fowler's
4. Trendelenburg's

81. A client had a laxative prescribed that acts by causing stool to absorb water and swell. Which of the following terms describes this type of laxative?
1. Bulk forming
2. Emollient
3. Lubricant
4. Stimulant

82. A nurse encourages a client to avoid foods that have a laxative effect, including which of the following?
1. Alcohol
2. Cheese
3. Eggs
4. Pasta

83. A client is complaining of moderate pain. Which of the following findings indicates a physiological response to pain?
1. Restlessness
2. Decreased pulse rate
3. Increased blood pressure
4. Protection of the painful area

84. A nurse notes crackles in the lung bases and pedal edema. Which of the following is a common cause of excess fluid volume?
1. Prolonged fever
2. Hyperventilation
3. Excessive I.V. infusion
4. Fluid volume shifts secondary to vomiting

85. A client is given instructions for a low-sodium diet. Which of the following statements best shows the client understands the diet instruction?
1. "Meat, fish, and chicken are high in sodium."
2. "I'll miss eating fruits."
3. "I'll enjoy eating at restaurants more often now."
4. "I'll avoid dairy products, potato chips, and carrots."

I knew you could do it! Super job! You're well on your way to total confidence for the NCLEX.

80. 2. Because gas rises, the knee-chest position facilitates the passage of flatus. In Trendelenburg's position the client lies flat with his head lower than his feet. Semi-Fowler's and Fowler's positions inhibit gas passage.
NP: Implementation; CN: Physiological integrity; CNS: Basic care and comfort; CL: Knowledge

81. 1. Bulk-forming laxatives cause stool to absorb water and swell. Emollients lubricate the stool; lubricants soften the stool, making it easier to pass; and stimulants promote peristalsis by irritating the intestinal mucosa or stimulating nerve endings in the intestinal wall.
NP: Data collection; CN: Physiological integrity; CNS: Pharmacological therapies; CL: Knowledge

82. 1. All the foods listed except alcohol have a constipating effect.
NP: Implementation; CN: Physiological integrity; CNS: Basic care and comfort; CL: Comprehension

83. 3. Increased blood pressure is a physiological, or involuntary, response to moderate pain. Restlessness and protection of the painful area are behavioral responses. Decreased pulse rate occurs when pain is severe and deep.
NP: Data collection; CN: Physiological integrity; CNS: Physiological adaptation; CL: Comprehension

84. 3. Excess fluid volume can result from excess I.V. fluids, especially in a compromised client. Vomiting, fever, and hyperventilation will result in loss of body fluids, leading to deficient fluid volume.
NP: Data collection; CN: Physiological integrity; CNS: Basic care and comfort; CL: Application

85. 4. Dairy products, potato chips, carrots, and restaurant food are all high in sodium. Meat, fish, chicken, and fruits aren't.
NP: Evaluation; CN: Physiological integrity; CN: Basic care and comfort; CL: Knowledge

Take this challenging comprehensive test that includes a variety of questions like a real NCLEX exam. Have a go at it!

COMPREHENSIVE
Test 3

1. A client in the postoperative phase of abdominal surgery is to advance his diet as tolerated. The client has tolerated ice chips and a clear liquid diet. Which diet would the nurse anticipate giving next?
  1. Fluid restricted
  2. Full liquids
  3. House
  4. Soft

2. The following information is recorded on an intake and output record: milk, 180 ml; orange juice, 60 ml; 1 serving scrambled eggs; 1 slice toast; 1 can Ensure oral nutritional supplement, 240 ml; I.V. dextrose 5% in water at 100 ml/hour; 50 ml water after twice daily medications. Medications are given at 9 a.m. and 9 p.m. What's the client's total intake for the 7 a.m. to 3 p.m. shift?
  1. 1,000 ml
  2. 1,250 ml
  3. 1,330 ml
  4. 1,380 ml

3. A nurse is witnessing consent from a client before a cardiac catheterization. Which of the following factors is a component of informed consent?
  1. Freedom from coercion
  2. Durable power of attorney
  3. Private insurance coverage
  4. Disclosure of previous answers given by the client

1. 2. Clear liquid diets are nutritionally inadequate but minimally irritating to the stomach. Clients are advanced to the full liquid diet next, adding bland and protein foods. A soft diet comes next, which omits foods that are hard to chew or digest. A regular, or house, diet has no limitations. A fluid restriction is ordered in addition to the diet order for clients in renal failure or heart failure.
NP: Implementation; CN: Physiological integrity; CNS: Basic care and comfort; CL: Comprehension

2. 3. 180 + 60 + 240 + 800 + 50 = 1,330 ml.
NP: Evaluation; CN: Physiological integrity; CNS: Basic care and comfort; CL: Knowledge

3. 1. The client must give consent voluntarily without any type of outside influences from persons involved with the procedure or research. The client must also be of sound mind and not under the influence of types of medications that may interfere with reasoning. Private insurance coverage shouldn't be a factor in informed consent. All clients (or another appointed individual) have the right to make their own decisions regardless of type of insurance. A durable power of attorney may be indicated if a client is unable to make decisions for himself. Previous answers given by the individual shouldn't be an influencing factor in the informed consent process.
NP: Implementation; CN: Safe, effective care environment; CNS: Coordinated care; CL: Knowledge

NP: Nursing process   CN: Client needs category   CNS: Client needs subcategory   CL: Cognitive level

4. In checking a client's chart, the nurse notes that there's no record of a narcotic being given to the client even though the previous nurse signed for one. The client denies receiving anything for pain since the previous night. Which of the following actions should be taken next?
1. Notify the physician that a narcotic is missing.
2. Notify the supervisor that the client didn't receive the prescribed pain medication.
3. Notify the pharmacist that the client didn't receive the prescribed pain medication.
4. Approach the nurse who signed out the narcotic to seek clarification about the missing drug.

4. 4. The nurse needs to seek clarification in a nonthreatening manner. If the nurse who signed out the narcotic can't give a plausible explanation, the nurse who discovered the error must then notify the supervisor. The nurse who signed out the narcotic may have a drug problem. The appropriate line of communication is to the hospital supervisor. The physician needs to be notified if the client didn't receive the prescribed medication. The pharmacist needs to be notified of discrepancies in the narcotic count.
NP: Implementation; CN: Safe, effective care environment; CNS: Coordinated care; CL: Application

5. A client is seen in the emergency department with bruises on her face and back. She has the signs of a battered wife. Which of the following community resources could provide assistance to the client?
1. Alcoholics Anonymous (AA)
2. Crime Task Force
3. Lifeline Emergency Aid
4. Women's shelter

5. 4. A women's shelter can house women and children who need protection from an abusive partner or parent. AA is a support group for alcoholics and their families. The Crime Task Force and Lifeline Emergency Aid don't provide housing for women or children who want to leave an abusive relationship.
NP: Planning; CN: Safe, effective care environment; CNS: Coordinated care; CL: Application

6. Multidisciplinary team meetings are used frequently as a method of communication among health care disciplines. Which of the following units uses this method of communication?
1. Critical care units
2. Home health care services
3. Labor and delivery units
4. Outpatient surgical units

6. 2. Home health care services and restorative care services (such as rehabilitation units) that use different disciplines are required by the Joint Commission on Accreditation of Healthcare Organizations or Medicare to hold multidisciplinary team meetings. This serves as a means of communicating the client's diagnosis, care plan, and discharge needs using all disciplines for input. Critical care units, outpatient surgical units, and labor and delivery units use between-shift reporting as a method of communicating.
NP: Implementation; CN: Safe, effective care environment; CNS: Coordinated care; CL: Application

7. A nurse finds a client crying after she was told hemodialysis is needed due to the development of acute renal failure. Which of the following interventions is best?
1. Sitting quietly with the client
2. Referring the client to the hemodialysis team
3. Reminding the client this is a temporary situation
4. Discussing with the client the other abilities she has

7. 1. Sitting with the client shows compassion and concern, and may help the nurse establish therapeutic communication. Making a referral doesn't allow the client to explore feelings with the nurse. The nurse can't guarantee the acute renal failure is temporary. Discussing the client's other abilities is diverting the emphasis away from the primary issue for this client.
NP: Implementation; CN: Psychosocial integrity; CNS: Coping and adaptation; CL: Knowledge

NP: Nursing process   CN: Client needs category   CNS: Client needs subcategory   CL: Cognitive level

8. A client was admitted to a mental health unit for hyperexcitability, increasing agitation, and distractibility. Which of the following nursing interventions has priority?
1. Involving the client in a group activity
2. Being direct and firm and setting rules for the client
3. Using a quiet room for the client away from others
4. Channeling the client's energy toward a planned activity

9. Which of the following controllable risk factors identified on a client history may contribute to heart disease?
1. Race
2. Prostate cancer
3. Diabetes mellitus
4. Previous myocardial infarction (MI)

10. A public health nurse visiting a new postpartum client notices that the client has two children under age 4. The nurse notices one infant playing in the cabinet under the sink. Which of the following instructions should the public health nurse give the client?
1. Cover the infant's hands with gloves.
2. Make sure all liquid cleaners are labeled.
3. Tighten all cap tops on the bottles under the sink.
4. Remove all liquid cleaners that could be ingested orally.

11. A nurse arrives at an automobile accident involving a school bus and a large truck. The school bus is lying on its side. Several people had been thrown from the windows of the school bus. Which of the following victims needs priority care?
1. A girl crying hysterically
2. A boy who's unconscious
3. A boy with a laceration of the scalp
4. A girl with an obvious open fracture

You've finished 10 questions already! Good job!

8. 3. Being in a quiet environment away from stimuli facilitates helping the client regain a sense of control. If the nurse attempts to be firm and set rules for this client, it will most likely heighten the agitation. The client is too excited to focus at this time and group activities may worsen the client's situation. The client is unable to focus on activity.
NP: Implementation.; CN: Psychosocial integrity; CNS: Psychosocial adaptation; CL: Knowledge

9. 3. Diabetes mellitus, if uncontrolled, can lead to heart disease. Race is a factor that can't be controlled. Previous MI and history of prostate cancer aren't risk factors.
NP: Data collection; CN: Health promotion and maintenance; CNS: Prevention and early detection of disease; CL: Knowledge

10. 4. All liquid cleaners must be removed to reduce the risk of poisoning. Safety locks should be placed on cabinets to prevent young children from opening the cabinets or the bottles. Infants can't read danger labels.
NP: Implementation; CN: Safe, effective care environment; CNS: Safety and infection control; CL: Application

11. 2. An unconscious or unresponsive client always needs assistance first. The client's breathing and circulation status should be checked. Once help arrives, the girl's fracture can be stabilized, pressure can be applied to the laceration of the scalp to stop the bleeding, and emotional support can be given to the girl crying hysterically.
NP: Data collection; CN: Safe, effective care environment; CNS: Safety and infection control; CL: Application

12. Which of the following statements from a newly diagnosed client with diabetes indicates more instruction is needed?
1. "I need to check my feet daily for sores."
2. "I need to store my insulin in the refrigerator."
3. "I can use my plastic insulin syringe more than once."
4. "I need to see my physician for follow-up examinations."

12. 2. Insulin only needs to be stored in the refrigerator if it won't be used within 6 weeks after being opened; it should be at room temperature when given to decrease pain and prevent lipodystrophy. According to a poll by the Juvenile Diabetes Foundation, a high percentage of diabetics reuse their insulin syringes. However, it's recommended they be carefully recapped and placed in the refrigerator to prevent bacterial growth. The remaining statements show that the client understands his condition and the importance of preventing complications.
NP: Evaluation; CN: Safe, effective care environment; CNS: Safety and infection control; CL: Analysis

13. A client with terminal cancer is receiving large doses of narcotics for pain control. He becomes agitated and continues trying to get out of bed but can't stand without a two-person assistance. To reduce the risk of falling, which of the following types of restraint is the most beneficial?
1. Leg restraints
2. Chemical restraints
3. Mechanical restraints
4. Tying him in bed with a sheet

13. 2. Chemical restraints are effective, especially with clients who are highly agitated and receiving large doses of narcotics. For example, anxiety medications can be used to calm the client. Other forms of restraint will only increase the client's agitation and hostility, thus increasing the safety risk.
NP: Implementation; CN: Safe, effective care environment; CNS: Safety and infection control; CL: Application

14. A client who had a heart transplant is in protective isolation postoperatively. Which of the following explanations for this is correct?
1. To protect the client from his own bacteria
2. To protect the hospital staff from the client
3. To protect the other clients on the nursing unit
4. To protect the client from outside infections from others

14. 4. Immunosuppressed clients need to be protected from infections from others. Infections can occur if strict hand-washing techniques aren't observed, especially with hospital staff going from one room to the next. Protective isolation isn't used to protect the hospital staff and other clients from an infected client.
NP: Planning; CN: Safe, effective care environment; CNS: Safety and infection control; CL: Application

15. A physician ordered a sterile dressing tray set up in a client's room to insert a subclavian central venous catheter. Which of the following steps is done first to set up the sterile field?
1. Open the tray toward the nurse.
2. Use correct hand-washing technique.
3. Put on sterile gloves before opening the tray.
4. Place the sterile dressing tray on an overbed table.

15. 2. Use appropriate hand-washing technique before participating in a sterile procedure. Clean the area with an appropriate antiseptic, place the tray in the center of the clean area, and open it away from the nurse. After the dressing tray is opened, put on sterile gloves to assist the physician.
NP: Implementation; CN: Safe, effective care environment; CNS: Safety and infection control; CL: Application

NP: Nursing process   CN: Client needs category   CNS: Client needs subcategory   CL: Cognitive level

16. The nurse is instructing a homebound client on safety precautions for chemotherapy. Which color plastic bag is universally used for handling chemotherapy supplies?
1. Blue
2. Purple
3. Red
4. White

17. A public health nurse is interviewing a young Pakistani client in her home. The nurse notices the client and the infant wear long skirts and coverings over their heads. The home isn't air-conditioned and the room is very warm. The dress code is recognized as part of which of the following characteristics?
1. Culture
2. Economic status
3. Race
4. Socialization

18. Which of the following actions is included in the assessment step of the nursing process?
1. Identifying actual or potential health problems specific to the individual client
2. Judging the effectiveness of nursing interventions that have been implemented
3. Identifying goals and interventions specific to the individualized needs of the client
4. Systematically collecting subjective and objective data with the goal of making a clinical nursing judgment

19. During an interdepartmental team meeting at a hospice, a nurse who practices Catholicism verbalizes concern for the spiritual needs of a terminally ill infant and her non-Catholic family. She suggests the infant be baptized before death. Which of the following recommendations of the multidisciplinary team is most likely?
1. Insist the infant obtain baptism before death occurs.
2. Bathe the infant with special oil to prepare for death.
3. Schedule an appointment with a Catholic priest to see the family.
4. Recognize that not all religions practice infant baptism.

16. 3. Biohazardous waste products are placed in red biohazard bags. White bags are used for normal trash products and blue bags are used for recycling plastics. Purple bags aren't used for biohazardous waste products.
NP: Planning; CN: Safe, effective care environment; CNS: Safety and infection control; CL: Knowledge

17. 1. Many cultures have specific dress codes. The client's dress, as described, doesn't indicate economic status. Race refers to a group of people with similar physical characteristics such as skin color. Socialization is the process by which individuals learn the ways of a given society to function within that group.
NP: Data collection; CN: Health promotion and maintenance; CNS: Growth and development through the life span; CL: Application

18. 4. Assessment involves data collection, organization, and validation. Evaluation involves judging the effectiveness of nursing interventions and whether the goals of the care plan have been achieved. The nurse and client work together to identify goals, outcomes, and intervention strategies that will reduce identified client problems in the planning step. The diagnosis step of the nursing process involves the identification of actual or potential health problems.
NP: Data collection; CN: Safe, effective care environment; CNS: Coordinated care; CL: Knowledge

19. 4. Many religious organizations (for example, Baptist, Adventist, Buddhist, Quaker) don't practice baptism or only baptize an individual when he is an adult. Hospice organizations use the family's religious leader as a choice for spiritual directions. Seventh Day Adventists believe in divine healing and anointing with oil. Deciding whether to baptize the infant isn't the nurse's responsibility. It's important to honor all customs and religious beliefs of families.
NP: Data collection; CN: Psychosocial integrity; CNS: Coping and adaptation; CL: Comprehension

20. A home health client asks a nurse for information on sources of financial support. The client's elderly parent is blind and living with her. To which of the following programs would it be appropriate for the nurse to refer the client?
1. Medicare
2. Meals On Wheels
3. Supplemental Security Income
4. Aid to Families with Dependent Children

Stay focused, now. You're doing great.

20. 3. Supplemental Security Income is a governmental subsidy assisting the poor and medically disabled. Aid to Families with Dependent Children is a state subsidy given to poor families with dependent children. Meals On Wheels is a nonprofit organization that delivers food to the poor. Medicare is available to elderly individuals age 65 and older and individuals younger than age 65 with long-term disabilities or end-stage renal disease.
NP: Implementation; CN: Health promotion and maintenance; CNS: Growth and development through the life span; CL: Application

21. Giving hearing and vision screening to elementary school children is an example of which of the following types of prevention strategy?
1. Primary
2. Secondary
3. Tertiary
4. None of the above

21. 2. Screening is a major secondary prevention strategy. Secondary prevention is aimed at early detection and treatment of illness. Primary prevention strategies are aimed at preventing the disease from beginning by avoiding or modifying risk factors. Tertiary prevention strategies focus on rehabilitation and prevention of complications arising from advanced disease.
NP: Data collection; CN: Health promotion and maintenance; CNS: Prevention and early detection of disease; CL: Application

22. Which of the following nursing actions is most appropriate in relieving pain related to cancer?
1. Using heat or cold on painful areas
2. Keeping a hard bedroll behind the client's back
3. Allowing the client to stay in one position to prevent pain
4. Keeping bright lights on in the room so the nurse can assess the client more quickly

22. 1. Using either heat or cold can reduce inflammatory responses, which will reduce pain. Avoid pressure (such as bedrolls) on painful areas. Change position frequently. Coordinate activity with pain medication. Reduce bright lights and noise to prevent anxiety, which can increase pain.
NP: Implementation; CN: Physiological integrity; CNS: Physiological adaptation; CL: Application

23. A new graduate is assigned to a nursing unit. A nurse manager assesses that the graduate's skills are deficient. Which action is most appropriate for the nurse manager to take?
1. Talk with the supervisor about terminating the new graduate.
2. Discuss with the graduate that a transfer to another unit is necessary.
3. Work with the graduate and develop a plan to improve the graduate's deficiencies.
4. Counsel the graduate that, if performance doesn't improve, the graduate will be terminated.

23. 3. A principle of leadership involves mastery over ignorance by working with people. The leader needs to work with the new graduate and provide opportunities for the graduate to grow and develop. The other responses wouldn't give the new graduate the opportunity and support needed for improvement.
NP: Implementation; CN: Safe, effective care environment; CNS: Coordinated care; CL: Application

NP: Nursing process   CN: Client needs category   CNS: Client needs subcategory   CL: Cognitive level

24. A local community health nurse is asked to speak to a group of adolescent girls on the topic of preventing pregnancy. Which of the following statements indicates the adolescents need more information on this topic?

    1. "I can get pregnant even the first time we have sex."

    2. "I can get pregnant even though I don't have sex regularly."

    3. "I can't get pregnant because my menstrual cycle isn't regular yet."

    4. "I can get pregnant even if my boyfriend withdraws before he comes."

24. 3. Many adolescents have misunderstandings related to risk periods and timing, including periods of susceptibility during the menstrual cycle, age-related susceptibility, and timing of male ejaculation.

NP: Evaluation; CN: Health promotion and maintenance; CNS: Growth and development through the life span; CL: Comprehension

25. Which of the following nursing actions is most appropriate in stimulating the appetite of a child with cancer?

    1. Using food as a reward system

    2. Serving large meals frequently

    3. Preparing foods appropriate to the age of the child

    4. Placing the child on a rigid time schedule for eating

25. 3. It's important to prepare foods appropriate to children in certain age groups. Involve the child in food preparation and selection. Encourage parents to relax pressures placed on eating by stressing the legitimate nature of loss of appetite. Let the child eat all food that can be tolerated. Assess the family's beliefs about food habits. Take advantage of a hungry period and serve small snacks.

NP: Implementation; CN: Physiological integrity; CNS: Physiological adaptation; CL: Application

26. A nurse working in a public health clinic is planning tuberculosis (TB) screening. Screening is indicated for which of the following groups?

    1. All clients coming into the clinic

    2. People living in a homeless shelter

    3. Clients who haven't received the TB vaccine

    4. Clients suspected of having human immunodeficiency virus (HIV)

26. 4. Clients with HIV infection or suspected of having HIV are at greater risk for developing TB. A screening test should be done and, if positive, treatment with isoniazid (Nydrazid) given. Clients coming to the clinic don't need to be tested unless they're at high risk; for example, living with someone infected with TB, abusing I.V. drugs, or suffering from chronic health conditions, such as diabetes mellitus and end-stage renal disease. Clients living in a homeless shelter aren't necessarily at greater risk unless other residents in the shelter have TB. The TB vaccine isn't widely used in the United States.

NP: Data collection; CN: Health promotion and maintenance; CNS: Prevention and early detection of disease; CL: Application

27. A professional nurse should report positive tuberculosis (TB) smears or cultures to the health department within which of the following time periods?

    1. 12 hours

    2. 48 hours

    3. 1 week

    4. 10 to 14 days

27. 2. A client is considered contagious if he has a positive TB smear or culture, so the results must be reported within 24 to 48 hours. One week or 10 to 14 days is too long to wait. The smear or culture may not have grown an organism in 12 hours.

NP: Implementation; CN: Health promotion and maintenance; CNS: Prevention and early detection of disease; CL: Application

28. Which of the following methods is used for the Mantoux test, tuberculin skin testing used to identify clients infected with *Mycobacterium tuberculosis?*
   1. Intradermal injection
   2. Multiple puncture test
   3. I.V. infusion
   4. I.M. injection

29. A nurse must have expertise in many roles. In which of the following nurse-client interactions is the nurse showing a secondary intervention as an advocate?
   1. Contacting the local church to borrow a walker for the client to use
   2. Listening to a client express feelings of frustration over the limitations imposed by his condition
   3. Giving I.V. antibiotic therapy every 12 hours with attention to sterile technique and prevention of complications
   4. Teaching a client with chronic obstructive pulmonary disease the effect of abdominal distention on breathing and ways to help bowel function

30. Which of the following times would be ideal to begin discharge planning for a client admitted with an exacerbation of asthma?
   1. At the time of admission
   2. The day before discharge
   3. After the acute episode is resolved
   4. When the discharge order is written

31. A young pregnant client attending prenatal classes is concerned about her alcohol intake. Which of the following statements indicates the client's child is at high risk for fetal alcohol syndrome (FAS)?
   1. "I just snort once or twice a day."
   2. "I don't feel like anyone loves me."
   3. "I drink a six pack of beer daily to settle my nerves."
   4. "I smoke marijuana with my boyfriend and his friends."

28. 1. A Mantoux test is given intradermally, not I.M., I.V., or by multiple puncture test.
NP: Implementation; CN: Health promotion and maintenance; CNS: Prevention and early detection of disease; CL: Knowledge

29. 1. Referral to community agencies is an advocacy role for nurses. The role of the advocate implies the nurse is able to advise clients how to find alternative sources of care. Instructing clients about disease processes, giving emotional support, and giving therapies to clients are direct care activities.
NP: Implementation; CN: Safe, effective care environment; CNS: Coordinated care; CL: Application

30. 1. Discharge planning should begin as soon as the client is admitted. Client stays are increasingly shorter, giving the interdisciplinary team less time to accomplish care goals. Beginning discharge planning as early as possible will give staff the greatest amount of time and make appropriate resources available. Waiting until the acute episode is resolved will greatly diminish the time available for discharge planning. Also, waiting for a discharge order or deferring planning until the end of the client's stay doesn't allow sufficient time.
NP: Planning; CN: Safe, effective care environment; CNS: Coordinated care; CL: Comprehension

31. 3. Ingestion of alcohol on a daily basis increases the risk of FAS. Other forms of addictive behavior, such as the ingestion of cocaine and smoking marijuana, increase the risk of fetal abuse — not FAS.
NP: Evaluation; CN: Health promotion and maintenance; CNS: Prevention and early detection of disease; CL: Comprehension

NP: Nursing process   CN: Client needs category   CNS: Client needs subcategory   CL: Cognitive level

32. According to the Centers for Disease Control and Prevention, which of the following groups doesn't need preventive therapy for tuberculosis (TB)?
1. Clients with human immunodeficiency virus (HIV) infection
2. Recent tuberculin skin test converters
3. People with no contact with infectious TB cases
4. Previously untreated or inadequately treated clients with abnormal chest X-rays

33. Which of the following psychosocial approaches should the emergency department nurse use when dealing with suspected family violence?
1. Punitive
2. Supportive treatment
3. Disgust and avoidance
4. Get the facts at all costs

34. A concerned client called the school asking that the nurse assess her 13-year-old son for signs of depression. Which of the following symptoms would the nurse expect to see?
1. Becoming angry at peers easily
2. Seeking out support from peers
3. Eating several small meals daily
4. Feeling he can control everything in his life

35. The nurse is speaking to a 56-year-old client who has recently lost his 82-year-old father to lung cancer. Which of the following signs of grief would the nurse expect to find when speaking with this client?
1. Decreased libido
2. Absence of anger and hostility
3. Difficulty crying or controlling crying
4. Clear dreams and imagery of the deceased

Keep up the good work! I'm so impressed!

32. 3. Clients with no contact with infectious TB cases aren't at high risk for developing TB. Preventive therapy should be initiated for recent tuberculin skin test converters, previously untreated or inadequately treated clients with abnormal chest X-rays, and clients infected with HIV.
NP: Implementation; CN: Health promotion and maintenance; CNS: Prevention and early detection of disease; CL: Application

33. 2. Emotional support is a nonthreatening approach when dealing with suspected family violence. Aggressive, punitive, and disdainful approaches can increase the anxiety of the perpetrator, increasing the risk of more violence.
NP: Implementation; CN: Psychosocial integrity; CNS: Coping and adaptation; CL: Application

34. 1. Adolescents experiencing depression may experience and express anger at peers. Adolescents feel a lack of control over their current situation, so they isolate from their peers. The adolescent often has an intake of nutrients insufficient to meet metabolic needs.
NP: Data collection; CN: Psychosocial integrity; CNS: Coping and adaptation; CL: Analysis

35. 4. A grieving client usually has vivid, clear dreams and fantasies. He also has a good capacity for imagery, particularly involving the loss. Difficulty crying or controlling crying, absence of anger and hostility, and decreased libido are signs of depression.
NP: Data collection; CN: Psychosocial integrity; CNS: Coping and adaptation; CL: Analysis

36. A 72-year-old client experienced the death of her husband 1 year ago. She now needs home health services due to severe osteo-arthritis. Which of the following statements indicates the client will need further bereavement counseling?
1. "I'm lucky my children live so close."
2. "I really don't have anything to live for."
3. "My health isn't very good, but I can live with it."
4. "I've always had trouble remembering where I placed things."

36. 2. Wishing for death is a sign of depression. Usually after a year, most individuals accept the death of their loved ones and begin restoring their life. Being grateful for good health and close family ties is a sign of acceptance of a new life that one experiences after the loss of a loved one. Memory loss can be a sign of dementia or depression.
NP: Data collection; CN: Psychosocial integrity; CNS: Coping and adaptation; CL: Analysis

37. Which of the following drugs and routes of administration is best to treat secondary syphilis?
1. Penicillin G orally
2. Penicillin G rectally
3. Cephalexin (Keflex) I.V.
4. Penicillin G I.M.

37. 4. Penicillin is the drug of choice to treat syphilis. Because of the long-term consequences of inadequate treatment, penicillin is usually given either I.M. or I.V., especially for syphilis of the nervous system or secondary syphilis. Keflex isn't the drug of choice for syphilis.
NP: Implementation; CN: Physiological integrity; CNS: Pharmacological therapies; CL: Knowledge

38. Your client must undergo an obstetric sonography. This test is typically indicated during the third trimester to rule out which of the following disorders?
1. Adnexal mass
2. Blighted ovum
3. Molar pregnancy
4. Breech presentation

38. 4. An obstetric sonography is indicated in the third trimester to assess presentation of the fetus. If the fetus is in breech position, external cephalic version or cesarean delivery may be considered. An adnexal mass or blighted ovum would have been ruled out during the first trimester. A molar pregnancy would have been ruled out during the first or second trimester.
NP: Data collection; CN: Health promotion and maintenance; CNS: Prevention and early detection of disease; CL: Analysis

39. A client developed oral ulcerations secondary to chemotherapy agents. Which of the following nursing actions is most appropriate for reducing pain and irritation in the mouth?
1. Serving a high-fiber diet
2. Using a toothbrush to clean teeth
3. Avoiding oral temperatures
4. Rinsing the mouth with hydrogen peroxide and water

39. 3. If oral ulcers are present, taking oral temperatures will be painful. Use the axillary region, rectum, or ear as sites for temperature readings. Use a soft-sponge toothbrush, cotton-tipped applicator, or gauze-wrapped finger to clean teeth. Give normal saline solution mouthwashes and rinses to reduce pain and inflammation. Hydrogen peroxide mixed with water is too irritating if oral ulcers are present.
NP: Implementation; CN: Physiological integrity; CNS: Physiological adaptation; CL: Application

40. A client is admitted to the emergency department after being sexually assaulted. A policewoman accompanies her. The nurse realizes that several important tasks should be done in a sexual assault case. Which of the following nursing interventions receives priority?
1. Assisting with medical treatment
2. Collecting and preparing evidence for the police
3. Attempting to reduce the client's anxiety from a panic to a moderate level
4. Providing anticipatory guidance to the client about normal responses to sexual assault

41. The nurse is caring for an 8-year-old boy diagnosed with attention deficit hyperactivity disorder (ADHD). Which of the following behaviors is most common in children with ADHD?
1. Lethargy
2. Long attention span
3. Short attention span
4. Preoccupation with body parts

42. The second stage of labor is characterized by strong urges to bear down. This reflex is known as:
1. Babinksi's reflex.
2. Ferguson's reflex.
3. Moro's reflex.
4. Myerson's reflex.

43. A 35-year-old professional woman is admitted to an inpatient substance abuse unit with a diagnosis of alcohol dependence. Which of the following comments by the client indicates that she's using rationalization to deal with her alcohol problem?
1. "I don't drink more than two beers when I'm out."
2. "I always remember what happens the next day."
3. "I always ask a friend to drive me home when I'm drinking."
4. "I've had four tickets for driving while intoxicated in the last month."

Excellent! You're nearly halfway to the end. You should feel proud—and motivated!

40. 3. Reducing anxiety will help the client participate in medical, forensic, and legal follow-up activities. Medical treatment should begin as soon as the client's anxiety decreases below the panic level. Collecting and preparing evidence and providing anticipatory guidance aren't high-priority interventions.
NP: Implementation; CN: Psychosocial integrity; CNS: Coping and adaptation; CL: Comprehension

41. 3. Short attention span is a common characteristic of ADHD due to difficulty concentrating. These children show hyperexcitability, not lethargy. Children with this disorder are distracted by environmental stimuli, so they won't be concentrating on their body parts.
NP: Data collection; CN: Psychosocial integrity; CNS: Psychosocial adaptation; CL: Comprehension

42. 2. Ferguson's reflex is characterized by the strong urge to bear down during the second stage of labor. Babinski's reflex results in dorsiflexion of the big toe and fanning of the other toes when the sole of the client's foot is scraped, Moro's reflex is a normal generalized reflex in an infant when he reacts to a sudden noise such as when a table is struck next to him. Myerson's reflex results in blinking when the client's forehead, bridge of the nose, or maxilla is tapped.
NP: Data collection; CN: Physiological integrity; CNS: Physiological adaptation; CL: Analysis

43. 4. Driving while intoxicated can be seen as a symptom of alcohol dependence. Designating drivers and limiting alcohol consumption are self-responsible actions, but don't address the underlying problem. By asking someone to drive them home, clients with alcohol dependence rationalize that it's okay to drink if they're responsible. The amount one drinks doesn't matter. An alcoholic experiences blackouts, which are periods of amnesia about experiences while intoxicated.
NP: Evaluation; CN: Psychosocial integrity; CNS: Psychosocial adaptation; CL: Analysis

44. A nurse is working with a client with alcoholism in an acute care mental health unit. The client has been referred to Alcoholics Anonymous (AA). Which of the following statements by the client indicates that the client is ready to begin the AA program?

    1. "I know I'm powerless over alcohol and need help."

    2. "I think it will be interesting and helpful to join AA."

    3. "I'd like to sponsor another alcoholic with this same problem."

    4. "My family is very supportive and will attend meetings with me."

44. 1. In step 1 of AA, a person admits their powerlessness over alcohol and is ready to accept help. This should occur before they begin AA. A supportive family and a desire to help others with the same problem are good for the client, but don't necessarily indicate readiness to participate in the program.

NP: Evaluation; CN: Psychosocial integrity; CNS: Psychosocial adaptation; CL: Application

45. In planning care for a client diagnosed with paranoid schizophrenia, which of the following actions is correct for the psychiatric nurse?

    1. Confront the client about her hallucinations.

    2. Ask the minister to provide spiritual direction.

    3. Instruct family members to discourage delusions.

    4. Affirm when client's perceptions and thinking are in touch with reality.

45. 4. The nursing care plan focuses on reinforcing perceptions and thinking that are in touch with reality. Confronting a client about her hallucinations and delusions isn't effective or therapeutic. Using family members could create distrust between the client and the family. Spiritual direction is important, but a client with paranoid schizophrenia may have issues surrounding her religious or spiritual orientation. Therefore, asking a minister to provide spiritual direction may not be effective or therapeutic.

NP: Planning; CN: Psychosocial integrity; CNS: Psychosocial adaptation; CL: Application

46. A psychiatric nurse finds a client with bipolar disorder sitting in the dayroom. The client is wearing a red polka dot dress, large yellow hat, and heavy makeup with large gold jewelry. Which of the following phases of the illness is the client most likely in?

    1. Delusional

    2. Depressive

    3. Manic

    4. Suspicious

46. 3. Extreme labile moods are characteristic of clients in the manic phase of bipolar disorder. Hyperactivity, verbosity, and drawing attention to oneself through dress are typical of the manic phase. Delusions and suspiciousness may be seen in bipolar disorder, but are more commonly seen in schizophrenia. In the depressive phase, clients are withdrawn, cry, and may not eat. Visual or auditory hallucinations, delusional thoughts, and extreme suspiciousness are behaviors seen in clients diagnosed with paranoid schizophrenia.

NP: Data collection; CN: Psychosocial integrity; CNS: Psychosocial adaptation; CL: Analysis

NP: Nursing process   CN: Client needs category   CNS: Client needs subcategory   CL: Cognitive level

47. Which of the following signs would the nurse expect to observe in a 4-week-old neonate in acute pain?
　1. Whimpering
　2. Eyes opened wide
　3. Limp body posture
　4. Desire to breast-feed frequently

48. Which of the following positions or equipment is needed to maintain posture for a child with juvenile rheumatoid arthritis?
　1. Soft mattress
　2. Prone position
　3. Large fluffy pillows
　4. Semi-Fowler's position

49. There are four main types of breech presentation. A complete breech occurs when:
　1. a foot extends below the buttocks.
　2. a knee extends below the buttocks.
　3. the buttocks are the presenting part, and thighs and knees are flexed.
　4. the buttocks are the presenting part, thighs are flexed, and knees are extended.

You've finished 50 questions! Wow! Way to fly high!

50. While putting an elderly client with an indwelling urinary catheter in bed, a nurse notices the tubing hanging below the bed. She places the tubing in a loop on the bed with the client and makes sure he won't lie on the tubing. Which rationale explains the nurse's action?
　1. To inhibit drainage
　2. To allow drainage to occur
　3. To allow the urine to collect in the tubing
　4. To have the client check the tubing for urine

51. A client complains of severe burning on urination. Which of the following instructions is best to give the client?
　1. Wear only nylon panties.
　2. Drink coffee to increase urination.
　3. Soak in warm water with bubble bath.
　4. Drink 2,500 to 3,000 ml of water per day.

47. 1. Crying, whimpering, and groaning are vocal expressions of acute pain in the neonate. Eyes tightly closed, decreased appetite, and fist clenching with rigidity are also signs of acute pain in the neonate.
NP: Data collection; CN: Health promotion and maintenance; CNS: Growth and development through the life span; CL: Application

48. 2. Lying in the prone position is encouraged to straighten hips and knees. A firm mattress is needed to maintain good alignment of spine, hips, and knees, and no pillow or a very thin pillow should be used. Semi-Fowler's position increases pressure on the hip joints and should be avoided.
NP: Implementation; CN: Physiological integrity; CNS: Basic care and comfort; CL: Application

49. 3. In a complete breech, the buttocks are the presenting part and the thighs and knees are flexed. Two types of incomplete breech are footling breech, when a foot extends below the buttocks, and knee presentation, when a knee extends below the buttocks. A frank breech happens when the buttocks are the presenting part, the thighs are flexed, and the knees are extended.
NP: Data collection; CN: Physiological integrity; CNS: Physiological adaptation; CL: Analysis

50. 2. Catheter tubing shouldn't be allowed to develop dependent loops or kinks because this inhibits proper drainage by requiring the urine to travel against gravity to empty into the bag. Permitting the urine to collect in the tubing increases the risk of infection. Observing the catheter and tubing is the responsibility of the nurse.
NP: Implementation; CN: Physiological integrity; CNS: Reduction of risk potential; CL: Application

51. 4. Drinking large amounts of water will help flush bacteria from the urinary tract. Avoid nylon underwear; wear only cotton undergarments to decrease the warm, moist environment. Avoid using bubble baths, perfumed soaps, or bath powders in the perineal area. The scent in toiletries can be irritating to the urinary meatus. Avoid tea, coffee, carbonated drinks, and alcoholic beverages because of bladder irritation.
NP: Implementation; CN: Physiological integrity; CNS: Basic care and comfort; CL: Application

52. Which of the following instructions is given to a client with a hearing aid?
1. Clean the hearing aid with baby oil.
2. Wear the hearing aid while sleeping.
3. Keep the hearing aid out of direct sunlight.
4. Leave the hearing aid in place while showering.

52. 3. The hearing aid should be kept out of direct sunlight and away from high temperatures. Solvents or lubricants shouldn't be used on the aid. If there's a detachable ear mold, it can be washed in warm, soapy water and dried with a soft cloth. The hearing aid should be left in place while the client is awake, except when showering.
NP: Implementation; CN: Physiological integrity; CNS: Basic care and comfort; CL: Application

I'm impressed with your progress! Keep it up!

53. A 72-year-old client is being discharged from same-day surgery after having a cataract removed from her right eye. Which of the following discharge instructions does the nurse give the client?
1. Sleep on the operative side.
2. Resume all activities as before.
3. Don't rub or place pressure on the eyes.
4. Wear an eye shield all day and remove it at night.

53. 3. Rubbing or placing pressure on the eyes increases the risk of accidental injury to ocular structures. The nurse should also caution against lifting objects, straining, strenuous exercise, and sexual activity because such activities can increase intraocular pressure. Glasses or shaded lenses should be worn to protect the eye during waking hours after the eye dressing is removed. An eye shield should be worn at night. Caution against sleeping on the operative side to reduce the risk of accidental injury to ocular structures.
NP: Implementation; CN: Physiological integrity; CNS: Reduction of risk potential; CL: Application

54. To ensure the safe administration of medications, which of the following actions is correct?
1. Make sure the client is in the right room.
2. Check the name band of the client.
3. Have the client repeat his name.
4. Note the physician's name.

54. 2. The five rights for proper administration of medications and therapies are the right client, right dose, right time, right route, and right medication. Checking the name band is the most reliable way to verify client identification.
NP: Implementation; CN: Physiological integrity; CNS: Pharmacological therapies; CL: Knowledge

55. Which of the following instructions is correct for a client taking nortriptyline (Pamelor) for depression?
1. Be aware that this drug can cause a slow heart rate.
2. This drug will work immediately to treat depression.
3. Take this drug in the morning because it causes drowsiness.
4. Wear protective clothing and sunscreen when out in the sun.

55. 4. A common adverse effect of this drug is sensitivity to the sun. Protective clothing and sunscreen are worn in the sun. This drug can cause an irregular heart rate. It doesn't work immediately, but takes 2 to 3 weeks to achieve the desired effect. Take this drug at bedtime if it causes drowsiness.
NP: Implementation; CN: Physiological integrity; CNS: Pharmacological therapies; CL: Application

NP: Nursing process   CN: Client needs category   CNS: Client needs subcategory   CL: Cognitive level

56. Which of the following instructions is correct for a client receiving lithium for bipolar disorder?
1. Avoid drugs containing ibuprofen.
2. Drink at least two cups of coffee daily.
3. Be aware that you may experience increased alertness.
4. It isn't necessary to monitor the blood level of this drug.

57. Which of the following nursing interventions is correct for clients receiving I.V. therapy?
1. Change the tubing every 8 hours.
2. Monitor the flow rate at least every hour.
3. Change the I.V. catheter and entry site daily.
4. Increase the rate to catch up if the correct amount hasn't been infused at the end of the shift.

58. Which of the following nursing interventions is correct for a client receiving total parenteral nutrition (TPN)?
1. Discard TPN solutions after 24 hours.
2. Discard lipid emulsions after 20 hours.
3. Inspect the TPN solution for clearness and visibility.
4. Teach the client to blow out during expiration when the tubing is disconnected.

59. Which of the following drug forms can be administered through a nasogastric (NG) tube?
1. Enteric coated
2. Oral
3. Parenteral
4. Sublingual

56. 1. Avoid drugs that alter the effect of lithium, such as ibuprofen (Advil, Motrin, Nuprin) and sodium bicarbonate or other antacids containing sodium. Lithium can decrease alertness and coordination. Avoid beverages with caffeine because they increase urination, which may alter the effect of lithium. Lithium levels may need to be monitored every 2 weeks, especially if adverse effects occur. The dose may need to be regulated.
NP: Implementation; CN: Physiological integrity; CNS: Pharmacological therapies; CL: Application

57. 2. Closely observing the rate of infusion prevents underhydration and overhydration. Tubing is changed according to facility policy but not at the frequency of every 8 hours. The I.V. catheter and entry site should be changed every 48 to 72 hours in most situations. Increasing the rate may lead to fluid overload.
NP: Implementation; CN: Physiological integrity; CNS: Pharmacological therapies; CL: Comprehension

58. 1. TPN solutions are good media for fungi, so they should be discarded after 24 hours. The nurse should teach the client to perform Valsalva's maneuver, taking a deep breath and holding it, when the tubing is disconnected. Valsalva's maneuver increases intrathoracic pressure, which prevents air entry. Inspect TPN solutions for cloudiness, and the container for cracks or leaks before hanging. Lipid emulsions are also good media for fungi and should be discarded after 12 hours.
NP: Implementation; CN: Physiological integrity; CNS: Pharmacological therapies; CL: Application

59. 2. Most oral medications can be given via an NG tube because they're intended for passage into the stomach. Some oral drugs have special coatings intended to keep the pill intact until it passes into the small intestine; these enteric-coated pills shouldn't be crushed and put through an NG tube. Sublingual means under the tongue and the medication requires placement there for adequate absorption. Parenteral means I.V., I.M., and subcutaneous. Some parenteral medications, such as insulin, may be destroyed by gastric juices.
NP: Implementation; CN: Physiological integrity; CNS: Pharmacological therapies; CL: Comprehension

60. Which of the following diagnostic tests is used to diagnose bacterial endocarditis?
1. Electrolytes
2. Blood cultures
3. Prothrombin time (PT)
4. Venereal Disease Research Laboratory (VDRL)

60. 2. Blood cultures are crucial in diagnosing bacterial endocarditis. Electrolyte levels indicate abnormalities that occur with drug therapy as well as with complications associated with heart failure. A positive VDRL may be evidence of syphilitic heart disease. PT values are useful in monitoring anticoagulant therapy.

NP: Data collection; CN: Physiological integrity; CNS: Reduction of risk potential; CL: Application

61. In preparing a client for cardiac catheterization, which of the following statements or questions is most appropriate?
1. "Are you allergic to contrast dyes or shellfish?"
2. "Have you ever had this kind of procedure before?"
3. "You'll need to fast 24 hours before the procedure."
4. "You'll be given medication to help you sleep during the procedure."

61. 1. The nurse must assess for allergies to iodine before the procedure because the dye used during catheterization contains iodine. The client needs to stay awake during the procedure to follow directions, such as taking a deep breath and holding it during injection of the dye, and to report chest, neck, or jaw discomfort. The client is instructed to fast for 6 hours before the procedure. The client will be asked to empty his bladder before the procedure. Knowing the client's history and prior experience with this procedure would be helpful, but knowing the client's allergies is more important.

NP: Implementation; CN: Physiological integrity; CNS: Reduction of risk potential; CL: Application

62. Which of the following techniques is considered a noninvasive diagnostic method to evaluate cardiac changes?
1. Cardiac biopsy
2. Cardiac catheterization
3. Magnetic resonance imaging (MRI)
4. Pericardiocentesis

62. 3. MRI is a noninvasive procedure that aids in the diagnosis and detection of thoracic aortic aneurysm and evaluation of coronary artery disease, pericardial disease, and cardiac masses. Cardiac biopsy, cardiac catheterization, and pericardiocentesis are invasive techniques used to evaluate cardiac changes.

NP: Data collection; CN: Health promotion and maintenance; CNS: Prevention and early detection of disease; CL: Knowledge

63. A client recovering from surgery tells a nurse, "I feel like I have to urinate more often than usual, and it burns when I urinate." The nurse should plan to obtain a urinary specimen for which of the following factors?
1. Culture
2. Glucose
3. Ketones
4. Specific gravity

63. 1. The signs and symptoms of this client are those of a urinary tract infection, and a culture will probably show a bacterial infection. Ketones and glucose in the urine don't cause the client to complain of burning and frequency. Specific gravity is a measure of the concentration of the urine. A high or low specific gravity won't cause these symptoms.

NP: Implementation; CN: Physiological integrity; CNS: Reduction of risk potential; CL: Comprehension

---

NP: Nursing process   CN: Client needs category   CNS: Client needs subcategory   CL: Cognitive level

64. Which of the following foods can alter results when stool is checked for occult blood?
1. Red meat
2. Dairy products, canned fruit, and pretzels
3. Horseradish, raw fruits, and vegetables
4. Potatoes, orange juice, and decaffeinated coffee

65. In caring for a client with arterial insufficiency, which of the following instructions is most appropriate?
1. "You may leave your feet open to the air."
2. "Sit and rest for several hours per day."
3. "Avoid crossing your legs at the knees or ankles."
4. "Wear tight socks instead of no socks."

66. Which of the following instructions would be appropriate for a client taking oral anticoagulants?
1. "Shave with a standard razor."
2. "Take ibuprofen or aspirin for pain."
3. "Take the anticoagulant at the same time each day."
4. "Eat a large quantity of green leafy vegetables."

67. Which of the following actions is most appropriate to reduce sensory deprivation for a visually impaired elderly client in the hospital?
1. Keep the lights dimmed.
2. Close the curtains or blinds on windows to reduce glare.
3. Open the hospital door so bright light can shine in the room.
4. Open the curtains during the day so the sun can shine brightly.

Only 20 left. Hang in there!

64. 1. Consumption of red meat has caused false-positive readings. Avoid foods that are high in iron. The other foods don't cause false-positive readings.
NP: Data collection; CN: Physiological integrity; CNS: Reduction of risk potential; CL: Comprehension

65. 3. Leg crossing should be avoided because it compresses the vessels in the legs. Feet and extremities must be protected to reduce the risk of trauma. Avoid constrictive clothing, such as tight elastic on socks, to prevent compression of vessels in the legs. Injury to the extremity will require more blood to heal than to keep the tissue intact; an extremity with compromised circulation may not be able to provide the extra blood required.
NP: Implementation; CN: Physiological integrity; CNS: Reduction of risk potential; CL: Application

66. 3. It's important to take the anticoagulant at the same time each day to maintain an adequate blood level. Avoid taking aspirin or ibuprofen because these drugs decrease clotting time. Eating a large amount of green leafy vegetables, which contain vitamin K, increases the clotting time, thus requiring more anticoagulants. An electric razor reduces the risk of cutting the skin. Avoid the use of standard razors.
NP: Implementation; CN: Physiological integrity; CNS: Reduction of risk potential; CL: Application

67. 2. Closing curtains or blinds on windows can reduce glare and improve vision for the older client. Controlled lighting can help the older client see better in the hospital. Adequate background lighting helps the older client decrease visual accommodation when moving from brightly lit to dimly lit rooms and hallways.
NP: Implementation; CN: Physiological integrity; CNS: Reduction of risk potential; CL: Application

68. Which of the following actions is most appropriate to reduce sensory overload for a hearing-impaired elderly client?
    1. Keep the overhead light on continuously.
    2. Discuss the client's condition at the bedside.
    3. Allow all family members to stay with the client.
    4. Limit bedside conversation to that directed to the client.

68. 4. Limiting bedside conversation to that directed to the client creates fewer disturbances, thus reducing sensory overload. Turning off or dimming the overhead lights further reduces visual stimulation and facilitates day and night light fluctuations. Although fostering family interaction with the client is necessary, only one or two family members should be allowed to visit with the client at one time. Crowding of people in the client's room may precipitate a loss of privacy and control for the client.
NP: Implementation; CN: Physiological integrity; CNS: Reduction of risk potential; CL: Application

69. Which of the following instructions is most appropriate for a client with osteoarthritis of the left knee?
    1. Use cold on joints.
    2. Keep the knee extended.
    3. Develop a weight-reduction plan.
    4. Have someone help the client in activities of daily living.

69. 3. Reducing weight decreases joint stress. Local moist heat provides pain relief and will decrease stiffness. The client should perform muscle-strengthening exercises, which help prevent joint stiffness. The nurse should allow the client to perform activities of daily living as independently as possible.
NP: Implementation; CN: Physiological integrity; CNS: Reduction of risk potential; CL: Application

70. A newly diagnosed client with diabetes is found semi-comatose with a rapid heart rate and low blood pressure. The client's skin is warm and dry. Which of the following conditions would the nurse suspect?
    1. Hypoglycemia
    2. Cardiogenic shock
    3. Diabetic ketoacidosis (DKA)
    4. Hyperosmolar hypoglycemic nonketotic syndrome (HHNS)

70. 3. DKA develops as a result of severe insulin deficiency. The incidence of DKA generally results from undiagnosed diabetes and inadequacy of prescribed medication and dietary therapies. Hypoglycemia involves episodes of low blood glucose levels caused by erratic or altered absorption of insulin. In cardiogenic shock, the client has pale, cool, and moist skin. HHNS is a deadly complication of diabetes distinguished by severe hyperglycemia, dehydration, and changed mental status.
NP: Implementation; CN: Physiological integrity; CNS: Physiological adaptation; CL: Application

71. A nurse is standing next to a person eating fried shrimp at a parade. Suddenly, the man clutches at his throat and can't speak, cough, or breathe. The nurse asks the man if he's choking and he nods yes. Which of the following responses is most appropriate?
    1. Attempt rescue breathing.
    2. Perform the Heimlich maneuver.
    3. Deliver external chest compressions.
    4. Use the head tilt-chin lift maneuver to establish the airway.

71. 2. If a conscious victim acknowledges that he's choking, the best response is to perform the Heimlich maneuver to relieve the airway obstruction. The other options are used for an unresponsive victim with absent heart rate and breathing.
NP: Implementation; CN: Physiological integrity; CNS: Physiological adaptation; CL: Application

NP: Nursing process   CN: Client needs category   CNS: Client needs subcategory   CL: Cognitive level

72. Preparation is the key to successful resuscitation. Which of the following responses is most appropriate to prepare for a cardiopulmonary emergency?
　　1. Have nasal oxygen ready when needed.
　　2. Place an oropharyngeal airway at the bedside.
　　3. Keep the medication cart locked up for safety.
　　4. Don't start an I.V. line unless necessary.

73. Obtaining the dorsalis pedis and posterior or tibial pulses are helpful in establishing which of the following data?
　　1. Heart rate
　　2. Pulse rate
　　3. Lower extremity circulation
　　4. Evidence of tachycardia

74. A young client is admitted to the emergency department unconscious from an overdose of salicylates. The physician orders dialysis. Which of the following dialysis methods is most appropriate?
　　1. Hemodialysis
　　2. Peritoneal dialysis
　　3. Continuous hemofiltration
　　4. Continuous ambulatory peritoneal dialysis

75. Which of the following nursing actions is most appropriate for a low birth-weight neonate?
　　1. Keeping the temperature cool
　　2. Gavage feeding the neonate if he has a weak sucking reflex
　　3. Keeping the neonate uncovered in the humidified incubator
　　4. Keeping the I.V. infusion at a keep-vein-open rate

72. 2. A nurse should learn to anticipate clinical deterioration before overt signs and symptoms are apparent. If a client is having breathing difficulties, the nurse should place an oropharyngeal airway at the bedside while the client is monitored for deterioration. The emergency cart should be placed outside the client's room for easy access. If breathing stops, the client will need to be intubated and placed on a respirator, if necessary. The client should have a stable I.V. line for administration of emergency drugs.
NP: Implementation; CN: Physiological integrity; CNS: Physiological adaptation; CL: Analysis

73. 3. These pulses represent peripheral circulation and are rated by their quality from 0 (absent) to 4 (bounding). Heart rate and pulse rate are used interchangeably and established through examination of the radial or apical pulses. Tachycardia is a fast pulse rate best established by examination of the apical or radial locations.
NP: Data collection; CN: Physiological integrity; CNS: Physiological adaptation; CL: Knowledge

74. 1. Hemodialysis is a rapid method to correct fluid and electrolyte problems. It's also a fast way to treat accidental or intentional poisonings as a means of clearing drugs or toxins from the body. Peritoneal dialysis, continuous ambulatory peritoneal dialysis, and continuous hemofiltration are slow methods of removing toxins.
NP: Implementation; CN: Physiological integrity; CNS: Physiological adaptation; CL: Application

75. 2. To maintain adequate nutrition, a low-birth-weight neonate may need to be gavage fed. The neonate should be kept warm. The nurse should avoid situations that might predispose the neonate to chilling, such as exposure to cool air. Maintain adequate parenteral fluids to prevent dehydration.
NP: Implementation; CN: Physiological integrity; CNS: Physiological adaptation; CL: Application

76. A client's laboratory values are: calcium, 18 mg/dl; potassium, 3 mEq/L; and magnesium, 4 mEq/L. These values put this client most at risk for which of the following complications?
1. Bleeding
2. Renal failure
3. Cardiac arrhythmia
4. Respiratory distress

77. A client is 2 days postoperative from a femoral popliteal bypass. The nurse's assessment finds the client's left leg cold and pale. Which of the following actions has priority?
1. Checking distal pulses
2. Notifying the physician
3. Elevating the foot of the bed
4. Wrapping the leg in a warm blanket

78. A 40-year-old client is scheduled to have elective facial surgery later in the morning. The nurse notes that the client's pulse rate is 130 beats/minute. Which of the following reasons best explains the tachycardia?
1. Age
2. Anxiety
3. Exercise
4. Pain

79. Which of the following statements is an example of a key element in the nursing care plan?
1. Advance diet to regular as tolerated.
2. Ambulate 30′ (9.1 m) with walker by discharge.
3. Give furosemide (Lasix) 40 mg I.V. now.
4. Discontinue I.V. fluids when tolerating oral fluids.

76. 3. Low potassium, magnesium, and calcium levels cause the heart muscle to become irritable, resulting in arrhythmias. Respiratory weakness may occur with a low potassium level. Bleeding is seen with a low calcium level. Renal failure isn't caused by low electrolyte levels.
NP: Evaluation; CN: Health promotion and maintenance; CNS: Prevention and early detection of disease; CL: Comprehension

77. 1. The client has arterial disease and had vascular surgery. The nurse must assess the client for complications. A potential problem would be a clot at the surgical site, so the nurse must assess circulation by checking for distal pulses. Before the physician is notified, the nurse should determine if distal pulses are present. Elevating the foot of the bed would promote venous return but decrease arterial blood flow and shouldn't be done. The leg can be covered lightly after circulation is assessed.
NP: Implementation; CN: Physiological integrity; CNS: Physiological adaptation; CL: Application

78. 2. Anxiety tends to increase heart rate, temperature, and respirations. Exercise will also increase the heart rate but most likely won't occur preoperatively. The normal heart rate for a client this age is 60 to 100 beats/minute. The client shouldn't be in any pain preoperatively.
NP: Data collection; CN: Physiological integrity; CNS: Physiological adaptation; CL: Comprehension

79. 2. This is a measurable expected outcome or goal, a key element of a nursing care plan. Other key elements include nursing diagnoses and interventions. The other options are physician's orders, not key elements of care plans.
NP: Planning; CN: Safe, effective care environment; CNS: Coordinated care; CL: Application

NP: Nursing process   CN: Client needs category   CNS: Client needs subcategory   TL: Cognitive level

80. A client had an appendectomy 24 hours ago. Which of the following goals is appropriate for this client?
1. The client will be able to walk in the hallway.
2. The client will be able to attend physical therapy.
3. The client will be able to accomplish all activities of daily living (ADLs).
4. The client will be able to state the rationale for all postoperative medications.

81. Which of the following actions is included in the principles of asepsis?
1. Maintaining a sterile environment
2. Keeping the environment as clean as possible
3. Testing for microorganisms in the environment
4. Cleaning an environment until it's free from germs

82. Two days after admission to a psychiatric unit, a client with bipolar disorder becomes verbally aggressive during a group therapy session. Which nursing response is best?
1. "You're behaving in an unacceptable manner, and you need to control yourself."
2. "You're scaring everyone in the group. You can't participate until you know how to act."
3. "This is why you have relationship troubles; you don't know how to talk to people."
4. "You're disturbing the other clients. Let's go to the exercise room to help you release some of your energy."

83. A client receiving phenothiazine has become restless and fidgety and has been pacing the hallway continuously for the past hour. This behavior suggests that the client may be experiencing which of the following adverse reactions to phenothiazine?
1. Dystonia
2. Akathisia
3. Parkinsonian effects
4. Tardive dyskinesia

80. 1. A 24-hour postoperative client is expected to be able to walk in the hallway. A client who just had an appendectomy shouldn't need physical therapy unless deconditioning was evident. At 24 hours, a client should begin to assume responsibility for ADLs, but shouldn't necessarily be responsible for all activities. It's too early to expect a client to state the rationale for all postoperative medications, especially if the client is elderly.
NP: Planning; CN: Physiological integrity; CNS: Basic care and comfort; CL: Application

81. 2. Asepsis is the process of avoiding contamination from outside sources by keeping the environment clean. A clean environment has a reduced number of microorganisms, but isn't necessarily sterile (the absence of all microorganisms). Testing for microorganisms or culturing isn't indicated in the promotion of asepsis.
NP: Implementation; CN: Safe, effective care environment; CNS: Safety and infection control; CL: Comprehension

82. 4. This response shows the nurse finds the client's behavior unacceptable, yet still regards the client as worthy of help. The third option is judgmental and not therapeutic. The other options give the false impression that the client is in control of the behavior; the client hasn't been in treatment long enough to control the behavior.
NP: Implementation; CN: Psychosocial integrity; CNS: Coping and adaptation; CL: Analysis

83. 2. The client's behavior suggests akathisia — an adverse effect of phenothiazines. Dystonia appears as excessive salivation, difficulty speaking, and involuntary movements of the face, neck, arms and legs. Parkinsonian effects include a shuffling gait, hand tremors, drooling, rigidity, and loose arm movements. Tardive dyskinesia is characterized by odd facial and tongue movements.
NP: Data collection; CN: Physiological integrity; CNS: Reduction of risk potential; CL: Comprehension

84. A client who had her gallbladder removed 2 days ago now complains of pain in her right calf. Which of the following responses has priority?

1. Assess the client for Homans' sign.
2. Instruct the client to flex her knee and hip.
3. Apply a warm compress and call the physician.
4. Gently massage the calf and notify the physician.

85. Which of the following statements by a client with chronic arterial disease indicates further teaching is needed?

1. "I'm going to stop smoking."
2. "I'm going to have the podiatrist check my feet."
3. "I'm going to keep the heat in my house at 80° F."
4. "I'm going to walk short distances every morning."

84. 1. Pain in the calf is a symptom of possible deep vein thrombosis. The nurse must assess further. Assessing the client for Homans' sign would be the next intervention. Making the client flex her knee and hip won't help assess for the presence of a clot. Warm compresses may be ordered after a diagnosis of deep vein thrombosis is made. Never massage the calf muscle as the clot could be dislodged.

NP: Implementation; CN: Physiological integrity; CNS: Reduction of risk potential; CL: Comprehension

85. 3. Clients with peripheral vascular disease need to be at a comfortable temperature because of impaired circulation. Having the heat at 80° F is too warm. The other choices are all appropriate interventions for a client with peripheral vascular disease.

NP: Evaluation; CN: Physiological integrity; CNS: Reduction of risk potential; CL: Application

Congratulations! You did a great job, and I'm totally proud of you! Yaaaaay!

NP: Nursing process   CN: Client needs category   CNS: Client needs subcategory   CL: Cognitive level

Here's the final comprehensive test to practice for the NCLEX. Check it out!

COMPREHENSIVE
Test 4

1. You're caring for a neonate at 10 minutes after birth. Which finding would require you to notify the physician?
1. Crackling noises during auscultation of breath sounds
2. Respiratory rate of 50 breaths/minute
3. Bluish gray pigmented nevi on the buttocks
4. Shrill, high pitched cry that doesn't cease

2. A team leader notes increasing unrest among the staff members. Which of the following actions is best for the team leader to take?
1. Discuss the problem with a coworker.
2. Report the problem to the nurse-manager.
3. Discuss the problem with the staff.
4. Ignore the problem and hope it won't interfere with the functioning of the floor.

3. A physician orders a urine specimen for culture and sensitivity stat. Which approach is best for the nurse to use in delegating this task?
1. "We need a stat urine culture on the client in room 101."
2. "Please get the urine for culture for the client in room 101."
3. "A stat urine was ordered for the client in room 101. Would you get it?"
4. "We need a urine for culture stat on the client in room 101. Tell me when you send it to the lab."

4. A 67-year-old widow is visiting a physician for an annual check up. The client asks the nurse, "Do you think it's wrong to masturbate?" Which response by the nurse is best?
1. "How do you feel about that?"
2. "Do you really want to do that?"
3. "I think you're a little too old for that."
4. "Why don't you ask your physician?"

1. 4. A shrill, high pitched cry that doesn't cease may indicate a central nervous system injury or abnormality. It's often found in neonates with genetic abnormalities, drug withdrawal, or fetal alcohol syndrome. Crackling noises heard during auscultation of breath sounds and a respiratory rate of 50 breaths/minute are normal. Bluish gray nevi on the buttocks (Mongolian spots) are also common.
NP: Data collection; CN: Health promotion and maintenance; CNS: Prevention and early detection of disease; CL: Analysis

2. 3. The leader should comment to the group on the observed behavior. This is a firm approach but one that shows concern. Ignoring problems or discussing them with someone else doesn't confront the issue at hand.
NP: Implementation; CN: Safe, effective care environment; CNS: Coordinated care; CL: Application

3. 4. This option not only delegates the task, but also provides a checkpoint. To effectively delegate, you need to follow up on what someone else is doing. The other options don't provide for feedback, which is essential for communication and delegation.
NP: Implementation; CN: Safe, effective care environment; CNS: Coordinated care; CL: Application

4. 1. It's essential in communication to find out how the client thinks and feels. The second and third options are biased and put the client down. The last option tells the client the nurse isn't interested. The client might be too uncomfortable to discuss this topic with the physician.
NP: Implementation; CN: Psychosocial integrity; CNS: Coping and adaptation; CL: Comprehension

NP: Nursing process   CN: Client needs category   CNS: Client needs subcategory   CL: Cognitive level

5. Which of the following characteristics of the client goal in the care plan is correct?
1. Nurse-focused, flexible, measurable, and realistic
2. Client-focused, flexible, realistic, and measurable
3. Nurse-focused, time-limited, realistic, and measurable
4. Client-focused, time-limited, realistic, and measurable

5. 4. All goals should be client focused, allowing the client to understand what needs to be accomplished. Specify a time limit for when this task should be achieved. Be realistic, so the client may be successful in reaching the goal. The goal must be measurable so all staff can evaluate the client's progress. Nurse flexibility is an important attribute and necessary for reassessing needs and approaches for the client's optimal recovery. However, in the actual goal, specific criteria must be identified to allow all staff to work from the same data for achieving client goals.
NP: Planning; CN: Safe, effective care environment; CNS: Coordinated care; CL: Knowledge

6. Which of the following medications should a nurse withhold from a client 6 hours before a series of pulmonary function tests (PFTs)?
1. Antibiotics
2. Antitussives
3. Bronchodilators
4. Corticosteroids

6. 3. PFTs measure the volume and capacity of air. If a bronchodilator is given, it will improve the bronchial airflow and alter the test results. The other drugs would have no effect on the bronchial tree and PFT results.
NP: Planning; CN: Physiological integrity; CNS: Pharmacological therapies; CL: Comprehension

7. A client must choose a meal that follows a high-calorie, high-protein, decreased sodium, and low-potassium diet. Which of the following choices indicates the client understands these dietary guidelines?
1. Halibut, salad, rice, and instant coffee
2. Crab, beets, spinach, and baked potato
3. Salmon, rice, green beans, sourdough bread, coffee, and ice cream
4. Sirloin steak, salad, baked potato with butter, and chocolate ice cream

7. 3. The best choice of these meals is salmon with rice and green beans, which is high in protein, and the sourdough bread and ice cream add calories. Halibut, instant coffee, and potatoes are high in potassium, and beets are high in sodium.
NP: Evaluation; CN: Health promotion and maintenance; CNS: Prevention and early detection of disease; CL: Comprehension

You're on a roll! Keep going!

8. On the first day after undergoing a thoracotomy, a client exhibits a temperature of 100° F (37.8° C); heart rate, 96 beats/minute; blood pressure, 136/86 mm Hg; and shallow respirations at 24 breaths/minute with rhonchi at the bases. The client complains of incisional pain. Which of the following nursing actions has priority?
1. Medicate the client for pain.
2. Help the client get out of bed.
3. Give ibuprofen (Motrin) as ordered to reduce the fever.
4. Encourage the client to cough and deep-breathe.

8. 1. Although all the interventions are incorporated in this client's care plan, the priority is to relieve pain and make the client comfortable. This would give the client the energy and stamina to achieve the other objectives.
NP: Data collection; CN: Physiological integrity; CNS: Basic care and comfort; CL: Application

NP: Nursing process   CN: Client needs category   CNS: Client needs subcategory   CL: Cognitive level

9. Which of the following interventions is most important to include in a care plan for a client with atelectasis?
1. Giving oxygen continuously at 3 L/min
2. Coughing and deep-breathing every 4 hours
3. Using the incentive spirometer every hour
4. Getting the client out of bed to a chair every day

9. 3. Incentive spirometry is used to prevent or treat atelectasis. Done every hour, it will produce deep inhalations that help open the collapsed alveoli. Oxygen use doesn't encourage deep inhalation. Coughing and deep breathing is a good intervention, but rarely results in as deep an inspiratory effort as using an incentive spirometer. Getting the client out of bed will also help expand the lungs and stimulate deep breathing, but it's done less frequently than incentive spirometry.
NP: Implementation; CN: Physiological integrity; CNS: Reduction of risk potential; CL: Application

10. Two hours after nasal surgery, the client's nostrils are packed and a drip pad is anchored under his nose. Which of the following assessments alerts the nurse that the surgical site is bleeding?
1. Frequent swallowing
2. Dry mucous membranes
3. Decrease in urinary output
4. Temperature elevation

10. 1. Frequent swallowing is a sign of hemorrhage in nasal surgery. Decreased urine output, dry mucous membranes, and temperature elevation are usually signs of dehydration.
NP: Data collection; CN: Physiological integrity; CNS: Reduction of risk potential; CL: Analysis

11. Which of the following interventions is most important to reduce the risk of disuse osteoporosis in a bedridden client?
1. Turning, coughing, and deep breathing
2. Increasing fluids to 3,000 ml daily
3. Promoting venous return by elevating the legs
4. Providing active and passive range-of-motion (ROM) exercise

11. 4. All the interventions listed are good for a bedridden client. However, active and passive ROM exercises provide the mechanical stresses of weight bearing that are absent and lead to disuse osteoporosis.
NP: Implementation; CN: Health promotion and maintenance; CNS: Prevention and early detection of disease; CL: Application

12. An elderly client on bed rest for a week after a bout of pneumonia is in a negative nitrogen balance. Which of the following complications has highest priority?
1. Constipation
2. Renal calculi
3. Muscle wasting
4. Vitamin B_6 deficiency

12. 3. Negative nitrogen balance leads to muscle wasting. The body breaks down muscle tissue to use as energy. Renal calculi can be a complication of bed rest and demineralization of the bone but treating a negative nitrogen balance takes priority. Constipation and vitamin B_6 deficiency also need to be corrected but aren't of the highest priority.
NP: Implementation; CN: Physiological integrity; CNS: Physiological adaptation; CL: Comprehension

13. The Pao_2 of a client with asthma gives information on which of the following factors?
1. Respiratory status
2. Degree of dyspnea
3. Efficiency of gas exchange
4. Effectiveness of ventilation

13. 3. The Pao_2 reflects the gas exchange ventilation and perfusion. It doesn't measure the degree of dyspnea, effectiveness of ventilation, or respiratory status.
NP: Data collection; CN: Health promotion and maintenance; CNS: Prevention and early detection of disease; CL: Knowledge

14. A client with an acute myocardial infarction is given morphine for which of the following reasons?

1. To decrease cardiac output
2. To increase preload and afterload
3. To increase myocardial oxygen demand
4. To decrease myocardial oxygen demand

14. 4. Morphine will calm and relax the client and decrease respiratory rate, anxiety, and stress, decreasing the energy and oxygen demand of the heart.

NP: Implementation; CN: Physiological integrity; CNS: Pharmacological therapies; CL: Comprehension

15. A client with a wound dehiscence is started on ascorbic acid (vitamin C). Vitamin C is essential to wound healing for which of the following reasons?

1. Vitamin C reduces edema.
2. Vitamin C enhances oxygen transport.
3. Vitamin C enhances protein synthesis.
4. Vitamin C restores the inflammatory process.

15. 3. Vitamin C enhances the synthesis of protein and is essential for the maturation and repair of tissue. It doesn't affect the inflammatory process or oxygen transport or reduce edema.

NP: Implementation; CN: Physiological integrity; CNS: Pharmacological therapies; CL: Comprehension

16. A thin client is sitting up in bed talking on the phone and has a blood pressure of 96/56 mm Hg. Which of the following nursing actions is correct?

1. Increase fluids.
2. Call the physician.
3. Consider this a normal variation.
4. Suspect orthostatic hypotension.

16. 3. A thin client can have a blood pressure as low as 88/46 mm Hg and remain asymptomatic. Calling the physician with this information is inappropriate, as is increasing fluids. Orthostatic hypotension is a decrease in blood pressure and increase in heart rate that occur with a sudden change in position from lying to sitting. This condition could indicate some dehydration, but this client had been sitting up without symptoms for a while.

NP: Data collection; CN: Health promotion and maintenance; CNS: Prevention and early detection of disease; CL: Application

17. Which of the following injuries is least likely to become infected?

1. Contusion of the ankle in an 18-year old client
2. Laceration from glass in a 6-year-old client
3. Stab wound in the leg of a 37-year-old client
4. Cat bite to the left hand of an elderly client

17. 1. A contusion doesn't involve a break in the skin. The other options all involve a break in the skin that could lead to infection.

NP: Data collection; CN: Physiological integrity; CNS: Reduction of risk potential; CL: Comprehension

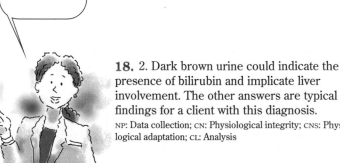

I'm impressed. Keep up the good work!

18. A client is hospitalized for 5 days with mononucleosis. Which of the following findings indicates a possibly serious consequence?

1. Vomiting
2. Dark brown urine
3. Temperature of 101° F (38.3° C)
4. Cervical lymphadenopathy

18. 2. Dark brown urine could indicate the presence of bilirubin and implicate liver involvement. The other answers are typical findings for a client with this diagnosis.

NP: Data collection; CN: Physiological integrity; CNS: Physiological adaptation; CL: Analysis

NP: Nursing process   CN: Client needs category   CNS: Client needs subcategory   CL: Cognitive level

19. A nurse is teaching a client about lifestyle changes that need to be made after a myocardial infarction (MI). The diagnosis of *Ineffective coping* is supported when the client is observed in which of the following actions?
 1. Reading a book about meal planning
 2. Pacing the floor of his room on occasion
 3. Sitting quietly in his room for a short time
 4. Telling his family he didn't have an MI

20. A client with a history of myasthenia gravis is admitted to the emergency department with complaints of respiratory distress. The client's condition worsens and arterial blood gases are drawn. Which of the following conditions is expected?
 1. Metabolic acidosis
 2. Metabolic alkalosis
 3. Respiratory acidosis
 4. Respiratory alkalosis

21. Which of the following foods is considered part of a full liquid diet?
 1. Gelatin dessert
 2. Milkshake
 3. Ice pops
 4. Tea

22. Which of the following conditions causes heart failure after a myocardial infarction (MI)?
 1. Increased workload of the heart
 2. Increased oxygen demands of the heart
 3. Inability of the heart chambers to adequately fill
 4. Impairment of contractile function of the damaged myocardium

19. 4. When the client tells his family he didn't have an MI, he's showing the defense mechanism of denial. Reading a book on meal planning is a positive intervention. Pacing the floor on occasion is a form of anxiety that's normal for the client to experience. Sitting quietly is a normal behavior. The client needs time to come to terms with his diagnosis.
NP: Evaluation; CN: Psychosocial integrity; CNS: Coping and adaptation; CL: Comprehension

20. 3. The client has a restrictive lung problem because of myasthenia gravis. This is aggravated by respiratory distress. Because of the restrictive problem, the client won't be able to exhale efficiently and carbon dioxide will build up, causing respiratory acidosis.
NP: Data collection; CN: Physiological integrity; CNS: Physiological adaptation; CL: Analysis

21. 2. Full liquid diets contain milk, cereal, gruel, clear liquids, and plain frozen desserts. The clear liquid diet contains only foods that are clear and liquid at room or body temperature, such as gelatin, fat-free broth, bouillon, ice pops and similar frozen desserts, tea, and regular or caffeinated coffee.
NP: Planning; CN: Physiological integrity; CNS: Basic care and comfort; CL: Knowledge

22. 4. After an MI, the injured myocardium is replaced by scar tissue. This scar tissue causes the ventricle to pump less efficiently. After an MI has resolved, oxygen and workload demands should normalize and the heart's chambers should fill adequately.
NP: Data collection; CN: Physiological integrity; CNS: Physiological adaptation; CL: Comprehension

23. A client is admitted to the emergency department with severe epistaxis. The physician inserts posterior packing. Later, the client is anxious and says he doesn't feel he's breathing right. Which of the following nursing actions is appropriate?
1. Cutting the packing strings and removing the packing
2. Reassuring the client that what he's experiencing is normal
3. Asking the client to fully explain what he means by "right"
4. Using a flashlight and inspecting the posterior oral cavity of the client

23. 4. The nurse must assess the patency of the airway. The packing might have become dislodged. The nurse shouldn't remove the packing or give the client false reassurance. The client is too anxious to explain what he means.

NP: Implementation; CN: Physiological integrity; CNS: Reduction of risk potential; CL: Analysis

24. While performing nasopharyngeal suction, a nurse hears a client's pulse oximeter alarm. The pulse oximeter indicates the client's oxygen saturation reading is 86%. Which of the following actions should the nurse take?
1. Stop suctioning and give oxygen to the client.
2. Withdraw the suction catheter and tell the client to cough several times.
3. Continue suctioning for 10 to 15 more seconds and then withdraw the suction catheter.
4. Keep the suction catheter inserted and wait a few seconds before beginning suctioning.

24. 1. The pulse oximeter reading indicates the client isn't oxygenating well. The normal range for oxygen saturation is 90% to 100%. Suctioning draws air as well as secretions from the lungs, reducing oxygen saturation in the blood. The nurse must stop suctioning and give oxygen to increase the saturation. Withdrawing the suction catheter will stop the removal of oxygen, but coughing will delay an increase in saturation. The suction catheter occupies space in the airway, making it harder for the client to breathe when it's left in place. Further suctioning will reduce the oxygen level even further.

NP: Implementation; CN: Physiological integrity; CNS: Physiological adaptation; CL: Application

25. Which of the following actions correctly explains the purpose of diaphragmatic breathing exercises for a client with chronic obstructive pulmonary disease (COPD)?
1. Dilate the bronchioles.
2. Decrease vital capacity.
3. Increase residual volume.
4. Decrease alveolar ventilation.

25. 1. In COPD, the bronchioles constrict during exhalation due to pressure changes in the lungs. Diaphragmatic breathing exercises keep the bronchioles open during exhalation. These exercises aren't performed for the other reasons stated.

NP: Data collection; CN: Physiological integrity; CNS: Reduction of risk potential; CL: Comprehension

26. A client with chronic obstructive pulmonary disease (COPD) is being discharged from the hospital. The nurse provided teaching on medications, diet, and exercise. Which of the following statements by the client indicates more teaching is needed?
1. "I'll eat six small meals per day."
2. "I'll get a flu shot every winter."
3. "I'll walk every morning before breakfast."
4. "I'll call my physician if I get cold symptoms."

26. 3. The worst time of the day for a client with COPD is morning. Exercise is important, but should be done later in the day. All other choices are appropriate for the client with COPD.

NP: Evaluation; CN: Physiological integrity; CNS: Basic care and comfort; CL: Comprehension

NP: Nursing process   CN: Client needs category   CNS: Client needs subcategory   CL: Cognitive level

27. A client is showing symptoms of bronchial obstruction. Which of the following findings is expected?
1. Hacking cough
2. Diminished breath sounds
3. Production of rust-colored sputum
4. Decreased use of accessory muscles

28. A client has just started treatment for tuberculosis. Rifampin is the drug ordered by the physician. Which of the following statements indicates the client has a good understanding of his medication?
1. "I won't go to family gatherings for 6 months."
2. "My urine will look orange because of the medication."
3. "Now I don't need to cover my mouth or nose when I sneeze or cough."
4. "I told my wife to throw away all the spoons and forks before I come home."

29. Which of the following actions should be initiated before feeding a client with Parkinson's disease?
1. Sit the client upright.
2. Have suction available.
3. Order a clear liquid diet.
4. Have a speech therapist evaluate the client.

30. Management of a pregnant client with cardiovascular disease focuses on which of the following treatments?
1. Rest
2. Hospitalization
3. Therapeutic abortion
4. Continuous cardiac monitoring

27. 2. Bronchial obstruction means no passage of air through the bronchi, so diminished or no breath sounds would be heard. A hacking cough is usually associated with upper respiratory infection and dryness in the upper airways. Rust-colored sputum is a sign of pneumococcal pneumonia. There would be increased use of accessory muscles.
NP: Data collection; CN: Physiological integrity; CNS: Physiological adaptation; CL: Knowledge

28. 2. Rifampin discolors body fluids, such as urine and tears. The client can go to family functions and eat with normal utensils. The client should cover his mouth and nose when coughing and sneezing until he has been on the medication for at least 2 weeks.
NP: Implementation; CN: Physiological integrity; CNS: Pharmacological therapies; CL: Application

29. 4. A speech therapist can evaluate the client's swallowing and make recommendations before the client is fed. Aspiration due to involuntary movement is common. Sitting the client upright and having suction available are helpful when feeding the client, but evaluation of the client's swallowing ability should come first. Clear liquids may be too difficult for the client; semisoft foods may be easier to swallow.
NP: Implementation; CN: Physiological integrity; CNS: Reduction of risk potential; CL: Knowledge

30. 1. The goal of antepartum management is to prevent complications and minimize the strain on the client. This is done with rest. Hospitalization may be required in older women or those with previous decompensation. Therapeutic abortion is considered in severe dysfunction, especially in the first trimester. Continuous cardiac monitoring isn't necessary.
NP: Implementation; CN: Physiological integrity; CNS: Reduction of risk potential; CL: Knowledge

31. Which of the following findings most likely indicates a urinary tract infection (UTI) in a 5-year-old child?

1. Incontinence
2. Lack of thirst
3. Concentrated urine
4. Subnormal temperature

31. 1. Incontinence in a toilet-trained child is associated with UTI. Subnormal temperature isn't a sign of UTI. Concentrated urine is a sign of dehydration. Lack of thirst wouldn't be expected in a child with UTI.

NP: Data collection; CN: Health promotion and maintenance; CNS: Prevention and early detection of disease; CL: Application

32. A client has two prescriptions for fluid retention. One prescription reads "Lasix, 40 mg, one tablet daily." The other one reads "Furosemide, 40 mg, one tablet daily." Which of the following instructions is given to the client?

1. Take both medications as ordered.
2. Lasix and furosemide are the same drug.
3. Use Lasix one day and furosemide the next day.
4. Throw away one of the drugs to not confuse the client.

32. 2. Using generic names for medications is common. It's the responsibility of the nurse to teach the client both brand and generic names of drugs. Setting up medications in a medication tray, using only one pharmacy to dispense medications, and using all medications until the bottle is emptied will reduce medication errors.

NP: Implementation; CN: Physiological integrity; CNS: Pharmacological therapies; CL: Analysis

33. A teacher tells the nurse that a preadolescent Vietnamese girl who is attending a new school in an affluent district sits in the back of the class and won't speak when spoken to, although her parents confirmed that the girl speaks English. Which of the following findings is most likely?

1. The student is experiencing cultural shock.
2. The student is developing a peer support system.
3. The student is going through a socialization period.
4. The student is becoming acculturated to the new school.

You're doing sensationally! I mean, I should do so well, eh? My, my, my.

33. 1. Cultural shock is a feeling of helplessness, discomfort, and a state of disorientation when an outsider attempts to comprehend or adapt to a new cultural situation. Acculturation occurs when there is a blending of cultural or ethnic backgrounds. This process takes time to develop. Peer groups usually develop based on the background, interests, and capabilities of its members. Developing peer cultures is part of the socialization process.

NP: Data collection; CN: Health promotion and maintenance; CNS: Growth and development through the life span; CL: Application

34. A school nurse is screening for hearing and vision at a local middle school. Which of the following techniques is used to communicate effectively with this age group?

1. Give undivided attention to each student.
2. Have the parents present during the screening.
3. Have several adolescents listen to each other's health histories.
4. Use puppets or dolls to show how the screening is going to take place.

34. 1. Give undivided attention to communicate effectively with adolescents. Respect their privacy. The presence of parents and use of puppets or dolls can be used to effectively communicate with younger children.

NP: Data collection; CN: Health promotion and maintenance; CNS: Growth and development through the life span; CL: Application

---

NP: Nursing process   CN: Client needs category   CNS: Client needs subcategory   CL: Cognitive level

35. An 18-month-old infant is screened for developmental problems. Which of the following screening tests would the nurse expect to be used?

1. Goodenough Draw-a-Person Test
2. Denver Developmental Screening Test (DDST)
3. McCarthy Scales of Children's Abilities (MSCA)
4. Preschool readiness screening scales

36. In preparing educational intervention for college students, the nurse understands that drinking alcoholic beverages may be a behavior commonly associated with the relief of which of the following problems?

1. Fatigue
2. Anxiety
3. Headache
4. Stomach pain

37. An educational program about relaxation techniques is provided for college students preparing for their final exams. Which of the following relaxation techniques is used to counteract anxiety?

1. Meditation
2. Music therapy
3. Dance therapy
4. Reality orientation

38. During the interview of a parent of a 12-year-old child, a teacher discovers the child has a history of conduct disorder. Which of the following behaviors is characteristic of this disorder?

1. Ability to play well with peers
2. Absence of aggressive behavior
3. Obsession with making fires
4. Competitiveness in sports

35. 2. The DDST is applicable for children from birth through age 6. Preschool readiness screening scales are designed for screening 5-year-old children's readiness for school. The Goodenough Draw-a-Person test is used to assess intellectual ability in children ages 3 to 10. The MSCA is a developmental tool for children ages 2⅓ to 8½.

NP: Planning; CN: Health promotion and maintenance; CNS: Growth and development through the life span; CL: Application

36. 2. Drinking alcoholic beverages is commonly thought to alleviate anxiety. These beverages aren't commonly used to relieve fatigue, headache, or stomach pain.

NP: Planning; CN: Psychosocial integrity; CNS: Coping and adaptation; CL: Application

37. 1. Meditation is a relaxation therapy used to counteract anxiety related to stress-inducing internal and external stimuli. Music therapy, dance therapy, and reality orientation are used as adjuncts to psychiatric care.

NP: Planning; CN: Psychosocial integrity; CNS: Coping and adaptation; CL: Application

38. 3. Adolescents with conduct disorder commonly become obsessed with making fires. They also have difficulty establishing relationships. These children aren't just competitive, but physically aggressive in sports, violating the rights of others.

NP: Data collection; CN: Psychosocial integrity; CNS: Psychosocial adaptation; CL: Comprehension

39. A 40-year-old client is admitted to the local women's shelter after being raped by her estranged husband. The client describes the traumatic event. Which of the following responses by the nurse is best?
1. Change the subject to prevent the client from crying.
2. Listen attentively while the client describes the event.
3. Arrange for the client to tell her story in group therapy.
4. Medicate the client with a tranquilizer to prevent hysteria.

39. 2. Retelling the event is part of the healing process. Giving medication and changing the subject don't allow the client to integrate the experience into her life. Group therapy may be helpful, but the best nursing response is to listen and convey empathy.
NP: Implementation; CN: Psychosocial integrity; CNS: Psychosocial adaptation; CL: Application

40. The client with bipolar disorder tells the nurse that her family physician prescribed lithium. Which of the following symptoms indicates the client is developing lithium toxicity?
1. Lethargy
2. Hypertension
3. Hyperexcitability
4. Low urine output

40. 1. Nausea, vomiting, diarrhea, thirst, polyuria, lethargy, slurred speech, hypotension, muscle weakness, and fine hand tremors are signs of lithium toxicity.
NP: Data collection; CN: Physiological integrity; CNS: Pharmacological therapies; CL: Application

41. Which of the following interventions is used when examining a pediatric client?
1. Ask the parents to leave the room during the examination.
2. Position the client on an examination table or bed at all times.
3. Organize the examination in the same way for every infant or child.
4. Identify the source (child, parent, caregiver, guardian) and indicate the reliability of the information obtained.

You're making great progress! Keep going!

41. 4. Document the source of information and its reliability. Depending on the child's age, parents may help position and hold the child, facilitating examination. Separation from the parent may cause anxiety and increase the child's fear and distrust. Organization of the examination is changed to accommodate the individual child's age and development.
NP: Data collection; CN: Health promotion and maintenance; CNS: Growth and development through the life span; CL: Comprehension

42. Which of the following interventions is used when examining an elderly client?
1. Ask the client to change positions quickly.
2. Keep the room temperature cool during health assessment.
3. Speak loudly and quickly to facilitate understanding of directions.
4. Change the height of the examination table or modify the client's position.

42. 4. Physiologically, an older client is prone to falls and dizziness due to their decreased ability to respond to sudden movements and position changes. The room temperature should be warm because older clients become hypothermic easily. Speak in a slow, normal tone of voice to facilitate communication.
NP: Data collection; CN: Health promotion and maintenance; CNS: Growth and development through the life span; CL: Comprehension

NP: Nursing process   CN: Client needs category   CNS: Client needs subcategory   CL: Cognitive level

43. Which of the following nursing diagnoses is appropriate for a client with chronic obstructive pulmonary disease who's anxious, dyspneic, and hypoxic?
1. Ineffective breathing pattern related to anxiety
2. Risk for aspiration related to absence of protective mechanisms
3. Impaired gas exchange related to altered oxygen-carrying capacity of the blood
4. Ineffective airway clearance related to presence of tracheobronchial obstruction or secretions

44. Which of the following statements is a wellness nursing diagnosis?
1. Readiness for enhanced spiritual well-being
2. Risk for activity intolerance related to prolonged bed rest
3. Grooming self-care deficit related to fatigue and muscle weakness
4. Constipation related to decreased activity and fluid intake as manifested by hard, formed stool every 3 days

45. Which of the following definitions is correct for collaborative problems?
1. Unusual or unexpected human responses to pharmacological agents
2. Pathophysiological responses of body organs or systems for which physicians have ultimate responsibility
3. Human responses for which registered nurses are capable of intervening legally and independently of physicians
4. Certain physiological responses or complications of body organs or systems for which nurses intervene in association with physicians and other disciplines

46. An intake nurse at a local mental health facility is admitting a client with psychosis. Which of the following techniques is most valuable to use when planning this client's care?
1. Rorschach test
2. Interview with the client
3. Mental Status Examination (MSE)
4. Review old records of the client

43. 3. The correct nursing diagnosis for this client is based on the impaired oxygenation at the cellular level. The first option applies to a client whose inhalation or exhalation pattern doesn't enable adequate pulmonary inflation or emptying. The second option applies if the client is at risk for aspirating gastric or pharyngeal secretions, food, or fluids into the tracheobronchial passages. The last option is appropriate for a client who is unable to clear secretions or obstructions from the respiratory tract.
NP: Planning; CN: Physiological integrity; CNS: Reduction of risk potential; CL: Application

44. 1. Wellness diagnoses are one-part statements containing the label only and begin with "Readiness for enhanced," followed by the higher level of wellness desired for the individual or group. The second option is a "risk for" nursing diagnosis. The third option describes an existing problem for which additional data are needed for confirmation. The last option describes a manifested health problem validated by identifiable major defining characteristics.
NP: Planning; CN: Psychosocial integrity; CNS: Psychosocial adaptation; CL: Application

45. 4. This option describes a collaborative problem that requires physician and nurse prescribed actions. The first option describes an adverse reaction to a pharmacological agent. The second option describes the focus of a medical diagnosis. The third option describes a nursing diagnosis; interventions for nursing diagnoses fall within the scope of practice of registered nurses.
NP: Planning; CN: Safe, effective care environment; CNS: Coordinated care; CL: Comprehension

46. 3. The MSE is a basis for planning care with a mental health client, especially one who is psychotic. An interview with a client with psychosis would be unreliable. The Rorschach test is used for depression. Review of old records won't assess the current state on which interventions are planned.
NP: Planning; CN: Psychosocial integrity; CNS: Psychosocial adaptation; CL: Comprehension

47. Which of the following actions best describes the planning step of the nursing process?

1. Collecting client health data
2. Implementing the interventions identified in the care plan
3. Evaluating the client's progress toward attainment of the outcomes
4. Identifying the expected goals or outcomes individualized for the client and family

47. 4. Client goals or outcomes are identified in the planning step of the nursing process. Health data are collected in the assessment step of the nursing process. Planned interventions are done in the implementation step of the nursing process. A client's achievement of outcomes or goals is determined in the evaluation step of the nursing process.

NP: Planning; CN: Safe, effective care environment; CNS: Coordinated care; CL: Knowledge

48. Which of the following client outcomes or goals is correct for a client with the nursing diagnosis *Risk for disuse syndrome*?

1. The client will be free of musculoskeletal complications.
2. The client will experience shorter periods of immobility and inactivity.
3. The nurse will stress the importance of maintaining adequate fluid intake.
4. The nurse will provide holistic care by collaborating with the health care team.

48. 2. This is an appropriate outcome for a client with this nursing diagnosis. Disuse syndrome, a result of prolonged or unavoidable immobility or inactivity, can be prevented. Musculoskeletal complications indicate actual disuse or complications of immobility. The last two options describe a nursing goal, not a client outcome.

NP: Planning; CN: Physiological integrity; CNS: Reduction of risk potential; CL: Application

49. A 42-year-old client who underwent a right modified mastectomy with insertion of a Hemovac drain will be hospitalized overnight because of minor complications. Which of the following goal statements is correct?

1. Teach proper care of the incision site and drain by October 12.
2. The client will know how to care for the incision site and drain by October 12.
3. The client will show the proper care of the incision site and drain by October 12.
4. The client will care for the incision site and contend with psychological loss by October 12.

You're doing well! Keep at it!

49. 3. This statement contains a specific measurable verb, clearly identifies the client behavior, and includes a date. The first option is a nursing goal as written, not a client-centered goal. The client goal of the second option isn't measurable as stated. The fourth option includes two goals that need to be addressed separately under the appropriate nursing diagnosis and it contains nonmeasurable verbs.

NP: Planning; CN: Physiological integrity; CNS: Basic care and comfort; CL: Application

50. Which of the following expected outcomes or goals is correct for a client with the nursing diagnosis *Risk for injury* related to lack of awareness of environmental hazards?

1. Encourage the client to discuss safety rules with children.
2. Help the client learn safety precautions to take in the home.
3. The client will eliminate safety hazards in their surroundings.
4. Refer the client to community resources for more information.

50. 3. This goal is appropriate and measurable as written and focuses on the client. The other options are nursing interventions as written.

NP: Planning; CN: Health promotion and maintenance; CNS: Prevention and early detection of disease; CL: Application

NP: Nursing process   CN: Client needs category   CNS: Client needs subcategory   CL: Cognitive level

51. Which of the following activities is performed in the planning step of the nursing process?
1. Collection, validation, and organization of data
2. Establishment of priorities, client goals, and outcomes
3. Setting priorities and nursing interventions and recording actions
4. Problem identification and synthesis from data collected

52. A client with heart failure is given furosemide (Lasix) 40 mg I.V. daily. The morning serum potassium level is 2.8 mEq/L. Which of the following actions is the most appropriate?
1. Question the physician about the dosage.
2. Give 20 mg of the ordered dose and recheck the laboratory test results.
3. Notify the physician, repeat the potassium as ordered, then give the furosemide.
4. Give the furosemide and get an order for sodium polystyrene sulfonate.

53. Which of the following clients is at greatest risk for developing respiratory alkalosis?
1. A client in labor
2. A client with diabetes
3. A client with renal failure
4. An immediate postoperative client

54. Which of the following conditions indicates that a sterile field has been contaminated?
1. Sterile objects are held above the waist of the nurse.
2. Sterile packages are opened with the first edge away from the nurse.
3. The outer inch of the sterile towel hangs over the side of the table.
4. Wetness on the sterile cloth on top of the nonsterile table has been noted.

51. 2. Establishing priorities, client goals, and expected outcomes are activities of the planning step of the nursing process. The first and fourth options describe activities of the assessment step of the nursing process. The third option describes activities of the implementation step of the nursing process.
NP: Planning; CN: Safe, effective care environment; CNS: Coordinated care; CL: Comprehension

52. 3. Furosemide is a diuretic. As water is lost, so is potassium. Diuresis is a treatment for heart failure. Notifying the physician of the low potassium level and getting an order for potassium chloride is the appropriate action before giving the furosemide. Furosemide, 40 mg, is an appropriate dose for the treatment of heart failure. The nurse shouldn't give half the dose without an order. Giving furosemide and sodium polystyrene sulfonate together would further lower the potassium level.
NP: Implementation; CN: Physiological integrity; CNS: Pharmacological therapies; CL: Application

53. 1. A client's respirations at certain stages of labor increase in volume, causing the $Paco_2$ to decrease, increasing the pH. Diabetes often causes a metabolic imbalance, resulting in metabolic acidosis. The respirations of a postoperative client are usually shallow after anesthesia and, because of pain, often cause respiratory acidosis. In renal failure, the inability of the kidneys to eliminate wastes increases the risk of developing metabolic acidosis.
NP: Data collection; CN: Physiological integrity; CNS: Physiological adaptation; CL: Application

54. 4. Moisture outside the sterile package and field contaminates it because fluid can be wicked into the sterile field. The outer inch of the drape is considered contaminated but doesn't indicate that the sterile field itself has been contaminated. Bacteria tend to settle, so there is less contamination above waist level and away from the nurse.
NP: Data collection; CN: Safe, effective care environment; CNS: Safety and infection control; CL: Knowledge

55. Which of the following interventions is helpful for a client to retain carbon dioxide?
1. Having the client breathe into a paper bag
2. Giving one ampule of bicarbonate as ordered
3. Giving oxygen at 3 L/minute through a nasal cannula
4. Repositioning the client and giving 75 mg of meperidine (Demerol) I.M. for pain

55. 1. By breathing into a paper bag, the client will rebreathe some of his own exhaled carbon dioxide and increase the carbon dioxide in his blood, which will correct his respiratory alkalosis. Giving oxygen won't increase the carbon dioxide to correct the imbalance. Giving one ampule of bicarbonate will worsen the alkalosis. Repositioning and giving Demerol are appropriate interventions for pain that may cause hyperventilation leading to respiratory alkalosis, but these interventions don't directly correct the problem.

NP: Implementation; CN: Physiological integrity; CNS: Physiological adaptation; CL: Comprehension

56. A nurse is checking the results of urine studies done on a client with dehydration. Which of the following results reflects an alteration in fluid balance?
1. Protein levels
2. Ketone levels
3. Specific gravity levels
4. Culture and sensitivity levels

56. 3. Specific gravity measures the concentration of urine. A dehydrated client will concentrate urine and decrease urine output to prevent further dehydration. Culture results indicate the presence or absence of an infection and aren't affected by fluid status. Protein may or may not be present in the urine and doesn't fluctuate with changes in fluid status. Ketones aren't present in the urine, and these levels don't fluctuate with fluid status.

NP: Data collection; CN: Physiological integrity; CN: Physiological adaptation; CL: Knowledge

57. A client admitted with hypoparathyroidism is being monitored for hypocalcemia. Which of the following signs is used to check for hypocalcemia?
1. Battle's sign
2. Brudzinski's sign
3. Chvostek's sign
4. Homan's sign

57. 3. Hypocalcemia can cause Chvostek's sign, abnormal facial muscle and nerve spasms elicited when the facial nerve is tapped. A positive Homans' sign indicates deep vein thrombosis. Brudzinski's sign is the flexion of the hips and knees in response to flexion of the head and neck toward the chest, indicating meningeal irritation. Battle's sign is bruising over the temporal bone in the presence of a basilar skull fracture.

NP: Data collection; CN: Physiological integrity; CNS: Reduction of risk potential; CL: Application

58. A client is complaining of pain 1 day after a colostomy. The nurse gives meperidine (Demerol) I.M. and, 30 minutes later, finds the respiratory rate at 8 breaths/minute, with the nasal cannula on the floor. Arterial blood gas (ABG) results are pH, 7.23; Pao_2, 58 mm Hg; $Paco_2$, 61 mm Hg; HCO_3^-, 24 mEq/L. Which of the following groups of factors contributes most to this client's ABG results?
1. Colostomy, pain, and Demerol
2. Meperidine, the nasal cannula on the floor, and the colostomy
3. Meperidine, respiratory rate of 8 breaths/minute, and the nasal cannula on the floor
4. Pain, respiratory rate of 8 breaths/minute, and the nasal cannula on the floor

59. Which of the following arterial blood gas (ABG) results is typical for a client with emphysema?
1. pH, 7.52; $Paco_2$, 18 mm Hg; HCO_3^-, 22 mEq/L
2. pH, 7.50; $Paco_2$, 38 mm Hg; HCO_3^-, 38 mEq/L
3. pH, 7.30; $Paco_2$, 52 mm Hg; HCO_3^-, 30 mEq/L
4. pH, 7.30; $Paco_2$, 40 mm Hg; HCO_3^-, 18 mEq/L

60. Which of the following factors is a major cause of metabolic alkalosis in a client who had a colon resection?
1. Hyperventilation
2. Pain management
3. Nasogastric suction
4. I.V. therapy

61. Which of the following reasons is correct for using a paper towel to turn off the faucet after hand washing?
1. To clean the faucet after use
2. To maintain sterility of the washed hands
3. To prevent contamination of the faucet handle
4. To prevent transmission of microorganisms from the faucet handle to the hands

You'll be finished soon! Hang in there!

58. 3. This client has respiratory acidosis. Narcotics can suppress respirations, causing retention of carbon dioxide. A Pao_2 of 58 mm Hg indicates hypoxemia, which is caused by the removal of the client's supplementary oxygen and the decreased respiratory rate. Pain increases — not decreases — the respiratory rate, which causes a decrease in $Paco_2$. Colostomy drainage doesn't start until 2 to 3 days postoperatively, and this drainage would contribute to metabolic alkalosis.
NP: Data collection; CN: Physiological integrity; CNS: Physiological adaptation; CL: Analysis

59. 3. Clients with emphysema retain carbon dioxide due to air trapping, causing an elevated $Paco_2$ and respiratory acidosis. Because emphysema is a chronic disease, the kidneys compensate over time for the increased $Paco_2$ by retaining HCO_3^-, thus attempting to normalize the pH. The other ABG results aren't consistent with results found in a client with emphysema.
NP: Data collection; CN: Physiological integrity; CNS: Physiological adaptation; CL: Comprehension

60. 3. Removing acidic gastric secretions from the stomach is a metabolic cause of alkalinization of the blood pH. Pain management may further decrease the respiratory rate. Most I.V. fluids don't influence pH. Hyperventilation decreases carbon dioxide and increases the pH, causing respiratory alkalosis.
NP: Data collection; CN: Physiological integrity; CNS: Physiological adaptation; CL: Analysis

61. 4. The faucet has microorganisms on it put there by all the "dirty" hands that touched it when turning it on. Although a paper towel isn't sterile, it's clean and protects the hands from becoming contaminated from the faucet after washing. Hand washing doesn't make the hands sterile. Wiping the faucet after use may make it look cleaner but isn't the reason for using a paper towel to turn off the faucet.
NP: Implementation; CN: Safe, effective care environment; CNS: Safety and infection control; CL: Comprehension

62. The nurse delegates to an aide the task of making an occupied bed for a 69-year-old client with dementia. Which of the following actions would the nurse stress to the aide as the priority when the task is completed?

1. Placing the bed in the lowest position
2. Putting the call light within the client's reach
3. Raising the top side rails to the upright position
4. Discarding soiled linen in a hamper or biohazard bag

62. 1. To reduce the risk of injury due to falls, the bed should be placed in the lowest position. The call button should be in reach of the client, but the immediate safety of the client comes first. All four side rails should be up to prevent accidental falls and to remind clients to stay in bed. Soiled linens should be placed in a hamper or biohazard bag, but client safety is a priority.

NP: Implementation; CN: Safe, effective care environment; CNS: Coordinated care; CL: Comprehension

63. After emptying urine from the bedpan of a client whose urinary output is being monitored, which of the following steps is done next?

1. Wash hands thoroughly.
2. Apply a clean pair of gloves.
3. Report the amount of urine to the nurse in charge right away.
4. Document the amount and characteristics of urine in the chart.

63. 1. After any procedure is completed, the nurse must wash her hands to prevent transmission of microorganisms. The application of gloves is only necessary if the nurse must attend to another item of personal care before documenting urinary output; even so, hands should be washed first. Crucial information is reported to the charge nurse, not routine intake and output.

NP: Implementation; CN: Safe, effective care environment; CNS: Safety and infection control; CL: Knowledge

64. Which of the following clients should be placed in an orthopneic position?

1. A client with edema of the lower legs and ankles
2. A client with a pressure ulcer on the coccyx and buttocks
3. An immobilized client with calf tenderness due to a thrombus
4. An elderly client with difficulty breathing unless in a sitting position

64. 4. The orthopneic position is a sitting position with the arms leaning on a bedside table. Sitting with the legs elevated to decrease edema is appropriate for clients with ankle and lower leg swelling. A client with a pressure ulcer will need to be positioned on his side and turned every 2 hours. Fowler's or semi-Fowler's positions are most appropriate for a client on complete bed rest.

NP: Planning; CN: Physiological integrity; CNS: Physiological adaptation; CL: Application

65. Two team members are helping a client transfer from the bed to a wheelchair. As the client rises from a supine to a sitting position, he complains of feeling light-headed and dizzy. Which of the following actions should the team members take next?

1. Lift the client quickly into the wheelchair.
2. Return the client to the supine position and apply a safety vest.
3. Ask the client to dangle at the bedside while leaving the room for a few seconds to get additional assistance.
4. Have the client sit at the side of the bed for a few minutes while supporting his back and shoulders.

65. 4. A quick change in position will decrease blood pressure, causing momentary light-headedness and dizziness. An additional change in position may further reduce the client's blood pressure to a level that may require emergency assistance. This can be avoided by waiting with the client in the sitting position until the blood pressure stabilizes. Leaving the room may put the client in danger if the blood pressure decreases further and the client needs emergency assistance. If the client continues to complain of dizziness and light-headedness, then return him to bed. A safety vest isn't necessary.

NP: Implementation; CN: Safe, effective care environment; CNS: Coordinated care; CL: Application

NP: Nursing process   CN: Client needs category   CNS: Client needs subcategory   CL: Cognitive level

66. Which of the following actions is the most effective infection-control measure for preventing the transmission of microorganisms?
1. Changing a client's bed linen daily
2. Washing hands before and after client contact
3. Wearing sterile gloves when touching a client's skin
4. Wearing a mask when in direct contact with infected clients

67. Which of the following actions represents a correct implementation of standard precautions?
1. Wearing eye goggles while giving a complete bed bath
2. Recapping a needle used for an injection before disposal
3. Disposing of blood-contaminated materials in a biohazard container
4. Using alcohol to decontaminate blood-contaminated steel instruments

68. A team leader would advise team members to wear a mask and protective eyewear or a face shield in which of the following situations?
1. When strong odors are emitted from an infected wound
2. When the client has an oral temperature greater than 101° F (38.3° C)
3. If needles or other sharp instruments are to be used in the procedure
4. During a procedure where splashing of blood or body fluid is anticipated

69. A nurse is caring for a client on neutropenic precautions. At which of the following times should the nurse remove the barrier protection when leaving the room?
1. Within the client's room, just inside the doorway
2. Out of the client's room, just outside the doorway
3. In the hallway, a significant distance from the client's room
4. At the bedside, immediately after completing work with the client

66. 2. Most often, the transmission of microorganisms occurs when health care personnel don't wash their hands before and after touching a client or contaminated object. A daily linen change isn't the most effective method of controlling infection. Sterile gloves and a mask aren't needed during routine client care.
NP: Implementation; CN: Safe, effective care environment; CNS: Safety and infection control; CL: Comprehension

67. 3. Blood-contaminated materials are disposed of in a biohazard container. Recapping needles puts the health care provider at risk for sticking herself. Standard precautions need not be observed during a bath because of the low risk for exposure to blood. Blood-contaminated steel instruments are decontaminated in an autoclave.
NP: Implementation; CN: Safe, effective care environment; CNS: Safety and infection control; CL: Comprehension

68. 4. Wearing eye goggles or face shields prevents blood or body-fluid splashes into the eyes. Odors don't transmit microorganisms. The use of needles or other sharp instruments doesn't mandate eye protection. A client with a fever won't transmit microorganisms into the eyes any more frequently than a client without a fever.
NP: Implementation; CN: Safe, effective care environment; CNS: Coordinated care; CL: Comprehension

69. 2. Disposing of gowns and gloves just outside the doorway provides sufficient distance from the client for all but airborne microorganisms. The client is protected from infection by airborne pathogens by keeping the door shut as much as possible to decrease the chance of exposure. Disposal of barriers at the bedside or inside the door negates the effectiveness of wearing barriers in the first place. It's unnecessary to wear the barriers away from the doorway as the door should remain closed.
NP: Planning; CN: Safe, effective care environment; CNS: Safety and infection control; CL: Knowledge

70. Which of the following time frames is most appropriate for completing client teaching for a client undergoing an open cholecystectomy?
1. The day of discharge
2. A few weeks before the surgery
3. The first 12 hours after surgery
4. Before discharge, 1 to 2 days after the surgery

70. 4. Pain levels should have sufficiently subsided 1 to 2 days after the surgical procedure, allowing the client to concentrate on the information. A few weeks before surgery is generally too early to retain information, and teaching within the first 12 hours after surgery isn't likely to produce retention of information, either. The day of discharge is too late, because it doesn't give the client time to ask questions or practice procedures (such as syringe preparation) that may be necessary. Also, the individual may be anxious about returning home, which may interfere with learning.

NP: Planning; CN: Physiological integrity; CNS: Basic care and comfort; CL: Application

You're nearing the finish! Only 14 questions left!

71. For which of the following individuals is the use of an oral thermometer contraindicated?
1. A school-age child
2. An unconscious client
3. An alert, oriented client
4. An elderly client in no acute distress

71. 2. Oral thermometers can be broken by an unconscious client. Only a client who can follow directions to hold the thermometer in his mouth should be allowed to do so. The clients in the other three options should have no difficulty.

NP: Implementation; CN: Safe, effective care environment; CNS: Safety and infection control; CL: Knowledge

72. Which of the following techniques is correct for tracheostomy care of a client at home?
1. Septic
2. Antiseptic
3. Medically aseptic
4. Surgically aseptic

72. 3. At home, the procedure of tracheostomy care is clean or medically aseptic because of the cost and difficulty of performing a completely sterile procedure. Septic means infected or dirty. An antiseptic is a solution used to prevent infection. Surgical asepsis can only realistically be practiced in the hospital.

NP: Implementation; CN: Safe, effective care environment; CNS: Safety and infection control; CL: Comprehension

73. A 46-year-old single mother was concerned about her 15-year-old son's behavior. He suddenly decided his mother shouldn't date or have men in the house. He told his mother he was the "man of the house." Which of the following disturbances was occurring in the internal dynamics of the family?
1. Age-appropriate behavior is occurring.
2. The son is powerful in the family system.
3. The son is trying to establish a role reversal.
4. It's culturally acceptable to be the man of the house at age 15.

73. 3. Role reversal occurs when the patterns of expected behavior aren't appropriate to age and ability. Males ages 13 to 17 are developing their identities, and separation from parents becomes necessary for individuation to occur. Males have a better understanding of their roles in relationships and families if they're raised around strong male role models. In healthy families, power is shared appropriate to age until the children are independent.

NP: Data collection; CN: Psychosocial integrity; CNS: Coping and adaptation; CL: Analysis

74. Which of the following behaviors in a preschooler is a cause for concern?
   1. Has nightmares
   2. Cries and holds tightly to parents
   3. Takes toys away from other children
   4. Sits quietly and doesn't participate in play activities

75. Which of the following techniques is appropriate for promoting proper breathing in a client experiencing pain or anxiety?
   1. Rapid, light respirations
   2. Rapid, deep respirations
   3. In through the mouth and out through the nose
   4. In through the nose and out through the mouth

76. A client is hospitalized with an acute sinus infection. Which of the following observations made by the nurse indicates serious complications?
   1. Orbital edema
   2. Nuchal rigidity
   3. Fever of 102° F (38.9° C)
   4. Frontal headache

77. Parents of a toddler are having problems putting him to bed at night. Which of the following interventions is most appropriate?
   1. Stop the afternoon naps.
   2. Allow the toddler to have a tantrum for ½ hour.
   3. Encourage the parents to develop night-time rituals.
   4. Allow the toddler to have some control over bedtime.

74. 4. This could be a sign of despair or hopelessness if observed in a preschooler. During times of stress, it's typical for the preschooler to be frightened at night. It's normal for a child at this stage to show separation anxiety and play typical of preschoolers.
NP: Data collection; CN: Psychosocial integrity; CNS: Psychosocial adaptation; CL: Knowledge

75. 4. Air inhaled through the nose is warmed, humidified, and filtered for large particles with the nasal hairs, conditioning the air for delivery to the lungs. Exhaling through the mouth after inhaling through the nose requires some concentration and provides a focus to distract a client experiencing pain and anxiety. This method is used to control respiratory rates when clients are anxious or in pain and optimizes air exchange. Rapid, light, or deep respirations cause the client to lose oxygen exchange time while continuing to blow off carbon dioxide. This leads to hypoxemia and respiratory alkalosis.
NP: Implementation; CN: Physiological integrity; CNS: Physiological adaptation; CL: Application

76. 2. Nuchal rigidity indicates neurologic involvement and possibly meningitis. Orbital edema, fever of 102° F, and a frontal headache are typical of a sinus infection.
NP: Data collection; CN: Physiological integrity; CNS: Physiological adaptation; CL: Comprehension

77. 3. Rituals are extremely important for toddlers to feel secure and relaxed. Allowing a toddler to make small decisions, such as choosing the order of the ritual and color of pajamas, will give him the feeling of some control. The toddler must clearly understand that tantrums won't get him what he wants. Stopping the naps may be helpful, depending on the toddler's needs.
NP: Implementation; CN: Psychosocial integrity; CNS: Coping and adaptation; CL: Knowledge

78. After abdominal surgery for repair of an aortic aneurysm, a client may show maladaptive coping behavior in response to body changes related to the surgery. Which of the following nursing interventions is best?
1. Let the client express his feelings.
2. Explain that a psychological referral would be beneficial.
3. Instruct the client on how to use positive coping strategies.
4. Encourage the client to participate in diversional activities.

79. A nurse is reviewing treatment of hypercyanotic spells (tet spells) with the parents of a 4-month-old client being discharged from the hospital. Which of the following discharge instructions is correct?
1. "Calm the baby down by holding her and placing her knees up to her chest."
2. "Call 911 immediately and begin cardiopulmonary resuscitation (CPR) on the baby."
3. "You'll need to administer four back blows to the baby if she begins having a tet spell."
4. "You don't need to worry about these spells yet because the baby is too young. You'll need to watch for them when she becomes more mobile."

80. Which of the following nursing interventions is used to treat gout?
1. Antibiotics, high fluid intake, narcotics
2. Antihyperuricemic drugs, low-purine diet, narcotics
3. High fluid intake, low-purine diet, antihyperuricemic drugs
4. High purine diet, nonsteroidal anti-inflammatory drugs, prednisone

81. Which of the following symptoms is a late sign of compartment syndrome?
1. Sudden decrease in pain
2. Swelling of toes or fingers
3. Inability to move fingers or toes
4. Change in skin color and diminished distal pulses

78. 1. Allowing verbalization of feelings is the most therapeutic nursing intervention. Making a referral may help, but initially the client should be allowed to express feelings. Giving advice may stop therapeutic communication. Providing diversional activities doesn't foster effective coping.
NP: Implementation; CN: Psychosocial integrity; CNS: Coping and adaptation; CL: Application

79. 1. Tet spells are acute episodes of cyanosis and hypoxia that occur when the infant's oxygen demand exceeds the available supply. They may occur when the infant is crying or eating. Tet spells are emergent situations that require immediate intervention. Begin by calming the infant down and placing the infant in the knee-chest position; this increases systemic vascular resistance by limiting venous return. This also decreases the right to left shunting and improves oxygenation. Back blows are given to infants who have something lodged in their trachea. CPR won't calm the infant down or improve oxygenation.
NP: Implementation; CN: Physiological integrity; CNS: Reduction of risk potential; CL: Application

80. 3. A low-purine diet decreases uric acid formation, and the high fluid intake increases urinary output to flush out the uric acid. Drugs, such as antihyperuricemics, are used to reduce serum urate concentrations. Anti-inflammatory medications are used during acute phases, but because this is a long-term condition, narcotics aren't generally given.
NP: Implementation; CN: Physiological integrity; CNS: Reduction of risk potential; CL: Application

81. 4. Compartment syndrome is a complication of a cast that places pressure on the blood vessels and nerves to the extremity. Symptoms include pain not relieved by analgesics and swelling of the extremity. A late symptom is a change in skin color with diminished distal pulses. After a fracture, some swelling and pain result, but pulses need to be monitored as well as color, sensation, and movement.
NP: Evaluation; CN: Physiological integrity; CNS: Reduction of risk potential; CL: Comprehension

82. A client with an arm cast complains of severe pain in the affected extremity and decreased sensation and motion are noted. Swelling in the fingers is also increased. Which of the following interventions has priority?

1. Elevate the arm.
2. Remove the cast.
3. Give an analgesic.
4. Call the physician.

83. Which of the following characteristics is correct for preoperative teaching of a 4-year-old child scheduled for a cardiac catheterization?

1. Basic and performed close to the implementation of the procedure
2. Done several days before the procedure so the child will have time to prepare
3. Detailed in regard to the actual procedure so the child will know exactly what to expect
4. Directed at the child's parents because the child is too young to understand the procedure

84. The nurse has been instructed to help with a morning blood specimen collection from a 4-year-old child in the hospital. The procedure is to be done in the treatment room. Which of the following rationales is correct?

1. The procedure will be faster in the treatment room.
2. The child won't fear painful procedures done while he's in his bed.
3. The child can only be restrained on the examination table.
4. The parents won't observe the procedure and upset the child.

85. A client tells a nurse, "My medical illness is the result of something bad I did to someone in the past." Which of the following responses by the nurse is the most appropriate?

1. "What did you do wrong?"
2. "Let's talk about your concerns."
3. "That's silly! Don't believe that!"
4. "You're suffering from a psychiatric delusion. Relax, it will end soon."

You made it and you did a wonderful job!

82. 4. The cast may be too tight and may need to be split or removed by the physician. The arm should already be elevated. Notify the physician when circulation, sensation, or motion is impaired. Giving analgesics wouldn't be the first step as it may mask the signs of a serious problem.
NP: Implementation; CN: Physiological integrity; CNS: Reduction of risk potential; CL: Application

83. 1. Four-year-old children are in Piaget's cognitive stage of preoperational thought. Their thinking is concrete and tangible, and can't make deductions or generalizations and are egocentric. They need simple explanations of procedures in relationship to how the procedure will affect them. They don't have a concept for the future so explanations need to be done close to the time of the procedure, not days in advance.
NP: Planning; CN: Health promotion and maintenance; CNS: Growth and development though the life span; CL: Comprehension

84 2. This implementation is based on the concept of "atraumatic care" and growth and development principles. Small children need to have a safe zone in their beds to relax and rest in their rooms. The treatment room is used instead. Parental support is important and needs to be encouraged during stressful and painful procedures. It won't be faster to draw blood in the treatment room; it would take the same amount of time regardless of where it's done. The child could be restrained in his room, but it isn't appropriate.
NP: Implementation; CN: Psychosocial integrity; CNS: Coping and adaptation; CL: Comprehension

85. 2. The nurse must be aware of cultural variations that relate to illness so she can appropriately address the client's needs. Calling the client silly or asking the client what he did wrong would likely escalate the client's concerns. Telling the client it will end soon gives false reassurance.
NP: Data collection; CN: Psychosocial integrity; CNS: Psychosocial adaptation; CL: Application

Index

A

Abdominal aortic aneurysm, 41, 42, 43, 44
Abdominal incision, 242
Abdominal pain, acute, 179
Abdominal surgery, diet after, 733
ABO blood incompatibility, 434
Abruptio placentae, 69, 392
Absolute granulocyte count, 501
Abstract thinking, assessment of, 286
Accelerations, during labor, 416
Acetabular dysplasia, 572
Acetaminophen, 476
Acetaminophen poisoning, 596, 715
Acetazolamide, 138
Achilles tendon release, 569
Acne, 663
Acquired immunodeficiency syndrome, 63, 80, 81, 82, 641, 693, 694
Acrocyanosis, 431
Acromegaly, 621
Active immunity, 487
Acute distress disorder, 315
Acute lymphoblastic leukemia, 68
Acute lymphocytic leukemia, 77
Acute renal failure, 224, 233
Acute schizophrenic reaction, 325
Acute spasmodic laryngitis, 512
Acyanotic heart defects, 461
Addisonian crisis, 195
Addison's disease, 192, 195, 196, 198
Adolescence, 373, 451, 761
Adrenal gland, 191
Adrenal insufficiency, 198
Adrenocortical hormone, 196
Advocate, nurse as, 740
Aerosol inhaler, 538
Affect, 318, 327
Afterload, 60
Afterpains, 422
Aging, and physiologic changes, 217, 293
Agnosia, 287
Agoraphobia, 264, 265
Airborne precautions, 719
Airway clearance, ineffective, 513

Airway obstruction, 515, 521, 702
Airway patency, 760
Akathisia, 753
Albuterol, 527
Alcohol abuse, 178, 180, 333, 334, 335, 336, 337, 338, 339, 340, 341, 342, 743, 744
Alcohol detoxification, 338, 714
Alcohol withdrawal, 183, 334, 335, 338, 703, 704
Alcoholics Anonymous, 727, 744
Aldosterone, 204
Allergic reactions, 77
Allergies, 525
Alopecia, 238, 506, 670
Alpha-adrenergic blockers, 216
Alphafetoprotein, 221, 555
Altered level of consciousness, 50, 140
Alzheimer's disease, 285, 286, 287, 288, 290, 294, 295
Ambivalence, 322
Amenorrhea, 384
Aminophylline, 527
Amnesia, 294, 295, 296, 359, 360
Amniotomy, 404
Amobarbital, 360
Amoxicillin, 517, 542
Amphetamine use, 343
Amphetamine withdrawal, 343, 348
Ampicillin, 519
Analgesic administration, 717
Anaphylaxis, 77, 95, 658
Anemia, 66, 67, 68, 69, 499. *See also specific type.*
Aneurysm,
    abdominal aortic, 41, 42, 43, 44
    cerebral, 116
Angina, 47, 48, 49, 190
Angiotensin II, 53
Angioplasty, percutaneous transluminal coronary, 49
Angiotensin-converting enzyme inhibitors, 60, 460
Ankle swelling, 399
Annular rash, 670
Anogenital warts, 638
Anorexia nervosa, 381, 382, 383, 384, 385, 386, 712
Antepartum care, 389-402
Antianxiety medication overdose, 345
Antianxiety medications, 264, 274, 282
Antibiotic therapy, 658

Anticholinergic drugs, 480
Anticoagulants, oral, 749
Antidiabetic medications, 188, 210
Antidiuretic hormone, 54
Antiembolism stockings, 574
Antihypertensive medications, 62, 367
Anti-infective medications, 642
Anti-inflammatory medications, 149
Antimotivational syndrome, 346
Antipsychotic medications, 319, 329, 330
Antiseizure medications, 543, 553, 554
Antisocial personality disorder, 303, 304, 305, 306, 307
Antistreptolysin-O test, 630
Anxiety, 254, 709, 763
Anxiety disorders, 263-284
Aortic stenosis, 471, 472
Apgar score, 416, 723
Aphakia, 138
Apical pulse, 35, 39
Appendectomy, 69, 753
Appendicitis, 166, 167, 599
Appetite stimulation, 739
Arrhythmias, 38, 190, 376, 386, 752
Arterial blood gases, 51, 94, 117
Arterial insufficiency, 749
Arteriogram, 42
Arthralgias, 182
Arthritis, 246. *See also specific type.*
Asepsis, 244, 753
Aspirin, 476, 485, 506, 594
Assertive behavior, 315
Assessment. *See specific type.*
Assessment, nursing process and, 737
Assistive walking devices, 149
Asthma, 89, 90, 91, 92, 525, 526, 527, 528, 528, 529, 713, 740
Asthma triggers, 90
Atelectasis, 93, 527, 757
Atherosclerosis, 33, 41
Atopic dermatitis, 670, 671
Atraumatic care, 775
Atresia
    esophageal, 584, 585, 586, 587
    tricuspid, 466, 467
Atrial fibrillation, 113
Atrial septal defects, 464
Atropine, 120, 137, 184, 480
Attention deficit hyperactivity disorder, 541, 546, 743
Aura, 124
Autologous transplantation, 68
Autonomic dysreflexia, 127, 133, 134
Autosomal-recessive disorders, 537
Avascular necrosis, 566
Azotemia, 218

B
Babinski's sign, 433, 552
Back, protecting, 151
Back pain, 42, 118
Baclofen, 134
Barium enema, 169, 179
Barium swallow, 166
Barking cough, 512
Barlow's test, 565
Basal cell epithelioma, 243
Baths, 241, 243, 433, 442, 671, 727
Behavior modification therapy, 728
Behavioral difficulties, 281
Benign prostatic hyperplasia, 216, 230
Benzodiazepine withdrawal, 275
Benztropine mesylate, 323, 330
Beta-adrenergic blockers, 38, 46, 93, 482
Betamethasone, 416
Beta thalassemia, 401
Beta-thalassemia trait, 496
Bilevel positive airway pressure, 112
Biliary cirrhosis, 180
Binge-purge cycle, 375, 377, 378, 379, 380
Biofeedback, 255
Bipolar cycling, 345
Bipolar disorder, 249, 276, 277, 278, 279, 744, 753, 764, 730
Bladder cancer, 225
Bladder catheterization, 136
Bladder infections, 721
Bladder, neurogenic, 136
Bladder spasms, 218, 228
Bleeding time, 71
Blood clot formation, preventing, 472
Blood cultures, 748
Blood glucose test, capillary, 189
Blood pressure, 52, 54, 408, 758
Blood pressure medication. *See* Antihypertensive medications.
Blood transfusions, 73, 74, 75, 414, 700
Blood-injection-injury phobia, 266
Blue bloater, 91
Blurred vision, in third trimester, 391
Body image, 365, 369, 706
Body-fluid splashes, 771
Bone formation, 143
Bone marrow cancer, 506

Bone marrow depression, 343
Bone marrow transplantation, 504
Borderline personality disorder, 298, 307, 308, 309, 310, 311, 312, 681, 729
Braces, 153
Brachial artery, 54
Bradypnea, 114
Brain stem infarction, 114
Braxton Hicks contractions, 410
Breast cancer, 223, 716
Breast cancer metastasis, 129
Breast engorgement, 418, 425, 725
Breast-feeding, 418, 425, 426, 428, 431, 491, 724
Breast-milk jaundice, 445
Breast milk supply, 417
Breast self-examination, 724
Breast surgery, 216. *See also* Mastectomy.
Breathing techniques, 773
Breath sounds, 85, 99, 701
Breech presentation, 410, 742, 745
Bronchial obstruction, 761
Bronchiolitis, 529, 531, 713
Bronchitis, 91, 92
Bronchodilators, 89, 756
Bronchopulmonary dysplasia, 432, 522, 523, 524, 525
Bronchoscopy, 534, 695
Bruises, 662
Bruit, 43
Bryant's traction, 158
Buck's extension traction, 158
Bulimia nervosa, 375, 376, 377, 378, 379, 380
Bulk-forming laxatives, 732
Burns, 235, 236, 240, 246, 653, 654, 661, 662, 664, 666, 675, 676

C

Calcium channel blockers, 39
Calcium deficiency, 598
Calcium gluconate, 207, 396, 402, 414
Calculus formation, 205, 229, 233, 705
Calf muscle cramps, 55
Cancer. *See specific type.*
Cancer, childhood, 507. *See also specific type.*
Cancer pain, 738
Candidiasis, 214, 658, 673
Capillary blood glucose test, 189
Captopril, 460
Caput succedaneum, 438, 672
Carbamazepine, 140
Carbon dioxide, 107, 768
Carbon monoxide poisoning, 239, 673

Carcinoembryonic antigen level, 102
Cardiac anomalies, 465, 610
Cardiac catheterization, 47, 456, 461, 463, 466, 471, 748, 775
Cardiac enzyme analysis, 50
Cardiac glycosides, 460, 699
Cardiac index, 50
Cardiac monitoring, 731
Cardiac output, 59, 457
Cardiac surgery, 457, 458
Cardiac tamponade, 457, 691, 692
Cardiogenic shock, 49, 50, 51, 454, 691
Cardiomegaly, 45
Cardiomyopathy, 44, 45, 46, 47, 716
Cardiopulmonary resuscitation, 515, 751
Cardiovascular disease, during pregnancy, 698, 699, 761
Cardiovascular disorders, 33-62, 453-482
Care plan, 752, 756, 757
Casts, 154, 155, 157, 160, 161, 162, 575, 775
Cataracts, 115, 131, 138, 746
Catatonic schizophrenia, 327
Cat-scratch disease, 661
Ceftazidime, 537
Celiac disease, 579, 580, 602, 713
Cellulitis, 669
Central nervous system injury, 755
Central venous catheter complications, 100
Cerebral aneurysm, 116
Cerebral palsy, 541, 544, 546, 547, 563, 671
Cervical cancer, 219, 638
Cervical dilation, 411
Cervical laceration, 424
Cervical polyps, 219
Cesarean delivery, 413, 419, 425
Chancres, 674
Chelation therapy, 597
Chemoreceptors, 53
Chemotherapy, 364, 503, 505, 506, 648, 716, 737, 742
Chest compressions, 515, 516
Chest drainage, 101
Chest pain, 49, 107, 109
Chest physiotherapy, 536, 537
Chest splinting, 85
Chest trauma, 691
Chest tubes, 100, 101, 103
Chest tubes, 458, 538
Chest X-rays, 87, 99, 101, 691, 712
Chiari II malformation, 557
Chickenpox, 489, 490, 491
Child abuse, 571, 659, 662

Children, growth and development in, 451-452
Chlamydial infections, 212, 219, 638, 640, 717
Chlorpromazine, 323, 329
Choking, 516, 750
Cholecystectomy, 772
Cholecystitis, 184, 185
Cholesterol levels, excessive, 34
Cholinergic crisis, 120
Chordee, 632
Chronic arterial disease, 754
Chronic lymphocytic leukemia, 76
Chronic obstructive pulmonary disease, 84, 90, 92,
        364, 760, 765
Chronic renal failure, 212, 231, 717
Chvostek's sign, 768
Chymopapain, 119
Cimetidine, 186
Circumcision, 435, 442, 444, 633
Cirrhosis, 180, 181, 183, 342, 704, 705
Clavicular fracture, 155, 447
Clean-catch urine specimen, 630
Cleft lip and palate, 580, 581, 582, 583, 600, 724
Client needs categories, 4, 5
Closed fracture, 156
Closed spinal surgery, 150
Clubbing, 462, 465
Clubfoot, 556, 559, 576, 577
Clues, covert, 699
Coarctation of aorta, 464, 465
Cocaine use, 342, 343, 344, 345, 346, 347, 348, 350
Cocaine withdrawal, in neonate, 446
Cognitive development, 452
Cognitive disorders, 285-298
Colchicine, 145
Colds, 525
Cold therapy, 164, 738
Collaborative problems, 765
Colon cancer, 173, 174
Colonoscopy, 172
Colostomy, 369, 706, 707, 719
Colostrum, 408
Compartment syndrome, 152, 153, 774
Compression garments, 672
Compulsive drug use, 342
Concussion, 548
Condom catheter, 217, 698
Conduct disorder, 717, 763
Confidentiality, 709
Congenital hip displacement, 565, 566, 572, 573, 577
Contraception, 429, 739
Contraceptive teaching, 639

Contractions
  Braxton Hicks, 410
  labor, 403, 404
Contraction stress test, 394
Contusion, 758, 662
Convection heat loss, 437
Conversion disorder, 256, 257, 258, 259, 260
Coombs' test, in neonate, 423
Coronary arteries, 33
Coronary artery disease, 33, 34, 36, 188, 692
Corticosteroids, 81, 473
Co-trimoxazole, 216, 504, 642, 647
Cough, barking, 512
Covert clues, 699
Crackles, 37, 39, 96, 732
Cramps, calf muscle, 55
Cranial nerve III, 140
Craniectomy, 486
Craniotomy, 544, 547
Creatine kinase isoenzymes, 36
Crises, client response to, 79
Crohn's disease, 170, 171, 172, 593
Croup, 512, 513, 514
Crowning, 411
Crush injury, 660
Crutch walking, 164
Cryosurgery, 219
Cryptorchidism, 717
Cultural differences, 737, 775
Cultural shock, 762
Cushing's syndrome, 192, 195, 197, 198, 204
Cyclosporine therapy, 227
Cystectomy, 225
Cystic fibrosis, 447, 534, 535, 536, 537, 538
Cystitis, 225
Cystoscopy, 232

D
Decelerations, during labor, 415
Deep vein thrombophlebitis, postpartum, 424
Deep vein thrombosis, 57, 58, 159, 160, 574, 575, 754
DEET, 668
Degenerative joint disease. See Osteoarthritis.
Dehydration, 530, 531, 663, 768
Delirium, 291, 292, 293
Delusion, 317
Delusional disorder, 323, 326, 327, 331
Delusional disorders, 317-332
Delusional thinking, 316
Dementia, 285, 288, 289, 290
Denial, 269, 722

Denver Developmental Screening Test, 763
Dependent edema, 40
Dependent personality disorder, 306, 312, 313, 314, 315, 316
Depersonalization disorder, 321, 361, 362
Depression, 250, 279, 280, 281, 282, 283, 310, 316, 350, 357, 718, 741
Dermatitis, atopic, 670, 671
Desmopressin acetate, 194, 614, 616, 623
Desquamative rash, 669
Detached retina, 138
Developmentally based care, 731
Dexamethasone suppression test, 197
Diabetes insipidus, 116, 193, 194, 202, 612, 613, 614, 615, 616, 622, 623, 624
Diabetes mellitus, 187, 188, 203, 209, 210, 380, 398, 400, 412, 420, 422, 428, 432, 437, 440, 447, 607, 608, 609, 610, 611, 612, 626, 707, 708, 723, 726, 736, 750
Diabetic ketoacidosis, 198, 199, 202, 209, 420, 608, 609, 750
Dialysis, 226, 227
Diaper rash, 658
Diaphragmatic breathing exercises, 760
Diaphragmatic hernia, 587, 588
Diarrhea, 64
Diastolic blood pressure, 52
Digoxin, 39, 41, 460, 463, 467, 469
Digoxin toxicity, 460
Dilatation and curettage, 217
Diphtheria-pertussis-tetanus immunization, 712
Discharge planning, 740
Disorganized schizophrenia, 318
Disposable needles, 607
Disseminated intravascular coagulation, 70
Dissociative amnesia, 359, 360
Dissociative disorders, 351-362
Dissociative fugue, 358, 359
Dissociative identity disorder, 351, 352, 353, 354, 355, 356, 357, 358
Disulfiram, 349
Disuse osteoporosis, 757
Diuretics, 205, 229, 469
Diverticulitis, 169
Diverticulosis, 168, 169
Dizygotic twins, 393
Dog bite, 660
Domestic abuse, 411, 687, 734, 741
Donuts, 244
Dopamine, 51, 141, 463

Dosage calculations, 113, 121, 131, 139, 403, 454, 460, 473, 474, 475, 480, 481, 517, 519, 523, 524, 527, 536, 537, 546, 547, 656, 674, 715
Douche administration, 215
Dressing changes, 241, 245
Drug addiction, 348, 349
Drug intoxication, 292
Drug overdose, 95, 96
Drug withdrawal, 436, 440
DTP immunization, 487
Duchenne's muscular dystrophy, 567, 571, 577
Duodenal ulcer, 185
Dyspareunia, 408
Dysplasia
  acetabular, 572
  bronchopulmonary, 432, 522, 523, 524, 525
Dysreflexia, autonomic, 127, 133, 134
Dysthymic disorder, 731

E

Ear canal irrigation, 139
Ear tubes, 517
Eardrops, administration of, 131, 517
Eardrum perforation, 518
Eating disorders, 375-386
Echocardiogram, 455, 462, 469
Echopraxia, 319
Eclampsia, 395
Ectopic pregnancy, 391
Eczema, 237, 665, 666
Effacement, 406, 410
Effleurage, 402
Ejaculation, retrograde, 363
Elastic bandages, 719
Elastic leg compression, 56
Elderly client, examination of, 764
Electronic fetal monitoring, 412
Electrocardiogram, 37, 48, 67, 455
Electrocautery, 666
Electroconvulsive therapy, 249, 276, 730
Electroencephalogram, 141
Electrolyte imbalance, 285, 376, 377, 394
Embolectomy, 110
Embolism
  fat, 97, 157
  pulmonary, 57, 105, 106, 107, 108, 109, 110
Emergency situations, nurse manager role in, 680
Emphysema, 91, 92, 727, 769
Endocarditis, 473, 474, 748
Endocrine disorders, 187-210, 603-627
Endotracheal tubes, 695

Enucleation, of eye, 139
Enzyme-linked immunosorbent assay, 64
Epidural anesthesia, 406, 408, 414
Epidural hematoma, 132
Epiglottiditis, 514, 519, 520, 701, 702, 703
Epinephrine, 191
Episiotomy, 408
Epispadias, 636
Epistaxis, 71, 501, 760
Erectile dysfunction, 363, 371
Erik Erikson, 373
Erratic behavior, substance abuse and, 345
Erythema infectiosum, 654, 655
Erythema toxicum, 442
Erythromycin, 474
Erythromycin ointment, 434, 440
Escharotomy, 675
Esophageal atresia, 584, 585, 586, 587
Esophageal reflux, 165
Esophageal varices, 181
Esophagogastroduodenoscopy, 185
Estrogen therapy, 372
Ethical and legal issues, 683-687
Eustachian tubes, 518
Evaporative heat loss, 437, 443
Evisceration, 215
Exercise, 92
Exophthalmos, 606
Eye trauma, 130

F
Familial short stature, 617
Family violence. *See* Domestic abuse.
Family-centered care, 512
Fasciotomy, 153
Fat embolism, 97, 157
Fecal occult blood test, 749
Femoral popliteal bypass, 752
Ferguson's reflex, 743
Fetal alcohol syndrome, 446, 447, 740
Fetal distress, 404, 405
Fetal heart rate, 407
Fetal heart tones, 413, 699
Fetal lung maturity, 416
Fetal monitoring, 404, 405, 407, 412
Fetal movement, 399
Fetal position, 406
Fetal station, 412
Fetus, postdate, 435
Fibrocystic disease, 223
Fifth disease. *See* Erythema infectiosum.

Finasteride, 218
Fish oil, 64
Flatulence, excessive, 725, 732
Fluid and electrolyte balance, postoperative, 457
Fluid deprivation test, 194
Fluid volume deficit, 470
Fluid volume excess, 732
Fluoxetine, 385
Folic acid, 558
Foot drop, 117, 161, 576
Foot fracture, 153
Foot swelling, 399
Foot ulcer, 203
Foreign body, 130
Foreign body aspiration, 533, 534, 516
Formula, diluting, 540
Fracture reduction, 162, 574
Fractures, 154, 155, 156, 157, 574. *See also specific type.*
Frejka splint, 578
Frontal lobe, 141
Frostbite, 656
Fundus assessment, 419, 420, 422, 429, 726
Furosemide, 36, 54, 213, 523, 524, 767

G
Gag reflex, 114
Gallbladder disease, and pancreatitis, 208
Gallbladder removal, 754
Gallop murmur, 458
Gallstones, and pancreatitis, 207, 208
Gardner-Wells tongs, 128, 134
Gastric cancer, 174, 175
Gastric resection, 167, 175
Gastric ulcer perforation, 180
Gastritis, 167, 168
Gastroenteritis, 602
Gastrointestinal bleeding, 185
Gastrointestinal disorders, 165-186, 579-602
Gastroschisis, 587, 589
Gastroscopy, 174, 178
Gender, ascribing, 371
Gender identity disorders, 367, 373, 363-371
Generalized anxiety disorder, 272, 273, 274, 275
Generic versus brand names, 762
Genital herpes, 212, 213, 222, 223, 413
Genitourinary disorders, 211-235, 627-652
German measles. *See* Rubella.
Gestational diabetes, 394
Giardiasis, 661
Gigantism, 206

Glaucoma, 120, 121, 138
Glipizide, 210
Glomerulonephritis, 627, 628, 629, 630, 631, 632
Glucose level, normal, 707
Glucose monitoring, 210, 609
Gonococcal infections, 211
Gonorrhea, 211, 637, 639, 640
Gout, 144, 145, 146, 163, 774
Gower's sign, 570
Granulocyte count, absolute, 501
Graves' disease, 201
Grieving, 425, 686, 721, 741, 742
Group B streptococci, 439
Group living, 83
Group therapy, 250, 266, 271, 340, 348, 386
Growth, intense activity and, 617
Growth and development, 451-452
Growth hormone, 206, 620
Growth hormone deficiency, 616, 617, 618, 619, 622
Growth hormone replacement, 613, 621
GTPAL formula, 428
Guillain-Barré syndrome, 94, 119
Gum disease, 669

H

Haemophilus influenzae, 514, 519
Haemophilus influenzae vaccination, 519, 521
Hallucinations, 291, 292, 293, 318, 320, 321, 322, 324,
    325, 330, 338
Halo vest, 128
Haloperidol, 288, 320, 326, 329
Hand washing, 769, 770, 771
Hand-foot syndrome, 495
Harrington rods, 554, 564
Head trauma, 115, 116, 126, 131, 132, 140, 545, 548, 552
Headache, 53, 549, 261
Health care delivery systems, 681
Hearing aids, 746
Hearing loss, 130, 582, 750
Heart defects, 454, 461
Heart disease, 380, 735
Heart failure, 37, 38, 39, 40, 41, 45, 46, 197, 335, 458,
    459, 460, 471, 759
Heart murmurs, 395, 402, 453, 455, 459
Heart sounds, 37, 402, 453
Heart transplantation, 47, 473, 736
Heartburn, 399
Heat loss, in neonate, 437, 443, 448
Heat therapy, 164, 738
Heelstick hematocrit test, 447
Heimlich maneuver, 727, 750

Helicobacter pylori infection, 168
Hemianopia, homonymous, 113
Hematologic and immune disorders, 63-82, 483-508
Hematoma, subdural, 116, 117
Hemodialysis, 212, 227, 234
Hemoglobin, 111
Hemoglobin electrophoresis, 401
Hemophilia, 500, 501, 508, 714
Hemorrhagic cystitis, 506
Hemorrhagic disease of neonate, 437
Hemorrhagic stroke, 133
Hemorrhoids, 176, 177
Hemothorax, 100
Heparin, 108, 113, 129, 69
Hepatic coma, 704
Hepatitis, 181, 182
Hepatitis B, 434
Hepatitis B immunization, 82
Hernia
    diaphragmatic, 587, 588
    hiatal, 165, 166
Herniated disk. *See* Herniated nucleus pulposus.
Herniated nucleus pulposus, 118, 134, 150, 151, 152
Heroin use, 347, 349
Herpes zoster, 667
Hiatal hernia, 165, 166
High Fowler's position, 59, 89
Hip displacement. *See* Congenital hip displacement.
Hip fracture, 160
Hip replacement, 158, 159, 574
Hip-spica cast, 571, 572, 573, 576
Hirschsprung's disease, 601
Histamine-2 receptor antagonists, 536
Histrionic personality disorder, 300
Hodgkin's disease, 74, 76, 498
Homans' sign, 58, 418, 575
Homonymous hemianopia, 113
Human chorionic gonadotropin, 392
Human immunodeficiency virus infection, 63, 64, 65,
    80, 491, 492, 498, 641, 739, 693
Human papilloma virus, 213, 368, 638, 671
Humeral fracture, 155
Hydatidiform mole, 391
Hydration, in neonate, 432
Hydrocarbons, poisonous, 594
Hydrocephalus, 539, 543, 557
Hydrocodone with acetaminophen, 220
Hydrocortisone, 195, 246, 665
Hydrops fetalis, 654
Hyperaldosteronism, 204
Hyperbilirubinemia, 434, 438, 439, 444

Hypercalcemia, 78, 79
Hypercoagulability, 57
Hypercyanotic spells, 466, 774
Hyperemesis gravidarum, 390, 394
Hyperglycemia, 187, 202, 609, 611
Hyperkalemia, 226, 240
Hyperparathyroidism, 205
Hyperphosphatemia, 206
Hyperpigmentation, 675
Hypersensitivity skin tests, 245
Hypersensitivity, drug, 95
Hypertension, 43, 52, 53, 54, 133, 135, 197, 380, 468
Hyperthyroidism, 195, 200, 201, 207, 606
Hypertrophic scarring, 672
Hypnosis, 355
Hypocalcemia, 206, 768
Hypocapnia, 59
Hypochondriasis, 252, 253, 254, 258
Hypoglycemia, 187, 203, 432, 440, 443, 608, 610, 611, 622, 726
Hypokalemia, 203, 375
Hyponatremia, 240
Hypoparathyroidism, 206
Hypophysectomy, 117
Hypopituitarism, 616, 617, 619, 620, 621
Hypoplastic left heart syndrome, 472
Hypospadias, 632, 633, 634, 635, 636, 637
Hypostatic pneumonia, 118
Hypotension, 228, 454
Hypothalamus, 54, 140
Hypothermia, 443
Hypothyroidism, 190, 191, 192, 603, 604, 605, 606, 624, 625
Hypovolemic shock, 495
Hypoxia, 465, 521
Hypoxic drive, 92
Hysterectomy, 143, 215, 365
Hysterosalpingography, 220

I
Ibuprofen, 64
Idiopathic thrombocytopenic purpura, 497
Ileal conduit, 225, 228
Ileostomy, and sexual dysfunction, 364
Illusion, 293, 324
Immune disorders, 63-82, 483-508
Immune response, 80
Immune status, altered, 81
Immune system, 71, 72
Immunizations, 487, 488, 521, 676, 701, 712
Immunoglobulin G, 446

Impetigo, 661
Impotence, 363, 371
Inactivated polio virus vaccine, 487
Incentive spirometry, 720
Incident reports, 684, 722
Increased intracranial pressure, 131, 484, 485, 502, 549
Indwelling urinary catheter, 218, 226, 228, 718, 721, 728, 745
Ineffective coping, 759
Infant, growth and development in, 545, 549, 553
Infection, neonatal, 432, 436
Infertility, 367, 368, 371
Influenza vaccinations, 91, 496
Informed consent, 685, 722, 733
Inhalants, 343
Inhaler, aerosol, 538
Inotropic medications, 40
Insulin, 187, 189, 202, 203, 207, 209, 210, 422, 428, 607, 608, 612, 622
Intake and output, calculating, 733
Integumentary disorders, 235-246, 653-676
Intermittent claudication, 58
Intervertebral disks, 151
Intracranial pressure, 118. *See also* Increased intracranial pressure.
Intraluminal valvular incompetence, 55
Intraocular pressure, 137
Intrapartum care, 403-416
Intubation, 693
Iodine-containing medications, 191, 231
Iron deficiency anemia, 71, 401 497, 498, 499, 508
Iron dextran injection, 499
Iron replacement therapy, 69
Iron supplements, oral, 500
Irritable bowel syndrome, 172, 173, 186
Isoniazid, 88
Isoproterenol, 481, 482
Isotretinoin, 237, 663
Itching sensation, reducing, 236
I.V. dosage calculations, 414
I.V. therapy, 747

J
Jaundice. *See* Hyperbilirubinemia.
Joint degeneration, hemophilia and, 501
Joint pain, 716
Jones criteria, 478
Jugular vein distention, 38
Juvenile rheumatoid arthritis, 745

K

Kaposi's sarcoma, 82
Kawasaki disease, 475, 476, 477, 492, 655, 656, 665
Kegel exercises, 232, 418, 569
Keloids, 673
Kernig's sign, 549
Kidney stone analysis, 234
Kidney transplantation, 227
Knee splints, 676
Korotkoff's sounds, 52
Kussmaul's respirations, 209
Kyphoscoliosis, thoracic, 98

L

La belle indifference, 258
Labor, 389, 403, 404, 405, 406, 407, 408, 409, 410, 411,
    412, 413, 414, 415, 416, 743
Lacerations, 80, 667, 715
Laënnec's cirrhosis, 180
Lamaze method, 401
Laminectomy, 118, 133
Large-for-gestational-age neonate, 447
Laryngitis, acute spasmodic, 512
Laryngotracheobronchitis, 521
Laxative abuse, 377
Laxatives, 732
Lead poisoning, 596, 597, 598
Leadership principles, 738
Learning disability, 141, 541
Lecithin to sphingomyelin ratio, 441
Leg contractures, preventing, 240
Leukapheresis, 78
Leukemia, 78, 497, 501, 502, 503, 504, 505, 506, 507. *See
    also specific type.*
Leukocytosis, 478
Level of consciousness, 94
  altered, 50, 140
Levodopa-carbidopa, 122
Levothyroxine, 190, 605, 606
Lice, 236, 659
Licensed Practical Nurse, 687
Life review, 719
Ligation and stripping, 56
Limb circumference, 154
Lindane shampoo, 659
Lipodystrophy, 201
Lithium, 277, 278, 282, 283, 284, 730, 747
Lithium toxicity, 764
Liver biopsy, 183
Liver cancer, 183
Liver disorders, assessment if, 183

Living wills, 687, 718
Loading dose, 124
Lobectomy, 103
Lochia serosa, 423
Logan bow, 581
Long bone fractures, 157
Loose associations, 296, 326
Lorazepam, 256, 334, 335
Low-birth-weight neonate, 401, 726, 751
Lucid interval, 132
Lumbar epidural block, 726
Lumbar lordosis, 567
Lumbar puncture, 116, 550, 558
Lung cancer, 101, 102, 103, 104, 105
Lyme disease, 246, 655
Lysergic acid diethylamide intoxication, 346

M

Maalox, 186
Macrosomia, 420, 437
Mafenide, 240, 676
Magnesium sulfate, 394, 395, 396, 402, 405, 414, 427,
    436
Magnetic resonance imaging, 131, 748
Maladaptive coping behavior, 774
Malignant lymphoma, 72
Malingering, 261
Management and supervision concepts, 679-682
Manic episodes, 275, 276, 278, 279, 284, 729, 744
Mannitol, 116, 117
Mantoux skin test, 87, 236, 655, 740
Marfan's syndrome, 44
Marijuana use, 342, 344, 346
Maslow's hierarchy, 12, 13
Mastectomy, 215, 766
Mastitis, 420, 421, 426
Maternal hormone withdrawal, 448
Maturational crisis, 728
Measles vaccine, 487
Mechanical ventilation, 135, 432, 527
Meckel's diverticulum, 600
Meconium, 438, 601
Meconium ileus, 447, 534
Medication administration, five rights of, 679, 746, 752
Medication errors, 685
Meditation, 763
Memory impairment, 141, 293, 296
Ménière's disease, 129, 130, 139
Meningitis, 439, 521, 540, 542, 543, 544, 545, 547, 550,
    551, 552
Meningococcemia, 550

Menorrhagia, 365
Menstruation, 378
Mental Status Examination, 765
Metabolic alkalosis, 769
Methylprednisolone, 127
Metronidazole, 219, 640
Microcephaly, 446
Microophthalmia, 446
Midnight croup, 512
Milwaukee brace, 561, 577
Mist tent, 512
Mongolian spots, 663, 665
Moniliasis, oral, 441
Monoamine oxidase inhibitors, 270
Mononucleosis, 758
Mood disorders, 263-284
Moro's embrace reflex, 543
Morphine, 36, 94, 758
Mouth ulcers, 505, 742
Multidisciplinary team meetings, 734
Multiple myeloma, 77, 78, 79
Multiple sclerosis, 123, 124, 142
Mumps, 220
Muscle cramps, 55
Muscular dystrophy, 563, 565, 566, 567, 568, 569, 570,
    571, 577
Musculoskeletal disorders, 143-164, 559-578
Musculoskeletal pain, 261
Myasthenia gravis, 119, 120, 759
Mycoplasmal pneumonia, 532
Myelography, 118
Myelomeningocele, 539, 557, 558
Myocardial infarction, 34, 35, 36, 37, 38, 43, 46, 48, 50,
    365, 758, 759
Myringotomy, 517
Myxedema coma, 190

N

Nägele's rule, 416
Naltrexone, 347
Narcolepsy, 255
Narcotics, 94, 115
Nasal surgery, 730, 757
Nasogastric tubes, 125, 184, 588, 710, 711, 747
Nasopharyngeal suctioning, 760
Nausea and vomiting, 505, 555, 591, 602
NCLEX, preparing for, 3-18, 19-30
Nebulizers, 513
Necrosis, avascular, 566
Necrotizing enterocolitis, 431, 601
Needles, disposable, 607

Negligence, 683
Neologisms, 328
Neonatal care, 431-448
Neonatal sepsis, 439
Nephritis, 75, 213
Nephrotic syndrome, 649, 650, 652
Neural tube defects, 555, 558
Neurogenic bladder, 136
Neurogenic shock, 127, 135
Neuromuscular blockers, 97
Neuropathy, 188
Neurosensory disorders, 113-142, 539-558
Neurovascular assessment, 154
Neutropenia, 78
Neutropenic precautions, 771
Night blindness, 136
Nitrates, 46
Nitrogen balance, negative, 757
Nitroglycerin, 48
Nodules, 661
Nonsteroidal anti-inflammatory drugs, 64, 66, 146, 167
Nonstress test, 400
Norepinephrine, 191
Normal saline solution, 75
Nortriptyline, 746
Novolin NPH, 189
Nuchal rigidity, 549, 773
Nurse Practice Act, 684
Nursing assistants, 679
Nursing diagnoses, 766
Nursing process, 765, 766, 767
Nystagmus, 124
Nystatin oral solution, 237, 669

O

Obesity, 184, 400
Obstetric sonography, 742
Occurrence policy, 685
Oculogyric crisis, 329, 330
Omnibus Reconciliation Act of 1986, 686
Omphalocele, 588
Ophthalmia neonatorum, 434
Ophthalmic medications, 130
Optic nerve function, 720
OPV immunization, 487
Organ donation, 686
Organ rejection, 233
Orthopneic position, 770
Orthostatic hypotension, 275, 427, 770
Osteoarthritis, 146, 147, 148, 149, 163, 750
Osteoblast activity, 143

Osteomyelitis, 163, 164
Osteoporosis, 143, 144, 367, 716, 757
Ostomy care, 173
Otitis media, 516, 517, 518
Otolaryngologist, 583
Oxycodone, 216
Oxygen administration, 84, 96, 109, 111, 112, 459, 535, 694
Oxygen supply, 48, 49
Oxygen therapy, 109
Oxygen toxicity, 98
Oxygenation, improving, 111
Oxytocin, 398, 403, 404

P

Pain
  cancer, 738
  chest, 49, 107, 109
  joint, 716
  pleuritic, 107
  pulmonary, 35
Pancreatic enzymes, 535
Pancreatitis, 178, 179, 180, 186, 207, 208, 209, 715
Pancuronium, 485, 486
Panic disorder, 263, 264, 265, 266, 272, 273, 374, 729
Papanicolaou test, 213
Papules, 237, 661
Paranoia, amphetamine withdrawal and, 348
Paranoid personality disorder, 297, 298, 299, 300, 301, 302, 315
Paranoid schizophrenia, 323, 326, 332, 744
Paraphilias, 371
Paresthesia, 153
Parietal lobe, 141
Parkinson's disease, 121, 122, 142, 697, 761
Paronychia, 657
Partial pressure of arterial oxygen, 757
Patent ductus arteriosus, 455, 460, 461
Patent foramen ovale, 470
Peak expiratory flowmeter, 713
Pediatric examination, 764
Pediculosis, 664
Pedophilia, 369
Pelvic inflammatory disease, 637, 641
Penicillin, 479, 490
Pentoxifylline, 58
Peptic ulcer disease, 176, 186
Percutaneous transluminal coronary angioplasty, 49
Pericardiocentesis, 692
Perinatal loss, 425
Peripheral vascular disease, 754

Peritoneal dialysis, 227
Peritonitis, 174, 177, 178
Permethrin, 657
Pernicious anemia, 168
Personality disorders, 297-316
Pertussis, 490, 496, 532
Petechiae, 237, 662, 670
Pharyngitis, 491
Phencyclidine use, 343, 346, 349
Phenobarbital, 554
Phenothiazine, 753
Phenytoin, 124, 125, 139, 553
Phimosis, 221
Phobias, 266, 267
Phototherapy, 438, 443, 444
Physiologic anemia, 393
Piaget's theory of cognitive development, 452
Pica, 598
Pilocarpine, 121
Pineal gland, 140
Pink puffer, 91
Pinworm infestation, 660
Pituitary gland, 194, 200
Pituitary growth hormone, 620
Pituitary hypofunction, posterior, 615
Placenta previa, 390, 403, 699, 700
Placental separation, 410
Plantar warts, 238
Plasmapheresis, 120, 142
Platelets, 72, 73
Pleural effusions, 102, 112
Pleur-evac, 101
Pleuritic pain, 107
Pneumococcal pneumonia, 532
Pneumocystis carinii infection, 693
Pneumocystis carinii pneumonia, 504
Pneumonectomy, 104
Pneumonia, 83, 84, 85, 118, 228, 504. 532. *See also specific type.*
Pneumothorax, 99, 100, 101, 527
Pneumovax, 91
Poisoning, 592, 593, 594, 595, 596, 597, 598, 735
Polio vaccine, 487
Polycythemia, 435
Polyhydramnios, as complication of delivery, 420
Polysubstance abuse, 342, 345
Positive end-expiratory pressure, 97, 111
Postcoke depression, 345
Posterior pituitary hypofunction, 615
Postoperative care, 172
Postoperative complications, 93

Postoperative wound care, 245
Postpartum assessment, 417, 421
Postpartum bleeding, 421, 422
Postpartum blues, 418, 424
Postpartum care, 417-430
Postpartum depression, 426
Postpartum hemorrhage, 420, 421, 426, 427, 429, 430
Posttraumatic stress disorder, 250, 268, 269, 270, 271, 282
Potassium, 36, 61, 203, 600
Potassium-rich foods, 524
Prednisone, 234, 473
Preeclampsia, 394, 395, 396, 400, 405, 413, 415, 427
Pregnancy, 55, 65, 69, 123, 344, 698, 699, 724, 727
Pregnancy test, 392
Pregnancy-induced hypertension. *See* Preeclampsia.
Premature atrial contractions, 455
Premature birth, 435, 509, 587, 722
Premature labor, 397, 401, 709, 725
Premature rupture of membranes, 396
Prenatal history, 397
Preschool behavior, 773
Prescription drug abuse, 344
Pressure dressings, 80
Pressure ulcers, 241, 242, 243, 671
Prevalence, 82
Progressive memory deficit, 287, 288
Progressive muscle relaxation, 257
Prone position, 98, 720
Propantheline, 635
Prostate biopsy, 222
Prostatectomy, 222, 363
Prostatitis, 216, 218
Pseudohypertrophy, 571
Pseudoseizures, 260
Psoriasis, 239, 672
Psychiatric care, essentials of, 249-250
Psychomotor retardation, 728
Psychopsychological disorder, 255
Psychosis, 765
Psychotropic drugs, low-potency, 331
Puerperal infection, 423
Pulmonary angiogram, 108
Pulmonary artery catheter, 51
Pulmonary edema, 41, 59, 60, 61, 398
Pulmonary embolism, 57, 105, 106, 107, 108, 109, 110
Pulmonary function tests, 756
Pulmonary pain, 35
Pulmonary venous congestion, 459
Pulmonic stenosis, 470, 471
Pulse oximetry, 94, 110, 760

Pulses, 154, 751
  apical, 35, 39
Pupil accommodation, 136
Pupil constriction, 140
Purified protein derivative test, 86, 396
Purine-rich foods, 144
Pustules, 661
Pyelonephritis, 224, 233, 392
Pyloric stenosis, 589, 590, 591, 592, 602
Pyrantel, 660

Q
QRS complex, 455
Quadriplegia, 126, 127, 133, 135

R
Radiation implant, 211, 219, 225
Radiation therapy, 174, 364, 505, 507, 648
Range-of-motion exercises, 164
Ranitidine, 186, 536
Rape prevention, 372
Rape-trauma syndrome, 370
Rashes, 238
  annular, 670
  desquamative, 669
  diaper, 658
Recessive disorders, 537
Rectal bleeding, 172, 175, 385
Rectal cancer, 175, 176
Rectal examination, 173, 176
Red blood cells, 73
Referential thinking, 328
Reflux, esophageal, 165
Regression, 322
Relaxation techniques, 729, 763
Religious beliefs, 649, 737
Renal biopsy, 232
Renal failure, 233
  acute, 224, 233
  chronic, 212, 231, 717
Renal injury, urinary tract infections and, 643, 646
Renal insufficiency, 228
Rescue breathing, 515
Respiratory acidosis, 94, 759, 769
Respiratory alkalosis, 107, 714, 767
Respiratory arrest, 94, 515
Respiratory changes, pregnancy and, 392
Respiratory disorders, 83-112, 509-538
Respiratory distress, 436, 513, 520, 521, 531
Respiratory distress syndrome, 96, 98, 435, 441, 445
Respiratory failure, 94, 95, 695

Respiratory syncytial virus, 529, 530, 531, 713
Restraints, 721, 736
Retina, 54
    detached, 138
Retinopathy, 188
Retrograde ejaculation, 363
Reye's syndrome, 483, 484, 485, 486
Rh isoimmunization, 397
Rh status, 425
Rheumatic fever, 471, 477, 478, 479
Rheumatoid arthritis, 64, 65, 66, 67, 69, 147, 716
Ribavirin, 531
Rickettsia rickettsii, 499
Rifampin, 533, 761
Right to refuse treatment, 684
Rings, 244
Risk for disuse syndrome, 766
Risk nursing diagnosis, 697
Risk-taking behavior, 639
Ritodrine, 725
Rocky Mountain spotted fever, 499, 670
Role reversal, 772
Roseola, 489, 490
Roundworm infection, 598
Rubella, 490, 498
Rubella vaccine, 427
Rule of Nines, 236, 244
Russell's traction, 155

S

Sacral edema, 40
Salicylate poisoning, 593, 594, 595, 751
Salmon patches, 662, 668
Scabies, 657
Scarlet fever, 490, 491
Schizoaffective disorder, 331
Schizophrenia, 317, 318, 319, 321, 322, 323, 324, 325,
    326, 327, 328, 329, 330, 331, 332, 346, 744
Schizophrenic disorders, 317-332
Schizophrenic reaction, acute, 325
Scissoring positioning, 546
Scoliosis, 560, 561, 562, 563, 564, 568, 574, 577
Screening, of school-age children, 738, 762
Sedative-hypnotic medications, 254
Segmented spine instrumentation, 560
Seizures, 124, 125, 529, 540, 541, 543, 544, 552,
    553, 554, 555
Self-esteem, 259, 281, 306, 309, 352, 377, 384, 568
Self-help groups, 320
Semi-Fowler's position, 177
Sensory-perceptual alterations, 292

Separation anxiety, 530
Serotonin reuptake inhibitors, 281
Severe combined immunodeficiency disease, 488
Sex education, 638, 639, 640
Sexual abuse, 335, 367
Sexual and gender identity disorders, 363-374
Sexual assault, 366, 368, 370, 743, 764
Sexual offenders, 373
Sexual reassignment surgery, 372
Sexually transmitted diseases, 637, 639, 640
Shaken baby syndrome, 551
Shock, 454, 594. *See also specific type.*
Short stature, 616, 617, 621
Shortness of breath, 99
Shoulder subluxation, 136
Sickle cell anemia, 492, 493, 494, 495, 496, 497
Sickle cell crisis, 395, 495, 714
Sickle cell trait, 493
Sinus arrest, 481
Sinus arrhythmia, 481
Sinus bradycardia, 480, 481
Sinus infection, 773
Sinus pause, 481
Skeletal muscle relaxants, 152
Skin breakdown, 118, 204, 693, 698
Skin cancer, 717
Skin grafts, 243
Skin integrity, maintaining, 238, 241, 244
Skin turgor, 243
Skull fracture, 115
Sleep disorders, 251-262
Sleep strategies, 254, 255, 720
Sleeping position, for infant, 511
Small-for-gestational-age neonate, 435
Smoke inhalation, 96, 664
Smoking, 101
Snellen chart, 137, 720
Social phobia, 266
Social theory of causation, 712
Somatic delusion, 324
Somatization disorder, 252
Somatoform disorders, 251-262
Somatoform pain disorders, 261, 262
Spasticity, 134
Specific gravity, 768
Sperm production, 217
Spider bite, 235
Spina bifida, 548, 556, 558
Spinal cord injury, 126, 127, 128, 129, 134, 135,
    363, 364, 705
Spinal fusion surgery, 128, 574

Spinal shock, 133
Spinal surgery, closed, 150
Spinal tap, 503
Spiral fracture, 156
Splenic sequestration crisis, 495
Splint
   Frejka, 578
   Thomas, 159
Splinting, of incision, 731
Sports-related injuries, 576
Sputum culture, 60, 84, 87, 520, 532
Squamous cell carcinoma, 102
Standard precautions, 771
Stapedectomy, 129
Status asthmaticus, 92, 526, 527
Status epilepticus, 126
Statute of limitation, 683
Steatorrhea, 579
Stenosis. *See specific type.*
Sterile field, 736, 767
Steroids, 173, 234, 507
Stool softeners, 116
Stork bites, 662, 668
Strawberry hemangioma, 666
Streptococcal infection, 479
Streptococcal pharyngitis, 491
Streptococcus pneumoniae, 83
Stress management, 81, 260, 692
Stress ulcers, 185
Striae, 409
Stroke, 113, 114, 133, 135, 136
Stroke volume, 454
Subarachnoid hemorrhage, 116, 117
Subdural hematoma, 116, 117
Subjective data, 699
Substance abuse, 265, 269, 378, 680, 727
Substance abuse disorders, 333-351
Sucking reflex, 541
Suctioning, 85
Sudden infant death syndrome, 509, 510, 511
Suicidal client, 249, 279, 280, 281, 283, 307, 308, 337,
   355, 356, 360, 384, 679, 718
Sunburn, 717
Sunscreen, 658
Supine position, 98
Supplemental Security Income, 738
Supraventricular tachycardia, 482
Surfactant, 431
Surgery preparation, 105, 106, 137, 156, 166
Survivor guilt, 270
Sutures, 667

Sydenham's chorea, 478, 479
Sympathetic nervous system stimulation, 40
Syndrome of inappropriate diuretic hormone secre-
   tion, 192
Syphilis, 223, 638, 642, 674, 742
Syrup of ipecac, 593
Systematic desensitization, 267
Systemic lupus erythematosus, 74, 75, 76

T

Tachycardia, 40
Tacrine, 287
Taking-hold phase, 430
Taking-in phase, 423, 429
Talipes equinovarus, 559, 561, 564, 577
Talipes varus, 559
Tangentiality, 328
Tap water, and burns, 664
Tardive dyskinesia, 320, 332
Tattoos, 673
T cells, 72
Team leader, responsibility of, 755
Telangiectasis, 393
Temporal lobe, 141
Tensilon test, 120
Terbutaline, 397, 405
Testes, undescended, 632, 717
Testicular cancer, 221, 717
Tet spells, 466, 774
Tetanus immunization, 676
Tetracycline, 664
Tetralogy of Fallot, 462, 465, 466, 470
Tetraplegia. *See* Quadriplegia.
Thalamus, 140
Thalassemia, 494
   beta, 401, 496
Theophylline, 522, 523
Therapeutic communication, 721
Therapeutic community, 727
Therapeutic nurse-client relationship, 729
Therapy groups, 321
Thermal neutral zone, 444
Thermometers, oral, 772
Thiazide diuretics, 205
Thioridazine, 62
Thomas leg splint, 159
Thoracentesis, 112, 714
Thoracic water-seal drainage, 528
Thoracotomy, 756
Thrombocytopenia, 69, 73
Thrombolytic medications, 43

Thrombophlebitis, 58
Thrombosis, 57, 59
Thrush, 441, 673
Thumb sucking, 451
Thymus gland, 72
Thyroid crisis, 196
Thyroid gland, 199
Thyroid storm, 200
Thyroidectomy, 196, 206, 207
Thyroid-stimulating hormone, 191, 200
Thyroxine, 189, 191, 195
Tick removal, 674
Ticlopidine, 113
Tinea capitis, 239
Tinea corporis, 239, 674
Tinea pedis, 672
Tissue donation, 686
Tocolytic medications, 398
Toddler, growth and development in, 451, 558, 773
Tonometer, 137
Tonsillectomy, 712
Tooth decay, 669
Tophi, 144, 145
TORCH, 424
Torticollis, 560, 561, 562, 563
Total anomalous venous return, 467, 468
Total parenteral nutrition, 579, 747
Touch-down weight bearing, 576
Toxic shock syndrome, 217
Tracheal suctioning, 710
Tracheoesophageal fistula, 448, 584, 585, 586, 587, 600
Tracheostomy, 132, 524, 525, 772
Traction, 155, 156, 159, 573, 575. *See also specific type.*
Transferring, body mechanics for, 723
Transfusion reaction, 496
Transient global amnesia, 296
Transient ischemic attack, 129, 135
Transient tachypnea of neonate, 445
Transitional milk, 425
Transplantation
  autologous, 68
  bone marrow, 504
  heart, 47, 473, 736
  kidney, 227
  xenogeneic, 72
Transposition of great arteries, 469, 470, 712
Transsexualism, 373, 374
Transverse fracture, 156
Transvestic fetishism, 369
Trendelenburg's test, 56

Triazolam, 254
Trichomoniasis, 214, 640
Tricuspid atresia, 466, 467
Tricyclic antidepressant medications, 268
Trihexyphenidyl, 122, 330
Triiodothyronine, 189, 195
Trophoblastic disease, 390
Truncus arteriosus, 468, 469
Tuberculin converter, 87
Tuberculin skin test, 696
Tuberculosis, 85, 86, 87, 88, 89, 396, 532, 533, 695, 696, 697, 739, 741
Tuberculosis screening, 739
Tumors, 661
Turning schedule, 242, 2445
Twins, 393, 430
Twin-to-twin transfusion syndrome, 389

U

Ulcer. *See specific type.*
Ulcerative colitis, 170, 171, 172, 586, 594
Umbilical cord, 440, 434, 700
Umbrella filter, 110
Urethra, 643
Urinary catheter
  condom, 217, 698
  indwelling, 218, 226, 228, 718, 721, 728, 745
  straight, 231
Urinary elimination, impaired, 627
Urinary tract infections, 232, 642, 643, 644, 645, 646, 647, 745, 748, 652, 705, 762
Urination, painful, 229
Urine retention, 118, 230, 651
Urine specimen, 226, 630, 643, 645, 718
Urine toxicology screen, 344
Urolithiasis, 224
Urticaria, 658
Uterine atony, 430
Uterine enlargement, 389
Uterine subinvolution, 426
Uteroplacental circulation, 400

V

Vaccinations, 91, 487, 488, 496, 519, 521, 701
Vagal maneuvers, 482
Vaginal birth, 419
Vaginal discharge, normal, 214
Vaginal irrigation, 220
Vagotomy, 176, 177
Valproic acid, 544

Valsalva's maneuver, 481
Varicella, 656
Varices, esophageal, 181
Varicose veins, 55, 56
Varicosities, 61
Vascular dementia, 289, 290
Vascular resistance, 52
Vasectomy, 221
Vaso-occlusive crisis, 493, 494, 495
Vasopressin, 117, 192, 194, 614, 616
Venous stasis, 574
Ventilation-perfusion scan, 108
Ventricular fibrillation, 50
Ventricular septal defect, 462, 463
Ventriculoperitoneal shunt, 540, 543, 546
Vernix, 440
Vesicoureteral reflux, 646, 652
Vision impairment, 121, 749
Vital signs, 125, 709, 710
Vitamin A deficiency, 136
Vitamin C, 758
Vitamin D, 206
Vitamin K, 433, 437
Vomiting. *See* Nausea and vomiting.
von Willebrand's disease, 501
Voyeurism, 370

W

Warfarin, 59, 109
Warts, 668, 671
    anogenital, 638
    plantar, 238
Wedge resection, 103
Weight, determining, 442, 566
Wellness diagnoses, 765
Wet-to-dry dressings, 203
Wheelchairs, 568, 570
Wheezes, 89, 526
White blood cells, 504, 508
Wilms' tumor, 646, 647, 648, 649, 650, 651, 652
Wound care, 245, 457
Wound culture, 242
Wound healing, 242, 758, 80
Wound infection, 667

XY

Xenogeneic transplantation, 72

Z

Zinc deficiency, 598

Notes